THE ACRONYM BOOK
Acronyms in Aerospace and Defense

Compiled by

Peter B. Landecker, Ph.D.
Senior Staff Scientist
TRW Space and Electronics, Redondo Beach, California

and

Fernando B. Morinigo, Ph.D.
Professor Emeritus
California State University, Los Angeles, California

AIAA

American Institute of Aeronautics and Astronautics
1801 Alexander Bell Drive
Reston, VA 20191–4344

Publishers since 1930

ISBN 1-56347-536-7

PREFACE

This book consists of more than 51,000 entries. Most entries in this list are abbreviations constructed from the initial letters or syllables of the principal elements of the word or phrase that is represented. Such abbreviations are today referred to as acronyms, even though the traditional meaning of acronym does not quite fit. In compiling this list, the assumption has been made that any grouping of all capital letters and the symbols -, /, and & that occurs within text that is not all in capitals deserves to be included, unless the capitalization appears to denote emphasis of a complete word of ordinary English.

All the entries listed have been found in publications and documents in active use at major aerospace and defense companies or U.S. government agencies, or in press coverage of these industries, through the end of 1999. No entries have been included only because of an individual's oral or electronic report or opinion.

According to procedures that would be used in compiling language dictionaries, the only acronyms worthy of listing would be those that appear embedded in ordinary English text without an immediate expansion shown in parentheses. That has been the method for a large fraction of the entries. As a next best method, we have accepted lists appearing as appendices for proposals and similar documents circulated within companies and government agencies active in the aerospace or defense industries. Those lists are normally prepared after examination of a document and include acronyms in fact used within the document, so the standard linguistic criterion will usually have been met.

All-purpose lists of acronyms prepared by aerospace and defense professionals have also been used, but principally for comparison and corroboration, that is, to verify the entries.

We are confident that this list will be useful to you, the user, as an aid to the interpretation of other people's writings. If you are about to create a new acronym for your own purposes, do check this list to see what kind of competition you may have for the same set of capital letters.

Peter B. Landecker
Fernando B. Morinigo

CRITERIA AND CONVENTIONS

Omitted from the list are most groupings of capitals that include periods, numbers, lowercase letters, calendar information, variable or command names from computer programs, groupings from classified advertisements, stock market lists, and TELEX codes. The first few letters of a single English word have not been held to constitute an acronym, even if appearing in all capitals.

One problem that has become more noticeable since the publication of the 1990 and 1992 versions of this book is the rapid creation of humorous or "joke" acronyms that usually stand for a complete sentence or grammatically complete phrase rather than a concept or entity. Early examples that are in widespread use today are TGIF for Thank God It's Friday, ASAP for As Soon As Possible, MYOB for Mind Your Own Business, and BYOB for Bring Your Own Bottle. In compiling this list, if an acronym has been identified as this type, the authors have included it only if in their opinion it has already gone through the process of becoming part of American culture. Without such a criterion, this list would have been significantly padded by "garbage" acronyms not meriting the effort of being disseminated in a working tool for professionals. The indiscriminate inclusion of any reported acronym is a characteristic of lists disseminated principally through the World Wide Web.

There are many instances of a concept or entity that is represented by a group of all capital letters that is totally unrelated to any descriptive phrase or any recognizable set of initials of words. A famous example is SOS, the distress signal devised in the early days of Morse code radio communications. Although such groupings are not abbreviations, they are included in this list if information about the concept or entity could be reasonably inferred from the context.

The usage of acronyms is grammatically haphazard. Some acronyms are, in the grammatical sense, nouns, others are adjectives or verbs, some are proper nouns, and so forth. The following are recommended:

The plural of an acronym that is a noun is indicated by appending a lowercase "s" to the singular (singular, BAPTA; plural, BAPTAs).

The genitive case (possessive) of an acronym noun is indicated by an apostrophe and "s" (The NCO's means it belongs to the NCO).

The slash / is used most often as a replacement for "or," although frequently it stands for "per." In many cases it is used as a sort of punctuation mark, separating acronyms that have a distinct meaning in their own right. It is recommended that it should never stand for "and," which has its own symbol, &. In a few dozen instances in this list, the Arabic numerals 2 or 4 are used as abbreviations for the prepositions "to" and "for," respectively.

Some acronyms are often written in an algebraic notation of powers, C to the second power stands for CC, C to the third power stands for CCC, and so on. In this list, these acronyms are listed as if they were written CC or CCC.

When two acronyms are placed in juxtaposition and there is a clear conceptual significance to the individual parts, a slash or hyphen is often used to separate the parts. It is clear from the usage that often there is an evolution in the manner of writing such acronyms, so that the slash or hyphen tends to disappear.

Direct quotations from text found are given between parentheses ( ). Square brackets [ ] give optional material and are mostly used to denote context, to complete the definition, and to contain translations of acronyms derived from words in languages other than English. Ordinary parentheses are used to denote possible variants, as when the same acronym may stand for an adjective or a noun, or when the same letter in an acronym may stand for closely related concepts. Curly braces { } are used to indicate that the explanation given is a possible inference from the context and that a substantial degree of doubt exists. Braces enclose a question mark {?} when the explanation given has been extracted accurately from a text, but a suspicion exists that the explanation may be garbled or contain misprints.

When acronyms really catch on because they represent often useful concepts, consciousness of the precise original words that led to the acronym is blurred, and alternative explanations of the initial letters are given by different writers. A representative list of alternative explanations of what various letters stand for follows.

Air = Aerial	Operations = Operating
Authority = Authorization	Procedure = Plan
Authorized = Approved	Program = Project
Automated = Automatic	Quote = Quotation
Description = Descriptor	Record = Report
Effective = Equivalent	Repair = Restore
Electrical = Electronic	Report = Response
Improved = Increased	Reprint = Reprinting
Inner = Internal	Resonator = Resonating
Least = Lowest	Shipborne = Shipboard
Log = List	Silicon = Semiconductor

Measure = Measurement	Sonic = Sound
Notice = Notification	System = Subsystem
Observing = Observation	Talks = Treaty
Office = Officer	

The only method available to check the accuracy of the information in this list is cross correlation of various sources. The very earliest lists were kept without a record of where the acronyms or their explanations were found. Later versions listed only one source, the one believed most authoritative. For the most recent versions, multiple sources are listed. The larger the number of sources given, the more likely it is that the acronym is useful and that the explanation for its meaning is firmly based on professional usage.

The matter in parentheses after the colon serves to identify the document or publication in which the acronym has been found, either used or explained. The explanation of the codes is given in the list of citation references.

No more than one citation per identifying code is listed.

In general, different meanings of the same set of capital letters are listed separately. However, where the difference in meaning is simply due to grammatical function (adjective vs noun, for example) leading to differences in suffixes, the variants are listed in parentheses under the same entry.

Often, just listing the words that originated an acronym will be insufficient to define what the acronym means or stands for. For this reason, in many cases we have expanded beyond giving the list of words represented in the acronym and have given additional material, making this list one step closer to a dictionary of acronyms.

One warning: In many acronym explanations in the documents and sources, and in many acronym lists, the probability of finding misprints is very high. Where a misprint is obvious, the corrections have been made. However, it is not possible to guarantee that all such misprints have been caught, nor, indeed, is it possible to have a clear idea of how many of the entries may have deficient expansions.

This work is available both in print-on-demand format and online. The online version is available at http://www.aiaa.org/research/index.hfm?res=4. Both versions of the book are updated periodically. This is in contrast to previous editions of *The Acronym Book,* which were published in hardcopy format only by AIAA in 1990 and 1992. This edition also includes a significant expansion in the number of acronym entries.

The authors welcome comments as well as suggested additions and corrections to this work at landecker@cyberdude.com or addressed to the authors' attention at AIAA.

Finally, we wish to acknowledge that this work would have been impossible to prepare without the active support of literally hundreds of colleagues and coworkers who have sent us the citations and their own acronym collections for inclusion here and given us much encouragement throughout the years. It is their expressed interest and gratitude that is the truly rewarding part of preparing this list.

CITATION REFERENCE CODES

0 = Abbreviations and Acronyms, Hughes Aircraft Co. IDC 2255.1/1400

1 = Eastport Study Group, Report to the Director, SDIO, Summer Study, 1985

2 = NASA Goddard Space Flight Center, Tracking and Data Relay Satellite System (TDRSS) Users' Guide, September 1984, STDN No. 101.2

3 = Intelsat VI ADCS Summary, Hughes IDC 4142.10/1307

4 = W.L. Woodson, TDRSS Glossary, TRW IOC, 29 July 1980, TDRSS-80-200-084T

5 = U.S. Department of the Air Force, Space-Based Kinetic Energy Weapon System Request for Proposal, 1986

6 = Semiconductor and Computer Glossary, Design News, /3-22-76/, p. 101 ff

7 = Glossary of Technical Terms for Satellite Communication System Engineering, Indonesian Communication System Engineering Training Program

8 = Term and Acronym List, Hughes Aircraft Co., from MIS

9 = Acronym List, AUSSAT Spacecraft Training Manual, Hughes, #836609WP

10 = James Bamford, "The Puzzle Palace"

11 = Robert M. Bowman, "Star Wars," Jeremy P. Tarcher, Los Angeles, CA, 1986

12 = W. M. Arkin, R. W. Fieldhouse, "Nuclear Battlefields," Ballinger, 1985

13 = Walter A. McDougall, "The Heavens and the Earth," Basic Books, NY, 1985

14 = J. Rotblat and S. Hellman, Nuclear Strategy and World Security, Macmillan, 1985

15 = Acronyms, List circulated at Hughes Aircraft Co., 1987

16 = Acronyms/Glossary, Proposal for Defense Meteorological Satellite Program, Hughes Aircraft Co., circa 1988

17 = Acronyms, ISBR/ACM Proposal, Hughes Aircraft Co., 1988

18 = Space and Naval Warfare Systems Command, UHF F/O Communications Satellite RFP, 1988

19 = Satellite Communications Applications Research Program, NASA Research Announcement Preproposal Conference, Holiday Inn Capitol, Washington, DC, 25 August 1989

aa = Aerospace America (ISSN 0740-722X), published by AIAA, Reston, VA (vol.no.page)

ab = A.H. Bogen (page)

ac = L.K. Acheson (ed.), SHREWD, Hughes Aircraft Co., 1980

ad = miscellaneous advertising leaflet or flyer, period 1985–1992

ae = Aerospace Daily (ISSN 0193-4546), Washington, DC (vol.no.page)

af = Air Force Magazine (ISSN 0730-6784) Air Force Association, Arlington, VA

ag = Austin H. Gale, List of Acronyms, Hughes Aircraft Co., circa 1985

ah = "Short List of SEAO HSC Acronyms," 248 lines, file acroshrt.doc by Peter Landecker, forwarded October 1997

aj = Armed Forces Journal International (year.month.page)

an = Astro News, Los Angeles AFB, CA (vol.no.page)

as = System Specification for Anti Satellite System (ASAT), ESD, 21 September 1990

at = ATEP flyer, Hughes Aircraft Co., circa 1991

au = Aura Systems, Inc., press release, 2 October 1997

aw = [AW&ST] Aviation Week and Space Technology, McGraw Hill Publishing Co. (ISSN 0005-2175)

bf = BFV-T2SS Acronyms List (Bradley Fighting Vehicle), HAC, circa 1983

bkr = Barney Krinsky, Hughes Aircraft Co. senior scientist, circa 1990

bl = Z. Bleviss (ed), Geosynchronous Spacecraft Case Histories, Hughes Aircraft Co., 1981

bm = Bate, Mueller, Orbital Mechanics, <?>, circa 1980

bo = Boeing Aircraft Co., Inertial Upper Stage Users' Guide D290 10652-1

ca = Ashton B. Carter, Directed Energy Ballistic Missile Defense in Space, OTA-BP-ISC 26, Washington, DC, 1984

cb = Commerce Business Daily (month.day.year)

cca = Charles Campbell, formerly at Hughes Aircraft Co., later professor at USC, circa 1989

cd = C&DP Booklet or C&DP flyer, Hughes Aircraft Co., circa 1985

ce = W. L. Pritchard and J. A. Sciulli, Satellite Communication Systems Engineering, Prentice–Hall, Upper Saddle River, NJ, 1986

ch = International Organization for Standardization (ISO) Online Help from http://www.iso.ch/infoe/guide.html, downloaded 29 April 1998

ci = U.S. Civilian Space Programs, 1958–1978, Commission on Space and Technology

cl = C.M. Leary, AUSSAT program environmental requirements acronyms, 29 October 1990 cm = Computer (vol.no.page)

cp = Capabilities to Support ASAT System Development, SAIC publication, 9 August 1989

cr = Chemical Rubber Co., CRC Mathematical Tables, 1986

crl = Acronymes/Acronyms, AFCRL Air Force Cambridge Research Laboratory, http://www.univ-perp.fr/fuseurop/acronym.htm, downloaded 17 May 1999

cs = CSSC, Proceedings of Communications Satellite System Conference, 7 March 1982

ct = Heath D. Albert, "Computer and Telecommunications Acronyms," 1243 entries, comm.htm file, forwarded October, 1997

cu = "Commonly Used Acronyms," 83 entries, space.htm file, forwarded October 1997

dh = DOD-HAC Acronyms, Department of Defense and Hughes Aircraft Co. list, circa 1986

dm = DMSP Block 6 Risk Reduction (Phase 2) Request for Proposal, 1 November 1990 (SSD/SDMP Control No: S-00705)

dmi = Don Miller, Hughes Aircraft Co., circa 1992

dn = David Newlands, Acronyms List, Hughes Aircraft Co., circa 1985

do = D.R. Doll, Data Communication Facilities, Networks, Systems, J. Wiley, 1978

dt = U.S. Department of Defense Telephone Directory, August 1985 (Superintendent of Documents)

du = Internet For Dummies, 4th ed., Levine et al., IDG Books, 1997

ecm = Engineering Command Media Handbook, Hughes Space and Communications, August 1993

em = "The Acronym Emporium," emporium.htm file, 20 October 1997, 1283 entries

eo = Science Exploration Opportunities for Manned Missions, NASA, 30 June 1989

ep = Space and Communications Group, HAC, Engineering Procedure Acronyms List, 11-07-88

es = ESA Bulletin, European Space Agency, Noordwijk, Netherlands (no.page)

ett = Ed T. Tamura, Hughes Missile System Group, memorandum, circa 1992

fq = Franklin Quest, Writing Winning Proposals Manual, 1996

fr = Fletcher Report (a secret document on the feasibility of SDI) circa 1985

fu = Acronyms List, Hughes Fullerton, received 1997

gb = GETS Booklet, Hughes Aircraft Co., circa 1983

gc = Glossary of Computer Security Terms and Acronyms (copy from John W. Pope of HAC)

gd = GOES DEF Data Book, Hughes Aircraft Co., August 1980

ge = Acronyms for Global Environmental Monitoring (received 1997)

gg = Hughes Aircraft Co., Proposal to NASA Goddard Space Flight Center for GOES GHI, Part III

go = D.V. Goode, Anti-Satellite System Submittal of Contractor Capabilities, 89/1L06.10-3453-JO478, 19 July 1989

gp = GOES Proposal Acronyms List, GOES Statement of Work Acronyms List, 1987

gr = Daniel O. Graham, "High Frontier," circa 1983

gs = Z. Bleviss (ed.), IV Geosynchronous Spacecraft Case Histories, Hughes Aircraft Co., circa 1982

hc = Hughes Aircraft Co. Corporate Marketing Acronyms, CT/C1/MS A142, 16 June 1988

hd = Hughes Aircraft Co., Internal Telephones Directory, circa 1984

he = author HSC/EDS, 176 entries, file hs702.doc or hs702.htm

heh = Hughes Electronics Herald (vol.no.page)

hf = HAC form 27 CS

hg = Donald R. Merritt, Hughes Global Environmental Monitoring, December 1991

hi = Harfang Internationale, Inc., Charlesbourg, Quebec, Canada, AcronymBuster database, copyright 1996

hj = Hughes Space and Communications Journal (vol.no.page)

hm = Hughes Aircraft Co.: Miscellaneous (unarchived) Communication (flyer), period 1982–1992

hn = Hughes News biweekly, Hughes Aircraft Co. (vol.no.page)

hq = Hughes News Supplier Quarterly Edition (vol.no.page)

hs = Hughes Aircraft Co., publications quality archived viewgraph, period 1982–1992

hy = Highlights of the Year 1989, NASA Earth Science and Applications Division, Appendix C

ic = Innovative Concepts for Space Based Radar, PRDA No. 88-I-PKRZ, 11 December 1987

id = Hughes Aircraft Co., Inter Departmental Correspondence, archived, period 1982–1992

ie = Inside EDSG, Hughes Aircraft Co. (vol.no.page)

ig = Industrial Engineering, Institute of Industrial Engineers, Norcross, GA (vol.no.page)

in = Intelsat Document

is = IEEE Spectrum (ISSN 0018-9235), IEEE, Piscataway (vol.no.page)

jb = Joy F. Berman, ASAT/BMCCC, GSG, Hughes Aircraft Co., 2 November 1989 jdr = James D. Richardson, Hughes Milstar and UHF Follow-On Programs, September 1996

jgo = J. Godwin, Hughes Aircraft Co., circa 1990

jhm = Jon H. Myer, Hughes Aircraft Co., circa 1990

jj = J. T. Johnson, Abbreviations and Acronyms List, Field Engineering Technical Letter, Hughes Aircraft Co., Phoenix Tomcat Program, 20 February 1979

jk = Jerry D. Kendrick, GeoMobile Contracts, Hughes Space and Communications, 1997

jmj = Joint KE ASAT Program Briefing to Industry, J. Morgan Jellett, U.S. Army Strategic Defense Command, 7 September 1989

jn = Jay Nam, Orion 3 Program, Hughes Space and Communications, 2 October 1997

jo = Joh. (a document having pages numbered iv and v)

jp = JPL Galileo program Blueprints, NASA, circa 1990

jr = James D. Richardson, Hughes Space Communications Co., 55 Additions to Acronym List of 4

jrm = Jack R. Murphy, Hughes Aircraft Co. senior scientist, circa 1990

js = Journal of Spacecraft and Rockets, AIAA, Reston, VA

kb = Kirk Bretney, 15th DARPA Systems and Technology Symposium Trip Report, 16 October 1989

kf = Kathleen A. Fricks, Hughes Space Based Infrared Program, July 1996

kg = Irving Kind, "BABEL: A glossary of computer related (oriented) abbreviations and acronyms," 3879 entries, file babel.htm, September 1997

kgl = Keith G. Lyon, Hughes Aircraft Co., circa 1988

kja = K. Jarett, Hughes Aircraft Co., circa 1990

lat = Los Angeles Times

lc = Labor Collection Network publication, Hughes Aircraft Co.

le = Donald P. LeGalley (ed.), Space Physics, John Wiley and Sons, Inc., circa 1982

lka = Louis K. Acheson, chief scientist, Hughes Aircraft Co., circa 1992

ls = Launchspace, Magazine of the Space Industry (vol.no.page)

lt = LEASAT Technical Proposal, Hughes Aircraft Co., circa 1984

lyg = Lynn Grasshoff, Hughes Aircraft Co., circa 1990

ma = Hughes Aircraft Co., Mars Orbiter Study, Vol. 1, 18 February 1982

mb = Mars Observer Proposal Information Package, NASA-JPL, 3 May 1985

mc = MAPS Catalog, Hughes Aircraft Co.

md = MCD Digest, a publication of Microelectronics Circuits Division of IEG, Hughes (vol.no.page)

mi = Massachusetts Institute of Technology, Astro/Space Frequently Seen Acronyms, 11th ed., 10 August 1994

mj = Microwave Journal (ISSN 0192-6225), Horizon House-Microwave, Inc., Norwood, MA

mo = Mars Orbiter Proposal, Hughes Aircraft Co., circa 1986

mp = SMP Handbook, published by Computer Mathematics Corp., circa 1985

ms = MS Newsletter, Hughes Aircraft Co. (vol.no.page)

msm = Michael Smith, Hughes Aircraft Co., circa 1991

mt = Megaterms, by Mike Bandor et al., a database of 14,167 military terms and acronyms, maintained at the Software Technology Support Center, Hill AFB, UT, July 1997

muc = M.U. Chang, Hughes Aircraft Co., circa 1990

nd = National Defense, Journal American Defense Preparedness Association, Arlington, VA (ISSN 0092-1491)

ne = Hughes Aircraft Co., Newport Design Center (flyer or other publication), 1982–1992

no = NOAA, SESC Glossary of Solar-Terrestrial Terms, Boulder, CO, August 1988

np = National Polar Orbiting Operational Environmental Satellite System Integrated Program Office Interface Control Document, 1 September 1997

ns = Goddard Space Flight Center, Guide to the National Space Science Center, 1990

nt = Near-Term Kinetic Energy Antisatellite Program, Draft RFP, U.S. Army Strategic Defense Command, 6 October 1989

nu = Nunan, Telecommunications Acronyms, Hughes Aircraft Co., 1997

oe = Geostationary Operational Environmental Satellite, NASA Code 415, Rev O, 31 March 1997

pbl = Peter B. Landecker, Hacronyms List (Hughes Space and Communications Acronyms), 1997

pf = Performance, a magazine published by Defense Systems unit of Unisys Corp. (issue.page)

ph = Photonics Spectra, 10/84, pp. 88–89, Acronyms List

pk = Panos Kotsianas, Aura Systems, Inc., Defense/Computer/Software Industry Acronyms Glossary, 1996

pl = Peter Landecker, Technical Acronyms, HAC Internal Memorandum, 11 January 1989 pm = Pat Molloy, MSAMS Baseline System Concept Description Report, HADC, circa 1990

pr = Products and Services, SESC (NOAA), Boulder, CO, May 1990

ps = Paul B. Stares, Space and National Security, Brookings Institution, 1987

re = A Robotic Exploration Program, JPL, Glossary of Acronyms, 1 December 1989

rf = Reliability and Maintainability Forum (vol.no.page)

rg = R.J. Goldstein, "IR&D Technology Development," Defense Initiatives Department, Hughes Aircraft Co., 5 May 1989

ric = Jim Richardson, Acronym Delta Update, 1997 (cf. jdr)

rih = Jim Richardson, Acronym Delta Update, 1999 (cf. jdr)

rir = Richard Rommel, Hughes Aircraft Co., circa 1989

rm = Managing for Reliability and Maintainability, Hughes Aircraft Co. newsletter (vol.no.page)

ro = "Glossary of Acronyms," 151 entries, file romelab.htm, file forwarded October 1997

rva = Richard E. Van Allen, Hughes Aircraft Co., circa 1992

sa = Systems Analysis Laboratories, Hughes Aircraft Co., archived IDC, period 1982–1992

sb = SBKEWS Proposal Acronyms List, Hughes Aircraft Co., Phase II Proposal, 23 January 1987

sc = The SCG Journal, Hughes Aircraft Co., El Segundo, CA (vol. no. page)

sd = R. Goldstein (ed.), Hughes Aircraft Co., Space Defense and Asset Protection Study, 1984

se = Security Briefs, Hughes Aircraft Co. internal newsletter (date.page)

sf = Spaceflight, British Interplanetary Society Ltd., London (monthly) (vol.no.page)

sh = HS 350-F6 Fuel Slosh Test Results, SAL 14104, Hughes Aircraft Co., circa 1988

sm = Satellite Manual and Reference for Telecommunications, Hughes Aircraft Co., December 1980

sn = Synergy, a publication of Hughes Aircraft Co., a subsidiary of GM (no.page)

so = Cassini Mission: Saturn Orbiter, NASA Announcement of Opportunity, 10 October 1989

sp = J.J. Spilker Jr., Digital Communications by Satellite, Prentice–Hall, Upper Saddle River, NJ

spi = Sam Pierce, Hughes Aircraft Co., circa 1992

sr = Space Defense Review (a periodical)

ss = Space Defense Study, Hughes Aircraft Co., September 1982

st = Glossary of Strategic Defense Command Abbreviations and Acronyms, 18 November 1986

su = "Acronyms and Abbreviations," HAC/HSC, Security, 147 entries, forwarded October 1997

sv = Security Views, Hughes Aircraft Co. (a periodic publication)

sy = You and Hughes Security, Hughes Aircraft Co. Security Training Manual

tb = NTB Acronym List (from NTB Training) Roger Cornell, 3rd ed., circa 1987

tc = T. Cohen, Hughes IDC, 13 January 1984 (an acronyms list)

te = NASA Tech Briefs, Associated Business Publications (vol.no.page)

th = Technology Center Highlights, Hughes Microelectronics Center Newsletter

tk = Thiokol Corp. catalog, circa 1984

tm = Tomislav Matic, Hughes Engineering Procedures, April 1996

tr = TRW Acronyms List, circa 1984

ts = Technical Symposium on SDI/ADI Simulation and Modeling, USAF Electronic Systems Division, Hanscom AFB, MA, 23–24 March 1988

tt = Common Telecommunication Terms and Acronyms from http://www.hughesglobal.com/glossary.htm, downloaded 29 April 1998

uc = McDaniel/Lee: UCLA, Glossary of Acronyms and Abbreviations, 15 December 1982

uh = Marty G. Backe, "Acronym Definitions," 455 entries, file uhf-fo.htm, last updated 26 August 1997

un = unrecorded (citation source not recorded or source record lost)

vc = Vectors (a periodical published by Hughes Aircraft Co.)

vv = Tracking and Data Relay Satellite, Hughes file TDRSBD08.DOC, sent 2 October 1996 by V. Vaultz

w9 = Webster's Ninth New Collegiate Dictionary, Merriam-Webster, Inc., Springfield, MA, 1988

wl = WYLE Laboratories, Acronyms for Space Station Program, September 1985

wp = Wright–Patterson Air Force Base, OH, ACROFIND database, 1 April 1990

ws = World Spaceflight News (ISSN 0737-8548) (vol.no.)

y2k = Investigating the Impact of the Year 2000 Problem, U.S. Senate Special Committee on the Year 2000 Technology Problem, S. Prt. 106-10, Appendix II, 24 February 1999

A

A [designator for Thunderbolt II (A-10), Skyhawk (A-4), Intruder (A-6), Corsair II (A-7), USA DOD] *mt*
A&A Astronomy and Astrophysics *mi*
A&AEE Aerospace and Armament Experimental Establishment [Boscombe Down, UK, formerly Aeroplane and Armament Experimental Establishment] *mt*
A&C Arithmetic and Control *jj*
A&CO Assembly and Check Out *tb, jjA&A, jj, jjling wp, fu, mt*
A&CO Astronomy and Astrophysics *mi*
A&DCP Analysis and Data Collection Plan *mt*
A&E Architectural Architecture and Engineering *tb, fu, mt*
A&M Aeronautical and Meteorological *fu*
A&R Alert and Resolution *fu*
A&R Automation and Robotics *aa27.4.b54*
A&RC Attitude and Reaction Control *dm*
A&VT Angle and Velocity Tracker *jj*
A&W Alert and Warning *fu*
A-A Air to Air *hc, jj*
A-AFCWF Army Air Force Civilian Welfare Fund *dt*
A-D Analog to Digital [also A/D] *jj*
A-E Architect-Engineer *mt*
A-HOUR Alert-Hour *mt*
A-NORS Anticipated Not Operationally Ready Supply *jj*
A-SAR Advanced SAR *hg, ge*
A-SCAT Advanced Scatterometer *hg, ge*
A-TIDE Automatic TOW Infantry Depot Equipment *hc*
A-V Audio Visual [also AV] *jj, fu*
A-XAWS A-X Close Air Support A/C Weapon System *hc*
A/A Air to Air *jj, fu, mt*
A/ASTR Air to Air Search and Track *hc*
A/B After Burner *jj, kf*
A/B Air Borne *hc, jj, mt*
A/C Absolute Ceiling *mt*
A/C Air Conditioner *hn49.3.2, jdr, fu*
A/C Aircraft *sd, hc, aa26.11.43, jb, tb, jj, kf, fu, mt*
A/C Attitude Control *0*
A/CA Air Craft Attrition *mt*
A/D Alphanumeric Display *fu*
A/D Analog to Digital *bl, 0, 3, hc, aa27.4.57, lt, st, jj, kf, fu, oe, hi, mt, ai1*
A/D Antenna Driven *tb*
A/DC Analog to Digital Converter *0*
A/E Architect/Engineer *tb, kf, fu*
A/EQ Amplitude Equalizer *bl2-9.15*
A/G Air / Ground *mt*
A/G Air to Ground *jj, fu, mt*
A/L Air Lock *jdr*
A/L Airborne Launch *jj*
A/M Aeronautical and Meteorological *fu*
A/M Ampere per Meter *jdr*
A/N Alpha Numeric *jj, fu, mt*
A/N Army / Navy [standard] *hc, jj*
A/P AutoPilot *jj*
A/R Accept or Reject *gp*
A/S Air Speed *jj*
A/S Air to Service *jj*
A/SPP Antenna Solar Panel Position *hc*
A/T Angle Tracking *hc, jj*
A/T Assembly and Test *kf*
A/T/RM Antenna Transmit Receive Module *hc*
AA Adaptive Array *hc*
AA Air to Air *wp, mt*
AA Angle of Attack *wp*
AA Anti Aircraft *w9, wp*
AA Arithmetic Average *fu*
AA Atomic Absorption *wp*
AA Attack Assessment *tb, mt*
AA Automatic Approach *wp*
AA/SLBM Airborne Anti Sea Launched Ballistic Missile *hc*
AAA Alternative Antenna Array *fu*
AAA Anti Aircraft Armament *hc*
AAA Anti Aircraft Artillery *fu, hc, wp, jj, mt*
AAA Assessment of Aircraft Attrition [USAFE] [USA] *mt*
AAA Automated Airlift Analysis System [AMC*] [USA] *mt*
AAA Automated Analyst Aides *hc*
AAAA Army Aeronautical Activity at Ames *hc*
AAAA Army Aviation Association of America *hc*
AAAAWS All Arms Anti Armor Weapon System *hc*
AAAD Airborne Anti Tank [Armor] Air Defense *wp*
AAAD All Arms Air Defense [UK] *mt*
AAADOC AAA Division Operations Center [also AAA DIV OC *fu*
AAAE American Association of Airport Executives *aw129.14.112*
AAAM Advanced Air to Air Missile [replacement for AIM*-54 Phoenix] [USA] *mt, wp, fu, hc, aw129.14.34, kb*

AAAM Air to Air Attack Management *hc, wp*
AAAMA Air to Air Mission Analysis *wp*
AAAMLE Advanced Air to Air Missile Launch Envelope *hc*
AAAMS Advanced Anti Armor Missile System *wp*
AAAOB Anti Aircraft Artillery Order of Battle *mt*
AAAS Advanced Aircraft Armament System *hc*
AAAS American Association for the Advancement of Science *wp, 0, le42, w9, aa26.11.b7*
AAAS Armored Anti Aircraft System *wp*
AAAS Aural Aided Acquisition Signal *hc*
AAAS Automated Antenna Alignment System *hc*
AAAWS Advanced Anti Armor Weapon System *wp*
AAB Aircraft Accident Board *wp*
AAB All to All Broadcast *pk, kg*
AAB Army Air Base *hc, wp, jj*
AABCP Advanced Airborne Command Post [also AACP] *wp*
AABL Advanced Atmospheric Burst Locator *mt*
AABM Airborne Anti Ballistic Missile *wp*
AABNCP Advanced Airborne Command Post *ro*
AABNCP Advanced Airborne National Command Post [also AACP] *kf, hc, tb, fu, ro*
AAC Adaptive Antenna Control *hc, wp*
AAC Aircraft Aviation Armament Change *wp, jj*
AAC Alaskan Air Command [USA] *hn44.9.10, 12, hc, af72.5.47, wp, fu, oe, mt*
AAC Army Air Corps [forerunner of USAF] *af72.5.45*
AAC Army Air Corps [UK] *mt*
AAC Automatic Answering Capability *mt*
AAC Automatic Approach Control *wp*
AAC Aviation Armament Change *wp*
AACB Aeronautics and Astronautics Coordinating Board [group] *ci.566, wp*
AACC Airborne Alternate Command Center *tb*
AACC Airport Associations Coordinating Council *aw124.21*
AACC Army Airborne Command and Control [USA] *mt*
AACC Army Airspace Coordination Center *fu*
AACE Aircraft Alerting Communications EMP *wp, mt*
AACF Advanced Aircraft Concept Formulation *wp*
AACG Arrival Airfield Control Group *mt*
AACISP AAC Information Systems Plan *mt*
AACISS AAC Intelligence Support System *mt*
AACM Active Access Control Module *mt*
AACO Assault Airlift Control Officer *wp*
AACONV ASAT-to-AIDES Conversion Program *fu*
AACP Advanced Airborne Command Post [also AABCP, AABNCP] *hc, wp*
AACS Airborne Astrographic Camera System *wp*
AACS Airways and Air Communications Service *hc*
AACS Anti Aircraft Control Station *wp*
AACS Army Command and Control System *mt*
AACS Attitude and Articulation Control System *aa27.4.60*
AACS Auxiliary Attitude Control System *wp*
AACSENT Air to Air Covert Sensor Technology *hc*
AAC_CCCC AAC Command Control Communications Computers *mt*
AAD Airborne Assault Division *wp*
AAD Aircraft and Armament Development *hc*
AAD Analog Alignment Diskette *ct, em*
AAD Anniston Army Depot *hc*
AAD Anti Aircraft Defense *kf*
AADC Advanced Avionics Digital Computer *hc, wp*
AADC Alaska Aerospace Development Corporation *ai2*
AADC Area Air Defense Commander [USA DOD] *mt, kf*
AADC Army Air Defense Command *wp*
AADCOM Army Air Defense Command *pm*
AADCP Antiaircraft Defense Command Post *fu*
AADCP Army Air Defense Command Post *fu, hc, mt*
AADE Army Air Defense Element *mt*
AADEOS Advanced Air Defense Electro Optical Sensor *hc*
AADGE Allied Command Europe Air Defense Ground Environment [NATO] *mt*
AADHS Advanced Avionics Data Handling System *wp*
AADOC Army Air Defense Operations Center *fu*
AADR Alaskan Air Defense Region *mt*
AADS Anti Aircraft Defense System [Army] [also called Missile Monitor and AN/MSG-4 *fu, hc, wp*
AADS Area Air Defense System *wp*
AADS Army Air Defense School *jj*
AADS Automated Air Defense System [renamed SAM-D *fu*
AADSS Access Area Digital Switching System Structure *hc*
AADT Avionics Advanced Development Task *hc*
AADWS Alternate Air Defense Weapon System *wp*
AAE Aerospace Auxiliary Ancillary Equipment *hc, wp, jj*
AAE American Association of Engineers *wp*
AAE Army Acquisition Executive *st*
AAED Advanced Airborne Expendable Decoy *mt*
AAED Airborne Active Expendable Decoy *hc*

AAEOFCS Air to Air Electro Optical Fire Control System *hc*
AAES Advanced Aircraft Electrical System *wp*
AAES Automated Aeromedical Evacuation System *mt*
AAEWR Advanced Aircraft Early Warning Radar *wp*
AAF Advanced Airburst Fuze *hc*
AAF Army Air Field [USA] *mt, wp*
AAF Army Air Forces *13, af72.5.45, wp*
AAF Auxiliary Air Force *mt*
AAFA Army Aviation Flight Activity [USA] *mt*
AAFBD Army and Air Force exchange service Board of Directors *dt*
AAFC Anti Aircraft Fire Control *wp*
AAFCE Allied Air Forces Central Europe [NATO Command] *mt, 12, af72.5.97, ts17-1, fu*
AAFCS Advanced Automatic Flight Control System *wp*
AAFCS Anti Aircraft Fire Control System *hc*
AAFE Advanced Application Flight Experiment *ac9.2-9, hc*
AAFES Army and Air Force Exchange Service *dt, wp*
AAFES Auto Alert Force Exercise Schedule [SAC*] *mt*
AAFIR Army Air Force Intelligence Report *wp*
AAFIS Advanced Avionics Fault Isolation System *hc, wp*
AAFMAA Army and Air Force Mutual Aid Association *af72.5.150*
AAFPS Army and Air Force Postal Service *dt*
AAFS Advanced Airburst Fuze System *wp*
AAFSS Advanced Aerial Fire Support System *hc, wp*
AAFTC Air Force Flight Test Center [Edwards AFB, USA] *mt*
AAG Aeromedical Airlift Group [AMC*] *mt, af72.5.121*
AAGR Air to Air Gunnery Range *wp*
AAGS Aircraft Adapter Groups *mt*
AAGS Army Air Ground System *mt*
AAH Advanced Attack Helicopter (the AH-64) [USA] *mt, ph, hc, wp*
AAHTME Advanced Attack Helicopter TOW Missile Equipment *hc*
AAI (already available are AAI's Pioneer and) (AAI Corporation, Huntsville, AL 35801) *aa26.11.50, af72.5.32, hc, jmj*
AAI Air to Air Identification Interrogation *hc, jj*
AAI Air to Air Intercept Interceptor *wp, wp*
AAI Angle of Approach Indicator *mt*
AAI Application/ Application Interface *fu*
AAIB Air Accident Investigation Branch [UK] *aw130.13.33, mt*
AAIFF Air to Air Identification Friend or Foe *wp*
AAIP Advanced Avionics Integration Program *wp*
AAIPS Automated Air Information Production System *wp*
AAIS Atmospheric Attack Indications System *mt*
AAL American Airlines *wp*
AAL ATM {Asynchronous Transfer Mode} Adaptation Layer [collection of standardized protocols] *nu, kg, ct, em, hi*
AALAAW Advanced Air Launched Anti Armour Weapon *mt*
AALC Advanced Airborne Launch Center *wp*
AALCC Alternate Airlift Control Center *fu*
AALPS Automated Aircraft Load Planning System [FORSCOM] *mt*
AAM Adaptive Array Module *fu*
AAM Air to Air Missile [e.g. Falcon, Sidewinder, Sparrow] *mt, fu, uc, hc, wp, jj, kf*
AAM Airborne Armament Maintenance *hc*
AAM Aircraft Availability Model *mt*
AAM Airfield Attack Munition *wp*
AAM Apogee Adjust Maneuver *oe*
AAMA Area Airspace Management Authority *mt*
AAME Airspace Management Element [AAC] *mt*
AAMP Army Automation Master Plan *mt*
AAMRL Armstrong Aerospace Medical Research Laboratory *wp*
AAMS Advanced AFSATCOM Monitoring System *hc, mt*
AAMS Advanced Automatic Monitoring System *wp*
AAMS Air To Air Missile Seeker *wp*
AANCP Advanced Airborne National Command Post [USA] *mt*
AAO Airborne Area of Operation *mt*
AAO Anglo Australian Observatory *mi*
AAO Army Authorized Acquisition Objective *st, jj*
AAOC After Award of Contract *fu*
AAOC Anti Aircraft Operations Center *fu, hc*
AAODS Airborne Autonomous Orbit Determination System *tb*
AAOE Airborne Antarctic Ozone Experiment *hg, ge*
AAOFA Army Aviation Operating Facility [USA] *mt*
AAP Advance Acquisition Plan *tb*
AAP Advanced Automation Program *fu*
AAP Air Augmented Propulsion *wp*
AAP Allied Administrative Procedure *mt*
AAP Allied Administrative Publication *mt*
AAP Apollo Applications Program *pl, hc*
AAP Applications Access Point [DEC] *pk, kg*
AAP Army Ammunition Plant *-12*
AAP Associative Array PA&A
AAP Astronomy and Astrophysics *mi*
AAP/SHORAD Air Augmented Propulsion for Short Range Air Defense *hc*
AAR Aircraft Accident Report *wp, jj*
AAR Airport Acceptance Rate *fu*

AAR Automated Aircraft Reporting *ag*
AAR Automatic Alternate Routing *mt*
AARAD Attack Assessment Radar *wp*
AARB Advanced Air Aerial Refuelling Boom *wp, mt*
AARC Automatic Adaptive Radar Control *fu*
AARM Advanced Anti Radiation Missile *wp*
AARMS Authorized / Assigned Resource Management System [AFLC] *mt*
AARPS Air Augmented Rocket Propulsion System *hc, wp*
AARS Advanced Airborne Radar System *wp*
AARS Air Force Rescue and Recovery Service *wp*
AARS Ammunition Asset Reporting System *mt*
AARTS Automated Audio Remote Test System *mt*
AAS Advanced Antenna System *0*
AAS Advanced Automation System [FAA] *mt, hn42.4.8, aw129.21.102, is26.1.62, wp, tb, fu*
AAS Aeromedical Airlift Squadron [USA] *af72.5.121, mt*
AAS Air to Air Search *hc*
AAS All to All Scatter *pk, kg*
AAS American Astronautical Society *ci.1245, 0, aa26.11.b55, pbl, mb, mi, cu*
AAS Army Air Service [forerunner of USAF] *af72.5.45*
AAS Attack Assessment System *mt*
AAS Automatic Addressing System [of DOD] *hc*
AAS Automatic alignment Angle Sensor *hc*
AASA Advanced Airborne Surveillance Antenna *hc, mt*
AASC Army Area Signal Center *hc*
AASF Army Aviation Support Facility [USA] *mt*
AASI Advanced Atmospheric Sounder Imaging *hc*
AASIR Advanced Atmospheric Sounder and Imaging Radiometer *gp*
AASLBM Airborne Anti-SLBM *fu*
AASM Advanced Air to Surface Missile *wp, dh, nd74.445.39, mt*
AASMS Advanced Anti Ship Missile Seeker *hc*
AASO Army Aeronautical Service Office *dt*
AASP Army Automation Standardization Program *mt*
AASR Advanced Airborne Surveillance Radar [RADC] *mt, hc, wp, fu*
AASRS Advanced Airborne Signal [intelligence] Reconnaissance System *uc*
AASRT Alternate Air Support Radar Team *fu*
AASS Advanced Airborne Surveillance Sensor *wp*
AASW Airborne Anti Submarine Warfare *hc, jj*
AASWA Advanced Aircraft Anti Submarine Avionics *wp*
AAT Air to Air Track *hc*
AAT Army Assault Team *mt*
AAT Average Access Time *hi, kg*
AATC Automatic Air Traffic Control *wp*
AATH Automatic Approach To Hover *mt*
AATIDS Air to Air Target Identification System *hc*
AATMS Advanced Air Traffic Management System *wp*
AATPG Analog Automatic Test Program Generation *hc*
AATRI Army Air Traffic Regulation and Identification *hc*
AATS Advanced Automation Training System *fu*
AATS Aircraft Avionics Tradeoff Study *hc*
AATS Alternate Aircraft Takeoff System *wp*
AATSR Advanced Along Track Scanning Radiometer *hg, ge*
AATW Advanced Anti Tank Weapon *wp*
AAU Acoustic Add on Unit *wp*
AAV Advanced Aerospace Vehicle *tb*
AAV Airborne Assault Vehicle *wp*
AAV Amphibious Assault Vehicle [USMC] *mt, wp*
AAV Autonomous Air Vehicle *wp*
AAVA Audio Visual Activity, Army *dt*
AAVCS Automatic Aircraft Vectoring Control System *wp*
AAVRS All Altitude Vertical Reference System *wp*
AAVS Aerospace Audio Visual Service [MAC] *mt, hc, an31.7.9*
AAVSO American Association of Variable Star Observers *mi*
AAW Aeromedical Airlift Wing [MAC*] *mt, wp*
AAW Airfield Attack Weapon *wp*
AAW Anti Air Warfare [US Navy] *mt, wp, fu, uc, hc, ts10-3, jj, kf*
AAWC Anti Air Warfare Commander [US Navy] *mt, jj, fu*
AAWEX Air to Air Weapons Exercise *jj*
AAWS Advanced Antiarmor Weapon System *hc, wp*
AAWS Advanced Antitank Weapon System *hc*
AAWS Airborne Alert Weapon System [a battlefield missile system] *17, hc*
AAWS Anti Aircraft Warning System *wp*
AAWS Automated Attack Warning System *wp, mt*
AAWS-H Advanced Antiarmor Weapon System-Heavy *hc*
AAWS-M Advanced Antiarmor Weapon System-Medium *aw130.13.67, hc*
AB Able Bodied [seaman] *w9*
AB Address Bus *hi*
AB After Burner *wp, jj*
AB Agile Beam antenna *jdr*
AB Air Base *wp, fu, 12, af72.5.14, mt*
AB Air Blast *wp*

AB Airborne *wp, w9*
AB Airman Basic *w9*
AB As Built *oe*
AB Assault Breaker *hc*
AB Avionics Bay *wp*
AB/FS Air Base / Force Survivability *mt*
AB/TC Afterbody/Tailcone *fu*
ABA Aminobutyric Acid *wp*
ABACUS Air Battle Analysis Center Utility System *wp*
ABAD Air Base Air Defense *wp*
ABB As Built Baseline *tb*
ABC Advance Base Components *jj*
ABC Advanced Ballistics Concepts *wp*
ABC Air Borne Cigar [codename for equipment used in WWII to jam German communications] [UK] *mt*
ABC As Built Configuration *fu*
ABC Atanasoff Berry Computer [first digital calculating machine to use vacuum tubes] *kg, ct, em*
ABC Atomic, Biological, Chemical [chemical warfare] *wp*
ABC Automatic Brightness Control *jj*
ABC Automatic Buying Code *fu*
ABCA American, British, Canadian, and Australian armies *mt*
ABCBS Air Base Commander's Battle Staff *fu*
ABCC Airborne Command Center *hc, wp, mt*
ABCCC Airborne Battlefield Command and Control Center [USA DOD] *mt, wp, hc, af72.5.138, pf.f88.3, fu*
ABCD Advanced Beam weapon Concept Definition *tb*
ABCI Advanced Business Communications, Incorporated *sc3.5.4*
ABCMR Army Board for the Correction of Military Records *dt*
ABCP Acid Base Characteristics Phototropism *hc*
ABCS Airborne Beam Control System *17, hc*
ABDIS Area B Data Interchange System *fu*
ABDL Automatic Binary Data Link *wp*
ABDR Air Battle Damage Report *mt*
ABDR Aircraft Battle Damage Repair *wp, mt*
ABDS Accounting Budget and Distribution System [USAF] *wp, mt*
ABDW Air Base Defense Weapon *wp*
ABE Agent Building Environment *ct, em*
ABEL Air Breathing Electric Laser *wp*
ABEL Army Baseline Electric Laser *hc*
ABEND Abnormal Ending End *fu, kg*
ABERT Auto Bit Error Rate Test *wp*
ABF Adaptive Beam Forming *fu*
ABF Advanced Beam Former *fu*
ABFC ABEA Better Environment [DARPA] {?} *mt*
ABFC Advanced Base Functional Component system [US Navy] *mt, jj*
ABG Air Base Group [USA] *wp, af72.5.127, mt*
ABGD Air Base Ground Defense *wp, af72.5.10, mt*
ABI Application Binary Interface *pk, kg*
ABIC Army Battlefield Interface Concept *fu, mt*
ABIC Automated Battlefield Interface Concept *mt*
ABICS ADA Based Integrated Control System *wp*
ABIOL Advanced Base Initial Outfitting List [USA] *mt, jj*
ABIOS Advanced Basic Input / Output System *kg, ct, em, hi*
ABISL Advance Base Initial Support List *jj*
ABIST Automatic Built In Self Test [IBM] *pk, hi, kg*
ABIT Airborne Image Imagery Transmission *hc, wp*
ABKK Autonomous Onboard Computing Complex *ac3.1-43*
ABL Air Borne Laser *sd, wp, tb*
ABL Allegany Ballistics Laboratory *ci.150*
ABL Allocated Baseline [also AB/L *fu*
ABL As Built List *fu*
ABL Automated Biological Laboratory *hc*
ABLE Adaptive Battery Life Extender *pk, kg*
ABLE Advanced Building Block for Large Area Exploitation *wp*
ABLE Atmospheric Boundary Layer Experiment *hy, hg, ge*
ABLR Agile Beam Laser Radar *sd*
ABM Advance Bill of Material *jj, hc*
ABM Advanced Battery Management *hi*
ABM Air Battle Management *fu*
ABM Air Breathing Missile *wp*
ABM Anti Ballistic Missile *mt, fu, wp, sb, 11, 12, 13, 14, 17, ca, hc, ps202, nt, tb, st*
ABM Apogee Boost Motor *gd2-0, sf31.1.20*
ABM {Army Ballistic Missile} *ci.204*
ABMA Army Ballistic Missile Agency [later ASDC] *ci.44, 0, 13, hc, jj*
ABMC American Battle Monuments Commission *dt*
ABMD Advanced Ballistic Missile Defense *wp*
ABMD Air Battle Management Demonstration *hc*
ABMD Army Ballistic Missile Division *tb*
ABMDA Advanced Ballistic Missile Defense Agency [of US Army, later ASDC] *hc, dt*
ABMDA Antiballistic Missile Defense Agency *fu*
ABMIS Airborne Ballistic Missile Intercept System *wp*
ABML Aesop Basic Machine Language *hc*

ABMOC Air Battle Management Operations Center *wp, pm, fu*
ABMT Anti Ballistic Missile Treaty *tb*
ABMT Automated Bus Module Test *jdr, ric*
ABNCP Air Borne [National] Command Post [USA] *mt*
ABNFE Air Borne Force Element *mt*
ABNOC Airborne Nuclear Operations Center [SHAPE] *mt, wp*
ABO After Burn Out *fu*
ABO Air Base Operability *wp, mt*
ABO Air Borne Order *mt*
ABOC Air Base Operations Center *fu*
ABOM ADA Bit Oriented Message *mt*
ABP Auxiliary Beam Positioning *hc*
ABPS Air Breathing Propulsion System *wp*
ABR Available Bit Rate *ct, em, kg, hi*
ABRES Advanced Ballistic Re-Entry Systems Studies [later ASMS] *0, 17, sb, hc, wp, tb, st*
ABRES Advanced Ballistic [missile] Reentry System *ro, ai2*
ABRV Advanced Ballistic Reentry Vehicle *hc, wp*
ABS Agile Beam System *tb*
ABS Air Base Simulator *fu*
ABS Air Base Squadron [USA] *wp, mt*
ABS Air Base Survivability *wp, mt*
ABS Antenna Bridge Structure *sc4.9.2*
ABS Anti lock Brake System *hc*
ABSLOA Approved Basic Stock List Of Ammunition *mt*
ABSS Airborne Subsystem *hc*
ABT ADGE Bus Technology *hc*
ABT Air Breathing Threat *kf*
ABTI Agile Beam Tracker/Illuminator *hc*
ABTS ASCII Block Terminal Services *pk, kg*
ABU Adapter Building Units *hc*
ABW Air Base Wing [USA] *mt, wp, fu*
AC Access Communications *fu*
AC Accomplishment Criteria *kf*
AC Account Control *wp*
AC Acoustic Coupler *wp*
AC Action Code *fu*
AC Active Component *mt*
AC Adjacent Channel *fu*
AC Aerodynamic Center *fu, mt, ai1*
AC Air Conditioning *w9, wp, mt*
AC Air Crew *jj*
AC Aircraft *aw130.11.75, tb*
AC Aircraft Commander [USA] *mt, tb*
AC Aircraft Controller *fu*
AC Allied Command *wp, mt*
AC Alternating Current *oe, ct, em, wp, rm1.1.3, w9, tb, jj, kg, fu, mt*
AC Antenna Control *jj*
AC Antenna Coupler *fu*
AC Area Code *w9, hi*
AC Area Coverage *kf, mt*
AC Arithmetic Capability *fu*
AC Assistant Controller *fu*
AC Atlantic Council [NATO] *mt*
AC Attack Characterization *mt*
AC Attack Control *fu*
AC Automatic Computer *pk, kg*
AC [chemical warfare codename for hydrogen cyanide] *wp*
AC [USA DOD designator for Hercules (AC-130), Spectre (AC-130A/H)] *mt*
AC&W Aircraft Control and Warning *hc, jj, fu*
ACA Accuracy Control Analysis *hc*
ACA Advanced Combat Aircraft *wp*
ACA After Contract Award *fu*
ACA Agile Combat Aircraft *mt*
ACA Air Canada *wp*
ACA Air Clearance Authority *mt*
ACA Airlift Clearance Authority *wp, mt*
ACA Airspace Control Authority *kf, mt*
ACA Airspace Coordination Area *fu, mt*
ACA Alarm Correlation and Assessment *kf*
ACA All Composite Aircraft *wp*
ACA Approval of Contract Actions *fu*
ACA Associate Contractor Agreement *fu, kf*
ACA Asynchronous Communications Adapter *hi*
ACA Attitude Control Actuator *kf*
ACA Attitude Control Assembly *wl*
ACA Automatic Circuit Analysis *jj*
ACAB Air Cavalry Attack Brigade *fu, mt*
ACADA Automatic Chemical Agent Detector and Alarm [chemical warfare] *wp*
ACAF Advanced Counter Air Fighter *hc, wp, mt*
ACALS Airborne Command And Launch Subsystem *hc*
ACAMPS Automatic Communications And Message Processing System *wp*
ACAMS Automated Continuous Air Monitoring System *hc, wp*

ACAP Advanced Composite Aircraft Airframe Program [a helicopter design study] *mt, hc, wp*
ACAP Application Configuration Access Protocol *pk, kg*
ACAP Assembly Computer Aided Planning *fu*
ACAPS Advanced Composite Aircraft Primary Structure *aw129.21.40*
ACAPS Automated Chemical Analysis for Process Solutions *hc, aw129.10.12*
ACARS ARINC Communication, Addressing and Reporting System *pl*
ACAS Advanced Counter Air System *mt*
ACAS Airborne Collision Avoidance System *mt*
ACAS Aircraft Collision Avoidance System *wp*
ACAS Army Crisis Action System *mt*
ACAS Assistant Chief of Air Staff [UK] *mt*
ACAT Automated Computer Assisted Test *st*
ACATT Advanced Combat Arms Team Trainer *hc*
ACATT Aviation Combined Arms Team Trainer *hc*
ACAW Aircraft Control And Warning *wp*
ACB Advanced Concepts Base *mt*
ACB Air Circuit Breaker *jj*
ACB Air Crew Bulletin *wp*
ACB Amphibious Construction Battalion [USA] *mt*
ACBA Allied Command Baltic Approaches *fu*
ACBACCIS Allied Command Baltic Approaches Command Control and Information System [NATO] *mt*
ACBMDA Automated CCCCM Battle Management Decision Aid [RADC] *mt*
ACBP Advanced Concepts Base Program *mt*
ACBT Air Combat Training *mt*
ACC Access Control Center *mt*
ACC Acoustic Contact Correlation *fu*
ACC Advanced Communications Corporation *sc4.2.3*
ACC Air Combat Command [created in 1992 by merging TAC* and SAC*] [USA] *kf, mt*
ACC Air Component Commander *mt*
ACC Air Control Center *fu, wp, mt*
ACC Airlift Coordination Center *wp, mt*
ACC Airspace Control Center *mt*
ACC American Control Conference *aa26.8.b34*
ACC Area Control Center [VAFB] *fu, pl*
ACC Armament Control Console *jj*
ACC Armored Column Cover [continuous presence of tactical aircraft to support advancing armored forces] [Allied] *mt*
ACC Army Command Center *mt*
ACC Army Communications Command *dt, mt*
ACC Assistant Chief Controller *fu*
ACC Attack Control Console *fu*
ACC Automated Command Center *pm*
ACC {accelerometer} *bl1-5.127*
ACC-C Army Communications Command - CONUS *dt*
ACC/MASS Airlift Coordination Center / Military Airlift Support Squadron [AMC*] *mt*
ACCAT Advanced Command and Control Architectural Testbed *tb, fu, mt*
ACCB Aircraft Change Control Board *jj*
ACCC Antisatellite Command and Control Center *sd*
ACCC Area Control Computer Complex *fu*
ACCCCSL Advanced CCC Systems Laboratory *tb*
ACCDS Air Control Computer Display System *fu*
ACCE Aircraft Control Equipment *hc*
ACCES Automated Command and Control Evaluation System [WWM-CCS] *mt, wp*
ACCESS Automated Catalog of Computer Equipment and Software System [USA] *mt*
ACCESS Automated Command and Control Evaluation Systems [also ACCES] *mt*
ACCESS Automated Command and Control Executive Support System *wp*
ACCHAN Allied Command Channel [NATO Command] *12, mt*
ACCIS Automated Command and Control Information System [NATO] *wp, mt*

ACCLAIMS Army COMSEC Commodity Log and Accounting MIS *mt*
ACCM Advanced Charged Coupled Mosaics *hc*
ACCM Asbestos Containing Construction Material *hn49.12.1*
ACCMP Army Command and Control Master Plan *mt, fu*
ACCMPS Automated CC Message Processing System [NATO] *mt*
ACCNET Army Command and Control Network *mt*
ACCORDS Active Acoustic Correlation Detection System *hc*
ACCP Anti Char Courte PortÈe [French anti tank short range missile] *mt*
ACCP Army Correspondence Course Program *mt*
ACCP Automated Communications Control Processor *fu*
ACCS Adaptive Communications and Control Subsystem *hc*
ACCS Advanced Combat Control System *fu*
ACCS Air Command and Control System [NATO] *fu, aw130.22.61, wp, pm, mt*
ACCS Airborne Command and Control Squadron [USA] *pf.f88.1, mt*
ACCS Allied Command and Control System *hc, mt*

ACCS Army Command and Control System *fu, wp, hc, aw129.14.103, st, mt*
ACCSA Allied Communications and Computer Security Agency *mt*
ACCT Account Code [SARAH*] *mt*
ACCTP Attitude Control, Command, and Telemetry Program *-3*
ACD Administrative Commitment Document *wp*
ACD Aircraft Change Directive *jj*
ACD Airlift Communications Division *mt*
ACD Alert Condition Display *fu*
ACD Assistant Command Director *mt*
ACD Auxiliary Command Display *fu*
ACDA Arms Control and Disarmament Agency [US] *su, fu, wp, ci.876, 11, 13, hc, tb*
ACDAR Advanced Chemical Defense Aircrew Respirator [chemical warfare] *wp*
ACDB Army Corporate Data Base *mt*
ACDB Army Corps of Engineers {?} *mt*
ACDI Asynchronous Communications Device Interface *hi, kg*
ACDP Alternating Current Discharge Plasma *wp*
ACDS Advanced Combat Direction System [US Navy Program] *fu, hn44.12.1, hc, aa27.4.9, pf1.4*
ACDS Attitude Control and Determination System *oe*
ACDS Automatic Comprehensive Display System *hc*
ACE Acceptance Checkout Equipment *hc*
ACE Access Control Entry *pk, kg*
ACE Acquisition Cost Evaluation *wp*
ACE Advance Compound Engine *hc*
ACE Advanced Composition Explorer *mi*
ACE Advanced Computing Environment *pk, kg*
ACE Advanced CRT Editor *fu*
ACE Adverse Channel Enhancements *pk, kg*
ACE Aerospace Control Environment *wp*
ACE Agile Control Experiment *hc*
ACE Air Combat Engagement *wp*
ACE Airborne Command Element *mt*
ACE Aircraft Configuration Equipment *mt*
ACE Airspace Control Element *fu*
ACE Allied Civil Engineers *mt*
ACE Allied Command Europe [NATO Command] *fu, 12, 14, mt**
ACE Alternate Command Element *mt*
ACE Army Corps of Engineers [USA] *mt, wp*
ACE Array Computing Element *hc*
ACE ASIT Communication
ACEF ADA Compiler Evaluation Facility *mt*
ACEM Air Combat Evaluation Model *hc*
ACEORBAT ACE* Order of Battle [NATO] *mt*
ACEREP Allied Command Europe Reporting System [NATO] *mt*
ACES Action Cards for Engineers and Scientists *aa26.12.77*
ACES Adaptation Controlled Environment System *fu*
ACES ADP Command and Control Evaluation System *mt*
ACES Advanced Beam weapon Concept Evaluation Study *tb*
ACES Advanced Crew Escape System *wp*
ACES Aerial Combat Engagement Simulator *wp*
ACES Automated Command and Control Evaluation System [DSSO] *mt*
ACES Automated Cost Evaluation System *mt*
ACES Automatic Characterization and Extraction System *ne*
ACES Automatic Checkout and Evaluation System *wp*
ACES-I Advanced Control Evaluation for Structures [facility] *aa26.12.63*
ACESD Antenna Circular Equivalent Standoff Distance *jdr*
ACESS Automated Command and Control Executive Support System *wp*
ACESTRIKE Allied Command Europe Strike Command [NATO] *mt*
ACET Automatic Cancellation of Extended Targets *wp*
ACETEF Air Combat Environment Test and Evaluation Facility [USA] *mt*
ACEVAL Air Combat Evaluation [NATO] *mt, wp, jj*
ACF Access Control Facility *tb*
ACF Access Control Field *pk, kg*
ACF Advanced Communications Function *kg, ct, em*
ACF Advanced Computational Facility *tb*
ACF Air Combat Fighter *hc, mt*
ACF Alternate Command Facility *mt*
ACF Area Control Facility *fu*
ACF Asynchronous Communications Facility [Honeywell] *mt*
ACF Autocorrelation Function *fu*
ACF Automated Command Function *jn*
ACFPS Advanced Computer Flight Planning System [AFGWC] *mt*
ACG Acoustic Channel Gain *fu*
ACG Air Cargo Glider *wp*
ACGE Analog Command Generation Equipment *oe*
ACH Attempts per Circuit Hour *mt*
ACI Adjacent Channel Interference *jdr*
ACI Air Combat Intercept *mt*
ACI Airborne Controlled Intercept *wp*
ACI Alarm Control Interface [tech control] *mt*
ACI Allocated Configuration Identification *fu, hc, tb, mt*

ACI Altered Commercial Item *hc, jj*
ACI Analytic Condition Inspection *hc*
ACI Automated Carrier Interface [MTMC] *mt*
ACIA Asynchronous Communications Interface Adapter *hi*
ACIAS Automated Calibration Interval Analysis System *pk, kg*
ACIC ADCOM Current Intelligence Center *mt*
ACIC Aeronautical Chart and Information Center *hc*
ACIC Aeronautical Charting Intelligence Center *hc*
ACIMD Advanced Common Intercept Missile Demo *dh, hc*
ACINT Acoustic Intelligence *wp, mt*
ACIO Air Combat Intelligence Office *wp*
ACIS Advanced Cabin Interphone System *hd1.89.y27*
ACIS American Committee for Interoperable Systems *pk, kg*
ACIS Automated Claim Information System [USMC] *mt*
ACIS Avionics, Control and Information System *wp*
ACIT Air Combat Interference Techniques *hc*
ACIU Army Information Unit [U. S.] *dt*
ACL Access Control List *ct, wp, kg, em, mt*
ACL ADA Command Language *mt*
ACL Aircraft Circular Letter *jj*
ACL Allowable Cabin Load *fu, mt*
ACL Allowable Cargo Load *wp*
ACL Authorized Configuration List *hc, jj*
ACL Automatic Carrier Landing *jj*
ACL Automatic Clutter Mapping *wp*
ACLANT Allied Command, Atlantic [NATO Command] *fu, 12, mt*
ACLC Attack Center Launch Console *fu*
ACLDB Army Control Logistics Data Bank *mt*
ACLG Air Cushion Landing Gear *wp*
ACLICS Airborne Communications Location Identification and Collection System *wp*
ACLIS Airborne Communications Location Identification and Collection System [USA] *mt*
ACLM AUTODIN Communication Line Module *mt*
ACLS Air Cushion Landing System *fu*
ACLS Air Cushion Landing System [tested on a De Havilland Canada XC-8A Buffalo] *mt, fu*
ACLS Automatic All weather Carrier Landing System *fu, wp, jj, mt*
ACM Acoustic Counter Measures *fu*
ACM Active Counter Measures *fu, jj*
ACM Additional Crew Member *wp, mt*
ACM Advanced Cruise Missile [the AGM-129] *mt, wp, dh, aw129.10.4, af72.5.31*
ACM Air Chief Marshal [UK] *mt*
ACM Air Combat Maneuvering *fu, mt*
ACM Air Combat Model *hc, wp, jj, mt*
ACM Air Conditioning Module *fu*
ACM Anti Armor Cluster Munition *wp, uc, hc*
ACM Association for Computing Machinery *do46, 0, hc, is26.1.71, kg, fu, mt*
ACM Attitude Control Motor *st*
ACM Audio Compression Manager *pk, kg*
ACM Automated Communications Monitor *mt*
ACM Automatic Clutter Mapper *fu*
ACM Auxiliary Crew Member *mt*
ACMA Advanced Civil / Military Aircraft *wp*
ACMA Army Class Managing Activity *jj*
ACME Advanced Core Military Engine [Rolls-Royce] *mt*
ACME Air Crew Combat Mission Enhancement *wp*
ACMFS Automated Combat Mission Folder System *hc*
ACMI Air Combat Maneuvering Instrumentation [USA] *mt, wp, 12, hc*
ACMIS Automated Configuration Management/Integration System [WWMCCS] *mt*
ACMR Air Combat Maneuvering Range [USA] *mt, hc, jj*
ACMR/I Air Combat Maneuvering Range/Instrumentation *hc*
ACMS Air Combat Maneuver Simulator *wp*
ACMS Aircraft Condition Monitoring System *mt*
ACMS Application Control [and] Management System *pk, kg, mt*
ACMSC Airborne Communications Mode Selector Control *hc*
ACMT Advanced Cruise Missile Technology [DARPA] *wp, hc*
ACN Activity Control Number *jj*
ACN Advanced Change Notice *wp, fu, mt*
ACN Aircraft Classification Number *mt*
ACN Approval Control Number *fu*
ACNET Acquisition Computer Network, Federal [OMB] *mt*
ACNZ Airways Corporation of New Zealand *fu*
ACO Administrative Contracting Officer Office *su, fu, wp, 18, hc, nt, tb, st, jj, kf, mt*
ACO Airspace Control Order [USA DOD] *mt*
ACO Airspace Coordination Order *mt*
ACO Aspect Cutoff *jj*
ACO Asset Control Operator *fu*
ACOC Air Command Operations Center *fu, mt*
ACOC Airborne Combat Operations Center *mt*
ACOC Allied Combat Operations Center [NATO] *mt*
ACOC Alternate Alternative Command Operations Center [NATO] *hc, mt*

ACOC Area Communications Operations Centers [DISA] *mt*
ACOC Area Control Operations Center *mt*
ACOC AUTODIN Communication Operations Center *mt*
ACOM Asian Communist [operational division of NSA's] *-10*
ACOM Automated Centralized Operations and Maintenance centers *mt*
ACOMMW Aerospace Communications Squadron *mt*
ACORD Assembly Configuration, Operation and Routing Document *fu*
ACOSS Active Control of Space Structures *hc*
ACOUSID Acoustic Seismic Intrusion Detector [deployed from air] *wp*
ACP Adaptation Calculation Program *fu*
ACP Advance Change Proposal *fu*
ACP Advanced Communications Processor *fu, hc*
ACP Aerosol Collector and Pyrolyzer *es55.28*
ACP Air [ADCOM] Command Post *sd, hs*
ACP Airborne Command Post *tb, mt*
ACP Airlift Command Post *mt*
ACP Airspace Control Plan [USA DOD] *mt*
ACP Airspace Control Procedures *fu*
ACP Allied Communications Procedures *mt*
ACP Allied Communications Publication *fu, jj, mt*
ACP Alternate Command Post *wp, mt*
ACP Ancillary Control Program *pk, kg*
ACP Area Coordination Paper *wp*
ACP Arithmetic Control Processor *fu*
ACP Armament Control Panel *hc, jj*
ACP Army Capabilities Plan *mt*
ACP Army Cost Position *nt*
ACP Attitude Control Processor *gp, 8, ep*
ACP Automatic Communications Processor *mt*
ACP Auxiliary Control Process *pk, kg*
ACP Azimuth Change Pulse *fu*
ACPA Airborne Adaptive Controlled Phased Array [antenna system] *wp*
ACPB Advanced Concepts Base Program *mt*
ACPC Avionics Corrosion Prevention Control *jj*
ACPERS Army Civilian Personnel System *mt*
ACPI Advanced Configuration [and] Power Interface *ct, kg, em*
ACPL Atmospheric Cloud Physics Laboratory *ci.xxiii*
ACPS Advanced Collective Protection Shelter *wp*
ACPS Automated Contract Preparation System *mt*
ACPTT Air Combat Part Task Trainer *hc*
ACQ/TRK Acquisition/Tracking *kf*
ACQUOT Acquisition On Target *jj*
ACR Active Cavity Radiometer *hg, ge*
ACR Air Control Radar *wp, mt*
ACR Anti Circular Run *fu*
ACR Approach Control Radar *wp*
ACR Armored Cavalry Regiment [USA] *mt, fu, 12*
ACR Automatic Call Recording *hi*
ACRA Airlift Concepts and Requirements Agency *wp, mt*
ACRB Army Council of Review Boards *dt*
ACRC Area Calibration Repair Center *fu*
ACRD Assistant Chief of Staff [for] Research and Development [Army] *fu, hc*
ACRE Automatic Checkout of Readiness Equipment *hc*
ACRES Airborne Communications Relay Station *wp, mt*
ACRFAET Aircraft Crash Rescue Filed Assistance and Evaluation Team *wp*
ACRIM Active Cavity Radiometer Irradiance Monitor [EOS, UARS, JPL] *hy, ge, hg, oe*
ACROSS Automated Cargo Release and Operations Service System *pk, kg*
ACRV Assured Crew Return Rescue Vehicle *mi*
ACS Access Control Set *pk, kg*
ACS Access Control System *pk, kg*
ACS Acoustic Correlation System *fu*
ACS Acquisition and Command Support *wp*
ACS Active Control System *aa26.8.20, jdr*
ACS ADA Compilation System *mt, fu*
ACS Advanced Communications Satellite *sc2.8.4, hc*
ACS Advanced Communications Services *hi*
ACS Advanced Computer System *pk, kg*
ACS Afloat Correlation System *fu, mt*
ACS Air Commando Squadron [USA] *mt*
ACS Air Control System *mt*
ACS Airborne Control System *mt*
ACS Aircraft Control Surveillance *wp*
ACS ALADNS Control Station *fu*
ACS Altitude Control System *fu*
ACS Altitude Control System {misprint ? for Attitude Control System} *hc*
ACS American Chemical Society *wp, 0, w9*
ACS Antenna Coverage Subsystem *jdr*
ACS Armament Control System *hc, jj*
ACS Array Controller/Sequencer *fu*
ACS Artillery Computer System *wp*
ACS Asset Control System [USA] *mt*

ACS Assistant Chief of Staff *mt, wp*
ACS Asynchronous Communication Server *pk, hi, kg*
ACS Attitude Control [Sub]System *wp, id4142-10/831, 0, 4, 9, 16, 17, sb, lt, ce, re, tb, wl, kf, vv, jn, oe, he, uh, ai1, ai2*
ACS Automatic Class Selection *ct, em*
ACS Automatic Control System *hc*
ACS Auxiliary Crane Ship [US Navy] *mt*
ACS/I Assistant Chief of Staff for Intelligence *af72.5.107*
ACSA Allied Communications Security Agency *mt*
ACSA Army Combat Surveillance Agency, [U. S.] *dt*
ACSAS Advanced Conformal Submarine Acoustic Sensor *uc*
ACSASALCM Configuration Status and Accounting System {?} *mt*
ACSC Air Command and Staff College *af72.5.72*
ACSC Army Computer Systems Command, [U. S.] *dt*
ACSC Association Control Service Element *mt*
ACSD Automatic Color Scanned Device [Rome ADC] *wp*
ACSD Aydin Computer Systems Division *fu*
ACSFOD Assistant Chief of Staff, Force Development [Army] *fu*
ACSI Assistant Chief of Staff for Intelligence [Department of the Army] *hc, fu, tb, st, jj*
ACSIM Assistant Chief of Staff for Information Management [USA] *mt*
ACSL Advanced Continuous Simulation Language *wp, mt*
ACSL Assistant Chief of Staff, Logistics [Army] *fu*
ACSM Advanced Conventional Stand off Missile *mt, hc*
ACSN Advance Change [or] Study Notice *hc, fu, pl, wp, tb, mt*
ACSO Assistant Chief of Staff for Operations [US Army] *fu, hc*
ACSQ Airborne Communications Squadron *mt*
ACSS Acquisition Sun Sensor *cl, jdr, jn, uh*
ACSS Augmented Contact Support Set *hc*
ACSSS Allied Cooperative Support Sharing System [JCS] *mt*
ACSW Air Communications Service Wing *fu*
ACT Acoustic Charge Transport [DARPA] *wp, fu, mt*
ACT Activation and Characterization Test *oe*
ACT Active Cleaning Techniques *hc*
ACT Active Control Technology *mt*
ACT Advance Circuit Technology *fu*
ACT Advanced Capability Tanker *wp*
ACT Advanced Computer Technology *wp*
ACT Advanced Concepts Technology *hc*
ACT Aerial Combat Tactics *mt*
ACT Air Cargo Terminal *mt*
ACT Air Combat Tactics [USA] *mt*
ACT Air Correlation Tracker *wp*
ACT Aircraft Controller Technician *fu*
ACT Armored Cavalry Troop [USA] *mt*
ACT Automatic Capacitor Tester *hc*
ACTC Asynchronous Chirp Transform Correlator *fu*
ACTD Acoustic Charge Transport Device *wp*
ACTD Advanced Concept and Technology Demonstration *kf, mt*
ACTDS Automatically Cued Target Detecting System *wp*
ACTGEN Automatic Generation *hc*
ACTIS Advanced Circular Scan Thermal Imaging System *wp*
ACTIVE Advanced Control Technology for Integrated Vehicles [application of fully thrust-vectoring nozzles to the F-15 SMTD] [USA] *mt*
ACTS Activated Carbon Wastewater Treatment System *ci.1021*
ACTS Adaptive Computerized Training System *hc*
ACTS Advanced CNI Technology Study *hc*
ACTS Advanced Communications Technology Satellite [NASA] *wp, sc3.2.7, 19, mt, ls4, 5, 12*
ACTS AEGIS Combat Training System *hc*
ACTS Armament Control Test Set *jj*
ACTS Automated Circuit Test System *mt*
ACTS Automated Computer Time Service *wp, kg*
ACTT Advanced Communication and Timekeeping Technology *pk, kg*
ACU Access Control Unit *mt*
ACU Air Control Unit *fu*
ACU Antenna Control Unit *jdr*
ACU Antenna Coupler Unit *fu*
ACU Arithmetic [and] Control Unit *fu, wp*
ACU Armament Control Unit *hc, jj*
ACU Assault Craft Unit [US Navy] *mt*
ACU Automatic Call Calling Unit *fu, kg*
ACU Avionics Control Unit *mt*
ACUREX (ACUREX Corporation . . . Huntsville, Alabama 35806) *jmj*
ACV Air Cushion Vehicle [USA DOD] *mt, hc, w9, aa27.4.b12, tb, jj*
ACV Armored Command Vehicle *mt*
ACV Auxiliary Carrier Vehicle [aircraft carriers later renamed CVE] [USA] *mt*
ACVD Aircraft Chart Viewing Device *wp*
ACW Air Control Wing [USA] *mt*
ACW Aircraft Control and Warning *mt, wp, jj*
ACW Anti Carrier Warfare [US Navy] *fu, mt*
ACWAR Advanced Continuous Wave Acquisition Radar *hc*
ACWP Actual Cost of Work Performed *mt, wp, fu, 8, 17, sb, tb, jdr, oe*
ACWS Actual Cost of Work Scheduled *wp, oe*
ACWS Aircraft Control and Warning Squadron *mt, 12*

AD Active Duty [USA DOD] *w9, mt*
AD Advanced Deployability posture *mt*
AD Advanced Development *hc, mt*
AD Air Defense *fu, wp, ic2-6, pm, tb, st, mt*
AD Air Directive *mt*
AD Air Division [USAF] *wp, mt*
AD Air Drop *mt*
AD Airworthiness Directive *wp, aw129.10.20, mt*
AD Analog [to] Digital *hc, wp*
AD Ancillary Document *fu*
AD Armament Division [AFSC*] *mt*
AD Army Dispensary [U. S.] *dt*
AD Artillery Detector *fu*
AD Authorized Delay *wp*
AD Destroyer Tender [US Navy] *mt*
AD&D Accidental Death and Dismemberment [insurance] *hn43.19.1*
AD/CAC Automated Detection/Computer Assisted Classification *hc*
AD/EXJAM Artillery-Delivered Expendable Jammer *mt*
AD/T Air Detector Tracker *jj*
ADA Aerial Delivery Adapter *mt*
ADA Air Defense Aircraft *mt*
ADA Air Defense Area *fu, mt*
ADA Air Defense Artillery *wp, fu, 12, hc, pm, jj, mt*
ADA Airborne Data Automation *mt*
ADA AJPO Director's Advisory *fu*
ADA Analog Digital Analog *jj*
ADA Angular Differentiating Accelerometer *hc*
ADA Automatic Data Acquisition *wp, kg*
ADA [military standard programming language, named for Augusta Ada Byron, "the world's first computer programmer"] *ro, 17, aw130.11.56, wp, wl, kg, fu, hi, mt*
ADABECS A/B Demonstrator Advance Beam Extraction and Control System *hc*
ADAC Acoustic Data Analysis Center [USA] *mt*
ADAC Attitude Determination And Control *gp*
ADACP Air Defense Artillery Command Post *mt*
ADACS Attitude Determination And Control Subsystem *16, dm, oe*
ADACS Automated Data Acquisition and Control System *wp*
ADAD After Date of Award Document *fu, tb*
ADADS Army Depot Automatic Diagnostic System *hc*
ADAF Air Defense Alert Facility *mt*
ADALT Advanced Radar Altimeter *hg, ge*
ADAM Administrative, Documentation, and Management System *mt*
ADAM Advanced Aircrew Mask [chemical warfare] *wp*
ADAM Advanced Digital Avionics Map *hc, wp*
ADAM Advanced Dynamic Anthropomorphic Mannequin *wp*
ADAM Aerial Port Documentation and Management System [MAC*] (ADAM I [OVERSEAS], ADAM II [CONUS], ADAM III [CAPS]) *mt*
ADAM Air Defense Analysis Model *ts14-11*
ADAM Air Deflection And Modulation [concept for STOL] *wp*
ADAM Air Delivered Attack Marker *wp*
ADAM Air Force Dispersion Assessment Model *wp*
ADAM Airbase Damage Assessment Model *wp*
ADAM Architecture Development Analysis Model *fu*
ADAM Area Denial Artillery Munition [scatterable mine] *wp, dh*
ADAM Artillery Delivered Antipersonnel Mine *wp*
ADAM Automated Design and Maintenance System *mt*
ADAM Automated Design and Manufacturing *fu*
ADAM Automated Drafting and Manufacturing System *hn44.13.1, hc*
ADAMS Airborne Data Analysis and Monitoring System *wp*
ADANS Airlift Deployment Analysis System [AMC*] *mt*
ADAO Air Defense Artillery Officer *fu*
ADAP Airborne Development Aid Program [FAA] *wp*
ADAPP Active Damping and Precision Pointing *hc*
ADAPS Advanced Data Processing System *fu*
ADAPT Adaptive Digital AMTI Processing Technique *hc*
ADAPT Analog-Digital Analysis Processing Techniques *hc*
ADAPT ARPA Database Access and Presentation Terminal *mt*
ADAR Advanced Design Array Radar [for ICBM identification] *wp, hc, fu, ro*
ADARS Adaptive Antenna Receive System *ro*
ADARS Army Defense Acquisition Regulation Supplement *st*
ADARTS ADA based Design Approach for Real Time Systems *kf*
ADARTS ADA for Real Time Systems *kf*
ADAS Airborne Damage Assessment System *wp*
ADAS Airborne Data Acquisition System *hc, wp, mt*
ADAS Airborne Dynamic Alignment System *hc*
ADAS Airfield Damage Assessment System *mt*
ADAS Architecture Design and Assessment System *wp*
ADAS Army Digital Avionics System *hc*
ADAS Automatic Data Acquisition System *fu*
ADASCC ADA Support and Control Capability [NATO] *mt*
ADAT Air Defense Anti-Tank *fu*
ADATA Army Drug and Alcohol Technical Activity, [U. S.] *dt*
ADATP Allied Data Publication [NATO] *mt*
ADATS Air Defense Anti Tank System *hc, aw130.22.19, wp*

ADATS Air Defense Antitank System *mt*
ADATS Airborne Digital Avionics Test System [USAF] *wp, mt*
ADA_FDC Air Defense Artillery Fire Control Direction Center *mt*
ADA_JUG ADA Joint Users Group *mt*
ADB Air Defense Board *fu, hc*
ADB Asset Database *fu*
ADBFLIR Advanced Dual Band Forward Looking Infrared *hc*
ADBMT Air Defense Battle Management Technology [ADI] *mt*
ADC Acoustic Display Converter Console *fu*
ADC Actual Date of Contract *fu*
ADC Actual Direct Costs *-4*
ADC Adaptive Data Compression [protocol] [Hayes] *pk, kg*
ADC Add with Carry *pk, kg*
ADC Advanced Data Connector *em*
ADC Aerospace Air Defense Command *w9, hc, st, jj*
ADC Aerospace Defense Command [USA] *fu, mt*
ADC Affiliated Data Center *hg, ge*
ADC Aide de Camp *w9, wp*
ADC Air Data Computer *mt, fu, jj*
ADC Air Defense Command Commander *fu*
ADC Air Defense Console *fu*
ADC Air Development Center *wp*
ADC Airborne Digital Computer *hc*
ADC Analog to Digital Coding *ce*
ADC Analog to Digital Converter *mt, wp, gp, 6, 8, 17, hc, pl, st, jj, jdr, kf, kg, oe, fu, em, ct, hi, ai1*
ADC Architecture Design Contractor *mt*
ADC Assistant Division Commander *w9*
ADC Assistant Duty Controller *fu*
ADC Astronomical Data Center *ns*
ADC Attitude Director Converter *jj*
ADC AUTODIN Display Console *mt*
ADC Automated Data Collection *wp, mt*
ADCAP Advanced Capability [a torpedo program] *fu, hn42.4.8, hc, md2.1.1, rm8.1.1*
ADCC Air Defense Command and Control *fu, wp*
ADCC Air Defense Control Center *fu, hc, mt*
ADCC Air Defense Coordination Center *mt*
ADCC Airborne Defense Control Center *hc, tb*
ADCCP Advanced Data Communications Control Procedure *ct, em, do369, sd, jb, ce, tb, kf, kg, fu, hi*
ADCCP Advanced Data Communications Control Procedures *mt*
ADCCS Air Defense Command and Control System [Missile Command] *hc, fu, mt*
ADCOLE [a digital sun sensor with wide lateral field of view] *sa122052-115*
ADCOM Administrative Command *mt*
ADCOM Aerospace Defense Command *fu*
ADCOM Air Defense Command *fu, wp, sr9-82, hc, tb. st*
ADCON Administrative Control Code *mt*
ADCON Advance Concepts *hc*
ADCON Analog to Digital Converter *hc, jj*
ADCOS Air Defense Combat Operations Staff [ACC] *mt*
ADCP Acquisition and Distribution of Commercial Products *hc*
ADCP Advance [Flight] Control Programmer *hc*
ADCP Air Defense Command Post *hc, fu, jj*
ADCS ACCC Data Control Subnet *fu*
ADCS Activity Data Control System *wp*
ADCS Air Data Computing Computer [Set/]System *hc, jj*
ADCS Air Deployment and Control Squadron [USA] *mt*
ADCS Attitude Determination and Control Subsystem *id4142-10/831, 8, ep*
ADCSP Advance Defense Communication Satellite Project *hc*
ADCSS Advanced Defense Communications Satellite System *hc*
ADD Acoustic Discrimination of Decoys *hc*
ADD Aerospace Deployment Device *hc*
ADD Airstream Direction Detector [to measure the AoA] *mt, jj*
ADD ["long range air arm", replaced the DBA-GK in 1942] [USSR] *mt*
ADDA Analog to Digital, Digital to Analog *hc*
ADDC Air Defense Direction Center *hc, fu*
ADDCM Adaptive Differential Pulse Code Modulation [also ADPCM] *wp*
ADDCS Aircraft Decontamination/Deicing/Cleaning System [chemical warfare] *wp*
ADDD Australian Defense Data Dictionary *hc*
ADDDS Automatic Direct Distance Dialing System *hi*
ADDER Algorithm, Description, Development, Evaluation and Refinement *mt*
ADDISS Advanced Deployable Digital Imagery Support System [PACAF] *wp, hc, mt*
ADDJUST Automatic Determination and Dissemination of Just Updated Steering Terms *oe*
ADDO Assistant Deputy Director for Operations *mt*
ADDR Address Register *hi*
ADDS Acquisition, Discrimination and Designation System *fu*
ADDS Advanced Data Delivery System *fu, hc*
ADDS Air Defense Demonstration System *hc*

ADDS Airborne Detection and Discrimination Sensor *hc*
ADDS Alternate Attitude {? possibly AADS} Determination System *hc*
ADDS Army Data Distribution System [later called PJH] *fu, wp, mt*
ADDS Atmospheric Distributed Data System *ag*
ADDS Attitude Determination/Despin Subassembly *jdr*
ADDS Automatic Data Distribution System [Navy, part of NTDSA] *fu*
ADE Ada Development Environment *fu*
ADE Air Defense Emergency *fu, wp, mt*
ADE Authorized Data Element *mt*
ADE Autotrack Detector Equipment *-4*
ADEA Army Development and Employment Agency *mt*
ADEA Army Development and Engineering Agency *fu*
ADEC Airborne Defensive Electronic Countermeasures *wp*
ADEEM Air Defense Effectiveness Evaluation Model *fu*
ADEOL Air Defense Electro Optical Laboratory *hc*
ADEOS Advanced Earth Observing Satellite [Japan] *hy, ge, hg*
ADEOS Air Defense Electro Optical Support *hc*
ADEPT Analysis based Data Exchange for Publications and Training *mt*
ADES Advanced Diagnostic Executive System *fu*
ADES Automatic Digital Encoding System *hi*
ADESS Analog Data Equipment Switching System *hc*
ADEW Airborne Directed Energy Weapons *wp, mt*
ADEWS Air Defense Electronic Warfare System *hc, fu, mt*
ADF Adaptive Doppler Filtering *fu*
ADF Advanced Direction Finder *wp*
ADF Aeronutronics Division, Ford [Motor Company] *0*
ADF Aerospace Data Facility *sc4.9.center*
ADF Air Defense Fighter *wp, af72.5.32*
ADF Application Development Facility *fu*
ADF Astrophysics Data Facility [GSFC, Code 631] *cu*
ADF Automated Automatic Design Facility *fu, hc, jj, hc, w9, wp, jj*
ADF Automatic Direction Finding Finder *fu, mt*
ADF Automatic Document Feeder *pk, kg*
ADF Automatic Drafting Facility *fu*
ADF Automatically Defined Function *pk, kg*
ADFB Adaptive Doppler Filter Bank *fu*
ADFDS Air Defense Fire Distribution System *fu*
ADFG Automatic Diode Function Generator *hc*
ADFLIR Advanced Forward Looking Infra Red [also AFLIR] *hc, jj*
ADFRF Ames-Dryden Flight Research Facility [now DFRC, NASA] *mi*
ADG Acoustic Display Generator *fu*
ADG Adaptation Data Group *fu*
ADG Aeronautical Development Group *jj*
ADG Air Defense Group [Canada] *mt, wp*
ADG Application Design Guide *fu*
ADGA (ADGA Systems International Limited of Toronto, Ontario) *fu, hq6.2.7*
ADGB Air Defense of Great Britain [UK] *mt*
ADGE Air Defense Ground Environment Equipment *fu, hn47.7.8, 11, hc, jj, mt*
ADH Advanced Development Hardware *tb, wl*
ADHEL Advanced High Energy Laser *hc*
ADHI Area Defense Homing Interceptor *hc*
ADHP Advanced Development Hardware Phase *hc*
ADI Advanced Development Implementation *hc*
ADI Air Defense Initiative [USAF] [USA] *fu, wp, ro, aw126.20.126, ic2-19, hc, tb, mt*
ADI Air Defense Integration *fu*
ADI Air Defense Intercept *wp*
ADI Allied Defense Initiative *st*
ADI Allowable Daily Intake *wp*
ADI Alternating Direction Implicit *ai1*
ADI Analog Digital Interface *mt*
ADI Analog Display Indicator *jj*
ADI Application to Database Interface *fu*
ADI Architecture Definition Interface *fu*
ADI ASPRO/DB Interface *fu*
ADI Attitude Directional Indicator *wp, hc, mt*
ADI AutoCad AutoDesk Device Interface [driver] *pk, kg*
ADI AUTODIN DNN Interface *fu*
ADI/SIS Air Defense Initiative/System Integration Study *fu*
ADIC ADCOM Intelligence Center *sr9-82, hc, mt*
ADIC Aerospace Defense Command Intelligence Center *mt*
ADIC Air Aerospace Defense Intelligence Center *hs, tb*
ADIC Automatic Data Interface Console *mt*
ADICT Automatic Digital Circuit Tester *fu*
ADIRS Aid Data and Inertial Reference System *mt*
ADIS Acquisition and Due In System *mt*
ADIS Air Defense Integrated System *wp*
ADIS Airborne Digital Instrumentation System *wp, hc*
ADIS Automated Data Interchange System *fu, wp, mt*
ADISCC Air Defense Initiative Simulation for Command and Control *ts.6*
ADISS Advanced Deployable Digital Imagery Support System *hc*
ADIT Analysis and Design Interface Transform *fu*
ADIT Automatic Digital Test *wp*

ADIU Aircraft Data Interface Unit *fu*
ADIZ Air Defense Identification Zone *wp, hc, w9, fu, jj, mt*
ADL ADA Based Design Language *tb*
ADL ADA Design Language *mt*
ADL Address Data Latch *pk, kg*
ADL Addressable Latch *fu*
ADL Airborne Data Link *wp*
ADL Aircraft Data Link *hc*
ADL Armament Datum Line *jj*
ADL Armament Development Laboratory [Eglin AFB] *wp*
ADL Asynchronous Data Link Control *pk, kg*
ADL Authorized Data Link *hc*
ADL Authorized Data List *fu, wp, mt*
ADL Automatic Data Link [USAF] *mt, fu, hc, jj*
ADL Avionics Development Laboratory [NASA] *wp*
ADLA Assistant Director for Legal and Legislative Affairs [NSA] *-10*
ADLAT Adaptive Lattice Filter *pk, kg*
ADLC Adaptive Lossless Data Compression [IBM] *pk, kg*
ADLC Asynchronous Data Link Control *hi*
ADLI Advanced Deck Launched Interceptor *hc*
ADLIPS Advanced Data Link Processing System *fu*
ADLK Aircraft Data Link Kit *hc*
ADLO Air Defense Liaison Officer *wp, fu, mt*
ADLS Automated Decoy Launch System *fu*
ADM (ADM-20 Quail) *mt*
ADM Acquisition Decision Memorandum *wp, nt, jmj, kf*
ADM Adaptive Delta Modulation *ce*
ADM Add / Drop Multiplexers *ct, em*
ADM Advanced Development Model *15, hc, jj, fu, mt*
ADM Air Decoy Missile *wp*
ADM Air Launched Decoy Missile [USA] *mt*
ADM Antenna Deployment Actuator *he*
ADM Antenna Deployment Mechanism *pl, uh*
ADM Applications Development Machine *fu*
ADM Archive Data Management *oe*
ADM Area Denial Munition *wp*
ADM Atomic Demolition Munition *uc, 12*
ADM Attitude Data Multplexer *oe*
ADM Automatic Drafting Machine *jj, hc*
ADM [a Xenon ion thruster manufactured by Hughes] *aa26.12.61*
ADMA Acoustic Direct Memory Access *fu*
ADMAC Air Defense Monitor and Control *hc, fu*
ADMAPS Automated Document Management and Publishing System *mt*
ADMATT Advanced Mobile Acoustic Torpedo Target *fu*
ADMD Administrative Management Domain [X.400] *pk, kg*
ADMIS Administrator Management Information System *mt*
ADMISREP Air Defense Mission Report *fu*
ADMP Air Defense Master Plan [NORAD] *mt*
ADMRL Application Data Material Readiness List *hc, jj*
ADMS Advanced Developmental Model Subsystem *hc*
ADMS Automatic Data Monitoring System *hc, jj*
ADMS Automatic Digital Message Switch *mt*
ADMS AWIS Data Management System [USA] *mt*
ADMU Attitude Data Multiplexer Unit *oe*
ADN Air Defense Notification *fu*
ADNAC Air Defense of the North American Continent *wp*
ADNC Air Defense Notification Center *fu*
ADO Active Data Objects *kg, em*
ADO ActiveX Data Objects *em*
ADO Advanced Development Objective *wp, hc, fu*
ADO Air Defense Operations *wp*
ADO Airline Dispatch Office *fu*
ADOA Air Defense Operations Area *mt*
ADOC Air Defense Operations Center [SPACECMD] *mt*
ADOC Area Aerospace Defense Operations Center *wp, fu, sd, hc, hs, tb*
ADOCC Alternative Defense Operations Control Center *tb*
ADOCS Advanced Digital Optical Control System *aw118.18.74*
ADOP Advanced Distributed Onboard Processor *wp, hc, tb*
ADOUR [Rolls-Royce turbofan engine] *wp*
ADP Acceptance Data Package *4, oe*
ADP Adaptive Digital Processor *fu*
ADP Adenosine Diphosphate *wp*
ADP Advanced Development Program *hc, aw129.14.27, tb, wl, jj*
ADP Advanced Digital Processor *tb*
ADP Advanced Ducted Propeller [P&W] *mt*
ADP Allied Defense Publication [NATO] *mt*
ADP Array Data Processor *fu*
ADP Attitude Data Processor *gp, 9, 8, lt*
ADP Automatic Automated Data Processing *mt, ct, em, wp, ph, hc, pl, w9, af72.5.124, nt, tb, st, jj, jdr, kg, oe, hi, fu*
ADP Automatic Display Processor *mt*
ADP Autopilot Data Processor *fu*
ADP {Air Defense Plan} (NCMC ADP System Architecture) *sd*
ADP/CE Automated Data Processing and Communications Equipment *fu*
ADPA American Defense Preparedness Association *wp, fu, hn43.17.8, aw130.11.78, aa26.10.b3, nd74.445.3*

ADPC Automatic Data Processing Center *wp*
ADPCM Adaptive Differential Pulse Code Modulation [also ADDCM] *hc, ct, em, jgo, ce, kg, hi*
ADPCS Advanced Data Processing Control Subsystem *hc*
ADPE Automated Automatic Data Processing Equipment *mt, wp, hn44.20.3, 2, 4, fu, hc, nt, tb, jj, kf, hi*
ADPEP Automated Data Preparation Evaluation Program *fu*
ADPESO [Hughes Aircraft Company customer for RIXT remote information exchange terminal] *hc*
ADPF Automatic Data Processing Facility *mt*
ADPG Air Defense Planning Group *hc*
ADPL Assistant Director for Policy and Liaison [NSA] *-10*
ADPLO Automatic Data Processing Liaison Officer *mt*
ADPM Automatic Data Processing Machine *jj*
ADPMIS Automatic Data Processing Management Information System *mt*
ADPMP Automated Data Processing Master Plan *mt*
ADPO Advanced Development Program Office *wp*
ADPP Automatic Data Processing Plan *hc*
ADPR Assistant Director for Plans and Resources [NSA] *-10*
ADPR Automatic Data Processing Resources *mt*
ADPREP Army Automated Data Processing Resource Estimating Procedures *mt*
ADPREPS Automated Documentation and Preparation System *fu*
ADPS ASARS Deployable Processing System *wp*
ADPS Automated Automatic Data Processing System [USAF] *mt, 0, hc, fu, tb, jj*
ADPSEC Automated Data Processing Security *mt*
ADPSI Advanced Data Processing System Investigation *st*
ADPSO Automatic Data Processing Selection Office *hc, dt, mt*
ADR Acceptance Data Review *jdr*
ADR Accident Data Recorder *mt*
ADR Achievable Data Rate *2, 4*
ADR Air Defense Region [UK] *mt*
ADR Airbase Damage Repair *mt*
ADR Airborne Digital Recorder *hc, wp*
ADR Aircraft Discrepancy Report *jj*
ADR Aircraft/Aeronautical Directive Requirements *jj*
ADR Altitude Difference Ranging *jj*
ADR Ammunition Disposition Report *jj*
ADR Automatic Data Relay *mt*
ADR Auxiliary Data Record *ac2.7-14*
ADRES Advanced Reconnaissance Sensor System *hc*
ADRES Army Data Retrieval Engineering System *hc*
ADRS ADCAP Data Reduction System *fu*
ADRSS Automated Data Reports Submission System [USAF] *mt*
ADRTS Applied Dynamics Real Time System *kf*
ADRV Auxiliary Display - Remote View *hc, jj*
ADS Advanced Deployable System *mt*
ADS AEGIS Display System *fu*
ADS Aerial Delivery System *mt*
ADS Aerospace Data Systems *gp, 2, 4, oe*
ADS Aerospace Defense Squadron *-12*
ADS Air Data System *mt*
ADS Air Defense Sector [USA] *wp, mt*
ADS Air Defense Suppression *mt*
ADS Air Defense System *fu*
ADS Air Deployable Deception Device *wp*
ADS Airborne Data System *hc*
ADS Airborne Decommutation System *hc*
ADS Angular Displacement Sensor *oe*
ADS Antenna Design Freeze *jn*
ADS Application Development Solutions [AT&T] *pk, kg*
ADS Application Development System *pk, kg*
ADS Area Detection System [Cw; By Lidar] *wp*
ADS Astrophysics Data System *cu*
ADS Audio Distribution System *hn44.11.3, hc*
ADS Automated Automatic Data System *fu, mt*
ADS Automatic Dependent Surveillance *aw132.10.9*
ADS Automatic Distribution System *pk, kg*
ADS Automation of Design for Supportability *mt*
ADSACS Advanced Strategic Aerospace Crew Systems *wp*
ADSAMS Advanced Surface to Air Missile System *mt*
ADSC Active Duty Service Commitment *an31.13.4*
ADSC Air Defense Systems Command *hc, jj*
ADSC Army Data Services and Administrative Systems Command, U. S. *dt*
ADSC Automated Data Source Collection *mt*
ADSCEPS Advanced Stored Chemical Energy Propulsion System *wp*
ADSCS Advanced Defense Satellite Communications System *hc*
ADSG Airborne Digital Symbology Generator *hc*
ADSHIPDA Advise Shipping Date *jj*
ADSI Active Directory Services Interface *ct, em*
ADSI Air Defense System Integration *fu*
ADSIA Allied Data Systems Interoperability Agency [NATO] *mt*
ADSID Air Defense Systems Integration Division [USAF *fu, hc*
ADSID Air Delivered Seismic Intrusion Device *wp*

ADSIM Advanced Simulation Missions *hc*
ADSIM Applied Dynamics Simulation Language *jdr*
ADSL Asymmetric Digital Subscriber Line *du, rih*
ADSL Asymmetrical Digital Subscriber Loop Line [analog voice + data] *nu, kg, ct*
ADSL Asynchronous Digital Subscriber Line *em*
ADSLT Asymmetric Digital Subscriber Line Technology *hi*
ADSM Air Defense Suppression Missile [an air launched, anti radiation development of Stinger] [USA] *mt, wp*
ADSM Automated Data Systems Management *fu*
ADSMO Air Defense Systems Management Office [USAF *fu, hc*
ADSP Advanced Digital Signal Processor *tb, st*
ADSR Attack, Decay, Sustain, Release *pk, kg*
ADSS Aerospace Data Systems Standards *2, 4*
ADSS Air Defense Suppression System *wp*
ADSS Army Decision Support System *mt*
ADSTAR Automated Document Storage and Retrieval *fu*
ADSU Air Data Sensor Unit *mt*
ADSW Anti Diesel Submarine Warfare *mt*
ADT Airborne Data Terminal *fu, wp*
ADT Application Data Types *pk, kg*
ADT Application Design Tool *fu*
ADT ASAT Data Terminal *jb*
ADT Assistant Director for Training [NSA] *-10*
ADT Automatic Detect and Track Detection and Tracking *fu, wp*
ADTAC Air Defense Tactical Air Command [later ADTAC was renamed "1st Air Force"] [USA] *mt*
ADTAM Air Delivered Target Activated Munition *wp*
ADTC Armament Development and Test Center [Eglin AFB] [USAF] *mt, hc, wp, fu*
ADTOC Air Defense Tactical Operations Center *kf*
ADTR Acquisition Phase Demonstration and Test Requirements *fu*
ADTRAC Air Defense Tactical Replacement Aircraft *hc*
ADTS Anomaly Data Tracking System *jdr*
ADTS Automated Data and Telecommunications Service [GSA] *mt*
ADTS Automated Data Test Station *oe*
ADTS Avionics Depot Test Station *hc, wp*
ADTTT Air Defense Tactical Training Theater *mt*
ADTU Automatic Detection and Tracking Unit *fu*
ADTWCN Air Defense Tactical Weather Communication Network *fu*
ADU Accumulation and Distribution Unit *mt*
ADU Air Data Unit *mt*
ADU Airborne Digitizing Unit *ws4.10*
ADU Automatic Dialing Unit *pk, kg*
ADU Auxiliary Display Unit *mt*
ADV Advanced Development Vehicle *hc, jj*
ADV Air Defense Variant [the interceptor version of the Panavia Tornado] [UK] *af72.5.28, mt*
ADVA Advanced Soviet [part of NSA's Production Organization A Group] *-10*
ADVANCE Airborne Doppler Velocity Altitude Navigation Compass Equipment *wp*
ADVANCE Army Data Validation And Netting Capability Establishment *mt*
ADVCAP Advanced Capability *mt*
ADVON Advanced Cadre *mt*
ADVON Advanced Echelon *mt*
ADW Air Defense Warning *mt, wp, tb, fu*
ADW Area Denial Weapon *mt*
ADWAO Air Defense Weapons Assignment Officer *fu*
ADWAR Advanced Directional Warhead *wp*
ADWC Air Defense Weapons Center *jj*
ADWC Air Division Warning Center *mt*
ADWS Air Defense Weapon System *wp*
ADWS Automated Digital Weather Switch *mt, wp, ag, hc*
ADX Automatic Data Exchange Extract *fu, hi*
ADZ Air Defense Zone *wp*
AE Above or Equal *pk, kg*
AE Action Entry *fu*
AE Active Emission *fu*
AE Aeromedical Evacuation *mt, wp*
AE American Ephemeris and Nautical Almanac *bm367*
AE Ammunition ship [US Navy] *jj, mt*
AE Analog Electronics *hg, ge*
AE Anti Earth *oe*
AE Arithmetic Element *fu*
AE Assault Echelon *mt*
AE Associate Engineer *jj*
AE Atmosphere Atmospheric Explorer *ge, hc, hg*
AE Autumnal Equinox *kf, oe*
AE Auxiliary Equipment *tb*
AE Aviation Electrician's mate *jj*
AEA American Electronics Association *is26.1.25*
AEA Association of European Airlines *aw129.21.3*
AEAF Allied Expeditionary Air Force [the air support force for 'Overlord'] [Allied] *mt*
AEAO Airborne Emergency Action Officer *mt*

AEARC Army Equipment Authorization Review Center [U. S.] *dt*
AEAS ASW Anti Submarine Warfare Environmental Aquatic Support [US Navy] *mt*
AEB Aerial Exploitation Battalion [USA] *mt*
AEB Agencia Espacial Brasileira [Brazilian Space Agency] *ai2*
AEB Aircrew Evaluation Board *mt, wp*
AEB Area Exploration Battalion [USA] *mt*
AEBLE Advanced Electron Beam Lithography Equipment *fu*
AEC Air Event Conference *mt*
AEC Arithmetic Element Controller *fu*
AEC Atomic Energy Commission [later DOE] *mt, fu, su, wp, ac1.7-14, 13, hc, w9, jj*
AEC Automatic Engine Control *mt*
AECA Arms Export Control Act *wp, fu*
AECC Aeromedical Evacuation Control Center *mt*
AECE Aeromedical Evacuation Control Element *mt*
AECIP Aircraft Engine Component Improvement Program *wp*
AECL Aircraft and Equipment Configuration List *jj*
AECM Active Electronic Countermeasures *fu*
AECS Advanced Entry Control System *mt*
AECS Advanced Environmental Control System *wp*
AECS Aerospace Engineering Conference and Exhibits *aa26.12.77*
AED Active Electronic Decoy *fu*
AED Aerospace Electrical Division [Westinghouse] *ci.973*
AED Allocation Engineering Division *mt, wp*
AED Automated Engineering Design *fu*
AEDAS Automated ESM Data Analysis System *mt*
AEDC Arnold Engineering Development Center [TN] [AFMC] *mt, fu, wp, js7.4, 0, 17, sb, hc, aa26.12.b26, tb, st*
AEDIF Avionics Engineering Development and Integration Facility *hc*
AEDOC After Effective Date Of Contract *-18*
AEDPS Automated Engineering Document Preparation System *hc, fu*
AEDS Atomic Energy Detection System *mt, wp*
AEE Absolutely Essential Equipment *fu*
AEEC Airline Electrical Engineering Council *hc*
AEELS Automated ELINT Emitter Emission Location System *wp, fu*
AEELS Automatic Airborne ELINT Emitter Location System *mt*
AEF Aerospace Education Foundation *af72.5.24*
AEF Air Experience Flight [UK] *mt*
AEF American Expeditionary Force *w9, af72.5.58*
AEFA Advanced Engineering Flight Activity [Army Organization at Edwards AFB] *wp*
AEFA Aviation Engineering Flight Activity *hc*
AEFS [Fleet replenishment ship, US Navy] *mt*
AEG Active Element Group *jj*
AEG Aeromedical Evacuation Group *mt, wp*
AEG Airport Equipment Gateway *fu*
AEG Allgemeine Elekrizit%ots Gesellschaft [German "general electricity company"] (pioneer German electrical company) *sc2.8.2, aa26.11.12, wp*
AEGIS Airborne Early warning / Ground environment Integration System [an advanced anti missile and anti aircraft system fitted to a number of US Navy ships] *mt*
AEGIS Airborne Early Warning Ground Integration System *fu, hn41.18.4, 17, hc, ts.5*
AEHM Aeromedical Evacuation and Hospitalization Model [3CS] *mt*
AEHP Atmospheric Electrical Hazards Protection *wp*
AEI American Electronics, Inc. *oe*
AEI American Enterprise Institute *nd74.445.9*
AEI Average Error Integration *oe*
AEIA Airport Equipment Interface Adapter *fu*
AEICC Alternate Emergency Information and Coordination Center [FEMA] *mt*
AEISF Advanced Extendible Integration Support Facility *hc*
AEK Action Entry Keyboard *fu*
AEL Allowance Equipage List *fu*
AEL American Electronics Laboratory *fu*
AEL Aviation Electric, Limited [Canada] *hc, aw130.22.49*
AELA Airport Equipment Line Adapter *fu*
AELASS Automatically Exploiting the Large Area Surveillance Sensors *fu*
AELP Allied Electrical Publication [NATO] *mt*
AELT Aeromedical Evacuation Liaison Team *mt*
AEM Advanced Engineering Model *jj*
AEM Ancillary Equipment Module *fu*
AEM Applications Explorer Mission [USA] *hg, ge*
AEM Arsenal Exchange Model *wp*
AEM Atmospheric Explorer Missions *ac2-ii, ns*
AEMO Advance Engineering Material Order *wp*
AEMS Aircraft Engine Management System [US Navy] *mt*
AEMS Ammunition Executive Management System [USA] *mt*
AEO Air Electronics Officer [UK] *mt*
AEO Armed Escort Operations *hc*
AEOO Aeromedical Evacuation Operations Officers *mt*
AEOQ Advanced Economic Order Quantity *-18*
AEOS Advanced Electro Optical System *hc*
AEOSS Advanced Electro Optical Sensor System *wp, hc*

AEOT Advanced Electro Optical Tracker *hc*
AEP Action Entry Panel *fu*
AEP Allied Engineering Publication [NATO] *mt*
AEP Altitude Error Probable *fu*
AEP Anti Earth Panel *oe*
AEP Applied Engineering Products [electronics firm, New Haven, CT] *mj31.5.6*
AEPA Airport Environmental Protection Areas *aw130.6.60*
AEPDS Automatic Emergency Action Message {?} *mt*
AEPDS Processing and Dissemination System {?} *mt*
AEPF Action Entry Processing Function *fu*
AEPG Army Electronic Proving Ground *hc, jj*
AEPI Aerospace Engineering Process Instruction *hc*
AEPRT All Equipment Product Production Reliability Test *fu*
AEPS Airlift Execution Planning System [AMC*] *wp, mt*
AEPS Automated Environmental Prediction System *ag*
AEPS Average Expenditure Per Sortie *mt*
AEQUARE [air-launched mini RPV for target acquisition Lockheed] *wp*
AER Army Emergency Relief [office of] *dt*
AER Atmospheric and Environmental Research *hg, ge*
AER-TOW Airborne Extended Range TOW *hc, jj*
AERA Advanced Automated En-Route Air Traffic Control *wp, fu*
AERB Army Educational Requirements Board *fu*
AERDC Army Electronics Research and Development Command *mt*
AERITALIA [Italian aerospace company] *aw129.21.118*
AERL AVCO Everett Research Laboratory [Massachusetts] *hc*
AERLAB Aerolab Development Company *hc*
AERMIP Aircraft Equipment Reliability and Maintainability Program *jj*
AERNO Aeronautical Equipment Reference Number *fu*
AEROJ Aerojet-General Corporation *hc*
AEROMED Aeromedical [evacuation system] [AMC*] *mt*
AERONT Aeronautronics [Division, Philco Corporation] *hc*
AEROP Aerodynamics Measurements Project *hc*
AEROS Advanced Earth Resources Observations Satellite *hg, ge*
AERP Advanced Equipment Repair Program *fu*
AERP Aircrew Eye and Respiratory Protection [System] *wp, hc*
AES Advanced Engineering Staff *sn2.4, rm8.1.2*
AES Aeromedical Evacuation Squadron *mt*
AES Aerospace Electrical Society *0*
AES Apollo Extension System *hc*
AES ATCCS Experimentation Site *fu*
AES Atmospheric Environment Service *fu*
AES Atomic Emission Spectroscopy *wp*
AES Auger Electron Spectroscopy *wp*
AESAT Avionics and Electrical System Advanced Trainer *wp*
AESC Aerojet Electro Systems Company *pl, hc, pl*
AESC Aerospace Environmental Support Center *fu*
AESC Automatic Electronic Switching Center [AUTODIN] *wp, mt*
AESD Aerojet Electro Systems Division *kf*
AESI Advanced Electronics Systems International *hc, fu*
AESIR Aerospace Instrumentation Range [Hughes Aircraft Company GSG] *hc*
AESP Auxiliary Engineering Signal Processor *hc*
AEST Advanced EHF SATCOM Terminal *hc*
AEST All Equipment Screening Test *fu*
AET Aircrew Egress Trainer *wp*
AET Architecture Enterprise Team *pbl, jn, ah*
AETACS Airborne Elements of the TACS *mt*
AETC Air Education and Training Command [UK] [USA] *kf, mt*
AETI Alternative EOCM Techniques Investigation *hc*
AETL Army Engineer Topographic Laboratories [U. S.] *tc*
AETMS Airborne Electronic Terrain Map System *wp, hn44.4.8, hc*
AETP Allied Electronic Publication *hc*
AETRON (AETRON Division, Aerojet-General Corporation) *hc*
AEU Airborne Electronics Unit *mt*
AEU Auxiliary Electronics Unit *kf*
AEV Aerothermodynamic Environmental Vehicle *0*
AEW Advance Early Warning *hc*
AEW Airborne Early Warning [US Navy] [NATO] *mt, fu, wp, aw118.14.65, hc, jj*
AEW&C Airborne Early Warning and Control *mt, aw130.6.34*
AEWC Airborne Early Warning and Control *wp, jj*
AEWEDS Advanced Electronic Warfare Evaluation Display System *wp*
AEWIC Army Electronic Warfare Intelligence Committee [DCSOPS DA] *mt*
AEWICS Airborne Early Warning and Interceptor Control System *wp*
AEWINGPAC Airborne Early Warning Wing, Pacific Fleet [Miramar] *jj*
AEWIS Army Electronic Warfare Information System *hc*
AEWMISREP AEW Mission Report *fu*
AEWS Advanced Electronic Warfare System *wp, fu, hc, mt*
AEWTF Aircrew Electronic Warfare Tactics Facility *hc, wp*
AEWTS Advanced Electronic Warfare Trainer System *hc*
AF Air Force *mt, fu, ci.891, 12, hc, w9, ep, tb, st, jj, kf, oe*
AF Air Frame *jj*
AF Airborne Fixed [Wing *fu*
AF Airway Facilities *fu*

AF Assembly Fixture *oe*
AF Audio Frequency *fu, nu, 0, w9, wp, jj, hi, ai1*
AF Auxiliary carry Flag *kg, ct*
AF Award Fee *kf*
AF [stores ship, US Navy] *mt*
AF-RF Audio Frequency-Radio Frequency *mt*
AF/IG Air Force Inspector General [U. S.] *dt*
AF/IN {Air Force Intelligence Network} *mt*
AFA Accelerated File Access *pk, kg*
AFA Air Force Academy *mt, wp*
AFA Air Force Association *wp, an31.7.1, af72.5.24, ts.4*
AFA Audio Frequency Amplifier *wp*
AFAA Air Force Audit Agency *mt, af72.5.47, dt, wp*
AFAADS Advanced Forward Area Air Defense System *hc*
AFAARS Air Force After-Action Reporting System *mt*
AFAC Air Force Armament Center *hc, wp, jj, kf*
AFAC Airborne Forward Air Controller *mt, hc, wp, kf*
AFAD Air Force Acquisition Document *tb, fu*
AFADPP Air Force Automation Data Processing Plan *mt*
AFAE Advanced Fuel Air Explosives *wp*
AFAE Air Force Acquisition Executive *wp, kf*
AFAF Air Force Assistance Fund *mt, an31.7.9*
AFAFC Air Force Accounting and Finance Center *wp, af72.5.47*
AFAISRC Air Force Automated Information Systems Review Council *mt*
AFAITC Armed Forces Air Intelligence Training Center *mt*
AFAL Air Force Armaments Laboratory [Eglin AFB] *tc, 5, 17, sb*
AFAL Air Force Avionics Laboratory [Wright Patterson AFB] *hc, fu*
AFALC Air Force Acquisition Logistics Center [AFMC] *hc, wp, mt*
AFALCIIS AFALC Integrated Information System [AFLIC] *mt*
AFALD Air Force Acquisition Logistics Division [WPAFB] *mt, fu, tc, wp*
AFALT Air Force Alternate Command Post *mt*
AFAM Air Force Acquisition Model *kf*
AFAMPE Air Force Automated Message Processing Exchange *mt*
AFAMRL Air Force Aerospace Medical Research Laboratory [AFMC] *mt*
AFAMS Automated Flying and Maintenance Scheduling *mt*
AFAO Approved Force Acquisition Objective *wp*
AFAP Artillery Fired Atomic Projectiles *ph, wp*
AFAPG Air Force ADP Planning Guide *mt, wp*
AFAPL Air Force Aero Propulsion Laboratory [WPAFB] *hc*
AFAR Advanced Field Array Radar *wp*
AFAR Airborne Fixed Array Radar [Hughes Aircraft Company] *hc, wp*
AFAR Automatic False Alarm Rate *hc, jj*
AFAR Azore Fixed Acoustic Range [US Navy] *mt*
AFARC Air Force [Systems Acquisition Review Council *fu*
AFARN Air Force Air Request Net *mt*
AFAS Air Force Aid Society *wp*
AFAS Articulated Flexible Appendage Subroutine *hc*
AFAS Automated Fuels Accounting System *mt*
AFASPO Air Force Automated System Project Office *dt*
AFATDS Advanced Field Artillery Tactical Data System [USA] *hc, fu, mt*
AFATDS Advanced Field Artillery Target Designation System [USA] *mt*
AFATL Air Force Advanced Armament Technology Laboratory [Eglin AFB] *wp, tb*
AFATL Air Force Armament Laboratory *mt*
AFB Air Force Base [USA] *mt, aw130.13.6, sc2.7.1, 12, 13, hc, w9, aa26.11.1, jb, wp, no, pr, tb, st, jj, kf, oe, fu, ro, uh, ai1*
AFBMA Anti Friction Bearing Manufacturers Association, Inc. *fu*
AFBMC Air Force Ballistic Missile Center *hc*
AFBMD Air Force Ballistic Missile Division [USA] *hc, mt*
AFBMO Air Force Ballistic Missile Office *tc*
AFBS Air Force Board Structure *wp*
AFBSD Air Force Ballistic Systems Division *hc*
AFC Air Force Cleaner "C" [chemical warfare decontaminant] *wp*
AFC Air Force Commander *fu*
AFC Air Force Council *wp*
AFC Air Frame Change *jj*
AFC Area Frequency Coordinator *mt*
AFC Assault Fire Console *hc*
AFC Automatic Flight Control *wp*
AFC Automatic Frequency Control *mt, wp, 0, ct, 7, hc, w9, ce, jj, jdr, kg, fu, hi, ai1*
AFCAA Air Force Cost Analysis Agency *kf*
AFCAC Air Force Computer Acquisitions Command *wp, hc*
AFCALS Air Force Computer Aided and Acquisition Logistics Support *mt*
AFCAN Analogue Factor Calibration Network *hc*
AFCAS Automatic Flight Control Augmentation System *mt*
AFCC Air Force Chief of Staff [U. S.] *dt*
AFCC Air Force Combat Command *wp*
AFCC Air Force Communications Command [USA] * *mt, 12, hc, af72.5.47, aw130.11.72, jb, tb*
AFCC Air Force Component Commander *mt*
AFCCCMSDB Air Force CCCCM Support Data Base [also AFC3CMSDB] *mt*

AFCCCS ASW Force Command, Control, and Communications System *fu*

AFCCDC Air Force Command and Control Development Command *hc*

AFCCDD Air Force Command and Control Development Division *hc, fu*

AFCCMM Air Force Command and Control Modernization Methodology [also AFC2M2] *mt*

AFCCMMH Air Force Command and Control Modernization Methodology Handbook [also AFC2M2H] *mt*

AFCCP Air Force Component Command Post *hc*

AFCCP Award Fee and Corporate Commitment Plan *kf*

AFCCPC Air Force Communications Computer Programming Center *mt*

AFCCR Air Force Communications Command Regulation *mt*

AFCCS Air Force Cargo Coordination Support System [AMC] *mt*

AFCCS Air Force Chief of Staff *hc, jj*

AFCCS Air Force Command and Control Systems *mt*

AFCCSSPO Air Force Command and Control Systems Systems Program Office [also AFC2SSPO] *mt*

AFCD Advanced Fighter Capability Demonstrator *hn42.1.1, hc*

AFCD Air Force Cryptologic Depot [Kelly AFB] *mt, hc*

AFCE Automatic Flight Control {Equipment} [System] *jj*

AFCEA Armed Forces Communications and Electronics Association *aw126.20.7, tb, fu, mt*

AFCENT Allied Forces Central Europe [NATO Command] *12, fu, mt*

AFCH Air Force Component Headquarters *mt*

AFCIMS Air Force Claims Information Management System *mt*

AFCIPS Air Force Communications-Computer Integration and Planning System *mt*

AFCMC Air Force Contract Maintenance Center *mt*

AFCMD Air Force Contract Management Director *jdr*

AFCMD Air Force Contract Management Division [Kirtland AFB] [AFMC] *mt, wp, hn44.16.1, hc, hq6.2.2, tb*

AFCMO Air Force Contract Management Office *hc*

AFCMR Air Force Court of Military Review *af72.5.108*

AFCNF Air Force Central NOTAM Facility [AFCC*] *mt*

AFCO Air Force Contracting Officer *hc*

AFCOA Air Force Contracting Office Approval *hc*

AFCOLR Air Force Coordinating Office for Logistics Research *tc, mt*

AFCOM (military synchronous communications satellite) *hc*

AFCOMAC Air Force Combat Ammunition Center *mt*

AFCOMPMET Air Force Comptroller Management Engineering Team *af72.5.109*

AFCOMS Air Force Commissary Commission *wp*

AFCOMS Air Force Commissary Service *mt, af72.5.47*

AFCOS Air Force Combat Operations Staff *af72.5.47*

AFCOS Air Force Contingency Operations Staff [HQ USAF] *mt*

AFCP Advanced Flight Control Programmer *hc*

AFCPMC Air Force Civilian Personnel Management Center *af72.5.47*

AFCRC Air Force Cambridge Research Center *0, hc*

AFCRI Air Force Computer Resource Inventory *jdr*

AFCRL Air Force Cambridge Research Laboratories [later AFGL] *fu, le30, hc, crl*

AFCS Air Force Communications Service System *fu*

AFCS Army Facility Components System *st, mt*

AFCS Automatic Fire Control System *wp*

AFCS Automatic Flight Control System *mt, fu, hc, jj, ai1*

AFCSA Air Force Center for Studies and Analyses *mt, af72.5.47*

AFCSA Air Force Communications-Computer Systems Architecture *mt*

AFCSC Air Force Cryptologic Support Center [AFIC] *wp, tb, kf, mt*

AFCSDO Air Force Communications-Computer Systems Doctrine Office *mt*

AFCSM Air Force Communication Security Manual *wp*

AFCSP Air Force Computer Security Program *mt*

AFCSPO Air Force Computer Security Program Office *mt*

AFCSTC Air Force Cost Center *af72.5.124*

AFCUISPD Air Force Common User Information Services Planning Document *mt*

AFCW Accept Forced Control Word *fu*

AFD Aft Flight Deck *gp, jdr, oe*

AFD Air Force Depot *hc, jj*

AFD Arming Firing Device *wp*

AFD Automatic File Distribution *pk, kg*

AFDA Air Force Director of Administration [U. S.] *dt*

AFDAA Air Force Data Automation Agency *hc*

AFDAP Air Force Designated Acquisition Program *mt*

AFDATACOM Air Force Data Communications *mt*

AFDCO Air Force Distribution Control Office *wp*

AFDD Air Force Data Dictionary *mt*

AFDIGS Air Force Digital Graphics System *mt*

AFDR Air Force Deficiency Report *fu*

AFDR Automatic Fault Detection and Recovery *jdr, uh*

AFDS Amphibious Flagship Data System *fu*

AFDS Automatic Flight Director System *mt*

AFDS Automatic Frequency Deconfliction System *mt*

AFDSDC Air Force Data Systems Design Center *hc, mt*

AFDTAMS Air Force Data Transmission and Message System *mt*

AFDW Active Framework for Data Warehousing [Microsoft] *pk, kg*

AFDW Air Force District of Washington *mt, af72.5.47*

AFE Aeroassist Flight Experiment [NASA] *aa26.12.57*

AFE Aerospace Facilities Engineer *hc*

AFE Alternate Fighter Engine *wp, af72.5.28*

AFEB Award Fee Evaluation Board *fu*

AFECO Air Force Electronic Combat Office *aw129.14.103*

AFEDSC Air Force Engineering Data Service Center [Scott AFB] *mt*

AFEE Airborne Forces Experimental Establishment [Beaulieu, Hampshire, UK, closed in 1950] *mt*

AFEEIP Air Force Exercise Evaluation and Improvement Program *mt*

AFEKDS Air Force Electronic Key Distribution System *mt*

AFELIS Air Force Engineering and Logistic Information System *wp*

AFEMMIS Air Force Equipment Maintenance Management Information System *mt*

AFEMS Air Force Equipment Management System *wp, mt*

AFEOC Air Force Emergency Operations Center *mt, wp*

AFEP Air Force Exercise Program *mt*

AFER Air Force Eastern Range *oe*

AFERC Air Force Edwards Research Center *hc*

AFES Automatic Feature Extraction System [RADC] *wp*

AFES Automatic Fire Extinguishing System *wp*

AFESC Air Force Electronic Security Systems Command *10, tb*

AFESC Air Force Engineering [and] Services Center *mt, wp, tc, af72.5.47*

AFESD Air Force Electronics Systems Division *hc*

AFESMC Air Force Eastern Space and Missile Center *oe*

AFESMET Air Force Engineering and Services Management Team *af72.5.109*

AFETAC Air Force Environmental Technical Applications Center *mt*

AFETR Air Force Eastern Test Range [Patrick AFB, Florida] *fu, wp, hc, jdr, kf*

AFETRM Air Force Eastern Test Range Manual *bo*

AFETRR Air Force Eastern Test Range Regulation *oe*

AFETS Air Force Engineering Technical Services *wp, kf*

AFEWC Air Force Electronic Warfare Center [ESC*] *jb, wp, mt*

AFEWES Air Force Electronic Warfare Evaluation Simulator [operated by GD] *wp*

AFEWL [Hughes Aircraft Company customer for OWL/D] *hc*

AFFARS Air Force Federal Acquisition Regulations Supplement *kf*

AFFDL Air Force Flight Dynamics Laboratory *hc, wp*

AFFE Airborne Fire Fighting Equipment *wp*

AFFIF Air Field Facilities Information File [WWMCCS] *mt*

AFFIS Air Field Facilities Information System [WWMCCS] *mt*

AFFMC Air Force Frequency Management Center [AFCC*] *mt*

AFFOR Air Force Forces *mt*

AFFOR Air Force Theater Command {?} *mt*

AFFSIM Airborne Formation Flight Simulator *wp*

AFFTC Air Force Flight Test Center [Edwards AFB, California, USA] *mt, wp, 12, hc, aw129.14.120, jj*

AFFTIC Air Force Foreign Technology Intelligence Center *mt*

AFFTIS Air Force Flight Test Instrumentation System *hc*

AFGE American Federation of Government Employees *mt, ci.72*

AFGL Air Force Geophysics Geophysical Laboratory [formerly AFCRL] [Hanscom AFB, MA] [AFMC] *mt, wp, ci.894, 12, pl, no, dm, hy, tb, kf*

AFGS Automatic Flight Guidance System *mt*

AFGWC Air Force Global Weather Central Center [Omaha, NE] *fu, wp, ac7.5-9, 12, 15, hc, pl, pf1.27, jb, no, dm, kf*

AFGWC Air Force Global Weather Central [AWS] *mt*

AFGWIS Air Force Global Weather Information System *mt*

AFHIS Air Force Headquarters Information System *mt*

AFHRL Air Force Human Resources Laboratory *mt, wp, tc, fu*

AFI Air Force Instruction *kf*

AFI Air Forces Iceland [LANTCOM] *mt*

AFI Allied Forces Italy [NATO] *mt*

AFI Automatic Fault Isolation *hc, jj*

AFI Award Fee Incentive *jdr*

AFIA Air Force Intelligence Agency *af72.5.47, jmj*

AFIAMS Air Force Industrial Assistance Management Survey *hc*

AFIC Air Force Intelligence Center *wp*

AFIC Air Force Intelligence Command *mt, kf*

AFICP Air Force Intelligence Communications Plan *mt*

AFIE Abnormal Fluctuations In the Economy *hc*

AFIF Air Force Industrial Fund *mt, wp*

AFIFIO Air Force Information for Industry Office [AFMC] *mt*

AFII Association for Font Information Interchange *pk, kg*

AFIISP Air Force Intelligence Information Systems Plan *mt, wp*

AFIMMSS Air Force Integrated Maintenance Management Support System *mt*

AFIN Assistant Chief of Staff, Intelligence *dt*

AFINTELMET Air Force Intelligence Management Engineering Team *af72.5.109*

AFIP Air Force Intelligence Plan *mt*

AFIP Armed Forces Institute of Pathology *dt*

AFIPS American Federation of Information Processing Societies *fu, do32*

AFIRM Automated Fingerprint Image Reporting and Match *pk, kg*

AFIRST Advanced Far Infra Red Search Track *hc*

AFIS Air Force Intelligence Service *dt*

AFIS Automated Fingerprint Identification System *pk, kg*
AFIS Automatic Fault Insertion System *fu*
AFIS/IND Air Force Intelligence Service/Intelligence Data Management *mt*
AFISA Air Force Information Systems Architecture *mt*
AFISA Air Force Integrated Systems Architecture *mt*
AFISA Air Force Intelligence Support Agency *mt*
AFISC Air Force Inspection and Safety Center *mt, wp, af72.5.47*
AFISP Air Force Information Systems Plan *mt*
AFIT Air Force Institute of Technology *mt, wp, af72.5.72, tb*
AFIT Automatic Fault Isolation Test *jj*
AFIT/LS Air Force Institute of Technology School of Systems and Logistics *mt*
AFJIMS Air Force Justice Information System *mt*
AFJPO Air Force AF JINTACCS Program Office *mt*
AFK Air Forces Korea *mt*
AFL American Federation of Labor *wp*
AFL Automatic Line Feed *wp*
AFL-CIO American Federation of Labor and Congress of Industrial Organizations *wp, w9, aw130.22.108*
AFLAN Air Force Local Area Net *mt*
AFLANSPO Air Force Local Area Network Systems Program Office *mt*
AFLANT Air Force, Atlantic Command [ACC] *mt*
AFLAS Aviation Fuels Logistical Area Summary *mt*
AFLC Air Force Logistics Center *hc*
AFLC Air Force Logistics Command [USA] *mt, fu, ro, wp, rm6.4.4, 12, hc, af72.5.35, tb, jj, jdr*
AFLCM Air Force Logistics Command Manual *wp, hc*
AFLCP (requirements as described in AFSCP/AFLCP 173-5) *nt*
AFLCR Air Force Logistic Command Regulation *wp*
AFLETS Air Force Law Enforcement Terminal System *mt*
AFLIR Advanced Forward Looking Infra Red [also ADFLIR] *hc*
AFLMC Air Force Logistics Management Center [AU] *mt*
AFLOGMET Air Force Logistics Management Engineering Team *af72.5.109*
AFLSC Air Force Legal Services Center *af72.5.47*
AFM Air Force Manual *mt, fu, wp, tb, kf*
AFM Arrival Flow Management *fu*
AFMAG Air Force Management Analysis Group *mt*
AFMBP Air Force Model Base for Communications-Computer Systems Program *mt*
AFMC Air Force Materiel Command [USA] *kf, fu, mt*
AFMC/LMSC AFLC {?} / Logistics Management Support Center *mt*
AFMCBEAMS AFLC {?} Base Engineering Automated Management System *mt*
AFMCO Army Force Modernization Coordination Office *mt*
AFMCSCE Air Force Materiel Command Support Center, Europe *mt*
AFMDC Air Force Missile Development Center [Holloman AFB] *hc, wp*
AFMEA Air Force Management Engineering Agency *mt, wp, af72.5.47*
AFMEDICS Air Force Medical Information and Control System *mt*
AFMEDMET Air Force Medical Management Engineering Team *af72.5.109*
AFMER Aircraft Field Material Evaluation Report *jj*
AFMIC Armed Forces Medical Intelligence Center *mt*
AFMIG Air Force Management Improvement Group *wp*
AFMIS Air Force Major Command Information System *mt*
AFMIS Air Force Management Information System [Project 6000] *mt*
AFMIS Army Food Management Information System *mt*
AFML Air Force Materials Laboratory [Wright Patterson AFB] [AFSC*] *mt, wp, hc, tb*
AFMLO Air Force Medical Logistics Office *mt*
AFMMFO Air Force Medical Material Field Office *mt*
AFMPC Air Force Manpower and Personnel Center [Randolph AFB] *mt, fu*
AFMPC Air Force Military Personnel Center *mt, af72.5.47*
AFMPMET Air Force Manpower and Personnel Management Engineering Team *af72.5.109*
AFMR Anti Ferro Magnetic Resonance *wp*
AFMS Advanced Factory Management System *fu*
AFMS Automatic Flight Management System *mt*
AFMSC Air Force Medical Service Center *mt*
AFMSISP Air Force Medical Services Information System Plan *mt*
AFMT Air Force Manufacturing Technology *fu*
AFMTC Air Force Missile Test Center *wp, 0, jj*
AFN Armed Forces Network *wp*
AFNAG Air Force Manuals of National Technical Criteria of Compromising Emanations *tb*
AFNORTH Allied Forces Northern Europe [NATO Command] *mt, fu, 12, hc*
AFNS {Air Force News Service} *an31.13.1*
AFO Accounting and Finance Office *wp*
AFO Air Forward Observer *hc, jj*
AFO Announced Flight Opportunity *pl*
AFO Assault Fire Platoon *hc*
AFOC Air Force Operations Center [HQ USAF] *hc, fu, mt*
AFOC Army Flight Operations Center *mt*
AFOCAL A Spherical Optics {?} *tb*

AFOE Assault Follow On Echelon *wp, mt*
AFOEA Air Force Organizational Excellence Award *ro, mt*
AFOG Air Force Orientation Group *af72.5.119*
AFOIS Air Force Office of Special Investigations [also AFOSI] *dt*
AFOLDS Air Force On Line Data Systems *mt*
AFOLTA Air Force Office For Logistic Technology Applications *wp*
AFOMS Air Force Office of Medical Support [formerly AFMSC] *mt, af72.5.47*
AFOOTS Air Force Obligation / Outlay Tracking System *mt*
AFOREP Air Force Operational Reporting System *mt*
AFORMS Air Force Operations Resources Management System *mt*
AFOS Automated Forecasting System *fu*
AFOS Automation of Field Operations and Services *pl, oe*
AFOSH Air Force Occupational Safety and Health *mt, kf, wp*
AFOSI Air Force Office of Special Investigations [also AFOIS] *mt, wp, af72.5.47, tb*
AFOSP Air Force Office of Security Police *af72.5.47, mt*
AFOSR Air Force Office of Scientific Research *mt, 0, hc, aa27.4.b30, wp*
AFOT Air Force Office of Technology *tb*
AFOTEC Air Force Operational Test and Evaluation Center *mt, hc, af72.5.47, jb, wp, tb, kf*
AFOUA Air Force Outstanding Unit Award *ro*
AFP Afloat Pre Positioning Force *wp*
AFP Air Force Pamphlet *mt, tb, kf*
AFP Armed Forces, Philippines *fu*
AFP Assault Fire Platoon *fu*
AFP Award Fee Plan *kf*
AFPC Air Force Procurement Circular *hc*
AFPCP Air Force Potential Contractor Program *mt*
AFPD Air Force Policy Directive *kf*
AFPE Air Force Preliminary Evaluation *hc*
AFPEO/SP Air Force Program Executive Officer For Space *kf*
AFPG Air Force Planning Guide [also APG] *mt, wp, af72.5.107*
AFPI Air Force Procurement Instruction *hc, fu*
AFPLA Air Force of the People's Liberation Army [China] *mt*
AFPMB Armed Forces Pest Management Board *dt*
AFPR Air Force Plant Representative *hc, jj, fu*
AFPRO Air Force Plant Representative Office *mt*
AFPRO Air Force Plant Representative Office Organization *su, fu, hn43.23.2, 8, 17, sc7.4.1, hc, pl, hq6.2.2, wp, ep, tb, jj, kf, oe*
AFPTS Air Flow and Pressure Test Station *jj*
AFQA Air Force Quality Assurance *gp, fu*
AFQAR Air Force Quality Assurance Representative *hc, fu, jj*
AFQCR Air Force Quality Control Representative *hc, jj*
AFR Air Force Regulation *mt, wp, fu, 17, hc, aw129.10.72, tb, jj, kf*
AFR Air Force Reserve *mt*
AFRAMS Air Force Recoverable Assembly Management System *wp*
AFRAMS Air Force Recoverable Asset Management System *mt*
AFRAP Air Force Remedial Action Program *mt*
AFRAPT Air Force Research in Aero Propulsion Technology *aa26.11.55*
AFRB Award Fee Review Board *kf*
AFRBO Air Force Reviews Boards Office *af72.5.47*
AFRCC Air Force Rescue Coordination Center [Scott AFB] *mt, wp*
AFRCE Air Force Regional Civil Engineer Engineering *kf, hc, mt*
AFRCS Air Force Reports Control System *mt*
AFRES Air Force Reserve [USA] *mt, wp, 12, af72.5.5, kf*
AFRL Air Force Research Laboratory [U. S.] *ai2*
AFRLC Air Force Regional Civil Engineers *mt*
AFRO Air Force Recruiting Office *mt*
AFROTC Air Force Reserve Officer Training Corps *mt, af72.5.24*
AFRP Air Force Recurring Publication *mt*
AFRPL Air Force Rocket Propulsion Laboratory [AFMC] *tc, 5, 17, tb, mt*
AFRRI Armed Forces Radiobiology Research Institute *dt*
AFRS Advanced Fighter Radar System *hc*
AFRS Air Force Recruiting Service *mt*
AFRS Automated Fitness Reports System [USMC] *mt*
AFRS Auxiliary Flight Reference System *wp, hc, jj*
AFRTS Armed Forces Radio and Television Service *mt*
AFS Aeronautical Fixed Service *fu*
AFS Air Force Specialties *mt*
AFS Air Force Standard *wp*
AFS Air Force Station *mt, 12, af72.5.103, jb, tb, jj*
AFS Andrew File System [protocols] *pk, kg*
AFS Automatic Flight System *mt*
AFS Fleet Stores Ship [US Navy] *mt*
AFSA Armed Forces Security Agency *-10*
AFSAB Air Force Science Advisory Board *wp*
AFSAC Air Force Special Activities Center *mt*
AFSAC Armed Forces Security Agency Council *-10*
AFSAC/IRC AFSAC Intelligence Requirements Committee *-10*
AFSARC Air Force System Acquisition Review Council *mt, wp, fu, hq*
AFSAS Advanced Fire Support Avionics System *hq*
AFSAT Air Force Satellite *mt, jb, fu*
AFSATCOM Air Force Satellite Communications [system] *mt, fu, wp, sc2.8.2, 12, 17, hc, jb, tb, jdr, kf*

AFSC Air Force Skill Specialty Code *hc, kf, wp, jj*
AFSC Air Force Specialty Code *mt, fu*
AFSC Air Force Supply Code Catalog *wp*
AFSC Air Force Systems Command [Andrews AFB, Maryland, USA] *
mt, wp, hc, sc2.7.1, rm6.4.1, 12, 13, 17, an31.7.9, af72.5.47, jb, dt, tb, jj, fu, ro
AFSC Armed Forces Staff College [NDU]
AFSCAPS Air Force Standard Civilian Automated Pay System *mt*
AFSCC Air Force Special Communications Center *hc*
AFSCCCC Air Force Systems for Command, Control, Communications, and Computers *mt*
AFSCCCCI Air Force Systems for Command, Control, Communications, and Computers Intelligence *mt*
AFSCCISA Air Force Standard Command and Control Information Systems Architecture *mt*
AFSCF Air Force Satellite Control Facility *mt, wp, 4, 5, 11, 16, bo, sb, hc, hs, jb, dm, tb, jdr, kf, oe, fu*
AFSCM Air Force Systems Command Manual *hc, fu*
AFSCN Air Force Satellite Space Control Network *mt, uh, 16, hc, af72.5.67, dm, jdr, kf, oe*
AFSCOASO Air Force Small Computer / Office Automation Service Organization *mt*
AFSCP Air Force Service Cost Position *kf, nt*
AFSCR Air Force Systems Command Regulation *wp, fu, dm*
AFSCS Air Force Satellite Communications System *wp*
AFSCSSD Air Force Systems Command, Space Systems Division *0*
AFSD Air Force Space Division [El Segundo, California] *tc, 16, hc, pl, tb*
AFSD Air Force Supply Directive *wp*
AFSF Air Force Stock Fund *wp*
AFSFC Air Force Space Forecast Center *kf*
AFSFRS Air Force Stock Fund Reporting System *mt*
AFSG Amphibious Force Support Group [US Navy] *mt*
AFSHRC Albert F. Simpson Historical Research Center *mt*
AFSIG Air Force Weapons Support Improvement Group *mt*
AFSINC Air Force Service Information and News Center *mt, af72.5.47*
AFSIP Air Force Signals Intelligence Plan *mt*
AFSIP Air Force Space Intelligence Plan Program *hc, wp*
AFSISC Air Force Standard Information Systems Center *mt*
AFSLV Air Force Small Launch Vehicle [U. S.] *ai2*
AFSME Air Force Systems Maintenance Engineering *hc*
AFSMO Air Force Space and Missile Organization *mt*
AFSOB Air Force Special Operations Base [USA, DOD, that base, airstrip, or other appropriate facility that provides physical support to Air Force special operations forces] *mt*
AFSOC Air Force Special Operations Command [USA] *mt*
AFSOC Air Force Special Operations Component [USA DOD] [the Air Force component of a joint force special operations component] *mt*
AFSOD Air Force Special Operations Detachment [USA DOD] *mt*
AFSOE Air Force Special Operations Element [USA DOD] *mt*
AFSOF Air Force Special Operations Forces [USA DOD] *mt*
AFSOS Automated Food Service Operations System *mt*
AFSOUTH Allied Forces Southern Europe [NATO Command] *mt, 12, fu*
AFSPACECOM Air Force Space Command [USA] *mt, uh, af72.5.31, go2.4, jb, cp, jmj, tb, st, as*
AFSPC Air Force Space Command *kf*
AFSPCR Air Force Space Command Regulation *kf*
AFSPOC Air Force Space Operations Center *jb, cp*
AFSS Air Force Security Service [Kelly AFB, Texas, USA] *mt, hc*
AFSS Automated Flight Service Station *fu*
AFSS Automated Fuels Service Station *mt*
AFSSD Air Force Space Systems Division *hc*
AFSSMET Air Force Special Staff Management Engineering Team *af72.5.109*
AFSSO Air Force Special Security Office *mt*
AFSTC Air Force Satellite Test Center *sb*
AFSTC Air Force Space Technology Center [AFSC*] *mt, tc, 16, ts4-14, cp, tb*
AFSWC Air Force Special Weapons Center [Kirtland AFB, New Mexico] *wp, hc, fu, jj*
AFSWF Air Force Software Facility [WIS] *mt*
AFSWP Armed Forces Special Weapons Project *hc*
AFT Automated Funds Transfer *hi*
AFTA Avionics Fault Tree Analyzer *hc, ws4.10*
AFTAC Air Force Tactical Applications Center *kf*
AFTAC Air Force Technical Applications Center [Patrick AFB] *mt, 12, hc, af72.5.47, hs, dt, tb*
AFTACIE Air Force Tactical Imagery Exploitation *wp*
AFTACIES Air Force Tactical Imagery Exploitation System [TAF] *wp, mt*
AFTCO Air Force Telecommunications Certification Office *mt*
AFTD Air Force Test Director *kf*
AFTEC Air Force Test and Evaluation Center [later AFOTEC] *mt, fu, aw118.18.42, hc*
AFTI Advanced Fighter Technology Integration [USA] *mt, wp, aw118.14.62, hc, aa26.11.b9*
AFTI Automated Interchange of Technical Information {?} *mt*
AFTIP Air Force Theater Intelligence Plan *mt*
AFTN Aeronautical Fixed Telecommunications Network *fu, mt*

AFTO Air Force Technical Order *mt, hc, jj, jdr, fu*
AFTP Anonymous File Transfer Protocol *hi, kg*
AFTPC Air Force Tele Processing Center *mt*
AFTRCC Aerospace and Flight Test Range Coordinating Company *hc*
AFTT Air Force Test Team *hc*
AFTTL Air Force TACS/TADS Test Location *fu*
AFU Airborne Fixed-wing Unit *fu*
AFU Alaska Forecast Unit *mt*
AFU Assault Fire Unit *fu*
AFV Armored Family Vehicles *hc*
AFV Armored Fighting Vehicle *mt*
AFWAL Air Force Wright Aeronautical Laboratories [WPAFB] *mt, fu, wp, tc, aa27.4.b29, hc, tb*
AFWAM Air Force WWMCCS ADP Modernization Program *mt*
AFWC Air Force Wargaming Center [AU] *mt*
AFWIS Air Force WWMCCS Information System *mt*
AFWL Air Force Weapons Laboratory [Kirtland AFB, USA] *mt, wp, tc, hc, pl, ie26.1.16, tb, st*
AFWMPRT Air Force Wartime Manpower and Personnel Readiness Team *mt, af72.5.109*
AFWR Air Force Western Range *oe*
AFWR Atlantic Fleet Weapons Range *hc, jj*
AFWSIG Air Force Weapon System Improvement Group *mt*
AFWSMC Air Force Western Space and Missile Center *oe*
AFWTF Atlantic Fleet Weapons Training Facility *fu*
AFWTP Air Force Weapons Training Facility *jj*
AFWTR Air Force Western Test Range [Vandenberg AFB] *wp, hc*
AFWWMCCS Air Force Worldwide Military Command and Control System *mt*
AFXP Advance Feature Extractor Processor *hc*
AG Adjutant General *w9, mt*
AG Aerospace Groups [Hughes Aircraft Company] *fu, jj*
AG Air Gunner *mt*
AG Air to Ground *wp, jj*
AG Airlift Group [USA] *mt*
AG Aktien Gesellschaft [German {Incorporated}] *aw130.13.63, hn43.22.4, es55.3*
AG Angle Gate *jj*
AG Army Garrison *dt*
AG Army Group [NATO] *mt*
AG Arresting Gear *jj*
AG Auxiliary Ground *fu*
AGA Air to Ground to Air *fu*
AGA Airborne Gas Accumulator [Sweden] *hc*
AGA American Gas Association *y2k*
AGACS Automatic Ground to Air Communications System [USAF/FAA] *wp, hc*
AGAD Awaiting Government Approval/Direction *oe*
AGADS Advanced Graphics Avionics Display System *hc, mt*
AGARD Advisory Group for Aerospace Research and Development [NATO] *mt, 0, ac7.2-5, wp, st*
AGAT Automatic General Analog Tester *fu*
AGATE Aerial Gunnery Advanced Technology Evaluation *wp*
AGATE Air to Ground Acquisition and Tracking Equipment *wp*
AGATE Air to Ground Automatic Tracking Equipment *hc, jj*
AGB Autonomous Guided Bomb *wp*
AGBT Advanced Gear Box Technology *aa26.10.20*
AGC Automatic Automated Gain Control *ct, em, uh, fu, kg, wp, ac2.7-5, hc, 0, 7, 17, sb, mj31.5.26, lt, jj, jdr, oe, ai1*
AGC Automatic Gain Control *mt*
AGCA Altitude Gyro Control Assembly *wp*
AGCA Automatic Ground Control[led] Approach *hc, jj*
AGCL Automatic Ground Controlled Landing *hc*
AGCM Aerospace Group Configuration Management *hc*
AGCMM Aerospace Group Configuration Management Manual *hc*
AGCS Automatic Ground Control Station *wp*
AGCW Autonomous Guidance For Conventional Weapons *wp*
AGD Attack Geometry Display *jj*
AGDDS Aerospace Group Directive Document System *hc*
AGDS [launch area support ship, US Navy] *mt*
AGE Acquisition Gate Enable *jj*
AGE Aerospace Ground Equipment *mt, fu, ah, pbl, gp, wp, hc, ep, tb, jj, kf, tm, jn, oe, ai2*
AGE Aircraft Ground Equipment *mt*
AGE Auxiliary Ground Equipment *mt, hc, jj, jdr*
AGE/SE Aerospace Ground Equipment/Support Equipment *wp*
AGEI Aerospace Ground Equipment Illustration *hc*
AGENORS Aerospace Ground Equipment Not Operationally Ready Supply *jj*
AGEP Advisory Group On Electronic Parts *wp*
AGEP Aerospace Ground Equipment Plan *hc, jj*
AGERD Aerospace Ground Equipment Recommended Recommendation Data *hc, jj, fu*
AGETS Automated Ground Engine Test Set *wp*
AGF Auxiliary Command Ship [US Navy ship classification] *mt, fu*
AGI Annee Geophysique Internationale *crl*
AGI Auxiliary General Intelligence *mt*

AGI [CIS intelligence collecting ship] *mt*
AGIK Auxiliary Ground Installation Kit *fu*
AGIR Advanced Growth Infrared *fu*
AGIS Air to Ground Integration System *hc*
AGIS Apex Global Information Services *ct, em*
AGISS Advanced Gigahertz Input Snap Shot *hc*
AGL Above Ground Level *mt, wp, fu, aw129.10.73*
AGM Air to Ground Missile [system] [USA] *fu, hc, wp, jj, mt*
AGM [missile range instrumentation ship, US Navy] *mt*
AGMC Aerospace Guidance and Meteorology Center *wp, kf, mt*
AGMD Aerospace Group Management Directive *hc, jj*
AGMTI Air to Ground Moving Target Indicator *hc, jj*
AGMX Air to Ground Missile Excalibur *hc*
AGN Active Galactic Nucleus *es55.18, mb, mi*
AGOR Oceanographic Research Ship [US Navy] *mt*
AGOS Air Ground Operations School [USAF] *mt*
AGOS Air Ground Operations System *mt*
AGOS [ocean surveillance ship, US Navy] *mt*
AGPA Aerospace Group Property Administrator *hc, jj*
AGPB Advanced General Purpose Bomb *wp*
AGPE Aerospace Group Property Engineering *hc*
AGPLAN Army General Purpose Local Area Net *mt*
AGPM Aerospace Group Property Management *hc*
AGR Air to Ground Ranging *hc, jj*
AGR Annual Growth Rate *hi*
AGRAS Antiglare Antireflective Antistatic *hi, kg*
AGREE Advisory Group for Reliable Electronic Equipment *hc*
AGRISTARS Agriculture and Resource Inventory Survey Through Aerospace Remote Sensing *un*
AGS Abort Guidance Section *ci.392*
AGS Abort Guidance System *wp*
AGS Administrator of General Services *wp*
AGS Advanced Guidance Sensor *hc*
AGS Advanced Gun System *hc*
AGS Aircraft Generation Squadron *mt, fu*
AGS Alliance Ground Surveillance *rih*
AGS Alternate Guidance Section *hc*
AGS Alternating Gradient Synchrotron [at BNL] *oe*
AGS Armored Gun System [Army] *mt*
AGS [surveying ship, US Navy] *mt*
AGSC Air Force System Command *fu*
AGSP Automatic Graphic Science Program *fu*
AGSS Attitude Ground Support System *oe*
AGSS Auxiliary Research / Experimental Submarine [conventional] [US Navy] *mt*
AGSSS Airborne Graphics Software Support System [ASD] *mt*
AGT Above Ground Test *mt*
AGT Above Ground [Nuclear] Testing *st*
AGT Advanced Guidance Technology *wp*
AGT Advanced Gun Technology [USA] *mt*
AGT Automotive Gas Turbine *aw129.10.63*
AGTC Advanced Airport Ground Traffic Control System *wp*
AGTELIS Army Ground Transportable Emitter Location and Identification System *fu*
AGTR Auxiliary Technical Reconnaissance Ship [US Navy] *mt*
AGU Address Generation Unit *pk, kg*
AGU Aerospace Ground Unit *wp*
AGU American Geophysical Union *le42, mb, hg, ge, mi*
AGU Auxiliary Ground Unit *fu*
AGUIK Auxiliary Ground Unit Installation Kit *fu*
AGV Automated Automatic Guided Vehicle *hc, wp*
AGV Avion ‡ Grande Vitesse [a design for a Mach 5 airliner, that could enter service in 2015] [France] *mt*
AGVT Advanced Ground Vehicle Technology [USA] *mt*
AGVT Advanced Guided Vehicle Technology *hc*
AGW Actual Gross Weight *wp*
AGW Allowable Gross Weight *mt*
AGZ Actual Ground Zero *mt*
AH Attack Helicopter *fu*
AH Authentication Header *em*
AH [designator, (AH-64A) Apache Helicopter, (AH-1J) Sea Cobra] [USA DOD] *mt*
AH [hospital ship, US Navy] *mt, wp*
AH/ERS Alternate Headquarters / Emergency Relocation Sites *mt*
AHA Antenna Hoisting Handling Adapter *hc, jj*
AHAAMS Advanced Heavy Anti Armor Missile System *hc*
AHACS Attack Helicopter Armament Control System *hc*
AHAMS Advanced Heavy Antitank Missile System [USA] *hc, mt*
AHARS Airborne Heading-Attitude Reference System *wp*
AHARS Altitude Heading Aircraft Reference System *fu*
AHB Air Historical Branch [of the Ministry of Defense, UK] *mt*
AHC Aerospatiale Helicopter Corporation *aw130.6.54*
AHC Aircraft Handling Characteristics *mt*
AHC Army Health Clinic *dt*
AHE Automatic Height Extractor *fu*
AHEAPS Automated Hazard Abatement Program System [USAF] *mt*

AHEP Advanced Hardware Exploitation Program *hc*
AHEP Atmospheric Electrical Hazards Protection *wp, mt*
AHF Adaptive High Frequency *mt*
AHGC Advanced Hardened Guidance Computer *hc*
AHGS Altitude Heading Gyroscope System *wp*
AHHQ Alternate Hardened Headquarters *mt, wp*
AHIP Armed Helicopter Improvement Program. [modification program for the OH-58 Kiowa helicopters of the US Army, initiated in 1981] [USA] *mt*
AHIP Army Advanced Helicopter Improvement Program *ph, hc*
AHOS Automated Hydrologic Observing System *ag*
AHR Adsorptive Heat Recovery *wp*
AHRA Advanced Helmet Sight Reticule Assembly *wp*
AHRS Advanced Heat Rejection System *hc*
AHRS Altitude and Heading Reference System *fu, wp, hc, tb, mt*
AHRS Automatic Heading Reference System *jj*
AHS American Helicopter Society *aw118.18.3, aa26.10.b12*
AHSSS Adaptive High Speed Signal Sorter *fu*
AHU Air Handling Unit *fu*
AHUD Austere Head-Up Display *wp*
AHWT Advanced Heavy Weight Torpedo *uc*
AI Acquisition Instruction *tb*
AI Action Item *4, jdr, jn*
AI Air Intercept Interception *fu, w9*
AI Air Interception *mt*
AI Air Interdiction *wp, mt*
AI Airborne Intercept Interceptor *mt, wp, hc, w9, jj*
AI Analog Input *jj, kg*
AI Artificial Intelligence *ct, em, hn44.9.3, hc, ic2-13, 17, sb, w9, wp, st, kg, fu, mt, ai1*
AI Assembly Instruction *fu*
AI Atomics International *ci.961*
AI Automatic Initiation *fu*
AI&T Assembly, Integration and Test *kf*
AIA Advanced Integrated Avionics *kf*
AIA Aerospace Industries Industry Association [of America] *hc, nd74.445.26, aw129.21.73, hq6.2.1, jj, heh2.21.5, fu*
AIA Air Intelligence Agency *kf*
AIA Application Installation Aid *mt*
AIA Applications Integration Architecture [DEC] *pk, kg*
AIA Army Information Architecture *mt*
AIA Army Intelligence Agency *mt*
AIA Auxiliary Integration Assembly *kf*
AIA Auxiliary Interface Adapter *hc*
AIAA American Institute of Aeronautics and Astronautics *mt, sc2.8.4, 0, aw130.13.104, hc, aa26.12.2, ns, tb, kf, hg, mb, oe, fu, mi, ai*
AIAAM Advanced Intercept Air to Air Missile [a proposed successor for AIM-54 Phoenix, USA] *uc, hc, wp, mt*
AIAD Accountable Inventory Adjustment Document *wp*
AIC Aerodynamic Influence Coefficient *hc*
AIC Air Intercept Controller Controlled *mt, wp, jj, fu*
AIC Airborne Information Correlation Program *wp*
AIC AIXwindows Interface Composer [IBM] *hi, kg*
AICARS Airlift Interim CAMS and REMIS System *mt*
AICBM Anti Intercontinental Ballistic Missile *wp, jj*
AICC Airborne Intercept Command Control *mt*
AICC Airborne Interceptor Common Channel *fu*
AICCM Air Intercept [fire] Control ECCM Study *hc*
AICCS Autopilot Interface Circuit Card Set *fu*
AICE American Institute of Chemical Engineers [also AICHE] *hc, jj*
AICHE American Institute of Chemical Engineers [also AICE] *0, kf*
AICS Advanced Integrated Control System *fu*
AICS Air Induction Control System *jj*
AICS Airborne Inlet Control System *jj*
AICS Airborne Intercommunication System *jj*
AICS Avionics Interim Contractor Support *wp*
AID Act for International Development *-13*
AID Aeronautical Inspection Directorate [UK] *mt*
AID Agency for International Development *mt, ac1.2-21, wp, w9, hc*
AID Agile Interceptor Development *tb, st*
AID Alphanumeric Information Display *fu*
AID Altered Item Drawing *fu*
AID Artificial Intelligence Diagnostics *wp*
AID Authorized Interchangeability Drawing *fu*
AID AUTODIN Interface Device *mt*
AID Automated Information Dictionary *fu, hc*
AID Automatic Identification *fu*
AIDA Airbase Damage Assessment [HQ USAF] *wp, mt*
AIDA Aircraft Identification Assistant *fu*
AIDA Automatic Identification by Area *fu*
AIDAPS Automatic Integrated Diagnostic and Prognostic Subsystem *hc*
AIDAT Automatic Integrated Dynamic Avionics Tester *hc*
AIDAY Airborne Interceptor DAY *jj*
AIDC Aero Industry Development Center [Taiwan] *mt*
AIDE Automatic Inspection and Diagnostic Equipment *hc, jj*
AIDES Airborne Infrared Decoy Evaluation System *hc*

AIDES Analyst's Intelligence Display and Exploitation System [JCS] *mt*
AIDES Automated Image Data Extraction System *wp*
AIDES Automated Interactive Design and Evaluation System *fu*
AIDR Acceleration Insensitive Drift Rate *oe*
AIDS Acoustic Intelligence Data System *wp*
AIDS Acquired Immune Deficiency Syndrome *hn43.19.5, wp, af72.5.119, is26.1.69*
AIDS Advanced Airborne Integrated Display System *hc, jj*
AIDS Advanced Identification System *hc*
AIDS Advanced Integrated Data System *wp*
AIDS Advanced Integrated Display System [USA] *mt*
AIDS Air Force Intelligence Data System *wp*
AIDS Airborne Integrated Data System *wp*
AIDS Aircraft Identification System *fu*
AIDS Aircraft Integrated Data System *mt, pl*
AIDS Authorized Independent Demand System *fu*
AIDS Automated Information Distribution System [TRADOC] *wp, mt*
AIDS Automated Intelligence Data System *mt*
AIDS Automatic Infrared Diagnostic System *hc*
AIDS Automatic Interactive Debugging System *hd1.89.y13, fu*
AIDSC Army Information and Data Systems Command *dt*
AIE Acceptance and Inspection Equipment *hc, jj*
AIE ADA Integrated Environment *fu, mt*
AIE Airborne Interceptor Equipment *wp*
AIE Antenna Interface Equipment *kf*
AIEE American Institute of Electrical Engineers *0, hc, jj*
AIEEE American Institute of Electronic and Electrical Engineers *hc, jj*
AIES Airborne Intruder Engagement System *hc*
AIF Army Industrial Fund *jj, mt*
AIF Army Installation File *mt*
AIF Atomic Industrial Forum *hc*
AIF Audio Interchange Format *pk, kg*
AIF Automated Installation Intelligence File *mt*
AIF Avionics Integrated Fuzing *wp*
AIFF Audio Interchange File Format *hi*
AIFS Advanced Indirect Fire System *uc, hc*
AIFTDS A/B Integrated Flight Test Data System *hc*
AIFV Armored Infantry Fighting Vehicle *hc*
AIG Address Indicator Group *jj, mt*
AIG Air Intelligence Group *dt*
AIGS Acoustic Information Gathering System *fu*
AII Automated Item Identification *mt*
AIIM Association for Information and Image Management *pk, kg*
AIL (the AIL subsidiary of Eaton Corp.) [Hughes Aircraft Company customer] *aw130.6.67, hc*
AIL Administrative/ Intelligence/Logistics *fu*
AIL Airborne Instruments Laboratory [Long Island, NY] *mt, ac2.3-6, hc*
AIL AMRAAM International Licensing *aa26.11.12*
AIL Automated Interface List *fu*
AILAS Automatic Instrument Low Approach System *hc*
AILC AMRAAM International Licensing Company *hn49.12.1*
AILS Advanced Integrated Landing System *wp*
AILS Automatic Instrument Landing System [ATLAS] *hc, jj*
AIM Activity Indication Monitor *hc*
AIM Aerial Intercept Mobile *hc*
AIM AFSATCOM / IEMATS Microprocessor *mt*
AIM AIDES Interactive Metrics *fu*
AIM Air Aerial Airborne Intercept Missile [formerly GAR] * *wp, hc, jj, mt, fu, ph, wp*
AIM Air Force Office of Aerospace Industry Modernization *mt*
AIM Airman's Information Manual *af72.5.169*
AIM American Institute of Management *wp*
AIM Antenna Induced Modulation *-2*
AIM Architectural Implications of Mission *tb*
AIM Automated Input Methodology *mt*
AIMACO Air Materiel Computer *mt*
AIMD Aircraft Intermediate Maintenance Depot Department *hc, jj*
AIMD Avionic Integrated Maintenance Department [USA] *mt*
AIMDAPS Automatic Inspection, Diagnostic, and Prognosis System *hc*
AIME American Institute of Mechanical Engineers *0*
AIME American Institute of Mining, Metallurgical, and Petroleum Engineers *0*
AIMEVAL Airborne Intercept Missile Evaluation *wp, jj*
AIMI Aviation Intensive Managed Items [USA] *mt*
AIMILO Army Industrial Materiel Information Liaison Office *hc*
AIMIS Advanced Integrated Modular Instrumentation System *hc*
AIMS Ada Information Management System *hc*
AIMS Advanced Interceptor Missile Subsystem *tb, st*
AIMS Air Launched Intercept Missile Record System [USAF] *mt*
AIMS Airborne IFF Mark II System *hc*
AIMS Aircraft Identification Military System *hc*
AIMS Aircraft Identification Monitoring System *mt*
AIMS Aircraft Integrated Munition System *wp*
AIMS Airlift Implementation and Monitoring System [MAC/PACAF] *mt*
AIMS ATCRBS IFF Mk XII System *fu*
AIMS Audit Information Management System *fu*

AIMS Automated Information Management System *mt, fu, st*
AIMSS Advanced Integrated Maintenance Support System *kf*
AIMU Advanced Inertial Measurement Unit *af72.5.81*
AIMUX ATM Inverse Multiplexing *em*
AIMVAL Air Intercept Missile Evaluation *mt, hc, wp*
AIN Advanced Intelligent Network *kg, em*
AIN Army Interoperability Network *fu*
AIO Air Intelligence Officer *wp*
AIP Acquisition Improvement Program *ph*
AIP Airborne Instrumentation Platform *hc, jj*
AIP Airport Improvement Program *fu*
AIP Allied Intelligence Publication *mt*
AIP American Institute of Physics *aa26.10.b5, hc*
AIP Annual Incentive Program Plan *pbl, kf, jn, ah*
AIP Application Input Processor *mt*
AIP ATOC Implementation Plan *mt*
AIP Australian Industrial Participation *hc*
AIP AUTODIN Improvement Program *mt*
AIP Autotrack Interface Processor *-4*
AIP Avionics Integration Plan Program *wp, hc, jj*
AIP Avionics Integrity Program *wp*
AIPM Assembly Instructions, Parts and Materials *fu*
AIPPS Army Integrated Publishing and Printing Service *mt*
AIPS Advanced Intelligence Processing System *wp*
AIPS Aeronautical Information Processing System *fu*
AIPS Aircrew Integrated Protection System *wp*
AIPS Army Information Processing Standards [USA] *mt*
AIPS Astronomical Image Processing System *mi*
AIR Aerospace Information Report *fu*
AIR Air Inflatable Retarder [bomb decelerator] *mt*
AIR Air Intercept Radar *mt*
AIR Air to Air rocket [USA] (AIR-2 Genie) *mt*
AIR Airborne Infrared Sounding System *kf*
AIR Airborne Intercept Radar *wp*
AIR Aircraft Incident Report *jj*
AIR Aircraft Installation Report *jj*
AIR Aircraft Inventory Record *jj*
AIR American Institute of Research *hc, jj*
AIR Automatic Image Refinement [Canon] *hi*
AIR Aviation Infrastructure Roundtable *aw126.20.35*
AIR [Airborne data processing system] [Canada] *ac3.5-6*
AIRARM Aircraft Armament, Incorporated [Maryland] *hc*
AIRBM Anti Intermediate Range Ballistic Missile *fr, hc*
AIRCAP Air Flying Capability Model *mt*
AIRCAP Airframe Capability Model *mt*
AIRCENT Allied Air Forces Central Region *fu*
AIRCLOP Airborne Interceptor Radar Counter Low Observables [HAVE TARP] *hc*
AIRCOM Aerospace Communications Complex *mt*
AIRCOM [an ITT program] *hc*
AIRCOMNET [USAF Communications Network] *wp*
AIREP Automated Incident Report System [WWMCCS] *mt*
AIRES Advanced Imagery Requirements and Exploitation System *mt, hc, wp, ro*
AIRES Advanced Infrared Entertainment System *hn44.23.1, hc*
AIRES Automated Intelligence Reconnaissance Exploitation System *mt*
AIREVAC Aeromedical Evacuation *mt*
AIREVAC Air Evacuation Report *mt*
AIREW Airborne Infrared Early Warning [system] *wp, hc*
AIRGLO Airborne Infrared Gunfire Locator *hc*
AIRL Aeronautical Icing Research Laboratory *wp*
AIRLANT [Naval Air Force, Atlantic Fleet] *jj*
AIRLOGSITREP Air Logistics Situation Report *mt*
AIRMATCOMDK Air Materiel Command, Denmark *fu*
AIRMATCOMNOR Air Materiel Command, Norway *fu*
AIRMET Airman's Meteorological [Information *fu*
AIRMICS Army Institute for Research in Management Information, Communications, and Computer Science *mt*
AIRMOD Air Freight Module *mt*
AIRMS Airborne Infrared Measurement System *kf*
AIROPS Air Operations *mt*
AIROPS Air Operations Report System *mt*
AIRP Airborne Instrumentation Research Program [a NASA project, USA] *mt*
AIRPAC [Naval Air Force, Pacific Fleet] *jj*
AIRS Advanced Inertial Reference Sphere *mt*
AIRS Advanced Inertial Reference System *dh*
AIRS Advanced InfraRed Seeker *hn49.4.8*
AIRS Airborne Infrared Radiometer System *wp*
AIRS Airborne Integrated Reconnaissance System *hc, wp*
AIRS Aircraft Inventory Reporting System *wp*
AIRS AMDF Integration and Retrieval System [USA] *mt*
AIRS Atmospheric InfraRed Sounder [EOS] *hy, hg, ge*
AIRS Automated Automatic Image Retrieval System *hc, wp*
AIRS Autonomous InfraRed Sensor *hc*
AIRSOUTH Air Forces, Southern Region [NATO] *mt*

AIRSOUTH Allied Air Forces Southern Region *fu*
AIRSTAT Air Defense Status *mt*
AIRTAS Air Deployed Towed Array Surveillance System *wp*
AIRTASS Airborne Towed Array Experimental Modules *hc*
AIRTRAC Air Transportable Armament Control [system] *hc, fu*
AIR_CAD/CAM Aviation Computer Aided Design / Computer Aided Manufacturing *mt*
AIR_SEC Air Section / Sector *mt*
AIS Adaptive Interface Suppression *hc*
AIS Advanced Indication System *mt*
AIS Advanced Instructional System *hc*
AIS Aeronautical Information Service *fu*
AIS AFWAL Information System [ASD] *mt*
AIS Air Intelligence System *mt*
AIS Airborne Imaging Spectrometer *wp, hg, ge*
AIS Airborne Instrumentation System *hc, jj*
AIS Aircraft Instrumentation System *hc*
AIS Aircrew Instruction System *hc*
AIS American Interplanetary Society *-13*
AIS Anti Icing System *wp*
AIS AUTODIN Information System *mt*
AIS Automated Identification System *pk, kg*
AIS Automated Information System *mt, wp, tb, jdr, kf, su*
AIS Automatic Intercept System *wp*
AIS Avionics Intermediate Shop *mt, hc, wp*
AIS Avionics Intermediate Supporter Station *mt*
AISAF Artificial Intelligence Situation Assessment Focus *hc*
AISARC Automated Information Systems Acquisition Review Council *mt, wp*
AISB Association of Imaging Service Bureaus *pk, kg*
AISC Advanced Information Systems Company *hc*
AISC American Institute of Steel Construction *0*
AISC Assessment and Information Services Center *ag*
AISD Airlift Information Systems Division *wp*
AISE Artificial Intelligence Support Environment *fu*
AISE Association of Iron and Steel Engineers *aa26.8.b34*
AISE Automated Information System Equipment *su*
AISF Avionics Integration Integrated Support Facility *hc, wp*
AISG AUTODIN Intersite Gateway *wp*
AISI Army Information System Integration *mt*
AISIM Automated Interactive Simulation Model [also AISM] *17, tb, fu*
AISM Automated Interactive Simulation Model [also AISIM] *hc*
AISP Association of Information Systems Professionals *pk, kg*
AISRC Automated Information Systems Review Council *mt*
AISS Airborne Information Systems Squadron *mt*
AISS Automated Information System Security *tb*
AISSAA [Hughes Aircraft Company customer for EISS] *hc*
AIST Air Intelligence Support Team *mt, wp*
AIST Avionics Integration Support Technology *hc*
AISTE Aircraft Integration Special Test Equipment *hc*
AISWG Air Interface Standards Working Group [NATO] *mt*
AIT American Institute of Technology *wp*
AIT Assembly and Integration Trailer *ai2*
AIT Assembly, Integration and Test *oe*
AIT-SAEC Analysis Input To SAEC *jj*
AITC Army Interoperability Test Center *fu*
AITI Automated Interchange of Information *hc*
AITS Action Item Tracking System *kf*
AIU ADCOM IDHS Upgrade *mt*
AIU Antenna Interface Unit *fu*
AIU Array Interface Unit *fu*
AIU AUTODIN Interface Unit *mt*
AIWS Advanced Interdiction Weapon System [USA] *mt, hc*
AIWX Airborne Interceptor All Weather *jj*
AIX Advanced Interactive Executive [IBM] *mt, hi, kg*
AJ Anti Jam Jamming *mt, sa, ic2-12, hc, aw130.22.86, wp, ep, tb, jj, jdr, kf, uh, fu*
AJ Astronomical Journal *mi*
AJARS Airborne Radar Jamming System *wp*
AJC Anti Jam Command *uh*
AJCC Alternate Joint Communications Center *mt, wp*
AJCON Automated Job Control *mt*
AJE Anti Jam Equipment *wp*
AJEM Army Joint Exercise Manual *mt*
AJFH Anti Jam Frequency Hopper *wp*
AJH Anti Jam Hopper *wp*
AJO Anti-Jam Operator *fu*
AJP Anti-Jam Processor *fu*
AJPO ADA Joint Program Office *mt, wp*
AJRCC Alternate Joint Rescue Coordination Center *fu*
AJT Anti-Jam Technician *fu*
AK Adaptation Kit *jj*
AK Aviation Storekeeper *jj*
AK [general cargo ship, US Navy] *mt*
AKA AK cargo ship, Attack [US Navy] *mt*
AKDC Automatic Key Distribution Center *mt*

AKG (AKG 51 'Immelmann') *mt*
AKG Aufkl%orungs Geschwader [reconnaissance wing] [Germany] *mt*
AKM Apogee Kick Module Motor *bl1-2.162, 9, 16, hc, lt, ep, ce, dm, jdr, oe*
AKO Aviatsionno Kosmicheskoye Ob'edinenie [Aviation Space Organization] *ai2*
AKOH Alcoholic KOH [chemical warfare decontaminant] *wp*
AKR [vehicle cargo ship, US Navy] *mt*
AKROCC Alaskan Region Operations Control Center *mt*
AKS Apogee Kick Stage *ai2*
AKV [aircraft ferry] [often old carriers] [USA] *mt*
AL Acquisition Letter *nt*
AL Activity Leader *pbl, jn, ah*
AL Air Land *mt, wp*
AL Assembly Language *wp, hi*
ALA Allowable Lead Angle *jj*
ALABINT Air Launched Boost Intercept *hc*
ALABM Air Launched Anti Ballistic Missile *hc, jj*
ALACS A La Carte System *mt*
ALAD Automatic Liquid Agent Detector [chemical warfare] *wp*
ALADNS Automated Location And Data Netting System *hc, fu*
ALANG Alabama Air National Guard *-12*
ALAP Appletalk Link Access Protocol *hi*
ALAP Associative Linear Array Processor *hc*
ALARM Air Launched Anti Radiation Missile [UK] *mt*
ALARM Airborne Laser Receiver Module *wp*
ALARM Airborne LIDAR for Agent Remote Measurement *wp*
ALARM Alert, Locate And Report Missiles *kf*
ALARM Alerting Long range Airborne Radar for MIT *hc*
ALARM Automatic Light A/C Readiness Monitor *hc*
ALARR Air Launched Recoverable Rocket *hc*
ALASAT Air Launched Anti Satellite [weapon] *go3.2, jb, jmj*
ALAT Aviation LegÈre de l' ArmÈe de Terre ["light Air Force of the land army"] [France] *mt*
ALB Air-Land Battle *uc, tb*
ALBATROSS [a design display and analysis system] *hc*
ALBIS Aircraft Launched Boost Intercept Study *hc*
ALBM Air Land Battle Management System *kb*
ALBM Air Land Battle Management [USA/USAF/DARPA] *mt*
ALBM Air Launched Ballistic Missile *mt, 17, sb, hc*
ALC Air Logistics Center [AFMC] *mt, fu, 12, hc, af72.5.31, aw130.22.29, wp, tb*
ALC Arithmetic and Logic Circuits *pk, kg*
ALC Assured Logistics Communications *mt*
ALC Automatic Launch Control *jj*
ALC Automatic Level Control *fu, gp, uh, 4, jdr, kg, oe*
ALC Automatic Light Control *hc*
ALC Automatic Load Control *jj*
ALCA Advanced Link Cluster Algorithm *fu*
ALCAM Air Launched Conventional Attack Missile *mt*
ALCC Air Launch Control Center [USA] *mt, wp, sr9-82, hc, hs, go3.2, jb, tb*
ALCC Air Lift Control Center *hc, fu, wp*
ALCC Airborne Launch Control Center *mt*
ALCC Airlift Control Center *mt*
ALCE Airlift Control Element *fu, mt*
ALCEP AUTOSEVOCOM Life Cycle Extension Program [DISA] *mt*
ALCM Air Launched Cruise Missile [AGM-86 cruise missile carried by the B-52 and B-1] [USA] *mt, aw118.18.72, fu, hc, 12, 14, ts12-6, wp, tb, st, jj, kf*
ALCOA Aluminum Company Of America *dn*
ALCOM Alaskan Command *hc, wp, jj, mt*
ALCOM Algebraic Compiler *wp*
ALCON [AWG-9 shipboard system four weather tests] *hc*
ALCOP Alternate Command Post *-12*
ALCOR ARPA Lincoln C-Band Observables Radar *bm140, hc, ps204, go2.3.2.1, jb, tb*
ALCP Alternate Command Post *mt*
ALCS Airborne Launch Control System *wp, mt*
ALCS Airlift Control Squadron *mt*
ALCS Augmented Longitudinal Control System [USA] *mt*
ALD Accounting Line Designator [USA, DOD] *mt*
ALD Airborne Laser Densitometer *hc*
ALD At a Later Date *jj*
ALD Available to Load Date [at APOE] [USA DOD] *mt*
ALDH Aldehyde Dehydrogenase *wp*
ALDP Advanced Laser Designator Pod *hc*
ALDPS Automated Logistic Data Processing System *wp*
ALDT Administrative and Logistics Delay Time *mt*
ALDT Administrative Logistics Down Time *fu*
ALE Address Latch Enable *pk, kg*
ALE Annual Loss Expectancy *mt*
ALE Annual Loss Exposure *fu*
ALE Army Liaison Element *mt*
ALE Avionics Logistics Effects *hc*
ALEC Assembly Line Effectiveness Center *sn2.9*
ALERT Active Laser Resonator Techniques *hc*

ALERT Acute Launch Emergency Reliability Tip *pl*
ALERT Air Logistics Early Requirements Technique *mt*
ALERT Aircraft Low Energy Retrieval Technique *mt*
ALERT Attack and Launch Early Reporting to Theater *kf*
ALERTCONS Alert Conditions *wp*
ALERTS Airborne Early Warning and RV Tracking System *mt*
ALERTS Airborne Laser Equipment Real Time Surveillance *hc*
ALESA American League For Exports and Security Assistance *wp*
ALEXIS Array of Low Energy X-ray Imaging Sensors *mi*
ALF Absorption Limiting Frequency *wp*
ALF Air Lift Flight [USA] *mt*
ALF Air Lift Forces *mt*
ALF Auxiliary Landing Field *jj*
ALFA Air Land Force Application Agency [USA / USAF] *mt*
ALFA Air Land Forces Agency *wp*
ALFGL Automatic Low Frequency Gain Limiting Circuit *hc*
ALFI Air Land Force Interface *mt*
ALFS Airborne Low Frequency Sonar *af72.5.15, fu*
ALFSE Allied Land Forces South East [NATO] *mt*
ALFSEA Allied Land Forces South East Asia [a central command for these forces, created in November 1944] *mt*
ALFT Annual Long Range Airlift Requirements *mt*
ALG Aircraft Landing Gear *wp*
ALGOL Algorithmic Algebraic Oriented Language *fu, kg, wp, hi*
ALH Active Laser Homing *mt*
ALH Advanced Light Helicopter *mt*
ALI Acer Laboratories, Inc. *ct*
ALIADS Alaskan Integrated Air Defense System *hc*
ALIBI Adaptive Location of Internetworked Bases of Information *pk, kg*
ALIC Aircraft Launcher Interface Computer *mt*
ALICAT Advanced Long Wave Infrared Circuit Array Development *hc*
ALICE Automated Logistics Integrated Computer Extracts [AFMC] *mt*
ALICTS Automatic Linear Integrated Circuit Test Set *fu*
ALIM Air Launched Intercept Missile *wp*
ALIM Automatic Line Integration Medium *fu*
ALIMS Automatic Laser Instrumentation Measurement System *hc*
ALINK Active Link [HTML] *pk, kg*
ALIRATS Airborne Laser Illuminator Ranging and Tracking Scheme *hc*
ALIT Automatic Line Integration Thresholded *fu*
ALIU Automatic Line Integration Unthresholded *fu*
ALIVE Artificial Life Interactive Video Environment *ct, em*
ALIWEB Archie Like Indexing in the Web [Internet] *hi, kg*
ALL Airborne Laser Laboratory [USA] *mt, 14, wp, hc, af72.5.138, tb, st*
ALL Assembly Loading List *fu*
ALLA Allied Long Lines Agency [NATO] *mt*
ALLCE All Light Load Carrying Equipment *fu*
ALLD Airborne Laser Locator Designator *hc, wp*
ALLDS Airborne Laser Locator Designator System *hc*
ALLNAV All Navy [message to all navy personnel] [USA] *mt*
ALLWACS All Weather Attack Control System *hc*
ALM Advanced List of Materials *fu, hc, jj*
ALM Air Launched Missile *fu, jj*
ALM Airlift Loading Model [Army Data Base Generator] *mt*
ALM Amplified Modulated Link *hc*
ALM Attack Laydown Model *st*
ALM Automatic Landing Mode *jj*
ALMB Air Launched Missile Ballistics *jj*
ALMC Army Logistics Management Center *mt*
ALMEDS Alaskan Meteorological Data System *no, mt*
ALMIDS Army Logistics Management Integration Data System *mt*
ALMIS Acquisition Logistics Management Information System *mt*
ALMS Air Launched Missile Systems *jj*
ALMV Air Launched Miniature Vehicle *sd, wp, tb*
ALN Active Launch Normal *jj*
ALN Asynchronous Learning Network *pk, kg*
ALNF Application of Limited Nuclear Forces *mt*
ALNOT Alert Notice *mt*
ALO Advanced Logistics Order *fu*
ALO Air Liaison Office Officer *mt, fu, hc, jj*
ALO Airlift Liaison Officer *mt*
ALO Army Liaison Officer *mt*
ALO Authorized Level of Organization *mt*
ALOC (Cold Hunt Missions in support of the ALOC project.) [USA] *mt*
ALOC Air Lines Of Communication *mt*
ALOC Airborne visual Laser Optical Communication [USA] *mt*
ALOFT A Language Oriented to Flight engineering and Testing *wp*
ALOFT Airborne Light Optical Fiber Technology *wp*
ALOREP Airlift Operational Reporting System [USAFE] *mt*
ALOT Adaptive Laser Oscillator Technique *hc*
ALOT Airborne Lightweight Optical Tracking *wp*
ALOTS Advanced Light Optical Tracking System [USA] *mt*
ALOTS Airborne Lightweight Optical Tracking System *hc*
ALP Advanced Language Program *-10*
ALP Allied Logistics Publication *mt*
ALP Arithmetic Logic Processor *wp*
ALP Assembly Language Program *wp*

ALP Automated Library Program *wp*
ALP Automatic License Plate *hc*
ALP Automatic Logistic Process *kf*
ALPA Air Line Pilots' Association [USA] *mt, aw118.14.31, hc, jj*
ALPHA AMC Logistic Program Hardcore Automated *jj*
ALPHA Automatic Literature Processing, Handling and Analysis *wp*
ALPO Association of Lunar and Planetary Observers *mi*
ALPS Advanced Linear Programming System [Burroughs] *mt*
ALPS Advanced Liquid Propulsion System *0, wp*
ALPS Air Launched Probe System *tb, st*
ALPS Air Launched Propulsion System *hc, jj*
ALPS Alternative Low level Primitive Structure *fu*
ALPS Automated Lecture and Presentation System *fu*
ALPS Automated Logistics Planning System *mt*
ALR Advance Linebacker Radar *hc*
ALR Advanced Logic Research, Inc. *pk, kg*
ALR Airborne Laser Rangefinder *hc, wp*
ALR Airlift Requirement *mt*
ALR Artillery Locating Radar *fu*
ALRA Advanced Laser Requirements Assessment *fu*
ALRAAM Advanced Air launched Long Range Air to Air Missile *hc, jj, wp*
ALRAD Airborne Laser Rangefinder And Designator *mt*
ALRAR Annual Long Range Airlift Requirements *mt*
ALRC Airlift Logistics Readiness Center *mt*
ALRI Advanced Airborne Long Range Interceptor *hc, wp, jj*
ALRS Advanced Laser Reconnaissance Set *hc*
ALRS Alternate Landing Strip *mt*
ALS Active Laser Seeker *wp*
ALS ADA Language System *mt, fu*
ALS Advanced Landing System *wp*
ALS Advanced Laser System *hc*
ALS Advanced Launch System [vehicle] *wp, 16, ic2-63, aa26.8.26, af72.5.150, tb, mb, mi*
ALS Advanced Life Support System [also ALSS] *wp*
ALS Advanced Limb Scanner *hc*
ALS Advanced Logistics System *mt*
ALS Air Launched Segment *jb*
ALS Aircraft Landing System *wp*
ALS Airlift Squadron [USA] *mt*
ALS Approach Lighting System *fu*
ALS Augmented Logistic Support *wp*
ALSAM Advanced Strategic Air Launched Missile [also ASALM] *hc*
ALSAM Air Launched Ship/Surface Attack Missile *hc*
ALSD Advanced Large Screen Display *fu*
ALSEP Apollo Lunar Surface Experiment Package *ci.395, hc*
ALSF Approach Lighting System with Sequence Flashers *fu*
ALSIP Approach Lighting System Improvement Program *fu*
ALSRC Apollo Lunar Sample Return Container *hc*
ALSS Advanced Life Support System [also ALS] *wp*
ALSS Advanced Location Strike System *hc*
ALSS Advanced Logistic Support Site [US Navy] *mt*
ALSS Apollo Logistics Support System *hc*
ALSTS Advanced Laser Search and Track Set *hc*
ALT Accelerated Life Testing *fu*
ALT Administrative Lead Time *wp, jj*
ALT Airborne Laser Technology *hc*
ALT Airborne Laser Tracker *hc, wp*
ALT Approach and Landing Test [of STS] *ci.ix*
ALTAC Algebraic Translator And Compiler *wp*
ALTAIR ARPA Long Range Tracking And Instrumentation Radar *bm140, ps202, jb, tb*
ALTC Automatic Laser Target Classification *wp*
ALTCOM Alternate Command *wp*
ALTE Altitude Transmitting Equipment *wp*
ALTF Air Lift Task Force *wp*
ALTHQ-BS Alternate Headquarters Battle Staff *mt*
ALTRAN Assembly Language Translator *wp*
ALTS Automatic Laser Test Set *hc, tb*
ALU Air Lift Unit *mt*
ALU Arithmetic [and] Logic Unit *3, 6, wp, jj, oe, fu, kg, hi, ai1*
ALUSNA US Naval AttachÉ {?} [USA] *mt*
ALV Air Launched Vehicle *wp*
ALV Autonomous Land Vehicle *hn47.7.5, tb*
ALVRJ Air Launched Low Volume Ram Jet *hc*
ALW Advanced Laser Warning *hc*
ALW Air Launched Weapon *wp*
ALWS Air Launched Weapon School *jj*
ALWSIM Army Laser Weapon Simulation *hc*
ALWT Advanced Light Weight Torpedo *ph, fu, hc*
AM Acoustic Module *fu*
AM Active Matrix *hi*
AM Air Manifest *mt*
AM Air Mass *jj*
AM Air Medal *w9*
AM Air Movement Message *mt*

AM Airlock Module *ac2.3-3*
AM Amendment Message *fu*
AM Amplitude Modulation Modulated *mt, fu, ct, uh, bl1-5.1, wp, 0, 2, 4, hc, ce, jj, jdr, kf, vv, oe, hi, tt, ai1*
AM Angle Memory *jj*
AM Antenna Module *-4*
AM Apogee Motor *oe*
AM Area Manager *fu*
AM Arithmetic Mean *cr*
AM Asynchronous Mode *hi*
AM Avionics Maintenance *jj*
AM-CW Amplitude Modulated Continuous Wave *mt*
AM/AM Amplitude Modulation to Amplitude Modulation [conversion] *vv*
AM/LRAAM Advanced Medium or Long Range Air to Air Missile *hc*
AM/PM Amplitude Modulation to Phase Modulation [conversion] *vv, 4*
AMA Actuated Mirror Array *pk*
AMA Adaptive Multifunction Antenna *hc*
AMA Advanced Minuteman Accelerometer *hc*
AMA Air Materiel Area [later called Air Logistics Center] *fu*
AMA Angle Angular Measurement Accuracy *st, kf*
AMA Assistant Manager for Automation *fu*
AMA Automatic Message Accounting *kg, fu*
AMA Automobile Manufacturers Association *wp*
AMAA Army Mutual Aid Association *dt*
AMAC Aircraft Airborne Monitor[ing] And Control *hc, wp, jj*
AMAD Airframe Mounted Accessory Drive *fu*
AMAG Advanced Microprocessor Analysis Group *fu*
AMAMS Advanced Medium Anti Missile System *hc*
AMANDA Automatic MET and NBC Data Processing System *fu*
AMANDDA Automated Messaging and Directory Assistance *pk, kg*
AMAO Advanced Missions Analysis Office *oe*
AMAP Anti Mast Armor Program *hc*
AMAR Anti Missile Array Radar *hc*
AMARC Aerospace Maintenance And Regeneration Center *wp*
AMARC Army Materiel Acquisition Review Committee *hc*
AMARS Automatic Message Address Routing System *wp*
AMARV Advanced Maneuvering Reentry Vehicle *wp, tb, st, ro*
AMAS Advanced Manned Aerial Scout *wp*
AMAS Advanced Midcourse Active System *hc*
AMAS Advanced Multipurpose Armor System *hc*
AMAS Automated Maneuvering Attack System *wp*
AMAS Automatic Mobile Aerostation *mt*
AMAS AVFUEL Management Accounting System [USAF] *mt*
AMAS Aviation Management Accounting System *mt*
AMASM Air Materiel Area System Management *wp*
AMASS Advanced Marine Airborne SIGINT System [USMC] *mt*
AMAWS Advanced Medium Assault Weapons System *hc*
AMB Aircraft Maintenance Branch *mt*
AMB Airways Modernization Board *hc*
AMB Angular Measurement Bias *kf*
AMBA Automatic Main Beam Acquisition *kf*
AMBCS Alaska Meteor Burst Communications System *mt*
AMBIT Augmented Built In Test *hc, jj*
AMC Advanced Minuteman Computer *hc*
AMC Advanced Monopulse Countermeasures *af72.5.117*
AMC Aerodynamic Mean Chord *ai1*
AMC Aerospace Maintenance Cell *jj*
AMC Air Mobility Command [created in 1992, combining the transport part of MAC, TAC and SAC] [USA DOD] * *mt*
AMC Army Materiel Command [formerly DARCOM] [USA] *mt, hc, wp, dt, tb, st, fu, jj*
AMC Automatic Mixture Control *jj*
AMCC Ashore Mobile Contingency Communications *mt*
AMCCOM Armaments, Munitions, and Chemical Command [USA] *hc, mt*
AMCD Active Matrix Color Display *hi*
AMCDK Air Material Command, Denmark *fu*
AMCHIPOT Advanced Mode Control and High Power Optics Techniques *hc*
AMCHQ Air Material Command Headquarters *jj*
AMCM Advanced Mine Counter Measures *hc, mt*
AMCNO Air Material Command, Norway *fu*
AMCOM Armament Munitions and Chemical Command [US Army] *hc, tc*
AMCR Army Materiel Command Regulation *hc*
AMCRC Alternate Master Control and Reporting Center *fu*
AMCS Air Materiel Command System *hc*
AMCS Airborne Master Control Station *fu*
AMCS Airborne Missile Control System *hc, jj*
AMCS Assets Management and Calibration System *fu*
AMCS/SS Automated Management Control System/Supply Support *fu*
AMCSA Army Material Command Support Activity *dt*
AMCSS AFSAT II Modulation Capability Subsystem *kf*
AMCSS Airborne Missile Control Subsystem *hc*
AMCU Auto/Manual Cutover Unit *fu*
AMD Active Matrix Display *hi, kg*

AMD Advanced Micro Devices, Inc. [Sunnyvale, California] *hq5.3.3, kg, fu, ct*
AMD Air Movement Data *fu*
AMD Air Movement Designator *mt*
AMD Aircraft Maintenance Department *jj*
AMD Angular Measurement Drift *kf*
AMD Assembly Module Drawing *fu*
AMDAHL [an avionics and military electronics company] *ie5.11.9, aw130.22.77*
AMDAS Army Military Damage Assessment System *hc*
AMDEL Army Military Damage Data Element List *hc*
AMDF Army Master Data File *jj, mt*
AMDP Aircraft Maintenance Delayed for Parts *mt*
AMDS Alternate Media Delivery System *fu*
AME Airspace Management Element [USA] *fu, pm, mt*
AME Alternate Mission Equipment *jj*
AME Amplitude Modulation Equivalent *fu*
AME Angle Measuring Equipment *hc, tb, jj*
AME Average Magnitude of Error *sp659*
AMEDD Army Medical Department *mt*
AMEDP Allied Medical Publication *mt*
AMEL Active Matrix Electro Luminescent *hi*
AMELT Advanced Missile Ejector Launch Technology *hc*
AMEM Army Multiple Engagement Model *st*
AMES Automatic Message Entry System *mt*
AMETA Army Management Engineering Training Activity *mt*
AMETA Army Management Engineering Training Agency *jj*
AME_LNO Airspace Management Element Liaison Officer *fu*
AMF ACE Mobile Force [NATO] *mt*
AMF Allied Mobile Force *wp*
AMF Apogee Maneuver Firing *oe*
AMF Apogee Motor Firing *bl1.3.2, 9, lt, jn, oe, uh*
AMF Army Array Maintenance Facility *fu*
AMF Automatic Maneuvering Flaps [fitted to the A-7D] [USA] *mt*
AMF-80 [Air Force chemical warfare Shelter] *wp*
AMFA Apogee Motor Firing Attitude *id4142-10/831, lt*
AMG Air Maintenance Group [RNAS Yeovilton, UK] *mt*
AMG Air Movements Group [Royal Navy, UK] *mt*
AMG Allied Military Government *w9*
AMGAP Aerodynamic Missile Guidance Analysis Program *hc*
AMGOT Allied Military Government Occupied Territories [Allied] *mt*
AMH Automated Message Handler/Handling *mt*
AMHFS Advanced Miniature High Frequency System *hc*
AMHS Advanced Mine Hunting Sonar *hc*
AMHS Automated Message Handling System *wp, mt*
AMH_MROC AMH Multicommand ROC *mt*
AMI Active Microwave Instrumentation *ac3.7-23*
AMI Advanced Manned Interceptor *hc, wp, jj*
AMI Aeronautica Militare Italiana [Italian Air Force] *mt*
AMI Alternate Alternate Mark Inversion *ct, kf, em*
AMI American Megatrends, Inc. *kg, ct*
AMI Angle Memory Initiate *jj*
AMICUS Automated Management Information Civil Users System *wp*
AMIDS Airborne Mine Minefield Detection System *wp, hc*
AMIE AUTODIN Management Index *mt*
AMILO Army/Industry Material Information Liaison Office *hc*
AMIM Army Material Information Memorandum *mt*
AMIS Acquisition Management Information System *wp, kf*
AMIS ADCAP Management Information System *fu*
AMIS Aircraft Movement Information Center *wp*
AMIS Aircraft Movement Information System *fu*
AMIS Automated Management Information System *mt*
AMJAMS Automated Military Justice Analysis and Management System *mt*
AML Adaptive Maneuvering Logic *wp*
AML Amplitude Modulated Link *hc, fu*
AML Any Modulation Link *hd*
AML Approved Materials List *oe*
AMLCD Active Matrix LCD *hi*
AMM Advanced Missile Model *tb*
AMM Anti Missile Missile *hc, wp, jj*
AMM Avionics Maintenance Modifier *jj*
AMMA Advanced Memory Management Architecture [Everex Systems] *hi, kg*
AMMC Army Maintenance Management Center *fu*
AMME Army Multiple Message Exchange *mt*
AMME Automated Multiple Media Exchange [USA] *mt*
AMMH Annual Maintenance Man-Hours *fu*
AMMIS Aerospace Maintenance Manpower Information System [HQ USAF] *mt*
AMMIS Aircraft Maintenance Manpower Information System *mt, wp*
AMMP Approved Modification and Maintenance Program *wp*
AMMPCR Aircraft and Missile Maintenance Production Compression Report *mt*
AMMPS Advanced Mission Materials Preparation Program *wp*
AMMPTS Aircraft and Missile Maintenance Production Total System [PACAF] *mt*

AMMR Advanced Multimode Multi-Mission Radar *hc, jj*
AMMRC Army Materiel and Mechanics Research Center *hc, wp*
AMMRL Aircraft Maintenance Material Readiness List *jj*
AMMS Advanced Microwave Moisture Sensor *hy*
AMMS Allowable Material Management System *mt*
AMMS Automated Material Management System *ep*
AMMS Automated Message Management System [NATO] *mt*
AMMUS Air Force Minicomputer Multiple User System [Wang] *mt*
AMMUSNET Air Force Minicomputer Multiple User System Network *mt*
AMMWS Advanced Millimeter-Wave Seeker *fu*
AMO (cells of..2x2 cm with an AMO efficiency in the range of) *sm16-1*
AMO Advanced Materiel Order *hc, jj, fu*
AMO Air Movement Designator *mt*
AMO Automation Management Office *mt*
AMO Aviation Material Office [Norfolk, San Diego] *jj*
AMON ATM Monitoring *ct, em*
AMOP Amplitude Modulation On Pulse *wp*
AMOPS Army Mobilization and Operations Planning System *mt*
AMOS Air Force ARPA Maui Optical Station [site calibration test] [RADC] *mt, ws4.10, ps205, jb, cp, tb, ro, hc*
AMOS Army Main Optical Site *st*
AMOS Automatic Message Operating Scheduling program *mt*
AMP (AMP Federal Systems Business Group, Harrisburg, PA) *aw129.10.7*
AMP (AMP Incorporated) [an avionics/military electronics company] *aw130.22.49*
AMP Adenosinemonophosphate *wp*
AMP Advanced Management Program *tb*
AMP Advanced Manned Penetrator *wp*
AMP Advanced Meteoroid Program *hc*
AMP Advanced Minuteman Platform *hc*
AMP Aircraft Modernization Program *wp*
AMP Airways Modernization Project *fu*
AMP Amplification Message *mt*
AMP Analytical Maintenance Program *hc*
AMP Angular Measurement Precision *tb*
AMP Army Material Plan *jj*
AMP Automated Data Processing System Master Plan *mt*
AMP Avionics Master Plan *wp*
AMP Avionics Modernization Program [USAF] *hc, mt*
AMP/E Advanced Minuteman Platform/Electronic *hc*
AMPA Adaptive Multibeam Phased Array *hc*
AMPAC Airborne Multi Purpose Avionics Computer *hc*
AMPE Automated Message Processing Equipment *fu, mt*
AMPE Automated Message Processing Exchange *mt*
AMPL Approved Materials Parts List *he*
AMPL Authorized Approved Materials and Processes List *pbl, gp, ep, jdr, tm, kf, vv, jn, ric*
AMPOA Army Military Police Operations Agency *dt*
AMPR Advanced Microwave Precipitation Radiometer *hy*
AMPS Active Microwave Pressure Sounder [UK/France] *hg, ge*
AMPS Advanced Manned Penetrator System [later AMPSS] *hc*
AMPS Advanced Mobile Phone System Service [Motorola] *ric, jdr, fu, kg, hi*
AMPS Airlift Mission Planning and Scheduling System [MAC] *mt*
AMPS Atmospheric, Magnetospheric, and Plasma in Space *hc*
AMPS Automated Message Processing System *mt, wp, tb*
AMPSA Advanced Manned Precision Strategic Aircraft. Requirement, 1965 [USA] *mt*
AMPSS Advanced Manned Precision Strike System *hc*
AMPTE Active Magnetospheric Particle Tracer Explorer *sc4.9.3*
AMR (AMR Corporation, parent company of American Airlines) *aw129.1.69*
AMR Advance Material Request *jj*
AMR Advanced Microscan Receiver *fu*
AMR Advanced Modular Radar *hc, wp*
AMR Advanced Multifunctions Radar *hc*
AMR Air Movement Request *fu*
AMR Aircraft Microwave Refractometer *ag*
AMR Altitude Marking Radar *0, hc*
AMR Angular Measurement Random *kf*
AMR Area Management Region *fu*
AMR Atlantic Missile Range [later ETR] *ci.86, 0, jj*
AMR Automated Management Reports *wp*
AMR/IT Atlantic Missile Range Integrated Telemetry *hc*
AMRA Army Materiel Research Agency *hc*
AMRAAM Advanced Medium Range Air to Air Missile [also known as AIM-120, the replacement for AIM-7 Sparrow, similar in layout but smaller, and with an active seeker, USA] *mt, hn41.16.4, 17, aw130.13.71, sb, hc, aa26.11.12, af72.5.135, wp, pm, tb, jj, kf, fu*
AMRAC Anti Missile Research Advisory Council *hc, fu*
AMRAD Armament/Munitions Requirements And Development committee *dt*
AMRARM Advanced Medium Range Anti Radiation Missile [USA] *mt*
AMRBA Army Military Review Boards Agency *dt*
AMRCA Advanced Multi Role Combat Aircraft *wp*

AMREP Aircraft and Missile Maintenance Production Compression Report [USAF / PACAF / AFLC] *mt*
AMRIR Advanced Medium Resolution Imaging Radiometer *hc, hy*
AMRL Aerospace Medical Research Laboratory *wp*
AMRO Army Medical Regulating Office *mt*
AMROC American Rocket, Incorporated *aa26.8.30*
AMROO Atlantic Missile Range Operations Office *ci.86*
AMRPV Advanced Multi Mission RPV *wp*
AMRR ARPAT Medium Range Radar *hc*
AMRS Assets Management Reporting System *hc*
AMS Academy of Military Science *af72.5.32*
AMS Access Method Services *ct, em*
AMS Advanced Minuteman System *hc*
AMS Aerial Measuring System *mt*
AMS Aeronautical Material Specification *jj*
AMS Aerospace Materials Specification *fu*
AMS Air Maintenance Squadron [USA] *mt*
AMS Air Monitor System *hc*
AMS Aircraft Maintenance Squadron *mt*
AMS Alternate Master Station *fu*
AMS American Meteorological Society *pbl, ge, hg, oe*
AMS Ammunition Maintenance Squadron *mt*
AMS Antitank Missile System *hc*
AMS Apogee and Maneuvering State *sc4.4.4*
AMS Asymmetric Multiprocessing System *hi*
AMS Auto Microbic System *ci.1013*
AMS Automated Maintenance System [C5] *mt*
AMS Automated Management System [ASD] *wp, mt*
AMS Automated Measurement Subsystem *fu*
AMS Automatic Monitoring System *fu*
AMS Auxiliary Master Station *fu*
AMS Avionics Maintenance Squadron *mt*
AMS-H Advanced Missile Systems - Heavy *hq6.2.7*
AMSA Advanced Manned Strategic Aircraft [requirement, 1965] [USA] *hc, wp, mt*
AMSA Association of Metropolitan Sewerage Agencies *y2k*
AMSAR Airborne Multi role Solid-state Active array Radar [development began in 1993, with planned service introduction in 2005] [UK / France] *mt*
AMSAT (Ariane will have aboard AMSAT's ..space vehicle) *sc2.12.3*
AMSC Advanced Military Spaceflight Capability [study] *wp*
AMSC Aeronautical and Maritime Satellite Communications *-19*
AMSC American Mobile Satellite Corporation *heh2.21.4*
AMSDL Acquisition Management System and Data [requirements control] List *mt, hc, wp, tb, kf*
AMSDS Automated Management Supporting Data System *mt*
AMSE Advanced Manufacturing Systems Engineering *fu*
AMSE Aircraft Maintenance Support Equipment *hc*
AMSF Area Maintenance Support Facility *jj*
AMSL Above Mean Sea Level *mt, wp*
AMSL Armament Mode Selection Logic *hc*
AMSOE Automated Mobility Schedule of Events *mt*
AMSP Allied Military Security Publication *hc*
AMSR Advanced Microwave Scanning Radiometer [Japan] *hy, ge, hg*
AMSRC Acquisition Management Systems Review Committee *mt*
AMSS Advanced Minehunting Sonar System *fu*
AMSS Advanced Mission Sensor System *hc*
AMSS Advanced Multi Spectral Scanner *hc*
AMSS Advanced Multi Spectral Sensor *hc*
AMSS Advanced Multiple Mission Sensor System [US Navy] *fu, wp, mt*
AMSS Aeronautical Mobile Satellite for Service and Safety *wp*
AMSS Aeronautical Mobile Satellite Service *pbl*
AMSSA Army Management Systems Support Agency *dt*
AMSSA Assured Mission Support Space Architecture *mt*
AMSSB Amplitude Modulation - Single Sideband *mt*
AMST Advanced Medium STOL Transport [USA] *hc, wp, mt*
AMST Air Missile Space Threat *hc*
AMST Assembly Master Schedule Transmittal *fu*
AMSU Advanced Microwave Sounder Sounding Unit [NOAA] *ge, sc4.11.4, pl, hy, hg*
AMT Active Matrix Technology *hi*
AMT Advanced Manufacturing Technology *wp*
AMT Air Mail Terminal *wp*
AMT Assistant Manager for Training *fu*
AMTA Advanced Multicolor Tracker for AMOS *hc*
AMTA Advanced Multispectral Tracking Adjunct *-17*
AMTAF All Mobile Tactical Air Force [model] *wp*
AMTBF Attack Management Test Bed Facility *mt*
AMTD (Draw on successful AMTD POC accomplishments) *-19*
AMTEC (AMTEC Corporation . . . Huntsville, AL 35816) *jmj*
AMTEC Alkaline Metal Thermoelectric Converter *re, tb*
AMTESS Army Maintenance Test Evaluation and Simulation System *hc*
AMTEX Air Mass Transformation Experiment *pl*
AMTF Airmobile Mechanized Task Force *mt*
AMTF Average Modulation Transfer Function *hi*
AMTFT Active Matrix Thin Film Transistor *hi*

AMTI Advanced Missile Technology Installation *wp*
AMTI Airborne Moving Target Indicator *mt, hc, wp, fu, jj*
AMTRAC Airlift Monitoring and Tracking System [USAFE] *mt*
AMTS Advanced Moisture and Temperature Sounder *ge, hg*
AMTS Array Module Test Set *fu*
AMU Add on Memory Unit *fu*
AMU Aircraft Maintenance Unit *mt*
AMU Alarm Monitoring Unit *fu*
AMU Alternate Master Unit *fu*
AMU Analog Monitor Unit *jj*
AMU Astronaut Maneuvering Unit *ci.375*
AMU Atomic Mass Unit *wp, le32, w9, ai1*
AMU Autonomous Maintenance Unit *jj*
AMU Auxiliary Memory Unit *hc, jj*
AMUC Auxiliary Memory Unit Controller *hc*
AMUS Automated Manpower Utilization System [USA] *mt*
AMUX Avionics Multiple Bus *wp*
AMUX Avionics Multiplexing [system] *hc*
AMVER Automated Mutual Assistance Vessel Rescue System [USCG] *mt*
AMVETS American Veterans [of World War II] *w9*
AMW Air Mobility Wing [USA] *mt*
AMW Anti Missile Warfare *wp*
AMX [a short take off aircraft] *aw130.22.89*
AN Access Node *hi*
AN Airman, Navy *w9*
AN Alteration Notice *pl*
AN Analysis Note *fu*
AN Army/Navy *fu, hc, nt*
AN Ascending Node *oe*
AN/APS Army-Navy / Airborne Pulse Search equipment [a radar set carried by aircraft] [USA] *mt*
AN/GYQ {Army-Navy} GYQ [military designation for PDP-11-Based Computers] *mt*
ANA Air Force-Navy Aeronautical *jj*
ANA All Nippon Airways *aw129.21.127*
ANA Antinuclear Antibody *wp*
ANA Army Network Architecture *mt*
ANA Army Nuclear Agency *dt*
ANA Army-Navy-Air Force *hc, jj*
ANA Automatic Network Analyzer *jdr*
ANA Automatic Number Announcement *ct*
ANAC Analog Card *fu*
ANAD Army Navy Aeronautical Design *hc*
ANAF Army Navy Air Force *jj*
ANAF AUTOSEVOCOM Network Assessment Facility *mt*
ANC Access Node Control *mt*
ANC Active Nutation Control *gd2-46, hc, 8, 9, lt, pbl, oe*
ANC Adaptive Noise Canceller *fu*
ANC American National Standard *fu*
ANC Ashore Navigation Center *fu*
ANC Automatic Nautical Charting *hc*
ANCA Allied Naval Communications Agency [NATO] *mt*
ANCIB Army Navy Communications Intelligence Board *-10*
ANCICC Army Navy Communications Intelligence Coordinating Committee *-10*
ANCOVA Analysis of Covariance *pk, kg*
ANCP Adaptive Network Cognitive Processor *hc*
ANCS AUTOVON Network Control Subsystem *mt*
AND (Coordinated Military standards . . . including . . . AND's) *nt*
AND Active Nutation Damper *gs2-8.2*
ANDE Active Nutation Damper Electronics *gs2-8.28*
ANDF Architecture-Neutral Distribution Format *pk, kg*
ANDIP American National Dictionary of Information Processing *fu*
ANDVT Advanced Narrowband Digital Voice Terminal *mt, hc, fu, wp, kf*
ANEW Airborne Navy Early Warning *fu*
ANFE Aircraft Not Fully Equipped *mt, hc, jj*
ANG Air National Guard [USA] *fu, mt, hn44.9.10, wp, hc, af72.5.5, jj, kf*
ANG Army National Guard [USA] *mt*
ANG Atlantique Nouvelle Generation [French anti-submarine aircraft] *mt, fu*
ANG/WARS Air National Guard Workday Accounting and Reporting System *mt*
ANGB Air National Guard Base [USA] *mt, 12, af72.5.55, kf*
ANGLICO Air / Naval Gunfire Liaison Company [USA DOD, an organization composed of Marine and Navy personnel specially qualified for shore control of naval gunfire and close air support] *mt*
ANGOSTA Air National Guard Operational Support Transport Aircraft [USA] *mt, af72.5.141, wp*
ANGSC Air National Guard Support Center *mt*
ANHR Access Node Hub Router *hi*
ANI Automatic Number Identification *kg, mt, ct*
ANIK [Eskimo brother] [Canadian Communications Satellite] *hc, pl, lt*
ANIP Army Navy Instrumentation Program *hc*
ANK Alphanumeric Keyboard *fu*
ANL Automatic Noise Limiting Limiter *mt, wp, jj*
ANMC American National Metric Council *fu*

ANMCC Alternate National Military Command Center *mt, fu, 12, wp, hc, jb, tb, kf*
ANN Artificial Neural Network *pk, kg*
ANNA Army, Navy, NASA, and Air Force *le37*
ANO Air Navigation Order *mt*
ANORS Anticipated Not Operationally Ready Supply *jj, mt*
ANOVA Analysis Of Variance *w9, wp, kg*
ANP Aircraft Nuclear Propulsion [an attempt to build a nuclear-powered aircraft, primarily a strategic bomber, USA] *mt*
ANP Allied Navigation Publication *mt*
ANPO Aircraft Nuclear Propulsion Office *dt*
ANR Access Node Router *hi*
ANR Alaskan NORAD Region *mt*
ANRCC Alaskan NORAD Region Control Center *mt*
ANROCC Alaskan NORAD Region Operations Control Center *mt*
ANRS Advanced Night Reconnaissance Sensor *hc*
ANS Advanced Network Services *hi*
ANS Airborne Navigation System *mt*
ANS American Nuclear Society *fu*
ANS Astronomical Netherlands Satellite {1974} *ci.736*
ANS Automatic Navigation System *mt*
ANSCS Andean Nations Satellite Communication System [for education] *hc*
ANSD (a comprehensive investigation into cluster and ANSD problems) *ie5.11.7*
ANSI American National Standards Institute *do369, hc, aa26.11.b4, aw130.6.70, nt, wp, ns, dm, tb, jj, as, jdr, kf, mb, oe, ct, em, fu, kg, nu, mi, hi, mt*
ANSP Adaptive Network Sensor Processor *hc, wp*
ANSS Active Network Sensitivity Study *hc*
ANSYS Analysis System [a program for finite element analysis] *md2.1.1, aw130.22.41, ep*
ANT Abstract of New Technology *hc, jdr, fu*
ANT Air Navigational Terminal *hc*
ANT Armament NuclÉaire Tactique [nuclear armed part of the Air Force] [France] *mt*
ANT Automatic Night Tracking [system] *hc*
ANT [German telecommunications company, member of Eurosatellite consortium] *sc4.3.3, es55.4*
ANTP Automatic Navy Training Plan *fu*
ANTS Airborne Night Television System *hc, wp*
ANTS Alternate Net Time Station *fu*
ANTSAM Anti Surface to Air Missile [missile] *hc*
ANUTC Adjusted Negotiated Unit Target Cost *mt*
ANVIL Artificial Neural Vision Learning System *hc*
ANVIS Aviator Night Vision Imaging System [USA] *mt, hc, aw130.11.14, hd1.89.y22*
ANWC Alternate National Warning Center *mt*
ANWD Alpha Numeric Wall Display *mt*
ANWSG Army Nuclear Weapons Surety Group *dt*
ANX Automotive Network Exchange [Chrysler/Ford/General Motors] *pk, kg*
ANZUK Australia, New Zealand and UK [military alliance] *mt*
ANZUS Australia, New Zealand, and United States [a treaty] *mt, wp*
AO Action Officer *mt, wp*
AO Adaptive Optics *dh*
AO Aerosol Obscurant *wp*
AO Air Order *mt*
AO Airborne Optics *tb, st*
AO Allotment Order *mt*
AO Analog Output *jj, kg*
AO Announcement of Opportunity *ge, ci.718, es55.63, so, hy, re, hg*
AO Area of Operations [USA DOD] *fu, mt*
AO Artillery Observer *fu*
AO Assembly Order *fu*
AO Assisting Organization *hc*
AO Atomic Orbital *wp, ai1*
AO Aviation Ordnanceman *jj*
AO Oiler ship [US Navy] *mt*
AO Operational Availability *fu*
AO-N Administrative Office, Navy *dt*
AOA Abort Once Around [Shuttle abort plan] *ci.608, ws4.10, mb, mi*
AOA Airborne Observation Adjunct *tb*
AOA Airborne Optical Adjunct *mt, fu, uc, wp, hc, 17, sb, th4, jmj. tb, st, kf*
AOA Amphibious Objective Area [US Navy] *mt*
AOA Amphibious Operations Area *mt, fu, hc, jj*
AOA Angle Of Arrival *fu, hc, wp*
AOA Angle Of Attack *mt, hc, aw130.22.28, wp, jj, mb, mi, ai1*
AOA Atlantic Ocean Area *jj*
AOAP Army Oil Analysis Program Standard Data System *mt*
AOAT Allowed Off Aircraft Time *jj*
AOB Advanced Operations Base [USA DOD] *mt*
AOB Air Order of Battle *mt, wp*
AOB Alternate Operating Base *mt, wp*
AOB Automatic Optical Bench *hc*
AOB Average On Board *fu*

AOBHIST AOB History File *mt*
AOC Advanced Operational Capability *mt*
AOC Air Operations Center [USAF] *mt, fu, hc, tb, kf*
AOC Airlift Operations Center *mt*
AOC Amphibious Operations Center *fu*
AOC Area Operations Center *mt*
AOC Army Operations Center *mt*
AOC Attitude and Orbit Control *ggIII.1.55, oe*
AOC Automated Operations Center *fu*
AOC Automated Operations Control *hc*
AOC Automatic Overload Control *fu*
AOC Aviation Officer Cadet *mt*
AOCC Air Operations Coordination Center *mt*
AOCC Alternate Operations Control Center [USAF] *tb*
AOCE Attitude and Orbit Control Electronics *oe*
AOCI Airport Operators Council International *aw126.20.32*
AOCM Advanced Optical Counter Measures *hc*
AOCP Airborne Operation Computer Program *fu*
AOCP Aircraft Out of Commission For Parts *mt, wp, jj*
AOCP Attitude/Orbit Control Program *oe*
AOCS Attitude and Orbit Control Sub System *gd2-28, sf31.1.20, es55.60, aw130.6.68, mb, mi, oe*
AOD Aeronautical Operations Division *fu*
AOD Airlift Operations Directive *mt*
AODS All Ordnance Destruct System *-17*
AOE Application Operating Environment [AT&T] *hi, kg*
AOE Army Of Excellence *mt*
AOE [fast combat support ship, US Navy] *mt*
AOE [fast combat support ship] *jj*
AOG Aircraft On the Ground *mt*
AOG [gasoline carrier / tanker, US Navy] *mt*
AOH Available Operating Hours *mt*
AOHR Airborne Over the Horizon Radar *wp*
AOI Area Of Interest *mt, wp, kf, fu*
AOJ Acquisition On Jam *jj*
AOL Active Oxygen Loss *hc*
AOLO Advanced Orbital Launch Operations *hc*
AOM/NM Administration, Operations and Maintenance / Network Management *mt*
AON Special Purpose Air Arm [strategic force created in 1936] [USSR] *mt*
AOO Aviation Ordnance Officer *wp*
AOP Additive Operational Project *fu*
AOP Aerial Observation Post *mt*
AOP Airborne Observation Post *mt*
AOP Airborne Optical Platform *hc, wp, tb, st*
AOP Allied Ordnance Publication *mt*
AOP Application Output Processor *mt*
AOP Assembly and Operations Plan *0*
AOPA Aircraft Owners and Pilots' Association *aw132.10.9*
AOPF Active Optical Proximity Fuze *hc*
AOPL Association of Oil Pipelines *y2k*
AOR Airborne Optical Radar *mt*
AOR Atlantic Ocean Region *lyg*
AOR Replenishment Oiler [US Navy] *mt*
AOS Acquisition Of Signal *4, tb, oe, uh*
AOS Add Or Subtract *pk, kg*
AOS Airborne Optical Sensor *fu, hc, wp, tb*
AOS Airborne Optical Surveillance *fu, 17, hc, wp, tb*
AOS Airborne Optical System *fr, tb, st*
AOS Archive and Operations System *hy*
AOS-M AOS-Midcourse *tb, fu*
AOS-S AOS-Submarine Launched Ballistic Missile *tb*
AOS-T AOS-Terminal *tb*
AOS/VS Advanced Operating System/Virtual Storage *oe, fu*
AOSC Alternate Operational Support Center [USAFE] *mt*
AOSO Advanced Orbiting Solar Observatory *le25, hc, 0*
AOSP Advanced On board Signal Processor [RADC] *1, 17, hc, tb, mt*
AOSP Advanced Orbiting Solar Panel *hc*
AOSS AOS SLBM *fu*
AOSS Assembly Order Status System *fu*
AOST AOS Terminal *fu*
AOT Acquisition Active On Target *hc, jj*
AOT Alignment Optical Telescope *hc*
AOT Angle Only Track *tb, st*
AOT [transport oiler / tanker, US Navy] *mt*
AOTDD Active Optical Target Detecting Device *hc*
AOTH Acquisition On Target Held *jj*
AOTN ACE Operational Teletype Network [NATO] *mt*
AOTS Advanced On-the-job Training System *hc*
AOTS ASOC Optical Transceiver System *mt*
AOTV (Impact Ionization on Martian Return AOTV Flowfields) *aa27.4.b23*
AOU Area Of Uncertainty *kf*
AOU Arithmetic Output Unit *hi*
AOV Area Outline Video *fu*

AP Acquisition Phase *fu*
AP Acquisition Plan *mt, tb, st, jj, kf*
AP Active/Passive *tb*
AP Adjunct Processor *pk, kg*
AP Aerial Port *fu*
AP Air Patrol *mt*
AP Airborne Platforms *mt*
AP Airspace Probe *fu*
AP Alert Processing *fu*
AP Allied Publication [NATO] *mt*
AP Ammonium Perchlorate *aa26.12.38, ai2*
AP Analysis Package [SPACECMD] *mt*
AP Antenna Pattern *mt*
AP Anti Personnel *hc, w9*
AP Anticipation Point *jj*
AP Applications Processor Processing *mt, hc, kg, oe*
AP Arithmetic Processor *fu*
AP Armor Piercing *w9, mt*
AP Array Processor *muc, wp, tb, oe*
AP Assembly Planning *fu*
AP Associated Press *sc21.12.4, w9*
AP Attack Planning *fu*
AP Auto Pilot *jj*
AP Automatic Programming *mt*
AP Automation Programmer *fu*
AP Auxiliary Processor *fu*
AP [troop ship, non-landing, US Navy] *mt*
AP-S Antennas and Propagation Society *mj31.5.15*
APA Adaptive Packet Assembly *hi, kg*
APA Airborne Power Adapter *fu*
APA All Points Addressable *ct, kg, em*
APA Appropriations Purchase Account *jj*
APA Arithmetic Processing Accelerator *pk, kg*
APA Avionics Planning Activity *wp*
APA [attack transport ship, US Navy] *mt*
APADE Automation of Procurement and Accounting Data Entry System *mt*
APAFO Advanced Particles and Fields Observer *ge, hg*
APAG Atlantic Policy Advisory Group [NATO] *mt*
APAL AP Assembly Language *fu*
APAM Anti Personnel and Anti Materiel [USA] *mt*
APAMS Automated Pilot Aptitude Measurement System *wp*
APANA Australian Public Access Network Association *em*
APAR Authorized Program Analysis Report [IBM] *pk, kg*
APAREN Address Parity Enable [IBM] *pk, kg*
APAS Automatic Performance Analysis *wp*
APAT Automated Prediction Analysis Technique *jj*
APATS Acquisition Planning and Tracking System [USA] *mt*
APATS ARIA Phased Array Telemetry System *hc, wp*
APB Acquisition Program Baseline *wp*
APB Avionics Planning Baseline *wp*
APC Accelerated Provisioning Concept *hc*
APC Adaptive Predictive Coding *mt*
APC Adequate Price Competition *kf*
APC Aim Point Correlator *hc*
APC Air Pollution Control *wp*
APC Airborne Particulate Cleanliness *he*
APC American Power Conversion *ct*
APC Antenna Pattern Correction *ac7.4-11, pl*
APC Approach Power Control/Compensator *jj*
APC Armament Practice Camp *mt*
APC Armored Personnel Carrier *mt, hc, fu, w9, wp*
APC Automatic Phase Control *0*
APC Automatic Power Compensator *jj*
APC Automatic Power Control *jdr*
APCA Audio Peak Clipping Amplifier *hc*
APCC Aerial Port Control Center *mt*
APCC Alternate Processing and Correlation Center *kf*
APCE Antenna Advanced Positioner Control Electronics *hc, jj, jdr*
APCI Armor Piercing Capped Incendiary *wp*
APCI Atmospheric Pressure Chemical Ionization *wp*
APCM Adaptive Pulse Code Modulation *sp56*
APCM Aerial Port Capability / Manpower Simulation Model [MAC*] *mt*
APCO Association of Public Safety Communications Officials *y2k*
APCP Army Potential Contractor Program *mt*
APCS Action Paper Control System [OJCS] *mt*
APCVD Atmospheric Pressure Chemical Vapor Deposition *wp*
APD Advanced Program Development *cca, hc, jj*
APD Advanced Programs Division [of RSG at Hughes Aircraft Company] *44.12.3*
APD Advanced Projects Department *fu*
APD All Purpose Decontaminant [chemical warfare] [Monoethanolamine + Isopropanolamine + Lioh] *wp*
APD Attitude Pulse Digitizer *-9*
APD Automated Problem Detection *fu*
APD Avalanche Photo Diode *fu, wp, jdr*

APD Fast Destroyer Transport [US Navy] *mt*
APDS Advanced Personnel Data System [USAF] *mt*
APDS Antenna Pattern Data System *hc*
APDS Automated Personnel Data System *wp*
APDS Automated Postal Directory Service [USMC] *mt*
APDSMS Advance Point Defense Surface Missile system *hc*
APE Advanced Production Engineering *hc*
APE Advanced Production Engineering [Army] *fu*
APE Analog Processing Electronics *-4*
APE Antenna Positioning Electronics *bl2-7a.16, 9, lt, ep, jdr*
APE Automatic Plot Extractor *fu*
APEB Army Physical Evaluation Board *dt*
APECS Aerospace Planning, Execution, and control System *aw129.14.94*
APES Aerial Port Exploitation Squadron *mt*
APES Army Profit Evaluation System *fu*
APES Automated Patient Evacuation System [MAC*] *mt*
APEX Ariane Passenger Experiments *ac3.7-5*
APF Afloat Pre-Positioning Force [also AFP] *wp*
APFO Aerial Photography Field Office *ci.877*
APFSDS-DU Armor Penetrating, Fin-Stabilized, Discarding Sabot-Depleted Uranium *mt*
APG Aberdeen Proving Ground [Aberdeen, Maryland] *mt, fu, wp, hc, jj*
APG Air Force Planning Guide *wp*
APG Air Proving Ground *hc*
APG Antenna Positioner Group *-17*
APG Automatic Priority Group *fu*
APGC Air Proving Ground Center [Elgin AFB, Florida] *jj*
APGM Autonomous Precision Guided Munition *hc*
APHIDS Advanced Panoramic Helmet Interface Designator System *mt*
APHIDS Airborne Positive Hostile Identification System *hc*
API Advanced Performance Interceptor *hc*
API Air Position Indicator *w9, jj*
API Aircraft Performance Indicator *hc*
API American Petroleum Institute *fu*
API Antecedent Precipitation Index *ac7.5-21*
API Aperture Processed Imagers *sa830182P.ii*
API Application Program Interface *kg, ct, em*
API Armor Piercing Incendiary *mt, wp*
API Assembly Planning Instruction *fu*
APIC Advanced Programmable Interrupt Controller *em, kg*
APIC Arithmetic Processor Interface Control *fu*
APICS American Production and Inventory Control Society *hn49.11.12*
APIIS Airborne Passive Infrared Imaging System *hc*
APIS Advanced Passenger Information System *pk, kg*
APIS Airborne Passive Identification System *hc*
APIS Application Program and Information System [AFLC] *mt*
APIT Armor Piercing Incendiary Tracer *wp*
APIU Adaptive Programmable Interface Unit *fu*
APIU Analog Processor Interface Unit *hc*
APJ Astrophysical Journal *pl, mb*
APL A Programming Language *mt, fu, kg, gb, wp, hi*
APL Acceptance Performance Level *fu*
APL Advance Parts List *fu*
APL Advanced Programming Language *fu*
APL Aero Propulsion Laboratory [of AFWAL] [AFSC*] *tc, mt*
APL Algorithmic Programming Language *hi*
APL Allowance Parts List *fu, jj*
APL Applied Physics Laboratory [Navy] *ci.173, fu, uh, kf, oe*
APL Approved Parts List *mt*
APL Approved Processes List *tb*
APL Assembled Assembly Program Listing *fu, jj*
APL Assembly Parts List *hc, jj*
APL Authorized Approved Parts List [a document] *ah, bl2-9.50, 8, 17, sb, ep, tb, pbl, tm, kf, jn, oe, he, jdr*
APL Automatic Phase Lock *fu*
APL Auxiliary Personnel Lighter [US Navy] *mt*
APL Average Picture Level *2, 4*
APL/AEL Allowance Parts/Equipage List *fu*
APLU Automatic Programming Loading Unit *jj*
APM Acquisition Program Project Manager *fu, mt*
APM Advanced Penetration Model [model] *wp*
APM Advanced Power Management *ct, em, kg, hi*
APM Aircraft Porous Media, Incorporated *hq5.2.3*
APM Angle Position Memory *jj*
APM Antenna Positioner Mechanism *bl2-6.36, hc, lt, jj*
APM Assistant Program Project Manager *fu, 4, kf*
APM Associate Project Manager *fu*
APM Associate Proposal Manager *fu*
APM Atmospheric Pressure Mass Spectroscopy *wp*
APM Attached Pressurized Module [a.k.a. Columbus] *mi, es55.46, mb*
APM Automated Performance Measurement *wp*
APMAL Aerial Port Management Level *mt*
APMD Automatic Pitch Momentum Dumping *jdr, uh*
APME Associative Processor Microelectronic Element *hc*
APMIS Automated Project Management Information System *wp*

APML Assistant Program Project Manager, Logistics *jj, mt*
APMO Area Petroleum Management Office *mt*
APMPL Approved Parts, Materials, and Processes List *ep*
APMT Asia Pacific Mobile Telecommunications Company *kf*
APN Aircraft Procurement, Navy *jj*
APN Army Part Number *hc, jj, fu*
APO Antarctic Projects Office *dt*
APO Army Air Force Post Office *wp, tb, su, hc, w9, wp, af72.5.154*
APO Automatic Purchase Order *mt*
APOD Aerial Port Of Debarkation *mt, wp*
APOE Aerial Port Airport Of Embarkation *mt, 17, wp*
APOJI Automatic Processing of Jezebel Information *fu*
APORT Aerial Port *mt*
APORTS Aerial Port Capability / Manpower Simulation Model *mt*
APORTS Aerial Ports File [WWMCCS] *mt*
APOSTLE Analysts' Program for Open Source Literature Exploitation *fu*
APP Acoustic Performance Prediction *fu*
APP Advanced Planetary Probe *hc*
APP Advanced Procurement Plan *fu*
APP Allied Procedures Publication *mt, fu*
APP Army Procurement Procedure *hc*
APPA American Public Power Association *y2k*
APPC Advanced Program to Program Communications *ct, em, kg, hi*
APPCP Advanced Program to Program Communications Protocol [IBM] *hi*
APPI Advance Planning Procurement Information [Army] *hc*
APPL Authorized Program Parts List *fu*
APPLE Apollo Pay Load Exploration *hc*
APPLE Ariane Passenger Payload Experiment *ac3.2-3*
APPM Application Project Manager *fu*
APPN Advanced Peer to Peer Networking [IBM] *mt, hi, kg*
APPS Agency Procurement Requests *mt*
APPS Analytic Photogrammetric Positioning System [DMA] *mt*
APPS Automated Personnel Procurement System *wp*
APPS Automated Publications and Printing System [USA] *mt*
APPS Automated Publications Production System *mt*
APPS Automatic Pattern Recognition *mt*
APPS Automatic Proposal Preparation System *fu*
APQ Available Position Quality *fu*
APR Agency Procurement Request *mt*
APR Air Priority *0*
APR Airmen Performance Report *mt*
APR Annual Percentage Rate *hn41.22.1, w9*
APR Atmospheric Profile Record *pl*
APR Automated Problem Resolution *fu*
APR Automated Power Reserve *fu*
APRC Army Physical Review Council *dt*
APREP AMRAAM Producibility Enhancement Program *hc*
APREQ Altitude Reservation Approval Request *fu*
APRO Army Procurement Research Office *hc*
APRP Adaptive Pattern Recognition Processing *pk, kg*
APS ADCAP Propulsion System *fu*
APS Advanced Photo System *pk, kg*
APS Advanced Photon Source *is26.1.76mm*
APS Advanced Planning System [USAF] *mt, pf1.27, ro*
APS Advanced Printing Service [IBM] *pk, kg*
APS Advanced Propulsion System [AFSC*] *mt*
APS AFLC Provisioning System [AFSC*] *mt*
APS Aft Provision Stage *kgl*
APS Aircraft Pilot Simulator *fu*
APS Aircraft Prepared for Service *mt*
APS Airport Site *jj*
APS Airway Planning Standard *fu*
APS Alternate Path Selection *fu*
APS American Physical Society *aw126.20.104*
APS Antenna Positioning System *jdr*
APS Antennas and Propagation Symposium *fu*
APS Application Software *fu*
APS Applications Set *fu*
APS Army Packaging System *mt*
APS ASARS Processing Segment *hc*
APS Asynchronous Protocol Specification *pk, kg*
APS Automated Publication System *fu*
APS Auxiliary Power Supply *jdr*
APS Auxiliary Propulsion SubSystem *jdr, oe*
APS Transport Submarine [US Navy] *mt*
APSE ADA Programming Support Environment *mt, kg, at, tb, fu*
APSI Advanced Propulsion System Integration *wp*
APSIN Alaskan Public Safety Information Network *mt*
APSL Advanced Programs Support Laboratory [Hughes Support Systems] *17, sb*
APSM Automated Power System Management *hc*
APSS [vehicle transport submarine, US Navy] *mt*
APT Acquisition Pointing and Tracking *tb, st*
APT Address Pass Through *pk, kg*

APT Advanced Parallel Technology *pk, kg*
APT Air Passenger Terminal *mt*
APT Airborne Pointer/Tracker Pointing and Tracking *hc, wp, st*
APT Army Pointing and Tracking [program] *hc*
APT Automatic Picture Terminal *oe*
APT Automatic Picture Transmission *ge, ac3.1-10, 14, hc, jj, hg, oe, ai1*
APT Automatically Programmed Tool *fu, kg*
APTA American Public Transit Association *y2k*
APTAAM Advanced Processor Technology Air to Air Missile *hc*
APTAS Automatic Programming Technologies for Avionics Software *hc*
APTS Airborne Pointing and Tracking System *hc*
APTS Automated Package Test Set *fu*
APTU Army Participating Test Unit *fu*
APU Analog Processing Unit [GOES S/DB] *pl*
APU Autopilot Unit *jj*
APU Auxiliary Pointing Unit *hc*
APU Auxiliary Power Unit *mt, fu, mi, ci.343, wp, 0, 9, aw129.21.38, jj, mb*
APU Auxiliary Processing Unit *oe*
APUP Analog Programmable Microprocessor *hc*
APV Automatic Problem Verification *jj*
APV-F [plasma and wave instrument on Phobos mission] *es55.98*
APVL Approved Parts and Vendors List *hc, jj*
APVO [Soviet Air Defense Aviation] *wp*
APWM Adaptive Pulse Width Modulation *-17*
APXS Alpha, Proton, X-ray Spectrometer *re*
AQ Aviation Fire Control Technician *jj*
AQAP Allied Quality Assurance Publication *fu*
AQCESS Automated Quality of Care Evaluation for Support Systems *mt*
AQF Advanced Quick Fix [USA] *mt*
AQL Acceptable Acceptance Quality Level *gp, hc, jj, jdr, fu*
AQMD Air Quality Management District *fu*
AQPSK Asymmetrical Quadra-Phase Shift Keying *oe*
AQR Assembly Quality Record *jj*
AQT Acceptable Acceptance Quality Test *fu, jj*
AQUILA [reconnaissance drone] *wp*
AR Acceptance Requirements *jj*
AR Acceptance Review *tb*
AR Accuracy Ratio *fu*
AR Acquisition Radar *fu*
AR Active Radar *mt*
AR Aeronautical Requirements *jj*
AR Air Refueling *mt, wp*
AR Airborne Radar *fu*
AR Airborne Rotary wing *fu*
AR Altitude Return *jj*
AR Amplifier Research [electronics firm, Souderton, Pennsylvania] *mj31.5.25*
AR Angle Ranging *hc, jj*
AR Anti Radar *wp*
AR Anti Reflection *wp*
AR Apportionment Request *mt*
AR Army *mt*
AR Army Regulation *mt, w9, nt, tb, st, jj*
AR Aspect Ratio *wp, tb, jj*
AR Axial Ratio *jdr*
AR [repair ship, US Navy] *mt*
AR [training ship, USA] *mt*
ARA Aerial Rocket Artillery *jj*
ARA Airborne Radar Approach *mt, wp*
ARA Aircraft Replaceable Assembly *hc, jj*
ARA Assigned Responsible Agency *mt, wp, jj*
ARA Autoresonant Accelerator *wp*
ARAAV Armored Reconnaissance Airborne Assault Vehicle *wp*
ARAC Aerospace Research Application Center *ci.1005*
ARAC Atmospheric Release Advisory Capability *mt*
ARADCOM Armament Research and Development Command [later AMCCOM] [US Army] *hc*
ARADCOM Army Air Defense Command *wp*
ARAG Anti Reflective Anti Glare *hi, kg*
ARAMIS Advanced Repeater for Aeronautic and Maritime Integrated Services *sf31.1.23*
ARAMIS Automations Robotics and Machine Intelligence Systems *hc*
ARANG Arkansas Air National Guard *-12*
ARAP Army Remedial Action Project *mt*
ARAPS Area Requirement and Product Status *mt, hc*
ARAR Accident Risk Assessment Report *17, sb, jdr, kf*
ARAS Antireflective-Antistatic *pk, kg*
ARATE Autofocus Requirements Analysis and Technique Evaluation *hc*
ARB Airport Rotating Beacon *fu*
ARB Anomaly Review Board *oe*
ARB [repair ship, battle damage, US Navy] *mt*
ARBAT Application of Radar for Ballistic Ammunition Acceptance Testing *fu, hc*
ARBF Active Receive Beamformer *fu*
ARBS Angle Rate Bombing System Set [Hughes AN/ASB-19, V] *mt, fu, hn41.17.1, hq5.2.2, hc, jj*

ARC (Conserve the ARC/Spectrum Resource Through Technology Innovations) *-19*
ARC Accessory Record Card *jj*
ARC Acquisition Reportback Crosslink *jr*
ARC Acquisition Review Council Committee [DCA] *mt, wp, jdr*
ARC Advanced Raster Console *fu*
ARC Advanced Research Center [Huntsville, Alabama 35816, USA] *fu, mt, nt, jmj, tb, st*
ARC Advanced RISC Computing *em*
ARC After Receipt of Contract *fu*
ARC Airborne Radar Control *mt*
ARC Airlift Readiness Center *mt*
ARC Airlift Requirements Center [USAFE] *mt*
ARC Airlift Requirements Collector *mt*
ARC Ames Research Center [California] *sm18-1, 0, wp, ns, tb, wl, jdr, mb, mi, oe*
ARC Application Review Committee *fu*
ARC Area Reprogramming Capability *mt, hc, wp*
ARC Aspect Ratio Corrected *oe*
ARC Assembly Record Card *fu*
ARC Astrophysical Research Consortium *mi*
ARC Atlantic Research Company *jn*
ARC Atlantic Richfield Corporation *ai2*
ARC Attached Resources Computing *em, ct*
ARC Automatic Range Control *wp*
ARC Automatic Reaction Control *wp*
ARC Automatic Responsibility/Responsivity Control *hc, jj*
ARC Average Response Computer *wp*
ARC [cable repair / laying ship, US Navy] *mt*
ARCA Advanced RISC Computing Architecture *hi, kg*
ARCADE Audits/Reviews Corrective Action Database *fu*
ARCC Airborne RPV Control Center *fu*
ARCC cyclic Redundancy Check Character *hi*
ARCCNET Army Command and Control Network *mt*
ARCEA Armed Forces Communications and Electronics Association *wp*
ARCENT US Army Forces, Central Command [Third US Army] *mt*
ARCM Atlas Roll Control Module *ai2*
ARCNET Attached Advanced Resource Computer Network *hi, kg*
ARCOM Army Reserve Command *mt*
ARCONET Army Command and Control Communications Network *wp*
ARCP Air Refueling Contact Point *fu*
ARCP Air Refueling Control Point *fu, mt, wp*
ARCS Achievement Rewards for College Scientists [Foundation] *hn44.13.5*
ARCS Acquisition Radar and Control System *fu, hn44.3.1, hc*
ARCT Air Refueling Contact Time *fu*
ARCT Air Refuelling Control Time *mt*
ARD Acceptance Requirements Document *fu, hc, ep, jj, jdr*
ARD Airborne Radar Development *fu*
ARD Amplitude Ratio Detector *jj*
ARD Asset Reporting Data System *mt*
ARD Automatic Release Date *jj*
ARD [floating dry dock, US Navy] *mt*
ARDB Associated Report Data Base *fu*
ARDC Air Research and Development Command [USAF] *fu, ro, 13, ac7.1-34, tb, jj*
ARDC Armament Research and Development Center [of AMCOM] *tc*
ARDCOM Army Air Defense Command *jj*
ARDE (ARDE Inc.. . Norwood, NJ 07648) *jmj*
ARDEC Armaments Research, Development, and Engineering Center Command *nd74.445.57, hc, st*
ARDEMS Artillery Delivered Multi Purpose Submunitions *hc*
ARDF Airborne Radio Direction Finding *10, mt*
ARDIS Airlift Requirement Data Integrated Systems *mt*
ARDIS Airlift Requirements Data Information System [USAFE] *mt*
ARDIS Army RDTE Information System *tb*
ARDME Automatic Range Detector and Measuring Equipment *hc*
ARDU Advanced RADEC Data Unit *kf*
ARDU Aircraft Research and Development Unit [of RAAF] *aw130.6.41*
ARE Aircraft Recovery Equipment *wp*
ARE Arab Republic of Egypt *fu*
ARE Atmospheric Research Equipment *wp*
AREFG Air Refueling Group [USA] *mt, af72.5.121*
AREFS Air Refueling Squadron [SAC. USA] *mt, af72.5.121*
AREFW Air Refueling Wing [SAC, USA] *mt, af72.5.121*
ARENTS ARPA Environmental Test Satellite *hc*
AREP Air Refueling Egress Point *wp*
ARES (has flown its ARES special mission support prototype aircraft) *aw132.10.260*
ARES Advanced Railroad Electronics System *is26.1.63*
ARES Advanced Research EMP Simulator *wp*
ARES Airborne Radar Effectiveness Simulation *fu*
ARES Amateur Radio Emergency Services *mt*
ARESTEM A Recording Stray Energy Monitor *hc*
ARE_OC ARE Operations Centers *fu*
ARF Air Reserve Facility *mt, af72.5.52*
ARF Air Reserve Forces [AF Reserves and Air National Guard] *mt*

ARF Airport Reservation Function *fu*
ARF Anomaly Report Forms *hc*
ARF-BLMPS Air Force Reserve - Base Level Military Personnel System *mt*
ARFA Allied Radio Frequency Agency [NATO] *mt*
ARFCOS Armed Forces Courier Service *mt, 10, fu, su, hc, tb*
ARFF Air Reserve Forces Facility *af72.5.120*
ARFOCS Armed Forces Courier Service [headquarters office] *dt*
ARFOR Army Forces *mt*
ARFOR-AK Army Forces-Alaska *mt*
ARFORM Atomic Resonance Filter Optical Receiver Module *hc*
ARFTG (The ARFTG Conference reception and banquet) *mj31.5.40*
ARFTIC Autonomous Radio Frequency Target Identification/Classification *hc*
ARG Accident Response Group *mt*
ARG AFCAP Review Group *mt*
ARG Air Refueling Group [SAC, USA] *12, mt*
ARG Air Reserve Group [Canada] *mt*
ARG Amphibious Ready Group *mt*
ARG [repair ship, landing craft, US Navy] *mt*
ARGADS Army Radar Gun Air Defense System *hc*
ARGOS [space based data collection and position location system, ESA] *ge, ac3.6-1, hg*
ARH Active Radar Homing *mt*
ARH Advanced Reconnaissance Helicopter *wp*
ARH Anti Radiation Homing [device] Homer *fu, mt, wp, tb*
ARH [repair ship, hull repairs, US Navy] *mt*
ARHP Anti Radiation Homing Projectile *hc*
ARI Active Relay Indicator *fu*
ARI Activity Routing Identifier *jj*
ARI Air Research Institute *fu*
ARI Applied Research, Incorporated *tb*
ARI Army Research Institute [for Behavioral, Social Sciences] *mt, fu, hc, dt, wp, tb*
ARIA Advanced Range Instrumentation Instrumented Aircraft *mt, 4, aa27.4.27, af72.5.138, wp, tb, oe*
ARIA Apollo Radar [Range] Instrumentation Aircraft *ci.131, hc*
ARIA Apollo Range Instrumentation Aircraft [later Advanced Range Instrumentation Aircraft, an EC-135 / EC-18] *mt*
ARIEL-V [a UK X-ray mission, also known as UK-5] *cu*
ARIES Airborne Reconnaissance Integrated Electronic System *wp*
ARIES ATM Research and Industrial Enterprise Study *em, ct*
ARIMA Auto Regressive Integrated Moving Average *wp*
ARIMS Airborne Radar Inflight Monitoring System *hc*
ARIN Added Radar Integration Netherlands *hc*
ARINC Aeronautical Radio Incorporated (ARINC Research Corporation . . . Colorado Springs, CO 80916) *jmj, aw114.14.25, fu*
ARIP Air Refueling Initial Point *mt*
ARIS Advanced Range Instrumentation Ship *0, hc, tb, jj, fu*
ARIS Automatic Recording Infrared Spectrometer *wp*
ARISTOTELES Applications and Research Involving Space Technologies Observing the Earth's Field From Low Earth Orbiting Satellite [ESA] *ge*
ARJS Airborne Radar Jamming System *mt*
ARKIVE Archive Subsystem *mt*
ARL Acceptable Reliability Level *fu*
ARL Adjusted Ring Length *pk, kg*
ARL Applied Research Laboratory *fu, wp*
ARL Armament Ready Legend *jj*
ARL [landing craft repair ship, USA] *mt*
ARLANT Army Forces, Atlantic [USA, FORSCOM] *mt*
ARLL Advanced Run Length Limited *ct, kg, em, hi*
ARLO Air Reconnaissance Liaison Officer *mt*
ARLOC Army Location Code *mt*
ARM Advanced RISC Machine [processor] *em, kg, hi*
ARM Angle Rate Memory *jj*
ARM Annotated Reference Manual *pk, kg*
ARM Anti Radar Missile *0, aw130.22.106*
ARM Anti Radiation Missile *mt, fu, uc, wp, hc, pm, tb, jj, kf*
ARM ARPANET Reference Model *mt*
ARM Asynchronous Response Mode *pk, kg*
ARM Attitude Reference Module *jdr*
ARMA Association of Records Managers and Administrators *pk, kg*
ARMA Autoregressive Moving Average *wp, fu*
ARMCOM Armament Command *jj*
ARMMAT [Hughes Aircraft Company customer for INFANT] *hc*
ARMMS Automatic Reliability and Maintainability Measurement System *jj*
ARMMS Automatically Reconfigurable Modular Multiprocessor System *fu, hc*
ARMP Aerial Refueling Master Plan *wp*
ARMPREP Army Manpower and Personnel Requirements Process *hc*
ARMS Aerial Radiological Measurements Study *hc*
ARMS Air Resource Management System [USAFE] *mt*
ARMS Aircraft Readiness Maintainability Simulator *wp*
ARMS Aircrew Resource Management System *wp*
ARMS Ammunition Reporting and Management System [USAF] *wp, mt*

ARMS Annual Reliability and Maintainability Symposium *fu*
ARMS Army River Mine System *hc*
ARMS Automated Records Management System [AFMPC] *mt*
ARMS Automated Recruit Management System [USMC] *mt*
ARMS Automated Retrieval Mail Service [a system used by the NDADS to manage retrieval requests] *cu*
ARMS Automatic Requisition and Monitoring System *mt*
ARMTA Angular Rate Matching Transfer Alignment *17, sb*
ARMTE Army Material Test Evaluation Directorate *jj*
ARN Additional Reference Number *fu*
ARN Airborne Radio Navigation *hc, jj*
ARNG Army National Guard [USA] *jj, mt*
ARNO Air Request Net Operator *mt*
ARO After Receipt of Order *gp, wp, 16, hc, lt, jj, jdr, kf, oe, fu*
ARO Airport Reservations Office *fu*
ARO Army Research Office *mt, fu, hc, wp*
ARO Auxiliary Readout *fu*
AROD Airborne Range and Orbit Determination *hc*
AROD Airborne Remotely Operated Device *mt*
AROF A Register Occupancy Flip-flop *0*
AROH Automatic Reporting Of Height *fu*
AROS Airborne Radar Optical System *tb*
AROW Acquisition Report Back Order Wire *uh*
AROZ Army Restricted Operations Zones *mt*
ARP Abnormal Release - Provider *fu*
ARP Address Resolution Protocol *ct, em, kg*
ARP Aero Rifle Platoon [infantrymen carried by helicopters, USA] *mt*
ARP Aeronautical Recommended Practice *hc, jj*
ARP Aerospace Recommended Practice *fu*
ARP Air Raid Precaution *w9*
ARP Airborne Relay Platform *fu*
ARP Aircrew Respiratory Protection *wp*
ARP Altitude Reservation Playback *fu*
ARP Anti Radiation Projectile *wp, dh, hc*
ARP Automatic Reporting Post *wp*
ARP Azimuth Reference Pulse *fu*
ARPA Advanced Research Projects Agency [DOD] *mt, ci.46, 0, 13, 14, hc, jb, wp, tb, jdr, kf, ct, em, ro, fu, hi*
ARPANET Advanced Research Project Agency Network [prototype of the Internet] *mt*
ARPANET Advanced Research Projects Agency [communications] Network *1, wp, ns, tb, ct, em, fu, kg, hi*
ARPAS Air Reserve Pay and Allowance System [USAF] *mt*
ARPAT Advanced Research Project Agency, Terminal *17, sb, hc*
ARPC Air Reserve Personnel Center *mt, hc, af72.5.32*
ARPC Automated Radar Prediction Capability *mt*
ARPCA AUTOVON Remote Reprogrammable Conference Arranger *mt*
ARPERCEN Army Reserve Personnel Center *mt*
ARPL Adjust Requested Privilege Level *pk, kg*
ARPLA Alert Response Plans *mt*
ARPS Advanced Radar Processing System [for the E-2C Hawkeye, USA] *wp, mt*
ARPS Automated Resource Projection System *mt*
ARPSE Aerospace Research Pilot School, Edwards [AFB] *hc*
ARPTT Aerial Refueling Part Task Trainer *wp*
ARPV Advanced Remotely Piloted Vehicle [RPV development program, USA] *mt*
ARQ Automatic Repeat Request *ct, em, fu, kg, bl, 0*
ARR Advance Rocket Ramjet *hc*
ARR Aeronautica Regala Romana [Royal Rumanian Air Force] *mt*
ARR Airborne Radio Relay *mt*
ARR Airborne Reference Radar *wp*
ARR Andoya Rocket Range *crl*
ARR Anti Repeat Relay *fu*
ARRADCOM Armament Research and Development Command [Army] *fu, hc*
ARRC ACE Rapid Reaction Corps [NATO] *mt*
ARRCOM Armament Materiel Readiness Command [later AMCCOM] [US Army] *hc*
ARRDATE Arrival Date *mt*
ARRED US Army Forces, US Readiness Command [FORSCOM] *mt*
ARRES Automatic Radar Reconnaissance Exploitation System *wp*
ARRG Aerospace Rescue and Recovery Group [MAC*] *mt, af72.5.121*
ARRI Automation and Robotics Research Institute [Texas] *aa26.11.53*
ARRL American Radio Relay League *mt*
ARRS Aerospace Rescue and Recovery Service [USA] *mt*
ARRS Aerospace Rescue and Recovery Squadron *mt, af72.5.121, ag*
ARRW Aerospace Rescue and Recovery Wing [MAC*] *mt*
ARS Active Radar Seeker *wp*
ARS Active Repeat Satellite *fu*
ARS Advanced Record System *mt*
ARS Air Refueling Squadron [SAC] *wp, mt*
ARS Air Rescue Squadron [USA] *mt*
ARS Airborne Receiving System *fu*
ARS Airborne Relay Station *fu, wp*
ARS Airborne Relay System *hc*

ARS Airport Surveillance Radar *fu*
ARS American Rocket Society [later AIAA] *fu, 13*
ARS Angular Rate Sensor *wp*
ARS Atmosphere Revitalization System *-9*
ARS Attack Radar Set [FB-111] *wp, mt*
ARS Automated Relay System *mt*
ARS Automatic Registration System *fu*
ARS Automatic Restart Switch *jj*
ARS Automation Research Systems, Ltd. *fu*
ARS [salvage vessel / ship, US Navy] *mt*
ARS [submarine rescue ship, US Navy] *mt*
ARSA Airport Radar Service Area *fu*
ARSAG Aerial Refueling Systems Advisory Group *mt, wp*
ARSC Army Recruiting Support Center *dt*
ARSDP Aerial Refueling Systems Development Plan *wp*
ARSOC Army Special Operations Component [USA DOD] *mt*
ARSOF Army Special Operations Forces [USA DOD] *mt*
ARSP Airborne Reconnaissance Support Program *mt*
ARSPC Army Space Command *kf*
ARSPOC Army Space Command Operations Center *jb, cp, jmj, as*
ARSR Air Route Surveillance Radar *fu*
ARSTAF Army Staff Agency *mt*
ARSTU Auto Remote STUIII *mt*
ART Advanced Radiation Technology *wp*
ART Automated Advanced Reasoning Tool *fu*
ART Automatic Radar Tracking *hc, fu*
ART Automatic Radiating Tester *hc*
ART Avionics Reconfiguration Technology *hc*
ARTAC Advanced Reconnaissance and Target Acquisition Capability *wp*
ARTADS Army Tactical Data System *mt, hc, fu, dt*
ARTB Advanced Radar Test Bed *mt, hc, hn49.11.8, wp*
ARTBASS Army Training Battlefield Simulation System *hc*
ARTC Adaptive Rapid Turnaround Communication *wp*
ARTC Air Route Traffic Control *fu*
ARTCC Air Route Traffic Control Center [FAA] *mt, ag, fu, aw129.21.143*
ARTCP Air Route Traffic Computer Program *fu*
ARTCS Advanced Radar Traffic Control System *hc*
ARTEMIS Advanced Relay TEchnology MISsion *mi*
ARTEMIS Automated Scheduling System *hc*
ARTEP Army Training and Evaluation Program *hc*
ARTEWG ADA Runtime Environment Working Group *mt*
ARTF/TA Automated Radar Terrain Follow/Terrain Avoidance *hc*
ARTI Advanced Rotorcraft Technology Integration [a research effort to evaluate systems for new combat helicopters] [USA] *hc, mt*
ARTI Automatic Radar Target Identification *hc, wp*
ARTIS Advanced Radar Target Identification System *wp*
ARTRAC Advanced Range Testing, Reporting and Control *hc*
ARTS Airborne Radar Target Simulator *wp*
ARTS Arkansas Research and Test Station *wp*
ARTS Asynchronous Remote Takeover Server *hi, kg*
ARTS Automated Remote Tracking Stations *kf, mt*
ARTS Automatic Radar Tracking System *hc, fu*
ARTS Automatic Return To Search *jj*
ARTT Asynchronous Remote Takeover Terminal *hi, kg*
ARTU Automatic Range Tracking Unit *0*
ARTW Aerospace Reconnaissance Technical Wing *wp*
ARU Abnormal Release-User *fu*
ARU Airborne Rotary wing Unit *fu*
ARU Airspace Altitude Reservation Unit *fu*
ARU Altitude Reference Unit *hc, 17, sb*
ARU Antenna/RF Unit *fu*
ARU Audio Response Unit *pk, kg*
ARU Auxiliary Readout Unit *fu*
ARV Airborne Relay Vehicle *wp*
ARV Aviation Repair Vessel [US Navy] *mt*
ARVN Army of the Republic of Viet Nam [South Viet Nam] *w9*
ARW Aerial Refueling Wing *12, wp*
ARW Air Raid Warning *wp*
ARW Air Refuelling Wing [USA] *mt*
ARWR Advanced Radar Warning Receiver *ae151.52.460*
ARWS Aerial Reconnaissance Weather System *wp*
ARX (ARX Incorporated) [an avionics/military electronics company] *aw130.22.49*
ARX AutoCAD Runtime eXtension *em*
AS Access Switch *mt*
AS Access Systems *mt*
AS Acquisition Strategy *tb, st*
AS Aeronautical Systems *jj*
AS Aerospace Standard *fu*
AS After Sight *w9*
AS Air Sortie *mt*
AS Air Sovereignty *fu*
AS Air Speed *jj*
AS Air Station *mt*
AS Air Superiority *hc, jj*

AS Air to Surface *mt*
AS Ammeter Switch *fu*
AS Anti Submarine *fu, w9, mt*
AS Aperture Stop *wp*
AS Application System *kg, tb*
AS Area Air Surveillance *wp, tb*
AS Assembly Specification Document *tm*
AS Augmented Support *jj*
AS Automation Specialist *fu*
AS Autonomous System *ct, em, kg*
AS Avionics Specification *jj*
AS Submarine Tender [US Navy] *mt*
AS {Air Station} *af72.5.103*
AS&E American Science and Engineering, [Incorporated] *pl*
AS&T American Science and Technology [Corporation] *sc2.8.5*
AS&TC American Science and Technology Corporation *sc3.10.3*
AS/IDO Air Surveillance/ID Officer *fu*
AS/IR Antenna Servo/Infrared *jj*
AS/RICMO Air Surveillance/Radar Inputs Countermeasure Officer *fu*
AS/RS Automated Storage and Retrieval System *fu*
AS/U Advanced Server for Unix *pk, kg*
ASA Acoustic Signal Analysis *fu*
ASA Acoustical Society of America *fu*
ASA Advanced SCSI Architecture *hi*
ASA Advanced System Architecture *hc*
ASA Advanced System Avionics *wp*
ASA American Standards Association *mt, hc, w9, jj*
ASA American Statistical Association *wp*
ASA Army Security Agency [Fort Meade] *mt, hc, 10, hc, tb, jj, fu*
ASA Army Security Service [USA] {?} *mt*
ASA Assistant Secretary of the Army *tb, st*
ASA Astronomical Society of the Atlantic *mi*
ASA Automated Separation Assurance *fu*
ASA Automatic Spectrum Analyzer *hc*
ASAC All Source Analysis Center *fu, mt*
ASAC Allied Signal Aerospace Company *af72.5.105*
ASAC Anti Submarine Air Controller *jj*
ASAC Automated System for Army Commissaries [USA] *mt*
ASACS Airborne Surveillance and Control System *hc*
ASAE American Society of Association Executives *hc*
ASAF Assistant Secretary of the Air Force *kf*
ASAFA (Cadet *USMA, ASAFA, or ROTC Contract/Scholarship > af72.5.151*
ASAI Applied Science Associates, Incorporated *hc*
ASALM Advanced Strategic Air Launched Missile [a Mach 4 air breathing missile, canceled in 1980] [USA] *mt, hc*
ASALM Advanced Supersonic Air Launch Missile *fu*
ASAM Advanced Scanning Array Module *wp*
ASAM Advanced Surface to Air Missile *dh, hc, fu*
ASAMS Austere Surface to Air Missile System *hc*
ASAP Accelerated Strike Aircraft Program *jj*
ASAP Advanced Sensor Application Program *hc*
ASAP Advanced Systematic Analysis Production *hc*
ASAP Aerospace Safety Advisory Panel *ci.605*
ASAP Any Service / Any Port *ct, em*
ASAP Ariane Strucure for Auxiliary Payloads *ai2*
ASAP Automated Statistical Analysis Program *wp*
ASAP Automatic Sourcing of Assets Program [AFCC*] *mt*
ASAP Automatic Switching And Processing *pk, kg*
ASAP Avionics Sensor Adaptive Planner *hc, wp*
ASAR Advanced Surface to Air Rocket *hc*
ASAR Advanced Synthetic Array Radar [System] *fu*
ASARC Army Systems Acquisition Review Council *mt, fu, hc, wp, tb, st, jj*
ASARDA Assistant Secretary of the Army for Research, Development and Acquisition *hc*
ASARG Autonomous Synthetic Aperture Radar Guidance *wp*
ASARS Advanced Airborne Synthetic Array Radar System *fu*
ASARS Advanced Synthetic Aperture Radar System [TR-1] [USA] *mt, hn43.7.5, wp, hc, nd74.445.29*
ASARS Airborne Surveillance and Reconnaissance System *hc*
ASAS Advanced Solid State Array Spectrometer *hy*
ASAS All Source Analysis System [USA] *mt, ph, wp, hc, fu*
ASAT Anti Satellite Analysis And Tracking *kf*
ASAT Anti Satellite missile [launched by an F-15 at high altitude] [USA] *mt*
ASAT Anti Satellite [weapon or system] *mt, fu, sb, 11, 14, 17, ic2-8, hc, af72.5.18, jb, nt, ae151.52.459, wp, tb, st, as, jdr, kf, ci.576, 5, 11, sb*
ASAT Automated Specification software Analysis Tool *fu*
ASB Air Staff Board *wp*
ASB Army Science Board *tb*
ASB As Built *kf*
ASBCA Armed Services Board of Contract Appeals *hc, wp, dt*
ASBD Advanced Sea Based Deterrent *hc*
ASBM Air to Surface Ballistic Missile *-14*
ASBTB Ammunition Storage Bay Terminal Box *hc*
ASC ADGE Standard Controller *fu*

ASC Advanced Satellite Control *kf*
ASC Advanced Scientific Computing *aa26.12.95*
ASC Aeronautical Systems Center *jj*
ASC Aircraft Scheduling Cell *mt*
ASC Allowable Stock Concentrations *hc*
ASC Allowance Source Code *mt*
ASC Allowance Support Code *fu*
ASC American Satellite Company *sc2.8.4, hd1.89.y2*
ASC Analog Signal Conditioner/Converter *jj*
ASC Analog Signal Converter *fu*
ASC Army Space Council *mt*
ASC AUTODIN Switching Center *mt, jb, tb*
ASC Automatic Saturation Control *jj*
ASC Automatic Sensitivity Control *jj, fu*
ASC Automatic Switching Center [AUTODIN] *mt*
ASC Auxiliary Switch Breaker Closed *fu*
ASCA [Japanese Asuka spacecraft, formerly ASTRO-D] *cu*
ASCAC Anti Submarine Contact Classification Analysis Center *fu, hc*
ASCC Air Standards Coordinating Committee [responsible for allocating codenames to aircraft of the potential enemy] *mt*
ASCC Automatic Sequence Controlled Calculator [IBM] *pk, kg*
ASCD Aircraft Sensor Correlation Device *hc*
ASCE American Society of Civil Engineers *0, aw, aa26.11.b3*
ASCEVAL Airspace Control Evaluation *fu*
ASCI Accelerated Strategic Computing Initiative *em*
ASCII American Standard Code for Information Interchange *mt, do263, wp, no, tb, jdr, kf, oe, nu, kg, fu, ct, em, hi*
ASCM Advanced Spaceborne Computer Module *ro, kf*
ASCM American Society of Computing Machinery *kf*
ASCM Antenna Surface Contour Measurements *hc*
ASCM Anti Ship Cruise Missile *mt, fu*
ASCN Advance Study Change Notice *fu*
ASCOD Area Systems Coordinating Document *hc*
ASCON Associate Contractor *mt, tb*
ASCOT Adaptive Scan Centroid Optical Tracker [a DARPA project on ASAT] *hc, kb, jmj*
ASCP AJCC Subnet Communications Processor *mt*
ASCP Army Strategic Capabilities Plan *mt*
ASCS Adaptive Sidelobe Cancellation Study *fu*
ASCS Automatic Stabilization and Control System *ci.349, hc*
ASCSI Advanced SCSI *hi*
ASCU Armament Station Control Unit *hc*
ASD Action Sequence Diagram *fu*
ASD Activity Support Date *jj*
ASD Advanced Safe Device [program] *hc*
ASD Advanced Sensor Demonstration *uc, fu*
ASD Aeronautical Systems Division [Wright Patterson AFB] [USAF] *mt, hc, rm6.4.4, tc, aa26.12.b20, af72.5.14, wp, tb, jj, fu*
ASD Aerospace Systems Division [of Bendix Corporation] *ci.940*
ASD Air Aircraft Situation Display *fu, aw129.14.113*
ASD Air Situation Display *mt*
ASD Air Sortie Duration *mt*
ASD Aircraft Statistical Data *jj*
ASD Ammunition Supply Depot *wp*
ASD Army Software Development *mt*
ASD Assistant Secretary of Defense [Manpower, MRA&L, Reserve Affairs, and Logistics] [OSD] *mt, fu, hc, wp, tb, kf*
ASD Attitude Sensor Digitizer *hc*
ASD Average Sortie Duration *mt*
ASD Aviation Supply Depot *wp*
ASD Aydin Systems Division *fu*
ASD/ENASC (Our monolithic solutions . . . are ASD/ENASC certified.) *aw130.22.115*
ASDAC Anti Submarine Data Analysis Center *hc*
ASDACS Active Sonar Data Analysis and Conversion System *hc*
ASDAR Aircraft to Satellite Data Relay *pl*
ASDC Army Strategic Defense Command [Huntsville] *hc*
ASDC Assistant Secretary of Defense Comptroller *su*
ASDC Automated Source Data Collection *mt*
ASDCS Automated Source Data Collection System *mt*
ASDE Airport Surface Detection Equipment *hc, fu, ro*
ASDI Assistant Secretary of Defense Intelligence *tb*
ASDIC Allied Submarine Detection Investigation Committee [committee that developed the device named after it, ASDIC, also known as SONAR] [UK] *mt*
ASDM Automated System Design Methodology *mt, wp*
ASDO Air Situation Display Operator *fu*
ASDP Advanced Sensors Demonstration Program *hc*
ASDR Acceleration Sensitive Drift Rate *oe*
ASDR Air Situation Display Room *fu*
ASDR Airport Surface Detection Radar *wp, fu*
ASDS Air Situation Display System *mt*
ASDS Antenna Signal Distribution System *hc*
ASD[CCCI] Assistant Secretary of Defense for Communications, Command and Control, and Intelligence *mt*
ASD[C] Assistant Secretary of Defense [Comptroller] *mt*
ASE Advanced Sensor Exploitation *mt, fu, hc, wp*

ASE Aerospace Support Equipment *wp*
ASE Airborne Search Equipment *wp*
ASE Airborne Support Equipment *gp, 4, 9, 16, hc, mb, ep, dm, tb, jdr, kf, ai2*
ASE Aircraft Serviceability Equipment [USA] *mt*
ASE Aircraft Survivability Equipment [USA] *fu, mt*
ASE Allowable Steering Error *hc, jj*
ASE And may Simultaneously Execute *fu*
ASE Army Support Engineering *jj*
ASE Assistant System Engineer *fu*
ASE Auxiliary State Estimator *-3*
ASEA/TOW Advanced Sea TOW *hc*
ASEAN Association of Southeast Asian Nations *fu, hc, wp, w9*
ASEB Aeronautics and Space Engineering Board *ci.874*
ASEE American Society for Engineering Education *0, aa26.10.b12*
ASET Advanced Sensor Evaluation and Test *hc, wp*
ASET Advanced Sensor Exploitation Test *fu*
ASETS A/B Suher Evaluation Test System *hc*
ASETS Airborne Sensor Evaluation Test Set *fu*
ASF Advanced Striking Force [RAF units sent to France in 1939, UK] *mt*
ASF Aeromedical Staging Facility *mt*
ASF Air Superiority Fighter *wp*
ASF Alaska SAR Facility [University of Alaska, USA] *hg, ge*
ASF Authorized Strength File [USA] *mt*
ASFIR Active Swept Frequency Interferometer Radar *ro*
ASFISS Advanced Simulation Facility Interconnection and Setup Subsystem *hc*
ASFM Administrative Support (Comptroller) *dt*
ASG Administrative Support Group *dt*
ASG Advanced Systems Group *mt, ct, em*
ASG Aeronautical Standards Group *fu*
ASG Aerospace Systems Groups [of Hughes Aircraft Company] *hn41.26.6, jj, fu*
ASG Airborne Spiral Gun [laying fire control system] *hc*
ASG Airborne Systems Group *jj*
ASG American Standards Group *jj*
ASG Area Support Group [USA] *mt*
ASH Advanced Surveillance Helicopter *hc*
ASH Alternate Support Headquarters *mt*
ASHRAE American Society of Heating, Refrigerating and Air Conditioning Engineers *fu*
ASI Additional Skills Identifier *fu, mt*
ASI ADI System Integration *fu*
ASI Aerospace Studies Institute *wp*
ASI Agenzia Spaziale Italiana *mi, crl*
ASI Air Speed Indicator *mt, w9, jj*
ASI Alpha Scattering Instrument *hc*
ASI Army Space Institute *lka*
ASI Associated Software Inquiries *fu*
ASI Atmospheric Structure Instrument *es55.28*
ASI August Systems Incorporated *fu*
ASI Azimuth Speed Indicator *fu*
ASI/TDP Advanced Sensor Integration/Tactical Data Processing *fu*
ASIA Advanced Sensor Integration Analysis *fu*
ASIAC Aerospace Structures Information and Analysis Center *wp*
ASIC All Source Intelligence Center *mt, fu, kf*
ASIC Application Specific Integrated Circuit *fu, 17, sb, pl, ct, em, ah, he, kg, aa27.4.35, pf1.26, is26.1.4, aw130.22.78, rm8.2.5, wp, pbl, jdr, tm, kf, jn, hi*
ASIC Army Intelligence and Security Command [U. S.] *dt*
ASIC Avionics Subsystem Interface Contract Contractor *hc, wp*
ASID Address Space IDentifier *em*
ASID Advanced System Integration Demonstration *hc, wp*
ASIDS Abbreviated Status Information Display System *mt*
ASIDS Airborne Surveillance and Intercept Defense System *fu*
ASIF Airlift Service Industrial Fund [MAC] *mt*
ASIF Airlift Services Industrial Found *wp*
ASIM Aircraft Stores Interface Manual *fu*
ASIMIS Aircraft Structural Integrity Management Information System *mt*
ASIMS Air Staff Information Management System *mt*
ASIMS Army Standard Information Management System *mt, wp, fu*
ASIO Air Surveillance/Identification Officer *fu*
ASIOE Associated Support Items Of Equipment *mt, wp*
ASIP AFGWC Software Improvement Program [MAC] *mt*
ASIP Aircraft Structural Integrity Program [USA] *mt*
ASIP Aircraft Structures Structural Integrity Program *wp, aw130.22.29*
ASIP All Source Imagery Processor [USMC] *mt*
ASIP All Source Imagery Program *hc*
ASIS Advanced Sensor Integration Study *fu*
ASIS Alpha Scattering Instrument Sensor/Surveyor *hc*
ASIS Amphibious Ship Information System [MSC] *mt*
ASIS Army Space Initiative Study *tb*
ASIS Automotive Sensor Instrumentation System *hc*
ASISH Alpha Scattering Instrument Sensor Head *hc*
ASISP Air Staff Information Systems Plan [USAF] *wp, mt*

ASIST A Simple Integration of Statistical Techniques *mt*
ASIT Adaptable Surface Interface Terminal [JTIDS] *hc, fu, mt*
ASIT Advanced Security and Identification Technology *pk, kg*
ASK Amplitude Shift Keying *wp, sm3-6, fu, hi*
ASL Above Sea Level *fu, wp*
ASL Adaptive Speed Levelling *pk, kg*
ASL Airborne Systems Laboratory *jj*
ASL Alternate Storage Location *mt*
ASL Approved Source Suppliers List *fu*
ASL Atmospheric Science Laboratory [of ERADCOM] *tc*
ASL Authorized Stockage List / Level *mt*
ASL Average Sortie Length *mt*
ASL Azimuth Scan Limit *jj*
ASLAR Aircraft Surge Launch and Recovery *mt*
ASLBM Anti Submarine Launched Ballistic Missile *fr, fu, hc*
ASLBMS Anti Submarine Launched Ballistic Missile System *hc*
ASLC Adaptive Side-Lobe Cancellation Canceller *fu, wp*
ASLE American Society of Lubricating Engineers *0, fu*
ASLT Advanced Submarine-Launched Torpedo *fu*
ASLV [Indian Launch Vehicle] *ac3.2-3*
ASM (a trademark of ASM International) *aa26.11.45*
ASM (teammate Thompson Sintra ASM will demonstrate) *af72.5.15*
ASM ActivitÈs Sous Marines [French under sea activities] *fu*
ASM Adaptive Signal Masking *wp*
ASM Advanced Surface Missile *hc*
ASM Air to Surface Missile [earlier used by the US Navy to label its missiles] *mt, dh, fu, wp, hc, 12, jj, kf*
ASM Airfield Survival Measures *mt*
ASM All Sky Monitor *cu*
ASM Anti Ship Missile *tb, fu, mt*
ASM Automated Scheduling Message *mt*
ASM Automated Scheduling Module *mt*
ASM Automatic Switch Monitor *mt*
ASM Available Seat Mile *aw129.21.95*
ASM Aviation School of Medicine *jj*
ASMA Aero Space Medical Association *ci.645, wp*
ASMA Air Staff Management Aid *mt*
ASMC AAS System Monitor and Control *fu*
ASMC Alternate System Monitoring Center [DDN] *mt*
ASMCOMM Airfield Survival Measures Communications *mt*
ASMD Anti Ship Missile Defense [US Navy] *un, fu, mt*
ASME American Society of Mechanical Engineers *hc, sc2.8.4, 0, aw130.13.104, aa26.8.b34, ct, fu*
ASMIRES Air Staff Manpower Information Requirement System *mt*
ASMMIC Application Specific Microwave Integrated Circuit *wp*
ASMOD Anti Submarine Warfare Module [CV/CVN] *mt*
ASMP Air-Sol Moyenne PortÈe [medium range Air to Surface missile, supersonic, air-breathing and nuclear-armed] [France] *mt*
ASMP Asymmetric Multiprocessing *hi, kg*
ASMRO Armed Forces Medical Regulating Office [USA DOD] {?} *mt*
ASMRO Armed Services Medical Regulating Office *mt, dt*
ASMS Advanced Strategic Missile System [formerly ABRES] [BMO] *mt, uc, 17, sb, wp, tb*
ASMS Advanced Surface Missile System [became AEGIS] [USA] *hc, fu, mt*
ASMS Air Space Management System *fu*
ASMS American Society For Mass Spectroscopy *wp*
ASN Abstract Syntax Notation *fu, kg*
ASN Access Stack Node *hi*
ASN Autonomous System Number *pk, kg*
ASN Average Sample Number *fu*
ASNE American Society of Naval Engineers *fu*
ASNM Advance Stack Network Management [HP] *hi*
ASNS Automatic Switching and Networking System *tb*
ASO Acoustic Sensor Operator *fu, jj*
ASO Air Situation Officer *fu*
ASO Air Staff Office *hc*
ASO Air Surveillance Officer *mt, wp, fu*
ASO Ammunition Supply Officer *wp*
ASO Army Space Office *mt*
ASO Astrotech Spacecraft Space Operations *jn, oe*
ASO Aviation Supply Office *mt*
ASO Aviation Supply Officer *jj*
ASOA Army School Of the Air [USA] *mt*
ASOAS Air Staff Office Automation System *mt*
ASOC Air Sector Operations Center *mt*
ASOC Air Support Operations Center [USAF/NATO] *wp, fu, mt*
ASOC Allied Sector Operations Center [NATO] *mt*
ASOC Alternate Sector Operations Center *fu*
ASOC_OTS Air Support Operations Center Optical Transceiver System *ro*
ASOD Assistant Secretary Of Defense *wp*
ASOJ Airborne Advanced Standoff Jammer *hc*
ASOJ Anti Standoff Jammer *uc*
ASOP Army Strategic Objectives Plan *mt*
ASOP Automatic Structural Optimization Program *wp*

ASOS Army Secure Operating System *fu*
ASOS Automated Surface Observation System *fu, oe*
ASOT Air Surveillance Officer Technician *fu*
ASP (ACSM/ASP Plenary Session, 10 March 1980, St. Louis, MO) *ac2.2-27*
ASP AAS Sector Processor *fu*
ASP Accelerometer Signal Processing *-3*
ASP Acquisition Strategy Plan *wp*
ASP Advanced Signal Processor *fu*
ASP Advanced Strategic Penetrator *wp*
ASP Advanced Study Program *wp*
ASP Advanced System Planning *wp*
ASP Aerial Scout Program *hc*
ASP Aerial Superiority Program *hc*
ASP Affordable System Performance *kf*
ASP Agile Seeker Program *hc*
ASP Air Spotting Pool [Allied] *mt*
ASP Air Superiority Program *jj*
ASP Airborne Sensor Platform *wp, tb, st*
ASP All Service Processor *mt*
ASP All Source Production [USA] *mt*
ASP Ammunition Supply Point *hc, jj*
ASP Analog Signal Processor *kf, oe*
ASP Annual Service Practice *jj*
ASP Annual Service Practice [live firing of a 'Nike Hercules' SAM on NAMFI] [Belgium] *mt*
ASP Annular Suspension Pointing *hc*
ASP Army Strategic Plan *mt*
ASP Association of Shareware Professionals *kg, ct, em*
ASP Association-Storing Processor *hc*
ASP Astronomical Society of the Pacific *mi*
ASP Augmented Support Period *hc, jj*
ASP Authorized Service Provider *pk, kg*
ASP Autonomous Strike Platform *wp*
ASP Avionics Status Panel *wp*
ASPA Air Strike Planning Assistant *mt*
ASPA Armed Services Procurement Act *wp*
ASPA [signal processor architecture, generic] *hc*
ASPAB Armed Services Patent Advisory Board *-10*
ASPADOC Alternate Space Defense Operations Center *mt, jb, cp, jmj*
ASPC Analysis of Spare Parts Change *hc, jj*
ASPG Automatic Data Processing Standards Policy Group *mt*
ASPI Advanced SCSI Programming Interface [Adaptec] *kg, ct, em, hi*
ASPJ Airborne Self Protection Protected Jammer *mt, wp, aw130.13.81, uc, hc, af72.5.101, jj, fu*
ASPK Associated Spare Parts Kit *fu*
ASPM Armed Service Pricing Manual *mt*
ASPO Army Space Program Office *mt*
ASPO Automated Systems Program Office *mt*
ASPP Antenna Solar Panel Positioner *hc*
ASPPO Armed Services Production Planning Officer *hc, fu, mt*
ASPR Armed Services Procurement Regulations [replaced by DAR, FAR] *mt, mc4-83, hc, tb, jj, fu, su*
ASPRO Associative Processor *aw130.13.56, fu*
ASPS Adaptable Space Propulsion System *hc*
ASPS Advanced Signal Processing System *pk, kg*
ASPS Advanced Spaceborne Processor System *oe*
ASPS Advanced Speech Processing Station *hc*
ASPS Annular Suspension Pointing System *hc*
ASPT Advanced Simulator For Pilot Training *wp*
ASPU Automatic Signal Processor Unit *hc*
ASPUR Automated System for Processing Unit Requirements [MTMC] *mt*
ASQ Assignment Queue *fu*
ASQC American Society for of Quality Control *fu, 0, jj, kf, oe*
ASR Address Space Register *pk, kg*
ASR Air Sea Rescue *w9, wp, jj, mt*
ASR Air Search Radar *fu*
ASR Air Staff Requirement *hc, mt*
ASR Airborne Air Surveillance Radar *fu, wp, jj*
ASR Airport Surveillance Radar *mt, fu, w9*
ASR Alternate Source Request *fu*
ASR Asynchronous Sender/Reader *pl*
ASR Atomic Strike Recording *hc*
ASR Automated Automatic Speech Recognition *kg, wp, hi*
ASR Automatic Send and Receive *mt, fu, 0, kg, ct, em*
ASRA-PLUS Automated Storage and Retrieval System [USA] *mt*
ASRAAM Advanced Short Range Air to Air Missile [a Sidewinder replacement] [USA] *mt, hn43.21.1, wp, hc, aa26.11.12*
ASRATS Advanced Strike Radar Technology System *hc*
ASRC Alabama Space and Rocket Center *ci.626*
ASRM Advanced Solid Rocket Motor [Shuttle] *aa26.12.38, mb, mi*
ASRMO Armed Services Medical Regulatory Organization Reporting System [MAC] *mt*
ASROC Anti Submarine Rocket [USA DOD] *mt, fu, aw130.13.56, wp, 11, ph, hc, jj*
ASRS Aircraft Suitability Reporting System [MAC] *mt*
ASRS Automatic Storage and Retrieval System *15, fu*

ASRT Air Support Radar Team [USAF/USMC] *fu, hc, wp, mt*
ASSA Automated Support System Architecture *mt*
ASSAULT_BREAKER [DARPA program for smart submunitions] *wp*
ASSAW Aerospace Surveillance and Warning *hc*
ASSB Avionics Subsystem for Strategic Bombers *hc*
ASSC Alternate Space Surveillance Center *jb, cp*
ASSD Autonetics Strategic Systems Division [Rockwell] *ci.961*
ASSESS Airborne Science Spacelab Experiments Systems Simulation *ci.494*
ASSET Aerothermodynamic / Elastic Structural Systems Environmental Tests [this was a lifting body reentry vehicle, USA] *0, mt*
ASSET Airborne System for Evaluation and Testing *hc*
ASSG Army Space Study Group *mt*
ASSH Advanced Space Systems Hardening *hc*
ASSIST ADPE System Sizing and Simulation Test *fu*
ASSIST Advanced System Support and Information System Technology [AFLC] *mt*
ASSIST Army System for Intelligence Support Terminals *mt*
ASSIST Automated Special Security Information Systems Terminal *mt*
ASSIST [a simulation program] *tb*
ASSM Abstract System Semantic Model *fu*
ASSOCX Association [a software program] *jb*
ASSOTW Airfield and Seaplane Stations Of The World *mt*
ASSP Auxiliary Transport Submarine [US Navy] *mt*
ASSR Autonomous Soviet Socialist Republic *w9*
ASSS Automated Service Station Systems [USAF] *mt*
ASSSAP ANSI SQL Standard Scalable and Portable *hi, kg*
ASSSC Automated Self Service Supply System [USA] *mt*
ASSW Anti Surface Ship Warfare *hc*
AST (AST Research, Inc.) [from first initials of the founders *Albert Wong, Safi Qureshey, Thomas Yuen*] *pk, kg*
AST Advanced Simulation Technology *tb*
AST Advanced Supersonic Transport *hc, wp, jj*
AST Advanced Surveillance Technology *wp*
AST Advanced Surveillance Testbed *kf*
AST Aerothermal Structures {?} *st*
AST Air Staff Target [an operational requirement] [UK] *mt*
AST Air Support Threat *tb*
AST Airborne Surveillance Testbed *kf*
AST Army Staff Table *fu*
AST Asynchronous System Trap *ct, em*
AST Atlantic Standard Time *wp, kg*
AST Atmospheric Surveillance Technology [ADI] *wp, mt*
AST Auto Shop Tester *hc*
AST/AR Advanced Surveillance and Tracking/Airborne Radar *go2.2*
AST/AR Atmospheric Surveillance Technology/Airborne Radar *fu*
ASTA Advanced Strategic Transport Aircraft *wp*
ASTA Aerial Surveillance and Target Acquisition *wp*
ASTA Aero Space Technologies of Australia *hc, aw130.6.43*
ASTAAM Advanced Seeker Technology Air to Air Missile *hc*
ASTAB Automated Status Board *fu*
ASTACS ASW Tactical Support *fu*
ASTAM Automated Station Test and Monitor *hc*
ASTAP Advanced Survival Techniques and Analysis *hc*
ASTAR Advanced Spacecraft Thermal Analysis Routine *hc*
ASTAR Advanced Surveillance and Target Acquisition Radar *wp*
ASTAR Airborne Search Target Attack Radar *mt*
ASTAR Atomic Strike Approval Request *mt*
ASTC Adaptive Sensitivity Time Control *hc, jj*
ASTCM Astrodynamics Common *uh*
ASTD Anti Ship Torpedo Defense *fu, uc, mt*
ASTDS Air Supported Threat Defensive Systems *hc*
ASTE Advanced Surveillance Technology Experiment [SDI] *mt*
ASTEI Air-Surface Technology Evaluation and Integration [program to reduce radar cross-sections of equipment, a stealth technology] [USA] *mt*
ASTER Advanced Spaceborne Thermal Emission and Reflection Radiometer *hg, ge*
ASTERIX Airborne Stereo Rain Radar *ge, hg*
ASTEX Atlantic Stratocumulus Transition Experiment *hy*
ASTF Aeropropulsion Systems Test Facility *wp*
ASTIA Armed Services Technical Information Agency [former acronym for Defense Documentation Center] *fu, hc*
ASTL Approved Supplier Tab List *hc, jj*
ASTM American Society of for Testing Materials *fu, wp, aw118.18.3, 0, w9, jj, oe*
ASTM Apollo Soyuz Test Mission *-11*
ASTO Aerospace and Strategic Technology Office [DARPA] *kb*
ASTOR Airborne Stand Off Radar *mt*
ASTOR Anti Submarine Torpedo [USA DOD] *fu, mt*
ASTOVL Advanced Short Take Off and Vertical Landing [aircraft] [NASA] [USA] *mt, wp, aa26.12.34, aw129.10.4, ail*
ASTP Apollo Soyuz Test Project *ac2.5-1, 13, wl*
ASTP Army Specialized Training Program *w9*
ASTR Authorized Strength *mt*
ASTRA Advanced System Training Aircraft [a modified Hawk Mk.1] [UK] *mt*

ASTREC Atomic Strike Recording Complex [Sperry-Rand] *hc*
ASTROMAST [deployable 60 or 100-foot antenna] *ro*
ASTS Advanced Surveillance and Tracking System *wp*
ASTS Airborne Survivable Tracking System *hc*
ASTSP Adaptive Space-Time Signal Processing *fu*
ASTT ARD Atmospheric Surveillance and Tracking Technology, Airborne Radar Demonstration *fu*
ASTV (Seasat-A ASTV *Commercial Demonstration Experiment* > *ac2.7-25*
ASU Aeromedical Staging Unit *mt*
ASU Approved Approval for Service Use *fu*
ASU Auto Switch Unit *-9*
ASUPS Ammunition Supply Points *mt*
ASURES Advanced Surveillance and Recon System *mt*
ASUT Adapter Sub Unit Tester *hc*
ASUW Anti Surface Warfare [US Navy] *ph, ts10-2, mt*
ASUW Antisurface Ship Undersea Warfare *fu*
ASUWC AntiSurface Warfare Commander [US Navy] *fu, mt*
ASV Aerothermodynamic Structural Vehicle *0*
ASV Analog Secure Voice *mt*
ASV Anti Surface Vessel [radar] *mt*
ASV-OA Audio Visual Services Office, [U. S.] Army *dt*
ASVAB Armed Services Vocational Aptitude Battery *hc*
ASVP ADA Simulated Validation Program [USAF] *mt*
ASW Air to Surface Weapon *hc*
ASW Antenna Safety Switch *fu*
ASW Anti Submarine Warfare *mt, fu, ca, wp, aw129.14.1, 11, 12, 14, aa27.4.b13, af72.5.15, kb, tb, st, jj, jdr*
ASW Applications Software *mt*
ASW Atmospheric Surveillance and Warning *tb*
ASW Auxiliary Switch *fu*
ASWA Adjustable Solar Wing Actuator *jn, he*
ASWAC Aerospace Warning and Control *hc, fu, jj*
ASWC Anti Submarine Warfare Center *tb*
ASWC Anti Submarine Warfare Commander [US Navy] *fu, mt*
ASWC Average Strobe Width Criterion *fu*
ASWCCS Anti Submarine Warfare Command and Control System *tb, fu, mt*
ASWCS Anti Submarine Warfare Control System *fu*
ASWDS Anti Submarine Warfare Data System *hc*
ASWEPS Anti Submarine Weapon Electronic Prediction System *hc*
ASWG American Steel Wire Gage *fu*
ASWG Army Space Working Group *mt*
ASWM Anti Submarine Warfare Module *fu, hc, pf1.4*
ASWMC {antisubmarine warfare equipment on board destroyers} *pf1.4*
ASWO Air [tactical action control] Officer *jj*
ASWOC Anti Submarine Warfare Operations Center [US Navy] *mt, fu, 12, hc, pf1.4*
ASWS Advanced Strike Weapon System *hc*
ASWS Aerospace Surveillance Warning System *fu, hc*
ASWSOW ASW Standoff Weapon *fu*
ASWSPO Anti Submarine Warfare Systems Project Office *dt*
ASWTDA Anti Submarine Warfare Tactical Decision Aid *pf1.10*
AT Acceptance Temperature *oe*
AT Acceptance Test *mt, tb, oe, fu*
AT Advanced Technology *kg, em, hi*
AT Air Tasking *mt*
AT Air Traffic *fu*
AT Air Transport *fu*
AT Alert Team *tb*
AT Alert Velocity Track *jj*
AT Ampere Turn *fu, w9*
AT Anti Terrorism [USA DOD] *mt*
AT Apogee Thruster *oe*
AT Automated Telecommunications Program *mt*
AT Automatic Testing *mt*
AT Automatic Transmission *w9, kg*
AT Aviation [electronics] Technician *jj*
AT Awaiting Transportation *wp*
AT [tug ship, US Navy] *mt*
AT&E Advanced Technology and Engineering *fu*
AT&T American Telephone and Telegraph [a communications corporation] *mt, sc2.7.2, aa26.12.65, aw130.11.57, jb, ce, kf, kg, ct*
ATA Actual Time of Arrival *mt, wp, fu*
ATA Advanced Tactical Aircraft [USA, US Navy] *mt, fu, wp, aw129.14.26, pf.f88.27*
ATA Advanced Technology Aircraft [US Navy] *aa26.12.70*
ATA Advanced Technology Attachment *kg, em, ct, hi*
ATA Advanced Test Accelerator *hc, wp, tb*
ATA AEW Target Area *fu*
ATA Air Transport Transportation Association [USA] *mt, aw129.14.115, y2k*
ATA Airborne Target Area *go3.1-1*
ATA Airborne Target Augmentor *hc*
ATA Alimentary Toxic Aleukia *wp*
ATA Antenna Train Angle *fu*
ATA Antenna Turn Around *jj*

ATA Auto Terrain Avoidance [USAF] *mt*
ATA Automatic Terrain Avoidance *wp*
ATA Automatic Threshold Adapter *oe*
ATA Automatic Threshold Adjust *oe*
ATA Auxiliary Tug [ship] [US Navy] *mt*
ATAAD Anti Tank Assault Air Defense [weapon system] *hc*
ATAAM Advanced Tactical Air to Air Missile *hc*
ATABCS Advanced Tactical Airborne Battle Control System *hc*
ATABOI Auto Target Acquisition By Observation Integration *hc*
ATAC Advanced Tactical *hc*
ATAC Army Tank and Automotive Command *fu*
ATACC Advanced Tactical Air Command Central [US Navy] *mt, hc, aw130.6.32*
ATACC Advanced Tactical Aircraft Control Capability *hc*
ATACC Alternate Tactical Air Control Center *fu, mt*
ATACC Army Tactical Air Control Center *mt*
ATACCS Advanced Tactical Air Command and Control System *hc*
ATACCS Allied Tactical Command and Control System [NATO] *mt*
ATACM Army Tactical Missile *nd74.445.39*
ATACMS Army Tactical CC Information System {?} *mt*
ATACMS Army Tactical Missile System *mt, wp, aw126.20.18, aa26.12.72, af72.5.28, nd74.445.30*
ATACO Air Tactical [Action] Control Operator Officer *hc, jj*
ATACS Advanced Tactical Air Combat Simulation *hc*
ATACS Advanced Tactical Air Control System *hc*
ATACS Advanced Tactical Cruise System *hc*
ATACS Airborne Tactical Air Control System *mt*
ATACS Army Tactical Communications System *mt*
ATADS Advanced Tactical Air Defense System *mt*
ATAF Advanced Tactical FLIR *hc*
ATAF Allied Tactical Air Force [NATO] *mt, fu, 12*
ATAFCS Airborne Target Acquisition and Fire Control System *mt, wp*
ATAL (ATAL airborne television surveillance system with a coded down-link) *aw129.21.13*
ATAL Automatic Target Acquisition Logic *17, sb*
ATAM Automated Threat Assessment Model *mt*
ATAP Anti Tank, Anti Personnel *mt*
ATAP Apollo Telemetry Aircraft Project *hc*
ATAP ATOC Target Analysis and Planning *mt*
ATAPI Advanced Technology Attachment Packet Interface *em, ct, hi*
ATAPS Advanced Tactical Aircraft Protection System *wp*
ATAR Acquisition Tracking And Recognition *wp*
ATAR Air To Air Recovery *mt, af72.5.142*
ATAR Automatic Target Recognition [NASA] *hc*
ATARS Advanced Tactical Air Airborne Reconnaissance System [USA] *mt, uc, aw, hc, af72.5.138, pf.f88.26, wp, kf*
ATARS Air To Air Recognition System [later TISEO] *hc*
ATARS Army Tactical Airspace Regulation System *fu*
ATARS Automatic Traffic Advisory and Resolution Service *fu*
ATAS Advanced Target Acquisition Sensor *mt*
ATAS Advanced Target Acquisition System *wp*
ATAS Aerodynamic Threat Analysis System *hc*
ATAS Airborne Target Assessment System *hc*
ATAS Automatic Terrain Avoidance System *wp*
ATASPI Advanced Technology Attachment Software Programming Interface *hi*
ATAST Airborne Tactical Air Support Team *hc*
ATAWS Advanced Tactical All Weather Strike *hc*
ATAWS Autonomous Tactical All Weather Strike *wp*
ATB Active Transfer Bias *jj*
ATB Advanced Technology Bomber [development program for the stealth Northrop B-2, USA] *mt, wp, 11, ph, af72.5.133*
ATB Advanced Test Bed *kf*
ATB Airborne Test Bed *hc, wp, jj*
ATB Assembly and Test Building [Plesetsk, Russia] *ai2*
ATBCB Architectural and Transportation Barriers Compliance Board *fu*
ATBM Advanced Tactical Ballistic Missile *wp, fu*
ATBM Advanced Tactical Battle Management *mt*
ATBM Anti Tactical Ballistic Missile *mt, fr, wp, hc, nd74.445.18, tb, st*
ATBN Amine-Terminated Butadiene / Acrylonitrile *aa26.12.76*
ATC Active Transfer Command *jj*
ATC Advanced Technology Center *tb, st*
ATC AEGIS Training Center *fu*
ATC Aerojet Techsystems Company *-17*
ATC Air Threat Conference *mt*
ATC Air Traffic Control Controller *mt, fu, aa26.12.42, aw118.18.77, ic2-31, wp, tb, jj, y2k*
ATC Air Training Command [Randolph Air Force Base] [USAF] *mt, hc, fu, af72.5.47, wp, tb, jj*
ATC Air Training Corps [UK] *mt*
ATC Analog Technology Corporation *pl*
ATC Army Training Command *mt*
ATC Automated Technical Control *fu*
ATC Automatic Target Cueing *wp*
ATC Automatic Test Console *fu*
ATC Automatic Threshold Control *jj*
ATC Automatic Traffic Control *fu*

ATCA Advanced Tanker / Cargo Aircraft *mt, wp, af72.5.143*
ATCA Air Traffic Controllers Association *fu*
ATCA Allied Tactical Communications Agency [NATO] *mt*
ATCALS Air Traffic Control And Landing Systems [USA DOD] *mt*
ATCAP Army Telecommunication Automation Program [USA] *mt*
ATCAS Airborne Tactical Control And Surveillance *hc*
ATCBI Air Traffic Control Beacon Interrogator *fu*
ATCC Air Traffic Control Center *mt, hc, wp, fu*
ATCC Airborne Transmitter Control Center *mt*
ATCC Automated Telecommunications Center *mt*
ATCCC Advanced Tactical Command and Control Capacity *hc*
ATCCC Air Traffic Control Command Center *fu*
ATCCS Army Tactical Command and Control System *mt, pm, fu, kf*
ATCCSS Air Traffic Control Communications Switching System *hc*
ATCE Advanced Threat to Central Europe *fu*
ATCE Automatic Test and Checkout Equipment *fu*
ATCF Air Traffic Control Flight [AFCC*] *mt*
ATCG AFCAP Technical Coordinating Group *mt*
ATCM Advanced Technology Cruise Missile *uc, wp*
ATCM Air Training Command Manual *tb*
ATCMD Advanced Transportation Control [and] Movement Document *wp, mt*
ATCO Air Traffic Control Officer *wp*
ATCO Air Traffic Coordinating Officer *mt*
ATCOS Air Traffic Control Simulator *fu*
ATCOS Atmospheric Composition Satellite *wp*
ATCRABS Air Traffic Control Radar Beacon System *mt*
ATCRBS Air Airspace Traffic Control Radar Beacon System *fu, hc, jj, wp*
ATCS Active Thermal Control [Sub] System *9, wl*
ATCS Advanced Train Control System *is26.1.62*
ATCS Air Traffic Control Simulator Specialist *fu*
ATCSF Air Traffic Control Simulation Facility *fu*
ATCT Air Traffic Control Transponder *wp*
ATCT Airport Traffic Control Tower *fu*
ATCU Air Transportable Communication Unit *wp, mt*
ATCV Air Transportable Communication Van *mt*
ATD Actual Time of Departure *mt, fu*
ATD Advanced Technology Demonstrator [US Navy] [USA] *hc, mt*
ATD Advanced Technology Developments *hc*
ATD Advanced Technology Directorate *st*
ATD Aerospace Technology Directorate [NASA] *-19*
ATD Air Tasking Directive *mt*
ATD Aircrew Training Devices *mt*
ATD Automatic Target Detection *fu*
ATD Automatic Target Detector *wp*
ATDE Advanced Turboshaft Development Engine *wp*
ATDL Adaptive Targeting Data Link [USAF] *mt*
ATDL Army Tactical Data Link [USA] *kf, fu, mt*
ATDM Asynchronous TDM [concentrator, statistical multiplexing] *nu*
ATDMA Advanced Time Division Multiple Access *fu*
ATDMX Asynchronous Time Division Multiplexing *hi*
ATDP Attention Dial Pulse *pk, kg*
ATDPS Airborne Tactical Data Processing System *hc, wp*
ATDRS Advanced Tracking and Data Relay Satellite *mi*
ATDRSS {Advanced} TDRSS *rg, 19*
ATDS Airborne Tactical Data System [data processing system fitted to the E-2C Hawkeye] [USA] *hc, fu, jj, mt*
ATDS Automated Target Data System [USEUCOM] *mt*
ATDT Attention Dial Tone *hi, kg*
ATE Acceptance Test Equipment *fu*
ATE Actual Time Enroute *mt*
ATE Advanced Technology Engine [USA] *wp, mt*
ATE Aluminum Triethyl *wp*
ATE Asynchronous Terminal Emulation *fu*
ATE Automatic Automated Test Equipment *mt, fu, rm1.2.3, wp, hc, 17, sb, aa26.11.51, af72.5.185, tb, jj, kf, kg, oe*
ATE Automatic Threat Evaluation *fu*
ATE/CPL Automatic Test Equipment/ Commercial Project Line *fu*
ATEC Analyse des Travaux En Cours [French analysis of work in progress] *fu*
ATEC Automated Technical Control [Program] *mt, ro*
ATEGG Advanced Turbine Engine Gas Generator *wp*
ATEP Advanced Technical Education Program [Hughes Aircraft Company] *pbl, 8, hc, ep, jj, jdr, tm, kf, jn, fu, ah*
ATER Automatic Testing, Evaluation and Reporting *wp*
ATERS Automatic Test Equipment Reporting System *mt*
ATES AEGIS Tactical Executive System *fu*
ATEWS Advanced Tactical Electronic Warfare System *wp*
ATF Advanced Tactical Fighter *fu, aw129.14.26, wp, ph, hc, aa26.12.70, af72.5.39, jj*
ATF Advanced Technology Fighter [F-22 and F-23 program] [USA] *mt*
ATF Air Transport Force [UK] *mt*
ATF Amphibious Task Force [includes Navy forces and a landing force, with their organic aviation] [USA DOD] *mt, fu, hc, jj*
ATF Automatic Target Following *fu*
ATF Automatic Terrain Following *hc, wp, jj, fu*

ATF Automatic Test Function *fu*
ATF Automatic Track Following *fu*
ATF Oceangoing Fleet Tug [US Navy] *mt*
ATFE Advanced Tactical Fighter Engine *wp*
ATFES Army Tactical Frequency Engineering System *mt*
ATFP Airlift Total Force Plan *mt*
ATFR Automatic Terrain Following Radar *fu*
ATFS Advanced Technology Frequency Synthesizer *fu*
ATFS Automatic Terrain Following System *fu*
ATFTS Automatic TOW Field Test Sets *hn44.11.8, hc*
ATG Advanced Technology Group *pk, kg*
ATG Air Transport Group [Canada] *mt*
ATG Antenna Transceiver Group *fu*
ATG Applications and Test Group *fu*
ATG Automatic Tank Gauging *mt*
ATG Avionics Technical Group *oe*
ATGM Anti Tank Guided Missile *mt, hc, wp, jj*
ATGR Anti Tank Guided Rocket *hc*
ATGS Advanced Terminally Guided Submunitions *wp*
ATGW Anti Tank Guided Weapon *mt, fu, ph, hc*
ATH Above The Horizon *mt, fu, 5, 17, sb, hc, tb, kf*
ATH Acceleration Time History *jdr, ric*
ATH Air Transportable Hospital *wp*
ATH Attention Hang Up [a modem command] *pk, kg*
ATH Autonomous Terminal Homing *tb*
ATHOC Automatic Target Hand Off Controller *hc*
ATHP Autonomous Thermal Homing Program *hc*
ATHRS Air Transportable Hydrant Refuelers *mt*
ATHS Airborne Target Handoff Handover System *fu, hc, aw129.14.109, wp*
ATI Active Transfer Initiate *jj*
ATI Advanced Technologies Incorporated *mt*
ATI Application Transport Interface [Honeywell] *mt*
ATI Automated Tactical Intelligence *wp*
ATI Automated Technical Information *mt*
ATI Automation of Technical Information *mt*
ATIC Air Technical Intelligence Center *0, hc, jj*
ATIC ASOC/TACP Improved Communications *mt*
ATIDS Army Tactical Information Distribution System *hc*
ATIFPA Advanced Thermal Imaging Focal Plane Array *hc*
ATIGS Advanced Tactical Inertial Guidance System *hc, wp*
ATIM Advanced Technology Insertion Module *kf*
ATIPS Army Threat and Intelligence Production System *mt*
ATIRCM Advanced Threat Infra Red CounterMeasures *mt*
ATIS Advanced Thermal Imaging Scanner *hc*
ATIS Aerodrome Terminal Information Service *mt*
ATIS Alliance for Telecommunications Industry Solutions *y2k*
ATIS Automatic Terminal Information Service *fu*
ATIS Automatic Traffic Information System *wp*
ATISD Air Training Information Systems Division [ATC] *mt*
ATJS Advanced Tactical Jamming System *hc, wp, jj*
ATK Anti Tank *mt*
ATL Active Target List *tb*
ATL Active Template Library *kg, em*
ATL Advanced Technical, Limited *hc*
ATL Antenna Turn Left *jj*
ATL Applied Technology Laboratory [of AVSCOM] *tc*
ATL/CPL Automatic Test Equipment {?}/ Commercial Product Line *fu*
ATLANTIC Airborne Targeting Low Altitude Navigation Thermal Imaging and Cueing *mt*
ATLAS Abbreviated Test Language for All Systems [US Navy] *mt*
ATLAS Abbreviated Test Language for Avionics Systems *fu, hc, wp*
ATLAS Advanced Tactical Lightweight Avionics System *hc*
ATLAS Advanced Target Location And Strike *wp*
ATLAS Atmospheric Laboratory for Applications and Science *hy, hg, mi, ge*
ATLAS [a manufacturer of personal computers] *hn43.25.2*
ATLID Atmospheric Lidar [ESA] *hy*
ATLIS Airborne Tracking Laser Identification System *wp*
ATLO Assembly, Test, and Launch Operations *pl*
ATLS Automated Tape Library Support *mt*
ATLSS Advanced Technology For Large Structural Systems *wp*
ATM Adobe Typeface Manager *em, ct, kg*
ATM Advanced Tactical Missile *uc*
ATM Advanced Test Module *wp*
ATM Aft Termination Module *fu*
ATM AGM Training Missile [USA] *mt*
ATM Air Tactical Missile *hc*
ATM Air Tasking Message *mt*
ATM Air Terminal Management *mt*
ATM Air Traffic Management *hc, fu*
ATM Air Training Missile *hc*
ATM Amateur Telescope Maker *mi*
ATM Analog Telemetry *he*
ATM Anti Tactical Missile *hc, tb, st*
ATM Anti Tank Missile (ATM-12 Bullpup Trainer) *fu, mt*

ATM Apollo Telescope Mount *ac2.2-27, hc*
ATM Articulating Thermal Membrane *jdr*
ATM Asynchronous Transfer Mode *hi*
ATM Asynchronous Transfer Mode [cell relay, hundreds of MBPS] *is26.1.41, jdr, kf, heh2.21.7, em, ct, nu, kg*
ATM Attack Laydown Model *st*
ATM Automatic Automated Teller Machine *ct, sc2.11.2, kg, hi*
ATM AWCAP Transmittal Memorandum *jj*
ATMAC Air Traffic Management And Control *fu*
ATMAC Air Traffic Management Automated Center *fu*
ATMAC Air Traffic Management Automated Control *hc*
ATMCT Air Terminal Management Control Team *mt*
ATMCT Air Terminal Movements Control Team *mt*
ATME Atmospheric Transmission Measurement Equipment *hc*
ATMF ATM forum [recommends and defines ATM standards] *nu*
ATMH Automatic Text Message Handling *mt*
ATMOS Atmospheric Trace Molecules Observed by Spectroscopy *hy, hg, ge*
ATMOS High Resolution Atmosphere Infrared Spectrometer [Spacelab] *ge*
ATMOS-UFS ATMOS-Umwelt Forschungs Satellit [FRG] *ge*
ATMS Air Traffic Management System [USACC] *fu, mt*
ATMS Analog Telemetry Monitor Software *oe*
ATMS Anti Tank Missile Study *hc*
ATMS Army Traffic Management System *mt*
ATMS Automatic Track Management System *fu*
ATN Adaptive Tactical Navigation *hc*
ATN Advanced TIROS-N [NOAA] *hg, ge*
ATNVS Airborne TOW Night Vision System *hc*
ATO Abort To Orbit [Shuttle abort plan] *ci.608, mb, mi*
ATO ADP Terminal Operator *mt*
ATO Air Targets Officer *mt*
ATO Air Tasking Officer *mt*
ATO Air Tasking Order *mt, fu, wp, pbl, kf*
ATO Air Training Officer *mt*
ATO Aircraft Transfer Order *jj*
ATO Airlift Tasking Orders *mt*
ATO Alternate Technical Officer *fu*
ATO Assisted Takeoff *mt, wp*
ATO Old Tug [ship] [US Navy] *mt*
ATOC Air Tactical Operations Center *mt, wp*
ATOC Air Tasking Order Center *mt*
ATOC Air Terminal Operations Center *mt, wp*
ATOC Allied Tactical Operations Center [Sembach] [NATO] *mt, af72.5.98*
ATOC Automated Tactical Operations Center *fu*
ATOL Assisted Takeoff and Landing *wp*
ATON ACE Operational Telegraph Network *mt*
ATOS Air Tasking Order System [PACAF] *mt*
ATOS Assisted Takeoff System *wp, jj*
ATOS Automated Tasking Order System [USAF] *mt*
ATOS Automated Technical Order System [LFSMS] *mt, fu, hc, wp, kf*
ATOT Actual Time Over Target *fu*
ATOT Angle Track On Target *hc, jj*
ATOW Acquisition Timing Orderwire *jdr*
ATP Acceptance Test Procedure Plan *wp, gp, 16, 17, sb, hc, tb, jj, jdr, kf, jn, oe, kg*
ATP Acquisition, Tracking and Pointing *17, wp, sb, tb, kf*
ATP Adenosine Triphosphate *wp*
ATP Advanced Technical Plan Program *hc, jj, fu*
ATP Advanced Test Program *fu*
ATP Advanced Tracker Program *hc*
ATP Advanced Tracking Prototype *mt*
ATP Advanced Turbo Prop [BAe program] *mt, aa26.10.15, aw129.21.68*
ATP Agreement To Proceed *jdr*
ATP Allied Tactical Publication *mt, fu*
ATP Authorization Authorized To Proceed *5, 8, sb, re, tb, st, wl, jdr, kf, fu*
ATP Automated Test Plan *wp*
ATP-FC Acquisition, Tracking, Pointing and Fire Control [RADC] *mt*
ATP-OD Acceptance Test Procedure - Ordnance Data *fu*
ATP/ALTP (ATP/ALTP or FE certificate . . . required) *aw130.11.76*
ATPD Ambient Temperature and Pressure Dry *wp*
ATPG Armament Technology and Procurement Group *hc*
ATPG Automatic Test Pattern Generator *rm8.2.1, fu*
ATPG Automatic Test Program Generator *hc, wp*
ATR (3/4 ATR military enclosure) *hd1.89.y13*
ATR Acceptance Test Report *fu, hc, jj*
ATR Advanced Tactical Radar *mt, fu, uc, wp, hc*
ATR Aided Target Recognition *ie5.11.2*
ATR Air Tasking Request *mt*
ATR Air Transport Rack *fu, hc*
ATR Air Transport Racking [standard electronics racks] *mt*
ATR Air Transport Regulations *mt*
ATR Airborne Transit Requirements *fu*
ATR Airborne Transport Rack *mt*

ATR Aircrew Training System *mt*
ATR Antenna Turn Right *jj*
ATR Anti Tactical Radar *hc*
ATR Anti Transmit Receive [device] *fu, hc, jj, jdr*
ATR Attenuated Total Reflectance *fu*
ATR Automatic Target Recognition *mt, wp, st*
ATR Automatic Terminal Recognition *pk, kg*
ATR Avions de Transport Regional [regional transport aircraft, built by French / Italian manufacturers] *mt, aw130.13.5*
ATR Rescue Tug [ship] [US Navy] *mt*
ATRA Advanced Technology Reconnaissance Aircraft *wp*
ATRAN Automatic Terrain Recognition And Navigation *fu, wp, jj*
ATRAS Automated Travel Record Accounting System *mt*
ATRC Air Traffic Regulation Center *hc*
ATRCCM Automatic Target Recognition CCM *hc*
ATREQ Atomic Supporting Request *mt*
ATRID Automatic Target Recognition Identification and Detection *wp*
ATRK Antenna Track *jj*
ATRR Automatic Target Recognition and Reporting *wp*
ATRS Advanced Teleoperator Retrieval System *hc*
ATRS Aerial Training Squadron [USA] *mt*
ATRS Automated Travel Record System *mt*
ATRS Aviator Training Research Simulator *hc*
ATS Acceptance Test Specification *hc*
ATS Acquisition and Tracking Seeker *hc*
ATS Acquisition and Tracking System [target] *sd*
ATS Active Test Sequence *oe*
ATS Administrative Telephone System *tb*
ATS Administrative Terminal System *fu*
ATS Administrative Terminal System [Hughes Aircraft Company] *hc, jj, kg*
ATS Advanced Target Study *hc*
ATS Advanced Technology Satellite *mt, ge, sc3.2.5, wp, 11, ep, hg, oe*
ATS Advanced Training System [ATC] *mt*
ATS Aerothermal Structure *st*
ATS Air Traffic Services *fu, mt*
ATS Air Traffic System *y2k*
ATS Air Transportable Sonar *fu*
ATS Aircrew Training System *mt, wp, hn49.14.8*
ATS Airmanship Training Squadron [USA] *mt*
ATS ALFS Target Simulator *fu*
ATS Alternate Thermal Sight *hn49.3.8*
ATS Anomaly Tracking System *oe*
ATS Applications Technology Satellite [laser communications experiment] *fu, ac7.1-10, hc, ac2.2-7, hc, lt, ep, oe*
ATS Armament Test Set *jj*
ATS Automated Telemetry System *aw129.21.5*
ATS Automated Telephone Switch *fu*
ATS Automated Terminal Service *fu*
ATS Automated Terminal System *mt*
ATS Automated Test Set *fu*
ATS Automated Travel System [USAF] *mt*
ATS Automatic Test System *wp, dm*
ATS Automatic Time Sharing *fu*
ATS Avionics Test Station *fu*
ATS Salvage and Rescue Ship [US Navy] *mt*
ATSA Automatic Time Slot Assignment *fu*
ATSAR Atomic Strike Approval Request *mt*
ATSC Automated Telecommunications Systems Center [USA] *mt*
ATSD Airborne Traffic Situation Display *wp*
ATSD Assistant To the Secretary of Defense *dt*
ATSE Army TPS Support Environment *fu*
ATSGMS Augmented Test Set, Guided Missile System *hc*
ATSIM Acquisition and Track Simulation *tb*
ATSR Advanced Tactical Surveillance Radar [RADC] *fu, hc, mt*
ATSR Along Track Scanning Radiometer [ERS, ESA] *ge, hg*
ATSR ATS Ranging Station *ac2.5-17*
ATSR/M Along Track Scanning Radiometer Microwave Sounder *ge, hg*
ATSS Acquisition and Tracking Seeker Subsystem *hc*
ATSS Advanced Tactical Surveillance System *mt, hc*
ATSS Alaskan Telephone Switching System *fu*
ATSS Attitude Sun Sensor *gp*
ATSS Automatic Test Support System *aw118.18.6, hc*
ATST Air Traffic Services Tower *fu*
ATSTAT Airlift Status *mt*
ATT Acquisition Track Techniques *hc*
ATT Advanced Tactical Transport [MAC] *mt*
ATT Advanced Theater Transport [USA] *wp, mt*
ATT American Telephone and Telegraph *fu*
ATT Army Troop Test *fu, hc*
ATT Automatic Target Tracking *wp*
ATT Automatic Track Transfer *jj*
ATTAK Advanced Tank Target Auto Kill *hc*
ATTAP Advanced Turbine Technology Program *aw129.10.63*
ATTD Advanced Technology Transition Demonstrator *wp*
ATTDET Attitude Determination [Hughes Aircraft Company] *hc*

ATTG Adversary Threat Training Group *mt*
ATTG Automated Tactical Target Graphics *mt*
ATTIP Advanced Technology for Thermal Imaging Program *hc*
ATTMA Advanced Transport Technology Mission Analysis *wp*
ATU Acoustical Transmitter Unit *fu*
ATU Adaptive Tracker Unit *ac10.4.4*
ATU Advanced Timing Unit *jdr*
ATU Airborne Transceiver Unit *fu*
ATU Antenna Tuning Unit *fu*
ATU Arab Telecommunications Unit *fu*
ATU Automatic Tracking Unit *fu*
ATURF Airborne TOW USAREUR Repair Facility *hc, jj*
ATV Aerodynamic Test Vehicles *hc*
ATV All Terrain Vehicle *wp*
ATV Automated Transfer Vehicle *ls4, 5, 39, ai2*
ATV Automatic Threshold Variation *fu*
ATVC Automatic Target Voltage Control *hc*
ATVI Advanced Technology Validation and Integration *hc*
ATVS Ada Test and Verification System *wp, ro, hc*
ATW Aerospace Test Wing *hc*
ATW Anti Torpedo Weapon *fu*
ATWC Atmospheric Tactical Warning Connectivity *mt, fu*
ATWS Automatic Track While Scan *hc, fu, jj*
ATZ Attention Restore configuration profile from nonvolatile RAM [Modem command] *pk, kg*
AU Air University [USAF] *mt, hc, wp, af72.5.47*
AU Airborne Unit *fu*
AU Analyzer Unit *fu*
AU Angstrom Unit *w9*
AU Arithmetic Unit *fu, hc, jj*
AU Astronomical Unit *ma23, aa26.10.37, w9, sf31.1.12, es55.36, so, mb, re, mb, mi, ai1*
AU Atomic Unit *ai1*
AUASAM Automatic Aim Point Selection and Maintenance [ground-based] *hc, pl*
AUCV Advanced Unmanned Cargo Vehicle *17, sb*
AUDIT Army Uniform Data Inquiry Technique *mt*
AUDIT Automated Data Input Terminal *pk, kg*
AUEL Automated Unit Equipment List [USA] *mt*
AUF Airborne-Uninhabited Fighter *fu*
AUFIS Automated Ultrasonic Flaw Imaging System *wp*
AUG Add-ons and Upgrades *wp*
AUG AFCAP Users Group *mt*
AUI Analysis User Interface *fu*
AUI Attachment Unit Interface *kg, ct*
AUI Attachment User Interface *em*
AUIK Airborne Unit Installation Kit *fu*
AUL Air University Library *af72.5.72*
AUL All Up Round *mt*
AUM Air to Underwater Missile *hc, fu, jj*
AUP Acceptable Use Policy [Internet] *em, kg*
AUP AUTODIN Upgrade Project *mt*
AUPCU Arithmetic Unit Program Control Unit *hc*
AUR All Up Round [missile] *jj*
AURA Association of Universities for Research in Astronomy *mi*
AURIO Auroral Imaging Observatory *ge, hg*
AURORA Automatic Recovery Of Remotely piloted Aircraft *mt*
AURT All Up Round Table/Tester [missile GSE] *hc, jj*
AUS Army of the United States *w9*
AUSA Association of the US Army *hc*
AUSCANUCUS Australia, Canada, United Kingdom, United States *wp*
AUSMS Automated Unbundled Software Management System [JDSSC] *mt*
AUSS Advanced Underwater Search System *hc*
AUSS Autonomous Universal Seeker Study *hc*
AUSSAT Australian [National Telecommunications] Satellite [Company] *sc2.8.4, 17, hn49.9.2, kf*
AUSTACCS Australian [Army] Tactical Command Control System *fu, hc*
AUT Antenna Under Test *R. Rommel, jdr*
AUT Aperture Uncertainty Time *hc*
AUTEC Acoustic Range Project *hc*
AUTEC Atlantic Undersea Test and Evaluation Center *mt, fu, 12*
AUTO-ICPL Automatic Interrupted Circuit Path Locator *fu*
AUTOBOOK Automated Booking and Distribution System [AFLC] *mt*
AUTOCAT Automatic Communications Airborne Transfer *mt*
AUTOCONET AUTODIN Conference Network [CINCPAC] *mt*
AUTODIN Automatic Digital [Information] Network [DCA] *mt, fu, wp, 12, ph, hc, jb, no, tb, jj, kf*
AUTOEXEC Automatic Execution *pk, kg*
AUTOFAC AUTODIN Test Facility [AUTO KC135] *mt*
AUTOFAC [KC-135 Quarterly Schedule, SAC] *mt*
AUTOMATE Automatic Manufacturing Test Equipment *hc*
AUTOPAT Automatic Panel and Assembly Tester Hughes *fu*
AUTOPATH Automated Flight Path [SAC/DARPA] *mt*
AUTOROS Automated Retail Outlet System [USA] *mt*
AUTOSEV Automatic Secure Voice Communications *tb*

AUTOSEVOCO Automatic Secure Voice Communications *wp*
AUTOSEVOCOM Automatic Secure Voice Communications *mt, hc, wp, tb*
AUTOSPEC Automated Sensor Processor Evaluation Center *hc*
AUTOSTRAD Automated System for Transportation Data [MTMC] *mt*
AUTOVON Automatic Voice [Switching] Network [USA DOD] *mt, wp, 12, jb, an31.13.4, tb, jj, fu*
AUV Administration Use Vehicle *jj*
AUV Autonomous Underwater Vehicle *fu*
AUW Advanced Underwater Weapons [naval nuclear depth bombs] *12, fu*
AUW All Up Weight *mt*
AUW Anti Underwater Warfare *fu*
AUWMP Air University War and Mobilization Plan *mt*
AUXCP Auxiliary Command Post *mt*
AUX_MEA Auxiliary Maintenance Engineering Analysis *jj*
AV Air Vehicle *fu*
AV Aircraft [seaplane] Tender (AV-8 is Harrier) [USA DOD] [USA] *mt*
AV Allan Variance [RMO] *jdr*
AV Authenticity Verification *pk, kg*
AV-MF Aviatsiya Voenno-Morsko Flota [Naval Air Force] [USSR] *mt*
AVA A Verifiable ADA *fu*
AVA Activity Value Analysis *wp*
AVA Audio Visual Authoring [IBM] *pk, kg*
AVA Azimuth Versus Amplitude *fu*
AVARS Audio Visual Analysis and Retrieval System *mt*
AVAS Aided Visual Attack System *hc, jj*
AVAS Automatic Vfr Advisory Service *wp*
AVAWOS Aviation Automated Weather Observation System *hc*
AVC Advanced Vidicon Camera *ac2.8-2*
AVC Audio Visual Connection [IBM] *pk, kg*
AVC Automatic Volume Control *mt, 0, wp, hc, w9, jj, jdr, hi, ai1*
AVC Avionics Change *jj*
AVCAL Aviation Consolidated Allowance List *jj*
AVCDP Analog Video Common Data Point *fu*
AVCO Active Voltage Controlled Oscillator *hc, jj*
AVCO American Viscose Company *0, hc*
AVCOM Aviation Material Command [US Army] *jj*
AVCRAD Aviation Classification Repair Activity Depot [USA] *mt*
AVCS Advanced Vidicon Camera [Sub] System *sd, hc, ge, hg*
AVD Advanced Vehicle Detection *hc*
AVD Alternating Voice and Data *pk, kg*
AVDS Audio Visual Display System *mt*
AVE Aerospace Vehicle Equipment *17, wp, sb, hc, pl, jj, jdr, kf*
AVE Analog Video Even *jj*
AVEN Axisymmetric Vectoring Exhaust Nozzle [a vectoring nozzle for the General Electric F-101 engine, applied to MATV] [USA] *mt*
AVETV Aided Visual Element TV Freeze *hc*
AVF All Volunteer Force *wp*
AVG American Volunteer Group [aka the 'Flying Tigers' unit] [USA/China] *mt*
AVG [Aircraft Escort Vessel, later renamed ACV, USA] *mt*
AVGAS Aviation Gas Gasoline *wp, mt*
AVHRR Advanced Very High Resolution Radiometer [NOAA] *hc, ac1.6.-13, aw129.14.43, hy, hg, oe, ge*
AVI Airborne Vehicle Identification *wp*
AVI Audio Video Interleaved Interlaced *ct, kg, hi*
AVID Airborne Vehicle Identification [System] *fu, hc*
AVIM Aviation Intermediate Maintenance [facility] [USA] *hc, jj, mt*
AVIONICS Aviation Electronics *wp, jj*
AVIP Avionics [/Electronics] Integrity Program *mt, wp, aa26.8.b25*
AVIRIS Airborne Visible Infrared Imaging Spectrometer[NASA] *ge, wp, hy, hg*
AVIS Airborne Audio Visual Instrumentation System *wp*
AVISH Air Vehicle Inventory Status and History *mt*
AVISURS Aerospace Vehicle Inventory Status and Utilization Reporting System [USAF] *mt*
AVL Automatic Volume Control *fu*
AVL Avionics Laboratory [of AFWAL] *tc*
AVLIS Atomic Vapor Laser Isotope Separation *hc*
AVM Airborne Vibration Monitoring *wp*
AVN Ameritech Virtual Network *ct, em*
AVNIR Advanced Visible and Near-Infrared Radiometer *hg, ge*
AVNL Automatic Video Noise Levelling *fu*
AVO ADA Validation Office [DOD] *mt*
AVO Air Vehicle Operator *fu*
AVO Analog Video Odd *jj*
AVO Avoid Verbal Orders [a document] *ah, pbl, sc4.5.2, 8, 15, hc, ep, jj, jdr, tm, jn*
AVP Acoustic Adaptive Video Processor *fu*
AVP Adaptive Video Processing *wp*
AVP AUTOVON Precedence *mt*
AVP Small Seaplane Tender [USA] *mt*
AVR Automatic Voice Recognition *hi, kg*
AVR Automatic Voltage Regulation *hi*
AVRADA Avionics Research and Development Activity [of AVSCOM] *tc*

AVRADCOM [Hughes Aircraft Company customer for AIMDAPS] *hc*
AVRE Armoured Vehicle, Royal Engineers [UK] *mt*
AVS Air Vehicle System *st*
AVS Architecture Validation Software *fu*
AVS Audio Visual Service [MAC] *dt*
AVS Aviation Supply Ship [US Navy] *mt*
AVS Avionics Visual System *hn49.9.8*
AVSAD A-new VS Airborne Drum [NADC] *hc*
AVSCOM Aviation Systems Command [US Army] *mt, fu, tc, hc, jj*
AVSDU Analog Video Signal Display Unit *fu*
AVSS Aided Visual Sensor Set *hc*
AVT Anti Virus Technology *hi*
AVT Automatic Video Tracker *hc*
AVT Auxiliary Aircraft Transport [includes some former CV's of the Essex class and CVL's of Saipan and Independence class, AVT-6 Saipan, AVT-7 Wright] [US Navy] *mt*
AVTA [Hughes Aircraft Company customer for SAVA Standard Army Vectronics Architecture] *hc*
AVTR Airborne Video Tape Recorder *wp*
AVTR Avionics Technology Roadmap *wp*
AVTS Advanced Visual Technology System *mt*
AVTS Automated Visual Technology System *wp*
AVUIK Aircraft Vehicle Unit Installation Kit *fu*
AVUM Aviation Unit Maintenance *hc, jj*
AV_BN Aviation Battalion *mt*
AW Above Water *fu*
AW Acoustic Warfare *fu*
AW Administrative Wing *fu*
AW Advanced Westar *-4*
AW Aircraft Warning *w9*
AW Airlift Wing [USA] *mt*
AW All Water *w9*
AW All Weather *wp, jj*
AW Alternate Weapon *jj*
AW Articles of War *w9*
AW Attack Warning *tb*
AW Automatic Weapon *w9*
AW&ST Aviation Week and Space Technology [McGraw-Hill Publishing Company] *aw118.14.11, 14, mb, mi*
AW-SHORADS All Weather Short Range Air Defense System *hc*
AW/AA Attack Warning/Attack Assessment *mt, tb, kf, fu*
AW/PN Assist Work/Part Numbers *fu*
AWA Alternate Weapon Attack *jj*
AWA Amalgamated Wireless Australasia *sc4.7.1, lt*
AWA Assist Work Authorization *hc, fu, su*
AWA Aviation [and Space] Writers' Association *hn44.16.8*
AWAC Airborne Weapon and Control *hc*
AWACD Airborne Warning and Control Division *mt*
AWACS Adverse Weather Close Air Support *sb*
AWACS Airborne Warning and Control Squadron *mt*
AWACS Airborne Warning And Control System *aa26.12.20, wp, aw130.13.5, hn41.19.1, hc, 12, af72.5.49, pm, tb, jj, kf, fu, ro*
AWACS Airborne Warning And Control System [the Boeing E-3A Sentry] [USA] *mt*
AWACTS Airborne Warning And Control Training Squadron [USA] *mt*
AWACW Airborne Warning And Control Wing *wp, af72.5.139*
AWADS Adverse Weather Aerial Delivery System [C-130] *mt*
AWADS Adverse Weather Air Aerial Delivery System *hc, af72.5.142, wp*
AWANS Aviation Weather And NOTAM System *fu*
AWAPS Advanced Weather Analysis and Prediction System *mt, hc*
AWARE Advanced Weapon/Aircraft Requirements Evaluation *hc*
AWARES Assessment of Wholesale and Retail System [RAND] *mt*
AWARS All Weather Reconnaissance System *hc, wp*
AWAY [France, rocket extraction system for the pilot; made by Hurel-Dubois] *mt*
AWB Aglets Workbench *ct*
AWBS Aircraft Weight and Balance Requirements System [PACAF] *mt*
AWC Air War College [AU] *mt, af72.5.72*
AWC Air Warfare Control *wp*
AWC Army War College *mt*
AWCAP Airborne Weapons Corrective Action Program *jj*
AWCAS Adverse Weather Close Air Support Study *hc*
AWCCV Advanced Weapon Carriage Configuration Vehicle *hc*
AWCIS Aircraft Weapons Control Intercept Interceptor System *hc, wp, jj*
AWCL All Weather Carrier Landing *jj*
AWCLS All Weather Carrier Landing System *mt, jj*
AWCO Area Wage and Classification Office *dt*
AWCS Air Weapons Control System *mt*
AWCS Air [Airborne] [Aircraft] Weapons Control System *fu*
AWCS Automatic Weapons Control System *hc, wp, jj*
AWCW Airborne Warning Composite Wing [USA] *mt*
AWDR Advanced Weapon Delivery Radar *hc, wp*
AWDS Automated Weather Distribution System [MAC/AWS] *ag, fu, pf1.27, mt*
AWE Atomic Weapons Establishment [UK] *mt*
AWES All Weather Electronic System *hc*
AWG Activation Working Group *kf*

AWG Airborne Weapon Guided Guidance *hc, hd1.89.y3, jj*
AWG American Wire Gauge [wire standard] *ep, jj, jdr, he, fu, kg, hi*
AWG Attack Working Group *st*
AWGN Additive Average White Gaussian Noise *2, fu, 4, jdr*
AWI All Weather Interceptor *hc, jj*
AWI Awaiting Installation *jj*
AWIP [a weather information system] *hc*
AWIPS Advanced Weather Information Processing System *oe*
AWIPS Advanced Weather Interactive Processing System [NOAA] *mt*
AWIS Advanced Wafer Imaging System *hc*
AWIS Aircraft Warning Interface System *hc*
AWIS All Weather Identification Sensor *wp*
AWIS Army WWMCCS Information System *mt, fu, mt*
AWIS Automated Weather Information System *fu, ag*
AWK [Unix language named after its authors... Al Aho, Peter Weinberger and Brian Kernighan] *hi, kg*
AWL All Weather Landing *wp, jj*
AWL Automated Wire List *fu*
AWLAR All Weather Low Altitude Route *wp*
AWLS All Weather Landing System *hc, wp, jj*
AWM Advanced Waveform Modulation *fu*
AWM Adverse Weather Maverick *hc*
AWM Airborne Weapon Maintenance *jj*
AWM Awaiting Maintenance *wp, jj*
AWN Automated Weather Network [USAF worldwide weather information] *mt, pl*
AWO Aircraft Work Order *jj*
AWOL Absent Without [Official] Leave [USA] *wp, mt*
AWOP Automated Weaponeering Optimization Program *mt*
AWOS Automated Weather Observation Observing System *ag, fu*
AWP Aviation Weather Processor *fu*
AWP Awaiting Parts *mt, wp, jj*
AWPDS Attack Warning Processing and Display System[NORAD] *mt*
AWPR Advanced Weapon Delivery Radar *hc*
AWR A-Word Register *jj*
AWR Alteration Work Requirements *hc*
AWR Army War Room *hc, jj*
AWRS Airborne Weather and Reconnaissance System *wp*
AWRS All Weather Reconnaissance System *hc, wp*
AWS Acoustic Warfare System *fu*
AWS Advanced Warning System *ph, wp, kf*
AWS Advanced Weapons System *hc*
AWS Advanced Workstations and Systems [group] [IBM] *pk, kg*
AWS Air Weather Service *mt, wp, sc3.1.312, hc, no, jj, pbl*
AWS American Welding Society *jj*
AWS Area Weapon Subsystem *hc*
AWS Automated Warehouse System [AFLC] *hc, wp, mt*
AWSACS All Weather Stand Off Attack Control System *hc*
AWSC Agricultural Weather Service Center *ag*
AWSD Antisubmarine Warfare Systems Division [Hughes GSG] *hq5.2.4*
AWSMIS Air Weather Service Management Information System [MAC] *mt*
AWST Automated Weapon System Trainer *hc, wp*
AWSWO Air Weather Service Washington Office *dt*
AWT Abstract Windows Windowing Toolkit *kg, em*
AWTBS All Weather Tactical Bombing System *wp*
AWTS (the testing phase of AWTS-Reporters) *hn49.11.8*
AWTSS All Weather Tactical Strike System *hc, wp*
AWTT Aircraft Weapons Tactics Trainer *jj*
AWU Atomic Weight Unit *wp*
AX Architecture Extended *hi, kg*
AX {Advanced Experimental} [concept for long range advanced missile] *wp*
AXAF Advanced X-Ray Astrophysics Facility *sc4.10.3, hc, es55.61, hn51.16.1, mb, mi*
AXE [surface launched ballistic missile concept] *dh*
AXLE Adaptive Emitter Location Equipment *hc*
AXP Allied Exercise Publication *mt*
AXP Almost Exactly Prism *ct*
AXP Axial Pitch *fu*
AYACC Ada YACC [Ada program compiler compiler] *fu*
AYC Automatic Yaw Control *wp*
AZ Assault Zone *mt*
AZ AZ Engineering Company [Hughes Subsidiary in Santa Ana] *fu*
AZ/EL Azimuth/Elevation *uh, hd*
AZA Azusa Range Safety System Cape Kennedy *0*
AZANG Arizona Air National Guard *-12*
AZEL Azimuth Elevation *fu*
AZM Azimuth Memory *jj*
AZSCAN Azimuth [frequency] Scan Radar *hc, jj*
AZT Azinothymidine *wp*
AZUR [solar physics spacecraft of ESA] *ci.735*
A_SPEC [top level system] Specification *fu*

B

B Bombardirovshkchik [bomber] [USSR] *mt*
B Bomber [B-1B is a long range penetrating bomber, B-52 is Stratofortress] [USA DOD] *mt*
B Burroughs [computer designator, as in B3500 for Burroughs 3500 computer, B3700 for Burroughs 3700, etc.] *mt*
B&L Buildings and Lands *fu*
B&P Bidding Bid and Proposal [budget] *mt, fu, ah, pbl, tm, ep, tb, jdr, kf*
B-B Bar to Bar *jj*
B-CDMA Broadband Code Division Multiple Access *ct*
B-CHANNEL Bearer Channel *pk, kg*
B-ICI Broadband - ISDN Inter Carrier Interface *ct*
B-ISDN Broadband - Integrated Services Digital Network *ct*
B-LEVEL Development Specification Level *tb*
B-R Boost Regulator *0*
B-SCAN Range [Rate] vs. Azimuth Display *jj*
B/A Booster/Adaptor *st*
B/B Breadboard, Brassboard *17, sb*
B/D Binary to Decimal *fu*
B/F Background/Foreground *pk, kg*
B/G {Brigadier General} *af72.5.28*
B/H Base to Height [ratio] *ac2.5-17*
B/H Bulkhead *jdr*
B/I Burn-In *jdr*
B/L Baseline *16, fu, oe*
B/L Bill of Lading *fu, hc, jj*
B/M Bill of Material *hc, jj, fu*
B/M Buffer Multiplexer *fu*
B/N Bombing/Navigation *jj*
B/O Back Order *fu*
B/O Bake Out *jdr*
B/O Burn Out *jdr*
B/PB Boost/Post Boost *hc, fu*
B/S Beam Splitter *oe*
B/U Back-up Unit *tb*
B/V Boost Booster Vehicle *sb*
B/W Black and on White *ac2.8-14, hc, jj*
B3ZS Bipolar, 3 Zero Substitution *fu*
B8ZS Binary 8-Zero Substitution *ct, em*
BA Base Aerea [Air Force Base] [Portugal] *mt*
BA Basic Agreement [DECCO] *hc, mt*
BA Best Accuracy *fu*
BA British Aerospace *sc3.2.7*
BA Bromo Acetone [chemical warfare codename] *wp*
BA Budget Authority *mt*
BA Budget Authorization *mt, fu*
BA Buffer Amplifier *fu*
BA&E British Aerospace and Electronics *fu*
BAA Backup Aircraft Authorization *wp*
BAA British Airport Authority *aw130.13.11*
BAA Broad Agency Announcement *mt, hc, nt, jdr*
BAAARS Base Access Area Automated Reporting System *mt*
BAAE Bachelor of Aeronautical and Astronautical Engineering *w9*
BAAP Beam And Add Program *hc*
BAAS Broadband Acoustic Array Section *fu*
BABS Base Automated Budget System *mt, wp*
BABS Blind Approach Beacon System *jj*
BAC Bearing Area Curve *fu*
BAC Billing Adjustment Code *wp*
BAC Boeing Aircraft Aerospace Company Corporation *bo, hc, 4, tb, jj, fu*
BAC British Aircraft Corporation Company *15, jj*
BAC Budget at Completion *4, 17, sb, tb, jdr, fu*
BAC {a type of telemetry mode} *gs2-13.21*
BACE Basic Automatic Checkout Equipment *hc, jj*
BACKFIRE [TU-26 Swing-Wing] *wp*
BACP Bandwidth Allocation Control Protocol *pk, kg*
BAD BLACKER / AUTODIN / DDN *mt*
BAD Budget Allocation and Documentation *fu*
BADG Battle Group AAW Display Group *fu*
BADGE Base Basic Air Defense Ground Environment [Japan] *mt, ro, wp, hc, fu*
BADGE Budget and Accounting Data General purpose Extract [PACAF] *mt*
BADGE-X Next Generation BADGE *fu*
BAE Bachelor of Aeronautical Engineering *w9*
BAE Battlefield Area Evaluation *wp*
BAE Beacon Antenna Equipment *fu*
BAE British Aerospace, [Limited] *pl, mb, mi*
BAF Batiment d'Assemblage Final [Ariane final assembly building] *ai2*
BAFB Bolling Air Force Base *dt*
BAFO Base Accounting and Finance Office *wp*
BAFO Best and Final Offer *mt, pbl, fu, ah, hc, wp, jdr, kf, jn, oe*
BAFO British Air Force of Occupation [created in Germany in 1945] [UK] *mt*
BAH Booz, Allen, Hamilton *uh*

BAI Backup Aircraft Inventory *mt, wp*
BAI Barometric Altitude Indicator *wp*
BAI Battlefield Air Area Interdiction *mt, wp, aw124.21, aa26.12.84*
BAJ Bristol Aerojet Ltd. *crl*
BAK Binary Adaptation Kit [Microsoft] *pk, kg*
BAKER_NUNN [ground based optical system] *mt*
BAL Basic Assembler Assembly Language *fu, kg*
BAL British Anti Lewisite [Dimercaprol] *wp*
BALFRAM Balanced Force Requirements Analysis Methodology *mt*
BALFRAM Balanced Force Requirements Attrition Model *mt*
BALLAST Balanced Loading via Automated Stability and Trim *wp*
BALLUTE Balloon Parachute [bomb retarder] *mt*
BALUN Balanced / Unbalanced [device] *kg, jj, ct, em, hi*
BAM Binary Angular Measurement *fu*
BAM Binary Azimuth Measurement *fu*
BAM Boyan Action Module *pk, kg*
BAM Bus Analog Module *hn47.7.8, hc, fu*
BAMBI Ballistic Advanced Missile Boost Interceptor [system] *17, sb, hc, tb*
BAMIRAC Ballistic Missile Radiation Analysis Center *hc*
BAMP Battlefield Automation Management Plan [USA] *mt*
BAMS Base Automated Mobility System [TAC] *wp, mt*
BAMS Binary Angle Measurement System *fu*
BAN Basic Access Node *mt*
BANCS Bell Application Network Control System *ct*
BANGB Buckley Air National Guard Base *kf*
BANM Bell Atlantic Nynex Mobil *ct*
BAO Basic Attack Options *mt*
BAOF Brassboard Active Optical Fuze *st*
BAPC British Aircraft Preservation Council [UK] *mt*
BAPTA Bearing And Power Transfer Assembly *sc3.6.3, 8, 9, 15, hc, pl, es55.23, lt, ep, re, pbl, jdr, jn*
BAQ Basic Allowance for Quarters [USAF] *af72.5.49, an31.13.1*
BAQ/SITW Basic Allowance Quarters / State Income Tax Withholdings *mt*
BAR Browning Automatic Rifle *w9*
BARB Beacon Aided Radar Bombing *wp*
BARCAP Barrier Combat Air Patrol *mt, jj*
BARCS Battlefield Area Reconnaissance System *wp*
BARD Baseline Allocation Requirements Document *tb*
BARE_BASE [to equip and support a tactical unit moving from a home base to bare] *wp*
BARN Bombing And Reconnaissance and Navigation *hc*
BARRNET Bay Area Research Network *ct*
BARSTUR Barking Sands Tactical Underwater Range [Hawaii] *12, hc*
BARTS Bell Atlantic Regional Timesharing *pk, kg*
BAS Basic Allowance for Subsistence [USA DOD] *mt, af72.5.49*
BAS Battlefield Automated Systems *mt, fu*
BAS Beam Alignment System *hc*
BAS Biological Agent Simulant [chemical warfare] *wp*
BAS Breathing Air System *jj*
BAS Budget Allocation Summary *wp*
BASAT Ballistic ASAT *sd*
BASCOP Base Communications Plan [USA] *mt*
BASD Ball Aerospace Systems Division *ggIII.1.43, oe*
BASE Basic Army Strategic Estimates *mt*
BASE Boeing Automated Software Engineering *fu*
BASE British Aerospace Systems and Equipment *aa26.12.79, aw130.11.41*
BASES Beam Approach Seeker Evaluation System *hc*
BASF Badische Anilin Und Sodafabric [a German corporation] *wp*
BASIC Basic Algebraic Symbolic Interpretive Compiler *wp*
BASIC Battle Area Surveillance and Integrated Communications *wp*
BASIC Beginner's All purpose Symbolic Instruction Code *fu, kg, ct, em, wp, hi*
BASICS Battle Area Surveillance Intelligence Communications System *wp*
BASM Built in Assembler *hi, kg*
BASO Base Accountable Supply Officer *wp*
BASOPS Base Operating System *wp*
BASOPS Base Operations *fu*
BASS Backup Avionics Subsystem Software *wp*
BASS Base Augmentation Support Set *mt*
BASS Base Automated Service Store *mt*
BASS Basic Saville System *hc*
BAT Best Available Treatment *wp*
BAT Block Address Translation *pk, kg*
BAT [TU-2 bomber] *wp*
BATCAP Battery Capacitor *hc*
BATES Battlefield Artillery Target Engagement System *fu*
BATO Balloon Assisted Take-Off *wp*
BATS Ballistic Aerial Target System *hc, wp, jj*
BATS Broadband Antenna Test System *mt*
BATSE Burst And Transient Source Experiment [on CGRO] *mi*
BAW Bulk Acoustic Wave *jr*
BB Bar to Bar *hc*
BB Bare Base *mt*

BB Base Band *-7*
BB Battle Ship [US Navy ship classification] *mt, fu, wp*
BB Black Body *oe*
BB Bottom Bounce *fu*
BB Break Bulk *mt*
BB Broad Band *jj*
BBAS Base Budget Automation System [USAF] *wp, mt*
BBAU Base Band Assembly Unit *jdr, ric*
BBB Blood Brain Barrier *wp*
BBBG Battleship Battle Group *mt*
BBC Back up Bus Controller *he*
BBC Bromobenzyl Cyanide [chemical warfare] *wp*
BBCAL Blackbody Calibration *oe*
BBG British [Aerospace] and Bodenseewerk Geraetetechnik *aa26.11.12*
BBI Base-Band Interface *fu*
BBM Blocking and Bracing Materials [USA] *mt*
BBM Blue Ballistic Missile *tb*
BBMA Buffer Bulk Memory Address *jj*
BBN Bolt, Beranek, and Newman [Incorporated] (BBN Software Products Co.) (BBN Systems and Technologies) *fu, aw130.13.42, csIII.96, 19*
BBO Boost Burnout *kf*
BBPT Base-Band Plain Text *fu*
BBRC Ball Brothers Research Corporation *pl*
BBS Bulletin Board Service System *ct, em, wp, kf, kg, hi, du*
BBSP Bare Base Support Package *wp*
BBT Booster Burn Time *tb*
BBU Base Band Unit *jdr, uh*
BBU Bit Buffer Unit *jdr*
BBXRT Broad Band X-Ray Telescope [on ASTRO-1 shuttle flight, Dec. 1990] *cu, mi*
BC Bar Code *hi*
BC Basic Channel *fu*
BC Battle Command *kf*
BC Battle Cruiser [US Navy] *mt*
BC Battlespace Characterization *kf*
BC Binary Code *hi*
BC Binary Counter *fu*
BC Biological Chemical [chemical warfare] *wp*
BC Block Check *do366*
BC Body Current *jj*
BC Broadcast Controller *fu*
BC Bus Controller *he, fu*
BC/CC Basic Channels per Composite Channel *fu*
BC/SD Battle Commander/Senior Director *fu*
BCA Base Contracting Automated System *mt*
BCA Baseline Change Authorization *fu*
BCA Beacon Code Assignment *fu*
BCA Best Cruise Altitude *hc*
BCA Board of Contract Appeal *kf*
BCA Broadcast Control Authority *mt*
BCAG Boeing Commercial Aircraft Group *y2k*
BCAR British Civil Airworthiness Requirement [UK] *mt*
BCAS Base Contracting Automated System [USAF] *wp, mt*
BCC Base Communications Center *mt*
BCC Battery Charge Controller *he, kgl, cl, jdr, jn, uh*
BCC Battery Control Center [USA] *mt, hc, fu, nt, jmj, as*
BCC Binary Check Code *jdr*
BCC Block Check Character *do366, jdr, kg*
BCC Body Centered Cubic *ai1*
BCCSWG Base Communications Center Strategy Working Group *mt*
BCD Bad Conduct Discharge *mt, wp*
BCD Baseline Concept Description *mt*
BCD Baseline Configuration Description *tb*
BCD Binary Coded Decimal *mt, 4, wp, ct, em, fu, kg, 6, hc, w9, jj, oe, hi, ai1*
BCDMA Broadband Code Division Multiple Access *em*
BCDU Bias Capacitive Discharge Unit *uh*
BCDU Bias Charge Discharge Unit *jdr*
BCE BAPTA Control Electronics *hc*
BCE Base Civil Engineering [USA] *wp, jj, mt*
BCE Baseline Cost Estimate *mt, nt, wp, tb, st, jj*
BCE Battlefield Coordination Element [USA DOD] *mt*
BCE Bench Checkout Equipment *hc, jj*
BCE-IWP Base Civil Engineer In-service Work Plan *mt*
BCEP Base Communications-Electronics Plan *mt*
BCF Bit Control Field *do293*
BCFI Build CFI *kf*
BCFSK Binary Code Frequency Shift Keying *fu*
BCFX Bidding Category Codes *hc*
BCG Battalion Control Group *fu*
BCGN Battle Cruiser, Guided missile armed, Nuclear powered [US Navy] *mt*
BCH Binary Coded Hexadecimal *uh*
BCH Bose, Chadhuri, Hocquenghem [a type of cyclic block code] *do293, kf, fu, he*

BCI Binary Coded Information *jj, fu*
BCI Bit Count Integrity *tb*
BCI Broadcast Control Intercept *mt*
BCI Broadcast Interference *fu*
BCI Budgetary Cost Information *wp*
BCL Batch Command Language *pk, kg*
BCM Baseline Correlation Matrix *mt*
BCM Beyond Capability of Maintenance *fu, hc, jj*
BCMR BCM Rate *jj*
BCMV Battery Cell Voltage Monitor *-4*
BCN Backbone Concentrator Node *hi*
BCO Base Contracting Office *mt*
BCO Binary Coded Octal *fu*
BCOC Backup COC *fu*
BCOC Base Cluster Operations Centers *mt*
BCP Base Communications Processor *mt*
BCP Battery Command Post *fu, hc*
BCPL Basic Computer Programming Language *hi, kg*
BCPSM Binary Continuous Phase Shift Modulation *fu*
BCR Baseline Change Request [AF form 1773] *mt, wp, jdr, kf*
BCR Business Communications Review *ct*
BCR Byte Count Register *pk, kg*
BCRT Binary Coded Range Time *fu*
BCRT Bus Controller Remote Terminal *aw130.22.115*
BCRU Battery Charge and Reconditioning Unit *gp, jdr*
BCS Backup Channel Switch *fu*
BCS Bar Code Sorter *pk, kg*
BCS Bardeen Cooper Schrieffer *ai1*
BCS Baseline Comparison System *fu, wp, tb*
BCS Battery Computer System *fu*
BCS Battery Control System *hc*
BCS Battlefield Computer System *fu*
BCS Beam Control System *st*
BCS Bench Check Serviceable *wp*
BCS Block Check Sequence *jdr*
BCS Boeing Computer Services *fu*
BCS Broadcast Control Station *mt*
BCSA Base Communications-Computer System Assessment *mt*
BCST Broadcast *uh*
BCT Battle Coordination Team [USA] *mt*
BCTF Base Central Technical Facility *mt*
BCTF Base Central Test Facility *mt*
BCTF Base Control Test Facility *mt*
BCTF Battle Commanders Training Facility *mt*
BCU Bench Check Unit *oe*
BCU Buffer Control Unit *mt, fu*
BCV Boost Glide Vehicle *fu*
BCVM Battery Cell Voltage Monitor *lt, he, cl, jn*
BD Back Deck Supervisor *fu*
BD Battle Director *mt, fu*
BD Binary Decoder *fu*
BD Bomb Disposal *w9*
BD Broadcast Driver *fu*
BD Brought Down *w9*
BDA Battle Damage Assessment [USA DOD] *pbl, mt*
BDA Battle Damaged Aircraft *mt*
BDA Bermuda STDN *oe*
BDA BIOS Data Area *pk, kg*
BDA Bomb Damage Assessment *hc, fu, jj, kf*
BDA Bomb Damage Assessment [USA DOD] *mt*
BDAS Battle Damage Assessment System *mt, wp*
BDB Big Dumb Booster *mi*
BDB Blue Data Base *tb*
BDC Backup Domain Controller *ct, kg, em*
BDC Battery Discharge Controller Control *he, uh, gp, 9, cl, jdr, kf, jn*
BDC Binary Decimal Counter *fu*
BDD Binary To Decimal Decoder *wp*
BDE Block Diagram Editor *jdr*
BDE Borland Database Engine *em, kg*
BDE Brigade [USA] *wp, pm, mt*
BDHI Bearing, Distance, Heading Indicator *wp, fu*
BDI (capability to merge SDI, ADI, CDI, and BDI concerns) *ic2-19*
BDI Bearing Deviation Indicator *jj, fu*
BDIC Battelle Defender Information Center *hc*
BDK Beans Development Kit *em*
BDK Bol'shoy Desantniy Korabl' ["large landing ship"] [USSR] *mt*
BDL Battery Data Link *fu*
BDM (BDM International, Incorporated) [a defense electronics firm] *aa27.4.45, aw130.22.69, jmj, st*
BDM Base Defense Measure *wp*
BDM Bomber Defense Missile *mt, hc, wp*
BDM Braddock, Dunn and McDonald *wp*
BDMS Baseline Data Management System *mt*
BDN Bulk Data Network *mt, wp*
BDO Basic Delivery Order *mt*
BDOC Base Defense Operations Center *mt*

BDOS Basic Disk Operating System *pk, kg*
BDP Base Development Planning *mt*
BDP Battlefield Development Plan *mt*
BDPA Black Data Processing Associates *hn49.8.3*
BDPI Base Data Processing Installation *mt*
BDR Base Destination Requirements [HQ USAF] *mt*
BDR Baseline Design Review *jdr*
BDR Battle Damage Repair *mt*
BDR Bus Device Request *pk, kg*
BDRI Bearing Distance Heading Indicator *jj*
BDRT Battle Damage Repair Training [UK] *mt*
BDS Battle Dressing Station *mt*
BDS Battlefield Data Systems *mt, fu, hc*
BDS Budget Development System [USMC] *mt*
BDS Business Development Specialist *jdr*
BDSL Broadband ADSL *pk, kg*
BDSR Base Dental Service Report [USAF] *mt*
BDT Bulk Data Transfer *mt*
BDTS Boost Detection and Track System *fu*
BDU Basic Display Unit *fu*
BDU Battery Display Unit *fu*
BDU Battle Dress Uniform [USA] *wp, mt*
BDU [practice versions of nuclear bombs, USA] *mt*
BDV Binary Divide *fu*
BDW Block Data Word *fu*
BDZ Base Defense Zone [USA DOD] *mt*
BE Below or Equal *pk, kg*
BE Bidding Expense *hc, jj*
BE Bottoms up Estimate *kf*
BE Brilliant Eyes *kf*
BE Bulk Encryption *mt*
BEAB Bofors Electronic AB [Sweden] *hc*
BEACONS Bid Estimating and Control System *fu*
BEAGLE [NATO code name for Soviet Il-28 Bomber] *wp*
BEAM BE Additional Mission *kf*
BEAMS Base Engineering Automated Management System *mt, wp*
BEAR [TU-20 Strategic Bomber] *wp*
BEAST [IL-10] {Ilyushin-10 Bomber} *wp*
BEC Best Estimate Constrained [with SALT II] *st*
BEC Bit Error Comparator *-4*
BEC Board Error Correction *fu*
BECN Backward Explicit Congestion Notation Notification *kg, ct, em*
BECO Booster Engine Cut Off *fu, uh, sd, hc, jj, oe*
BECS Battlefield Electronic Communications System *mt*
BED Block Error Detector Detection [NASA] *4, oe*
BEDO Burst Extended Data Out *pk, kg*
BEE Battlefield Electromagnetic Environments [USA] *dt*
BEEF Base Emergency Engineering Force *mt*
BEEP Bus Express Employee Program *sc2.11.1*
BEF British Expeditionary Force *mt, w9*
BEF Buffer Emitter Follower *fu*
BEG BMEWS Exercise Generator [SAC / SPACE CMD] [began in 1946] *mt*
BEHP Bis-2-Ethylhexyl Phosphite chemical warfare Simulant *wp*
BEL Bell *pk, kg*
BELLCORE Bell Communications Research *kg, ct*
BEMO Base Equipment Management Office *mt, wp, jj*
BENELUX Belgium, Netherlands, Luxembourg [precursor of EEC] *mt*
BEP Bit Error Probability *fu*
BEP Break Even Point *jj*
BEP Budget Expenditure Plan *-4*
BER Basic Encoding Rules *pk, kg*
BER Beyond Economical Repair *jj*
BER Bit Error Rate *mt, fu, kg, gg73, 2, 4, 16, 17, sb, 19, wp, ce, tb, st, jdr, kf, vv, oe, uh, hi*
BER Bit Evaluation Report *jj*
BER Budget Estimate Review *jj*
BERH Boards of Engineers for Rivers and Harbors *dt*
BERP Bit Error Rate Postprocessor *fu*
BERP British Experimental Rotor Program [UK] *mt*
BERT Bit Error Rate Test Tester Testing *fu, kg, hi*
BERTS Bit Error Rate Test Set *-4*
BES Budget Estimate Submittal Submission *mt, wp, tb*
BESEP Base Electronics System Engineering Plan *mt, fu*
BESIM Brilliant Eyes Simulation *kf*
BESS Binary Electromagnetic Signal Signature *fu*
BESS Biomedical Experiments Scientific Satellite *hc*
BEST Base Engineering Support, Technical [US Navy] *mt*
BEST Battery Energy Storage Test *hc*
BEST Bilan d'Energie du SystÉme Tropical [France] *hg, ge*
BEST Built in Electronic System Test *tb*
BET Best Estimate of Track Trajectory *17, sb, tb*
BET Brunauer Emmet Teller *wp*
BETA Battlefield Elevated Target Acquisition *hc*
BETA Battlefield Exploitation and Target Acquisition [system] *mt, fu, 17, hc, dt, wp, kf*

BETA [ballistic coefficient] *tb*
BETAS Battlefield Exploitation and Target Acquistion System *hc*
BETS Building Energy Technical Survey *mt*
BEU Best Estimate Unconstrained [without SALT II] *st*
BEU Bypass Electronics Unit *jdr, uh*
BEX Broadband Exchange *hi*
BEZS Bandwidth Efficient Zero Support *em*
BEZS Bandwidth Efficient Zero Suppression *ct*
BF Bad Flag *pk, kg*
BF Baseline Finish *kf*
BF Battle Force *fu, jmj*
BF Beam Forming *ic2-11*
BF Beat Frequency *fu*
BF Bomber-Fighter *af72.5.71*
BFA Battlefield Functional Area *fu*
BFAP Blocking Factor Analysis Program *mt*
BFC Budget and/or Forecast Calendarization *hc*
BFCC Battle Forces Command and Control [US Navy] *mt*
BFCO Band Filter Cutoff *fu*
BFD Back Focal Distance *fu*
BFD Bistatic Fleet Defense *hc*
BFD Bookform Drawing *fu*
BFE Beam Forming Electrode *fu*
BFE Blacker Front End *mt*
BFE Buyer Furnished Equipment *wp*
BFEC Bendix Field Engineering Corporation *fu, ci.940, oe*
BFF Binary File Format [IBM] *hi, kg*
BFI Battle Field Interdiction *wp*
BFI Bearing Frequency Intensity *fu*
BFIM Battle Force Information Management [US Navy] *fu, mt*
BFL Back Focal Length *fu*
BFL Bomb Fall Line *jj*
BFM Basic Fighter Maneuvers *wp*
BFM Basic Flight Maneuver *mt*
BFMDS Base Flight Management Data System *wp*
BFMO Base Fuel Management Officer *wp*
BFN Beam Forming Network *bkr, jdr*
BFO Beat Frequency Oscillator *mt, 0, fu, hc, wp, jj*
BFP Back Focal Plane *wp, ai1*
BFPC Bulk Flight Plan Conversion *fu*
BFR Blip Frame Ratio *fu*
BFRL Basic Facilities Requirements List *jj*
BFS Bottom of Free Space *fu*
BFS Broadcast Frequency Stability Selectability *jdr*
BFT Beam Forming Trigger *fu*
BFT Binary File Transfer *ct, kg, em*
BFTF Battle Force Track File *fu*
BFTP Batch FTP *pk, kg*
BFTS Bomber Fighter Training System *wp*
BFTSC Brassboard Fault Tolerant Spaceborne Computer *hc*
BFTT Battle Force Tactical Trainer [US Navy] *mt*
BFV Bradley Fighting Vehicle [formerly FVS] *uc, hc, bf*
BFVS Bradley Fighting Vehicle System [USA] *mt, hq5.3.5, hn41.19.4*
BG Battle Group [US Navy] *fu, jmj, tb, mt*
BG Battlefield Geometry *fu*
BG Bomb Glider [USA] *mt*
BG Bombardment Group [USA] *mt*
BG Brigadier General *mt, fr, w9, nt, jmj, wp*
BG/BF Battle Group / Battle Force *fu*
BGA Ball Grid Array *ct, kg, em*
BGAAWC Battle Group AAW Coordination *fu*
BGE Branch if Greater or Equal *pk, kg*
BGI [aerospace company, France] *es55.44*
BGIGS Battle Group Interactive Gaming System *hc*
BGK Bomb Guidance Kit *hc*
BGM Ballistic Guided Missile (BGM-71 TOW, BGM-109 Tomahawk) [USA] *mt*
BGP Border Gateway Patrol *ct*
BGP Border Gateway Protocol *em, kg*
BGPHES Battle Group Passive Horizon Extension System *mt, fu*
BGRV Boost Glide Reentry Vehicle *hc, wp*
BGT Branch if Greater Than *pk, kg*
BGTT Battle Group Tactical Trainer *fu*
BGV Boost Glide Vehicle [AFSC*] *hc, wp, mt*
BH Black Hole *mi*
BH Brinell Hardness *w9, wp*
BH Buried Heterostructure *fu*
BHA Base Helix Angle *fu*
BHASKARA [India] *ge*
BHC Bell Helicopter Company *hc*
BHC Benzene Hexachloride *wp*
BHI Bell Helicopter International *hc*
BHI Branch if Higher *pk, kg*
BHIS Branch if Higher or Same *pk, kg*
BHN Brinell Hardness Number *w9, wp*
BHP (BHP Melbourne Research Laboratories) *is26.1.74*

BHP Brake Horse Power *wp, jj*
BHT Bell Helicopter Textron *hc, jj*
BHU Basic Hybrid Unit *fu*
BI Background Investigation [security] *10, wp, su, tb, mt*
BI Battlefield Interdiction *mt, hc, wp*
BI Binary Input *pk, kg*
BI Bus Interface *fu*
BIA Base Information Analysis [USAF] *mt*
BIAP Base Information Analysis Program [AFCC*] *mt*
BIAS Battlefield Illumination Airborne System *wp*
BIC Baseband Interface Circuit *fu*
BIC Battlefield Information Center [USA] *wp, mt*
BIC Bombardment Induced Conductivity *hc*
BIC Braduskill Interceptor Concept *fu, hc, tb, st*
BIC Broadcast Interface Controller *fu*
BIC Bus Interface Controller *lt, tb*
BICC Battlefield Information [and] Control Center *mt, fu*
BICES Battlefield Information Collection and Exploitation System *mt, hc, wp*
BICES Battlefield Information Correlation Enemy System *mt*
BICMOS BIpolar Complementary Metal Oxide Semiconductor *is26.1.48, 19, em*
BID Battlefield Information Distribution *mt*
BID Built In Diagnostics *mt*
BID Bus Interface Device *fu*
BIDDS Base Information Digital Distribution System [AFCC*] *wp, mt*
BIDE Basic Identity Information *mt*
BIDOPS Bipolar Doppler Scoring *jj*
BIEP Bases Integrated Electronics Package *hc*
BIFET Bipolar Field Effect Transistor *hc, kg*
BIFF Battlefield Identification Friend or Foe *hc, wp*
BIFF Binary Interchange File Format *pk, kg*
BIG_BEAR [classified USAF ELINT program] *wp*
BIG_BIRD [early reconnaissance satellite] *wp*
BIG_CROW [modified KC-135 for EW laboratory] *wp*
BIG_EYE [500 lb binary deep strike chemical bomb] [chemical warfare] *wp*
BIG_NOSE [Soviet radar on FIDDLER] *wp*
BIG_WING [laser guided bomb] *wp*
BII Basic Issue Item *wp, fu, jj*
BIIC Basic Issue Item Code *wp*
BIIL Basic Item Issue Issue Items List [USA] *hc, jj, mt*
BIIR Basic Imagery Interpretation Report *wp*
BIL Band Interleaved by Line *ct, bl3-3.12, em*
BIL Basic Impulse Level *hc*
BIL Batiment d'Integration Lanceur [Ariane booster integration building] *ai2*
BIL Bomb Impact Line *jj*
BIL Built In Logic *wp*
BIL Bulk Items List *hc, fu, jj*
BILBO Built-in Logic Block Observation *fu*
BILT Basic Impulse Level Tests *hc*
BIM Ballistic Intercept Missile *uc, wp, hc, tb*
BIM Beginning of Information Marker *pk, kg*
BIM Bus Interface Module *fu*
BIMA Berkeley Illinois Maryland Array *mi*
BIMAT [fast film processing method for reconnaissance] *wp*
BIML Battle Information Management Laboratory [RADC] *mt*
BIMOS Bipolar [plus] Metal Oxide Semiconductor *17, fu, sb, th4, hc*
BIMS Battle Staff Information Management System [AAC] *mt*
BIMS Battlefield Integration Management System *wp*
BINAC Binary Automatic Computer *pk, kg*
BINGEN Binary Generation [a software product] *fu*
BINHEX Binary Hexadecimal *pk, kg*
BIO Buffered Input/Output *fu*
BIONICS Biological Electronics *wp, jj*
BIOREP Biological Attack/Incident Report *fu*
BIOS Basic Input / Output System *fu, kg, wp, ct, em, hi, ai1*
BIOS Battlefield Information Distribution System *wp*
BIOS British Intelligence Objectives Subcommittee *wp*
BIOSPEX Biological Space Experiments *ci.641*
BIOSTAT Biostatistical Reporting System *mt*
BIOU Buffered Input/Output Unit *fu*
BIP Band Interleaved by Pixel *ct, bl3-3.12, em*
BIP Binary Image Processor *wp*
BIPS Billion Instructions Per Second *wp, ic2-66, tb, hi*
BIPS Brayton Cycle Isotope Power System *ci.898*
BIRDIE Battery Integration Radar Display Equipment *hc*
BIRP Bus Monitor Input Output/Interrupt Processor *hc*
BIRS Battlefield Information Reporting System *fu*
BIS (BIS follows market trends in the electronics industry) *aw130.13.6*
BIS Binary Synchronous *mt*
BIS Board Bureau of Inspection and Survey [USA] *hc, jj, mt*
BIS British Interplanetary Society *sf31.1.10*
BIS Business Information System *pk, kg*
BISAM Basic Indexed Sequential Access Method *wp*

BISDN Broadband Integrated Services Digital Network [Cell relay, ATM] *nu, kg, hi*
BISEPS Basic Interceptor Sortie Evaluation Program System *mt*
BISMC Base Information System Management Center *mt*
BISP Base Information Systems Plan *mt*
BISS Base and Installation Security System [DOD] *hc, ro, fu, wp, mt*
BISS Base Intrusion Security System *mt*
BISS Base Intrusion Surveillance System *hc*
BIST Backward Instability Shock Test *jr, jdr*
BIST Baseline Integrated System Test *jr*
BIST Built In Self Test *hc, fu, kg, is26.1.51, rm8.2.1, wp, jj, jdr, jr*
BISTAR Bistatic Thinned Array Radar System *hc*
BISYNC Binary Synchronous Communication protocol [also BSYNC] *kg, mt*
BIT Binary Digit *fu, wp, tb, jj, kg*
BIT Build Integration Test *kf*
BIT Built In Test *mt, he, uh, fu, rm1.1.1, 17, sb, hc, wp, tb, jj, kf, vv*
BITBLT BIT BLock Transfer *ct, em, hi*
BITC Base Information Transfer Center *mt*
BITE Built In Test Equipment [USA] *mt, fu, 4, sb, hc, aw129.21.120, wp, tb, jj, kf*
BITG Boeing Industrial Technology Group *aw130.6.20*
BITNIC BITNET Network Information Center *pk, kg*
BITS Base Information Transfer System *mt*
BITS Battlefield Information Targeting System *aw129.1.61*
BIU Basic Inteface Unit *wp*
BIU Bus Interface Unit *mt, fu, kg, ct, em*
BIV Ballistic Intercept Vehicle *fu*
BIVD Biocular and Indirect View Display *hc*
BIWAY [concurrent TADIL C data exchange via JTIDS and Conventional Link-4A] *fu*
BJSM British Joint Services Mission *hc, dt*
BJT Bipolar Junction Transistor *fu, wp, jdr*
BK Bord Kanone [Germany] *mt*
BKAV [Bureau of Commissars of Aviation and Aeronautics, created in 1917, USSR] *mt*
BKEP Boosted Kinetic Energy Penetrator *mt, dh, wp*
BKGD Background *17, sb*
BKS Broadcast Keying Station *mt*
BL Bachelor of Law *w9, wp*
BL Back Lit *hi, kg*
BL Bill of Lading *su, hc, w9, tb*
BL Bit Line *pk, kg*
BL Blue Lightning [Chip] *ct, em*
BL Butt Line *jj*
BL [Loop noise Bandwidth] *-2*
BLA Base Level Assessment *mt*
BLACKER [NSA multiple level security project for data communications] *mt*
BLADES BMD LWIR Advanced Exoatmospheric Sensor *tb*
BLAIDS Base Level AUTODIN Incoming Distribution *mt*
BLAMES Base Level AUTODIN Message Extract System *mt*
BLARS Base Level Accounting and Reporting System [USAF] *mt*
BLAST Blocked Asynchronous Transmission [protocol] [Communications Research Group] *hi, kg*
BLATS Built up Low cost Advanced Titanium Structures *wp*
BLC BLACK Line Conducted Conduction *fu, jdr*
BLC Boundary Layer Control [a flow of pressured engine bleed air over the flaps and other lift surfaces, to keep the stalling speed down] *mt*
BLC Boundary Layer Control [system] *hc*
BLCS Base Level Contingency Simulations *mt*
BLD Balanced Line Driver *fu*
BLD Battalion Level Device *mt*
BLDASP Base Level Data Automation Standardization Program *mt*
BLE Branch if Less or Equal *pk, kg*
BLED Blue Light Emitting Diode *hi*
BLEDE Backsattered Laser Energy Digitizing Equipment *hc*
BLER Block Error *pk, kg*
BLERT Block Error Rate Test *fu*
BLF Best Lock Frequency *oe*
BLIP Background Limited Infrared Photoconductive Photoconductor *hc, st, fu*
BLIP Background Limited Infrared Photography *wp*
BLIP Background Limited Performance *wp*
BLIP Battlefield Laser Implication Program Study *wp*
BLIP {ELINT program on Ec-121 to monitor SAM activity} *wp*
BLIS Base Level Information Services Program [AFLC] *mt*
BLIS Base Level Inquiry System *mt*
BLM Bilayer Lipid Membrane *wp*
BLMC Buried Logic Macrocell *pk, kg*
BLMPS Base Level Military Personnel System [USAF] *wp, mt*
BLMS Base Level Military [personnel] System *wp*
BLN Billing Log Number *fu, 8*
BLO Base Level Operations *wp*
BLOB Binary Large Object *hi, kg*
BLOS Beyond Line Of Sight *mt, hc, wp, tb*
BLOS Branch if Lower Or Same *pk, kg*

BLP Bypass Label Processing *fu*
BLPS Base Level Personnel System [USAF] *wp, mt*
BLPS-C Base Level Personnel System - Civilian [USAF] *mt*
BLPS-M Base Level Personnel System - Military [USAF] *mt*
BLR Broken Link Report *fu*
BLS Beach Landing Site [USA DOD] *mt*
BLS Bureau of Labor Statistics *bl1-2.180, 18, w9, wp*
BLSS Base Level Self Sufficiency Spares *wp, mt*
BLSS Baseline Standardized [avionics] Subsystem *hc*
BLT Battalion Landing Team [USMC] *mt*
BLU Bomb Live Unit [submunition for cluster bombs or dispensers, USA] *mt*
BLu (FAB/BLu Force AÈrienne Belge / Belgische Luchtmacht) [French / Flemish Belgian Air Force] *mt*
BLUE_FLAG [CCCI exercises, Eglin AFB, FL] *mt*
BLUE_FORCES [US and friendly forces, opposite of "red"] *mt*
BLUE_SPIKE [an infrared spectral band where aircraft engine emissions peak] *wp*
BLX Basic Launch Complex *mt*
BM Ballistic Missile *fu*
BM Battle Management Manager *mt, fu, wp, tb, st*
BM Bench Mark *fu, jj*
BM Bill of Material *fu, w9*
BM Bulk Memory *jj*
BM Bulk Message *fu*
BM/CC Battle Management/Command and Control *tb*
BM/CCC Battle Management / Command, Control, and Communications *mt, 1, 5, 17, sb, hc, nt, st, as, fu*
BM/CCCI Battle Management / Command, Control, Communications, and Intelligence *tb, st*
BMA Battle Management Algorithm *tb*
BMAC Battle Mode Access Control *fu*
BMAC Boeing Military Airplane Co. *fu, hc*
BMAR Backlog of Maintenance and Repair *mt*
BMAR Bulk Memory Address Register *jj*
BMC Bit Map [Memory] Controller *fu, hn50.8.4*
BMC Bulk Media Conversion *fu*
BMC Bulk Memory Controller *jj, jdr, ric*
BMC Burst Multiplexed Channel *oe*
BMCC Battle Management Control Center *fu*
BMCCC Battle Management Command, Control, and Communication *kf*
BMCR Bulk Memory C Register *jj*
BMD Ballistic Missile Defense *mt, ca, 1, 5, 11, 12, 14, ic2-8, 17, sb, hc, jb, nt, wp, tb, st, kf, fu*
BMD Ballistic Missile Division [USAF] *ci.157, fu*
BMD/STP Ballistic Missile Defense/Systems Technology Program *17, sb*
BMDATC Ballistic Missile Defense Advanced Technology Center *tc, hc, tb, st, fu*
BMDC Ballistic Missile Defense Center *mt, kf*
BMDCC Ballistic Missile Defense Control Center *hc, tb*
BMDO Ballistic Missile Defense Office [a less ambitious, lower cost follow up of the SDI project] [USA] *mt*
BMDO Ballistic Missile Defense Organization [was SDIO] *kf, mb, mi*
BMDO BMCCC Ballistic Missile Defense Organization Battle Management Command, Control and Communications *kf*
BMDOC Ballistic Missile Defense Operations Center *fu, tb*
BMDPM Ballistic Missile Defense Program Manager *fu, tb*
BMDPO Ballistic Missile Defense Program Office *dt, tb, fu*
BMDS Ballistic Missile Defense System *hc, tb, st*
BMDS Base Manpower Data System [USAF] *mt*
BMDSC Ballistic Missile Defense Systems Command *hc, wp, tb*
BMDSCOM Ballistic Missile Defense Systems Command [Army's] [later USASDC] *uc, tb*
BMDTEC Ballistic Missile Defense Test and Evaluation Center Command *hc, tb*
BME Bench Maintenance Equipment *fu*
BME [a phenomenology used by NORAD {?}] *sd*
BMEDS Base Management Engineering Data System [USAF] *wp, mt*
BMEP Brake Mean Effective Pressure *jj*
BMEWS Ballistic Missile Early Warning System [USA] *mt, bm132, 12, hc, is26.1.76cc, jb, cp, wp, tb, jj, kf, fu, ro*
BMFT Bundes Ministerium fur Forschung und Technologie [German Federal Ministry for Research and Technology] *ns, hg, ge*
BMI Branch if Minus *pk, kg*
BMIC Branch Module Interface Channel *oe*
BMIC Bus Master Interface Controller [Intel] *pk, kg*
BMIDE Bus Managed Integrated Drive Electronics *em*
BML Battle Management Laboratory *tb*
BML Bidder's Mailing List *fu*
BMM Bit Map Memory *fu*
BMO Ballistic Missile Office Organization [Air Force] [AFSC*] *mt, 17, hc, wp, tb, st, kf*
BMO Business Management Office *fu*
BMP Basic Mapping Support *pk, kg*
BMP Batch Message Processing Program *pk, kg*
BMP Batch Mode Message Processing *fu*
BMP Best Manufacturing Practices [program] *nd74.445.35*

BMP Bit Map *jr, ct, hi*
BMP Bronevaya Machina Piekhota [Soviet armored vehicles, infantry] *fu*
BMPS Battle Management Processing Simulation *mt*
BMPS Battle Management Processing System *tb*
BMR Basic Military Requirement [NATO] *mt*
BMR Battery Management Risk *jn*
BMR Battery Management Requirements *fu*
BMR Body Mounted Radiator *hc*
BMS Background Mapping Sensor *hc*
BMS Background Measurements Satellite *hc*
BMS Battle Battlefield Management System *1, wp, tb, hc, fu*
BMS Bus Module Subsystem *kf*
BMSO Base Medical Supply Officer *wp*
BMT Ballistic Missile Threat *tb*
BMT Basic Military Training *wp, mt*
BMT Bus Module Test *jn*
BMTB Battle Management Test Bed *st*
BMTD Ballistic Missile Terminal Defense *wp*
BMTST Bus Module Thermal Stress Test *uh*
BMU Basic Memory Unit *fu*
BMU Battery Monitor Unit *jdr*
BMUX Bus Multiplexer *hn47.7.8, fu*
BMW (contacted the 28 BMW director of operations) {Black Mills West} *aw129.10.73*
BN Binary Number *fu*
BNC Bayonet and Coaxial [a type of cable connector] *aw129.10.7*
BNC Bayonet N-type Compact *em*
BNC Bayonet Neill-Concelman [connector for coaxial cable invented by Mr. Neill-Concelman] *kg, fu*
BNC British National Connector {?} *ct*
BNCC Base Network Control Center *kf*
BNCSR British National Committee on Space Research *crl*
BNDD Bureau of Narcotics and Dangerous Drugs *10, w9*
BNE Branch if Not Equal *pk, kg*
BNEP Basic Naval Establishment Plan [USA] *mt*
BNF Backus-Naur Form [developed by John Backus and Peter Naur] *kg, wp, fu*
BNL Brookhaven National Laboratory *he, oe*
BNS Backbone Network Service *hi, kg*
BNS Backup North-South Stationkeeping *-3*
BNS Bomb Navigation System *wp*
BNSC British National Space Centre *mi, crl*
BO Beat Oscillator *fu*
BO Binary Output *pk, kg*
BO Binary to Octal *wp*
BO Blocking Oscillator *jj*
BO Body Odor *w9, wp*
BOA Basic Ordering Agreement *mt, 8, fu, hc, wp, jj, jdr*
BOA Broad Ocean Area *0*
BOA Buffer Output, ASIC *jdr*
BOAC British Overseas Airways Corporation *wp*
BOB Break Out Box *kg, oe, hi*
BOB Bureau Of the Budget *13, fu, jj*
BOC Base Operations Center *fu*
BOC Basic Operating Companies *mt*
BOC Basic Operator Console *pk, kg*
BOC Battalion Operations Center *wp, fu*
BOC Beginning Of Contract *kf*
BOC Bell Operating Companies *mt*
BOC Best Operational Capability *wp*
BOC Break Of Configuration [meeting] *oe*
BOCCA Board Of Coordination of Civil Aviation [NATO] *mt*
BOD Base Ordnance Depot *wp*
BOD Basic Operational Data *fu*
BOD Beneficial Occupancy Date *mt, fu, 4, 17, hc, tb, jj, kf*
BOD Biochemical Biological Oxygen Demand *w9, wp*
BOE Baseline Operational Evaluation *tb*
BOE Basis Of Estimate *4, tb, jdr, kf*
BOF Beginning Of File *wp, kg, hi*
BOF Bias Oscillator Frequency *0*
BOI Basis Of Issue *mt*
BOIP Basis Of Issue Plan *mt, st, fu*
BOJO (Flexible Reporting and Display System [SAC]) {?} *mt*
BOKKV Basic Operational Kinetic Kill Vehicle *17, sb*
BOL Bearing Only Launch *mt*
BOLT Beam Of Light Transistor *fu*
BOM Base Operation Manager *wp*
BOM Batch Optimization Method (liquid propulsion system) *jn*
BOM Beginning Of Message *kg, fu*
BOM Beginning Of Month *jdr*
BOM Bill Of Materials *pbl, fu, wp, ep, jdr, tm, jn, hi*
BOM Bit Oriented Message *fu*
BOM Bureau Of Mines *wp*
BOMEX Barbados Oceanographic and Meteorological Experiment *sm18-13, hc*

BOMP Bill Of Material Program *fu*
BOND Bandwidth On Demand *pk, kg*
BOOMER (Fleet Ballistic Missile Submarine [US Navy]) {?} *mt*
BOOMER [air refueling boom operator, SAC*] *mt*
BOOPP Boost Phase Prediction [a trajectory model] *-17*
BOOTP BOOT Bootstrap Protocol [Internet] *ct, em, kg, hi*
BOP Basic Operating Operations Plan *ah, pbl*
BOPS Billion Operations Per Second *wp, kg, hi*
BOQ Bachelor Officers' Quarters *mt, wp, w9, jj*
BOR Bespilotnye Orbitalnye Raketoplany [unmanned orbital rocketplane] *ai2*
BORAM Block Oriented Random Access Memory *hc, fu*
BORPQU Borland Pro Quattro *pk, kg*
BORQU Borland Quattro *pk, kg*
BOS Back Of Secondary [mirror] *oe*
BOS Back Order Suspense *fu*
BOS Bands Of Separation *mt*
BOS Base Operating Supplies *wp*
BOS Base Operating Support *mt, tb*
BOS Basic Base Operating System *fu, tb, kg*
BOS Brassboard Optical Seeker *hc, st*
BOSCH Boschma Ordinate Selectable Concept for Homing to Intercept *st*
BOSN Boatswain *jj*
BOSS B-1B Operational and Support Software *mt*
BOSS Basic Operating System Software *fu*
BOSS Biocular Sensor and Symbology *hc*
BOSS Block [diagram] Oriented Systems Simulator *19, ric, jdr*
BOSS Bomb Orbital Strategic System *hc*
BOSS Boss Operations Support Center *mt*
BOSS Business Open Shop Scheduling *cd*
BOT Beginning Of Transmission *fu*
BOTH Bombing Over The Horizon *hc, jj*
BOTOSS Bottom Topographic Survey Subsystem *hc*
BOW Burn Out Weight *oe*
BP Band Pass *0, jj*
BP Base Page *wp*
BP Base Pay *wp*
BP Base Plate *jdr*
BP Base Pointer *pk, kg*
BP Batch Processing *wp*
BP Battery Pack *cl*
BP Blue Print *w9, jj*
BP Boiling Point *w9, wp, ai1*
BP Boost Phase *17, sb, tb*
BP Break Point *jj*
BP Brilliant Pebbles *kf*
BP Budget Programs *mt*
BP Bulk Petroleum *mt*
BP Bulk Processing *mt*
BP Bus Power *jdr*
BP/MC Boost Phase/Midcourse *fu*
BPA Bent Pin Analysis *fu*
BPA Business Publications Audit *aa26.8.3*
BPA [Grumman Bethpage - Hughes Aircraft Company customer for metallic return transfer breaker] *hc, jj*
BPAC B-Loop Pointer Address Control *jj*
BPAC Budget Program Activity Code *mt, hc, jj*
BPB BIOS Parameter Block *ct, kg, em*
BPC Battery Power Controller *kf*
BPC Binary Phase Code *hc, jj*
BPC Bus Power Controller *he*
BPC Business Process Council *jdr*
BPD Basic Point Defense *fu*
BPD Bombrini-Parodi-Delfino *crl*
BPD Buy Per Drawing *fu*
BPDIU Basic Point Defense Interface Unit *fu*
BPDMS Basic Point Defense Missile System [US Navy] *jj, fu, mt*
BPDSMS Basic Point Defense Ship Surface Missile System *hc, fu*
BPDU Basic Power Distribution Unit *fu*
BPDU Bridge Protocol Data Unit *pk, kg*
BPDU Bus Power Distribution Unit *cl, kf, he, uh*
BPE Best Preliminary Estimate *wp*
BPF Band Pass Filter *fu, uh, bl2-9.15, 2, 4, 9, lt, jdr, oe, ai1*
BPF Berkeley Packet Filter *ct, em*
BPF Blood Profusion Monitor *is26.1.18*
BPHS Battlegroup Passive Horizon System *hc*
BPI Bits Per Inch *mt, pl, hc, jj, hg, oe, fu, kg, ge, hi*
BPI Boost Phase Intercept Interceptor *17, sb, tb, st*
BPI Bytes Per Inch *hi*
BPK Bol'shoy Protivolodochniy Korabl' [large Anti submarine ship] [USSR] *mt*
BPL Band Pass Limiter *ai1*
BPL Bits Per Line *em*
BPL Boost Phase Leakage *tb*
BPL Branch if Plus *pk, kg*

BPMS Buy Per Material Specification *fu*
BPO Base Post Office *wp*
BPO Base Procurement Office *wp*
BPOL Book Plan Outline *fu*
BPP Bits Per Pixel *em, hi*
BPP Black Patch Panel *mt*
BPP Boarding Pass Printer *mt*
BPPI Backorder by Project P/N Inquiry *fu*
BPPM (BPPM optical intersatellite link) *-19*
BPR Bits Per Record *pl*
BPR Bypass Ratio *mt*
BPS Battle Planning Simulation *tb*
BPS Beacon Present Signal *-3*
BPS Bits Per Second *mt, gp, 4, 17, pl, jb, wp. mb, tb, jj, jdr, kf, hg, vv, oe, ct, em, uh, fu, kg, ge, hi, du*
BPS Bus Propulsion System *kgl*
BPS Bytes Per Second *wp, kg, hi*
BPSK Binary Bi Phase Shift Key Keying *fu, uh, sp659, 2, 4, 17, mj31.5.30, ce, kf, vv, oe, tt*
BPSM Binary Phase Shift Modulation *fu*
BPSN Budget Program Sequence Number *hc*
BPT Bipropellant Thruster *he*
BPTF Black Patch and Test Facility *mt*
BPTS Boost Phase Tracking System *fr, fu, st*
BPU Basic Processing Unit *mt, fu*
BPU Battery Protection Unit *gd2-34*
BPV Boost Phase Vehicle *hc*
BPZ Below the Promotion Zone *an31.13.4*
BQ [unmanned aircraft carrying explosives, USA] *mt*
BQM [a designator] (BQM-34 Firebee) [USA DOD] *mt*
BR Bad Register *pk, kg*
BR Baud Rate *hi*
BR Bend Radius *fu*
BR Bit Rate *hi*
BR Buffer Register *fu*
BRA Bench Replaceable Assembly *hc, jj*
BRAA Base Recovery After Attack *wp*
BRAAT Base Recovery After Attack Team *mt*
BRAAT Base Recovery After Attack Transport *wp, af72.5.19*
BRASO Branch Aviation Supply Office *jj*
BRASS PRIME BEEF Reporting Analysis and Status System *mt*
BRAT Battle management Requirements Analysis Tool *hc*
BRAT Budget / Readiness Analysis Technique [AFLC] *mt*
BRAVE BRAZO Avionics Equipment *hc*
BRB Basic Reference Book *jj*
BRCCS Body Referenced Cartesian Coordinate System *hc*
BRCT Base Repair Cycle Time *wp*
BRD Baseline Requirements Document *mt*
BRD Booster Requirements Document *0*
BRDF Bidirectional Reflectance Distribution Function *17, hc, kf*
BRDM [floating dry dock, US Navy] *mt*
BRET Bistatic Radar Evaluation Techniques *hc*
BRGC Binary Reflected Gray Code *pk, kg*
BRI Basic Rate Interface *kg, ct, em, hi*
BRI Brain Response Interface *pk, kg*
BRITE Basic Research in Industrial Technologies for Europe *mt*
BRITE Bright Radar Indicator Tower Equipment *fu, hc*
BRITT Baseline Reduction Intelligent Target Tracker *hc*
BRK Bol'shoy Raketniy Korabl' [large rocket ship] [USSR] *mt*
BRL Ballistics Research Laboratory [of AMCOM] [USA] *mt, tc, hc, id4091.2/1111, wp*
BRL Bit Rate Loop [timing subsystem in GOES S/DB] *pl*
BRL Bomb Release Line *hc, jj*
BRM Bio Research Module *ci.969*
BROI Best Return on Investment *fu*
BROT Boresight Range On Target *jj*
BRP Beacon Ranging Pulse *fu*
BRP Beam Riding Platform *hc*
BRPA Basic Research and Proposal Announcement *mt*
BRPL Ballistichekaya Raketa Podvodnykh Lodok [SLBM] [USSR] *mt*
BRS Bibliographic Retrieval Service *fu*
BRS Broadcast Radiation Station *mt*
BRSL Boresight Launch *jj*
BRSU Battery reconditioning Switch unit *he*
BRT Base Repair Time *mt*
BRTS Bilateration Ranging Transponder System *-2*
BRU Battery Regulator Unit *jdr*
BRU Bomb Release Unit *jj*
BRUCE Basic Remote User's Content Editor *mt*
BRUSA British-United States Agreement *-10*
BRV Ballistic Reentry Vehicles *hc, st*
BRW Brake Release Weight [maximum permitted weight at the start of take off run] *mt*
BS Base Support *kf*
BS Baseline Start *kf*
BS Battle Short *jj*

BS Battle Staff *mt, fu*
BS Berezin Skorostrelnye [Berezin rapid-firing gun] [USSR] *mt*
BS Bit Sliced *fu*
BS Bomb Squadron [SAC*] [USA] *mt*
BS Bombardment Squadron *-12*
BS Bore Sight *jj*
BS Borescope *fu*
BS Breaking Strength *fu*
BS British Standard *w9, wp*
BS Broadcast Satellite *ce*
BS Business Systems Department *fu*
BSA Base Support Agreement *mt*
BSA Beach Support Area *mt*
BSA Body to Sun Angle *uh*
BSA Bovine Serum Albumin *wp*
BSAM Basic Sequential Access Method *fu, wp, kg*
BSAS Battle Surveillance Airship System *hc*
BSAS Beam Stabilization Auto-alignment Subsystem *hc*
BSB British Satellite Broadcasting *hq5.3.2, sc8.9.2*
BSC Battle Space Characterization *kf*
BSC Battle Staff Coordinator *mt*
BSC Beam Steering Computer *fu*
BSC Beam Steering Controller *fu*
BSC Bi Binary Synchronous Communication *ct, em, csIII.44, ce, kg, fu, hi*
BSC Billet Sequence Code *fu*
BSC Binary Symmetric Channel *mt, sp659*
BSC Binary Synchronous Communications *mt*
BSCS Bachelor of Science [degree] in Computer Sciences *kg, aw130.11.57*
BSCU Battery Signal Conditioning Unit *kgl*
BSD Battle Staff Directive *mt*
BSD Battle Staff Director *mt*
BSD Berkeley Software Distribution *ct, em, kg, hi*
BSD Berkeley Standard [a UNIX version] *fu, is26.1.37, kg*
BSD [Hughes Aircraft Company customer for solid propellant reentry test vehicle] *hc*
BSE Base Support Element *mt*
BSE Battle Staff Element *mt*
BSE Bit Systems Engineering *jj*
BSE Boresight Error *wp*
BSE Broadcast Satellite Experiment *ac3.3-10*
BSEC Broad Area Surveillance Executive Committee *tb*
BSEP Basic Skills Education Program *hc*
BSF Backspace File *jj*
BSF Basic Shelter Facility *hc, jj*
BSF Bit Scan Forward *pk, kg*
BSI British Standards Institution Institute *w9, kg*
BSIT Base Systems Integration Test Bed *mt*
BSL Basic System Listing *mt*
BSL Bench Stock List *jj*
BSM Business Systems Management *fu*
BSMO Base Supply Management Office *wp*
BSMS Battle Staff Management System *mt*
BSOC Backup Sector Operations Center *fu*
BSOD Blue Screen Of Death *em*
BSOO Battle Staff Operations Officer *pf.f88.4*
BSOT Background System Operability Test *fu*
BSP Bell Systems Practice *ct*
BSP Bulk Synchronous Parallelism *kg, em*
BSP Business Strategy Panel *mt*
BSQ Band Sequential *ct, em, bl3-3.12*
BSR Back Space Record *jj*
BSR Base Support Radio [USAF] *mt*
BSR Baseline Surveillance Review *oe*
BSR Battlefield Surveillance Radar *mt*
BSR Bi Static Radar *fu*
BSR Bit Scan Reverse *pk, kg*
BSR Bit Slippage Rate *2, 4*
BSR Blip/Scan Ratio *fu*
BSR Buffered Send/Receive *fu*
BSS Base Service Store *mt*
BSS Base Supply System *hc*
BSS Baseline Select Switch *fu*
BSS Beam Select Switch *jdr*
BSS Block Started by Symbol *pk, kg*
BSS Broadcast Satellite Services *dn, 19*
BSSC Battle Staff Support Center *mt, hc, wp*
BSSC Bit Synchronizer Signal Conditioner *hc*
BST Barium Strontium Titanate *jdr*
BST Base Shop Tester *hc*
BST Business Solution Team *ah, pbl*
BSTAR Battlefield Surveillance Target Acquisition Radar *hc*
BSTAT Base Status Report [Logistic] *mt*
BSTS Boost [phase] Surveillance [and] Tracking Targeting System [AFSC*] *mt, 17, sb, aa26.8.23, hc, aw130.11.15, hq6.2.7, cp, wp, tb, st, kf, fu*

BSU Beam Steering Unit *fu*
BSWS Boeing Software Standard *fu*
BSYNC Binary Synchronous Communications [protocol] *hi, kg*
BT Bathy Thermograph *mt, jj*
BT Bazoviy Tral'shchik [base minesweeper] [USSR] *mt*
BT Begin Transmission [start of message] *0*
BT Bit Test *pk, kg*
BT Breadboard Test *jdr*
BT Brightness Temperature *hg, ge*
BT British Telecom *ct*
BTA Best Technical Approach *jmj, fu, st*
BTA {Bulgarian news agency} (Sofia BTA, 68 Dec 20, 08 *15 GMT >
ci.384*
BTAM Basic Telecommunications Access Method [IBM] *do422, kg, hi*
BTB Branch Target Buffer *ct, kg, em*
BTC BAPTA Torque Command *-3*
BTC Bench Test Cooler *gp*
BTC Bit Test and Complement *pk, kg*
BTC Block Transfer Controller *jj*
BTCE Bench Test and Calibration Equipment *gp*
BTCF Base Technical Control Facility *mt*
BTD Basic Track Data *fu*
BTD Binary To Decimal *wp*
BTE Battery Terminal Equipment *hc, fu*
BTE Beacon Target Extractor *fu*
BTE Bench Test Equipment *gp, oe*
BTF Baseline Test Facilities *st*
BTF Battalion Task Force *mt*
BTF Bromine Trifluoride *wp*
BTG (BTG, Inc. . . . Vienna, VA 22182) *jmj*
BTH Below The Horizon *fu, 5, 17, sb, hc, tb, kf*
BTI Balanced Technology Initiative *aw126.20.23, wp*
BTI Bearing Time Intensity *fu*
BTK Bench Test Kit *jj*
BTL Beacon Tracking Level *fu*
BTL Bell Telephone Laboratories *st*
BTM Bi level Telemetry *he*
BTM Bunkered Target Munition *hc*
BTMS Battalion Training Management System *hc*
BTN Battalion [USA] *mt*
BTO Bombing Through Overcase *jj*
BTO Bombing Trough Overcast [a WWII bombing radar, USA] *mt*
BTOA Binary To ASCII *hi, kg*
BTOC Battalion Tactical Operations Center *pm*
BTOC Brigade Tactical Operations Center *fu*
BTP Batch Transfer Program *pk, kg*
BTP Bench Test Procedures *fu*
BTR Beacon Track Receiver *-3*
BTR Bearing Time Recorder *fu*
BTR Bench Test Repair *fu*
BTR Bit Test and Reset *pk, kg*
BTR Broken Tool Report *fu*
BTR Bronetransporter [Soviet armored personnel carrier] *fu*
BTS Base Transceiver Station *ct, em*
BTS Base [station] Telecommunications System *mt, rih*
BTS Beacon Tracking System *id4142-10/831, 8*
BTS Bit Test and Set *pk, kg*
BTT Bank To Turn *hc*
BTT Burst Transmit Timing *fu*
BTU Beamformer Test Unit *fu*
BTU British Thermal Unit *mt, aa26.12.22, w9, wp, jj, jdr, kf, oe*
BTU Bus Terminal Unit *jdr*
BTV Blast Test Vehicles *hc*
BTX Botulinal Toxin [possible chemical warfare biological agent] *wp*
BU Battery Unit *mt*
BU BM/C2 Utility *tb*
BU Branch Unit *pk, kg*
BU Business Unit [organization] *pbl, tm, jdr*
BUBL Bulletin Board for Libraries [Janet] *pk, kg*
BUC Back Up Computer *wp*
BUC Back Up Controller *wp*
BUCKET [DCS Station on Osan AB] *mt*
BUCOC Backup Combat Operations Center *fu*
BUCON Bureau Control Number *jj*
BUD Back Up Decoder *hc*
BUD Block Update Development *fu*
BUDOCK Bureau of Yards and Docks *jj*
BUEC Backup Emergency Communications *fu*
BUF Back Up Facility *mt, 12*
BUF Benign Undetected Failures *fu*
BUIC Backup Interceptor Control [system] [USA] *mt, fu, ro, hn41.18.1,
hc, wp*
BUICS Backup Interceptor Control System *hc*
BUL Business Unit Leader *pbl, ah, jdr*
BULK Bulk Cargo *mt*
BULLSEYE [high frequency / direction finding systems] *mt*

BUNO Bureau Number *jj*
BUP Burst Upload *uh*
BUPERS Bureau of Navy Personnel [US Navy] *0, fu, jj, mt*
BUPPIE Black Urban Professional *wp*
BURP Backup Radar Processing *fu*
BURP Basic Unit Readiness Program *mt*
BURT Bit Universal Readout Tester *wp*
BUS Backscatter Ultraviolet Spectrometer *wp*
BUS Broadcast and Unknown Server *ct, kg, em*
BUSHIPS [Hughes Aircraft Company customer for synchro azimuth con-
verter] *hc*
BUSOC Back-Up Sector Operations Center *fu*
BUST (Breakup Processing software program) *jb*
BUTE Basic Unit Test Equipment *hc*
BUU Basic User Unit *fu*
BUV Backscatter Ultraviolet [Spectrometer] *ac2.8-2*
BUVS Backscatter Ultraviolet Spectrometer *hc*
BUWEPINST Bureau of Weapons Instruction *jj*
BUWEPS Bureau of Weapons [US Navy] *mt*
BUWEPS Bureau of [Naval] Weapons [Hughes Aircraft Company cus-
tomer for airborne infrared search set] *fu, hc, jj*
BUYNONE [Operation Readiness Inspection Exercise, SAC] *mt*
BV (In Europe, contact *Hercules BV, P.O. Box > aa26.12.4*
BV Boost Vehicle *5, 17, sb*
BV Breakdown Voltage *fu*
BVC Bus voltage Control *he*
BVCL Boundary Value Channel Library *fu*
BVH Base Video Handler *pk, kg*
BVL Bus Voltage Limiter *cl, jdr, kf, uh, he*
BVP Beacon Video Processor *fu*
BVR Beyond Visual Range *mt, hc, fu, aw129.14.109, ts17-11, wp*
BVSP Beacon Video Signal Processor *fu*
BVW Binary Voltage Weighter *bl2-7b.25, hc, jj*
BW Backward Wave *jj*
BW Bacteriological Warfare *ph, hc, w9*
BW Band Width *mt, uh, fu, sm1-20, 2, 4, pl, wp, jj, jdr, kf, oe*
BW Below Water *fu*
BW Biological Bacteriological Warfare *mt, w9, jj*
BW Bomb Wing [SAC] *mt*
BW Bombardment Wing [USA] *12, mt*
BWA Backward Wave Amplifier *hc, wp, fu*
BWB [Federal office for military technology and procurement, West
Germany] *hn43.6.2*
BWC Biological Weapons Convention *wp*
BWG Birmingham Wire Gage *fu*
BWIA British West Indian Airways *aw118.18.11*
BWL Benet Weapons Laboratory [of AMCOM] *tc*
BWM Backward Wave Magnetron *fu*
BWM Block Write Mode *pk, kg*
BWO Backward Wave Oscillator *hc, fu, wp, jj*
BWOFS Battlefield Weather Observation and Forecast System *mt*
BWPA Backward Wave Power Amplifier *fu*
BWR B Word Register *jj*
BWR Band Width Ratio *fu*
BWS Base Workshop *fu*
BWS Buy With Structure *fu*
BX Base Exchange *w9, wp*
BXMTR Beacon Transmitter *-9*
BXNT Bit Transmitter Noise Test *jj*
BY Billion Years *eo*
BY Budget Year *mt, wp, fu, jj*
BYRON Program Development Language *fu*
BYTE [binary element string, group or word of eight or sixteen binary
digits] *wp, kg, hi*
BZ [chemical warfare codename for 3-Quinuclidylbenzylate, a psycho-
chemical] *wp*
BZL [designators, experimental] *jj*

C

C [designator, (C-130) Hercules, (C-141) Starlifter, (C-5) Galaxy] [USA
DOD] *mt*
C&C Command and Control [also CC and C2] *hc, jj, kf, fu*
C&CE Command and Control Electronics *th4*
C&CR Command and Control Reports *mt*
C&CS Command and Control Subsystem *jdr*
C&D Command and Decision *fu*
C&D Communications and Data *fu*
C&D Controls and Displays *hc, jj, fu*
C&D Cover and Detection *fu*
C&DH Command and Data Handling *hc, mb*
C&DH Communications and Data Handling [for a satellite] *mt, ci.947,
2, tb*
C&DM Configuration and Data Management *ep, fu*
C&DP Communications and Data Processing *hn44.13.8, 8, 15, fu*

C&DP Computing and Data Processing *hc*
C&E Communications and Electronics [also C-E and CE] *hc, fu*
C&I Compatibility and Interoperability [also C+I] *fu, mt*
C&I Control and Integration *fu*
C&J Collection and Jamming *mt*
C&J Cut & Jumper *fu*
C&M Calibration and Maintenance *hc, jj*
C&N Communications and Navigation *fu*
C&N Communications and Networking *hi*
C&O Concept and Objectives *mt*
C&P Contracts and Policy *hc*
C&PDO Configuration and Product Definition Office *hc*
C&RD Communications and Radar Division *fu*
C&S Configuration and Status *kf*
C&S Control and Status *kf*
C&T Chips and Technologies *pk, kg*
C&T Communications and Telemetry *fu*
C&T Communications and Tracking *sc4.11.4, 2, 4, wl, mb, mi*
C&W Caution and Warning *jdr, fu*
C-DAY [deployment date, USA DOD] *mt*
C-DITEG Common Digital Television Generator *fu*
C-E Communications [and] Electronics [also C&E and CE] *mt, jj, fu*
C-E_ESR Communications - Electronics Equipment Status Reporting [USAF] *mt*
C-FLIRTS Common Forward Looking Infrared Test Sets *hc*
C-LEVEL [produce specification level] *tb*
C-NET COMPESNET [USAF] *mt*
C-RET Color Resolution Enhancement Technology [HP] *hi*
C-SAM Special Airlift Mission [an aircraft program] *hc*
C-SCAN [azimuth versus elevation display] *jj*
C-SCANS Client-Systems Computer Access Networks *ct, em*
C-SPAN Cable Satellite Public Affairs Network *mt, hn43.25.3*
C-U Command Unique *mt*
C-WIFS Sea-viewing, Wide Field-of-view Sensor *-16*
C/A Computer Assisted *jj*
C/A Contract Administrator *fu*
C/A Corrective Action *fu*
C/A Cost Account *hc, fu*
C/A Course Acquisition *sd*
C/B Center of Balance *mt*
C/B Circuit Breaker *jj*
C/B Correlator/Buffer *fu*
C/B/U Computer/Buffer/ Utility [cabinet] *fu*
C/C Carbon/Carbon *hc*
C/CC Counts per Count Cycle *kf*
C/CS Channel/Configuration Switch *fu*
C/D Command Decoder *0*
C/D Control Data *pk, kg*
C/D Controls and Displays *jj*
C/DTS Computer Digital Test Station *hc*
C/E Components and Equipment *jj*
C/E Cost Effectiveness [ratio] *hc*
C/MH Cost per Man Hour *jj*
C/N Carrier to Noise [received power ratio] *sp659, 2, 4, 17, ce, fu,*
C/N Change Notice *jj, jdr, fu*
C/N Charge Number *-8*
C/O Call Out *jj*
C/O Check Out *pl, hc, wl, jj*
C/O Close Out *fu*
C/PC Cartridge/Printer Controller *fu*
C/PU Camera/Projection Unit *fu*
C/R Central Recorder *0*
C/S C or S-[band antenna] *ac3.5-2*
C/S Call Sign *mt, fu*
C/S Client/Server *pk, kg*
C/S Clutter to Signal [power ratio] *ic2-37*
C/S Cycles Per Second *w9*
C/S {Chief of Staff ?} *af72.5.31*
C/SCS Cost and Schedule Control System *mt, 17, fu, sb*
C/SCSC Cost Scheduling /Schedule Control System Criteria *mt, ms3.1.3, 17, ep, bf, st, jdr, kf, fu*
C/SPCS Cost / Schedule Planning and Control Specification *hc, fu*
C/SSR Cost / Schedule Status Report *mt*
C/SSR Cost/Schedule Status Report *sb, nt, st, fu*
C/T Carrier to Noise [power ratio] *csIII.69*
C/W Carrier Wave *fu*
C/W Complete With *fu*
C2D Character To Decimal [REXX] *pk, kg*
C2X Character To Hexadecimal [REXX] *pk, kg*
C5 Computer Program Product Specification *fu*
CA Cell Averaging *fu*
CA Channel Adapter *fu*
CA Clutter Amplitude *fu*
CA Clutter Attenuation *hc, fu*
CA Collision Avoidance *pk, kg*
CA Combat Assessment [USA DOD] *mt*

CA Commander AFDSDC [technical assistant to the] *mt*
CA Commitment Authorization *hc*
CA Computer Animation *ct, em*
CA Conditioned Analog (data) *he, ggIII.2.373*
CA Configuration Audits *mt*
CA Conflict Alert *fu*
CA Connection Approval *mt*
CA Contract Amendment *-4*
CA Contract Analysis *mt, wp*
CA Contract Article *fu*
CA Contract Authorization *oe*
CA Contract Award *mt, wp, fu*
CA Controller Aircraft [UK] *mt*
CA Corrective Action *wp*
CA Cost Account *fu, kf*
CA Cost Analysis *mt*
CA Cost Avoidance *fu*
CA Crisis Alerting *mt*
CA Current Awareness Bulletin *mt*
CA Gun Cruiser USN ship classification *fu*
CA [chemical warfare codename for bromobenzyl cyanide] *wp*
CA [heavy gun cruiser] [traditionally, 'heavy' cruisers are armored, and 'light' cruisers are not, relying on speed for protection.] [US Navy] *mt*
CA/VFR Conflict Alert/Visual Flight Rules *fu*
CAA Civil Aeronautics Administration *hc, jj*
CAA Civil Aviation Authority [UK] *mt, aw130.13.15, fu*
CAA Clear Aperture Antenna *re*
CAA Combined Armed Armies *hc*
CAA Combined Arms Army *mt*
CAA Concepts Analysis Agency [US Army] *mt, dt, st*
CAA Converter, Analog to Analog *jj*
CAAC Change Analysis And Coordination *fu*
CAAC Civil Aviation Administration of China *aw129.21.3*
CAAD Cost Account Authorization Document *-17*
CAALS Computer Aided Acquisition Logistic Support *wp*
CAANG California Air National Guard *-12*
CAANS Collision Avoidance And Navigation System *hc, jj*
CAAP Computer-Assisted Array Processing *fu*
CAATS Canadian Advanced Automated Air Traffic Transport System *hq6.2.7, hn50.8.1, hn48.13.1, fu*
CAB Captured Air Bubble *hc*
CAB Civil Aeronautics Board [USA] *mt, aw118.14.25, w9, wp, tb, jj, fu*
CAB Command Actuated Bit *jj*
CAB Communications Action Board *ah, pbl*
CAB Consolidated Agile Beam *jr*
CAB Contract Adjustment Board *wp*
CAB Control Advisory Board *jj*
CAB Corrective Action Board [an organization] *pbl, ep, jdr, tm, kf, jr, ah, fu*
CAB Cost Analysis Brief *st*
CAB Cost Audit Board *wp*
CABAPS Chemical And Biological Agent Protection System [chemical warfare] *wp*
CABERS Command Automated Budget Execution and Retrieval System *mt*
CABS Command Automated Budget System [USA] *mt*
CAC Cable Assembly Case *jj*
CAC Capabilities Assessment Center *mt*
CAC Carrier Aircraft Contractor *hc*
CAC Climate Analysis Center *ag*
CAC Combat Air Command *mt, wp*
CAC Combat Alert Center / Cell *mt*
CAC Combat Analysis Capability *mt*
CAC Combat Assessment Capability *wp, mt*
CAC Combined Arms Center [USA] *fu, mt*
CAC Computer Aided Classification *fu*
CAC Computer Assisted Classification *mt*
CAC Connection Admission Control *ct, em*
CAC Console Access Controller *fu*
CAC Control and Analysis Center *mt*
CAC Corrective Action Committee *fu, jj*
CAC Cost at Completion *fu, ep, 8*
CAC Current Actions Center *mt*
CACC Canadian Aerospace Control Center *fu*
CACC Commander, Air Component Command *mt*
CACDA Combined Arms Combat Developments Activity *mt, fu*
CACI (CACI, Incorporated - Federal) {?} *mt*
CACO Corporate Administrative Contracting Officer *hn50.8.2*
CACP Central Arbitration Control Point *ct, em*
CACRS Canadian Advisory Committee on Remote Sensing *ac3.5-2*
CAD Capital Air Defense *fu*
CAD Card Assembly Department *fu*
CAD Cartridge Activated Device *jj*
CAD Civil Aviation Division *fu*
CAD Coherent Amplitude Detector *gg154*
CAD Collective Address Designator *mt*

CAD Computer Aided Design *aa26.12.18, sc21.12.4, rm7.1.2, 8, 15, 17, hc, pl, w9, md2.1.1, aw130.11.75, wp, ep, tb, st, wl, jj, pbl, tm, kf, kg, ct, em, ah, ah, fu, hi, mt, ai1*
CAD Computer Aided Drafting *em*
CAD Computer Assisted Aided Dispatch *hc, y2k*
CAD Contract Award Date *fu*
CAD Corrective Action Document *fu*
CAD Counter Arm Decoy [US Navy] *mt*
CAD/CAE Computer Aided Design/ Computer Aided Engineering *fu*
CAD/CAM Computer Aided Design / Computer Aided Manufacturing *mt, aa26.12.2, 16, ai1*
CAD/M Computer Aided Design / Manufacturing *hc*
CADA Computer Aided Design and Analysis *wp*
CADA Computer Aided Design Applications *oe, wp*
CADAD Computer Aided Derating Analysis Documentation [system] *rm7.1.4*
CADAM Computer [graphics] Augmented /Aided Design And Manufacturing [software] *fu, hn43.22.8, hc, aa26.11.43, ep, jdr*
CADB Cost Analysis Data Base *tb*
CADBIT Computer Aided Design for Built In Test *mt*
CADC Capital Air Defense Center Controller *fu*
CADC Central Air Data Computer [USA] *mt, wp, hc, jj*
CADC Control Air Data Computer *hc*
CADCC Capital Air Defense Control Center *fu*
CADCP Capital Air Defense Computer Program *fu*
CADD Combat Air Delivery Division *wp*
CADD Computer Aided Design and Drafting *wp, kg, jdr, hi*
CADD Computer Aided Design Development *wp*
CADDA Computer Aided Design and Design Automation *wp*
CADDS [3D modeling software by Prime Computer Incorporated] *aw130.22.37*
CADE Client/server Application Development Environment *pk, kg*
CADE Combined Allied Defense Experiment / Effort *mt*
CADE Combined Common Allied Defense Experiment *wp, tb, st, fu, hc*
CADE Computer Aided Document Engineering *hi*
CADEF Computer Aided Design and Engineering Facility *fu*
CADENS CONUS Air Defense *hc*
CADESS Computer Aided Data Exchange Standards / Specifications *mt*
CADF Central Control Air Defense Force *hc, jj*
CADFISS Computation and Data Flow Integrated Sub System *hc*
CADGE Common Air Defense Ground Environment *fu*
CADH Communications And Data Handling *oe*
CADI Computer Aided Diagnostic Information *wp*
CADIN Continental [Canadian] Air Defense Integration North *12, hc*
CADIS Computer Aided Design Information System *fu*
CADIZ Canadian Air Defense Identification Zone *fu*
CADM Computer Aided Design and Manufacturing *kf*
CADM Configuration and Data Management *-4*
CADMAT Computer Aided Design, Manufacturing and TestTesting *fu, ai1*
CADNET Chemical Agent Detector Network [chemical warfare] *wp*
CADOSS Computer Aided Development of Standards / Specifications *mt*
CADQAD Computer Aided Development of Quality Assurance Data *mt*
CADRAA CONUS Air Defense Resource Allocation Advisor *ts15-23*
CADRE Center for Aerospace Doctrine, Research and Education [AU] *mt, af72.5.72*
CADS Command Arm-Disarm System *fu*
CADS Computer Assisted Display System *mt*
CADS Continental Air Defense Study *hc*
CADSAT Computer Aided Design and Specification Analysis Tool *fu*
CADSI Computer Aided Design Software, Incorporated *aa27.4.b57*
CAE (a contract with Canada's CAE Electronics, Ltd.,) *hn43.18.1*
CAE (Teledyne CAE also were competing for the [FOG-M] contract) *aw130.13.13, aa26.12.59, af72.5.150*
CAE Carrier Aircraft Equipment *hc*
CAE Client Application Enabler *pk, kg*
CAE Common Applications Environment *hi, kg*
CAE Computer Aided Assisted Engineering *mt, kg, fu, aa26.12.18, rm7.1.2, wp, ep, wl, jdr, hi, te23, 11, 24, ai1*
CAEDS Computer Aided Emulation Design System *mt*
CAEL Consolidated Aerospace Equipment List *fu*
CAEL Consolidated AGE Equipment List *hc*
CAEMS Computer Aided Embarkation Management System [USMC] *mt*
CAEO Commandement Air en ExtrËme Orient ["Far East Air Command"] [France] *mt*
CAEOS Common Aperture Electro Optical System *hc*
CAEWIS Computer Aided Electronic Warfare Information System *mt, ro*
CAEX Computer Aided Exploration *pk, kg*
CAF Canadian Armed Forces [Canada] *mt*
CAF Command Advisory Function [USAF] *mt*
CAF Commander Airlift Forces *mt*
CAF Counter Air Fighter *wp*
CAF Critical Assessment Factor *fu*
CAFDA Commandement Air des Forces de Defense AÈrienne ["Air Command, Aerial Defense Forces"] [France] *12, mt*

CAFES Computer Aided Function-allocation Evaluation System *hc*
CAFI Composante Air des Forces d'Intervention ["air component of the intervention forces"] [France] *mt*
CAFIT Computer Aided Fault Isolation Testability *mt*
CAFMS Computer Aided Assisted Force Management System [USAF] *mt, hc*
CAFSU Carried And Field Service Unit *jj*
CAFVIMS Command Air Force Vehicle Integrated Management System [USAF] *mt*
CAG Carrier Air Group [commander] [USA] *jj, mt*
CAG Carrier Arresting Gear *jj*
CAG Civilian Advisory Group *mt*
CAG Column Address Generator *pk, kg*
CAG Combat Analysis Group *mt*
CAG Combat Applications Group *hc, jj*
CAG Commander of Air Group [US Navy] *mt*
CAG Commercial Advocacy Group *wl*
CAG Commercial Airplane Group *hc*
CAG Communications Advisory Group *fu*
CAG Corporate Advisory Group *rm7.1.5*
CAG heavy cruiser CA, Guided missile [US Navy] *mt*
CAGC Coded Automatic Gain Control *mt*
CAGC Course Automatic Gain Control *jj*
CAGE Canadian Air-Ground Environment *hc*
CAGE Commercial And Government Entity Number [Formerly FSC] *su, 18, nt, kf*
CAGE Contractor And Government Entity *wp*
CAGEC Commercial And Government Entity Code [a designator, formerly FSCM] *ep, jdr, tm*
CAGEL Consolidated Aerospace Ground Equipment List *hc, jj, fu*
CAGEL Consolidated AGE List *mt, wp*
CAGR Compound Annual Growth Rate *hi*
CAGR Compounded Average Growth Rate *hi*
CAH [through-deck cruiser, i.e. a small carrier with helicopters and VTOL aircraft, like the British HMS Invincible] [USA] *mt*
CAHY Calcium Hydroxide, chemical warfare Decontaminant *wp*
CAI Code Address Identification *hc*
CAI Computer Assisted Aided Instruction *mt, hc, w9, wp, jj, kf, fu, kg, ai1*
CAI Configuration Acceptance Inspection *hc*
CAI [a division of Recon Optical] *hn41.19.3*
CAIA Conference on Artificial Intelligence Applications *is26.1.76*
CAIF Command Automated Installation File [REDCOM / JDA] *mt*
CAIG Cost Analysis Improvement Group [OSD] *ph, hc, jj, kf, mt*
CAIMS Computer Aided Instruction Management System *mt*
CAIMS Conventional Ammunition Integrated Management System [US Navy] *mt*
CAIN Code to Analyze mutual Interface from Nuclear bursts *st*
CAINS Carrier Airborne Inertial Navigation System [USA] *mt*
CAINS Carrier Aircraft Inertial Navigation System [USA] *hc, aw130.11.c2, jj, fu*
CAINS Central Avionics Information System *hc*
CAIR [Hughes Aircraft Company customer for PACER WEB] *hc*
CAIS Casualty Assistance Information System [USMC] *mt*
CAIS Common ADA Interface Set *mt*
CAIS Common APSE Interface Set *mt, fu*
CAISS Computer Assisted Information Support System [USAF] *mt*
CAIV Cost As an Independent Variable *kf*
CAL Canister Airlock *jdr*
CAL Columbia Astrophysics Laboratory *pl*
CAL Command Asset List *mt*
CAL Command Authorization List *wp*
CAL Common Assembler Language Code *mt*
CAL Common Assembly Language *hd1.89.y13, fu*
CAL Communications Application Language *mt*
CAL Computer Aided Learning *kg, wp, ai1*
CAL Computer Animation Language *wp*
CAL Configured Articles List *gp, oe*
CAL Cornell Aeronautical Laboratory *jj*
CAL Cray [Research, Inc.,] Assembly Language *tb*
CALC Cargo Acceptance and Load Control *wp*
CALDS Computer Aided Logistics Data System *fu*
CALFAB Computer Aided Layout and Fabrication *nd74.445.20*
CALL Computer Aided Language Learning *hi*
CALM Computer Assisted Load Manifest [USAF] *mt*
CALMA [CAD development system] {computer program used for chip layout design} *fu, ne*
CALMS Continuous Automatic Line Monitoring System *hc*
CALP Caution and Advisory Lamp Panel *jj*
CALPA Canadian Air Line Pilots' Association *aw130.13.33*
CALPHAD Calculation of Phase Diagrams [a technical journal] *hn43.15.5*
CALPS COMPES Automated Load Planning System *mt*
CALS Command And Launch Subsystem *hc*
CALS Computer Aided Logistic System *wp*
CALS Computer Aided Logistics Support [DOD] *fu, mt*
CALS Computer Aided [acquisition and in] Logistics Support *mt, rm7.1.1, hc, aa26.12.66, aw130.22.65, nt, wp, kg*

CALS Continuous Acquisition and Life Cycle Support *kf*
CALS/ATI Computer Aided Logistics Support / Automation of Technical Information *mt*
CALSA Computer Aided Integrated Logistics Life Cycle Support Analysis [CONGRESS] *mt*
CALSA Computer Aided Logical Support Analysis *mt*
CALSTAT Cargo Location and Status Program [MSC] *mt*
CALSU Combat Air Logistics Support Unit *hc, jj*
CALS_MIO CALS Management Information Office *mt*
CALS_TIMS CALS Technical Information Management System *mt*
CALT China Academy of Launch Vehicles *ai2*
CALTAI Calibration [and Metrology Program] Taiwan *hc*
CALTECH California Institute of Technology *jj*
CALVN Calibration [and Metrology Program] Viet Nam *hc*
CAM Checkout and Automatic Monitoring *fu*
CAM Chemical Agent Monitor [chemical warfare] *wp*
CAM Collision Avoidance Maneuver *bo, wp*
CAM Combat Airlift Mission *mt*
CAM Common Access Method *ct, kg, em, hi*
CAM Communications Assets Monitoring *jdr*
CAM Computer Aided Assisted Manufacturing *mt, sc21.12.4, 8, 15, 17, hc, pl, w9, aa26.11.53, wp, ep, tb, wl, jj, jdr, kf, kg, fu, hi, ai1*
CAM Computer Aided Machining *ct, em*
CAM Computer Assisted Maintenance *jj*
CAM Contents Addressable Memory *fu, kg, ai1*
CAM Contract Audit Manual *kf*
CAM Cost Account Manager *17, wp, tb, jdr, kf, fu*
CAM-RFTS Computer Assist Modification Radio Frequency Test *hc*
CAMA Centralized Automatic Message Accounting [Soviet] *fu, mt*
CAMAAFCS Clear Air Mass Anti Aircraft Fire Control System *hc*
CAMAC Computer Automated Aided Measurement and Control (KSC has supplied powerful CAMAC data acquisition systems) *aa27.4.57, fu, oe*
CAMAL Continuous Airborne alert, Missile-launching Aircraft with Low level penetration capability [USA] *mt*
CAMD Collect, Audit, Maintain, Display *mt*
CAMD Computer Aided Maintenance Device *fu*
CAMDS Chemical Agent Munitions Disposal System [chemical warfare] *hc, wp*
CAME Corps Air Space Management Element [USA] *mt*
CAMEL Critical Aeronautical Material Equipment List *jj*
CAMEO Covert Active Modular Electro Optical [system] *hc*
CAMESA Canadian Military Electronics Standards Agency *fu*
CAMGEN Computer Aided MAPL Generation *fu*
CAMI Civil Aero Medical Institute *aw129.14.112, fu*
CAMI Continued Action Maintenance Item *jj*
CAMIS Continental Army MIS [later RCAS] *mt*
CAMMIS Command Aircraft Maintenance and Manpower Information System [USAF] *mt*
CAMMP Capabilities Assessment Model for Munitions Production [USAF] *mt*
CAMMS Computer Aided Maintenance Management System *mt*
CAMMS Computer Aided Manufacturing Management System [USAF] *tb*
CAMMS Coscom Automated Maintenance Management System *fu*
CAMMU Cache Memory Management Unit *hi*
CAMP Cabin Air Manifold Pressure *wp*
CAMP Claim Administrative Management Program *mt*
CAMP Command ADP Modernization Program [USAF] *mt*
CAMP Command Automation Master Plan *mt*
CAMP Common ADA Missile Package *hc, wp, fu*
CAMP Computer Aided Maintainability Predictor *wp*
CAMP Computerized Aircraft Maintenance Program *wp*
CAMPAL Computer Aided Mission Preparation at Airbase Level *wp*
CAMPS Computer Aided Message Processing System *hc, fu*
CAMPS Computer Aided Mission Planning System *hc*
CAMPS Computer Assisted Message Processing System [NATO] *mt*
CAMPS Computer Assisted Mission Planning System *mt*
CAMPSIM Campaign Simulation Expert System [PACFLT] *mt*
CAMPUS Center Automated Manpower and Project Update System *mt*
CAMS Coastal Anti Missile System *hc, fu*
CAMS Collection Automated Management System *mt*
CAMS Combat Aviation Management System [FORSCOM] *mt*
CAMS Core Automated Maintenance System [USAF] *wp, kf, mt*
CAMS Crisis Action Management System [MTMC] *wp, mt*
CAMTI Coherent Airborne Moving Target Indication *hc, jj*
CAMTT Coherent Airborne Moving Target Track *hc*
CAN Campus Area Network *ct, kf*
CAN Canberra, Australian [deep space] Network *oe*
CAN Controller Area Network *em*
CANA Comando de Aviaciûn Naval Argentina [Argentine Naval Aviation Command] *mt*
CAND Controls and Displays *fu*
CANSAT Canadian [domestic communications] Satellite *hc*
CANU Computer Air Navigation Unit *jj*
CANUS Canada, United States *mt*

CAO Combat Air Operations *wp*
CAO Computer Aided Operation *fu*
CAO Contract Administration Officer *mt, fu, hc, wp, jj*
CAO Counter Air Operations *mt*
CAOC Combined Air Operations Center [NATO] *wp, mt*
CAOCC Combined Air Operations Coordination Center *mt*
CAOE Contractor Acquired Operational Equipment *hc, jj*
CAOSOP Standing Operating Procedures for the Coordination of Atomic Operations *mt*
CAP Cable Access Protocol *mt*
CAP Capabilities and Procedures Document *0*
CAP Carrier Air Patrol *fu*
CAP Carrierless Amplitude and Phase [Modulation] *ct, kg, em*
CAP Central Arbitration Point *pk, kg*
CAP Centralized Assignment Procedures (CAP_III) [USA] *mt*
CAP Civil Air Patrol *mt, w9, af72.5.72, jj*
CAP Codec Analysis Program *fu*
CAP Combat Air Patrol *mt, hc, wp, jj, fu*
CAP Command Acceptance Pattern *-4*
CAP Commander's Audit Program *mt*
CAP Communication Application Platform *pk, kg*
CAP Communications Access Processor *mt*
CAP Communications Alternative Provider *ct*
CAP Communications and Processor SMTS *kf*
CAP Communications Electronics Authorization Program *mt*
CAP Community Arbitration Program *fu*
CAP Computer Address Panel *jj*
CAP Computer Aided Planning [system] *8, ep*
CAP Computer Aided Programming *hi*
CAP Computer Aided Publishing *hi, kg*
CAP Conduct Air Patrol *fu*
CAP Consolidated Aerial Port *mt*
CAP Contract Administration Plan *mt*
CAP Contractor Acquired Property *hc*
CAP Corrective Action Plan Process *fu*
CAP Cost Account Plan *17, oe, fu*
CAP Countermeasures Advisory Panel *mt*
CAP Countermeasures Assessment Program *mt*
CAP Crew Activity Plan *wl*
CAP Crisis Action Planning [USA DOD] *mt*
CAP Customer Accommodation Plan *wl*
CAPCOM Capsule Communicator *pf1.24*
CAPD Computer Automated Performance Data *mj31.5.7*
CAPD Computing to Assist Persons with Disabilities [Johns Hopkins University] *pk, kg*
CAPER Cost and Planning Evaluation Reports *ms3.1.2*
CAPI Common API *hi*
CAPLAN Computer Aided Planning *fu*
CAPM Capital Asset Pricing Model *wp*
CAPMAR Cost Account Performance Measurement and Analysis Report *oe*
CAPP Computer Aided Process Planning *hc, fu*
CAPPI Constant Altitude Plan Position Indicator *hc*
CAPS Cassette Programming System *pk, kg*
CAPS Communication Antenna Pointing System *fu*
CAPS Computer Aided Planning System [a data base] *ep*
CAPS Computer Aided Plating System *fu*
CAPS Computer Assisted Acquisition System *mt*
CAPS Computerized Airlift Planning System *mt*
CAPS Configuration Analysis and Projection System *mt*
CAPS Conventional Armaments Planning System *wp*
CAPSL Canon Printing System Language *hi*
CAPSTONE [army program to align AC and RC units] *mt*
CAPTOR Capsulated Torpedo [antisubmarine mine] *aw130.13.56, uc, wp*
CAPW Corrective Action Problem Worksheet *fu*
CAQ Computer Aided Quality Control *wp*
CAR Capital Appropriations Request *fu*
CAR Chief, Army Reserve *jj*
CAR Civil Air Regulations *w9, wp*
CAR Civil Airworthiness Regulation *mt*
CAR Cloud top Altitude Radiometer *hc*
CAR Coherent Antistokes Raman *wp*
CAR Command Assessment Review *mt, wp*
CAR Configuration Analysis Report *fu*
CAR Configuration Audit Review *mt, hc, fu, wp, jj*
CAR Contract Appraisal Report *wp*
CAR Corrective Action Report *mt*
CAR Corrective Action Request [a document] *fu, oe, gp, 17, ep, jj, pbl, tm*
CAR Customer Account Representative *mt*
CARA Cargo And Rescue Aircraft *wp*
CARA Center for Astrophysical Research in Antarctica *mi*
CARA Civilian Appellate Review Agency [US Army, USAF] *dt*
CARA Combat Aircrew Rescue/Recovery Aircraft *hc*
CARA Combined Altitudes Radar Altimeter *wp*

CARA Computer Aided Requirements Analysis *fu*
CARC Chemical Agent Resistant Coating [chemical warfare] *wp*
CARCN Central American Regional Communications Network *mt*
CARCO [a company that makes 3-axis tables] *-17*
CARD Computer Aided Remote Driving *re*
CARD Cost Analyses Requirements Description *kf*
CARDA Continental Airborne Reconnaissance For Damage Assessment [US] *mt, wp, fu*
CARDS Carrier Dynamic Simulation Program [US Navy] *mt*
CARDS Catalogue of Approved Required Documents *st*
CARE Computer Aided Reliability Estimation *hc*
CARE Cooperative for American Relief Everywhere *w9, wp*
CARET Computer Aided Revising, Editing, and Translating system *fu*
CARF Central Altitude Reservation Function *fu*
CARGRU [Aircraft] Carrier Group *12, fu*
CARI Control Channel Active Relay Indicator *fu*
CARL Colorado Alliance of Research Libraries [Internet] *pk, kg*
CARL Configuration Accounting Requirements List *fu*
CARMISH Commander Army Mission *jj*
CAROPS Carrier Operations *jj*
CARP Computed Air Release Point *mt, wp*
CARP Contingency Alternate Routing Plan *mt*
CARP Contingency Alternate Routing Program *mt*
CARP Contract Award Release Package *mt*
CARP/HARP {a special use of GPS} *N75-27202-483*
CARPS Calculated/Computed Air Release Point Steps *hc*
CARQUAL Carrier Qualification [USA] *jj, mt*
CARRE Computerized Algorithm for the Radar Range Equation *st*
CARS Coherent Anti Stokes Raman Scattering *aa26.12.b9*
CART Central Automated Replenishment Technique *jj*
CART Conformal Array [Radar] Technology *wp, mt*
CART Corrective Action Review Team *jj*
CARTS Computer Aided Reconstruction Tool, Shipboard *fu*
CARTS Covert Aircraft Recovery Tracking System *hc*
CARVER Criticality, Accessibility, Recuperability, Vulnerability, Effect, and Recognizability [special operations forces acronym used to assess mission validity and requirements] [USA] *mt*
CAS (CAS, Inc. . . . Huntsville, AL 35814) *jmj*
CAS Calculated Air Speed *wp*
CAS Calibrated Air Speed [IAS corrected for system errors] *mt, aw130.13.38, jj*
CAS Chemical Abstracts Service *wp*
CAS Chief of Air Staff *mt*
CAS Chinese Academy of Sciences *ac3.4-6*
CAS Circuit Assignment Selection *fu*
CAS Close Air Support [USA DOD] *mt, fu, aw130.13.106, hc, aa26.12.83, wp, jj*
CAS Cockpit Avionics System *wp*
CAS Collision Avoidance System *mt, wp*
CAS Column Address Select *ct, em, kg*
CAS Column Address Strobe *fu*
CAS Combat Ammunition System [USAF] *mt*
CAS Combat Aviation Squadron *mt*
CAS Combined Antenna System *jj*
CAS Combined Arms Simulation *hc*
CAS Command Augmentation System *hc*
CAS Command Automated System [CINCOM] *mt*
CAS Command Automation System *mt*
CAS Communications Application Specification *pk, kg*
CAS Computer Aided Search *mt*
CAS Computer Aided Styling *pk, kg*
CAS Computer Aided Support *wp*
CAS Computer Aided System *hc*
CAS Contract Administration Services *mt, jj, fu*
CAS Control Assembly Set *hc*
CAS Cost Account Schedule *oe*
CAS Cost Accounting Standard *mt, fu, 8, 18, pl, nt, wp, tb, kf*
CAS Cost Allocation Application Summary *hc, jj*
CAS Crisis Action Support *mt*
CAS Crisis Action System *mt*
CAS-C (phase out of communications spacecraft role after CAS-C) *ci.114*
CAS/BAI Close Air Support/Battlefield Air Interdiction *aa26.11.17, af72.5.136*
CASA California Astronautics and Space Agency *sc3.10.3*
CASA Computer Aided Software Analysis *kf*
CASA Construcciones Aeronauticas, S. A. [Spain] *aa26.12.20, aw118.18.11, es55.34, af72.5.32, ai2*
CASA Cost Analysis Strategy Assessment *kf*
CASADA Close Air Support Alternative Design Analysis *wp*
CASAT Computer Assisted System Approach to Training [USMC] *mt*
CASB Canadian Aviation Safety Board *aw130.13.33*
CASB Cost Accounting Standards Board *hc, tb, fu*
CASC Capital Area Support Center *dt*
CASC Contract Administration Service Component *jj*
CASCADE Combined A/B Surveillance and Control Aerospace Defense *hc*

CASDE Computer Assisted Software Development Environment *mt*
CASE Central AISIM Support Equipment *fu*
CASE Complete Aim-4D Systems Equipment *hc*
CASE Computer Aided Assisted Software Engineering *mt, aw129.14.119, pl, is26.1.37, go1.5-1, wp, kf, kg, ct, em, hi, ai1*
CASE Computer Aided System Engineering *fu*
CASE Computer Aided System Evaluation *mt*
CASE Computer Automated Support Equipment *fu*
CASE Consolidated Aerospace Supplier Equipment *hc*
CASE Coordinated Aerospace Supplier Evaluation *fu*
CASE Coordinating Agency for Supplier Evaluation *fu, gp*
CASE/MIS Construction, Automation, and Specialized Equipment Management Information System [US Navy] *mt*
CASEE Comprehensive Aircraft Support Effectiveness Evaluation *mt*
CASES Capabilities Assessment Expert System [PACFLT] *mt*
CASES Capability Assessment System for Expert Systems *hc*
CASEVAC Casualty Evacuation *mt*
CASF Composite Air Strike Force [USA] *mt*
CASFDD Crisis Action System Force Deployment Data *mt*
CASIMS Calibrated Airborne Spatial Infrared Measurement System *hc*
CASM Close Air Support Missile *17, sb, hc, wp*
CASM Combined Arms Simulation Model [HQ USAF] *hc, mt*
CASMC Close Air Support Mission Card *hc*
CASMU [aerospace company, Italy] *aw130.22.76*
CASOFF Control And Surveillance Of Friendly Forces *hc*
CASOM Conventionally Armed Stand Off Missile [UK] *mt*
CASP Canadian Airspace Systems Plan *fu*
CASP Contracting and Acquisition Support Plan *mt*
CASP Control and Status Panel *fu*
CASPAR Coherent Advanced Solid State Phased Array System *hc*
CASPR Command Automated System for Procurement [USA] *mt*
CASPR Computer Assisted Planning-Reconnaissance [SAC] *mt*
CASR Canadian Air Surveillance Radar *fu*
CASREP Casualty Reporting [US Navy, USA] *mt*
CASREP Casualty [Summary] Report *jj, fu*
CASS Carrier Aircraft Support Study *hc*
CASS Coarse Analog Sun Sensor *oe*
CASS Command Active Sonobuoy System *fu, jj*
CASS Computer Assisted Search Service *pk, kg*
CASS Computer Automated Support System *wp*
CASS Consolidated Automated Automatic Support System [US Navy] *mt, hc, fu*
CASSA Coarse Analog Sun Sensor Assembly *oe*
CASSE Coarse Analog Sun Sensor Electronics *oe*
CASSE Cost Analysis Scheduling and Schedule Analysis *st*
CASSIS Classified and Search Support Information System *pk, kg*
CAST Change Authorization, Scheduling and Tracking *fu*
CAST Chinese Academy of Space Technology *ac3.4-6*
CAST Common Access Security Terminal *hc*
CAST Computer Aided Submode Training *fu*
CAST Corsair Avionics Subsystems Tester *hc*
CASTOR Corps Airborne Stand Off Radar *mt*
CASTS Canal Safe Transit System *hc*
CASW Close Air Support Weapon *hc*
CASWS Close Air Support Weapon System *wp*
CAT Capsule Ariane Technologique *ac3.4-6*
CAT Cartographic Advanced Terminal *hc*
CAT Clear Air Turbulence *hc, w9, jj, ai1*
CAT Clustered Air Technology *hc*
CAT Cockpit Automation Technology *mt, aa26.8.b24, hc, wp*
CAT Combined Acceptance Test *fu*
CAT Command Action Team *mt, wp*
CAT Common Air Tasking *mt*
CAT Computer Adaptive Test *pk, kg*
CAT Computer Aided Assisted Testing *kg, vc25.4.5, hc, wp, jj, hi*
CAT Computer Aided Assisted Training *mt, fu*
CAT Computer Aided Test *mt, fu*
CAT Computer Aided Tomography *pk, kg*
CAT Computer Aided Tracking *hc, jj*
CAT Computer Aided Transcription *pk, kg*
CAT Computerized Adaptive Testing [system] *hc*
CAT Computerized Axial Tomography *w9, wp*
CAT Configuration and Traceability *hc*
CAT Contour Center Adaptive Tracker *hc*
CAT Contract Assistance Team *wp*
CAT Controlled Atmosphere Technique *fu*
CAT Corrective Action Team *jj*
CAT Crisis Action Team *mt*
CAT/FCS Command Adjusted Trajectory Fire Control System *hc*
CATA Computer Aided Trouble Analysis *fu*
CATA Correlation Algorithm and Techniques Analysis *wp*
CATAC Commandement AÈrien Tactique ["tactical air command"] [France] *12, mt*
CATANET Concatenated Network *ct*
CATC Civil Air Traffic Controller *hc, fu*
CATC Computer Assist to Technical Control [NORAD] *mt*

CATCC Civil Air Traffic Control Center *fu*
CATCOMS Credibility Assessment Team Communications System *mt*
CATCS Central Air Traffic Control School [UK] *mt*
CATD Crisis Action Team Director *mt*
CATE Ceramic Applications for Turbine Engines *aw129.10.63*
CATE Computer Aided Test Equipment *fu*
CATF Commander, Amphibious Task Force [USA DOD] *fu, mt*
CATGEN Combinational Array Test Generator *hc*
CATGME Canadian Air Task Group Middle East *mt*
CATI Computer Assisted Telephone Inquiry *mt*
CATIA (the positions require CADAM/CATIA experience) *aa26.11.43*
CATIC China Aero Technology Import-Export Corporation *aw129.1.45*
CATIES Common Aperture Techniques Imaging Electro Sensor *hc*
CATIS Computer Aided Tactical Information System [AFIS] *mt*
CATIS Computer Aided Tactical Intelligence System *ro, wp*
CATIS Computer Aided Technical Information System [WWMCCS] *mt*
CATNBF Catalog Nuclear Burst File *tb*
CATREP Crisis Action Team Report *mt*
CATS Cheap Access To Space *mi*
CATS Coherent Acoustic Torpedo System *hc*
CATS Communications And Tracking System *jdr*
CATS Computer Aided Tolerance Selection *fu*
CATS Computer Assisted Training System *hi, kg*
CATS Computer Automated Test System *jj*
CATS Contained Armament Test Set *wp*
CATS Convert Acquisition Tracking System *hc*
CATSCAN Computerized Axial Tomography Scan *pk, kg*
CATSOP Computer Aided Tailoring of Software Programs *wp*
CATSS Cartographic Applications for Tactical and Strategic Systems [RADC] *hc, mt*
CATTS Combined Arms Tactical Training Simulator [Army] *fu*
CATV Community Antenna Television [later renamed Cable TV] *mt, sc2.10.6, hc, w9, lt, tb, jj, kg, ct, hi*
CAT_XSS Command Acquisition and Telemetry *jdr*
CAU Channel Adapter Unit *mt*
CAU Controlled Access Unit *pk, kg*
CAU Crypto Adapter Unit *mt*
CAU Crypto Ancillary Auxiliary Unit *mt, fu*
CAV Constant Angular Velocity *ct, kg, em, hi*
CAVISAT Center for Audio Visual Instruction via Satellite *hc*
CAVOK Ceiling And Visibility OK *mt*
CAVS COBOL Automated Verification System *ro*
CAVU Ceiling and Visibility Unlimited [USA DOD] *w9, jj, mt*
CAW Conflict Alert Warning *fu*
CAWG Calibration Advisory Working Group *hy*
CAWGS Covert All Weather Gun System *hc*
CAWSS Crisis Action Weather Support System [WWMCCS] *mt*
CAX Combined Arms Exercise *mt*
CAZ Cursor Azimuth *wp*
CB Capsule Bus *hc, jj*
CB Center of Balance *mt*
CB Chemical / Biological *mt*
CB Chemical and Biological *fu*
CB Circuit Breaker *fu, jj*
CB Citizens Band *ct, wp*
CB Cloud Break *kf*
CB Common Battery *mt*
CB Comparator Buffer *fu*
CB Component Board *fu*
CB Component Broker *ct, em*
CB Conditioned Binary (data) *ggIII.2.373*
CB Construction Battalion [also written "seabee"] [USA] *mt*
CB Contract Brief *8, hc, jj, fu*
CB Cost Benefit *-16*
CB [large cruiser, US Navy] *mt*
CB {Corrections Board} [executive secretary of the AF board for the correction of military records] *dt*
CB/CM Counter Battery/ Counter Mortar *fu*
CB/L Commercial Bill of Lading *fu*
CBA Component Board Assembly *fu*
CBA Cost Benefit Analysis *wp*
CBA [Brazilian Aerospace Company] *aw129.21.88*
CBAS Command Budget Automated System [USAF] *mt*
CBAT Computer Based Aid for Troubleshooting *mt*
CBB Contract Budget Base[line] *17, hc, fu*
CBC Cipher Block Chaining *pk, kg*
CBC Common Core Booster *ai2*
CBC/MIS Construction Battalion Center MIS [US Navy] *mt*
CBCR Channel Byte Count Register *pk, kg*
CBD Commerce Business Daily *mt, cb, tb, st, fu*
CBE Command Budget Estimate *jj*
CBEMA Computer and Business Equipment Manufacturers Association *pk, kg*
CBESS Computer Based Educational Software System *hc*
CBF Central Band Filter *jj*
CBFS Command Budget Formulation System *mt*

CBGA Ceramic Ball Grid Array *pk, kg*
CBI China, Burma, India [operational sector of Allied forces in WWII] *mt*
CBI Computer Based Instruction *mt, w9, tb, kf, fu, kg*
CBIAC Chemical Biological Information Analyis Center [chemical warfare] *wp*
CBIL Common [and] Bulk Items List *fu, jj*
CBIOS Compatibility BIOS *hi*
CBIS Computer Based Information System *mt, fu*
CBITS Computer Based Instruction Training System *wp*
CBL Commercial Bills of Lading *su, hc, tb, fu*
CBL Computer Based Learning *hi, kg*
CBL Configuration Baseline *fu*
CBL Configuration Breakdown Lists *jj*
CBLS Carrier, Bomb, Light Store [UK] *mt*
CBM Central Battle Manager *tb*
CBM Confidence Building Measures *-14*
CBMS Computer Based Mail Message System *hi, kg*
CBMU Construction Battalion, Maintenance Unit *mt*
CBMU Current Bit Monitor Unit *fu*
CBN Cubic Boron Nitride *17, sb*
CBNRC Communications Branch, National Research Council *-10*
CBO Circuit Board Outline *fu*
CBO Combined Bomber Offensive [Allied] *mt*
CBO Congressional Budget Office *mt, aa27.4.10, wp*
CBPO Consolidated Base Personnel Office *mt, af72.5.112*
CBPS Chemical-Biological Protective Shelter [chemical warfare] *wp*
CBR California Bearing Ratio Rate *wp, jj*
CBR California Bearing Ratio [measure for the wheel pressure an airfield surface will sustain] *mt*
CBR Chemical / Biological / Radioactive weapons [USA] *mt*
CBR Chemical, Biological, Radiological [chemical warfare] *fu, wp, jj*
CBR Common Bomb Rack *mt*
CBR Constant Bit Rate *ct, kg, em, hi*
CBRN Caribbean Basin Radar Network *mt, hc*
CBS Columbia Broadcasting System *sc2.8.4, w9, wp*
CBS Committee for Basic Services *ag*
CBS Controlled Blip Scan *fu*
CBS Cost Breakdown Structure *hc*
CBS-X Continuing Balance System - Expanded [USA] *mt*
CBSE Computer-Based System Engineering *wp*
CBSS [a commercial satellite] *sd*
CBT Computer Based Training *ct, sc7.4.3, hc, aw130.22.8, wp, tb, kf, fu, kg*
CBTAO Cost Benefit of Tactical Air Operations *mt*
CBTDEV Combat Development *st*
CBTOPS Combat Operations Division *mt*
CBTPLANS Combat Plans Division *mt*
CBTS Computer Based Training System *hc*
CBU Cluster Bomb Unit *mt, wp, jj*
CBU Conference Bridge Unit *mt*
CBU Customer Business Unit *jdr*
CBW Chemical [and] Biological Warfare *mt, 14, ph, w9, wp*
CBW Convert Byte to Word *pk, kg*
CBX Computer Controlled Branch Exchange *hi, kg*
CC Call Contract *hc*
CC Cape Canaveral [AFS, Florida] *sc4.1.5*
CC Card Column *hc*
CC Central Computer *fu*
CC Ceramic Cards *-17*
CC Change Control *hc, jj*
CC Channel Counter *hc*
CC Chief Controller *fu*
CC Cluster Controller *fu, kg*
CC Combat Coded *wp*
CC Command and Control [also C2 and C^2] [USA DOD] *mt, ro, ph, hc, jb, wp, pm.ii, tb, st, kf*
CC Command Cab *mt*
CC Command Center *mt, fu, lka, tb*
CC Command Compiler *pl*
CC Common Collector *fu*
CC Common Console *fu*
CC Communications Cabin *fu*
CC Communications Controller *oe*
CC Complex Conjugate *ai1*
CC Component Commander *mt*
CC Composite Channel *fu*
CC Computer Comment *-4*
CC Configuration Control *mt, wp*
CC Container Code *mt*
CC Control Center *mt, sd*
CC Conventional Conflict *mt*
CC Cost Center *fu*
CC Critical Capability *kf*
CC [tactical command ship, US Navy] [USA] *mt*
CC&CS Command, Control and Common Services *kf*

CC&D {a type of threat to satellite missions} *sd*
CC&M Command Control and Monitoring [SOCC portion of GIM-TACS] *oe*
CC&S Central Computer and Sequencer *0*
CC&S Command, Control, and Status *kf*
CC/DC Command Center and Direction Center colocated *fu*
CC/SM Control Center/Security Monitor *mt*
CCA Carrier Controlled Approach *mt, jj*
CCA Change Control Authorization *fu*
CCA Chroma and Analog Integrator *fu*
CCA Chromated Copper Arsenate *wp*
CCA Circuit Card Assembly *hn47.7.8, oe, fu*
CCA Combat Coded Aircraft *wp*
CCA Command and Control Agency *mt*
CCA Command Center Automation *mt*
CCA Common Communications Adaptor *hi*
CCA Communicant Circuit Assignment *fu*
CCA Communication Circuit Assignment *fu*
CCA Comparative Cost Analysis *hc*
CCA Computer and Control Abstracts *fu*
CCA Contamination Control Area [chemical warfare] *wp*
CCA Continental Control Area *wp*
CCA Core Current Address *jj*
CCA Critical Capability Area *kf*
CCACS Command Center Automation System [CINCLANT] *mt*
CCAEP Computer-Controlled Action Entry Panel *fu*
CCAF Command Center Analysis Facility *kf*
CCAFS Cape Canaveral Air Force Station *gp, 4, 9, jdr, kf, mi, mb, oe*
CCAM Collision and Contamination Avoidance Maneuver *kf, oe, ai2*
CCAPS Computer Controlled ATLAS Pressurization System *oe*
CCAR Core Current Address Register *hc*
CCAR Critical Clearance Analysis Report *jdr*
CCARS Change Control And Reporting System *hd1.89.y30, fu*
CCAS Cape Canaveral Air Station *jr, uh, kf*
CCAS Central Computer And Sequencer *jdr*
CCAS Command Center Automation System [PACAF] *mt*
CCAT Computer Controlled Automatic Test *fu*
CCB Change Control Board *pl, pl, jdr, kf*
CCB Command Control Block *ct, em*
CCB Communications Control Board *fu*
CCB Configuration Control Board [an organization] *mt, ah, gp, 4, 16, hc, wp, tb, st, jj, jdr, pbl, tm, kf, oe, fu*
CCBD Configuration Control Board Directive *mt, hc, tb, jj, fu*
CCBI Central Computer Bus Interface *fu*
CCBL Configuration Control Baseline *fu*
CCBM Command and Control Battle Management *mt*
CCC Calibrator Control Console *gp*
CCC CARDA Coordination Center *mt*
CCC Cargo Category Code *mt*
CCC Central Computer Center [CCAFS] *oe*
CCC Central Computer Complex *fu*
CCC Change and Configuration Control *mt*
CCC Change Control Center *jj*
CCC Cognizant Controlling Custodian *jj*
CCC Command and Control Center *mt, kf, fu*
CCC Command and Control Console *tb*
CCC Command Center Complex *mt*
CCC Command, Control, and Communications [also C3 and C^3] *mt, ro, tb, st, kf, fu, aw130.13.56, 5, 11, 16, 17, ph, sb, hc, jb, wp, dm*
CCC Commonality Change Center *jj*
CCC Communications Center Console *0*
CCC Component Change Control *fu*
CCC Computer Central Complex *hc*
CCC Correspondence Control Center *hc*
CCC Crisis Coordination Center *mt*
CCC Critical Control Circuit [JRCS] *mt*
CCC/DM Change Configuration Control/Development and Maintenance *aw130.11.57*
CCC/IRM Command, Communications and Control / Information Resource Management [Coast Guard] *mt*
CCCB Component Change Control Board *jj*
CCCB Contractors' Configuration Control Board *tb*
CCCC Centralized COMINT Communications Center *-10*
CCCC Command, Control and Communications Countermeasures [also C4] *mt*
CCCC Command, Control, Communications, and Computers [systems] [also C4] [USA DOD] *dh, hc, wp, mt*
CCCC Corporate Communications and Computing Center Hughes *fu*
CCCC Cross Channel Coordinating Committee *mt*
CCCCI Command, Control, Communications, Computers and Intelligence [also C^4I and C4I] *kf, mt*
CCCCM Command, Control and Communications Counter Measures [also C^3CM] *hc, dh, 12, tb, mt*
CCCCMSDB CCCCM Support Data Base *mt*
CCCCS Center for Command and Control Communication Systems [DCA] [also C4S] *mt*
CCCC_SMP CCCC C4 Systems Master Plan *mt*

CCCD Camouflage, Cover, Concealment and Deception *mt*
CCCE Command, Control, Communications Electronics *kf*
CCCI Command, Control, Communications, and Intelligence [also C3I and C^3I] *mt, tb, st, kf, aa26.12.2, hc, ph, 13, 14, af72.5.41, aw130.11.59, jb, ae151.52.459, dt, wp, pm, fu, ro*
CCCII Command, Control, Communications Intelligence and Interoperability [also CCCI2] *mt*
CCCL Cathode Current Control Loop *fu*
CCCM Command, Control, and Communications Countermeasures [also CCCCM] *fu*
CCCN CC Computer Network [USCENTCOM] *mt*
CCCOB Command, Control, and Communications Order of Battle *mt*
CCCP Contingency, Command, and Control Plans *kf*
CCCP Critical Clearance Control Plan *jdr*
CCCPP Command, Control, Communications Program Plan [also CCCP2] *mt*
CCCRIS CCCS Readiness Information System [OJCS] *mt*
CCCS Command and Control Coordinate System *fu*
CCCS Command, Control and Communications Systems *mt*
CCCS Command, Control and Communications Systems Directorate [of OJCS] *mt*
CCCS Command, Control, and Communications Segment *-16*
CCCS Customer Correspondence Control System *fu*
CCCSN Central Command and Control System Network *wp*
CCCSPPBS CCCS Planning Programming and Budgeting System *mt*
CCCT Command, Control, Communications and Track *lka*
CCD Camouflage, Concealment and Deception *wp, mt*
CCD Central Command Decoder *fu, hc*
CCD Charge Coupled Device *mt, fu, ac3.4-14, 6, 17, sb, es55.57, wp, so, ep, tb, jj, hg, mb, kf, oe, kg, ge, mi, ai1*
CCD Charge Coupled Display *hi*
CCD Communication Circuit Deassignment *fu*
CCD Conference of the Committee on Disarmament *-14*
CCD Consolidated Cab Display *fu*
CCD Continental Communications Division *mt*
CCD Contract Cost Data *tb*
CCD Core Current Drive *fu*
CCDD Charged Coupled Device Detector *hc*
CCDD Command Control Development Division *hc, fu*
CCDP Central Computer Diagnostic Program *fu*
CCDP Command and Control Display Processing *hc*
CCDR CAD/CAM Database Release *fu*
CCDR Contract Cost Data Report *mt*
CCDR Contractor Cost Data Report Reporting *wp, jdr, kf, fu*
CCDS Carrier Combat Direction System *fu*
CCDS Centers for the Commercial Development of Space *mi*
CCE Central Control Element *fu*
CCE Command and Control Element *kf*
CCE Command and Control Evaluation *mt*
CCE Commercial Construction Equipment *jj*
CCE Common Control Element *fu*
CCE Competitive Concept Evaluation *fu*
CCE Contamination Control Engineer *oe*
CCE Contingency Communications Elements *wp, mt*
CCE Continuous Comprehensive Evaluation *kf, fu*
CCE Cost to Complete Estimate *fu*
CCE [Executive Officer, AFDSDC {?}] *mt*
CCEB Combined Communications Electronics Board *mt*
CCEC Command and Control Engineering Center [DCA] *mt*
CCEF Command Center Evaluation Facility [ESD] *mt*
CCEL Command and Control Engineering Laboratory *fu*
CCEP Commercial Communications Security Endorsement Program [NSA] *mt*
CCEP Commercial COMSEC Endorsement Program *tb*
CCES Change Cost Estimate Summary *fu*
CCES Command Center Evaluation System [ESD] *mt*
CCEU Component Control Expediter Unit *jj*
CCEV Control Center Experimental Version *tb*
CCF Central Computing Facility *0, tb*
CCF Change Coordination Form *jj*
CCF Chinese Communist Forces *w9*
CCF Collection Coordination Facility *mt*
CCF Communications Control Facility *mt*
CCF Compartmented Communications Facility *mt*
CCF Computer Control Function *fu*
CCF Confidence Checking Function *fu*
CCF Configuration Control Form *jj*
CCF Consolidated Computer Facility *mt*
CCFB Central Computing Facility Backup *0*
CCFGEN Configuration Control File Generator *fu*
CCFS Closed Cycle Fluid Supply System *hc*
CCFSS Collection Coordination Facility Support System [DIA] *mt*
CCFT Cold Cathode Fluorescent Tube *hi, kg*
CCFT Controlled Current Feedback Transformer *fu*
CCG Combat Communications Group *fu*
CCG Constant Current Generator *fu*
CCG Crisis Coordinating Group [OSD] *mt*

CCH Crew Chief *kf*
CCHK Continuity Check *jj*
CCI Charge Coupled Imager *hd, hc*
CCI Command and Control Information *fu*
CCI Command, Control, and Intelligence *pm*
CCI Common Client Interface *pk, kg*
CCI Composite Configuration Item *mt*
CCI Configuration Control Identifier [a designator] *ep*
CCI Controlled Cryptographic Item *jdr*
CCI Critical Configuration Item *fu*
CCID Configuration Control Identification Code *fu*
CCID Cursor Computer Interface Data *hc*
CCIE Command and Control Information Exchange *mt*
CCII Compatibility Class II *fu*
CCIL Continuously Computed Impact Line *mt*
CCINC Cabinet Committee on International Narcotics Control *-10*
CCIP Command and Control Initiatives Program *mt*
CCIP Command Center Improvement Program *mt*
CCIP Command, Control, and Intelligence Processor *mt*
CCIP Continuously Computed Impact Point *mt, aw118.18.95, hc, wp, jj*
CCIPS Command and Control Information Processing System *mt*
CCIR ComitÈ Consultatif International de Radio [French Consultative Committee on International Radio] [later renamed ITU-R] *gp, 2, 4, 7, ce, fu, nu, oe*
CCIR Commander's Critical Information Requirement *mt*
CCIS Centrally Controlled Interconnection System *hc*
CCIS Command Center Information System [PACOM] *mt*
CCIS Command [and] Control [and] Information System [USAREUR] *mt, fu, hn42.4.8, 17, hc, dt, tb*
CCIS Common Channel Interoffice Signaling *mt, fu*
CCIS Comprehensive Communications and Information System *hc*
CCISM Command and Control Information Systems Manager *mt*
CCISMO Command and Control Information Systems Management Office *mt*
CCISS Command, Control and Intelligence Support Squadron [AFCC*] *mt*
CCISS Command, Control, Intelligence Support System [AAC] *mt*
CCITT ComitÈ Consultatif International TÉlÈgraphique et TÉlÈphonique [International Telegraph and Telephone Consultative Committee] [Renamed to ITU-T] *mt, fu, nu, do24, 7, is26.1.41, ce, tb, ct, hi*
CCITTX.409 International Telephone and Telegraph Consultative Committee X.409 *mt*
CCIU Central Computer Interface Unit *fu*
CCIU Component Control Issue Unit *jj*
CCJB Crew Chief Junction Box *hc, jj*
CCL Carrier Common Line *mt*
CCL Change Control Log *fu*
CCL Commodities Control List *un*
CCL Common Command Language *mt*
CCL Composite Configuration List *jj*
CCL Connection/Cursor Control Language *pk, kg*
CCLAW Close Combat Laser Weapon *uc*
CCM Call Control Module *mt*
CCM Charge Coupled Memory *hi*
CCM Configuration Control Model *fu*
CCM Controlled Carrier Modulation *fu*
CCM Conventional Cruise Missile *wp*
CCM Counter Counter Measures *uc, bf, tb, jj, fu*
CCMD Continuous Current Monitoring Device *fu*
CCMR Central Contract Management Region [USAF] *hc*
CCMS Checkout, Control, and Monitor Subsystem *tb*
CCMS Command and Control Modernization Methodology *mt*
CCMSF Compass Call Mission Support Facility *mt*
CCN Contract Change Notice Notification *8, hc, jj, fu, oe*
CCO Ceiling Cutoff *fu*
CCO Chief of Combat Operations *mt*
CCO Circuit Control Office *mt*
CCO Crystal Controlled Oscillator *wp, fu*
CCOC Consolidated Contingency Operations Center *mt*
CCOMS Correlation Center Output Message Set *kf*
CCOS Command Control Operating System *fu*
CCP Central Control Processor *oe*
CCP Certificate in Computer Programming *mt*
CCP Certificate of Current Pricing *fu*
CCP Command and Control Processor *hc*
CCP Common Users Communication Program [Canada] *hc*
CCP Communication Control Processor *fu*
CCP Configuration Control Panel *mt*
CCP Configuration Control Plan [TACS/TADS] *jj, mt*
CCP Configuration Control Processor *hc*
CCP Console Command Processor *pk, kg*
CCP Consolidated Control Point *mt*
CCP Consolidated Cryptologic Program *mt, 10*
CCP Consolidation / Containerization Program *mt*
CCP Contamination Control Plan *jdr, oe*
CCP Contract Change Proposal *mt, fu, wp, tb, jj, jdr*

CCP Conventional Contingency Plan *mt*
CCP Coordinate Conversion Processor *hc*
CCP Crystal Cryptologic Program *fu*
CCP Cubic Close Packed *ai1*
CCP Custom Color Palette *hi*
CCPC Communications Computer Programming Center *mt*
CCPC Critical Collection Problems Committee [U. S. Intelligence Board] *-10*
CCPDS Command Center Processing and Display System *mt, hc, 17, go3.1.1, tb*
CCPDS-R Command and Control Processing Display System Replacement *mt, dh, hc, tb, kf*
CCPO Consolidated Civilian Personnel Office *mt, dt*
CCPS Centralized Civilian Pay System *mt*
CCQ [Headquarters Squadron Section, AFDSDC] *mt*
CCR Circulation Control Rotor *hc*
CCR Configuration Change Request *gp, 16, oe*
CCR Contract Change Request *wp*
CCR Contract Compliance Review *tb*
CCR Cost to Cost Relationship *kf*
CCRA Circuit Card Rack Assembly *fu*
CCRA Command and Control Requirements Analysis *fu*
CCRB Commonality Change Review Board *jj*
CCRB Components Change Review Board *fu*
CCRC Communications Control and Restoral Center *mt*
CCRC Conditional Conventional Retaliation Capability *-14*
CCRF Consolidated Contractor Repair Facility *hc*
CCRS Canadian Centre for Remote Sensing CDEMR *ge, hg, ac3.5-2*
CCS Central Computer System *hc*
CCS Central Control System *mt*
CCS Clear Channel Sensing *hc, fu*
CCS Combat Control System *fu*
CCS Command and Control Segment *uh*
CCS Command and Control System *mt, kf*
CCS Command Control Status *kf*
CCS Committee of Chiefs of Staff [Allied] *mt*
CCS Common Channel Signaling *ct, em*
CCS Common Command Set *kg, ct, em*
CCS Common Communications Services *pk, kg*
CCS Common Communications Support *pk, kg*
CCS Communications Control System *hc*
CCS Component Control Section *jj*
CCS Computer Controlled Simulator *jj*
CCS Configuration Control System *fu*
CCS Continuous Composite Servo *pk, kg*
CCS Control Computer Subsystem *jdr*
CCS Crew-member Communications Selector *pf.f88.11*
CCS Cryptologic Combat System *fu*
CCS Hundreds of Call Seconds *do457*
CCSA Command and Control Support Agency [DAMO] *dt*
CCSA Command and Control Support Agency [US Army] *mt*
CCSA Common Control Switching Arrangement *do129*
CCSB Configuration Control Sub Board *mt*
CCSC Command and Control Systems Center *mt*
CCSC Cryptologic Combat Support Console [US Navy] *fu, mt*
CCSD Cellular Circuit-Switched Data *pk, kg*
CCSD Collins Communication Systems Division *fu*
CCSD Command and Control Systems Division *fu*
CCSD Command Communications Service Designator *mt*
CCSDS Consultative Committee for Space Data Systems *es55.72, ns, re*
CCSG Consolidated Contingency Steering Group [USAFE] *mt*
CCSIP Combat Control Systems Improvement Program *fu*
CCSK Cyclic Code Shift Keying *fu*
CCSL Compatible Current-Sinking Logic *fu*
CCSO Command and Control Systems Office [AFCC*] *mt*
CCSO Command and Control Systems Organization [DISA] *mt*
CCSPMO Command and Control Systems Program Management Office *mt*
CCSS Command and Control Subordinate System [USA] [also CCS2] *fu, mt*
CCSS Commodity Command Standard System [USA] *fu, mt*
CCSS Cryptologic Combat Support System *fu*
CCST Combat System Certification Test *fu*
CCT Chief Controller Technician *fu*
CCT Combat Control Team [USAF] *fu, hc, mt*
CCT Command Control and Telemetry *-16*
CCT Command Control Transmitter *hc*
CCT Computer Compatible Tape *ac2.2-15, ge, ep, hg*
CCT Constant Current Transformer *fu*
CCT Contingency Communications Team *mt*
CCT Critical Component Temperature *hc*
CCTC Command and Control Technical Center [DCA] *hc, dt, mt*
CCTF Combat Crew Training Facility *kf*
CCTG Command and Control Test Group *fu*
CCTS Combat Crew Training School *wp, kf, jj*
CCTS Combat Crew Training Squadron [USA] *mt*

CCTS Coolant Contamination Test Set *jj*
CCTT Command and Control Team trainer *hc*
CCTV Closed Circuit Television *mt, ct, csIV.3, 0, 9, hc, w9, tb, st, wl, jj, jdr, fu, oe, hi*
CCTWT Coupled Cavity Travelling Wave Tube *hc, tb, fu*
CCU Cable Connecting Unit *fu*
CCU Camera Control Unit *hc*
CCU Central Command Unit *sd*
CCU Central Control Unit *hc, jdr, kf*
CCU Channel Control Unit *fu*
CCU Command Center Upgrade *mt*
CCU Common Control Unit *mt*
CCU Communications Control Unit *wp, fu*
CCU Computer Control Unit *jj*
CCU Configuration Control Unit *wp*
CCUP Central Computer Utility Program *fu*
CCV Close Combat Vehicle *hc*
CCV Control Configured Vehicle *aa26.8.14, hc*
CCV Control Configured Vehicle [a test program that modified aircraft with additional control surfaces, involving a B-52, a F-104 and an F-16] [USA] *mt*
CCVCS Closed Cycle Vaporization Cooling System *hc*
CCVPS Computer Controlled Vent and Pressurization System *oe*
CCW Command and Control Warfare *mt*
CCWS Command Center Watch Station *mt*
CCWS Common Controller Workstation *fu*
CCWT Command Center Watch Teams *hc*
CCX [Phase IV Management Office, AFDSDC] *mt*
CC_FACRP CC Functional Analysis and Consolidation Review Panel *mt*
CC_MPPCS CC Manpower and Personnel Contingency Support System [HQ USAF] *mt*
CD Cable Duct *fu*
CD Capacitor Diode *fu*
CD Carrier Detect *ct, fu, em, kg*
CD Change Directory *pk, kg*
CD Channel Designator *mt*
CD Chemical Defense [chemical warfare] *wp*
CD Civil Defense *mt, fu, w9*
CD Classification of Defects *fu*
CD Clearance Directives *fu*
CD Collision Detection *fu, kg, hi*
CD Color Display *hi, kg*
CD Command Decoder *ggIII.1.52*
CD Command Destruct *tb*
CD Command Director *hc*
CD Commercial Equipment *jj*
CD Committee Draft *ch*
CD Committee on Disarmament *-14*
CD Common Digitizer *fu*
CD Community Designator *fu*
CD Compact Disk *ct, em, nd74.445.5, wp, kg, hi*
CD Components Department *fu*
CD Concept Development Definition *tb, jmj, fu, st*
CD Conceptual Design *fu*
CD Conference on Disarmament *wp*
CD Constant Drive *jj*
CD Contract Data *8, hc*
CD Contractor Definition *mt, hc, tb, jj, kf, fu*
CD Contractor Demonstration *hc, jj*
CD Control Document *fu*
CD Controls and Displays *jj*
CD Corps Diplomatique [French diplomatic corps] *w9, wp*
CD Counter Drug [USA DOD] *mt*
CD Critical Design *fu*
CD Current Dollars *mt*
CD+G Compact Disk plus Graphics *hi, kg*
CD-DA Compact Disc - Digital Audio *kg, ct, em, hi*
CD-E Compact Disk - Erasable *pk, kg*
CD-I Compact Disk - Interactive *hi, kg*
CD-MO Compact Disk - Magneto Optical *hi, kg*
CD-R Compact Disc - Recordable *kg, ct, em, hi*
CD-RDx Compact Disk - Read Only Memory [Data Exchange Standard] *kg, mi, fu, ge*
CD-ROM Compact Disc - Read Only Memory *mt, ns, hg, mb, ct, em, hi*
CD-ROMXA Compact Disk - Read Only Memory Extended Architecture *hi*
CD-RTOS Compact Disk - Real Time Operating System *hi, kg*
CD-RW Compact Disc, Re-Writable *ct, em*
CD-V Compact Disk - Video *hi, kg*
CD-WO Compact Disk - Write Once *hi, kg*
CD-XA Compact Disk - Extended Architecture *hi, kg*
CD/UAS Channel Divider/User Assignment Switch *-4*
CDA Central Design Activity *mt*
CDA Central Design Authority *es55.72*
CDA Combat Development Agency *fu*

CDA Command Data Administration *mt, wp*
CDA Command [and] Data Acquisition [system] *fu, gp, hc, hy, oe*
CDA Compound Document Architecture [DEC] *pk, kg*
CDA Critical Design Audit *kf*
CDA [Hughes Aircraft Company customer for TDM Time Division Multiplex] *hc*
CDAA Circularly Disposed Antenna Array *-10*
CDAP Centrally Directed Audit Program *mt*
CDAS Command and Data Acquisition Station [Wallops Island, Virginia] *bl3-1.59, pbl, oe*
CDAU Computer Data Adapter Unit *kf*
CDAW Coordinated Data Analysis Workshop *ns*
CDAY Commencement of Deployment Date *mt*
CDB Central Data Bank *mt*
CDB Central Data Buffer *fu*
CDB Chemical warfare Decontaminant Sodium Dichloroisocyanurate *wp*
CDB Clock Divider Buffer *jdr*
CDB Command Data Base [MSC/Navy] *mt*
CDB Command Data Buffer *wp*
CDB Configuration Baseline Document {?} *tb*
CDB Corporate Data Base *mt*
CDB Current Data Bit *fu*
CDBDD Computer Data Base Design Document *fu*
CDBS Configuration Data Base System *fu*
CDC Career Development Course *mt, jj*
CDC Center for Disease Control Atlanta *wp*
CDC Centre de Detection et de Control [French center for detection and control] *-12*
CDC Classified Document Control *se1.8.2, hc, jj*
CDC Coded Data Channel *fu*
CDC Combat Development Command [Army] *fu*
CDC Command and Data Handling Console *0*
CDC Command Data Channel *mt*
CDC Command Destruct Control *tb*
CDC Computer Display Channel *fu*
CDC Configuration Data Control *wp*
CDC Confined Detonating Cord *aw130.11.71*
CDC Consolidated Duplicating Center *mt*
CDC Control and Delay Channel *ce*
CDC Control Data Corporation *ct, kg, fu, hc, pl, jb, nt, tb, st, jj*
CDC/CE Common Display Console/ Commercial Emulator *fu*
CDC/DCC Computer Display Channel/Display Channel Complex *fu*
CDCA Central Data Collection Agency *jj*
CDCF Cost Data Collection Format *hc*
CDCN Command Document Control Number *mt*
CDCN Contract Document Change Notices *mt*
CDD Certificate of Disability for Discharge *w9*
CDD Command Demodulator Decoder *gg55, 16*
CDD Common Data Dictionary *mt, fu*
CDD Competitive Design / Development *mt*
CDD Computer Driven Display *fu*
CDD Concept and Development Definition *tb*
CDD Connecting Devices Division [of IEG at Hughes Aircraft Company] *fu, hn41.18.3, hq5.2.4, hc*
CDD Courseware Design Document *fu*
CDDAE COPAT Digital Data Analysis Equipment *hc*
CDDB Central Demand Data Base [USA] *mt*
CDDC Company Document Distribution Center *15, ep, fu*
CDDF Central Data Distribution Facility [Marlow Heights, Maryland] *gd1-4, oe*
CDDFTS Controls-Displays-Doppler Filter Test Station *jj*
CDDI Copper Distributed Data Interface *kg, ct, em, hi*
CDDIS Crustal Dynamics Data Information System *ns*
CDDV Copilot's Display, Direct View *hc*
CDE Chemical Defence Establishment [chemical warfare] *wp*
CDE Committee on Disarmament, Europe *-14*
CDE Common Desktop Environment *hi, kg*
CDE Complex Data Entry *kg, fu*
CDE Computer Design and Engineering *hi*
CDEC Combat Development Evaluation Center *hc*
CDEC Combat Developments Experimentation Command *hc*
CDEMR Canada Department of Energy, Mines and Resources *hg, ge*
CDEP Common Data Extraction Program *wp*
CDF Calibration Data Feedback *fu*
CDF Central Data Facility *oe*
CDF Channel Definition Format *kg, em*
CDF Combat Direction Finding *mt*
CDF Comma Delimited Format *hi*
CDF Command Data Formatter *oe*
CDF Commander of Defense Forces *fu*
CDF Common Data Format *ns*
CDF Confined Detonating Fuse *lt*
CDF Cumulative Distribution Function *aa26.11.37, wp, ai1*
CDFS CD-ROM Compact Disk File System [Microsoft] *ct, kg, em, hi*
CDG Capacitor Diode Gate *fu*

CDG Central Display Generator *hc*
CDG Coder Decoder Group *hc, fu*
CDG COMPOOL Data Generator *fu*
CDG Concept Development Group *tb, wl*
CDH Command and Data Handling *re*
CDHC Command and Data Handling Console *hc*
CDHF Central Data Handling Facility *hy, oe*
CDI California Disability Insurance *sa816391B*
CDI Compact Disc Interactive *ct, em*
CDI Computer to Disk Interface *fu*
CDI Connecting Devices, Incorporated [Long Beach, California] *mj31.5.15*
CDI Conventional Defense Initiative *mt, wp, ic2-19*
CDIA Certified Document Imaging Architect *kg, ct*
CDIF Component Development and Integration Facility [Butte, Montana] *aa26.12.25*
CDIP Consolidated Defense Intelligence Program *mt*
CDITEG Common Digital Television Generator *fu*
CDIU Cradle Driver Interface Unit *gp*
CDL Capacitor Diode Logic *fu*
CDL Centre De Lancement [Ariane launch control center] *ai2*
CDL Compliance Document List *kf*
CDL Computer Design Language *pk, kg*
CDM CIP Device Management *fu*
CDM Code Division Modulation *mt*
CDM Code Division Multiple Access *-17*
CDM Color Display Monitor *fu*
CDM Command Data Message *fu*
CDM Common Data Model *fu*
CDM Compliance Demonstration Matrix *oe*
CDM Concept Decision Model *wp*
CDM Configuration [and] Data Management *fu, pbl, ah*
CDM Contract Data Management *fu*
CDM Contractor Developed Material *hc, jj*
CDM Control Data Message *fu*
CDMA Code Division Multiple Access *fu, ct, em, gs1-5.90, ce, jdr, kg, hi*
CDMA Configuration and Data Management Administrator *fu*
CDMD Configuration and Data Management Department *jj, fu*
CDMF Commercial Data Masking Facility [IBM] *pk, kg*
CDMM Chemical Defense Multipurpose Mask [chemical warfare] *wp*
CDMO Configuration and Data Management Operations *fu*
CDMPCS Contract Depot Maintenance Production and Cost System *mt*
CDMS Contracting Data Management System *mt, wp*
CDMS Control and Data Management System *gp, oe*
CDMS Crisis Deployment Management System [USA] *mt*
CDNATS FEMA {Civil Defense ?}National Teletypewriter System *mt*
CDNAVS FEMA {Civil Defense ?}National Voice System *mt*
CDNTS Civil Defense National Telephone System *mt*
CDO Command Center Duty Officer [watch supervisor] *mt*
CDOS Concurrent Disk Operating System *pk, kg*
CDOS Customer Data and Operations System *hc, hy, tb*
CDOVHL Crash Damage Overhaul *jj*
CDP Central Data Processor *mt, wp*
CDP Certificate in Data Processing *mt, w9*
CDP Communications and Data Processing *hn49.3.5*
CDP Communications Data Processor *mt*
CDP Compressor Discharge Pressure *wp*
CDP Computer and Data Processing *fu*
CDP Computer Diagnostic Program *fu*
CDP Conditioned Diphase *fu*
CDP Configuration Definition Phase *hc*
CDP Contract Definition Phase *fu, hc*
CDP Control and Display Panel *fu*
CDP Convening Date Planned *fu*
CDP Crustal Dynamics Project *hy*
CDPC Course Data Processing Code *fu*
CDPD Cellular Digital Packet Data Delivery *kg, jdr, ct, em, jr, hi*
CDPD Chemical Defense Planning Document *mt*
CDPF Central Data Processing Facility *ac2.2-13, hc, jj*
CDPIR Crash Data Position Indicator Recorder *wp*
CDPS Central Data Processing System *-15*
CDPS Corporate Data Processing Service *fu*
CDR Call Detail Record Recording *pk, kg*
CDR Central Data Repository *fu*
CDR Classified Document Receipt *fu*
CDR Combat Deployable Radio [USAF] *mt*
CDR Command Destruct Receiver *fu, ai2*
CDR Commander *mt, w9, ws4.10, wp, jj*
CDR Conceptual Design Review *pbl, tm, jdr, ah*
CDR Contract Data Requirement *wp*
CDR Contractor Deficiency Report *jdr, tm*
CDR Crash Data Recorder *aw130.11.71*
CDR Critical Design Review *mt, fu, ah, gp, 4, 8, 15, 16, 17, sb, hc, nt, lt, wp, ep, bf, re, tb, st, wl, jj, pbl, jdr, tm, tm, kf, oe*
CDR Current Directional Relay *fu*

CDRAM Cached Dynamic RAM *hi*
CDRL Contract Data Deliverable Requirements List [a document] *17, sb, H, hc, pl, nt, lt, wp, ep, dm, tb, st, jj, jdr, pbl, tm, kf, fu, kg, ah, uh*
CDRL Contract Data Requirements List *mt*
CDRL Contract Documents Requirements List *gp, 16, oe*
CDRL Contractor Data Requirement List *vv*
CDROM Compact Disk Read Only Memory *wp, ai1*
CDRP Contract Data Requirements Plan *hc*
CDRS Container Design Retrieval System *fu*
CDRS Control and Data Retrieval System *fu*
CDRS Crash Data Recorder Subsystem *hc*
CDRT Combat Damage Repair Time *hc*
CDRT Component Defect Review Team *fu*
CDRV Copilot's Display, Remote View *hc*
CDRWS Critical Design Review Work Sheets *hc, jj*
CDS Canadian Defense Sector Staff *mt*
CDS Capability Design Specification *mt*
CDS Central Data Subsystem [Shuttle] *mt*
CDS Centralized Data Systems [F-16] *mt*
CDS Centre de DonnÈes astronomiques, Strasbourg [French astronomical data center, Strasbourg] *es55.66, ns*
CDS Combat Data Systems *fu*
CDS Combat Direction System [US Navy] *fu, mt*
CDS Command Data System [on VRM and VOIR] *pl*
CDS Command Direct Support *mt*
CDS Commander's Data Summary *mt*
CDS Communications Deception System *hc*
CDS Computer Data System *hc, jj*
CDS Container Delivery System *mt*
CDS Control Diagnostic System *oe*
CDS Correlated Double Sampling *hg, ge*
CDS Countermeasure Dispenser Systems *mt*
CDS Coupled to a Differential Synchro *hc*
CDS Current Directory Structure *pk, kg*
CDS Current Document Status *fu*
CDSA Common Data Security Architecture *pk, kg*
CDSAR Course Development and Student Administration / Registrar Keeping System [USAF] *mt*
CDSF Combat Developer Support Facility *fu*
CDSL Current Document Status List *fu*
CDSP Communications and Digital Signal Processing *is26.1.86*
CDSR Cost Data Summary Report *fu*
CDSS Canadian Department of Supply and Services *hc, tb*
CDSS Communications Digital Switching System *hc*
CDSS Compressed Data Storage System *fu*
CDSV Composite Data Service Vendor *do34*
CDT Calibrated Data Tape *tb*
CDT Central Data Terminal *mt*
CDT Central Daylight Time *w9, wp, kf, kg*
CDT Color Display Terminal *fu*
CDT Command Destruct Transmitter *fu*
CDT Consecutive Duty Tour *wp*
CDT Contract Definitization Team *kf*
CDT Contractor Demonstration Test *tb, jj*
CDT Contractor Development Test *jj*
CDT Control Data Terminal *fu*
CDT Coordinate Data Terminal *fu, hc*
CDT Crew Duty Time *wp*
CDT&E Contractor Development, Test and Evaluation *fu*
CDTS Computer Directed Training System [WWMCCS] *mt*
CDTS Controls and Displays Test Station *jj*
CDTUC Conventional Data Terrain Unit Cartridge *mt*
CDTV Compact Disk Television *hi*
CDU Central Distribution Unit *fu*
CDU Command Decoder Detector Decoding Unit *gg52, uh, mb, re, cl, jdr, kf, oe*
CDU Control Data Unit *hc*
CDU Control [and] Display Unit *mt, fu, hc, wp*
CDV Cell Delay Variation *ct, em*
CDV Comma Delimited Values *hi*
CDV Correlation Decision Verification *fu*
CDVM Capacitor Diode Voltage Multiplier *hc*
CDVT Cell Delay Variation Tolerance *ct, em*
CDW Command Data Word *fu*
CDWO Civil Defense Warning Operator *fu*
CDWS Civil Defense Warning System *mt*
CDX Control Differential Transmitter *fu*
CD_CHRDY Card Channel Ready [IBM] *pk, kg*
CE Cache Enable *pk, kg*
CE Calibration Equipment *-4*
CE Capital Equipment *jj*
CE Carbon Equivalent *wp*
CE Cargo Element *jdr*
CE Chemical Energy *mt*
CE Chip Enable *pk, kg*
CE Civil Engineering *fu, mt, wp, w9*

CE   Collision Elimination   *hi, kg*
CE   Command Element [USMC]   *fu, mt*
CE   Command Executioner   *pl*
CE   Commercial Emulator   *fu*
CE   Common Emitter   *fu*
CE   Common Equipment   *fu*
CE   Communications Electronics   *fu, tb, mt*
CE   Communications Emulator   *fu*
CE   Communications Equipment   *fu, jdr*
CE   Commutator End   *fu*
CE   Component Engineering   *fu*
CE   Concept Exploration   *wp*
CE   Concurrent Engineering [process]   *jdr, tm*
CE   Conducted Emission   *fu*
CE   Conducted Emissions   *oe*
CE   Consumer Electronics   *ct, em, hi*
CE   Control Electronics   *jdr*
CE   Control Element   *fu*
CE   Convert Enable   *pk, kg*
CE   Corps of Engineers [US Army]   *wp, jj*
CE   Current Estimate   *hc*
CE   Customer Engineer   *mt*
CEA   Canadian Electric Association   *y2k*
CEA   Circular Circuit {?} Error Average   *hc, fu, jj*
CEA   Commisiariat ‡ l'Energie Atomique [atomic energy commission] [France]   *-12*
CEA   Complete Effort Authorized   *hc, jj*
CEA   Core End Address   *jj*
CEAA   Commandement des Ecoles de l'ArmÈe de l'Air [Air training command of the Air Force] [France]   *mt*
CEAC   Comprehensive Estimate At Completion   *jdr*
CEAM   Centre d'Experimentations AÈriennes Militaires [military aeronautical experiment center, at Mont-de-Marsan] [France]   *mt*
CEAP   Corps of Engineers Automation Plan   *mt*
CEAR   Core End Address Register   *hc*
CEAS   Center for Environmental Assessment Services   *ag*
CEAT   Aeronautic Test Center of Toulouse   *fu*
CEAT   Centre d'Essais Aeronautiques de Toulouse [aeronautical trials center in Toulouse] [France]   *mt*
CEB   Change Engineering Board   *jdr*
CEB   CNO Executive Board   *jdr*
CEB   Combined Effects Bomblet   *wp*
CEB   Communications Electronics Board [NATO]   *mt*
CEB   Consumer Electronic Bus   *em*
CEBM   Corona, Eddy Current, Beta Ray, Microwave   *0*
CEBZ   Central European Buffer Zone   *mt*
CEC   Central Electronics Complex   *fu*
CEC   Components Evaluation Center   *fu*
CEC   Consolidated Electrodynamics Corporation   *0*
CEC   Cooperative Engagement Capability   *kf*
CECAS   Civil Engineer Combat AFSC Simulation [USAF]   *mt*
CECAT   Combat Enhancing Capability Aviation Team [USA]   *mt*
CECM   Contractor Executive Committee Meeting   *fu*
CECOM   Communications [and] Electronics Command [Fort Monmouth] [US Army]   *mt, tc, hc, fu*
CECORS   Civil Engineering Contract Reporting System [USAF]   *mt*
CECP   Compatibilty Engineering Change Proposal   *wp*
CED   Components Engineering Department   *jr*
CED   Console Electronics Drawer   *fu*
CED   Continuing Engineering Development   *wp*
CEDARS   Communications Electronics Delivery and Retrieval System   *hc*
CEDE   Critical Event Discrimination Experiment   *hc*
CEE   Commercial Equivalent Equipment   *fu*
CEE   Conducted Electromagnetic Emissions   *jj*
CEEE   Common Electronic Equipment Enclosure   *fu*
CEEIA   Communications Electronics Engineering Installation Agency   *dt*
CEERS   Command Excess Equipment Redistribution System   *mt*
CEES   Chloroethyl Ethylsulfide [half mustard; chemical warfare simulant]   *wp*
CEESIM   Combat Electromagnetic Simulator   *wp*
CEF   Career Executive Force   *wp*
CEF   Carrier Elimination Filter   *fu*
CEF   Centralized Environmental Facility   *mt*
CEF   Civil Engineering Files [JRS]   *mt*
CEFD   Clutter Equivalent Flux Density   *kf*
CEFR   Communications Electronics Facility Records   *mt*
CEFSR   Committee for Evaluating the Feasibility of Space Rocketry   *ci.37*
CEG   Combat Evaluation Group [USAF]   *wp, mt*
CEG   Continuous Edge Graphics   *pk, kg*
CEG   Control Equipment Group   *fu*
CEG   Converter Equipment Group   *fu*
CEGE   Combat Equipment Group, Europe [USA]   *mt*
CEI   Coherent Error Integrator   *oe*
CEI   Communications Electronic Instructions   *0*

CEI   Component End Item   *jj*
CEI   Components Engineering Instruction   *fu*
CEI   Conducted Electromagnetic Interference   *pk, kg*
CEI   Configuration End Item   *mt*
CEI   Contract End Item   *fu, gp, ah, 8, hc, pl, jj, jdr, pbl, tm, oe*
CEI   Critical End Item   *jj*
CEID   Communications Electronics Implementation Document   *mt*
CEIP   Communications Electronics Implementation Plan   *mt*
CEIPA   Communications Electronics Implementation Plan Amendment   *wp*
CEIS   Communications Equipment List {?}   *mt*
CEIS   Contract End Item Specification   *hc, jj*
CEIS   Cost and Economic Information System [later CIR]   *hc*
CEIT   Cost Engineering Integration Tool   *kf*
CEL   California Eastern Laboratories [Santa Clara, California]   *mj31.5.51*
CEL   Centre d'Essais des Landes [Landes missile testing center] [France]   *mt, aw130.13.34, 12, crl*
CEL   Colorado Engineering Laboratories [of Hughes Aircraft Company]   *pl*
CEL   Components Engineering Laboratory   *fu*
CEL-MP   Chief Engineer, Logistics - Manpower and Personnel   *fu*
CELM   Contract ELM Logistics Readiness Support   *mt*
CELS   Controlled Environment and Life Support System   *wl*
CELTS   Coherent Emitter Location Test Targeting System   *fu, wp*
CELV   Complementary Expendable Launch Vehicle   *mt, tb, ai2*
CEM   Captured Enemy Material   *wp*
CEM   Centre d'Essais en Mediterranee   *crl*
CEM   Combined Effects Munition   *mt, aw118.14.11, wp*
CEM   Communications Electronics Meteorological [equipment status and inventory system] [USAF]   *mt*
CEM   Communications Equipment Maintenance   *fu*
CEM   Communications, Electronics Meteorological   *hc*
CEM   Concepts Evaluation Model [USA]   *wp, mt*
CEMARS   COMSEC Equipment Modification Application Reporting System   *mt*
CEMAS   Civil Engineering Materials Acquisition System [USAF]   *wp, mt*
CEMES   Communication, Electronic and Meteorological Equipment Status   *mt*
CEMM   Compaq Expanded Memory Manager   *hi*
CEMO   Command Equipment Management Office [USAF]   *wp, mt*
CEMO   Communications Electronics Mission Order   *mt*
CEMPDS   Combined Engineering Manufacturing Product Definition System   *fu*
CEMS   (if you are using Photodiodes, PM tubes, CEMS or MCPs)   *es55.6*
CEMS   Central Electronic Management Systems [SST]   *hc*
CEMS   Chloroethyl Methyl Sulfide [chemical warfare Simulant]   *wp*
CEMS   Comprehensive Engine Management System [USAF]   *mt*
CEMS   Constituent Electronic Mail System   *pk, kg*
CEMT   Command Equipment Management Team   *mt, wp*
CENEN   Corps of Engineers National Energy Network [USA]   *mt*
CENPAC   Central Processor and Controller [unit]   *hc*
CENSEI   Center for System Engineering and Integration [USA]   *mt*
CENTACS   Center Tactical Computer Sciences [Army]   *hc*
CENTAF   Central Command, Air Forces [US 9th AF]   *mt*
CENTAG   Central Army Group [Central Europe NATO Command]   *mt, 12, wp, fu*
CENTCOM   Central Command [McDill AFB, Florida] [USA]   *mt, 12, jb*
CENTO   Central Treaty Organization [later defunct]   *mt, hc, su, w9, wp, tb, fu*
CENTREX   [telephone switch]   *mt*
CENTRO   Central Treaty Authorization   *jj*
CEO   Chief Executive Officer   *ah, aw130.13.41, sc3.5.1, 17, w9, hn49.9.4, af72.5.175, wp, pbl, kf, fu, hi, y2k*
CEO   Communications Electronics Officer   *mt*
CEOI   Communication Electronics Operational Operating Instruction   *fu, mt*
CEOP   Conditional End Of Page   *pk, kg*
CEOPL   Civil Engineer Operational Project List [MAC]   *mt*
CEOS   Committee on Earth Observing Satellites   *hy*
CEP   Circular Error Probability Probable [the radius within which 50% of the launched weapons hit]   *mt, fu, 5, 11, 13, 14, 17, ph, sb, hc, nd74.445.30, wp, tb, st, jj, kf, vv, oe*
CEP   Civil Engineering Plan   *mt*
CEP   Clock Enable Pulse   *jj*
CEP   Combat Entry Point   *mt*
CEP   Combination Entry Panel   *fu*
CEP   COMSEC Equipment Program   *mt*
CEP   Concept Exploration Phase   *fu*
CEP   Contract Execution Plan   *mt*
CEPDR   Communications Electronics Post Deployment Requirement   *mt*
CEPG   Command Execution Planning Group   *mt*
CEPIS   Civil Engineering Project Information System [USAFE]   *mt*
CEPRO   CEP Reporting Official   *mt*
CEPS   Center Exploration Program Scientist [NASA]   *hy*

CEPS Central European Pipeline System *mt*
CEPS CITS Expert Parameter System *wp*
CER Canonical Encoding Rules *pk, kg*
CER Character Error Rate *mt, fu*
CER Chip Error Rate *jdr*
CER Civil Engineering Report *wp*
CER Complete Engine Repair *jj*
CER Component Engineering Request *fu, hc, wp*
CER Cost Effectiveness Ratio [also C/E] *ic2-17*
CER Cost Estimating Estimate Relationships *fu, 16, sb, hc, go3.5.1, wp, tb, st, wl, kf*
CER Cost Exchange Ratio *17, sb*
CERC Coastal Engineering Research Center [US Army] *tc*
CERCOM Communications and Electronics Materiel Readiness Command [Army] *fu, hc*
CERCS Central Engineering Records Control System *fu*
CERES Clouds and Earth's Radiant Energy System [EOS] *hy, ge, hg*
CERF Complementary Error Function *wp*
CERFNET California Educational Research Network *ct*
CERIS European Center for International Relations and Strategy *tb*
CERL Construction Engineering Research Laboratory [US Army] *hc, tc*
CERN Conseil EuropÈen pour la Recherche NuclÈaire [French European Council for Nuclear Research] *wp, ac3.7-1, kg, hi*
CERR Coherence Enhanced Radiation Rejection *17, sb*
CERS Carrier Evaluation and Reporting System *mt*
CERS {Centre EuropÈen de Recherche Spatiale} [European space research organization, also known as ESRO] *es55.91*
CERT Character Error Rate Test *fu*
CERT Combined Electrical Readiness Test *kf*
CERT Combined Environmental [and] Reliability Test Testing *fu, wp, hc*
CERT Communications-Electronics Readiness Team *mt*
CERT Component Evaluation and Reliability Testing *nd74.445.27*
CERT Composite Electrical Readiness Test *oe*
CERT Computer Emergency Response Team *kg, ct, kb, em*
CERV Crew Emergency Reentry Vehicle *aa26.12.53*
CES Candidate Evaluation System *fu*
CES Capacity Engineering Support *hc*
CES Circuit Emulation Services *ct, em*
CES Civil Engineering Squadron *mt*
CES Committee on Earth Sciences *hy*
CES Common Elements System *fu*
CES Communications Electronics Subsystem *vv*
CES Consumer Electronics Show *hi*
CES Contractor Engineering Support *hc*
CES Control Electronics Section *ci.392*
CES Cost Estimate System *hc*
CES/MIS Civil Engineer Support/Management Information System [US Navy] *mt*
CESAC Communications Electronics Scheme Accounting and Distribution Control system *mt*
CESD Circular Equivalent Standoff Distance *jdr*
CESE Communications Equipment Support Element *fu*
CESL Component Engineering and Support Laboratory [Hughes] *hm, ep*
CESM Cryptologic Electronic [Warfare] Support Measures [US Navy] *fu, mt*
CESP Civil Engineering Support Plan *mt, wp*
CESP Civil Engineering Support Program *mt*
CESP Command Element Specific Program *mt*
CESPG Civil Engineering Support Planning Generator [JOPS] *mt*
CESPS Civil Engineering Operational Planning System *mt*
CET Central European Time *wp*
CET Clutter Equivalent Target *kf*
CET Cumulative Elapsed Time *wp, jj*
CETA Comprehensive Employment and Training Act *sb, w9*
CETACS Cost Estimating and Tracking System *fu*
CETD Concept Exploration and Technology Demonstration [SDI] *mt*
CETI Communication with Extra-terrestrial Intelligence *fu*
CETS Contractor Engineering [and] Technical Services *mt, hc, fu, wp, jj*
CETU Computer Emulator Transfer Unit *jj*
CEU Camera Electronics Unit *hc*
CEU Common Electronic Unit *hc*
CEU Control Electronics Unit *kf, ric, jdr*
CEV Centre d'Essais en Vol [Flight testing center, Istres, France] [France] *mt*
CEV Corona Extinction Voltage *wp*
CEVG Combat Evaluation Group *mt, 12*
CEVM Consumable Electrode Vacuum Melt *gs2-9.40*
CEVR Circle of Equivalent Vulnerability Radius *sd*
CEVSV Centre d'Entrainement en Vol Sans VisibilitÈ [IFR training center] [France] *mt*
CEWI Combat Electronic Warfare and Intelligence [USA] *wp, fu, mt*
CEWS Cost Estimated Work Sheets *hc, jj*
CEXP Control Exercise Processor *fu*
CE_ROCC Canada East Regional Operational Control Center [SPACECMD] *mt*

CF Carrier Frequency *fu*
CF Carry Flag *ct*
CF Cathode Follower *jj, fu*
CF Cellular Frequency *hi*
CF Centrifugal Force *fu, w9, wp, jj*
CF Communications Flight *mt*
CF Complementary Function *cr*
CF Compression Filter *fu*
CF Control Facility *mt*
CF Control Frame *fu*
CF Conventional Forces *dh*
CF Counter Force *mt, tb*
CF Crew Factor *mt*
CF&I Contractor Furnish and Install *fu*
CFA Center For Astrophysics [Harvard Smithsonian] *ns, mi*
CFA Chartered Financial Analyst *hn44.23.2*
CFA Cognizant Field Activity *jj*
CFA Color Filter Array *hi*
CFA Combined Field Army [US/ROK] *fu, mt*
CFA Contract Funds Available *fu*
CFA Contractor Furnished Accessories *jj*
CFA Contractual Funding and Authorization *fu*
CFA Crossed Cross Field Amplifier *fu, hc, jj*
CFAE Contractor Furnished Accessory Equipment *fu*
CFAE Contractor Furnished Aeronautical Aerospace Equipment *hc, wp, jj, fu*
CFAR Constant False Alarm Rate [/Receiver {?}] *sa830182P.10, hc, jj, fu*
CFAS Chaplain Fund Accounting System *mt*
CFAT Captive Flight Acceptance Test *hc, jj*
CFAW Comfair Airwing Pacific *jj*
CFB Canadian Forces Base *mt, 12, aw*
CFB Cipher Feedback *pk, kg*
CFB Configurable Function Block *pk, kg*
CFC Central Fire Control *fu*
CFC Central Flow Control *fu*
CFC Chloro Fluoro Carbon *wp, mi, ge, hn50.8.i, ns, hy, mb, hg, kf*
CFC Combined Federal Campaign *wp*
CFC Combined Forces Command [Korea] [US / ROK] *wp, fu, mt*
CFCC Central Flow Control Complex *fu*
CFCF Central Flow Control Function *fu*
CFCS Command Fund Control System *mt*
CFD Captive Flight Demonstration *hc*
CFD Chaff Flare Dispenser *jj*
CFD Computational Fluid Dynamics *aa26.12.b8, aw129.10.74, wp, re, st, te23, 11, 28, ai1*
CFD Computer Fire Detection *fu*
CFD Coordinated Firing Doctrine *tb*
CFDS Composite Force Requirements and Development System *mt*
CFDSS Combat Follow on Spares Support *mt*
CFE Central Fighter Establishment [UK] *mt*
CFE Consolidated Front End *mt, 17*
CFE Contractor Furnished Equipment *mt, fu, gp, hc, wp, tb, jj, kf, oe*
CFE Conventional armed Forces in Europe treaty [a force reduction treaty] *mt*
CFE Conventional Forces In Europe *wp*
CFE Customer Furnished Equipment *fu, gp, jdr*
CFEN Contractor Furnished Equipment Notice *hc, jj, kf*
CFES Continuous Flow Electrophoresis Separation *sc4.6.6*
CFF Call For Fire *fu*
CFF Columbus Free Flyer *mi*
CFF Critical Flicker Frequency *fu*
CFF Critical Fusion Frequency *hc*
CFHT Canada-France-Hawaii Telescope *mi*
CFI Call For Improvements *kf*
CFI Card Format Identifier *fu*
CFI Certified Flight Instructor [certificate] *w9, aw130.11.74*
CFI Chief Flying Instructor *w9*
CFI Contract to Furnish Information *tb*
CFI Contractor Furnished Information *wp, tb*
CFI Contractor Furnished Item *fu*
CFI Customer Furnished Information *fu*
CFIT Controlled Flight Into Terrain *mt*
CFL Calibrated Focal Length *fu*
CFL Courant Friedrichs Lewy *ai1*
CFLOS Cloud Free Line Of Sight *hc, jj, kf*
CFM (CFM International, a joint company of GE . . . and SNECMA) *aw132.10.6*
CFM Captive Flight Model/Missile *jj*
CFM Cathode Follower Mixer *fu*
CFM Center Frequency Modulation *fu*
CFM Code Fragment Manager [Macintosh] *pk, kg*
CFM Contractor Furnished Material *fu*
CFM Cubic Feet per Minute *wp, jdr, hi, fu*
CFM Customer Furnished Material *fu*
CFMCON [US] Freight Management [Program] [MTMC] *mt*

CFMI CFM International [a joint company of GE and SNECMA] *aa26.12.43*
CFMR Coherent Frequency Multiplexed Radar *hc, fu*
CFMS Centralized Fuels Management System *mt*
CFMS Combat Fuels Management system [USAF] *mt*
CFO Chief Financial Officer *hn49.8.8, fu*
CFO Consolidated Front End *sb*
CFOSS Combat Follow On Supply System *mt*
CFP Computer Freedom and Privacy *ct, hi*
CFP Computerized Flight Plan *mt*
CFP Concept Formulation Plan Package *hc, fu, tb, st, jj*
CFP Contractor Furnished Property *fu, wp, jj*
CFP Customer Furnished Property *fu, jdr*
CFP/FSM Computer Flight Plan / Flight Simulation Model [AWS] *mt*
CFPD Command Flight Path Display *hc*
CFPE Conventional / Special Force Planning and Execution *mt*
CFPM Contract Financial Reporting Manual *hc, jj*
CFPS Cubic Feet Per Second *hi*
CFPTS Coolant Flow and Pressure Test Station *jj*
CFR Carbon Fiber Reinforced *wp*
CFR Carbon Film Resistor *wp, fu*
CFR Code of Federal Regulations *mt, fu, hc, nt, tb, jj*
CFR Computerized Facial Recognition *pk, kg*
CFR Computerized Fault Reporting *mt*
CFR Contract Funds Reserve *kf*
CFR Cumulative Failure Rate *wp*
CFRC Continuous Fiber Reinforced Composites *wp*
CFRDS Composite Force Requirements and Deployment System *mt*
CFRE Contractor Financial Requirements Estimate *hc*
CFRP Carbon Fiber Reinforced Plastic *mt, es55.23, ai2*
CFS Caching File System *pk, kg*
CFS Canadian Forces Station *-12*
CFS Carrier Frequency Shift *fu*
CFS Center Frequency Stabilization *fu*
CFS Central Flying School [UK] *mt*
CFS Collection and Forwarding System *mt*
CFS Common File System *tb, kg*
CFS Contractor Furnished Software *tb*
CFS Contract[or] Field Service *hc, jj*
CFS Conventional Fighter Simulator *fu*
CFS Critical Field Strength *wp*
CFS Cubic Feet per Second *wp*
CFS Customer Field Service *jj*
CFS Customer Furnished Services *fu*
CFS Customer Furnished Software *kf*
CFSE Customer Furnished Support Equipment *jj*
CFSR Contract Field Service Representative *hc*
CFSR Contract Funds Status Report *mt, 17, nt, dm, tb, st, jj, jdr, kf, fu*
CFT Captive Flight Training *jj*
CFT Charge Flow Transistor *wp*
CFT Conformal Fuel Tank *af72.5.135, wp*
CFT Cray FORTRAN Compiler *tb*
CFT Crystal Field Theory *wp*
CFTE Contractor-Furnished Test Equipment *fu*
CFTMATH Cray Mathematical Library *tb*
CFU Captive Flight Unit *hc*
CFV Cavalry Fighting Vehicle *hc*
CFWP Central Flow Weather Processor *fu*
CFWSU Central Flow Weather Service Unit *fu*
CFY Current Fiscal Year *mt, wp, tb*
CG Center of Gravity *mt, gp, 4, 9, aa26.10.25, hc, w9, aw129.21.41, re, st, jj, jdr, kf, mb, mi, vv, oe, he, uh, fu, ai1*
CG Coast Guard *mt, w9*
CG Command Guidance *wp*
CG Commanding General *mt, fu, wp*
CG Communications Gateway *mt, fu*
CG Comptroller General *wp*
CG Consolidated Ground *kf*
CG Consolidated Guidance *mt, wp*
CG Control Gate *pk, kg*
CG Cruiser, Guided missile [US Navy ship classification] *mt, wp, jj, fu*
CG [chemical warfare codename for phosgene] *wp*
CGA Color Graphics Adapter *wp, ct, em, kg, hi*
CGA Computer Graphics Adapter *fu*
CGA Configurable Gate Array *ne, fu, rm31.1.3*
CGA Controlled Gain Amplifier *fu*
CGAD Collins Government Avionics Division *fu*
CGAS Coast Guard Air Station [USA] *mt*
CGC Coast Guard Cutter *mt*
CGCC Command, Ground Component Command *mt*
CGCO Cincinnati Gear Company [Cincinnati, Ohio] *nd74.445.2*
CGE Common Graphics Environment *pk, kg*
CGE Computer Generated Images *pk, kg*
CGFMF Commanding General, Fleet Marine Force *jj*
CGH [Light aircraft carrier / aviation cruiser] [a cruiser carrying missiles, helicopters and possible VSTOL aircraft] [USA] *mt*

CGI Cognizant Government Inspector *hc, jj*
CGI Common Gateway Interface *kg, ct, em*
CGI Computer Generated Imagery *wp*
CGI Computer Graphics Interface *fu, kg, em, hi*
CGIDS Commander's Graphics and Information Display System *mt*
CGLNO Coast Guard Liaison Officer *mt*
CGM (Missile, CGM-13 Mace) [USA] *mt*
CGM Computer Graphics MetaFile *mt, ct, em, aw130.22.65, hi*
CGMAW Commanding General, Marine Air Wing *jj*
CGMTI Coherent Ground Moving Target Indication *hc, jj*
CGMTT Coherent Ground Moving Target Tracking *hc*
CGN Cruiser, Guided Missile, Nuclear powered [USN ship classification] *mt, hc, jj, fu*
CGO Company Grade Officer *an31.13.4*
CGOC Company Grade Officer Council *an31.13.4*
CGP Central Ground Point *jj*
CGP CITS Ground Processor *wp*
CGP Common Ground Point *fu*
CGPM Conference GÈnÈrale des Poids et Mesures [French General Conference on Weights and Measures] *fu*
CGRD Consolidated Ground Requirements Document *kf*
CGRO [Arthur Holley] Compton Gamma Ray Observatory [formerly GRO] *mi, cu*
CGRT Carbon Glass Resistance Thermometer *wp*
CGS Centimeter, Gram, Second [a system of units] *wp, fu, ai1*
CGS CONUS Ground Station [DSP] *mt, kf, wp*
CGSE Common Consolidated Ground Support Equipment *hc, jj, fu*
CGSEL Common Consolidated Ground Support Equipment List *wp, hc, jj*
CGSP Conventional Geometry Smart Projectile *hc*
CGT Consolidated Ground Terminal *fu*
CGU Countermeasures Generator Unit *fu*
CGWIC China Great Wall Industrial Corporation *ai2*
CG[N] Cruiser, Guided Missile [Nuclear] *mt*
CH ConfÈdÈration Helvetique [Switzerland] *es55.22*
CH Corporate Headquarters *hc*
CH [designator, CH-53A is Sea Stallion] [USA DOD] *mt*
CHAD Combined Health Agencies Drive *hn43.21.8, hc*
CHAMP Competitive Health And Medical Plan *wp*
CHAMP Computerized HARVEST_BARE Asset Management Prototype *mt*
CHAP Challenge Handshake Authentication Protocol *kg, ct, em*
CHARA Center for High Angular Resolution Astronomy *mi*
CHARM Countermeasure Hardened FLIR Sensor *hc*
CHAS Chemical Hazard Assessment System [chemical warfare] *wp*
CHASE Combat History Analysis Study Effort [USA] *mt*
CHASIS Change Handling and Statusing Information System *fu*
CHASM COMCOR's Hybrid Adaptation of Sigma Monitor *jj*
CHAT Conversational Hypertext Access Technology [Internet] *pk, kg, hi*
CHB Cargo Handling Battalion, Navy [USA DOD] *mt*
CHB Circular History Buffer *jdr*
CHCK Channel Check *pk, kg*
CHCP Change Code Page *pk, kg*
CHCS Composite Health Care System [TRIMIS] *mt*
CHDIR Change Directory *pk, kg*
CHE Collimating Holographic Experiment *hc*
CHECKMATE Compact High Energy Capacitor Modular Advanced Technology Experiment *wp*
CHECMATE Compact High Energy Capacitor Module Advanced Technology Experiment *wp*
CHEMFET Chemical Field Effect Transistor *wp*
CHEMREP Chemical Attack/Incident Report *fu*
CHEOPS Chemical Operations System Study *hc*
CHESDIV Chesapeake Division [Naval Facilities Engineering Command] *dt*
CHEU Cradle Heater Electronics Unit *gp*
CHFN Change Finger [Unix] *pk, kg*
CHG Cruiser Helicopter with Guided missile armament [US Navy] *mt*
CHI Cloud Height Indicator *hc*
CHI Computer-Human Interface *fu*
CHIC Compact High-Intensity Cooler *fu*
CHIRP Configuration Handling and Information Retrieval Program *fu*
CHN Carbon, Hydrogen, Nitrogen *wp*
CHNAVMAT Chief of Naval Material *jj*
CHOD Chief Of Defense *fu*
CHODNORWAY Chief Of Defense, Norway *fu*
CHOP Change of Operational Control [USA DOD] *0, wp, mt*
CHOP Change of Operational Procedures *mt*
CHOWN Change Owner *pk, kg*
CHRP Common Hardware Reference Platform *hi, kg*
CHS Combat Support Hospital *mt*
CHS Commercial High Speed *kf*
CHS Common Hardware and Software *pm, fu*
CHS Composite Hospital System [TRIMIS] *mt*
CHS Cylinder Head Sector *hi, kg*
CHS Cylinders / Heads / Sectors *ct, em*

CHSF Cargo Hazardous Servicing Facility *jdr*
CHSTR Characteristics of Transportation Resources File [JOPS] *mt*
CHT Cylinder Head Temperature *aw130.13.39*
CHU CATIS Host Upgrade *mt*
CHX Cycloheximide *wp*
CI Card Input *fu*
CI Channel Interrupt *fu*
CI Chemical Ionization *wp*
CI Command Interpreter *fu*
CI Communication Identification *hc*
CI Communications Interface *mt*
CI Component Interface *pk, kg*
CI Computer Interface *pl*
CI Configuration Identification *st*
CI Configuration Item *mt, fu*
CI Configuration Item Index *16, 17, hc, wp, ep, dm, tb, jj, jdr, kf, oe, mt, fu*
CI Contract Identifier *8, hc*
CI Contract Item *fu, hc, wp*
CI Control Indicator *jj*
CI Control Item [hardware or software] *he, fu, gp, ep, pbl, tm*
CI Cost Improvement *sc9.6.1*
CI Counter Intelligence [USA DOD] *fu, mt*
CI Counterinsurgency *mt*
CI Critical Item *tb*
CI Customer Integration *wl*
CI/CPCI Configuration Item / Computer Program Configuration Item *mt*
CIA Central Intelligence Agency *mt, fu, ci.159, 0, 11, 13, hc, w9, aa26.11.b3, aw129.10.17. jb, wp, tb, st, jdr*
CIA Change Impact Analysis *fu*
CIA Chemiluminescence Immuno Assay *wp*
CIA Commander's Information Analysis *mt*
CIA Communications Interface Adapter *fu*
CIA Compagnia Industriale Aerospaziale [Italy] *hc*
CIA Computer Interface Adapter *hi*
CIA Current Instruction Address *pk, kg*
CIAC Career Information And Counseling *wp*
CIAC Computer Incident Advisory Capability *pk, kg*
CIACS Coded Integrated Armament Control System *jj*
CIAP Communication Improvement Augmentation Program *hc*
CIAPS Customer Integrated Automated Procurement System *hc*
CIAPS Customer Integrated Automated Purchasing System [USAF] *mt*
CIAT Crew Initiated Automatic Test *hc, jj*
CIBA Chemical Industry in Basle *wp*
CIC Cockpit Interface Computer *wp*
CIC Combat Information Center [USA DOD] *fu, hc, jj, mt*
CIC Combat Intelligence Center *mt, wp*
CIC Combined Intelligence Center *mt, kf*
CIC Computing Information Center *tb*
CIC Content Indicator Code [AUTODIN] *mt*
CIC Control Interface Controller *fu*
CIC Counter Intelligence Corps *fu*
CIC Cover Integrated Cell *oe*
CIC Cover Interconnect Cell *uh*
CIC Custom Integrated Circuit *fu*
CIC Customer Identification Code *wp*
CICA Competition In Contracting Act *mt, wp*
CICERO Communications Integrated Control, Engineering, Reporting, and Operations *hc, fu*
CICM Communications Interface Control Module *mt*
CICP Critical Items Control Plan *jdr*
CICS Combat Information Center System *fu*
CICS Commercial / Industrial Activities and Contract Services [USAF] *mt*
CICS Commercial Industrial Contract System *mt*
CICS Computer Information Control System *em*
CICS Customer Information Control System [IBM data base access system] *mt, do424, ct*
CICS/VS Customer Information Control System/ Virtual Storage [IBM] *hi, kg*
CICU Central Interface Converter Unit [later CSDC] *jj*
CICWG Commonality Interface Control Working Group *fu*
CID Charge Injection Device *fu, pl, 17, wp, kg*
CID Collision Induced Dissociation *wp*
CID Combat Intelligence Division *mt*
CID Combined Immunodeficiency Desease *wp*
CID Communications electronics Implementation Directive *mt*
CID Component Identification *fu*
CID Comprehensive Interior Design *mt*
CID Computer Identification *fu*
CID Configuration Identification *mt*
CID Configuration Identification Documentation *fu*
CID Configuration/Installation/Distribution *pk, kg*
CID Critical Issues Document *fu*
CID Critical Item Demonstration *fu*
CID Cryptologic Interface Device *fu*

CIDB CINCPACAF Integrated Data Base *mt*
CIDC Criminal Investigation Command [US Army] *mt, dt*
CIDEM Combat ID Evaluation Model *fu*
CIDIN Common ICAO Data Interchange Network *fu*
CIDR Classless Inter Domain Routing *kg, ct, em*
CIDR Configuration Item Development Record *fu*
CIDS Central ICAM Development System *fu*
CIDS Configuration Critical Item Development Specification *fu*
CIDS Contracting Information Data Base System *mt*
CIDSS CINCPACAF Integrated Display Support System *mt*
CIE Commission Internationale de l'Eclairage [French International Commission on Illumination] *fu, ch*
CIE Common Inspection Equipment *fu*
CIEES Centre Interarmees d'Essais d'Engins Speciaux *crl*
CIEH Centre d'Instruction des Equipages d'HelicoptÈres [center for instruction of helicopter personnel] [France] *mt*
CIEMS Catalog of Information Exchange and Message Standards *mt*
CIF Cal Tech Intermediate Format *fu*
CIF Cells In Frame *ct*
CIF Common Interchange Format *hi*
CIF Common Interchange/Intermediate Format *pk, kg*
CIF Convert Initialization Function *fu*
CIF Cost of Insurance and Freight *fu*
CIF Cost, Insurance, and Freight *w9, wp, jj*
CIF Crystallographic Information File *pk, kg*
CIF Customer Information Feed *pk, kg*
CIFAS (prime contractor, Symphonie communication satellites) *sc3.2.7*
CIFAS Centre d'Instruction de Force AÈrienne Strategique [training center of the strategic Air Force] [France] *mt*
CIFE Control Index File Europe [DOD] *hc*
CIFS Common Internet File System *em, kg*
CIG Communications Interface Group *fu*
CIG Computer Image Generator Generation Generated *mt, hn44.2.1, wp, fu*
CIGARS Console Internally Generated And Refreshed Symbols *hc, fu*
CIGS Computer Image Generation System *hc, hq6.2.7*
CIGTF Central Inertial Guidance Test Facility *hc, wp*
CIGUD Computer Image Generation Unifying Device *wp*
CII Compatibility, Interoperability and Integration *kf*
CII Configuration Identification Index *fu*
CIIL Configuration Item Identification List *oe*
CIIL Control Interface Intermediate Language *fu*
CIIWG Communications, Implementation, and Installation Working Group *mt*
CIL Clear Indicating Light *fu*
CIL Configuration Identification List *pl, oe*
CIL Critical Item List *fu, jdr, oe*
CILOP Conversion In Lieu Of Procurement *mt, hc, jj*
CIM Chemical Ionization Mass Spectroscopy *wp*
CIM Coffin launched Intercept Missile [a surface-to-Air missile that is fired from a horizontally stored box] [CIM-10 Bomarc] [USA] *mt*
CIM Common Information Model *ct, kg, em*
CIM Communications Improvement Memorandum [US Navy] *mt*
CIM Communications Interface Module *fu*
CIM Component Item Manager *wp*
CIM CompuServe Information Manager *hi, kg*
CIM Computer Input Microform *fu*
CIM Computer Integrated Manufacturing *fu, kg, mt, aw130.13.77, hn48.19.3, aa26.11.53, is26.1.52, wp, jdr, ai1*
CIM Configuration Identification Manual *fu*
CIM Control Interface Module *fu*
CIM Corporate Information Management *mt, wp*
CIM Critical Item Monitor *mt*
CIMEX Civil Military Exercise *mt*
CIMF Centralized Intermediate Maintenance Facilities *fu*
CIMS Chemical Ionization Mass Spectrometry *wp*
CIMS Command Information Management System *fu*
CIMS Coordination and Interference Management System *hc*
CIMSS CAATS Integrated Management Support System *fu*
CIMSS Cooperative Institute for Meteorological Satellite Studies [NOAA/NESDIS, Univ. Wisconsin] *oe*
CIMT Computer Integrated Manufacturing Techniques *hc*
CIN Cargo Increment Number *mt*
CIN Center Information Network *tb*
CIN Code Identification Number *fu*
CIN Command Information Network *mt*
CIN Commander's Information Network *wp*
CIN Component Configuration Identification Number *fu*
CINC Commander In Chief [USA] *mt, 11, ph, hc, aw129.1.47, w9, af72.5.31, jb, wp, tb, as, jdr, kf, fu*
CINC1_NET (Pacom Priority UHF SATCOM Jtf Voice Net) *mt*
CINCAD Commander In Chief, Aerospace Defense Command *mt*
CINCAD Commander In Chief, Air Defense *fu*
CINCAIR Commander In Chief, Air Base [USA] *mt*
CINCANT CINC Atlantic *kf*
CINCAREUR Commander In Chief, US Army Forces - Europe Command *mt*

CINCARLANT Commander In Chief, US Army Forces - Atlantic *mt*
CINCARRED Commander In Chief, US Army Forces - Readiness *mt*
CINCCENT Commander In Chief, Central Command *kf, mt*
CINCCFC Commander In Chief, Combined Forces Command *mt*
CINCCHAN Commander In Chief, Channel Command [NATO] *mt*
CINCEASTLANT Commander In Chief, Eastern Atlantic *fu*
CINCENT Commander In Chief, Allied Forces Central Europe [NATO] *fu, mt*
CINCENT Commander In Chief, USCENTCOM [USCINCCENT] *mt*
CINCEUR Commander In Chief, Europe [USA/USAF/USN] *mt, hc, fu, jj*
CINCHAN Commander In Chief, Channel [NATO Command] *fu, 12*
CINCLANT Commander In Chief, Atlantic [Norfolk, Virginia, USA] *mt, 12, hc, tb, fu*
CINCLANT/PAC Commander In Chief, Atlantic/Pacific *jj*
CINCLANTFLT Commander In Chief, Atlantic Fleet [US Navy] *jj, fu, mt*
CINCNAVBASE Commander In Chief, Naval Base [USA] *mt*
CINCNAVEUR Commander In Chief US Naval Forces Europe *mt*
CINCNAVSURFLANT Commander In Chief, Naval Surface Fleet, Atlantic [USA] *mt*
CINCNET CINC Network *kf*
CINCNET Commander In Chief, Internet *fu*
CINCNORAD Commander In Chief, North American Aerospace Defense Command *mt, jb*
CINCNORNAVEUR Commander In Chief, Northern Naval Forces in Europe [USA] *mt*
CINCNORTH Commander In Chief, Northern Europe [NATO] *fu, mt*
CINCPAC Commander In Chief, Pacific [USA/USAF/USN] *mt, kf, 10, 12, hc, tb, fu*
CINCPACAF Commander In Chief, Pacific Air Force [USA] *mt*
CINCPACFLT Commander In Chief, Pacific Fleet [US Navy] [USA] *jj, fu, mt*
CINCPT CINC Report System *mt*
CINCRED Commander In Chief Readiness Command *mt*
CINCSAC Commander In Chief, Strategic Air Command *mt, tb, fu*
CINCSD Commander In Chief, Strategic Defense *tb*
CINCSEURCOM Commander In Chief, Southern Europe Command [USA] *mt*
CINCSOUTH Commander In Chief US Southern Command *mt*
CINCSOUTH Commander In Chief, Southern Europe [NATO] *fu, mt*
CINCSPACE Commander In Chief, US Space Command *kf, fu*
CINCSTRAT Commander In Chief, US Strategic Command *mt*
CINCUKAIR Commander In Chief, United Kingdom Air Forces *fu*
CINCUNC Commander In Chief United Nations Command *mt*
CINCUSAFE Commander In Chief, United States Air Force, Europe *mt, hc, jj*
CINCUSAREUR Commander In Chief, United States Army, Europe *aj89.9.40*
CINCUSNAVEUR Commander In Chief, US Navy in Europe [USA] *fu, mt*
CINCWESTLANT Commander In Chief, Western Atlantic *fu*
CINDA Chrysler Improved Numerical Difference Differencing Analyzer *gp, 16, kf, fu*
CINFARS Command Integrated Financial Accounting and Reporting System *mt*
CINS Command Information Network Supplement/System *mt*
CINS Commander's Information Network System [AFLC] *mt*
CIO Chief Information Officer *kg, ct, y2k*
CIO Customer Integration Office *wl*
CIOC Communications I/O Controller *fu*
CIOC Current Intelligence Operations Center *mt*
CIOCS Communication Input/Output Control System *pk, kg*
CIP Carbonyl Iron Powder *wp*
CIP Combat Intelligence Processor *mt*
CIP Combined Interoperability Program *mt*
CIP Command Interface Port *pk, kg*
CIP Commercial Instruction Processor [Honeywell] *mt*
CIP Common Integrated Processor *hc, rm8.2.5, wp*
CIP Communication Improvement Plan *mt*
CIP Communications Implementation Plan *mt*
CIP Communications Interface Processor *mt, fu*
CIP Component Improvement Program *mt, wp*
CIP Compressor Inlet Pressure *wp*
CIP Computed Impact Point *hc, jj*
CIP Configuration In Process *jdr*
CIP Contract Implementation Plan *oe*
CIP Control Item Planners *hc*
CIP Cost Improvement Program Proposal *fu, hn43.22.9, 8, 15, sc8.9.6, hc, ig21.12.17, ep, jj, jdr*
CIP Critical Item Program *mt*
CIPFS Configuration Item Product Fabrication Specification *fu*
CIPM ComitE International des Poids et Mesures [French International Committee on Weights and Measures] *fu*
CIPP Communications Individual Positioning Program *mt*
CIPS Computer based Information Processing System *fu*
CIPT Cost Integrated Product Team *kf*
CIR Cargo Integration Review *gp, 4, oe*

CIR Central Intelligence Report [CIA] *wp, mt*
CIR Circuit Information Release *fu*
CIR Committed Information Rate *ct, kg, em, hi*
CIR Cost Information Report [formerly CEIS] *fu, hc, wp, jj*
CIR Critical Information Requirement *mt*
CIRA COSPAR International Reference Atmosphere *ns*
CIRA {Italian Aeronautical Research Center} *aa26.8.12*
CIRC Central Information Reference and Control *hc, wp*
CIRC Circular Reference *pk, kg*
CIRC Cross Interleaved Reed Solomon Code *pk, kg*
CIRCUIT Trunk and Circuit Listing [REDCOM/JDA] *mt*
CIRF Centralized Intermediate Repair Facility *mt*
CIRIS CC Integrated Resources Information System [USAFE] *mt*
CIRIS Consolidated Intelligence Resources Information system [AF/IN] *wp, mt*
CIRM Celestial Infrared Measurement [study] *hc*
CIRRIS Cryogenic InfraRed Radiance Instrument for Shuttle *mb, mi, tb*
CIRRUS (a sensor known as CIRRUS, failed) *sf31.1.29*
CIRS Composite Infrared Spectrometer *es55.26*
CIRVIS Communications Instructions for Reporting Vital Intelligence Sightings [USA DOD] *mt*
CIS Card Information Structure *pk, kg*
CIS Client Customer Information System *pk, kg*
CIS Combat Identification System *mt, ph, hc, wp*
CIS Command and Information System *tb*
CIS Communications and Information Systems *mt*
CIS Communications Interface Subsystem *mt*
CIS Compensated Imaging System *ro*
CIS Computer Information Systems *pk, kg*
CIS Computer Interface Serial *fu*
CIS Contact Image Sensor *pk, kg*
CIS Control Indicator Set *fu*
CIS Cooled Infrared Spectrometer *hg, ge*
CIS Cost Information System *fu, sc2.10.2, 8, 15, 16, 17, sb, hc, hn49.11.8, lt, ep, jj, jdr*
CIS-ISS Combat Identification System - Indirect Sub System *mt*
CISC Complex Instruction Set Code *em*
CISC Complex Instruction Set Computing Computer *ct, em, is26.1.53, kg, hi*
CISC CSOC Integration Support Contractor *hc*
CISD Communications and Information Systems Division *hy*
CISF Computer Integration SW Facility *jdr*
CISF Consolidated Integration Support Facility *kf*
CISG Combat Information Systems Group [AFCC*] *mt*
CISM Conseil International du Sport Militaire *dt*
CISPD Consolidated Information Services Planning Document *mt*
CISRO Cost Information System Report Order *-8*
CISS Combat Information Systems Squadron [AFCC*] *mt*
CISS Computer Interface Sub System *jj*
CIST Combined Interim System Test *hc*
CIT California Institute of Technology *le32, fu fu*
CIT Center for Innovative Technology [Reston, VA] *mt*
CIT Circumstellar Imaging Telescope *mi*
CIT Combined Interrogator-Transponder *mt*
CIT Communications Interface Terminal *tb*
CIT Compressor Inlet Temperature *wp*
CIT Computer Integrated Telephony *hi, kg*
CIT Contractor Independent Testing *fu*
CIT Critical Item Test *fu*
CIT Current Injection Test Testing *jdr, kf*
CITA Commercial / Industrial Type Activities *mt*
CITA Controller Interaction and Task Analysis *fu*
CITE CADE Integration, Test, and Evaluation *mt, st*
CITE Cargo Integration Test Equipment *ci.628, 4, 9, oe*
CITE Chemical Instrumentation Test and Evaluation *hy*
CITE Contractor Independent Technical Effort [a budget item] *mc4-83, hc, ep, pbl, tm, fu, jdr, ah*
CITE Contractor Independent Technical Evaluation *sc4.9.6*
CITEC Contractor Independent Technical Effort *wp*
CITIS Contractor Integrated Technical Information Services *kf*
CITS Centrally Integrated Test System Station [B-1B] *mt, hc, fu, aw130.11.71, wp*
CITSMUX Central Integrated Test System Multiplex *hc*
CITV Commander's Independent Thermal Viewer *hc, wp*
CIU Central Interface Unit [Honeywell] *wp, jj, mt*
CIU Communications Interface Unit *mt, fu, 4*
CIU Computer Interface Unit *fu, hc*
CIU Console Intercom Unit *fu*
CIU Control Interface Unit *fu*
CIU Customer Interface Unit *fu*
CIUS Corps Interim Upgrade System [USA] *mt*
CIV Corona Inspection Voltage *gp*
CIVR Computer and Interactive Voice Response *pk, kg*
CIVR Configuration Item Verification Review *fu*
CIVT CONUS Integration and Verification Test *st*
CIWC Cheyenne Mountain Intelligence War Center *kf*

CIWS Close-In Weapon System Support *mt, wp, fu, jj*
CIX Commercial Internet Exchange *hi, kg*
CIX Compulink Information Exchange *pk, kg*
CJCS Chairman [of the] Joint Chiefs of Staff [OJCS] [USA DOD] *hc, mt*
CJCSI Chairman of the Joint Chiefs of Staff Instruction [USA DOD] *mt*
CJLI Command Job Language Interpreter *pk, kg*
CJTF Combined Joint Task Force [USA] *mt*
CJTF Commander, Joint Task Force *mt*
CK [chemical warfare codename for cyanogen chloride] *wp*
CKD Count Key Data [device] *pk, kg*
CKV Conventional Kill Vehicle *tb, st*
CL Center Line *fu, w9, jj*
CL Centerline Lights *fu*
CL Change Class *fu*
CL Chemical Laser *wp*
CL Communication Laboratory *jr*
CL Communications Link *fu*
CL Components Library *fu*
CL Configuration Listing *fu*
CL Control Limit *fu*
CL Cruiser, Light [US Navy] *mt*
CLA Calibration Lead Assembly *pl*
CLA Centerline Average *fu*
CLA Centro de Lan amento de Alcantara [Alcantara Launch Center, Brazil] *ai2*
CLA Collector Lens Assembly *jj*
CLA Communications Line Adapter *fu*
CLA Control Logic Assembly *hc, jj*
CLA Coupled Load Analysis *gp, he, 16, jdr, oe*
CLA Cross Link Assignment *fu*
CLA Custom Logic Array *hi*
CLAA Cruiser, Light, Anti Aircraft [US Navy] *mt*
CLADE Chemical Laser Advanced Diffuser/Ejector *hc*
CLAES Cryogenic Limb Array Etalon Spectrometer [UARS] *hg, ge*
CLAM Chemical Low Altitude Missile *hc*
CLAM Clear Air Mass Systems *jj*
CLAM Completed Leasing Action Message *mt*
CLAMDOC CLAMS Library Documentation *tb*
CLAMP Closed Loop Aerospace Aeronautical Management Program *hc, wp, jj*
CLAMS Clear Air Mass Systems *hc*
CLAMS Cleared Lane Marking System *mt*
CLAMS Common Los Alamos Mathematical Software *tb*
CLAMSRC CLAMS Library Source *tb*
CLAMTI Clutter-Locked Moving Target Indicator *wp*
CLANS Computerized Link Analysis System *fu*
CLAR Channel Local Address Register *pk, kg*
CLAS Clean Air Station *aa26.8.42*
CLAS Combat Logistics Assessment System *mt*
CLASP Computer Language Aeronautics and Space Program *hc*
CLASS Carrier Landing Aid Stabilization System *hc*
CLASS Client Access to Systems and Services *pk, kg*
CLASS Close Air Support System *hc, wp*
CLASS Communications Link Analysis and Simulation System *2, 19*
CLASS Computing Load Assessment Simulation System *hc*
CLASS Cooperative Library Agency for Systems and Services *pk, kg*
CLASS Custom Local Area Signaling Services *mt, pk, kg*
CLAW Close Air Support Weapon *wp*
CLAW Concept for Low Cost Air to Air Weapon *hc*
CLBRP Cannon Launched Beam Rider Projectile [USA] *hc, mt*
CLC Clear Carry Flag *pk, kg*
CLC Command Launch Console *fu*
CLC [task fleet command ship, US Navy] *mt*
CLD Chemoluminescence Detector *wp*
CLD Clear Direction Flag *pk, kg*
CLD Current Limited Device *hc*
CLDSIM Cloud Simulation *kf*
CLE Console Local Equipment *jj*
CLE Customer Located Equipment *ct*
CLEC Competitive Local Exchange Carrier *ct*
CLEO Conference on Lasers and Electro-Optics *aw118.18.4*
CLEP College Level Examination Program *w9, an31.13.4*
CLF Catalog of Films *fu*
CLF Commander, Landing Force [USA DOD] *mt*
CLFC Closed Loop Fire Control *hc, jj*
CLFCS Closed Loop Fire Control System *hc, jj*
CLG Cruiser, Light, Guided Missile [US Navy] *mt*
CLGN Cruise, Light, Guided Missile, Nuclear powered [US Navy] *mt*
CLGP Cannon Launched Guided Projectile *fu, hc*
CLI Call Level Interface *hi, kg*
CLI Clear Interrupt Flag *pk, kg*
CLI Client Library Interface *pk, kg*
CLI Coherent Laser Illumination *hc*
CLI Command Language Interpreter *fu*

CLI Command Line Interface *pk, kg*
CLI Command Line Interpreter *fu, hi*
CLI Communications Line / Link Interface *mt*
CLI Compression Laboratories, Incorporated *csIV.1*
CLI Contract Line Item *jj*
CLI Control Level Item [hardware or software] *fu, 8, lt, ep, pbl, tm, jdr*
CLI Cost of Living Increase *jj*
CLI Critical Line Item *mt*
CLIC Center for Low Intensity Conflict *mt*
CLID Calling Line Identification *pk, kg*
CLIFFS Cost, Life, Interchangeability, Function, Fit *hc*
CLIFT Closed Loop Infrared Countermeasures Test *hc*
CLIIP Command Logistics Information Improvement Program *mt*
CLIN Contract Line Item Number *mt, hc, ts4-11, nt, wp, tb, st, jj, jdr, fu, kf*
CLIP Compiler Language For Information Processing *wp*
CLIR Limb Emission Spectrometer/Radiometer [Shuttle] *hg, ge*
CLIRCM Closed Loop Infrared Countermeasures *hc, wp*
CLIST Command List [in a computer program] *8, kg*
CLK Hunter-Killer Cruiser [US Navy] *mt*
CLL Consolidated Load List *jj*
CLM Communications Load Module *fu*
CLM Compact LIST Machine *tb*
CLMA Channel Linearity Measurement Assembly *-4*
CLNP Connectionless Network Protocol *kg, nu, hi*
CLO Communications Liaison Officer *mt*
CLO Computer Lock On *jj*
CLOC Central Logic Oscillator Control *fu*
CLOS Clear Line Of Sight *wp*
CLOS Common Lisp Object System *hi, kg*
CLP Captive Line Parts *hc*
CLP Cell Loss Priority *kg, ct*
CLP Central Line Power *jj*
CLP Constraint Logic Programming *pk, kg*
CLR Cell Loss Ratio *ct*
CLR Current Limiting Resistor *fu*
CLRCs Country Logistics Readiness Centers [USAFE] *mt*
CLS Clear Screen *hi, kg*
CLS Cloud Lidar System *hy*
CLS Combat Logistics System [HQ USAF] *mt*
CLS Command Level Schedule *oe*
CLS Component Level Schedule *oe*
CLS Constant Level Signals *mt*
CLS Contingency Landing Site *9, oe*
CLS Contractor Logistics Support *mt, wp, kf, fu, hc*
CLS/CALM Combat Logistics System Computer Aided Load Manifesting *mt*
CLSA Control Logic and Switching Assembly *hc*
CLSA Cooperative Logistics Support Arrangement *jj*
CLSC Contractor Logistics Support Commitment *kf*
CLSDT Colored Large-Screen Display Terminal *fu*
CLSF COMSEC Logistics Support Facility *jj*
CLSI Custom Large Scale Integration *hc*
CLSID CLasS Identifier *em*
CLSS Combat Logistic Support Squadron *wp*
CLSS Combat Logistics Support System *mt*
CLSS Combined Logistics Support System *mt*
CLT Cargo Load Transporter *hn44.5.3*
CLT Closed Loop Targeting *fu*
CLTC China [satellite] Launch and Tracking Control *ai2*
CLTP Communications Line Termination Panel *fu*
CLTP Connectionless Transport Protocol *pk, kg*
CLTS Clear Task Switch Flag *pk, kg*
CLU Central Logic Unit *fu*
CLU Command and Launch Unit *wp*
CLU Command Logic Unit *hc, jj*
CLUI Command Line User Interface *pk, kg*
CLV Constant Linear Velocity *ct, kg, em, hi*
CM Cache Memory *fu*
CM Calibrated Magnification *fu*
CM Candidate Material *hc, tb, su*
CM Center of Mass *sa12474.9, 17, jdr, mi, mb, oe, ai1*
CM Charge Module *jdr*
CM Circuit Management *fu*
CM Circuit Master *fu*
CM Clock Module *fu*
CM Coding Modulation *jj*
CM Color Monitor *fu*
CM Combat Management *fu*
CM Command Message *fu*
CM Command Module [Apollo spacecraft] *fu, ci.388, mi, mb*
CM Common Mobile *kf*
CM Conceptual Model *mt*
CM Configuration Management *mt, fu, mc, 16, hc, wp, ep, tb, st, jj, jdr, kf, oe*
CM Continuous Monitoring *hc, jj*

CM Contract Memoranda *fu*
CM Contract Modification *oe*
CM Contract Monitor *mt, kf*
CM Control Mark *pk, kg*
CM Core Memory *fu*
CM Corrective Maintenance *mt, fu, kg, wp, tb*
CM Counter Measure *mt, sd, hc, wp, tb, st, jj, fu*
CM Crisis Management *mt*
CM Cruise Missile *tb, st, kf, fu*
CM&D Collection, Management and Dissemination Section [USA] *mt*
CM&D Countermeasures and Deception *fu*
CM/CAI Computer Management / Computer Assisted Instruction *un*
CM/CB Countermortar/ Counterbattery *fu*
CM/DM Configuration Management and Data Management *ep, ful*
CM/PM Corrective Maintenance/Preventive Maintenance *tb*
CMA Chemical Manufacturers' Association *hy, y2k*
CMA Communications Module Assembly *fu*
CMA Composition Modulated Alloy *tb*
CMA Concert Multi thread Architecture *pk, kg*
CMA Configuration Management Administrator Activity *fu*
CMA Contact Making Ammeter *fu*
CMA Control, Monitor and Alarm *mt, kf*
CMA Court of Military Appeals *dt*
CMAC Complex Multiplier-Accumulator Controller *fu*
CMAFB Cheyenne Mountain Air Force Base *jb, jmj, as, kf*
CMAFS Cheyenne Mountain Air Force Station *wp*
CMAG Cruise Missile Advanced Guidance *mt, hc, wp*
CMAH CINC Mobile Alternate Headquarters *mt, kf, ric, jdr*
CMARP Contingency Mating and Ranging Program [SAC] *mt*
CMAS Cheyenne Mountain Air Station *kf*
CMAS Computer based Maintenance Aids System *mt, wp*
CMAS Countermeasures Assessment System *fu*
CMAS Crisis Management Automation System *wp*
CMB Central Meteorological Bureau [China] *ac3.4-9*
CMB Collection Management Branch *mt*
CMB Combat Mobility Branch *mt*
CMB Configuration Management Board *fu*
CMBDB Configuration Management Baselines Data Base *fu*
CMBR Cosmic Microwave Background Radiation *mi*
CMBS Central Missile Support Base *tb*
CMBTC/F/S/G Combat Communications Flight / Squadron / Group *mt*
CMC Call Management Center [WIS] *mt*
CMC Ceramic Matrix Composite *aa26.12.30, wp*
CMC Change Management Center *ep*
CMC Cheyenne Mountain Complex [SPACECOM] *mt, jo.v, hc, go3.3-1, wp, tb, kf, fu*
CMC Collection Management Center *mt*
CMC Commandant, Marine Corps [USA] *tc, jj, mt*
CMC Common Mail Calls *pk, kg*
CMC Common Messaging Calls *hi, kg*
CMC Communication Management Configuration *pk, kg*
CMC Communication Master Control *-9*
CMC Complement Carry Flag *pk, kg*
CMC Computer Mathematics Corporation *ad*
CMC Computer Mediated Communication [Internet] *hi, kg*
CMC Configuration Management Capability *tb*
CMC Corrective Maintenance Card *fu*
CMC Crisis Management Council [OSD] *mt*
CMCA Cruise Missile Carrier Aircraft [USA] *hc, wp, mt*
CMCC Central Mission Control Centre [ESA] *mi*
CMCC CINC Mobile Command Center *mt*
CMCC Combined Movements Control Center *mt*
CMCD COMSEC Mode Control Device *hc*
CMCHS Civilian Military Contingency Hospital System *mt*
CMCM Contingency Management Computer Model [also CM2] *mt*
CMCO Classified Material Control Officer *mt, wp, jj*
CMCS Central European Movements Control System *mt*
CMD Circuit Mode Data *pk, kg*
CMD Coherent Magnitude Detection *fu*
CMD Color Matrix Display *hc*
CMD Components and Materials Department *fu*
CMD Configuration Management Directive *fu, jj*
CMD Contract Management District *fu, hc*
CMD Core Memory Driver *fu*
CMD Counter Measures Dispenser *hc*
CMD Cruise Missile Defense *wp*
CMDAC Current Mode Digital-to-Analog Converter *fu*
CMDH Command and Data Handling *jdr*
CMDR Coherent Monopulse Doppler Radar *wp*
CMDR Crew Commander *kf*
CMDS Command Manpower Data System [USAF] *mt*
CMER Components and Materials Engineering Request *hc, fu, jj*
CMER Controlled Material Evaluation Request *jj*
CMEST Cruise Missile Engagement System Technology [ADI] *mt*
CMET Command Management Engineering Team *mt, af72.5.109*
CMF Central Maintenance Facility *fu*

CMF Coherent Memory Filter *fu*
CMF Combat Mission Folders *mt*
CMF Complementary Multirole Fighter *wp*
CMF Creative Music Format *hi, kg*
CMF Crisis Management Facility *mt*
CMFE Civilian Manpower and Funding Extract *mt*
CMFESIS Civilian Manpower and Funding Extract {?} [PACAF] *mt*
CMFS Coherent Multiple Frequency Signature *hc*
CMG Condensed Maintenance Guide *fu*
CMG Control Moment Gyro Gyroscope [system] *ah, ci.432, 17, sb, wl, pbl, ai1*
CMGS Cruise Missile Guidance System *mt*
CMH Collapsible Maintenance Hangar *jj*
CMH Countermeasures Homing *fu*
CMI Cable Microcell Integrator *ct, em*
CMI Computer Managed Instruction *mt, wp, tb, kf, fu, ai1*
CMI Configuration Management Implementation Plan *fu*
CMI Continuous Measurable Improvement *ah, uh, fu, sc9.6.1, hn50.9.8, pbl, tm, kf*
CMI Cruise Missile Integration [B-52] *mt*
CMI [Hughes Aircraft Company customer Belgium for armored infantry fighting vehicle] *hc*
CMI/HIC Cable Microcell Integrator / Headend Interface Converter *ct, em*
CMICS Corporate Manufacturing Information Control System *fu*
CMIP Common Management Information Protocol *kg, ct, em, hi*
CMIP Cost Methods Improvement Program [AFSC*] *mt*
CMIS CCS Mk2 Management Information System *fu*
CMIS Command Management Information System [AIA] *wp, mt*
CMIS Commissary Management Information system [USMC] *mt*
CMIS Common Management Information Services System *pk, kg*
CMIS Configuration Management Information System *kf*
CMIS Conical Scanning Microwave Imager/sounder *jr*
CMIT Coherent Moving Target Indicator *wp*
CML Components and Materials Laboratory [Hughes Aircraft Company] *gp, 15, hc, lt, jj*
CML Computer Managed Learning *hi, kg*
CML Conceptual Modelling Language *pk, kg*
CML Configuration Management Library *fu*
CML Current Mode Logic *mt, fu, kg*
CMLC Civilian Military Liaison Committee *ci.888*
CMLC Computer Analysis of Analog and Logic Circuit Cards *fu*
CMLS Conventional Mine Laying System *wp*
CMM Capability Maturity Model *mt, pbl, jdr, kf, kg, ah*
CMM Cargo Movement Module *mt*
CMM Configuration Management Manual *gp, hc, jj*
CMM Coordinate Measurement Machine *nd74.445.27*
CMMC Contractor Major Milestone Chart *fu*
CMMCA Cruise Missile Mission Control Aircraft *hc, af72.5.140, hn49.11.9, wp*
CMMF Circuit Message and Monitoring Facility *kf*
CMMR Common Modular Multi-Mode Radar *hc, wp*
CMMS Chloromethyl Methylsulfide *wp*
CMMS Command Munitions Management System [USAFE] *mt*
CMMS Computerized Maintenance Management Software *pk, kg*
CMMS Congressionally Mandated Mobility Study *mt*
CMMU Cache Memory Management Unit [Motorola] *hi, kg*
CMN Common Mode Noise *oe*
CMN Constructions MÉcaniques de Normandie *aa27.4.b10*
CMNCC Cheyenne Mountain National Command Center *tb*
CMO Central Measurement [and Signature Intelligence NASINT] Office *kf*
CMO Chief Maintenance Officer *wp*
CMO Civil Military Operations *mt*
CMO Collection Management Office *mt*
CMO Configuration Management Office Officer [an organization] *pbl, fu, tm, ah, gp, 4, 15, 17, sb, hc, ep, jj, jdr, oe*
CMO Configuration Management Organization *mt*
CMO Contract Management Office *wp*
CMOC Cheyenne Mountain Operations Center *kf*
CMOM Consumable Maintenance Overhaul Materials *jj*
CMOML Consumable Maintenance and Overhaul Material List *jj*
CMOP Configuration Management Operating Plan *fu, jj*
CMOR Components Material Organization Release *hc*
CMOS Cargo Movement Operations System [USAF] *mt*
CMOS Complementary Metal Oxide Semiconductor Silicon *mt, pl, aw129.1.61, gp, 6, 16, 17, th4, hn43.11.1, hc, wp, st, jj, kf, fu, kg, oe, ct, em, uh, hi*
CMOS-SOS Complementary Metal Oxide Semiconductor - Silicon On Sapphire *16, 17, sb, tb*
CMOS/SOI Complementary Metal Oxide Semiconductor/Silicon On Insulator *kf*
CMOV Conditional Move *pk, kg*
CMP Capabilities Master Plan [of AWS] *pl*
CMP Celestial Mapping Program *hc*
CMP Celestial Measurements Program *hc*
CMP Circuit Management Processor *fu*

CMP Circuit Message Processor *fu*
CMP Communications Management Processor *fu*
CMP Communications Plenum Cable *ct*
CMP Company Materiel Practices *fu*
CMP Component Material and Process *fu*
CMP Computational Map Processor *wp*
CMP Configuration Management Plan *mt, fu, gp, hc, nt, ep, tb, st, jj, kf, oe*
CMP Controlled Material Plan *hc*
CMP Counter Military Potential *fu, wp*
CMPB Configuration Management Policy Board *fu*
CMPCO Conventional Mission Planning Concept of Operations [SAC] *mt*
CMPNTEXP Component Exception *mt*
CMPS Compare word String *pk, kg*
CMR Common Module Retrofit *hc*
CMR Communications Riser Cable *ct*
CMR Configuration Management Requirements *pl*
CMR Contract Management Region *hc, fu*
CMR Contract Management Review *wp*
CMR Corrective Maintenance Report *jj*
CMR Corrective Maintenance Request [GAC] *jj*
CMR Counter Measure Receiver *hc*
CMR Critical Milestone Review *fu*
CMR Customer Material Return *hc, jj*
CMR Customer's Management Room *oe*
CMRA Current Maximum Reimbursable Amount *hc*
CMRB Contract Maintenance Review Board *wp*
CMRF Central Maintenance Repair Facility *fu*
CMRF Complementary Multi Role Fighter *wp*
CMRS Calibration Measurements Requirements Summary *hc, tb, fu, jj*
CMS Cabin Management System *aw126.20.35*
CMS Cargo Movement System *mt*
CMS Cockpit Management System *mt*
CMS Code Management System *fu, ct, kg, oe, em*
CMS Collapsible Maintenance Shelter *jj*
CMS Color Management Settings *hi*
CMS Combat Maintenance System [SOUTHCOM] *mt*
CMS Combat Mission Simulator *hc*
CMS Command Management System *-4*
CMS Command Memory System *oe*
CMS Communications Management System *mt*
CMS Communications Module Subsystem *kf*
CMS Communications Security Equipment Inventory *mt*
CMS Compiler Monitor System *fu, kg*
CMS Configuration Management Strategy *fu*
CMS Configuration Management System *fu, mt*
CMS Constellation Maintenance System *jr*
CMS Container Management System [MTMC] *mt*
CMS Contractor Maintenance Services *fu, wp, jj*
CMS Contractor Maintenance Support *fu*
CMS Controlled Mode Security *mt*
CMS Conventional Munitions System [USAFE] *mt*
CMS Conversational Monitoring System *mt, fu, kg, wp*
CMS Crisis Management System [OSD] *mt*
CMS CRITICOM Multiplexer Unit *tb*
CMSAS Configuration Management Status Accounting System *fu*
CMSB Central Missile Support Base *tb*
CMSC Consolidated Mission Support Center *mt*
CMSDG Configuration Management System Development Group *mt*
CMSEP Contractor Management System Evaluation Program *mt, fu*
CMSGT Chief Master Sergeant *mt*
CMSIJ Crisis Management System Interface to JOPES [OJCS] *mt*
CMSP Configuration Management Support Plan *mt, wp*
CMSQ Communications Maintenance Squadron *mt*
CMSS Calibration and Measurement Standards System *hc*
CMSS Configuration Management System Services *fu*
CMSS Crisis Management Support System [MSC] *mt*
CMSSO Contract Maintenance and Supply Support Operations *hc*
CMST Cruise Missile Surveillance Technology [ADI] *hc, wp, mt*
CMT CAMS Mobile Terminal *mt*
CMT Change Management Term Team *hc, jj*
CMT Combat Mission Trainer *wp*
CMT Common Module Tester *fu*
CMT Computer Maintenance Technician *kf*
CMT Computer Managed Training *hc, jj*
CMT Configuration Management Team *fu*
CMT Connection Management *fu*
CMT Critical Military Target *fu, tb, st*
CMT Critical Mobile Target *kf*
CMTI Coherent Moving Target Indication *hc*
CMTRK Configuration Management Tracking System *mt*
CMTU Cartridge Magnetic Tape Unit *hc, fu*
CMU Cheyenne Mountain Upgrade *mt*
CMU Command Memory Unit *oe*
CMU Computer Maintenance Unit *fu*

CMU Mobile Command Unit *mt*
CMUX Communications Multiplexer *fu*
CMVC Configuration Management and Version Control [IBM] *tb, fu, kg*
CMVM Contact Making Voltmeter *fu*
CMW Compartmented Mode Workstation *pk, kg*
CMWG Configuration Management Working Group *mt, fu*
CMY Cyan, Magenta, Yellow [a color model] *pk, kg*
CMYK Cyan Magenta Yellow Black [a color model] *hi, kg*
CN Change Notice *hc, tb, jj, fu, jdr*
CN Communications Network *hi*
CN [chemical warfare codename for teargas] *wp*
CN&I Communications, Navigation and Identification *hc*
CNA Camp New Amsterdam *mt*
CNA Center for Naval Analysis *mt, tb*
CNA Certified NetWare Network Administrator [Novell] *ct, kg, hi*
CNA Copper Nickel Alloy *fu*
CNABATRA Chief of Naval Air Basic Training *jj*
CNAD Committee of National Armament Directors [NATO] *mt*
CNAD Conference of National Armaments Directors [NATO] *mt, ph, aa26.20.10*
CNAPS Co-Processing Node Architecture for Parallel Systems *pk, kg*
CNARESTRA Chief, Naval Air Reserve Training *jj*
CNATECHTRA Chief, Naval Air Technical Training *jj*
CNATRA Chief of Naval Air Training [USA] *jj, mt*
CNAVANTRA Chief, Naval Air Advanced Training *jj*
CNAVRES Commander, Naval Reserve *mt*
CNC Computer Numeric Controlled Computerized Numerical Control *hn43.22.8, hd1.89.y26, ct, em, kg*
CNC Computer Numerically Numerical Controlled *kf, fu*
CNC [chemical warfare codename for chloroacetophenone/chloroform mixture] *wp*
CNCC COINS Network Control Center *mt*
CNCC Country Name to Country Code [JCS] *mt*
CNCE Communications Network Control Element [TRI-TAC] *mt*
CNCE Communications Nodal Control Element *mt, fu*
CND Chief of Naval Development *fu*
CND Could Can Not Duplicate *oe, fu, rm2.3.4, wp, jj*
CNDO Complete Neglect of Differential Overlap *wp*
CNDWI Critical Nuclear Defense Weapon Information *mt*
CNE Certified NetWare Engineer [Novell] *hi, kg*
CNE Certified Network Engineer *ct*
CNES Centre National d'Etudes Spatiales [France] [National Center for Space Studies] *sc2.8.4, aa26.8.12, sf31.1.11, es55.36, aw130.11.29, hy, re, mb, hg, ge, mi, oe, crl, ai2*
CNET Chief of Naval Education and Training *mt, fu*
CNFE Composite Network Front End *mt*
CNGB Chief, National Guard Bureau *mt*
CNI Communication, Navigation, and Identification system [USAF] *mt*
CNI Communication/Navigation/Instrument *jj*
CNI Communications, Navigation, [and] Identification *ro, fu, hc, mj31.5.14, wp*
CNIDR Clearinghouse for Network Networked Information Discovery and Retrieval [Internet] *pk, kg*
CNIE National Space Research Commission [Argentina] {ComisiÛn Nacional de Investigaciones Espaciales} *aw118.14.20, crl*
CNIF Clutter to Noise Improvement Factor *fu*
CNIN Composite NFE Internal Network [DIA] *mt*
CNISI Communication, Navigation, Identification, Shipboard Integrated *fu*
CNL Communication, Navigation and Landing *hc*
CNL Communications Architecture Laydown *fu*
CNM Chief of Naval Materiel *12, fu, jj*
CNM Communications Network Management *hi*
CNN Cable News Network *sc3.9.3, jb*
CNN Composite Network Node *pk, kg*
CNO Carbon Nitrogen Oxygen *mi*
CNO Chief of Naval Operations [US Navy] [USA] *mt, 12, hc, w9, jj, fu, jdr*
CNOC Commander, Naval Oceanography Command *tc*
CNP Chief of Naval Personnel *fu*
CNR Canadian National Railway *wp*
CNR Carrier to Noise [power] Ratio *bl1-5.9, kg*
CNR Centro Nazionale delle Recerche *crl*
CNR Chief of Naval Research [US Navy] *tc, fu, mt*
CNR Combat Net Radio *mt*
CNR Consiglio Nazionale delle Recerche [National Research Council Italy] *hy*
CNR Critical Node Recognizer *wp*
CNRS Centre National de la Recherche Scientifique [French National Scientific Research Center] *aa27.4.b32, crl*
CNS Certified Novell Salesperson *ct*
CNS Compagnia Nazionale Satellini [Italy] [National Satellites Company] *sc2.10.3*
CNS Compressible Navier-Stokes *aa27.4.b31*
CNS Consolidated NOTAM System *fu*
CNS Constant Watch Network Support Subsystem [PACAF] *mt*

CNSR Comet Nucleus Sample Return [mission] *es55.8, mi, mb*
CNSS Core Nodal Switching Subsystem [Internet] *hi, kg*
CNTC Commander, Naval Telecommunications Command *dt*
CNU Container Unit *jj*
CNVEO [Hughes Aircraft Company customer for air defense electro-optical support] *hc*
CNWDI Critical Nuclear Weapon Design Information *mt, su, fu, sy9, hc, nt, tb, jdr, kf*
CNX Certified Network Expert *pk, kg*
CO Call Out *jj*
CO Central Office *mt, do146, ct, kg*
CO Change Order *fu, hc, wp, jj*
CO Co-Orbital *sd*
CO Command Output *pk, kg*
CO Commanding Officer *mt, fu, w9, wp, jj*
CO Communications Orbiter *re*
CO Computer Operator *fu*
CO Contracting Officer Office *mt, fu, hc, tb, wl, jj, kf, oe*
CO Convert Out *pk, kg*
CO-ED Composition and Editing Display *hc*
CO2 Carbon Dioxide *tb, jdr*
COA Central Operating Authority *mt*
COA Certificate Of Authenticity *ct*
COA Commonwealth Of Australia *fu*
COA Comptroller Of the Army *mt, st*
COA Constant Output Amplifier *fu*
COA Contract Administration Office *tb*
COA Course Of Action [USA DOD] *fu, as, mt*
COAA Confirmation Of Assigned Altitude *fu*
COADS Comprehensive Ocean Atmosphere Data Set *ns*
COAN Comptroller Office Automation Network [HQ USAF] *mt*
COANG Colorado Air National Guard *12, kf*
COAP Cost Analysis of Maintenance Policies *jj*
COAR Coherent Array Radar *fu*
COARE Coupled Oceans Atmospheric Responses Experiment *hy*
COARS Command On Line Accounting and Reporting System *mt*
COAS (DTO [is] . . . HUD backup to COAS) [Shuttle] *ws4.10*
COAST Cache On A Stick *ct, em, hi*
COAST Configuration Operational and Support Tape *fu*
COAT Coherent Optical Adaptive Technology *fu, hc*
COATS CIACS Operational Assessment Timing System *jj*
COB Chip On Board *wp, kg, fu*
COB Close Of Business *mt, wp, oe, ah, fu*
COB Collocated Operating Base [USAF] *wp, tb, mt*
COB Command Operating Budget *mt, jj*
COB Currently On Board *mt*
COBAE Commissio Brasileira de Atividades Espaciais [Brazilian Space Activities Commission] *ai2*
COBE Cosmic Background Explorer *ci.716, hc, mi, mb*
COBI Coded Biphase *fu*
COBIM Collocated Operating Base Information Model [USAFE] *mt*
COBIS Command Operating Budget Information System *mt*
COBOL COmmon Business Oriented Language [a computer programming language] *mt, ro, em, ct, hc, wp, fu, kg, jj, hi*
COBRAH Coin Beacon Ranging And Homing *hc*
COBRA_DANE [space ICBM warning phased array radar system, Sheyma, now Eareckson AFB, AK] *wp, fu, mt*
COBRA_JUDY [airborne / shipborne phased array radar study for USAF] *fu, mt*
COBS Command Operating Budget System *mt*
COC Center Of Coverage *jdr*
COC Certificate Of Compliance *fu*
COC Certificate Of Conformance *fu*
COC Circle Of Containment *fu*
COC Combat Operations Center [NORAD] *mt, fu, 12, hc, tb*
COC Command Operation Center *mt*
COC Command Operations Central *fu, hc, kf*
COCAM Coordinating Committee for Multilateral Export Controls {?} [OSD] *mt*
COCC Central Operations/Command Center *mt*
COCC Contractor's Operational Control Center *sc4.1.center, lt*
COCDAS CONARC Class One Automated System Ground System *hc*
COCMAGTF Combat Operations Center Marine Air Ground Task Force *mt*
COCOM Combat Combatant Command [command authority]. [USA DOD] *mt, as, kf*
COCOM Coordinating Committee [for export restrictions] *ph, mt*
COCOMO Constructive Cost Model *kf, fu*
COCORADS Covert Communications Radio System *hc*
COCPFF Conventional Ordinance Consumption Plan Factors File *mt*
COD Carrier Onboard Delivery [aircraft designed to deliver goods to carriers, not necessarily carrier-based] *jj, mt*
COD Change Operations Directive *tb*
COD Code Of the Day *fu*
COD Combat Operations Division *mt*
COD Correction Of Deficiencies *jj*
COD Current Operations Division *mt*

CODAN Carrier Operated Device, Anti Noise *mt*
CODAR Communications Detection And Recognition *fu*
CODAR Correlation Display Analyzing and Recording *fu*
CODAS Control and Data Acquisition System *wp*
CODASYL Conference on Data System Languages [group that designed COBOL] *mt, kg, fu, hi*
CODD Central On-line Data Directory *ns*
CODE Client-Server Open Development Environment *pk, kg*
CODEC COMpression / DECompression *kg, ct, em, hi*
CODEC Encoder Coder and Decoder *mt, em, ct, csIII.30, tb, kg, fu*
CODEM Coded Modulator Demodulator system *hc*
CODES Common Digital Exploitation System *wp*
CODES Computerized Deployment System [MTMC] *mt*
CODES Computing Development System *hc, tb*
CODIAK Concurrent Development, Integration, and Application of Knowledge *hi*
CODIT Computer Direct to Telegraph *fu*
CODS Classified Output Distribution System *mt*
CODS CONUS Output Distribution System [MAC] *mt*
CODSIA Council Of Defense and Space Industries Association *wp*
CODSTAT Combat Distribution Team Status Report *mt*
COE Chief Of Engineering Engineers *st, jj*
COE Common Operating Environment [GCCS] [USA] *mt*
COE Concept Of Engineering *mt*
COE Corps Of Engineers [US Army] *mt, 0, hc, st, kf*
COEA Cost and of Operational Effectiveness Analysis *mt, wp, fu, st, jj, kf*
COEA Cost Effectiveness Analysis *mt*
COED Computer Operated Electronic Display *fu, jj*
COEI Component End Item *wp*
COEM Commercial Original Equipment Manufacturer *pk, kg*
COEMIS Corps Of Engineers Management Information System [USA] *mt*
COES Committee for Operational Environmental Satellites *ag*
COF Conduct-of-Fire *fu*
COFA Colocation Flutter Analysis *hc*
COFDEN Chief of Operational Forces, Denmark *fu*
COFF Common Object File Format [Unix] *hi, kg*
COFT Conduct Of Fire Trainer *hc*
COG Candidate Optimal Group *uh*
COG Center Of Gravity *gp, wp*
COG Compilation Order Generator *fu*
COG Continuity Of Government *mt, 12*
COGO Coordinate Geometry [Programming Language] *pk, kg*
COGS Chief Of the General Staff *hc, jj*
COGSTA Cognizant Shore Station *fu*
COHO Coherent Oscillator *hc, fu, wp, jj*
COHP/CARE Computerized Occupational Health Program Coronary Artery Risk Evaluation *mt*
COI Communications Operation Instruction *mt*
COI Community Of Interest *mt, fu*
COI Conflict Of Interest *kf*
COI Course Of Instruction *mt, fu*
COI Critical Objects of Interest *kf*
COI Critical Operational Issue *kf*
COIC Combat Operations Intelligence Center [USAFE] *hc, ro, wp, mt*
COID Combat Operations Intelligence Division [USAF] *mt*
COIL Chemical Oxygen-Iodine Laser *wp*
COIN Counter Insurgency [anti guerrilla warfare] *fu, mt*
COIN Counterinsurgency Information Analysis Center *hc*
COINS Commercial Operation Integrated System *mt*
COINS Community On Line Intelligence Network System [NSA] *mt*
COINS Community On Line System *ro*
COINS Consolidated On Line Intelligence System *wp*
COIR Commanders Operational Intelligence Requirements [NATO] *mt*
COL Communications Oriented Language *fu*
COL Computer Oriented Language *kg, wp*
COLAR Carbon di-Oxide Laser Radiometer *hc*
COLD Coherent Light Detector *hc*
COLD Computer Output to Laser Disk *pk, kg*
COLDS Common Optoelectronic Laser Detection System *wp*
COLDSAT (cryogenic . . . NASA . . . experiment . . . 1995) *aa27.4.32*
COLIDAR Coherent Light Detection And Ranging *hc, wp, jj, fu*
COLIDS Coherent Light Detector System *hc*
COLO Coherent Local Oscillator *hc*
COLOC Change Of Location Of Command *mt*
COLREGS International Nautical rules *mt*
COLT Carbon di-Oxide Laser Technology *hc*
COLTS Contrast Optical Laser Tracking System *hc*
COLUP Collection and Update *hn44.18.2*
COLV Cost Optimized Launch Vehicle *ci.xviii*
COM Character Oriented Message *fu*
COM Collection Management [the process of converting intelligence requirements into collection requirements] [USA DOD] *mt*
COM Collection Operations Management [USA DOD] *mt*
COM Component Object Model *em*

COM Computer Operation Manual *mt*
COM Computer Output [to] Microfilm *w9, fu, mt*
COM Concept Of Maintenance *mt*
COM Cost of Ownership Model *hc*
COM/ELEC Communications / Electronics *mt*
COMAAC Commander, Alaskan Air Command *mt*
COMAAFCE Commander, Allied Air Forces Central Europe [NATO] *mt, af72.5.97, fu*
COMAC Commander, Military Airlift Command *mt*
COMAFFOR Commander, Air Force Forces *mt*
COMAFFOR-AK Commander, Air Force Forces - Alaska *mt*
COMAFK Commander, Air Forces - Korea *mt*
COMAFSOC Commander, Air Force Special Operations Cmd *mt*
COMAIRBASE Commander of Air Base [USA] *mt*
COMAIRPAC Commander, Air Forces, Pacific Fleet [USA] *mt*
COMAL Common Algorithmic Language *hi*
COMALF Commander, Airlift Forces *mt*
COMALF-E Commander, Airlift Forces - Europe *mt*
COMANR Commander, Alaskan Norad Region *mt*
COMAO Combined Air Operations [USA] *mt*
COMARFOR Commander, Army Forces *mt*
COMARFOR-AK Commander, Army Forces - Alaska *mt*
COMARSPACE Commander, Army Space Command *as*
COMBALTAP Commander, Baltic Approaches *fu*
COMBAT_GRANDE [USAF project name for automation of Spain's Air Defense System] *fu*
COMBAT_TALON [AAQ-15 FLIR] *wp*
COMBO Computation Of Miss Between Orbits *mt*
COMBOX Computation of Miss Between Orbits [a software program] *jb*
COMBS Contractor Operated and Maintained Base Supply *wp*
COMBS Contractor Operated and Managed Base Supply *mt*
COMCEN Communications Center *mt*
COMCENTAG Commander, Central Army Group [NATO] *mt*
COMCO [Multicorporate team CECSA/HAC, Spain, for COMBAT_GRANDE] *hc, fu*
COMCONS Command and Control Reports *mt*
COMCUWTF Commander, Combined Unconventional Warfare Task Force *mt*
COMDEG Communications Degradation *mt*
COMDEV (Canada's COMDEV Limited will design the..filters) *sc2.8.2*
COMDEX Communications Development Exposition *ct, em*
COMDEX Computer Dealers Exposition *kg, wp, hi*
COMDIS Computerized Display of Deployment Considerations Subsystem [USA] *mt*
COMECON Council for Mutual Economic Assistance Aid *w9, wp*
COMED CONUS Meteorological Data [System] *fu*
COMEDS CONUS Meteorological Data System *mt, ag, no*
COMEDS CONUS Meteorological Environmental Distribution System *mt*
COMELSUM Communications-Electronics Summary *mt*
COMET Command Experience and Training *wp*
COMET Commercial Experiment Transporter *ai2*
COMET Cornell Macintosh Terminal Emulator *pk, kg*
COMFAC Communications Facility *hc*
COMFAIR Commander, Fleet Air *jj*
COMFAIRMIR Commander, Fleet Air, Miramar *jj*
COMFEP Communications Front-End Processor *fu*
COMFID Compressed Fault-Isolation Data *fu*
COMFITWINGONE Commander Fighter Wing One [Oceana] *jj*
COMFY_BANNER [USAF automatic processing system] *wp*
COMFY_BEE [high altitude drone for ELINT and IMINT] *wp*
COMFY_BOX [installing COMPASS_EARS hardware on USAF sites] *wp*
COMFY_BRIDLE [secure communications network in sea] *wp*
COMFY_CARD [air transportable mobile huts for radio communication] *wp*
COMFY_CHALLENGE [modular jamming and deception for tactical CCC] *wp*
COMFY_COAT [low altitude drone ? for ELINT] *wp*
COMFY_DISC [secure communications network in sea] *wp*
COMFY_DISH [SATCOM Jammer For Training, USAF] *mt*
COMFY_DOZEN [air transportable ground stations for communication] *wp*
COMFY_FLUFF [Reports In EW *wp*
COMFY_FOX [Mobile Signal Security Assessment Capability System] *wp*
COMFY_HARVEST [real time reporting and processing system] *wp*
COMFY_LEVI [Airborne COMINT Collection System, a psychological warfare version of the C-130 Hercules, EC-130E] [USAF] *mt*
COMFY_SHIRE [program to improve CCC countermeasures] *wp*
COMFY_SMOG [Spectrum Occupancy Study] *wp*
COMFY_SWORD [communications jammer for training, USAF] *wp, mt*
COMFY_TIP [Weekly Abstract In EW] *wp*
COMFY_VELVET [intercom system] *wp*
COMI Communications Interface *mt*

COMINT Communications Intelligence *mt, 10, 14, hc, af72.5.22, aw129.21.48, wp, tb, jj, fu, jdr*
COMIREX Committee on Imagery Requirements and Exploitation [sets specifications for spy satellites and the like, USA] *10, mt*
COMIT Communications Intelligence *hc*
COMJAM Communications Jamming *mt, tb, fu*
COMJSOTF Commander, Joint Special Operations Task Force *mt*
COMJTF Commander Joint Task Force *mt*
COMJTF-AK Commander, Joint Task Force - Alaska *mt*
COMJUWTF Commander, Joint Unconventional Warfare Task Force *mt*
COMLANDFOR Commander, Land Forces *mt*
COMLNOFAC Communication Liaison Facility *mt*
COMLOGNET Combat Logistics Network *jj*
COMLOGNET Communications Logistics Network *mt*
COMM Commercial Missions *wl*
COMM Communication[s] *mt*
COMM-HIL Communication Hardware In the Loop [simulation - emulation] *-17*
COMMA Composite Maneuver Augmentation *hc*
COMMAC Communications MACRO [Computer Language] *mt*
COMMANDO_ESCORT [code name for PACAF HF/SSB communications] *mt*
COMMARDEZ Commander, Maritime Defense Zone *mt*
COMMARFOR Commander, Marine Forces *mt*
COMMERMARLANT Commander, Merchant Marine Atlantic [USA] *mt*
COMMO Communications Officer *mt*
COMMZ Communications Zone [Theater] *mt*
COMM_CC Communications CC system [USA] *mt*
COMNAB Commander, Naval Air Base *jj*
COMNATONAVFORLANT Commander, NATO Naval Baltic Forces [USA] *mt*
COMNAVAIRLANT Commander, Naval Air Force, Atlantic Fleet *jj*
COMNAVAIRPAC Commander, Naval Air Force, Pacific Fleet *jj*
COMNAVAIRSTA Commander, Naval Air Station [USA] *mt*
COMNAVBASE Commander, Naval Base [USA] *mt*
COMNAVFOR Commander, Naval Forces *mt*
COMNAVOCEANCOM Commander Naval Oceanography Command *ag*
COMNAVSPACECOM Commander, Naval Space Command *uh*
COMNAVSURFLANT Commander of the Naval Surface Force, Atlantic fleet [USA] *mt*
COMNON Commander Allied Forces Northern Norway *fu*
COMNORTHAG Commander, Northern Army Group [NATO] *mt*
COMO Coherent Master Oscillator *fu*
COMO Combat Oriented Maintenance Organization *mt, wp*
COMOD Core Modification *fu*
COMOPS Communications Operating Summary *mt*
COMOPTEVOR Commander, Operational Test and Evaluation Force *jj*
COMOR Committee on Overhead Reconnaissance *10, jdr*
COMOSY Component Mode Synthesis [a computer program] *gg20, hc*
COMPASS CDC Assembly Language *tb*
COMPASS Commander Third Fleet Computer Assisted Search [PACFLT] *mt*
COMPASS Computer Alerted Surveillance System [USAF programs involving battle field surveillance] *wp, fu*
COMPASS Computerized Movement Planning and Status System [FORSCOM] *mt*
COMPASS COPE [USAF project name for a type of RPV] *fu*
COMPASS_ARROW [High Altitude IMINT Surveillance Flights] *wp*
COMPASS_BIN [Low Altitude IMINT Air Launched Reconnaissance Drone] *wp*
COMPASS_BRAKE [Installation of Ka-51 Cameras On Rf-4c] *wp*
COMPASS_BRIGHT [cryptological program dealing with compass dwell recovery] *wp*
COMPASS_CALL [communication jammer on C-130] *wp*
COMPASS_CAPE [reconnaissance program with the TR-1] *wp*
COMPASS_CLEAR [ECM integration program] *wp*
COMPASS_COOL [infrared countermeasures program] *wp*
COMPASS_COPE [high altitude IMINT reconnaissance drone] *wp*
COMPASS_COPE [high altitude, long endurance RPV for reconnaissance] *wp*
COMPASS_COUNT [the AN/AVD-2 laser line scanner camera RF-4C] *wp*
COMPASS_DART [comint with jamming capability] *wp*
COMPASS_DWELL [program to develop reconnaissance drones 24 hr loitering] *wp*
COMPASS_EARS [Emergency Backup For Intell Communications With An RPV] *wp*
COMPASS_FLAG [Sort of Reconnaissance Drone Involving the Qu-22b Aircraft ???] *wp*
COMPASS_FOG [system to detect and classify high energy sources] *wp*
COMPASS_FROST [high altitude reconnaissance exploitation test program] *wp*
COMPASS_GHOST [EO CM program on tactical aircraft] *wp*

COMPASS_GOLF [Installation of Ks-87 Cameras On Rf-4c] *wp*
COMPASS_HAMMER [pod mounted EO countermeasures] *wp*
COMPASS_HOME [radar to develop search and destroy missile] *wp*
COMPASS_LOOK [Airborne Laser/Infrared Transmission For Navigation] *wp*
COMPASS_NICKEL [ELINT program along the East German border] *wp*
COMPASS_ORGAN [Program To Demonstrate ECM and EW Capabilities of RPV TAC] *wp*
COMPASS_PASS [SAC signal analysis program] *wp*
COMPASS_POINTER [former program drone with foliage penetration radar] *wp*
COMPASS_PREVIEW [Single Console Analysis for IMINT Reconnaissance] *wp*
COMPASS_QUEEN [improved infrared reconnaissance sensor] *wp*
COMPASS_QUICK [Fine ELINT Analyzer On Rc-135] *wp*
COMPASS_ROBIN [Expendable COMINT Sensors From RPV-S] *wp*
COMPASS_ROSE [RPV Research Effort] *wp*
COMPASS_ROYAL [Advanced Collection Capability] *wp*
COMPASS_SEE [B-52 ECM upgrade to navigate using hostile chemical warfare and pulsed sign.] *wp*
COMPASS_SEVEN [special EO collection system on the RC-135] *wp*
COMPASS_SIGHT [Data Link To Transmit Infrared IMINT From Rf-4c] *wp*
COMPASS_WIDGET [communications jammer on C-130] *wp*
COMPASS_WORLD [Reconnaissance Drone By Sperry Univac] *wp*
COMPATWINGSLANT Commander, Patrol WIngs Atlantic [USA] *mt*
COMPCAMP Commentated Phase Comparison Amplitude Monopulse *hc*
COMPES Contingency Operation / Mobility Planning and Execution System [USAF] *mt*
COMPG Composite Group [USA] *mt*
COMPLEX Committee on Planetary and Lunar Exploration *so*
COMPOOL Communications Pool *fu*
COMPTEL COMPton TELescope [on CGRO] *mi*
COMPUSEC Computer Security *mt, af72.5.74, nt, wp, fu, tb*
COMPW (24th COMPW, Howard AFB, Panama) *af72.5.55*
COMRAD Computer Aided Design Environment *wp*
COMRATS Commuted Rations [USA] *mt*
COMREP Communications Report *mt*
COMS Collection Objectives Management System *mt*
COMS Communication Satellite *jdr*
COMS Contractor Operation / Maintenance of Simulators *hc*
COMS Contractor Operation and Maintenance Support *fu, hc*
COMS Conventional Munitions Management System *mt*
COMSAT Commercial Satellite *kf*
COMSAT Communications Satellite [Corporation] *mt, blIII.iii, fu, 16, kg, hc, jb, 19, tb, jj*
COMSATCOM Commercial Satellite Communications *mt*
COMSC Commander, Military Sealift Command *mt, fu*
COMSEC Communications Security *mt, su, uh, rm3.1.1, 4, 16, 17, th4, fu, hc, af72.5.74, nt, wp, dm, tb, st, jj, jdr, kf*
COMSECONDFLT Commander of the Second Fleet [USA] *mt*
COMSOC Commander, Special Operations Command *mt*
COMSONOR Commander Allied Forces Southern Norway *fu*
COMSPOT Communication Spot Report *mt*
COMSR Communications Support Requirements *mt, fu*
COMSS Coastal Ocean Monitoring Satellite System *ac3.7-21*
COMSTAC Commercial Space Transportation Advisory Committee [FAA] *ai2*
COMSTAR COMSTAT Global Telecommunications Satellite *mt*
COMSTAR [US domestic telephone communications satellite network] *hc, lt*
COMSTARC Commander, State Area Command *mt*
COMSTAS Communications Stations *fu*
COMSTAT Communications Status Report *mt*
COMSTRKFORLANT Commander, Strike Force Atlantic [USA] *mt*
COMSUBGRP Commander, Submarine Group [USA] *mt*
COMSUBLANT Commander, Submarine Forces, Atlantic Fleet [USA] *mt*
COMSUBPAC Commander, Submarine Forces, Pacific Fleet [USA] *mt*
COMSUBRON Commander, Submarine Squadron [USA] *mt*
COMSURFPAC Commander, Surface Forces, Pacific Fleet [USA] *mt*
COMTAC Commander, Tactical Air Command *mt*
COMTACWINGSLANT Commander, Tactical Wings, Atlantic Fleet *jj*
COMTAF Commander, Tactical Air Force *mt*
COMTEX Communications Oriented Multiple Terminal Executive *do422*
COMTHIRDFLT Command Third Fleet *fu*
COMTUEX Competitive Training Unit Exercise *jj*
COMUKADR Commander, UK Air Defense Region [NATO] *mt*
COMUSJ Commander, United States Forces, Japan *mt*
COMUSK Commander, United States Forces, Korea *mt*
COMUSMARFORK Commander, USMC Forces, Korea *mt*
COMVAT Combat Vehicle Armament Technology *hc*
CONAD Continental Aerospace Defense Command *mt*
CONAD Continental Air Defense [Command] *dh, hc, fu, jj*
CONARC Continental Army Command [USA] *hc, fu, jj, mt*

CONCRETE_SKY [USAF Development of Tactical Aitcraft Shelters] *wp*
CONDIGS Continental United States Digital Graphics System *mt*
CONDOR (a CONDOR like aircraft for an endurance mission) {an UAV} *kb*
CONELRAD Control of Electromagnetic Radiation *wp, jj*
CONEVAL Concurrent Evaluation *jj*
CONFAC [a computer program for cryogenic design] *hc*
CONFICS Cobra Night Fire Control System *hc*
CONIN Contamination Investigation *hc*
CONOCO (a participating organization) *ac2.7-19*
CONOP Concept of Operation *wp*
CONOPS Concept of Operations *mt, jb, jmj, wp, kf, uh*
CONOPS Continental United States Operations *wp*
CONOPS Continuity of Operations *pm, fu*
CONPLAN Concept Plan [operation Plan in Concept form] [USA DOD] *mt*
CONPLAN Contingency Plan *mt, wp*
CONRAD Computerized National Range Documentation *hc*
CONREP Contingency Construction Report *mt*
CONS Connection Oriented Network Service *hi, kg*
CONSCAN Conical Scan *hc*
CONSCARDS Constant Cards *mt*
CONSIDO Consolidated Special Information Dissemination Office *-10*
CONSTANT [programs connected with operational test and evaluation USAF] *wp*
CONSTANT_ANGEL [expendable tactical radar jamming drone] *wp*
CONSTANT_CARIBU [air-to-surface weapon system evaluation for anti aircraft] *wp*
CONSTANT_CLUSTER [initial operational test and evaluation of improved cluster munition] *wp*
CONSTANT_CONE [operational evaluation of the AIM-7F missile] *wp*
CONSTANT_DATA [initial operational test and evaluation of in-flight data transmission] *wp*
CONSTANT_FALL [comparison of glass and aluminum chaff] *wp*
CONSTANT_GUARD [aircraft development program] *wp*
CONSTANT_HIT [improved air to air missile reliability study] *wp*
CONSTANT_IMAGE [Modified AN/AAAS-18 Infrared System Operational Test] *wp*
CONSTANT_QUALITY [test and evaluation of a concept to improve reconnaissance system performance] *wp*
CONSTANT_STATIC [air to ground communication test] *wp*
CONSTANT_TRACK [test and evaluation of FLIR in the RF-4C] *wp*
CONSTEL [BMD] Constellation Analyzer [an SDI battle simulation program] *17, sb*
CONSTOCS CONUS Contingency Stocks [USA] *mt*
CONS_MODE Console Mode *mt*
CONT Confidence Test *jj*
CONT Container Ship [US Navy] *mt*
CONTAD Concealed Target Detection *hc, wp*
CONTONE Continuous Tone *pk, kg*
CONTRAIL Condensation Trail *wp*
CONUS Contiguous Continental United States [i. e., excluding Alaska] *mt, sm18-35, 5, sb, hc, af72.5.10, wp, no, tb, st, jj, jdr, kf, oe, su, uh*
CONUSA Continental US Army *mt*
CONVEX [a computer manufacturer] *nt*
COO Chief Operating Officer *kf, fu*
COOF Comptroller Office Of the Future [USAF] *mt*
COOL [programs connected with Alaskan Air Command] *wp*
COOP Continuity Of Operations Plan *mt, 12*
COOP Craft Of Opportunity Program [USA] *mt*
COORDWG Coordination Working Group *mt*
COOT Soviet Il-18 Aircraft *wp*
COP Career Opportunity Program *sc2.8.6, 15, ie11.5.2*
COP Central Operating Program *fu*
COP Combat Operations Processor *mt*
COP Command [and] Observation Post *mt, wp*
COP Contingency Operation Plan *mt*
COP Contingency Operation Procedure *oe, pl*
COP Craft of Opportunity Program [CSCG] *mt*
COP Current Operating Plan Procedure *fu, 8, 15, hc, lt, ep, jj, jdr, kf*
COPA Crisis Operations Procedures Analysis [JDSSC] *mt*
COPAN Command Post Alerting Network [USAF] *wp, mt*
COPASS Covert Programmable Authentication Signaling System *hc*
COPE Computer Operated Peripheral Equipment *hc*
COPERNICUS Naval SEW, CCC Architecture *mt*
COPERS COmmission Preparatoire Europeenne de Recherces Spatiales *crl*
COPES Cooperating Expert Systems [RADC] *mt*
COPH Contingency Operation Procedure Handbook *oe*
COPPER Contracting Improved Computer Support ICOMPS *mt*
COPPER_IMPACT USAF program to Improve Pricing And Costing Techniques *wp*
COPR Concept of Operations Review *mt*
COPRAM Contingency Personnel Resource Availability Model [USAF] *mt*
COPREC Command Post Record Capability Network [USAF] *mt*

COPS   Command Post Record Capability   *mt*
COPS   Communicators Of Problem Solutions   *rm3.1.4*
COPS   Concept Oriented Programming System   *ct, em*
COPS   Current Operations Support [OJCS]   *mt*
COPTR   Controllability, Observability, Predictability, and Testability Report   *fu*
COPTRAN   Computer Communication Optimization Program   *hc*
COPUOS   Committee on Peaceful Uses of Outer Space   *ci.1043, 13*
COR   Central Office of Record   *fu, hc, tb, su*
COR   Coherent On Receive   *fu*
COR   Continental Operating Range   *fu, wp*
COR   Contract Contractor Operations Review   *hn49.11.12, ie5.11.12*
COR   Contracting Officer's Representative   *mt, fu, tb*
COR   Contractor Operational Review   *jdr*
CORA   Coherent Radar Array   *fu, hc*
CORA   Conditioned Response Analog   *fu*
CORAD   Correlation Radar   *wp, fu*
CORADCOM   Communications Research and Development Command [US Army]   *mt, fu, hc*
CORAL   Command Radio Link   *fu*
CORAL   RADC Optical Radar System   *hc*
CORAPRAN   COBELDA Radar Automatic Preflight Analyzer   *hc*
CORBA   Common Object Request Broker Architecture   *ct, kf, kg, em, hi*
CORCENS   USAF Correlation Centers   *wp*
CORDIC   Coordinate Rotation Digital Computer   *fu*
CORDS   Coherent On Receive Doppler System [Look-down radar system, to be incorporated in the AN/APQ-120 radar, canceled in 1968] [USA]   *mt, hc, wp*
CORE   Coherent On Receiver   *hc*
CORE   Common Operating Research Equipment   *ci.xxiii*
CORE   Contingency Response [program]   *mt, wp*
CoREN   Corporation for Research and Enterprise Network   *pk, kg*
CORF   Committee on Radio Frequencies   *fu*
CORL   Complete Operational Recording List   *fu*
CORONA_ACE   [Review of Air To Air Combat Capabilities and Skills]   *wp*
CORONA_LOCK   [USAF arm and munitions security study]   *wp*
CORONET   [programs usually connected with the tactical air command]   *wp*
CORONET_BARE   [TAC Bare Base Demo]   *wp*
CORONET_BRIEF   [EW Training Briefing Kits]   *wp*
CORONET_CLAYMORE   [TAC HF single sideband network for TAC commanders]   *wp*
CORONET_COMMAND   [CCC in broad front operations study]   *wp*
CORONET_DRAGON   [TAC operations security program]   *wp*
CORONET_ECHO   [RPV/Drone Aerial Target Program]   *wp*
CORONET_OPTIC   [reconnaissance support for the Alaskan Air Command]   *wp*
CORONET_PRINCE   [electro-optical countermeasures program]   *wp*
CORONET_STEAM   [training program in communications electronics]   *wp*
CORONET_STEED   [concepts for ground sensor technology]   *wp*
CORONET_VOX   [single side band communications test]   *wp*
CORSPERS   Committee on Remote Sensing Programs for Earth Resource Surveys   *ac1-1.1*
COS   Change Of Status   *-15*
COS   Checkout Station   *bo*
COS   Chief Of Staff   *w9, jj, mt*
COS   Chief Of Station   *-10*
COS   Class Of Service   *ct, fu*
COS   Co Orbiting Satellite   *ac3.7-3, wl*
COS   Compatible Operating System   *pk, kg*
COS   Computer Operator Station   *fu*
COS   Conceptual Operational System   *hc*
COS   Critical Occupational Specialty [USA DOD]   *mt*
COSA   Carrier Onboard Spare Aircraft   *jj*
COSA   Combat Operational Support Aircraft   *jj*
COSAL   Coordinated Ships Allowance List   *fu*
COSAR   Compression Scanning Array Radar   *hc, wp*
COSATI   Committee on Scientific and Technical Information   *fu*
COSBAL   Coordination Shorebased Allowance List   *fu*
COSCOM   Corps Support Command [USA]   *fu, mt*
COSDIF   Cost Differential   *mt*
COSE   Combined Office Standard Environment   *pk, kg*
COSE   Common Open Software Systems Environment   *hi, kg*
COSG   Combat Operational Support Group   *mt*
COSG   Consolidated Contingency Steering Group   *mt*
COSG   Consolidated Operations Steering Group   *mt*
COSIN   Control Staff Instruction [EXERCISES]   *mt*
COSL/COSP   Atlantic, P
COSL/COSP   Commander, Ocean Systems L
COSL/COSP   Pacific   *fu*
COSMIC   Computer Service [Software] Management [and] Information Center [NASA]   *hc, ci.930, wp, kg*
COSMIC   [TOP SECRET information which is the property of NATO]   *su, tb*
COSMOS   Computer System for Mainframe Operations   *ct, kg, em*

COSMOS   Cost/Schedule Management Oriented System   *fu*
COSMOS   Norsk Data Intercomputer Communications   *fu*
COSMOS   [European Space Consortium]   *hc*
COSO   Combat Oriented Supply Organization   *mt*
COSPAR   Committee On Space Research [NATO, of ICSU]   *ac2.7-19, 0, 13, aa26.22.b4, ns, hg, ge, ai2*
COSPAS   [search and rescue satellite project, USSR]   *sc3.4.2, 19*
COSPOTS   Computer Simulation Pointing and Tracking System   *hc*
COSQ   Communications Operations Squadron   *mt*
COSRO   Conical Scan Receive Only   *fu, wp*
COSRO   Conical Scanning On Receive Only   *mt*
COSS   Common Object Services Specification   *pk, kg*
COSS   Conventional Ordnance Status System   *mt*
COSSAC   Chief Of Staff Allied Supreme Headquarters   *mt*
COSSAC   Cooled Spectral Shared Aperture Concepts   *hc*
COSSC   The Compton Observatory Science Support Center   *cu*
COSSTA   Computer for Special Small Tactical Application   *hc*
COSTAR   Corrective Optics Space Telescope Axial Replacement   *mi*
COSTARS   Combine Sensors Target Acquisition, Recognition, Strike   *hc*
COSTI   Committee on Scientific and Technical Information   *dt*
CoSysOp   Co-Systems Operator   *ct*
COT   Certificate Of Test   *fu*
COT   Commanding Officer of Troops [USA DOD]   *mt*
COT   Consolidated Operability Testing   *fu*
COTAM   Commandement du Transport AErien Militaire [military air transport command] [France]   *mt*
COTAR   Correlated Orientation, Tracking And Range Ranging [System]   *0, ful, hc, wp*
COTAWS   Collision and Obstacle Terrain Avoidance Warning System   *hc*
COTD   Commanding Officer's Tactical Display   *fu*
COTD   Critical Optics Technology Development   *hc*
COTES   Combat Orders, Training and Evaluation System [USACGSC]   *mt*
COTR   Contracting Officer's Technical Representative   *mt, 18, lt, tb, st, wl, fu, jdr, oe*
COTS   Commercial Off The Shelf [equipment, software]   *mt, go1.5, wp, tb, as, pbl, fu, jdr, kg, kf, oe, ah, y2k*
COTS   Contractor Off The Shelf   *kf*
COTSS   Classified Operational Telecommunications Switching System   *mt*
COTV   Cargo Orbital Transfer Vehicle   *ci.1056*
COVERT_STRIKE   [USAF program to develop bistatic radar technology]   *wp*
COVMAP   Coverage Map   *fu*
COZI   Communications Zone Indicator   *jj, fu, ro*
CP   Center of Pressure   *w9, tb, st, ai1*
CP   Central Processor   *fu, jj*
CP   Centrally Procured   *mt*
CP   Change Proposal   *mt*
CP   Channel Processor   *fu*
CP   Chemically Pure   *w9, wp*
CP   Circuit Package   *fu*
CP   Circular Polarization   *fu, sm9-17*
CP   Circularly Polarized   *wp*
CP   Civilian Personnel   *tb*
CP   Clock Pulse   *fu, bl2-7b.42, hc, jj*
CP   Cluster Process   *tb*
CP   Coefficient of Performance   *fu*
CP   Collective Protection [chemical warfare]   *wp*
CP   Command Post   *mt, fu, sd, w9, hs, wp, pm, tb, jj*
CP   Command Preparation [group]   *0*
CP   Command Procedure   *oe*
CP   Command Pulse   *fu*
CP   Commercial Plane   *wp*
CP   Communications Processor   *mt, fu*
CP   Company Policy [directive]   *pbl, tm, jdr, ah*
CP   Company Practice   *fu*
CP   Company Private   *hf, hc*
CP   Computer Program[ming]   *fu, sc4.1.6, 4, tb*
CP   Cone Point   *fu*
CP   Connect Presentation   *fu*
CP   Constant Pressure   *fu*
CP   Contact Point   *fu*
CP   Control Panel   *fu*
CP   Control Platform [see WCP]   *17, sb*
CP   Control Point   *fu*
CP   Control Procedure   *tb*
CP   Control Processor   *fu*
CP   Control Program   *mt, hi*
CP   Copy Protected   *pk, kg*
CP   Correlation Processor   *fu*
CP   Cost Plus   *tb*
CP   Cost Proposal   *hc*
CP   Cruise Power   *jj*
CP&DC   Corrosion Prevention and Deterioration Control   *nt*
CP-DBFN   Control Platform Digital Beam Forming Network   *-17*

CP/M   Computer Program / Microprocessor   *mt*
CP/M   Control Program / [for] Microcomputers   *ct, wp*
CP/M   Control Program for Microcomputers Microprocessors [Digital Research]   *kg, fu*
CP/VC   Centre de Perfectionnement / Vervolmakingscentrum [advanced training center] [Belgium]   *mt*
CPA   Cash Purchasing Agent   *wp*
CPA   Central Pulse Amplifier   *fu*
CPA   Certified Public Accountant   *fu, hn44.23.2, w9, kg*
CPA   Civilian Personnel Authorized and Assigned Analyzer   *mt*
CPA   Closest Point of Approach   *fu*
CPA   Command - Complement A   *jj*
CPA   Concurrent Photon Amplification [increased film speed for RECCE]   *wp*
CPA   Connect Presentation Accept   *fu*
CPA   Continuous Patrol Aircraft   *uc*
CPA   Corporate Purchase Agreement   *fu*
CPA   Cost Performance Analysis   *wp*
CPA   Cost Planning and Appraisal   *hc, wp, jj*
CPA   Cost Plus Award   *wp*
CPA   Czechoslovak People's Army   *fu*
CPAC   Coded Pulse Anticlutter System   *fu*
CPADS   Covert Passive Air Defense Sensors   *wp*
CPAE   Contractor Furnished Aeronautical Equipment   *fu*
CPAF   Cost Plus Award Fee [a type of contract]   *mt, ci.926, 8, 17, sb, hc, ts4-9, wp, ep, tb, jj, jdr, fu, kf, oe*
CPAO   Calibration Property Action Order   *hc, jj*
CPAR   Contractor Performance Assessment Report   *kf*
CPARS   Contractor Performance Assessment Reporting System   *kf*
CPAS   Central Procurement Accounting System   *mt*
CPASS   Command Post Automated Staff Support System [TRADOC]   *mt*
CPAT   Critical Process Assessment Tool   *kf*
CPB   Capabilities, Programming and Budgeting   *mt*
CPB   Charged Particle Beam   *uc, wp, tb, st*
CPB   Combined Parts Buy   *jdr*
CPB-N   Council of Personnel Boards - Navy   *dt*
CPBW   Charged Particle Beam Weapon   *hc*
CPC   Clock Pulsed Control   *fu*
CPC   Computer Process Control   *fu*
CPC   Computer Program Component   *mt, fu, ac7.5-9, pl, jb, tb, jdr*
CPC   Constant Point Calculation   *pk, kg*
CPCI   Computer Program Configuration Item   *mt, ac7.5-9, 17, hc, hn49.11.7, fu, wp, tb, jj, jdr, oe, uh*
CPCIT   Computer Program Configuration Item Test   *fu*
CPCMP   Computer Program Configuration Management Plan   *fu*
CPCP   Contamination Prevention and Control Plan   *jdr*
CPCP   Contractor Performance Certification Program   *nd74.445.55*
CPCS   Check Processing Control System [IBM]   *pk, kg*
CPCS   Combat Personnel Control system [USAF]   *wp, mt*
CPCSB   Computer Program Configuration Sub Board   *mt*
CPD   Center for Professional Development   *af72.5.72*
CPD   Central Processing and Distribution System [TRIMIS]   *mt*
CPD   Clutter Pattern Detector   *fu*
CPD   Combat Plans Division   *mt*
CPD   Contact Potential Difference   *ai1*
CPDA   Computer Program Development Activity   *fu*
CPDF   Central Personnel Data File [OPM]   *mt*
CPDM   Computer Program Development Manager   *fu*
CPDP   Command Post Display Processor   *mt*
CPDP   Computer Program Development Plan   *mt, wp, tb, jdr, fu, kf*
CPDR   Computer Program Design Requirements   *jj*
CPDS   Computer Program Design Development Specification   *mt, jj, fu, jdr*
CPE   Central Processing Element   *pk, kg*
CPE   Circular Probable Error [Cfr. CEP]   *wp*
CPE   Collective Protection Equipment [chemical warfare]   *wp, hc*
CPE   Communications Processor Element   *mt*
CPE   Computer Performance Evaluation   *mt*
CPE   Contractor Performance Evaluation   *hc, jj*
CPE   Conventional Planning and Execution   *mt*
CPE   Customer Premises Equipment   *mt, kg, ct, y2k*
CPE   Customer Provided Equipment   *pk, kg*
CPEG   Contract Performance Evaluation Group   *hc, jj*
CPET   Contractor Performance Evaluation Team   *mt*
CPF   Canadian Patrol Frigate   *pf1.9*
CPFA   Computer Program Functional Area   *fu*
CPFB   Chloropentafluorobenzene   *wp*
CPFF   Cost Plus Fixed Fee   *mt, bl2-3.20, 8, hc, w9, ie11.5.2, wp, ep, tb, st, jj, jdr, fu, kf, oe*
CPFS   Computer Program Functional Specification   *fu*
CPFSK   Continuous Phase Frequency Shift Keying   *fu*
CPG   Clock Pulse Generator   *kg, wp*
CPG   Communications Products Group   *fu*
CPG   Corporate Planning Group   *wp*
CPG   Custom Products Group   *fu*
CPI   Channel Protocol Interpreter   *mt*

CPI   Characters Per Inch   *mt, fu, hi, kg*
CPI   Chemical Process Industries   *y2k*
CPI   Clock [pulse cycles] Per Instruction   *hi, kg*
CPI   Common Programming Interface [IBM]   *hi, kg*
CPI   Comparability Pay Increases   *tb*
CPI   Consumer Price Index   *w9, wp*
CPI   Continuous Process Improvement   *kf*
CPI   Contractor Productivity Improvement   *nd74.445.55*
CPI   Cost Performance Index [efficiency]   *fu, jdr, kf, mt*
CPI   Crash Position Indicator   *wp*
CPI-C   Common Programming Interface for Communications [IBM]   *pk, kg*
CPIF   Cost Plus Incentive Fee [a type of contract]   *mt, ci.926, 8, hc, wp, ep, tb, jj, fu, jdr, kf*
CPIN   Computer Program Identification Number   *mt, hc, wp, fu, tb, jdr*
CPIO   Central Processor Input Output   *jj*
CPIP   Cooperative Propulsion Integration Program   *wp*
CPIR   Centre de Prediction et Instruction Radar [France]   *mt*
CPIR   Crash Position Indicator Recorder   *hc*
CPIR   Critical Path Items Report   *fu*
CPIS   Contingency Planning Information System   *mt*
CPKM   C Perigee Kick Motor [Long March 2C]   *ai2*
CPL   Certified Professional Logistician   *wp*
CPL   Characters Per Line   *hi*
CPL   Commercial Pilot License   *mt, wp*
CPL   Commercial Price List   *mt*
CPL   Communications Payload   *kf*
CPL   Computer Program Library   *fu*
CPL   Computer Programming Laboratory   *fu*
CPL   Console Programming Language   *fu*
CPL   Consolidated Parts List   *tb*
CPL   Control Panel Layout   *em*
CPL   Critical Path List   *fu*
CPL   Current Privilege Level   *pk, kg*
CPLF   Centralized Program Library Facility   *fu*
CPLR   Coupler   *bl2-9.15*
CPM   Cards Per Minute   *jj*
CPM   Characters Per Minute   *mt, wp, hi*
CPM   Chinese Procurement Mission   *hc*
CPM   Command Post Modem   *fu*
CPM   Company Property Manual   *fu*
CPM   Complex Phasor Modulator   *fu*
CPM   Computer Performance Management   *mt*
CPM   Computer Programming Manual   *jj*
CPM   Contractor Performance Measurement   *mt, fu*
CPM   Controlled Product Management   *wp*
CPM   Counts Per Minute   *wp, hi*
CPM   Critical Path Method   *mt, fu, 8, hc, wp, jj, pbl, jdr, ah*
CPM   Cycles Per Minute   *w9, wp, jj*
CPM/P   Command Post Modem/Processor   *fu*
CPMF   Computer Program Maintenance Facility   *mt, fu*
CPMF   Computer Program Management Facility   *mt*
CPMI   Computer Program Mnemonic Identifiers   *fu*
CPMIEC   China Precision Machinery Import and Export Corporation   *aw130.6.19*
CPMIS   Civilian Personnel Management Information System   *mt*
CPMM   Computer Program Maintenance Manual   *fu*
CPMP   Communications Panel Master Processor   *fu*
CPMP   Computer Program Management Plan   *mt*
CPMR   Computer Program Management Review   *mt*
CPMS   Career Planning Management system [USMC]   *mt*
CPMS   Computer Performance Management System   *mt*
CPMT   Command Procedure Message Text   *oe*
CPMU   Command Post Modem Unit   *fu*
CPNF   Cost Plus No Fee   *wp*
CPO   CAATS Project Office   *fu*
CPO   Canadian Program Office   *fu*
CPO   Cash Purchase Order   *fu*
CPO   Chief Petty Officer   *w9, jj*
CPO   Civilian Personnel Office   *mt, wp*
CPO   Command Pulse Output   *fu*
CPO   Company Policy [a directive]   *ep*
CPO   Component Pilot Overhaul   *jj*
CPODA   {a model for packet transmission demand assignment}   *csll.85*
CPOM   Computer Program Operator's Manual   *fu*
CPOM   Master Chief Petty Officer   *w9*
CPOP   Computer Operator   *fu*
CPOS   Senior Chief Petty Officer   *w9*
CPP   Civilian Personnel Policy   *wp*
CPP   Command Post Processor   *fu*
CPP   Communications Panel Processor   *fu*
CPP   Component Procurement Plan   *jj*
CPP   Critical Performance Parameter   *kf*
CPP   Cumulative Pulse Priority   *fu*
CPPC   Cost Plus Percent of Cost   *hc, jj*
CPPL   Contractor Preferred Parts List   *hc*

CPPS Computer Program Product Performance Procurement Specification *tb, jj, jdr, fu*
CPPSO Cargo Projects Payload Support Office *oe*
CPPSO Consolidated Personal Property Shipping Office *mt*
CPPU Command Post Processor Unit *fu*
CPQ Command - Complement Q *jj*
CPR Cardio Pulmonary Resuscitation *sc3.9.3, w9, hn49.12.3, wp*
CPR Change Proposal Request *tb*
CPR Changing Pulse Radar *hc*
CPR Circuit Package Schematic *fu*
CPR Company Practice [a directive] *ep, jdr, tm*
CPR Component Pilot Rework *hc, jj*
CPR Connect Presentation Reject *fu*
CPR Contract Performance Report *fu, oe*
CPR Contract Procurement Request *jj*
CPR Contract Procurement Requirement *fu*
CPR Cost Performance Report *mt, fu, 17, sb, nt, wp, dm, tb, st, jdr, kf*
CPR Critical Program Review [General Dynamics] *oe*
CPRC Calibration Problem Referral Card *jj*
CPRS Command Procurement Reporting System [USAF] *mt*
CPS Cards Per Second *w9*
CPS Central Processing System *fu*
CPS Central Propulsion System *kgl*
CPS Ceramics Process Systems [Corporation] *aw129.14.103*
CPS Characters Per Second *mt, ct, em, kg, w9, wp, tb, hi*
CPS Circuit Package Schematic *fu*
CPS Civilian Public Service *w9*
CPS Commander's Panoramic Sight *hc*
CPS Computer Power Supply *fu*
CPS Computerized Presentation System *fu*
CPS Contractor Plant Services [Personnel] *hc, jj*
CPS Control Power Supply *hc*
CPS Core Processing System *mt*
CPS Corporate Pricing System *jj, fu*
CPS Cycles Per Second *mt, w9, wp, tb, jj, kg, hi*
CPSC Computer Services Center *mt*
CPSC Contingency Planning Support Capability *mt*
CPSD Computer Programming Standards Document *fu*
CPSD Cursor Positioning/ Selection Device *fu*
CPSE Complementary Pair Switch Element *hc, fu*
CPSK Coherent Detection Phase Shift Keying *csIII.33*
CPSL Computer Program Support Library *fu*
CPSM Continuous Phase Shift Modulation *fu*
CPSM Critical Path Scheduling Method *wp*
CPSO Computer Program Support Organization *fu*
CPSR Contractor Procurement Purchasing System Review *fu, tb*
CPSS Central Processing Sub System *hc*
CPSS Contract Parts Status System *fu*
CPT Captive Pod Trainer *hc, jj*
CPT Channel Parameter Table *fu*
CPT Cockpit Procedures Trainer *mt, wp, jj*
CPT Command Pass Through *pk, kg*
CPT Comprehensive Performance Test *oe*
CPT Control Power Transformer *fu*
CPT Critical Path Technique *fu*
CPT&E Computer Program Test and Evaluation *mt, tb*
CPTC Central Processor Test Console *hc*
CPTC Computer Performance Technical Center *mt*
CPTC Consolidated Pentagon Telecommunications Center *mt*
CPTE Computer Program Test Equipment *mt*
CPTP Computer Program Test Plan *fu*
CPTPR Computer Program Test Procedures and Results *fu*
CPTS Computer Program Test Specification *fu*
CPTSD Computer Program Timing and Sizing Data *jdr*
CPU Central Processing Unit *0, 6, 16, 17, bo, sb, aw129.14.103, hc, pl, aa26.11.44, nt, wp, tb, st, jj, jdr, pbl, tm, kf, oe, kg, 8, hc, wp, wp, jj, pbl, jdr, ah, fu, hi, mt, ai1*
CPU Collective Protection Unit [chemical warfare] *wp*
CPU Communications Processor Unit *mt, fu*
CPU Control Processing Unit *jdr*
CPUM Computer Program Users' Manual *jj*
CPUSS Computer Support Squadron *mt*
CPV Correlation Priority Value *fu*
CPVC Critical Pigment Volume Concentration *hc*
CPWG Center Projects Working Group [DMS] *mt*
CPX Command Personnel Exercise *fu*
CPX Command Post Exchange *fu*
CPX Command Post Exercise *mt, wp*
CQ Call to Quarters *w9*
CQ Charge of Quarters *w9*
CQ Correlation Quality *fu*
CQMINT Communications Intelligence *su*
CQP Company Quality Practice *gp*
CQV Crew Quarters Vehicle MGS *kf*
CR Capability Release *st*
CR Card Reader *fu*

CR Cathode Ray *w9, jj*
CR Central Region *fu*
CR Change Request *mt, 4, kf*
CR Channel Register *hc*
CR Character Reader *mt*
CR Clarification Request *mt, kf*
CR Combat Rescue *mt, wp*
CR Communications Relay *tb*
CR Company Records *fu*
CR Computer Resource Acquisition Course [USAF] *mt*
CR Continuous Rod *hc, jj*
CR Contrast Ratio *fu*
CR Control and Reporting [USA] *mt*
CR Control Relay *fu*
CR Correlation Routine *jj*
CR Cost Reimbursement *hc*
CR [chemical warfare codename for dibenz-1-4-oxazepine an irritant] *wp*
CR&D Contract Research and Development *kf*
CRA Catalog Recovery Area *fu*
CRA Central Repair Activity *fu*
CRA Centro Ricerche Aerospaziali *crl*
CRA Change Review Authorization Action *17, hc*
CRA Conflict Resolution Alert *fu*
CRA Continuing Resolution Authority *mt, tb*
CRAB Caging Retainer And Boresighter *hc, jj*
CRAB [Aircraft Discrepancy - GAC] *jj*
CRAC Computer Resource Acquisition Course [USAF] *mt*
CRAD Chief Or Research And Development *wp*
CRAD Composite Radar Data Processing *wp*
CRAD Composite Requirements Allocation Document *kf*
CRAD Contract Research And Development *ah, aa26.11.31, wp, pbl*
CRAF Civil Reserve Air Fleet [a program in which the US Department of Defense uses aircraft owned by a US entity or citizen] *mt, aw118.14.31, hc, af72.5.77, wp*
CRAF Comet Rendezvous Asteroid Flyby [mission] *es55.29, aa27.4.10, so, eo, re, mi, mb*
CRAFTS Civil Reserve Auxiliary Fleet Ships *wp*
CRAG Contractor Risk Assessment Guide *wp*
CRALCC Central Region Airlift Control Center *mt*
CRALD Central Region Airlift Division *mt*
CRAM Card Random Access Memory *em*
CRAM Centro de RetransmisiÛn Autom-tico de Madrid [automatic retransmission center] *fu*
CRAM Combat Resource Allocation Model *wp*
CRAM Computational Random Access Memory *em*
CRAM Computerized Reliability Analysis Method *wp*
CRAM Cyberspatial Reality Advancement Movement *pk, kg*
CRAOC Central Region Air Operations Center [NATO] *mt*
CRASH Computer Remediation and Shareholder *y2k*
CRATE [NATO code name for Soviet IL-14 transport aircraft] *wp*
CRAW Carrier Replacement Air Wing [later RCVW] *hc, jj*
CRAY [a computer manufacturer] *nt*
CRB Change Review Board *gp, fu, 15, 17, hc, lt, wp, ep, jj, jdr, kf*
CRB Configuration Review Board *mt*
CRB Contract Budget Baselines *jj*
CRB Contract Review Board [DCA] *mt*
CRB Cramer-Rao Bound *fu*
CRC Carlsbad Research Center Hughes *fu*
CRC Chemical Rubber Company *cr*
CRC Coleman Research Corporation *fu, st, ls4, 5, 5*
CRC Column Reference Code *mt*
CRC Combat Readiness Center *fu*
CRC Combat Reporting Center *mt, wp*
CRC Command and Reporting Center *fu, mt*
CRC Command Review Council *mt*
CRC Communications Relay Center *mt, wp*
CRC Company Records Center *fu, hd1.89.y19*
CRC Computer Resources Council *mt*
CRC Contingency / Regional Conflict *mt*
CRC Contractor Recommended Code *hc, jj*
CRC Control and Reporting Center [NATO] *fu, mt*
CRC Control [and] Report and Reporting Center *hc, pf.f88.25, wp, pm, kf*
CRC Cost Reduction Curve *fu*
CRC Cyclic Redundancy Check Checking *mt, kg, nu, em, uh, ct, csII.14, wp, jdr, oe, fu, hi*
CRC Cyclical Redundancy Code *kf*
CRCC Combined Rescue Coordination Center *mt*
CRCC Cyclic Redundancy Check Character *oe*
CRCS Change Request Control System *hd1.89.y16*
CRD Capstone Requirements Document *kf*
CRD CINC's Required Date [the original date relative to C-day, specified by the combatant commander for arrival of forces or cargo at the destination] [USA DOD] *mt*
CRD Computer Readout Devices *fu*
CRD Concept Requirements Document *fu*

CRD   Contractor Repair Depot   *fu*
CRD   {Corporate Research and Development}   *-19*
CRDB   Contingencies Resources Data Base [US Navy]   *mt*
CRDC   Chemical Research and Development Center [of AMCOM]   *tc*
CRDEC   Chemical Research Development and Engineering Center [chemical warfare]   *wp*
CRDF   Cathode Ray Direction Finding Finder   *hc, fu*
CRDL   Chemical Research and Development Laboratories [Army]   *fu*
CRDS   Component Repair Data Sheet   *jj*
CRDS   Component Requirement Data Sheet   *hc, jj*
CRE   Combat Readiness Evaluation   *wp*
CRE   Computer Reprogramming Equipment   *mt*
CRE   Console Remote Equipment   *jj*
CRE   Crisis Response Element   *kf*
CREATE   Combat Readiness Analysis Team Effort   *wp*
CREATE   Computational Resources for Engineering And [Simulation] Training and Education   *wp, mt*
CREDATA   Communications Resources Data System   *hc*
CREDIT   CINC_NORAD Remote Display Information Terminal   *mt*
CREDIT   Cost Reduction by Early Decision Information Technology   *hc*
CREDO   Comprehensive Report Examination and Display Option   *mt*
CREEK   [NATO code name for Soviet YAK-12 aircraft]   *wp*
CREEK   [USAFE Programs]   *wp*
CREEK_BOY   [USAFE night/day photo reconnaissance program]   *wp*
CREEK_DOOR   [intelligence project]   *wp*
CREEK_MAGPIE   [intelligence project ??]   *wp*
CREEK_YUCCA   [intelligence collection program]   *wp*
CREF   Civil Reserve Air Fleet   *wp*
CREN   Computer Research Education Network   *pk, kg*
CREN   Corporation for Research and Education Networking   *pk, kg*
CRES   Command Readiness Exercise System [AU]   *mt*
CRES   Computer Readability Editing System   *fu*
CRES   Corrosion Resistant Stainless Steel   *uh, fu, hc, aw129.21.46, jj, jdr, oe*
CRESS   Combat Readiness Evaluation Simulation System   *st*
CREST   Consolidated Reporting and Evaluating System, Tactical   *wp*
CREST   Covert Reconnaissance/Strike   *wp*
CREST   Crew Escape Technology   *wp*
CRESTED   [nickname for HQ USAF programs]   *wp*
CRESTED_DRAGON   [USAF chemical warfare munitions capability study]   *wp*
CRESTED_FALCON   [non-nuclear munitions capability study]   *wp*
CRESTED_ROOSTER   [telemetry program]   *wp*
CREW_CHIEF   [Air Force maintenance model, computer simulation]   *aa26.12.16*
CRF   Cable Retransmission Facility   *pk, kg*
CRF   Central Repair Facility   *wp, fu*
CRF   Clutter Rejection Factor   *sa830182P.10*
CRF   Contractor Repair Facility   *hn44.10.1*
CRF   Cross Reference File   *kg, jj*
CRF   Cryptographic Repair Facility   *wp*
CRFA   Contract Requirements Flowdown Analysis   *fu*
CRG   Change Review Group   *wp*
CRHS   Component Reliability History Surveys   *hc*
CRI   Collective Routing Indicator   *mt*
CRI   Color Reproduction Indices   *pk, kg*
CRI   Cray Research, Incorporated   *tb*
CRICS   Central Region Integrated Communications System [NATO]   *mt*
CRIMREP   Crisis Management Information Reporting System   *mt*
CRIMS   Cost Risk Identification and Management System   *kf*
CRIP   Collective Routing Indicator Processing   *mt*
CRIS   Center for Rectification of Images from Space [Toulouse]   *ac3.6-23*
CRIS   Coastal Radar Integration Segment   *hc*
CRIS   Command Retrieval Information System   *jj*
CRIS   Computer / Radar Interface System   *fu*
CRIS   Current Research Information System   *hc*
CRISC   Complex Reduced Instruction Set Computer   *hi*
CRISD   Computer Resources Integrated Support Data Document   *mt, fu, kf*
CRISL   Contract Repair Initial Support List   *wp*
CRISP   Computer Resources Integrated Support Plan   *mt, wp, fu*
CRISP   Cost, Risk, Installation, Supportability, and Price   *jdr*
CRISPR   CRISP Review   *mt*
CRISYS   Computer Related Information Symposium   *mt*
CRITIC   Critical Intelligence Report   *mt*
CRITIC   Critical Intelligence [message]   *-10*
CRITIC   Critical Report   *mt*
CRITICOM   Critical Communication   *mt*
CRITICOM   Critical Intelligence Communications [network] [system]   *mt, 10, tb*
CRITICOM   Cryptologic Communications   *wp*
CRL   Common Rail Launcher   *mt*
CRL   COMSEC Requirements Lists   *mt*
CRL   Conflict Resolution Logic   *fu*
CRLCMP   Computer Resources Life Cycle Management Plan   *mt, fu*
CRM   Collection Requirements Management [USA DOD]   *mt*

CRM   Communications Resource Manager   *mt*
CRM   Cover and Removal Mechanism   *sd*
CRM   Cradle Release Mechanism   *gp, jdr*
CRM   Crew Resource Management   *kf*
CRMA   Combat Rescue Mission Analysis   *mt*
CRMO   Corporate Records Management Office   *hd1.89.y19*
CRMP   Computer Resources Management Plan   *wp, fu, st*
CRMP   Computer Resources Management Program   *mt*
CRMT   Computer Resources Management Technology   *mt*
CRMTS   Cryogenic Refrigerator Maintenance Test Set   *hc*
CRN   Carriage Reference Number   *jj*
CRN   Computer Reseller News   *ct*
CRO   Cathode Ray Oscilloscope   *fu, 0, jj*
CRO   Character Readout   *fu*
CRO   Civilian Repair Organization [set up during WWII to repair aircraft damaged in combat, UK]   *mt*
CRO   CRT Readout   *fu*
CROB   Construction Resources Data Base {?}   *mt*
CROC   Combat Required Operational Capability   *wp*
CROH   Control Room Operations Handbook   *oe*
CROM   Control Read Only Memory   *kg, fu, hi*
CRONUS   [advanced computer distributed operating system]   *ro*
CROPP   Clutter Rejection Optical Pre-Processing   *hc*
CROW   [NATO code name for Soviet YAK-12 trainer]   *wp*
CROWN   [HMX1 helicopter for presidential movement]   *mt*
CRP   Capacity Requirements Planning   *fu*
CRP   Celestial Relays and Power   *0*
CRP   Central Repair Point   *fu*
CRP   Change Request Process   *jdr*
CRP   Combat Reporting Post   *fu*
CRP   Communications/Radar Processor   *fu*
CRP   Completion Report   *mt*
CRP   Component Repair Program   *jj*
CRP   COMSEC Resources Program / Plan   *mt*
CRP   Contract Requirements Package   *tb*
CRP   Control and Reporting Post [USA, USAF]   *fu, mt*
CRP   Control [and] Reporting Post   *hc, wp*
CRP   Coordinated Reconnaissance Plan   *mt*
CRP   Cost Reduction Program   *fu, jj*
CRP   Modular Control Equipment {?}   *ts17-1*
CRPL   Central Radio Propagation Laboratory   *hc*
CRPO   Consolidated Reserve Personnel Office [USAF]   *mt*
CRPSP   Contractor Repair Parts Stock Point   *fu*
CRR   Communicant RAPID Report   *fu*
CRRD   Cost Reimbursement Research and Development   *hc*
CRREL   Cold Regions Research and Engineering Laboratory [US Army]   *tc*
CRRES   Combined Release and Radiation Effects Satellite [US Navy]   *mt, sc3.8.3, hc, ns, hy, mi, mb*
CRS   Calibration Recall System [USA]   *mt*
CRS   Calibration Requirements Summary   *fu, hc, jj*
CRS   Capability Reporting System [US Navy]   *mt*
CRS   Chip Requirements Specification   *fu*
CRS   Command Readout Station [of DMSP]   *-16*
CRS   Component Repair Squadron   *mt, fu*
CRS   Comprehensive Retrieval System   *mt*
CRS   Computer Reservation System   *aw130.13.33*
CRS   Computer Resources Support   *mt, wp*
CRS   Congressional Research Service Center   *ci.v, wp, tb*
CRS   Control and Reporting Center   *wp*
CRS   Control Reset   *jj*
CRS   Cooperative Reporting System   *mt*
CRSP   Contractor Recommended Support Plan   *jj*
CRSTADL   Communications, Reconnaissance, Survivability Target Acquisition Data Link   *hc*
CRT   Calibrated Radiance Tape   *oe*
CRT   Cathode Ray Tube   *mt, aw130.13.107, sc4.1.center, 6, 15, hc, aa26.12.78, w9, af72.5.135, lt, dm, kg, jj, wp, tb, oe, ct, em, ro, su, fu, hi, ai1*
CRT   Circuit Requirement Table   *fu*
CRT   Controller/ Receiver/ Transmitter   *fu*
CRT   Cyclic Redundancy Test   *oe*
CRTC   CRT Controller   *pk, kg*
CRTS   Controllable Radar Target Simulator   *fu*
CRTS   Cryogenic Refrigerator Test Station   *jj*
CRU   Command Response Unit   *fu*
CRUD   Corrosive, Radioactive, Undetermined Deposit [USA] "crud"   *mt*
CRUDESGRU   Cruiser-Destroyer Group   *-12*
CRUS   Create / Retrieve / Update / Delete   *em*
CRUSK   Center for Research on Utilization of Scientific Knowledge   *fu*
CRV   Canadian Rocket Vehicle   *hc*
CRW   Carrier Wave   *wp*
CRWG   Computer Resources Working Group   *mt, hc, fu, wp*
CRYPTA   Cryptoanalysis   *wp*
CRZ   Controlled Reaction Zone   *fu*
CS   Calibration Switch   *fu, hc, w9, jj*

CS Calm Sea *ac7.1-37*
CS Camera System *re*
CS Celestial Sensor *hc*
CS Central Secretarial *ch*
CS Chief of Staff *fu*
CS Chip Select *pk, kg*
CS Circuit Switching *mt*
CS Civil Service *mt, w9, wp*
CS Clear to Send *hi, kg*
CS Code Segment *fu, wp, kg*
CS Color Scanner *hy*
CS Combat Support *mt, fu*
CS Command Select *-9*
CS Commercial Standard *fu*
CSI Commisaryman [USA] *mt*
CS Common Subscriber *fu*
CS Communications Computer Systems *mt*
CS Communications Satellite *sc2.8.4, ce*
CS Communications Segment *hc*
CS Communications Systems *fu*
CS Company Standard *fu*
CS Competitive-Sensitive *wp*
CS Computer Science *hi*
CS Computing Station *fu*
CS Conducted Susceptibility *oe, jj*
CS Constant Source *mt*
CS Control Store memory *fu*
CS Control Switch *fu*
CS Cross Scan *mt*
CS Crossover Switch *mt*
CS Customer Service *hi*
CS Cycle Shift *fu*
CS Seaplane Tender [WW2] [USA] *mt*
CS Strike cruiser [post WW2] [USA] *mt*
CS [chemical warfare codename for o-chlorobezylidene malonic nitrile an irritant] *wp*
CS&C Communications Switching and Control [CDA portion of GIM-TACS] *oe*
CS/DL Combined Sensor/Data Link *hc*
CS/SS Card Service/Socket Service *pk, kg*
CS/TA Combat Surveillance/Target Acquisition *fu*
CSA Calendaring and Scheduling API [IBM] *pk, kg*
CSA Canadian Space Agency *hg, ge*
CSA Canadian Standards Association *fu*
CSA Celestial Sensor Assembly *pl*
CSA Chief of Staff, US Army *mt*
CSA Combat Support Aircraft *mt*
CSA Command Support Aircraft [USA] *mt*
CSA Communications Security Agency *fu*
CSA Communications Service Authorization *mt*
CSA Computer Sciences of Australia *hn43.17.8*
CSA Configuration Status Accounting *fu, mt, wp, kf*
CSA Continuous Semi Active *jj*
CSA Control Store Assembler *fu*
CSA {a high level organization within USASDC} *jmj*
CSAF Chief of Staff of the Air Force *mt, wp*
CSAF Czechoslovakian Air Force *mt*
CSAG Combat System Advisory Group *fu*
CSAGI Special Committee of the International Geophysical Year *ci.41, 13*
CSAM Counter Soviet AWACS [SUAWACS] Missile *uc*
CSAR Channel System Address Register *pk, kg*
CSAR Coherent Synthetic Aperture Radar *ci.955*
CSAR Combat Search And Rescue [USA DOD] *wp, mt*
CSAR Configuration Status Accounting Reports *mt, hc, fu, jj*
CSAR Control Store Address Register *hc*
CSAR/PR Combat Search and Rescue/Personnel Recovery *mt*
CSARC Command Systems Acquisition Review Council *fu*
CSAS Configuration Status Accounting System *mt, wp, fu*
CSAT Components Selected At Test *fu*
CSAT Crew Station Automation Automated Technology *hc, mt*
CSAW Close Support Assault Weapon *wp*
CSB Chemical Safety and Hazard Investigation Board *y2k*
CSB Closely Spaced Basing [also called dense pack] *mt, uc, st*
CSB Computer Support Base *mt*
CSBM Confidence and Security Building Measures *14, wp*
CSBSA Computer Support Battle Staff Actions *mt*
CSC Central Screening Control *fu*
CSC Civil Service Commission *hc, w9, wp*
CSC Coherent Sidelobe Cancellation *hc*
CSC Communications Satellite Corporation *hc*
CSC Communications System Control *fu*
CSC Computer Sciences Corporation *mt, fu, ah, hc, sc3.2.7, aa26.10.55, pbl, oe*
CSC Computer Security Center [DOD] *fu, mt*
CSC Computer Software Component *mt, fu, uh, ah, hc, pl, ep, tb, pbl, jdr, tm, kf*

CSC Computer Software Configuration *fu*
CSC Computer System Center *mt*
CSC Computer Systems Command [USA] *mt*
CSC Conventional Standoff Capability *hc*
CSC Core Support Capability *tb*
CSC Cryptologic Support Center [USAF] *mt*
CSC-STD Computer Sciences Corporation- Standard *kf*
CSC-STD Computer Security Center Standard *fu*
CSC/AC Computer Systems Command/Army Communication *fu*
CSCC Communications System Category Code *mt*
CSCE Communications System Control Element [TRITAC] *mt, fu, aw129.1.57*
CSCE Conference on Security and Cooperation in Europe *-14*
CSCFM Chief of Staff Commander's Flash Message *mt*
CSCI Computer Software Configuration Item *mt, he, uh, fu, pl, hc, ep, dm, tb, pbl, tm, jdr, kf, oe*
CSCMP Computer System Capacity Management Plan *fu*
CSCS Component Support Cost System *mt*
CSCS Contract Security Classification Specification [DD Form 254] *su*
CSCS Contractor Satellite Control Site [LEASAT] *mt, sc4.1.center, lt*
CSCSC Cost Schedule Control System Criteria *-4*
CSCSGE Computer System Command Support Group Europe [USA] *mt*
CSCU Crew Station Control Unit *fu*
CSCW Computer Supported Cooperative Work *pk, kg*
CSD (two CSD . . . solid propellant booster rocket motors) *af72.5.149*
CSD Canonical Signal Digit *fu*
CSD Channel Selection Device *fu*
CSD Chemical Systems Division [of Pratt and Whitney] *ai2*
CSD Chemical Systems Division [of United Technologies] *ep*
CSD Circuit Switched Data *pk, kg*
CSD Color Status Display *fu*
CSD Combat Support Display Description *fu*
CSD Commercial Systems Division *hn43.6.2*
CSD Committee for Systems Development *ag*
CSD Communications Computer System Directive [USAF] *mt*
CSD Communications Systems Division *fu*
CSD Computer Systems Division *mt, fu*
CSD Constant Speed Drive *mt, jj*
CSD Contract Start Date *tb, wl*
CSD Contract Support Detachment *hc*
CSD Contracted Sortie Duration *mt*
CSD Contractor Support Depot *fu*
CSD Corrective Service Diskette *ct, kg, em*
CSDC Central Signal Data Converter *fu*
CSDC Combat System Display Console *fu*
CSDC Computer Signal Data Converter *jj*
CSDD Computer Subprogram Design Document *fu*
CSDDD Combat System Design Data Document *fu*
CSDF Crew Station Design Facility [AFSC*] *wp, mt*
CSDL Charles Stark Draper Laboratory *hc*
CSDM Computer System Diagnostic Manual *mt, fu*
CSDM Continuous Slope Delta Modulation *sp660*
CSDP Communications Software Developmental Package *hc*
CSDS CCC Software Development and Support *hc*
CSDS Packet Switched Data Service *pk, kg*
CSE Calibration Support Equipment *hc, jj*
CSE Candidate System Evaluation *fu*
CSE Center for Software Engineering *fu*
CSE Certified System Engineer *hi, kg*
CSE Chief Software Engineer *fu*
CSE Common Support Element *mt*
CSE Common Support Equipment *mt, fu, hc, dm, kf*
CSE Communications Support Element *mt, fu*
CSE Computer Security Evaluation [Center] *tb*
CSE Computer Support Equipment *jj*
CSE Contingent Security Element *af72.5.115*
CSE Cryogenic Space Experiment *kf*
CSEA Combat System Engineering Agent *hc*
CSEASP Computer Security Evaluation and Assistance Support Program *mt*
CSEC Combat System Engineering Center *fu*
CSEC Computer Security Evaluation Center *mt*
CSED Combat System Engineering Development *fu*
CSED Coordinated Ship Electronic Design *hc*
CSEDS Combat Systems Engineering Development Site *hc, ts10-7*
CSEE {Computer Sciences and Electrical Engineering} *-19*
CSEF Central Software Engineering Facility *tb*
CSEF Combat System Engineering Facility *fu*
CSEF Committee for Space Environment Forecasting *ag*
CSEL Consolidated Support Equipment List *wp*
CSEL Contractor Support Equipment List *fu*
CSEP Communications System Engineering Plan *mt*
CSES Combat Systems Electronic Space *fu*
CSES Communications Satellite Earth Station *hc*
CSESD Communications Security Equipment System Documentation *fu*

CSETWG Computer Security Education and Training Working Group *mt*
CSF Casualty Staging Facility *fu*
CSF Central Supply Facility *fu*
CSF Central Support Facility *mt*
CSF Close Supportive Fire *hc*
CSF Communications Sans Fil [French wireless communications] *wp, sc2.8.2, aw130.11.29*
CSF Contractor Support Facility *wp, fu*
CSF Cut Sheet Feeder *fu*
CSFDR Crash Survivable Flight Data Recorder *wp*
CSFI Communication Subsystem For Interconnection *pk, kg*
CSG Centre Spatial Guyanais Center for Space, French Guiana [CNES] *ac3.6-23, crl, ai2*
CSG CINCPAC Support Group *mt*
CSG Combat Support Group *mt*
CSG Command Signal Generator *hc*
CSG Command Subsystem Group *hc*
CSG Composite Support Group *mt*
CSG Constructive Solid Geometry *pk, kg*
CSG Consulting Services Group [Lotus] *pk, kg*
CSG Cryptologic Support Group *mt, fu, wp*
CSGP Computer Systems Group *mt*
CSGU Control Signal Generator Unit *hc*
CSI Command Sequence Introducer *pk, kg*
CSI Commonalty, Standardization, And Interoperability *kf*
CSI Component Source Inspection *jdr*
CSI CompuServe Incorporated *pk, kg*
CSI Computer System Integration *mt, wp*
CSI Construction Specification Institute *fu*
CSI Contractor Source Inspection *oe*
CSI Critical Safety Item [USA DOD] *mt*
CSID Call Subscriber ID *pk, kg*
CSID Calling Station Identification *ct*
CSIF Communications Service Industrial Funds *mt*
CSII-RCME Collins System International, Inc-Rockwell Collins Middle East *fu*
CSIO Command Systems Integration Office [US Army] *af72.5.64, dt*
CSIP Component Sponsored Investment Program *mt*
CSIP Crew-Station Integration Program *hc*
CSIR Committee for Scientific and Industrial Research [South Africa] *0, ci.134*
CSIRO Commonwealth Scientific and Industrial Research Organization *wp*
CSIRS Covert Survivable In Weather Reconnaissance / Strike [claimed development project that created the TR-3A stealth aircraft and the F-117] *mt*
CSIS Center for Strategic and International Studies *mt, aw129.1.14, wp*
CSIS Central Secondary Item Stratification *mt*
CSISM COMSEC Supplement to the Industrial Security Manual *fu, hc, tb*
CSIST Chung Shan Institute of Science and Technology *fu*
CSIT Combat System Interface Testing *fu*
CSIT CONUS System Integration Testing *fu*
CSK Code Shift Keying *fu*
CSKO Counter Sink Other Side *jj*
CSL Common Switch Logic *fu*
CSL Component Save List *jj*
CSL Computer Science Library *mt*
CSL Computer Sensitive Language *pk, kg*
CSL Configuration Summary List *ep*
CSL Control and Simulation Language *hc, wp*
CSL Customer Supplied Link *fu*
CSLB Common Switch Logic Board *fu*
CSLC Coherent Sidelobe Cancellation *fu*
CSLIP Compressed Serial Line Internet Protocol *ct, em, kg, hi*
CSLS California Space Launch Complex [VAFB] *pbl*
CSM Color Shadow Mask *fu*
CSM Command and Service Module [Apollo spacecraft] *ac2.5-12, mi, hc, mb, fu*
CSM Command Sergeant Major *w9*
CSM Communicant Status Message *fu*
CSM Communication System Monitor *jo.iv, 9, lt, ep, jdr*
CSM Communications Services Manager *pk, kg*
CSM Computer Systems Manual *mt*
CSM Computerized Structural Mechanics *wp*
CSM Cost and Schedule Management *oe*
CSMA Carrier Sense Multiple Access *mt, ct, em, fu, tb, hi*
CSMA Common Spectrum Multiple Access *hc*
CSMA/CA Carrier Sense Multiple Access/with Collision Avoidance *pk, kg*
CSMA/CD Carrier Sense Sensing Multiple Access with Collision Detection *mt, jj, ct, em, wp, fu, nu, kg, oe, hi*
CSMC Central System Management Computer *hc*
CSMP Command Surgeon's Microcomputer Program [USAFE] *mt*
CSMP Continuous System Modeling Program *fu*
CSMP Current Ship's Maintenance Project [US Navy] *mt*

CSMS Central Systems Management Segment *tb*
CSMS Combat Supply Management System [USAF] *mt*
CSMS Corps Support Missile System *hc*
CSMS Cost/Schedule Management System *jdr, kf*
CSMS Customer Support Management System *pk, kg*
CSMT Circuit Switching Magnetic Tape *fu*
CSN Card Sequence Number *fu*
CSN Card-Select Number *pk, kg*
CSN Circuit Sequence Number *fu*
CSN Command Sequence Number *fu*
CSN Customer Contractor Serial Number *fu*
CSNET Computer Science Network *ct, kg*
CSNL Customer Serial Number Log *fu*
CSO Central Services Organization *pk, kg*
CSO Chief Satellite Officer *jb*
CSO Closely Spaced Objects *5, 17, sb, hn48.19.5, tb, st, kf*
CSO Cognizant Security Office *hc, tb, jdr, su*
CSO Contract Settlements Office *hc*
CSO Contractor Self Oversight *jdr, kf*
CSOC Combined Sector Operations Center *fu*
CSOC Consolidated Space Operations Center [Falcon AFS, Colorado] *mt, hc, 11, 16, 17, wp, tb, jdr, kf, uh, fu*
CSOC Consolidated Space Operations Contract *te23, 11, 18*
CSOD Command Systems Operations Division *mt*
CSOM Computer System Operator's Manual *mt, fu*
CSOP Consolidated Space Operations Plan *kf*
CSORO Conical Scan on Receive Only *fu*
CSP Central Signal Processor *hc, jj*
CSP Central Supply Point *mt*
CSP Certified Systems Professional *mt, kg*
CSP Coder Sequential Pulse *fu*
CSP Command Support Plans *uh*
CSP Commercial Subroutine Package *pk, kg*
CSP Common Signal Processor *mt, fu, aw129.1.59, wp*
CSP Communicating Sequential Processes *pk, kg*
CSP Communications Support Pallet *kf*
CSP Communications Support Processor [DODIIS] [USA] *mt*
CSP Communications System Processor *mt*
CSP Company Standard Practice [Hughes Aircraft Company] *hc, jj*
CSP CompuCom Speed Protocol *ct, em, pk, kg*
CSP Computer Support Program *wp*
CSP Computer Switch Panel *fu*
CSP Contact Support Processor *uh*
CSP Contract Strategy Paper *mt*
CSP Control Station Processor *fu*
CSP Crisis Staffing Procedures [OJCS] *mt*
CSP Cross System Product [IBM] *pk, kg*
CSPA Charge Sensitive Pre-Amplifier *oe*
CSPAR CINC'S Preparedness Assessment Report *mt*
CSPDN Circuit Switched Public Data Network *hi, kg*
CSPE Communications System Planning Element *fu*
CSPL Commercial Spacecraft Product Line *jdr*
CSPL Control Store Programming Language *fu*
CSPO Communications Satellite Project Office [DCA] *mt*
CSPO Computer Security Program Office *mt*
CSPP Communications Computer [System] Program Plan [USAF] *mt, wp*
CSPP Crew Scheduling and Phasing Plan *fu*
CSPPSP Communications-Computer Program Plan Support Plan [USAF] *mt*
CSPU Central System Processor Unit *oe*
CSQ Company Source Qualification *fu*
CSQA Computer Software Quality Assurance *oe, fu*
CSQAO Corporate Supplier Quality Assurance Office *fu*
CSQPP Computer Software Quality Program Plan *fu*
CSR Central Security Room *fu*
CSR Change Status Report *mt*
CSR Clutter to Signal Ratio *ic2-41*
CSR Coastal Surveillance Radar *wp*
CSR Combat Search and Rescue *wp*
CSR Combat Surveillance Radar *wp*
CSR Combat System Requirement *fu*
CSR Computer System Resources *fu*
CSR Conically Scanning Radiometer *hg, ge*
CSR Contract Status Report *8, hc, jj*
CSR Control Shift Register *fu*
CSR Current Situation Room *mt, wp*
CSRB Computer Systems Requirements Board *mt*
CSRB Consent to Ship Review Board *fu*
CSRB-WG Computer Systems Requirements Board Working Group *mt*
CSRC Combat System Readiness Control *fu*
CSRD Communications Computer Systems Requirement Document *mt, wp, kf*
CSRL Common Strategic Rotary Launcher [launchers that can accept both ALCM and SRAM] [USA] *mt, dh, wp, af72.5.133*
CSRM Combat System Readiness Management *fu*
CSRM Common Short Range Missile *hc*

CSRP Computer and Software Review Panel *wp*
CSRR Combat Systems Readiness Review *hc*
CSRS Civil Service Retirement System *wp*
CSRT Combat Systems Readiness Test *hc, fu*
CSS Cascading Style Sheets [Microsoft] *ct, em, kg*
CSS Central Security Service [NSA] *10, mt*
CSS Central Simulation System *fu*
CSS Combat Service Support *fu, mt*
CSS Combat Skyspot *mt*
CSS Combat Supply System [USAF] *wp, mt*
CSS Combat Support Service *mt*
CSS Combat Support Subsystem *mt*
CSS Combat System Simulation *fu*
CSS Commercial Satellite Survivability *mt*
CSS Communication System Simulator *muc*
CSS Communications Subset *mt*
CSS Communications Support Segment [SPACECMD] *mt*
CSS Communications Support System [US Navy] *mt, fu*
CSS Communications System Segment *mt, hc, rva, hs, jb*
CSS Complex Combat System *fu*
CSS Consolidated Support System *hc*
CSS Constant Surveillance Service *su*
CSS Contact Start Stop *pk, kg*
CSS Contingency Support Staff *mt*
CSS Continuous System Simulator [Language] *pk, kg*
CSS Contract Support Set *jj*
CSS Controller Sector Suite *fu*
CSS Controls Subsystem *jdr*
CSS Cooperating Space System *tb*
CSS Cooperative Security System *is26.1.70*
CSS Cooperative Spacecraft System [SDI] *mt*
CSS-R Communications System Segment Replacement [SPACECMD] *mt*
CSSA Central Supply Support Activity *hc, jj*
CSSA Computer Software Support Agency *mt*
CSSA Computer Systems Selection and Acquisition [Agency, US Army] *dt*
CSSA Cryptologic Shore Support Activity *mt, fu*
CSSC Computer System Security Officer *mt*
CSSC Computer System Support Center *mt*
CSSCS Combat Service Support Control System [USA] *fu, mt*
CSSE Combat Service Support Element [USMC] *mt*
CSSF Contractor Software Support Facility *fu*
CSSM Client-Server Systems Management [IBM] *pk, kg*
CSSO Computer System Security Officer *mt*
CSSO Contractor Special Security Officer *tb, kf*
CSSQT Combat Systems Ship's Qualification Test *fu*
CSSR Communications System Segment Replacement *fu*
CSSR Communications System Segment Replacement Requirement [program] *hc, as*
CSSR Cost and Schedule Reporting System *ep*
CSSR Cost Schedule Status Report *mt*
CSSR Cost Schedule Status Review *wp, bf*
CSSR Cost/Schedule Status Reporting Report *fu, kf*
CSSR Czechoslovakian Soviet Socialist Republic *sb*
CSSS Combat Service Support System *fu, hc*
CSSS Combat Service Support System [also CS3] [USA] *mt*
CSSS Conceptual Satellite Surveillance System *hc*
CSSS Conceptual Satellite System Studies *hc*
CSSS Cross Spin Stabilization System *hc*
CSSTM Configurable Spread Spectrum Test Module *hc*
CSSTS Configurable Spread Spectrum Test Set *fu*
CSSTV Color Slow Scan Television *hc*
CST Center for Space, Toulouse [CNES] *ac3.6-2*
CST Central Standard Time *0, wp, w9, kf, kg*
CST Combined Systems Testing *hc, 0, ai2*
CST Comprehensive System Testing *4, fu*
CST Control Section Tester *jj*
CST Critical Sea Test *fu*
CST Customer Support Team *mt*
CSTA Combat Systems Test Activity [USA] *mt*
CSTA Computer-Supported Telephony Applications *pk, kg*
CSTAL Combat Surveillance and Target Acquisition Laboratory [of ERADCOM] *T.Co*
CSTC Consolidated Satellite Test Center [USAF] *mi, oe*
CSTC Consolidated Space Test Center *uh, jb*
CSTC Consolidated Space Tracking Center *oe*
CSTI Civil Space Technology Initiative [NASA] *mt, aa27.4.31*
CSTI Control Surface Tie-In *hc*
CSTL Combat Surveillance and Tracking Labs *fu*
CSTOL Cargo Short Take Off and Landing *wp*
CSTS Combat Support Test Set *hc*
CSTSF Combat Systems Test and Support Facility *pf1.9*
CSTSR Computer Software Timing and Sizing Report *fu*
CSU California State University [Los Angeles] *ie5.11.10*
CSU Central Switching Unit *fu*

CSU Channel Service Unit *ct, em, do114, kg, hi*
CSU Circuit Switching Unit *mt, fu*
CSU Cleveland State University *is26.1.76z*
CSU Clock Synchronization Unit *fu*
CSU Communications Switching Unit *fu*
CSU Computer Software Unit *fu, uh, pl, pbl, tm, jdr*
CSU Control Sector Unit *fu*
CSU Conventional Signals Upgrade *mt*
CSU Crew Station Unit *hc*
CSU Customer Service Unit *jj*
CSUF California State University, Fresno *is26.1.87*
CSUN California State University, Northridge *hn43.11.3*
CSUS Containership Strike Up System *hc*
CSV Circuit-Switched Voice *pk, kg*
CSV Corona Start Voltage *wp*
CSV Crew Support Vehicle MGS *kf*
CSVRP Computer Security Vulnerability Reporting Program *tb*
CSW Central Signals Workshop *fu*
CSW Control Power Switch *fu*
CSW Conventional Standoff Weapon [USA] *hc, ph, mt*
CSWBS Contract Summary Work Breakdown Structure *fu*
CSWG Communication-Computer System Working Group *mt*
CSWP Common Software Processes *jdr, kf*
CSWS Conventional Standoff Weapon System *hc, un*
CSWS Corps Support Weapon System *fu, hc, uc*
CT (CT39 Automated Schedules) [MAC] *mt*
CT Case-Telescoped [a round that contains the bullet within the cartridge, to have a compact, cylindrical round] *mt*
CT Center Central Tap *jj*
CT Center Tray *jdr*
CT Cipher Text *fu*
CT Clock Time *hc, jj*
CT Collective Table *fu*
CT Command Table *pl*
CT Communications Technician *fu, jj*
CT Computed Tomography *aw118.18.71, is26.1.68*
CT Computer Telephony *ct, em*
CT Confidence Test *jj*
CT Corporate Towers [a Hughes site] *fu, hn42.1.8, 15*
CT Counter Terrorism [USA DOD] *mt*
CT Current Transformer *fu*
CT Cypher Text *mt*
CT&CU Central Telemetry and Command Unit *ma78*
CT&E Contractor [Development] Test and Evaluation *mt, fu*
CT-SURV Communications/Tracking Survivability [simulation] *-17*
CTA Calculated Time of Arrival *fu*
CTA CAPS Test Adapter *fu*
CTA Case Telescoped Ammunition *mt*
CTA Center Tray Assembly *jdr*
CTA Centro Tecnico Aeroespacial [Aerospace Technical Center, Brazil] *sc3.10.3, ai2*
CTA Cognizant Technical Activity *fu*
CTA Companion Trainer Aircraft *hc*
CTA Computer Technology Associates, Inc. *fu, is26.1.45, aw132.10.7, tb*
CTA Contractual Technical Assistance *mt*
CTA Course Time Adjust *jr*
CTA Critical Task Analysis *kf*
CTA-ISO Communication Terminal Asynchronous-International Standards Organization *mt*
CTA-STD Communication Terminal Asynchronous-Standard *mt*
CTAK Cipher Text Auto Key *fu*
CTAN CINCPAC Teletype Alert Network *mt*
CTAPS Contingency TACS Advanced Planning System [TAC] *mt*
CTASC Corps and Theater ADP Service Center [USA] *mt*
CTB Comprehensive Test Ban *14, wp*
CTB Cypher Type Byte *pk, kg*
CTBT Comprehensive Test Ban Treaty *hc*
CTC Carbon Tetrachloride *wp*
CTC Centralized Traffic Control *fu, w9*
CTC Channel To Channel *pk, kg*
CTC Chirp Transform Correlator *fu*
CTC Chlortetracycline *wp*
CTC Combat Targeting Center *mt*
CTC Combat Theater Communications *mt*
CTC Computer Test Console *hc*
CTC Computing Technology Center *hd1.89.y15*
CTC Comsec-Transec Controller *jdr*
CTC Contract Target Cost *jdr, fu*
CTC Cost To Complete *gb, 8, jj, jdr*
CTC Cross Track Contiguous [scanning mode] *ac7.3-1*
CTCF Channel Technical Control Facility *mt*
CTCP Client To Client Protocol *pk, kg*
CTCS CDA Telemetry and Command System [CDA portion of GIM-TACS] *oe*
CTCU Central Telemetry and Command Unit *kf, trds*

CTCU   Crosslink Central Telemetry and Command Unit   *jr, he*
CTCU   Crosslink Telemetry and Command Unit   *jdr*
CTD   Cell Transfer Delay   *ct, em*
CTD   Charge Transfer Device   *hc, fu, ai1*
CTD   Color Tactical Display   *mt*
CTD   Composite Tactical Display   *kf*
CTDC   Company Technical Document Data Center [Hughes Aircraft Company]   *fu, ah, hc, pl, ep, pbl, tm, jdr*
CTDF   Contracting Technical Data File [DLA]   *mt*
CTDP   Communications and Tracking Data Processor   *mt*
CTDR   Commercial Training Devices Requirements   *st*
CTE   Cable Termination Equipment   *fu*
CTE   Central Timing Equipment   *hc*
CTE   Coefficient of Thermal Expansion   *hc, gp, 16, aa26.10.42, jdr*
CTE   Commander, Task Element [USA]   *mt*
CTE   Commercial Test Equipment   *fu, jj*
CTE   Communications Test Equipment   *oe*
CTE   Contractor Technical Evaluation   *wp*
CTE   Contractor Test and Evaluation   *fu*
CTE   Crew Training Emulator   *kf*
CTE   Cross Track Error   *hc*
CTEC   Corporate Technical Education Center   *hn44.18.4*
CTEI   Contractor Trainer End Item   *hc, jj*
CTF   Carrier Task Force   *fu*
CTF   Central Training Facility   *fu*
CTF   Chlorine Trifluoride [chemical warfare]   *wp*
CTF   Clear To Fire   *jb*
CTF   Combined Task Force   *mt*
CTF   Combined Test Force   *aw118.14.62, aw*
CTF   Commander, Task Force [GCCS] [USA]   *dt, mt*
CTF   Common Transmission Format   *fu*
CTFE   ChloroTriFluoroEthylene   *aw130.22.35, wp*
CTFM   Continuous Transmission Frequency Modulation   *fu*
CTFS   Common Time and Frequency Standard System   *4, vv*
CTG   Cartridge Tape Generator   *fu*
CTG   Commander, Task Group [USA]   *mt*
CTG   Communications Transmission Group   *mt*
CTHOUS   (Catalog Breakup Pieces Software Program)   *jb*
CTI   Circuit Technology Incorporated   *-17*
CTI   Computer [and] Telephony Integration   *ct, kg, em, hi*
CTI   Critical Technical Issue   *nt*
CTIA   Capacitive-feedback Trans-Impedance Amplifier   *kf*
CTIC   Contractor Technical Information Code   *fu*
CTIO   Cerro Tololo Interamerican Observatory   *mb, mi, ns*
CTIS   Command Tactical Info System [11TH AF]   *mt*
CTL   Communications Test Laboratory   *tb, wl*
CTL   Complementary Transistor Logic   *jj, fu*
CTL   Core Transistor Logic   *fu*
CTLD   Command and Training Launch Demonstration   *-17*
CTLED   Clear Track Light Emitting Diode   *oe*
CTM   Captive Training Missile   *hc, jj*
CTM   Cargo Traffic Message   *mt*
CTM   Communications Terminal Module   *mt*
CTM   Constellation Time Management   *jr*
CTM   Cycle Time Management   *sn2.14, ig21.12.19*
CTMC   Communications Terminal Module Controller   *mt*
CTMP   Channel Traffic Movement Plan   *mt*
CTN   Commercial Telecommunications Network   *mt*
CTNC   Cross Track Non-Contiguous [scanning mode]   *ac7.3-4*
CTNE   CompaÒla TelefÚnica Nacional de EspaÒa [National Telephone Company of Spain]   *fu*
CTO   Cognizant Transportation Office   *mt*
CTO   Contractor Technical Organization [Calverton]   *jj*
CTO   Contracts Termination Office   *hc, jj*
CTO   Corporate Telecommunications Office [Hughes]   *fu*
CTO   Corps Transportation Officer   *mt*
CTOC   Corps Tactical Operations Center [USA]   *pm, mt*
CTOCSE   Corps Tactical Operations Center Support Element   *mt*
CTOL   Coarse Time Ordered List   *fu*
CTOL   Conventional Take Off and Landing   *mt, ph, wp, ai1*
CTOS   Computerized Tomography Operating System   *pk, kg*
CTOS   Convergent Technologies Operating System   *em, kg*
CTP   Card Test Procedure Program   *fu*
CTP   Channel Trap Processor   *fu*
CTP   Charge Transfer Photography   *wp*
CTP   Chief Test Pilot   *mt*
CTP   Command and Telemetry Processor   *oe*
CTP   Comprehensive Test Plan   *fu*
CTP   Consolidated Target Program   *st*
CTP   Contract Technical Personnel   *hc, jj*
CTP   Coordinated Test Program   *jj*
CTP   Critical Technology Program   *fu*
CTP   Port Captain   *mt*
CTPA   Coaxial to Twisted Pair Adapter   *hi, kg*
CTPE   Central Tactical Processing Element   *kf*
CTPP/TS   Central Tactical Processing Program/Talon Shield   *kf*

CTPS   Console, Telemetry, Pulse, and Serial command   *jdr*
CTR   Center for Turbulence Research   *aa26.11.56*
CTR   Command Track Receiver   *id4142-10/831, 9*
CTR   Configuration and Traceability Report   *hc*
CTR   Consolidated Training Record   *fu*
CTR   Controlled Thermonuclear Reactor   *wp*
CTRCO   Calculating, Tabulating, Recording Company [later renamed International Business Machines]   *pk, kg*
CTRI   Crash Test Rating Index   *wp*
CTRL   Control   *wp, kg, jj*
CTS   Card Test Set   *fu*
CTS   Central Tactical System   *mt*
CTS   Central Training Section   *mt*
CTS   Clear to Send [signal]   *mt, do202, kg, ric, jdr, ct, em, fu, hi*
CTS   Cloud Top Scanner   *hy*
CTS   Colorado Tracking Station   *hc*
CTS   Command, Track Station   *jdr*
CTS   Communications Technology Satellite [Canada]   *hc, tk, 19*
CTS   Communications Test System   *fu*
CTS   Composite Training Squadron [USA]   *mt*
CTS   Computer Test Station [F-15]   *jj, hc, fu, mt*
CTS   Computerized Training System   *jj*
CTS   Comsec Transec Subsystem   *jdr*
CTS   Configuration Tracking System   *fu*
CTS   Contact Test Set   *mt, fu*
CTS   Contract Tooling System   *fu*
CTS   Contractor Technical Services   *jj, fu*
CTS   Control Test Set   *-4*
CTS   Conversation Time Sharing   *mt*
CTS   Cosmic Top Secret   *mt*
CTS   Customer Telephone System   *pk, kg*
CTSA   Cosmic Top Secret Atomal [NATO]   *mt*
CTSP   Contract Technical Service Personnel   *hc, jj*
CTSR   Consent To Ship Review   *kf, fu*
CTSS   Compatible Time Sharing System   *pk, kg*
CTSS   Contractor Technical Support Services   *fu*
CTSS   Cray Time Sharing System   *m, tb*
CTT   Command Tactical Trainer   *hc*
CTT   Commander's Tactical Terminal [TAC]   *kf, wp, mt*
CTT   [electronics firm, Santa Clara, California]   *mj31.5.13*
CTT/T   Commander's Tactical Terminal / TEREC   *mt*
CTTO   Central Tactics and Trials Organization [UK]   *mt*
CTTP   Centralized Theater Processing Program   *kf*
CTTY   Console Teletype   *ct, em*
CTU   Cable Transmitter Unit   *fu*
CTU   Calibration Test Unit   *jj*
CTU   Cartridge Tape Unit   *fu*
CTU   Central Telemetry Unit   *gp, 8*
CTU   Central Timing Unit   *fu, pl*
CTU   Command Telemetry Unit   *-4*
CTU   Commander, Task Unit [USA]   *mt*
CTV   Cable Television   *sc3.9.5, hi*
CTV   Compatibility Test Van   *gp, 2, 4, oe*
CTV   Control Test Vehicle   *st*
CTV   Corona Test Voltage   *gp*
CTVS   Cockpit TV Sensor   *wp*
CU   Call Up   *kf*
CU   Central Unit   *gp*
CU   Common User   *fu, mt*
CU   Control Unit   *fu*
CU   Conversion Unit   *fu*
CU   Cryptographic Unit   *mt*
CUA   Common User Access [IBM]   *pk, kg*
CUBE   Cooperating Users of Burroughs Equipment   *mt*
CUBIC   Common User Baseline for the Intelligence Community   *mt*
CUC   Common User Console   *kf*
CUC   Common User Contract [WIS]   *wp, mt*
CUCV   Cargo Utility Commercial Vehicle [USA]   *fu, hc*
CUCV   Commercial Utility and Cargo Vehicle [USA]   *mt*
CUDG   Compilation Unit Dependency Graph   *fu*
CUDIXS   Common User Digital Information Exchange Subsystem [US Navy]   *mt*
CUE   Common User Element   *jdr, kf*
CUE   Common User Equipment   *uh*
CUE   Computer Updating Update Equipment [of GSG at Hughes Aircraft Company]   *hc, fu*
CUE   Custom Updates and Extras [card] [Egghead Software]   *pk, kg*
CUES   Combat Underwater Exploitation System [US Navy]   *mt*
CUF   Common User Follow-on   *mt*
CUF   Compromising Undetected Failures   *fu*
CUI   Centre Universitaire d'Informatique [French University Computing Science Center]   *ct*
CUI   Character Oriented User Interface   *pk, kg*
CUI   Common User Interface [IBM]   *kg, tb, hi*
CUINCU   Copper-Invar-Copper   *wp*
CUJT   Complementary Unijunction Transistor   *fu*

CUL Common User Language *mt*
CULPRIT [data processing language for data extraction and report writing] *hm*
CUM Reprogram Expenditures and Commitments *jj*
CUO Coupon Outline *fu*
CUP Common User Products *mt*
CUP Computer Utility Program *fu*
CUP COMSEC Utility Package *mt*
CUP Console Upgrade Program *fu*
CUPID Completely Universal Processor I/O Design [AST] *pk, kg*
CUPS Council On Uniform Procurement System *wp*
CURT Combined Unit Record Tape service *hc*
CUS Common User Software *uh*
CUS Common User Subsystem [WIS] *mt*
CUS Common User Support *mt*
CUS Common User System *mt*
CUS Customer Unique Software *jdr*
CUS {Central User Services} *es55.44*
CUSFO Capability for US Firms Overseas [JCS] *mt*
CUSIP Committee for Uniform Security Identification Procedures [US Treasury] *pk, kg*
CUSR Center U. S. Registry *dt*
CUSRPG Canada-U. S. Regional Planning Group *dt*
CUSV Commercial Utility Cargo Vehicle {?} *mt*
CUT Coding and Unit Testing *fu*
CUT Control Unit Terminal *kg, ct, em*
CUT Coordinated Universal Time *fu*
CUT Cross Utilization Training *mt*
CUTE Clarkston University Terminal Emulator *pk, kg*
CUV Common User Voice *fu*
CUWG Configuration Update Working Group *wp*
CUWTF Combined Unconventional Warfare Task Force *mt*
CV Cargo Variant *wp*
CV Carrier Vehicle [short for SBICV] *gr, ts7-14, tb*
CV Cataclysmic Variable *mi*
CV Check Valve *jj, uh, oe*
CV Common Vision *fu*
CV Computer Vision *hn43.9.4, ep*
CV Computervision *fu*
CV ComputervisionTM [Computer Software Co.] *tm*
CV Configuration Verification *fu*
CV Continuously Variable *fu*
CV Cost Variance *mt, jdr*
CV Counter Voltage *fu*
CV Cryptovariable *fu*
CV Vice Commander *mt*
CV [multipurpose, conventionally powered aircraft carrier USN ship classification, as in CV-22 for Osprey] *fu, uc, af72.5.15, pf1.4, wp, jj*
CV {Carrier Vehicle} [aircraft carrier of conventional power, numbers such as CV-5 or CV-12 are unique for ships] [USA] *mt*
CV-ASWM Carrier Anti-Submarine Warfare Module *fu*
CV-IC Carrier Intelligence Center *fu*
CV-SES Carrier-Surface Effect Ship *hc*
CV-TSC Carrier [Vehicle] - Tactical Support Center *fu, he*
CVA Attack Carrier *hc*
CVA Carrier Vehicle for Aircraft [e.g. aircraft carrier] *un*
CVA Central Verifications Activity *su*
CVA Clandestine Vulnerability Analysis *mt*
CVA Communications Vulnerability Analysis *wp*
CVA Communications Vulnerability Assessment *mt*
CVA {Carrier Vehicle, Attack} [attack carrier, USA] *mt*
CVA-WSIP Aircraft Carrier Weapon System Improvement Program *hc*
CVAC Continuously Variable Amplitude Carrier *hc*
CVALIN Complete Value In *hi*
CVAN CINCPAC Voice Alerting Network *mt*
CVAN {Carrier Vehicle, Attack, Nuclear} [attack aircraft carrier, nuclear powered, e.g. CVAN-65 Enterprise] [USA] *mt*
CVANCIN CPAC Voice Alerting Network *mt*
CVAST Combat Vehicle Armament System Technology *hc*
CVB Large Aircraft Carrier [USA] *mt*
CVBF Carrier Battle Force *mt*
CVBG Carrier Vehicle Battle Group [US Navy battle group designation] *mt, uc, fu*
CVC Combat Vehicle Crewman *jj*
CVC Cryogenic Vacuum Calorimeter *hc*
CVC-MTI Coherent Video Cancellation-Moving Target Indicator *mt*
CVCM Collected Volatile Condensable Material *jr, he, oe*
CVD Chemical Vapor Deposition *aa26.11.42, fu, hd1.89.y20, is26.1.76v, wp*
CVD Compact Video Disk *wp*
CVE Code Validity Elements *fu*
CVE {Carrier Vehicle, Escort} [aircraft carrier of Casablanca class, also known as Anzio class] [USA] *mt*
CVF Compressed Volume File *ct, kg, em, hi*
CVFCS Combat Vehicle Fire Control System *hc*
CVFR Controlled Visual Flight Rules *fu*
CVGA Color Video Graphics Array *hi, kg*

CVH Helicopter Carrier [US Navy] *mt*
CVHA Assault Helicopter Transport [e.g. CVHA-1 Thetis Bay] [US Navy] *mt*
CVHE Carrier Vehicle, Escort Helicopter [an escort helicopter carrier] [US Navy] *mt*
CVHG Carrier Vehicle, Helicopter, with Guided missile armament [US Navy] *mt*
CVI Code Validity Interval *fu*
CVI Configuration Verification Index *fu*
CVIA Computer Virus Industry Association *pk, kg*
CVIC Aircraft Carrier Intelligence Center *fu*
CVIS Combat Vehicle [VHSIC] Integrated System *hc*
CVL Carrier Vehicle, Light [an aircraft carrier] [US Navy] *mt*
CVL Configuration Verification List *fu*
CVLL Crypto Variable Logic Label *fu*
CVM Configuration Verification Multirequest *fu*
CVMS Crow Valley Microwave System *hc*
CVN Carrier Vehicle, Nuclear [an aircraft carrier] [US Navy] *mt*
CVN [multipurpose aircraft carrier,] Nuclear [USN ship classification] *fu, uc, wp, jj*
CVP Component Verification Program *hc*
CVP Cryptographic Verification Plan *fu*
CVR Cockpit Voice Recorder *mt, hc, aw129.14.17, jj*
CVR Configuration Verification Request *fu*
CVR Contract Value Report *jdr*
CVR Cost Variance Report *fu*
CVRAM Cached VRAM *hi*
CVRP Communication/Computer Vulnerability Program *wp*
CVRS Central Vehicle Registration System [USA] *mt*
CVRT Combat Vehicle Reconnaissance Tank *hc*
CVS Carrier Suitability *jj*
CVS Carrier Vehicle, Support [an ASW aircraft carrier] [includes some former CVs of the Essex class] [USA] *mt*
CVS [ASW aircraft carrier, US Navy ship classification] *fu*
CVSD Continuously Variable Slope Delta [modulation] *mt, fu*
CVSPS Combat Vehicle Self Protection System *hc*
CVT Carrier Vehicle, Training [a training aircraft carrier] [US Navy] *mt*
CVT Combine Verification Transmission [program] *0*
CVT Cryptographic Verification Test *fu*
CVU Carrier Vehicle, Utility [a utility aircraft carrier] [USA] *mt*
CVW Aircraft Carrier Air Wing *12, jj*
CVW Carrier Vehicle [Air] Wing [USA] *mt*
CVW Code View for Windows *pk, kg*
CV[N] Aircraft Carrier [Nuclear] *mt*
CV_EWIIN Carrier Electronic Warfare/ Intelligence Information Network *fu*
CW Air Force Satellite and Launch Control Systems Program Office {?} *uh*
CW Carrier Wave *wp, jj*
CW Catalog Writing *jj*
CW Chemical Warfare *mt, hc, ph, fu, w9, wp, jj*
CW Chief Warrant Office *w9*
CW Civil Works *dt*
CW Composite Wing [USA] *mt*
CW Constant Watch [PACAF] *mt*
CW Continuous Wave *mt, fu, hc, 4, 17, sb, aw, w9, 19, wp, tb, st, jj, oe, ct, em, uh, ai1*
CW Control Word *fu*
CW/BW Chemical Warfare / Biological Warfare *ph*
CWAN Contractor Wide Area Network *pbl*
CWAR Continuous Wave Acquisition Radar *hc, wp, jj, fu*
CWAS Contractors' Weighted Average Share [of risk] *hc*
CWB Ceramic Wiring Board *fu*
CWBS Contract Work Breakdown Structure *mt, fu, hc, 8, 16, 17, sb, nt, wp, tb, jj, jdr, kf, oe*
CWC Cable and Wireless Communication *ls4, 5, 62*
CWC Calibrating Work Center *wp*
CWC Clear Write Condition *fu*
CWC Composite Warfare Commander [US Navy] *fu, mt*
CWCPR Compute Weights Constant Percent Resolution *fu*
CWD Change Working Directory [Internet] *pk, kg*
CWD Chemical Warfare Defense *wp*
CWDA Cooperative Weapon Delivery Aircraft *hc*
CWFM Continuous Wave Frequency Modulation *fu*
CWFZ Chemical Weapon Free Zone *-14*
CWG Communication Working Group *mt, 17*
CWG COMSEC Working Group *mt*
CWI Continuous Wave Illuminator Illumination [radar] *jj, fu, wp, hc*
CWIF Continuous Wave Intermediate Frequency *fu*
CWIS Campus Wide Information Service/System [Internet] *pk, kg*
CWIS Community Wide Information Service / System [Internet] *hi, kg*
CWL Cloud Water [over] Land *ac7.5-12*
CWNET Constant Watch Network [PACAF] *mt*
CWO Chief Warrant Officer *hn44.9.10, w9, jj*
CWO Cloud Water [over] Ocean *ac7.5-12*
CWO Communications Watch Officer *mt*
CWO Contract Contractor Work Order *hc, jj*

CWO/IA Constant Watch Operations and Intelligence Automation *mt*
CWP Central Weather Processor *fu*
CWPTI Conventional Weapons Proficiency Technical Inspection *jj*
CWS Compiler Writing System *fu*
CWS Container Weapon System [Germany] *mt*
CWSA Chemical Warning Scanning Alarm *hc*
CWSU Center Weather Service Unit [FAA] *fu, ag*
CWT Hundredweight *jj*
CWTDC Continuous-Wave Target Detection Console *fu*
CWTN Continuous Wave Tone Jamming *fu*
CWTSAR Chemical Warfare Tsar chemical warfare Model *wp*
CWW Cruciform Wing Weapon [the GBU-15 guided bomb] [USA] *hc, mt*
CW_ROCC Canada West Regional Operational Control Center [SPACECMD] *mt*
CX Control Transmitter *fu*
CX [chemical warfare codename for phosgene oxime] *wp*
CX-HLS Cargo Transport - Heavy Logistics Supply *hc*
CX-X [requirement for an utility aircraft for US embassies; the Beech C-12F was selected] [USA] *mt*
CXBR Cosmic X-ray Background Radiation *mi*
CY Calendar Year *mt, hc, ph, 0, w9, jmj, wp, tb, jj, fu, jdr, oe, ai2*
CY Cubic Yard *wp*
CY Current [fiscal] Year *mt, tb*
CYBORG Cybernetic Organism *ct, em*
CYI Canary Islands [STDN site] *ac2.7-11*
CYMK Cyan Yellow Magenta Key *em*
CYSA Cape York Space Agency *ai2*
CZ Canal Zone *w9, jj*
CZ Communications Zone *0*
CZ Convergence Zone *fu*
CZAF Czechoslovakian Air Force *-12*
CZCS Coastal Zone Color Scanner *hc, sc2.7.3, ns, hy, ge, hg*

D

D Democrat *hn41.22.5, w9, aa27.4.10*
D Demonstration *tb, kf*
D Density *jj*
D Detectivity [radiometric] *oe*
D Deutschland [(German) Germany] *es55.35*
D Dextro [right rotatory] *wp*
D Discriminator *tb*
D {Dimension} (3D) *aa26.11.44, hi*
D&D Design and Development *kf*
D&DE Display and Data Entry *fu*
D&EC Defense and Electronics Center *oe*
D&F Determination(s) and Finding(s) *mt, hc, tb, jj*
D&I Development and Implementation [Plan] *mt*
D&O Deployment and Operations *fu*
D&P Design and Performance *jj*
D&V Demonstrations and Validation *fu*
D&V Design and Validation *mt*
D&W Detection and Warning *mt*
D-AMPS Digital Advanced Mobile Phone Service *hi*
D-BRITE Digital Bright Radar Indicator Tower Equipment *fu*
D-CHANNEL Data Channel *pk, kg*
D-CLS Depot Contractor Logistics Support *kf*
D-ECL Dielectric-Isolation Emitter Coupled Logic (process) *fu*
D-PAT Drum-Programmed Automatic Tester *hc*
D-S (D-S multiplex mode on) *ggIII.2.404*
D/A Digital to Analog *hc, sp8, 3, aa27.4.57, lt, kg, fu, tb, jj, oe, mt, ai1*
D/AC Digital to Analog Converter *-17*
D/AC Digital/Analog Card *fu*
D/AM Digital/Analog Mixer *hi*
D/C Down Converter *jo.iv, jdr*
D/D Data Display *0*
D/DD Display/Display Driver *hc*
D/DIRNSA Deputy Director, National Security Agency *-10*
D/E Data Encoder *0*
D/E Depression/Elevation *fu*
D/F Directional Finder *hc*
D/I Design Indication *fu*
D/I Disassembly and Inspection *jj*
D/L Data Link (San Marco D/L Satellite (Italy)) [also DL] *sc4.3.6, 17, jj*
D/L Down / Load *hi*
D/L Downlink *jdr, kf*
D/N Day or Night *pl*
D/R Direct or Reverse *pk, kg*
D/S Dhrystone Per Second *pk, kg*
D/V Demonstration / Validation *mt*
D2C Decimal To Character [REXX] *pk, kg*
D2X Decimal To Hexadecimal [REXX] *pk, kg*
DA Dalnaya Aviatsiya [Strategic Aviation] [USSR] *mt*

DA Damage Assessment *mt, tb*
DA Data Administrator *mt*
DA Date Accepted *jj*
DA Days After *fu*
DA Decision Analysis *mt*
DA Demand Assignment *hc*
DA Department of the Army [also DOA] *mt, fu, nt, wp, tb, st*
DA Design Assistance *mt*
DA Design Authorization *hc*
DA Design Automation *wp*
DA Development Agent *fu*
DA Dimension Array [as in 3DA for three dimension array] *hi*
DA Direct Access *fu, mt*
DA Direct Action [USA DOD] *mt*
DA Direct Ascent *tb*
DA Direct Attack *fu*
DA Directorate of Administration *mt*
DA Dispersal Airfield *mt*
DA Double Amplitude *fu*
DA Dust Analyzer *es55.26*
DA [chemical warfare codename for diphenyl chloroarsine] *wp*
DA/TS Data Accumulation/Transmittal Sheet *hc*
DAA Data Access Arrangement *mt, do37, kg*
DAA Days After Approval *fu*
DAA Decimal Adjust for Addition *pk, kg*
DAA Designated Approval (Approving) Authority *mt, jj, kf*
DAA Direct Access Acknowledge *hc*
DAA Divisional Assistance Authorization *hc*
DAAA {Department of the Army Audit Agency} [US Army] *dt*
DAAC Automation and Communications Assistant Chief of Staff *dt*
DAAC Data Acquisition and Analysis Center *fu*
DAAC Days After Award of Contract *fu*
DAACM Direct Airfield Attack Combined Munition [CBU-98/B cluster bomb] [USA] *mt*
DAAG Office of the Adjutant General *dt*
DAAH Office of the Chief of Chaplains [also DACH] *dt*
DAAR Day Air to Air Refueling *mt*
DAAR Office of the Army Reserve *dt*
DAARC Demonstration of Automatic Adaptive Radar Control *fu*
DAAS Defense Automated Addressing System *mt, wp*
DAASO Defense Automated Addressing System Office *mt*
DAATS Digital Analog Automatic Test System *hc*
DAB Data Announcement Bulletin *ns*
DAB Defense Acquisition Board [USA] *mt, fu, wp, aw129.14.25, jb, nt, ae151.52.459, tb, jdr, kf*
DAB Deployable Aft Blanket *kf, uh*
DABM Defense Against Ballistic Missiles [study] *dh, 14, tb, st*
DABS Director, Air Battle Staff *pf.f88.4*
DABS Discrete Address Beacon System *fu, hc*
DABS Dynamic Airblast Simulator *wp*
DAC Data Acquisition *wp*
DAC Data Acquisition and Control *pk, kg*
DAC Data Acquisition Center *hc*
DAC Data Acquisition Computer *st*
DAC Days After Contract [award] *fu, hc, wp, tb, jj*
DAC Defense Acquisition Circular *kf*
DAC Design Analysis and Commonality *fu*
DAC Diazo Aperture Card *jj*
DAC Digital to Analog Converter *fu, ct, em, hc, 6, 8, th4, wp, jj, kg, jdr, oe, hi, ai1*
DAC Direct Access program *mt*
DAC Disaster Assistance Center *mt*
DAC Discretionary Access Control *tb*
DAC Document Availability Code *fu*
DAC Douglas Aircraft Company *0*
DAC Dual Attachment Concentrator *pk, kg*
DAC Duplicate Aperture Card *fu, ep, jdr, tm*
DACA Days After Contract Award *fu, wp*
DACAFO Field Operating Agencies, US Army *dt*
DACAR Damage Assessment and Casualty Report *wp*
DACC Direct Access Communications Channel *hc, wp*
DACCC Defense Area Communications Control Center *hc, wp, jj*
DACCS Department of the Army Command and Control System *mt*
DACE Data and Command Equipment *hc*
DACG Departure Airfield Control Group *mt*
DACH Office of the Chief of Chaplains [also DACA] *dt*
DACI Drafting and Checking Instruction *fu*
DACM Days After Contract Modification *fu*
DACM Dissimilar Air Combat Maneuvering [i.e. training in air combat with 'aggressor' aircraft] [USA] *jj, mt*
DACOM Digital Data Communication System *hc*
DACON Digital to Analog Converter (Conversion) *wp, fu*
DACOR Data Comparator *hc, kg*
DACOR Data Correlator *wp*
DACS Data Acquisition and Control Subsystem *ag, oe*
DACS Data and Analysis Center for Software [RADC] *mt*

DACS Design and Commonality Study *fu*
DACS Despin and Altitude Control Subsystem *hc, jdr*
DACSS Deployable AWACS Computer Support Systems *mt*
DACT Dissimilar Aerial Combat Tactics *mt*
DACT Dissimilar Air Combat Training *mt*
DAD Data Description language *mt*
DAD Decision Action Diagram *fu*
DAD Design Approval Data *wp*
DAD Desktop Application Director [Borland] *pk, kg*
DAD Digital to Analog to Digital *jj*
DAD Draft Addendum *fu*
DAD {Differential Air Density} [mission of Explorer spacecraft] *ci.729*
DADA Design Analysis and Diagramming Aid *fu*
DADCAP Dawn And Dusk Combat Air Patrol [USA DOD] *mt*
DADE Decision Aid Development Environment *fu*
DADE Differential Absolute Delay Equalization *fu*
DADEM Dynamic Air Defense Effectiveness Model *fu*
DADM Detailed Air Defense Model [JDSSC] *mt*
DADMS Defense Automated Document Management System *wp*
DADS Data Access and Drive System *jj*
DADS Data Archive and Distribution System *ns, hy*
DADS Digital Audio Distribution System *hc*
DADS Discrimination Algorithm Development Study *fu*
DADS Document Availability and Distribution Services *ns*
DADS Dosimetry Acquisition and Display System *hc*
DADS Dynamic Analysis and Design System *aa26.10.44*
DAE Defense Acquisition Executive *mt, hc, wp, kf*
DAE Department of Atomic Energy [India] *ac3.2-1*
DAE Design, Analysis, and Engineering *cp*
DAED Diisopropyl Amino Ethyl Disulfide [chemical warfare mustard simulant] *wp*
DAED Directorate of Aerospace Engineering and Development *hc*
DAEN Office of the Chief for Engineers *dt*
DAES Defense Acquisition Executive Summary *mt*
DAF Data Acquisition Facility *hc*
DAF Department of the Air Force *mt, tb*
DAF Discard At Failure *fu*
DAFCCS Department of the Air Force Command and Control System *mt*
DAFCS Digital Automatic Flight Control System *mt, aw129.10.61*
DAFI Director of Air Force Intelligence *hc, jj*
DAFP Data Acquisition and Processing Facility *oe*
DAG Digital Address Group *jj*
DAG Division Advisory Group *hc, wp*
DAGC Delayed Automatic Gain Control *fu*
DAGC Digital Automated Gain Control *hc, jj*
DAHAP Data Acquisition, Handling and Analysis Plan *17, sb, nt*
DAI Days After Installation *mt*
DAI Distributed Artificial Intelligence *hi, kg*
DAIC Defense Automation Resources Information Center *dt*
DAIG Office of the Inspector General [also OIG] *dt*
DAILY_PLANIT [a workload management system] *mt*
DAIM Dual ASAS Interface Module *fu*
DAIN/ACDB Department of the Army Information Network/Army Corporate Data Base *mt*
DAIO Office of the Chief of Information *dt*
DAIP Defense Acquisition Improvement Program *wp*
DAIPR Department of the Army In Process Review *st*
DAIS Digital Avionics Information System *wp, jj, fu*
DAIS Digital Avionics Information System/Instruction Set *hc, 17, aw129.1.61*
DAISS Digital Airborne Intercommunication and Switching System *wp*
DAISY Disposal Automated Information System *mt*
DAL Data Access Language [Apple Computer] *hi, kg*
DAL Data Access List *wp*
DAL Data Accessions List *fu*
DAL Data Analysis Laboratory *0*
DAL Defended Assets List *mt*
DAL Design Analysis Language *fu*
DAL Disk Access Lockout *pk, kg*
DALFA Directorate of Air Land Forces Application *mt*
DALO Office of the Deputy Chief of Staff for Logistics *dt*
DAM Data Acquisition and Monitoring *pk, kg*
DAM Data Addressed Memory *fu*
DAM Device Adapter Modules *mt*
DAM Diagnostic Acceptability Measure *fu*
DAM Direct Access Memory *wp*
DAM Direct Action Mission *mt*
DAMA Demand Access Multiple Assignment [US Navy] *mt*
DAMA Demand Assignment (Assigned) Multiple Access *sp9, hc, jdr, fu, ct, em, tt*
DAMA Deputy Chief of Staff Research, Development and Acquisition *dt*
DAME Damage Assessment Methods *wp*
DAME Division Airspace Management Element *mt, fu*
DAMES Demonstration of Avionics Module Exchangeability via Simulation *aw129.21.135*
DAMES Division Airspace Management Element System *mt*

DAMES Dynamic Airlift Management and Execution System [MAC*] *mt*
DAMH={Department of the Army} [center for] Military History :(dt) Dynamic Airlift Management and Execution System [MAC*]
DAMI Office of the Assistant Chief of Staff for Intelligence *dt*
DAMIS Defense Analysis Modeling Information System [NATO] *mt*
DAMMH Direct Annual Maintenance Man Hours *jj*
DAMMS Department of Army Movements Management System *mt*
DAMO Department of Army Military Operations [DSCOPS] *mt*
DAMP Distribution Amplifier *fu*
DAMP Downrange Antiballistic Measurement Program *0*
DAMPL Department of Army Master Priority List *mt*
DAMPS DDN Automated Message Processing System [USAF] *mt*
DAMPS Digital Advanced Mobile Phone Service [also D-AMPS] *hi, kg*
DAMS Deployable Automated Maintenance System *mt*
DAMS Depot-Atelier Munitions SpÈciales [(French) special munitions depot and workshop] *-12*
DAMWO Department of the Army Modification Work Order *jj, fu*
DAN Data Analysis *fu*
DAN Data Analysis Network [Canada] *ns*
DAN Delivery Adjustment Notice *fu*
DANA Delay And Network Analysis *ne*
DANASAT Direct Ascent Nuclear Antisatellite [weapon or system] *5, 17, sb, tb*
DANC Despin Active Nutation Control *-3*
DANC [chemical warfare] Decontaminating Agent, Non Corrosive [Dimethyldichlorohydantoin] *wp*
DAND Despin Active Nutation Damper (Damping) *3, 8, uh*
DANDE Despin Active Nutation Damping Electronics *3, bl2-5.26, 9, lt, ep, jdr*
DAO Defense Attache Office *hc*
DAOD Days after the As Of Date *tb*
DAOI Data Area of Interest *fu*
DAOMIS Division Ammunition Office Management Information System [USA] *mt*
DAP (the DAP is built by Active Memory Technology, Inc.) *is26.1.30*
DAP Data Access Processor *mt*
DAP Data Access Protocol [DEC] *pk, kg*
DAP Data Automation Panel *mt*
DAP Data Automation Proposal *fu*
DAP Day of Arrival at POE *mt*
DAP Depot Availability Period *fu*
DAP Designated Acquisition Program *st*
DAP Developer Assistance Program *pk, kg*
DAP Digital Auto Pilot [card] *rm8.1.3*
DAP Diode Attach Process *jdr*
DAPG Data Analysis Processing Group *fu*
DAPIE Developers Application Programming Interface Extensions *pk, kg*
DAPIL Digital Assembly Parts Identification List *fu*
DAPR Digital Automatic Pattern Recognition *wp*
DAPS Data Preparation System *fu*
DAPS Deployable Aircraft Planning System [SAC] *mt*
DAPS Deployable Automated Planning System *wp*
DAQ Digital Angle Quantizer *jdr*
DAR Daily Activity Report *kf*
DAR Data Acquisition Requirements *jj*
DAR Data Assembly Register *jj*
DAR Data Automation Requirement *mt, wp*
DAR Days Awaiting Repair *jj*
DAR Defense Acquisition Radar *fu, hc, wp*
DAR Defense Acquisition Regulation [earlier ASPR, later replaced by FAR] *fu, su, hc, wp, ph, 8, ig21.12.16, tb, jdr, mt*
DAR Design Analysis Report *kf*
DAR Design Approval Request *fu, hc*
DAR Direct Access Request *hc*
DAR Distributed Array Radar *wp*
DARA Deutsche Agentur fur Raumfahrt *crl*
DARC Days After Receipt of Contract *fu*
DARC Direct Access Radar Channel *aw118.18.76, fu*
DARC Office of Reserve Components *dt*
DARCOM Development And Readiness Command [for Department of Army Materiel] *hn41.17.1, 12, nt, wp, st, jj, fu, mt*
DARD Depot Acceptance Requirements Document *fu*
DARI Database Application Remote Interface [IBM] *pk, kg*
DARK_SPOT [techniques to detect concealed weapons] *wp*
DARMS Data Automation Resource Management System [PACAF] *mt*
DARO Days After Receipt of Order *fu*
DARO Defense Airborne Reconnaissance Office *kf*
DARP DAAS ADPE Replacement Program *mt*
DARP Days After Report Period *fu*
DARP Direct Access Radar Processor *fu*
DARPA Defense Advanced Research Projects Agency [a research institute of the DOD, USA] *1, 11, 12, 14, 17, hc, aw130.13.24, sc3.3.2, hn47.7.5, aa26.12.26, af72.5.38, wp, tb, st, fu, ct, mt, ai2*
DARPANET Defense ARPANET *hi*
DARS Data Acquisition and Reduction (Recording) System *hc, st*
DART Data Analysis Recording Tape *fu*

DART Dedicated Advanced Reprogramming Testbed *oe*
DART Defect Awareness Review Team *rm3.1.1*
DART Defense Acquisition Review Team *tb*
DART Demonstration of Advanced Radar Technology (Techniques) *hc, wp*
DART Digital Audio Reconstruction Technology *pk, kg*
DART Director of Army Research and Technology *mt*
DART Duplex Army Radio/Radar Targeting *wp*
DART Dynamic Analysis and Replanning Tool [GCCS] [USA] *ro, mt*
DART Dynamic Analytic Replanning Tool [TRANSCOM] *mt*
DART Dynamically Adaptive Receiver Transmitter *mt*
DARTS Design Approach for Real Time Systems *fu*
DAS Damage Assessment Strike *wp*
DAS Data Acquisition System *gp, hc, tb, wl, jj, oe, ai2*
DAS Data Analysis System *ci.909*
DAS Data Approval Sheet *fu*
DAS Data Automation [Sub] System *0, hc, wp*
DAS Decimal Adjust for Subtraction *pk, kg*
DAS Defense Audit Service *wp*
DAS Defensive Avionics System *wp*
DAS Design Analysis Simulation (System) *fu*
DAS Despun Atmospheric Sounder *gp*
DAS Direct Access Service *fu*
DAS Direct Air Support *fu, mt, wp*
DAS Direct Analog Synthesizer *fu*
DAS Directorate of Administrative Services *dt*
DAS Distant Air Superiority *hc*
DAS Dual Attached Station *pk, kg*
DASA Defense Atomic Support Agency *hc, fu*
DASC Defense Logistics Agency Administrative Support Center *dt*
DASC Department of the Army System Coordination *st*
DASC Direct Air Support Center [USMC] *hc, fu, jj, mt*
DASC DLA's Administrative Support Center *mt*
DASD Department of Army Strategic Defense [center] *-17*
DASD Direct Access Storage Device *mt, wp, kg, fu, tb, ct, em*
DASFA Dynamic Analysis of Spacecraft with Flexible Appendage *hc*
DASG Data Acquisition Subsystem Group *hc*
DASH Drone Anti Submarine Helicopter *mt, hc, jj*
DASI Digital Altimeter Setting Indicator *fu*
DASIAC DOD Nuclear Information and Analysis Center *tb*
DASM Data Acquisition System Manual *jj*
DASO Demonstration and Shakedown Operation *fu*
DASP Direct Air Support Center *wp*
DASPS-E Department of the Army Standard Port Systems - Enhanced [USA] *mt*
DASS Defensive Aids Sub System [EW system for the Eurofighter 2000] *mt*
DASS Direct Air Support System *hc*
DASSS Decentralized Automated Service Support System [also DAS3] [USA] *mt*
DAST Design, Architecture, Software, Test *fu*
DAST DOD AUTODIN Subscriber Terminal *mt*
DASTARD Destroyer Antisubmarine Transportable Array Detection *fu*
DAT Daily Activity Team *mt*
DAT Deployment Action Team [JDA] *mt*
DAT Design Acceptance Test *wp*
DAT Design Analysis Tool *fu*
DAT Development Acceptance Test *mt*
DAT Digital Assembly Tester *fu*
DAT Digital Audio Tape *ct, em, wp, kg, hi, ai1*
DAT Direct Ascent Threat *5, sb*
DAT Disk Array Technology *pk, kg*
DAT Documentation Analysis Technique *tb*
DATA Decision Aids for Target Aggregation *mt*
DATA Design (Drafting) Aids Techniques Advice *fu, hn43.19.2*
DATACOM Data Communications *fu*
DATAEASE [commercial name for a fourth generation language data base] *mt*
DATAMAN Data Management [also DM] *wp*
DATAR Distributed Aperture of Tethered Array Radar *wp*
DATATRAC Data Traceability, Retrievability, Accountability, and Control *fu*
DATB Diaminotrinitrobenzene *wp*
DATC Days After Test Completion *fu*
DATD Digital Audio Tape Drive *hi*
DATE Dash Automatic Test Equipment *hc*
DATE Decision Aids Test Environment *hc*
DATE Digital Angular Torquing Equipment *hc*
DATE Dual Axis Test Equipment *hc*
DATIS Direct Access Technical Information System *hc*
DATIX Defended Area Threat Index *fu*
DATM Department of the Army Technical Manual *hc, jj*
DATOM Data Aids to Training, Operation, and Maintenance *fu*
DATOS Detection And Tracking Of Satellites *wp*
DATPA Decision Aid For Threat Penetration Analysis *wp*
DATPG Digital Automatic Test Pattern Generator *mt*

DATPG Digital Automatic Test Program Generation *hc*
DATR Design Approval Test Report *wp*
DATS Dynamic Accuracy Test Set *hc*
DATSA Depot Automatic Test System for Avionics *fu, hc*
DATU Direct Access Testing Unit *ct, em*
DAU Digital Adapter Unit *kf*
DAU Distribution Amplifier Unit *kf*
DAUPHIN [X-ray laser test by the Energy Department] *wp*
DAV Digital Audio-Video *hi, kg*
DAVC Delayed Automatic Volume Control *0, jj, fu*
DAVIC Digital Audio-Visual Council *pk, kg*
DAVID Digital Audio / Video Interactive Decoder *ct, em*
DAVID Distributed Access View Integrated Database [NASA] *ns, ge, hg*
DAW Data Address Word *wp*
DAWCLM Data Automation Workload Control and Library Maintenance *mt*
DAWNS Digital Automatic Weather and NOTAM System *mt*
DAX Digital Access Exchange *hc, hd1.89.y19*
DAX Digital Cross Connect *fu*
DAZD Double Anode Zener Diode *fu*
DB Data Base *mt, do422, 8, pl, tb, kg, jdr, kg, fu, hi, oe*
DB Data Block *fu*
DB Data Buffer *pk, kg*
DB Depth Bomb *mt*
DB Digital Bridge *do338*
DB Display Buffer *fu*
DB Double Biased *fu*
DB-TBS Deceptively Based-Terminal Defense System {?} *st*
DB/K Decibels [per degree] Kelvin *mb, vv, oe*
DB/L Development Baseline *fu*
DBA Data Base Administrator *mt, fu*
DBA Decibel, Adjusted *pk, kg*
DBA Demonstration and Briefing Auditorium *tb*
DBA Despin Bearing Assembly *gd2-0, 15, hc*
DBA Direct Budget Authority *wp*
DBA Dual Band Assembly *fu*
DBA-GK [USSR, Supreme Command's Long Range Bomber Arm; replaced the AON in 1940] *mt*
DBASI Digital Barometer Altimeter Setting Indicator *mt*
DBB {Design and Breadboard} *pl*
DBC Data Base Control *fu*
DBC Data Base Controller *mt*
DBC Data Bus Control *wp*
DBC Decibel(s) [referenced to Carrier] *vv, uh, oe*
DBC Digital to Binary Converter *fu*
DBCCR Data Base of Configuration Change Request *oe*
DBCP Data Base Control Processor *fu*
DBCR Data Base Change Request [AF Form 1776] *mt*
DBCS Delivery Bar Code Sorter *pk, kg*
DBCS Double Byte Character Set *pk, kg*
DBD Data Base Design *fu*
DBD Data Base Design Document [also DBDD] *tb*
DBD Data Base Document *mt*
DBD Differential Broadcast Driver *fu*
DBDD Data Base Design Document [also DBD] *mt, fu, ep*
DBDL Data Base Definition Language *hi*
DBE Data By Example *hi*
DBF Data Base Fomat *hi*
DBF Demodulator Band Filter *fu*
DBF Digital Beam Former *hc*
DBF Digital Beamforming Function *ic2-56, rg*
DBFN Digital Beam Forming Network *17, ic2-12, sb*
DBG Data Base Generator *fu*
DBGEN Data Base Generator *fu*
DBI Data Bus Interface *fu*
DBI Decibels relative to Isotropic *jj, oe*
DBID Data Block Identifier *oe*
DBIDI Data Base (DB) of Imagery Derived Information [DOD] [DIA] *mt*
DBIS Data Base Information System *sc4.2.4, 8*
DBIU Data Bus Interface Unit *hc*
DBL Direct Burdenable Labor *-8*
DBL DOS Batch Language *em*
DBM Data Base Manager (Management) *fu, kg, mt*
DBM Decibels [relative to 1 MilliWatt] *uh, ct, mb, jj, kf, vv, oe*
DBM Double Balanced Mixer *fu*
DBM Dynamic Balance Mechanism *gp, hc, pbl*
DBMF Data Base Management Function *fu*
DBMI Decibel(s) [referenced to 1 milliWatt isotropically received power] *vv*
DBMM Database Maintenance and Monitoring *fu*
DBMP Data Base Management Plan *mt*
DBMS Data Base Management Software *fu*
DBMS Data Base Management System *cu, ct, em, ge, kg, w9, wp, hc, hd1.89.y29, ns, tb, st, wl, hg, kf, fu, hi, oe, mt, ai1*
DBN Data Bus Network *-9*

DBNK {Data Bank} [Command Supply Management Data Bank, USAF] *mt*
DBNS Doppler Bombing / Navigation System *mt*
DBO Date Block Zero *jj*
DBP Database Processor *fu*
DBP Dibutylphtalate *wp*
DBPSK Differential BPSK *kf*
DBR Data Buffer Register *jj*
DBR Detail Billing Record *wp*
DBR DOS Boot Record *ct, em*
DBRB Data Base Review Board *-4*
DBRITE Digital Bright Radar {Indicator} Tower Equipment [display] *fu, mt*
DBS Data Base Server *hi, kg*
DBS Data Base Specification *mt*
DBS Data Base System *wp*
DBS Database Specification (Services) *fu*
DBS Demand Broadcast System *ct*
DBS Dibutylsulfide [chemical warfare Mustard simulant] *wp*
DBS Direct Broadcast by Satellite [system] *ct, hc, sc2.8.4, 19, ce, fu, tt, ai1*
DBS Direct Broadcast System *kf*
DBS Doppler Beam Sampling *wp*
DBS Doppler Beam Shaping *jj*
DBS Doppler Beam Sharpened (Sharpening) (Sharpener) *mt, hc, wp, fu, aw118.14.5*
DBSC Direct Broadcast Satellite Company *sc4.5.3*
DBSFTT Doppler Beam Sharpening Fixed Target Tracking *hc*
DBSGM Doppler Beam Sharpened Ground Map *hc*
DBSP Doppler Beam Sharpening Patch *jj*
DBSS Doppler Beam Sharpening Sector *jj*
DBT Data Base Transmission *mt*
DBTC Data Bus Technology Concept *fu*
DBU Database Update *fu*
DBU Days Before Use *jj*
DBUE Data Base Utilization Expert *fu*
DBV Data Base Viewer *fu*
DBV Digital Broadcast Video *pk, kg*
DBW Decibel(s) [relative to 1 Watt] *vv, kf, uh*
DBW Design Bandwidth *fu*
DBW/M2 Decibel(s) [relative to 1 Watt per meter squared] *vv*
DBWI Decibel(s) [referenced to 1 Watt isotropic radiated] *vv*
DBX Digital Branch Exchange *fu*
DBX Distributed Branch Exchange *mt*
DC Damage Control *jj*
DC Data Center *fu*
DC Data Collection *gg23, kg*
DC Data Communication *pk, kg*
DC Data Compression *hi*
DC Data Concentrator *mt*
DC Data Control *pk, kg*
DC Dedicated Circuit *mt*
DC Definite Contract *hc*
DC Delta Clipper *mi*
DC Depth Charge *mt, fu*
DC Design Confidence (Code) *fu*
DC Device Control *kg, sp5, hi*
DC Device Controller *fu*
DC Direct Current *mt, gg140, wp, 4, w9, es55.72, aw129.21.7, mj31.5.3, tb, jj, kf, ct, kg, em, fu, oe, ai1*
DC Direction Center *fu*
DC Disable Communications *fu*
DC Disarmament Conference *-14*
DC Display Controller *fu*
DC Display Coordinator *fu*
DC Distribution Code *fu*
DC [chemical warfare codename for diphenylcyanoarsine] *wp*
DC [Methylphosphonic Dichloride (chemical warfare nerve agent precursor)] *wp*
DC&P Data Collection and Processing *mt*
DC&T Detection, Classification and Targeting *mt*
DC-A Development Center - Atlanta [USA] *mt*
DC/SR Display and Control/Storage and Retrieval [System] [USAF] *fu, mt*
DCA Data Correlator Array *fu*
DCA Defense Communications Agency * *mt, fu, hc, wp, gs1-5.A5, 5, 12, sb, is26.1.3, jb, dt, tb, jj, jdr*
DCA Defensive Counter Air *mt*
DCA Digital Communications Associates *pk, kg*
DCA Distributed Communications Architecture *hi*
DCA Dual Capable Aircraft *wp*
DCAA Defense Contract Audit Agency [an organization] *mt, fu, hc, ma24, 8, aw130.11.24, ie5.11.12, nt, dt, tb, jj, pbl, tm, jdr, wp, ah, kf, oe*
DCAAO Defense Contract Audit Agency Office *st*
DCAC DCA* {Defense Communications Agency} Circular *mt*
DCACAS Data Collection, Analysis, and Corrective Action System *fu*
DCAEUR Defense Communications Agency Europe *mt*

DCAF Distributed Console Access Facility [IBM] *pk, kg*
DCAM Direct Chip Attach Module *pk, kg*
DCAMPS Deployable Computer Aided Mission Planning System [SAC*] *mt*
DCAMS Deployable Core Automated Maintenance System [CAMS] [USAF] *mt*
DCANG District of Columbia Air National Guard *12, af72.5.27*
DCAOC Defense Communications Agency Operations Center *mt, fu*
DCAP Deficiency Corrective Action Program *fu*
DCAP Digital Communications Access Processing *fu*
DCAP Document Change Analysis Program *fu*
DCAPAC Defense Communications Agency Pacific *mt*
DCAS Defense Contract Administrative (Administration) Service *mt, fu, su, hc, 17, sb, wp, tb, st, jj, oe*
DCAS Defense Contract Audit Service *mt*
DCAS Defense Control Administration Services *hc*
DCAS Digital Core Avionics System *aw118.18.78*
DCASMA Defense Contract Administration Services Management Area *mt, fu, wp*
DCASO Defense Contract Administration Services Office *fu*
DCASPRO Defense Contract Administration Services Plant Representative Office *su, mt, hm, wp, fu*
DCASR Defense Contract Administration Service(s) Region *hc, tb, jj, fu, su, mt*
DCB Data Control Block *ct, em*
DCB Design Change Board *-15*
DCB Device Control Block *pk, kg*
DCB Digital Communications Buffer *fu*
DCB Disk Coprocessor Board [Novell] *pk, kg*
DCC Damage Control Center *mt*
DCC Data Change Code [designator] *ep, pbl, jdr, tm, ah*
DCC Data Communications Channel *0*
DCC Data Computation Complex *ci.944*
DCC Data Condition Code *0*
DCC Defense Command Center *hc*
DCC Design Change Coordination *tb*
DCC Detection Criteria Control *fu*
DCC Device Control Character *wp*
DCC Digital Command Control *pk, kg*
DCC Digital Compact Cassette *hi, kg*
DCC Direct Cable (Client) Connection *ct, em, kg*
DCC Display Channel Complex *fu*
DCC Display Combination Code *pk, kg*
DCC Display Control Console *fu*
DCC Document Control Center [organization] *ah, ep, jj, pbl, jdr, tm, oe*
DCC Drone Control Center *hc*
DCCA Design Change Cost Analysis *wp*
DCCO Defense Commercial Communications Office *mt*
DCCS Data Communications Control System *mt*
DCCS Defense Communications Control System *hc, wp, jj*
DCCS Digital Cross Connect Switching *hi*
DCCS Distributed Command and Control System [USA] *kf, mt*
DCCU Digital Communications Control Unit *fu*
DCD Data Carrier Detect *pk, kg*
DCD Design Change Document *wp*
DCD Design Control Document *hc*
DCD Design Criteria Document *tb*
DCD Desired Completion Date *mt*
DCD Diagnostic Control Device *fu*
DCDPC Defense Command and Data Processing Center *hc*
DCDRS Drone Control and Data Retrieval System *fu, hc*
DCDT Data Collection and Drawing Tool *fu*
DCE Data Carrier Equipment *hi*
DCE Data Circuit terminating Equipment *hi, mt, tb, kg, nu, fu*
DCE Data Communications Equipment *mt, em, ct, do248, kg, tb, hi*
DCE Data Control Equipment *-4*
DCE Deployment Control Electronics *oe*
DCE Despin Control Electronics [also DECEL] *gs2-9.14, jdr*
DCE Digital Communications Equipment *fu, wp*
DCE Distributed Computing Environment *ct, em, kg, hi*
DCE Distributed Computing Equipment *pk, kg*
DCE Drive Command Electronics *-4*
DCEC Defense Communications Engineering Center [DCA*] *mt, dt, tb*
DCED Distributed Computing Environment Daemon *pk, kg*
DCEM Digital Communications Efficiency Model *ro*
DCEO Defense Communications Engineering Office *wp*
DCF Data Collection Form *mt*
DCF Data Communication Facility [IBM] *pk, kg*
DCF Data Compression Facility *pk, kg*
DCF Data Count Field [IBM] *pk, kg*
DCF DCEC Computer Facility [DCA] *mt*
DCF Document Composition Facility *fu*
DCF Driver Configuration File [Lotus] *pk, kg*
DCF Drone Control Facility *tb, fu*
DCFEM Dynamic Crossed Field Electron Multiplication *fu*
DCFO Defense Communications Field Office *mt*

DCFS Data Communications Functional Support *mt*
DCG Diode Capacitor Gate *fu*
DCG {Deputy Commanding General} *jmj*
DCGE Digital Command Guidance Electronics *bf*
DCI Defense Computer Institute *mt*
DCI Development Configuration Identification *fu*
DCI Direct Computer Interface *fu*
DCI Director of Central Intelligence [USA] *mt, ph, wp, 10*
DCI Director of Combat Intelligence *mt*
DCI Display Control Interface *fu, kg, hi*
DCIB Data Communications Input Buffer *fu*
DCID Director of Central Intelligence Directive *mt, kf*
DCII Defense Central Index of Investigations *hc, tb*
DCIM Data Control and Interface Module *fu*
DCL Data Communications Link *jmj*
DCL Data Control Language *pk, kg*
DCL DEC Command Language [DEC] *mt, hi, kg*
DCL DEC Control Language *fu*
DCL Declaration *pk, kg*
DCL Defense Capability Level *fu*
DCL Device Clear *pk, kg*
DCL Digital Command Language [Digital] *kg, hd1.89.y17, oe*
DCL Digital Control Logic *pk, kg*
DCL Direct Communications Link [the Washington to Moscow "hot line"] *mt, hc, 12*
DCLD Digital Communications Section [Rome Air Development Center] *mt*
DCLSO DLA CALS Support Office *mt*
DCLZ Data Compression Lempel-Ziv *hi*
DCM Data Communications Manager *hi*
DCM Data Concentrator Multiplexer *fu*
DCM Data Configuration Management *oe*
DCM Deputy Commander for Maintenance *mt*
DCMC Defense Contract Management Command *jdr, kf, pbl, oe*
DCMO DLA Civilian Personnel Management Support Office *mt*
DCMU Digital Colour Map Unit [UK] *mt*
DCN Data Change Notice *wp*
DCN Design Change Notice *fu, kf*
DCN Development Change Notice *15, hc, jj, fu*
DCN Document (Drawing, Documentation) Change Notice *fu, wp, tb*
DCN Document Control Number *jj, fu*
DCNM (shall require approval of the DCNM (RM&QA)) *rm1.1.3*
DCO Data Category Option *jj*
DCO Depth Cutoff *fu*
DCO Deputy Commander for Operations *mt, tb, fu*
DCO Dial Central Office *mt, fu*
DCO Dial Control Office *mt*
DCOM Distributed Component Object Model *em, kg*
DCOSS Defense Communications Operational Support System [DCA] *mt*
DCP Data and Command Processor *sc4.3.8*
DCP Data Collection Platform *gd1-4, 16, pbl, oe*
DCP Data Communications Processor *mt*
DCP Data Control Point *fu*
DCP Database (Data) Control Processor *fu*
DCP Decision Coordinating Plan *mt*
DCP Decision Coordination (Coordinating) Paper *mt, fu, hc, wp, tb, st, jj, jdr*
DCP Design Change Proposal *hc, jj*
DCP Design Competition (Concept) Phase *fu*
DCP Despin Control Processor *-8*
DCP Development Concept Paper *mt, hc*
DCP Development Concept Plan *hc, jj*
DCP Diagnostic Computer Program *fu*
DCP Diagnostic Control Panel *fu*
DCP Digital Light Processing [TI] *pk, kg*
DCP Disaster Control Plan *wp*
DCP Display Control Processor *fu*
DCP Distributed Communications Processor [as in DCP-40, Model 40] *mt*
DCP Division Control Point *fu*
DCP Document Control Point *fu*
DCPC Direct Current Power Converter *fu*
DCPC Dual Channel Port Controller *fu*
DCPG Defense Communications Planning Group *mt, hc*
DCPI Data Collection Platform Interrogation *gp, 16, oe*
DCPI Data Collection Platform Interrupt [signals] *bl3-1.26*
DCPLS Data Collection and Platform Location System *ag*
DCPR Data Collection Platform Radio *gp, oe*
DCPR Data Collection Platform Response (Report, Reply) [temperature, pressure] *bl3-1.26, gp, 16, oe*
DCPR Defense Contractors Planning Report *hc*
DCPS Data Collection Platform System *ac1.7-10*
DCPS Dynamic Crew Procedure Simulator *ci.971*
DCPSK Differentially encoded Coherent Phase Shift Keying *csIII.11*
DCR Data Conversion Receiver *fu*

DCR Deficiency/ Concern Report *fu*
DCR Design Change Recommendation *fu*
DCR Design Change Request *wp*
DCR Digital Conversion Receiver *fu*
DCR Direct Current Restorer *fu*
DCR Directorate of Collateral Responsibility *mt*
DCR Document (Drawing) Change Request *fu*
DCR Document Change Report *fu*
DCS Data Collection System *ci.881, 16, hy, hg, ge, kg, oe*
DCS Data Communications System *mt*
DCS Data Conditioning System [Science] *0*
DCS Data Control System *pk, kg*
DCS Decision Communications System *tb*
DCS Defense Communications System [DCA*] *mt, 12, wp, ph, hc, tb, st, jj, kf, fu*
DCS Defense Courier Service *jdr, kf*
DCS Deployment Control System *fu*
DCS Deputy Chief of Staff *mt, af72.5.31, jj, ro, fu*
DCS Design Change Schedule *hc*
DCS Desktop Color Separation *pk, kg*
DCS Despin Control [Sub] System *hc, gs2-8.2*
DCS Digital (Data) Communications System *fu*
DCS Digital Channel Simulator *fu*
DCS Digital Command System *0*
DCS Digital Communications Services *jdr, ric, ct*
DCS Digital Communications System *hc*
DCS Digital Control System *y2k*
DCS Digital Controller Subsystem *fu*
DCS Display and Control Station (System) *fu*
DCS Display Coordinator Subsystem *fu*
DCS Distributed Computing Services *jdr*
DCS Distributed Computing System *fu*
DCS Distributed Control System *pk, kg*
DCS/C Deputy Chief of Staff/Controller *hc*
DCS/D Deputy Chief of Staff/Development *hc*
DCS/I Deputy Chief of Staff/Intelligence *hc*
DCS/IN Deputy Chief of Staff for Intelligence *fu*
DCS/L Deputy Chief of Staff/Logistics *hc*
DCS/LOG Deputy Chief of Staff/Logistics *fu*
DCS/M Deputy Chief of Staff/Material *hc*
DCS/O Deputy Chief of Staff/Operations *hc*
DCS/P Deputy Chief of Staff/Personnel *hc, af72.5.31*
DCS/RD Deputy Chief of Staff for Research and Development *fu, hc*
DCSC Defense Construction Supply Center *mt, wp, jj*
DCSG Data Computation Subsystem Group *hc*
DCSIRM Deputy Chief of Staff for Information Resource Management *mt*
DCSLOG Deputy Chief of Staff for Logistics [USA] *st, jj, mt*
DCSO Defense Communications System Office *mt*
DCSO Defense Communications System Organization [DCA*] *mt*
DCSO Deputy Commander for Space Operations *sf31.1.27*
DCSOPS Deputy Chief of Staff for Operations and Plans *mt, st*
DCSP Defense Communications Satellite Program *mt, hc*
DCSPER Deputy Chief of Staff for Personnel *mt*
DCSRDA Deputy Chief of Staff for Research, Development and Acquisition *hc, st*
DCSS Deployable Combat Support System [USAF] *mt*
DCSS Digital Communications Subsystem *hc*
DCT Data Communications Terminal *mt*
DCT Depot Chassis Testers *hc*
DCT Design Certification Test *fu*
DCT Detached Console Trainer *fu*
DCT Diagnostic Computer Terminal *fu*
DCT Digital Communications Terminal *fu*
DCT Digital Communications Terminals [USMC / USAF] *mt*
DCT Discrete Cosine Transform *pk, kg*
DCT DND Controller Terminal *fu*
DCTL Direct Coupled Transistor Logic *hc, jj, fu*
DCTN Defense Communications Telecommunications Network [DCA] *mt*
DCTV Despun Control [Compartment] Thermal Vacuum *8, 9, lt, ep*
DCU Data Cache Unit *pk, kg*
DCU Data Conversion Unit *fu*
DCU Decade Counting Unit *fu*
DCU Demodulator/Computer Unit *hc*
DCU Digital Communications Unit *fu*
DCU Digital Computer Unit *hc*
DCU Digital Conference Unit *mt*
DCU Digital Control Unit *fu, oe*
DCU Digital Counting Unit *fu*
DCU Display and Control Unit *fu*
DCU Display Controller Unit *fu*
DCUTD DOD Controlled Unclassified Technical Documents *fu*
DCV Direct Current Voltmeter [multipurpose voltmeter] *hc*
DCWG Data Collection Working Group *-17*
DCWP DC Warm Power *fu*

DCWV Direct Current Working Voltage *jj*
DCX Delta Clipper eXperimental *mi*
DCXO {a type of Crystal Oscillator (?)} *hd*
DD Data Definition *fu, ct*
DD Data Dictionary *fu, tb*
DD Data Dictionary and Directory [team] *mt*
DD Day [used in DD-MM-YY date format definitions] *hi, kg*
DD Defense Department [usually DOD] *mt, hn42.3.6*
DD Definition Description (HTML) *ct*
DD Deny and Deception *wp*
DD Destroyer [US Navy ship classification] *mt, fu, wp*
DD Detailed Design *fu*
DD Digital Display *jj, kg*
DD Disconnecting Device *fu*
DD Dishonorable Discharge *w9*
DD Double Density *ct, em, kg, hi*
DD Duplex Drive [designation for Sherman tanks for amphibious operations, fitted with twin propellers and a buoyancy screen] [Allied] *mt*
DD&E DOD Research and Engineering office *-4*
DD&P Detection, Diagnosis and Prognosis *rf84.3.26*
DD/D Data Dictionary / Directory *mt*
DD/DDG Destroyer/Destroyer Guided Missile *jj*
DD/DS Data Dictionary Directory System *mt*
DDA Deputy Director for Administration *-10*
DDA Designated Development Agency *mt*
DDA Destroyer, Attack *hc, jj*
DDA Device Driver Architecture *hi*
DDA Digital Differential Analyzer *fu, 0, hc, jj*
DDA Digitally Directed Analog *fu*
DDA Distributed Data Access *pk, kg*
DDA Domain Defined Attribute *pk, kg*
DDAA Department Of Defense Audit Agency *jj*
DDAS Digital Data Acquisition System *hc*
DDAT Depot Design Assistance Task *jj*
DDAY [date of commencement of hostilities] [also D-Day] *mt*
DDB Decision Data Base *mt*
DDB Design Decision Board *fu*
DDB Device Dependent Bitmap *hi, kg*
DDB Device Descriptor Block *pk, kg*
DDB Digital Data Base *wp*
DDB Digital Data Bus *kf, he, vv*
DDB Distributed Data Base *wp*
DDBMS Distributed DBMS
DDBS Display Data Base Subs *mt*
DDC Dansk Datamatik Center *fu*
DDC Data Display Center *mt*
DDC Data Display Console *fu*
DDC Data Distribution Center *fu, mt, wp, kf*
DDC Defense Documentation Center [formerly ASTIA, later DTIC] *mt, su, hc, wp, tb, jj, fu*
DDC Design and Development Contract *mt*
DDC Design Development Center *hd1.89.y20*
DDC Digital Data Channel [VESA] *hi, kg*
DDC Digital Data Converter *fu*
DDC Digital Data Corporation *fu*
DDC Digital Downconverter *fu*
DDC Digitally Directed Control *fu*
DDC Direct Data Command *0*
DDC Direct Data Connection *0*
DDC Direct Digital Control *hc, jj*
DDC Display Data Channel (DDC1 for display data channel #1) *hi, kg*
DDC Display Data Controller *fu*
DDC Distributed Data Channel *fu*
DDC Document Distribution Center *jdr*
DDC Dual Diversity Comparator *fu*
DDCMP Digital Data Communications Message Protocol [DEC] *ct, em, kg, do286, hi*
DDCP Definitive Design Change Proposal *hc*
DDCP Document Distribution Control Point *fu*
DDCS Distributed Database Connection Services [IBM] *pk, kg*
DDCU DC to DC Converter Unit *mi*
DDD Data Definition Language *pk, kg*
DDD Data Description Document *mt*
DDD Design Data Dictionary *fu*
DDD Design Disclosure Document *fu*
DDD Design, Development and Delivery *hc*
DDD Desired Delivery Date *wp*
DDD Detailed Data Display *hc, jj*
DDD Detailed Design Document *fu*
DDD Detect, Discriminate and Designate *fu*
DDD Detection, Discrimination, and Designation [also D3] *hc, tb, st*
DDD Digital Data Display *hc*
DDD Digital Density Detector *fu*
DDD Digital Diagnostic Diskette *ct, em*
DDDD Design Data Dictionary and Directory [also D4] [AWIS] *mt*
DDDP Detail Data Display Processor *jj*

DDDS Danish Defense Data Service *fu*
DDE DD (Destroyer), Escort [USA] *mt*
DDE Digital Data Extractor *fu*
DDE Direct Data Entry *pk, kg*
DDE Dynamic Data Evaluation *hc*
DDE Dynamic Data Exchange *ct, kg, em, hi*
DDEML Dynamic Data Exchange Manager Library [Microsoft] *pk, kg*
DDEP Defense Development Exchange Program *ph*
DDESB Department Of Defense Explosives Safety Board *dt*
DDEU Digital Data Entry Unit *fu*
DDEW Database Design and Evaluation Workbench *fu*
DDF Display Data Channel *pk, kg*
DDF Dynamic Data Formatting [IBM] *pk, kg*
DDFD Detailed Data Flow Document *fu*
DDFD Digital Delay Frequency Discrimination *fu*
DDG Data Distribution Group *fu*
DDG Destroyer, Guided Missile armed [USA DOD] *mt*
DDG Destroyer, Guided Missile [US Navy ship classification] *uc, wp, fu*
DDG Dynamic Data Generator *fu*
DDH Destroyer, Helicopter *hc, jj*
DDH Destroyer, Helicopter carrying [US Navy] *mt*
DDI Data Display Indicator *fu*
DDI Delivery Distribution Indicator *mt*
DDI Depth Deviation Indicator *fu*
DDI Detailed Departmental Instructions *oe*
DDI Device Driver Interface *pk, kg*
DDI Digital Data Indicator *jj, fu*
DDI Digital Diagnostics Inc. *fu*
DDI Digital Display Indicator *mt, jj, fu*
DDI Direct Dial In *pk, kg*
DDI Display Definition Interface *fu*
DDial Diversi-Dial *ct*
DDIS Danish Defense Intelligence Service *fu*
DDIU Display Dedicated Interface Unit *fu*
DDK Destroyer, Hunter-Killer [US Navy] *mt*
DDK Device Driver Kit [Microsoft Windows] *pk, kg*
DDL Data Definition Language *mt, kg, fu*
DDL Data Description Language [Honeywell] *kg, mt*
DDL Database Description Language *em*
DDL Delegation of Disclosure authority Letter *mt*
DDL Digital Data Link *mt, wp*
DDL Digital Delay Line *hc, jj*
DDL Direct Data Link *kf*
DDL Direct Downlink *kf*
DDL Dispersive Delay Line *hc, pl, jj*
DDL Document Description Language *hi*
DDL Document Distribution List *jj*
DDLC Data Description Language Committee *mt*
DDLT Direct Detector Laser Transceiver *hc*
DDM Distributed Data Management *pk, kg*
DDMO Defense Data Management Office *mt*
DDMO Divisions Data Management Office *hc*
DDMS Distributed Data Management System *tb*
DDN Defense Data Network [DCA*] *mt, is26.1.76ee, pf.f88.27, jb, tb, kg, kf*
DDN Defense Department Network *ct*
DDN Defense Digital Network *wp*
DDN Digital Data Network *tb*
DDN Distributed Data Node (Network) *tb*
DDNIA Defense Data Network Interface Adapter *mt*
DDNS Dynamic Domain Naming System *pk, kg*
DDO Defensive Duty Officer *mt*
DDO Deputy Director for Operations [NMCC] *10, mt*
DDP Defense Dissemination Program *mt*
DDP Department Of Defense Production [Canada] *hc*
DDP Deployment Detailed Planning *mt*
DDP Digital Data Processing (Processor) *mt, wp, hc, jj*
DDP Distributed Data Processor (Processing) *mt, tb, kg*
DDP Distribution Data Processing *st*
DDPP Deputy Director for Plans and Policy *-10*
DDPR Deputy Director for Programs and Resources *-10*
DDPR Digital Data Processing Request *hc*
DDPS Digital Data Processing System (Subsystem) *mt, ac2.8-14, 4, oe*
DDR Daily Demand Rate *mt, wp*
DDR Decoy Discrimination Radar *hc*
DDR Destroyer, Radar *hc*
DDR Detail Design Review *jj*
DDR Deutsche Demokratische Republik [German Democratic Republic, a.k.a. "East Germany"] *mt, fu*
DDR Digital Display and Recorder *hc*
DDR Radar Picket Destroyer [US Navy] *mt*
DDR&E Deputy Director, Research and Engineering *hc*
DDR&E Director of (Directorate for) Defense Research and Engineering [DOD office] *mt, 2, 13, tb*
DDRB Design Data Review Board *wp*
DDRR Directional Discontinuity Ring Radiator *hc*

DDRS Defense Data Repository System [GCCS, DOD] [USA] *mt*
DDRS Digital Data Recording System *hc*
DDS Data Display System (Station) *mt, wp, jj*
DDS Dataphone Digital Service [AT&T] *do44, mt*
DDS Defense Dissemination System *mt*
DDS Design Data Sheet *pk, kg*
DDS Digital Data Storage *hi, kg*
DDS Digital Data System *mt*
DDS Digital Dataphone Service [AT&T] *mt, kg*
DDS Direct Digital Synthesizer *hc, uh, jdr*
DDS Direct Distribution [of Scientific Data] Satellite *-19*
DDS Display Design System *fu*
DDS Distributed Database Services *pk, kg*
DDS Dry Deck Shelter [USA DOD] *mt*
DDS Dust Detector Subsystem *jp*
DDSC Design and Drafting Standards Committee *ep, jj, pbl, tm*
DDSP Digital Doppler Signal Processor *hc*
DDSS DS Decision Support System [AFLC] *mt*
DDSSJ Drone Deceptive Self Screening Jammer *hc*
DDT Data Description Table *wp*
DDT Dichloro Diphenyl Trichloroethane [an insecticide] *wp*
DDT Downlink Destination Table *jdr*
DDT Dynamic Debugging Tool *ct, em*
DDT&E Design, Development, Test, and Evaluation *mt, hc, ci.591, tb, wl, jdr*
DDT&E Director, Defense Test and Evaluation *ph*
DDTE Design, Development, Test, and Evaluation *mi*
DDTS Distributed Data and Telecommunications System *mt*
DDTT Dye Diffusion Thermal Transfer [printing] *ct, kg*
DDTU Discrimination, Designation and Track Unit *hc*
DDU Data Detector Unit *jdr*
DDU Data Distribution Unit *oe*
DDU Despun Driver Unit *gp*
DDU Digital Display Unit *jj*
DDU Digital Distribution Unit *mt*
DDU Disk Drive Unit *hi*
DDVN Direct Dedicated Voice Network *mt*
DDVP Dimethyl Dichloro Vinyl Phosphate *wp*
DDW Discrete Data Word *jj*
DDX Digital Data Exchange *pk, kg*
DE Damage Estimation *mt*
DE Damage Expectancy *wp, tb*
DE Data Element *mt, wp*
DE Data Entry *mt, jj*
DE Delco Electronics [Corporation] *heh2, 21, 1*
DE Deputy Chief of Staff / Engineering and Services *mt*
DE Development Engineering *fu, hc, ep, jj*
DE Device End *fu, pk, kg*
DE Diesel Electric *wp*
DE Digital Encoder *fu*
DE Direct Escort *sd*
DE Directed Energy [USA DOD] *jb, jmj, as, mt*
DE Discard Eligible *ct, em*
DE Dynamic Exercise *fu*
DE Dynamics Explorer *4, hy*
DE Escort Destroyer [USA] *mt*
DE&D Data Entry and Display *fu*
DE&DD Data Entry and Display Device *fu*
DEA Data Encryption Algorithm *tb*
DEA Data Exchange Agreement *hc, wp*
DEA Defense Exchange Agreement *ph*
DEA Di Ethyl Amine *wp*
DEA Drug Enforcement Administration (Agency) *mt, wp*
DEAL Decision Evaluation Logic *wp*
DEARAS Defense Emergency Authorities Retrieval and Analysis System *mt*
DEB Data Electronics Box *oe*
DEB Digital European Backbone *mt, wp*
DEBI DMA Extended Bus Interface *em*
DEBS Digital Electron Beam Scanner *fu*
DEC Data Entry Console *jj*
DEC Declination *0*
DEC Decoder *jdr*
DEC Decrement *pk, kg*
DEC Delco Electronics Corp. *fu*
DEC Device Clear *pk, kg*
DEC Digital Equipment Corporation *mt, fu, kg, ct, hc, sc4.5.2, aa26.11.44, aw130.11.56, nt, ns, tb, st, jdr, hi, oe*
DEC Dual Engine Centaur *ai2*
DEC [control escort ship] [US Navy] *mt*
DECAN Distance measuring Equipment Command And Navigation *hc*
DECC Department Of Defense Concessions Committee *dt*
DECCO Defense Commercial Communications Office [DISA] *mt*
DECCO Defense Engineering Commercial Communications Office *dt, tb*
DECEL Despin Control Electronics [also DCE] *hc*
DECL Declaration of War / National Emergency {?} *mt*

DECL Dielectrically Isolated Emitter Coupled Logic *fu*
DECM Deceptive Electronic Counter Measures [AN/ALQ-126] [USA] *wp, jj, fu, mt*
DECM Defense Electronic Countermeasure *hc*
DECM Digital Electronic Countermeasure *wp*
DECNET Digital Equipment Corporation Network (Networking) [products, generic family name] *ct, fu, ns, do395, kg*
DECON Defensive Electronic Countermeasures *mt*
DECON Distributed Execution Control for Online Networks *fu*
DECON [chemical warfare] Decontaminant (DECON40 is trichloroisocyanuric acid] *wp*
DECS Defense Electronics Control Service *wp*
DECS DSCS ECCM Control Subsystem *mt*
DECU Digital Engine Control Unit *mt*
DED Data Element Dictionary (Definition) *mt, wp, jj, fu*
DED Dual Element Drive *fu*
DEDD Data Entry and Display Device *fu*
DEDS Data Element Dictionary Subsystem *fu*
DEDS Data Entry and Display Station *fu*
DEDS Data Entry Device Set *fu*
DEDS Data Entry Display System *wp*
DEE Data Encryption Equipment *tb*
DEEC Digital Electronics Engine Control [USA] *mt, aw118.14.11*
DEEP Di Ethyl Ethyl Phosphonate [chemical warfare simulant] *wp*
DEEPOB Deep Space Objects [a software program] *jb*
DEEPT Diethyl Ethyl Phosphono Thioate [chemical warfare Simulant] *wp*
DEER Directed Energy Experimental Range *tb*
DEERS Defense Eligibility and Enrollment Reporting System [DOD] *mt, af72.5.119, wp*
DEES Dynamic Electronic Environment Simulator *fu*
DEF Data Exchange Format *pl*
DEF Denver Engineering Facility [a part of Hughes Aircraft Company] *pl*
DEF Development [and] Evaluation Facility [WIS] *tb, mt*
DEF Display Evaluation Form *mt*
DEFCOMNON Defense Command North Norway (Bodo, NO) *fu*
DEFCOMSONOR Defense Command South Norway (Oslo, NO) *fu*
DEFCON Defense [readiness] Condition *af72.5.119, wp, jb, fu, wp, tb, kf*
DEFCORD Defense Coordination Network *mt*
DEFIL Data Edit File *mt*
DEFPLAN Defense Plan *mt*
DEFREP Defense Response Status *mt*
DEFSIM Defense Simulation *tb*
DEFSMAC Defense Special Missile and Astronautics Center *10, jb, kf*
DEFSMAC Defense Status Military Alert Center *hs*
DEFSMAC Department Special Missile and Astronautic Center *mt*
DEG Destroyer Escort, Guided-missile *jj*
DEG [Guided missile Escort ship] [US Navy] *mt*
DEG/DB Degree(s) per Decibel *vv*
DEG/SEC Degree(s) per Second *vv, kf*
DEGM Dynamic End Game Model *hc*
DEI Design (Development) Engineering Inspection *hc, fu, jj*
DEIIS Detailed Experiment Integration Interface Specification *pl*
DEIMOS Discrete Element Idealization Model of Solar *hc*
DEIMS Defense Economic Impact Modeling System *ph, wp*
DEIOP Display Extended Input/Output Processor *fu*
DEIS Defense Energy Information System [DFSC] *mt*
DEK Data Encryption Key *pk, kg*
DEK Data Entry Keyboard *fu*
DEL Data Evaluation Laboratory *-4*
DEL Data Export License *fu*
DEL Denver Engineering Laboratories *hc, sc3.1.4, hn49.11.7*
DELEX Data Elements Lexicon *mt*
DELPHI Discriminating Electrons By Laser Photon Ionization *wp*
DELSTR Delete String [REXX] *pk, kg*
DEM Data Energy Modernization {?} *mt*
DEM Data Entry/ Management *fu*
DEM Diethyl Malonate [chemical warfare Agent Simulant] *wp*
DEM Digital Elevation Model *fu*
DEM/VAL Demonstration and Validation [program phase] [also DEM-VAL] *go4.1, nt, kf*
DEMA Distributed Emission Magnetron Amplifier *fu*
DEMANDS Depicts Each Month's Averages and New Demand *mt*
DEMC Defense Electronics Management Center *jj*
DEMON Demodulated Noise *fu*
DEMON Discrimination Effectiveness Model for Orbital Networks *kf*
DEMOS Demonstrations *tb*
DEMP Dispersed Electro-Magnetic Pulse *jdr, kf*
DEMPS Dispersed Electro-Magnetic Pulse Simulator *hc*
DEMS Data Entry Management System *fu*
DEMSTAT Deployment/Employment/Mobilization Status [FORSCOM] *mt*
DEMUX Demultiplexer [DMX is also seen] *hc, jdr, fu, uh*
DEMVAL Demonstration and Evaluation [USA] *mt*

DEMVAL Demonstration and Validation [also DEM/VAL] *nt*
DEN Data Element Number *jj*
DEN Document Enabled Networking [Novell-Xerox] *pk, kg*
DENIM Detector Enhancement, Integration and Multiplexing *hc*
DEO Directed Energy Office *tb*
DEOB Deck Edge Outlet Box *jj*
DEP Data Entry Panel *fu*
DEP Data Exchange Program *hc, jj*
DEP Delayed Entry Program *mt*
DEP Depot / Data Element Profile *mt*
DEP Diethylphtalate *wp*
DEP Draft Equipment Publication *fu, jj*
DEPACK Deployment Package system *mt*
DEPDA Deployment Data file [OJCS] [replaced by TPFDD] *mt*
DEPDATE Departure Date from POE *mt*
DEPECH Deployment Echelon *mt*
DEPEST Deployment Package Estimation System *mt*
DEPGUIDE Deployment Guide *mt*
DEPID Deployment Indicator Code *mt*
DEPLAN Deployment Plan *mt*
DEPMAS Deployment Management System [FORSCOM] *mt*
DEPOPSDEP Deputy Operations Deputies *mt*
DEPOS USAREUR Standard Depot System *mt*
DEPRA Defense European and Pacific Redistribution Activity *wp*
DEPREP Deployment Reporting System [JCS] *mt*
DEPS Development Engineering Prototype Site *wp*
DEPSECDEF Deputy Secretary of Defense [OSD] *mt, hc, wp, tb*
DEPSO Navy Department Standardization Office *dt*
DEPUS Departmental User *mt*
DER Distinguished Encoding Rules *pk, kg*
DER Double Ended Receiver *fu*
DER [Radar picket Escort ship] [US Navy] *mt*
DERA Defense Evaluation and Research Agency *crl*
DEROS Date of Expected Return from Overseas *mt*
DES Data Encoding System *0*
DES Data Encryption Standard(s) *mt, 10, wp, tb, em, kg, nu, fu, hi, tt*
DES Data Entry Service *fu*
DES Data Entry Sheet *pk, kg*
DES Deployed Encryption Standard *ct*
DES Digital Encryption Standard *ct*
DES Doppler Error Sensor *hc*
DESA Dimensional Electronic Scan Antenna *hc*
DESC Defense Electronics Supply Center [Dayton, Ohio] [DLA] *mt, fu, wp, hc, nt. jj, jdr*
DESC Double Ended Scan Converter *hc*
DESCOM Depot Systems (Support) Command [USA] *mt, wp, jj*
DESIRE Direct English Statement Information Retrieval Extraction [USAF] *mt*
DESMO Data Element Standardization and Management Office [Department Of Defense Logistics] *dt*
DESO Defense Export Services Organization [UK] *mt*
DESRON Destroyer Squadron *-12*
DESSIM Design Simulator *hc*
DET Data Entry Terminal *fu*
DET Device Execute Trigger *pk, kg*
DETA Dielectric Thermal Analyzer *wp*
DETA Diethylenetriamine *wp*
DETD Drift Equivalent Temperature Difference [frequency] *pl*
DETEC Defensive (Defense) Technology Evaluation Code *ts4-17, tb*
DETECT (Detects (Software Program)) *jb*
DETIR Defense Technology Information Repository *tb*
DETPA Diethylenetriaminepentacetic Acid *wp*
DEU Detector Electronics Unit *pl*
DEU Display Electronics Unit *hc*
DEV Design Evaluation Vehicle *0*
DEVAT Depot Vehicle Automatic Tester [Hughes Aircraft Company - DSD] *hc*
DEVR Distortion Eliminating Voltage Regulator *fu*
DEW Directed Energy Warfare [USA DOD] *mt*
DEW Directed Energy Weapon *mt, fu, wp, aw130.13.104, dh, 1, 14, 17, ic2-77, sb, hc, tb, st, jdr, kf*
DEW Distant Early Warning [a line of radar sites on the northern coast of Canada and on Greenland, replaced by the Northern Warning System NWS] *mt, fu, ro, 0, 12, 13, 15, aw, ph, hc, w9, af72.5.142, wp, tb, jj, kf*
DEWG Directed Energy Weapon Ground *tb*
DEWIZ Distant Early Warning Identification Zone *mt, fu*
DEWL Directed Energy Weapon, Laser *tb*
DEWO Directed Energy Weapon, Orbital *tb*
DEWP Directed Energy Weapon, Particle Beam *tb*
DF Data Field *pk, kg*
DF Destination Field *pk, kg*
DF Deuterium Fluoride *fu, wp, tb, st*
DF Device Flag *pk, kg*
DF Dicke-Fix *fu*
DF Direct Frequency *0*
DF Direction Finder (Finding) *mt, fu, wp, 0, w9, tb, jj*
DF Direction Flag *fu*

DF Distribution Free *fu*
DF Dog Fight *jj*
DF Double Flag *pk, kg*
DF Methylphosphonyldifluoride [chemical warfare nerve agent precursor] *wp*
DFA Data Flow Analyzer *hc*
DFA Designated Field Activity *jj*
DFA Domestic Field Allowance *fu, jj*
DFAD Digital Feature Analysis Data *mt, fu*
DFAE Director of Facilities Engineering *jj*
DFAMS Defense Fuel Automated Management Systems *mt*
DFAR DOD Federal Acquisition Regulation [a document] *ah, pbl, tm*
DFARS Department Of Defense Federal Acquisition Regulation Supplement *18, jdr*
DFASC Deployable Force Automated Service Center [USMC] *mt*
DFB Digital Filter Bank *ac10.4.4*
DFB Doppler Filter Bank *fu*
DFBW Digital Fly By Wire *wp*
DFC Data Flow Control [OSI Layer 5] *kg, nu*
DFC Diagnostic Flow Chart *wp, jj*
DFC Distinguished Flying Cross *w9*
DFCC Digital Fire Control Computer *wp*
DFCD Data Format Control Document *oe*
DFCS Defensive Fire Control System *hc*
DFCS Deployment Flow Computer System [MAC*] *mt*
DFCS Digital Fire Control System *mt*
DFCS Digital Flight Control System *hc, fu, wp, aw130.22.34*
DFD Data Flow Diagram *mt, gp, tb, kf, fu, oe*
DFD Digital Frequency Discrimination (Discriminator) *fu*
DFDR Digital Flight Data Recorder *mt*
DFDSS Data Facility Dataset Services *ct, em*
DFDT Di Fluoro Diphenyl Trochloroethane {?} *wp*
DFE Decision Feedback Equalizer *sp660*
DFE Derivative Fighter Engine *wp*
DFHSM Data Facility Hierarchical Storage Manager *ct, em*
DFI Data Field Identifier *mt*
DFI Design, Fabricate, and Install *mt*
DFI Development Flight Instruments *sc4.6.6*
DFL Deployed Fixed Location *kf*
DFLE Dog Fight Lock-on Enable *jj*
DFM Data Fax Modem *hi*
DFM Departure Flow Management *fu*
DFM Design For Manufacturability *pk, kg*
DFM Direct Function Mechanization *fu*
DFM Distinguished Flying Medal *w9*
DFM Dog Fight Mode *jj*
DFMA Design For Manufacturing and Assembly *jdr, kf*
DFMSR Directorate of Flight and Missile Safety *wp*
DFO Disaster Field Office *mt*
DFOLS Depth of Flash Optical Landing System *wp*
DFP Davidon, Fletcher, Powell *wp*
DFP Detailed Flight Plan *fu*
DFP Diisopropyl Fluoro Phosphate *wp*
DFPA (Fire Protection Conference, sponsored by DFPA) *aw130.13.104*
DFR-E Defense Fuels Region, Europe *mt*
DFRC Dryden Flight Research Center *ci.70, mi, mb*
DFRF Dryden Flight Research Facility [was ADFRF, now DFRC] *mi*
DFRIF Defense Freight Railway Interchange Fleet *mt*
DFRL Daily Failure Reporting Log *fu*
DFRN Data File Reference Number *jj*
DFS Development Flight System *jdr*
DFS Digital Formatting System *hc*
DFS Display Formatter Subunit *fu*
DFS Distributed File System *pk, kg*
DFS Drain and Fill Stand *jj*
DFS [German telecommunications satellite system] *sc3.12.4, es55.4*
DFSC Defense Fuel Supply Center [DLA] *dt, wp, mt*
DFSK Double Frequency Shift Keying *fu*
DFSM Detailed Functional System Requirements *jj*
DFSMS Data Facility Storage Management Subsystem *ct, em, kg*
DFSP Defense Fuel Supply Point [DLA] *mt*
DFT Design For Test (Testability) *wp, kf, kg, fu*
DFT Diagnostic Function Test *pk, kg*
DFT Discrete Fourier Transform *ep, kg, fu*
DFT Distributed Function Terminal *pk, kg*
DFTE Depot/Factory Test Equipment *fu*
DFTS Doppler Filter Test Station *hc, jj*
DFU Data File Utility *pk, kg*
DFVLR Deutsche Forschungs und Versuchsanstalt f̦r Luft und Raumfahrt [German research and test institute for air and space travel] *mt, ac3.7-15, aa26.8.31, crl*
DFWMAC Distributed Foundation Wireless Media Access Control *hi*
DFXP Digital Feature Extractor Processor *hc*
DG Data Group *2, 4, vv*
DG Deception Generator *fu*

DG Defense Guidance *mt, wp, tb, jdr*
DG Diesel Generator *fu*
DG Directional Gyro *jj*
DG Director General [CNES] *ac3.6-2, w9*
DG Display Generator *fu*
DG-UX Data General Unix *ct, em*
DGA [French General Directorate for Armament] *wp*
DGAC Direction GÉnÉrale de l'Aviation Civile [General Directorate of Civil Aviation] [France] *mt, aw129.1.28*
DGB Doppler Gravity Bias *jj*
DGBC Digital Geoballistic Computer *hc*
DGIS Direct Graphics Interface Standard *pk, kg*
DGIS DOD Gateway Information System *mt*
DGM Digital Group Multiplex *mt*
DGP Department of Programs and Industrial Policy [CNES] *ac3.6-2*
DGS Data Gathering Systems *fu*
DGS Degaussing System *fu*
DGS Domestic Ground Station *kf*
DGSA Diego Garcia Station [A-antenna] *uh*
DGSC Defense General Supply (Support) Center [DLA] *hc, tb, jj, mt*
DGSC Defense General Support Center [DLA] *mt*
DGSE Development Ground Support Equipment *jj*
DGSG Defense General Supply Center {?} {probably DGSC} *tb*
DGTB Data Generating Technology Base *wp*
DGTS Dynamic Ground Target Simulator *wp*
DGTS Dynamic Ground Truth Simulator *fu*
DGU Display Generator Unit *fu*
DGZ Designated Ground Zero [nuclear target] *fu, wp, tb, mt*
DH Digital Hierarchy *hi*
DH {an organization within Hughes Aircraft Company} *hn50.8.5*
DHA Detachement d'HelicoptÉres ArmÉs [armed helicopter detachment] [France] *mt*
DHARS Doppler Heading, Attitude and Reference System *hc*
DHB Defended Hard Basing [for ICBMs] *hc*
DHCP Dynamic Host Configuration Protocol *ct, kg, em, acp*
DHD Double Heatsink Diode *fu*
DHE Data Handling Equipment *pl*
DHEP Detailed Human Engineering Plan *fu*
DHF Demand History File *jj*
DHIT Development Hardware Integration Test *he*
DHL Dynamic Head Loading *pk, kg*
DHMI Devlet Hava Mcydanlari Isletmesi [General Directorate of State Airfields] *fu*
DHR Dehumidifier - Regulator *fu*
DHR [Phoenix N-3 Serial Number Designators] *jj*
DHRS Data Handling and Recording System *hc, wp*
DHS Data Handling Subsystem (System) *hc, fu, hy, jdr*
DHS Despun Heat Shield *hc*
DHU Data Handling Unit *pl*
DI (Data Input (In)) *fu, kg*
DI Data Item *mt, fu, tb, jj, kf*
DI Delivery Incentive [fee] *jdr*
DI Design Implementation (Group) *fu*
DI Design Integration *hc, jj, jdr*
DI Destination Index *ct, wp, kg, em*
DI Device Interface *fu*
DI Digital (Digitswitch) Input *hc, jj*
DI Direct Impact *mt*
DI Directivity Index *fu*
DI Division Instruction [a directive] *ep*
DI Document Imaging *pk, kg*
DI Driving Issue *17, sb*
DI/O Data Input/Output *fu*
DI/SR Descent Imager/Spectral Radiometer *es55.28*
DIA Defense Information Analysis *fu*
DIA Defense Intelligence Agency [USA] *mt, hc, wp, fu, su, 12, 17, ph, jb, jmj, dt, tb, st, jj, kf*
DIA Deficiency In Allowance *jj*
DIA Document Interchange Architecture [IBM] *hi, kg*
DIAC Defense Industry Advisory Council *hc*
DIAC Defense Intelligence Analysis Center *mt*
DIACS Documentation Information and Control System *fu*
DIAG Device Interface Address Generator *fu*
DIAL Deficiency In Allowance List *jj*
DIAL Design Issues Analysis Log *fu*
DIAL Differential Absorption LIDAR [or Ladar, UV to IR] *16, wp, hy, hg, ge*
DIAL Display Interactive Assembly Language *wp*
DIAL-A-LOG [Air Force on line bulletin board systems] *mt*
DIALCOM Dialed Communications *mt*
DIAM Defense Intelligence Agency Memorandum/Manual *mt*
DIAM Defense Investigative Agency Manual *kf*
DIAN Doppler Inertial Airdata Navigation *hc*
DIANA [space surveillance radar] *wp*
DIANE Digital Integrated Attack and Navigation Equipment [USA] *mt*
DIAOBS DIA Order of Battle *mt*

DIAOLS Defense Intelligence Agency On Line System *ro*
DIAOLS DIA On Line Intelligence System *mt*
DIB Defense Intelligence Board *wp*
DIB Device Independent Bitmap *hi, kg*
DIB Directory Information Base *pk, kg*
DIB Dual Independent Bus [Intel] *pk, kg*
DIBA Di Iso Butyl Amine *wp*
DIBOL Digital Business Oriented Language [DEC] *hi, kg*
DIC Defense Industrial Cooperation *ph*
DIC Device Interface Controller *fu*
DIC Digital Interface Converter *hc*
DIC Document Identifier Code *wp, fu*
DICASS Directional Command Active Sonobuoy System *fu*
DICBMS Defended ICBMs *hc*
DICE Dichroic Caladioptic Element *hc*
DICE Digital Interface Countermeasures Equipment *wp*
DICE Dimensional Centering *hc*
DICE Direct Course Error *fu*
DICIFER Digital Image Complex for Image Feature Extraction and Recognition *wp*
DICO Data Information and Coordination Office [OJCS] *mt*
DICOMNET Defense Intelligence Teletypewriter Network *mt*
DICOR Directional Control Rocket *hc*
DICOSE Digital Communications System Evaluator *ro*
DICP Developmental Interface Change Proposal *fu*
DICS Digital Image Correction System *ac3.5-4*
DICS Double Image Concentric System *hc*
DID Data Item Description (Descriptor) [a document] *mt, fu, ah, wp, 4, 16, 17, sb, ep, dm, tb, st, jj, pbl, tm, jdr, kf, oe*
DID Defense In Depth *17, sb, hc, tb, st, fu*
DID Direct Inward (Internal) Dialing (Dial) *kg, fu*
DID Display Interface Device *fu*
DID Dynamic Interaction Diagnostics [spacecraft] *oe*
DID/S Data Item Description / System *mt*
DIDAC Digital Data Computer *hc*
DIDB Defense Integrated Data {Base ?} System *mt*
DIDB DLA Inventory Data Base [DLA] *mt*
DIDDS Dynamic Integrated Data Display System *hc*
DIDHS Deployable Intelligence Handling System *ro*
DIDHS Deployed IDHS [USRECOM] *mt*
DIDO Data Input, Data Output *fu*
DIDP Display Data Processor *fu*
DIDS Data Item Description System *wp*
DIDS Decision Information Distribution System *mt*
DIDS Defense Integrated Data System [DSA] *wp, mt*
DIDS Distributed Intrusion Detection System *pk, kg*
DIDSIM Defense In Depth Simulation *tb, st*
DIDSRS Defense Intelligence Dissemination, Storage and Retrieval System *wp*
DIE (Designs are available in both DIE and Packaged Configurations) *mj31.5.73*
DIE Digital Interface Electronics *oe*
DIE Digital Interface Equipment *oe*
DIEOB Electronic Order of Battle Files *mt*
DIEU DIRA Interface Extension Unit *oe*
DIF Data Interchange Format *tb, kg, wl, hi*
DIF Data Interface Facility *hy*
DIF Directory Interchange Format *hg, ge*
DIF Document Interchange Format *mt*
DIFA Digital Interface Assembly *hc*
DIFAD DIG Tactical Airborne Computer *hc*
DIFAR Direction Finding And Ranging *wp, jj*
DIFAR Directional Frequency Analysis and Recording *fu*
DIFAR Directional Frequency Analysis Reporting [buoy] *pf1.9*
DIFAX Defense Intelligence Facsimile Network *mt*
DIFFSENS Differential Sense *pk, kg*
DIFM Due In From Maintenance *mt, wp, jj*
DIFMOS Double Injection Floating-gate Metal Oxide Semiconductor *em*
DIFOT Duty Involving Flying, Operational and Flight Training *jj*
DIFP Diisopropyl Fluoro Phosphate *wp*
DIFS Deployable Intelligence Fusion System *mt*
DIFU Digital Interface Unit *hc, jj*
DIG Data Interface Group *fu*
DIG Digital electronics unit *pl*
DIGICOM Digital Communications *fu, jj*
DIGIT Digital Unit *hi*
DIGITAIR Advanced Digital Airborne Computer *hc*
DIGS Delta Inertial Guidance System *ai2*
DIGS Digital Image Generation System *wp*
DII Defense Information Infrastructure *jdr*
DIIG Digital Information Infrastructure Guide *ct, em*
DIIP Direct Interrupt Identification Port *pk, kg*
DIIVS Defense Intransit Item Visibility System [USA] *mt*
DIL Design Integration Lab [of Hughes Aircraft Company] *msm, ep*
DIL Doppler-Inertial LORAN *hc*
DILAG Differential Laser Gyro *wp*

DILC Dedicated Intelligence Loop Circuits *mt*
DILC Defense Intelligence Loop Circuit *mt*
DILC Display Interface Processor {?} *mt*
DIM Digital Image Model *re*
DIMAP Digital Modular Avionics Program *wp*
DIMCP Defense Item Management Coding Program *wp*
DIME Dynamic Infrared Missile Evaluator *hc, wp*
DIMES Defense Integrated Management Engineering System *mt, wp*
DIMES Digital Image Manipulation and Enhancement System *wp*
DIMM Dual In Line Memory Module *ct, kg, em, hi*
DIMODE Discontinuity Modulation Effect *hc*
DIMOTE Digital Module Tester *hc*
DIMP Di Isopropyl Methyl Phosphate [chemical warfare nerve agent simulant] *wp*
DIMS Dynamic Inertial Measurement Systems *hc*
DIMUS Digital Multibeam Steering *hc, jj, fu*
DIN AUTODIN Data Identification Number *mt*
DIN Defense Intelligence Notice *mt, wp*
DIN Deutsche Industrie Normen (Norm) [(German) German industry standards] *mt, ct, wp, kg, hi*
DIN/DSSCS Digital Information Network / Defense Special Security Communications System *10, mt*
DINA Direct Noise Amplification *hc, wp, jj, fu*
DINAH Desktop Interface to AUTODIN Host *mt*
DINAS Digital Inertial Navigation / Attack System *mt*
DINCS Distributed Intelligent Numerical-Control System *fu*
DINDAC Digital Access Direct Access *mt*
DINK Double Income No Kid [childless yuppie couple] *wp*
DINOB Naval Order of Battle {?} *mt*
DINS Digital Inertial Navigation System *mt*
DINS Dormant Inertial Navigation System *hc*
DIO Data Input Output *pk, kg*
DIO Defense Intelligence Officer *wp*
DIO Direct Input/Output *jj*
DIO Director Of Industrial Operations *jj*
DIO Disk Input/Output *jj*
DIOBS DIA Order of Battle *mt*
DIOC Displayed Independent Of Computer *jj*
DIOD Digital Input / Output Device *fu, hc*
DIODS Diagram-Oriented Documentation System *fu*
DIOLAMINE Diethanolamine *wp*
DIOP Defense Intelligence Objectives and Priorities *mt*
DIP (an ideal . . . image source for DIP systems) *aw130.11.51*
DIP D-BRITE Interface Processor *fu*
DIP Dialup Internet Protocol [Internet] *hi, kg*
DIP Digital Imaging Processing *pk, kg*
DIP Diisopropyl Phosphate (chemical warfare Simulant) *wp*
DIP Dual In line Package *fu, wp, 6, hc, kg, jj, ct, em, hi*
DIP Dual In line Pin *wp, kg, jdr, hi*
DIPA Diisopropanolamine *wp*
DIPEC Defense Industrial Plant (Plan) Equipment Center *mt, hc, fu, wp, jj*
DIPF Di Isopropyl Phospho Fluoridate *wp*
DIPL Dipole *jdr*
DIPOLES Defense Intelligence Photoreconnaissance On-Line Exploitation System *wp*
DIPP Defense Intelligence Projections for Planning *mt, af72.5.107, tb*
DIPPA Digital Parallel Processing Array *hc*
DIPS Display Information Processor System *mt*
DIPS Dynamic Isotope Power System (Subsystem) *mt, hc, tb*
DIPT Development Integrated Product Development Team *kf*
DIR Defense Intelligence Report *wp*
DIR Depot Inspection and Repair *wp*
DIR Design Information Release *fu*
DIR Directed Investigation Report *jj*
DIR Disassembly and Inspection Report *jj*
DIRA Digital Integrating Rate Assembly *oe*
DIRC Dithered IR Configuration *hc*
DIREP Difficulty Report *mt*
DIRK Dual Independent Ranging Kit *hc*
DIRM Director, Information Resource Management *mt*
DIRNSA Director, National Security Agency *mt, 10*
DIRS Damage Information Recording System [AFSC/USAFE/PACAF] *wp, mt*
DIRSO Defense Industrial Resources Support Office *dt*
DIRT Defense InfraRed Test *wp*
DIRT Delayed InfraRed Track *hc, jj*
DIS Data and Information Service *sc4.1.3*
DIS Data Input System *ac2.2-13*
DIS Days In Shop *jj*
DIS Defense Intelligence Service *mt, wp*
DIS Defense Investigative Service *fu, hc, sc3.11.7, 10, 18, nt, dt, tb, st, jdr, kf, su*
DIS Design Implementation Specification *fu*
DIS Diagnostic Set *fu*
DIS Digital Integrating (Integration) System *hc, wp*
DIS Disabled In Service *mt*

DIS Distributed Interactive Simulation *kf*
DIS DOD Investigative Service *mt*
DIS Draft Interface Specification (Standard) *fu*
DIS Draft International Standard *pk, kg, ch*
DIS Dynamic Impedance Stabilization *ct, em, kg*
DISA Data Interchange Standards Association *pk, kg*
DISA Defense Industrial Security Agency *fu*
DISA Defense Information Systems Agency *mt, jdr, kf*
DISA Deployable Information Systems Architecture *mt*
DISAM Defense Institute of Security Assistance Management *mt*
DISC Data Index For Software Configuration *wp*
DISC Defense Industrial Supply Center *mt, hc, fu, wp, jj*
DISC Differential Scattering *wp*
DISCASS Dynamic Interaction of Solar Cell Arrays with Space *hc*
DISCO Defense Industrial Security Clearance Office *hc, fu, su, wp, tb*
DISCOID Direct Scan Operating with Integrating Delays *hc*
DISCOM Division Support Center [USA] *mt*
DISCOM Division Support Command *fu, wp*
DISCOS Disturbance Compensation System *-16*
DISCR Director for Industrial Security Clearance Review *hc, tb*
DISD Defense and Information Systems Division *fu*
DISI Defense Industrial Security Institute *mt, hc*
DISIDS Display and Information Distribution System *mt*
DISMAC Digital Scene Matching Area Correlation *rih*
DISMS Defense Integrated Subsistence Management System *mt*
DISN Defense Information Systems Network [GCCS] [USA] *kf, ric, jdr, mt*
DISNET Defense Integrated Secure Network *mt*
DISOSS Distributed Office Support System [IBM] *ct, em, mt*
DISP Defense (DOD) Industrial Security Program *su, hc, tb*
DISP Draft International Standardized Profiles *ch*
DISREP Discrepancy In Shipment Report *wp*
DISREP Discrepancy Report *fu*
DISS Digital Ionospheric Sounding System *mt*
DISTAFF Directing Staff [for exercises] *mt*
DISTAN Distributed Interactive Secure Telecommunications Area Network *wp*
DISTRA Distribution Authority *wp*
DISUM Daily Intelligence Summary [USA DOD] *mt*
DIT Days In Transit *jj*
DIT Directory Information Tree *pk, kg*
DIT Draft Initiation Time *mt*
DIT Dynamic Interaction Test *oe*
DITEG Digital Television Generator *fu, hc*
DITS Digital Imagery Transmission System *mt*
DITS Digital Television Spectrometer *fu*
DITV Digital Television *fu*
DIU Data Interface Unit *-4*
DIU Dedicated Interface Unit *fu*
DIU Digital Interface Unit *fu*
DIU Display Interface Unit *fu, hc, jj*
DIV-H Delta IV Heavy *ai2*
DIV-M Delta IV Medium *ai2*
DIVAD Division Air Defense *mt, hc, hn43.6.5, fu*
DIVADS Division (Divisional) Air Defense System *hc, fu*
DIVARTY Division Artillery *mt, hc, fu*
DIVE Direct Interface Video Extension [OS/2 Warp] *hi, kg*
DIVTAC Division Tactical Alternate Command *fu*
DIW D-Inside Wire [AT&T] *pk, kg*
DIW Dead In the Water *fu*
DIXIE Dynamic Intersite Switching Is Easy *hn41.25.5*
DJB Diode Junction Box *jdr*
DJC Dependent Job Control *fu*
DJET Delayed Jam Exceeds Threshold *jj*
DJOT Delayed Jam On Target *jj*
DJS Director, Joint Staff *mt*
DJSM Director, Joint Staff Memorandum *mt*
DK Denmark *es55.82*
DK Desantniy Korabl' [landing ship] [USSR] *mt*
DK Routine - Data K *jj*
DKO Delayed Key On *jj*
DL Damage Limiting *tb*
DL Data Link [also D/L] *mt, fu, hc, 17, jj*
DL Data List *wp, fu*
DL Definition List *ct*
DL Delay Line *jj*
DL Design Life *kf*
DL Diffraction Limited *wp*
DL Direct Labor *fu*
DL Display Line *fu*
DL Documentation Library *mt*
DL Down Link *tb*
DL Down Load [also D/L] *uh, kg, jdr*
DL Drawing List *fu*
DL Dummy Load *jj*
DL [frigate, US Navy] *mt*

DL/1 Data manipulation Language 1 [IBM] *pk, kg*
DLA Defense Logistics Agency [Headquarters] *mt, su, fu, ph, hc, dt, wp, tb, jj, kf*
DLA Department for Launch Vehicles [CNES] *ac3.6-3*
DLA-N Defense Logistics Agency, Executive Directory, Indiana *hc*
DLAM Defense Logistic Agency Manual *wp*
DLANET Defense Logistics Agency Network *mt*
DLAR Defense Logistics Agency Regulation *mt, wp*
DLAT Delayed Alert Velocity Tracker *jj*
DLC Data Link Command *jj*
DLC Data Link Control [protocol] *em, kg, nu*
DLC Digital Loop Carrier *ct, em*
DLC Direct Lift Control *mt, jj*
DLC Distributed Loop Carrier *em*
DLCI Data Link Connection Identification (Identifier) *ct, kg, em*
DLCS Data Link Communications System *mt, fu*
DLCS Data Link Control System *fu*
DLD Display List Driver *pk, kg*
DLDED Division Level Data Entry Device *mt, hc*
DLE Data Link Escape *do242, fu, kg*
DLE Down Link Expansion *-8*
DLED Dedicated Loop Encryption Device *fu*
DLED Digital Loop Encryption Device *mt*
DLG Digital Line Graph *fu*
DLG Dynamic Lead Guidance *hc*
DLG [guided missile frigate, US Navy] *mt*
DLGH Frigate, Guided missile, Helicopter [US Navy] *mt*
DLGN Frigate, Guided missile, Nuclear-powered [US Navy] *mt*
DLH Direct Labor Hour *fu, jj*
DLI Data Link Interface *jj*
DLI Deck Launched Interceptor *jj*
DLI Down Link Interface *oe*
DLI Dummy Load In *jj*
DLIFC Defense Language Institute, Foreign language Center *mt*
DLIM Data Link Interface Module *fu*
DLIR Down Looking Infra Red *wp*
DLIS Downward Looking Infrared System *wp*
DLJ Downlink Jamming *hc*
DLK Downbank Telemetry Processing *mt*
DLL Delay Lock Loop *sp660, fu*
DLL Dynamic Link Library [Microsoft] *ct, kg, em, hi*
DLM Data Line Monitor *oe*
DLM Depot Level Maintenance *mt, wp, fu*
DLM Designated Location Move *wp*
DLM Digital Logic Module *hc*
DLM Direct Logistics Maintenance *mt*
DLM Distributed Lock Manager *kg, em*
DLM Dynamic Link Module *pk, kg*
DLMF Depot Level Maintenance Facility *hc*
DLMS Digital Level Mass Simulator *hc*
DLMS Down Link Monitoring System *hc, 2*
DLMSFDS Digital Landmass System Feature Display System [JDSSC] *mt*
DLO Defense Liaison Office *wp*
DLO Difference of Longitude *wp*
DLOC Delayed Lock-On Command *jj*
DLOGS Division Logistics System [USA] *mt*
DLP Digital Light Processing [TI] *pk, kg*
DLP Downlink Processor *jdr*
DLPGSE Depot Level Peculiar Ground Support Equipment *jj*
DLPP Data Link Preprocessor *fu*
DLR Deutsches Zentrum fur Luft- und Raumfahrt (e. V., after integration of DARA into DLR) *crl*
DLR Deutsches [Forschungs Anstalt f̦r] Luft und Raumfahrt [(German) German Aerospace Research Institute] *ns, crl*
DLR DOS LAN Requester *pk, kg*
DLRN Data Link Reference Number *fu*
DLRP Data Link Reference Point *mt, fu*
DLRP Depot Level Maintenance Plan *jj*
DLRS Depot Level Repairables *mt*
DLRV Dual mode Lunar Roving Vehicle *hc*
DLS Data Link Set *fu*
DLS Data Link Simulator *fu*
DLS Data Link Subsystem *jj*
DLS Data Link Support *mt*
DLS Data Link System *mt*
DLS Decision Learning System *wp*
DLS Decoy Launching System *fu*
DLS Deep Look Surveillance *wp*
DLS Depot Level Services *mt*
DLS Dynamic Limb Sounder [EOS] *hy*
DLSA Defense Legal Service Agency *dt*
DLSC Defense Logistics Services (Support) Center *mt, wp, hc, jj*
DLSC Defense Logistics Supply Center *fu*
DLSDC Data Link Signal Data Converter *hc*
DLSIE Defense Logistics Studies Information Exchange *mt*

DLSS Defense Logistic Standard Service *wp*
DLSSO Defense Logistics [agency] Standard Systems Office *mt, dt, wp*
DLSW Data Link Switching *ct, em*
DLT Data Line Translator *fu*
DLT Data Link Terminal *wp*
DLT Decision Logic Table *mt*
DLT Di Lauryl Thiopropionate *wp*
DLT Digital Linear Tape *kg, ct, em, hi*
DLT Display List Traverser *fu*
DLTM Data Link Test Message *fu*
DLTS Deep Level Transient Spectroscopic *fu*
DLU Discharge Load Unit *gp*
DLU Display Logic Unit *jj*
DLV Demonstration Launch Vehicle *ai2*
DLVA Detector Logarithmic Video Amplifier *ct, em*
DLYSUM Daily Summary Spare Engines [USAF] *mt*
DLYSUMM Daily Summary Spare Engines [PACAF] *mt*
DLZ Dynamic Launch Zone *mt*
DM Data Management [also DATAMAN] *fu, hc, mc, 17, ep, jdr, kf, fu*
DM Data Manager (Management) *fu, mt*
DM Data Mark *mt*
DM Data Memory *jj, mt*
DM Decision Making *mt*
DM Decision Memorandum *wp*
DM Defensive Missile *fr*
DM Delay Modulation *sm3-12*
DM Delta Modulation *csIII.6, fu*
DM Departure Message *fu*
DM Depot Manufacture (Maintenance) *hc, jj*
DM Development Model *fu*
DM Development Module *hc*
DM Digital Modulation *do214*
DM Digital Multiplexer *gp, oe*
DM Directorate of ADPS Management *mt*
DM Discard Message *mt, fu*
DM Disconnect Mode *fu*
DM Disk Memory *fu*
DM Display Manager *fu*
DM Display Memory *fu*
DM Display Monitor *gp, oe*
DM Distributed Memory *pk, kg*
DM DITEG Module *fu*
DM Docking Module *ac2.5-12*
DM Dopolnitelny Motor [auxiliary motor] [ramjet engines, usually fitted underwing to WWII fighters, in an attempt to increase performance] [USSR] *mt*
DM Dual Mode *wp*
DM Light Minelayer [US Navy] *mt*
DM [chemical warfare codename for adamsite, diphenylamine chloroarsine] *wp*
DMA Data Management Administrator (Assistant) *mt, fu*
DMA Defense Mapping Agency [USA] *mt, 12, wp, dt, jdr, kf, ro, fu*
DMA Degraded [Mission/]Mode Assessment *hc, jj*
DMA Deployed Mechanical Assembly *hn43.25.5*
DMA Depot Maintenance Activity *jj*
DMA Designated Maintenance Agency *mt, wp*
DMA Direct Memory Access (Addressing) *mt, ct, em, fu, hc, wp, 4, 6, hd1.89.y13, is26.1.33, jj, jdr, kg, hi, oe, ai1*
DMA-AF Defense Message System - Air Force *mt*
DMAAC Defense Mapping Agency Aerospace (Aeronautical) Center [St. Louis, Missouri] *12, wp, dt*
DMAAC Defense Mapping Agency Aerospace Center *mt*
DMAB Direct Memory Access Buffer *fu*
DMAC Direct Memory Access Channel *fu, hi*
DMAC DMA Controller *pk, kg*
DMAC Dual Multipliers/ Accumulator *fu*
DMACS Distributed Manufacturing Automation and Control Software *pk, kg*
DMACTL Direct Memory Access Control *wp*
DMAHTC Defense Maping Agency Hydrographic and Topographic Center [Washington, D.C.] *12, dt*
DMAIAGS Defense Mapping Agency Inter American Geodetic Survey *dt*
DMAODS Defense Mapping Agency Office of Distribution Service *dt*
DMAP Depot Maintenance Activation Planning *fu*
DMAS Digital Modular Avionics System *wp*
DMAS Digital Multiplexed Audio System *hc*
DMAT Digital Module Automatic Tester *hc*
DMATC [Hughes Aircraft Company customer] *hc*
DMATS Defense Metropolitan Area Telephone Service/System *mt*
DMC Data Management Computer *hc*
DMC Data Management Coordinator *hc*
DMC Data Multiplexer Control *mt*
DMC Degraded Mission Capability *jj*
DMC Depot Maintenance Concept *mt*
DMC Disk Memory Controller *fu*
DMC Dynamic Management Center *fu*

DMCC Data Monitoring and Control Center *fu*
DMCO Delta Mission Check Out *ai2*
DMD Data Message Definition *kf*
DMD Deployment Management System *mt*
DMD Deployment Manning Document *mt*
DMD Digital Message Device *fu, hc*
DMD Digital Micromirror Device [TI] *pk, kg*
DMD Digital Multisensor Display *jj*
DMD Dual Mode Display *hc, jj*
DMDC Defense Manpower Data Center [of DLA] *mt, dt*
DMDL Dot Matrix Display Legibility *hc*
DMDM Data Management Directives Manual *hc*
DMDS Data Management Display System *mt*
DME Design Margin Evaluation *jj*
DME Despin Motor Electronics *hc*
DME Direct Maintenance effort *jj*
DME Distance Measuring Equipment *mt, 0, fu, wp, hc, aw129.10.73, tb, jj*
DME Distributed Management Environment *hi, kg*
DME/P Precision Distance Measuring Equipment *fu*
DMED Digital Message Entry Device *hc, fu, ro*
DMES Deployable Mobility Execution System [HQ USAF] *mt*
DMES Digital Message Entry Subsystem *hc*
DMF Data Matched Filters *jdr*
DMF Di Methyl Formamide *wp*
DMF Digital Matched Filter *fu*
DMF Distribution Media Floppy *ct, em*
DMF Distribution Media Format [Microsoft] *pk, kg*
DMFCP Digital Multimode Flight Control Program *fu*
DMFD Digital Matched-Filter Device *fu*
DMGS Digital Map Generation System *mt*
DMGS Digital Missile Guidance Set *dm*
DMGU Dual Mode Guidance Unit *hc*
DMH Di Methyl Hydrazine *wp*
DMH Direct Man Hours *hc*
DMHP Dimethyl Hydorogen Phosphite [chemical warfare nerve agent simulant and precursor] *wp*
DMI Depot Maintenance Interservicing *mt, fu*
DMI Desktop Management Interface *ct, kg, em, hi*
DMI Diagnostic Monitor Interface *gp*
DMI Direct Matrix Inversion *fu*
DMI Director of Military Intelligence *fu*
DMI Drill Master Image *fu*
DMI Dwell Mode Inhibit *jj*
DMIC Defense Metals Information Center *fu, hc*
DMICS Distributed Management Information and Control System *mt*
DMIF Depot Maintenance Industrial Fund *mt, hc, jj*
DMII Descriptive Method Item Identification *jj*
DMINS DLA's Distributed Minicomputer System *mt*
DMIP Defense Mediterranean Improvement Program *mt*
DMISA Depot Maintenance Interservice Support Agreement *fu*
DMISA Depot Maintenance Interservice Support Management {?} *wp*
DMJO Defense Management Journal Office *dt*
DML Data Manipulation Language *mt, kg, fu*
DML Digitized Message Link *hc*
DMM Diagnostic Maintenance Manual *fu*
DMM Digital Multi Meter *hc, jj, kg, em, ct, jdr, fu, oe*
DMM Director of Materiel Management *wp*
DMM DynaMetric Model [USAF] *mt*
DMMBF Delta Modulated (Modulation) Multibeam Beamformer *fu, hc*
DMMC Division Materiel Management Center *fu*
DMMH/FH Direct Maintenance Man Hours per Flight Hours *hc, jj*
DMMIS Depot Maintenance Management Information System [AFLC] *mt*
DMMIS Depot Management Information System *wp*
DMMM Direct Maintenance Man Minutes *hc, jj*
DMMP Dimethylmethylphosphonate (A chemical warfare Simulant) *wp*
DMMPA Dimethylmorpholinophosphoramidate (chemical warfare Simulant) *wp*
DMMS Depot Maintenance Management System [USMC] *mt*
DMMS Dynamic Memory Management System *hi, kg*
DMMSS Depot Maintenance Management Support System [AFLC] *mt*
DMNA Distributed Microprocessor Network For Avionics *wp*
DMO Contract Data Management Officer *mt*
DMO Data Management Office (Officer) *hc, wp, tb*
DMO Data Management Operations *fu*
DMO Data Management Organization *-17*
DMO Diode Microwave Oscillator *fu*
DMO Documentation Management Officer *mt*
DMOD Demodulator *jdr*
DMOP Data Management Operating Plan *fu, hc*
DMOS Diffused (Diffusion) Metal Oxide Semiconductor *is26.1.55, em, wp*
DMOS Double diffused Metal Oxide Semiconductor *pk, kg*
DMP Data Management Plan *fu, hc, wp*
DMP Depot Modernization Program *fu*

DMP Digital Modification Program *jj*
DMP Dimercaptopropanol *wp*
DMP Dimethylphtalate *wp*
DMP Director of Materiel Procurement [Canada] *hc*
DMP Dot Matrix Printer *hi, kg*
DMP Drive Motor Protector *fu*
DMPC Distributed Memory Parallel Computer *pk, kg*
DMPE Depot Maintenance Plant Equipment *hc*
DMPI Desired Mean Point of Impact *mt*
DMPI Dimethylphosphite *wp*
DMQS Display Mode Query and Set [IBM] *pk, kg*
DMR Data Management Routines *mt*
DMR Date Materiel Required *jj*
DMR Defective Material Report *hc, jj*
DMR Defense Management Review *wp*
DMR Discrepant Material Report *oe*
DMR Division Management Review *-8*
DMR Dual Mode Radar *hc*
DMR Dual Mode Recognizer (Recognition) *hc, jj*
DMRIS Defense Medical Regulating Information System *mt*
DMS Data Management Software *pk, kg*
DMS Data Management System (Subsystem) *mt, fu, bo, wp, sc4.11.4, hc, es55.71, no, tb, kg, wl, jj, hi, ai2*
DMS Data Multiplexing Subsystem [DMSP] *pl*
DMS Database Management System [as in DMS-1100 for S1100 UNISYS computer systems] *mt*
DMS Defense Mapping Schools *dt*
DMS Defense Materiel (Materials) Systems *hc, fu*
DMS Defense Message System *mt, kf, uh*
DMS Defense Meteorological Satellite *ci.904*
DMS Defensive Management System *aw130.13.85*
DMS Dependent Milestone Slip *kf*
DMS Depot Maintenance Support *wp*
DMS Detailed Module Specification *fu*
DMS Digital Matrix Switch *-2*
DMS Digital Microwave System *-17*
DMS Digital Missile Simulator *fu*
DMS Digital Multiplex Switch *ct*
DMS Digital Multiplex System *ct*
DMS Diminishing Manufacturing Sources *mt, tb, fu*
DMS Documentation Management System *mt*
DMS Dual Mode Seeker *hc*
DMS High Speed Minelayer [USA] *mt*
DMSD Digital Multi Sensor Display *hc, jj*
DMSD Digital Multistandard Decoding *pk, kg*
DMSIG DMS Information Group *mt*
DMSO Dimethyl Sulfoxide *wp*
DMSP Defense Meteorological Satellite Program *hc, wp, sc3.2.7, 12, 14, 16, pl, af72.5.68, cp, no, ep, dm, hy, re, tb, hg, mb, kf, ge, mi, oe, mt, ls4, 5, 16*
DMSP Depot Maintenance Support Plan *wp*
DMSPA Defense Mobilization and Support Planning Agency *mt*
DMSS Defense Meteorological Satellite System *16, pl*
DMSS Distributed Mass Storage System *pk, kg*
DMSS Dual Mode Surveillance System *hc*
DMSSC Defense Medical Systems Support Center *mt*
DMSSO Defense Materiel Specifications and Standards Office [DLA] *mt, fu, rf84.3.13, dt*
DMT Depot Module Tester *hc*
DMT Design Maturity Testing *fu*
DMT Digital Multi Tone *pk, kg*
DMT Discrete Module Tester *hc, jj*
DMT Discrete Multi Tone *ct, kg*
DMT Doppler Modulated Target *st*
DMT Dual Mode Tracker *hc*
DMT Technology Division of the Directorate of ADPS Management *mt*
DMTDL Doppler Modulated Target Delay Line *st*
DMTF Desktop Management Task Force *kg, ct, em*
DMTI Digital (Digitized) Moving Target Indicator *mt, fu, hc*
DMU Data Management Unit *mt*
DMU Disk Memory Unit *fu*
DMU Display Multiplexer Unit *fu*
DMU Distribution and Measurement Unit *fu*
DMU Downlink Modulator Unit *jdr, uh*
DMV Department of Motor Vehicles *wp*
DMW [a telecommunications corporation] *do289*
DMWR Depot Maintenance Work Requirements (Request) *mt, fu, hc, jj*
DMWRO Depot Maintenance Work Requirements Order *hc*
DMX Data Multiplex *fu*
DMY Day Month Year *hi, kg*
DMZ Demilitarized Zone *mt, w9, fu, wp*
DM_IV Data Management, Version IV [Honeywell] *mt*
DN Data Net [as in Honeywell's DN355 or DN8] *mt*
DN Data Number *0*
DN Departure Notice *oe*
DN Descending Node *oe*

DN Discrepancy Notice *oe*
DN Domain Name *ct, em*
DNA DEC Network Architecture *ct*
DNA Defense Nuclear Agency *mt, 5, fu, wp, 12, 17, ph, sb, hc, dt, tb, st, jdr, kf*
DNA Deoxyribo Nucleic Acid [molecular structure that carries genetic information] *wp, ci.811*
DNA Distributed Network Architecture *hi*
DNA Do Not Attack *jj*
DNACC Defense National Agency Check Center *hc, tb*
DNAS DOD IIS Network Access System *hc, mt*
DNC Day-Night Capability *wp*
DNC Dinitrocellulose *wp*
DNC Direct Numerical Control *kg, fu*
DNC Distributed Network Computing *wp*
DNCCC Defense National Communications Control Center *hc, dt*
DNCG Digital Null Command Generator *hc*
DNCONV Downconverter *jdr*
DND Department of National Defense [Canada] *fu, mt*
DNFYP Department of the Navy Five-Year Program *fu*
DNG District of Columbia National Guard *dt*
DNIC Data Network Identification Code *pk, kg*
DNIF Duty Not Involving Flying *mt, wp*
DNIS Dialed Number Identification Service *kg, ct*
DNJ Drone Noise Jammers *wp*
DNMP Domestic Net Material Product *wp*
DNMR Document Number Master Record *wp*
DNOS Distributed Network Operating System *fu*
DNP Reoxyribonucleoprotein *wp*
DNR Digital Number Recorder *ct*
DNS Data Network Service *hc*
DNS Data Network System *sc2.7.3*
DNS Distributed Network Supervisor *mt*
DNS Domain Name Server *du*
DNS Domain Naming (Name) System (Service) [translation of internet domain names and addresses] *kg, nu, ct, em*
DNSDP Defense Navigation Satellite Demonstration Program *hc*
DNSIX DOD IIS Network Security for Information Exchange *mt*
DNSS Defense Navigation Satellite System *mt*
DNSVT Digital Non Secure Voice Terminal *hc*
DNV Detector Number Valid *jj*
DNVT Digital Nonsecure Voice Terminal [TRI-TAC] *hc, fu, mt*
DNW (the German-Netherlands Wind Tunnel (DNW)) *aa26.12.36*
DO Daily Operations *mt*
DO Data Ordering *fu*
DO Data Out *pk, kg*
DO Data Output *fu*
DO Defense Order *w9*
DO Deputy Commander for Operations *mt*
DO Director of Operations *kf*
DO Directorate, Operations *af72.5.100*
DO Distributed Object *em, kg*
DO Ditto [the same] *wp, jj*
DO Dual Operation *fu*
DO Due Out *jj*
DO Duty Officer *fu*
DO-IT Disabilities, Opportunities, Internetworking and Technology *pk, kg*
DOA Dead On Arrival *w9, wp*
DOA Delegation Of Authority *jj*
DOA Department Of Agriculture *wp*
DOA Department Of the Army [also DA] *wp*
DOA Difference Of Arrival *hc*
DOA Direction Of Arrival *mt, fu, wp*
DOAS Differential Absorption Spectroscopy *hg, ge*
DOB Dispersal Operations Base *fu*
DOB Dispersed Operating Base *wp, mt*
DOC Data Operations Center *0*
DOC Data Output Channel *fu*
DOC Date Of Change *wp*
DOC Defense Operations Center *tb*
DOC Department Of Commerce [USA] *mt, ge, gp, 16, hc, no, jj, jdr, hg, oe, y2k*
DOC Designated Operational Capability *wp*
DOC Designed Operational Capabilities *mt*
DOC Diagnostic Operations Controller *fu*
DOC Direct Operating Cost *hc, wp, jj*
DOC DTDMA Operating Characteristics *fu*
DOCC DCA Operations Control Complex *mt*
DOCPREP Document Preparation [in support of E-3A AWACS] *mt*
DOCS Designated Operational Capability Statement [USAF] *mt*
DOCS Display Operational Control Set *fu*
DOCS DSCS Operational Control System *mt*
DOCSV Data Over Circuit-Switched Voice *pk, kg*
DOD Department Of Defense [also DoD] [US] *aw130.13.77, hc, ci.867, 0, 1, 2, 4, 8, 12, 14, 15, 16, 17, sb, pl, aa26.12.15, w9, wp, no, ep, dm, hy, re, tb, tb, st, wl, jj, pbl, mb, tm, hg, kf, jdr, ge, mi, fu, ah, su, fu, oe, mt, ai2*

DOD Depth Of Discharge [of a battery] *gg11, 16, lt, kf, jr, uh, oe*
DOD Digital Optical Disk *hi*
DOD Direct Outward Dialing *fu, hi*
DOD-STD Department Of Defense Standard *mt, tb*
DODAAC Department Of Defense Activity Address Code *mt*
DODAAD Department Of Defense Activity Address Dictionary (Directory) *mt, fu, hc*
DODAC Department Of Defense Ammunition Code *fu*
DODADL Department Of Defense Authorized Data List *fu*
DODCI Defense Computer Institute *dt*
DODCI Department Of Defense Computer Institute *dt, mt, fu*
DODCSC Department Of Defense Computer Security Center *mt, tb*
DODD Department Of Defense Directive *fu, hc, wp, nt, st, jj, kf*
DODD DOD Directive *mt*
DODDS Department Of Defense Dependent Schools *dt*
DODE Developmental Optint Diagnostic Equipment *hc*
DODE Diagnostic Optical Demonstration Equipment *hc*
DODI Department Of Defense Instruction *fu, hc, wp, nt, st*
DODI DOD Instruction *mt*
DODIC Department Of Defense Identification Code *mt, fu*
DODIIS Department Of Defense Intelligence Information System [DIA] *hc, wp, mt*
DODINST Department Of Defense Instruction [also DODI] *mt, fu*
DODIP Department Of Defense Information Program *mt*
DODISM Department Of Defense Industrial Security Manual *fu*
DODISS Department Of Defense Index of Specifications and Standards *mt, fu, hc, nt, wp, kf*
DODM DOD Manual *mt*
DODMDS Department Of Defense Material Distribution System Study Group *fu*
DODMIS Department Of Defense Management Information System *fu*
DODNAF Department Of Defense Non Appropriated Fund *dt*
DODR Department Of Defense Regulation *tb, fu*
DODSASP DOD Small Arms Serialization Program [USA] *mt*
DODT Design Option Decision Tree *wp*
DODTP Deployment Observation Discrimination Technical Program *hc*
DODWHS Department Of Defense Washington Headquarters Services *dt*
DOE Date Of Enlistment *mt*
DOE Department Of Energy [formerly ERDA] *mt, ci.9, 12, hc, w9, hn49.9.3, wp, hy, re, tb, st, jj, jdr, mb, su, fu, mi, y2k*
DOE Design Of Experiments *rm7.1.1, ad, ig21.12.17, kf*
DOE Distributed Objects Everywhere *pk, kg*
DOE&E Director, Operational Test and Evaluation *kf*
DOEA Distributed Objects Everywhere Architecture *hi*
DOEE Directorate of Observation of the Earth and its Environment [ESA] *ge, hg*
DOES Defense Organization Entity System *hc*
DOF Degrees Of Freedom *gp, wp, 16, 17, sb, hc, cp, re, tb, st, jj, jdr, kf*
DOF Developmental Optics Facility *hc*
DOFL Diamond Ordnance Fuze Laboratories *hc*
DOGFIGHTER [short range tactical air to air missile, Aim-82] *wp*
DOHL Degree Of Hard Limiting *uh*
DOI Department Of the Interior *mt, ac1.2-21, hc, hy*
DOI Department Operating Instruction *jj*
DOI DSSCS Operating Instruction *jj*
DOIM Director Of Information Management *mt*
DOIP Dial Other Internet Providers [IBM] *pk, kg*
DOJ Department Of Justice *mt, y2k*
DOJ Directorate Of Combat Employment [USAFE] {?} *mt*
DOL Department Of Labor *wp, mt*
DOL Deputy Operations Leader *pbl, ah*
DOL Director Of Laboratories *wp*
DOL Dispersed Operating Locations *mt, wp*
DOLARS Departmental On Line Accounting and Reporting System *mt*
DOLCE Digital On Line Cryptographic Equipment *mt, fu*
DOLLARS Dedicated On Line Logistical Airlift Ratemaking System *mt*
DOLRAM Detection Of Laser, Radar, And Millimeter *wp*
DOM Description, Operation, Maintenance *wp*
DOM Digital Ohm Meter *wp*
DOMF Distributed Object Management Facility *hi, kg*
DOMS Director Of Military Support *mt*
DOMSAT Domestic [communications] Satellite *fu, 2, 4, 19, wl*
DONCS Director of Operations Narcotics Control Reports *-10*
DONOACS Department Of the Navy Office Automation and Communication System [US Navy] *mt*
DOO Daily Operations Order [NATO] *mt*
DOO Delivery On Orbit *kf*
DOORS Dynamic Object Oriented Requirements System *pbl, kf*
DOP Designated Overhaul Point *fu, jj*
DOP Development Options Paper *mt*
DOP Dilution Of Precision *jdr*
DOP Dioctyl Phalate *hc*
DOPA DihydrOxyPhenylAlanine *wp*
DOPAA Description Of the Proposed Actions and Alternatives *nt*
DOPCOM Doppler Command Missile Delivery System *hc*
DOPMA Defense Officer Personnel Management Act [DOD] *mt*

DOPS Director Of Operations *fu*
DOPS Documentation Production System *fu*
DOPSUM Daily Operations Summary [USAFE] *mt*
DOR Date Of Rank *wp, mt*
DOR Date Of Request *fu*
DOR Defense Operations Room *fu*
DOR Differential One way Ranging *mi*
DOR Due Out Release *jj*
DORA Dynamic Operator Response Apparatus *jj*
DORAN DLA's Operations Research Analysis Network *mt*
DORAN Doppler Range *jj*
DORAN Doppler Range and Navigation *fu*
DORIS Doppler Radiopositioning System [France] *ge, hg*
DOS Date Of Separation *mt*
DOS Days Of Supply *mt*
DOS Deep Ocean Survey *aa27.4.b14*
DOS Department Of State *mt, ci.837, jdr, fu, ac3.2-20*
DOS Disk Operating System *mt, hd, wp, 8, hc, w9, hd1.89.y13, kf, ct, kg, fu, em, hi, oe, ai1*
DOS Distributed Operating System {?} *mt*
DOSAAF Dobrovoln'oe Obshchestvo Sodeistviya Armii, Aviatsii i Flotu [Voluntary Association for the support of Army, Aviation and Fleet, a paramilitary sport organization that undertook basic flying training] [USSR] *mt*
DOSEM DOS Emulation *pk, kg*
DOSS DSCS Operational Support Systems *mt*
DOSTN Department Of State Telecommunication Network *wp*
DOT Delayed On Target *jj*
DOT Department Of Transportation *mt, su, wp, ac1.7-14, 16, aa26.8.8, hc, w9, 19, tb, mi, jj, mb, fu, oe*
DOT Designating Optical Tracker *hn43.10.8, 17, sb, hc, tb, st, kf, fu*
DOT Director Of Training *mt*
DOT Double Offset Tactic *fu*
DOTM Due Out To Maintenance *wp*
DOT_EO Department Of Transportation Emergency Organization *mt*
DOVAP Doppler Velocity And Pattern *fu*
DOVAP Doppler Velocity And Position *jj*
DOW Day Of Week *pk, kg*
DOW Direct Over Write *ct, em*
DOX Directorate of Operational Plans *mt*
DOY Day Of Year *uh*
DP Damage Potential *fu*
DP Dash Pot *fu*
DP Data Processing (Processor) *mt, fu, hf, hc, w9, tb, st, jj, 4, 17, sb, kg, tb, jdr*
DP Days Prior *tb*
DP Decision Package *tb, jdr*
DP Decision Point *jj*
DP Deep space Perturbations *jb*
DP Defense Plan *fu*
DP Defense Programs *mt*
DP Deliberate Planning *mt*
DP Demarcation Point *mt*
DP Departure Point *mt*
DP Description Pattern *hc*
DP Detailed Process [specification document] *fu, hc, ep, jj, pbl, tm, jdr*
DP Development Plan *jj, jdr, fu*
DP Dew Point *w9, jj*
DP Dial Pulse *fu*
DP Differential Pressure *fu*
DP Diffusion Pump *jdr*
DP DiPhosgene [chemical warfare codename] *wp*
DP Disk Processor *oe*
DP Display Processor *fu*
DP Double Pole *jj, fu*
DP Double Pulse *fu*
DP Draft Proposal *fu*
DP Drill Plate *fu*
DP&D Data Processing and Display *fu*
DP&P Data Processing and Peripherals *fu*
DP-RT Data Processor-Receiver-Transmitter *fu*
DP/MCT Data Processing/Maintenance Control Terminal *fu*
DPA Data Processing Activity *wp, jj*
DPA Data Processing Area *0*
DPA Data Processing Authorization *fu, hf, 8*
DPA Defense Production Act *wp*
DPA Delegation of Procurement Authority *mt*
DPA Demand Protocol Architecture *pk, kg*
DPA Destructive Physical (Parts) Analysis *ah, fu, gp, 16, pl, lt, ep, pbl, tm, jdr, oe*
DPA Division Property Administrator *hc*
DPA Divisional Procurement Agreement *fu*
DPA Double Precision Arithmetic *wp*
DPA&E Director Program Analysis And Evaluation *mt, st*
DPACT Defense Policy Advisory Committee on Trade *ph*
DPAF Dual Payload Attachment Fitting *ai2*
DPAM Demand Priority Access Method *ct, kg, em*

DPANS Draft Proposal American National Standards *mt*
DPAREN Data Parity Enable [IBM] *pk, kg*
DPAS Digital Patch and Access System [DCS] *hc, mt*
DPATS Detector Packing Assembly Test Station *hc, ie5.11.9*
DPB Device Parameter Block *wp, kg*
DPBAX Digital Private Branch Automatic Exchange *tb*
DPBC Double-Pole Back-Connected *fu*
DPC Data Processing Center [an organization] *mt, fu, 8, 15, ep, tm, pbl*
DPC Data Processing Control *0*
DPC Defense Planning Committee [NATO] *mt*
DPC Defense Procurement Circular *hc*
DPC Digital Pulse Compression *fu*
DPC Direct Program Control *pk, kg*
DPCA Dual Port Communications Adapter *mt*
DPCC Data Processing Control Console *0*
DPCCP Defective Parts and Components Control Program *hc*
DPCM Differential Pulse Code Modulation *fu, sp660, hi*
DPCM Digital Pulse Code Modulation *wp*
DPCMB Defense Procurement Career Management Board *mt*
DPCS Data Processing Control System *fu*
DPD Data Procurement Document *pl*
DPD Data Project Directive *mt, wp*
DPD Demonstration Procedure Document *fu*
DPD Detailed Procedures Description *mt*
DPD Digital Phase Difference *fu*
DPDFA Data Processing and Display Functional Area *fu*
DPDM DOD IIS Protocol Development and Maintenance *mt*
DPDO Defense Property Disposal Office *mt*
DPDP Data Processing Development Plan *hc*
DPDR-E Defense Property Disposal Region - Europe *mt*
DPDS Defense Property Disposal Service *wp*
DPDT Double-Pole, Double-Throw *wp, fu, jj, hi*
DPDTSW Double-Pole, Double-Throw Switch *fu*
DPE Data Processing Element *mt*
DPE Data Processing Equipment *mt, 16, wp*
DPE Demand Processing Unit {?} *mt*
DPE Dynamic Phase Error *gg158*
DPESO Defense Project Engineering Service Organization *fu*
DPFC Double-Pole Front-Connected *fu*
DPFG Data Processing Functional Group *fu*
DPG Data Processing Group *fu*
DPG Defense Planning Guidance [USA DOD] *wp, mt*
DPG Defense Policy Guidance *tb*
DPH DiPhenylHydantoin *wp*
DPI Data Preparation Instruction [a document] *ah, 17, ep, pbl, tm, jdr*
DPI Data Processing Installation *mt, wp, st*
DPI Detected Pulse Interference *fu*
DPI Distributed Protocol Interface *pk, kg*
DPI Dot Pitch Integer *ct, em*
DPI Dots Per Inch *mt, em, kg*
DPICM Dual Purpose Improved Conventional Munitions *mt*
DPL Descriptor Privilege Level *pk, kg*
DPL Detached Parts List *ep, tm*
DPL Display Programming Language *fu*
DPL Drawing Parts List *ms3.1.2, ep, jdr, tm*
DPL Dual Processor Logic *jdr*
DPLS Deck Projector Landing Sight *hc*
DPM Data Processing Machine *jj, fu*
DPM Data Processing Module *mt, fu*
DPM Defense Program Memorandum *tb*
DPM Deputy Program Manager *jmj, st, kf, fu, oe*
DPM Digital Panel Meter *pk, kg*
DPM Divert Propulsion Module *17, sb*
DPMA Data Processing Management Association *pk, kg*
DPMI DOS Protected Mode Interface [Microsoft] *kg, em, hi*
DPML Deputy Program Manager for Logistics *mt, wp, tb*
DPMO Defense Productivity Measurement Office *dt*
DPMS Display Power Management Signaling *ct, em*
DPMS Display Power Management Support *pk, kg*
DPMS Display Power Management System [VESA] *hi*
DPMS DOS Protected Mode Services *em, ct*
DPNL Distribution Panel *jj*
DPO Data Phase Optimization *pk, kg*
DPO Director Planning and Operations *fu, hc*
DPO Disaster Preparedness Office *mt, wp*
DPO Double Pulse Operation *fu*
DPOB Date and Place Of Birth *wp*
DPP Data Project Plan *mt*
DPP Defect Prevention Process *jdr*
DPP Defense Payload Program *jdr*
DPP Development and Production Prove-out *kf*
DPP Digital Post-Processor *kf*
DPP Digital Pre-Processor *kf*
DPPB Defense Intelligence Information Systems Products Priorities Board *mt*
DPPC Data Processor Power Conditioner *fu*

DPPD Data Processing Products Division *fu*
DPPG Defense (DOD) Policy and Planning Guidance *mt, tb*
DPPH Direct Program (Product) Person Hours *st, fu*
DPPO Defense Productivity Program Office *dt*
DPPP Discrete Parts Parameter Program *fu*
DPQ Defense Planning Questionnaire [DOD] *mt*
DPR Data Processing Request *8, hc, jj, fu*
DPR Design Problem Report [AF Form 1774] *mt, 4, wp, oe*
DPR Directed Procurement Request *fu*
DPR Document Problem Report *fu*
DPR Double Pulse Ranging *fu*
DPRB Defense Planning and Resources Board *wp*
DPRC Defense Program Review Committee *mt*
DPRO Defense Plant Representative Office *pbl, ah, tm, jdr, kf*
DPROC Data Processing Statement *jj*
DPRP Disaster Prevention and Recovery Plan *mt*
DPS (AAS-DPS Conference, Boston, 1977) *ma16*
DPS Data Present Signal *jj*
DPS Data Processing Set *fu*
DPS Data Processing [Sub] System *mt, 0, hc, st, jj*
DPS Decision Package Set *mt, wp*
DPS Defense Printing Service *dt*
DPS Defense Priorities System *fu*
DPS Delegated Production System *mt*
DPS Detail Process Specification *jdr*
DPS Digital Processing Subsystem *jdr*
DPS Display Processing System *mt*
DPS Display Programming System *fu*
DPS Distributed Processing System [DPS8/WWMCCS ADP] *mt, jb, kg*
DPSB Defense Program Strategy Board *mt*
DPSC Defense Personnel Support Center *mt, wp, jj*
DPSCPAC Data Processing Service Center Pacific Fleet *mt*
DPSK Differential Phase Shift Keying (Keyed) *mt, uh, gs1-5.60, 17, tb, kg, fu, jdr, hi*
DPSS Data Processing and Services Subsystem *ag*
DPSS Data Processing Switching System *0*
DPSS Data Product and Service System *oe*
DPST Double Pole Single Throw *lt, fu, jj*
DPSW Double Pole Switch *fu*
DPT Digital Picture Terminal *hc*
DPT Distributed Processing Technology *ct, em*
DPTC Days Prior to Test Completion *fu*
DPTE Data Processing Terminal Equipment *do19*
DPtoTP Display Coordinates to Tablet Coordinates [converting] *pk, kg*
DPTRAJ Double Precision Trajectory [program] *sf31.1.14*
DPTT Days Prior to Test *fu*
DPU Data Processing Unit *mt, fu, gp, wp, 16, oe*
DPU Defects Per Unit *jj, fu*
DPU Digital Processor Unit *sd*
DPU Dual Processor Unit *mt*
DPWS Dual Purpose Weapon System *hc*
DQ Draft Quality *fu*
DQDB Distributed Queue Dial (Dual) Bus *ct, em, kg*
DQL Data Query Language *pk, kg*
DQM Data Quality Monitor *fu*
DQPSK Differentially encoded Quadrature Phase Shift Keying *fu*
DQS Double Q Switch *hc*
DR Damping Ratio *hc*
DR Data Rate *2, 4*
DR Data Received *pk, kg*
DR Data Register *hc*
DR Data Requirement *tb, wl*
DR Data Router *jdr*
DR Dead Reckoning *mt, fu, w9, wp, jj*
DR Deficiency Report *mt, wp, kf, fu*
DR Design Reports *fu*
DR Design Requirement *0, tb*
DR Design Review *mt, 17, jdr, fu*
DR Detection Radar *mt*
DR Discrepancy Report *mt, 4, oe*
DR Discrimination Radar *st, fu*
DR Display Register *jj*
DR Dockside Reeler *fu*
DR Document Record *fu*
DR&A Data Reduction and Analysis *fu*
DR/BOND Dialup Router / Bandwidth On Demand [NEC] *hi*
DR/DR Data Recording and Data Reduction *fu*
DRA Data Reduction and Analysis *fu*
DRA Dead Reckoning Analyzer *fu, jj*
DRA Decision Risk Analysis *mt*
DRA Defense Research Agency [UK] *mt*
DRA Detection, Recognition and Acquisition *hc*
DRAC Digital Radar Azimuth Converter *fu*
DRADS Defense Radar Degradation System *hc*
DRAM Data Responsibility and Approval Matrix *fu*
DRAM Dimension Random Access Memory [as in 3DRAM] *hi*

DRAM Dynamic Random Access Memory [hardware] *mt, fu, ah, ct, em, wp, aw130.13.62, is26.1.4, kg, pbl, tm, hi, ai1*
DRAR Data Reduction And Reporting *jdr*
DRAR Depot Repair Activity Report *fu*
DRAS Data Requirements Authorization Sheet *hc*
DRAT Digital Range and Angle Tracker *hc*
DRAW Direct Read After Write *mt, kg*
DRB Data Review Board *fu*
DRB Defense Research Board [Canada] *hc*
DRB Defense Resources Board [OSD] *mt, ph, wp, tb, kf*
DRB Design Requirements Baseline *hc, wp, jj*
DRB Design Review Board *hc, ep, tb*
DRB Discrepancy Review Board *fu*
DRBOND Dial-up Router Bandwidth On Demand [NEC] *pk, kg*
DRC Data Reduction Center *mt, kf*
DRC Defense Research Committee *wp*
DRC Defense Research Corporation *hc*
DRC Design Rule Check *kf, fu*
DRC Director of Reserve Components [USA] *mt*
DRC Disposal Release Confirmation *wp*
DRC Document Record Card *hc, jj, fu*
DRC Dynamics Research Corporation *mt, jj*
DRCF Data Reduction Computer Facility *fu*
DRCP Data Reduction Computer Program *fu*
DRCP Data Routing and Control Processor [aka router] *jdr*
DRCP Defect Reduction Control Point *fu*
DRCU DEW Line Record Communications Upgrade *mt*
DRD Data Requirements Document *mt, fu*
DRD Data Resources Directory *hc*
DRD Description of Required Documents *oe*
DRD Device Reliability Data *hc*
DRD Document Requirement Description [NASA] *hc*
DRDA Data Reduction and Data Analysis *fu*
DRDA Distributed Relational Database Architecture *pk, kg*
DRDBA Distributed Relational Database Architecture [also DRDA] *hi*
DRDP Detection Radar Data Processing *fu*
DRDTO Detection Radar Data Take Off *mt, fu*
DRDW Direct Read During Write *pk, kg*
DRE Dead Reckoning Equipment *fu*
DRE Diversity Reception Equipment [ESD] *mt*
DREC Detection Radar Electronic Component *fu*
DRED Deferred Requisitioning of Engineering Drawings *mt*
DRET Direction des Recherches, Etudes, et Techniques [France] [Directorate for Research, Studies, and Technology] *aw126.20.104*
DRF Data Recording Function *fu*
DRF Data Request Form *wp*
DRF Depot Recovery Factor *jj*
DRF Disaster Response Force *mt*
DRF Dual Role Fighter *ph*
DRFM Digital Radio Frequency Memory *mt*
DRFM Digital RF Memory *hc, wp, fu*
DRG Data Receiver Group *fu*
DRG Data Recording Group *fu*
DRG Defense Research Group *hc*
DRG Directional Receiver Group *mt*
DRGS Direct Readout Ground System *hc, gp*
DRI Data Requirements Instructions *oe*
DRI Data Resources Incorporated *aa26.12.48*
DRI Dead Reckoning Indicator *fu*
DRI Defense Research Institute *hc, jj*
DRI Digital Research Incorporated *kg, hi*
DRI Double Rate Imager *oe*
DRIE Department of Regional Industrial Expansion [Canada] *fu, aw129.1.70*
DRIPS Dynamic Real Time Information Projection System *ro*
DRL Data Redefinition Language *fu*
DRL Data Reduction Laboratory *0*
DRL Data Requirements List [NASA] *hc, wp, jj*
DRL Date Required to Load *mt*
DRL Documents Requirements List *gp, oe*
DRLMS Digital Radar Land Mass Simulator *wp*
DRM Dependability, Reliability, Maintainability *kf*
DRM Deployable Radiator Mechanism *he*
DRM Design Reference Mission [IUS] *-4*
DRM Digital Rate Multiplier *fu*
DRM Disaster Resource Manager *mt*
DRM Drafting Room Manual [a document] *fu, hn42.1.6, hc, ep, jj, jdr*
DRMAP Data Responsibility Matrix/ Acceptance Plan *fu*
DRMC Defense Resources Management Course *mt*
DRMO Defense Reutilization and Marketing Office *wp*
DRMS Deficiency Report Management System *wp*
DRMWG Dependability, Reliability, and Maintainability Working Group *kf*
DRO Data Readout *fu*
DRO Data Request Output *pk, kg*
DRO Destructive Read Out *hc, kg, jj, fu*

DRO Dielectric Resonator [stabilized] Oscillator *mj31.5.26, wp, jdr, fu*
DROLS Defense RDT&E On Line [Retrieval] System *fu, mt*
DROT Delayed Range On Target *hc, jj, fu*
DRP Data Reduction Program *fu*
DRP Design Review Program *oe*
DRP Designated Rework Point *jj*
DRP Deutsches Reich Patent [German Empire patent] *wp*
DRP Direct Requisitioning Procedure *jj*
DRP Disaster Recovery Plan *mt*
DRP Drawing Release Point *hc, jj*
DRPD Divided-Time Pulse Distributor {?} *fu*
DRPS Data Recording and Playback Subsystem *fu*
DRPS Design Recovery Processing System *fu*
DRPS Display Rapid Prototyping (Prototype) System *fu, hc*
DRPS Display Recording and Playback Subsystem *fu*
DRR Digital Radar Relay *fu*
DRRA Data Recording, Reduction and Analysis *fu*
DRRB Data Requirements Review Board *mt, fu*
DRRL Digital Radar Relay *fu*
DRRP Data Requirements Review Board *wp*
DRRT Directorate of Requirements, Training Acquisition Branch *kf*
DRS Data Reduction Set *fu*
DRS Data Relay Satellite [an ESA program] *es55.41*
DRS Deficiency Reporting System [USA] *mt*
DRS Design Recovery System *hd1.89.y30, fu*
DRS Design Requirements Specifications *mt*
DRS Designator Ranging Subsystem *hc*
DRS Detecting and Ranging Set (System) *fu, hc, hn43.5.1, jj*
DRS Display Remoting System *fu*
DRS Document Registration System *pk, kg*
DRS Drawing Record Summary *hc, jj*
DRSEM Deployable Receive Segment Engineering Model *mt*
DRSN Down Range Station *ai2*
DRSP Defense Reconnaissance Support Program *mt*
DRSS Data Relay Satellite System *hc, sf31.1.21*
DRT Data Relay Terminal *hc*
DRT Dead Reckoning Tracker/Tracer *jj*
DRT Design Review Team *mt*
DRT Diagnostic Rhythm Test *fu*
DRTE Defense Research Telecommunications Establishment [Canada] *hc, le36*
DRTL Diode Resistor Transistor Logic *jj, fu*
DRTP Design Requirements Traceability Package *fu*
DRTS Digital Recording Technique Study *hc*
DRU Data Regenerator Unit *fu*
DRU Direct Reporting Unit *mt, af72.5.5*
DR_DOS Digital Research-Disk Operating System *hi, kg*
DS Data Segment *wp, kg*
DS Data Send *pk, kg*
DS Data Server *pk, kg*
DS Data Store *fu*
DS Database Specification [DOD Standard Documents] *mt*
DS Decade Scaler *fu*
DS Defense Suppression *mt, wp*
DS Deficiency System [REDCOM/JDA] *mt*
DS Design (Detail) Specification *hc, jj*
DS Design Study *fu*
DS Designated Subcontractor *tb*
DS Detection Sensitivity *jj*
DS Development Specification *fu*
DS Digital Signal level (DS- n) *nu, ct*
DS Digital Switch *fu*
DS Diode Switch *fu*
DS Direct Support *mt, fu, jj*
DS Disconnect Switch *fu*
DS Dispatch Sequencer *fu*
DS Display Subsystem (System) *fu*
DS Distributed System *fu*
DS Doppler Science *es55.29*
DS Dual Scan *hi*
DS Dual Spin *sd*
DS Dwell Sounding [data] *gd4-5*
DS&DH Data Switching and Data Handling *fu*
DS&H Document Status and History *fu*
DS&R Data Storage and Retrieval *fu*
DS&RS Document Storage and Retrieval System [OJCS] *mt*
DS/CC Display Subsystem/Command Console *fu*
DS/PM Data Sequencer/Performance Monitor *fu*
DS2 [chemical warfare decontaminating agent, 70% Diethylentriamine + 28% Cellosolve + 2% NaOh] *wp*
DSA Dataroute Serving Area *do180*
DSA Defense Shipping Authority *mt*
DSA Defense Supply Agency *mt, hc, wp, jj, kf, fu*
DSA Digital Spectrum Analyzer *fu*
DSA Direct Support Aircraft *hc, jj*
DSA Directory System Agent *pk, kg*

DSA Distributed Sparse Array *wp*
DSA Distributed System Architecture [Honeywell] *mt, ct, em, hi*
DSA Downselect Authority *kf*
DSA-SIM Digital Spectrum Analyzer Simulator *fu*
DSAA Defense Security Assistance Agency *mt, hc, fu, wp, dt*
DSAC DLA Systems Automation Center *mt*
DSAC Down Select Advisory Council *kf*
DSACS Defense Standard Ammunition Computer System *mt*
DSAD Data System Authorization Directory [USAF] *mt*
DSADAP Digital Synthetic Array Data Processor *hc*
DSAM Double Sideband Amplitude Modulation *fu*
DSAP Data Systems Automation Program *mt*
DSAP Defense Security Assistance Programs *jj*
DSAP Destination Service Access Point *pk, kg*
DSAR Defense Supply Agency Regulation *mt*
DSARC Defense System Acquisition Review Council [DOD] [OSD] *mt, hc, wp, fu, ph, tb, st, jj*
DSAT Defensive Satellite *5, 17, sb, wp, tb, fu*
DSATS DLA Standard Automation Transportation System *mt*
DSB Defense Science Board [DOD] [OSD] *mt, ph, wp, tb, fu*
DSB Direct Sounder Broadcast *pl*
DSB Double Side Band [AM modulation method] *mt, bl1-5.44, pl, wp, jdr, fu*
DSB-AM Double Sideband Amplitude Modulation *mt*
DSBL Disable *he*
DSC Data Synchronizer Channel *fu*
DSC Data Systems Coordinator *fu*
DSC Decommutator Synchronization Code [also DSL] *-16*
DSC Deep Submergence Computer *hc*
DSC Defense Supply Center *mt, jj*
DSC Depot Support Concept *mt*
DSC Differential Scanning Calorimetry (Calorimeter) *wp, eo, re, jdr*
DSC Digital Scan Converter *fu, hc, jj*
DSC Digital Synchro Converter *hc*
DSC Distinguished Service Cross *w9, af72.5.172*
DSCB Data Set Control Block *fu*
DSCC Deep Space Communications Complex *hc, re*
DSCD Dual Scan Color Display *hi*
DSCP DSC Position *fu*
DSCS Defense Satellite Communications System *fu, mi, hc, wp, ci.571, 5, 12, 14, 16, 17, sb, af72.5.68, jb, tb, mb, kf, ai2*
DSCS Defense Satellite Communications System [DCA] *mt*
DSCS Defense Service Communications Satellite *ce*
DSCSI Differential SCSI *hi*
DSCSOC DSCS Operations Center [DCA] *mt*
DSCT Double Secondary-Current Transformer *fu*
DSCX Digital Scan Converter - Translator *fu*
DSD Data Security Device *hc*
DSD Data System Designator *mt, wp*
DSD Data Systems Division *fu*
DSD Defense Support Division *mt*
DSD Defense Systems Division *sc3.1.4*
DSD Digital Sharing Device *fu*
DSD Drawing Status Diagram *fu*
DSD Dual Scan Display *hi*
DSD Dynamic Situation Display *fu*
DSDCS Dynamic Sensor Display and Control Simulator *hc*
DSDD Double Sided, Double Density [diskette] *hi, kg*
DSDIO Director, Strategic Defense Initiatives Organization *st*
DSDO Data System Design Office *mt*
DSDP Deputy [under] Secretary of Defense for Policy *tb*
DSDS Dataphone Switched Digital Service *do183*
DSDS Decision Support Display System [also DS2] [FEMA] *mt*
DSDT Deformographic Storage Display Tube *hc*
DSE Data Storage Equipment *kg, fu*
DSE Depot Support Equipment *wp*
DSE Development Supportability Engineering *wp*
DSE Distributed Systems Environment *mt*
DSEA Display Station Emulation Adapter *pk, kg*
DSEB Down Select Evaluation Board *kf*
DSECT Dummy Control Section *pk, kg*
DSEL Data Systems, Designators Exchange List *mt*
DSES DSP Software Engineering Support *kf*
DSESTS Direct Support Electrical System Test Sets *af72.5.185*
DSF Danish Simulation Facility *hc*
DSF Development Support Facility *fu*
DSFC Direct Side Force Control *mt*
DSFH Direct Sequence Frequency Hopping *fu*
DSFM Direct System Function Mechanization *fu*
DSG Defense Systems Group [Hughes Aircraft Company] *sc2.8.un, fu*
DSG Density Signal Generator *fu*
DSG Down Select Guide *kf*
DSGMM Detailed Sensor Geometric Math Model *pl*
DSGS Densely Spaced Geodetic Systems *hy*
DSHD Double Sided, High Density [diskette] *hi, kg*
DSI Digital Speech Interpolation *hc, jgo, ct, em*

DSI Digital Speech Interruption *gs2-10.54*
DSI Dissimilar Source Integration *fu*
DSID Data Stream Identifier *oe*
DSID Disposable Seismic Intrusion Detector *wp*
DSIF Deep Space Instrumentation Facility [at JPL] *fu, hc, sm18-2, 0*
DSIOC Data Slic, Input/Output Controller *fu*
DSIPS Digital Satellite Image Processing System *wp*
DSIS Defense Satellite Interface System *hc*
DSIS Defense System Interaction Study *hc*
DSIS Digital Software Integration System *hn48.13.4*
DSIS Distributed Support Information Standard *pk, kg*
DSISR Delinquent Supply Item Status Report *hc, jj*
DSISS AMC Standard Installation Supply System [USA] *mt*
DSL Data Structure Language *wp*
DSL Decommutator Synchronization Code [also DSC] *gp*
DSL Deep Scattering Layer *fu*
DSL Depot Supply Level *wp*
DSL Digital Subscriber Line *mt, ct, kg, em*
DSL Drawing Status List *hc, jj, fu*
DSL Dynamic Simulation Language *pk, kg*
DSLAM Digital Subscriber Line Access Multiplexer *ct, kg, em*
DSLOC Delivered Source Lines of Code *kf*
DSM Data System Manager *oe*
DSM Data System Modernization *16, tb*
DSM Defense Suppression Missile *wp, hc*
DSM Design Standards Manual *jdr*
DSM Direct Support Maintenance *mt*
DSM Display Switch Module *fu*
DSM Display System, Multipurpose *jj*
DSM Distinguished Service Medal *w9, wp*
DSM Drafting Standards Manual [a document] *pbl, tm, ah*
DSM Dynamic Scattering Mode *wp*
DSMA Defense Supply Management Agency *hc*
DSMA Digital Sense Multiple Access *pk, kg*
DSMC Defense Systems Management College *mt, fu, wp, dt, hc*
DSMC Direct Simulation Monte Carlo [method] *aa26.12.b8, ai1*
DSMDPS Deployable Strategic Mission Data Preparation System [KC-135] *mt*
DSN Deep Space Network *wp, sc3.12.4, 4, 16, hc, es55.28, so, mb, re, mb, mi, pbl, vv, oe*
DSN Defense Switchboard (Switched) Network [USA DOD] *mt, kf, ph, is26.1.76ee*
DSN Digital Services Network *mt*
DSN Digital Switched Network *hi*
DSN Document Serial Number *jj*
DSNET Distributed Systems Network *hi*
DSO Data Systems Operator *kf*
DSO Defense Sciences Office [DARPA] *kb*
DSO Delco Systems Operations [of Santa Barbara] *hn49.14.6*
DSO Dependents Schooling Office [Atlantic] *dt*
DSO Detailed Supplementary Objective *ws4.10*
DSO Dielectrically Stabilized Oscillator *pl*
DSO Digital Storage Oscilloscope *is26.1.24*
DSO Distinguished Service Order *w9*
DSOC Defense Space Operations Committee *mt*
DSOM Distributed System Object Model *hi, kg*
DSOT Daily System Operability (Operational) Test *st, fu*
DSP Data Systems Products [Department] *fu*
DSP Deep Space Probe *0*
DSP Defense Satellite Program *tb, st*
DSP Defense Science Program *hc*
DSP Defense Support Program [USAF/NRO] [a network of early warning satellites, USA] *mt, fu, hc, wp, ci.571, 12, 14, 17, sb, af72.5.67, aw130.11.15, mi, hn49.11.8, jb, tb, jdr, mb, kf, ls4, 5, 16*
DSP Digital Signal Processing (Processor) *fu, kg, ct, em, hc, wp, jj, jdr, hi*
DSP Digital Speckle Pattern Interferometry *wp*
DSP Director for Security Plans and Programs *hc, tb*
DSP Directory Synchronization Protocol [Lotus] *pk, kg*
DSP Display System Protocol *fu*
DSP Dynamic Support Program *fu*
DSP/C DSP Consolidation *kf*
DSPAUG Defense Support Program Augmentation *tb*
DSPCD Dual Scan Passive Color Display *hi*
DSPFO Defense Support Program Follow On *hc*
DSPG Defense Special Projects Group *dt*
DSPMO SAMMS Program Management Office *dt*
DSPN Direct Sequence Pseudonoise *fu*
DSPO Defense Support Project Office *mt*
DSPT Digital Signal Processing Techniques *hc*
DSQD Double Sided, Quad Density [diskette] *hi, kg*
DSR Data Scanning and Routing *fu*
DSR Data Status Report *fu*
DSR Date Set Ready *ct, kg, em*
DSR Device Status Register (Report) *pk, kg*
DSR Direct Scope Radar *hc*
DSR Directorate Status Review *oe*

DSR Document Status Record *fu*
DSRC Data Systems Requirements Committee *2, 4*
DSRC Defense Systems Acquisition Council *fu*
DSRD Data Systems R&D Program [MMES] *mt*
DSRD Depot Support Requirement Document *wp*
DSRE Defense Subsistence Region, Europe *mt*
DSREDS Digital Storage and Retrieval Engineering Data System [USA] *mt*
DSRI Danish Science Research Institute *cu*
DSRI Destination Station Routing Indicator *mt*
DSRS Disk Storage and Retrieval System *fu*
DSRV Deep Submergence Rescue Vehicle [US Navy] *hc, mt*
DSS Data Storage Set *tb*
DSS Data System Supervisor *kf*
DSS Decision Support System *mt, kg, fu*
DSS Deep Space Station *hc, 0, mb, re*
DSS Defense Supply Services *wp*
DSS Defensive Satellite System *sd*
DSS Department of Supply [and] Services [Canada] *hc, fu*
DSS Digital (Dipping) Sonar Subsystem *fu*
DSS Digital Satellite System *heh2.21.6*
DSS Digital Signature Standard *kg, em, hi*
DSS Digital Subsystem *jj, jdr*
DSS Digital Sun Sensor *jdr, oe*
DSS Direct Satellite System *jdr*
DSS Directed Stationing System *mt*
DSS Disk Storage System *hi*
DSS Distribution and Switching System *-2*
DSS Dual Sector Suite *fu*
DSS Dynamic Satellite Simulator *jdr*
DSS Dynamic Systems Simulator *wp*
DSS-W Defense Supply Service-Washington *dt*
DSSBD Double Single Sideband *fu*
DSSBD Dual Solar Sail Boom Drive *oe*
DSSC Double Sideband Suppressed Carrier *hc*
DSSCS Defense Special Secure Communications System *mt*
DSSD Double Sided Single Density *hi*
DSSE Daily Summary Spare Engines [PACAF] *mt*
DSSE Development Software Support Environment *fu*
DSSE Digital Sun Sensor Electronics *oe*
DSSI Digital Standard Systems Interconnect [DEC] *pk, kg*
DSSM Downselect Security Monitor *kf*
DSSO Defense System Support Office [Formerly JDSSC/DISA] *mt*
DSSP Deep Submergence System Program *hc*
DSSP Defense Standardization and Specification Program *mt*
DSSR Deep Space Surveillance Radar *hc*
DSSS Deep Space Surveillance System *hc*
DSSS Direct Sequencing Spread Spectrum *pk, kg*
DSSSS Direct Support Unit Standard Supply System [also DS4] [USA] *mt*
DSSV Deep Submergence Search Vehicle *fu*
DSSW Defense Supply Services Washington *mt, wp*
DST Data Source Terminal *fu*
DST Daylight Saving Time *w9, wp, jj*
DST Defense Suppression Threat *mt, 5, 17, sb, cp, tb, kf*
DST Direct Support Team *mt*
DST Display Storage Tube *fu*
DSTB (ground terminals . . . fabricated by DSTB personnel) *-19*
DSTE Digital Subscriber Terminal Equipment *mt*
DSTF Delta Spin Test Facility *9, jdr, ai2*
DSTMM Detailed Sensor Thermal Math Model *pl*
DSTN Double (Dual Scan) Super Twisted Nematic [display] *hi, kg*
DSTP Director, Strategic Target Planning *mt*
DSTR Dual Stage Target Recognizer *hc*
DSTU Digital Signal Transfer Unit *hc*
DSTUMS Digital Signal Transfer Unit and Multi Sensor *hc*
DSTV Direct Support Team Vehicle *fu*
DSU Data Service Unit *do114, kg, ct, em*
DSU Data Switching Unit *tb*
DSU Data Synchronizer Unit *fu*
DSU Deployment Sensor Unit *pl*
DSU Digital Service Unit [DDS] *kg, mt*
DSU Direct Support Unit *10, jj*
DSU Direct Support Unit [USA] *fu, mt*
DSU Disk Storage Unit *mt, jj, fu*
DSU Drum Storage Unit *fu*
DSU/CSU Data Service Unit/Customer Service Unit *kf*
DSUWG Data Systems Users Working Group *ns*
DSV Deep Submergence Vehicle *jj*
DSVD Digital Simultaneous Voice and Data *kg, ric, jdr, hi*
DSVT Digital Secure Voice Terminal *mt, fu*
DSW Data Status Word *pk, kg*
DSW Defense Suppression Weapon *hc, st*
DSW Device Status Word *pk, kg*
DSWA Defense Special Weapons Agency *jdr*
DSWR Deep Space Warning Radar [ESD] *mt*

DSWS Division (Divisional) Support Weapons System *fu, uc*
DSX Digital Signals Cross-Connect *pk, kg*
DT Data Transmission *wp*
DT Day Tag *fu*
DT Decay Time *fu*
DT Definition Term *ct*
DT Delayed Time *ac3.7-30*
DT Deployment Transaction *mt*
DT Development Test[ing] *mt, st, jj, kf, fu*
DT Dial Tone
DT Display Test *jj*
DT Documentation Tree *fu*
DT Double Throw *fu*
DT Double Time *w9*
DT Down Time *0, fu*
DT&E Development, Test, and Evaluation [equipment] *mt, fu, wp, hc, 16, jb, jmj, dm, tb, st, jj, jdr, kf*
DT&IOT Developmental Test and Initial Operational Test *kf*
DT&OT Developmental Test and Operational Test *kf*
DT/OT Development Testing/Operational Testing (Test) *kf, fu, nt, jmj, jj*
DT/OT&E Developmental Test / Operational Test and Evaluation *kf*
DTA Data Transfer Area *wp*
DTA Differential Thermal Analysis *fu, wp, ai1*
DTA Disk Transfer Area *ct, em, wp, kg*
DTACC Deployable Tactical Air Control Center *mt, wp*
DTACC Deployed Tanker / Airlift Control Center *mt*
DTACCS Director, Telecommunications And Command and Control Systems *mt*
DTACK Data Transfer Acknowledge *fu*
DTADS Deutschland Trans Air Defense System *hc*
DTAE Depot Test and Acceptance Equipment *hc, jj*
DTAMS Data Transmission and Message System *mt*
DTAR Decision Theoretic Adaptive Radar *hc*
DTAR Dwell Time Adaptive Radar *fu*
DTAS Data Transmission and Switching [system for USMC] *fu*
DTAS Digital Transmission and Switching System *hc*
DTBS Duration of Time Between Slots *uh*
DTC Data Transmission Channel *hi*
DTC Design To Cost *mt, hc, ah, wp, 17, sb, nt, tb, st, jj, pbl, jdr, kf, hj16.9.3, fu*
DTC Desk Top Computer *fu*
DTC Desktop Conferencing *pk, kg*
DTC Desktop Tactical Computer *hc*
DTC Distributed Transaction Coordinator *em*
DTC Dynamic Test Chamber *pl*
DTCAM Design to Cost Analysis Model *fu*
DTCR Design To Cost Report *fu*
DTCS Digital Telephone Circuit Switch *fu*
DTCS Digital Telephone Communication Switch *fu*
DTCS Drone Tracking and Control System *af72.5.150*
DTD Data Transport Device *mt*
DTD Digital Television Display *hi*
DTD Document Type Definition *kg, em*
DTDMA Distributed (Distribution) Time Division Multiple Access *mt, fu*
DTE Data Terminal Equipment *mt, do19, wp, 4, tb, kf, kg, nu, ct, em, fu, hi*
DTE Data Transmitting Equipment *wp*
DTE Development, Test, and Evaluation *wp, hc, kf*
DTE Digital Target Extractor *fu*
DTE Dumb Terminal Emulator *pk, kg*
DTED Digital Terrain Elevation Data [DMA] *fu, mt*
DTED Digital Terrain Evaluation {misprint for elevation?} Data *wp*
DTER Data Transmittal and Evaluation Request *fu*
DTF Development Test Facility *jr*
DTF Discipline, Time, and Frequency source *jdr*
DTF Distributed Test Facility *pk, kg*
DTG Date Time Group [Julian Day and Zulu Time] *jj, fu, mt*
DTG Diagnostic Test Group *fu*
DTG Distance To Go *mt*
DTG Dry Tuned Gyro *17, ai2*
DTG Dynamically Tuned Gyro *17, sb*
DTH Direct To Home *jdr, hi*
DTI Design Team Incorporated [Ontario, Canada] *aw130.6.68*
DTI Digital Telesat, Incorporated *sc4.2.3*
DTIC Defense Technical Information Center [formerly Defense Documentation Center, Alexandria, VA] *mt, su, fu, hc, hn44.7.4, dt, wp, tb, st, kf*
DTID Disposal Turn In Document *wp*
DTIR Defense Technical Intelligence Report *wp*
DTL Dialog Tag Language [IBM] *pk, kg*
DTL Diode Transfer Logic *mt*
DTL Diode Transistor Logic *6, wp, hc, kg, fu, jj*
DTLCC Design to Life Cycle Cost *hc, fu, kf*
DTLS Descriptive Top Level Specifications *mt, tb*
DTM Data Test Message *fu*
DTM Data Transfer Module *wp, fu*

DTM Data Transmission Medium *mt*
DTM Developmental Telemetry *hc, jj*
DTM Digital Test and Monitor *fu*
DTM Digital Test Message *fu*
DTM Digital Topographic Model *re*
DTM DIXIE Toll Management *cd*
DTM Draft Technical Manual *fu, hc, jj*
DTM Dual Thruster Module *-4*
DTM Dual Tone Multifrequency *pk, kg*
DTMF Data Tone Multi(ple) Frequency *fu, kg*
DTMF Digital Tone Multi Frequency *hi*
DTMF Dual Tone Modulated Frequency *ct*
DTMF Dual Tone Multiple Frequency *mt*
DTMRP Detailed Tactical Mission Route Plan *fu*
DTMS Data Transmission and Multiplexing System *hc*
DTN Data Transmission Network [DCS] *mt*
DTNSRDC David Taylor Naval Ship Research and Development Center *mt, fu, 12*
DTO Daily Tasking Order *mt*
DTO Data Takeoff *fu*
DTO Delayed Test Objective *0, tb*
DTO Detailed Test Objective *wl*
DTO Development Test Objective *ws4.10*
DTO Direct Turn Over *jj*
DTO Division Transportation Officer *mt*
DTOA Differential Time Of Arrival *fu, jdr*
DTOC Division Tactical Operations Center *mt, fu, hc*
DTOCSE Division Tactical Operations Center Support Element [USA] *mt*
DTOS Division Tactical Operational System *fu*
DTOT Desired Time Over Target *fu*
DTP Data Transfer Process *mt*
DTP Design to Price *hc, fu*
DTP Detailed Planning *mt*
DTP Development Test Plan *wp*
DTP Differential Twisted Pair *jj*
DTP Distributed Transaction Process *pk, kg*
DTP Doppler Techniques Proposal [Hughes Aircraft Company] *hc*
DTP-EW Design to Price Electronic Warfare [system] [AN/SLQ-31 built by Hughes for Navy] *fu*
DTR Data Terminal Ready *mt, ct, kg, em, fu*
DTR Data Transfer Rate *pk, kg*
DTR Dedicated Token Ring *hi*
DTR Definite Time Relay *fu*
DTR Demand Totalizing Relay *fu*
DTR Detailed Rest Requirement *jj*
DTR Digital Tape Recorder *re*
DTR Digital Test Report [Line] *jj*
DTRA Defense Technical Review Activity *wp*
DTRA Defense Threat Reduction Agency *rih*
DTRG Dry Tuned Rate Gyro *hc*
DTRM Dual Thrust Rocket Motor *hc*
DTS Data Terminal Set *fu*
DTS Data Transfer Switchboard *fu*
DTS Data Transfer System *mt*
DTS Data Transmission Subsystem *mt*
DTS Dedicated Tracker System *fu*
DTS Defense Transportation System *mt, wp*
DTS Defensive Technology Study *mt, fr, tb*
DTS Diagnostic Test Set (Station) *fu*
DTS Digital Termination Service *cd, ce, fu*
DTS Diplomatic Telecommunications Service *mt*
DTS Dive - Toss *jj*
DTS Dock-To-Stress *fu*
DTS-W Defense Telephone Service-Washington *dt*
DTSA Defense Technology Security Administration *wp*
DTSCE Distributed Time-Slot Comparison Error *fu*
DTSE&E Director, Test, Systems Engineering and Evaluation *kf*
DTSHA Distributed Time-Slot Hardware Alarm *fu*
DTSS (includes a computer for DTSS processing) *hd*
DTSS Dartmouth Time Sharing System *fu*
DTSS Digital Topographic Support System [USA] *mt*
DTST Defensive Technologies Study Team *fr, tb, 10, fu*
DTSTP Derivative Truncated Sequential Test Program *wp*
DTT Digital Test Target *fu*
DTTL Data Transition Tracking Loop *-2*
DTTN Distributed Tactical Test Network *hc*
DTTS Defense Transportation Tracking System *mt*
DTTY Disadvantaged Teletype *uh*
DTU Data Transfer Unit *hc, wp, jj, fu*
DTU Digital Telemetry Unit *hc, jj, kf*
DTU Digital Test Unit *hc, jj*
DTUC Design To Unit Cost *fu*
DTUPC Design To Unit Production Cost *hc, fu, wp, nt, st, jj*
DTV Design Test Vehicle *hc, jj*
DTV Desk Top Video *ct, kg, em, hi*

DTV Development Test Vehicle *jdr*
DTV Digital Television *fu*
DTV Driver Thermal Viewer *hc, hn49.3.3*
DTVC Desktop Video Conferencing *pk, kg*
DTVC Digital Transmission and Verification Converter *fu*
DTVE Digital Television Element *mt*
DTVM Differential Thermocouple Voltmeter *fu*
DTWA Dual Trailing Wire Antenna *mt*
DU Defense Unit *fu*
DU Disk Usage *pk, kg*
DU Display Unit *hc, jj, fu*
DU Driver Unit *gg52*
DUA Directory User Agent *pk, kg*
DUAT Direct User Access Terminal *aw132.10.5, kg, fu*
DUATS Direct User Access Terminal Service *aw130.11.61*
DUB Deep Underground Basing *uc*
DUC Document Usage Card *hc, jj*
DUCA Distributed User Coverage Antenna *jdr*
DUCC Deep Underground Command Center *mt*
DUCE Distributed User Coverage Electronics *jdr*
DUCM Distributed User Coverage Mechanism *jdr*
DUEL Data Update Edit Language *mt*
DUER Design Unit Engineering Report *mt*
DUI Data Use Identifier *mt, fu, jdr*
DUI Design User's Interface *fu*
DUMD Deep Underwater Measuring Device *fu*
DUMR Display Unit Mounting Rack *hc*
DUN Dial-Up Networking *ct, kg, em*
DUNCE Dial Up Network Connection Enhancement *pk, kg*
DUNS Data Universal Numbering System *mt, hc, fu, 18, nt, tb*
DUPC Delayed Under Program Control *jj*
DUPO Dutch Powder [chemical warfare Decontaminant] *wp*
DURANDAL [air launched airfield attack weapon] *uc*
DUS Data Utilization Station *oe*
DUS Data Utilization System *gp, pl, 16*
DUS Design Unit Specification *mt*
DUS Design Utility System *fu*
DUSC Defense Underground Support Center *mt*
DUSD Deputy Under Secretary of Defense *mt, hc, jdr*
DUSD(P) Deputy Under Secretary of Defense for Policy *su*
DUSDRE Deputy Under Secretary of Defense for Research and Engineering *mt*
DUSD_CCCI Deputy Undersecretary of Defense for Communications, Command, Control, and Intelligence *mt*
DUSN Deputy Under Secretary of the Navy *dt*
DUSTER Dual Stage Target Recognition *hc*
DUSTWUN duty status - whereabouts unknown [USA DOD] *mt*
DUT Demand Usage Time *fu*
DUT Depot Unit Tester *hc*
DUT Device Under Test *fu, jdr*
DUT Document Update Transmittal *oe*
DUV Data Under Voice *do115*
DUV Deep Ultra Violet *fu*
DUWCAL Duluth Weapons Calibration Systems *hc*
DV Delta Velocity [a change in velocity] *uh*
DV Digital Video *ct, em*
DV Direct View *jj*
DVAL Data Link Vulnerability Joint Task Force *mt*
DVAL Demonstration Validation *wp*
DVARS Doppler Velocity Altimeter Radar Set *wp*
DVB Digital Video Broadcasting (Broadcast) *ct, jdr, em, tt*
DVC Desktop Video Conferencing *pk, kg*
DVC Digital Video Conference *ct, em*
DVCP Direct View Control Panel *hc, jj*
DVD Digital Versatile Disk *em*
DVD Digital Video Disk *ct, kg, em*
DVE Digital Video Effect *pk, kg*
DVESO Dod Value Engineering Services Office *wp*
DVF Divide Fractional - Command *jj*
DVFR Defense VFR *fu*
DVI Digital Video Interactive *ct, kg, em*
DVI Direct Voice Input *mt*
DVIIS Direct View Image Intensifier System *hc*
DVITS Digital Video Imagery Transfer System *mt*
DVL Direct Voice Line *fu*
DVM Design Verification Model (or Matrix) *4, jdr, fu*
DVM Digital Volt Meter *hc, wp, gp, ep, jj, fu, oe, ai1*
DVM Display Video Monitor *fu*
DVMRP Distance Vector Multicast Routing Protocol [Internet] *em, kg*
DVMST Design Verification Mixed Simulation Test *kf*
DVOM Digital Volt-Ohm Meter *wp*
DVOR Doppler VHF Omni Range *hc*
DVOT Delayed Velocity On Target *jj*
DVP Design Validation (Verification) Phase *fu*
DVP Design Verification Plan *jdr*
DVPPI Daylight View Plan-Position Indicator *fu*

DVPR Drawing Verification/Produc-ibility Review *fu*
DVR Design Verification Report *jdr*
DVR Digital Video Recorder *st*
DVRMP Distance Vector Multicast Routing Protocol [Internet] *hi*
DVRRE Digital Video Record - Reproduce Equipment *hc*
DVS Digital Versatility Service *fu*
DVS Digital Voltage Source *jj*
DVST Direct View Storage Tube *fu, hc, jj*
DVT Design Verification Test *8, fu*
DW Defensive Weapons *tb*
DWA (companies like DWA Composites Specialties) *aa26.11.14*
DWASP DLA Standard Warehousing and Shipping Automated System *mt*
DWC Douglas World Cruiser *af72.5.28*
DWCCCC Deployable WWMCCS Command, Control and Communications Capability [also DWC4] *mt*
DWD Deutsche Wetter Dienst [German Weather Service] *mt*
DWE Doppler Wind Experiment *es55.28*
DWG Design Working Group *fu*
DWG Digital Word Generator *fu*
DWL Dominant Wave Length *fu*
DWMSTD Defense Work Measurement Standard Time Date *mt*
DWMT Discrete Wavelet Multitone *pk, kg*
DWPS Deployable War Planning System *mt*
DWRIA Died of Wounds Received In Action [USA DOD] *mt*
DWS Defensive Weapons System *fu*
DWS Design Work Study *fu*
DWS Digital Weapon Simulator *fu*
DWS Disaster Warning System *wp*
DWS Dispenser Weapon System [Germany] *mt*
DWS Doppler Wind Sounder *hg, ge*
DWS/CS Defensive Weapon Subsystem/ Command Subsystem *fu*
DX Data Extraction *fu*
DX Direct Exchange *fu, jj*
DX/DR Data Extraction and Recording System *fu*
DXC Data Exchange Control *pk, kg*
DXE Data Transmitting Equipment *fu*
DXF Drawing Exchange Format *hi, kg*
DXI Data Exchange Interface *ct, em*
DXRB Diffuse X-Ray Background *es55.61*
DY Dysprosium *wp*
DYLAN Dynamic Language *ct, em*
DYM Day Year Month *hi*
DYNA-METRIC Dynamic Multiple Echelon Technique for Repairable Item Control *mt*
DYNACORDS Digital Dynamics Range Enhanced Cords *hc*
DYNAMETRICS [combat capabilities assessment tool] *mt*
DYNAMO Dynamic Action Management Operation *hc*
DYNOPT Dynamic Optimization [program] *hc*
DYOB Dynamic (Ground) Order of Battle *fu*
DYSIM Dynamic Simulation *fu*
DZ Drop Zone *mt, fu*
DZCO Drop Zone Control Officer *mt*
DZSO Drop Zone Safety Officer *mt*

E

E (Mode E 2 battery depth of discharge) {Eclipse} *gg26*
E [a designator, E-2 or E2 for US Navy Hawkeye, E-3A for AWACS Sentry]
E&A Evaluation and Assistance Visits *mt*
E&C Electronics and Control *hc, jj*
E&DCP Evaluation and Data Collection Plan *mt*
E&E Evasion and Escape *mt*
E&ERFTS Elementary and Reserve Training School [UK] *mt*
E&I Engineering and Installation *mt*
E&I Equip and Install *fu*
E&O [Eastern Launch Site (ETS) building] *bo*
E&R Extensions and Repairs *fu*
E&SP Equipment and Spare Parts *fu*
E&V Evaluation and Validation *mt*
E-4B {?} [National Airborne Operations Center [NAOC], formerly known as the National Emergency Airborne Command Post [NEACP]. [JCS]] *mt*
E-DARC Enhanced Direct Access Radar Channel *fu*
E-FORM Electronic Form *pk, kg*
E-I Engineering and Installation *mt*
E-O Electro-Optical (-Optics) *hc, uc, st, jj*
E-OCMS Electro-Optical Countermeasures System *hc*
E-OWDS Electro-Optical Weapon Delivery System *hc*
E-R Entity-Relationship *fu*
E-SPAN European SPAN *ns*
E-TDMA Extended TDMA *jdr*
E-W East-West *vv, oe*
E-ZINE Electronic Magazine *pk, kg*

E/A_CD Estimated or Actual Code *fu*
E/C Echo Cancelling *jgo*
E/E Electrical/Electronic *sn2.8*
E/EEC Extendable and Expandable Exit Cone *ai2*
E/FTE Eject / Fail To Eject *jj*
E/L Early/Late *fu, jdr*
E/M Electromagnetic *oe*
E/NE Effective/Non-Effective *mt*
E/O {Electro Optical} [also E-O] *af72.5.171*
E/P Equipment/Program *fu*
E/S Earth Station *csI.15*
E/W East West *gp, oe*
EA Earth Acquisition *0, oe*
EA Ecole d' Application [France] *mt*
EA Economic Analysis *mt, wp*
EA Effective Address *kg, fu*
EA Electronic Attack *mt*
EA Electronics Assembly *jj, fu*
EA Emergency Action *mt, tb*
EA Engineering Analysis *mt, tb*
EA Environmental Agency *ac3.3-23*
EA Environmental(al) Assessment *mt, wp, st, kf*
EA Equal Angle *pl*
EA Evolutionary Acquisition *mt*
EA Experimental Agent [chemical warfare] *wp*
EA Extended Attribute [OS/2] *pk, kg*
EA [a designator, EA-3 for Sky Warrior US Navy , EA-6A for Intruder, EA-6B for Prowler, USA DOD] *mt*
EAA Export Administration Act *ph*
EAARS Exercise After Action Reporting System [EUCOM] *mt*
EAASE ESM-Augmented Active Surveillance Evaluation *fu*
EAAT Electronic Aircraft Air Temperature *jj*
EAATS Eastern ARNG Aviation Training Site [USA] *mt*
EAB Emergency Air Breathing [mask] *mt*
EAB Extended Avionics Bus *fu*
EAC (EAC Industries) [an avionics/military electronics company] *aw130.22.54*
EAC Early Analysis Capabilities *tb*
EAC Echelons Above Corps [USA] *pm, fu, mt*
EAC Emergency Action Center (Cell) *mt*
EAC Emergency Action Center [CELL] *mt*
EAC Emergency Action Console *mt, jj*
EAC Emergency Action Coordinator *mt*
EAC Estimate At Completion *fu, 8, 17, sb, hc, tb, jj, jdr, kf, oe*
EAC Estimate of Completion [Government Estimate] *mt*
EAC Extended Area Coverage *oe*
EACB Engineering Analysis Control Boards *kf*
EACC Electronic Asset Control Center *mt, wp*
EACIC Echelons Above Corps Intelligence Center *mt*
EACP European Area Communication Plan *mt, hc*
EACT Emergency Action and Coordination Team *mt*
EAD Earliest Arrival Date [USA DOD] *mt*
EAD Eastern Australia Daylight *pk, kg*
EAD Echelons Above Division *mt*
EAD Estimated Availability (Arrival) Date *jj, wp, jj*
EAD Extended Active Duty *mt*
EADC Egyptian Air Defense Command *fu*
EADI Electronic Attitude Direction Indicator *hc*
EADS Egyptian Air Defense System *fu*
EADS Engineering Analysis and Data System *wp*
EADSI Egyptian Air Defense System Integration *fu*
EAE Emergency Action Element *mt*
EAF Emergency Action File *mt*
EAFB Edwards Air Force Base *mi, ai2*
EAFB Eglin Air Force Base *wp*
EAFE Europe, Australia and the Far East *aw130.13.11, hc*
EAFW Electronics and Auxiliary Fresh Water *fu*
EAG Extended Active Gate *hc*
EAGE Electrical Aerospace Ground Equipment *ah, gp, pbl, tm, jdr, kf, oe*
EAGLE Enhanced Automated Graphical Logistics Environment *kf*
EAGLE_EYE [EO sensor providing radar display of approaching aircraft] *wp*
EAI Electronic Associates, Incorporated *jj*
EAI Experimental/Early ARPAT Interceptor *hc*
EAID Equipment Authorization Inventory Date *wp, jj*
EAIM End Article Item Manager *hc, wp*
EAIR Enterprise Analysis Integration Review *mt*
EAL Equipment Airlock *jdr*
EALT Earliest Anticipated Launch Time [USA DOD] *mt*
EAM Economic Attrition Model *fu*
EAM Electronic Accounting Machine *mt, hc, wp, jj*
EAM Electronic Assembly Module *jdr*
EAM Emergency Action Message *mt, sd, tb, jdr, fu*
EAM Equipment Assembly Module *jdr*
EAMR Engineering Advance Material Requirements *fu*
EANS Emergency Action Notification System *mt*

EAO Emergency Action Officer *mt*
EAOSY Emergency Action Officer System *mt*
EAP Embedded Avionic Processor *hc*
EAP Emergency Action Procedures *mt*
EAP Employees Assistance Program *pbl, ah*
EAP Engineering Analysis Plan *oe*
EAP Equity Adjustment Program *hc*
EAP Etage d' Appoint Pourdre [Ariane 5 solid boosters] *ai2*
EAP Experimental Aircraft Program [technology demonstrator for the EFA] [UK] *mt, wp, aw129.21.6*
EAP-JCS Emergency Action Procedures of the Joint Chiefs of Staff *mt*
EAPL Engineering Assembly Parts List *hc, jj*
EAPROM Electrically Alterable Programmable Read-Only Memory *fu*
EAR Electric Alignment Reticle *dm*
EAR Electromagnetic Activity Receiver *hc*
EAR Electronically Agile Radar *hc, wp, jj, fu*
EAR Engineering Analysis Request *hc, jj*
EAR Enterprise Analysis Review *mt*
EAR Expenditure Authorization Request *fu*
EAR Export Administration Regulation *su*
EAR External Access Register *pk, kg*
EARFLAP Emergency Action Reporting for Logistical Analysis and Plans *mt*
EARLO Enhanced Airlift Reporting for Logistics and Operations *mt*
EARN European Academic Research Network *kg, es55.51*
EAROM Electronically (Electrically) Alterable Read Only Memory *mt, hc, 6, jj, kg, fu, hi*
EARP ECOS/ACOS Review Panel *fu*
EARS Electronic Access to Reference Services *pk, kg*
EARS Emergency Airborne Reaction System *hc, ro*
EARS Emergency AUTODIN Release System *mt*
EARS Explicit Archive and Retrieval System [Langley Research] *pk, kg*
EARTS Enroute Automated Radar Tracking System *mt, fu*
EAS Emergency Action System [REDCOM / JDA] *mt*
EAS En Route Automation Supervisor *fu*
EAS Engineering Analysis System *oe*
EAS Equivalent Air Speed [corrected for compressibility] *wp, jj, mt*
EAS Estimated Air Speed *wp*
EAS Expense Assignment System *mt*
EAS Experimental Antenna System *-19*
EAS [Aerospatiale subsidiary, France] *aa27.4.14, aw130.6.7*
EASE Electronic Analog Simulation Equipment *hc*
EASL Engineering Approved Supplier List *fu*
EAST Electronics And Space Technicians [Union] *fu, hc, hn43.5.4, jj*
EAST Experimental Army Satellite Tactical Terminal *hc*
EASTLANT Eastern Atlantic [NATO Command] *fu, 12*
EASTPAC Eastern Pacific *mt, tb*
EASY Automated Economic Analysis System [DLA] *mt*
EASY Efficient Assembly System [Honeywell] *hc*
EASY Emergency Action System [JCS] *mt*
EASY Engine Analyzer System *wp*
EAT Earliest Arriving Time *0*
EAT Engineering Assurance Test *fu*
EAT Environmental Acceptance Test *fu*
EATA Enhanced AT Bus Attachment *hi, kg*
EATS Extended Area Test System *fu, hc*
EATWG Executive Agency Transfer Working Group *mt*
EAU Extended Arithmetic Unit *jj*
EAUXCP East Auxiliary Airborne Command Post [SAC] *mt*
EAVS Emergency Action Voice System *mt*
EAW Electronic Assembly Warhead *fu*
EAX Electronic Automatic Interchange *fu*
EB Electron[ic] Beam *ph, st*
EB Engineering Bulletin *fu*
EB English Bias *jj*
EB Escadre de Bombardement [(French) Bombardment Squadron] *mt, 12*
EB&LP Engineering Breadboards and Laboratory Prototypes *fu*
EB/LSI Electron Beam LSI (process) *fu*
EB/NO Bit Energy to Noise density ratio *jdr*
EBA Electron Beam Accelerator *wp*
EBA Extendible Boom Assembly *oe*
EBACK End Block Acknowledge *fu*
EBC EISA Bus Controller *hi, kg*
EBCAD Electron Beam CAD *vc25.4.21*
EBCDIC Extended Binary Coded Decimal Information (Interchange) Code *wp, ct, em, do24I, jdr, fu, kg, hi, mt, ai1*
EBCT Electron Beam Computed Tomography *pk, kg*
EBG Electron Beam Generator *hc, jj*
EBI Electric Battery Initialization *fu*
EBI Equivalent Background Input *pk, kg*
EBI Exploding Bridgewire Initiator *tb*
EBI Extended Background Investigation *jdr, kg*
EBIC Electron Beam Induced Current *hc*
EBIC Electron(ic) Bombardment Induced Conductivity *hc, fu, jj*
EBL Electron Beam Lithography *hn44.14.3, fu*
EBM Electronic Battle Management *fu*

EBM Engagement Battle Manager *tb*
EBM Engineering Breadboard Model *fu*
EBMSC Enduring Battle Management Support Center *mt*
EBO Evaluation By Objective *wp*
EBOM Engineering Bill Of Material [a document] *ah, pbl, tm, jdr*
EBPA Electron Beam Parametric Amplifier *fu*
EBR Electron Beam Recording *fu*
EBR Extended partition Boot Record *ct, em*
EBS Electron Beam Switch *fu*
EBS Electron Bombarded Semiconductor (Silicon) *wp, hc, fu*
EBS Emergency Broadcast System [FEMA] *hc, wp, jj, mt*
EBS-AB [Swedish communications company] *sc4.10.3*
EBTS Electron Beam Test System *hc*
EBU European Broadcast Union *hc*
EBV Epstein-Barr Virus *wp*
EBW Electron Beam Welding *hc, jj*
EBW Explosive Bridgewire *nt*
EC (EC-130H Compass Call) [USAF] *mt*
EC (EC-18 BARIA Advanced Range Instrumentation Aircraft) [AFSC*] *mt*
EC Earth Coverage *mt, gp, 16, jdr, kf, uh*
EC Electronic Combat *mt, wp*
EC Electronic Commerce *kf, kg, y2k*
EC Electronics Card *17, sb*
EC Elevation Center *jj*
EC Elevation Control *jj*
EC Embedded Computer *mt*
EC Emitter Characterizer *fu*
EC Enable Communications *fu*
EC Engine Change *mt*
EC Engineering Change *fu, hc, oe*
EC Equipment Controller *fu*
EC Error Control *pk, kg*
EC Escadre de Chasse [(French) Aircraft Attack Wing] *12, mt*
EC Essentiality Code *fu*
EC Etched Circuit *fu*
EC European Community *aw130.13.91*
EC Evaluation Center *fu*
EC Event Controller *fu*
EC Executive Clock [Routine] *jj*
EC Executive Controller *fu*
EC Exercise Command *tb*
EC Exercise Controller *fu*
ECA Earth Central Angle *mt, kf*
ECA Earth Coverage Antenna *uh*
ECA Electrical Control Assembly *9, lt*
ECA Electromagnetic Compatibility {Analysis} *tb*
ECA Electronic Control Amplifier *jj*
ECA Enable Communications Acknowledge *fu*
ECA Engineering Change Analysis [a document] *fu, ah, hc, 17, pl, lt, ep, jj, pbl, tm, jdr*
ECA Engineering Change Authorization *mt, wp, hc, 17, pl*
ECA Equipment Condition Analysis *fu*
ECA European Communications Area [Wiesbaden, GE] *mt*
ECAB Enterprise Corrective Action Board *jdr*
ECAC Electromagnetic (Electronic) Compatibility Analysis Center *mt, fu, wp, sr9-82, hc, tb*
ECAD (Mentor . . . acquired . . . Tektronix's ECAD) *aa26.12.19*
ECAD Electrical Computer Aided Design *pbl, ah*
ECAE Engineering Change Analysis Evaluation [a document] *hc, jj, pbl, tm*
ECAL Enjoy Computing And Learn *pk, kg*
ECAM Electronic(ally) Computer Aided Manufacturing *hc, fu*
ECAMPS Environmental Compliance, Assessment and Management Programs *kf*
ECAP Electronic Circuit (Service) Analysis Program [Hughes Aircraft Company] *ful, hc, jj)*
ECAP Electronic Combat Action Plan *mt*
ECAP European Conflict Analysis Project *st*
ECAPB Engineering Unit Capability *mt*
ECAR ECA Revision *jj*
ECAT Electronic Card Assembly and Test [IBM] *pk, kg*
ECB Echelons Corps and Below *mt*
ECB Electronic Codebook *pk, kg*
ECB Event Control Block *pk, kg*
ECC (ECC International, Incorporated) [an avionics/military electronics company] *aw129.1.11*
ECC Elliptic Curve Crypto *pk, kg*
ECC Engagement Control Center *jmj, pm*
ECC Engineering Change Center *hc, lt, jj, jdr*
ECC Engineering Change Control (Coordination) [an organization] *ah, fu, 15, pl, hc, ep, tb, pbl, tm, jdr*
ECC Engineering Control Center *ep*
ECC Engineering Critical Component *hc*
ECC Error Check Code *hi*
ECC Error Checking and Correcting (Correction) *mt, em, kg, fu, hi*
ECC Error Correcting (Correction) Code *ct, em, wp, kg, fu, hi*

ECC Error Correction Circuitry *mt*
ECC Extended Core Configuration *mt*
ECCB Engineering Change Control Board *tb*
ECCB Engineering Configuration Control Board *fu*
ECCC Engineering Change Classification Concurrence *jj*
ECCCS Electronic Command, Communication, and Control System *hc*
ECCCS Emergency Command Control Communications System *hc*
ECCCS European Command and Control Communications System *mt*
ECCCS European Command and Control Console System *mt*
ECCCSE Enhanced CCC Survivability and Endurance *mt*
ECCD Electronic Cockpit Control Device *mt*
ECCFFA Escadrille de Commandement Central des Forces FranÁaises, Allemagne [central command flight squadron of the French forces in Germany] [France] *mt*
ECCL ESD Critical Checklist *jdr*
ECCM Electronic Counter Counter Measures *mt, fu, wp, hn43.14.3, 1, 17, ic1-1, hc, af72.5.135, mj31.5.15, tb, st, jj, kf, ro*
ECCMO ECCM Operator *fu*
ECCP Eielson Consolidated Command Post *mt*
ECCSL Emitter Coupled Current Steered Logic *fu*
ECCT Eros Inventory Computer Compatible Tape *hc*
ECD Electron Capture Detector *wp*
ECD Electronic Cash Disbursements *ct, em*
ECD Electronic Command Division *mt*
ECD Engineering Change Directive *-4*
ECD Engineering Computing Department *fu*
ECD Enhanced Color Display *pk, kg*
ECD Enhanced Compact Disk *pk, kg*
ECD Estimated (Expected) Completion Date *fu, wp, jj, jdr, oe*
ECD European Communications Division *mt*
ECDC Electrochemical Diffused Collector *fu*
ECDF Equipment Characteristics Data File [USA] *mt*
ECE Electrical and Computer Engineering *is26.1.86*
ECE Electrical Conversion Electronics *fu, hc*
ECE Electronic Control Enable *jj*
ECE Escadron de Chasse et d'Entrainment [fighter and trainer squadron] [France] *mt*
ECE Experimental Checkout Equipment *pl*
ECEF Earth Centered Earth Fixed *kf*
ECEMP Electron Caused Electromagnetic Pulse *17, sb*
ECEMP Electron Charging EMP *jdr*
ECET Electrical and Computer Engineering Technology *is26.1.87*
ECF Earth Center Finding *3, 8*
ECF Earth Centered Fixed *kf*
ECF Etched Circuit Fabrication *fu*
ECF Executive Control Function *fu*
ECG Electrocardiogram *wp, ac3.2-7, w9, es55.73*
ECG Emergency Coordination Group *mt*
ECG Exercise Control Group *mt*
ECG Exercise Controller Guide *fu*
ECH Earth Coverage Horn *uh*
ECHO European Commission Host Organization [Internet] *pk, kg*
ECHO EW test range at China Lake *wp*
ECI Earth Centered Inertial [coordinate system] *mt, fu, uh, ac3.2-20*
ECI Electronic Communications Instrument *hc*
ECI Engineering Change Information *hc*
ECI Engineering Change Instruction *fu*
ECI Equipment Configuration Item *mt*
ECI Extension Course Institute [AU] *mt, af72.5.72*
ECI External Call Interface *pk, kg*
ECI [a division of E-Systems, Incorporated, St. Petersburg, FL] *af72.5.32, jmj*
ECIL Electronics Corporation of India, Limited *ac3.2-20*
ECL Eddy Current Loss *fu*
ECL Emitter Characteristics Library *fu*
ECL Emitter Coupled Logic *mt, 6, hc, is26.1.29, kg, jj, fu*
ECL Equipment Configuration List *jj*
ECL Equipment Control List *wp*
ECL Executive Control Language *mt*
ECLIPSE Eclectic Capability for Logistic Information Support for Electronics *wp*
ECLSS Environmental Control and Life Support System *wp, dm, tb, wl, ai2*
ECM Electronic Control Module *pk, kg*
ECM Electronic Counter Measure *mt, hc, uc, 1, 12, 17, ic1-11, wp, sb, aw129.14.30, aa26.11.4, af72.5.8, pm, tb, st, jj, jdr, kf, fu, ro*
ECM Error Correction Mode *ct, em, 3, hi*
ECM-D Engineering Change Management-Development *hc*
ECMA European Computer Manufacturers' Association *tb, kg, ct*
ECMC Emergency Crisis Management Center *mt*
ECMC Enhanced Crisis Management Capability [PACOM / DNA] *mt*
ECMD ECM Display *jj*
ECME Electronic Countermeasures Environment *fu*
ECME Etylene Glycol Monomethylether (Methyl Cellosolve) *wp*
ECMF Electronic Combat Mission Folder *mt*
ECMJ Electronic Counter Measures Jammer *hc, jj*
ECMO Electronic Countermeasures Officer *mt*

ECMO Electronic Countermeasures Operator *fu*
ECMO Extra Corporal Membrane Oxygenation *wp*
ECMP Electronic Counter Measures Program *jj, fu*
ECMP&R EC Master Plan and Roadmap [USAF] *mt*
ECMR Eastern Contract Management Region [USAF] *hc*
ECMRITS Electronic Counter Measures Resistant Information Transmission System *aw130.6.7*
ECMWF European Center for Midrange Weather Forecasting *ns*
ECMWF [Tropical Wind Study] *ge, hg*
ECN Engineering Change Notice *mt, gp, 16, hc, jj, fu, oe*
ECNE Enterprise Certified NetWare Engineer [Novell] *hi, kg*
ECO Electron Coupled Oscillator
ECO ELINT Collection Outstation [USMC] *mt*
ECO Engineering Change Order *mt, pl, wp, hc, tb, jj, fu, oe*
ECO Engineering Checkout [phase] *re*
ECO Engineering Cognizant Office *jj*
ECO Equipment Control Officer *mt*
ECOD Estimated Cost Of Damage *wp*
ECOM Army Electronics Command *mt*
ECOM Electronic Computer Originated Mail *10, wp*
ECOM Electronics Command [US Army] *hc, jj, fu*
ECON (ECON Incorporated) *ac2.7.25*
ECOS EMSP Common Operational Software *fu*
ECP Electric Power Converter *lt*
ECP Emergency Change Package *mt*
ECP Emergency Command Precedence *mt*
ECP Engineering Change Proposal (Procedure) [a document] *mt, 0, wp, 4, 8, 17, pl, hc, ep, tb, st, jj, pbl, jdr, tm, kf, ah, fu, oe*
ECP Enhanced (Extended) Capabilities Port [Microsoft] *kg, ct, em, hi*
ECP Estimated Completion [date] *jj*
ECP External Communications Processor *mt*
ECPO Engineering Computer Processing Operation *0*
ECPS Enhanced Chemical Protection Suit [chemical warfare] *wp*
ECPTT Electronic Combat Part Task Trainer *hc*
ECR Earth Centered Rotational [coordinate system] *tb, kf*
ECR Electronic Combat and Reconnaissance [version of the Panavia Tornado] *mt*
ECR Embedded Computer Resource *hc, wp*
ECR Engineering Change Request [a document] *mt, ah, fu, hc, gp, 8, 16, 17, pl, lt, ep, tb, jj, pbl, tm, jdr, kf, oe*
ECR Engineering Change Requirement *0*
ECR Environmental Criteria Report *jj*
ECR Error Cause Removal *jj*
ECRA Electrical Contact Ring Assembly *uh, hc, pl, gp, 9, lt*
ECRA Engineering Change Request / Authorization *fu*
ECRB Engineering Change Review Board *-4*
ECRC ECR Coordination [No.] *fu*
ECRC Electron {?} Component Reliability Committee *hc*
ECRC Electronics Component Reliability Center *hc*
ECRS Embedded Computer Resource Standards *mt*
ECRU Enhanced Command Response Unit *fu*
ECS Electronic Combat Squadron [USA] *mt*
ECS Electronic Control System *mt*
ECS Emergency Coolant System *jj, fu*
ECS Emission Control System *wp*
ECS Enable Control System [Minuteman] *hc*
ECS Energy Conversion Subsystem *tb, wl*
ECS Engagement Control Station *pm*
ECS Engagement Controller Subsystem *tb*
ECS Engineering Contract Services *aw130.13.104*
ECS Environment(al) Control System *fu, mi, ci.406, wp, hc, tb, jj, jdr, ai2*
ECS Environmental Conditioning Subsystem *fu*
ECS European Communication System [Earth Station] *hc*
ECS European Communication [system] Satellite *sc2.8.2, sf31.1.17, es55.99, ce*
ECS Experimental Communication Satellite [Japanese] *oe*
ECS Experimental Communications Satellite *ac3.3-8*
ECS Exterior Communication System [US Navy] *mt*
ECS/CMS Embedded Computer System / Configuration {?} *mt*
ECSC European Coal and Steel Community *wp*
ECSS Executive Control Subordinate System [USA] *mt*
ECSS Extendable Computer Simulation System (System Simulation) *17, fu*
ECT Eddy Current Testing *wp*
ECT Escadron de Chasse et de Transformation [fighter and trainer conversion squadron] [France] *mt*
ECT Evaporative Cooling Technique *hc*
ECT Experiment Control Team *tb*
ECT Extended Coverage Telling *fu*
ECTD Electronics Command Technical Description *hc*
ECTED Electronic Combat Threat Environment Description *mt*
ECTF East Coast Test Facility *jb*
ECTL Electronic Communal Temporal Lobe *pk, kg*
ECTT Escadre de Chasse Tous Temps [All-weather fighter wing] [France] *mt*
ECU EISA Configuration Utility *pk, kg*

ECU Electrical (Electronic) Conversion Unit *0, 3, 17, sb, hc, jj*
ECU Electronic Control Unit *mt, jj*
ECU Engine Change Unit *mt*
ECU Entry Control Unit *fu*
ECU Environmental Control Unit *mt, fu, wp, ci.406, hc, jj*
ECV Electronic Combat Vehicle *hc*
ECWDS Enhanced COBRA Weapons Direction System *hc*
ECX EC-type aircraft, Experimental [follow-on aircraft for alerting strategic submarine forces] *uc, dh*
ED Edge Distance *jdr*
ED Effective Date [unit enters federal active duty] *mt*
ED Engineering Development *mt, fu, hc, wp, st*
ED Envelope Drawing *fu*
ED Erase Display *pk, kg*
ED Error Detector (Detecting) *hc, jj, fu*
ED EUCOM Directive *mt*
ED Event Dispatcher *mt*
ED Exercise Director *fu*
ED Exploratory Development *mt*
ED Extra High Density *ct, em*
ED&C Event Detection and Characterization *kf*
ED&T Experimentation, Demonstration and Test *fu*
EDA Electronic Design Automation *wp, kg*
EDA Electronic Digital Analyzer *fu*
EDA Elevation Drive Assembly *sm18-54*
EDA Embedded Document Architecture [Go Corporation] *pk, kg*
EDA Excess Defense Articles *wp*
EDAC Error Detection And Correction *mt, wp, st, jdr, kf, vv, uh, fu, hi, oe*
EDARC Enhanced Direct Access Radar Channel *fu*
EDAS Enterprise Data Access Services *jdr*
EDASRE Engineering Drawing Automated Storage and Retrieval Equipment *mt*
EDATE Effective Date *mt*
EDAX Energy Dispersive X-ray Analyzer *hc, hd*
EDB Energy Data Base [DOE] *mt*
EDB Engineering Data Base *oe*
EDB Experience Data Base *mt*
EDC Educational Development Center *mt*
EDC Electrical Discharge *wp*
EDC Electronic Digital Computer *pk, kg*
EDC Electronic Document Distribution {?} *hi*
EDC Engineering Data Center *fu*
EDC Engineering Data Control [an organization, later Product Data Services] *fu, hc, ep, jj, jdr, tm*
EDC Engineering Design Change *-8*
EDC Engineering Document Center *ep*
EDC Engineering Document Control *pl*
EDC Enhanced Data Correction *pk, kg*
EDC Eros Data Center [Sioux Falls, South Dakota] *ac2.2-13*
EDC Error Detection and Correction *wp, kg, fu, hi*
EDC Error Detection Coding (Code) *fu, hi*
EDC Estimated Date of Completion of Loading *mt*
EDC Estimated Date of Contract *jk*
EDC European Defense Community *wp*
EDC European Disarmament Conference *-14*
EDCARS Engineering Data Computer Assisted (Automated) Retrieval System [AFLC] *mt, wp*
EDCAS Equipment Designer's Cost Analysis System *fu*
EDCC Engineering Data Control Center *fu*
EDCC Enhanced Display Control Console *fu*
EDCC Error Detection and Correction Code *fu*
EDCD Engineering Data Control Department *jj*
EDCP Engineering Design Change Proposal *hc*
EDCS Engineering Data Control System *jdr*
EDCS Enhanced Display Control Station *fu*
EDCS European Defense Communications System *wp*
EDCT Expected Departure Clearance Time *fu*
EDCTP Environmental Design Criteria and Test Plan *fu*
EDD Earliest Departure Date *mt*
EDD Electron Dynamics Division [Hughes Aircraft Company] *fu, hc, hn43.12.1, lt*
EDD Electronic Data Display *fu*
EDD Estimated (Expected) Delivery Date *mt, fu, wp, jj*
EDD Estimated Departure Date *mt, jj*
EDD Experimental Destroyer [US Navy] *mt*
EDDC Extended Distance Data Cable *pk, kg*
EDDD Expanded Direct Distance Dialing *fu*
EDDF Error Detection and Decision Feedback *fu*
EDDS Electronic Data Distribution System *oe*
EDE Electronic Defense Evaluator *wp*
EDE Emergency Decelerating *fu*
EDE Environmental Data Entry *fu*
EDE Experimental Escort Ship [US Navy] *mt*
EDEE Equipment Design and Environmental Engineering *fu*
EDEN Engineering Design Network *hd1.89.y29*
EDF Elmendorf Air Force Base *mt*

EDF Engineering and Development Facility *mt*
EDF Engineering Data File *wp, jj*
EDF Engineering Design Format *wp*
EDF European Development Fund *wp*
EDG Exercise Display Group *fu*
EDG Exploratory Development Goal *fu*
EDGAR Electronic Data Gathering Analysis and Retrieval [system] *wp, kg, hi*
EDGE Electronic Data Gathering Equipment *fu*
EDI Electrical Design Integration *kf*
EDI Electronic Data Information *kf*
EDI Electronic Data Interchange [process] *mt, ah, ct, em, pbl, tm, jdr, kg, kf, hi, tt, y2k*
EDI Electronic Document Interchange [DEC] *hi, kg*
EDI Engineering Data Identifier [a designator] *ep*
EDI Engineering Data Interchange *hc*
EDI Engineering Design Identifier *jdr*
EDI Engineering Direction/Information *oe*
EDI Engineering Drawing Identifier *jdr*
EDI European Defense Initiative *mt, st*
EDIF Electronic Design Interchange Format *is26.1.37, mj31.5.358*
EDIFACT Electronic Data Interchange For Administration, Commerce and Transport *kf, kg*
EDIP European Defense Improvement Program *mt*
EDIPS EROS Digital Image Processing System *hc*
EDIS Engineering Data and Information System *hc*
EDIS Environmental Data and Information Service [NOAA] *ac2.4-9*
EDIS Executive Directorate, Industrial Security [formerly HQ DIS] *hc*
EDISILATE 1,2 Ethanedisulfonate *wp*
EDITAR Electronic Digital Tracking And Ranging *fu, hc*
EDL Edit Decision List *pk, kg*
EDL Electric(al) Discharge Laser *mt, hc, wp*
EDL Electro Dynamic Laser *hc*
EDL Expendable Data Link *fu*
EDLC Ethernet Data Link Control *hi, kg*
EDM Electrical Discharge Machining (Machine) *fu, hd1.89.y26*
EDM Emergency Defense Message *tb*
EDM Engineering Design Memorandum *hc, jj*
EDM Engineering Development Model *mt, fu, hc, hn44.19.8, jdr*
EDM Environmental Observation Mission *sc4.7.3*
EDM Exploratory Development Model *fu*
EDMICS Engineering Data Management Information and Control System *mt*
EDMIX Engineering Data Management Information Control System *hc*
EDMP Engineering Data Management Plan *mt*
EDMS Electronic Design (Data) Management System *fu*
EDMS Electronic Document Management System *em, kg*
EDMS Engineering Data Microreproduction System *fu*
EDN Emergency Data Network [AFCC*] *mt*
EDN Exercise Development Notebook *fu*
EDNET Educational Network [IBM] *jgo*
EDO (EDO Corporation) [an avionics/military electronics company] *aw130.22.49*
EDO Engineering Duty Officer [USA] *mt*
EDO Extended Data Out *ct, em, kg, hi*
EDO Extended Duration Orbiter *mi, ai2*
EDOC Effective Date Of Change *mt*
EDODRAM Extended Data Out DRAM *hi*
EDOR Early Development (Definition) Of Requirements *pbl, jdr, ah*
EDORAM Extended Data Out Random Access Memory *em, hi*
EDOS Enhanced DOS [for Windows] *pk, kg*
EDP Electron Density Profile *pbl, dm*
EDP Electronic Data Package *fu*
EDP Electronic Data Processing (Processor) *mt, fu, 0, ct, em, hc, w9, jj, kg, jdr*
EDP Emergency Defense Plan *mt*
EDP Engineering Development Program *wp*
EDP Engineering Drawing Procedure *hc, jj*
EDP European Digitization Program *mt*
EDP Event Display Process *mt*
EDPCS Engineering Design (Data) Product Control System [a data base] *17, ep*
EDPE Electronic Data Processing Equipment *mt, hc, 0, jj*
EDPF Experimental Distributed Processing Facility *hc*
EDPM Electronic Data Processing Machine *jj, kg*
EDPS Electronic Data Processing System *0, wp*
EDPS Exploratory Development Program Summary *hc*
EDPT Electronic Data Processing Test *mt*
EDR Electronic Data Requirements *kf*
EDR Engineering Design Review *sd, wp, hc*
EDR Engineering Document Record *fu*
EDR Environmental Data Record *ac7.5-9, 16, pl, dm, pbl*
EDR Equivalent Direct Radiation *fu*
EDR Exo Defense Regime *st*
EDR Experimental Data Record *ac2.7-11*
EDR External Deficiency Report *fu*
EDRA Engineering Drawing Release Authorization *hc, jj*

EDRAM Enhanced Dynamic Random Access Memory *hi*
EDRAM Erasable Dynamic Random Access Memory *em, kg*
EDRC External Deficiency Report Control *fu*
EDRCP External Deficiency Report Control Point *fu*
EDRD Electronic Data Requirements Document *kf*
EDRE Emergency Deployment Readiness Exercise *mt*
EDRIS Engineering Data Requisition and Index System *mt*
EDRL Engineering Data Records and Lists [a data base] *fu, hc, ep, jj*
EDRL Engineering Document Requirements List *fu*
EDRLS Engineering Data Records and Lists System *hc*
EDRP Evaluation Data Reduction Program *fu*
EDRS Engineering Data Release System [a data base] *ah, pl, ep, pbl, tm, jdr*
EDS Electronic Data Service *kf*
EDS Electronic Data Systems [a corporation] *mt, kg, hn49.14.1, fu*
EDS Electronic Display System *fu*
EDS Engine Diagnostic System *wp*
EDS Engineering Data System [a data base] *wp, ep*
EDS European Distribution System *mt, wp, af72.5.141*
EDS Executive Display Subsystem *fu*
EDS-FS European Distribution System Forward Stockage *mt*
EDS-LOG-CCC EDS Logistics Command, Control, and Communications *mt*
EDSA European Distribution System Aircraft [MAC] *mt*
EDSAC Electronic Delay Storage Automatic Computer *pk, kg*
EDSG Electro-optical and Data Systems Group [Hughes] *fu, hc, sc2.8.4, 8, 15, 16, 17, sb, hq5.3.5, hn49.9.1, ep, kf*
EDSGMD Electro-optical and Data Systems Group Manufacturing Division *hn41.17.1*
EDSI Enhanced Small Device Interface *pk, kg*
EDSL Extended Digital Subscriber Line *mt*
EDSM Electro-Optical Data Systems Manufacturing *-8*
EDSP Electronic Discrimination Signal Processor *hc*
EDST Eastern Daylight Saving Time *wp*
EDT Eastern Daylight Time *sc4.10.1, aw129.14.18, w9, ws4.10, kg, kf*
EDT Electrical Discharge Tube *fu*
EDT Electronic Transfer Device *fu*
EDT Engineering Design Test *bf, jj, fu*
EDT Engineering Development Test *fu*
EDT Enumerated Data Type *wp*
EDT Estimated Departure Time *wp*
EDTA EthyleneDiamineTetraacetic Acid *wp*
EDTCC Electronic Data Transmission Control Center [SAC] *mt*
EDTP Engineer Design Test Plan *jhm*
EDTR Experimental Developmental Test Research *hc*
EDTV Enhanced Definition Television *is26.1.59*
EDU Electronic Display Unit *mt, fu*
EDU Encoder-Decoder Unit *hc*
EDU Engine Diagnostic Unit *mt*
EDU Engineering Demonstration Unit *pbl*
EDU Engineering Development Unit *dm*
EDUSAT Educational Satellite *hc*
EDVAC Electronic Discrete Variable Automatic Computer [von Neumann's first stored-program digital computer] *kg, sf31.1.14*
EDVM Enhanced Design Verification Model *jdr*
EDX Energy Dispersive X-ray (analysis) *jdr*
EDX Event Driven Executive Operating System [IBM] *mt*
EE Eject Enable *jj*
EE Electrical Engineer *ac2.2-1, w9, hi*
EE Escadre Electronique [Electronic Squadron] [France] *mt*
EE Extended Edition [IBM] *pk, kg*
EEA Essential Elements of Analysis *mt*
EEA Exercise Electronics Assembly *fu*
EEC ECCM Executive Controller *fu*
EEC Electron Energy Corporation *hq5.2.2*
EEC Electronic Engine Controls *mt, aw124*
EEC Emergency Essential Civilians *mt*
EEC Engineering Executive Council *pbl, ah*
EEC European Economic Community *mt, 14, wp, aw129.14.110, w9*
EEC Extendable Exit Cone [IUS] *bo, 4, ai2*
EEC Extended Error Correction *hi, kg*
EECF Earthnet ERS-1 Central Facility *es55.44*
EECO (EECO Incorporated) [an avionics/military electronics company] *aw130.22.54*
EED Electro Explosive Device *fu, he, hc, gp, 4, 5, 16, 17, sb, nt, jj, jdr, oe*
EED Essential Elements of Data *mt*
EEE Electrical, Electronic, and Electromechanical *gp, nt, oe*
EEE Electromagnetic Environment Experiment *hc*
EEE Electromagnetic Environment(al) Effect *fu, hc*
EEE Electronic Entertainment Exposition [also E3] *hi*
EEE End to End Encryption *mt*
EEE Essential Elements of Evaluation *mt*
EEFI Essential Elements of Friendly Information [USA DOD] *su, tb, mt*
EEG Electromagnetic Environment Generator *hc, jj*
EEG Electronics Engineering Group *mt*
EEGS (improvements . . . include installation of GPS, EEGS) *af72.5.135*

EEI Edison Electric Institute *y2k*
EEI Essential Elements of Information [USA DOD] *mt*
EEIC Elements of Expense Investment Codes *mt, wp*
EEL Epsilon Extension Language *pk, kg*
EELC Electronic Equipment Liquid Cooler *jj*
EELS Electronic Emitter Location System *wp*
EELV Evolved Expendable Launch Vehicle *jdr, kf, ai2*
EEM Electron Emission Mass Spectroscopy *wp*
EEM Engineering Evaluation Module *fu*
EEM Ensembladores ElectrÛnicos de MÈxico *fu*
EEM Extended Memory Management *pk, kg*
EEMS Enhanced Expanded Memory Specification *hi, kg*
EEMSA Ensembladores ElectrÛnicos de MÈxico, S.A. [Hughes Aircraft Company subsidiary] *hd1.89.y27, fu*
EEO Elliptical Earth Orbit *ai2*
EEO Emergency Engineering Order *jdr*
EEOB Enemy Electronic Order Of Battle *mt*
EEP Ellipse of Equal Probability *fu*
EEP Elliptical Error Probable *wp*
EEPLRS Enhanced, Enhanced PLRS (VHSIC) *fu*
EEPROM Electronically Erasable Programmable Read Only Memory *ct, em, kf, hc, hd, 17, sb, af72.5.147, kg, fu, is26.1.27, hi, oe*
EER (EER Systems . . . has been awarded . . . contract) *aa26.10.55*
EER Electronic/Electrical Equipment Rack *hc, jj*
EER Energy Efficiency Rating *fu*
EEROM Electrically Erasable ROM [cf. EAROM] *fu, hi, ai1*
EES Electro Explosive Subsystem *hc*
EES Enlisted Evaluation System *mt*
EES Escrow Encryption Standard *pk, kg*
EES External Environment Simulator *hc*
EESIM End to End [kinetic kill vehicle endgame)] Simulation [END-SIM] *17, sb*
EESLC Electronic Equipment Shop Liquid Cooler *jj*
EEST Enhanced Ethernet Serial Transceiver *ct, em*
EET End to End Test *kf*
EETS Engineering Evaluation Test Station *hc*
EEU Exercise Electronic Unit *fu*
EEZ Electronic Exclusion Zone *mt*
EF (EF-111 Raven) *mt*
EF Early Finish *jdr, kf*
EF Emitter File *fu*
EF Emitter Follower *0, fu*
EF Emitter Frequency *fu*
EF Equivalent Focal length *wp*
EF External Function *fu*
EF&I Engineer Furnish and Install *fu*
EF&P Engineering, Fabrication and Planning *fu*
EF/EI External Function/External Interrupt *fu*
EFA Entry Feedback Area *fu*
EFA European Fighter Aircraft [also known as Eurofighter 2000] *mt, wp, hc, aa26.10.1, aw129.14.34*
EFA Extended File Attribute *pk, kg*
EFAMS Enhanced Fuel and Armament Management System [for the RAH-66 Comanche. Non-lifting, detachable stubs with additional attachment points] [USA] *mt*
EFAS En Route Flight Advisory Service *fu*
EFAS Engine Failure Assist System *aw118.18.42*
EFAS Enhance Fault Alarm System *mt*
EFC Effective Full Charge *hc*
EFC Electromagnetic Force Gun {?} *dh*
EFC Engineering Field Change *fu*
EFC Establish Flight Coordination *fu*
EFC Expect Further Clearance (Time) *fu*
EFCI Explicit Forward Congestion Indication *ct, em*
EFCS Electronic Flight Control System *mt*
EFCS Engineer Fuel Control System *jj*
EFD Engineering Field Divisions [US Navy] *jj, mt*
EFD {one of aeronomy science instruments on Mars orbiter} *ma40*
EFF Electronic Frontier Foundation *kg, ct, em, hi*
EFH Engine Flight Hours *wp*
EFI Electromechanical Frequency Interference *pk, kg*
EFI Electronics For Imaging *pk, kg*
EFI Exploding Foil Initiator *wp*
EFIS Electronic Flight Instrument System *mt, fu, wp, aw129.10.53*
EFISM Enhanced Footprint Improved Sensing Munition *hc*
EFIT Enhanced Fault Isolation Tester *fu*
EFL Effective Focal Length *hc, jj, kf, fu*
EFL Emitter Follower Logic *pk, kg*
EFL Equipment Focal Length *fu*
EFLM Extended Flight Line Maintenance *hc, jj*
EFM Enhanced Fighter Maneuverability [the X-31 test aircraft, USA] *mt*
EFM Environment File Maintenance *fu*
EFMP Extended File Management Package *hc*
EFO EIFEL Follow On *mt*
EFO Engineering Fabrication Order *fu*
EFOCL Enhanced Fiber Optic Communications Link *hc*

EFOV Extended Field Of View *oe*
EFP Electronic Front Panel *fu*
EFP Engineering Furnished Parts *fu*
EFPH Equivalent Full Power Hours [USA] *mt*
EFPR EF&P Stores Receiver *fu*
EFR Equipment (Electronic) Failure Report *hc, jj*
EFR External Function Request *fu*
EFS Enhanced Flight Screener [USA] *mt*
EFSE Electrical Flight Support Equipment *oe*
EFSM Ethernet / FDDI Switching Module *hi*
EFT Enhanced File Transfer *mt*
EFTA European Free Trade Association *wp*
EFTO Encoded (Encrypt(ed)) For Transmission Only *mt, hc, jj, kf*
EFTS Elementary Flying Training School [UK] *mt*
EFU European Forecast Unit *mt*
EFVS Electronic Fighting Vehicle System *pm*
EFW External Function Word *fu*
EFX Emitter File Index *fu*
EG Early Gate *jj*
EG Earth Gate *0*
EG Engineering General *pbl, ah*
EG Execution Guidance *mt*
EG Exercise Group *fu*
EG&G (EG&G Incorporated) [an avionics/military electronics company] *aw130.22.52, jmj*
EGA East German Army *-12*
EGA Engineering Assistance *hc*
EGA Enhanced Graphics Adapter *ct, em, wp, is26.1.33, kg, hi*
EGA Exhaust (Evolved) Gas Analyzer *wp, eo, re*
EGAD Electronic Ground Automatic Destruct *wp*
EGADS Electronic Ground Automatic Destruct Sequencer *0*
EGAF East German Air Force *-12*
EGAMS Evolved Gas Analysis Mass Spectroscopy *wp*
EGATR End Game Automatic Target Recognizer *tb*
EGATS Eurocontrol Guild of Air Traffic Services *aw129.21.140*
EGCS Environmental Generation Control System *hc*
EGDS Engineering Data Systems *kf*
EGDS Equipment Group Design Specifications *fu*
EGME Elevation Gimbal Mounted Electronics *hc*
EGO Eccentric Orbiting Geophysical Observatory *le27*
EGP Emergency Gear Down *jj*
EGP Exercise Generation Program *fu*
EGP Extensive Gateway Protocol *mt*
EGP Exterior (External) Gateway Protocol *ct, kg, em, hi*
EGR Exhaust Gas Recirculation *wp*
EGR Exploded Gross Requirement *fu*
EGREP Extended Global Regular Expression Print [Unix] *pk, kg*
EGRESS Emergency Global Rescue, Escape and Survival System *wp*
EGRET Energetic Gamma Ray Experiment Telescope [on CGRO] *mi*
EGS Emergency Generator System *jj*
EGS European Ground Station *kf*
EGSE Electronic (Electrical) Ground Support Equipment *kf, 4, 16, es55.7, oe*
EGT Estimated Ground Time *wp*
EGT Exhaust Gas Temperature *mt, wp, jj*
EGW Equipment Ground Wire *fu*
EGWS Enhanced General War System [OJCS] *mt*
EGYPTIAN_GOOSE [DARPA study of balloon suspended surveillance radar] *wp*
EH (EH Industries has selected the [Plessey receiver] system) *aw130.13.13*
EH Equivalent Hour *tb*
EH Escadre d'HelicoptÈres [helicopter wing units] *mt*
EH&S Environmental Health and Safety *fu, jj*
EHA Electro-Hydraulic Actuator *wp*
EHAC Escadrille d'HelicoptÈres Anti Chars [anti-tank helicopter flight squad] *mt*
EHATS Electro Hydraulic Actuator Test Set *jj*
EHD Elastohydrodynamic *gs2-9.44*
EHD Electrohydrodynamics *wp*
EHDC EMP Hardened Dispersal Communications *mt*
EHDT EROS Compatible High Density Tape *hc*
EHF Extremely High Frequency [30 - 300 GHz] *nu, fu, ah, ro, uh, wp, sc3.3.1, 0, 12, 15, 17, hc, w9, rg, go1.4.1, tb, st, jj, pbl, tm, jdr, kf, him, mt, ai1*
EHL Escadrille d'HelicoptÈres LÈgers [light helicopter flight squad] *mt*
EHLLAPI Emulator High Level Language Application Programming Interface [IBM] *pk, kg*
EHLM Eenheid voor de Herstelling van het Luchtvaart Materieel [repair unit for air equipment] [Belgium] *mt*
EHP Effective Horsepower *wp, jj, w9*
EHP Electric Horsepower *fu*
EHP Equivalent Horse Power *mt*
EHR Events History Recorder *wp*
EHS (must be labelled as acceptable by EHS) *id.82Aug04*
EHS Extremely Hazardous Substances *kf*
EHSS External High Speed Storage *fu*

EHT Electrothermal Hydrozine Thruster *ls4, 5, 28*
EHT Electrothermally Heated Thrusters *sd*
EHT Emitter Homing Technology *wp*
EHT Eye Head Tracker *hc*
EHV Electro Hydraulic Valve *wp*
EHV Extra High Voltage *fu, w9*
EI Effectiveness Index *wp*
EI Electron Impact Ionization *wp*
EI Electronic Information *kf*
EI Electronic Intelligence *tb*
EI Electronic Interface *fu*
EI End Item *hc, jj, fu*
EI External Input - Routine *jj*
EI External Interconnect *mt*
EI External Interface (Interrupt) *fu*
EI&T Emplacement, Installation and Test *fu*
EIA Electrical Interface Adapter *jr*
EIA Electrical Interference or Integration Assembly *jdr*
EIA Electronics Industry Association [formerly called RETMA] (Electronic Industries Association) *mt, fu, kf, kg, nu, ct, em, hn44.11.3, 4, hc, is26.1.59, tb, jj, hi*
EIA Energy Information Administration *mt*
EIA Environmental Impact Assessment *st*
EIA Enzyme Immuno Assay *wp*
EIA Experiment Integration Analysis *st*
EIA Extended Interaction Amplifier *hc, fu*
EIAP Environmental Impact Analysis Process *mt*
EIAP Environmental Impact Assessment Process *kf*
EIB Electronics Information Bulletin *fu*
EIC Eidal International Corporation *fu*
EIC Engineering and Integration Contractor *mt*
EIC Engineering Installation Center *fu*
EIC Equipment Identification Code *fu*
EIC External Interface Controller *fu*
EICAS Engine Indication and Crew Alert(ing) System *mt, aw129.1.69*
EICC Emergency Information and Coordination Center [FEMA] *mt*
EICD Electrical Interface Control Document (Drawing) *pl, jdr*
EICO External Instruction, Computer Operand *hc*
EID Electronic Identification Device [UK] *mt*
EID Emitter Identification Data *wp*
EID End Item Demand *fu*
EID Engineering Installation Division [AFCC*] *mt*
EIDE Enhanced Integrated Drive Electronics *ct, kg, em, hi*
EIDP End Item Data Package *jdr, ric, oe*
EIDS Electronic Information Delivery System *mt, hc, wp*
EIE Evaluation Instrumentation Exercise *fu*
EIEO External Instruction, External Operand *hc*
EIES Electron Impact Emission Spectroscopy *wp*
EIES Evaluation Instrumentation Exercise Section *fu*
EIF EIU Interface Controller *fu*
EIFEL Electronic Information CC System For the Luftwaffe [German Air Force type of CCIS] *fu, mt*
EIG Engineering and Installation Group *mt*
EIH Engineering Information Handbook *ep*
EIK Extended Interaction Klystron *fu*
EIL Electron Injector Laser *hc*
EIM End Item Manager *wp*
EIMC Electronic Image Motion Compensation *hc, jj*
EIMS Electron Ionization Mass Spectroscopy *wp*
EIMS Electronic Image Motion Stabilization *hc*
EIO Extended Interaction Oscillator *fu, hc*
EIP (EIP Microwave, Inc. . . . donated $1000 to MTT-S) *mj31.5.45*
EIP ELINT Improvement Program *wp*
EIP Emitter Identification Program (Radar) *wp*
EIP Engineering and Installation Plan *mt*
EIP Evolutionary Implementation Plan *mt*
EIP Exoatmospheric Interceptor Propulsion *st*
EIP External Input *jj*
EIP External Interface Processor *fu*
EIP/FGRT End-Item Presentation/ Finished Goods Routing Tag *fu*
EIR Electrostatic Image Reproducer *wp*
EIR Enable Interrupt Request *fu*
EIR Engineering Investigation Request *jj*
EIR Environmental Impact Review *wp*
EIR Equipment Improvement Recommendations *hc, jj*
EIR Equipment Improvement Report *fu*
EIR ERCS Interface Rack *fu*
EIR Expanded Infrared *hc, jj, fu*
EIR Explosive Incident Report *jj*
EIRP Effective (Equivalent) Isotropic Radiated Power *sc4.5.3, 2, 4, 5, 7, 9, 16, sb, hc, lt, mb, ep, ce, tb, st, pbl, tm, pbl, jdr, kf, vv, ct, uh, ah, fu, oe, mt, tt*
EIRR External Independent Readiness Review *oe*
EIRT Executive Independent Review Team *wp*
EIS Electromagnetic Intelligence System *wp*
EIS Electronic Information Services *kf*

EIS Electronic Instrument System *mt*
EIS Electronics / Engineering Installation Squadron [AFCC*] *mt*
EIS Engineering Information System *aa26.12.13, hc, tb*
EIS Environmental Impact Statement *mt, wp, st*
EIS Evaluation Information System *mt*
EIS Executive Information System *mt, wp, kg*
EIS Experiment/Economic Information System *hc*
EIS Extended Instruction Set *mt*
EISA Enhanced (Extended) Industry Standard Architecture *ct, em, wp, is26.1.32, kg, hi*
EISA {a corporation in Spain} *hn43.13.6*
EISC EUCOM Intelligence Summary Cable *mt*
EISD European Information Systems Division [USAFE] *mt*
EISDPAXRIV Engineering and Integrated Support Division, Patuxent River *jj*
EISM Electromagnetic Interference Safety Margin *oe*
EISN Experimental Integrated Switch Network *hc*
EISO Engineering and Integration Supply Office *jj*
EISP Engineering and Installation Support Plan *mt*
EISS European Intelligence System Support [EUCOM] *hc, mt*
EITT Employee Involvement Task Team *nd74.445.27*
EIU External Interface Unit *fu*
EIWG External Interface Working Group *kf*
EJ Electronic Jamming *fu*
EJ Expendable Jammer *wp*
EJASA Electronic Journal of the Astronomical Society of the Atlantic *mi*
EJRS Emergency Joint Reporting Structure *mt*
EJS Enhanced JTIDS System *mt, ph, wp, tb*
EJT Extended Joint Tests *jj*
EKA Elektronische Kampff,hrung und Aufkl‰orung [(German) Electronic Combat and Reconnaissance] [Germany] *mt*
EKD Electronic Key Distribution *mt*
EKF Extended Kalman Filter *ic2-10*
EKMS Electronic Key Management System *mt*
EKV Electromagnetic Kill Vehicle *tb, st*
EKW Equivalent Kilowatt *mt*
EL Emitter Library *fu*
EL Emitter Locator *mt*
EL Equipment Log *fu*
EL Erase Line *pk, kg*
EL Erector Launcher *kf*
ELA Ensemble de Lancement Ariane [Ariane launch complex] *ai2*
ELA Escadre de Liaison AÈriennes [France] *mt*
ELADS Early Launch Air Defense System *hc*
ELAN Emulated Local Area Network *kg, nu, ct, em, hi*
ELAN Error Logging Analysis [Honeywell] *hc*
ELAS Escadre de Liaison AÈriennes et de Sauvetage [France] *mt*
ELC Embedded Linking and Control *pk, kg*
ELD ECCM Load Device *fu*
ELD Electro Luminescent Display *hi*
ELDEC (ELDEC corporation, Lynwood, Washington) *ci.1025*
ELDO European Launcher Development Organization *hc, 13, ac3.7-1, es55.c, crl, ai2*
ELDP Executive Leadership Development Program *wp*
ELE Earth Landing Edge *uh, 3, 8*
ELE Engine Life Expectancy *wp, jj*
ELECSS Electronic Combat Satellite System *ic2-26*
ELECTRO-OPTINT Electro-Optical Intelligence [USA DOD] *mt*
ELEED Elastic Low Energy Electron Diffraction *wp*
ELEKLUFT Elektronik und Luftfahrger‰ote GMBH [Germany] *hc*
ELES Expanded Liquid Engine System [program] *cp*
ELF Ejected Lunar Flare *hc*
ELF Electronic Location Finder *wp*
ELF Element Lifting Fixture *hn51.16.1*
ELF Executable and Linking Format *pk, kg*
ELF Extremely Low Frequency [30-300HZ, frequency band used to communicate with submarines, USA] *mt, hc, ct, wp, 12, ph, w9, is26.1.76cc, kg, fu, tb, jj, hi*
ELG Exercise List Group *fu*
ELIAS Earth Limb Infrared Atomic Structure *tb*
ELIN Exhibit Line Item Number *wp, fu*
ELINT Airborne Electronic Emissions *hc*
ELINT Electromagnetic Information *mt*
ELINT Electromagnetic Intelligence *mt*
ELINT Electronics Intelligence [system] [USA DOD] *mt, ro, bm132, 11, 14, hc, af72.5.22, aw129.21.48, wp, tb, jj, kf, fu*
ELISA Electronic Intelligence Search and Analysis *wp*
ELISA Enzyme Linked Immunosorbent Assay *wp*
ELITE Executive Level Interactive Terminal *wp*
ELITE Extended Long Range Integrated Technology Evaluation *wp*
ELLPAC Elliptical Power Accumulation Technique *hc*
ELLSE Electronic Sky Screen Equipment *tb*
ELM Error Logging Memory *fu*
ELM Extended Length Message *fu*
ELMAIN Element Maintenance [a software program] *jb*
ELMS Earth Limit Measurement Satellite *hc*
ELOCAR [Hughes program name] *hc*

ELOCARS Electro Optical Collection and Analysis Reporting System *wp*

ELOP Electro Optical [Industry, Limited, Israel] *hc*

ELOS Experimental Land Observing System *hc*

ELOS Extended Line Of Sight *wp*

ELOTARSLOC Electro Optical Target Locating System *wp*

ELP Electronic Line Printer *fu*

ELPC Electroluminescent Photoconductive *fu*

ELS Earth Limb Sensor *hc, tb*

ELS Eastern Launch Site [Cape Canaveral, Florida, also known as ETR, KSC, ESMC] *uh, 4, 16, bo, mb, jdr, kf, oe*

ELS Emitter Location System *mt, wp*

ELS Entry Level System *pk, kg*

ELSA Emergency Life Support Apparatus *oe*

ELSCAN Elevation Frequency Scan Radar *hc*

ELSEC Electronic Security *mt, wp, tb, fu*

ELSET Element Set *mt, jb, jmj, ns*

ELSS Emergency Life Support System *wp*

ELSS Extravehicular Life Support System *w9*

ELT Electronic Technician *hc*

ELT Emergency Locator Transmitter *mt, ge, gp, wp, hg, oe*

ELT Enter Loop Test *fu*

ELT Enterprise Leadership Team *jr*

ELT Executive Leadership Team *ah*

ELTD Engineering Level Test Document *jj*

ELTD English Language Test Design / Documents *hc*

ELTD English Language Test Document *fu*

ELTI Elapsed Time Indicator *wp*

ELTP English Language Test Program *jj*

ELTRO (ELTRO Strahlungstechnik GMBH, Germany) *hc*

ELTRUGHES (ELTRUGHES Strahlungs Technik GMBH, Germany) *hc*

ELTS Emergency Locator Transmitters *mt*

ELV Electrically Operated Valve *fu*

ELV Expendable Launch Vehicle [NASA] *mt, uh, wp, ci.17, 16, 17, 18, sb, aa26.10.31, pl, dm, re, tb, mi, mb, oe, ai2*

ELXSI Electronics X System Integration (Silicon) [a computer manufacturer] *17, ts4-41, sb*

EM Early Midcourse [phase] *tb*

EM Effectiveness Monitor *fu*

EM Electromagnetic *5, sb, w9, rg, eo, tb, st, jj, jdr, fu, ai1*

EM Electronic Mail *mt, kg*

EM Electronics Module *fu, oe*

EM Emergency Mode *fu*

EM End of Medium *kg, fu*

EM Energy Management *mt*

EM Engineering Memorandum *jdr*

EM Engineering Model *gp, pl, hc, es55.44, jdr, kf, oe*

EM Enlisted Man *w9, jj*

EM Enterprise Model *mt*

EM Equipment Maintenance and Performance Reporting System [USA] *mt*

EM Eskadrenniy Minonosets [destroyer] [USSR] *mt*

EM Euromissile *hc*

EM Evaluation Missile *jj*

EM Expanded Memory *pk, kg*

EM-SMP Expanded Memory Signal Message Processor *fu*

EMA Electromagnetic Accelerator *st*

EMA Electromagnetic Accelerometer *hc*

EMA Electronic Mail (Messaging) Association *kg, ct, em*

EMA Electronic Missile Acquisition *fu*

EMA Engineering Method Analysis *wp*

EMA Enterprise Management Architecture *ct, kg, em*

EMACS Editing Macros [Unix] *hi, kg*

EMAPI Extended MAPI *hi*

EMAR Equipment Master Action Request *fu*

EMAS Equipment Marking Accounting System *mt*

EMAS Exercise Message Analysis System [OJCS] *mt*

EMATS Electro Mechanical Actuator Test Set *jj*

EMATS Emergency Message Automatic Transmission System *mt*

EMB Extended Memory Block [LIM/AST] *pk, kg*

EMBARC Electronic Mail Broadcast to a Roaming Computer [Motorola] *pk, kg*

EMBRAER {Empresa Brasileira de Aeronautica} (Aircraft Corp.) *af72.5.32*

EMBRATEL Empresa Brasileira de TelecomunicaÁoes [(Portuguese) Brazilian Telecommunications Enterprise] *sc2.8.1*

EMC (EMC Technology, Inc. . . . Cherry Hill, New Jersey) *mj31.5.38*

EMC E-Mail Connection *pk, kg*

EMC Early Midcourse *hc, tb, fu*

EMC Electromagnetic Compatibility *hc, wp, rm3.2.4, 4, 16, 17, sb, pl, aa26.11.31, aw130.22.87, kg, nt, mb, ep, dm, tb, st, jj, pbl, tm, jdr, kf, vv, ah, he, fu, oe, mt*

EMC Electromagnetic Coupling *mt*

EMC Electromagnetic [field] Control *gg14, 17*

EMC Electronic Material Change *jj*

EMC Electronic Mode Control *pl*

EMC Enhanced Memory Chip *pk, kg*

EMC Equipment Management Center, TRW *kf*

EMC Equivalent Mission Cycle *hc, jj*

EMC Extended Math Coprocessor *pk, kg*

EMCAB EMC Advisory Board *fu*

EMCE Electronic Mission Control Element *jdr*

EMCM Electromagnetic Counter Measures *sd*

EMCON Emissions Control (Controlled) [of electromagnetic radiations] [USA DOD] *mt, fu, sd, wp, hc, pm, tb, jj*

EMCS Electromagnetic Compatibility Simulation (Simulator) *fu*

EMCS Energy Monitoring and Control System *mt*

EMCTP EMC Test Plan *fu*

EMD Effective Miss Distance *hc, jj*

EMD Electric Motor Driven *fu*

EMD Electronic Map Display *hc, jj*

EMD Electronic Measurement and Display *hc*

EMD Electronique Marcel Dassault [France] *hc*

EMD Engine Model Derivative *aw118.14.11*

EMD Engineering and Manufacturing Development *jdr, kf*

EMD Equipment Maintenance Directive *-4*

EMD Equipment Manufacturer Design *hc*

EMDAS Expanded MINUTEMAN Data Analysis System *mt*

EMDB Equipment Maintenance Data Base *mt*

EMDEF Electronic Map Display Experimental Facility *hc*

EMDG Euro Missile Dynamics Group *mt, aa26.20.12*

EMDI Engineering Manufacturing Division Instruction *hc*

EMDP Engine Model Derivative Program *wp*

EMDS European Meteorological Data System *mt*

EME Electro Magnetic Effect *hc*

EME Electromagnetic Environment *fu*

EME Electromechanical Engineering *st*

EME Environmental Measurements Experiment *hc*

EMERCOM Federation State Committee of Emergencies [Russia] *mt*

EMERGCON Emergency Condition[s] *mt*

EMF Electro Motive Force [a line integral of electric fields on a closed loop, having units of volts] *0, wp, jj, ct, uh, fu, jdr, ai1*

EMF Electromagnetic Force [may refer to a gun] *wp, uc*

EMF Enhanced Meta file Format *ct, em, hi*

EMFA (Brazilian armed forces general staff) *aw130.22.35*

EMFM Electromagnetic Flow Meter *0*

EMG Electro Magnetic Gun *tb*

EMGRF EMG Research Facility *tb*

EMH External Message Handler *mt*

EMI Electric and Musical Industries, Limited [UK] *hc, hn44.4.4*

EMI Electro Magnetic Interface *tb, st*

EMI Electro Magnetic Interference [a phenomenon] *mt, fu, gg25, aw129.14.5, 4, 16, 17, sb, pl, hc, aa26.12.76, kg, nt, wp, mb, ep, dm, wl, jj, pbl, tm, jdr, kf, vv, ah, he, hi, oe, ai1*

EMI Electronic Memories, Incorporated [California] *hc*

EMI/EMC Electromagnetic Interference/Electromagnetic Compatibility *vv*

EMI/EQT EMI/Environmental Qualification Test *fu*

EMIC Electromagnetic Interference [and] Compatibility *wp, jj*

EMIC Electromagnetic Interference [and] Control *hc, jj, fu*

EMIDS Experiment for the Management of Information Data System *mt*

EMIEL EMI Electronics, Limited [UK] *hc*

EMIHUS (EMIHUS Limited, [UK]) *hc*

EMINT Electromagnetic Intelligence *mt, fu, jj*

EMIR Electromagnetic Interference Reduction *fu*

EMIS Electronic Mail Integration Services *hn49.14.2*

EMIS Executive (Equipment) Management Information System *wp, jj*

EMIS Exercise Message Intercept System *mt*

EMISM Electromagnetic Interference Safety Margin *hc, jj, kf*

EMIT Bit Moving Target *jj*

EMIT Emergency Message Initiation Terminal *hc, wp*

EML Electromagnetic Launcher *fu, 17, wp, sb, hc, aa26.11.6, tb, st*

EML Electromagnetic Levitation *wp*

EML EMIHUS Microcomponents, Limited *hc*

EML Engineering Mechanics Laboratory [Hughes] *ep*

EMLOC Emitter Location *fu*

EMM Expanded (Extended) Memory Manager *kg, ct, em, hi*

EMM Extended Memory Module *mt*

EMMA Engineering Maintenance Mockup Aid *jj*

EMMA Expert Missile Maintenance Aid *wp*

EMMADS Electronic Master Monitor and Display System *hc*

EMMU Extended Memory Management Unit *mt*

EMO Electronic Material Office *fu*

EMO Equipment Management Office *mt, wp, jj*

EMOC EOS Mission Operations Center *hy*

EMP Electromagnetic Pulse *mt, fu, ca.15, 1, 11, wp, 12, 14, 17, sb, hc, w9, aw130.6.6, nt, tb, st, jj, jdr, kf*

EMP Environmental Measure Package *tb*

EMPAR European Multifunction Phased Array Radar *mt*

EMPIRE Early Manned Planetary - Interplanetary Round Experiment *hc*

EMPP Electromagnetic Pulse Protection *fu*

EMPRESS Electromagnetic Pulse Radiation Environment Simulator For Ships *mt*

EMPRESS Expert Mission Planning and Replanning Scheduling System *tb*

EMPS Electromagnetic Pulse Simulator *wp*
EMPS Electronics Maintenance Publication System [USA] *mt*
EMPS Emergency Power Supply *fu*
EMPS En Route Maintenance Processor Subsystem *fu*
EMPSKD Employment Schedule [US Navy] *mt*
EMQ Electromagnetic Quiet *fu*
EMR Eastern Missile Range *5, sb*
EMR Electro Magnetic Radiation *mt, tb, kg, ct, fu, ai1*
EMR Electro Magnetic Resonance *wp*
EMR Electro Mechanical Research *hc*
EMR Electronic combat Multifunction Radar *hc, wp*
EMR Enhanced Metafile Record *hi, kg*
EMR Executive Management Responsibility *hc*
EMR Executive Management Review *fu*
EMR Explosive Mishap Report *jj*
EMRAAT Extended Medium Range Air to Air Technology *hc*
EMREL Emission Release *mt*
EMRG Electro Magnetic Rail Gun *tb*
EMRLD Excimer Moderate-power Raman-shifted Laser Device *aa26.12.25*
EMRO Electromagnetic Radiation Operation *fu*
EMRS Emergency Management Requirements Study *mt*
EMS Electromagnetic Sciences [Inc.] *oe*
EMS Electromagnetic Sounding (Sounder) *re*
EMS Electromagnetic Surveillance *wp, fu*
EMS Electronic Mail System *mt, kg, wp*
EMS Electronic Message Service *pk, kg*
EMS Emergency Management System *mt*
EMS Emergency Medical Service(s) *wp, aw130.6.54, y2k*
EMS Emergency Mission Support [system] *hc*
EMS Engine Monitoring System [US Navy] *mt*
EMS Engineering Master Schedule *wl*
EMS Equipment Maintenance Squadron *mt, wp, fu*
EMS Equipment Management System *mt*
EMS Expanded Memory Specification *kg, wp, ct, hi*
EMS Expanded Memory System *mt*
EMS Experiment Management System *st*
EMS Extended Memory Service *em*
EMSEC Electromagnetic Security *mt*
EMSEC Emanations Security *mt, tb, st*
EMSL Environmental Monitoring and Support Laboratory *ci.909*
EMSP Electromagnetic Pulse Survivability Military Standardization Program [AFSC*] *mt*
EMSP Enhanced (Embedded) Modular Signal Processor *hc, wp, uc, dh, fu*
EMSPGN EMSP Graph Notation *fu*
EMSS Emergency Mission Support System *mt*
EMSS Experimental Manned Space Station *0*
EMSTB Elevated Multi Sensor Test Bed *hc*
EMT Eastern Mediterranean *pk, kg*
EMT Elapsed Maintenance Time *jj*
EMT Electrical Metallic Tubing *fu*
EMT Electron Microscope Tomography *wp*
EMT Emergency Management Team *mt*
EMT Emergency Medical Technician *sc4.10.2, w9*
EMT Equivalent Mega Ton *tb*
EMT Executive Management Team *ah, pbl, jdr, kf*
EMTB Electro Mechanical Test Building *-9*
EMTC Electronic Module Test Console *jj*
EMTE Electromagnetic Test Environment *wp*
EMTS Electronic Module Tester System *fu*
EMU Electro Magnetic Unit [cgs] *fu, sm1-2, wp, jj, ai1*
EMU Environmental Measurement Unit *bo*
EMU Extravehicular Mobility Unit [NASA] *mt, aa26.11.29, wl, mi, mb*
EMUX Electrical Multiplexer Unit *mt*
EMUX Electronic Multiple Bus *wp*
EMUX {Electric} Power System Multiplexing *hc*
EMV Electromagnetic Vulnerability *fu, wp*
EMW Electromagnetic Window *wp*
EMW Equipment Manufacture Workmanship *hc*
EMWAC European Microsoft Windows NT Academic Centre *pk, kg*
EMWDS Expanded Missile Warning Display System *mt*
EMWG Electronic Module Working Group *hc*
EMWIN Emergency Managers Weather Information Network *oe*
EMX Electronic Support Measures *jj*
EMX Enterprise Messaging Exchange *hi*
EM_COMP Electromagnetic Compatibility *mt*
EN Estimation Number *fu*
ENAC Expanded National Agency Check *hc, tb, su*
ENACEOS Energetic Neutron Atom Camera for EOS *hy*
ENB Engineering Notebook *fu*
ENBC Engineering Notebook Coordinator *fu*
ENBL Enable *he*
ENC Emergency News Center *mt*
ENCE Enemy Situation Correlation Element *mt*
END European Nuclear Disarmament *wp*

ENDC Eighteen Nation Disarmament Committee *-14*
ENDEC Encoder / Decoder *hi, kg*
ENDEX End of Exercise *mt*
ENDO Endoatmospheric *hc, tb, st*
ENDO-NNK Endoatmospheric Non-Nuclear Kill [also ENNK] *hc, 17, sb*
ENDOSIM Endoatmospheric Non Nuclear [Kill] Simulation *hc, st*
ENDPLAN Endurability Plan *mt*
ENDS Ends Segment *pk, kg*
ENDSIM [Interceptor] End to End Simulation [also known as EESIM] *-17*
ENEL [Italian national utility company, Rome] *is26.1.58*
ENEWS Effectiveness of Navy Electronic Warfare Systems *fu*
ENFE Experimental Network Front End *mt*
ENG Electronic News Gathering *sd*
ENG Engineering-only [telemetry data stream] *mi*
ENG/EFP {type of video link} *hd*
ENIAC Electronic Numerical Integrator And Calculator [first fully electronic digital computer] *kg, fu, 10, wp, hc, sf31.1.14, nd74.445.46, ct, em*
ENJJPT Euro NATO Joint Jet Pilot Training *mt*
ENMAN Engine Management *mt*
ENMCC Expanded National Military Command Center *mt*
ENMES Engines Not Mission Capable Supply *mt*
ENNK Endo Atmospheric Non-Nuclear Kill *uc, hc, st*
ENOH Endoatmospheric Non Nuclear Optical Homing *st*
ENORS Engine Not Operationally Ready - Supply *jj*
ENOSA {a corporation in Spain} *hn43.13.6*
ENR Equivalent Noise Resistance *wp*
ENS Ensign *w9, jj*
ENSCE Enemy Situation Correlation Element [TAC] *mt, ph*
ENSCO (ENSCO, Inc. . . Springfield, VA 22151) *jmj*
ENSI Equivalent Noise Sideband Input *fu*
ENSIP Engine Structural Integrity Program [AFSC*] *wp, mt*
ENSO El NiÒo Southern Oscillation *hg, ge*
ENSS Exterior Nodal Switching Subsystem [Internet] *hi, kg*
ENSURE Emergency Non-Standard Urgent Requirements *hc*
ENTEL Empresa Nacional de Telecomunicaciones [Argentina] *hc*
ENWGS Enhanced Naval Warfare Gaming System [US Navy] *mt*
EO Electro-Optically [said of missile guidance systems and similar equipment having sensors that detect light (including infrared and ultraviolet) in their operation] *mt*
EO Electro-Optics (Electro-Optical) *dh, fu, wp, 16, 17, hc, dm, tb, kf*
EO Engineering Operations [organization] *pbl, tm, ah*
EO Engineering Order *8, hc, ep. tb, jj, pbl, tm, jdr, fu*
EO Evaluation and Operations [system IPT identifier] *kf*
EO Executive Order *mt, hc, wp, w9, tb, su*
EO-IR Electro Optical InfraRed *mt*
EO-RN Engineering Order - Revision Notice [a document] *fu, ep, tm*
EO/AA {Equal Opportunity and Affirmative Action} *aa26.11.56*
EOA Early Operational Assessment *nt*
EOA End Of Address *kg, fu, sp5*
EOAM End Of the Accounting Month *hc, jj, jdr*
EOAQ End Of the Accounting Quarter *hc, jj*
EOAR European Office of Aerospace Research *wp*
EOARD European Office of Aerospace Research and Development [AFSC*] *wp, mt*
EOB Edge (End) Of Beam *jdr, fu*
EOB Electronic Order Of Battle *mt, wp, fu*
EOB End Of Battle *sd*
EOB End Of Block *kg, fu*
EOB End Of Boost *st*
EOB Enemy Order of Battle *mt, tb, fu*
EOB Executive Office Building *mt*
EOC Early Operational Capability *mt, tb, kf*
EOC Earth Observation Center *ac3.3-13*
EOC Edge Of Clutter *fu*
EOC Edge Of Coverage *uh, jdr, oe*
EOC Element Operations Center (TT&C Centers For BE) *kf*
EOC Emergency Operations Center *mt*
EOC End Of Contract *wp, tb, st, jj, kf, fu*
EOC End Of Conversion *pk, kg*
EOC End Of Count *jj*
EOC Equivalent Operational Capability *mt, jb*
EOC Extended Operating Cycles *mt*
EOCC Experiment and Operations Control Center *tb*
EOCM Electro Optical Counter Measures *mt, hc, tb*
EOCTS Electro Optical Contact Test Set *hc*
EOD Electro Optical Division *hc, jj*
EOD Elements Of Data *fu*
EOD End Of Data *hi*
EOD End Of Day *mt*
EOD Every Other Day *wp*
EOD Explosive Ordnance Disposal [USA] *mt, hc, wp, 12, tb, st, jj*
EODAD End of Data Address *fu*
EODG Explosive Ordnance Disposal Group *wp*
EODSG Electro-Optical and Data Systems Groups *fu*

EOE Early Operational Experiments *jr*
EOE Expected Operational Environment *fu*
EOF Emergency Operating Facility *mt*
EOF End Of Field *fu*
EOF End Of File *mt, ct, em, kg, fu, wp, jj, hi, oe, ai1*
EOF End Of Fire *st*
EOF End Of Flight *st*
EOF End Of Frame *wp, jj*
EOF Enterprise Objects Framework [Next Computer] *pk, kg*
EOFCS Electro Optical Fire Control System *hc, jj*
EOG Electro-Oculogram *ci.667, is26.1.18*
EOGB Early Optical(ly) Guided Bomb *mt, wp*
EOH Engine Overhaul *mt*
EOHY End Of the Half Year (Calendar) *hc, jj*
EOI Electro Optical Instrumentation [division of SBRC] *gp, 16*
EOI End Of Interrupt *pk, kg*
EOI End Or Identify *pk, kg*
EOI Equipment Operating Instruction *fu*
EOI Expression Of Interest *mt*
EOIATS Electro Optical Identification And Tracking System *hc*
EOJ End Of Job *pk, kg*
EOL Electro Optical Laboratory *hc*
EOL End Of Life *ah, uh, he, gg4, 3, 4, 8, 9, 16, pl, lt, re, jdr, kf, oe*
EOL End Of Line *mt, kg, fu*
EOL End Of List *pk, kg*
EOL Expected Operating Life *wp*
EOLC End Of Line Code (words are added) *gd3-6*
EOLE Earth Orbiting Lab Equipment *hc*
EOLM Electro Optical Light Modulator *wp, fu*
EOM Electro Optical Modulator *fu*
EOM End Of Message *mt, sp5, wp, 0, jj, kg, fu*
EOM End Of Mission *ma20, ws4.10, re, jdr*
EOM End Of Month *w9, wp, jj, fu*
EOM Equations Of Motion *wp*
EOMS End Of Message Sequence *fu*
EOM_RATGT End Of Mission, Record as Target *fu*
EOOW Engineering Officer Of the Watch [US Navy] *mt*
EOP Emergency Operating Procedure *mt*
EOP Emergency Oxygen Pack *wp*
EOP End Of Program [computer] *hi*
EOP End Of Project *pl*
EOP Executive Office of the President *ci.868*
EOP Experiment of Opportunity Payload *ci.627*
EOP External Output *fu*
EOPAP Earth and Ocean Physics Applications Program *ac2.4-1*
EOPP Earth Observation Preparatory Programme *es55.41*
EOQ Economic Order Quantity *mt, 18, wp, fu*
EOQ End Of Quarter *hc, tb, jj, jdr*
EOR Earth Orbit Rendezvous *ci.386, 12*
EOR Emergency Override *fu*
EOR End Of Record *jj*
EOR End Of Run *fu*
EOR End Of Runway *mt*
EOR Environmental Operational Routines *fu*
EOR Estimates Of Recuperability *fu*
EOR Exclusive OR [Also XOR] *gp, kg, hi*
EOR Explosive Ordnance Reconnaissance *wp*
EORA Explosive Ordnance Reconnaissance Agency *wp*
EORL End-of-Run Locator *fu*
EORSAT Electronic (ELINT) Ocean Reconnaissance Satellite [USSR] *mt, fu, 14, sd, aj89.9.40, wp, tb*
EOS Earth Observation Satellite (Observing System) *ac7.4-1, hc, aa26.12.44, aw130.11.37, no, tb*
EOS Earth Observing System [NASA, ESA, NASDA, CSA] *ge, kg, mi, wp, ns, hy, wl. jdr, hg, mb, kf*
EOS Electrical Over Stress *hc, is26.1.76, jdr*
EOS Electro Optical Systems *wp, kf*
EOS Electrophoresis Operations in Space *wl*
EOS Eligible for Overseas *mt*
EOS End Of Scan *oe*
EOS End Of String *pk, kg*
EOS Equation Of State *st*
EOS/DIS Earth Orbiting Satellite Distribution System *hc*
EOSA Explosive Ordnance Safety Arming *jj*
EOSAEL Electro Optic Sensors Atmospheric Effects Library *wp*
EOSAT Earth Observation Satellite [Company] *hn44.12.1, hy, kf*
EOSDIS Earth Observing System Data and Information System [NASA] *hy, hg, ge, kg*
EOSP Earth Observing Scanning Polarimeter *hy*
EOSPC Electro Optical Signal Processing Computer *hc*
EOSR Electro Optical Solar Reflectors *oe*
EOSS Electro Optical Sensor System *wp, aw126.20.26, hc*
EOSS Electro Optical Surveillance System *hc*
EOSS Engineering Operational Sequencing System *hc*
EOSSRP EOS Scientific Research Program *hg, ge*
EOT Early On orbit Test *kf*

EOT End Of Table *pk, kg*
EOT End Of Tape [a marker] *16, jj, fu, kg, hi*
EOT End Of Task *mt*
EOT End Of Text *pk, kg*
EOT End Of Track *st*
EOT End Of Transfer *ct, em*
EOT End Of Transmission *mt, sp5, jdr, kg, fu, hi*
EOT Environmental Operational Test *fu*
EOT Equation Of Time *gp*
EOTADS (the mission processor, the EOTADS) *aw129.10.57*
EOTC Electro Optic Test Chamber *wp*
EOTF Early On-orbit Test Facility *kf*
EOTS Electro Optical Threat Sensor *hc, wp*
EOTV Expendable Orbital Transfer Vehicle *wp*
EOVAC Electro Optical Vulnerability Assessment Code [model] *wp*
EOW End Of Week *wp*
EOW End Of Word *jr*
EP Earth Panel *oe*
EP Earth Penetrator *wp*
EP Electric Panel *fu*
EP Electric Propulsion *ls4, 5, 28*
EP Electrical Performance *oe*
EP Electrically Polarized *fu*
EP Electronic Protection *fu*
EP Electrophotographic Engine *pk, kg*
EP Engagement Planning *jb*
EP Engagement Point *fu*
EP Engineering Procedure (Practice) [a directive] *fu, jr, ah, fu, hc, 16, ep, jj, pbl, tm, jdr*
EP Engineering Prototype *hc*
EP Entry Protected *jj*
EP Equivalent Part *pl*
EP Equivalent Person *jr*
EP Error Probable *hc*
EP Estimated Position *mt, w9*
EP Exceedance Processing *kf*
EP Execution Planning *mt*
EP Exercise Plane *fu*
EP Explorer Platform *kf*
EP&D Electric Power and Distribution *fu*
EP-L Electro Photographic-L cartridge *hi*
EPA Economic Price Adjustment *hc, 18, fu*
EPA Environmental Protection Agency *mt, fu, hc, wp, ac1.5-15, w9, nt, hn50.8.i, hy, tb, su, y2k*
EPA Evasive Plan of Action *mt*
EPA Exoathmospheric Penetration Aid *wp*
EPA Extended Planning Annex *wp, st*
EPA Extended Power Aging [Burn-in] *bl2-9.50*
EPA/ES Environmental Protection Agency / Energy Star compliant *hi*
EPA/VVS Ecole de Pilotage AvancÈ / Voortgezette Vliegopleidings School [(French / Flemish) advanced flying school] [Belgium] *mt*
EPAC East Pacific *mt*
EPADS Engineering Product Assurance Data Support *fu*
EPARCS Enhanced Perimeter Acquisition Radar Attack Characterization System *mt*
EPAS Exercise Production and Analysis System [OJCS] *mt*
EPAT Earliest Predicted Arrival Time *fu*
EPAT Earliest Probable Arrival Time *mt*
EPAW Enhanced Post Attack WWMCCS *mt*
EPB Engineering Process Bulletin *hc, jj*
EPBX Electronic Private Branch Exchange *fu*
EPC Electr(on)ic Power Conditioner *uh, sm18-23, lt, ep, jdr, kf*
EPC Engagement Planning Control *jb*
EPC Error Protection Code *pl*
EPC Etage Propergol Cryogenique [Ariane 5 core stage] *ai2*
EPC Experiment Package Console *pl, hc*
EPCRA Emergency Planning and Community Right-to-know Act [of 1986] *kf*
EPCU [payload preparation complex] *ai2*
EPD Earliest Possible Date *0*
EPD Electric Power Distribution *fu*
EPD Energetic Particle Detector *jp*
EPD Engineering Planning Document *0*
EPD Environmental Protection Devices *hc*
EPD European Programming Document [USAFE] *mt*
EPDB Extended Projects Data Base (Program) *fu*
EPDG Execution Plan Data Generation *mt*
EPDM Ethylene Propylene Diene Monomer *lt*
EPDS Electrical Power Distribution Subsystem *jdr*
EPDS ELINT Processing and Dissemination System *mt*
EPDU Extended Power Distribution Units *fu*
EPE Energetic Particle Explorer *ci.724*
EPE Enhanced Performance Engine *wp*
EPE Environmental Proof Engineering *fu*
EPE/EVS [Belgium] Ecole de Pilotage Elementaire / Elementaire Vliegschool [(French / Flemish) elementary flying school] *mt*

EPEC Electric Programmer, Evaluator, Controller *hc*
EPES Emergency Procedures Expert System *tb*
EPF Extended Payload Fairing *ai2*
EPG Electronics Proving Ground [US Army] *12, mt*
EPG Emergency Power Generator *wp*
EPG European Participating Governments *wp, af72.5.136*
EPG Exercise Planning Guidance [JCS] *mt*
EPG Exercise Preparation Group *fu*
EPG [United Video] {Cable TV Network} *sc3.9.3*
EPI Earth Path Indicator *ci.348*
EPI Electronic Position Indicator *fu*
EPI European Participating Industries *wp*
EPI Expanded Position Indicator *fu*
EPIC Electronic Privacy Information Center *ct, em*
EPIC Electronic Properties Information Center *fu, hc*
EPIC Extended Performance and Increased Capability *hc*
EPIC [a northern coast based environmental group] *hn50.8.i*
EPICS Evaluation of Payoff for Infrared Countermeasures Suppression (System) [model] *hc, wp*
EPIRB Emergency Position Indicating Radio Beacon *mt, gp, 16, ge, hg, oe*
EPIRSS Extended Performance Infrared Search (Surveillance) System *fu, hc*
EPITS Essential Program Information, Technologies and Systems *kf*
EPKM [Long March 2]E Perigee Kick Motor *ai2*
EPL Effective Privilege Level *pk, kg*
EPL Electronic Intelligence Parameter Limits List *mt*
EPL Emitter Parameter List *fu*
EPL Equipment (Engineering) Parts Lists *hc, jj*
EPLD Electrically Programmable Logic Device *hi, kg*
EPLRS Enhanced (Extended) Position Locating [and] Reporting System [USMC] *mt, fu, wp, aa26.12.27, hc, aw129.10.59, hn49.11.5, pm, kf*
EPM Earth Pointing Mode *jdr, uh*
EPM Electromagnetic Protection Modification *fu*
EPM Electron Probe Microanalysis *wp*
EPM Electronic Parts Manual *hc*
EPM Engineering Procedures (Processes) Manual *ep, jj*
EPM Enhanced Editor for Presentation Manager [IBM] *pk, kg*
EPM Enterprise Process Management [IBM] *pk, kg*
EPMIS Emergency Preparedness Management Information System [DCA] *mt*
EPMS Enlisted Personal Management System *hc*
EPMU Enhanced Precision Measurement Unit *fu*
EPO Emergency Power Off *fu*
EPOCH [A COTS software product] *kf*
EPOP European Polar Orbiting Platform *hg, hy, ge*
EPOS Electronic Point Of Sale *mt*
EPOW Enemy Prisoners Of War *mt*
EPP Emergency Power Package *wp*
EPP Enhanced Parallel Port *ct, kg, em, hi*
EPP European Polar Platform [ESA] *hg, ge*
EPP Executive Plan Program *hc*
EPPS Electronic Page Printing System *mt*
EPQ Economical Procurement Quantity *fu*
EPR Economic Production Rate *un*
EPR Electron Paramagnetic Resonance *fu, wp, ai1*
EPR Engine Pressure Ratio *mt, aw118.18.42, jj*
EPR Enlisted Performance Report [USAF] [USA] *mt*
EPR Equipment Performance Report *fu*
EPR Equipment Problem Report *fu*
EPR Ethylene-Propylene Rubber *oe*
EPR Experimental Packet Radio *fu*
EPRA Electronics Production Resources Agency *hc*
EPRI Electric(al) Power Research Institute *is26.1.57, y2k*
EPROM Electrically Programmable Read Only Memory [hardware] *kg, tm, hi*
EPROM Erasable Programmable Read Only Memory [hardware] *kg, ct, em, ah, fu, hc, wp, is26.1.47, ep, tb, jj, pbl, jdr, hi, mt, ai1*
EPS Electrical Power (Sub) System *hc, bo, wl, vv, he, fu, oe*
EPS Electrical Power Source (Supply) *wp, dm, jj, fu*
EPS Electronic Power System/Supply *hc, jj, kf*
EPS Encapsulated Post Script *jr, ct, em, hi*
EPS Energetic Particle Sensor *ggIII.2.69, 16, oe*
EPS Engineering Performance Standards *mt, jj*
EPS Engineering Process System *pbl, ah*
EPS Environmental Protection System *wp*
EPS Etage Propergol Stockable [Ariane 5 upper stage] *ai2*
EPS Exercise Production System *mt*
EPSE Experimental Packet Switched Service *hi*
EPSF Encapsulated PostScript File *hi, kg*
EPSF Expenditure Per Sortie Factor *mt*
EPSPL Encapsulated PostScript Programming Language *hi*
EPSS Experimental Packet Switching System [UK] *do132*
EPT Egress Procedures Trainer *wp*
EPT End to end Performance Testbed *kf*
EPTB Environmental Product Test Bed *pbl*
EPTS Enhanced PLRS Test Set *fu*

EPTS Enhanced Portable Test Set *fu*
EPTU Enhanced Portable Test Unit *fu*
EPU EAROM Programming Unit [TID] *jj*
EPU Electrical Power Unit *jj, fu*
EPU Emergency Power Unit *mt*
EPU Executive Processing Unit *fu*
EPUU Enhanced PLRS User Unit *mt, fu*
EPW Earth Penetrating Weapon *wp*
EPW Electrical Pulse Width *uh*
EPW Enemy Prisoner of War *mt, wp*
EPWR Emergency Power *jj*
EPY Expanded Planning Yard *mt*
EQ Example Query *mt*
EQBM Engineering/Qualification Back-up Model *hc*
EQE Equivalent Quantum Efficiency *wp*
EQL Environmental Quality Laboratory [Caltech] *cca*
EQL Extended Query Language *fu*
EQM Engineering Qualification Model *jdr, kf*
EQNCON Equation and PIN List Document Program *hc*
EQS Equality Search *jj*
EQSE Engineering Quality Support Engineer *fu*
EQT Environmental Qualification Test *fu*
EQUATE US Army Standard Automatic Test Equipment *fu*
ER Early Release *pl*
ER Eastern Range [was ETR] *mb, mi, kf, oe*
ER Echo Ranging *fu*
ER Electric Radiation *fu*
ER Electrical Resistance *fu*
ER Electro Rheological *aa26.10.42*
ER Electronic Reconnaissance *wp, fu*
ER Emergency Receiver *mt*
ER Employee Requisition *fu*
ER Engineering Report *hc*
ER Engineering Request *hc, fu*
ER Engineering Review *hc, gp, 16, fu*
ER Enhanced Radiation *mt*
ER Enhancement Request *fu*
ER Entity Relationship *mt, em*
ER Ephemeris Report *uh*
ER Equipment (Electronic) Requirements *jj*
ER Equipment Repair *mt*
ER Escadre de Reconnaissance [reconnaissance wing units] [France] *mt*
ER Established Reliability *gp, 16, 17, jj, fu*
ER Exchange Ratio *wp*
ER Expeditious Repair *jj*
ER Expense Report *15, jj, jdr*
ER Extended Range *mt, hc, aw129.21.129*
ER/RC Extended Result / Response Code *hi, kg*
ERA Engineering Release Authorization *hc, jj*
ERA Engineering Research Associates *-10*
ERA Entity Relation Attribute *fu*
ERA Extended Registry Attributes *pk, kg*
ERADCOM Electronics Research And Development Command [US Army] *mt, fu, hn43.11.4, 12, hc, tb*
ERAFSSO Emergency Reaction Air Force Special Security Office *fu*
ERAM Extended Range Anti armor (Antitank) Mine *uc, hc*
ERAM Extended Range Antiarmor Munition *wp*
ERAM Extended Range Aviation Munition *wp*
ERAM Extended Range Mine *wp*
ERAP Error Recording and Analysis Procedure *wp*
ERAPS Expendable Reliable Acoustic Path Sonobuoy *hc*
ERAS Electronic Routing and Approval System [Hughes Aircraft] *pk, kg*
ERASE Electromagnetic (Electronic) Radiation Source Elimination (Eliminator) *hc, wp*
ERB Earth Radiation Budget [instrument] *ge, ac2.8-3, hy, hg, oe*
ERB Engineering Review Board *ah, fu*
ERB Entry Request Block *fu*
ERB Executive Review Board *wp*
ERBE Earth Radiation Budget Experiment [NOAA, NASA] *ge, ci.634, hc, ns, hy, hg*
ERBM Extended Range Ballistic Missile *hc, fu, jj*
ERBS Earth Radiation Budget Satellite *hc, ns, hy*
ERBSS Earth Radiation Budget Satellite System [USA] *ge, plm, hg*
ERC Electronics Research Center [NASA] *ci.xv*
ERC Engineering Readiness Center *mt*
ERC Evaluation Research Corporation *mt*
ERCC Error Checking and Correction *fu*
ERCS Earth Referenced Cartesian coordinate System *hc*
ERCS ECM Resistant Communications System [NATO] *mt, go3.1-1, fu*
ERCS Emergency Rocket Communications System *mt, 12, tb*
ERD Entity Relationship Diagram *mt*
ERD Environmental Requirements Document *jdr, kf*
ERDA Energy Research and Development Agency [formerly AEC, later DOE] *mt, su, wp, ci.895, hc, jj*
EREP Earth Resources Experiment Package *ac2-i*
EREVS Extended Requirements Engineering and Validation System *fu*

ERF Error Function *wp*
ERF European Redistribution Facility *mt*
ERFM En Route Flow Management *fu*
ERGS Eastern Relay Ground Station *kf*
ERI Extended Range Interceptor *hc, jj*
ERIC Educational Resources Information Center *hc, jj, kg*
ERICS Emergency Rocket Communications Subsystem *tb*
ERILCO Exchange of RFI In Lieu of Concurrent Overhaul *jj*
ERIM Environmental Research Institute of Michigan *pl*
ERINT Extended Range Intercept Missile *wp*
ERINT Extended Range Interceptor Technology *mt, st*
ERIS Exoatmospheric Reentry [-vehicle] Intercept[or] [Sub-]System *hc, wp, 5, 17, sb, aw129.14.25, ae151.52.459, tb, st, fu*
ERISA Employee Retirement Income Security Act *pl, y2k*
ERL Environmental Research Laboratory [Boulder, Colorado] *gd1-4, aa26.11.b23, no, oe*
ERL Extended Relational Language *mt*
ERLL Enhanced Run Length Limited *hi, kg*
ERM Earth Return Module *hc*
ERM Ejection Restraint Mechanism *gp, 16*
ERM En Route Metering *fu*
ERM Error Reference Manual *fu*
ERMA Electronic Recording Method, Accounting [General Electric] *pk, kg*
ERMC Europe Regional Monitoring Center [DDN] *mt*
ERMIL Established Reliability Military *fu*
ERMU Experimental Remote Maneuvering Unit *ci.970*
ERM_II En Route Metering II *fu*
ERN Effectivity Revision Notice *jj*
ERN Engineering Release Notice *wp, jj, oe*
ERNO Entwicklungsring Nord [part of the MESH consortium] *hc, sc2.8.4, es55.45*
ERO Engine Running On / Off Load *mt*
EROM Erasable Read Only Memory *kg, fu, hi*
EROPS Extended Range Operations *mt*
EROS (emergency oxygen systems . . . by EROS) [a subsidiary of Intertechnique] *aw130.13.108*
EROS Earth Resources Observation System [US Geological Survey] *kg, ac1.2-5, ns, hy, oe*
EROS Electronic Retail Online Services *hi*
ERP ECM Resistant Communications System [NATO] *mt*
ERP Effective Radiated Power *mt, hc, fu, wp, 17, sb, tb, st, jj, jdr, kf*
ERP Emergency Requirements Plan *mt*
ERP Error Recovery Procedure *wp, hi*
ERP Exercise Reference Position *fu*
ERPA Evader Replica Penetration Aids *hc*
ERPC Emerson Radio and Phono Corporation *hc*
ERPD Effective Radiated Power Density *fu*
ERPL Engineering Requirements Parts List *fu*
ERPSL Essential Repair Parts Stockage List *fu*
ERR Engineering Release Record *hc, jj, kf, fu*
ERR Equipment Release Record *fu*
ERRC Expendability, Recoverability, Repairability Capability (Category) [code] *hc, wp, jj*
ERRC Expendability, Recoverability, Repairability Code *mt*
ERRC Expendability, Recoverability, Repairability Cost *fu*
ERS Earth Reference System *jj*
ERS Earth Remote Sensing Satellite [ESA] *hg, ge*
ERS Earth Resources Satellite [Japan] [as in ERS-1] *ge, mi, ac3.33-1, hc, wp, hg, mb*
ERS Emergency Relocation Site *mt*
ERS Environmental Research Satellite *aa29.10.10, pl*
ERS Equipment Requirement Specification *fu*
ERS ESA Remote Sensing [satellite] *es55.2, hy*
ERS Evaluation Record Sheet *jj*
ERS Expanded Radar Service *fu*
ERS Experimental Radar System [USAF] *mt*
ERSA Extended Range Strike Aircraft [USA] *mt*
ERSIR Earth Resources Shuttle Imaging Radar *hc*
ERSS Extended Range Surveillance System *hc*
ERT Emergency Response Team [FEMA] *mt*
ERT Engineering Release Ticket [a document] *fu, hc, ep, jj, pbl, tm, jdr*
ERT Equipment Repair Time *fu*
ERT Execution Reference Time *mt*
ERTS Earth Resources Technology Satellite [later called Landsat] *fu, hc, ac1.1-1, jj*
ERU Emergency Recovery Utility *kg, ct, em*
ERU Engineering Release Unit *oe*
ERU Equipment Replaceable Unit *hc, jj*
ERV Earth Return Vehicle *re*
ERV ECM-Resistant Voice *fu*
ERV Escadre de Ravitaillement en Vol [Squadron for Refueling in Flight, France] *12, mt*
ERVIS Exoatmospheric Reentry Vehicle Inteceptor Subsystem *tb*
ERVS Emergency Routine Verification SHO *-4*
ERW Explosive Radial Warhead *tb*
ERWTS Enhanced Return Wave Tracker System *hc*

ES (IP, ES, PSI, EX) Earth Sensor *bl1-5.127*
ES Early Start *jdr*
ES Earth Sensor *uh, oe*
ES Earth Station *mt, 2*
ES Echo Sounding *fu*
ES Ecole de Specialisation [Specialization School, France] *mt*
ES Electronic Support *mt*
ES Electronic Switching *fu*
ES Energy Star compliant *hi*
ES Engineering Standards [an organization] *ah, pbl, tm*
ES Engineering Studies *fu*
ES Equipment Specification *hc*
ES Ethernet Station *hi*
ES Exercise Scenario *mt*
ES Exercise Specialist *mt*
ES Exercise Supervisor *mt*
ES Exit Subroutine *fu*
ES Expert System *mt, 19, hi*
ES Extended Segment *wp*
ES Extensibility Scenario *fu*
ES Extra Segment *fu, kg*
ES-IS End System - Intermediate System *mt*
ESA Earth Scanner Assembly *hc*
ESA Earth Sensor Assembly *he, uh, id4142-10/831, 4, 16, pl, cl, jdr, oe*
ESA Electrical Surge Arrester *fu*
ESA Electronic Shelter Assembly *fu*
ESA Electronic Surge Arrestors *mt*
ESA Electronic Switch Assembly *jdr*
ESA Electronically Scanned Antenna *hc, hq6.2.7, jj, fu*
ESA Electronically Scanned Array *17, wp*
ESA Electronically Steerable Array [Radar] *hc, jj*
ESA Engineering Support Activity [USA] *mt, jj, fu*
ESA Enterprise Systems Architecture *ct, kg, em, hi*
ESA European Space Agency *hc, sc2.8.3, 13, 14, aa26.12.44, sf31.1.9, aw129.10.27, es55.1, so, ns, no, cea, hy, wl, hg, mb, ge, mi, kg, fu, oe, mt, crl, ai2*
ESA European Switching Center *mt*
ESA Explosive Safe Area *bo, 0, 9, hc, jj*
ESA/HCI Earth Sensor Assembly/horizon Crossing Indicator *he*
ESAAP EHF Satellite Adaptive Array Processor *hc, ro*
ESACCS Earth Stabilized Aircraft Centered Coordinate System *hc*
ESAD Earth Science and Applications Division *hy*
ESADS Earth Science and Applications Data System *hy*
ESAIRA Electronically Scanning Airborne Intercept Radar Antenna *hc, fu*
ESAM [name for a Short to Medium Air Defense System] *dh*
ESAMS Enhanced Surface to Air Missile Simulation [model] *wp*
ESANET ESA [communications] Network *es55.51*
ESAR Economic Synthetic Aperture Radar *wp*
ESAR Electronically Steerable (Scanned) Array Radar *fu, ro, hc, wp*
ESAS Electronic Solar Array Simulator *hc*
ESAU Expert System AOB (Air Order of Battle) Update [RADC] *hc, mt*
ESBM Equipment Section Boost Motor *ai2*
ESC Earth Sensor Compensation *oe*
ESC EISA System Component *pk, kg*
ESC Electronic Security Command [Kelly AFB] [USAF] * *mt, fu, hc, af72.5.47, wp, tb*
ESC Electronic Switching Center *mt*
ESC Electronic Systems Center *jdr*
ESC Engineering Service Circuit *ct, em*
ESC Engineers Study Center [US Army] *dt*
ESC Etage Superier Cryogenique [Ariane 5 cryogenic upper stages] *ai2*
ESC Executive Steering Committee *aa26.12.34*
ESC/P Epson Standard Code for Printers *hi, kg*
ESCA Electron Spectroscopy for Chemical Analysis *wp, jdr*
ESCAM Enhanced SORTS Capability Assessment Module *mt*
ESCAT Emergency Security Control Air Traffic *mt*
ESCES Experimental Satellite Communication Earth Station [India] *hc, ac3.2-2*
ESCM Extended Services Communications Manager [IBM] *hi, kg*
ESCON Enterprise System Connection [Architecture] [IBM] *pk, kg*
ESCORT An Airborne ELINT System *wp*
ESCP ECM Status and Control Panel *fu*
ESCS Emergency Satellite Communications System [services] *mt, hc*
ESCT ECM Status and Control Terminal *fu*
ESD Electronic Software Distribution *kg, ct, em*
ESD Electronics Systems Division [Hanscom Air Force Base] [AFSC*] *mt, fu, ro, hc, tc, 5, 12, 17, sb, ts.4, pf.f88.23, wp, tb, as*
ESD Electronics Systems Division [of Singer Corporation] *aw129.21.26*
ESD Electrostatic Discharge [phenomenon] *ah, he, uh, gp, 5, 8, 16, 17, sb, hn47.7.2, pl, hc, is26.1.76, jdr, wp, ep, st, kf, kg, fu, oe*
ESD Electrostatic Sensitive Device [hardware] *ah, fu, pbl, tm, jj, oe*
ESD Energy Systems Division [Westinghouse] *ci.972*
ESD Environmental Service Division *dt*
ESD Equipment Statistical Data *jj*
ESDA Electrostatically Deployed Antenna *hc*
ESDC Electrostatic Discharge Control *fu*

ESDI Enhanced Small (System) Device Interface *kg, ct, em*
ESDP Emergency ShutDown Procedure *jdr*
ESDP Emergency Shutdown Processor *ric*
ESDS Electrostatic Discharge Sensitive *fu*
ESE Earth Sensor Electronics *3, lt*
ESE Electronic Security Europe *mt*
ESE Electronic Support Equipment *tb*
ESEA El Segundo Employers' Association *hn43.14.6*
ESES Earth Moon Space Exploration Study *hc*
ESF Electronic Support Fund *wp*
ESF Electrostatic Focusing *fu*
ESF Explosive Safe Facility *ggIII.1.31, 16, hc, jj, oe*
ESF Extended Superframe [Format] *kf, kg*
ESFPS Earth Stabilized Fixed Point System *hc, jj*
ESG (ESG Aviation Services monitors the industry) *aw130.11.18*
ESG Electronic Scene Generator *hc*
ESG Electronic Security Group [ESC*] *mt*
ESG Elektronik Systems Gesellschaft [Germany] *hc*
ESG Eltrughes Strahlungstechnik GMBH [Germany] *hc*
ESG Energy Storage Group *kf*
ESG Engineering Support Gated *hc*
ESG Expanded Sweep Generator *fu*
ESGN Electrostatically Suspended Gyro Navigator *fu*
ESGP Earth Science Geostationary Platform *hy*
ESHP Equivalent Shaft Horse Power *jj*
ESI (ESI Publications, . . . , Paris, France) *aw130.13.104*
ESI Earth Station Interface *csII.96*
ESI Electrical Specialties, Incorporated *hc*
ESI Electromagnetic Susceptibility to Interference *fu*
ESI Electrospace Systems, Inc *fu*
ESI Enhanced Serial Interface [specification] [Hayes] *pk, kg*
ESI Essential Sustainment Items *mt*
ESI Extremely Sensitive Information *mt, wp*
ESIAC Electronic Satellite Image Analysis Center *wp*
ESIAS (a Hughes developed shell ESIAS) *go1.5.3*
ESICCS Earth Stabilized Interceptor Centered Coordinate System *hc*
ESID European Information System Division *mt*
ESIMS Engineering Services Information Management System *mt*
ESIP Embedded computer resources Support Improvement Program *mt, hc*
ESIS European Space Information System *es55.64, mi, mb*
ESJ Escort Jamming *fu*
ESJ Escort Screening Jammer *pm*
ESKE Enhanced Station Keeping Equipment *af72.5.142*
ESL Engineering Stop Lift *fu*
ESL Equipment Status List *jj*
ESLOC Equivalent Source Lines Of Code *kf*
ESM Edible Structural Material [proposed for the Lunar Module of the Apollo missions, but not adopted—ESM would have been made of powdered milk, corn starch, wheat flour, hominy grits, and banana flakes, baked at high pressure, USA] *mt*
ESM Electromagnetic (Electronic) Support Measures *fu, ic2-30, uc, aw129.14.109, hc, pf1.10, wp, tb, jj, hc*
ESM Electronic Signal Monitoring *mt*
ESM Electronic Support Measures [NATO] *mt*
ESM Electronic Surveillance Measures *fu, pm, mt*
ESM Energy Storage Module *jr*
ESM Engineering Services Memorandum *fu, hc, jj*
ESM Equipment Support Module *pl*
ESM Experiment Support Module *5, 17, sb*
ESMA Expert System Maintenance Aid *wp*
ESMC Eastern Space and Missile Center [also known as ETR] [AFSC*] *mt, uh, bl1-2.236, 12, 16, an31.7.94, jb, wp, jdr, oe*
ESMC Eastern Space and Missile Command *re*
ESMCR Eastern Space and Missile Center Regulation *oe*
ESMD El Segundo Manufacturing Division *fu, jj*
ESMDC Expandable Shielded Mild Detonating Cord *ai2*
ESMR Electr[on]ically Scanned Microwave Radiometer [NIMBUS] *ge, hc, ac1.5-17, hy, hg*
ESMR Enhanced (Extended) Specialized Mobile Radio *hi, kg*
ESN Earth Science Net *hy*
ESN Electronic Security Number *pk, kg*
ESN Electronic Serial Number *ct, em*
ESNG Engineering Subworking Group *mt*
ESO Electronic Supply Office [US Navy] *hc, fu, jj*
ESO Electronic Support Measure *mt*
ESO Emergency Security Option *mt*
ESO Engineering Stop Order [a document] *fu, ah, hc, ep, jj, pbl, tm, jdr*
ESO Entry Server Offering *ct, em*
ESO European Southern Observatory *es55.50, ns, mi, mb*
ESOC European Space Operations Centre [Darmstadt, Germany] *pl, sf31.1.14, es55.c, ns*
ESOH Equipment and Supplies on Hand *mt*
ESOP Employee Stock Ownership Plan *aw126.20.29*
ESP Electron Stream Potential *fu*
ESP Electronic Security Pacific [ESC*] *mt*
ESP Electronic Subsystem Project *hc*

ESP Electrostatic Signal Processor *jdr*
ESP Employee Suggestion Program [document] *pbl, tm, jdr, ah*
ESP Emulation Sensing Processor *pk, kg*
ESP Encapsulating Security Payload *em, kg*
ESP Energy Star Program *hi*
ESP Engineering Service Project *wp*
ESP Engineering Signal Processor *hc*
ESP Enhanced Serial Port *ct, em, kg*
ESPA Electronically Steered Phased Array *wp*
ESPAR Electronically Steerable Phased Array Radar *fu, jj*
ESPD El Segundo Police Department *sc2.11.1*
ESPE Earth Sensor Processing Electronics *-3*
ESPE Emergency Support Period Extension *-4*
ESPI Electronic Space Products, Incorporated *pl*
ESPIF Electro-Optical Sensor Performance Integration Facility *tb*
ESPIN Emergency Spare Parts Information Network *mt*
ESPRIT European Strategic Program for Research and Development In Information Technology *mt*
ESR Electron Spin Resonance *wp, fu, ai1*
ESR Electronic Scanning Radar *fu, wp*
ESR Equipment Status Report *4, jj*
ESR Equipment Status Reporting [USAF] *mt*
ESR Equivalent Series Resistance *pl, wp, fu*
ESR Equivalent Service Round [USA] *mt*
ESR Event Service Routine *pk, kg*
ESR Executive Service Request *fu*
ESR Grouped Comm-Electronics and Meteorological Reporting System [USAF] {?} *mt*
ESRA Extended SRA *jj*
ESRC Engineering and Services Readiness Center [USAFE] *mt*
ESRF Electronic System Repair Facility *hc*
ESRI Environmental Systems Research Institute *ct*
ESRIN European Space Records Information [Conference Center, Frascati] *ac3.7-2, es55.44*
ESRIS E-3A System Radar Identification System *hc*
ESRO European Space Research Organization [later called ESA] *ge, hc, 13, es55.c, hg, crl, ai2*
ESRP European Supersonic Research Program [for development of a successor to Concorde] *mt*
ESS Earth Sensor Sanity *uh*
ESS Electronic Security Squadron [ESC*] *mt*
ESS Electronic Switching System *mt, kg, fu*
ESS ELINT Support System [USMC] *mt*
ESS Emergency Ship Service *jj*
ESS EMSP Software System *fu*
ESS Energy Storage Subsystem *wl*
ESS Enterprise (Engineering) Scheduling System *pbl*
ESS Enterprise Scheduling System *jdr, kf, ah*
ESS Environmental Stress Screening *mt, rm2.1.2, fu, rm7.1.3, hc, nt, oe*
ESS Environmental Support System [JCS] *wp, mt*
ESS Environmental Survey Satellite *wp*
ESS Equipment Support Section *bo, ai2*
ESS Equivalent Sensor [Sub] System *hc*
ESS Executive Support Subsystem *mt*
ESS Executive Systems Software [MAC] *mt*
ESS Exercise Support system [OJCS] *mt*
ESS Expendable Signal System *wp*
ESS External Sound Speed *fu*
ESSA Environmental Science Services Administration [DOC] *ge, fu, ci.880, hc, jj, hg*
ESSC Earth System Sciences Committee *hy*
ESSC Emergency Support Schedule Changes *-4*
ESSC Environmental Studies Service Center *ag*
ESSEX Enhanced Satellite Survivability Experiment *hc*
ESSFNR Exercise Simulation System for Flexible Nuclear Response *mt*
ESSLR Eye-Soft System Laser Rangefinder *hc*
ESSO (ESSO Resources Canada, Limited) *ac2.7-19*
ESSP Earth Sensor Sun Presence *oe*
ESSPO Electronic Subsystem Support Project Office *hc*
ESSU Electronic Selective Switching Unit *fu*
EST Eastern Standard Time *kg, ci.350, 0, w9, aw129.10.48, wp, kf*
EST Electrostatic Storage Tube *jj, fu*
EST Emergency Support Team [FEMA] *mt*
EST Engineering String Test *kf*
EST Enhanced Simulation Training *fu*
EST Enroute Support Teams *mt*
EST Environmental System Test *mt*
EST Equipment Status Telemetry *pl*
EST ESM [mounted KKV] Seeker/Tracker *17, sb*
EST Expanded Service Test *mt*
EST-I Environmental System Test - Phase I *mt*
EST-II Environmental System Test - Phase II *mt*
ESTAR Electronically Scanned Thin Array Radiometer *hc*
ESTEC European Space Research and Technology Center [Netherlands] *ac3.7-2, sf31.1.16, es55.c, mj31.5.15, so, oe*
ESTO Equipment and System Test Objective *fu*
ESTP Equipment and System Test Procedure *fu*

ESTR Equipment and System Test Report *fu*
ESTRACK [network of ESA ground {tracking} stations] *sf31.1.15*
ESTS Electronic Systems Test Set *wp*
ESTV Electronic Shelf Thermal Vacuum *gs2-13.14*
ESU Electronic Switching Unit *fu*
ESU Electronics Support Unit *fu*
ESU Electrostatic Unit [cgs] *kg, sm1-2, wp, jj, fu, ai1*
ESU Environmental Simulation Unit *fu*
ESV Earth Satellite Vehicle *mt, 0, w9*
ESW Engagement Software *st*
ESWS Earth Satellite Weapon System *hc*
ESYLATE Ethanesulfonate *wp*
ET Earth Terminal *mt*
ET Eastern Time *jp, w9, hi*
ET Elapsed Time *jj, fu*
ET Electrical Time *fu*
ET Electro Thermal *id4133.10/922, st*
ET Electronics Technician *fu*
ET Embedded Training *wp*
ET Emerging Technology *nd74.445.29*
ET Encapsulation Technology *hi*
ET End Terminal *do24*
ET Engineering Task *fu*
ET Engineering Test *fu*
ET Enhancement Technology *pk, kg*
ET Ephemeris Time *mi, oe*
ET Escadre de Transport [transport wing units] [France] *mt*
ET Extended Test *mt*
ET Extended Threat *mt*
ET Extra Terrestrial *hn43.25.1, wp*
ET [Space Shuttle] External Tank *mi, bl1-2.250, 9, ws4.10, wl, jdr, mb, ai2*
ET&E Engineering Test and Evaluation *mt, hc, fu, tb*
ET/ST Engineering Test/Service Test *hc, jj*
ETA Earth Tangential Angle *tb*
ETA Engineering Task Assignment *hc, jj*
ETA Entity-Relation-Attribute *fu*
ETA Estimated Time of Arrival *mt, 0, wp, hc, w9, lt, fu, jj, oe*
ETA Exception Time Accounting *jj*
ETA Explosive Transfer Assembly *tk, 9, lt, oe*
ETA {a subsidiary of Control Data Corporation} *is26.1.31*
ETABS Electronic Tabular (Display) System *fu*
ETAC Enlisted Terminal Attack Controller [USA] *mt*
ETAC Environmental Technical Applications Center [USAF] *mt, aw130.11.61*
ETACCS European Theater Air Command and Control Study *hc*
ETACCS European Theater Air Command and Control System *mt*
ETACS Extended Total Access Communication System [IBM] *pk, kg*
ETADS Enhanced Transportation Automated Data System [AFLC] *mt*
ETANN Electrically Trainable Analog Neural Network [chip] [Intel] *pk, kg*
ETAP Elevated Temperature Aluminum Program *hc*
ETAS Elevated Target Acquisition Sensor *hc*
ETAS Elevated Target Acquisition System [USA] *mt, hc*
ETAS Escort Towed-Array Sensor *fu*
ETB End of Transmission (Transmitted) Block *kg, fu, do242, hi*
ETB Engineering Technical Bulletin *jj*
ETB Engineering Test Base *hc*
ETB Engineering Test Bed *kf*
ETC Earth Terminal Complex *mt*
ETC Electro Thermal Chemical [Gun] *mt*
ETC Electronic Toll Collection *pk, kg*
ETC Engineering Training Center [STDN Station, Maryland, USA] *ac2.9-10, 4*
ETC Enhanced Throughput Cellular [modem protocol] [AT&T] *hi, kg*
ETC Estimate To Complete *8, fu, 17, sb, hc, jdr, kf, oe*
ETC Estimated Time of Completion *wp, jj*
ETCA [Belgian {Company}, member of Eurosatellite consortium] *sc4.3.3*
ETCG Elapsed-Time Code Generator *fu*
ETCS External Threshold Clutter Select *jj*
ETD Effective Transfer Date *wp*
ETD Electronic Tactical Display *fu*
ETD Ensemble Threshold Detector *jj*
ETD Estimated Time of Departure *mt, w9, fu, jj*
ETD Experiment Test Document *hc*
ETDI Eurasian Target Data Inventory *mt*
ETDL Electronics Technology and Devices Laboratory [of ERADCOM] [Army] *fu, tc*
ETDMA Enhanced Time Division Multiple Access *hi*
ETE Earth Trailing Edge *3, 8, uh*
ETE End To End *kf, oe*
ETE Engineering Test Equipment *wp*
ETE Estimated (Expected) Time Enroute *mt*
ETE Estimated Time En Route *fu*
ETE Expendable Turbine Engine *hc*

ETE External Test Equipment *fu*
ETEC Expendable Turbine Engine Concept *wp*
ETER Estimated Time en Route *0*
ETES Early Training Estimation System *hc*
ETF Engine Test Facility *wp*
ETF Error Threshold Firing *-3*
ETFE Ethylene Tetra Fluoro Ethylene *he*
ETG Embedded Timing Generator *uh*
ETG Enhanced Target Generator [ATC] *fu, mt*
ETHERNET Xerox Local Area Network System [also ethernet] "ethernet" *mt*
ETI Elapsed Time Indicator *hc, wp, fu, jj*
ETIBS Enhanced Tactical Information Broadcast System [KOREA] *mt*
ETIC Estimated Time in Commission *mt, fu, jj*
ETIS Elapsed Time Indicator System *mt*
ETK Expected Time To Kill *wp*
ETL Environmental Test Laboratory *0*
ETL Equipment Test Laboratory *0*
ETLA Extended Three Letter Acronym *mi*
ETM Elapsed Time Meter *jj, fu*
ETM Electronic Tactical Map *hc, fu*
ETM Electronically Transmitted Message *wp*
ETM Engineering Test Model *bl1-9.9, 16, fu*
ETM Enhanced Thematic Mapper *16, kf*
ETM Extended Training Material *fu*
ETM+ Enhanced Thematic Mapper Plus *kf*
ETMA Engineering Tooling Manufacturing Aid *hc, jj*
ETN External Track Number *fu*
ETNF Estimated Time of Next Failure *wp*
ETO Electronic Trading Opportunity *ct, em*
ETO Emergency Time Out *vv*
ETO Estimated Take Off *wp*
ETO Estimated Time Outage *oe*
ETO European Theater of Operations *w9*
ETO Extensive Time Observations *hy*
ETOM Electron-Trapping Optical Memory *pk, kg*
ETOM Escadre de Transport d'Outre Mer [overseas transport wing] [France] *mt*
ETOP (twin engine extended range operations (ETOP)) *aw130.13.69*
ETOPS Extended Range Twin-Engine Operations *mt*
ETOT Estimated Time Over Target *mt, fu*
ETP Engineering Test Plan *hc, jj, fu*
ETP Equipment Test Plan (Procedure) *fu, wp*
ETP European Trunking Plan *mt*
ETP Experiment Test Plan *hc, jj*
ETP External Tracking Processor *hc*
ETP {aeronomy scientific instrument planned for Mars orbiter} *ma40*
ETPD Emerging Technology Program Database *mt*
ETPL Endorsed Tempest Products List *kg, wp*
ETPS Empire Test Pilots School. [at Cranfield until 1946, at Farnborough until 1968, later at Boscombe Down, UK] *mt*
ETR Eastern Test Range [Cape Canaveral] *fu, mi, uh, hc, gd2-44, 4, 12, 16, 17, lt, tb, jdr, mb, kf, vv, oe*
ETR Effective Transmission Rate *mt*
ETR Electronically Tuned Radio *hc*
ETR Estimated Time of Repair *wp*
ETR Estimated Time of Return *mt, wp, jj*
ETRO Expected Time for Return to Operation *mt, fu*
ETRP Engineering Test and Rework Plan (Repair Planning) *jr, jdr, ep*
ETRTB Estimated Time to Return to Base *fu*
ETS Econometric Time Series *pk, kg*
ETS Edwards Test Station *0*
ETS Embedded Training System *fu*
ETS Engineering and Technical Services *jj*
ETS Engineering Test Satellite *ac3.3-9*
ETS European Telephone System *mt, hc*
ETS European Troops Strength *wp*
ETS Evaluation Testbed System *fu*
ETS Evaluator Trainer System *hc*
ETS Experimental Technology Satellite [Japanese] *oe*
ETSREQ Engineering and Technical Services Request *jj*
ETT Elapsed Time Totalizer *jj*
ETTM Expanded Tactical Telemetry *jj*
ETU Environmental Transmitter Unit *fu*
ETUT Enhanced Tactical Users Terminals *mt*
ETV Educational Television *hc, w9, jj*
ETVM Electrostatic Transistorized Voltmeter *fu*
ETVS Educational Television Satellite *hc*
ETW Enemy Throw Weight *hc*
ETX End of Text *do242, jdr, kg*
ET_AL Et Alii [(Latin) and others] *wp*
ET_SEQ Et Sequens [(Latin) and the following] *wp*
EU Electron Unit *ai1*
EU Electronics Unit *17, sb, hc, jj, jdr*
EU Engineering Unit *oe*
EU Exciter Unit *jdr*

EU Execution Unit *kg, fu*
EUC End User Computing *mt, kg*
EUC Extended Unix Code [IBM] *pk, kg*
EUCE End User Computer Equipment *fu*
EUCLID European Co-Operations for the Long-term In Defense *mt*
EUCOM European Command [US forces in Europe] *mt, hc, ph, jb, jj, kf*
EUCOM AIDES European Theater Air Command and Control System *mt*
EUDAC European Command Defense Analysis Center [ELINT] *mt*
EUI End User Interface *pk, kg*
EULA End-User Licensing Agreement *ct, em*
EUMD Extended Unit Manning Document *mt*
EUMEAF Europe, Middle East and Africa *mt*
EUMETSAT European Meteorological Satellite [Organization] *mt, ge, sc3.8.3, hy, hg*
EUMR Emergency Unsatisfactory Material Report *jj*
EUNIE EUCOM Nuclear Interface Element [US] *mt*
EUNIE-F EUCOM Nuclear Interface Element-Fastbreak *mt*
EUP Engineering Unit Pages *oe*
EURAAM (the Anglo-German EURAAM consortium) {European AMRAAM} *aa26.11.12*
EURASEP [an archival center for Nimbus data] *ac2.8-14*
EURATOM European Atomic Energy Community *wp*
EURECA European Retrievable Carrier *ge, sf31.1.16*
EUREL European Electrical Engineers' National Associations Convention *hn42.4.5*
EURMEDS European Meteorological Data systems [USAFE] *mt*
EUROCAP European Communication Analysis Program *hc*
EUROCON European Conference on Electrotechnology *hn42.4.5*
EUROFAR [Europe] European Future Advanced Rotorcraft *mt*
EUROHEPNET European High Energy Physics Network *ns*
EUSC Effective US Controlled *mt*
EUSEC European Communications Security Agency *mt*
EUT Electronics Under Test *jdr, fu, wp*
EUT&E Early Users Test and Experimentation *fu*
EUTE Early Use Test and Evaluation *fu*
EUTELSAT European Telecommunications Satellite [Organization] *sc3.10.3, hc*
EUV Extreme Ultra Violet *fu, mi, pl, wp, es55.14, eo, tb, mb, oe, ai1*
EUVE Extreme Ultra Violet Explorer *ci.716, hc, mb, mi, kf*
EUVITA EUV telescope *cu*
EV Earned Value *8, fu, 17, sb, ep, jdr*
EV Electron Volt *wp, tb, oe*
EV Error Volume *st*
EV Experimental Version *ts7-30, tb*
EV Extravehicular *wp*
EVA Electromagnetic Valve Actuator *pk*
EVA Extra Vehicular Activity *mi, aa26.12.26, ac2.3-4, 8, 9, 17, hc, w9, ws4.10, wp, mb, eo, wl, jdr, mb, ai1*
EVAC Evacuation File [OJCS] *mt*
EVAS Extra Vehicular Activity System *wl*
EVCF Eastern Vehicle Checkout Facility *uh*
EVCF Eastern Vehicle Control Facility *oe*
EVDU Editing Video Display Unit *fu*
EVE Extensible VAX Editor *hi, kg*
EVE Extensive VAX Editor *oe*
EVE External Vernier Engine *ai2*
EVEA Extra Vehicular Engineering Activities *hc*
EVGA Extended Video Graphics Adapter *hi, kg*
EVGA Extended Video Graphics Array *pk, kg*
EVIL Extensible Video Interactive Language *hi*
EVK ['Flotilla a flying ships', the first unit of four-engined bombers, created on 10 December 1914. The aircraft were 'Ilya Mourometz' biplanes designed by Sikorsky] [Russia] *mt*
EVM Earned Value Measurement *jdr*
EVM Electronic Voltmeter *fu*
EVN European VLBI Network *es55.20*
EVOM Electronic Volt Ohm Meter *fu, jj*
EVP Estimated Vector Processor *fu*
EVP Experimental Version Prototype *ts7-11, tb*
EVRS Electro Optical Viewing and Ranging Set *hc*
EVS Electro-Optical Viewing System [system fitted in the turrets under the nose of the B-52] [USA] *mt, hc, wp, af72.5.134*
EVSD Energy Variant Sequential Detection *fu*
EW Early Warning *mt, fu, hc, wp, tb, st, jj*
EW East to West *in*
EW Electronic Warfare *mt, fu, wp, ic2-13, 16, ph, aw129.14.99, jb, tb, st, jj, kf*
EW Empty Weight *wp*
EW Enlisted Woman *w9*
EW C&R Electronic Warfare Coordination and Reporting *fu*
EW/AA Early Warning/Attack Assessment *tb, st*
EW/CCCCM Electronic Warfare Command and Control Communications Countermeasures *mt*
EW/DE Electronic Warfare Directed Energy *mt*
EW/GCI Early Warning/Ground Control Intercept *wp*

EW/RSTA Electronic Warfare / Reconnaissance and Target Acquisition Center [USA] *mt*
EWA Early Warning Augmentation *st*
EWAC Electronic Warfare Anechoic Chamber *wp*
EWACS Electronic Wide Angle Camera System *wp*
EWACS Electronics Warfare Close Air Support *un*
EWARS Electronics Warfare Assets Reporting Systems *mt*
EWASER Electromagnetic Wave Amplified by Stimulated Emission of Radiation *fu*
EWB Embedded Wiring Board *fu*
EWBN Early Warning Broadcast Net *fu*
EWC Electronic Warfare Center [USAF] *mt*
EWC Electronic Warfare Commander [US Navy] *mt*
EWC Electronic Warfare Coordinator [US Navy] *fu, mt*
EWCM Electronic Warfare Control Module *fu, mt*
EWCM Electronic Warfare Coordination Module *mt*
EWCS Electronic Warfare Combat System *hc*
EWCS Electronic Warfare Control Ship [US Navy] *mt*
EWCS Electronic Warfare Control System *fu*
EWCS European Wideband Communications System *mt, hc*
EWD Electronic Warfare Division *mt*
EWDG Electronic Warfare Display Group *fu*
EWDTC Electronic Warfare Design to Cost *hc*
EWE Electronic Warfare Element *fu*
EWEP Electronic Warfare Evaluation Program *mt*
EWES Electronic Warfare Environmental Simulator *hc, wp*
EWES Electronic Warfare Evaluation Simulator *hc*
EWGS European Weather Graphics System *mt*
EWI Education With Industry [U. S. Air Force program] *mt, hn41.16.8*
EWI Electronic Warfare Information *fu*
EWI Electronic Warfare Intelligence *mt*
EWICG EW Integration and Control Group *fu*
EWIOC Electronic Warfare and Intelligence Operation Center [Army] *fu*
EWIP Electronic Warfare Integrated Programming *mt*
EWIR Electronic Warfare Integrated Reprogramming [USAF] *wp, mt*
EWIRC Electronic Warfare Integrated Repograming Concept *wp*
EWIS Electronic Warfare Information system [WWMCCS] *mt*
EWIS European WWMCCS Information System *mt*
EWITS Early Warning Identification and Transmission System *hc*
EWJT Electronic Warfare Joint Test *hc*
EWL Electronic Warfare Laboratory [of ERADCOM] *tc*
EWMIS Electronic Warfare Management Information System *wp*
EWO Early Warning Operator *fu*
EWO Electronic Warfare Officer [USA] *mt, hc, wp, jj, fu*
EWO Emergency War Order *mt, 12, wp, fu*
EWO Engineering Work Order *wp, jj*
EWOC Electronic Warfare Operations Center *fu*
EWOPS Electronic Warfare Operations *wp*
EWP Emergency War Plan *wp*
EWP Escort Weapon Platform *sb*
EWQRC Electronic Warfare Quick Reaction Capability *wp*
EWR Early Warning Radar *wp, fu*
EWR Eastern and Western Range *kf, vv*
EWRL Electronic Warfare Reprogrammable Library [US Navy] *mt*
EWS Early Warning System *mt, hc, tb, kf*
EWS East-West Stationkeeping *-3*
EWS Electronic Warfare Supervisor *fu*
EWS Electronic Warfare System *wp, fu, jj*
EWS Employee Written Software [IBM] *pk, kg*
EWS Engineering Workstation *fu*
EWS Estimate Work Sheet *hc, jj*
EWS Extended Work Store *fu*
EWSIP Electronic Warfare Standardization and Improvement Program *wp*
EWSK East-West Stationkeeping *oe*
EWSM Early Warning Support Measures *mt*
EWSM Electronic Warfare Support Measures *wp*
EWSO Electronic Warfare Staff Officer *mt*
EWSS Electronic Warfare Support System *wp*
EWSUMS EW Spectrum Use Management System *fu*
EWTAP Electronic Warfare Tactics Analysis Program *wp*
EWTD Electronic Warfare Training Device *wp*
EWTES Electronic Warfare Threat Environmental Simulator *hc*
EWU Electronic Warfare Unit *fu*
EWVA Electronic Warfare Vulnerability Assessment *wp*
EWWS Electronic Warfare Warning System *mt, hc, wp, ts17-3, jj*
EW_ARC Electronic Warfare Area Reprogramming Capability *mt*
EX (IP, ES, PSI, EX) Executive Switching [routine] *bl1-5.127, jj*
EXACT Exchange of Authenticated Electronic Component *hc*
EXAMETNET Experimental Inter-American Meteorological Rocket Network *ci.102*
EXCA Exchangeable Card Architecture [Intel] *pk, kg*
EXCALIBUR [advanced forward area air defense gun] *wp*
EXCELS Expanded Communication Electronics System *hc*
EXCIMER Excited Dimer *wp*
EXCLASS Extended CLASS *mt*

EXCLOT Executive Committee For Low Observable Technology *wp*
EXCM Expendable Countermeasures *fu*
EXCODOP Externally Coherent Doppler *hc*
EXCOM Executive Committee *-10*
EXCON Executive Component *-4*
EXCRET Extended Cost Rebudgeting Tool *kf*
EXDM Exploratory Development Model *fu*
EXDRONE Expendable Drone *wp*
EXDRONE Expendable Drone Jammer [USMC] *mt*
EXE2BIN Executable To Binary [conversion program] *hi*
EXF External Function [command] *fu, jj*
EXJAM Expendable Jammer *mt*
EXM Enterprise Messaging Exchange [Lotus] *pk, kg*
EXO Exoatmospheric *hc, tb, st*
EXO-D Exoatmospheric Discrimination *st*
EXO-NNK Exoatmospheric Non-Nuclear kill *17, sb*
EXOCET [surface to surface missile] *wp*
EXOCHAFF {Exoatmospheric Chaff} [program to dispense chaff from ICBM before re-entry] *wp*
EXOKV Exoatmospheric Kill Vehicle *st*
EXOSAT European [Space Agency's] X-ray Observation (Observatory) Satellite *cu, ns*
EXSUM Exercise Summary *mt*
EXSWG Exploration Science Working Group *eo, re*
EXT Exercise Training *fu*
EXTOSS Extended Range Transfer Orbit Sun Sensor *cl, uh*
EXTRN External Reference *pk, kg*
EXU Execution Unit *fu*
EZ Extraction Zone *mt, fu*
EZCO Extraction Zone Control Officer *mt*

F

F (F-hour) [USA DOD] *mt*
F Fighter [US Air Force designator for fighter aircraft, such as F-4, F-8, F-15, F-20, F-111, etc.] *mt*
F {Fairing} (F-0 countdown baseline test) *ggIII.1.31*
F&A Fabrication and Assembly *jj*
F&E Facilities and Equipment *fu*
F&F Fire and Forget *dh*
F&FP Force and Financial Plan *mt*
F&PED Facilities and Plant Engineering Department *fu*
F-3D Fixed 3-Dimension *fu*
F-ATA Fast Advanced Technology Attachment *hi*
F-DADS Fault-tolerant Digital Airborne Data System *fu, hc*
F-F Flip-Flop also F/F] *fu*
F-O Follow-On *uc*
F-SCSI Fast SCSI *hi*
F-TOSS Fighter Time Ordered Spread Spectrum *hc*
F-TOSS Full Capability Time Ordered Spread Spectrum *fu*
F/A (F/A-18 Hornet) [USA DOD] *mt*
F/A Fabrication and Assembly *kf*
F/A Flight Acceptance *0*
F/A/T Fabricate, Assemble, Test *fu*
F/B Fleet Broadcast *jdr, uh*
F/C Fire Control [also FC] *5, sb*
F/D Fill and Drain [line, valve] *jdr, oe*
F/EM Flight/Engineering Model *es55.44*
F/F Far Field *jdr*
F/F Flip Flop *fu, jr*
F/G Flag/General Officer *mt*
F/O Fiber Optic *fu*
F/O Follow On *jdr, uh*
F/R Failure Rate *fu*
F/RD Fixed and Roll Down [panels] *kgl*
F/T Full Time *pk, kg*
FA Factored Analogy *kf*
FA False Alarm *jj, kf*
FA Feasibility Assessment [USA DOD] *mt*
FA Field Artillery *12, w9, fu*
FA Final Assembly *jj*
FA First Article *fu*
FA Flight Acceptance *jdr, jr*
FA Flight Anomaly *jr*
FA Focused Area *kf*
FA Forced Air *wp*
FA Forward Acquisition *tb*
FA Frequency Agility (Agile) *hc, wp, tb, jj, fu*
FA Frontovaya Aviatsiya [tactical Air Force] [USSR] *mt*
FA Full Action *jj*
FA Functional Area *mt, kf, fu*
FA&CO Fabrication, Assembly, and Checkout [also FACO] *hc*
FA-FR Focused Area-Fast Revisit [mode] *kf*
FA-HS Focused Area-High Sensitivity [mode] *kf*
FA-TE Focused Area-Transient Event [mode] *kf*

FA/EO First Article Engineering Order *fu*
FAA Federal Aviation Administration [Agency] [USA] *fu, hc, hn42.4.8, 2, 4, 12, aw129.14.112, aa26.12.5, w9, af72.5.169, nt, wp, hy, tb, jj, oe, mt, y2k, ai2*
FAA First Article Audit *fu*
FAA Fleet Air Arm [UK] *mt*
FAAA Final Acquisition Action Approach *mt*
FAAAC FAA Aeronautical Center *fu*
FAAB Frequency Allocation Advisory Board *wp*
FAAD Forward Area Air Defense *mt, fu, hc, wp, pm*
FAADCCI Forward Area Air Defense Command and Control and Intelligence *mt*
FAADS Field Army Air Defense System *fu*
FAADS Forward Area Air Defense System *mt, aa26.12.27, hn49.11.5, hq6.2.7, wp, kf*
FAALS Field Artillery Acoustic Locator System *hc*
FAALS Forward Area Artillery Locator System *hc*
FAAO Federal Aviation Accounting Office *jj*
FAAR Forward Air Acquisition Radar *hc*
FAAR Forward Area Alerting Radar *fu, hc, wp*
FAAS Family of Army Aircraft Systems (FAAS - 85) *fu*
FAASV Field Artillery Ammunition Support Vehicle *wp*
FAAT Fighter Attack Avionic Technology *hc*
FAATC Federal Aviation Administration Technical Center *fu*
FAAWC Fleet Anti Air Warfare Commander *hc, jj*
FAB (FAB/BLU) Force Aerienne Belge / Belgische Luchtmacht [(French / Flemish) Belgian Air Force] *mt*
FAB Fabricate *17, jj*
FAB Fabrication Plant [computer chip] *pk, kg*
FAB Fast Atom Bombardment *wp*
FAB Field Assistance Branch *mt*
FAB Fixed Action Button *fu*
FABMDS Field Army Ballistic Missile Defense System *hc*
FABMS Fast Atom Bombardment Mass Spectroscopy *wp*
FABPAC Fabry-Perot Power Accumulator *hc*
FAC Facilities Contract [special] *hc*
FAC Fast Attack Aircraft *wp*
FAC Fast Attack Craft *mt*
FAC Federal Acquisition Circular *kf*
FAC File Access Code *pk, kg*
FAC Filter Address Correction *jj*
FAC Final Acceptance Test [also FAT] *sb*
FAC First Alert Capability *wp*
FAC First Article Calibration *fu*
FAC Ford Aerospace Corporation [presently Space System/Loral] *oe*
FAC Forward Air Control(ler) [USA] *mt, fu, hc, af72.5.137, wp, jj*
FAC Free Address Count *jj*
FAC Full Action [Radar Switch] *jj*
FAC Fully Allocated Cost *do35*
FAC Functional Account Code *mt*
FAC Functional Area Chief *mt*
FAC Functional Audit Configuration *fu*
FACA Future Aircraft for Combat and Attack *hn41.16.1*
FACAR Failure Analysis and Corrective Action Report *fu, jdr*
FACC Ford Aerospace Communications Corporation *ag*
FACCE Family Concept of Computer Elements *hc*
FACE/CARD Forecasting Ammunition Consumption Expenditure / Critical Assets Reporting Data combined subsystem *mt*
FACI First Article Configuration (Compliance) Inspection *fu, hc, jj*
FACILE Flexibility And Control Interaction Linearized Equations *bl1-6.28*
FACMS Flexible Automated Composites Manufacturing Systems *kf*
FACO Fabrication, Assembly, and Checkout [also FA&CO] *hc, hn49.11.6*
FACO Final Acceptance and Checkout Facility *-17*
FACO Final Assembly [and] Checkout *hc, jj*
FACOR Failure Analysis and Close Out Report [document] *ep, pbl, tm, jdr*
FACP Forward Air (Area) Control Post [USAF] [USA] *mt, fu, hc, wp, jj*
FACP Forward Air Control Party [USMC] [USA] *mt*
FACR First Article Configuration Review *fu, hc, jj*
FACRP Functional Analysis and Consolidation Review Panel *mt*
FACSFAC Fleet Area Control Surveillance Facilities *fu*
FACSFAT Fleet Area Control and Surveillance Facility *mt*
FACT Fast Acting Control Transfer *wp*
FACT Field Test Analysis Correlation of Thermal Images *hc*
FACT Fleet Afloat Correlation Test *fu*
FACT Flexible Automatic Circuit Tester [Hughes Aircraft Company] *kf, rm1.2.3, hc, hd1.89.y4, lt, ep, jj, jdr, fu*
FACT Fully Automatic Compiler Translator *wp*
FACTS FLIR Augmented Cobra TOW System *uc, hc*
FAD Failure Analysis Diagnostic *jj*
FAD Feasible Arrival Date *mt*
FAD First Article Demonstration *fu, hc, jj*
FAD Fleet Air Defense [USA] *jj, mt*
FAD Flexible Automatic Depot *hc, jj*
FAD Force Activity Designator (Designation) *mt, jj*

FAD Fuel Advisory Departure *fu*
FADAC Field Artillery Digital Automatic Computer *hc, fu*
FADE Fully Automatic Depot Equipment *hc, jj*
FADEC Full Authority Digital Electronic (Engine) Control *mt, wp, hc, aw126.20.40*
FADM Fleet Admiral *w9, jj*
FADR Forward Area Defense Radar *fu*
FADS Fleet Air Defense System *fu, jj*
FAE Forward Acquisition Experiment *st*
FAE Fuel Air Explosive *mt, hc, wp*
FAEC Fuerza AÈrea del EjÈrcito Cubano [Cuban Army Air Force] *mt*
FAEL First Article Equipment Log *fu*
FAEO First Article Engineering Order *fu*
FAETU Fleet Airborne Electronics Training Unit *jj*
FAEW Fuel Air Explosive Weapon *wp*
FAFB Falcon Air Force Base *jr, kf, uh*
FAFS Falcon Air Force Station *tb*
FAFT First Article Factory Test *fu*
FAFTEEC Full Authority Fault Tolerant Electronic Engine Control *wp*
FAG Field Artillery Group *hc, jj*
FAGC Fast Automatic Gain Control *fu, hc, wp, jj*
FAHV Force AÈrienne de Haute Volta [(French) Air Force of Upper Volta] *mt*
FAI Fast Automatic Initiate *fu*
FAI Federal Acquisition Institute *dt*
FAI FÈdÈration AÈronautique Internationale [(French) International Aeronautics Federation] *mt, aw130.22.34*
FAI First Article Inspection *fu*
FAI Fly As Is *wp*
FAIA Functional Area Information Analysis *mt*
FAIP First Assignment Instructor Pilot *mt*
FAIR Fighter, Attack, Intercept, Reconnaissance *wp*
FAIR Fly Along Infra Red *hc, go4.2*
FAIR Forward Acquisition Infra Red *-17*
FAIR Functional Analysis Integration Review *mt*
FAIRS Focal Plane Array Infrared Seeker *hc*
FAISR File Archival Image Storage and Retrieval *wp*
FAIT First Article Inspection and Test *fu*
FAITE Final Acceptance Inspection Test Equipment *hc*
FAK Fly Away Kit *jj*
FALCON [an air to air missile] *wp*
FALCON_EYE [F-16 FLIR] *wp*
FALT FADAC Automatic Logic Tester *hc*
FALW-D Forward Area Laser Weapon Demonstration (Demonstrator) *hc, uc*
FAM Fat Array Module *fu*
FAM Flight Assurance Manager *oe*
FAMAS Field Artillery Meteorological Acquisition System *hc, fu*
FAMG Field Artillery Missile Group [USA] *mt*
FAMIS Factory Management Information System *fu*
FAMOS Floating gate (Avalanche) Metal Oxide Semiconductor *fu, kg, em*
FAMS Field Artillery Meteorological System *fu*
FAMS Fine Attitude Measurement System *ac3.7-29*
FAMS Fuels Automated Management System *mt*
FAMSIM Family of Automated Simulation *mt*
FAMST Flight Acceptance Mixed Simulation Test *kf*
FAMU Florida A and M University *aa26.12.92*
FAN Flaming Arrow Net [USEUCOM COMM NET] *mt*
FAN Forward Air Navigator [USA] *mt*
FAN Fuerza AÈrea Naval [Naval Air Force] [Cuba] *mt*
FAN Functional Area Network *mt*
FANG Flechette Area Neutralizing Gun *hc*
FANG [Soviet La-11 Fighter] *wp*
FANS Future Air Navigation System *mt, kg, wp*
FANSY Fan System *fu*
FANT Flight And Navigation Trainer *hc*
FANTAC Fighter Analysis - Tactical Air Combat *hc*
FANTAIL La-15 (Soviet) *wp*
FANTASY Foreign Antenna Assessment Study *hc*
FANX Friendship Airport Annex *-10*
FAO Fabrication / Assembly Order *fu*
FAO False Alarm Opportunities *fu*
FAO Food and Agriculture Organization [of the United Nations] *pl, w9, wp*
FAP False Alarm Probability *jdr*
FAP File Access Protocol *pk, kg*
FAP Flight Activity Performance [ANG] *mt*
FAP Fly Along Probe *tb*
FAP ForÁa Aerea Portuguesa [Portuguese Air Force] [Portugal] *mt*
FAPA Future Aviation Professionals of America *aw130.11.13*
FAPES Force Augmentation Planning and Execution System [GCCS] [USA] *mt*
FAPI Family Application Program Interface *pk, kg*
FAQ Frequently Asked Question *ct, em, kg, mi, bm, jdr, hi, du*
FAR Failure Analysis Report [document] *fu, ah, pbl, tm, gp, wp, 17, sb, jdr*

FAR False Alarm Rate *fu, sd, wp, hc, jj*
FAR Federal Acquisition Regulation [formerly DAR] *mt, fu, su, ah, hc, wp, 16, 18, ph, nt, ig21.12.16, ep, tb, st, jdr. pbl, tm, kf, oe*
FAR Federal Air Regulation *aw130.13.46, aa26.10.17, wp*
FAR Federal Aviation Regulation [USA] *mt, fu, ci.1154, af72.5.169*
FAR Field Assessment Review *wp*
FAR Final Acceptance Review *tb, wl, oe*
FAR First Alarm Register *fu*
FAR Fixed Array Radar *hc, fu, wp*
FAR Forward Acquisition Radar *fu*
FAR Fully Automatic Response *hi*
FAR Functional Analysis Review [AWIS] *mt*
FAR Functional Area Requirement *mt*
FAR [French rapid action force] *mt*
FAR/DAR Functional Area Requirement / Data Automation Requirement [PACAF] *mt*
FARA French Atmospheric Research Aircraft [France] *ge, hg*
FARADA Failure Rate Data [Program] *fu, hc*
FARADS Fast Response Air Defense System *hc*
FARGO Artillery Fire Control System *hc*
FARM Functional Area Record Manager *mt*
FARMAR Fighter Attack Reconnaissance Modular Adaptive Radar *hc*
FARMER [Soviet MIG-19 aircraft] *wp*
FARNET Federation of American Research Networks [Internet] *pk, kg*
FARP Forward Area Rearm / Refuel Point *mt*
FARP Fuel And Rearming Point *mt*
FARR FAA/Air Force Radar Replacement [ESD] *mt*
FARS Fighter Attack Reconnaissance System *hc*
FARS Financial Accounting and Reporting System *mt*
FAS Fault Alarm System *mt*
FAS Field Alert Status *mt*
FAS File Access System *fu*
FAS Fin Actuator System *ai2*
FAS Fire Support Aerial System *hc*
FAS Force Accounting System [USA] *mt*
FAS Forces AÈriennes Strategiques [(French) Strategic Air Command] *mt, 12*
FAS Forward Acquisition Sensor (System) *fu, hc, st, kf*
FAS Free Alongside Ship *w9*
FAS Fuel Advisory System *mt*
FAS Fueling At Sea *jj*
FAS Fuerza AÈrea SalvadoreÒa [Air Force of El Salvador] *mt*
FAS Functional Address System *mt*
FASA Flow And Structure Analyzer *oe*
FASAC Foreign Applied Science Assessment Center *mt*
FASC Fire and Air Support Center [USMC] *mt*
FASC Forward Area Signal Center [USA] *mt*
FASCAM Family of Scatterable Mines [US Army] *fu, hc, uc*
FASCAP Fast Payback Capital Investment Program *mt, af72.5.108*
FASCO Forward Area Support Company *fu*
FASCODE Fast Atmospheric Signature Code *wp*
FASCOM Field Army Support Command *hc, jj*
FASCS Federated Antisubmarine Combat System *fu*
FASCT Forward Acquisition System Center Technology *hc*
FASDPS Forward Acquisition System Data Processing *hc*
FASEX Fleet Air Superiority Exercise *jj*
FASH Full Action Switch Held *jj*
FASIC Function and Algorithm-Specific Integrated Circuit *pk, kg*
FASINEX Frontal Air-Sea Interaction Experiment *hy*
FASMS Forecast / Allocation Submission Management System [7115] [USAFE] *mt*
FASP Future Acoustic Signal Processor *hc*
FASS Frequency Agile Signal Simulation *wp*
FASSIM Forward Acquisition System Simulation *fu*
FASSP Flexible Adaptive Spatial Signal Processing *hc, fu*
FASST Forum for the Advancement of Students in Science and Technology *ci.1245*
FASST Forward Acquisition System Sensor Technology *hc*
FAST Facility for Automated Simulation Test *hc*
FAST Fan And Supersonic Turbine *wp*
FAST Fast Access Storage Technology *hi*
FAST Fast Acquisition Search and Track *hc*
FAST Fast Auroral SnapshoT [explorer] *mi*
FAST Fast Automatic Shuttle Transfer system [USA] *jj, mt*
FAST Field Artillery System Training *hc*
FAST Fighter Airborne Supply Tank [low drag conformal tanks attached to the sides of F-15s] [USA] *mt*
FAST Flight Analyses System *mt*
FAST Flow and Analysis System for TRANSCOM *mt*
FAST FMIS Applications Support Technique *mt*
FAST Force and Supply Tracking System [USAREUR] *mt*
FAST Forward Air Strike *wp*
FAST Forward Airborne Surveillance and Tracking *fu, hc*
FAST Forward Area Support Team *mt*
FAST Freight Automated System for Traffic Management [MTMC] *mt*
FAST Functional Automated Subsystem Test *fu*

FAST Future Aircraft Supersonic Transport *aw129.21.35*
FASTALS Force Analyses Simulation of Theater Administration and Logistic Support [Army] *mt*
FASTAR Frequency Angle Scanning, Tracking and Ranging *fu*
FASTCALS System Calibration Field Assist Support Team *hc*
FASTRAC EO Tracker Using LLTV *wp*
FAST_PACK [conformal tanks for fuel and sensors] *wp*
FASU Fleet Aviation Supply Unit *jj*
FASW Functional Application Software *fu*
FASWTC Fleet ASW Training Center *fu*
FAT Factory Acceptance Test (Testing) *fu, hc, tb, jj*
FAT Fighter Allocation Technician *fu*
FAT File Allocation Table *kg, wp, ct, em, hi*
FAT Final Acceptance Test (Testing) [also FAC] *hc, 17, oe*
FAT Final Acceptance Trial [US Navy] *fu, hc, jj*
FAT First Article Test *hc, fu*
FAT Flight Acceptance Test (Testing) *0, wp, jj, jdr*
FAT Fluorescent Antibody Test *wp*
FAT Formal Acceptance Test *jdr*
FAT Functional Area Testing *fu*
FATA Fast ATA [also F-ATA] *hi*
FATAC Force AÈrienne Tactique [French Tactical Air Command] *-12*
FATAC Force AÈrienne Tactique [tactical Air Force] [Air Force of the Congo Republic] *mt*
FATAFCS FLIR Airborne Target Acquisition/Fire Control System *hc*
FATCAT Film And Television Correlation Assessment Technique *wp*
FATDL Frequency and Time Division Link *fu*
FATOC Field Army Tactical Operations Center *hc*
FATP Functional Acceptance Test Procedure *fu*
FATS Focal Plane Array Test Station *hc*
FATS Forward Area Target Surveillance System *wp*
FAU Florida Atlantic University *aa26.11.56*
FAV Fuerzas AÈreas Venezolanas [Venezuelan Air Forces] *mt*
FAVS Fighter Attack Visual System *wp*
FAW Fleet Air Wing *wp*
FAWLA Fighter Aircraft Wing Lift Augmentation *wp*
FAWPS Flight And Weapons Planning System *mt*
FAX Facsimile [transmission] *mt, kg, ct, fu, aa26.22.39, aw129.21.6, wp, pr, tb, jj, hi, oe*
FAX Fuel Air Explosives *hc*
FB Feed Back *jr*
FB Fighter [/] Bomber [FB-111 is a medium range strategic bomber] *wp, mt*
FB/L Functional Baseline (also FBL) *fu*
FBB Fast Burn Booster *mt, tb, st*
FBC Fluidized Bed Combustion *wp*
FBFM Feedback FM *hc*
FBH Force Beach Head *mt*
FBI Federal Bureau of Investigation [USA] *mt, su, hc, sc3.2.1, w9, aw129.10.18, wp, tb*
FBIOS Flash BIOS *hi*
FBIS Foreign Broadcast Information Service *mt, jb, ns*
FBL Fly By Light *mt*
FBM (Transitions Using an FBM Monopole) *mj31.5.64*
FBM Fighter Battle Management *wp*
FBM Fleet Ballistic Missile [Polaris] [US Navy] *mt, fu, hc, 12, ph, tb, st, jj*
FBOE Frequency Band of Emission *fu*
FBR Fast Breeder Reactor *wp*
FBR Feedback Report *fu*
FBR Feedback Resistance *fu*
FBS (Earth Limb - 4 degrees FBS) *jmj*
FBS F-16 Bilgi Sistemi *mt*
FBS Forward Based Surveillance (system) *fu*
FBS Forward Based System *mt*
FBS Frame Buffer Slice *jdr*
FBUS Flight Back Up System *jj*
FBW Fly By Wire *mt, ci.349, wp, hc, jj, mt*
FC Facilities Contract *hc*
FC Fiber Channel *hi*
FC Field Change *fu*
FC Final Capability *kf*
FC Fire Control [also F/C] *17, wp, sb, w9, jj*
FC Fire Controlman *w9*
FC Firing Console *fu*
FC Force Control *mt*
FC Frequency Converter *jr*
FC Frequency Counter *0*
FC Fuel Cell *wp*
FC Function Code *jj*
FC-AL Fiber Channel Arbitrated Loop *hi*
FC/AL Fiber Channel/Arbitrated Loop *kg, hi*
FC/EL Fiber Channel/Enhanced Loop *pk, kg*
FCA Flow Control Assembly *wl*
FCA Functional Configuration Audit *mt, fu, hc, wp, pl, tb, jj, jdr, kf, oe*
FCAC Forward Control and Analysis Center *mt, fu*

FCB Field Change Bulletin *fu*
FCB File Control Block *kg, wp, ct, em*
FCB Forms Control Buffer *fu*
FCC Face Centered Cubic *ai1*
FCC Facility Configuration Console [formerly MACC] *fu*
FCC Federal Communications Commission [US] *mt, fu, ct, kg, hc, wp, sc2.7.3, 13, 15, w9, ce, tb, jj, kf, y2k, ai2*
FCC Fire Control Computer *hc, wp, jj*
FCC Flat Conductor Cable *hc*
FCC Fleet Command and Control *tb*
FCC Fleet Command Center [US Navy] *mt, kf, fu*
FCC Fleet Control Center *hc*
FCC Flight Clinical Coordination *mt*
FCC Flight Control Center *fu*
FCC Flight Control Computer *fu*
FCC Flight Coordinate (Coordination) Center *fu, hc*
FCC Fuels Control Center *mt*
FCCBMP Fleet Command Center Battle Management Program [PACFLT] *mt*
FCCM Facilities Capital Cost of Money *kf*
FCCS Flight Command Control System *fu*
FCCSET Federal Coordinating Council for Science Engineering and Technology *kg, ci.869, hy*
FCD Facility Completion Date *fu*
FCD Fire Detection Center *jj*
FCDA Fire Control Decision Aid *wp*
FCDNA Field Command Defense Nuclear Agency *mt*
FCDSSA Fleet Combat Direction Systems Support Activity *fu*
FCE Field Checkout Equipment *hc, jj*
FCE Field Control Element *hc*
FCE Fire Control Element *hc*
FCED Fire Control Engineering Description *hc*
FCEU Flight Control Engineering Description *hc*
FCF Functional Check Flight *mt, jj*
FCFS Frequency Coded Firing System *jj*
FCFT Fixed Cost, Fixed Time *wp*
FCG Foreign Clearance Guide *mt*
FCG Functional Coordinating Group *mt*
FCHP Full Coverage High Power *jdr*
FCHR Functional Cost Hour Report *fu, wp, jdr*
FCI Field Change Instruction *fu*
FCI Fluid Controls Institute *fu*
FCI Flux Changes per Inch *pk, kg*
FCI Functional Configuration Identification *mt, fu, wp*
FCIP Field Change Installation Program *hc*
FCIR Facility Change Initiation Request *hc*
FCIS Fire Control Interface Software *hc*
FCL Facility (Security) Clearance *su*
FCL Feedback Control Loop *wp*
FCLASS File Security Classification Codes *mt*
FCLP Field Carrier Landing Practice [USA] *jj, mt*
FCLS Factory Controlled Launch Support *kf*
FCM Flight Plan Conformance Monitoring *fu*
FCM Frequency Counter Measure *mc*
FCMS Functional Configuration Management System [AFLC] *wp, mt*
FCMSSR Federal Committee for Meteorological Services and Supporting Research *gr*
FCO Facility Control Office *mt*
FCO Federal Coordinating Officer *mt*
FCO Field Change Order *mt, tb*
FCO Fire Control Officer *fu, hc, jj*
FCO Firing Console Operator *fu*
FCOM Facilities Cost Of Money *kf*
FCP Facilities Criteria Plan *mt*
FCP Fire Control Panel *fu*
FCP Fire Control Pod *wp*
FCP Firm Cost Proposal *wp*
FCP Flight Control Processor *ai2*
FCP Frequency Control Processor *jdr*
FCPC Fleet Computer Programming Center *hc, fu*
FCPCP Fleet Computer Programming Center, Pacific *hc*
FCPS Fuel Consumption Projection System [PACOM] *mt*
FCR Facility Capability Report *hc*
FCR FIFO Control Register *pk, kg*
FCR File Change Report *kf, fu*
FCR File Change Request *fu*
FCR Fire Control Radar *mt, fu, hc, wp*
FCR Foreign Company Representative *jj*
FCR Functional Capability Requirement *wp*
FCR Fuse Current Rating *fu*
FCRC Federal Contract Research Center *mt, 12, fu, ts4-6, wp, tb, kf*
FCS Facility Checking Squadron [AFCC*] *mt*
FCS Federal Catalog System *mt*
FCS Fiber Channel Standard *pk, kg*
FCS Fire Control System *mt, hc, wp, fu, jj*
FCS Firmware Control System *fu*

FCS First Customer Release *ct, em*
FCS Fleet Communication Satellite *hc*
FCS Flight Control System *mt, hc, jj*
FCS Frame Check Sequence *mt, kg, fu*
FCS Fuel Control System *wp*
FCS Fuels Capabilities System [MAC] *mt*
FCS Full Communications Service *mt*
FCSC Federal Conversion Support Center *mt*
FCSC Fleet Command Support Center *fu, hc, kf*
FCSG Flight Control Sensor Group *0, hc*
FCSL Fire Control System Laboratory [Hughes Aircraft Company] *hc, jj*
FCST Federal Council for Science and Technology *ci.869*
FCT Field Communications Team *fu*
FCT Filament Center Tap *fu*
FCTC Fleet Combat Training Center *fu*
FCU Flight Control Unit *jj*
FCW Forced Control Word *fu*
FCW Frequency Control Word *uh, jdr*
FD Face of Drawing *jdr*
FD Fault Detection *mt, hc, wp, jj, fu*
FD File Description *mt*
FD Flex Density *-9*
FD Flexible Disk *fu*
FD Flight Data *fu*
FD Flight Director *kf, oe*
FD Floppy Disk (Drive) *kg, hi*
FD Force Direction *fu*
FD Frequency Diversity *fu*
FD Frequency shift, Doppler *jj*
FD Full Duplex *hc, kg, hi*
FD Functional Demonstration *st*
FD Functional Description *mt, fu*
FD&C Fault Detection and Correction *oe*
FD/FI Fault Detection/Fault Isolation *tb*
FD/FL Fault Detection/Fault Location *fu*
FDA Final Design Audit *-4*
FDA Food and Drug Administration *sc3.6.4, wp, w9, is26.1.69, y2k*
FDADS Fault Digital Airborne Data System *hc*
FDAT Final Development Acceptance Test *mt*
FDB Forced Draft Blower *fu*
FDB Full Data Block *fu*
FDBM Functional Data Base Manager [JDS] *mt*
FDC Facilities Design Criteria *fu*
FDC Fire Direction Center [USA] *mt, hc, fu, ts17-1*
FDC Fire Distribution Center *fu*
FDC Flight Director Computer *hc*
FDC Floppy Disk Controller *kg, ct, em, hi*
FDCACA Failure Data Collection Analysis and Corrective Action *fu*
FDCO Foreign Disclosure Coordinating Office [of DOD] *hc, dt*
FDD File Description Dictionary *fu*
FDD Floppy (Flexible) Disk Drive *fu, em, hi*
FDDI Fiber Digital Device Interface *pk, kg*
FDDI Fiber Distributed Interface *mt*
FDDI Fiber optic Digital Data Interface *go1.5.5*
FDDI Fiber optical Data Distribution Interface *tb*
FDDI Fiber [optic] Distributed Data Interface [100 Mbps LAN/MAN protocol] *is26.1.31, wp, pf.f88.27, kf, ct, em, nu, kg, fu, hi*
FDDL Frequency Division Data Link *fu, hc*
FDDS Flag Data Display System *mt, fu*
FDDS Frequency Division Discrimination Subsystem *hc, pf1.4*
FDE Field Decelerator *fu*
FDE Flight Data Entry *fu*
FDEN Flight Data Entry Notation *fu*
FDEP Flight Data Entry [and] Printout *mt, wp, fu*
FDESC Force Description *mt*
FDF Flight Data File *wl, oe*
FDF Flight Dynamics Facility *hy, oe*
FDG Fractional Doppler Gate *fu, jj*
FDHDD Floppy Disk High Density Drive *hi*
FDI Failure Detection and Isolation *ai2*
FDI Fault Detection and Isolation *kf*
FDIC Federal Deposit Insurance Corporation *y2k*
FDIO Flight Data Input / Output *hc, fu*
FDIO Flight Data Inputs Operator *fu*
FDIS Final Draft International Standard *ch*
FDISK Fixed Disk *pk, kg*
FDL Fast Deployment Logistics *dt, wp*
FDL File Definition Language *fu*
FDL Flight Dynamics Laboratory [of AFWAL] *wp, hc, tc, aa27.4.b31*
FDLP Final Design Load Cycle *ric, jdr*
FDLS Fast Deployment Logistic Ship *hc, jj*
FDM Feasibility Demonstration Model *hc, jj*
FDM Fill and Drain Module *-4*
FDM Fleet Demonstration Model *hc, jj*
FDM Flight Data Manager *wp*
FDM Formal Development Methodology *fu*

FDM Formatted Data Message *fu*
FDM Frequency Distribution Multiplexer *jdr*
FDM Frequency Division Multiplex(ing) (Multiplexed) *mt, kf, kg, hc, bl1-5.58, 7, ce, jj*
FDM Functional Development Model *fu*
FDMA Frequency Division Multiple Access *mt, fu, ct, em, uh, hc, bl1-5.74, rg, lt, ce, jdr, hi*
FDMG {a fluid dynamics organization at AFWAL} *aa27.4.b34*
FDMIS Force Development Management Information System [USA] *mt*
FDMS Flight Data Management System *mt*
FDMX Frequency Division Multiplexing [also FDM] *hi*
FDO Fee Determining Official *wp, tb, st, kf, oe*
FDO Fighter Duty Officer *mt, fu*
FDO Flight Deck Officer *mt*
FDO Fuse Delay Override *jj*
FDP Flight Data Processing (Position) *fu*
FDP Full Data Block *fu*
FDP Functionally Distributed Processing *fu*
FDPS Flight Data Processing System *fu*
FDR Final Design Review *mt, fu, id4142-10/831, 17, sb, hc, wp, ep, jj, jdr, oe*
FDR Flight Data Recorder (Record) *mt, fu, aw124.21.13, wp*
FDRP Fleet Data Reduction Program *fu*
FDRS Fleet Data Reduction System *fu*
FDS Firmware Development System *fu*
FDS Flash Detection Sensor *hc*
FDS Flight Demonstration System *kf*
FDS Flight Dynamics System *4, oe*
FDS Fuels Dispensing System *mt*
FDSP Flash Digital Signal Processor *hi*
FDSU Flight Data Storage Unit *wp*
FDT Fault Detection Tester *hc*
FDT Final Development Test *fu*
FDT Flexible Digital Terminal *mt, hc*
FDT Flight Data Terminal *fu*
FDT Foreign Technology Division {probably FTD ?} *tb*
FDT&E Force Development Test and Experimentation *fu*
FDTE Force Development Test and Experimentation *st*
FDTP Functional Demonstration and Test Program *fu*
FDTS Field Data Tracking system *mt*
FDU Frequency Distribution Unit *jdr*
FDU Frequency Doubling Unit *hc*
FDV Flight Development Vehicle [HS-601] *kf*
FDWG Flight Design Working Group *oe*
FDX Full Duplex *mt, do96, fu, nu, kg, hi*
FE Fast Ethernet *hi*
FE Ferro Electric *hc, jj*
FE Field Engineer *mt, hc, aw130.11.76, jj*
FE Flight Engineer *mt*
FE Flight Examiner *mt*
FE Flight Experiment *5, 17, sb, cp*
FE Force Element *fu*
FE Force Estimation *fu*
FE Force Execution *mt*
FE Forwarding Element *fu*
FE Free Electron *tb*
FE Frequency Error *jdr*
FE&D Facilities, Engineering, and Development *fu*
FEA Fast Ethernet Alliance *hi*
FEA Finite Element Analysis *wp*
FEA Fluid Experiments Apparatus *ws4.10*
FEADS Failure Effects and Data Synthesis *fu*
FEAF Far East Air Force [USA] *mt*
FEB Functional Electronic Block *fu*
FEBA Forward Edge of the Battle Area [USA DOD] *mt, fu, hc, dh, 14, aw129.1.47, wp, tb, kf*
FEBA Forward Engagement Battle Area *hc*
FEBI Front End Bus Interface *wp*
FEC Foreign Exchange Carrier *ct*
FEC Forward Error Control *csII.102, ce*
FEC Forward Error Correction [error correction at receiver without retransmission] *mt, 2, wp, 4, hc, kf, ful, nu, ct, em, hi*
FECN Forward Explicit Congestion Notification *kg, ct, em*
FECO Fourth Engine Cut Off *tb*
FECP Formal Engineering Change Proposal *fu*
FECS Front End Computer System *hc*
FED Field Effect Diode *hc, fu, jj*
FED Field Emission (Emitter) Display *kg, hi*
FED Formatted Event Data *kf*
FED-STD Federal Standard *mt, mi*
FEDAC Forward Error Detection and Correction *hc*
FEDB Failure Experience Data Bank *fu*
FEDD For Early Domestic Disassembly *fu*
FEDI Failure Experience Data Interchange *mt, hc*
FEDR Formatted Event Data Report *kf*
FEDRS Fleet Exercise/Data Reduction System *fu*

FEDSTD Federal Telecommunications Standard *mt*
FEDSTRIP Federal Standard Requisitioning and Issues Procedures *mt*
FEEDS Facilities Engineering Expert and Diagnostic System *mt*
FEEMS Facilities Engineering Equipment Maintenance System [USA] *mt*
FEES Front End Edit System [USA] *mt*
FEFA Future European Fighter Aircraft [later renamed EFA] *mt*
FEFO First Ended, First Out *kg, hi*
FEFT Fleet Exercise Fuel Tank *fu*
FEGLI Federal Employees' Group Life Insurance *mt, wp*
FEI Field Engineering Instructions *jj*
FEI Firing Error Indicator *hc*
FEI Fourth Engine Ignition *tb*
FEI Frequency Electronics, Incorporated *pl, jdr, oe*
FEIA Flight Engineers' International Association *hc*
FEIP Facility and Equipment Improvement Program *mt*
FEJE Facilities Engineer Job Estimating System *mt*
FEL Free Electron Laser *fu, hc, wp, dh, 14, 17, aw, ic1-2, sb, tb, st*
FEL-TIE Free Electron Laser/Technical Integration and Evaluation *-17*
FEM Finite Element Method *wp, kf, ai1*
FEM Finite Element Model *oe*
FEM Force Effectiveness Model *hc, jj*
FEM Front End Module *hc*
FEMA Federal Emergency Management Agency *mt, su, wp, sd, 12, hc, tb*
FEN Far East Network *mt*
FENCER [NATO code name for Soviet Su-19 air superiority aircraft] *wp*
FENP Front End Network Processor *mt*
FEO Field Engineering Order *fu, wp*
FEOC Field Emergency Operations Centers [States] *mt*
FEOTB Front End Of The Business *jdr*
FEP Fleet EHF Package [US Navy] *mt*
FEP Fleetsat (FLTSAT) EHF Program (Package) *uh, fuld*
FEP Flight Experiment Program *17, sb*
FEP Fluorinated Ethylene Propylene *he, fu*
FEP Forecast Expenditure Plan *-4*
FEP Front End Processor *mt, ct, kg, wp, em, fu, oe*
FEPI Front End Programming Interface *pk, kg*
FEPROM Flash Erasable PROM *kg, hi*
FER Flight Effectiveness Ratio *jj*
FER Flight Experiment Review *17, sb*
FER Force Exchange Ratio *wp*
FERAM Ferroelectric RAM *pk, kg*
FERC Federal Energy Regulatory Commission *mt, y2k*
FERF Far End Reporting Failure *ct, em*
FES Field Engineering Support *jj*
FES Fine Error Sensor *es55.42*
FES Fleet Exercise Section *fu*
FES Flywheel Energy Storage *pbl*
FES Functional Element Specification *fu*
FESD Flight Experiment Specification Document *17, sb*
FESDK Far East Software Development Kit *pk, kg*
FESE Field Enhanced Secondary Emission *fu*
FESS Facilities Engineering Supply System [USA] *mt*
FEST Fast Erase Storage Tubes *hc, jj*
FEST Field Engineering Support - Tucson *hc, jj*
FESTS Fleet Exercise Section Test Set *fu*
FET Field Effect Transistor [hardware] *mt, ah, kg, hc, wp, sc9.6.3, mj31.5.39, lt, ep, ce, jj, pbl, jdr, tm, kf, vv, fu, hi, oe, ai1*
FET Field Effort Transistor [(e.g., amplifier)] {sic} *4, 6*
FET Field Emission Transistor *hn43.25.4*
FET Fighter Evaluation Team *jj*
FET Functional Electrical Test *pl*
FETA Flexible Explosive Transfer Assembly *oe*
FETP Flight Experiment Test Plan *17, sb*
FETS Field Evaluation and Test System *hc*
FETS Fleet Exercise Test Set *fu*
FEWES Fleet Exercise Warhead Electronics Subsystem *fu*
FEWS Follow-on Early Warning System *kf*
FEWSG Fleet Electronic Warfare Support Group [US Navy] *fu, mt*
FEX Feature Extractor *17, sb*
FEZ Fighter Engagement Zone [USA DOD] *fu, mt*
FF Failure Factor *fu*
FF Fire and Forget *wp*
FF Fixed Format *fu*
FF Flip Flop *jj, kg, fu*
FF Fraction of Fill *sh*
FF Free Flyer *wl*
FF Frigate [US Navy ship classification] *mt, fu, wp*
FF/FFG Frigate/Frigate Guided Missile *jj*
FFA Fraction of False Alarms *fu*
FFAR Folding Fin Aerial Rocket *hc*
FFAR Folding Fin Aircraft Rocket [originally Forward Firing Aircraft Rocket] *mt*
FFAST Fire and Forget Anti Tank System Technology *hc*
FFBD Functional Flow Block Diagram *mt, fu, hc, tb, jj, jdr*

FFC Feed Forward Compensation *-3*
FFC Flat Flexible Cable *sd*
FFC Flip Flop Complementary *fu*
FFC French Flying Corps [World War I] *af72.5.58*
FFC Functional Flow Chart *hc, jj*
FFCC Forward Facing Crew Cockpit [the two-seat cockpit of the Airbus A300] [Europe] *mt*
FFD Fraction of Failures (Faults) Detected *fu, tb*
FFD Functional Flow Diagram *hc, dm, jj, fu*
FFDB Friendly Forces / Facilities Data Base [PACAF] *mt*
FFDC First Failure Data Capture [IBM] *pk, kg*
FFDS Fleet Flagship Data System *hc*
FFE Fire for Effect *fu*
FFEC Field-Free Emission Current *fu*
FFF FLIR Full Finder *fu*
FFG Fixed Frequency Generator *jdr*
FFG Frigate, Guided Missile armed [US Navy ship classification] *jj, mt, fu*
FFGH Frigate, Guided missile, Helicopter [US Navy] *mt*
FFH Fast Frequency Hopping (Hop) *hc, uc*
FFH Fixed Flight Hours *wp*
FFH Frigate, Helicopter [US Navy] *mt*
FFI Forsvaret Forskining Institut [defense research establishment, Norway] *mt*
FFI Fraction of Failures (Faults) Isolated *fu, tb*
FFIEC Federal Financial Institutions Examination Council *y2k*
FFK Fixed Function Key *mt, fu*
FFL Corvette [USA] [US Navy] *mt*
FFL Flip Flop Latch *fu*
FFL Front Focal Length *fu*
FFMED Fixed Format Message Entry Device *fu*
FFN Fleet Flash Net [US Navy] *mt*
FFO Factored Future Order *fu*
FFOB Flexible Fiber Optics Bundle *hc*
FFOP Failure Free Operating Period *wp*
FFOS Forward Flying Observation System *mt*
FFOT Fast Frequency On Target *hc, jj*
FFP Failure Free Period *rm1.2.2*
FFP Firm Fixed Price [contract] *mt, fu, hc, sm18-1, 17, 18, wp, ep, tb, jdr, kf, oe*
FFP Fixed Fee Procurement *wp*
FFPC Firm Fixed Price Contract *jj*
FFPI Firm Fixed Price Incentive *hc*
FFPLRM Firmed Fixed Price Labor Reimburseable Materials *hc*
FFPS Fuel Flow Power Supply *jj*
FFPVE Firm Fixed Price Value Engineering *hc*
FFR Flight Feasibility Review *bo*
FFR Frame-Frame Registration *oe*
FFRB Field Failure Review Board *jj*
FFRDC Federally Funded Research and Development Center *mt, kf*
FFRR Full Frequency Range Recording *fu*
FFS Fast File System *kg, hi*
FFS Formatted File System *mt*
FFST First Failure Support Technology [IBM] *pk, kg*
FFT Fast Formula Translation *jj*
FFT Fast Fourier Transform *kg, mi, fu, uh, hc, wp, sa830182P.24, is26.1.8a, rg, tb, st, jj, jdr, mb, hi, oe, ai1*
FFT File Format Table *mt*
FFT Final Form Text [IBM] *pk, kg*
FFT Fixture Functional Test *jj*
FFT For Further Transfer *jj*
FFT Formation Fighter Trainer *wp*
FFV Forenade Fabriksverken [Sweden] *hc, aw129.10.33*
FFW Failure Free Warranty *jj*
FG Fighter Group [USA] *mt*
FG Filament Ground *fu*
FG Floating Gate *pk, kg*
FGA Fighter Ground Attack [UK] *mt*
FGAS Flyable Generic ARM Seeker *hc*
FGC Functional Group Code *mt, fu, jj*
FGCS Flight Guidance and Control System *wp*
FGGE First GARP Global Experiment *pl, ns, hg, ge, oe*
FGM Floating Gear Mechanism *pbl*
FGM [man launched, surface-target missile, mainly anti tank missiles such as FGM-77 Dragon] [USA] *mt*
FGOS Flag and General Officers Seminar *mt*
FGR Fighter Ground Attack Reconnaissance [UK] *mt*
FGREP Fixed Global Regular Expression Print [Unix] *pk, kg*
FGS Fine Guidance Sensors [on HST] *mi*
FGSS Flexible Guidance Software System *hc*
FH Flight Hardware *jdr*
FH Flight Hours *wp, jj*
FH Frequency High *jdr*
FH Frequency Hopping *mt, fu, sd, jdr, kf*
FH/DSPN Frequency-Hopped Direct-Sequence Pseudo-Noise *fu*
FHC Fairchild Hiller Corporation *hc*

FHD Fixed Head Disk *mt, fu, hi*
FHDTC Fixed Head Disc Transfer Channel *mt*
FHE Forward Headquarters Element [CENTCOM] *mt*
FHSS Frequency Hopping Spread Spectrum *pk, kg*
FHST Fixed Head Star Trackers [on HST] *mi*
FHWA Federal Highway Administration *mt*
FI Fault Isolation *mt, hc, wp, jj, fu*
FI Field Intensity *fu*
FI Fixed Interface *hc, jj*
FI Flexible Interconnect (Intraconnect) *fu, mt*
FI Force Integration *mt*
FIA Financial Inventory Accounting *wp*
FIA Flame Ionization Analysis *wp*
FIA Fluoro Immuno Assay *wp*
FIA Functional Interoperability Architecture *mt*
FIA Future Imagery Architecture *rih*
FIAR Failed Item Analysis Report *fu*
FIAT First Installed Article Test *fu*
FIB File Interface Block *wp*
FIB Flight Information Bulletin *wp*
FIB Focused Ion Beam *hc, jdr*
FIB Foreground Initiated Background *fu*
FIC Fleet Intelligence (Information) Center [US Navy] *mt, fu*
FIC Force Indicator Code *mt*
FIC Frequency Interference Control *fu*
FIC Frequency Interval Counter *jj*
FICEURLANT Fleet Intelligence Center-Europe and Atlantic [US Navy] *mt*
FICON Fighter Conveyer [a GRB-36 carried a RF-84K reconnaissance aircraft in its bomb-bay] [USA] *mt*
FICPAC Fleet Intelligence Center, Pacific [US Navy] *mt*
FICS Fault Isolation Checkout System *hc*
FICS Forecasting and Inventory Control System *wp*
FID Fault Insertion Device *fu, hc*
FID Foreign Internal Defense [USA DOD] *mt*
FIDDLER [NATO code name for Soviet TU-28 Strike/RECCE aircraft] *wp*
FIDEL Focus Interactive Data Entry Language *fu*
FIDEL Full screen Integrated Data Entry Language *fu*
FIDI Forward Intra-Target Data Indicator *hc*
FIDO Fighter Interceptor Duty Officer *mt*
FIDO FOg DIspersal equipment {?} [UK] *mt*
FIDO Fog Investigation and Dispersal Operation *wp*
FIDO {Shuttle flight dynamics software} *pf1.24*
FIDOC Firing Doctrine *pm*
FIDRS Facilities Interface Data Requirement Sheets *jj*
FIDS Facility Intrusion Detection System *hc*
FIDU Filtered Input Data Unit *fu*
FIE Fleet Instrumentation / Exercise *fu*
FIET Field Integration Engineering Test *fu, hc*
FIF Fractal Image Format *kg, em*
FIFE First ISLSCP Field Experiment *hy, ge, hg*
FIFO First In First Out *mt, ma70, fu, kg, wp, hc, w9, ce, tb, ct, em, jj, hi, oe, ai1*
FIG Fighter-Interceptor Group (177th FIG 'Jersey Devils') [USA] *mt, af72.5.55, jj*
FIGAT Fiberglass Optical Target *jj*
FII Federal Item Identification *jj*
FII Release Item Identification *fu*
FIIG Federal Item Identification Guide *hc, fu, jj*
FIIN Federal Item Identification Number *mt, fu, hc, wp, jj*
FIIT Flash Internal Information Transfer *fu*
FIKS Forsvarets Integrerede Kommunikation System [(Danish) Defense Integrated Communications System] *mt, fu*
FILL Fleet Issue Load List *jj*
FILO First In, Last Out *kg, hi*
FILSYS File Handling Subsystem [Honeywell] *mt*
FIM Fabrication Instructions and Materials *fu*
FIM Fault Isolation Module *fu*
FIM Field Instruction Memorandum *0*
FIM Field Ion Microscopy *wp*
FIM Firing Instruction Manual *0*
FIM Flight Information Memorandum *0*
FIM [designator for man launched surface to air missiles, USA DOD] (FIM-43 Redeye, FIM-92A Stinger) *mt*
FIMS Fault Isolation Monitoring System *-2*
FIN Financial Inventory *fu*
FIND Fault Isolation By Nodal Dependency *wp*
FIND Federal Item Name Directory *fu*
FINDER Functionally Integrated Designating and Reference system *mt*
FINE Fixed Installation NAIAD Equipment [chemical warfare, UK] *wp*
FINRAE Ferranti Inertial Rapid Alignment Equipment. [equipment to initialize the INS of a BAe Harrier] [UK] *mt*
FINRS Far Infrared Noncoherent Radiating Systems *hc*
FIO Flight Information Officer *fu*
FIP Falcon Improvement Program *hc*
FIP Fault Isolation Plan *jj*

FIP File Processor [buffering] *pk, kg*
FIP Fleet Introduction Plan *fu*
FIP Fleet Introduction Program [USA] *hc, jj, mt*
FIPS Facilities Inventory and Planning System [US Navy] *mt*
FIPS Federal Information Processing Standards [a publication] *mt, kg, fu, csII.59, wp, tb, kf*
FIR Far Infra Red *hc, mi, wp, jj, mb*
FIR Fast Infrared *kg, em*
FIR Finite Impulse Response *kg, fu*
FIR Fixed Interface Ratio *jj*
FIR Flight Information Region *mt, fu*
FIR Full Indicator Reading *fu*
FIR Functional Item Replacement *fu*
FIRAMS Flight Incident Recorder and Aircraft Monitoring System *mt*
FIRAS Far Infra Red Absolute Spectrometer *hc*
FIRC Far Infra Red Camera *hc*
FIRE First ISCCP Regional Experiment *ge, ns, hy, hg*
FIREFINDER [AN/TPQ-36, 37 weapon locating radars] *fu*
FIREFLY Advanced Fighter Fire Control System *wp*
FIREFLY [reconnaissance drone] *wp*
FIRL Fleet Issue Requirement List *jj*
FIRL Franklin Institute Research Laboratories *hc*
FIRM Far Infrared Radiation Measurements *hc*
FIRM Fleet Introduction Replacement Model *jj*
FIRS Framing Infra Red Sensor *hc*
FIRST Far Infra Red Search and Track *go4.2*
FIRST Flexible Infra Red Search and Track *hc*
FIRST Forum of Incident Response and Security Teams *pk, kg*
FIRST Fourier Interferometer for Random Source Transient *hc*
FIRT FRACAS Incident Review Team *fu*
FIS Field Integration Services *gp, 16*
FIS Fighter Identification System *hc, jj*
FIS Fighter Interceptor Squadron (177th FIS 'Jersey Devils', 178th FIS 'Happy Hooligans', 87th FIS 'Red Bulls') [USA] *mt, 12, af72.5.55, jj*
FIS FIREFINDER Information System *fu*
FIS FORSCOM Information System *mt*
FIS {an instrument on Mars orbiter} *ma35*
FISA Flexible Integrated Solar Array *hc*
FISA Flight Information Services Automation *fu*
FISA Foreign Intelligence Surveillance Act *-10*
FISAR Fleet Information Storage and Retrieval *fu*
FISC Fleet Intelligence Support Center [US Navy] *mt*
FISCA Flexible Integrated Solar Cell Array [Hughes Aircraft Company] *hc*
FISHBED MIG-21 *wp*
FISI Fault Insertion Simulation [program] *hc*
FISINT Foreign Instrumentation Signals Intelligence [USA DOD] *mt*
FISK Fragment, Incendiary, Shaped Charge *hc*
FIST Fighter Interceptor Slaved Telescope *hc*
FIST Final Integration (Integrated) System Test *uh, gp, 17, sb, jdr, kf*
FIST Fire Support Team [also FST] *mt, fu, hc*
FIST Fleet Imagery Support Terminal *mt, fu*
FIST Forward Instability Shock Test *jdr*
FISTV Fire Support Team Vehicle *hc*
FIT Fault Isolation Test(er) *fu, hc, wp, jj*
FIT File Information Table *mt*
FIT Final Integration Test *pbl, ah*
FIT Flight Instrument Tester *pl*
FIT Flight Integrity Test *ah, pbl, jdr*
FIT [failure per billion hours, a characteristic] *tm*
FITD Far Infrared Target Detector *wp*
FITOW Future Improvement TOW *hc*
FITR Final Integrated Technology Reviews *-17*
FITR {Flight Instrument Test Report} *sb*
FITS Flexible Image Transport System *mi, ct, em*
FITS Flexible Interchange Transport Format [the IAU standard for astronomical data] *cu*
FITSIO [a portable suite of subroutines developed at the OGIP to provide convenient access to FITS files] *cu*
FITWING Fighter Wing *jj*
FIU Fault Isolation Unit *fu*
FIU Fingerprint Identification Unit *pk, kg*
FIU Flight Interim Unit *hc*
FIW Fighter Interceptor Wing *12, af72.5.55, jj*
FIX Federal Internet Exchange *kg, hi*
FJC Falcon Jet Corporation *aw130.6.72*
FJSRL Frank J. Seiler Research Laboratory *tc, wp*
FJX (FJX short-haul turbofan transport) *aw130.11.31*
FL Fault Location *fu*
FL Flight Level *mt, fu, aw129.10.72*
FL Focal Length *w9, fu*
FL Foot Lumen *mt*
FLA Field Logistics Activity *fu*
FLA Foreign Launch Assessment *mt*
FLA Future Large Airlifter [a proposed C-130 replacement] [Europe] *mt*
FLAC Final Load Analysis Cycle *kf*

FLAC Flutter Analysis by a Collocation Method *hc*
FLACT Forward Looking Active Classification Technology *wp*
FLACV Future Light Armor Combat Vehicle *hc*
FLAG Forward Looking Air to Ground *hc*
FLAGE Flexible Lightweight Agile Guided Experiment *mt, hc, wp, nd74.445.18, tb, st*
FLAGON [NATO code name for Soviet Su-15 Interceptor] *wp*
FLAIR Factory Liaison And Inspection Resources *nd74.445.27*
FLAM Flutter Analysis by a Model method *hc*
FLAMING_ARROW [code name for USEUCOM communications network] *mt*
FLAMR Forward Looking Advanced Multimode Radar *hc, jj, fu*
FLANG Florida Air National Guard *-12*
FLAP FLAMR Angle Processor *hc*
FLAP Forward Looking Angle Processor *fu*
FLAPS Force Level Automated Planning System [USAFE / German] *mt*
FLAPS Functional Layout and Presentation System *fu*
FLAR Forward Looking Airborne Radar *wp*
FLAS Fuels Logistical Area Summary *mt*
FLASH Folding Light Acoustic System for Helicopters *fu*
FLASH Force Level Alerting System *mt*
FLAT Flight Plan Aided Tracking *fu*
FLB Frame Lead Byte *fu*
FLC Federal Laboratory Consortium [for technology transfer] *ci.1007*
FLC Ferro electric Liquid Crystal *kg, hi*
FLC Final Loads Cycle *kf, ric, jdr*
FLC Force Level Control *fu*
FLCC Flight Control Computer *wp*
FLCS Force-Level Control System *fu*
FLDK Flight Deck *jj*
FLDR Flight Loads Data Recorder *wp*
FLEA Flux Logic Element Array *fu*
FLEEP Flying Lunar Excursion Experimental Platform *hc*
FLENUMOCEANCEN Fleet Numerical Oceanography Center *ag, dm*
FLEXAR FLexible Adaptive Radar *fu, hn41.22.6, hc*
FLEXREF Fleet Exercise Reconstruction Facility *fu*
FLICS Flight Command Simulation *hc*
FLIDAP Flight Data Position *fu*
FLIDIT Flight Line Detection and Isolation Technique *hc*
FLINN Fiducial Laboratory for an International Science Network *hy*
FLINT (Foreign Launch Processor (Software Program)) *jb*
FLINT Flow Field Interference Study *hc*
FLIP Flight Information Publication *mt, fu, wp*
FLIP Floating Instrument Platform *hc*
FLIPCO Flight Plan Coordination (System) *fu*
FLIPS Flight Information Processing System *mt*
FLIR Forward Looking Infrared (Radar) *mt, kf, fu, wp, pm, st, jj, aw130.13.53, hn42.4.3, hc, af72.5.134*
FLIRAS Forward Looking Infrared Attack Set *hc*
FLIRS Forward Looking Infrared Radar System *-17*
FLITE Federal Legal Information Through Electronics *mt*
FLL Flash Lamp Life *hc*
FLL FoxPro Link Library *pk, kg*
FLL Frequency Lock (Locked) Loop *fu, hc, lt, ep, jj*
FLLD Full Load *jj*
FLLLTV Forward Looking Low Light Television *hc*
FLM Flight Line Maintenance *hc, jj*
FLM Flow Management *fu*
FLML Flight Line Memory Loader *hc*
FLMMAR Forward Looking Multiple Mode Attack Radar *hc*
FLO File Operations Program *fu*
FLOGEN Flow Generator [MAC] *mt*
FLOGGER MIG-23 *wp*
FLOGRAP Fuels Logistics Readiness Assessment Program *mt*
FLOLS Fresnel Lens Optical Landing System *jj*
FLOP Floating Point Operation *em*
FLOP Front Line Of Troops *wp*
FLOPS Floating Point Operations Per Second *kg, wp, tb, hi*
FLOT Forward (Front) Line Of [Own] Troops *fu, 16, wp, pm*
FLOT Forward Line of Own Troops [USA DOD] *mt*
FLOX Fluorine-Liquid Oxygen [mixture] *ci.224, hc, wp*
FLP Fault Locate Program *fu*
FLP Flight Plan *fu*
FLQ Filtered (one-way) Link Quality *fu*
FLR Forward Looking Radar *mt, hc, wp, jj*
FLS Field Logistics Specialist *fu*
FLS Field Logistics System *mt*
FLS Forward Logistics Site [US Navy] *mt*
FLSC Flexible Linear Shaped Charge *wp, ai2*
FLSIP Fleet Logistics Support Improvement Program *fu*
FLSS Falcon Launching Saber System *hc*
FLST Final Launch Site Tests *kf*
FLT Flight Line Tester *hc, jj*
FLTAC Fleet Analysis Center *jj*
FLTBCST Fleet Broadcast [US Navy] *mt*
FLTBDCST Fleet Broadcast *jdr*

FLTCINC Fleet Commander In Chief *fu*
FLTCINCS Fleet Commanders in Chief [US Navy] *mt*
FLTCORGRU Fleet Composite Operational Readiness Group [US Navy] *fu, mt*
FLTDECGRU Fleet Deception Group *mt*
FLTF Full-Load Time Frame *fu*
FLTK Flight Line Test Kit *hc, jj*
FLTOPS Flight Test Oriented Pre-Compiler System *hc*
FLTSAT Fleet Satellite *mt, fu*
FLTSAT Fleet Satellite Communications *uh*
FLTSATCOM Fleet Satellite Communications [system] [US Navy] *mt, fu, sc2.8.2, 12, 17, af72.5.68, jb, kf*
FLTSEVO Fleet Secure Voice [US Navy] *mt*
FLU Front Line Units *jj*
FM Failure Management *fu*
FM Field Maintenance *hc, jj*
FM Field Manual *mt, hc, w9, jj, fu*
FM File Maintenance *mt*
FM File Mark *jj*
FM Flight Model *pl, es55.44, oe*
FM Force Modernization [USA] *mt*
FM Force Module [USA DOD] *mt*
FM Frequency Management [REDCOM / JDA] *mt*
FM Frequency Modulation (Modulated) *mt, ct, fu, hc, bo, bll-5.1, 0, 2, 4, 16, aa26.11.44, lt, wp, ce, jj, jdr, kf, hi, oe, tt, ai1*
FM Frequency Multiplex *fu*
FM Frequency Multiplier *0*
FM Functional Model *jdr*
FMA Failure Mode Analysis *hc, jj*
FMA Ferrite Modulator Assembly *3, 8, 9*
FMA Final Mission Analysis *oe*
FMA Fleet Maintenance Activity *fu*
FMA Foreign Media Analysis *mt*
FMA Foreign Military Assistance *mt*
FMA F·brica Materiales Aeroespaciales [CÛrdoba, Argentina] *aw129.10.4*
FMACCEL FM Accelerometer *-9*
FMAS Flush Mounted Antenna System *hc*
FMAT Frequency Modulation Anticipation Time *jj*
FMB Financial Management Board *wp, mt*
FMBS Frame Mode hearer Service {?} *nu*
FMC (FMC, BFVS prime contractor) *hn44.3.3, hc, aw130.22.58*
FMC Facilities Management Contract *fu*
FMC Federal Manufacturers' Code *hc, jj*
FMC Fighter Mode Command *jj*
FMC Full Mission Capability (Capable) [USA DOD] *mt, fu, wp, kf*
FMCCS Force Modernization Command and Control System [USA] *mt*
FMCS Flight Management Computer System *mt*
FMCS Flight Mission Control Study *hc*
FMCS Freight Movement Control system [MTMC] *mt*
FMCS Fuels Management Capabilities System [MAC] *mt*
FMCW Frequency Modulated Continuous Wave *mt, fu*
FMDS Failure Management Design System *gp*
FMDS Field Maintenance Data System *mt*
FMDS Flight Management Data (Display) System *hc, wp*
FMDS Flight Management Data System [USAF] *mt*
FMDS Flight Model Discharge System *hn44.11.1*
FME Field Modification Engineering *jj*
FMEA Failure Modes and Effects Analysis [a document] *fu, kf, 17, wp, hc, rm8.2.3, lt, ep, dm, tb, jj, pbl, tm, jdr, ah, he, oe*
FMEC Failure Modes, Effects, and Criticality [analysis] *gp, sb, pl*
FMECA Failure Mode(s), Effects(s) and Criticality Analysis *mt, fu, hc, wp, nt, st, jj, jdr, kf, oe*
FMET Functional Management Engineering Team *mt, af72.5.109*
FMF Fleet Marine Force]USMC] *hc, 12, jj, mt*
FMF-EUCE Fleet Marine Force - End User Computing Equipment [USMC] *mt*
FMFATL Fleet Marine Force Atlantic *hc*
FMFB Frequency Modulation Feedback *fu*
FMFEUR Fleet Marine Force, Europe [USMC] *mt*
FMFLANT Fleet Marine Force Atlantic [USMC] *mt, 12, jj*
FMFPAC Fleet Marine Force Pacific [USMC] *mt, 12, jj*
FMG Frequency Manager *-2*
FMH Free Molecular Heating *jdr, uh*
FMHS Formal Message Handling System *mt*
FMI Failure Mode Indicator *hc, jj*
FMI Finnish Meteorological Institute *ge, hg*
FMI Functional Management Inspection *mt, af72.5.107, wp*
FMI Future Manned Interceptor *hc*
FMIS Facilities Management Information System [OSD] *mt*
FMIS Financial Management Information System *mt*
FMIS Force Management Information System [SAC] *mt*
FMK Alternative form for AMC (FLYVEVAABNETS Material Kommando) *fu*
FML Feedback Multiple Loop *fu*
FML Major Force List *mt*
FMLS Force Module Logistics Sustainability [JCS] *mt*

FMMP Federal Master Mobilization Plan *mt*
FMO Fiscal Management Office *mt*
FMOCC Fleet Mobile Operational Command Center *mt*
FMOD File Modify *mt*
FMOF First Manned Orbital Flight *ci.620*
FMOMEM Failure Mode to Operating Modes by Effect Matrix *fu*
FMOMSM Failure Mode to Operating Mode by Severity Matrix *fu*
FMOP Frequency Modulation On Pulse *hc, fu*
FMOS Formatted Message Origination System *fu*
FMP Financial Management Plan *mt*
FMP Fleet Modernization Program [US Navy] *jj, fu, mt*
FMP Force Module Package [USA DOD] *mt*
FMP FORSCOM Mobilization Plan *mt*
FMPM Frequency Modulation Phase Modulation *fu*
FMPMIS Fleet Maintenance Program Management Information System *mt*
FMPO Fort Monmouth Procurement Office *fu*
FMPS FLIR Mission Payload Subsystem *hc*
FMR (General Skantze's FMR Data System) {a quality model} *nd74.445.26*
FMR Failure Maintenance Report *fu*
FMR Financial Management Review *mt, wp*
FMR Frequency Modulated Radar (Receiver) *fu*
FMR Frequency Modulation Ranging *hc, jj*
FMR Functional Management Review *mt*
FMRS Force Movement Requirements System [USA] *mt*
FMRS Forecast Movement Requirements System [EUCOM] *mt*
FMRS Fuels Management Requirements System *mt*
FMRT Frequency Modulation Real Time *9, lt*
FMS Field Maintenance Squadron *mt*
FMS Fighter Missile System *hc, jj*
FMS File (Forms) Management System *fu*
FMS File Maintenance System *mt*
FMS File Management Supervisor *mt*
FMS File Management System *mt*
FMS Financial Management Service *wp*
FMS Flexible Manufacturing System *fu*
FMS Flight Management System *mt, aw129.14.115*
FMS Force Management System [USAF] *mt*
FMS Foreign Military Sales [USA DOD] *mt, su, fu, hc, aw129.14.30, af72.5.28, wp, tb, jj*
FMS Formal Message Service *mt*
FMS Forms Management System *kg, hd1.89.y18*
FMS Fullerton Manufacturing Systems *fu*
FMSAEG Fleet Missile Systems Analysis and Evaluation Group *jj*
FMSCC Fullerton Manufacturing Systems Computer Center *fu*
FMSO Fleet Material Support Office [US Navy] *mt, dt, jj*
FMSO Foreign Military Sales Order *wp, jj*
FMSR Final Mission and Systems Review *re*
FMSS Force Module Subsystem [JOPS] *mt*
FMS_II Financial Management System [MTMC] *mt*
FMT Field Maintenance Technician *fu*
FMT Frequency Modulated Transmitter *fu*
FMTB Force Mobilization Troop Bases [USA] *mt*
FMTS Field Maintenance Test Set (Roland) *fu*
FMTS Field Maintenance Test Station/Sergeant (Set) *hc, jj*
FMU Fuze Munition Unit *wp*
FMV Forsvarets Materiel Verk [Sweden] *hc*
FMV Full Motion Video *pk, kg*
FMV-F [Defense Materials Administration] [Sweden] *hn42.4.1*
FMX FM Transmitter *wp*
FN Natural Frequency *vv*
FNAA Fast Neutron Activation Analysis *wp*
FNOC Fleet Numerical Oceanography Center [Monterey, California] *mt, ac7.5-9, pl, 16, ns*
FNP Front End Network Processor *mt*
FNR Flexible Nuclear Response *mt*
FNS Full NAEGIS Site *go3.1-1*
FNS Fully-implemented NADGE Site *fu*
FNWC Fleet Numerical Weather Central (Center) [Monterey, California] [US Navy] *mt, fu, ac2.7-11*
FO Fabrication Order *fu*
FO Fiber Optics *mt, tb, st, fu*
FO Field Officer *w9*
FO Field Order *w9*
FO Filtration Officer *fu*
FO Flash Override *fu*
FO Flight Observation *jr*
FO Flight Officer *w9*
FO Follow On *5, sb*
FO Forward Observer *mt, w9, fu*
FO Future Order *fu*
FOA Field Operating Agency *mt, fu*
FOA Fleet Operational Assets *hc*
FOA Forsvarets Forsknings Anstalt [Sweden] {Defense Research Institute} *hc*

FOAC Federal Office Automation Center *wp*
FOAD Field Operations Analysis Digest *jj*
FOB Fire-On-Board *fu*
FOB Forward Operating Base [USA DOD] *mt, hc, wp, fu, jj*
FOB Free On Board *mt, hc, wp, w9, jj*
FOB Freight On Board *gp, oe*
FOB Friendly Order of Battle *fu*
FOB Full Operational Base *mt*
FOBS Fractional Orbit (Orbital) Bombardment System *mt, fu, sd, hc, tb, ai2*
FOC Faint Object Camera [on HST] *mi, es55.42, mb*
FOC Final Operating (Operational) Capability *mt, pbl*
FOC First Operational Capability *hc, jj*
FOC Fleet Operations Command *fu*
FOC Flight Operations Center [USA] *mt, hc, fu, jj*
FOC Follow On Contract *hc, jj*
FOC Full (Final) Operational Capability *mt, 5, fu, 12, ic2-16, sb, hc, tb, kf, wp, hc, jb, st*
FOCA (the Swiss national aviation authority FOCA) *aw130.11.64*
FOCAP Fiber Optics Cost Analysis Program *hc*
FOCAS Fiber Optics Communications for Aerospace Systems *hc, wp*
FOCAS Force Capability Assessment System [TAF] *mt*
FOCI Foreign Ownership, Control, or Influence *mt, su, tb*
FOCIA Fiber Optics Coupled Image Amplifier *wp*
FOCR Final Operational Concept Review *17, sb*
FOCSI Fiber Optic Control System Integration *hc*
FOCUS Flight Operation Computer System *fu*
FOCUS Forum of Control Data Users *pk, kg*
FOCUS [programming language for report generation] [a commercial data base management system] *mt, mc, hd1.89.y14*
FOD Flag Officer, Denmark *fu*
FOD Foreign Object Damage [USA DOD] *mt, hc, wp, nd74.445.27, jj, fu*
FOD Foreign Object or Debris *oe*
FOD Function Operational Design *fu*
FODCCIS Flag Officer Denmark Command, Control and Information System *mt*
FODLM Fiber Optics Data Link Missile *hc*
FODS Fiber Optic Distributed System *wp*
FODU Fiber Optic Distribution Unit *fu*
FOE Figure Of Effectiveness *fu*
FOE Final Operational Evaluation *mt, wp*
FOE Follow On Engine *jj*
FOE Follow On Equipment *hc, st, jj*
FOE Follow On Evaluation *st, fu*
FOF Field Of Fire *wp*
FOF Field Operating Fund *fu*
FOF Flight Operations Facility *0*
FOFA Follow On Forces Attack *mt, 14, nd74.445.28, wp, tb*
FOG Fiber Optic Gyro *17, sb, st, ai2*
FOG Fiber Optics Guidance (Guided) *hc, wp*
FOG First Osborne Group *pk, kg*
FOG Flag Officer, Germany *fu*
FOG-M Fiber Optic Guided Missile [also FOGM] *mt, fu, aw130.13.13, hc, uc, 17, sb*
FOGD Fiber Optics Guidance Demonstration *hc*
FOGM Fiber Optic Guided Missile [also FOG-M] *wp*
FOI Fighter Officer Interceptor *mt*
FOI Follow On Interceptor *uc*
FOI Freedom Of Information *wp*
FOIA Freedom Of Information Act *mt, hc, jj, kf*
FOIM Field Office Information Management *ct, em*
FOIR Field Of Interest Register *hc*
FOIRL Fiber Optic Inter Repeater Link [IEEE] *kg, hi*
FOJ Fuze On Jam *wp*
FOL Fiber Optic Link *-17*
FOL Flight Over Land *jj*
FOL Forward Operating Location *mt*
FOLDOB Forward Operating Location Dispersed Operating Base *mt*
FOLDOC Free On Line Dictionary Of Computing *pk, kg*
FOLG Fiber Optic Laser Gyro *wp*
FOM Figure Of Merit *fu, wp, mb, tb*
FOM Flight Operations Memorandum *0*
FOMA Foreign Military Assistance file *mt*
FOMAE Follow On Management Application and Evaluation *mt*
FOMIS Fitting Out Management Information System *fu*
FOMM Functionally Oriented Maintenance Manual *wp, fu*
FOMMS Functionally Oriented Maintenance Manual [System?] *wp*
FOMS Functionally Oriented Maintenance [System?] *wp*
FON Fire Order Number *jj*
FOP Flight Operations Plan *sf31.1.20*
FOP Forward Observation Post *wp*
FOPEN Foliage Penetration [sensing or radar] *hc, dh, wp*
FOPT Fiber Optics Photon Transfer *hc*
FOR Fabrication Order Rework *fu*
FOR Field Of Regard *fu, 17, sb, wp, kf*
FOR Flight Operations Review *gp, 17, sb, oe*

FOR Formal Qualification Review *wp*
FORACS Fleet Operational Readiness Accuracy Check System *hc*
FORCEM Force Evaluation Model *wp*
FORD Fabrication Order *fu*
FORDAD Foreign Disclosure Automated Data *ro*
FORDIM Force Distribution Model *mt*
FORDTIS Foreign Disclosure and Technical Information System *mt, ph*
FORECAST Force Accounting System [USA] *mt*
FORECAST [airlift requirements forecast system, MAC*] *mt*
FORECAST [USAF Study "Vistas of Future Technologies And Science Areas"] *wp*
FORESEE Creativity, Concepts, Communications, Control [GSG program for improvement of productivity] [from pronunciation of "4C"] *fu*
FOREST Fast Order Radiation Effects Sampling Technique *hc*
FOREST_GREEN [program to develop nuclear weapon detonations] *wp*
FORGEN Force Generation report *mt*
FORMDEPS FORSCOM Mobilization and Deployment Planning System [USA] *mt*
FORMETS Formatting Message Text System [NATO, ADatP-3] *fu, mt*
FORRK Fiber Optic Radar Remoting Kit *mt*
FORSCAP Force Capability System [USAFE] *mt*
FORSCOM Forces Command [US Army] *mt, fu, 12, jj, kf*
FORSIZE Force Sizing exercise *mt*
FORSTAT Force Status and Identity Report System [WWMCCS [replaced by UNITREP] *mt*
FORSTAT Forces Status Report *jj*
FORSUM Force Summary System [USAFE] *mt*
FORTRAN Formula Translation (Translator) [a programming language] "fortran" *ct, em, ro, kg, hc, sf31.1.14, aw130.11.56, is26.1.89, wp, jj, oe, mt*
FORWARD_TALK [joint airborne command post] *wp*
FOS Faint Object Spectrograph [on HST] *mi, af72.5.3*
FOS Follow On Spares *wp*
FOS Fragmentary Order System *mt*
FOS Functional Operational Specification *fu*
FOSAMS Fleet Optical Scanning Ammunition Marking System [US Navy] *mt*
FOSC Flight Operations Support Complex *kf*
FOSE Federal Office Systems Exposition *pk, kg*
FOSI Format Option Specification Instance *pk, kg*
FOSIC Fleet Ocean Surveillance Information Center [US Navy] *mt, 12, go3.1.2*
FOSICS Fleet Ocean Surveillance Information Center *fu*
FOSIF Fleet Ocean Surveillance Information Facility [US Navy] *mt, fu, 12*
FOSIF Fleet Operational Support Intelligence Facility *mt*
FOSS Fabrication Order Status System *fu*
FOSS Fiber Optic Sensor System *wp*
FOSSIL Fido / Opus / Seadog Standard Interface Layer *kg, hi*
FOST Force Oceanique Strategique [(French) strategic nuclear submarine force] *-12*
FOT Final Operational Test *fu*
FOT Flight Operations Team *oe*
FOT Follow On Operational Test and Evaluation *hc*
FOT Follow On Testing *fu*
FOT Frequency of Optimum Transmission *fu*
FOT Frequency On Target *jj*
FOT Functional Operational Test *fu*
FOT&E Follow on Operational Test and Evaluation *mt, fu, kf*
FOT&E Full Operational Test and Evaluation *tb*
FOTACS Fleet Operational Telecommunications Automated Control System *fu, hc*
FOTAS Forward Observer Target Acquisition System *hc*
FOTC Force Over-the-Horizon Track Coordinator *fu*
FOTR Follow On Tactical Fighter *hc*
FOTRS Follow On Tactical Reconnaissance System *wp*
FOTS Fiber Optics Transmission System *hc, wp, hi*
FOUO For Official Use Only *fu, wp, jj, su, pbl, mt*
FOURATAF Fourth Allied Tactical Air Force [NATO] *mt*
FOV Field Of View *fu, ge, uh, he, hc, gd2-8, 2, 4, 8, 16, 17, sb, pl, rg, jb, jmj, lt, wp, mb, tb, wl, jj, pbl, hg, jdr, kf, vv, oe, mt, ai1*
FOW Family Of Weapons *ph*
FOWW Follow On Wild Weasel [F-15] *wp, mt*
FOXBAT MIG-25 *wp*
FP Fabrication Planning *fu*
FP Fast Path *tb*
FP Fault Protection *jdr, uh*
FP Federal Publication *mt*
FP Fixed Point *jj*
FP Fixed Price [contract] *mt, fu, hc, 8, st, jj, oe*
FP Flat Pack *hc, jj*
FP Flight Plan(ning) *fu*
FP Floating Point *hi*
FP Focal Plane *tb*
FP Fokker-Planck (equation) *sp660*
FP Freezing Point *wp, ai1*
FP Front Panel *fu*

FPA Final Power Amplifier *hc, jj*
FPA First Production Article *fu*
FPA Fixed Posting Area *fu*
FPA Fixed Price with [economic] Adjustment *ep*
FPA Flat Panel Assembly *fu*
FPA Flight Path Angle *kf*
FPA Flight Path Assurance *fu*
FPA Floating Point Accelerator *fu, oe*
FPA Focal Plane Array *fu, hc, wp, uc, 17, sb, th4, tb, st, jdr, kf*
FPA Focal Plane Assembly *gp, 16, kf*
FPAA Flight Path Analysis Area *0*
FPAC Flight Path Analysis and Command *0*
FPAF Fixed Price Award Fee [a type of contract] *wp, jj*
FPAG Flight Path Analysis Group *0*
FPB Fast Patrol Boat *mt*
FPC Fault Processor and Control (unit) *fu*
FPC Federal Power Commission *w9, wp*
FPC Filter Pin Connector *he*
FPC Filter Program Control *fu*
FPC Fire Permit Computer *jj*
FPC Firmware Program Component *fu*
FPC Fixed Price Contract *fu, jj*
FPC Flight Plan Checker *fu*
FPC Floating Point Calculation *pk, kg*
FPC/TP Flight Plan Conflict/Trial Plan *fu*
FPCE Floating-Point C Extension [a specification] *pk, kg*
FPCI Firmware Program Configuration Item *fu*
FPCP Flight Plan Conflict Probe *fu*
FPD Facilities Planning Document *jj*
FPD Flight Projects Directorate *oe*
FPDB Force Planning Data Base *mt*
FPDM Fault/Pattern Data Merger *rm8.2.5*
FPDMSS Flight Projects Directorate Multidiscipline Support Services *oe*
FPDS Federal Procurement Data System *mt, hc*
FPE Fixed Position Error *oe*
FPE Fixed Price [contract with] Escalation *hc*
FPE Flight Planning Element *hc*
FPE Force and Plan Execution *mt*
FPEM Force and Plan Execution Monitoring *mt*
FPF Final Protective Fire *fu*
FPF Fixed Price Firm *hc*
FPF Force Package File *mt*
FPFN Fast Pulse Forming Network *hc, jj*
FPGA Field Programmable Gate Array *kg, kf*
FPGM Four-Plane Graphics Memory *fu*
FPI Fixed Price Incentive [contract] *fu, hc, sm18-1, 8, 17, wp, ep, st, kf*
FPI Floating Point Interface *hi*
FPI Functional Process Improvement [GCCS] [USA] *mt*
FPI {instrument that is part of the Mars orbiter} *ma40*
FPIC Fixed Price Incentive Contract *hc, jj*
FPIF Fixed Price Incentive Fee *fu, hc, wp, tb, jj, jdr, kf*
FPIFP Fixed Price Incentive Fee Performance *jj*
FPIFV Fixed Price Incentive Fee Value *hc*
FPIS Force Planning Information System [NATO] *mt*
FPIS Forward Propagation Ionspheric Scatter *fu*
FPISP Fixed Price Incentive Successive Targets *hc*
FPJPA Fully-Proceduralized Job Performance Aid *fu*
FPL Flight Plan *fu*
FPL Fluorescent Penetrant Inspection *fu*
FPL Foreign Parts List *hc, jj*
FPL Frequency/Phase Lock *fu*
FPL Full Performance Level *fu*
FPLA Fair Packaging and Labeling Act *wp*
FPLA Field Programmable Logic Array *fu, kg, hc, hi*
FPLOE Fixed Price Level of Effort *mt*
FPM Facilities Planning Module [US Navy] *mt*
FPM Fast Page Mode *pk, kg*
FPM Federal Personnel Manual *mt*
FPM Federal Procurement Manual *mt*
FPM Feet Per Minute *mt, fu, pf1.22, wp, jdr, hi*
FPM Fill Pulse Map *fu*
FPM Floating Point Multiplier *fu*
FPM Focal Plane Module *kf*
FPM Functional Project Manager *mt*
FPMIS Federal Personnel Management Information System [OPM] *mt*
FPMR Fixed Price Materiel Reimbursable *hc*
FPMRAM Fast Page Mode RAM *hi*
FPN Fixed Pattern Noise *17, sb*
FPN Foreign Pendant Numbers Files *mt*
FPO Fleet Post Office *w9, wp, jj*
FPO Flight Projects Office *re*
FPOW Friendly Prisoner of War *mt*
FPP Fixed Path Protocol *pk, kg*
FPP Flight Plan Processing *fu*
FPP Flight Purpose Plans *jj*
FPP Floating Point Processor *tb, kg*

FPR Federal Procurement Regulation *mt, hc, fu*
FPR Field Problem Report *hc, jj, fu*
FPR Fixed Price Redeterminable [contract] *fu, hc, jdr*
FPR Flat Plate Radiometer *ge, hg*
FPR Flight Performance Reserve *oe, ai2*
FPR Floating Point Register *pk, kg*
FPR Foliage Penetration Radar *wp*
FPROM Field Programmable Read Only Memory *wp, hi*
FPRS Field Problem Resolution Support *fu*
FPS Fast Packet Switching *hi*
FPS Favorite Picture Selection *pk, kg*
FPS Feet Per Second *bo, wp, 4, w9, jj, jdr*
FPS Fighter Pilot Simulator *fu*
FPS Fixed Point Station *0*
FPS Flight Path Simulator *fu*
FPS Flight Planning Station *fu*
FPS Floating Point System *ct, em, fu*
FPS Focus Projection and Scanning *hc, fu*
FPS Foot, Pound, Second [a system of engineering units] *fu*
FPS Formatted Problem Statement *fu*
FPS Frames Per Second *kg, ct, em, hi, oe*
FPS Fuel Planning System [MAC] *mt*
FPS Functional Performance Structure *hc*
FPSO Flight Project Support Office *re*
FPSP Future Programmable Signal Processor *hc*
FPSR Field Problem Summary Report *jj*
FPSS Fine Pointing Sun Sensor *hc*
FPT Forced Perfect Termination *ct, kg, em, hi*
FPT Functional Performance Testing *fu*
FPT&E Formal Performance Test and Evaluation *fu*
FPTA Fully Proceduralized Troubleshooting Aids *hc*
FPTD Flight Plan Tabular Display *fu*
FPTD Focal Plane Technology Demonstration *hc*
FPTS Forward Propagation [by] Tropospheric Scatter *fu, mt*
FPU Field Processing Unit *fu*
FPU Floating Point Unit *kg, ct, em, hi*
FPU Forwarding Participating Unit [TACS / TADS] *fu, mt*
FPV Force Projection Vehicle *mt*
FPVIQMR Fixed Price Variable Indefinite Quantity Mate *hc*
FPVT Focal Plane Vector Table *kf*
FPVT Functional Performance Verification Test *fu*
FPWG Flight Planning Working Group *oe*
FQ Finish Queue *fu*
FQ Flight Quality *jj*
FQDN Fully Qualified Domain Name [Internet] *ct, kg, em, hi*
FQI Federal Quality Institute *wp*
FQM Four Quadrant Multiplier *fu*
FQR Formal Qualification Review *mt, hc, fu, tb*
FQR Functional Qualification Review *jdr*
FQS Finish Queue Sequencer *fu*
FQT Final Qualification Test *fu*
FQT Formal Qualification Test (Testing) *mt, fu, pl, wp, hc, tb, jdr, kf*
FQT Functional Qualification Test *jdr*
FQV Formal Qualification Verification *fu*
FR Fabrication Request *fu*
FR Failure Rate
FR Failure Report *gp, 15, 17, sb, ep*
FR Fast Release *fu*
FR Fast Revisit *kf*
FR Field Reversing *fu*
FR Flag Register *jj*
FR Flight Recorder *wp*
FR Flight Refuelling *mt*
FR Formal Report *fu*
FR Frame Relay [streamlined packet switching with variable frame size, T1 or E1transmission rates] *nu*
FR Frequency Received *-2*
FRA Federal Railroad Administration *mt*
FRACAS Failure Reporting, Analysis, and Corrective Action System *mt, fu, hc, nt, wp, oe*
FRAD Frame Relay Access Device *kg, ct, em*
FRAD Frame Relay Assembler/Disassembler *pk, kg*
FRAG Fragmentary Order *mt*
FRAG_II Fragmentary Order Processing System [PACAF] *mt*
FRAG_PREP Fragmentary Order Preparation System [WWMCCS] *mt*
FRAK Flak Radar Automatic Kanon *hc*
FRAM Ferroelectric Random Access Memory *kg, em, hi*
FRAM Flash Random Access Memory *hi*
FRAM Fleet Rehabilitation And Modernization [program] [USA] *fu, hc, jj, mt*
FRAMP Fleet Replacement Aviation Maintenance Program [Person] *hc, jj*
FRAN Framed Structure Analysis Program [IBM] *hc*
FRAND Fractionally Anded *jj*
FRAZ Frequency Azimuth *fu*
FRB Failure Review Board [an organization] *mt, fu, ah, gp, 4, 15, 17, pl, sb, hc, lt, ep, jj, pbl, tm, jdr, kf, oe*

FRB Federal Reserve Board *y2k*
FRBCAM Failure Review Board Corrective Action Matrix *fu*
FRBM Forward retractable Blanket mechanism *he*
FRC Facility Review Committee *wp*
FRC Federal Regional Center [FEMA] *12, mt*
FRC Federal Response Center *mt*
FRC Fiber Reinforced Composite *wp*
FRC Fixed Radio Communications *hc*
FRC Flight Research Center [NASA, Edwards AFB, old name for DFRC] *mi, tc, wp, jj, mb*
FRC Functional Redundancy Checking *pk, kg*
FRCB Financial Recap Contract Brief *fu, 8, hc, kf*
FRCP Federal Rules of Civil Procedure *hc*
FRCP Fiber Reinforced Composite Propellant *wp*
FRCT Fixed Record Communications Teletypewriter *mt*
FRD Facilities Requirement Document *jj*
FRD Flight Readiness Demonstration *0*
FRD Formerly Restricted Data *mt, sy9, tb, st, kf, su*
FRD Functional Requirements Description *mt*
FRD Functional Requirements Document *fu*
FRD {?}Two imaging X-ray CCD arrays on SXG *cu*
FRDS Flight Record Data System [AFISC] *mt*
FRE Flight Readiness Element *hc*
FRE Forward and Reverse Engineering *em*
FRE Functional Requirements Envelope *wl*
FRED Flexible Red [Attack Generator] *tb*
FRED Flexible Repository Engineering Data *hc*
FREJID Frequency Jumper Identification *hc*
FREQ Frequency Management [USREDCOM] *mt*
FRERP Federal Radiological Emergency Response Plan [FEMA] *mt*
FRESCAN Frequency Scan [radar] *fu*
FRESCANAR Free Scanning Radar (Frequency Scan Radar) [a Hughes trademark] *hc*
FRESH Force Readiness Expert System *mt*
FRESH Force Requirements Expert System [PACFLT] *hc, mt*
FRESTAR Frequency Scanned Typhon Array Radar *hc*
FRETT Fleet Readiness Emergency Travel Team *jj*
FRG (retropropulsion module (FRG)) *jp*
FRG Federal Republic of Germany [aka West Germany] *mt, su, fu, ge, sc3.8.3, hc, sf31.1.9, es55.86, aa27.4.b31, wp, so, ns, jj, hg*
FRG Force Requirements Generator [JOPS] *mt*
FRGATC Federal Republic of Germany ATC *fu*
FRI Frame Relay Internetworking *hi*
FRL Fuselage Reference Line *jj*
FRMAC Federal Radiological Monitoring and Assessment Center [DOE] *mt*
FRMAP Federal Radiological Monitoring and Assessment Plan [DOE] *mt*
FRMON Full R Monitor (Monitoring) *hi*
FRMS Frequency Record Management System *mt*
FRMS Frequency Resource Management System *mt*
FRN File Reference Number *jj*
FRN Force Requirement Number [USA DOD] *mt*
FRO Flexible Response Option *mt*
FRO Frequency Reference Oscillator *jj*
FRO Front Receiver On *jj*
FROG Free Rocket (Range) Over Ground [short range missile] *mt, fu, 12, wp*
FROKA First Republic of Korea Army *fu*
FROM Ferroelectric Read Only Memory *hi*
FROM Flash Read Only Memory *hi*
FROM Force Reception and Onward Movement system [USAREUR] *mt*
FROM Fusible Read Only Memory *hi*
FRP Facility Requirements Panel *mt*
FRP Fiber Reinforced Plastic *wp*
FRP For Record Purposes *jj*
FRP Force Rendezvous Point *mt*
FRPC Fast Reaction Procedure Card *mt*
FRPI Flux Reversals Per Inch *pk, kg*
FRR False Report Rate *kf*
FRR Flight Readiness Review *mi, gp, 4, 17, sb, tb, wl, jdr, mb, oe*
FRR Functional Recovery Routine *wp*
FRRS Frequency Resource Record System [PACAF] *mt*
FRS Failure Review System *oe*
FRS Federal Reserve System *mt, w9, tb, su*
FRS Field Repair Service *fu*
FRS Fighter, Reconnaissance, Strike [UK] *mt*
FRS Financial Reporting System *oe*
FRS Fleet Replacement Squadron [US Navy] *fu, mt*
FRS Fleet Replenishment Squadron [USA] *mt*
FRS [Hughes Aircraft Company customer for multiple diffraction gratings] *hc*
FRSI Flexible Reusable Surface Insulation *ci.xxii*
FRSM Force and Resource Status Monitoring *mt*
FRSM Future Radionavigation Systems Mix [DOD / DOT] *mt*
FRT Fast Reaction Team *mt*
FRT Functional Requirements Test *mt*

FRU Field Replaceable Unit *fu, hi*
FRU Forwarding Reporting Unit [TACS / TADS] *mt*
FRUSA Flexible Roll(ed) Up Solar Array *hc, sc4.3.6, 16, lt, ep*
FRV Force Rendezvous *mt*
FRXD Five-level Reperforator Transmitter Distributor *0*
FRXO Front Transmitter On *jj*
FS Fabrication Services *fu*
FS Factor of Safety *wl*
FS Fast Scan *oe*
FS Fast-Swept *fu*
FS Federal Specification *fu, jj*
FS Feedback Stabilized *fu*
FS Field Stop *wp*
FS Fighter Squadron (526th FS 'Black Knights') [USA] *mt*
FS File Separator [Information] *fu, kg, do242*
FS File System *tb*
FS Final Set *fu*
FS Final Site *fu*
FS Fire Support *mt, fu*
FS Flight Simulator *wp*
FS Flood - Setup *fu*
FS Forward Segment *-17*
FS Frame Synchronization *fu*
FS Framestore memory *sb*
FS Frequency Shift *mt*
FS Full Scale *wp, tb*
FS Full Service *mt, fu*
FS Function Switch *jj*
FS Fuselage Station *jj*
FS&S Field Service And Support Division [Division 28] *jj*
FS-X Fighter Support, Experimental [aircraft] *aa26.12.20, aw130.13.22*
FSA Family Separation Allowance *wp*
FSA Field Specification Analysis *jj*
FSA Field Support Activity *dt*
FSA Final Site Acceptance *fu*
FSA Final System Acceptance *fu*
FSA Flexible Spending Account *hn49.3.4*
FSA Free-Space Attenuation *fu*
FSA Front or Stern Advantage *fu*
FSA Functional System Analyzer *mt*
FSAGA First Sortie After Ground Alert *mt*
FSAR Failure Summary and Analysis Report *fu*
FSAS Flight Service Automation System *fu, ag*
FSAS Fuel Savings Advisory System *mt, wp*
FSAT Full Scale Aerial Target [radio controlled conversions of older aircraft, as the QF-100 and QF-106, USA] *mt, af72.5.150*
FSB Fleet Satellite Broadcast [US Navy] *mt*
FSB Forward Staging Base *mt*
FSBS Fixed Submarine Broadcast System [US Navy] *mt*
FSBS Fleet Satellite Broadcast System *mt*
FSC Facility Security Clearance *tb*
FSC Federal Supply Class(ification) *mt, fu, hc, jj*
FSC Federal Supply Code [later known as CAGE] *tb, su*
FSC Field Support Center *fu*
FSC Fire Support Center *hc*
FSC Fire Support Coordination *fu*
FSC Fire Support Coordinator [USMC] *mt*
FSC Flight Sensor Computer *wp*
FSC Flight Service Center *fu*
FSCA Full Systems Capable Aircraft *jj*
FSCAIS Family Services Center Automated Information System [USMC] *mt*
FSCC Fire Support Coordination (Control) Center *mt, fu, hc*
FSCE Fire Support Coordination Element *hc*
FSCIL Federal Supply Catalog Identification List *fu*
FSCL Fire Support Control Line *mt*
FSCL Fire Support Coordination Line *mt*
FSCM Federal Standard Codes of Manufacturers *tb*
FSCM Federal Stock Code, Manufacturers *hc*
FSCM Federal Supply Code for Manufacturers (Manufacturing) [a designator] *mt, fu, hc, wp, ep, jj, jdr, tm*
FSCOORD Fire Support Coordinator *mt*
FSCP Full Spectrum Color Projector *hc*
FSCR Final System Concept Review *17, sb*
FSCRC Follow on Small Computer Requirements Contract *mt*
FSCS Fleet Satellite Communications System *mt*
FSCT Floyd Satellite Communication Terminal *hc*
FSD Field Set Data *fu*
FSD File System Driver [OS/2] *pk, kg*
FSD Flying Spot Digitizer *fu*
FSD Forecast Support Date *fu*
FSD Front or Stern Disadvantage *fu*
FSD Full Scale Deflection *ai1*
FSD Full Scale Development *mt, fu, 4, 17, ph, hq5.3.1, hc, af72.5.136, ts4-48, nt, wp, tb, st, jj*
FSD Functional Sequence Diagram *fu*

FSD Functional Statement Document *mt*
FSD Functional System Description *mt*
FSD Functional System Design *tb*
FSDP Full Scale Development Phase *fu*
FSDPS Flight Service Data Processing System *fu*
FSDR Final Site Design Review *mt*
FSDT Functional System Design Team *mt*
FSDU Full Scale Demonstration Unit *tb*
FSDWS Fixed Site Detection and Warning System [chemical warfare] *mt, af72.5.19, wp*
FSE Facilities System Engineer *hc*
FSE Factory Support Equipment *wp*
FSE Field Service Engineer *fu*
FSE Field Support Equipment *jj*
FSE Fire Support Element [USA] *fu, mt*
FSE Fleet Supportability Evaluation *jj*
FSE Flight Support Equipment *gp, ep, fu, oe*
FSEC Fairchild Space and Electronics Company *ci.945*
FSED Full Scale Engineering Development *mt, fu, hc, wp, 17, sb, cb5.15.89, tb, st, jj, jdr*
FSEDM Full Scale Engineering Development Model *fu*
FSEP Federal Software Exchange Program *mt*
FSF Forward Space File *jj*
FSF Free Software Foundation [Internet] *kg, hi*
FSG Federal Supply Group *mt, fu, jj*
FSG Federal Systems Group *fu*
FSGM Forward Sector Ground Mapping *hc*
FSI Felec Services Incorporated *mt*
FSI Flight Safety International *af72.5.105*
FSI Flight Simulator Instructor *mt*
FSI Flight System Integration *wp*
FSI Frequency Sources Inc. *fu*
FSI Functionality Significant Item *fu*
FSIC Forward Sensor Interface and Control *fu*
FSIMM Flash SIMM *hi*
FSIPL Field Service Indentured Parts List *fu*
FSIS Flight Safety Information System [MAC*] *mt*
FSK Frequency Shift Key(ing) (Keyed) *mt, kg, ct, uh, bl1-5.114, hc, st, jj, jj, kf, fu, hi, oe*
FSL Forward Supply Locations *mt*
FSL Free Space Loss *fu, oe*
FSL Full System Listing *mt*
FSLSTE Flight Line System Level Special Test Equipment *hc, jj*
FSM Facility Security Manual *su*
FSM Field Strength Meter *fu*
FSM Firmware Support Manual *mt, fu, hc*
FSM Flight Simulator Model *mt*
FSM Formatted Status Message *fu*
FSM Frequency Shift Modulation *hc, jj, fu*
FSMSC Federal Software Management Support Center [formerly FCSC] [OIT] *mt*
FSMWO Field Service Modification Work Order *jj, fu*
FSN Federal Stock Number *mt, fu, hc, wp, tb, jj*
FSN Full Service Network *kg, ct, em*
FSO Facility Security Officer/Supervisor *su, tb*
FSO Field Support Office *hc, jj*
FSO Field Support Operations *jj*
FSO Fire Support Officer *fu*
FSO Flight Safety Officer *tb, fu*
FSO&S Flight Support Operations and Services *kf*
FSOC Fairchild Satellite Operation Center [Fairchild AFB, WA] *16, dm*
FSOF Forward Special Operations Facility *mt*
FSP Facility Security Plan *kf*
FSP Facility Support Plan *mt*
FSP File Service Protocol *kg, em*
FSP Flight Strip Printer *fu*
FSP Forward Supply Point *mt, jj*
FSP Frequency Standard Primary *fu*
FSP Full-Scale Production *fu*
FSPL First Spacelab Payload *ac3.7-15*
FSPP Flight (Progress) Strip Printing Program *fu*
FSR Field Service Report *wp*
FSR Field Service Representative *fu, hc, jj*
FSR Field Strength Ratio *fu*
FSR Forward Space Record *jj*
FSR Free System Resources *pk, kg*
FSR Frequency Scan Radar *hc, fu, jj*
FSR Frequency Selective Receiver *fu*
FSR Full System Requirement *fu*
FSR Functional Support Requirement *wp*
FSRT Firm Scheduled Return Time *mt*
FSS Fail Safe System [USAFE] *fu*
FSS Fast Symbol Synchronizer *fu*
FSS Federal Supply Schedule *tb, su*
FSS Federal Supply Service [GSA] *wp, mt*
FSS Field Storage Site *mt*

FSS Fine Sun Sensor *es55.42, oe*
FSS Fire Support System *hc*
FSS Fire Suppression System *fu*
FSS Firmware Support Station *fu*
FSS Fixed Satellite Service *sc4.5.3, 9, jdr*
FSS Fixed Service Structure *ai2*
FSS Flight Screening Squadron [USA] *mt*
FSS Flight Sensor System *hc*
FSS Flight Service Station *ag, wp, fu*
FSS Flying Spot Scanner *hc*
FSS Force Status System *mt*
FSS Forward Supply System *mt*
FSS Forward Support System *jj*
FSS Free Space Simulator *jj*
FSS Functional System Specification *mt*
FSSA Flexible Substrate Solar Array *jdr*
FSSE Foreign Service Sales Expense *hc*
FSSG Force Service Support Group [USMC] *12, mt*
FSSS Future Security Strategy Study [also FS3] *fr, tb*
FSSSF Flight Service Station Support Facility *fu*
FST Fail Safe Toggle *uh, jdr*
FST Field Service Tape *fu*
FST Field Support Team *fu*
FST Fire Support Team [also FIST] *hc*
FST Fire Support Terminal *hc*
FST Flat Square Tube [CRT monitor] *kg, hi*
FST Frequency Shift Transmission *fu*
FST Functional System Test *-4*
FST/FCT Field Support Team/Field Communications Team *fu*
FSTC Federal Software Testing Center *wp*
FSTC Foreign Science and Technology Center [USA] *wp, mt*
FSTDS Flight Station Tactical Display Set *hc*
FSTL Future Strategic Target List *mt*
FSTN Film Super Twist Nematics *hi*
FSTS Federal Secure Telephone System *mt*
FSU Factor of Safety Ultimate *sb*
FSU Florida State University *aa26.12.92*
FSU Former Soviet Union *kf, mt*
FSU Fuel Service Unit *oe*
FSVS Future Secure Voice System [NSA] *mt*
FSW Field Switch *fu*
FSW Flight Software *oe*
FSW Forward Swept Wing *mt, wp, af72.5.140*
FSW Frame Synchronization Word *fu*
FSX Fighter Support Experimental [see FS-X] *aw130.13.22*
FSY Factor of Safety Yield *sb*
FT Far Term [beyond 20 years] *fu*
FT Fault Tolerance *fu*
FT First year Thin [a type of sea ice] *ac7.1-17*
FT Flight Test *jj*
FT Free Text *fu*
FT Frequency Tracker *fu*
FT Frequency Translator *fu*
FT Functional Test *kf, fu*
FT/RF Frequency Translator and Recursive Filter *fu*
FTA Fast Time Analyzer *fu*
FTA Fault Tree Analysis *wp, fu*
FTACS Future Tactical Air Control System *hc*
FTAM File Transfer and Access Method *mt, kg, hi*
FTAM File Transfer And Management *es55.67*
FTAM File Transfer, Access and Management *mt, ct, em, kg, fu*
FTAS Fast Time Analyzer System *hc*
FTB Fort Bragg [range control project] *mt*
FTB Full Technology Validation *tb*
FTB Full Track Block *fu*
FTC Facilities Technical Criteria *mt*
FTC Fast Time Constant *mt, fu, hc, jj*
FTC Fast Time Control *hc*
FTC Federal Trade Commission *w9, ct*
FTC Fleet Training Center *jj*
FTC Forrestal Telecommunications Center *mt*
FTCA French Central Technical Armament Establishment *wp*
FTCC Flight Test Coordinating Committee *0*
FTCN Fleet Teletype Conferencing Network [US Navy] *mt*
FTCS Flight Training Control System *fu*
FTD Field Training Detachment *mt, jj*
FTD Fire Control Tracking and Designation *fr*
FTD Fire Training Device *fu*
FTD FIREFINDER Training Device *fu*
FTD Fix Time Determination *fu*
FTD Fixed Threshold Detector *jj*
FTD Flight Tabular Display *fu*
FTD Flight Test Directive *0*
FTD Flight Test Division *jj*
FTD Foreign Technology Division [AFSC*] *mt, jb, wp, dm, st*
FTD Formal Technical Documents *hc*

FTD Frequency Translation Distortion *fu*
FTD Functional Test Demonstration *mt*
FTD Functional Training Detachment *mt*
FTD Fuze Triggering Device *fu*
FTDMA Frequency and Time Division Multiple Access *mt*
FTE Factory Test Equipment *fu, hc, wp, jj*
FTE Frequency Tracks Error *hc*
FTE Full Time Equivalent *oe*
FTES Functional Test Expert System *fu*
FTF Factory Test Facility *-16*
FTF Fixed Time Firing *-3*
FTF Flared Tube Fitting *fu*
FTF Fullerton Test Facility *fu*
FTFCS Foreign Tank Fire Control System *hc*
FTFD Field Test Force Director *hc*
FTFGS Flared Tube Fitting Gasket Seal *fu*
FTG Fault Tolerant Group *fu*
FTGM Functional Task Group Manager *fu*
FTI Fast Technical Imagery *rih*
FTI First Flight Test in Flight Experiment Program *-17*
FTI Fixed Target Indicator *mt, fu, wp*
FTI Frequency Time Indicator *fu*
FTIC Fast Trial Intercept Calculation *fu*
FTIM Frequency and Time Interval Meter *jj*
FTIP Functional Test In Progress *jj*
FTIR Fourier Transform Infrared [spectrometer or spectroscopy] *hc, wp, jdr*
FTIRS Fourier Transform Infrared Spectroscopy *wp*
FTIS Flight Test Instrumentation System *jj*
FTIV Flight Test Instrumentation Van *bf*
FTL Federal Telecomm *0*
FTL Flash Transition Layer [Intel] *pk, kg*
FTL Formal Technical Literature *hc*
FTLS Final Top Level Statistics *mt*
FTLS Formal Top Level Specification *fu*
FTM Flat Tension Mask [Zenith] *kg, hi*
FTM Flexible, Taskable, Multimode *kf*
FTM Flight Test Model *hc, jj*
FTM Fourier Transform Mass [Spectroscopy] *wp*
FTM Functional Task Manager *fu*
FTMCC Flight Test Mission Control Complex *hc*
FTML Folded-Tape Meander Line *fu*
FTMP Foliage and Terrain Map Processing *fu*
FTMS Federal Test Method Standard *hc*
FTMS Fourier Transform Mass Spectroscopy *wp*
FTN FORTRAN Extended Compiler *tb*
FTNMR Fourier Transform NMR *wp*
FTNS Flight Track Navigation System *wp*
FTO Flight Test Objectives *17, sb*
FTOC Fleet Tactical Operations Center *mt*
FTOL Fine Time Ordered List *fu*
FTOOL FITS Tool [a suite of tools developed at the OGIP for general and mission-specific manipulation of FITS files] *cu*
FTP Factory Test Plan *jdr*
FTP File Transfer Program *fu*
FTP File Transfer Protocol [a widely available method for transferring files over the Internet] [MIL-STD-1780] *mt, cu, ct, em, csII.60, mb, jdr, ric, mi, kg, nu, hi, oe, du, ai1*
FTP File Transmission Protocol *mt*
FTP Functional Test Procedure *jj*
FTPD File Transfer Protocol Daemon *kg, ct, em*
FTPM Fundamental Topics in Program Management *heh2.21.7*
FTPP Factory Test Program Plan *jdr*
FTR Final Technical Report *fu*
FTR Functional Test Requirement *jj*
FTR Functional Throughput Rate *fu*
FTR Future Technology Requirements *wp*
FTRFLT Fighter Flight [USA] *mt*
FTS Factory Training School *hc, jj*
FTS Federal Telecommunications System *mt, kg, fu, ct, oe*
FTS Federal Telephone System [US] *mt, gp, 16, tb*
FTS File Transfer Service [WIN] *mt*
FTS Flexible Test Station *hc, jj*
FTS Flight Telerobotic Servicer *mi*
FTS Flight Termination System *17, hc, jj, oe, ai2*
FTS Flight Training (Target) Simulator *tb, jj*
FTS Flight Training School [UK] *mt*
FTS Flight Training Squadron [USA] *mt*
FTS Force Tracking System *mt*
FTS Fourier Transform Spectroscopy *wp, ai1*
FTS-IR Fourier Transform Spectroscopy Infrared *ai1*
FTSC Fault Tolerant Spacecraft Computer *hc*
FTSR Flight Termination System Report *hc*
FTT Field Test Telescope *hc, 17*
FTT Fixed Target Track *hc, jj*
FTTC Fiber To The Curb *kg, ct, em*

FTTH Fiber To The Home *ct, em*
FTU Field Test Unit *fu*
FTU Fuel Transfer Unit *oe*
FTV Flight Test Validation *st*
FTV Flight Test Vehicle *0, jmj, tb*
FTV Functional Test Vehicle *wp*
FTW Flying Training Wing [USA] *mt*
FTWG Flight Test Working Group (Ranger) *0*
FTX Fault Tolerant Unix *pk, kg*
FTX Field Training Exercise *mt, jj*
FU Fire Unit *hc, fu, jj*
FUE First Unit Equipped *fu*
FUELS [aviation fuels management system] *mt*
FUI File Update Information *pk, kg*
FUIF Fire Unit Integration Facility *hc*
FUJOPS FORSCOM Unique JOPS [FORSCOM] *mt*
FULCRUM [MIG-29 aircraft] *wp*
FULLDOC CLAMS Library Documentation *tb*
FULLER'S Fuller'S Earth [chemical warfare Decontaminant] *wp*
FUNI Frame User Network Interface *ct, em*
FUP Facility Utilization Plan *mt*
FUR Failure / Unsatisfactory Report *hc, jj*
FUSAG First US Army Group ['Ghost Army' created as part of the disinformation Allied campaign coded Quicksilver] *mt*
FUSE Far Ultraviolet Spectroscopic Explorer [NASA] *mi, ge, hg, mb*
FUT Function Under Test *jj*
FUV Fraction of Unexplained Variance *dm*
FUV {far ultraviolet} [120 - 200 nm] *es55.14*
FV Front View *jdr*
FVIPS First Virtual Internet Payment System *ct, em*
FVP Firefinder VHSIC Processor *fu*
FVR Fuse Voltage Rating *fu*
FVS Fighting Vehicle System [later BFV] *hc, 8*
FVS Flight Vehicle Simulation *pl*
FVSF Flight Vehicle Simulation Facility *-16*
FVT Full Video Translation *pk, kg*
FVT Functional Verification Testing *fu*
FW Fighter Wing [USA] *fu, mt*
FW Filter Wheel *oe*
FW Firmware *fu, 16*
FW Fixed Wing *fu*
FW Fixed Wiring *fu*
FW Fullwave *fu*
FW&A Fraud, Waste, and Abuse *mt*
FW-SCSI Fast/Wide SCSI *hi*
FW/DB Forward Warning and Deployment Base *mt*
FWAS Flight/Failure Warning and Analysis System *hc*
FWBR Full Wave Bridge Rectifier *jj*
FWC Filament Wound Case *hc*
FWC Force Weapons Coordinator [US Navy] *mt*
FWCC Fine Weave Carbon Carbon *hc*
FWCS Flight Watch Control Station *fu*
FWD Forward *sa12474.10, 4, wp, jj*
FWDP Force Command WWMCCS Development Plan [USA] *mt*
FWE Foreign Weapons Evaluation [DOD] *mt, wp, ph, hc*
FWF Fixed Word Format *fu*
FWG Facilities Working Group *hc*
FWG Financial Working Group *mt, wp*
FWHM Full Width at Half Maximum *mi, gp, kf, mb*
FWI Field Widened Interferometer *tb*
FWIT Fighter Weapons Instructors Training *mt*
FWOC Forward Wing Operations Centre [UK] *mt*
FWR Full Wave Rectifier *fu*
FWS Fighter Weapons School *mt, jj*
FWS Filter Wedge Spectrometer *hc, ac2.8-2, jj*
FWS Forward Swept Wing *mt*
FWW Fighter Weapons Wing (57th FWW, Nellis AFB, Nevada) *mt, af72.5.55*
FWWM Food, Water and Waste Management *wp*
FX Fighter, Experimental *wp*
FX Foreign Exchange *do128, w9*
FX Frequency Extension *fu*
FXR Flash X-Ray *jdr, kf*
FXR Fluence X-ray *fu*
FXS Foreign Exchange Service *hi*
FY First Year [a type of sea ice] *ac7.1-17*
FY Fiscal Year [1 October to 30 September] *mt, fu, hc, ac1.3-11, 0, 12, 13, sb, aa26.12.44, w9, af72.5.26, jb, wp, so, tb, st, jj, jdr, kf, oe, ai2*
FYCP Five Year Corporate Plan *mt*
FYDP Five Year Defense Program (Plan) *fu, mt, tb, hc, ph, af72.5.6, wp, st, jj*
FYDP Five Year Development Plan (Program) *fu*
FYDP Future Years Defense Program [formerly Five Year Defense Plan] *mt, kf*
FYMOP Five Year Master Objectives Program *mt*
FYP Five Year Plan (Program) *mt*

FYT Force Year Total Model [JDSSC] *mt*
FYTP Five Year Test Plan *jj*
FZG Flak Ziel Ger‰t [target for anti aircraft artillery, Germany] *mt*

G

G&A General and Accounting *fu*
G&A General and Administrative [expense] *fu, sc2.12.2, 8, 15, 17, sb, hc, nt, tb, jj, jdr, kf, oe*
G&C Guidance and Control [also G/C] *fu, uc, 16, 17, sb, hc, ep, tb, st*
G&N Guidance and Navigation *fu*
G&S General and Standard [also G/S] *hc, jj*
G&SSEL General and Standard Support Equipment List *jj*
G-M Geiger-Mueller *0*
G-MATE Generalized-Manual Adaptive Target Evaluation *fu*
G-S Gram-Schmidt *fu*
G-S/FFT Gram-Schmidt/Fast Fourier Transform *fu*
G/A Ground to Air *mt, fu*
G/A/G Ground to Air to Ground [radio] *fu, mt*
G/ADL Ground/Air Data Link *hc*
G/C Guidance and Control [also G&C] *jj*
G/G Ground to Ground *mt, fu*
G/L General Ledger *pk, kg*
G/N General Notes *jdr*
G/S General And Standard [also G&S] *hc, jj*
G/S Ground Speed *mt*
G/T Gain to Temperature [receive antenna gain to system noise temperature ratio, dB/K] *fu, gp, 4, 16, 17, lt, mb, ce, pbl, jdr, vv, kf, oe*
G/T Gain Transfer *uh*
G/T Ground Terminal [also GT] *-19*
G/VLLD Ground/Vehicle (Vehicular) Laser Locator Designator *hc, hn43.9.3*
GA Gate Array *wp*
GA General Arrangement [drawing] *mt*
GA General Availability *pk, kg*
GA General of the Army *w9*
GA Global Assessment *mt, wp*
GA Group Addressed *fu*
GA Guidance Assembly *jj*
GA Gun Assembly *fu*
GA [Organophosphate chemical warfare Nerve Agent (Tabun)] *wp*
GA {General Atomic} (GA Technologies, Incorporated) *aw130.11.55*
GAALN Gallium Aluminum Nitride *wp*
GAAS Gallium Arsenide *mt, dh, hc, tb, jdr, oe*
GAASFET Gallium Arsenide Field Effect Transistor *dn*
GAB Gates Aerospace Batteries *oe*
GABA Gamma Aminobutyric Acid *wp*
GAC Global Area Coverage *ge, hg*
GAC Goodyear Aerospace Corporation *ci.949*
GAC Grumman Aerospace Corporation *hc, jj*
GACC Ground Attack Control Capability [USAFE] *wp, mt*
GACC Ground Attack Control Center *hc, fu*
GACP Gunner Accuracy Control Panel *hc, jj*
GACT Graphic Analysis and Correlation Terminal *fu*
GAD Gallium Arsenide Diode *wp*
GADGES German Air Defense Ground Environment System *mt*
GADL Ground to Air Data Link *fu*
GAE Generate Airlift Requirement *mt*
GAEC Grumman Aeronautical and Engineering Company *hc, jj*
GAF German Air Force *mt, fu, hc, wp*
GAFS General Accounting and Finance System [Base Level] *mt*
GAG Gridding Accuracy Group [GOES] *spi*
GAG Ground to Air to Ground *fu, wp*
GAHRS Gyrocompassing Attitude and Heading Reference System *hc*
GAINS Gimballess Analytic Inertial Navigation System *hc*
GAIPS Graphic And Inter-Program Structure *tb*
GAL Generic Array Logic *pk, kg*
GALA Gate Array Layout Automation *ne*
GALCA Groupe d'Aviation LÈgÈre de Corps d'ArmÈe [army corps light aircraft group, France] *mt*
GALCIT Guggenheim Aeronautical Laboratory, California Institute of Technology *ci.152, 0, 13*
GALDiv Groupe d'Aviation LÈgÈre de Division [division light aircraft group, France] *mt*
GALE Generic Area Limitation Environment *kf*
GALOVAL Grapping And Lock-On Validation [facilities] *hc*
GALReg Groupe d'Aviation LÈgÈre de la Region [regional light aircraft group, France] *mt*
GALSTA Groupement ALAT de Section Technique de l'ArmÈe de Terre [France] *mt*
GAM Air to Surface missile {?}[USA] *mt*
GAM General Aeronautical Material *wp*
GAM Ground to Air Missile *wp*
GAM Group d'Activation Moteur [Ariane 5 main engine gimbal system] *ai2*
GAM Groupe AÈrienne Mixte [France] *mt*

GAM Guided Air (Aircraft) Missile *fu, hc*
GAMA General Aviation Manufacturers Association *aw126.20.13, aa27.4.8*
GAMD GÈnÈrale Aeronautique Marcel Dassault [France] *hc*
GAMM Generalized Air Mobility Model *wp*
GAMO Ground and Amphibious Military Operations [obsolete] *hc, fu, mt*
GAMP Guided Anti armor Mortar (Monitor) Projectile *hc, wp, dh*
GAN Gallium Nitride *hi*
GAO (GAO Associates. . . Northampton, MA 01060) *jmj*
GAO General Accounting Office [USA] *su, ah, fu, ci.xxiv, aw129.14.112, hc, w9, nd74.445.6, ae151.52.460, dt, tb, st, kf, mt, y2k*
GAP General Applications Program *fu*
GAP Generalized Availability Program *16, kf*
GAP Generic [Electro-Optical] Auto Processor *hc*
GAP Government Available Property *mt*
GAP Graphic Arts Purchasing *fu*
GAP Gun fired Antitank Projectile *wp*
GAPA Ground to Air Pilotless Aircraft *w9, jj*
GAPC Ground Attitude and Positioning Control *hc*
GAPEWS Graphic Air Picture Early Warning System [USAREUR] *mt*
GAPFILLER [Navy leased satellite communication service] *mt*
GAPI Gateway Application Programming Interface *pk, kg*
GAPL Group Assembly Parts (Provisioning) List *hc, jj, fu*
GAPP Geometric Arithmetic Parallel Processor *wp, tb*
GAPSAT Tactical Communications Satellite System *mt*
GAR Global Atmospheric Research [program] [also GARP] *-17*
GAR GOES Anomaly Report *oe*
GAR Government Analysis Report *mt*
GAR Grand Army of the Republic *w9*
GAR Guided Aircraft (Aerial) Rocket [later renamed AIM] [USA] *mt, fu, hc, wp, jj*
GAR Guided Antiarmor Rocket *wp*
GARB GOES Anomaly Review Board *oe*
GARB Guided Antiradiation Bomb *wp*
GARDEN/GARDEN PLOT [OPS plan for response to civil disturbances] *mt*
GARP Global Atmospheric Research Program [also GAR] [USA] *ge, bl, wp, hc, aw130.11.45, hy, hg, oe*
GARS GOES Archive and Retrieval System *oe*
GAS General Air Staff *wp*
GAS Get Away Special *mi, mb, jdr*
GASER Gamma Ray Laser *wp*
GASFET Gallium Arsenide Field Effect Transistor *wp*
GASL General Applied Science Laboratories [Ronkonkoma, New York] *aa27.4.60, aw130.22.17*
GASMS Ground Aircraft Services and Maintenance Support *hc, jj*
GASP General Activity Simulation Program *mt, fu, hc*
GASP General Aviation Synthesis Program *fu*
GASP Generalized Academic Simulation Program *wp*
GASS General Air and Surface Situation *fu*
GASSER Geographic Aerospace Search Radar *hc*
GAST German Auto Shop Tester *hc*
GAT General Analog Test(er) *fu*
GAT General Aviation Trainer *hc*
GAT Greenwich Apparent Time *fu*
GAT Ground to Air Transmitter *fu*
GAT Group d'Activation Tuyres [Ariane 5 booster nozzle gimbal system] *ai2*
GATE GARP Atlantic Tropical Experiment *pl*
GATE GBU-15 Automatic Test Equipment *hc*
GATE Generic Automatic Test Equipment *hc, wp*
GATE Graphical Analysts Tool Environment *kf*
GATE Ground Activity Target Elimination *hc*
GATERS Ground Air Telerobotic System [USMC] *wp, mt*
GATOR [air deliverable scatterable mine] *dh*
GATORS Ground Air Telerobotics System *hc*
GATR Ground [to] Air Transmitter [and] Receiver *fu, mt*
GATS Gbu-15 Automatic Test Station *wp*
GATX [aircraft conversion company] *aw129.21.115*
GATX/CL (a new aircraft leasing venture called GATX/CL Air Leasing) *aw130.22.27*
GAU [designation for aircraft guns, USA] *mt*
GAUGE Graphical Aids for the Users of GEMACS *ro*
GAV Gravity Accelerated Vehicle *wp*
GAVRS Ground Attitude Vertical Reference System *wp*
GAW Guided Atomic Warhead *hc, jj*
GB Generation Breakdown *tb*
GB Gravity Bomb *st*
GB Great Britain *w9, wp, es55.82*
GB Ground Based *fu, wp*
GB Guided Bomb [a standard bomb fitted with wings and a tail on twin booms, USA] *mt*
GB [organophosphate chemical warfare nerve agent, also called sarin] *wp*
GBASAT {Ground Based ASAT} *go3.2*
GBBCS Ground Based Beam Control System *hc*
GBBM Ground Based Battle Manager *tb*

GBCC Ground Based Control Center *hc, fu, tb*
GBCS Ground Based Common Sensor [USA] *mt*
GBD General Board *dt*
GBESM Ground Based Electronic Support Measures *hc*
GBFEL Ground Based Free Electron Laser *mt, fu, wp*
GBHRG Ground Based Hypervelocity Rail Gun *mt, wp, tb*
GBI Ground Based Interceptor *cp, kf*
GBJ Ground Based Jammer *hc*
GBL Government Bill of Lading *mt, hc, tb, jj, su, fu, oe*
GBL Ground Based Laser *fu, 5, wp, 16, 17, sb, hc, jmj, tb, st, kf*
GBL Ground Based Launcher *tb*
GBM Global Battle Manager *tb, fu*
GBM Ground Based Manager *tb*
GBM Ground Based Measurement *st*
GBMD Global Ballistic Missile Defense *11, wp*
GBMI Ground Based Midcourse Interceptor *tb*
GBPS Giga Bits Per Second *mt*
GBPS Giga Bytes Per Second *ct, em, hi*
GBR Ground Based Radar *mt, fu, jmj, tb, st, kf*
GBR Ground Based Remote *fu*
GBR Gun Boosted Rocket *wp*
GBRF Ground Based Radio Frequency *tb*
GBRG Ground Based Rail Gun *tb*
GBS Global Broadcast Service *jdr, pbl, kf, ah, uh*
GBS Ground Based Sensor *fu, hn49.11.5, pm*
GBS Ground Based System *mt*
GBS TT Ground Based Sensor Technical Test *fu*
GBSTS Ground Based Surveillance Tracking System *aw129.14.25*
GBT Green Bank Telescope *mi*
GBTEWS General Based Tactical Electronic Warfare System *hc*
GBTS Ground Based Surveillance and Tracking System *hq6.2.7*
GBTS Ground Based Training System [USA] *mt*
GBU General Buffer Unit *fu*
GBU Glide Bomb Unit *dh, st*
GBU Guided Bomb Unit [USA] *mt, wp, aw118.18.97, hc*
GC Gas Chromatography *wp*
GC Geographic Conversion *fu*
GC Graphic Controller *fu*
GC Guidance Computer *wp*
GC Guidance [and] Control *tb, fu, jj*
GC Guidance, Control and Autopilot *hc*
GC Gyro Caged *jj*
GC Gyro Control *0, wp*
GC&CS Government Code and Cipher School [UK] *mt*
GC/MS Gas Chromatography/Neutral-Mass Spectrometer *es55.28*
GCA Gain Control Attenuator *uh*
GCA Ground Control(led) Approach *mt, fu, ro, hc, w9, wp, jj, gs2-1.2*
GCAS Generic Configuration Accounting System [USAREUR] *mt*
GCAS Ground Collision Avoidance System *mt, wp*
GCB Graphic Control Block *fu*
GCC Global Command Center *tb*
GCC GNU C-Compiler [Unix] *pk, kg*
GCC Graduated Combat Capability *mt*
GCC Ground Command Count *uh, jdr*
GCC Ground Component Command *mt*
GCC Ground Control Center *fu*
GCC Guidance and Control Computer *hc, wp*
GCC Guidance Control Computer *ai2*
GCC Gulf Cooperation Council *wp*
GCCC Ground Communications, Command, and Control *-16*
GCCC&ME Ground Command, Control, Communication and Mission Equipment *kf*
GCCCCA Global Command, Control, Communications, and Computers Assessment [also GC4A] *mt*
GCCS Global Command and Control System [WWMCCS follow on] *mt*
GCCS Government Code and Cypher School *-10*
GCD Gain Control Driver *fu*
GCD Great Circle Distance calculator [WWMCCS] *mt*
GCDS Graphics Command and Data Set *fu*
GCE Gimbal Control Electronics *jdr*
GCE Global Change Encyclopedia *ge, hg*
GCE Ground Combat Element [USMC] *mt*
GCE Ground Communication Equipment *4, hc, jdr*
GCE Ground Control Equipment *hc, jj*
GCE Ground Crew Ensemble [chemical warfare] *wp*
GCF Ground Communications Facility *hc, re*
GCG Guidance Control Group *jj*
GCHQ Government Communications Headquarters [United Kingdom] *mt, 10, 12*
GCI Ground Control of Interception *mt*
GCI Ground Control(led) Intercept *mt, fu, hc, 12, wp, jj*
GCL Ground Controlled Landing *fu*
GCM General Circulation Model *ge, hy, hg*
GCM Guidance and Control Module [everything aft of KKV propulsion bulkhead] *17, sb*
GCMD Gaining Command *mt*

GCMS Gas Chromatography [Coupled] Mass Spectroscopy (Spectrometer) *wp, eo*
GCN Gage Control Number *fu*
GCN Ground Communications Network *mt, kf*
GCN Ground Control Network *kf*
GCN-X Ground Communications Network III Residual *mt*
GCN/SCIS Ground Control Network /Survivable Communication Integration System *kf*
GCO Ground Computer Change Out *kf*
GCOS General Comprehensive Operating System [Honeywell] *mt*
GCP Geodetic Control Points *hg, ge*
GCP Gimbal Control Processor *jdr*
GCP Global Change Program [ICSU] *ge, hg*
GCP Ground Control Point *bl3-3.20*
GCP Ground Control Processor *mt*
GCP Ground Correlator and Processor *pl*
GCPBL Ground Control Point Build Library *bl3-3.8*
GCR Galactic Cosmic Radiation *kf*
GCR Ground Controlled Radar *fu*
GCR Group Code Recording *pk, kg*
GCRA Generic Cell Rate Algorithm *ct*
GCS General Communications Subsystem *mt*
GCS Ground Communications System *hc, jj*
GCS Ground Control Squadron *wp*
GCS Ground Control Station *fu*
GCS Guidance and Control System (Set) *hc, tb, jj*
GCS Guidance Command Shutdown *ai2*
GCS Gun Control System *jj*
GCS/GTV Guidance and Control Simulator/Ground Test Verification *st*
GCSD Government Communications Systems Division [Harris Corporation] *kf*
GCT General Classification Test *jj, fu*
GCT Greenwich Civil Time *0, w9, fu*
GCTI Guidance and Control Technology Integration *hc*
GCTS Guidance and Control Test Set (Program) *fu*
GCU Generator Control Unit *jj*
GCU Guidance and Control Unit *hc, jj*
GCU Gunner's (Gun) Control Unit *hc, jj*
GCVS General Catalog of Variable Stars *mi*
GCVS Generic Crygenic Upper Stage *ai2*
GD General Discharge *wp*
GD General Dynamics [Corporation] *fu, wp, af72.5.25, nd74.445.26, hc, oe*
GD [organophosphate chemical warfare nerve agent (soman)] *wp*
GD/A General Dynamics/Astronautics *0*
GD/FW General Dynamics/Fort Worth *hc, jj*
GDA Gimbal Dish Antenna [DSCS] *mt*
GDA Gimbal Drive Assembly *-4*
GDA Global Data Area *pk, kg*
GDA Gross Distance Association *fu*
GDAP Generic Digital Autopilot *hc*
GDC General Dynamics Company [usually GD] *jj*
GDC Graphic Display Console *mt*
GDC Graphics Display Controller *fu*
GDC GRASP Data Centre *es55.35*
GDC Gross Distance Check *fu*
GDC Group Document Center *ep*
GDD Ground Display Device *fu*
GDDM Graphics Data Display Manager *pk, kg*
GDE General Data Entry *fu*
GDE Gibbs Duhem Equation *wp*
GDE Gimbal Drive Electronics *4, jdr*
GDE Ground Data Equipment *fu*
GDG Generation Data Group [IBM] *fu, ct, em, kg*
GDG Geography Data Generator *fu*
GDI General Data Interface *fu*
GDI Government of India [also GOI] *wp*
GDI Graphical (Graphics) Device Interface *kg, em, hi*
GDIP General Defense Intelligence Program *mt*
GDIS Gier-Dunkle Integrating Sphere *fu*
GDL Gas Dynamics Laboratory [Leningrad] *-13*
GDL Gas Dynamics Laser *hc, wp*
GDLC Generic Data Link Control [IBM] *pk, kg*
GDM Generalized Development Model *hc*
GDMO Group Data Management Office *hc*
GDOP Geometric Dilution Of Precision *ab.7, fu*
GDP General Defense Plan [NATO] *mt*
GDP Graphic Display Program *fu*
GDP Graphic Draw Primitive *pk, kg*
GDPS Global Data Processing System *pl*
GDR Generalized Data Retrieval System *mt*
GDR Geophysical Data Record *ge, hg*
GDR German Democratic Republic [a.k.a. "East Germany"] *mt, w9, wp, fu*
GDRL Government Data Requirement List *hc, fu, jj*
GDS Gas Deployed Skirt *tk*

GDS General Declassification Schedule *hc, su*
GDS Generation Dataset *ct, em*
GDS Graphic Data Set *mt*
GDS Great Dark Spot *mi*
GDS Gridlock Demonstration System *fu*
GDS Ground Data System *mt, hc, kf, oe*
GDS Ground Development System *mt*
GDS Group Downgrading Stamp *fu*
GDSO Global Data Systems Operators *kf*
GDSS Global Decision Support System [AMC] *wp, mt*
GDSTN (GDSTN supporting station, Merritt Island, Florida) *ac2.9-10, 2, 4*
GDT Geometric Dimensioning and Tolerancing *fu, kg*
GDT Global Descriptor Table *pk, kg*
GDT Graphic Display Terminal *fu*
GDT Graphics Development Toolkit *pk, kg*
GDT Ground Data Terminal *jdr, fu*
GDT&E Government Development Test and Evaluation *fu*
GDTA Aerospace Teledetection Development Group [France] *ac3.6-7*
GDU Geographic Display Unit *fu*
GDU Gimbal Drive Unit *jdr*
GDU Graphic Display Unit *fu*
GDU Ground Display Unit *fu*
GE General Electric [Corporation] *fu, aa26.12.28, aw130.13.4, hc, ac2.2-26, pl, af72.5.7, re, tb, jj, oe*
GE General Engineer *hc*
GE Graphic Engine *fu*
GE Green Energy *hi*
GE Ground Environment *fu, go3.1-1*
GE Groupement Ecole [France] *mt*
GE [ethyl sarin, chemical warfare nerve agent] *wp*
GEADGE German Air Defense Ground Environment [412L replacement] *mt, fu, hn43.6.1, go3.4.1, pm*
GEAEGIS AEGIS [at the four German air defense sites] *fu*
GEANS Gimbaled Electrostatic Gyro Aircraft Navigation System *hc, wp*
GEAS GOES Engineering Analysis System *oe*
GEBOS Generalized Exploratory Base Operations *wp*
GEBU Government Electronics Business Unit *ah, pbl, jdr*
GEC General Electric Company [U. K.] *aw130.13.11, sc3.2.7, aa26.11.12*
GECOS General Comprehensive Operating Supervisor *hc*
GECOS General Electric Comprehensive Operating System *pk, kg*
GED Gasoline Engine Driven *wp*
GED Generic Expendable Decoy *hc*
GED Global Event Detector *jdr*
GEE Grid [navigation system used until 1970, in which a chain of radio station emits synchronized pulses and aircraft calculate their position from the phase shift between pulses] [UK] *mt*
GEE Ground Electronics Engineering/Equipment *hc*
GEEIA Ground Electronics Engineering Installation Agency *fu*
GEF Ground Equipment Failure *fu*
GEG (write Motorola GEG . . . Scottsdale, Arizona) *aw130.6.16, 19*
GEIS General Electric Information Services [company] *sc8.9.4, kg*
GEIS General Electric Information Systems *ct*
GEIS General Experiment Interface Specification *pl*
GEIU Ground Environment Interface Unit *fu*
GEL General Emulation Language *wp*
GEL Generate Exception List *fu*
GEM Generalized Emulation Microcircuit *mt*
GEM Geo Mobile[satellite] *he*
GEM Gimbal Electronics Module *hc*
GEM Giotto Extended Mission *mi*
GEM Goddard Earth Model *ac10.1-15*
GEM Graphics Environment Manager [Digital Research Inc. program] *kg, hi*
GEM Graphite Epoxy Motor *af72.5.150, ai2*
GEM Ground Effect(s) Machine [MMRBM] *mt, hc, w9, tb, jj*
GEM Ground Electronics Maintenance *wp*
GEM Guidance Evaluation Missile *wp*
GEMACS General Electromagnetics Model for the Analysis of Complex Systems *ro*
GEMM Generalized Electronics Maintenance Model *hc, jj, fu*
GEMS General Energy Management System *5, 17, sb, tb*
GEMS Global Environment Monitoring System *hy*
GEMS Graphics Engineering and Mapping System *mt*
GEMS Gravity Environment Measurement System *hc*
GEMSS Ground Emplaced Mine Scattering System *fu, hc, dh*
GEN-X Generic Expendable [decoy] *hc*
GENIE General Electric Network for Information Exchange *pk, kg*
GENIE General Information Extractor *wp*
GENRAD General Radio *fu*
GENS General Soviet [part of NSA's Production Organization A Group] *-10*
GENSEL Generate System Engineering Language *fu*
GENSER General Service [communications] *fu, mt*
GENX Generic Expendable decoy *mt*
GEN_REL General Release *mt*

GEO Geosynchronous (Geostationary) Earth (Equatorial) Orbit *he, fu, mi, aa26.12.13, 14, 16, 17, hc, hy, tb, wl, mb, kf, kg, hc, ep, ce, jdr, vv, oe, ai2*
GEOALERT Geophysical Alert *pr*
GEODS Ground based Electro Optical Deep space Surveillance *tb*
GEODSS Ground based Electro Optical Deep Space Surveillance system [AFSPACECMD] [USA] *mt, tb, 12, wp, sc3.3.3, sb, hc, ps202, jb*
GEODSSS Ground based Electro Optical Deep Space Surveillance System *wp*
GEOFILE Geographic Locations Code File system [WWMCCS] *mt*
GEOLOC Geographic Location Code *mt*
GEOREF Geographic(al) Reference [Grid System] *fu, mt*
GEOS Geodynamics Experimental Ocean Satellite [USA] *ge, sc4.11.3, sf31.1.20, hy, hg*
GEOS Geosynchronous Earth Observation Satellite *wp*
GEOS Graphic Environment Operating System [Geoworks] *kg, hi*
GEOSAT Geodetic Satellite [US Navy] *ge, aa26.8.31, hy, hg*
GEOSIT Geographic Situation *fu*
GEP Graphics Entry Panel *fu*
GEP Ground Entity Point *mt, go3.1.1, jmj, wp, tb, kf*
GERT Graphical Evaluation and Review Technique *mt, wp*
GERTS General Electric Remote Terminal System *hc*
GES Global Enterprise Services *ct, em*
GES Ground Engineering System *hc*
GES Ground Entry Station *fu*
GEST Gemini Slow scan Television *hc*
GET Get Execute Trigger *pk, kg*
GET GOSIP Engineering Test Bed *mt*
GET Ground Elapsed Time *ci.371, 4*
GETOL Ground Effect Takeoff and Landing *wp*
GETS Government Emergency Telecommunication System *kf*
GETS Government Energy Telecommunications System *y2k*
GETS Group Estimating Techniques System *ms3.1.2, 8*
GEV Giga Electron Volt *eo, oe*
GEVS General Environmental Verification Specification *oe*
GEWCC Ground Environment Work and Calibration Center *fu*
GEWEX Global Energy and Water cycle Experiment *hg, ge*
GEWS Graphic Entry Work Station *fu*
GEX Government Employees Exchange *wp*
GF Gap Filler *fu*
GF Global Function *fu*
GF Government Furnished *kf*
GF Gravitational Force *fu*
GF Ground Forces *wp*
GF Guided Flight *st*
GF&CEM Group Facilities and Capital Equipment Management *fu*
GFA Government Furnished Accessory *jj*
GFA Government Furnished Assets *kf*
GFA Gross Failure Area *fu*
GFAC Ground Forward Air Controller *wp*
GFAE Government Furnished Aeronautical / Aerospace Equipment *fu, hc, jj*
GFAE Government Furnished Aircraft Equipment *fu*
GFB Government Furnished Baseline *hc*
GFC Gel Filtration Chromatography *wp*
GFC Generic Flow Control *ct, em*
GFC Graphite Fiber Composite *wp*
GFC Gun Fire Control *mt*
GFCES Glider Flight Control Electronic Subsystem *hc*
GFCS Gun Fire Control System *fu, hc, jj*
GFCSX Gun Fire Control System, Experimental *hc*
GFDL Geophysical Fluid Dynamics Laboratory [NOAA] *pl*
GFE Government Furnished Equipment *mt, su, uh, fu, wp, ggIII.1.46, hc, 5, 16, 17, sb, pl, w9, nt, lt, pm, dm, tb, st, wl, jj, jdr, kf, oe*
GFE Gross Feasibility Estimator [as in GFE III] [JOPS] *mt*
GFEL Government Furnished Equipment List *jj, fu*
GFF Government Furnished Facilities *kf*
GFFC Geophysical Fluid Flow Cell *hy*
GFI Government Furnished Information [item] *fu, wp, dm, tb, kf*
GFI Government Furnished Instructions [a document] *ep, hc*
GFI Ground Fault Interrupter *wp, fu*
GFL Government Furnished List *fu*
GFLOPS Giga Floating Point Operations Per Second *aa26.11.44, tb, hi*
GFM Gertsch Frequency Multiplier *0*
GFM Government Furnished Materiel *mt, hc, wp, tb, fu, jj*
GFM Graphics Function Manager *fu*
GFN Growth to Full NADGE [to convert IPG ADGE sites] *hc, fu*
GFO Ground Forward Observer *hc, jj*
GFP Government Furnished Property (Parts) *mt, fu, su, hc, wp, 17, sb, pl, nt, mb, tb, st, wl, jj, jdr, kf, oe*
GFP Group Finance Procedure *fu*
GFP&S Government Furnished Property and Services *fu*
GFP/E Government Furnished Property/Equipment *-17*
GFPL Government Furnished Property List *fu*
GFPP Guide for the Preparation of Proposals *hc, jj*
GFPT Government Furnished Property Tracking *fu*
GFPTS Government Furnished Property Tracking System *fu*

GFR General Flight Rules *wp*
GFR Generic Failure Rate *fu*
GFRP Glass Fiber Reinforced Plastic *mt, fu, wp*
GFRP Graphite Fiber Reinforced Plastic *hc, gg9, ep, jj, ric, jdr, oe*
GFS Government Furnished Service *tb*
GFS Government Furnished Software *mt, fu, wp, kf*
GFSE Government Furnished Support Equipment *hc, wp, jj*
GFSR General Functional Systems Requirements *jj*
GFTD Global Forces Trends Data Base [NATO] *mt*
GFW Gesellschaft f,r Weltraumforschung [(German) Society for Environmental Research] *hc*
GFY Government Fiscal Year *jdr, oe*
GG Global Gateway *fu*
GG Government Grade *-10*
GG Graphics Generator *fu*
GGAR Gas Guided Aircraft Rocket *hc*
GGI GPS Geoscience Instrument [EOS] *hy*
GGM Gravity Gradiometer Mission *hy*
GGN Global Geophysical Network *hy*
GGP Gateway to Gateway Protocol *mt, ct, kg, em*
GGS Global Geospace Studies (Science) [Program] *hc, ns, hy*
GGS Gyro Gun Sight *mt*
GGTS Gravity Gradient Test Satellite *pl*
GHA Greenwich Hour Angle *mt, hc*
GHC Guidance, Homing, Control *wp*
GHE Gaseous Helium *bo, oe, ai2*
GHE Ground Handling Equipment *hc, 0, jj*
GHIS Geosynchronous High-Resolution Interferometer Sounder *oe*
GHL Groupe d'HelicoptÈres LÈgers [light helicopter group] [France] *mt*
GHOST Global Horizontal Sounding Techniques *hc*
GHQ General Headquarters *w9*
GHRS Goddard High Resolution Spectrograph [on HST] *mi*
GI General Issue *w9, an31.13.4*
GI Government Issued *wp*
GI Groupement d'Instruction [training group, France] *mt*
GIA Government Inspection Agency [delegated] *oe*
GIANT_BEAR [worldwide SR-71 refueling support system] *wp*
GIANT_BLAZE [SAC flare program] *wp*
GIANT_CHERRY [terrain avoidance calibration program] *wp*
GIANT_ELK [an SR-71 related program (??)] *wp*
GIANT_FLASH [over-the-target motion picture documentation] *wp*
GIANT_LEAP [SR-71 Operation (??)] *wp*
GIANT_NAIL [U-2 Operational Employment In Sea] *wp*
GIANT_PLATE [SR-71 Contingency Plan (??)] *wp*
GIANT_QUEST [SR-71 Contingency Plan (??)] *wp*
GIANT_REACH [SR-71 Overseas Based Contingency Plan (??)] *wp*
GIANT_SCARE [SR-71 Contingency Plan (??)] *wp*
GIANT_TALK [code name for the USSTRATCOM HF/SSB Broadcast Network , SAC Commander In Chief's means to control strategic forces worldwide] *mt, wp*
GIANT_TASK [USAF reconnaissance special mission (??)] *wp*
GIAT Groupement Industriel des Armements Terrestres *aw130.22.97*
GIBP General Interface Bus Protocol *jdr*
GID Graphics Input Device *fu*
GIDEP Government / Industry Data Exchange Program *mt, ah, fu, 17, wp, sb, pl, hc, nt, ep, tb, jj, pbl, tm, jdr, oe*
GIE Government Inventory Equipment *fu*
GIE Gyro Interface Electronics *-4*
GIF Graphical (Graphics) Interchange Format *mi, ct, jr, hi, du*
GIF Graphics Image File *hi*
GIFAS (GIFAS Equipment Group, the union of French aeronautical and space industries) *aw130.22.32*
GIFT Ground and In Flight Training *jj*
GII Global Information Infrastructure *kg, hi*
GIIC Generic Item Indicator Code *wp*
GIK [State Test Cosmodrome] *ai2*
GIL German Infrared Laboratory *sc4.1.4*
GILS Government Information Locator Service *pk, kg*
GIM Gaining Inventory Manager *wp*
GIM Graphics Interface Module *fu*
GIMAD Generic Integrated Maintenance And Diagnostic *wp*
GIMADS Generic Integrated Maintenance And Diagnostics System *mt, hc, wp*
GIMS Geographic Information Management System *fu*
GIMS Ground Identification of Missions in Space *hc*
GIMSOT Gimbal System for Optical Tracker *hc*
GINGA [the third Japanese X-ray mission, also known as ASTRO-C] *cu*
GIOP General Interoperable ORB Protocol *em*
GIP Ground Instrumentation Plan *0*
GIP Gun Improvement Program *hc*
GIP Gunter Industrial Park *mt*
GIPD General Intelligence Production Division [INSCOM] *mt*
GIPSY GPS Inferred Positioning System *kf*
GIPSY Graphic Information Presentation System [WWMCCS] *mt*
GIR GOES Incident Report *oe*
GIRD Rocket Propulsion Study Group [formed in 1931] [USSR] *13, mt*

GIRLS Generalized Information Retrieval and Listing System *fu*
GIRTS Generic InfraRed Training System *mt, hc*
GIS General Instrument Specification *pl*
GIS Generalized Information System *hi*
GIS Geographic(al) Information System *ct, ge, kg, hg, ns, hi, ls4, 5, 40*
GIS Geometry Information System *fu*
GIS Global Information Solutions [AT&T] *kg, hi*
GIS Gopher Information System *hi*
GIS Government Information Services *hi*
GIS Graphical Information System *kf*
GISB Gas Industries Standard Board *y2k*
GISS Goddard Institute for Space Studies *hc, ci.78, hy*
GIT [Hughes Aircraft Company customer for TTR Target Tracking Radar] *hc*
GIUK Greenland, Iceland, United Kingdom *mt, pf1.5, wp*
GIX Global Internet Exchange [Internet] *kg, hi*
GJACS Gas Jet Attitude Control System *hc*
GK Germeticheskaya Kabina [pressured cabin] [USSR] *mt*
GKA Government Key Access *du*
GKS Graphical (Graphics) Kernel System *mt, kg, fu, wp*
GL Generation Language [as in 4GL for 4th Generation (procedural) Language] *hi*
GL Graphics Language *pk, kg*
GL Graphics Library *hi*
GL Grenade Launcher *wp*
GLA General Ledger Account *hf, 8, 15, hc, ep, jj, fu, he, jdr*
GLA/CI General Ledger Account/Contract Identifier [a designator] *ah, pbl, tm, kf*
GLAADS Gun Low Altitude Air Defense System *hc*
GLAC General Ledger Account Code *mt*
GLACI GLA Contract Identifier *jdr*
GLAD Gradient Light Analytical Detector *wp*
GLADIATOR Global Aerospace Defense Interceptor and Terrestrial Ordnance *fu*
GLADIS Ground Laser Attack Designator/Identification System *wp*
GLAM Groupe de Liaison Aeriennes Ministerielles [France] *mt*
GLARE Ground-Level Attack, Reconnaissance and Electronic Countermeasures *wp*
GLASS_BLADE [SAC tactical aircraft dispersal exercise] *wp*
GLASS_ORGAN [SAC KC-135 annual training exercise] *wp*
GLASS_SHIELD [SAC operational test and evaluation of the B-52 ECM] *wp*
GLAT Government Lot Acceptance Test *hc, jj*
GLAT Guidance Level Acceptance Test *jj*
Glaviaprom [USSR, Chief Directorate for the Aviation Industry] *mt*
GLC G Induced Loss of Consciousness *wp*
GLC Geographic Location Code *jj*
GLCM Ground Launched Cruise Missile [USA] *mt, fu, wp, hc, 12, 14, ph, af72.5.54, ts12-6, tb, st, kf*
GLD Ground Laser Designator *hc*
GLDS Ground Laser Designator System *hc*
GLDS Guidance Laser Designator Station *hc*
GLE Geolocation File *fu*
GLE Government Loaned Equipment *hc, fu, jj*
GLFS GOES Launch From SOCC *oe*
GLINT Gallium Arsenide Laser Illuminator for Night Tv *wp*
GLIS Global Land Information System [US Geological Survey] *pk, kg*
GLLD Ground Laser Locator Designator *mt, hc, wp, hn41.19.8, jj*
GLM General Linear Model *fu, st, kg*
GLMV Ground Launched Miniature Vehicle *sd*
GLO Ground Liaison Officer *mt*
GLO Gun Launched Orbiter *hc*
GLOB Glare Obstructor *-16*
GLOBE Global Backscatter Experiment *hy*
GLOBE Global Learning by Observations to Benefit the Environment [Internet] *pk, kg*
GLOBIXS Global Information Exchange Systems [COPERNICUS] *mt*
GLOCATER Fiberoptic Fault Locater *hc*
GLOMAGS Global Magnetic Survey Satellite *ci.969*
GLOMAR Global Maximum Array Radar *hc, fu*
GLOMB Glide Bomb [a conversion of a light plane with a 1814 kg warhead. 1942—1944] [USA, Pratt-Read LBE] *mt*
GLOMR Global Low Orbiting Message Relay satellite *mi, mt*
GLONASS Global Navigational (Navigation) Satellite System [USSR] *sc2.12.2, 14, ai1*
GLORIA Geographical Long Range Inclined ASDIC [USA] *mt*
GLP Government Loaned Property *hc, fu, jj*
GLRS Geoscience Laser Ranging System *hy*
GLSE Generalized (Linear) Least-Squares Estimation *fu*
GLV Gemini Launch Vehicle *ci.206*
GM General Manager *w9, hi*
GM General Motors [Corporation] *aw130.13.67, fu, 17, sb, hn49.9.1, kf*
GM Generalized Monitor *fu*
GM Geometric Mean *cr*
GM Global Memory *fu*
GM Global Mode *ac2.4-5*
GM Grid Modulation *fu*

GM Ground Mapping (Map) *hc, jj*
GM Ground Mode *0*
GM Group Modem *do214*
GM Guided Missile *0, wp, w9, jj, fu*
GM Gun Mode *jj*
GM&A General Management and Administration *mt*
GM/HE General Motors/Hughes Electronics Corporation [also GMHE] *17, sb*
GMA (Tektronix' GMA series ultra high resolution displays) *aw129.21.149*
GMAC {General Motors Acceptance Corporation} *hn49.8.3*
GMACC Ground Mobile Alternate Command Center *tb*
GMACS GOES Monitoring and Control System *gp, oe*
GMAP General Micro Assembler Processor [Honeywell] *mt*
GMAP Geometric Modeling Applications Project *mt*
GMAS General Management Assessment System *mt*
GMAS Group Material Administrative System *fu*
GMAT Greenwich Mean Astronomical Time *fu*
GMATS Government Metropolitan Area Telephone Service *mt*
GMB Graphic Model of Behavior *mt*
GMC General Motors Corporation *hc, jdr*
GMC Giant Molecular Cloud *mi*
GMC Ground Movement Controller *wp*
GMCC Ground Mobile Command Center *mt*
GMCM Guided Missile Counter Measure *jj, fu*
GMCP Ground Mobile Command Post *mt*
GMCPC General Motors Chevrolet Pontiac Canada *hc*
GMDD Generalized Multiple Disjoint Decomposition *hc*
GMDEP Guided Missile Data Exchange Program [Navy] *hc*
GMDL Guided Missile Destroyer Leader *hc*
GMDS Generic Miss Distance Simulation *fu*
GME Gimbal Mounted Electronics *hc, kf*
GME Guided Missile Evaluator [Hughes Aircraft Company developed] *hc*
GMEA Groupe de Manoeuvre de l'Ecole d'Applications [France] *mt*
GMF (solar cell assemblies are used in the . . . GMF . . . program) *hd1.89.y7*
GMF Ground Mobile Forces *mt, fu, kf*
GMFCS Guided Missile Fire Control System *wp, fu*
GMFSC Ground Mobile Force Satellite Communications [AFCC*] *mt*
GMG {Guidance} Control Moment Gyro *hc*
GMGO German Military Geophysical Office *mt*
GMHE General Motors Hughes Electronics [corporation] *fu, ah, hc, hq5.2.1, hn49.8.8, sn2.15, pbl, kf*
GMI General Military Intelligence [USA DOD] *mt*
GMI Goddard Management Instruction *gp, 4, oe*
GMI Ground Moving Indication [location techniques] *hc*
GMIB Graphical Management Information Base *hi*
GMIL [spacecraft tracking and data network station at KSC] *-4*
GML General Material List *jj*
GML Generalized Markup Language *fu, kg*
GML Guided Missile Laboratory *hc, jj*
GMLS Guided Missile Launching System *wp, fu*
GMLSS Guided Missile Launcher Sub System [TOW] *hc*
GMM Global Memory Manager *fu*
GMMIS GLCM Maintenance Management Information System *mt*
GMNC GWEN Maintenance Notification Center *mt*
GMP Generalized Message Processing *fu*
GMP Global Mobile Professional *pk, kg*
GMPA General Material and Petroleum Activity [USA] *mt*
GMR Giant Magneto Resistive (Resistance) *kg, hi*
GMR Ground Mobile Radar *wp*
GMRT Giant Meter-wave Radio Telescope *mi*
GMS Geostationary (Geosynchronous) Meteorological Satellite [Japan] *ge, fu, hc, wp, hn44.12.5, 16, 17, lt, no, ep, jdr, hg, kf, oe*
GMS Global Management System *pk, kg*
GMS Global Messaging Service [Novell] *kg, hi*
GMS Global {Mobile communications System} *hi*
GMS Ground Maintenance Support *wp*
GMS Guided Missile System *jj*
GMSS Geostationary Meteorological Satellite System *hc, pbl*
GMST Greenwich Mean Sidereal Time *0, ce, ai1*
GMT Graphic Multiple Terminal *fu*
GMT Greenwich Mean Time [a.k.a. UT, Zulu] *mt, mi, kg, uh, gp, wp, 0, 4, 15, hc, w9, sf31.1.1, ce, tb, jj, jdr, mb, kf, fu, hi, oe, ai1*
GMT Guided Missile Training *hc, jj*
GMTA General Motors Task Authorization *hc*
GMTB General Motors Truck and Body *hc*
GMTD Guided Missile Training Device *hc*
GMTI Ground Moving Target Indicator (Indication) *hc, jj*
GMTI/L Ground Moving Target Indication / Location *hc*
GMTS Guided Missile Test Set *hc, jj*
GMTT Ground Moving Target Tracker (Track) *hc, jj*
GMU Guided Missile Unit *jj*
GMW Gram-Molecular Weight *w9, wp*
GN (GN2 purge gas) Gaseous Nitrogen *9, 17, bo, jj*
GN General Navigation *jj*

GN Ground Network *oe*
GN&C Guidance, Navigation, and Control [also GNC] *hc, fu, ci.945, aa26.11.28, cp, tb, wl, jdr*
GN2 Gaseous Nitrogen *jdr, kf, oe, ai2*
GNATS General Noise and Tonal System *fu*
GNC Guidance, Navigation, and Control [also GN&C] *-9*
GNCS Guidance, Navigation, and Control System *ci.390*
GNDACF Ground Alternate Command Facility *mt*
GNDCP Ground Command Post *fu*
GNDFE Ground Force Element *fu*
GNE Government Nomenclature Equipment *fu*
GNE Gross National Expenditure *wp*
GNM Global Network Mission *re*
GNN Global Network Navigator [Spry Technologies] *kg, hi*
GNR Ground Netted Radar *fu*
GNS Global Navigation System *mt*
GNS Guidance and Navigation System *bo, ai2*
GNW General (Global) Nuclear War *kf, mt*
GO Guest Observer *cu*
GO&P General Operations and Plans *mt*
GO2 Gaseous Oxygen [LO2 boiloff] *oe*
GOALS Growth Opportunities and Leadership Seminars *fu*
GOASEX Gulf Of Alaska Seasat Experiment *ac2.7-13*
GOAT Goes Over All Terrain [mobilizer] *hc, jj*
GOB Ground Order of Battle *mt*
GOCC Government Operations Control Center *lt*
GOCC Ground Observation Control Center *fu*
GOCO Government Owned, Contractor (Commercially) Operated *mt, 12, tb, wp, fu, jj*
GOCOSAT Government Communication Satellite *hc*
GOCP Ground Observation Control Post *fu*
GOE Government Owned Equipment *fu*
GOE Ground Operational (Operation) Equipment *hc, jj*
GOES Geostationary (Geosynchronous) Operational Environmental Satellite [NOAA] *aa26.12.26, hc, hn43.21.4, 8, 17, aw130.11.29, lt, wp, ns, no, ep, hy, jdr, mb, hg, kf, ge, fu, mi, ct, oe, mt*
GOES-TAP Access to GOES imagery through commercial systems *gd3-12*
GOET Government Owned Expendable Tool *fu*
GOETS Ground Operations Estimating Techniques System *gb*
GOF Goodness Of Fit *wp*
GOFLAS Ground Fuel Logistical Summary *mt*
GOFS Global Ocean Flux Study *hy*
GOGW Ground Operations Working Group *-4*
GOI Government Of India *wp*
GOI Government Of Iran *hc, jj*
GOI Government Of Israel *hc*
GOLD Global On Line Data *hy*
GOLDS General On Line Display System [WWMCCS] *mt*
GOM GOES Operations Manager *oe*
GOM Government Of Mexico *wp*
GOMAC Government Conference on Microcircuits *hc*
GOMR Global Ozone Monitoring Radiometer *hy, ge, hg*
GOMS Geostationary Operational Meteorological Satellite [USSR] *sc3.2.3*
GOP Generated Options Plans *mt*
GOP GOES Operations Plan *oe*
GOP Group Operations Plan *mt*
GOPHER [a non-proprietary package for convenient access to anonymous FTP sites] *cu*
GOPS Giga Operations Per Second *ic2-63*
GOR General Operational Requirement [USA] *mt, fu, hc, wp, jj*
GOR Ground Operations Review *gp, oe*
GOS Geomagnetic Observing System [EOS] *hy*
GOS Global Observing System [part of WWW] *pl*
GOS Government Of Singapore *wp*
GOS Grade Of Service *fu, mt*
GOS Graphics Operating System *fu*
GOSG General Officer Steering Group *jb*
GOSIP Government Open Systems Interconnection Profile *mt, wp, kg, ct*
GOSPLAN [Soviet State Planning Organization] *ph, mt*
GOSS Ground Operational Support System [Apollo] *hc*
GOTS Government Off The Shelf *mt, kf*
GOWG Ground Operations Working Group *oe*
GOX Gaseous Oxygen *mi, 0, wp, jj, mb, ai2*
GP Gas Plasma *kg, hi*
GP General Perturbations *jb*
GP General Purpose *mt, kg, fu, wp, jj*
GP Giant Pulse *hc*
GP Global Processor *fu*
GP Graphics Processor *fu*
GP Gravity Probe *hy*
GP&RP Government Production and Research Property *hc*
GPA (GPA Jetprop is a joint venture of GPA Group and Canada's PWA) *aw130.13.90*
GPA Gas Processors Association *y2k*
GPA General Purpose Amplifier *fu*

GPA General Purpose Assembler *fu*
GPA Government Procurement Agency *fu*
GPA Grade Point Average *hn41.19.5, w9, fu*
GPAC Ground Positioning Attitude Control *hc*
GPADS General Purpose Airborne Data System *hc*
GPAL GOES Program Action List *oe*
GPALS Global Protection Against Limited Strikes *mt, kf*
GPAMS Ground Processing Automated Maintenance System *mt*
GPARS Generic Phased Array Radar Simulation *mt*
GPATE General Purpose Automatic Test Equipment *fu*
GPATS General Purpose Automatic Test System (Set) *hc, jj*
GPB GPS Position Beacon *ai2*
GPC General Purpose Computer *9, mi, wp, lt, mb, oe*
GPC Global Processor Complex *fu*
GPC Graph Process Controller *fu*
GPCSC General Purpose Computer Support Center [USA] *mt*
GPDC General Purpose Digital Computer *hc*
GPDV Generalized-Parameter Data Vector *fu*
GPE General Purpose Environment *fu*
GPE General Purpose Equipment *jj*
GPE Ground Processing Equipment *pl*
GPEE General Purpose Encryption Equipment *mt, fu*
GPES Ground Parachute Extraction System *hc*
GPETE General Purpose Electronic Test Equipment *hc, fu, wp*
GPF General Protection Fault *ct, kg, em, hi*
GPFS General Purpose Forces Satellite *fu*
GPFU Gas Particulate Filter Unit *fu*
GPHS General Power Heat Source *ci.898*
GPI General Packaging Instruction *fu*
GPI General Purpose Interface *fu*
GPI Glide Path Indicator *mt*
GPI Graphics Programming Interface *pk, kg*
GPI Ground Point Intercept *wp*
GPI Ground Position Indicator *hc, jj, fu, hi, ai1*
GPI/GPO General Purpose Input/General Purpose Output *fu*
GPIB General Purpose Information / Interface Bus (Board) [IEEE 488] *fu, wp, st, kg, oe*
GPIO General Purpose Input/Output *fu*
GPIOP General Purpose I/O Port *fu*
GPL General Precision Laboratory *hc*
GPL Generalized Programming Language *wp*
GPLA General Purpose Line Adapter *mt*
GPM Gallons Per Minute *w9, wp, jj, jdr*
GPM General Purpose Missile *hc*
GPM Generalized Poisson Model *fu*
GPM Government Program Manager *kf*
GPM Graphic Position Marker *fu*
GPM Gyro Pointing Mode *uh, jdr*
GPMG General Purpose Machine Gun *mt*
GPMR Government PMR *mt*
GPMS General Purpose Multiplexing System *hc*
GPNDS Global Positioning and Nuclear Detection System *wp*
GPO Government Printing Office *mt, fu, 0, wp, hc, w9, nd74.445.18, ps203, tb, jj*
GPOCC GOES Payload Operations Control Center *oe*
GPP General Purpose Processor *hc, fu, jdr*
GPR General Purpose Register [IBM] *kg, hc*
GPR Geodetic Position Reference *fu*
GPR Government Plant Representative *hc, fu*
GPR Group Practice [a document] *ep*
GPRS General Purpose Radar Simulator *un*
GPS Gallons Per Second *w9, wp*
GPS General Purpose Satellite *hc*
GPS Global Position(ing) [Satellite] System [NAVSTAR] *wp, hc, sc2.7.3, 5, 14, 16, 17, aw129.14.115, sb, aa26.12.64, af72.5.4, ns, dm, hy, tb, st, wl, jdr, mb, pbl, hg, kf, kg, ge, mi, ct, fu, hi, mt, ai2*
GPS Ground Power Subsystem *jdr*
GPS Ground Processing System [C-5] *mt*
GPS Group Pricing System *fu*
GPSCS General Purpose Satellite Communications System *mt*
GPSIU Global Positioning System Interface Unit *fu*
GPSPAC (the user set . . . flown on the Landsat D mission was GPSPAC) (sd)
GPSS General Purpose Simulation System *mt, fu*
GPSS General Purpose Systems Simulator [language] *hc, kg*
GPT GEC Plessey Telecommunications *aw129.21.26*
GPT Generic Principle Trainer *hc*
GPT Gross Provisions Tester *hc*
GPT&TE General Purpose Tools and Test Equipment *fu*
GPTE General Purpose Test Equipment *wp, jdr*
GPU Gas Pressurized Unit *jj*
GPU Ground Power Unit *mt, hc, jj*
GPU Guidance Power Unit *jj*
GPU Gun Pod Unit [e.g. GPU-2 is pod with the M197 20mm gun, GPU-5 is pod with a GAU-13 gun, USA] *mt*
GPV General Purpose Vehicle *mt*

GPWS General Purpose Work Station *fu*
GPWS General Purpose Work System *mt*
GPWS Ground Proximity Warning System *mt, wp, af72.5.142*
GPX Guided Projectile Experimental *dh*
GQ General Quarters [battle stations] *w9, jj, mt*
GR General Relativity *mi*
GR General Research *hc, ep*
GR Gravity Reference *jj*
GR Grid Resistor *fu*
GR Ground Attack, Reconnaissance [UK] *mt*
GR Gyro Rate *jdr, uh*
GR/XRS Gamma-Ray/X-Ray Spectrometers *re*
GRA Ground Resource Analyst *kf*
GRA Gyro Reference Assembly *-4*
GRABN WESTPAC TACAMO Broadcast {?} *mt*
GRADD Graphics Adapter Device Driver [IBM] *pk, kg*
GRADIO [French Gravity] Gradiometer *hy*
GRADS Ground Radar Aerial Delivery System *mt*
GRAM Generic Radar Analysis Model *st*
GRAMS Generalized Reliability And Maintainability Simulator [model] *wp*
GRAMS [quick look screening system for airborne data (Canada)] *ac3.5-6*
GRANDE Ground Active Nutation Damping Electronics *3, 8*
GRARR Goddard Range and Range Rate *hc, bl3-1.167*
GRAS Generally Recognized (Regarded) as Safe *w9, wp*
GRASER Gamma Ray Amplification by Stimulated Emission of Radiation *wp*
GRASP Gamma Ray Astronomy with Spectroscopy and Positioning *es55.1*
GRASP Generic Retrieve/Archive Services Protocol [an interface definition developed in the OGIP to insulate against archive vendor dependencies] *cu*
GRASP/OP Generalized Retrieval and Sort Processor/Output [OJCS] *mt*
GRASS Ground to Air Scanner Study/Surveillance *hc*
GRBM Global Range Ballistic Missile *hc*
GRBM Ground Regional Battle Manager *tb, fu*
GRBM Ground Regional Battle Manager/Control Center *hc*
GRBS Grating Rhomb Beam Sampler *hc*
GRC General Research Corporation *cp, tb, st, kf*
GRCA Ground Radar Coverage Area *fu*
GRCS Guard Rail Common Sensor [USA] *mt*
GRD Ground Resolved Distance *wp*
GRD Gruppe fur Rustungs Dienste [Switzerland] *mt*
GRE Graphics Engine *pk, kg*
GRE Ground Reconnaissance Equipment *wp*
GREEN_FLAG [electronic warfare exercises, Tyndall AFB, FL] *mt*
GREEN_PINE [unclassified code name for the STRATCOM Northern Area UHF System] *mt*
GREMLIN g-Resistant Electromagnetically Launched Interceptor *17, hc*
GREP Global Regular Expression Print *pk, kg*
GREP Graphite Epoxy *st*
GRF Group Repetition Frequency *fu*
GRFP Graphite Reinforced Fiber Plastic *-16*
GRI Gas Research Institute *y2k*
GRID Global Resources Information Database [UN] *hy, ge, hg*
GRID Ground Radio Interface Devices [ESD program] *wp*
GRIS Global Resources Information System *hy*
GRL Gross Requirements List *jj*
GRLD Graphic Remote Interface Display *mt*
GRM Geopotential Research Mission *hy*
GRM Ground Resource Management *kf*
GRNP General Remote Network Processor *mt*
GRO Gamma Ray Observatory [later CGRO] *mi, sc3.5.7, hc, aw132.10.5, mb, oe*
GROS Graphics Reconnaissance Operations System *mt*
GRP Glass Reinforced Plastic *mt*
GRP Greatest Response Probability *wp*
GRS Gamma Ray Spectrometer[on Mars Observer] *ma7, mi, mb, eo, re, mb*
GRS Great Red Spot *mi*
GRT Germanium Resistance Thermometer *wp*
GRT Gross Requirements Tapes *jj*
GRT Ground Receipt Time *oe*
GRTS General Remote Terminal System [Honeywell] *mt*
GRU Glavnoye Razvedivatelnoye Upravleniye [central military Intelligence Directorate of the General Staff, USSR] *su, fu, ph*
GRWT Gross Weight *jj*
GS General Schedule [government pay schedule] *mt, wp*
GS General Service *wp*
GS General Specification *jj*
GS General Staff *w9, fu*
GS General Support *mt, hc, jj, fu*
GS Gimbal, Stabilized *jj*
GS Glide Slope *fu*
GS Ground C3 and Mission Equipment [IPT Identifier] *kf*

GS Ground Segment *4, jdr, kf, vv*
GS Ground Speed *w9, wp, jj, mt*
GS Ground Station *oe*
GS Ground Support *fu*
GS Group Separator [Information] *kg, fu, do242*
GS Guidance System *jj*
GSA General Services Administration *mt, fu, su, hc, wp, ph, w9, nd74.445.11, tb, jj, kf, y2k*
GSA Ground Support Agency *mt*
GSABC General Services Administration Board of Contracting *mt*
GSAM Generalized Sequential Access Method *hi*
GSAT GTE Satellite [Corporation] *dn*
GSB Ground Systems Bulletin *fu*
GSBCA General Services [Administration] Board of Contract Appeals *wp, kf*
GSC General Staff Corps *w9*
GSC Guide Star Catalog [for HST] *mi*
GSCM Ground Station Control and Monitoring *oe*
GSCP Ground Support Computer Program *fu*
GSD General Support Division *wp*
GSD General Systems Description *jj*
GSD Ground Sample Distance *wp, kf*
GSDS Goldstone Duplicate Standard *0*
GSDSM Global Circulation Dust and Smoke Model [JDSSC] *mt*
GSE General Support Equipment *fu*
GSE Ground Support Equipment [hardware] *mt, ah, he, hc, wp, ggiii, 0, 4, 5, 9, 16, 17, sb, pl, jmj, so, mb, ep, tb, st, wl, jj, pbl, tm, jdr, kf, fu, oe, ai2*
GSE&I General Systems Engineering and Integration *tb, jdr, kf*
GSE/TD General Systems Engineering/Technical Director *hc*
GSEI Ground Support Equipment Illustrations *fu*
GSEP Ground Support Equipment Plan *fu*
GSERD Ground Support Equipment Recommendation (Requirement) Data *fu, hc, jj*
GSERS Ground Support Equipment Requirement Sheets *hc, jj*
GSESD Ground Support Equipment Statistical Data *jj*
GSET GEADGE Site Exercise Tape *fu*
GSF Ground Support Fighter *ph*
GSFC Goddard Space Flight Center [Greenbelt, Maryland] [NASA] *fu, cu, hc, gd1-8, 0, 4, 16, sf31.1.9, es55.49, jb, ns, no, hy, tb, wl, jdr, mb, hg, kf, ge, mi, oe, mt*
GSG Ground Systems Group [Hughes] *fu, rm1.1.1, 8, 15, 17, sb, th4, hq5.3.5, hc, md2.1.1, jj*
GSGMD Ground System Group Manufacturing Division *fu*
GSGP Ground Systems Group Practices *fu*
GSH Ground Segment Hardware [development, manufacture and integration process] *kf*
GSh Gryazev-Shipunov [gun designers, USSR] *mt*
GSI General Server Interface *pk, kg*
GSI Government Source Inspection *fu, 17, sb, hc, jj, jdr, oe*
GSIS Ground Safety Information System *mt*
GSL General Support Laboratory *fu*
GSL Ground Systems Laboratory [Hughes Aircraft Company] *hc, jj*
GSM General Support Maintenance *mt*
GSM Global Shared Memory *pk, kg*
GSM Global Summary Message *kf*
GSM Global System for Mobile [Communications network] *kg, em, jdr, ric*
GSM Ground Safety Monitor *wp*
GSM Ground Segment Manager *sf31.1.15*
GSM Ground Station Module *wp*
GSM Ground System Monitor *oe*
GSMC Global System for Mobile Communications [cf. GSM] *hi*
GSMST Ground Station Mixed Simulation Test (Testbed) *kf, uh*
GSNP [National Observing System for the State of the Environment and Climate, USSR] *hg, ge*
GSNR Government Source Not Required *fu*
GSO General Staff Officer *w9*
GSO Geo Stationary (Geosynchronous) [Earth] Orbit *kf, 14, jdr, ct, tt, ai2*
GSO Ground Safety Office *wp*
GSO Ground Staff Office *hc*
GSO Ground Support Officer *fu*
GSO Ground Support Operations *sc8.9.1*
GSO Ground System Operator *kf*
GSOIA General Security Of Information Agreement *tb*
GSORTS GCCS Status Of Resources and Training System [GCCS] [USA] *mt*
GSOSTATS Geosynchronous Satellite Orbital Statistics [NASA] *-19*
GSP General Strike Plan [NATO] *mt*
GSP Ground Systems (Group) Practices *fu*
GSPL Government Spacecraft Product Line *jdr*
GSPS Gamma Ray Spectrometer Penetrator System *hc*
GSQA Government Source Quality Assurance *fu*
GSR General Service Request *hc*
GSR General Support Reinforcing *mt*
GSR Government Spares Release *wp*

GSR Ground Surveillance Radar *mt, wp*
GSRS Ground Support Rocket System *hc*
GSS Generalized Simulation / Stimulation *fu*
GSS GOES Simulator System *oe*
GSS Graphic Software System *mt*
GSS Ground Segment Specification *kf*
GSS Ground Segment Subsystem [DMSP] *pl*
GSS Ground Support Station *fu*
GSS Ground Support System (Subsystem) *mt, hc, tb*
GSS/L General Staff Support / Large *mt*
GSS/M General Staff Support / Medium *mt*
GSSEL General and Standard Support Equipment List *hc*
GSSG Government System Safety Group *fu*
GSSR Ground System Support Requirements *hc*
GSST Ground Segment System Test *oe*
GST Greenwich Sidereal Time *0, w9, ce*
GST Ground Station Test *oe*
GST Guidance Section Tester *jj*
GSTA Ground Surveillance and Target Acquisition *wp*
GSTAR [a commercial satellite] *sd, hd1.89.y2*
GSTB Ground Segment Test Bed *kf*
GSTDN Goddard Space and Tracking Data Network *oe*
GSTDN Ground Spacecraft Tracking and Data Network *gp, 4*
GSTL Group System Test Log *fu*
GSTN General Switched Telephone Network *pk, kg*
GSTP Ground Segment Transition Plan *kf*
GSTP Ground Systems Test Plan *-4*
GSTRB Ground Station Test Review Board *-4*
GSTS Ground Based Surveillance and Tracking System *mt, jmj, tb, kf*
GSU General Support Unit *mt, jj, ful*
GSU Geographically Separated Units *mt*
GSU Ground Support Unit *hc*
GSV Guided Space Vehicle *0, wp, w9*
GT Generator Transmission *jj*
GT Government Test *fu*
GT Ground Terminal [also G/T] *4, vv*
GT Ground Test [aircraft] *hc, jj*
GT Ground Transmit *fu*
GT Ground Transportable *fu*
GT Group Technology *fu*
GTACS GOES Telemetry and Command System *oe*
GTAMS Ground to Air Measurements Systems *hc*
GTC Gain Time Control *fu, jj*
GTC Gas Turbine Compressor *jj*
GTC Gated Time Constant *fu*
GTC General Transformation Corporation *mt*
GTC Generic Type Code *fu*
GTCC Group Technology Characterization Code *hc*
GTCP Global Tropospheric Chemistry Program *hy*
GTD Geometrical Theory of Diffraction *fu*
GTDS Goddard Trajectory Determination System *oe*
GTE General Telephone and Electronics [corporation] *fu, kg, sc3.5.6, aw130.13.11, aa26.12.45, jb, tb*
GTE Global Tropospheric Experiment *hy*
GTEA Group Test Equipment Assembly *hc*
GTF Generalized Trace Facility *fu*
GTF Global Track File *fu*
GTF Ground Test Facility *fu, tb*
GTF GWEN Training Facility *mt*
GTGU Ground Test Guidance Unit *hc*
GTIC General Trial Intercept Calculation *fu*
GTL Gunning Transceiver Logic *kg, hi*
GTM Ground Test Missile *-17*
GTM Guidance Training Missile *hc, jj*
GTN Global Transportation System [USTRANSCOM] *mt*
GTNC German Territorial Northern Command *mt*
GTO Geostationary (Geosynchronous) Transfer Orbit *mi, sf31.1.19, uh, es55.14, ce, hy, mb, oe, ai2*
GTO Guide To Operations [IBM] *kg, hi*
GTOS Ground Terminal Operation and Support *-4*
GTP General Test Plan *nt, wp*
GTP Geometry Theorem Prover *pk, kg*
GTP Graphic Transform Package *fu*
GTPNet Global Trade Point Network *ct*
GTR Government Transport(ation) Request *fu, mt*
GTRI Georgia Technical Research Institute *fu, hc, st*
GTS Gas Turbine Starter *mt*
GTS GEODSS Test Site *mt*
GTS Global Telecommunications System *pl*
GTS Ground Tracking Station *wp*
GTS Guam Tracking Station *hc, uh*
GTSA German Telecommunications Statistics Agency *hc*
GTSB Guam Tracking Station (B-antenna) *uh*
GTSC German Territorial Southern Command *mt*
GTSS General Technology Support System *fu*
GTT Glucose Tolerance Test *wp*

GTU {an International Motorsports Association competition series} *hn50.9.2*
GTV General Test Vehicle *fu*
GTV Ground Transport Vehicle *un, oe*
GTV Guidance (Guided) Test Vehicle *17, hc, st*
GTW Gross Takeoff Weight *wp*
GTWT Gridded Travelling Wave Tube *hc, jj*
GU Guam *mt, w9, wp*
GU Guidance Unit *17, sb, hc, jj*
GU-RKKF Glavoce Upravlenie-Raboche-Krestyanskogo Krasnogo Vozdushnogo Flota [Chief directorate of the Workers and Peasants Military Air Fleet, created in 1918, USSR] *mt*
GUAP Chief Directorate of the Aviation Industry [USSR] *mt*
GUARD [Emergency Radio Channel] *mt*
GUARDRAIL [airborne COMINT DF HF/VHF/UHF intercept and location system, USA] *mt*
GUARDS General Unified Ammunition Reporting Data System [JCS] *mt*
GUCL General Use Consumables List *fu*
GUERAP General Unwanted Energy Rejection Analysis Program *hc*
GUI Graphical User Interface *pbl, jdr, kf, kg, ct, em, hi, oe*
GUID Global Universal Identifier *pk, kg*
GUID Globally Unique Identifier *kg, em*
GUIDE Graphical User Interface Design Editor *em*
GULS General Use Laser System *hc*
GUNK [chemical warfare decontaminant, Alcohol + Pine Oil + Soap + Naphta + Castor Oil] *wp*
GUS Guidance Update System *tb*
GUTS Guidance Unit Test Station *-17*
GVA Graphic Variable Address *fu*
GVHRR Geostationary Very High Resolution Radiometer *oe*
GVL Great Valley Labs *mt*
GVLLD Ground Vehicular Laser Locator Designator *hc*
GvOTBVP Gvardeyskii Otdelni'i Transportni'i Boevoi Vertoletni'i Polk [Guards Independent Combat Transport Helicopter Regiment, USSR] *mt*
GVS Glove Vane System *jj*
GVS Gypsy Verification Environment *tb*
GVSC Generic VHSIC Spaceborne Computer *17, tb, oe*
GVT Gated Video Tracker *hc, jj*
GVT Global Virtual Time *kg, hi*
GW GateWay *mt, tb*
GW Gross Weight *jj, mt*
GW Guided Weapon *mt, wp*
GW-BASIC Gee Whiz (General Work) - BASIC *kg, hi*
GWAM Get Well Analysis Module [WSMIS] *mt*
GWC Global Weather Center (Central) [AWS] *mt, fu, wp*
GWCLIBAF GWC Library [AWS] *mt*
GWDI Global Weather Dynamics, Inc. *fu*
GWE Global Weather Experiment *ag*
GWEN Ground Wave Emergency Network *mt, fu, hc, wp, 12, ph, is26.1.76cc, tb*
GWIP Global Weather Intercept Position *mt*
GWS General War System *mt*
GWS Graphics Work Station *mt*
GWT Gross Weight *wp*
GWU George Washington University *aa26.11.56*
GXP Gimbal XIPS Platform *he*
GZ Ground Zero *mt, wp, fu, jj*
GZT Greenwich Zone time *fu*

H

H [a designator, H-2 is Sea Sprite, H-3 is Sea King, H-46 is Sea Knight] [USA DOD]
H [chemical warfare codename for crude sulfur mustard] *wp*
H [code for time point in generalized operations plan, as in H-hour] [USA DOD] *mt*
H&K Homing and Kill *st*
H&MS Headquarters and Maintenance Station *jj*
H&R Hughes and Raytheon [Company] *hn47.7.1, hq6.2.7*
H&S Health and Status *tb*
H&V Hardening and Vulnerability *-17*
H-AMI Host - Automated Message Interface *mt*
H/O Hand Off *tb*
H/O Hand Over *tb*
H/P Hydrophone *fu*
H/V Horizontal/Vertical *pk, kg*
H/W Hardware *mt, 16, 17, sb, jb, tb, kg, oe*
HA Half Action *jj*
HA Head Aim *jj*
HA Health and Affairs *dt*
HA High Altitude *fu, wp, tb, jj*
HA Higher Authority *fu, tb*
HA Hour Angle *0, w9*
HA [US Navy attack helicopter squadron (HA[L]-3 Seawolves) USA] *mt*
HA-GO [Japanese plan for an attack in Burma, begun in February 1944, that failed] *mt*

HAA/SS High Altitude Active/Semiactive Seeker *st*
HAAC Helicopter Air to Air Combat *mt*
HAADS High Altitude Altitude Determination System *hc*
HAALS High Accuracy Airborne Location System *wp*
HAAM High Altitude Ablative Materials [Re-Entry ??] *wp*
HAARS High Altitude Airdrop Resupply System *mt*
HAARS High Altitude Altitude Reference System *hc*
HAB High Altitude Burst *tb*
HAC Heading Alignment Circle *aw129.14.20*
HAC HélicoptÈre Anti Char [anti tank helicopter, France] *mt*
HAC High Acceleration Cockpit *wp*
HAC House Appropriations Committee *mt, hc, wp*
HAC Hughes Aircraft Company *hc, sc2.7.1, 0, 8, 15, 16, sb, hq5.3.5, pl, lt, tb, pbl, tm, hg, kf, heh2.21.1, su, uh, ah, fu, ge, jn*
HACBSS Homestead And Community Broadcasting Satellite Service *sc4.5.3, 9*
HACMP High Availability Cluster Multi-Processing [IBM] *pk, kg*
HACNET [Hughes Aircraft Company wide information processing facility, a data communications computer network] *hn43.5.2, fu*
HACON {Hughes Aircraft Communications Network} [long range communications system] *hc*
HAD Heat Actuated Device *fu*
HADAM Hughes Air Defense Architecture Model *fu*
HADAS Helmet Airborne Display And Sight *mt*
HADB High Altitude Dive Bomb *wp*
HADES Helicopter Acoustic Detection System *hc*
HADES Hughes Analog Design Expert System *hc*
HADES Hybrid ASARS Demonstration Evaluation System *hc*
HADR Hughes Air Defense Radar *hn41.25.1, 17, hc, fu*
HADR-MR HADR-Medium Range *fu*
HADR-PT HADR-Precision Tracker *fu*
HADS Hawaii(an) Air Defense System [PACAF] *hc, wp, mt*
HADS Helicopter Air Data System *mt*
HADS High Altitude Defense System *tb, st*
HADS Hughes Advanced Development System *hc*
HAES High Altitude Effects Simulation *wp*
HAF Headquarters Air Force *mt*
HAF Hellenic Air Force [Greece] *hc, fu*
HAFB Holloman Air Force Base [in WSMR] *N75-27202-486, wp*
HAHO High Altitude High Opening *mt*
HAI Helicopter Association International [show] *aw130.6.54*
HAIDEX Hughes Artificial Intelligence Diagnostic Expert *hc, fu*
HAINS High Accuracy Inertial Navigation System *aw129.1.0*
HAIS Hawaiian Air Intelligence System [PACAF] *mt*
HAIS Hostile Aircraft Identification System *fu*
HAISC Hughes Aircraft International Service Company *hc, hn42.1.2, fu*
HAISS High Altitude Infrared Sensor System *wp*
HAJ Hydrazine Arc Jet *ls4, 5, 28*
HAL (HAL Incorporated) [an airlines/services company] *aw130.22.54*
HAL HAC ASH Laser *hc*
HAL Hard Array Logic *pk, kg*
HAL Hardware Abstraction Layer *kg, em, hi*
HAL High Order Assembly Language *wp*
HAL Highly Automated Logic *wp*
HAL House-programmed Array Logic *pk, kg*
HAL Hughes Authoring Language *fu*
HAL Hughes Automatic [Automated] Layout *hc, at*
HAL/S High-Order Programming Language for Spacelab Usage *wl*
HALAMINE Halogenated Amine *wp*
HALAP Hughes Associative Linear Array Processor *hc*
HALDIGS Howard And Lajes Digital Graphic System *mt*
HALE High Altitude Long Endurance [Aircraft] *mt, hc, wp, fu*
HALO High Altitude Large Optics *hc, wp, rva, tb*
HALO High Altitude Learjet Observatory *-17*
HALO High Altitude, Low Opening [a parachuting technique] *mt*
HALOE Halogen Occultation Experiment [UARS, NASA] *ac1.7-26, hy, hg, ge*
HALPRO (13 B-24s of HALPRO detachment .. flying from Egypt) *af72.5.60*
HAM Hard Acoustic Memory *jdr*
HAM Hold And Modify *hi*
HAM Home Amateur Mechanic *ct*
HAMI Hughes Aircraft, Mississippi *fu*
HAMMER Hughes Advanced Multi Mission Radar *hc*
HAMMER [Air Force air launched standoff (10-20 mi) munition . . . weapon] *uc*
HAMMER [nickname for USAF communication service] *wp*
HAMMER_ACE [rapid response secure communications] *wp*
HAMOTS High Accuracy, Multiple Object Tracking System *hc*
HAMP High Altitude Measurement Probe *hc*
HAMPS Host AUTODIN Message Processing System *mt*
HAMS Hardness Assurance, Maintenance and Surveillance *mt*
HAMS High Altitude Mapping System *wp*
HAMSTD Hamilton Standard [Division, United Aircraft] *hc*
HAMT Human Aided Machine Translation *wp*
HAN Home Area Network *ct, em*
HANE High Altitude Nuclear Effects *wp*

HANG Hawaii Air National Guard *mt*
HANGUL [Korean / US bilingual teletype] *mt*
HANSA Hanscom Satellite Analysis *jb*
HAO High Altitude Observatory [NCAR] *hc, no, mb, mi*
HAO Hughes Aeronautical Operations *hc*
HAOL Hybrid Array Optical Link *fu*
HAP Hazard Abatement Program *mt*
HAP HélicoptÈre d' Appui Protection [France] *mt*
HAP High Altitude Platform *wp*
HAP High Altitude Probe *tb*
HAP Host Access Protocol [research-oriented] *csII.60, kg*
HAP Hughes Assembly Program *fu*
HAP Hydraulic Actuator Package *17, sb*
HAP Hydroxylated Ammonium Perchlorate *17, sb*
HAP/ESP Hughes Aerobot Program/Elevated Sensor Program *hc*
HAPDAR Hard Point Defense (Demonstration) Array Radar *hc, fu*
HAPEX Hydrologic Atmospheric Pilot Experiment *hg, ge, hy*
HAPP Hardness Assurance Program Plan *jdr*
HAPP Hughes Aircraft Post Processor [computer] *hc*
HAPS Helicopter Acoustic Processing System *fu*
HAPS Hydrazine Auxiliary Propulsion System *hc, ai2*
HAPSAT High Altitude Probe Satellite *tb*
HAR Helicopter, search And Rescue [UK] *mt*
HARA High Altitude Radar Altimeter *hc, wp*
HARA Hughes Aircraft Retirees' Association *hn49.12.3*
HARAS Hughes Active Radar Augmentation System *hc*
HARC Hughes Acoustic Research Center *fu*
HARC Hybrid Array Radar Concepts *fu*
HARDMAN Military Manpower / Hardware Integration Program [USMC] *mt*
HARDS High Altitude Radiation Detection System *mt*
HARLG High Accuracy Ring Laser Gyro *wp*
HARM High Speed (Hypervelocity) Anti Radiation Missile [USA] *mt, aw130.13.7, ph, hc, af72.5.33, wp, fu*
HARP Halpern Anti-Radar Point *fu*
HARP High Altitude Reconnaissance Platform (Project) *wp, hc*
HARP High Altitude Relay Point *fu*
HARP High Altitude Release Point *mt*
HARP-TAP High Altitude Remotely Piloted-Target Acquisition Platform *st*
HARPS Hybrid AUTODIN Red Patch Service *mt*
HARPSS High Altitude Remote Platform Surveillance System *hc, wp, fu*
HARS Heading and Attitude Reference System *mt, hc*
HART (firms such as . . . HART Environmental) *hn50.8.i*
HART Hardened Amplifier for Radiation Transients *hc*
HART High Altitude Reconnaissance Technology *hc*
HARV High Alpha Research Vehicle [a modified F-18] [USA] *mt*
HARV High Altitude Reconnaissance Vehicle *wp*
HARVEST_BARE [bare base equipment program] *wp*
HAS Hardened Aircraft Shelter *mt, fu*
HAS High Altitude Search *st*
HAS Hydraulic Actuator System *17, hc, rm8.1.3*
HASC House Armed Services Committee [USA] *mt, hc, wp, af72.5.18*
HASC Hughes Aircraft South Carolina *hn47.7.4*
HASCL Hughes Aircraft Systems Canada Limited *fu*
HASCO Hughes Advanced Systems Company *fu*
HASEE Hughes Ada Software Engineering Environment *fu*
HASI Hughes Aircraft Systems International *hc, hn25.11.2, 15, fu*
HASP Heuristic Adaptive Surveillance Project *wp*
HASP High Accuracy Satellite Position *ac3.7-23*
HASP High Altitude Sampling Program [USA] *mt*
HASP High Altitude Space Program *hc*
HASP High Altitude Surveillance Platform *wp*
HASP Hollog-nosed Active Seeker Program {?} *st*
HASP Houston Automatic Spooling Priority (Program) [System] *kg, hc*
HASPA High Altitude Superpressure Powered Aerostat [a US Navy program for a high-altitude, remotely controlled airship for surveillance and data relay purposes] *mt*
HASPS Hardened Solar Power System *hc, fu*
HASS Hardware And Services Schedule *hc*
HASS High Accuracy Sun Sensor *oe*
HAST Harrier Avionics Systems Trainer [UK] *mt*
HAST High Altitude Supersonic Target [a project for a target drone that is to fly at Mach 4 and at altitudes up to 30500m, USA] *hc, mt*
HASTR Hughes Adaptive Search and Track Radar *fu*
HAT Hardened Antenna Technology *mt*
HAT Hardware Acceptance Test *fu*
HAT Height Above Touchdown *wp, aw130.11.72*
HAT High Angle Threat *hc, fu*
HATA Hardness Assurance Test Authorization *jdr*
HATLAS Hughes Automatic Test Language for Automated Systems *fu*
HATRAC Handover Transfer and Receiver Accept Change *fu*
HATS Helicopter Attack System *hc*
HATS Heuristic Automated Transportation mode System *mt*
HATS Huntsville Association of Technical Societies *aa27.4.44*
HATS Hybrid Automatic Test {System} *hc*

HATV High Altitude Test Vehicle *ci.37*
HAVE [nickname for AFSC programs] *wp*
HAVE_ACCENT [foreign technology collection program] *wp*
HAVE_ACORN [ASD support for lightweight fighter prototype] *wp*
HAVE_AGONY [foreign electronic research acquisition program] *wp*
HAVE_BAND [laser applications study] *wp*
HAVE_BELL [ASD support for advanced medium STOL prototype aircraft] *wp*
HAVE_Blue [precursor of the F-117, smaller, with inwards canted tailfins, USA] *mt*
HAVE_BOWL [Intelligence Collection System (?)] *wp*
HAVE_BRIDGE [AFSC Assistance of Tech. Spec. S, Equipment and Technology] *wp*
HAVE_CAB [phased array radar] *wp*
HAVE_CABLE [technique to jam Soviet infrared guidance] *wp, hc*
HAVE_CAN [AFSC Exploitation of Electronics] *wp*
HAVE_CAPABILITY [next generation trainer] *wp*
HAVE_CAR [AFCS Reposturing Actions] *wp*
HAVE_CHARCOAL [to provide infrared countermeasures for large aircraft] *wp*
HAVE_CURB [radar interference reports in the far eastern area] *wp*
HAVE_Dash [project for an active radar guided missile, USA] *mt*
HAVE_Dungeon [project to track cruise missiles and locate their launchers, USA] *mt*
HAVE_EDGE [integrated air weapons test range, ESD] *hc*
HAVE_EYES [AFSC Trade Fairs/Technical Conferences] *wp*
HAVE_FIDDLE [electronic components acquisition program] *wp*
HAVE_Flag [tactical missile project, USA] *mt*
HAVE_FLARE [exploitation of foreign radar systems] *wp*
HAVE_GARDEN [foreign aircraft exploitation program] *wp*
HAVE_GLANCE [electro optical countermeasures study, WPAFB] *hc*
HAVE_JEWEL [Near Real Time TAC Reconnaissance Test Program] *wp*
HAVE_KING [Aerial Intercept Photography] *wp*
HAVE_LACE HAVE Laser Airborne Communications Experiment [WPAFB] *hc, wp*
HAVE_LAW [electro optical countermeasures program, WPAFB] *hc*
HAVE_LEMON [WPAFB program] *hc*
HAVE_LIME [anti radiation and terminal guided missile program, Eglin AFB] *hc*
HAVE_NAP [conventionally armed stand off missile] *wp*
HAVE_Nap [the AGM-142 missile, a development of the Israli Popeye missile, USA] *mt*
HAVE_PLOT [chemical warfare Defense Improvement Program] *wp*
HAVE_QUICK [Improved TAC Air-Ground-Air Jam Resistant UHF Communications] *wp*
HAVE_Quick [jam resistant UHF radio, USA] *mt*
HAVE_RAKE [WPAFB program] *hc*
HAVE_RAM [WPAFB project] *hc*
HAVE_ROSE [Optical Equipment Program] *wp*
HAVE_RUST [WPAFB countermeasures system] *hc*
HAVE_SCREEN [LORAL Elec. program name] *hc*
HAVE_SLICK [air to surface weapon airframe under development] *wp*
HAVE_STING [Space Division program] *hc*
HAVE_SWALLOW [Hughes Aircraft Company SCG program] *hc*
HAVE_TARP [Airborne Interceptor Radar Counter Low Observables (AIRCLOP), WPAFB] *hc*
HAVE_TOWN [AFSC Collection Activities Program] *wp*
HAVE_WEDGE [dual mode guidance (missile) program, Eglin AFB] *hc*
HAW Hazard Analysis Worksheet *fu*
HAW Heavy Assault Weapon [TOW] *hc*
HAW Homing All the Way *wp*
HAWCS Hughes Aircraft Wireless Control Society *hn43.22.9*
HAWK Homing All the Way Killer [a mobile non-nuclear air defense artillery, surface to air missile] [USA DOD] *mt, pm, fu*
Hawkeye [a twin turboprop, multicrew airborne early warning and interceptor control aircraft designed to operate from aircraft carriers, designated as E-2] [USA DOD] *mt*
HAWS Hawaiian Area Wideband System *mt*
HAWTADS Helicopter Adverse Weather Target Acquisition and Detection System *hc*
HAZMAT Hazardous Material *kf*
HAZOP Hazards and Operations *kf*
HB Hazard Beacon *fu*
HB High Band *fu*
HB Homing Beacon *fu*
HB Horizontal Bridgeman *fu*
HBA Handbook Art *hc*
HBA Host Bus Adapter *kg, ct, em*
HBC High Band Centered *fu*
HBC Hughes Business Communications *pbl, tm, ah, jn*
HBH Hop-By-Hop *fu*
HBL Horizontal Blanking *hi*
HBM Homing Ballistic Missile *wp*
HBO Hyperbaric Oxygen *wp*
HBP Handbook Production *hc*
HBS Hardware Breakdown Structure *st*
HBS Hot Bench System *hc*

HBT Heterojunction Bipolar Transistor *wp, aw130.13.87, mj31.5.60*
HBTO Head Book Trade Off *fu*
HBW Half [power] Band Width *gp, oe*
HBWR High Band Warning Receiver *wp*
HC Hand Control *jj*
HC Handoff Coordinator *fu*
HC Hercules [a designator, as in HC-130] [USA DOD] *mt*
HC High Component *kf*
HC Hughes Communications [Incorporated, a subsidiary of Hughes Aircraft Company] *hn41.18.3*
HC [US Navy helicopter combat support squadron, USA] *mt*
HCA Half Cone Angle *mt*
HCA Hard Copy Channel *fu*
HCA Heading Crossing Angle *mt, fu*
HCC Hermetic Chip Carrier *hc*
HCCAS Hardware Configuration Statistical Accounting System *mt*
HCCB Hardware Change Control Board *fu*
HCCM Hardware Command and Control Manager *hc, fu*
HCDC House of Commons Defense Committee [UK] *mt*
HCE Heater Control Electronics *tr, oe*
HCG Horizontal (Location of) Center of Gravity *fu*
HCG Hughes Communications Galaxy [Incorporated] *sc4.2.1, heh2.21.1, fu*
HCI Hardness Critical Item *wp, tb, fu*
HCI Horizon Crossing Indicator *cl, jdr, uh, jn*
HCI Horizontal Center Line *fu*
HCI Hughes Communications [International], Incorporated *hc, hn42.3.1, 8, 17, sc8.5.1, pbl, kf, heh2.21.3, ah, jn*
HCI Hughes Communications, Inc. (Subsidiary) *fu*
HCI Human Computer Interface *kf, oe*
HCL Hardware Compatibility List *mt, kg, ct, em*
HCL Hardware Configuration List *mt*
HCM Hughes Computer Model *hc*
HCMM Heat Capacity Mapping Mission [USA] *hc, ac2.ii, hy, hg, ge*
HCMOS High-density (speed) Complementary Metal Oxide Semiconductor *ne, wp, fu*
HCMR Heat Capacity Mapping Radiometer *ac2.9-7*
HCMSS Hughes Calibration and Measurement Standard *hc*
HCMTS High Capacity Mobile Telecommunications System *hc*
HCOM Hughes Cost of Ownership Model *hc*
HCP Hardness Critical Process *fu, tb*
HCP Harpoon Control Panel *hc*
HCP Hexachlorophene *wp*
HCP Hexagonal Close Packed *ai1*
HCP Hypervelocity Countermeasures Program *wp*
HCPE Hybrid Collective Protection Equipment [chemical warfare] *wp*
HCR Hard Copy Record *fu*
HCRST Hardware Clip, Rotate, Scale, Translate *fu*
HCS Heliborne Common Sensor [USA] *mt*
HCS Helicopter Combat Support *mt*
HCS Host Computer System *tb, fu*
HCS Hughes Communication Services, Incorporated [Subsidiary of Hughes Aircraft Company] *hn44.5.8, lt*
HCS/MRR Helicopter Combat Support/Medium Range Recovery *hc*
HCSAS Hardware Configuration Status Accounting System *mt*
HCSE High Component Systems Engineering *kf*
HCSI Hughes Communication Systems, Incorporated *-15*
HCSS High Capacity Storage System *kg, em*
HCSS High Component System Specification *kf, ct*
HCT Hughes China Technology *heh2.21.1*
HCT Mercury Cadmium Telleruide *kf*
HCU Handheld Computer Unit *fu*
HCU Hard Copy Unit *fu*
HCU Home Computer User *pk, kg*
HCU HPA Control Unit *fu*
HD (HD Labs) [Hughes Aircraft Company customer for BETA] *hc*
HD Hard Disk *kg, wp, hi*
HD Hard Drive *ct, em*
HD Harmonic Distortion *fu*
HD Heavy Drop *mt*
HD Heavy Duty *w9, wp*
HD Henry Draper [catalog entry] *mi*
HD High Density *kg, em, ct*
HD Honorary Discharge *wp*
HD [chemical warfare codename for distilled (pure) sulfur mustard] *wp*
HDA Head Disk Assembly *mt, kg, ct, em, fu*
HDA Hybrid Detective Assembly *17, sb*
HDA Hydroxydopamine *wp*
HDAM Hierarchical Direct Access Method *hi*
HDB Hughes Data Bus *fu*
HDC Helicopter Direction Center *mt, fu*
HDC Hop Data Clock *jdr*
HDCC High Density Ceramic Card *17, sb*
HDCD High Definition Compatible Digital *pk, kg*
HDCS Hughes Developmental Correlation Sensor *hc*
HDD Hard Disk Drive *em*

HDD Head Down Display *mt*
HDD High Double Density *em*
HDDC Hard Disk Drive Controller *hi*
HDDR High Density Digital-[Tape] Recorder (Reproducer) *ac3.3-22, fu*
HDDT High Density Digital Tape *ac3.3-21*
HDE Hop Data Enable *jdr*
HDEP High-Density Electronic Packaging *fu*
HDF Hardness Design Feature *fu*
HDF Hierarchical Data Format [NCSA] *pk, kg*
HDFPT High Density Focal Plane Technology *hc*
HDGTS (NASA HDGTS) [Hughes Aircraft Company customer for geo-stationary earth orbiter] *hc*
HDH HDLC Distant Host *mt*
HDI Head to Disk Interference *pk, kg*
HDI Horizontal Data (Display) Indicator *hc, fu*
HDIB Hull-Dependent Information Base *fu*
HDIC High Density Integrated Circuit *wp*
HDIP Hazardous Duty Incentive Pay *af72.5.49*
HDL Hardware Description Language *fu, hc*
HDLC High level Data Link Control *mt, ct, em, kg, do348, ce, tb, kf, fu*
HDLCP High level Data Link Control Protocol *hi*
HDM Hardware Device Module *kg, em*
HDM Hierarchical Development Methodology *tb, fu*
HDM High Power Deformable Mirror *hc*
HDMI High Density Multichip Interconnect *aa27.4.9, af72.5.15, md2.1.1*
HDMIC High Density Microwave Integrated Circuit *sc9.6.3*
HDMS High Density Memory Set *mt*
HDOS Hughes Danbury Optical Systems *kf, jn*
HDP Hardware Development Plan *fu*
HDP Hughes [Large Screen] Display *fu*
HDPF Hughes Data Processing Facility *hc*
HDPP Hardware Development Project Plan *fu*
HDPS High Density Power Supply program *hc*
HDPWB High Density Printed Wiring Boards *fu*
HDR Hardware Discrepancy Reports *fu*
HDR High Data Rate *4, wl*
HDS Historical Document Status *fu*
HDS Horizontal Display System *hc*
HDS Hughes Driving Simulator *hn49.3.3*
HDSC High Density Signal Carrier [DEC] *kg, hi*
HDSL High-bit-(data)-rate Digital Subscriber Line (Link) [T1 over 4 pairs of twisted wires] *ct, em, nu, kg, hi*
HDSL Historic Drawing Status List *fu*
HDSS Hardpoint Defense System Study *hc*
HDSS Holographic Data Storage System *pk, kg*
HDT High Density Tape *bl3-3.9, hg, ge*
HDT Host Digital Terminal *kg, ct, em*
HDTR High Density Tape Recorder *hc*
HDTV High Definition Television *wp, sc3.6.4, md2.1.4, is26.1.25, 19, kg, ct, em, hi*
HDU Hard Disk Unit *fu*
HDU Hose Drum Unit *mt*
HDVD High Definition Volumetric Display *pk, kg*
HDVLWT High Delta V, Low Weight Thruster *st*
HDVS High Definition Video System *dn*
HE Heat Exchange *fu*
HE Heavy Equipment *mt*
HE Hercules Equivalent *fu*
HE High Energy *tb, st, kf*
HE High Explosive [on target] *mt, wp, hc, 5, sb, w9, tb, st, fu*
HE Highly Elliptic *sd*
HE Holographic Element *hc*
HE Hughes Electronics [Corporation] *pbl, tm, jdr, kf, heh2, 21, 1, ah, jn*
HE Human Engineering *mt, fu, wp, 0, hc, nt, kf, fu*
HE Hydrostatic Equilibrium *pl*
HE/EXJAM Hand Emplaced Expendable Jammer *mt*
HEA High-Efficiency Anti-reflective *fu*
HEAAMS Hughes ERT Anti Armor Missile System *hc*
HEADS High Endo atmospheric Defense System *hc, dh, sb, tb, fu*
HEADS Hughes Enhanced Anti Jam Data Link *hc*
HEAO High Energy Astronomy (Astronomical) Observatory *hc, ci.115, mb, mi, cu*
HEAP High Explosive Armor Piercing *wp*
HEASARC High Energy Astrophysics Science Archive Research Center *cu*
HEAT High Explosive Anti Tank *mt, wp*
HEAVY [Nickname for HQ USAF Programs] *wp*
HEAVY_GOLD [war consumables computation] *mt*
HEAVY_SMOKE [special operations support program] *wp*
HEAVY_WARRIOR [special technical evaluation prgram] *wp*
HEB High Energy Beam *tb*
HEC Header Error Control *ct, em*
HECKS Helicopter Close and Kill System *hc*
HECM Hughes Electronic Counter Measures *hc*
HECS Hughes-Developed Electronic Countermeasures *hc*

HED/S/I High Endoatmospherid Defense/System/Interceptor *st*
HEDA Human Engineering Design Approach *fu*
HEDAD Human Engineering Design Approach Document *tb, fu*
HEDB High Energy Density Battery *hc*
HEDI High Endoatmospheric Defense Interceptor *mt, hc, wp, 5, 17, sb, th4, aw129.14.25, nd74.445.18, ae151.52.459, tb, fu*
HEDL Hanford Engineering Development Laboratory *tb*
HEDS High Endo Designation Sensor *hc*
HEDTA Hydroxyethyl Ethylenediamine Triacetic Acid *wp*
HEED High Energy Electron Diffraction *wp*
HEF High Efficiency *re*
HEF High Energy Fuel *wp*
HEFR Human Engineering Final Report *hc*
HEFU High Energy Firing Units *-17*
HEL Hardware Emulation Layer *em*
HEL High Energy Laser [program] [USA] *mt, fu, 17, wp, uc, hc, af72.5.138, tb, st, kf*
HEL Human Engineering Laboratory [Aberdeen Proving Ground] *hc, tc, fu*
HELCM High Energy Laser Counter Measures *hc, wp*
HELDAF High Energy Laser Device and Facilities *hc*
HELETS High Energy Laser Experimental Test System *hc*
HELI-TOW Helicopter-mounted TOW *hc*
HELICAR Helicopter Radar *hc*
HELIP Hawk European Limited Improvement Program *hc*
HELIX [a simulator writing tool similar to ECSS] *-17*
HELLFIRE Heliborne, Laser, Fire-and-forget seeker *hc*
HELMAP High Energy Laser Mission Applications Project *wp*
HELMID Helmet Mounted Infantry Display *hc*
HELMS Helicopter Multifunction System *hc*
HELP Hazard Elimination is Loss Prevention *sc3.11.3*
HELP Helmet Position sensing system *hc*
HELP High Energy Laser Program *hc*
HELP High Energy Lightweight Propellant *wp*
HELP Hook Loads During Launch Program [also HOLD-UP] *hc*
HELP Hughes Emergency Locator Package *hc*
HELPEN High Energy Laser Penetration *hc*
HELRATS High Energy Laser Radar Acquisition and Tracking System *hc, wp*
HELSA High Energy Laser System Analysis *hc*
HELSTF High Energy Laser System Test Facility *17, ph, wp, tb*
HELTA High Endurance Lighter Than Air project [USN / USCG] *mt*
HELTADS High Energy Laser Tactical Air Defense System *hc*
HELTAS High Energy Laser Target Acquisition System *hc*
HELTAS High Energy Laser Technology Applications Study *wp*
HELVBES High Energy Laser Vacuum Beam Entry System *hc*
HELWS/TAS High Energy Laser Weapon System/Target Acquisition System *hc*
HEM High-Energy Microwave *fu*
HEMISEARCH Hemispherical Search [First Frequency Scanning] *hc*
HEMLAW Helicopter Mounted Laser Weapon *hc*
HEMP High Altitude Electromagnetic Pulse *mt, hc, tb, st, jdr, kf, fu*
HEMT High Electron Mobility Transfer *ep*
HEMT High Electron Mobility Transistor *wp, aa26.12.27, 17, hc, hn50.9.5, jdr, pbl, tm, fu*
HEMT High Electron Mobility Transition *mt*
HEMT {a nuclear effect} *un*
HEMTT Heavy Expanded Mobility Tactical Truck *pm*
HEMW Hybrid Electromagnetic Wave *fu*
HENEC High Endoatmospheric Nuclear Effects Code *tb*
HENILAS Helicopter Night Landing System *hc*
HEO High Earth Orbit [Greater Than 19,300 Nmi] *hs, wp, 5, sb, 16, tb, fu*
HEO Highly Eccentric (Elliptical) Orbit *es55.58, jdr, kf*
HEOEA Hughes Electro Optical Employees' Association *hn25.11.2*
HEOS Highly Eccentric Orbiting Satellite *hc, ac3.7-3*
HEP Held for Planning *fu*
HEP High Explosive Plastic *hc, wp*
HEP Human Engineering Program *fu*
HEPA High Efficiency Particle Air *he*
HEPA High Efficiency Particulate Air [filter] *pl, jdr, oe*
HEPAD High Energy Proton and Alpha-particle Detector *gp, 16, no, oe*
HEPC High energy proportional counter on SXG *cu*
HEPNET High Energy Physics Network *ns, ct*
HEPP Human Engineering Program Plan *fu*
HERA Hermes Robotic Arm *mi*
HERA High Explosive Rocket Assisted *wp*
HERCULES Hierarchical Editor and Router for Chips Using Logic Entry and Simulation *ep, fu*
HERD High Explosive Research and Development *wp*
HERF High Energy Radiation Field *wp*
HERO Hazards of Electromagnetic Radiation [to] Ordnance *hc, wp, fu*
HEROS Henry Rose Associates *fu*
HERT Headquarters Emergency Reconstitution Team *fu*
HERT Headquarters Emergency Relocation Team [SAC] *mt*
HESD Harris Electronics Systems Division *-4*
HESDEP Helicopter Sensor Development Program *hc*

HESH High Explosive Squash Head *mt*
HESI Hughes Environmental Systems, Incorporated *hn50.8.1*
HESP High Efficiency Solar Panel *hc*
HESP High Energy Solar Panel *sc3.8.3*
HESS Human Engineering System Simulator *hc*
HET Health/Education Telecommunications experiment *hc*
HETF Hill Engineering Test Facility *mt, wp*
HETP Height Equivalent To A Theoretical Plate *wp*
HETP Human Engineering Test Plan *fu*
HETR Human Engineering Test Report *fu*
HETS Hyper Environmental Test System [Scout] *hc, ci.198*
HEU High Estimate Unconstrained [W/O SALT II] *st*
HEUS High Energy Upper Stage *uc*
HEW Health, Education, and Welfare [U. S. Department of] *ac1.7-14, 13, w9, wp*
HEWH High Explosive Warhead *hc*
HEX Heat Exchanger [Large] *hc*
HF Harmonic Filter *bl2-9.15*
HF Height Finder (Finding) *w9, mt*
HF High Fidelity *tb*
HF High Frequency [3 to 30 MHz] *mt, hc, wp, gd4-8, 0, 12, aw129.14.109, w9, pf.f88.10, pr, tb, mb, fu, ro, nu, mi, hi, y2k, ai1*
HF Holding Fixture *fu*
HF Home Forces *w9*
HF Human Factors *mt, fu*
HF Hydrogen Fluoride *aa26.12.25, tb, st*
HF/DF High Frequency / Direction Finding *mt*
HF/E Human Factors / Engineering *mt*
HF/SSB High Frequency / Single Sideband *mt*
HFA Hardened Flexible Array *hc*
HFAJ High Frequency Anti Jam *mt, hc, wp*
HFB Height Finder Buffer *fu*
HFB Hughes Flying Boat *hc*
HFC Hop Frequency Control *jdr*
HFC Hybrid Fiber Coaxial *kg, nu, ct, em*
HFCM High Fast Cruise Missile *fu*
HFDF High Frequency Direction Finder [device that enabled Allies to obtain a quick fix on radio transmissions by German submarines, UK] *mt*
HFDF High Frequency Direction Finding *10, fu*
HFDP Human Factors Development Plan *fu*
HFE Helmholtz Free Energy *wp*
HFE Human Factors Engineering *hc, wp, tb, st, kf, fu*
HFEA Hughes Fullerton Employees' Association *hn42.3.5*
HFEA Human Factors Engineering Assessment *st*
HFECCM High Frequency ECCM [Applique] *hc*
HFEP Host Front End Processor *mt, fu*
HFEP Host Front End Protocol *tb*
HFIC High Frequency Intra-Task-Force Communications *hc*
HFK Heeres Flieger Kommando [Germany] *mt*
HFL Hydrogen Fluoride Laser *tb, st*
HFLLDF High Frequency Luneberg Lens Direction Finder *ro*
HFM Hyperbolic Frequency Modulation *fu*
HFMS Hospital Formulary Management System *mt*
HFO High Frequency Oscillator *mt, fu*
HFP Heat Flow Probe *re*
HFP Held For Planning *fu*
HFP Host to Front End Protocol *fu, mt*
HFPA Hardware Floating Point Accelerator *oe*
HFR Hardware Failure Rate *fu*
HFR Height Finding (Finder) Radar *fu, mt*
HFR Swedish Helicopter Association *aw130.22.77*
HFRB High Frequency Regional Broadcast *mt*
HFS Heeres Flieger Staffel [army air squadron] [Germany] *mt*
HFS High Frequency Synchronization *hc*
HFS Human Factors Society *fu, hc*
HFS Hyperfine Structure *hc*
HFSC Height-Finder Slew Control *fu*
HFSNAP High Frequency Steerable Null Antenna Processor *hc*
HFSSB High Frequency Single Side-Band [manpack radio] *hc*
HFT High Function Terminal [IBM] *pk, kg*
HFT&E Human Factors Test and Evaluation *mt, fu*
HFTB High Fidelity Test Bed *kf*
HFWS Heeres Flieger Waffen Schule [army air weapons school, Germany] *mt*
HFX High-Frequency Transceiver *fu*
HG/MD Hybrid Gun/Missile Demonstration *hc*
HGA High Gain Antenna *jp, so, re, mb, mi*
HGA Holographic Grating Axicon *hc*
HGAS High Gain Antenna System *hc*
HGB Half Bridge/Gateway *tb*
HGI Hughes-Georgia, Inc. *fu*
HGM [Silo-stored, guided ground-to-ground Missile. Silo stored means that the engines cannot be fired inside the silo. HGM-16 Atlas, HGM-25 Titan] [USA] *mt*
HGS HEO Ground Station *kf*
HGS Host Gateway Subsystem *fu*

HGS Hughes Gimbaled Sensor *kf*
HGS Hughes Global Services [Inc.] *tt*
HGS Hughes Government Services *pbl*
HGV Hypersonic Glide Vehicle *wp*
HH Half Height *hi*
HH Hawk Helicopter [HH-60A is Night Hawk Helicopter] [USAF] *mt*
HH Horizontal-Horizontal [polarization] *ac7.3-4*
HHA Hand-Held training Aid *hc*
HHB (HHB Systems Inc., Mahwah, New Jersey) *is26.1.36*
HHC Headquarters and Headquarters Company [USA] *mt*
HHD Headquarters and Headquarters Detachment *mt*
HHLAPI High Level Language Application Programming Interface *pk, kg*
HHLD Hand Held Laser Designator *hc*
HHLR Hand Held Laser Rangefinder *hc*
HHM Hand Held Module *hc*
HHMI Howard Hughes Medical Institute *hc, hn43.16.1, fu*
HHOC Headquarters, Headquarter and Operations Company *mt*
HHP Hand Held Product *hi*
HHQ Higher Headquarters *wp*
HHR High Hop Rate *jdr, kf, fu, uh*
HHRA Holographic Helmet Reticle Assembly *hc*
HHS Headquarters and Headquarters Squadron [USA] *mt*
HHS Health and Human Services [Department of] *mt, hc, w9, wp, tb*
HHS High High Star *hc*
HHS/BMS High High Star/Background Mapping by Satellite/Sensor *hc*
HHSRA Holographic Helmet Sight Reticle Assembly *hc*
HHT Hand Held Terminal *mt*
HHT Headquarters and Headquarters Troop [USA] *mt*
HHUD Holographic Head-Up Display *hc*
HI Human Interface *fu*
HI-CAMP Highly Calibrated Airborne Measurements Program [a support program for TEAL_RUBY] [USA] *mt*
HI-ROCC Hawaiian Regional Operational Control Center *fu*
HI-STAR High Speed Towed Array Research *hc, go4.2*
HIAC High Accuracy *wp*
HIAC High Altitude Camera [it weighed nearly 2000kg, and was carried in a side-looking configuration by the RB-47F] [USA] *mt*
HIAD Handbook of Instructions for Aircraft Designers (USAF) *fu*
HIAD High Altitude Defense *hc*
HIADS Hawaiian Integrated Air Defense System *hc*
HIAGED Handbook of Instructions for AGE Designers (USAF) *fu*
HIANG Hawaii Air National Guard *-12*
HIASC Hughes International Service Company *fu*
HIBEX High Boost Experiment [designated upstage] *hc*
HIBREL High Brightness Relay *tb*
HIC Hawkeye Interface Center *fu*
HIC Headend Interface Converter *ct, em*
HIC Height Input Converter *hc*
HIC Hybrid Integrated Circuit *hi*
HICAMP {Hughes inflight measurement program} (mosaic focal plane) *sb, go4.2*
HICAT High Altitude Clear Air Turbulence *wp*
HICCS Hardware Inventory Configuration Control System *mt*
HICLAS High-Power Closed Loop Adaptive System *hc*
HICLASS Hughes Integrated Classification System *hc, hn44.7.5*
HICOM High Command [USN HF/SSB Communication Network] *mt*
HID Hardware Integration Demonstration *fu*
HID Host Interface Device *fu*
HIDACZ High Density Airspace Control Zone [USA DOD] *mt*
HIDAD High Density Array Development *hc*
HIDAM Hierarchical Indexed Direct Access Method *hi*
HIDEC Highly Integrated Digital Electronic Control [USA] *wp, mt*
HIDEC Highly Integrated Digital Engine Control *wp*
HIDES Hardware Implant Detection Study *hc*
HIDSS Helmet Integrated Display (and) Sight (sub) System [USA] *mt, hc, aw129.10.58, hn49.3.4*
HIE Height Integration Equipment *hc*
HIEA {Hughes Irvine Employees' Association} *hn49.12.4*
HIEST Highest Image Enhancement Screen Technology *hi*
HIF Hyper-G Interchange Format *pk, kg*
HIFAR High Frequency Fixed Array Radar *hc, wp*
HIFD High Density Floppy Disk *kg, hi*
HIFI High Intensity Food Irradiator *hc*
HIFICT High Fidelity Color Technology *hi*
HIFR Helicopter In Flight Refueling *mt*
HIFR Hover In Flight Refuelling *mt*
HIFX High Intensity Flash X-Ray [a nuclear hardness test device] *-17*
HIG Honeywell Integrating Gyro *hc*
HIGFETS Heterostructure Isolated Gate Field Effect Transistor *wp*
HIGS Hypervelocity Intercept Guidance Simulator *hc*
HIHAT Height Resolution Hemispherical Reflector Antenna *hc*
HIHAT High Resolution Hemispherical Antenna Technique *fu*
HIHO High Insertion, High Opening *mt*
HIL Hardware In the Loop *17, sb, wp*
HIL Human Interface Link [HP] *pk, kg*

HILAC Heavy Ion Linear Accelerator *wp*
HILO High Insertion, Low Opening *mt*
HILS High Intensity Lightweight Searchlight *hc*
HIM Hardware Interface Module *wp*
HIMAD High Altitude Missile Air Defense *mt*
HIMAD High and Medium Altitude Air Defense *mt*
HIMAD High to (and) Medium Altitude Defense *wp, pm, fu*
HIMAD High to Medium Range Air Defense *mt*
HIMAG High Mobility Agility [vehicle] *hc*
HIMAT High Maneuverable Aircraft Technology *wp*
HiMAT Highly Maneuverable Aircraft Technology [a research RPV for investigation into the aerodynamics of highly maneuverable fighters] [USA] *mt*
HIMEM High Memory *kg, hi*
HIMES History Management and Retrieval System *mt*
HIMES Hughes IFF Mission Environment Simulation *fu*
HIMEZ High Altitude Missile Engagement Zone [USA DOD] *mt*
HIMOT High Mobility Tester *wp*
HIMSS High Resolution Microwave Spectrometer Sounder *hy*
HIOS Headquarters Integrated Office System [USA] *mt*
HIP Height Identification Plot *fu*
HIP Host Interface Processor *fu, hi*
HIP Host-IMP Protocol *mt*
HIP Host-IPLI Protocol *mt*
HIP Hot Isostatic Pressed (Pressure) *17, sb, jr*
HIP Howitzer Improvement Program *fu, hc*
HIPAAS High Performance Attack Avionics System *hc*
HIPAR High Power Acquisition Radar *hc, wp, fu*
HIPAR High Power Illumination Radar *mt*
HIPAS High Performance Attack System *hc*
HIPD Hughes Institute for Professional Development *pbl, tm, hj16.9.4, heh2.21.8, ah, jn*
HIPEG High Performance (External) Gun [Hughes system] *hc, wp*
HIPEHT High Performance Electrothermal Hydrazine Thruster *tr*
HIPERNAS High Performance Navigation System *hc*
HIPHAS High Power Phased Array Experiment *hc*
HIPIR High Power Illuminating (Illuminator) Radar *wp, fu*
HIPO Hierarchical (Hierarchy plus) Input, Process, Output *fu, hi*
HIPO Hierarchy Input Process Output *mt*
HIPO High Power *hc*
HIPOD High Performance Portable Discoid *hc*
HIPPI High Performance Parallel Interface *kg, em, hi*
HIPSAF High Performance Space Feed *hc*
HIR Horizontal Impulse Reaction *fu*
HIRAC High Random Access *fu*
HIRAM High Resolution Aerial Mapping *hc*
HIRAN High Precision Short Range Navigation *wp*
HIRDLS High Resolution Dynamics Limb Sounder [EOS] *pbl*
HIRE Hughes Infra Red Equipment *hc, hn43.15.1*
HIREL_HUD High Reliability Head Up Display *wp*
HIREP High Reliability Packaging *hc*
HIRES High Resolution *kf*
HIRES Hypersonic In-Flight Refueling System *wp*
HIRIS High Resolution Imaging Spectrometer [EOS] *aw130.11.49, hy*
HIRIS High Resolution Interferometer Spectrometer *wp*
HIRL High-Intensity Runway Lights *fu*
HIROCC Hawaii(an) Region Operations Control Center *hc, mt*
HIRPI High Resolution Pointable Images *hc*
HIRRLS High Resolution Research Limb Sounder [EOS] *hy*
HIRS High-resolution InfraRed (Radiation) Sounder [NIMBUS, NOAA] *hg, ge, ac, hy, oe*
HIRS Hughes IR Sensor *kf*
HIRSADAP High Resolution (Real Time) Synthetic Array (Aperture Data) Processor *hc, wp*
HIRTA High Intensity Radio Transmission Area *wp*
HIRU Hemispherical Inertial Reference Unit *kf, he*
HIRU Hughes Inertial Reference Unit *vv, hc*
HIS Helicopter Integrated System *hc*
HIS High-resolution Interferometer Sounder *hc, pl, hy*
HIS Honeywell Information System *mt*
HIS Hostile Intelligence Services *kf*
HIS House [of Representatives] Information System *wp*
HIS Hughes Information System *fu*
HISAM Hierarchical Indexed Sequential Access Method *hi*
HISS Hierarchical Interactive Schematic System *ne*
HIST Heavy Isotope Spectrometer Telescope *hc*
HISTAF High Sensitivity Tank FLIR System *hc*
HIT Hardware Integration Test *fu*
HIT Hawk Instrumentation Team *hc*
HIT Heterojunction Interface Trap *17, sb*
HIT Homing Interceptor Technology *hc, st*
HIT Hughes Aircraft Company Induced Turbulence study *hc*
HIT Hughes Improved Terminal (for JTIDS) *fu, hc*
HITL Hardware In The Loop *tb*
HITS Hardened INSB Technology for Scannners *hc*
HITS Hierarchical Integrated Test Simulator *mt, fu*

HITS Hughes Information Technology Systems *kf, jn*
HITS Human Intelligence Tasking System *wp*
HIU Host Interface Unit *tb, fu*
HIUK Hughes International, [Limited], UK *hc*
HIUS Host Interface Unit Simulator *fu*
HIWAS Hazardous In flight Weather Advisory Service *fu*
HIWI Hilfs Willige [Soviet volunteer in the Wehrmacht] [Germany] *mt*
HJ Helicopter Jammer *fu*
HJBT Heterojunction Bipolar Transistor *fu*
HJT Heterojunction (bipolar) Transistor *jdr*
HK-1 Hughes Flying Boat *fu*
HKV Homing Kill Vehicle *17, sb, tb*
HKV Hughes-Kongsberg Vaapenfabrikk [a joint venture, Hughes-NFT] *hn44.3.1*
HLA High Level Attack *fu*
HLC Heavy Lift Capability *mi*
HLCO High Low Close Open *pk, kg*
HLCU Hotline Control Unit *fu*
HLD Hardened Laser Designator *hc*
HLDC High Level Data Link Control [WIN] *mt*
HLF High Level Format (Formatting) *em, ct*
HLGS Hot Line Gun Sight *hc*
HLH Heavy Lift Helicopter *hc, mt, uc*
HLHS Heavy Lift Helicopter System *hc*
HLI Host Language Interface *mt*
HLL High Level Language *fu, kg*
HLLAPI High Level Language Application Programming Interface *kg, ct, em, hi*
HLLV Heavy Lift Launch Vehicle [SDI] *mt, hc, wp, ci.1055, aw129.21.160*
HLMS High Latitude Monitoring Station *no*
HLQ High Level Qualifier *pk, kg*
HLS Hue, Luminance, Saturation [a color model] *kg, hi*
HLSC Helicopter Logistic Support Center *mt*
HLSUA Honeywell Large Systems Users Association *mt*
HLTF High Level Task Force *wp*
HLTL High Level Transistor Logic *fu*
HLTTL High Level Transistor-Transistor Logic *fu*
HLV Heavy Lift Vehicle *sb, wp, tb, st, mb, fu, mi*
HLX Light Helicopter Experimental [also LHX] *wp*
HM Habitation Module *wl*
HM Hardness Margin *fu*
HM Helicopter, Mine sweeper squadron [US Navy] *mt*
HM&E Hull, Mechanical, and Electrical (HME also used) *fu*
HMA High Memory Area [Microsoft] *kg, ct, em, hi*
HMA Hub Management Architecture *kg, hi*
HMA Hybrid Microcircuit Assembly *hc*
HMAT High Mobility Avionics Tester *hc*
HMC (researchers visited HMC's new substrate processing facility) *md2.1.4*
HMC Halley Multicolor Camera [on Giotto] *mi*
HMCC Hughes Mission Control Center *vv, jn*
HMCT Highway Movements Control Team *mt*
HMD Hardware Monitoring Device *fu*
HMD Helmet (Head) Mounted Display *mt, hc, wp, aa26.11.30, kg*
HMD Hughes Maintenance Depot *hc*
HMD Hughes Microprogrammable Display *fu, hc*
HMDA Hexamethylenediamine *wp*
HMEI Hughes Missile Electronics, Incorporated [Eufala, AL] *-17*
HMEL Hughes Microelectronics Europa, Limited *heh2.21.5, kf*
HMI Host Message Interface *mt*
HMI Human Machine Interface *mt, kf, fu*
HMIC Hybrid Microwave Integrated Circuit *wp*
HMIS Hazardous Materials Information System *mt, wp*
HML Hard Mobile Launcher [USAF] *mt, wp, aw126.20.47, hc*
HMM Hidden Markov Model *pk, kg*
HMMCCC Highly Mobile Multimission CCC *fu*
HMMP Hyper Media Management Protocol *kg, ct*
HMMR High Resolution Multifrequency Microwave Radiometer *hy*
HMMS HELLFIRE Modular Missile System *fu*
HMMS Hughes Mission Management Software *kf*
HMMS HyperMedia Management Schema *ct*
HMMU Hardware Memory Management Unit *hi*
HMMWV High Mobility Multiple Purpose Wheeled Vehicle *mt, uc, wp, pm, kf, fu*
HMN Heptamethylnonane (Fuel) *wp*
HMOS High density Metal Oxide Semiconductor *kg, em*
HMOS High speed Metal Oxide Semiconductor *pk, kg*
HMP Heavy Machine Gun Pod [Belgium] *mt*
HMP Host Monitoring Protocol *kg, ct, em*
HMP Hughes Multiprocessor *hc*
HMP Hydrazine Monopropellant *wp*
HMPA Hexamethylpyrophosphoramide *wp*
HMPP Hazardous Material Program Plan *kf*
HMPS Hughes Mission Processing Software *kf*
HMS Helmet Mounted Sight *mt, hc, wp*

HMS Hughes Materials Specification *hc, gp, 8, ep, jdr, fu*
HMSC Hughes Missile Systems Company *heh2.21.2*
HMSS Helmet Mounted Sight System *hc*
HMU Hot Mock-up *fu*
HMVMA High Mach Vehicle Mission Applications *wp*
HMX High Melting Explosive (Cyclotetramethylenetetramine) *wp*
HN1 [chemical warfare codename for bis(2-chloroethyl)ethylamine, nitrogen mustard] *wp*
HNB Hexnitrobenzene *wp*
HNC Hughes NADGE Corporation *fu*
HNI Hierarchical Network Interface *hi*
HNIL High Noise Immunity Logic *6, fu*
HNL Helium Neon Laser *wp*
HNM Harris Network Manager *kf*
HNNS Hughes Night Navigation System *hc*
HNS Hughes Network Systems, [Incorporated] *sc8.9.4, hn49.8.1, pbl, heh2.21.4, ah, jn*
HNSMS Host Nation Support Management System [JCS] *mt*
HNV-IRD Helicopter Night Vision InfraRed Detector *hc*
HNVS Helicopter Night Vision Sight (System) *hc, ie5.11.6, kf, hn43.22.1, hq6.2.7*
HNVS Holographic Night Vision Goggle *hc*
HO Hand Over *tb*
HO Height Operator *fu*
HO History Office *mt*
HOB Height of Burst *mt, fu, tb, st, wp, kf*
HOB Homing on Offset Beacon *fu*
HOBO Homing Optical Bomb *wp*
HOBO Hornet Bomb [a PAVEWAY project] (HOBO modular weapon program) *hc, af72.5.148*
HOBOS Homing Bomb System [a large family of guided bombs, first introduced in combat in 1969, USA] *mt*
HOBS Homing Bombing System *hc, wp*
HOC Height Overlap Coverage *fu*
HOC High Orbit Component *kf*
HOC Hughes Operations Center *hc*
HOC Hughes Operations Chief *0*
HOE Holographic Optical Element *fu, hc*
HOE Homing Overlay Experiment (Equipment) *mt, 5, wp, 17, sb, hc, nt, tb, st, fu*
HOF Heat Of Formation *wp*
HOF Home Office Facility *hc, tb, su*
HOG Homing Optical Guidance *wp*
HOGHPOWS Holographic Gratings for High Power Outgoing Wavefront Samplings *-17*
HOGS Homing Optical Guidance System *wp*
HOI Headquarters Operating Instructions *mt*
HOIS Hostile Intelligence Services *jdr*
HOJ Home On Jam (Jamming) *mt, hc, wp, fu*
HOL High Order Language *mt*
HOL Higher (High) Order [computer programming] Language *hc, 16, 17, wp, tb, kf, fu*
HOLAB Holographic Alignment Brassboard *-17*
HOLD-UP Hook Loads During Launch Program [also HELP] *hc*
HOLWG High Order Language Working Group *mt*
HOM HAC Optical Module *fu*
HOMER Electromagnetic Homing Project *hc*
HOMI Handbook of Operator's and Maintenance Instructions *fu*
HOMR Human Oriented Mishap Reduction *hc*
HOMS Homing Overlay Missile Simulation *st*
HOP HAM Output Pulse *jdr*
HOPE HADR Operational Product Evolution [project] *hc*
HOPE Hands On Practical Exercise *fu*
HOPI Hughes Optical Products, Incorporated [Des Plaines, Illinois] *hn49.9.2, hd1.89.y22*
HOPS Hardware Performance Optimization System [also HPOS] *hc*
HOPS Helmet-mounted Optical Projection System *hc*
HOS Hardware Operating System *mt*
HOS Higher Order Software *mt, fu*
HOS Holographic Optical Sweden/SRA System *hc*
HOSAT Hands-On Stick And Throttle *wp*
HOSI Handbook Of Service Instructions *hc*
HOSM Host Operations Systems Monitor *mt*
HOSS Homing Systems Survey Study *hc*
HOSS Hydrogen-Oxygen Second Stage *hc*
HOST Headquarters On Line System For Transportation [MAC] *mt*
HOST [Nuclear] Hardened Optical Sensor Test Bed *wp, hc*
HOSTA Home Station *mt*
HOSV High Orbit Space Vehicle *kf*
HOT Hands On Training *fu*
HOT Haut subsonique, Optiquement TeleguidÈ, tirÈ d'un tube [(French) "high subsonic, optically guided, fired from a tube", a helicopter carried anti tank missile] *mt*
HOT High subsonic Optically Teleguided (anti-tank missile) *fu*
HOT Holographic One Tube [goggle] *hc*
HOT Homing Optics Technology *hc*
HOTA Human Operator Task Analysis *fu*

HOTAC Helicopter Optical Tracking and Control Unit *hc*
HOTAS Hands On Throttle And Stick *mt, wp*
HOTOL Horizontal Take Off and Landing [a proposed SSTO craft] *mt, mb, wp, mi*
HOTPHOTOREP Hot Photo Interpretation Report [USA DOD] *mt*
HOU [Hughes] Holographic Optical Unit *hc*
HOWLS Hostile Weapons Location System *hc, fu*
HOXM HOx Monitor *hg, ge*
HP Heat Pipe *jn*
HP Height Processor *fu*
HP Hewlett Packard [Incorporated] *mt, fu, kg, 17, aw130.11.57, is26.1.30, mj31.5.47, tb, ct, hi*
HP High Pass *jdr*
HP High Performance *wp*
HP High Power *ah, uc, hc, pbl, jn*
HP High Pressure *mt, w9, ai1*
HP Horizontal Polarization *fu*
HP&C Hardware Planning and Control *fu*
HP/EGL High Power/Energy Gas Laser [program] *hc*
HPA Handling and Positioning Aid *wl*
HPA High Performance Amplifier *hc*
HPA High Power (Powered) Amplifier *mt, cs1.7, 4, hc, w9, mj31.5.33, 19, lt, ep, ce, dm, jdr, kf, fu, uh, tt*
HPAPS High Power Amplifier Power Supply *fu*
HPB Hydrophone Positioning Band *fu*
HPBDA High Performance Bi-polar Device Array *hc*
HPBW Half Power Beam Width *jdr*
HPC High Power Converter *jgo*
HPC High Pressure Compressor *wp*
HPCC High Performance Computing and Communications *kg, hi*
HPCM Height Processor and Communications Module *fu*
HPD Hard Point Defense *fu*
HPDI Hard Point Defense Interceptor *hc*
HPDM High Performance Demonstration Motor *hc*
HPDS Hard Point Defense System *hc*
HPDU Heater Power Distribution Unit *cl, jdr, uh*
HPF Hazardous Processing Facility *9, jdr, kf, oe*
HPF High Pass Filter *fu, jdr*
HPF High Power Field *w9*
HPF Highest Possible Frequency *fu, w9*
HPF Horizontal Processing Facility *ai2*
HPFD Hughes Process Flow Diagram *sn2.8*
HPFE High-Performance Force Element *fu*
HPFS High Performance File System *kg, em, ct, hi*
HPG Hewlett-Packard Graphics *pk, kg*
HPGC Heading Per Gyro Compass *w9*
HPGL Hewlett Packard Graphics Language *kg, aw130.22.65, hi*
HPI High Performance Interceptor *hc*
HPI High Power Illuminator *fu*
HPI Hughes Process Instruction *fu, jdr*
HPIB HEPAD Interface Box *gp*
HPIB Hewlett Packard Interface Bus *kg, wp*
HPIC Host Platform Integrating Contractor *fu*
HPIEC High Performance Ion Exchange Chromatography *wp*
HPJ High Power Jammer *hc*
HPL Hewlett Packard Language *mt*
HPLC High Performance Liquid Chromatography *wp*
HPLDI High Performance Laser Designator/Illuminator *hc*
HPLJ Hewlett Packard Laser Jet *hi, wp*
HPLO High Power Laser Optics *hc*
HPM High Power Microwave *mt, hc, wp, jdr*
HPM High Power Model *fu*
HPMSK High Priority Mission Support Kit *mt, wp*
HPO Hourly Post Flight *mt*
HPOS Hardware Performance Optimization System [also HOPS] *hc*
HPOX High Pressure Oxygen *wp*
HPP High Pressure Panel *jn*
HPPA Hewlett-Packard Precision Architecture *pk, kg*
HPPI High Performance Parallel Interface *pk, kg*
HPPI High Performance Peripheral Interface *hi*
HPPS High-reliability Parts Pricing System *jdr*
HPR Hardware Problem Request *fu*
HPR High Performance Routing [IBM] *pk, kg*
HPR Hughes Process Requirement *fu, ep, jdr*
HPRF High Pulse-Repetition Frequency *fu*
HPRFICW High Pulse Repetition Frequency Interrupted Continuous Wave *hc*
HPS Human Reliability Program Status *mt*
HPS Hydraulic Power Supply *fu, hc*
HPSP Heat Pipe Sandwich Panel *hn44.2.3*
HPSP High Precision Scan Platform *so*
HPSSC Health Physics Society Standards Committee *fu*
HPT High Pressure Tin *ct*
HPT High Pressure Transducer *jn*
HPT High Pressure Turbine *wp*
HPTE High Performance Turbine Engine *wp*

HPTI High Performance Terminal Interceptor *hc*
HPTLC High Performance Thin Layer Chromatography *wp*
HPTS High Power Transmission System *wp*
HPW High Performance Workstation [Sun] *pk, kg*
HQ Hangar Queen [i.e., an x-configured nonflying craft] *fu*
HQ [chemical warfare codename for 25% sesquimustard (Q) + 75% sulfur mustard] *wp*
HQA Hardware Quality Audit (Assurance) *hc, tb*
HQAP Hardware Quality Assurance Plan *17, sb, fu*
HQDA Headquarters Department of the Army *fu, st*
HQMC Headquarters Marine Corps *mt*
HQPM Hughes Quality Practice Manual *17, sb*
HQ_AFCC Headquarters Air Force Communications Command *mt*
HQ_AFLC Headquarters Air Force Logistics Command *mt*
HQ_AFROTC Headquarters Air Force Reserve Officers {Training Corps} *mt*
HQ_ATC Headquarters Air Training Command *mt*
HQ_DA Headquarters, Department of the Army *mt*
HQ_MAC Headquarters, Military Airlift Command *mt*
HQ_PACAF Headquarters Pacific Air Force *mt*
HQ_SAC Headquarters, Strategic Air Command *mt*
HQ_TAC Headquarters, Tactical Air Command *mt*
HQ_USAF Headquarters, United States Air Force *mt*
HR Height Range *fu*
HR Hertzsprung-Russell [diagram] *mi*
HR High Rate *bl1-5.87*
HR History Record *jdr, pbl, tm, ah, jn*
HR Horizontal Rule *ct*
HR/IND Hard Item Indicator *fu*
HR/LL High Risk/Long Lead *jdr*
HRA Heavy Replaceable Assembly *hc*
HRB Hardware Review Board *fu*
HRBC High Rate Battery Charge *gg102*
HRBS High Rate Bit Synchronizer *-4*
HRC Hop Rate Control *jdr*
HRCS Hazard Response Computer System [Titan II] *mt*
HRD High Rate Demodulator *-4*
HRD High Resolution Diagnostic Diskette *ct, em*
HRDB Human Resource Data Base *tb*
HRDI High Resolution Doppler Interferometer [UARS] *hg, ge*
HREM High Resolution Electron Microscopy *wp*
HRF Height Range Finder *fu*
HRF High Reliability Fighter *wp*
HRFAX High Resolution Facsimile [for GMS] *pl*
HRFPA High Rate Focal Plane Assembly *hc*
HRG Hemispherical (Resonating) Resonator Gyro *17, sb*
HRG High Resolution Graphics *kg, hi*
HRGM High Resolution Ground Mapping *hc*
HRGM/TD High Resolution Ground Map/Target Designation *hc*
HRI Hardware RAID controller Interface *hi*
HRI Height Range Indicator *w9, fu*
HRI High Resolution Imager [on ROSAT] *mi, cu*
HRI High Resolution Interferometer *ns*
HRIR High Resolution InfraRed (Scanning) Radiometer *ac2.8-2, hg, ge*
HRIS High Resolution Imaging Spectrometer [ESA] *hy*
HRIS High Resolution InfraRed Sounder *ag*
HRL Horizontal Reference Line *fu*
HRL Hughes Research Laboratories (ful, sc2.7.3, 16, 17, sb, hc, md2.1.2, jn)
HRL Human Resources Laboratory *fu*
HRM High Resolution Mapping *hc*
HRMC Human Resource Management Center [Washington] *dt*
HRMS High Resolution Mass Spectroscopy *wp*
HRMS High Resolution Microwave Survey *mi*
HRMT Hughes Resident Management Team *fu*
HROAN Hardware Realization Of Adaptive Networks *wp*
HRP High Reliability Parts *jdr*
HRP Hovering Recoverable Probe *tb*
HRPD High Repetition-rate Pulsed Doppler *wp*
HRPD High Resolution Pulse Doppler [Radar] *wp*
HRPI High Resolution Pointable Imager *hc*
HRPT High Resolution Picture Transmission [NOAA] *pl, hg, ge*
HRPT High Resolution Pressure Transducer *cl, jn*
HRR High Range Resolution *fu*
HRR High Resolution Radar *hc, wp*
HRRL High Repetition Rate Laser *wp*
HRRM High Range Resolution Monopulse [Tracking Radar] *wp*
HRS (the HRS (microchannel plate) experiments) *es55.61*
HRS Hardware Requirements Section *ct*
HRSC High Resolution Stereo Camera *ct*
HRSI High Temperature Reusable Surface Insulation *ci.xxii*
HRT High Rate Terminal *kf*
HRT High Resolution Tracker *fu*
HRT Historical Record Tag *fu*
HRTF Head Related Transfer Function *pk, kg*
HRTV High-Resolution Television *fu*

HRTVM High-Resolution TV Monitor *fu*
HRU Heading Reference Unit *fu*
HRU High Risk Units *mt*
HRV High (Haute) Resolution Visible [SPOT] *ac3.6-8, hg, ge*
HRVIR High Resolution Visible and InfraRed *hg, ge*
HRWCM High Reliability Workmanship Criteria Manual *fu*
HS Harness Subsystem *vv*
HS Heat Shield *5, sb*
HS Helicopter anti submarine warfare Squadron (HS-9 Sea Griffens) [US Navy] *mt*
HS High Sensitivity *kf*
HS High Speed *hs, kg, jj, hi*
HS Horizon Sensor *fu*
HS Hughes Space [a prefix designator for programs] *hc, ep, pl, tm, ah, jn*
HS Hughes Spacecraft [bus, model] 601 *kf, fu*
HS Hughes Standard [a document] *ep*
HS Hybrid Simulator *fu*
HSAM Hierarchical Sequential Access Method *hi*
HSB High Speed Bus *mt, hc, wp*
HSB Hue, Saturation, Brightness [color model] *kg, hi*
HSBR High Speed Bombing Radar *hc*
HSC Hermetically Sealed Container *fu*
HSC Hierarchical Storage Controller *fu, kg, oe*
HSC High Speed Channel *pk, kg*
HSC Horizon Scanner *fu*
HSC Hughes Aircraft Systems Canada Limited [also HASCL] *fu*
HSC Hughes Space and Communications [Company] *pbl, jdr, tm, kf, vv, heh2.21.5, ah, he, uh, jn*
HSCDC Hughes Space and Communications [Company] Distribution Center [organization] *tm, pbl, jn*
HSCS High Speed Communications Subsystem *hc*
HSCT High Speed Civil Transport *aw130.13.74, aa26.12.b23*
HSCT High Speed Commercial Transport *mt*
HSCT Hughes Small Communication Terminal *hc*
HSD HAM Serial Data *jdr*
HSD Hard Site Defense *fu, hc, tb*
HSD High Speed Data *mt, ric, kf, jdr, fu, oe*
HSD Horizontal Situation Display [map] *fu, hc*
HSD Hughes Standard Design [a document] *ep*
HSD Human Systems Division *wp*
HSDB High Speed Data Bus *wp, fu*
HSDI High Speed Data Interface *oe*
HSDI Hughes Space Defense, Incorporated *17, sb*
HSDL High Speed Data Link *wp*
HSDS High Speed Digital Switch *fu*
HSE Hardware and Systems Engineering *fu*
HSE Headquarters Support Element *mt*
HSES Hughes Satellite Earth Station *hc*
HSETC Naval Health Sciences Education and Training Command *dt*
HSF Head Steerable FLIR *wp*
HSF Horizontal Scanning Frequency [display screen] *hi*
HSFB High Speed Fleet Broadcast *mt*
HSG Headquarters Support Group *dt*
HSI Hardware Software Integration *wp*
HSI Heading Select Indicator *hc*
HSI Horizontal Situation Indicator *mt, hc, wp, fu*
HSI Host Speed Interface *mt*
HSI Hue, Saturation, Intensity [color model] *kg, hi*
HSI Human Systems Integration *kf*
HSI Hyperspectral Imagery/Imaging *kf*
HSIBS Hull-Specific Information Base Subsystem *fu*
HSIM High Speed Interface Module *fu*
HSL High Speed Launch *w9*
HSL High Speed Logic *fu*
HSL Light Helicopter anti submarine warfare Squadron [US Navy] *mt*
HSLA High Speed Line Adapter *mt*
HSLCM High Speed Line Control Module *mt*
HSLP High Speed Line Printer *oe*
HSM Hard Structure Munitions *hc*
HSM Hardened Silo Missile *wp*
HSM Hierarchical Storage Management *kg, hi*
HSM High Speed Memory *fu, mt*
HSMP High Speed Message Processor *kf*
HSMT Hughes Site Management Team *fu*
HSN Hardened Surface Communications Network *fu*
HSN High Speed Net *fu*
HSP High Speed Photometer [on HST] *es55.26, mi*
HSP High Speed Printer *mt, kg, fu, tb*
HSP Hughes programmable Signal mini Processor [Minipro] *fu*
HSP Hughes Spare (Space) Part [a hardware specification] *ep, pbl, tm, ah, jn*
HSPL Hughes Space Parts List *jdr, jn*
HSPL Hughes Standard Parts List *pbl, jdr, ah*
HSR Hardware Support Resources *sd*
HSR High Speed Reader *fu*
HSRP Headquarters Systems Replacement Program [7CG] *mt*

HSRT Hydraulic System Response Test *hc*
HSS Helicopter Support Ship *mt*
HSS High Speed Serial [interface chip] *oe*
HSS High Speed Switch *jdr, uh*
HSS Hybrid Simulation System *hc*
HSSA Helmet Sight Sensor Assembly *hc*
HSSA High Speed Switch Assembly *jdr, uh*
HSSC Hughes Standard SEM Cabinet *fu*
HSSDB High Speed Special Data Buffer *hc*
HSSI High Speed Serial Interface *kg, ct, em*
HSSI Hughes Simulations Systems, Incorporated *hn49.14.3*
HSSS High Speed Signal Sorter *fu*
HSSS Hydraulic System Service Set *hc*
HST Hauldown, Securing, and Training [system] *hc*
HST High Speed Technology [US Robotics] *kg, ct, em*
HST Hubble Space Telescope *es55.13, no, mb, cu, mi*
HST Hughes Standard Tool *fu*
HST Hypersonic Transport *mt, w9, wp*
HSTISA Host / Tenant Interservice Support Agreement [PACAF] *mt*
HSTP Hardware and Software Turnover Plan *mt*
HSTS High Speed Text Search System *hc*
HSTSP Hardware and Software Turnover Support Plan *mt*
HSTT High Speed Test Track *st*
HSTTY High Speed Teletypewriter *mt*
HSTV High Survivability Test Vehicle *hc*
HSTVL High Survivability Test Vehicle Lightweight *hc*
HSU Heater Switch Units *jdr*
HSU Host Support Unit *fu*
HSV Hue Saturation Value *kg, hi*
HSVD Horizontal Situation Video Display *fu*
HSVP High Speed Vector Processor *wp*
HSWG High Speed Wire Guidance *wp*
HSWS High Speed Waveform Sampler *ac10.4-4*
HSYNC Horizontal Synchronization *hi*
HSZD Hermetically Sealed Zener Diode *fu*
HT Heavy [Satellite Ground] Terminal *mt*
HT Height Technician *fu*
HT Horizontal Tabulation *fu, do242*
HT [chemical warfare codename for 60% crude mustard and 40% t-mustard] *wp*
HT/MT Heavy Terminal/Medium Terminal *kf*
HTA Heavier Than Air *wp*
HTAC Hard Target Attack Capability *wp*
HTACC Hardened Tactical Air Control Center [OSAN] *mt, aw130.13.87*
HTAF Hellenic Tactical Air Force *mt*
HTB High Technology Brigade *wp*
HTB Hughes Test Bed *fu*
HTC Hop Timing Control *jdr*
HTC Hughes Tool Company *hc*
HTCS High Critical Temperature Superconductor *-19*
HTD Horizontal Tactics Display *hc*
HTDM Helicopter Team Defense Missile *hc*
HTEA Hughes Torrance Employees' Association *hn42.4.4*
HTEC High Technology *wp*
HTEST Hughes VLSI Test [system] *rm8.2.1*
HTF Hardware Test Facility *fu*
HTF Hyper-G Text Format *hi*
HTHL Horizontal Takeoff Horizontal Landing *mi*
HTI Horizontal Tactics Indicator *hc*
HTI Horizontal Technology Integration *kf*
HTI Hughes Training, Incorporated *heh2.21.5*
HTK Hit To Kill *17, tb*
HTKP Hard Target Kill Potential *wp*
HTL High Threshold Logic *6, fu*
HTLD High Technology Light Division [USA] *mt, fu*
HTML Hyper Text Markup Language [www presentation language] *kg, nu, ct, em, jr, hi, du*
HTOVL Horizontal Take Off, Vertical Landing *mt*
HTPB Hydroxyl Terminated Poly Butadiene [propellant] *17, lt, oe, ai2*
HTR High Technology Radar *fu*
HTR High Thermal Regime *st*
HTRB High Temperature Reverse Bias *bl2-9.50, mj31.5.7, jdr, fu*
HTS HARM Targeting System [USA] *mt*
HTS Hawaiian Tracking Station *16, hc*
HTS High Temperature Semiconductors *mt*
HTS High Temperature Superconductor *wp*
HTS Hughes Telecommunications and Space *pbl, ah, jn*
HTS Hughes Terminal System *fu*
HTS Hughes Text System *fu, hn44.18.2*
HTSA Host Tenant Support Agreement *mt*
HTSB Hawaii Tracking Station (B-antenna) *uh*
HTSC High Temperature Superconductivity *wp*
HTSC Hughes Technical Services Company (Corporation) *hn48.13.1, hq6.2.2, heh2.21.1, fu*
HTSI Hughes Training Systems, Incorporated *hn49.14.8*
HTT Hard Target Tracking *hc*

HTT High Technology Testbed *wp*
HTTB High Technology Test Bed *wp, hc, aw126.20.127*
HTTP Hyper Text Transfer (Transport) Protocol [underlying protocol in the www] *nu, kg, pbl, ah, ct, em, jr, hi, du*
HTTP-NG HTTP Next Generation *pk, kg*
HTTPD HyperText Transfer Protocol Daemon *em*
HTTPS Hyper Text Transfer Protocol Security *du*
HTTT High Temperature Turbine Technology *wp*
HTU Handheld Terminal Unit *fu*
HTU HEPAD Test Unit *gp*
HTVL Horizontal Takeoff Vertical Landing *mi*
HTW Hard Target Weapon *wp*
HTWI High Temperature Wire Insulation *hc*
HTX (Chemical Industries' HTX thermoplastics resin) *aa26.12.22*
HUCO Hughes Consortium *hc*
HUD Head(s) Up Display *mt, wp, hn42.4.1, aw130.13.75, hc, ws4.10, sn2.14, fu, kg*
HUD Housing and Urban Development, [Department of] *aw118.14.11, w9, aa27.4.10, wp, y2k*
HUD/WAC Head Up Display/Weapon Aiming Computer *hc*
HUDWAC Head-Up Display Weapon Aiming Computer *mt*
HUDWASS Head-Up Display Weapon Aiming Sub-System *mt*
HUI Hardcopy User Interface *fu*
HUK Hunter Killer [force] *mt, hc, wp*
HULTEC Hull To Emitter Correlation *mt*
HUMID Hughes Unit Malfunction Isolation Detector *hc*
HUMINT Human Intelligence *10, wp, tb, kf, fu*
HUMINT Human resources Intelligence [USA DOD] *mt*
HUMS Health and Usage Monitoring System *mt*
HUPS H6000 Utilization Project System *mt*
HUSAR Hughes System for Artillery *fu*
HUSFU Hard Summary Fault Unit *mt*
HUSINT {Hughes Switzerland International} [a Hughes associate company] *hc*
HUT Hopkins Ultraviolet Telescope [ASTRO package] *mi, kg*
HUTE Hard User Terminal Element *mt*
HV (on dual polarized (HH and HV) Ka-band aircraft images) *ac1.4-14*
HV High Value *t*
HV High Voltage *bl3-1.193, 9, w9, mb, fu, mi, hi, oe*
HV High Volume *tb*
HV Homing Vehicle *tb*
HVAC Heating, Ventilation, and Air Conditioning *hn48.19.3, hc, wp, tb, jdr, kf, fu, jn, y2k*
HVAM High Volume Automated Manufacturing *wp*
HVAP High Velocity Armor Piercing *wp*
HVAR High Velocity Aircraft Rocket [the 5 inch Holy Moses and later Zuni rockets] [USA] *mt*
HVC Hardened Voice Circuit *fu*
HVCH Hardened Voice Channel *fu*
HVD Helmet Visor Display *hc*
HVDC High Voltage Direct Current *wp, is26.1.56, fu*
HVG Hardware Vector Generator *fu*
HVG Hyper Velocity Gun *fu, wp, tb*
HVHMD Holographic Visor Helmet Mounted Display *hc*
HVL Hyper Velocity Launcher *wp, tb, fu*
HVM Hyper Velocity Missile [technology demonstration] [USA] *mt, hc, aw124.21, wp*
HVP High Velocity Projectile *hn48.20.1*
HVP Horizontal and Vertical Position *pk, kg*
HVPS High Voltage Power Supply *ggIII.2.403, hc, fu, oe*
HVR High Voltage Regulator *fu*
HVT High Value Target *hc, tb, st*
HVT Hypersonic Velocity Technology *wp*
HVTIP High Volume Transaction Interface Package *mt*
HVU High Value Unit *fu*
HVU High Voltage Unit *fu*
HVV Hyper Velocity Vehicle *wp*
HWC Hand Wire Check *fu*
HWCI Hardware Configuration Item *mt, hc, ep, dm, tb, kf, fu*
HWCI Hardware Control Item [hardware] *pbl, tm, jdr, ah, jn*
HWCP Hardware Code Page *pk, kg*
HWD Height-Width-Depth *kg, hi*
HWDM Hardware Development and Manufacturing *kf*
HWIL Hardware In the Loop *nt, jmj, tb, st, kf, fu*
HWLS Hostile Weapons Locating System *hc, fu*
HWM Human Workload Model *fu*
HWS Harassment Weapon System *hc*
HWS Harpoon Weapon System *hc*
HWSS Harassment Weapon System Sensor *hc*
HWT Hypersonic Wind Tunnels *0*
HWT Hypervelocity Wind Tunnel *wp*
HX Heat Exchanger *wp*
HYBMED Hybrid Microelectronic Device *fu*
HYCANS Hydrofoil Collision Avoidance and Navigation System *hc*
HYCOL Hybrid Computer Link *fu*
HYD (Pilot . . . responsible for APU, HYD, MPS, EPS, OMS, RCS) [Shuttle] *ws4.10*

HYDAPT Hybrid Digital Analog and Pulse Time *fu*
HYFES Hypersonic Flight Environmental Simulator *wp*
HYPER_ISC Hyperchannel Inter Systems Communication *mt*
HYSAS Hybrid Signal Analysis System *hc*
HYSAT Hybrid Satellite [System] *fu*
HYSAT Hybrid Sensor Experiment Satellite Study *hc*
HySTP Hypervelocity System Technology Program [USA] *mt*
HYTELNET Hypertext-browser for TELNET Network accessible sites *kg, hi*
HYTREC Hydrospace Target Recognition, Evaluation and Control *hc*
HYVIA Hyper-Velocity Interceptor Armament *hc*
HYWAYS Hybrid With Advance Yield Surveillance *hc, hn49.11.1, ie5.11.8*

I

I&C Installation and Checkout *hc, tb, kf, fu*
I&C/O Installation and Checkout [also I&C and I&CO] *fu*
I&CO Installation and Checkout *tb, kf, fu*
I&CS Information and Communications System *tb*
I&E Instrumentation and Exercise *fu*
I&EW Intelligence and Electronic Warfare [USA] *mt*
I&I Installation and Integration *fu*
I&KP Instructor and Key Personnel *fu*
I&L Installations and Logistics *hc, dt*
I&Q In-Phase and Quadrature *hc, fu*
I&RB Industrial and Regional Benefits *fu*
I&S Interchangeability and Substitution system [USA DOD] *fu, mt*
I&SI Information and Software Integration *mt*
I&T Identification and Traceability *fu*
I&T Installation and Test *tb*
I&T Integration and Test *gp, 17, sb, jdr, kf, fu, oe*
I&V Integration and Verification *kf*
I&W Indications and Warnings *mt, jb, tb, tb, fu*
I&W Intelligence and Warning *mt*
I&WS Indications and Warnings System *mt*
I-CASE Integrated Computer Aided Systems Engineering *kf*
I-HAWK Improved Hawk [USA] *mt*
I-IDNS In-garrison IDNS [USREDCOM] *mt*
I-RAMS Integrated Retractable Aircraft Munition System [for the RAH-66 Comanche, USA] *mt*
I-S/A_AMPE Inter-Service / Agency Automated Message Processing Exchange [program canceled] *mt*
I-SBR Innovative Space Based Radar *ic1-1*
I/C Implementation / Conversion *mt*
I/D Integrate and Dump *fu*
I/D Interceptor/Destroyer *fu*
I/E Instrumentation and Exercise *fu*
I/EW Intelligence and Electronic Warfare *fu*
I/F Image to Frame (Ratio) *fu*
I/F Interface (gp, 17, sb, jdr, kf, fu, kg, oe)
I/FOA Installation / Field Operating Activities [USA] *mt*
I/IP Installation / Implementation Plan *mt*
I/O Input / Output *mt, w9, gp, 0, 12, 16, 17, hc, aa27.4.52, jb, lt, tb, st, jdr, kf, kg, ct, uh, fu, oe, ai1*
I/OB Input Output Buffer *fu*
I/OE Input Output Element *fu*
I/OP Input Output Processor *fu*
I/OR Input Output Register *fu*
I/P Identification of Position *mt, fu*
I/P Input (power) *bl1-5.9, jdr*
I/P Inspection Plan *fu*
I/Q Inphase/Quadrature [phase] *oe*
I/S Imager/Sounder *oe*
I/T Information Technology *pbl, ah*
I/T Integration and Test *-17*
I/WAC Interface / Weapon Aiming Computer [fitted to USMC AV-8A Harriers in place of the more complicated British FE541 navigation system] [USA] *mt*
IA Imagery Analyst *mt, wp*
IA Implementing Agency *wp*
IA Indirect Access *fu*
IA Information Architecture *mt*
IA Initial Assessment [USA DOD] *mt*
IA Initial Authorization *hc*
IA Intel Architecture *ct, em*
IA Intersite Architecture [team] *mt*
IA&T Integration, Assembly and Test *tb, st*
IA/LP Ion Analyzer and Langmuir Probe *es55.26*
IA/SR Intelligence Analysis/Storage and Retrieval *fu*
IAA Interim ASOC Automation *mt*
IAA International Academy of Astronautics *ac3.6-4, sf31.1.11*
IAADS Integrated Anti Airborne Defense System *hc*
IAADS Interim Automated Air Defense System *hn48.19.1, fu*
IAANG Iowa Air National Guard *-12*

IAAWS Infantry Anti Armor Weapons Study *fu*
IAB Internal Auto Boresight *hc*
IAC Industrial Applications Center *ci.929*
IAC Information Analysis Center [of DOD] *pl, wp*
IAC Instructor Aircraft Commander *aw129.10.72*
IAC Integrated Analyses Center [USMC] *mt*
IAC Integrating Associate Contractor *fu*
IAC Integration, Assembly, and Checkout *wp, hc, fu*
IAC Intelligence Analysis Center [USMC] *fu, mt*
IAC Intelligent Asynchronous Controller *fu*
IAC Inter Application Communications [Macintosh] *kg, hi*
IAC Inter Array Communication *fu*
IAC Inter Array Correlator *fu*
IAC International Advisory Committee *mt*
IAC/ADP Interagency Committee on Automatic Data Processing *mt*
IACCR Industrial Activity Capability and Capacity Response *fu*
IACG Inter Agency Consultative Group *ns*
IACRS Interagency Committee on Remote Sensing [Canada] *ac3.5-2*
IACS Integrated Acoustic Communications System *fu*
IACS Integrated Armament Control System *hc, wp*
IACS Integrated Avionics Control System *hc, wp*
IACS Integrated Avionics Crew Station *hc*
IACS International Annealed Copper Standard *fu*
IAD Integrated Access Device *ct, em*
IAD Inventory Adjustment Document *wp*
IADB Inter American Defense Board *mt, dt, wp*
IADC Inter American Defense College *dt*
IADE Interactive Data Enhancement *hc*
IADL Indentured Assembly Drawing List *fu*
IADS Iceland Air Defense System *hn48.19.5*
IADS Integrated Air Defense System *mt*
IADS Interactive Authoring and Display System *mt*
IADS Italian Air Defense System *hc*
IAE Instituto de Aeronutica e Espa o [Institute of Aeronautics and Space, Brazil] *ai2*
IAE Integrated Architecture Experiment *st*
IAE International Aero Engines *aw129.1.125*
IAEA International Atomic Energy Agency *mt, 14, wp, w9, fu*
IAF Initial Approach Fix *fu*
IAF Interactive Applications Facility *mt*
IAF International Astronautical Federation *ci.154, aa26.10.b10, sf31.1.11*
IAF Israeli Air Force *wp*
IAF Italian Air Force *hc*
IAFU Integrated Assault Fire Unit *fu*
IAG Instruction Address Generation *pk, kg*
IAG Interactive Application Generation *mt*
IAGA International Association of Geomagnetism and Aeronomy *ns, no*
IAGC Instantaneous Automatic Gain Control *hc, wp, fu*
IAI Inactive Aircraft Inventory *wp*
IAI Israel Aerospace (Aircraft) Industries *aw126.20.27, aa26.11.8, af72.5.132, ai2*
IAIPS Integrated Automatic Intelligence Process System *hc*
IAIPS Intelligence Analyst Integrated Processing System *mt*
IAK Internet Access Kit [IBM] *pk, kg*
IAL International Algebraic Language [later renamed ALGOL] *kg, wp*
IALC Instrument Approach and Landing Chart *w9*
IALCE International Airlift Control Element *mt*
IAM Improved Aimpoint Maintenance *hc*
IAM Inertially Aided Munition *wp*
IAM Institute of Aviation Medicine [UK] *mt*
IAM International Association of Machinists [and aerospace workers] *w9, aw118.18.30*
IAM Inverse ATM multiplexing [Fractional T3 capability transmitting cells over multiple T1 circuits] *nu, ct*
IAMP Imagery Acquisition and Management Plan [OSD] *mt*
IAMS Individual Alert Measures *mt*
IAMUS Installation Automated Manpower Utilization System [USA] *mt*
IANA Internet Assigned Numbers Authority [Internet] *kg, em*
IAP Improved Accuracy Program *hc*
IAP Instrument Approach Procedure *fu*
IAP Inter Array Processing *fu*
IAP Interactive Application Processor *mt*
IAP Internet Access Provider [Internet] *kg, hi*
IAP Intrasystem Analysis Program *hc*
IAP Isdn Adjunct Processor *wp*
IAP Istrebitelnye Aviatsionnye Polki [Fighter Aviation Regiment] [USSR] *mt*
IAP-IASA International Airport - Integrated System Architecture *mt*
IAPPP International Amateur/Professional Photoelectric Photometry *mi*
IAPS Ion Auxiliary Propulsion System [Hughes Aircraft Company electric rocket engine] *hc, sc4.1.2, ep*
IAPSO International Association of Physical Sciences of the Ocean *ge, hg*
IAR Imagery Analysis Report *wp*
IAR Instrument Accommodation Review *re*
IAR Intelligence Automation Requirements *mt*

IAR Interface Analysis Report *fu*
IAR Internal Assessment Review *mt*
IAR Inventory Adjustment Report *wp*
IARMS Integrated Aircrew Resource Management System [MAC*] *mt*
IARN Immediate Air Request Net *mt*
IARS Improved Aerial Refueling System [KC-135] *wp, mt*
IARS Inertial Automatic Registration System *fu*
IAS Immediate Access Storage *mt, fu*
IAS Indicated Air Speed *mt, wp, aw130.13.39, w9, fu*
IAS Information Adoptive System *hc*
IAS Institute of Aeronautical Sciences *hc*
IAS Institute of the Aerospace Sciences *0*
IAS Instrumentation Acquisition System *17, sb*
IAS Integrated Air Surveillance *hc*
IAS Integrated AUTODIN System *mt*
IAS Integrated Avionics System *hc, wp*
IAS Intelligent Analyst System *hc*
IAS Interactive Application System *mt*
IAS Interface Applications Software *mt*
IASA Integrated AUTODIN System Architecture *mt*
IASN Intelligence Analysis Support Network *mt*
IASPG Integrated AUTODIN System Planning Group *mt*
IASPM Infrared Atmospheric Signature Prediction Model *wp*
IASRP Integrated AUTODIN System Requirements Panel *mt*
IAT Immediate Attitude Trim *-3*
IAT Import Address Table *pk, kg*
IAT Integrated Acceptance Test *pl*
IAT Integration, Assembly and Test *fu*
IAT Intelligence Analysis Team *mt*
IAT International Atomic Time *fu*
IAT&C Integration, Assembly, Test And Checkout *kf*
IATA International Air Transport Association *mt, hc, aw118.18.26, w9, fu, y2k*
IATACS Improved Army Tactical Communications System [USA] *mt*
IATACS Integrated Acquisition Tracking and Aimpoint Control System *hc*
IATC International Air Traffic Control *fu*
IATO Interim Authority To Operate *kf*
IATS Intermediate Avionics Test Station *fu*
IAU Infrastructure Accounting Unit *hc*
IAU Interface Adapter Unit *mt, fu*
IAU International Astronomical Union *pl, w9, sf31.1.12, hy, mb, cu, mi*
IAUC IAU Circular *mi*
IAUP Internet User Account Provider [Internet] *pk, kg*
IAV Inventory Adjustment Voucher *wp*
IAVC Instantaneous Automatic Volume Control *fu*
IAVMS Installation Automated Vehicle Management System [USA] *mt*
IAW In Accordance With *mt, wp, hc, nt, jmj, tb, st, jdr, kf, fu*
IAW/AA Integrated Attack Warning / Attack Assessment *mt*
IB (logic) Input Buffer *gg68*
IB Interface Builder *em*
IB Internet Bridge *tb*
IB Istrebitel Bombardirovschik [fighter bomber, as in SU-27IB] [USSR] *mt*
IBA Independent Broadcast Authority *dn*
IBACK Intermediate Block Acknowledge *fu*
IBAS Input Beam Alignment System *hc*
IBC Impurity Band Conduction *tb*
IBC Instrument Bus Computer *pk, kg*
IBCA Interior Board of Contract Appeals [Department of the Interior] *hc*
IBCC Improved Battery Control, Central *fu*
IBCS Integrated Battlefield Control System *fu*
IBCS Intel Binary Compatibility Specification *kg, hi*
IBECS Innovative Beam Control Study *hc*
IBERPAC (the Spanish packet-switching data network (IBERPAC)) *es55.51*
IBGTT Interim Battle Group Tactical Trainer [US Navy] *mt*
IBIS Imaging Background Limited Infrared System *wp*
IBIS Integrated Blade Inspection System *wp*
IBL Intercomputer (Inter) Bus Link *fu, oe*
IBL Internal Boundary Layer *fu*
IBM International Business Machines [a corporation] *mt, aw130.13.103, sc2.7.3, 0, 4, hc, es55.100, jb, ce, tb, hg, ct, em, fu, ge, kg, hi*
IBM-GL IBM Graphics Language *kg, hi*
IBMGNET IBM Global Network *hi*
IBP Inbound Processor *jdr*
IBR Implementation Baseline Review *tb*
IBR Initial Baseline Release *jdr*
IBR Integrated Baseline Review *jdr, kf*
IBR Internal Baseline Review *jdr*
IBRS Intra Base Radio System *mt*
IBS Image Buffer System *oe*
IBS Integrated Boresight Sensor *hc*
IBS IntelSat Business Services *ct, em*
IBS International Business Service *sc3.5.7*
IBS Internet Board Shop *hi*

IBS Ionospheric Beacon Satellite *hc*
IBSS Infrared Background Signature Survey *wp, sf31.1.33*
IBT Instructor-Based Training *kf*
IBX Integrated Branch Exchange *fu*
IC Ice Concentration *ac7.5-12*
IC Infrastructure Committee [NATO] *mt*
IC Initial Capability *kf*
IC Input Circuit *pk, kg*
IC Input Control *fu*
IC Inspection Code *fu*
IC Instruction Cache *fu*
IC Integrated Chip *mt*
IC Integrated Circuit *mt, fu, kg, bl2-3.34, 6, pl, hc, aa26.12.72, af72.5.15, md2.1.1, aw130.11.61, wp, st, jdr, pbl, tm, kf, ct, em, ah, ai1*
IC Integration Cell *jdr*
IC Integration Contractor *mt*
IC Intelligence Center *fu*
IC Intelligence Community *fu*
IC Intelligence Community / Center / Collection *mt*
IC Interceptor Controller *fu*
IC Interchangeability Code *fu*
IC Intercommunity *fu*
IC Interface Controller *fu*
IC Interim Change *fu*
IC Internal Combustion *wp*
IC Internal Communications *fu*
IC Interrupt Controller *pk, kg*
IC Inventory Code *fu*
IC Ion Chamber *fu*
IC&D Installation, Checkout and Demonstration *hc*
IC&FCD Interior Communication and Fire Control Distribution *fu*
IC/SPRI Integrated Circuits-Standard Parts Reserve Inventory *fu*
ICA Independent Cost Analysis *mt, wp*
ICA Intelligence Communications Architecture *hc*
ICA Intelligent Console Architecture *kg, em*
ICA Interface Connector Assembly *fu*
ICA Interfacility Communication Adapter *fu*
ICA Intra application Communications Area *pk, kg*
ICAAS Integrated Controls and Avionics for Air Superiority *mt, hc, wp*
ICAD Individual Chemical Agent Detector [chemical warfare] *wp*
ICAD Interactive Computer Aided Design *wp*
ICADS Integrated Correlation And Display System *kf*
ICADS Integrated Cover And Deception System *mt, fu*
ICAF Industrial College of the Armed Forces *mt, dt*
ICAF International Committee on Aeronautical Fatigue *aw130.22.136*
ICAI Intelligent Computer Aided Instruction *tb*
ICAM Improved COBRA Agility and Maneuverability *hc*
ICAM Integrated Communications Access Method *hi*
ICAM Integrated Computer Aided Manufacturing [program] [USAF] *mt, hc, wp, fu*
ICAM Inter Company Advance Memo *jdr*
ICAM International Civil Aircraft Markings *wp*
ICAMP Integrated Computer Aided Manufacturing Program *hi*
ICAO International Cargo Aircraft Organization *fu*
ICAO International Civil Aviation Organization *mt, fu, hc, 0, aw129.14.115, w9, aa27.4.19, wp, y2k*
ICAP Improved COBRA Armament Program *hc*
ICAP Inductively Coupled Argon Plasma *wp*
ICAP Integrated Circuit Application Program *ne*
ICAP Integrated Contractor Assessment Program *hm*
ICAP Integrated Corrective Action Program *hc*
ICAP Internet Calendar Access Protocol [Lotus] *pk, kg*
ICAPS Interactive Computer Aided Provisioning System *fu*
ICAPS Interim Carrier Acoustic Prediction System *fu*
ICAS Integrated Configuration Accounting System *mt*
ICAS Intel Communicating Applications Specifications *kg, hi*
ICAS International Council (Congress) of Aeronautical Sciences *aa26.8.b2*
ICAT Integrated Circuit Automatic Tester *fu*
ICAT Intercept Controller Air Traffic *fu*
ICATS Intermediate Capacity Automated Telecommunications System *mt*
ICAWS Improved Cannon Artillery Weapon System *hc*
ICB Improvement Control Board *fu*
ICB Input Circular Buffer *fu*
ICB Interface Connector Bracket *jdr*
ICB Interface Control Board *hc, st*
ICB International Competitive Bidding *mt, fu*
ICBAD Improved Chemical Biological Agent Decontaminant [chemical warfare] *wp*
ICBC Interagency Committee on Back Contamination *ci.684*
ICBM Inter Continental Ballistic Missile *mt, aw130.13.18, hc, ac3.1-5, 11, 12, 13, 14, 17, sb, aa26.12.38, w9, af72.5.54, wp, tb, st, kf, fu, crl*
ICC Improved Command Center *mt*
ICC Independent Communications Center *mt*
ICC Information and Coordination Center [Patriot] *mt*
ICC Information Coordination Control *mt*

ICC Information [and] Coordination, Central (Center) *fu, pm*
ICC Instrument Control Center [EOS] *hy*
ICC Interceptor Control Concept *st*
ICC Interface Control Contractor *jdr, fu*
ICC International Communications Corporation [Miami, Fla.] *dov*
ICC Interstate Commerce Commission *mt, 17, sb, hc, w9, tb, su*
ICC Item Category Code *wp*
ICCB Interim Configuration Control Board *tb*
ICCC International Conference on Computer Communications *fu*
ICCCCP Intertheater CCC COMSEC Package [also IC3CP] *mt*
ICCD Imaging Charge Coupled Devices *hc*
ICCIS Indonesian CCIS *fu*
ICCIS Interim Command and Control Information System *fu*
ICCIT In Circuit Component Inspection Tester [Hughes] *fu*
ICCP Institute for the Certification of Computing Professionals *mt, kg*
ICCP Integrated Communications Control Panel *fu*
ICCP Interface Control and Converter Processor *jb*
ICCS Integrated Communication and Control System *fu*
ICCS Integrated Communications Collection System *hc*
ICCT Image Computer Compatible Tape *hc*
ICCU Intercomputer Control Unit *mt*
ICCU Interface Control and Converter Unit *jb*
ICD Imitative Communication Deception *fu, wp*
ICD Initial Check Distance *fu*
ICD Initiative Communications Deception *mt*
ICD Installation Control Document *fu*
ICD Interconnection Device *fu*
ICD Interface (Installation) Control Document (Drawing) *mt, fu, ah, he, uh, ep, pbl, tm, jdr, hc, gp, 4, 16, 17, 18, pl, nt, lt, wp, mb, ep, tb, wl, jdr, pbl, kf, tm, vv, jn, oe*
ICD Internal Control Document *tb*
ICD International Code Designation *ct, em*
ICD Interoperability Control Document *mt*
ICDA Infrared Charge Coupled Detector Array *hc*
ICDD Intercomputer Data Device *fu*
ICDD Intercomputer Data Duplexer *fu*
ICDI Inter-Community Data Input *fu*
ICDL Integrated Circuit Design Language *fu*
ICDR Incremental Critical Design Review *fu*
ICDR Initial Critical Design Review *mt*
ICDR Internal Critical Design Review *jdr*
ICDS Integrated Chemical Defense System [chemical warfare] *wp*
ICDS Intelligence Communications Distribution System *wp*
ICDSI Independent Commission on Disarmament and Security Issues -*14*
ICDWS Integrated Chemical Detection and Warning System [chemical warfare] *wp*
ICE Improved Combat Efficiency [an upgrade of the F-4 Phantom II, Germany] *hc, mt*
ICE In Circuit Emulator [Intel] *kg, fu, oe*
ICE Increased Combat Effectiveness *hc*
ICE Independent Cost Estimate *mt, nt, wp, tb, st, kf*
ICE Initial Cost Estimate *mt*
ICE Integrated Computing Environment [Langley Research] *pk, kg*
ICE Integrated Conceptual Environments *mt*
ICE Internal Communications Element *mt*
ICE International Cometary Explorer *sc4.1.4, ns, no, mb, mi*
ICE Intrusion Countermeasure Electronics *ct, em*
ICECS Integrated Closed Environmental Control System *wp*
ICES Interference-Causing Equipment Standard *pk, kg*
ICES Intersociety Conference on Environmental Systems *aa26.10.b20*
ICET Integrated Cost Engineering Tool *kf*
ICF Inertial Confinement Fusion -*12*
ICF Inspection Check Fixture *fu*
ICF Interconnect Facilities *mt*
ICG Interactive Computer Graphics *wp*
ICG Interface Control Group -*4*
ICI Image Component Information *pk, kg*
ICI Imperial Chemical Industries (Ltd.) *crl, aw129.21.13*
ICI Incoming Call Indicator *fu*
ICI Instructional Communications Corporation *hc*
ICI Instrument Control Interface *oe*
ICI Integrated Configuration Index
ICI Interactive Communications Interface *mt*
ICI Interagency Committee on Intelligence -*10*
ICI Investment Company Institute *y2k*
ICIC {International} Commission for Interplanetary Communications -*13*
ICICLE Integrated Cryogenic Isotope Cooling Engine [System] *hc, wp*
ICIS Integrated Communication Identification System *hc*
ICL Indentured Control List *jdr*
ICL Integrated/Item Configuration List *hc*
ICL Interconnection List [a document] *gp, 8, ep, jdr, pbl, tm, ah, jn*
ICL Interface Clear *pk, kg*
ICLID Incoming Call Line Identification *kg, hi*
ICM Image Color Matching [Kodak] *kg, hi*

ICM Improved Capability Missile *wp*
ICM Improved Conventional Munition *ph, st, fu*
ICM Interface Communication Memorandum *hc*
ICM Intra Cavity Modulation *hc*
ICM Intra Contractor Correspondence Memorandum *hc*
ICM Inventory Control Manager *hc*
ICMA Intelligence Collection Management Architecture *hc*
ICMF Intercontinental Missile Facility equipment *hc*
ICMP Intelligence Collection Management Process *mt*
ICMP Internet Control Message Protocol [Novell] *mt, kg, ct, em, hi*
ICMSE Interagency Committee for Marine Science and Engineering (ac1.6-22)
ICN Installation Completion Notice *fu*
ICN Integrated Computing Network *tb*
ICN Interface Change Notice *fu*
ICN Inventory Control Number *jdr*
ICNI Integrated Communication, Navigation, and Identification *hc, fu*
ICNIA Integrated Communication, Navigation, and Identification Avionics system [ATF] [RADC] [USA] *mt, hc, aw129.10.59, wp, tb*
ICNS Integrated Communications and Navigation System *fu*
ICO Inclined Circular orbit (intermediate Circular orbit) *he*
ICO Installation and Checkout *mt*
ICO Intercept Control Operator *fu*
ICO Intermediate Circular Orbit *jdr, kf*
ICOM Internal Cost Of Money *kf*
ICOMP Intel Comparative Microprocessor Performance *kg, hi*
ICOMS Improved Conventional Mine System *dh, hc, fu*
ICON Integrated Acoustic Console *hc*
ICON Integration Concept *hc*
ICOP Interface Control Operator Position *fu*
ICP Incremental Change Package *fu*
ICP Indicator Control Panel *fu*
ICP Initial Contract Period *mt*
ICP Instructor Control Program *fu*
ICP Integrated Channel Processor *pk, kg*
ICP Intellectual Currency Protection *pbl, jdr, hj16.9.3, ah, jn*
ICP Inter Computer Processor *mt*
ICP Interactive Command Processor *fu*
ICP Interface Change Proposal *hc, kf*
ICP Interface Control Panel *fu, wp*
ICP Interface Control Plan *tb*
ICP Intertheater COMSEC Package *mt*
ICP Inventory Communications Processor *mt*
ICP Inventory Control Point *mt, fu*
ICPCN Intersite Command Post Communication Network [AFLC] *mt*
ICPL Interrupted Circuit-Path Locator *fu*
ICPNET Inventory Control Point Network *mt*
ICPP Interactive Computer Presentation Panel *hc*
ICR Inductance, Capacitance, Resistance *fu, 0*
ICR Integrated Cueing Requirements *hc*
ICR Intelligence Collection Requirements *mt*
ICR Intelligent Character Recognition *kg, ct, em*
ICR Interface Change Request *pl, hc*
ICR Interim Change Record *fu*
ICRC Inventory Control Record Card *fu*
ICRIS Integrated Customer Record Information System *ct, em*
ICRM Intercontinental Reconnaissance Missile *hc*
ICS (temperature adjustment factors for ICS and discrete parts) *smv*
ICS Image Correction System *bl3-3.8*
ICS Improved Combat System *fu, hc*
ICS Inertial Control System *mt*
ICS Inspection Check Sheet *fu*
ICS Integrated Card System *fu*
ICS Integrated Communications System *mt, fu, hc, af72.5.44*
ICS Interface Control Specification *fu, kf*
ICS Interim Contractor Support *mt, fu, hc, wp, kf*
ICS Interior Communication System *fu*
ICS Internal Calibration Sources *hc*
ICS Internal Countermeasure System *mt*
ICS Internal Countermeasures Set *hc*
ICS International Classification for Standards *ch*
ICS Interphone Control Station *fu*
ICS Interpretive Computer Simulator *hc*
ICS Intuitive Command Structure *pk, kg*
ICS Inventory Control System *fu, hc*
ICSA Infrared Charge Scanned Array *hc*
ICSAPI Internet Connection Services API *pk, kg*
ICSC Interim Communications Satellite Committee *hc*
ICSC Irvine Computer Sciences Corporation *hd1.89.y16*
ICSE Item Change/Schedule Estimate *tb*
ICSS Initial Conventional Support System [OJCS] *mt*
ICSS Interim Contractor Supply Support *fu*
ICST Institute for Computer Sciences and Technology [NBS] *mt*
ICST Integrated Control Subsystem Testing *kf*
ICST Integrated/Interim Combined System Test *hc*
ICSTF Integrated Combat Systems Test Facility *fu*

ICSU International Council of Scientific Unions *ci.720, ns, hy, hg, ge*
ICT Inspection Check Template *fu*
ICT Integrated Combat Turn *mt, wp*
ICT Integrated Cost Tool *kf*
ICT Internal Calibration Target [imager and sounder] *oe*
ICT International Critical Tables *wp*
ICT50 Incapacitating [vapor inhalation dosage, with 50% probability] *wp*
ICTP Individual and Collective Training Plan *fu*
ICTP Installation Compatibility Test Plan *fu*
ICTR Individual Consolidated Training Record *fu*
ICU Indicator Control Unit *hc*
ICU Infrared Command Unit *hc*
ICU Instruction Cache Unit *pk, kg*
ICU Interface and Control Unit *ac10.4-5*
ICU Interim Capacity Upgrade *pf.f88.26*
ICU Interleave Control Unit *fu*
ICU ISA Configuration Utility *pk, kg*
ICV Internal Correction Voltage *fu*
ICW Interrupted Continuous Wave *mt, wp, ac10.1-3, 0, fu*
ICWAR Improved Continuous Wave Acquisition Radar *hc*
ICWG Interface Control Working Group *mt, 4, wp, 18, pl, hc, ts7-32, nt, tb, st, jdr, kf, fu, oe*
ICWG International Coordination Working Group *hy*
ICWGA Interface Control Working Group Action [plan] *hc*
ICX IC Cross-Reference *fu*
ID Increased Deployability Posture *mt*
ID Infantry Division *mt, fu*
ID Inner (Internal) (Inside) Diameter *w9, wp, ai1*
ID Inside Dimensions *w9*
ID Instruction Decode *fu*
ID Integrated Diagnostics *wp*
ID Intelligence Department *w9*
ID Interactive Discrimination *tb, st*
ID Interconnection Device *hc*
ID Interface Device *fu*
IDA Input Data Assembler *fu*
IDA Institute for Defense Analyses *mt, fu, fr, wp, hc, tb*
IDA Integrated Digital Access *pk, kg*
IDA Integrated Digital Avionics *wp*
IDA Intelligent Disk (Drive) Array *pk, kg*
IDA International Dark-sky Association *mi*
IDA-CRD Institute for Defense Analysis, Communications Research Division *-10*
IDAC Interconnecting Digital-to- Analog Converter *fu*
IDAD Infrared Detector Array Development *hc*
IDAD Internal Defense And Development [USA DOD] *mt*
IDAPI Integrated Database Application Programming Interface [Borland] *kg, hi*
IDAPS Image Data Processing System [at NASA/MSFC] *pl*
IDAPS Inspection Diagnostic and Prognostic System *hc*
IDAS Integrated Design Automation System *hc, wp*
IDASP Infrared Detection and Acquisition Signal Processing *hc*
IDB Information Distribution Bus *fu*
IDB Integrated Data Base *mt*
IDB Integrating Double Buffer *mt*
IDB Intercept During Boost *hc*
IDB Interface Data Buffer *fu*
IDB Internal Data Bus *fu*
IDBMS ICAM Database Management System *fu*
IDBMS Integrated Database Management System *fu*
IDC Image Dissector Camera *hc, gp, 16*
IDC Information Distribution Capability *fu*
IDC Inspection Data Code *wp*
IDC Integrated Database Connector *pk, kg*
IDC Integrated Desktop Connector *pk, kg*
IDC Intelligent Disk Controller *fu*
IDC Inter Departmental Correspondence [Hughes Aircraft Company] *hc, mc, 8, 15, ep, pbl, tm, jdr, fu, su, ah, uh, jn*
IDC International Data Corporation *mt*
IDC Internet Database Connector [Microsoft] *pk, kg*
IDCL Information Design Change List *hc*
IDCM Interplant Debit and Credit Memo *hc, 8*
IDCPS Integrated DC Power Supply *0*
IDCS Image Dissector Camera [Sub] System *hc, ac2.8-2*
IDCS Interfacility Data Communications System *fu*
IDCSC Integrated Defense Communications System Control *mt*
IDCSP Initial Defense Communications Satellites Project *mt*
IDCSP Interim Defense Communication Satellite Program *14, hc*
IDCSS Initial Defense Communication Satellite System *hc*
IDD Information Display Division *fu*
IDD Integrated Data Dictionary *mt, fu*
IDD Interface Definition Document *oe*
IDD Interface Description Document *fu*
IDD Interface Design Document *mt, hc, ep, jdr, fu*
IDDD International Direct Distance Dialing *fu*

IDDE Integrated Development and Debugging Environment [Symantec] *kg, hi*
IDDP Interface Device Dependent Port *mt*
IDDS Information Distribution and Display System *mt*
IDE Imbedded Drive Electronics *kg, hi*
IDE Integrated Development Environment [Borland] *kg, em, hi*
IDE Integrated Device Electronics *em, hi*
IDE Integrated Drive Electronics *kg, ct, hi*
IDE Intelligent Drive Electronics *pk, kg*
IDE Interactive Design and Engineering *pk, kg*
IDE Interface Design Enhancement *pk, kg*
IDE Interim Data Element *mt*
IDE Internal Display Element *mt*
IDEA Integrated Dose Environment Analysis *st*
IDEA Intelligence Design Assistant *hc*
IDEA International Data Encryption Algorithm *kg, em*
IDEA Internet Design, Engineering, and Analysis *em*
IDEAL Integrated Design and Engineering Analysis Languages *wp*
IDEAS Integrated Design and Analysis System *wp*
IDEAS Integrated Design Engineering Analysis Software *fu*
IDEAS Intelligence Data Elements Authorization Standards *mt*
IDEF ICAM Definition Methodologies [IDEF-1 for ICAM definition method One] *wp, fu, mt*
IDEP Interagency (Interservice) Data Exchange Program *fu*
IDEP Interservice Data Exchange Program *hc, wp*
IDEPC IEMATS Data Entry Personal Computer *mt*
IDEPS IDHSS Deployable Segment *hc*
IDEX Imagery Digital Exploitation (IDEX_II System) [AF/IN] *wp, mt*
IDF Indigenous Defense Fighter [Taiwan] [the Ching-Kuo] *mt*
IDF Instantaneous Direction Finder (Finding) *fu, mt*
IDF Integrated Development Facility *fu*
IDF Intermediate Distribution (Distributing) Frame *mt, fu, kg, 0, hc*
IDF Internal Data Format [document] *-4*
IDF {Israel Defense Forces} *nd74.445.16*
IDF/AF Israeli Defense Force / Air Force *mt*
IDF/IFM (accurate 2 degree, 1.5MHz IDF/IFM receivers) *af72.5.22*
IDFT {Inverse ? Fourier Transform} *rg*
IDG Integrated Drive Generator *aw118.18.43*
IDHS Intelligence Data Handling System *mt, wp, sr9-82, hc, jb, tb, fu, ro*
IDHSC Intelligence Data Handling System Communications *mt*
IDHSS Intelligence Data Handling Support System *hc*
IDHTS Instrument Data Handling and Transmission System *ac3.7-23*
IDI Interrupt Digital Inputs *hc*
IDIP Intelligence Data Input Package *hc, wp*
IDIQ Indefinite Delivery, Indefinite Quantity *jdr*
IDIV Integer Divide *pk, kg*
IDL Improved Data Link *hc*
IDL Indentured Drawing List [a document; an obsolete designation replaced by IPL] *fu, 8, hc, ep, jdr, tm*
IDL Intelligent Data Language *fu*
IDL Interactive Data Language *kg, mi, ns, mb*
IDL Interface Definition Language *kg, kf, em*
IDL Internal Distribution List *fu*
IDL International Data Link *fu*
IDL Interoperable Data Link *fu*
IDL [a proprietary data analysis system of Research Systems International] *cu*
IDLH Immediately Dangerous to Life and Health *wp*
IDLHL Immediately Dangerous to Life or Health Limit *oe*
IDM Information Display Monitor *fu*
IDM Integrated Defence Model *-14*
IDM Integrated Diagnostic Model *wp*
IDM Intelligent Data Management *ns*
IDM Intelligent Database Machine *fu*
IDMP Integrated Diagnostic Master Plan *wp*
IDMR InterDomain Multicast Routing *em*
IDMS Information and Data Management System *tb, wl*
IDMS Integrated Database (Data) Management System *mt, kg, 8, 15, sc8.5.3, hc, ep, fu, hi*
IDMS Integrated Disposal Management System *wp*
IDMS Interim Deployable Maintenance System *mt*
IDN Integrated Data Network *mt, hn49.3.5*
IDN Integrated Digital Network *wp, hi*
IDNE Inertial Doppler Navigation Equipment *wp*
IDNS (I-IDNS ~ In-garrison IDNS [USREDCOM]) *mt*
IDNS Internet Domain Name System *hi*
IDNX Integrated Digital Network Exchange [IBM] *pk, kg*
IDO Identification Officer *fu*
IDOC Intrusion Detection Optical Cable *hc*
IDOCS Intrusion Detection Optical Communication System *hd1.89.y27, fu*
IDOE International Decade of Ocean Exploration *pl*
IDOP Identification Operator *fu*
IDP Image Data Processor *pbl*
IDP Initial Design Phase *fu*
IDP Input Data Processor *fu*

IDP Inquiry Dictionary Processor *fu*
IDP Integrated Data Processing *fu, kg, wp*
IDP Integrated Diagnostic Plan *wp*
IDP Integrated Digital Processor *hc*
IDP Intercept Deployment Plan *-10*
IDP Interface Design Plan *fu*
IDP Interface Development Plan *mt*
IDP Interface Diagnostic Program *fu*
IDPP Integrated Diagnostics Program Plan *wp*
IDPS Initial Deployable Processing Station *hc*
IDPS Interface Data Processor Segment *pbl*
IDPS Interim Deployable Process Station [PACAF] *mt*
IDR Initial Design Review *mt, fu, wp, jn*
IDR Inspection (Internal) Discrepancy Report *fu, oe*
IDR Intelligent Document Recognition *pk, kg*
IDR Inter Divisional Review *fu*
IDR Interim Design Review *hc, jdr*
IDR Intermediate Design Review *oe*
IDR Internal Design Review [a meeting] *pl, hc, ep, jdr, pbl, tm, kf, ah, jn*
IDR/MPR Internal Discrepancy Report/ Material Purge Report *fu*
IDRAS Interactive Data Reduction Analysis Station *hc*
IDRC Interim Depot Repair Capability *hc*
IDRC Internal Discrepancy Report Control *fu*
IDRO Industrial Development and Renovation Organization *hc*
IDS ICAM Documentation Standard *fu*
IDS Image Display System *wp*
IDS Immediate Drum Storage *fu*
IDS Implementation Design Specification *fu*
IDS Information Display Subsystem *fu*
IDS Infrared Detecting (Detection) Set *wp, hc*
IDS Instrument Data Subsystem *-17*
IDS Integrated Data System *mt*
IDS Integrated Design Support [System] *hc, wp*
IDS Integrated Display Set *hc*
IDS Inter Disciplinary Scientist *so*
IDS Interactive Display Station *tb*
IDS Interdictor / Strike [Panavia Tornado version] *mt*
IDS Interface Design Specification *fu*
IDS Intrusion Detection System *mt*
IDS Ionization Detector System *wp*
IDSCS Initial Defense Satellite Communication(s) System *mt, hc*
IDSD Individual Data Storage Device *mt*
IDSD Institutional Data Systems Divisions *ci.944*
IDSL ISDN Digital Subscriber Line *ct, em*
IDSO Instructor Defensive Systems Operator *aw129.10.72*
IDSRD Investigation Description and Science Requirements Document *re*
IDSS ICAM Decision Support System *fu*
IDSS Integrated Decision Support System *mt*
IDSS Integrated Design Support System *mt*
IDSS Integrated Diagnostic Support System *mt, wp, aw126.20.115, hc*
IDSTN Integrated Digital Switching and Transmission Network *hi*
IDT Identification Technician *fu*
IDT Inactive Duty Training [USA DOD] *mt*
IDT Indentured Document Tree *oe*
IDT Instrument Development Team [for facility instruments] *s0*
IDT Integrated Document Traceability *wp*
IDT Interface Design Tool *pk, kg*
IDT Interface Display Terminal *fu*
IDT Interrupt Descriptor Table *pk, kg*
IDTN Interim Data Transmission Network *mt*
IDTV Improved Definition Television *is26.1.60*
IDU Input Data Unit *fu*
IDU Instrumentation Data Unit *hc*
IDU Interactive Display Unit *mt*
IDVM Integrating Digital Voltmeter *oe*
IDW Input Data Word *fu*
IDWA Intra (Inter) Divisional Work Authorization *hc, fu*
IDWS Inter Divisional Work Structure *fu*
IDZ Inner Defense Zone *hc*
IE Ice Edge [location] *ac7.5-12*
IE Industrial Engineering (Engineer) *8, w9, fu*
IE Information Engineering *mt*
IE Infrared Emission *wp*
IE Inspection Equipment *fu*
IE Interface Electronic *hc*
IE Interface Equipment *hc*
IE Internet Explorer [Microsoft] *pk, kg*
IE/P/R Independent Evaluation/Plan/Report *st*
IEA Interface Electronics Assembly *hc*
IEA International Energy Agency *mt*
IEC Integrated Engine Control *wp*
IEC Intelligence Evaluation Committee *10, wp*
IEC Inter Exchange Carrier *mt, hi*
IEC International Electric Corporation *hc*
IEC International Electrotechnical Commission *mt, aa26.11.b7, nu, kg, fu, ch*

IECEC Intersociety Energy Conversion Engineering Conference *aa26.11.b53*
IECIRTS Improved Electrical Circuit Test Set *hc*
IED Imitative Electronic Deception *hc*
IEEE Institute of Electrical and Electronics Engineers *mt, hc, hn44.18.6, aa26.10.b2, w9, aw130.11.56, wp, ce, tb, st, kf, fu, kg, ci, em, oe*
IEF Information Engineering Facility *pk, kg*
IEG Imagery Exploitation Group *wp*
IEG Independent Evaluation Group *tb*
IEG Industrial Electronics Group [Hughes Aircraft Company] *hq5.3.5, hn44.10.3, 8, 16, 17, md2.1.1, fu, su*
IEG Information Exchange Group *hc*
IEI Industrial Education Institute *0*
IEI Iran Electronics Industries *hc*
IEIS Integrated Engine Instrument System *hc*
IEL Inspection Equipment List *hc*
IEM Initial Entry Message *fu*
IEMATS Improved Emergency Message Automatic Transmission System *mt, hc, jb, cp, fu*
IEMCAP Intrasystem Electromagnetic Compatibility Analysis Program *hc, fu*
IEMP Internal Electromagnetic Pulse *17, st, jdr, kf*
IEMS Integrated Emergency Management System [FEMA] *mt*
IEMSI Interactive Electronic Mail Standard Identification *pk, kg*
IEMSL Instantaneous Electromagnetic Mean Sea Level *ac3.7-23*
IEN Integrated Enterprise Network *ct, em*
IEN Internet Engineering Notes *pk, kg*
IEO Interim Engineering Order *fu*
IEOD Iranian Electro Optical Depot *hc*
IEP Independent Evaluation Plan *mt*
IEP Information Exchange Program *wp*
IEP Initial Engineering Phase *hc*
IEPG Independent European Program Group *hc, wp, aa26.10.10*
IER Institute of Engineering Research [University of California] *0*
IERS International Earth Rotation Service *hy*
IES Illuminating Engineering Society *fu*
IES Industrial Engineering Services *hc*
IES Institute of Environmental Sciences *rf84.3.8*
IESG Internet Engineering Steering Group [Internet] *kg, ct, em*
IESS Integrated Electromagnetic System Simulator *hc, wp*
IETF Internet Engineering Task Force [Internet] *kg, nu, ct, em, hi, du*
IETM Interactive Electronic Technical Manuals *kf*
IETS Individual Extension Training System *hc*
IEU Interface Electronics Unit *mt, hc*
IEW Intelligence and Electronic Warfare [USA] *fu, mt*
IEW-UAV Intelligence Electronic Warfare-Unattended Aerial Vehicle *hc*
IEW/UAV Intelligence Electronic Warfare/Unmanned Area Vehicle *hc*
IEW_UAV IEW Unmanned Aerial Vehicle [USA] *mt*
IF Imprest Fund *wp*
IF Industrial Fund *mt*
IF Intermediate Forward *fu*
IF Intermediate Frequency *mt, wp, sp661, 0, 4, 7, pl, w9, lt, ep, wl, jdr, uh, fu, hi, oe, ai1*
IF Intermittent Frequency {?} *hc*
IF Internal Filtering *he, fu*
IF Interrupt enable Flag *fu*
IF {Incentive Fee} *ts4-9*
IFA Information Flow Analysis *mt*
IFA Integrated Functional Audit *fu*
IFA Interface Format Adapter *fu*
IFA Interface Formatter Assembly *hc*
IFA Intermediate Frequency Amplifier *fu*
IFAAR Improved Forward Area Alerting Radar *fu*
IFAC (paper presented at the IFAC symposium, June 1973) *id4091.2/1111*
IFALPA International Federation of Air Line Pilots' Associations *aw118.18.29*
IFAR Interferometric Angle Resolver *hc*
IFAST Integrated Facility for Avionics Simulation and Testing *wp*
IFAST Integration Facility for Avionics Systems Testing *mt, ph, hc*
IFB Information For Bidders *fu*
IFB InterFace Box *ac9.4-7*
IFB Invitation For Bid *mt, hc, wp, tb, jdr, su, fu*
IFC In Flight Calibration [system] *gp, oe*
IFC Incrementally Funded Contract *st*
IFC Instantaneous Frequency Correlation *fu*
IFC Inter Facility Link Equipment *-9*
IFC Internet Foundation Classes *em*
IFCA Interim Functional Configuration Audit *kf*
IFCL Inter Facility Communication Link *mt*
IFD Image File Directory *kg, em*
IFD In Flight Disconnect *oe*
IFDAPS Integrated Flight Data Processing System *mt, wp*
IFDS Inertial Flight Data System *wp*
IFDS InterFacility Data Set *fu*
IFE In Flight Emergency *mt*
IFE Industrially Furnished Equipment *5, sb*

IFES Image Feature Extraction System *wp*

IFF Identification (Interrogation) Friend or Foe [USA DOD] *mt, uc, wp, aa26.10.9, hc, aw129.10.53, pm, tb, kf, ro, fu*

IFF Interchangeable File Format [Amiga] *pk, kg*

IFF/SIF Identification, Friend or Foe / Selective Identification Feature *fu, mt*

IFFA Independent Federation of Flight Attendants *aw120.10.67*

IFFC Integrated Flight and Fire Control *hc, wp, fu*

IFFN Identification Friend or Foe Network [NATO] *mt*

IFFN Identification Friend, Foe or Neutral *wp, aw130.13.10, hc, ph, ts4-2, tb, fu*

IFFN_JTF Identification Friend or Foe, or Neutral Joint Test Force *mt*

IFFS Identification, Friend or Foe Switching Circuit *fu*

IFFT Inverse Fast Fourier Transform *hc, rg*

IFG Incoming Fax Gateway *pk, kg*

IFH Image File Header *em*

IFHV In Flight Homing View *tb*

IFIS Integrated Flight Instruments System [UK] *mt*

IFL Improvement Factor Limit *fu*

IFL Interactive Fault Locator *fu*

IFL InterFacility Link *kf*

IFLCS Initial Force Level Control System *fu*

IFLIR Integrated Forward Looking InfraRed *hc*

IFM Instantaneous Frequency Measurement *mt, fu, wp*

IFM Integrated Flow Management *fu*

IFM Intelligent Flow Management *ct, em*

IFMIS Integrated Force Management Information System [USAFE] [now Constant Control] *mt*

IFMS In-Flight Maintenance Spares *fu*

IFMS Interactive File Management System *hc*

IFO Identified Flying Object *w9*

IFO Information For Offerers *mt*

IFO Intensive Field Observations *hy*

IFOG Interferometer Fiber Optic Gyro *17, sb, jmj*

IFORG Integrated Fiber Optics Resonator Gyro *st*

IFOV Instantaneous Field Of View *wp, bl3-2.10, 16, 17, sb, hc, so, tb, st, pbl, hg, kf, ge, fu, oe*

IFOV Instrument Field Of View *mi*

IFP Instruction Fetch Pipeline *pk, kg*

IFP Interface Processor *mt*

IFP Intermediate Frequency Patch *mt*

IFP Intermediate Frequency Processor *fu*

IFP Invitation For Proposal *wp*

IFPAD Integrated Focal Plane Array Dewar *17, sb*

IFPM In-Flight Performance Monitoring *hc*

IFPP Industrial Facilities Protection Program *mt*

IFPP Information (Instructions) For Proposal Preparation *tb, fu*

IFPP Instructions For Proposal Preparation *mt*

IFR In Flight Refueling *mt, wp*

IFR Increased Frequency Reporting *mt*

IFR Instant Failure Report *ep, jdr*

IFR Instrument Flight Rules *mt, w9, aw129.14.113, hc, fu*

IFRB International Frequency Registration Board *sc4.6.6, 7, 14*

IFRB International Frequency Regulation Board *fu*

IFRNA Inhibited Fuming Red Nitric Acid [cfr. IRFNA] *hc*

IFS Inshore Fire Support Ship [US Navy] *mt*

IFS Installable File System *pk, kg*

IFS Integrated Facilities System *mt*

IFS Intermediate Frequency Strip *fu*

IFSAL Integral Frequency Scan Approach and Landing *hc*

IFSED Integrated Full Scale Engineering Development *hc*

IFSM IF Switch Matrix *jdr*

IFSM Information Systems Management *pk, kg*

IFT In Flight Training *hc*

IFT Intermediate Frequency Transformer *fu*

IFTE Intermediate Field Test Equipment *hc*

IFTE Intermediate Forward Test Equipment *mt, fu*

IFTES Integrated Functional Test System *fu*

IFTRP Input File For Trajectory Reconstruction Program *kf*

IFTU In-Flight Target Update *kf*

IFTU Intensive Flying Trials Unit [UK] *mt*

IFTV In Flight Targeting Updates{?} *tb*

IFU Instruction Fetching Unit *fu*

IFU Interface Unit *fu, hc*

IFV Infantry Fighting Vehicle *hc*

IFV Interface Verification *-4*

IFVCGE Infantry Fighting Vehicle Command Guidance Equipment *hc*

IG Identification Group *fu*

IG Inertial Guidance *fu*

IG Input Generator *tb*

IG Inspector General *mt, w9, wp, af72.5.31, tb*

IG Instructor's Guide *fu*

IG Inverse Gain *fu*

IGA Integrated Graphics Array *pk, kg*

IGA Integrating Gyro Accelerometer *wp*

IGACS Integrated Guidance And Control System *hc*

IGB Inner (Inter) German Border *mt, fu*

IGBP International Geosphere-Biosphere Program *hy, hg, ge*

IGBT Insulated Gate Bipolar Transistor *is26.1.55*

IGC Integrated Graphics Controller *pk, kg*

IGCE Independent Government Cost Estimate *tb, st*

IGCP Intelligence Guidelines for COMINT Priorities *-10*

IGD Interactive Graphics Display *fu*

IGDS Interaction Graphics Display System *hc*

IGE In Ground Effect [noted by operational ceiling of helicopters; this is the ceiling if the helicopter is flown near to the earth's surface, i.e. in mountainous terrain] *mt*

IGES Initial Graphics Exchange Specification *mt, fu, hn43.22.8, is26.1.37, aw130.22.65*

IGES Initial Graphics Exchange Standard *mt, kg, kf*

IGES International Graphic Exchange System *fu*

IGFET Insulated Gate Field Effect Transistor *hc, fu*

IGFOV Instantaneous Geometric Field Of View *hc, bl3-1.32, oe*

IGI International Graphics, Incorporated *hn43.18.5*

IGL Interactive Graphics Library *fu*

IGM InterGalactic Medium *mi*

IGMIRS Inspector General Management Information and Reporting *mt*

IGMP Internet Group Management Protocol *em*

IGMP Internet Group Multicast Protocol *kg, ct, em, hi*

IGN Institut Geographique National [France] *ac3.6-4*

IGOSS Integrated Global Ocean Station System *pl*

IGP Inertial Guidance Package *wp*

IGP Intelligent Gateway Processor [component of LOGNET] *mt*

IGP Interior Gateway Protocol *pk, kg*

IGP Internet Gateway Protocol *ct, em*

IGRF International Geomagnetic Reference Field *ns*

IGRP Interior Gateway Routing Protocol [Cisco] *pk, kg*

IGRV Improved GUARDRAIL V *mt*

IGS Inertial Guidance System *fu*

IGS Integrated Graphics System *mt*

IGS Internal German Service *aw129.1.64*

IGS Internet Go Server [Internet] *pk, kg*

IGS IONDS Global Satellite *mt*

IGT Integrated Government Testing *mt*

IGT Interactive Graphics Terminal *fu*

IGTD Inertial Guidance Technology Demonstration *wp*

IGWA Inter Group Work Authorization *-4*

IGWO Integrated and Guided Wave Optics *is26.1.75*

IGY International Geophysical Year [1957] *ac3.2-1, 13, w9, no, mb, mi, crl*

IH International Harvester *hc*

IH/SR Integration (Integrated) Hardware / Software Review *gp, oe*

IHADSS Integrated Helmet And Display Sighting System *mt*

IHAP Image Handling and Processing *es55.50*

IHAS Integrated Helicopter Avionics System [Navy] *hc, fu*

IHAWK Improved HAWK [surface to air missile] *fu*

IHD Integrated Help Desk [IBM] *pk, kg*

IHE Improved High Explosive [mainly intended for nuclear bombs] [USA] *mt*

IHE Insensitive High Explosive *ph, wp*

IHFF Inhibit Halt Flip-Flop *fu*

IHFR Improved High Frequency Radio *mt, hn42.4.8, fu*

IHHS Improved Hawk Hardware Simulation *fu*

IHI Ishikawajima Harima Heavy Industries [Japan] *ci.192, aw129.21.47*

IHIPIR Improved High-Powered Illumination Radar *fu*

IHPTET Integrated High Performance Turbine Engine Technology [USA] *wp, mt*

IHR Infrared Heterodyne Radiometer *hc*

IHSBR Improved High Speed Bombing Radar *fu*

IHV Independent Hardware Vendor *pk, kg*

II Image Intensifier *wp*

II Imagery Interpretation (Image Interpreter) [USAF] [USMC] *mt, fu, wp*

II Item Identification *fu*

IIA Interim ASOC Automation *mt*

IIAATC Imperial Iranian Array Aviation Training Center *hc*

IIAH Interim Improved Armed Helicopter *hc*

IIC Information for Industry Committee *fu*

IIC Interceptor Identification Capability *hc*

IICF Interdisciplinary Investigator Computing Facilities *hy*

IICP IJMS Interface Computer Program *fu*

IICWG Interprogram Interface Control Working Group *tb*

IID Information Input Device *fu*

IID Integrated Information Display *mt*

IID Interface Identifier *em*

IIDB Integrated Intelligence Data Base *mt*

IIDP IJMS Interface Data Processor *fu*

IIF Imagery Interpretation Facility [HQ USMC] *wp, mt*

IIF Immediate Interface *pk, kg*

IIFS Intelligent Information Fusion System *ns*

IIGF Imperial Iranian Ground Forces *hc*

III Incapacitating Illness or Injury [USA DOD] *mt*

III Integrated Imaging Irradiance *hc*

III Interstate Identification Index [NCIC] *pk, kg*
III/C Interoperability, Integration, Immunity, Continuity *hc*
IIIDB International Interchangeability Interface Data Base *hc*
IIIL International Interchangeability Interface List *hc*
IIIWG Interservice Interoperability Implementation Working Group *mt*
IIL Integrated Injection Logic *fu, 6, hc*
IILS Image Interpretation Light Station *wp*
IIMAPS Incremental IMAPS *mt*
IINCOM Intratheater Intelligence Communications Network *mt*
IINCOMNET Intra Theater Communications Network *wp*
IINCOM_NET Intratheater Intelligence Communications Network [also IINCOM] *mt*
IIOP Internet Interoperable ORB Protocol *kg, ct, em*
IIP Implementation and Installation Plan *mt*
IIP Instantaneous (Initial) Impact Point *17, sb, oe*
IIP Integrated Information Presentation (Processing) *hc*
IIP Interceptor Improvement Program *hc*
IIR Imaging InfraRed *mt, fu, wp, hn42.4.3, 17, hc, af72.5.147, cp*
IIR Imaging Infrared Guidance *mt*
IIR Immediate Impulse Response *pk, kg*
IIR Infinite Impulse Response *fu*
IIR Initial Integration Review *mt*
IIR Integrated Instrumentation Radar *fu*
IIR Interim IEMATS Replacement *hc*
IIR Intermediate InfraRed *fu*
IIRA Integrated Inertial Reference Assembly *mt, wp*
IIRES Imagery Reporting / Exploit System [PACAF] *mt*
IIRGU Imaging InfraRed Guidance Unit *hc*
IIRN Inactive Item Review Notification *wp*
IIRS Imaging InfraRed System *hc*
IIRV Improved Inter Range Vector *4, oe*
IIS Imagery Interpretation Segment *mt*
IIS Indirect ID System *fu*
IIS Information Improvement Support Program [MAC] *mt*
IIS Innovative Information Systems [vendor for magnetic motion simulator and magnetic gimbal that evolved into Aura Systems] *17, sb*
IIS Inspection Instruction Sheet *hc*
IIS Institute of Industrial Sciences [Japan] *ac3.3-3*
IIS Integrated Information System *wp*
IIS Interactive Instructional System *fu*
IIS Internet Information Services (Server) [Microsoft] *kg, ct, em*
IIS Internetted Information System *mt*
IIS Inventory Item Specification *fu*
IISA Integrated Information System Architecture *wp*
IISC Intelligence Information Systems Committee [HQ USAF] *mt*
IISP Interim Inter Switch Protocol *ct, em*
IISS Integrated Information Support System *mt, fu*
IISS Intelligence Information Subsystem [USAREUR] *mt*
IISS International Institute for Strategic Studies *-14*
IIST Initial Integrated (Integration) (and) System Test *jdr, uh, jn*
IIST Initial Interface Systems Test *sb*
IIST Integrated Interface Systems Test *gp, 17, sb*
IIT Illinois Institute of Technology *aa26.11.52*
IIT Independent Integration Test *fu*
IIT Internal Information Transfer *mt*
IITF Information Infrastructure Task Force *pk, kg*
IITP Initial Integrated Test Plan *jn*
IITRI Illinois Institute of Technology Research Institute *hc, jdr*
IITS Intratheater Imagery Transmission System [ESD] *wp, mt*
IITV Interactive Instructional Television *sc2.10.2, fu*
IIW International Institute for Welding *ch*
IIWG IEMATS Implementation Working Group *mt*
IIWG Internal Interface Working Group *kf*
II_MAF Second Marine Amphibious Force *mt*
II_SEGMENT Imagery Interpretation Segment [TAC] *mt*
IJD Interim JOPES Dictionary *mt*
IJMS Interim Joint Message Standard *fu*
IJMS Interim JTIDS Message Specification *mt, fu*
IK Installation Kit *fu*
IKA Immediate Kill Assessable *-17*
IKI Space Research Institute [Russia] *cu*
IL Ilyushin [Soviet aircraft designer, IL-20 is Coot-A, IL-38 is May, IL-76 is AEW&C mainstay] *mt*
IL Index List *fu*
IL Intermediate Level *fu*
IL&FM Installations, Logistics, and Financial Management *dt*
ILA Image Light Amplifier *pk, kg*
ILA Incremental Linear Actuator *jdr, he, uh*
ILAADS Interim Low Altitude Air Defense System *hc*
ILAAS Integrated Light Attack Avionics System *hc*
ILAAT Inter Lab Air to Air Technology *hc*
ILAN Implementation of Local Area Networks *mt*
ILANG Illinois Air National Guard *-12*
ILAS Improved Limb Atmospheric Spectrometer [Japan, Env. Agency] *ge, hg*
ILAS Instrument Landing Approach System *fu, hc, wp*

ILAS Instrument Low Approach System *mt*
ILAS Interrelated Logic Accumulating Scanner *fu*
ILAW Improved Light Antitank Weapon system *hc*
ILB Inner-Lead Bonded *fu*
ILC Initial Launch Capability *pbl, ai2*
ILC International Licensed Carriers *-10*
ILC International Logistics Center [AFLC] *mt*
ILCO International Logistic Control Office *wp*
ILCS International Logistic Communication System *wp*
ILD Injection Laser Diode *mt, jdr*
ILD Integrated Logistics Data *mt*
ILDF Integrated Logistic Data File *wp*
ILDR Item Logistics Data Record *hc, fu*
ILE Interactive Load and Execute *fu*
ILF Indirect Labor Factor *wp*
ILF Infralow Frequency *fu*
ILFC International Lease Finance Corporation *aw129.10.67*
ILG International Leisure Group *aw130.11.32*
ILIOS In-Line-Infinity Optical System *fu*
ILIR In-house Laboratory Independent Research *wp*
ILIS Inner Layer Inspection System *vc25.3.20*
ILJS ISO Loading Jack System *mt*
ILLIAC Illinois [Large Scale Highly Parallel] Computer *hc*
ILLLTV Integrated Low Light Level Television *hc*
ILM Intermediate Level Maintenance *mt, hc*
ILMF Intermediate Level Maintenance Facility *hc*
ILMI Integrated Layer Management Interface *ct, em*
ILMR Independent Landing Monitor Radar [system] *hc*
ILMT Intermediate Level Maintenance Trainer *hc*
ILN International Logistic Negotiation *un*
ILO Injection Locked Oscillator *fu*
ILO Integrated Logistics Overhaul *mt*
ILOG [a COTS software company] *kf*
ILOSS Integrated Laser Optical Sight Set *hc*
ILP Intermediate Level Programs *hc*
ILP International Logistics Program *wp*
ILPFL Ideal Lowpass Filter *fu*
ILR Integrated Logistics Report *mt*
ILRIS Intermediate Long Range Interceptor System *hc*
ILRV Integrated Launch and Re-entry (Recovery) Vehicle *ci.447, hc, ai2*
ILS Input Laser System *hc*
ILS Instrument Landing System *mt, fu, hc, N75-27202-487, aw129.14.115, w9, af72.5.142, wp*
ILS Integrated Logistics Support [an organization] *mt, fu. hc, wp, hn44.17.12, 8, 16, 17, sb, nt, ep, tb, st, jdr, pbl, tm, kf, ah, jn, oe*
ILS International Language Support *pk, kg*
ILS International Launch Services *ai2*
ILS/MT/P Integrated Logistics Support/Management Team/Plan *st*
ILSD Integrated Logistics Support Division [Westinghouse] *gp*
ILSF Integrated Logistics Support Facility *mt*
ILSG Improved Linear Sequence Generator *fu*
ILSL Initial Logistics Support Letter *mt*
ILSM Integrated Logistics Support Manager *mt, fu*
ILSMIS Industrial Logistics MIS (Management Information System) [US Navy] *mt*
ILSMP Integrated Logistics Support Management Plan *hc*
ILSMR Integrated Logistics Support Management Review *fu*
ILSMR/T Integrated Logistics Support Management Review / Team *mt*
ILSMT Integrated Logistics Support Management Team *mt, hc, hn44.17.12, tb, fu*
ILSO Integrated Logistic Support Office *wp*
ILSP Integrated Logistics Support Plan(ning) *mt, hc, nt, tb, kf, fu*
ILSS Integrated Logistics Support Standards *kf*
ILST Initial Launch Site Test *kf*
ILST Integrated Logistics Support Team *hc*
ILST&E Integrated Logistics Support Test and Evaluation *mt*
ILSTD Integrated Logistics Support Technical Director *fu*
ILSWG Integrated Logistics Support Working Group *mt. tb*
ILT Imprecisely Located Target *st*
ILT Integration Level Test *oe*
ILT Interface Level Test *oe*
ILTO Integrated Logistic Technology Office *wp*
IM Impulse Modulation *fu, wp*
IM Industrial Manager *dt*
IM Information Management *mt*
IM Infrared Microwave *hi*
IM Ingot Metallurgy *wp*
IM Injection Module *jp*
IM Inner Modulator *-9*
IM Intercept(or) Missile [i.e. surface to air missile] [USA] *fu, mt*
IM Interface Module *fu*
IM Interim Memory *fu*
IM Intermediate Missile *fu*
IM Intermediate Modulation *fu*
IM Intermodulation [products] *fu, sc3.12.2, 7, hc, lt, jdr*
IM Inventory Manager *fu, mt*

IM Item Management *wp*
IM Item Manager [AFLC] *wp, mt*
IM/C Intermodulation per Carrier *dmi*
IMA Individual Mobilization Augmentee *mt, af72.5.119*
IMA Information Management Analyzer *fu*
IMA Information Management Architecture *mt*
IMA Information Management Area *mt*
IMA Information Mission Area [USA] *mt*
IMA Integrated Microwave Assembly *kf*
IMA Interface Module Assembly *fu*
IMA Intermediate Maintenance Activity (Action) *mt, hc, fu*
IMAAWS Infantry Manned portable Anti Armor Assault Weapon System *hc*
IMAC Integrated Material Accountability Control *hc*
IMAC Interim Message Automated Capability *mt*
IMACS Image Management And Communication System *kg, hi*
IMAD Integrated Multisensor Airborne Display *wp*
IMAGE Indonesian Mobile Aerospace Ground Environment (System) *fu*
IMAGE Innovative Management Achieves Greater Effectiveness *mt*
IMAP Initial Manufacturing Assignment Program *hn43.7.5*
IMAP Internet Message (Messaging) (Mail) Access Protocol [Internet] *kg, ct, em, hi*
IMAPS Integrated Military Airlift Planning System [MAC*] *mt*
IMAS Industrial Management Assistance Survey *hc, wp*
IMAT Integrated Modification and Trial *hc*
IMB Integration Management Board *tb*
IMBLMS Integrated Medical and Behavioral Lab Measurement System *ci.955*
IMC Image Motion Compensation *mt, hc, oe*
IMC Information Management Center *hy*
IMC Initial Microcode Load *ct, em*
IMC Instrument Meteorological Conditions [USA DOD] *mt, fu, aw130.13.39, wp*
IMC Integrated Management Control *wp*
IMC Intermodule Coupling *fu*
IMC Item Management Code *wp*
IMCC Initial Mobile Command Center *mt, fu*
IMCC Integrated Mission Control Center *fu, hc*
IMCO Intercontinental Manufacturing Company *hc*
IMCO Intergovernment(al) Maritime Consultative Organization *mt, hc, aa27.4.b12*
IMCS Integrated Management Control System *hc, kf*
IMD Implementation Management Document *mt*
IMD India Meteorological Department *ac*
IMD Instructional Materials Development *fu*
IMD Intelligent Missile Defense *wp*
IMD Interactive Minefield Display *hc*
IMD Intermodulation Distortion *fu*
IMD [San Diego company in joint venture with GSG] *hn44.22.1*
IMDS Image Data Stream [format] [IBM] *pk, kg*
IME Input Method Editor *pk, kg*
IME Institute of Makers of Explosives *fu*
IMET Infrared Mission Effectiveness Team *kf*
IMET International Military Education and Training *wp*
IMEWS Integrated Missile Early Warning System *ci.1249*
IMF Integrated Maintenance Facility *wp*
IMF Interconnection Master File *fu*
IMF Intermediate Maintenance Facility *mt*
IMF Interplanetary Magnetic Field *-16*
IMFRAD Integrated Multi Frequency Radar *hc*
IMG Interface Management Group *fu*
IMG Interferometric Monitor for Greenhouse Gases [MITI] *hg, ge*
IMG Intertheater Movement Generator *mt*
IMGS Integrated Mechanical Graphics System *fu*
IMH Interactive Message Handling *fu*
IMI Improved Manned Interceptor [USA] *hc, fu, mt*
IMI Individual Marketing Initiative *st*
IMINT Image(ry) Intelligence [USA DOD] *mt, ph, wp, kf*
IMIP Industrial Modernization Incentives (Improvement) Program [DOD] *mt, ph, tb, fu, hc*
IMIR Interceptor Missile Interrogation Radar *fu*
IMIS Integrated Maintenance Information System [USAF] *mt, af72.5.28, hc, wp*
IMIS Intelligent Management Information System *tb*
IML Initial Microcode Load [IBM] *pk, kg*
IML International Manufacturing Liaison *fu*
IML International Micrography Laboratory *es55.46, hy*
IMM Immunization Area *mt*
IMM Independent Manned Manipulator *ci.970*
IMM Integrated Maintenance (and) Management *fu, hc*
IMM Integrated Material Management *wp*
IMM Integrated Material Manager *mt*
IMM Integrated Memory Modules *fu, oe*
IMM Intermediate Maintenance Manual *hc*
IMMM Internal Monthly Management Meeting *-4*
IMMP Information Management Master Plan [USA] *mt*
IMMS Integrated Maintenance Management System *mt*

IMMS-RD Interim Maintenance Activity Management System [US Navy] *mt*
IMN Indicated Mach Number *mt, fu*
IMO Installation Mobility Officer *mt*
IMO International Maritime Organization *mt, fu*
IMO International Meteor Organization *mi*
IMOAS Information Management and Office Automation System [AFOTEC] *mt*
IMOM Improved Many On Many [USAFE] *mt*
IMOS Ion implanted Metal Oxide Semiconductor *fu*
IMP Implementation Management Plan *mt*
IMP Information (Interface) Message Processor *kg, ct, em, wp, tb*
IMP Information Management Plan (Program) [USA] *mt*
IMP Input Message Processing *fu*
IMP Instrument Mounting Panel *oe*
IMP Integrated Management Plan *jdr*
IMP Integrated Master Plan *pbl, kf, ah, jn*
IMP Integrated Memory Processor *hc*
IMP Integrated Mission Processor *fu*
IMP Integration Management Panel *oe*
IMP Integration Modification Package *fu*
IMP Intensity Measurement Program study *hc*
IMP Interface Management Plan (Processor) *fu*
IMP Interface Message Processor [ARPANET/WIN] *mt*
IMP Intermodulation Product *gp, 4*
IMP Interplanetary Monitoring Platform *ci.731, hc, ns, no*
IMP Inventory Management Plan [POL] *mt*
IMP/MON Implementation and Monitoring *mt*
IMPA Intelligent Multiple Port Adapter [DCA] *pk, kg*
IMPACTS Integrated Manpower, Personnel, And Comprehensive Training and Safety *kf*
IMPATT Impact (Ionization) Avalanche Transit Time *wp*
IMPATT Impact Avalanche (and) Transit Time *fu, hc, hd1.89.y3*
IMPATT Impact Avalanche and Transmit Time [diode] *17, sd, fu*
IMPIS Integrated Manufacturing Process Information System *aw129.14.91*
IMPRINT Image Processing Identification of Non-cooperative Targets *ie5.11.2*
IMPROVE Immediate Production Verification *aw129.14.91*
IMPS Integrated Mission Planning Station *hc*
IMPS Interaction Measurements Payload for Shuttle [RADC] *mt*
IMR Imaging Microwave Radiometer *hg, ge*
IMR Input Message Report *mt*
IMR Internal Management Review *tb*
IMR Internet Monthly Report *pk, kg*
IMR Inventory Management Record *wp*
IMRADS Information, Management, Retrieval, And Data System *hc*
IMRC Item Management Responsibility Code *hc*
IMRL Individual Material Readiness List *hc*
IMRO Interplant Material Requisition Order *hc*
IMRS Improved Munitions Requirements System [USAFE] *mt*
IMRS Input Mechanical Redundancy Switching *jdr*
IMS Information Management System *mt, kg, hd, tb, fu*
IMS Institutional Management Support *oe*
IMS Instructional Management System [USMC] *hc, mt*
IMS Integrated Management Schedule *jdr, kf*
IMS Integrated Master Schedule *jn*
IMS Integrated Measurement System *fu*
IMS Integrated Measurement Systems [Incorporated, Beaverton] *is26.1.51*
IMS Interceptor Missile Subsystem *tb*
IMS Interim Meteorological Satellite [NOMSS] *hc*
IMS Intermediate Maintenance Standards *pk, kg*
IMS International Magnetospheric Study *ns*
IMS International Military Staff [NATO] *mt*
IMS Interpretive Microinstruction Simulator *tb*
IMS Inventory Management Specialist *wp*
IMS Ion Mass Spectrometer *ns*
IMS Ion Mobility Spectroscopy *wp*
IMSE Intermediate Maintenance Support Equipment *fu*
IMSL Integrated Mathematics and Statistics Library *tb*
IMSL International Mathematical and Statistical Library *wp, hd1.89.y18, st*
IMSP Internet Message Support Protocol *pk, kg*
IMT Insertion Mount Technology *fu*
IMT Intelligent Mission Toolkit *jn*
IMT Intermachine Trunk *mt*
IMT Intermediate Maintenance Trainer *fu, hc*
IMTC International Multimedia Teleconferencing Consortium *pk, kg*
IMTS Improved Mobile Telephone Service *fu*
IMTV Interactive Multimedia Television *kg, hi*
IMU Inertial Measurement Unit *mt, bo, wp, 4, 5, 16, 17, sb, hc, tb, st, kf, fu, oe, ai2*
IMUL Integer Multiply *pk, kg*
IMUX Input Multiplexer *uh*
IMUX Inverse Multiplexing (Multiplexer) *kg, ct, em*
IMV Instrumented Measurement Vehicle *hn49.12.3*

IMVDS Item Management Vehicle Data System *mt*
IMWRP Item Manager Wholesale Requisition(ing) Process *wp, mt*
IN Intelligence Network *mt*
IN Intelligent Network *ct, em*
INAS Inertial Navigation Attack System *wp*
INAS Integrated Navigation / Attack System *mt*
INAS Integrated Night Attack Sensor system *hc*
INBATIM Integrated Battlefield Interactive Model *mt*
INC Intelligence Collection [Directorate] *mt*
INCA (INCA Engineering) [Hughes Aircraft Company customer for A/D conversion program LEMDE] *hc*
INCA Integrated Nuclear Communications Assessment *mt*
INCA Intelligence Communications Architecture [DIA] *mt*
INCE Institute of Noise Control Engineering *fu*
INCI Integrated Navigation, Communication and Identification *fu*
INCNR Increment Number *mt*
INCO Installation and Checkout *fu*
INCOFT Intelligent Conduct Of Fire Trainer *hc*
INCONREP Intra CONUS Movement Reports *mt*
INCONUS Intra Continental United States *mt*
INCOSE International Council On Systems Engineering *pbl, ah*
INCOSPAR Indian National Committee for Space Research *ac3.2-1*
IND Improvised Nuclear Device [USA DOD] *mt*
INDEX Indian Ocean Experiment *pl*
INDEX Intra NASA Data and Experience Exchange *hc*
INDICOM Indications Intelligence Communications Network *mt*
INE Inertial Navigation Equipment *wp*
INEWAM Integrated Electronic Warfare Analysis and Model *wp*
INEWS Integrated Electronic Warfare System [USA] *mt, aw118.18.97, hc, af72.5.95, fu*
INEWSS Integrated Electronic Warfare System Simulation [model] *wp*
INF Intermediate range Nuclear Force(s) *mt, hn48.13.1, wp, 11, 14, ph, aw129.1.47, aa26.12.72, af72.5.94, nt, tb*
INFANT Iroquois Night Fighter And Night Tracker *hc*
INFO Invitation For Offers *fu*
INFOCEN Information Central [ASD] *mt*
INFOSEC Information Security *mt, go1.1, tb, kf, fu*
INFRAD [a reconnaissance SAR to penetrate foliage] *wp*
ING Inactive National Guard *wp*
INGAA Interstate Natural Gas Association of America *y2k*
INGRES Integrated Graphics and Retrieval System *mt*
INGRES Interactive GRaphics and Retrieval System *hd1.89.y18, cu, fu*
INICOS Universal Compiler FORTRAN compatible {?} *hi*
INIS Information System Internetted [MAC] *mt*
INIU Interim Network Interface Unit *mt*
INLAW Infantry Laser Weapon *hc*
INM Integrated Network Management *pk, kg*
INM Internet Network Model *hi*
INMARSAT International Maritime Satellite [organization] *mt, sc2.8.4, hc, wp, kf*
INMS Ion Neutral Mass Spectrometer *es55.26, re*
INNS International Neural Network Society *hn48.19.8*
INOA Intelligence Operations Division *mt*
INOZ Indications and Warning Center {?} *mt*
INP Indium Phosphide *jdr*
INP Intelligence Processor *mt*
INP Iranian National Police *hc*
INPE Instituto National de Pesquisas Espaciaes (Brazilian National Institute for Space Research) *es55.87, aw130.22.35*
INPO Institute of Nuclear Power Operation *mt*
INQUEST Interim Query System [MAC] *mt*
INR Image Navigation / Registration *oe*
INR Inertial Reference *wp*
INRI (INRI . . . Yorktown, VA 23692) *jmj*
INS Inertial Navigation System *mt, hc, wp, aw129.1.0, af72.5.135, tb, fu*
INS Information Network Systems *mt*
INS Input String *pk, kg*
INS Insert Code *mt*
INS Integrated Network Server *pk, kg*
INSA Indian National {Science} Academy *sf31.1.11*
INSAC Integrated National Surveillance and Control *hc*
INSAT Indian National Satellite [system] *ge, hg, ac3.2-5, hc, hy, oe*
INSB (HITS
INSB Indium Antimonide *oe*
INSCOM Intelligence and Security Command [US Army] *mt, 10, dt, st*
INSCOM International Satellite Communication, [Limited] *aw130.22.4*
INSERT Insert Code *mt*
INSIT Intelligence Situation Report *mt*
INSMAT Inspector of [Naval] Material *fu, hc*
INSORD Office of Naval Inspector of Ordnance, Washington *dt*
INSP Interactive Network Simulation Program *fu*
INSS Information Network Satellite System *sc2.8.4*
INSTAR Inertialess Scanning, Tracking and Ranging *fu*
INSTOR Inventory and Storage *wp*
INSURV [Board of] Inspection and Survey [Navy] *fu, dt*
INT/WP Weapon Platforms and its Interceptors *-17*

INTA Instituto Nacional de Tecnica Aerospacial *crl*
INTA Interrupt Acknowledge *pk, kg*
INTAAS Integrated Aircraft Armament System *wp*
INTACS Integrated Tactical Communications System *mt, fu*
INTACVAL Intelligence Aid For Plan Evaluation *mt*
INTCPTS Intercepts *jb*
INTEL Intelligence *mt, tb*
INTELECT Interceptor Electronics [program] *17, tb*
INTELECT [an aimpoint algorithm testbed concentrating on boost phase] *sb*
INTELPOST International Electronic Post *-10*
INTELSAT International Telecommunications Satellite [organization] [a consortium] *mt, fu, wp, scJ2.8.4, 13, 16, hc, 19, ep, jdr, ct, em, oe*
INTERACT Integrated Research Aircraft Control Technology [USA] *mt*
INTERLOC Computerized Management Information System *hc*
InterNIC Internet Network Information Center [Internet] *kg, ct, hi*
INTERPOL International Police [organization] *wp*
INTF Interim National Test Facility [CSOC location] *tb*
INTIPS Integrated Information Processing System *ro*
INTLA Interfacility Line Adapter *fu*
INTO Interrupt if Overflow occurs *pk, kg*
INTRAN In Transit *wp*
INTREP Intelligence Report *mt, fu*
INTSUM Intelligence Summary *mt, fu*
INU Inertial Navigation Unit *wp, oe, ai2*
INUK-SAT [Greenland domestic satellite] *hc*
INVEST Integrated Vehicle Systems Technology *hc*
INW Isotropic Nuclear Weapon *wp*
INWG International Network Working Group *ct, em*
INXLTR Input Translator *fu*
IO Injection Opportunity *bl1-3.16*
IO Inoperability *fu*
IO Input / Output *em*
IO Issue Order *mt*
IOAU Input / Output Access Unit *mt*
IOB Input/Output Buffer *hc, wp*
IOC Index of Cooperation *hc*
IOC Initial Operational Capability *mt, fu, hc, wp, 1, 5, 12, 17, ic2-16, ph, sb, af72.5.133, jb, nt, dm, tb, st, wl, pbl, kf, uh*
IOC Input / Output Controller *fu, hc, jdr*
IOC Intelligence Operations Center *mt, as, fu*
IOC Inter Office Channel *pk, kg*
IOC Inter Office Correspondence *jdr, kf*
IOC Intergovernmental Oceanographic Commission *pl*
IOC Interim Operational Capability *mt*
IOC Interpretive Object Code *hc*
IOCC Input/Output Channel Converter *pk, kg*
IOCC Input/Output Controller Chip *pk, kg*
IOCE Input/Output Control Element *fu*
IOCL Input/Output Control List *oe*
IOCS Input / Output Computer Services *mt*
IOCS Input/Output Control System *kg, fu*
IOCTL Input/Output Control *pk, kg*
IOCU Input/Output Control Unit *hc*
IOCW Input/Output Control Words *hc*
IOD Increased Operational Data *fu*
IOD Infrared Observation Device *fu*
IOD Initial Operational Date *mt*
IOD Input/Output Device *fu*
IOD Intercept Opportunities Determination *jb*
IODC Input / Output Data Controller *mt*
IODE Integrated Development and Debugging Environment [Symantec] *pk, kg*
IODP Input / Output Digital Processor *mt*
IOE Input/Output Expander unit *jdr*
IOGA Industry Organized, Government Approved *wp*
IOHMD Interim Operational Helmet Mounted Display *hc*
IOIC Integrated Operational Intelligence Center *mt*
IOL Initial Outfitting List *hc*
IOL Instantaneous Overload *fu*
IOM Input / Output Multiplexers *mt*
IOM Input/Output Multiplexer *jb*
IOMODX (Initial Orbit Module [software program]) *jb*
ION (ION . . . Huntsville, AL 35806) *jmj*
ION {Ion} [Implantation System] *mt*
IONA A COTS software company *kf*
IONDDS Integrated Operational Nuclear Detonation Detection System *wp*
IONDS Integrated Onboard Nuclear Detonation Detection System *tb*
IONDS Integrated Operational NUDET (Nuclear Detonation) Detection System [USAF] *ph, mt*
IOP Initial Operational Period *mt*
IOP Input / Output Processor *mt, hc, tb, em, fu, oe*
IOP Interface Operational Procedures [TACS/TADS] *mt*
IOPL Input/Output Privilege Level *pk, kg*
IOR Immediate Operational Requirement *mt*

IORG Interoperability Review Group *fu*
IORP Interface Operational and Recording Program *hc, fu*
IOS Indian Ocean Station (Site) [Air Force] *uh, hc, oe*
IOS Initial Operations System *tb*
IOS Input / Output System *em*
IOS International Organization for Standardization *fu, hc*
IOS Internetwork Operating System *em, ct, hi*
IOSA Integrated Optic Spectrum Analyzer *hc*
IOSGA Input / Output Support Gate Array *pk, kg*
IOSPS Integrated Open Standards Profile Systems *kf*
IOT In Orbit Test *jdr, uh, jn*
IOT Initial Operational Test *mt, kf*
IOT Inter-Organizational Transfer *jdr, kf, jn*
IOT&E Initial Operational Test(ing) and Evaluation *mt, wp, hc, N75-27202-492, 16, jb, dm, tb, kf, fu*
IOT&E Installation Operational Test and Evaluation *mt*
IOT/ESVA In Orbit Test and Earth Station Verification and Assistance *hc*
IOTA Infrared Optical Telescope Array *mi*
IOTA Infrared Overhead Tasking Authority *kf*
IOTA International Occultation Timing Association *mi*
IOTB Infrared Overhead Tasking Board *kf*
IOTC Input/Output Test Console *hc*
IOTE In-Orbit Test Equipment *uh*
IOU Input/Output Unit *fu*
IOWG Instruments Operations Working Group *oe*
IOX Input / Output Transfer Unit *mt*
IP Identification of Position *fu*
IP Identification Peculiarity *hc*
IP Image (Imagery) Processing *17, sb, wp*
IP Impact Point *fu*
IP Implementation Phase *fu*
IP Implementation Procedures *mt*
IP Index Pointer *wp*
IP Index Pulse *gp, 8, pl*
IP Individual Protection [chemical warfare] *wp*
IP Information Provider *hi*
IP Initial Point *fu, mt, hc, w9, tb*
IP Initial Procurement *fu*
IP Installation Procedure *fu*
IP Instruction Pointer *pk, kg*
IP Instructor Pilot *mt*
IP Inter Phone *fu*
IP Interactive Processing *mt*
IP Intercept Point *fu*
IP Interface Processor *mt, fu*
IP Interface Program *wp*
IP Interlocks Package *ai2*
IP Intermediate Pressure *mt, w9*
IP Intermediate Processor *fu*
IP Internet Protocol [standard of DOD] *mt, fu, csII.60, aw130.11.56, hn49.11.8, tb5, tb, jdr, pbl, kg, nu, ct, em, ah, hi*
IP/MP In Phase/Mid Phase *spx*
IPA In Plant Adjustment *fu*
IPA Intelligent Program Assessment *kf*
IPA Intermediate Power Amplifier *fu, hc, jdr*
IPA Internet Protocol Address *hi*
IPAA Independent Petroleum Association of America *y2k*
IPAAACS Image Process Auto Acquisition and Aimpoint Control System *hc*
IPAC Intelligence Center PACOM *mt*
IPAC Intelligence Pacific *mt*
IPAD Integrated Programs for Aerospace Vehicle Design *hc*
IPADAE Integrated Passive-Action Detection and Acquisition Equipment *fu*
IPADS Integrated Passive Air Defense System *hc*
IPADS Integrated Processing And Display System *mt, wp*
IPAM Improved Point Analysis Model *wp*
IPAR Intercepted Photosynthetically Active Radiation *wp*
IPAS Inter Platform Alignment System *hc*
IPAT Improved PLRS/ADDS Terminal *fu*
IPB Illustrated Parts Breakdown *hc, fu*
IPB Illustrated Parts Brochure *hc*
IPB Installation Parts Breakdown *mt*
IPB Intelligence Preparation of the Battlefield (Battlespace) [USA DOD] *wp, kf*
IPB Intercept Priorities Board *-10*
IPB Interprocessor Bus *hc*
IPC Imaging Proportional Counter *cu*
IPC Information Processing Center *mt, wp*
IPC Initial Planning Conference *mt*
IPC Initial Production Center *fu*
IPC Institute of Printed Circuits *fu*
IPC Instructions Per Clock *kg, hi*
IPC Integrated Power Controller *kf, he, ah*
IPC Inter Process Communication *em*
IPC Inter-Process Communication *1, tb, kf, kg*

IPC Intermittent Positive Control *fu*
IPC Internet Proxy Cache *ct, em*
IPCA Interim Physical Configuration Audit *kf*
IPCE Independent Parametric Cost Estimate *st*
IPCE Interface Power Control Electronics *oe*
IPCL Inter-Process Communications Library *fu*
IPCP Improved Platoon Command Post *fu*
IPCP Internet Protocol Control Protocol *kg, jdr, ric*
IPCS Integrated Personnel Communication System *hc*
IPCS Interactive Problem Control System *fu*
IPD Implementation Difficulty *kf*
IPD Improved Point Defense *hc*
IPD Industrial Products Division [of IEG] *fu, hn43.6.2, hq6.2.7*
IPD Information Processing Division [at Goddard] *ac2.8-12, 4*
IPD Insertion Phase Delay *fu*
IPD Integrated Product Development [Process] *pbl, tm, jdr, kf, ah, jn*
IPD Issue Priority Designator *wp*
IPD/I Improved Point Defense Integrated *fu*
IPD/TAS Improved Point Defense/Target Acquisition System *fu*
IPDC Integrated Product Development Center *kf*
IPDL Indentured Parts and Document List *hc, ep*
IPDMS Improved Point Defense Missile System *fu*
IPDMS Integrated Product Development Management System *jdr*
IPDR Initial Program Design Review *mt*
IPDS Imagery Processing and Dissemination System *mt*
IPDSMS Improved Point Defense Surface Missile System *fu, hc*
IPDTAS Improved Point Defense Target Acquisition System *hc*
IPE Idle Industrial Plant Equipment *hc*
IPE Increased (Improved) Performance Engine *af72.5.28, aw130.11.32, wp*
IPE Individual Protection Equipment *wp*
IPE Individual Protective Equipment *fu*
IPE Industrial Plant Equipment *fu, hc*
IPE Information Processing Equipment *wp, mt*
IPE Integrated Protective Entrance *fu*
IPE/T Improved Protective Entrance/Tent *fu*
IPECO [a manufacturer of crew seats, Lawndale, California] *af72.5.32*
IPECS Integrated Power Environment Control System *fu*
IPEI Ionospheric Plasma and Electrodynamics Instruments [EOS] *hy*
IPF Image Processing Facility [NASA digital] *ac2.2-13*
IPF In Process Factor *hc*
IPF Initial Production Facilities (Facilitization) *fu, hc, st*
IPF Initial Production Funds *mt*
IPF Integration and Processing Facility *ai2*
IPF Interactive Processing Facility *mt*
IPF Interactive Productivity Facility *fu*
IPFC Information Presentation Facility Compiler [IBM] *pk, kg*
IPG Index Pulse Generator [on GOES] *-8*
IPG International Planning Group [designation for early ADGE equipment for Germany, Netherlands and Belgium] *fu*
IPG-GFN International Planning Group-Growth to Full NADGE *hc*
IPG/GFN IPG Growth-to-Full-NADGE *fu*
IPHIR [helioseismology instrument on Phobos mission] *es55.98*
IPI Intelligent Peripheral Interface *pk, kg*
IPI Internal Process Improvement *kf*
IPI&T In-Plant Integration and Test *fu*
IPICS Integrated Part Information Control System *fu*
IPIR Initial Photographic Interpretation Report *wp, fu, mt*
IPIR Initial Programmed Interpretation Report [USA DOD] *mt*
IPIS In Process Inspection Section *fu*
IPIU Instrument Power Interface Unit *ma37*
IPKF Indian Peace Keeping Force *wp*
IPL Indentured Parts List [document] *fu, pbl, tm, ah, jn, oe*
IPL Information Programming Language *pk, kg*
IPL Initial Program Load (Loader) *fu, ct, em, kg*
IPL Integrated Priority List [USA DOD] *mt*
IPL Ion Projection Lithography *pk, kg*
IPL/POR Initial Program Load/Procedure Oriented Language *fu*
IPLC Interferometer Position Location Concept *hc*
IPLE Index Pulse Leading Edge *-3*
IPLI Internet Private Line Interface *mt, tb*
IPM Implementation Program Manager *mt*
IPM Impulses Per Minute *wp*
IPM Inches Per Minute *hi*
IPM Incremental Phase Modulator *mt*
IPM Industrial Preparedness Measure *hc*
IPM Interpersonal Message *pk, kg*
IPM Interprocessor Multiplexer *-4*
IPM Interrupts Per Minute *fu*
IPMI Internet Provider Multicast Initiative *ct, em*
IPMO Implementation Program Management Office *mt*
IPMO International Project Management Office *hc*
IPMS Information Processing Management System *mt*
IPMS Integrated Planning and Monitoring System *fu*
IPMS Integrated Program Master Schedule *oe*
IPMS Integrated Property Management System *fu*

IPMS Integrated Provisioning Management System *fu*
IPN Internet Protocol Number *ct, em*
IPNG Internet Protocol, Next Generation *ct, kg, hi*
IPO Input, Process and Output *fu*
IPO Integrated Program Office *pbl*
IPO International Pact Organization *hc, su, tb*
IPO International Post Organization *mt*
IPO Iterative Planning Optimization [RADC] *mt*
IPOI Initial Plan Of Instruction *kf*
IPOMS International Polar Orbiting Meteorological Satellite *hy*
IPP Imaging Photopolarimeter [Telescope] *aw118.18.24*
IPP Impact Point Prediction *mt, wp, tb, st, kf*
IPP Impact Predictor Building [Atlantic Missile Range] *0*
IPP Industrial Preparedness Planning *hc, mt, fu*
IPP Integrated Payload Package *hc*
IPP Interactive Post Processor *muc*
IPP Intercept Planning and Prelaunch *tb*
IPP Interface Program Plan *fu*
IPP Interpulse Period *fu*
IPPBS Integrated Personnel Planning and Budgeting System *wp*
IPPL Indentured Parts Price List *hc*
IPPL Industrial Preparedness Planning List *mt*
IPR Improved Packet Radio *fu*
IPR In Process (Progress) Review *mt, fu, 8, hc, nt, tb, st, kf*
IPR Initial Program Review *fu*
IPR Inpulse Response *pbl*
IPR Intelligence Production Requirements *mt*
IPR Inter Page Reference *fu*
IPRA In Place Repairable Assembly *hc*
IPRD In Process Review Document *st*
IPRS Intelligence Production Requirements *tb*
IPRT In Process Review Team *fu*
IPS Image Processing System *ac3.5-4*
IPS In Plant Suborder *fu*
IPS In Plant Support *fu*
IPS Inches Per Second *mt, fu, pl, wp, hi, oe*
IPS Inertial Pointing System *mi*
IPS Inertial Positioning System *hc, wp*
IPS Information Processing (Sub)System [GMCC] [MAC] *wp, mt*
IPS Instructions Per Second *fu*
IPS Instrument Pointing System *hc, sc4.11.3, wl*
IPS Instrument Property Services *fu*
IPS Instrumentation Power Supply *fu*
IPS Integrated Procurement System *mt*
IPS Integrated Program Study *mt*
IPS Integrated Program Summary *mt, fu, hc, wp, tb, st, kf*
IPS Intelligence Processing Subsystem *hc*
IPS Intelligence Production Support *mt*
IPS Intelligence Production System [PACAF] *mt*
IPS Interceptor Pilot Simulator *fu*
IPS Interruptions Per Second *fu*
IPSAR Instrument and Property Services Action Request *fu*
IPSC Information Processing Standards for Computers *kf, fu, mt*
IPSD Integrated Plans and Scheduling Document *oe*
IPSE Integrated Programming (Project) Support Environment *wp, kg*
IPSM Improved Performance Space Motor *oe*
IPSP Intelligence Priorities for Strategic Planning *mt*
IPSS Initial Preplanned Supply Support *mt, wp*
IPSS International Packet Switched Service *hi*
IPT Identified Property Transfer *fu*
IPT Information Processing Terminal *fu*
IPT Infrared Plume Target *wp*
IPT Integrated Payload Test *jdr*
IPT Integrated Process Team *kf*
IPT Integrated Product Team [an organization] *ah, pbl, tm, jdr, kf, jn*
IPT Inter-system Performance Testing *fu*
IPT Interactive Programming Terminal *fu*
IPTC International Press Telecommunications Council *pk, kg*
IPTD Interim Provisioning Technical Documentation *fu*
IPTN [Indonesian Aircraft Industry] *aw129.1.32*
IPX Internetwork Packet Exchange [Novell] *kg, ct, em, hi*
IPX/SPX Internet Packet Exchange / Sequenced Packet Exchange *hi*
IPY International Polar Year *ci.720*
IQ In-phase and Quadrature *kf*
IQ Interactive Query *mt*
IQC Indefinite Quantity Contract *mt*
IQCS Inservice Quality Control System *mt*
IQI Inspection Quality Indicator *fu*
IQL Interactive Query Language *pk, kg*
IQS Inspection Quality System *fu*
IQSY International Quiet Sun (Solar) Years [1964-65] *hc, ci.735*
IQT Informal Qualification Test *fu*
IQT Initial Qualification Training *kf*
IQUE In-Plant Quality Evaluation *fu, oe*
IR Impound Request *fu*
IR In flight Refueling *mt*

IR Incident Report *mt, oe*
IR Industrial Relations *fu*
IR Information Retrieval *fu, w9, ai1*
IR Infra Red (InfraRed) *mt, aw130.13.21, gd1-0, 11, 16, 17, sb, th4, pl, w9, wp, dm, hy, eo, tb, st, pbl, jdr, hg, kf, mb, he, em, fu, nu, kg, ge, mi, jn, hi, oe, ai1*
IR Initial Rekey *fu*
IR Initial Requirements *fu*
IR Inside Radius *fu*
IR Inspection Report *oe*
IR Instrument Route *aw129.10.72*
IR Insulation Resistance *fu*
IR Internal Resistance *fu*
IR Interrogator Responder *fu*
IR&D Independent Research and Development *mt, fu, ro, ah, pl, gg60, 17, sb, th4, aw129.14.103, lt, hn50.8.3, ep, tb, wl, jdr, pbl, tm, kf, jn, oe*
IR&D Internal Research and Development *15, ph, hc, md2.1.2, hg, ge*
IR-NOR Interim Release, Notice of Revisions *fu*
IR/EME InfraRed/Electromagnetic Environment *hc*
IR/MMW/SAR InfraRed/Millimeter Wave/Synthetic Aperture Radar *fu*
IRA InfraRed Associates *oe*
IRA InfraRed [Detection and] Acquisition *hc*
IRA Intelligence Related Activity *mt, wp*
IRA Interface Requirement Agreement *mt*
IRA Interim Repair Activity *hc*
IRAC Information Resources Administration Council *wp, mt*
IRAC InfraRed Analysis Center *hc*
IRAC Interdepartmental Radio Advisory Committee *-4*
IRAC [International Relations Committee of ESA] *es55.94*
IRACQR InfraRed Acquisition Radar *fu*
IRAD Independent Research And Development *wp, tb, kf, pbl*
IRAD InfraRed Adaptive Discrimination *hc*
IRAD InfraRed Ambush Detection System *hc*
IRAD [an area of expertise at General Dynamics] (IRAD and CRAD) *aa26.11.31*
IRADDS InfraRed Air Defense Detection System *hc*
IRADS Infra Red Acquisition and Detection System [USA] *mt*
IRAF Image Reduction and Analysis Facility [a large astronomical analysis system developed at KPNO] *cu, mi*
IRAH InfraRed Active Homing *wp*
IRAH InfraRed Alternate Head [advanced Sidewinder] *hc*
IRAM Increased (Improved) (Integrated) Reliability and Maintenance *hc*
IRAMS InfraRed Automatic Mass Screening System *hc*
IRAN Inspect(ion) and Repair As Necessary [a repair facility] [USA] *wp, hc, tb, fu, mt*
IRAP Integrated Radome Antenna Platform Assembly *hc*
IRAP ISLSCP Retrospective Analysis Project (Program) *ns, hy*
IRAS InfraRed Acquisition Sensor *hc*
IRAS InfraRed Astronomy (Astronomical) Satellite *mi, sc3.1.5, hc, sf31.1.12, ns, mb*
IRAS InfraRed Attack [Sub] System *hc*
IRAS InfraRed Automatic System *fu*
IRAS Interdiction Reconnaissance Attack System *hc*
IRASP InfraRed Advanced Seeker Program *hc*
IRAT InfraRed Angle Track *hc*
IRATE Interim Remote Area Terminal Equipment *fu*
IRATS InfraRed Algorithm Test Simulator *hc*
IRB Instruction Reorder Buffer *hi*
IRBC Intermediate Rate Battery Charge *gg102*
IRBM Intermediate Range Ballistic Missile *mt, hc, wp, ac3.4-3, 0, 5, 12, 13, 14, 17, sb, w9, tb, st, kf, fu, crl, ai2*
IRBO InfraRed Bomb *wp*
IRBOSS InfraRed Beacon Offset Strike System *hc*
IRBS InfraRed Background Sensor *st*
IRBS Intermediate Range Booster System *5, sb*
IRC Information Recovery Capsule *hc*
IRC Information Retrieval Center *fu*
IRC Internet Relay Chat *ct, em, kg, hi, du*
IRCCD InfraRed Charge Coupled Device *hc, wp*
IRCCM Infra Red Counter Counter Measures *mt, hc, fu*
IRCM Infra Red Counter Measures *mt, hc, wp, fu*
IRCMIS Initial Requirements Computation and Management Information System [RDB] [AFLC] *mt*
IRCOS InfraRed Catadioptic Optical System *hc, wp*
IRCS Integrated Radar and Communications Subsystem *sc2.7.1*
IRD Independent Research and Development *pl, wp*
IRD Information Resource Dictionary *mt*
IRD Integrated Receiver / Decoder *tt*
IRD Integrated Receiver / Descrambler *kg, hi*
IRD Integrated Record Data System *mt*
IRD Integration Requirements Definition *fu*
IRD Interface Requirements Document *mt, fu, bo, tb, wl, kf, oe*
IRD Interoperability Requirements Documents *mt*
IRDA Infrared Data Association *pk, kg*
IRDA InfraRed Detection and Acquisition [system] *hc*
IRDA InfraRed Developers' Association *hi*
IRDA Innovative Research and Development Announcement [SDIO] *17, sb*

IRDA Integrated Reliability Design Assessment *wp*
IRDATS InfraRed Data Seeker [Redeye] *wp*
IRDDS InfraRed Decoy Discrimination System *hc*
IRDHS Imagery Related Data Handling System *wp*
IRDS Information Resource Dictionary System *mt, kg*
IRDS InfraRed Detection System *wp*
IRDU InfraRed Detection Unit *hc*
IRE InfraRed Electronics *hc*
IRE Institute of Radio Engineers [later IEEE] *mt, fu, hc, sm3-1, 0*
IRE Interrogation Reply Evaluator *fu*
IRED InfraRed Emitting Diode *wp*
IREECON Institute of Radio and Electronics Engineers Conference (SCGJ 3.9.8)
IREG InfraRed Environment Generator *hc*
IREPS Integrated Refractive Effects Prediction System [US Navy] *fu, mt*
IRES Imagery Reporting and Exploitation Station [TAC] *mt*
IRET Interrupt Return *pk, kg*
IRETS Infantry Remoted Target System *hc*
IREW InfraRed Electronic Warfare *wp*
IRF Infrared Reference Frame [DSP] *kf*
IRF Inherited (Inheritance) Rights Filter [Novell] *kg, ct, em*
IRF Interim Repair Facility *hc*
IRF Intermediate Routing Function *pk, kg*
IRFCS InfraRed Fire Control System *hc*
IRFNA Inhibited Red Fuming Nitric Acid [a propellant] *hc, ci.189, mb, mi*
IRFPA InfraRed Focal Plane Array *wp*
IRG Inertial Rate Gyro *wp*
IRG Interactive Report Generation *fu*
IRGD InfraRed Guidance Demo *hc*
IRGM InfraRed Guidance Module *wp*
IRGP InfraRed Guided Projectile *wp*
IRH Infra Red Homing *mt*
IRHVTA InfraRed High Value Target Acquisition *hc*
IRI International Reference Ionosphere *ns*
IRIA InfraRed Information and Analysis *fu*
IRIA Institut de Recherche d'Informatique et de Automatique [(French) Automation and Cybernetics Research Institute] *wp*
IRIAF Islamic Republic of Iran Air Force [Iran] *mt*
IRID InfraRed Identification *hc*
IRIDS InfraRed Identification System [SEA SPARROW] *hc*
IRIG Inertial Rate Integrating Gyro *pl, hc*
IRIG Inter Range Instrumentation Group *hc, bl1-5.122, 0, 4, 16, aa26.11.44, mj31.5.31, lt, tb, jdr, kf, fu, oe*
IRIG/TM Inter Range Instrumentation Group Telemetry *hc*
IRIS Image Retrieval and Information System *hi*
IRIS Inferential Retrieval Indexing System *mt*
IRIS InfraRed Imagery (Imaging) of Shuttle *hc, tb*
IRIS InfraRed Imaging Seeker *hc, wp, tb*
IRIS InfraRed Information Symposium *hc*
IRIS InfraRed Information System *wp*
IRIS InfraRed Interferometer Spectrometer *wp, ac2.8-2, 17*
IRIS Integrated Reconnaissance Intelligence System *wp*
IRIS Integrated Requirement Implementation System *tb*
IRIS Iran Radar Intercept System *hc*
IRIS Italian Research Interim Stage *hy*
IRISH InfraRed Imaging Seeker Hardware *hc*
IRISH InfraRed Imaging Seeker Head *hc*
IRISS InfraRed Radiometric Imager/Surrogate Seeker *hc*
IRJE InfraRed Jamming Equipment *wp*
IRL Inter-Repeater Link *pk, kg*
IRL Interactive Reader Language *pk, kg*
IRLAP InfraRed Link Access Protocol *pk, kg*
IRLAS InfraRed Laser Sensor System *hc*
IRLAS InfraRed Laser [Rangefinder] *hc*
IRLED InfraRed Light Emitting Diode *kg, hi*
IRLS Information, Retrieval, and Location System *hc*
IRLS Infra Red Line Scan(ner) *hc, mt*
IRLS Interrogation, Recording, and Location [Sub] System *ac2.8-2, hc*
IRM Information Resource Management (Manager) *mt, fu, 8, wp, jdr, kg*
IRM Information Resource Monitor subsystem [JDS] *mt*
IRM InfraRed Measurement *fu*
IRM Inherent Rights Mask *pk, kg*
IRM Initial Release Memorandum *fu, hc, ep, jdr*
IRM Integrated Resource Management [software] *pbl, tm, kf, ah, jn*
IRM Ion Release Module *hy*
IRMA Information Resource Management Agency *hi*
IRMC Information Resources Management College *wp*
IRMS Information Resource Management System *mt, fu, wp*
IRMS InfraRed Manual System *fu*
IRMX Intel Realtime Memory Executive *fu*
IRN Interface Revision Notice *hc, 4*
IRO Immediate Response Option *mt*
IROAN Initial Repair Only As Necessary *hc*
IROAN Inspect and Repair Only As Needed *hc, mt*
IROC Intrusion Resistant Optic(al) Communications *mt, hc, ro*

IROL Imagery Reconnaissance Objectives List *wp*
IROL Imagery Requirements Objective Listing *mt*
IROP InfraRed Optical Intelligence *wp*
IROR Improved Range Only Radar *fu*
IROR Inspection, Repair, Overhaul, and Rebuild *hc*
IROS Increased Reliability of Operational System *fu, hc*
IROT InfraRed On Target *hc*
IRP Improved Radar Program *wp*
IRP InfraRed Probe *jb*
IRP Interference Rejection Processing *fu*
IRP Interim Repair Parts (Kit) *fu*
IRPL InfraRed Payload *kf*
IRPL Interim Repair Parts List *fu*
IRPM InfraRed Payload Module *kf*
IRPPS Information Resource Planning and Projection System Survey *mt*
IRPRL Initial Repair Parts Requirements List *hc*
IRPSL Interim Repair Parts Support List *hc*
IRPU InfraRed Processing Unit *kf*
IRQ Interrupt Request *ct, em, kg, hi*
IRQC InfraRed Quantum Counter *wp*
IRR Imaging InfraRed {misprint for IIR?} *hc*
IRR Individual Ready Reserve *mt*
IRR Initial Requirements Review *mt, tb, wl*
IRR Integrated Radio Room *fu, hc*
IRR Intelligence Radar Reporting *fu*
IRR Interface Requirements Review *tb, wl*
IRR Interim Requirements Review *st*
IRR Internal Rate of Return *hi*
IRRAD InfraRed Ranging (Range) And Detection *wp, fu*
IRRMP Information Reports Requirement Management Program *mt*
IRRTI InfraRed Reconnaissance Target Imagery *hc*
IRS Identification and Reference Sheets *hc*
IRS Indian Remote Sensing [satellite] *ac3.2-1, aa26.10.10, sf31.1.8, ls4, 5, 40, ai2*
IRS Inertial Reference System *mt, wp, ai2*
IRS Information Retrieval System *hc, es55.51*
IRS InfraRed Spectrometer [IRSS] *hg, ge*
IRS InfraRed Surveillance [system] *fu*
IRS Installation Readiness System *mt*
IRS Interactive Retrieval Software [Exercise Schedule] [JCS] *mt*
IRS Interface Requirements Specification *mt, fu, hc, pl, jdr, kf*
IRS&GHL InfraRed Systems and Guidance Heads Laboratory *hc*
IRSF InfraRed Simulation Facility *hc*
IRSIGS Infra Red Signatures *mt*
IRSIM InfraRed Simulation *kf*
IRSS Indian Remote Sensing Satellite [India] *hg, ge*
IRSS InfraRed Search Sensor *hc*
IRSS InfraRed Search Set *hc*
IRSS InfraRed Search System *wp*
IRSS InfraRed Sensor Simulator *hc*
IRSS InfraRed Sensor System *fu*
IRSS InfraRed Subsystem *fu*
IRSS InfraRed Surveillance Sensor *fu*
IRSS InfraRed Surveillance [Sub] System *hc*
IRSSO InfraRed Search Set Operator *hc*
IRST Infrared Search and Track Sensor [ATF] *mt*
IRST InfraRed Search and Track [Set] [System] *mt, hc, wp, ph, aw129.14.103, pm, st, fu*
IRSTD InfraRed Search and Target Designation System (SAR-8) *fu*
IRSTS Infra Red Search and Track System (Set) *wp, hc*
IRT (IRT Corporation) [an avionics/military electronics company] *aw130.22.54*
IRT Independent Review Team *wp*
IRT InfraRed Technology *hi*
IRT InfraRed Telescope *wp*
IRT InfraRed Tracker *fu, hc*
IRT Initial Report Time *kf*
IRT Input Responsibility Table *fu*
IRT Integrated Risk Team *kf*
IRT Intermediate Range Technology *wp*
IRT Interrogator Responder Transducer *fu*
IRTAS InfraRed Target Simulator [also IRTS] *hc, wp*
IRTF Internet Research Task Force *ct, em, kg, hi*
IRTOS I2O Real Time Operating System *ct, em*
IRTS InfraRed Target Simulator [also IRTAS] *hc*
IRTVSU InfraRed Television Sight Unit *hc*
IRTW InfraRed Surveillance and Threat Warning System *hc*
IRTWS InfraRed Tail Warning System *hc*
IRU Inertial Reference Unit *sd, 0, hc, aw130.11.c2, re, cl, jdr, kf, uh, jn*
IRU Integrated Recovery Utility *mt*
IRWL Interchangeability and Reliability Working List *hc*
IRWR Infra Red Warning Receiver *hc, wp, mt*
IRWSS InfraRed Weapon System Simulation *hc*
IRX Information Retrieval Experiment *pk, kg*
IRX InfraRed Telescope System *hc*
IS Imaging Spectrometer *re*

IS Incomplete Sequence *fu*
IS Information Science *hi*
IS Information System *mt, kg*
IS Infrared Spectrophotometer *hc*
IS Initial Sequence *fu*
IS Input Signal *hc*
IS Input Slice *jdr*
IS Interface Specification *fu, 16*
IS Interface Structure *sd*
IS Interim Standard *kf*
IS International Staff [NATO] *mt*
IS Internet Society *hi*
IS Interrupt Status *pk, kg*
IS Inventory Schedule *fu*
IS&R Information Storage and Retrieval *mt*
IS&RP Initial Spares and Repair Parts *mt*
IS&T Innovative Science and Technology [also IST] *tb*
IS/A Interservice / Agency *mt*
ISA Imaging Sensor Autoprocessor *wp*
ISA Industry Standard Adapter *em*
ISA Industry Standard Architecture *ct, kg, hi*
ISA Inertial Sensor Assemblies *wl*
ISA Information Systems Architecture *mt*
ISA Input Switch Assembly *jdr*
ISA Instruction Set Architecture *mt, fu, kg*
ISA Instrument Society of America *fu, id4091.2/1111, 0, aa26.8.b34*
ISA Integrator Shaft Assembly *hc*
ISA Inter-Stage Adapter *oe*
ISA Interactive Services Association *em, ct*
ISA Intermediate Storage Area *oe*
ISA International Security Affairs *mt, dt, hc*
ISA International Security Agency *fu, hc*
ISA International Standard Atmosphere *mt*
ISA Interservice Support Agreement *nt*
ISABE International Symposium on Air Breathing Engines *aa26.8.b3*
ISABPS Integrated Submarine Automated Broadcast Processing System *mt*
ISACC Initial Satellite Control Center *hc*
ISACCC Initial Military Satellite Command and Control Center *hc*
ISACCC Interim Satellite Communications Control Center *mt*
ISAD Information Systems, Authorization Directory [AFP 700-19] *mt*
ISADC Interim Standard Airborne Digital Computer *hc*
ISAF Intelligent Sensor Assessment Facility *hc*
ISAF Israeli Air Force *wp*
ISAGE Imager / Sounder Analysis Ground Experiment *oe*
ISAGEX International Geodetic Satellite Experiment *ci.916*
ISAKMP Internet Security Association and Key Management Protocol *em*
ISAL Inverse Synthetic Aperture Laser *fr, wp*
ISAM Indexed Sequential Access Management (Method) *kg, wp, hi*
ISAM Information System Acquisition Methods *hc*
ISAMS Improved Stratospheric and Mesospheric Sounder [UARS] [United Kingdom] *ge, hy, hg*
ISAPI Internet Server Advanced Programming Interface *em*
ISAPI Internet Server Application Programming Interface *kg, em*
ISAR Inverse Synthetic Aperture Radar [USA] *mt, hc, pf1.10, hd1.89.y7, mj31.5.24, tb, fu*
ISAS Institute of Space and Aeronautical (Astronautical) Sciences [University of Tokyo] *aw129.14.45, ac3.3-3, hy, mb, mi, ai2*
ISAS Integrated Strike Avionics System [PAVE PILLAR] *hc*
ISASI International Society of Air Safety Investigators *aw129.14.120*
ISASM Intelligent Shelter Attack Submunition *hc*
ISAT Integrated Site Acceptance Test *jr*
ISB Independent Side Band *mt, fu*
ISB Industrial Security Bulletin *hc, tb*
ISB Intermediate Staging Base [PACOM] *mt*
ISBN Integrated Satellite Business Network *sc8.9.4*
ISBU Integrated Systems Business Unit *ah*
ISC (ISC Defense and Space . . . Arlington, VA 22202) *jmj*
ISC Image Stabilization Compensation *hc*
ISC Incremental Support Capability *hc*
ISC Information Systems Command [replaced ACC] [USA] *fu, wp, mt*
ISC Input Signal Conditioner *fu*
ISC Instruction Set Computer *pk, kg*
ISC Instructional Systems Consultant *fu*
ISC Integrated Satellite Control *kf*
ISC Integrated Support Center *fu*
ISC Intelligence Support Cells *mt*
ISC Intelligent Synchronous Controller *fu*
ISC Inter-Systems Communication [IBM] *pk, kg*
ISC Inter-Task System Communication (SW Function) *fu*
ISC Interactive Systems Corporation (Center) *fu*
ISC Intermediate Switching Center *mt*
ISCA Integrated Systems Control Architecture *mt*
ISCAMS Installation Standard Command Automatic Data Processing Management System [USA] *mt*

ISCC ISSS Support Computer Complex *fu*
ISCCP International Satellite Cloud Climatology Project *ns, hy, hg, ge, oe*
ISCI Input Signal Conditioner Interface *fu*
ISCM Instructor Station Controller Module *fu*
ISCNET Inter System Communications sub Network *mt*
ISCO Initial Systems Check-Out *hc*
ISCRO Industrial Security Clearance Review Office *hc, tb, su*
ISCS Integrated Sensor Control System *wp*
ISCT&M Integrated System Cost, Technical, and Management *kf*
ISD (time on the ISD and Prime computers) *id4091.2/1111*
ISD Image Section Descriptor *pk, kg*
ISD In Service Date *mt*
ISD Information System Design, [Incorporated] *un*
ISD Information Systems Directive *mt*
ISD Information Systems Division *mt*
ISD Infrared Suppression Device *wp*
ISD Instructional Systems Design *fu, kg*
ISD Instructional Systems Development *mt, fu, hc, kf*
ISD Integrated Strategic Defense *mt*
ISD Integrated System Dictionary *mt, fu*
ISD Integrated Systems Division *fu*
ISDB Integrated Satellite Data Base *mt*
ISDCP Integrated Strategic Defense Concept Plan *mt*
ISDDM Integrated System Data Design Methodology *fu*
ISDM Integrated Software Development Method *fu*
ISDN Integrated Services Digital Network *mt, wp, af72.5.64, is26.1.27, 19, kf, ct, em, hi, du, tt*
ISDN Integrated System Digital Networks *ric, kg, fu, nu, jdr*
ISDOS Information System Design Optimization System *fu*
ISDRP Interface Simulation and Data Reduction Program *hc, fu*
ISDS Imager / Sounder Dynamics Simulator *oe*
ISDS Inadvertent Separation Destruct System *ai2*
ISDS Integrated System Development System *fu*
ISDS Internet Society Domain Survey *hi*
ISDSMS Improved Self Defense Surface Missile System [US Navy] *hc, mt*
ISDU Interim Secure Data Unit *hc, fu*
ISE In Service Engineering *hc*
ISE Information Systems Engineering *mt*
ISE Initial Support Element *mt*
ISE Integrated SATKA Experiments *tb*
ISE Intelligence Support Element [part of EACIC] *mt*
ISE Interim Support Equipment *fu*
ISE Ion Specific Electrode *wp*
ISEC Information Systems Engineering Command [USA] *mt*
ISEE International Sun-Earth Explorer [ESA satellite] *hc, ci.734, sf31.1.20, es55.89, no, hy*
ISEIC Information Systems Engineering and Integration Center [USA] *mt*
ISEM Improved Standard Electronics Module *fu*
ISEP Integrated System Engineering Plan *st*
ISESA Information Systems Engineering Support Activity [USA] *mt*
ISET Independent Software Evaluation Test *mt*
ISET Integrated System Engineering Team *fu*
ISF Improved Support Facility *mt*
ISF Information Systems Flight [AFCC*] *mt*
ISF Integrated Satellite Factory *pbl, jdr, kf, ah, jn*
ISF Integrated Support Facility *fu*
ISF Intelligence Support Facility *mt*
ISF Intermediate Support Facility *hc*
ISFET Ion Sensitive Field Effect Transistor *wp*
ISG Information Systems Group [AFCC*] *mt*
ISG Intersite Gateway [component of LOGNET] *wp, mt*
ISH Information Super Highway *kg, hi*
ISHC Improved Standard Hardware Cabinet *fu*
ISHM International Society for Hybrid Microelectronics *hn41.22.2, aw129.14.120*
ISI Information Science Institute *fu*
ISI Initial Support Items *mt*
ISI Integral Systems, Incorporated *oe*
ISI Internally Specified Index *pk, kg*
ISI International Satellite, Incorporated *sc3.9.7*
ISIC Information and Software Integration Contractor *mt*
ISIC Interceptor System Integration Contractor *st*
ISICAD [A COTS software product] *kf*
ISID Improved Standard Information Display *hc, fu*
ISIL Interim Support Items List *fu*
ISIN Information Systems Internetting *mt*
ISIP Information Systems Planning Services [DLA] *mt*
ISIS Information System Integration Support *mt*
ISIS Integrated Satellite Information Service *ac3.5-4*
ISIS Integrated Satellite Interface System *jn*
ISIS Integrated Shipping Information System *fu*
ISIS Integrated Signals Intelligence System [USMC] *mt*
ISIS Integrated SPADOC IDHS SCC *jb*

ISIS Integrated Strike and Interceptor System *wp*
ISIS Integrated Systems and Information Services *pk, kg*
ISIS Internally Switched Interface System *em, ct*
ISIS International Satellite for Ionospheric Studies *hc, wp, ci.730*
ISIS International Science Information Service *wp*
ISIS Investigative Support Information System *em, ct*
ISIT Initial System Integrated Test *lt*
ISJIS Infrared Solder Joint Integrity System *hc*
ISL Industrial Security Letter *hc, tb*
ISL Inter Satellite Link *hc, in, 19*
ISL Inter Sensor Linking *fu*
ISL Interactive System Language *pk, kg*
ISLI Interrogation Sidelobe Inhibition *fu*
ISLR Integrated Side Lobe Ratio *hc*
ISLS Interrogation Sidelobe Suppression *fu*
ISLSCP International Satellite Land Surface Climatology Project *ns, ge, hy, hg*
ISM Igniter Safety Mechanism *hc*
ISM Industrial Security Manual [for safeguarding classified information , DOD Directive 5220.22-M] *mt, hc, tb, kf, fu, su*
ISM Information System Management *hi*
ISM Inter Stellar Medium *mi*
ISM Intermediate Service Module *em*
ISM Internet Service Manager *em, kg, ct*
ISMA International Satellite Monitoring Agency *11, 14*
ISMC Integrated Sheet Metal Center *fu*
ISMF Interactive Storage Management Facility *kg, ct, em*
ISMMS Integrated Stores Monitor and Management Set *hc*
ISMS Information Systems Maintenance Squadron [AFCC*] *mt*
ISN Internal Serial Number *fu*
ISNSA Independent Software Nuclear Safety Analysis [JCMPO] *fu, mt*
ISO Information Systems Office *ns, hy*
ISO Infrared Space Observatory *sf31.1.15, mi*
ISO Intelligence Support Office *mt*
ISO International Organization for Standardization [recommends and defines mostly OSI layers 7,6,5,4] *nu, mi, kg, ct, em, wp, csII.59, aa26.11.b4, es55.64, ns, tb, jdr, kf, mb, ch*
ISO International Satellite Organizations *jdr*
ISO International Standards (Standardization) Organization [cf. ISO] *mt, fu, hi*
ISO International Standards / Services Organization *mt*
ISO Interplant Shipping Order *hc*
ISO Intershop Order *fu*
ISOC Internet Society *kg, hi*
ISOD Interplanetary Satellite Orbit Determination *hc*
ISODE International Organization for Standarization Development Environment *mt*
ISOO Information Security Oversight Office *mt*
ISORF International Standards Organization Reference Model *hi*
ISOS Information Systems Operations Squadron [AFCC*] *mt*
ISOS International Southern Ocean Studies *pl*
ISP Implementation Support Plan *mt*
ISP Index Sequential Processor *mt*
ISP Information Security Program *mt*
ISP Information System Planning / Plan *mt*
ISP Infrared Spectrophotometer *wp*
ISP Instrumentation Signal Processor *sb*
ISP Integrated Software Process *fu, 17, sb, hc, tb, kf*
ISP Integrated Support Plan *mt, fu*
ISP Intelligence Support Plan *kf*
ISP Interim Support [later Augmented Support] *hc*
ISP International Standardized Profiles *ch*
ISP Internet Service Provider *ct, em, kg, nu, hi*
ISP Interrupt Stack Pointer *pk, kg*
ISP Interrupt Status Port *pk, kg*
ISP [Ion] Specific Impulse *16, 17, sb, re, tb, jdr, vv, uh, kf, jn, oe*
ISPA International Society of Parametric Analysts *fu*
ISPA Inverted Socket Process Architecture *ct, em*
ISPE Improved Sonar Processing Equipment *hc, fu*
ISPE Information Systems Processing Equipment *wp*
ISPE Interim Software Progress Emulation *mt*
ISPF Interactive System Programming Facility *kg, ct, em*
ISPF/PDF Interactive System Productivity Facility / Program Development Facility *ct, em*
ISPI Information Systems Processing Installation *mt*
ISPM International Solar Polar Mission [later renamed Ulysses] *hc, ci.717, mb, mi*
ISPO Instrumentation Ships Project Office *dt*
ISPP Information Systems Program Plan *mt*
ISPPS Item Support Plan Policy Statement *hc*
ISPS Instruction Set Processor Specification *fu*
ISPS Integrated Secondary Power System *wp*
ISPS Item Support Policy Statement *hc*
ISPTC Information Systems Performance Technical Center *mt*
ISQL Interactive SQL *hi*
ISR Industrial Security Regulation *mt, hc, tb*
ISR Information Storage and Retrieval *pk, kg, ai1*

ISR Instruction Set Register *kf*
ISR Instrument Service and Repair *fu*
ISR Integrated Safety Review *17, jdr*
ISR Interim System Review *tb, wl, kf*
ISR Interrupt Service Routine *wp, kg*
ISR Interrupt Status Register *pk, kg*
ISR {Indian Space Research} (the ISR 1-B earth resources satellite) *sf31.1.8*
ISRB Information Systems Requirements Board *mt*
ISRC Information Services Readiness Center *mt*
ISRD Information Systems Requirements Document *mt*
ISRL Indentured Special Request List *fu*
ISRO Indian Space Research Organization *ac2.5-15, ce, ai2*
ISS Imaging Science Subsystem [a candidate Facility Instrument] *so*
ISS Impact Surface Science *es55.29*
ISS Indirect Sighting System *hc*
ISS Indirect Subsystem *fu*
ISS Information Systems Security *mt*
ISS Information Systems Squadron [AFCC*] *mt*
ISS Infrared Surveillance Sensor *wp*
ISS Instrumentation System Support [Atlas launch vehicle] *oe*
ISS Integrated Source Sensor *hc*
ISS Integrated System Support *st*
ISS Integration and Structures Subsystem *jdr*
ISS Intelligence Support Staff *mt*
ISS Intelligence System Simulation *mt*
ISS Intelligent Support Systems *mt*
ISS Intercept Surveillance Stations *17, hc*
ISS Interim Standard Set [VATE] *hc*
ISS International Space Station *pl, ai1, ai2*
ISS Interrupt Safety System *oe*
ISS Inventory Status System *hc*
ISS Ionosphere Sounding Satellite *ac3.3-8*
ISSA Information Systems Security Architecture *wp*
ISSA Installation Supply Support Activity *hc*
ISSA Installation Support Site Activity *hc*
ISSA Inter Service Support Agreement *mt, tb*
ISSA Iranian Site Support Activity *hc*
ISSAA Information Systems Selection and Acquisition Agency (Activity) [Army] *mt, wp*
ISSB Information Systems Standards Board *mt*
ISSC Information Systems Steering Committee [USMC] *mt*
ISSE Imaging Science Subsystem Electronics *hc*
ISSE Interim System Support Equipment *fu*
ISSG Information Systems Support Group [AFCC*] *mt*
ISSL ICAM System Specification Language *fu*
ISSL Initial Spares Support List *mt, fu*
ISSL Initial Supplies Support List *mt*
ISSM Information System Security Manager [GCCS] [USA] *mt*
ISSN International Standard Serial Number *w9, sf31.1.1, aw129.10.5, kg, ch*
ISSO Information Security Officer *kf*
ISSO Information Systems Security Officer *mt*
ISSO Information Systems Staff Officer *mt*
ISSP Information Systems Standardization Program *mt*
ISSP-S Interim Single Source Processor-Signals Intelligence [DIA/NSA] *mt*
ISSS Information Systems Support Squadron [AFCC*] *mt*
ISSS Initial Sector Suite System *fu*
ISSS Integrated Support Software System *hc*
ISSSM Imaging Seeker Surface to Surface Missile *wp*
ISSSP Information System Security Support Plan *wp*
ISST ICBM SHF Satellite Terminal *mt*
ISST Independent System Status Test *fu*
IST Imagery Support Terminal *mt*
IST Implementation Systems Test *mt*
IST Incremental System Testing *fu*
IST Independent Software Test(ing) *fu*
IST Initial Support Team *mt*
IST Innovative Science and Technology [also IS&T] *wp, tb*
IST Inspection Status Tag *gp, hc, ep, tm, jdr, pbl, ah, jn*
IST Instrument Support Terminal *hy*
IST Integrated Status Tag *-17*
IST Integrated System Test(ing) *mt, gs2-13.10, 5, 9, 17, sb, hc, es55.43, lt, ep, st, jdr, pbl, tm, kf, ah, jn*
IST Inter Switch Trunk *mt*
IST International Standard Thread *fu*
IST Intersegment Terminal *jdr*
IST {India Standard Time} *ac3.2-7*
ISTA Intelligence, Surveillance and Target Acquisition *mt, wp, hc*
ISTAP Information Systems Technology Application Program *mt*
ISTAR Image Storage Translation and Reproduction *fu*
ISTC Integrated System Test Complex *oe*
ISTECH Information Systems Technology *mt*
ISTF Integrated Services and Test Facility [Canada] *wl*
ISTI International Space Technology, Inc. *ls4, 5, 29*

ISTM International Society For Testing Materials *wp*
ISTO Information Science and Technology Office [DARPA] *kb, wp*
ISTP Information Systems Tasking Plan *mt*
ISTP International Solar Terrestrial Physics (Program) *hc, ns, no*
ISTRAC [ISRO telemetry, tracking and telecommand network] *ai2*
ISU I/O Terminal Support *fu*
ISU Instructor Scoring Unit *hc*
ISU Integrated Sight Unit *hc, bf*
ISU Interface Simulator Unit *fu*
ISU International Space University [at MIT] *es55.100*
ISV Independent Software Vendor *kg, hi*
ISV Intelligence Secure Voice *mt*
ISVCS Improved Secure Voice Conferencing System *mt*
ISVN Interim Secure Voice Network *mt*
ISVP Integrated System Verification Plan *jdr*
ISW Information Systems Wing [AFCC*] *mt*
ISW Intermediate Scale Warfare *hc*
ISWG Information Systems Working Group [USMC] *mt*
ISWG Integrated Support Working Group *tb*
ISY International Space Year [1992] *hy, wl, mb, mi*
ISYSCON Integrated System Controller *fu*
IT Independent Test *fu*
IT Information Technology *ct, em, kg, kf, jn, hi, y2k, ai1*
IT Installation Technician *fu*
IT Insulating Transformer *fu*
IT Integrated Technology *5, 17, sb*
IT Integration Testing *mt*
IT Interactive Terminal *fu*
IT Intermediate Terminal *do24*
IT&E Integration Test and Evaluation *kf*
IT&M Integrated Technical and Management *kf*
IT&TEL Initial Tools and Test Equipment List *hc*
ITA Individual Task Authorization *hc*
ITA Infrared Tasking Authority *kf*
ITA Intelligent Task Automation *hc*
ITA Interceptor Training Area *fu*
ITA Interface Test Adapter *fu, wp*
ITA International Telegraph Alphabet *mt, fu*
ITA International Trade Administration [US DOC] *mt, fu*
ITAA Information Technology Association of America *y2k*
ITAADS Installation the Army Authorization Document System *mt*
ITAC Intelligence Threat Analysis Center [USA] *dt, mt*
ITACIES Interim Tactical Imagery Exploitation System *mt*
ITACS Integrated Tactical Air Control System *mt, wp, fu*
ITACS Interim Tactical Air Control System *hc*
ITAD Integrated Thermal Avionics Design *hc, wp*
ITAI Interceptor Technology Analysis and Integration *st*
ITALSS Integrated Test And Logistic Support System *hc, fu*
ITAR Integrated Terrain Access and Retrieval *hc*
ITAR International Traffic in Arms Regulations [Code of Federal Regulations, Title 22, Chapter 1, Parts 121-127] *mt, ph, hc, tb, jdr, fu, su*
ITARS Integrated Terrain Access and Retrieval System *hc, wp*
ITAS Indicated True Air Speed *wp*
ITAS Instantaneous Telephone Alerting System *mt*
ITAS Integrated Tactical Attack System *wp*
ITASC Interim Theater ADP Service Center [USA] *mt*
ITASS Interim Towed-Array Surveillance System *fu*
ITAWDS Integrated Tactical Amphibious Warfare Data System *mt, fu*
ITB Information Technology Branch *pk, kg*
ITB Integrated Test Bed *hc, st*
ITB Intermediate Text Block *pk, kg*
ITB Invitation To Bid *su*
ITC Ice Technology Conference *ac3.7-36, aw*
ITC In Track Contiguous [mode] *ac7.3-1*
ITC Increased TOW Capability *hc*
ITC Information Technology Center *mt*
ITC Integrated Telemetry and Command *id4142-10/831, gp, 16*
ITC Integrated Telemetry Complex *hc*
ITC Intelligence Training Consolidation *mt*
ITC International Trade Commission *mt*
ITC International Transducer Corporation *fu*
ITC International Typeface Corporation *pk, kg*
ITCAN Inspect, Test and Correct As Necessary *wp*
ITCC Intelligence Technical Coordinating Committee *mt*
ITCCCC Interim Transportable Command and Control Computer Center [USA] *mt*
ITCI Inadvertent Transmission of Classified Information
ITCP/IP Intercomputer Protocol; Transmission Control Protocol/Internal Protocol *mt*
ITCS Integrated Target Command (Control) System *hc*
ITCU International Telemetry and Command Unit *jn*
ITD Inception To Date *fu, 8, hc, lt, jdr, kf*
ITD Integrated Technology *fu*
ITD Integrated Technology Demonstration *mt, 17, sb, tb*
ITD Integration Test and Demonstration *wp*

ITD Interactive Text Display *fu*
ITD Ion Trap Detector *wp*
ITDA Interim Tactical Decision Aid *fu*
ITDM Intelligent Time Division Multiplexer *fu*
ITDN Integrated Tactical Data Network *wp*
ITDP Integrated Technology Demonstration Plan *17, sb*
ITDR Integrated Technology Design Review *17, sb*
ITDS Improved Technical Data System *mt, wp*
ITDSS Intelligent Target Development Support System *mt*
ITDT Improved Technical Documentation and Training *hc, fu*
ITDT Integrated Technical Documentation and Training *fu*
ITE Information Technology Equipment *pk, kg*
ITE Institute of Tool Engineering *0*
ITE Interim Test Equipment *hc, kf*
ITE Intersite Transportation Equipment *bo, kf*
ITEP Integrated Test and Evaluation Plan (Program) *kf, wp*
ITEP Interim Tactical ELINT Processor [USA] *wp, mt*
ITEWS Integrated Tactical Electronic Warfare System *hc, wp*
ITF Integrated Test Facility *hc*
ITF Integration Test Facility *fu*
ITF Integration Test Folders *wp*
ITF Intelligence Task Force *mt*
ITF Intensity Transfer Function *es55.53*
ITF Interactive Terminal Facility *fu*
ITF Interactive Test Facility *pk, kg*
ITFS Instructional Television Field Service *dn, hc*
ITG Information Transfer Group *fu*
ITGBL International Through Government Bill of Lading *mt*
ITI India Telephone Industries *ac3.2-7*
ITI Industrial Tectonics, Incorporated *oe*
ITI Instrument Technology, Incorporated [Westfield, Massachusetts] *aa26.11.39*
ITI Intermediate Time Integration *fu*
ITIC-PAC INSCOM Theater Intelligence Center-Pacific [USA] *mt*
ITIP Improved Transtage Injector Program *ai2*
ITIP International Technology Integration Panel *wl*
ITIR Intermediate and Thermal Infrared Radiometer [Japan] *hy*
ITIS Intelligent Target Imaging System *hc*
ITIS Intra Theater Imagery Transmission System *mt*
ITL Integrate, Transfer, and Launch [complex] *hc, ci.85, 0, jdr, ai2*
ITL Intent To Launch *hc*
ITM Integrated Test and Maintenance *wp*
ITMD Interim Theater Missile Defense *st*
ITMG Intra Theater Movement Generator [EUCOM] *mt*
ITMS Ion Trap Mass Spectroscopy *wp*
ITN Identification Tasking and Networking *pk, kg*
ITN Information Transfer Node *mt*
ITNC In Track Non Contiguous [mode] *ac7.3-1*
ITNS Integrated Tactical Navigation System *wp, fu*
ITO Independent Test Organization *hc, pl*
ITO Indium Tin Oxide *gp, jdr, kf, uh, jn*
ITO Installation Transportation Officer *mt*
ITO Instrument Take Off *mt*
ITO Interim Technical Order *wp*
ITO Intermediate Transfer Orbit *uh, jn*
ITO International Trade Organization *w9, wp*
ITOC Italian Operations Central *hc*
ITOPS Interim Terminal Overseas Processing System *mt*
ITOPS Interim Transportation Overseas Processing System [MAC] *mt*
ITOS Improved TIROS Operational Satellite *ac3.1-14, fu, oe*
ITOW Improved TOW *hc, vc25.3.10*
ITP Impact Time Prediction *mt, kf*
ITP Index of Technical Publications *fu*
ITP Industrial TEMPEST Program *fu*
ITP Instruction To Proceed *mt*
ITP Integrated Technology Plan *-17*
ITP Integrated Test Plan *mt, tb, jn*
ITP Integrated Training Program *fu*
ITP Integration and Test Plan *kf*
ITP Intent To Purchase *mt*
ITP Interactive Theorem Prover *fu*
ITP Istrebiteltyazhely Pushechny [heavy fighter, cannon-armed, USSR] *mt*
ITPR Infrared Temperature Profile Radiometer *ac2.8-3*
ITPT Integrated Technology and Product Development *pbl, ah, jn*
ITQ Information To Quoters *fu*
ITQ Invitation To Quote *hc, wp*
ITR In Transit Rendezvous *hc*
ITR Initial Technical Review (Report) *fu*
ITR Inspection Test Report *hc, kf*
ITR Inspection/Test Request *fu*
ITR Instrumentation Tape Recorder *hc*
ITR Internet Talk Radio *pk, kg*
ITR Intersatellite Timing Resolution *jdr*
ITRAM International Traffic Passenger Management System [MTMC] *mt*

ITRB Initial Test Review Board *mt*
ITRC Information Technology Requirements Council *is26.1.55*
ITRO Interservice Training Requirements Organization *mt*
ITRR Integrated Technology Requirements Review *17, sb*
ITS Individual Training Standard *hc*
ITS Information Transfer Satellite *hc*
ITS Information, Technologies and/or Systems *kf*
ITS Institute of Telecommunications Sciences *fu*
ITS Integrated Training Station *fu*
ITS Integrated Training System *hc, kf*
ITS Integration Test Set *fu*
ITS Interactive Terminal System *mt*
ITS Interface Tracking System *fu*
ITS Intermediate Tape Store *fu*
ITS Internet Telephony Server *ct, em*
ITS Item Tracking System *fu*
ITSEC Information Technology Security Evaluation Criteria *em*
ITSO International Telegraphy Society Organization *jb*
ITSOP Integrated Telecommunications Systems Operational Planning *mt*
ITSS Integrated Tactical Surveillance System *mt, wp, sc2.10.4.*
ITSTEC Integrated Transmission Switching and Technical Control *mt*
ITT Integration Test Terminal *fu*
ITT Intelligent Transform Tool *fu*
ITT International Telephone and Telegraph [a corporation] *mt, ct, ac2.9-7, aw129.14.34, 17, hc, af72.5.32, oe*
ITT Invitation To Tender [an offer] *mt, es55.42*
ITTA Information Technology Training Association *ct, em*
ITTAV International Telephone and Telegraph Avionics Division *hc*
ITTCSI International Telephone and Telegraph Communication System, Incorporated *hc*
ITTS Interrogator and Transponder Test Set *fu*
ITU Interface Transformation Unit *hc*
ITU International Telecommunications Union [Geneva] *mt, ct, em, jr, fu, kg, sc3.12.5, 7, 13, w9, ce, hi, y2k*
ITU-R International Telecommunications Union-Radio [Former CCIR, frequency and orbit allocations] *nu*
ITU-T International Telecommunications Union-Telecommunications [Former CCITT, recommends and defines mostly OSI layers 1,2,3] [Telecommunications standardization sector of ITU] *nu, hi, du*
ITU-TIES ITU-Telecom Information Exchange Services *pk, kg*
ITU-TSS ITU-Telecommunication Standards Section *pk, kg*
ITUE Integrated Technology Uplink Experiment *hc, 17, tb, st*
ITUG International Telecommunications User Group *pk, kg*
ITV Improved TOW Vehicle *hc*
ITV Instructional Television *mt, hc, w9*
ITV Instrumented Test Vehicle *mt, hc, 5, 17, sb, jb, cp, tb*
ITV Interactive Television *kg, jgo, ct, em, hi*
ITV Intercept Test Vehicle *jb*
ITV Interface Verification Test *sb*
ITV&V Independent Technology [and system] Verification and Validation *st*
ITVE Integrated Technology Validation Experiment *tb*
ITW&A Integrated Tactical Warning and Assessment *ts7-37, tb, mt*
ITW/AA Integrated Tactical Warning / Attack Assessment *kf, mt*
ITWG Interagency Technical Working Group *oe*
ITX Intermediate Text Block *pk, kg*
IU Identification Unit *fu*
IU Initial User *mt*
IU Instrument Unit *ac2.3-3*
IU Integer Unit *pk, kg*
IU Integration Unit *fu*
IU Interface Unit *mt, fu*
IUAP Internet User Account Provider [Internet] *kg, hi*
IUE International Ultraviolet Experiment *tk*
IUE International Ultraviolet Explorer [NASA and ESA satellite] *wp, hc, ci.737, es55.13, ns, mb, cu, mi*
IUESIPS IUE Spectral Image Processing System *ns*
IUGG International Union of Geodesy and Geophysics *no, hy*
IUGS International Union of Geophysical Sciences *hy*
IUI Integrated User Interface *tb*
IUI Intelligent User Interface *ns*
IUID Internal Unit Identification *mt*
IUKADGE Improved United Kingdom Air Defense Ground Environment [UK] *mt*
IULTCS International Union of Leather Technologists and Chemists Societies *ch*
IUM Interim Use Material *wp*
IUP Installed User Program *fu*
IUPAC International Union of Pure and Applied Chemistry *wp*
IUPAP International Union of Pure and Applied Physics *wp*
IURL Internet Uniform Resource Locator *hi*
IUS Inertial Upper Stage *mt, ci.568, 4, aw129.14.18, hc, sf31.1.29, ws4.10, af72.5.149, wp, ep, dm, tb, jdr, mb, kf, mi, ai2*
IUS Interim Upper Stage *ci.874, hc, ce*
IUS/ITB Interchange Unit Separator/Intermediate Transmission Block *pk, kg*

IUSS Integrated Undersea Surveillance System [NAVELEX] *hc, ph, fu, mt*
IUWDS International URSIGRAM and World Days Service *ns, no, pr*
IV Initial Velocity *fu*
IV Intercept Vehicle *tb*
IV Intermediate Voltage *fu*
IV&V Independent Validation and Verification *mt, hc, wp, 17, sb, aa27.4.63, tb, kf, kg, fu, oe*
IVA Intermediate Volatility Agent *-14*
IVA Intravehicle (Intravehicular) Activity *eo, wl*
IVCE Internal Voice Communications Equipment *mt*
IVCS Integrated Voice Communications System *fu*
IVCS Intrasite Voice Communication Subsystem *-4*
IVD Image Velocity Detector *fu*
IVD Indirect View Display *hc*
IVDCD Interactive Video Disc Coursewear Development *hc*
IVDS Interim Variable Depth Sonar *fu*
IVDW Integrated Voice / Data Workstations *mt*
IVEN Interactive Video Extension Network *wp*
IVES Information Vending Encryption System *hi*
IVIS Inter Vehicular Information System *hc, fu*
IVIS Interactive Video Information System *pk, kg*
IVL Independent Vendor League *pk, kg*
IVL Intel Verification Lab *pk, kg*
IVR Interactive Voice Response *kg, em*
IVRS Interim Voice Response System *fu*
IVS Indirect Viewing Substitution *hc*
IVS Interactive Videodisk System *pk, kg*
IVS Interactive Voice System *mt*
IVSAN Initial Voice Switching Network [NATO] *mt*
IVSN Initial Voice Switch Network [NATO] *mt*
IVT Interface Vacuum Test *-9*
IVT Interface Verification Test (Testing) [launch] *gp, 17, kf, jn, oe*
IVT Interrupt Vector Table *pk, kg*
IVT Iverson Technology [Corporation, McLean, Virginia] *aw130.22.98*
IVTP Installation Verification Test Procedure *fu*
IVTS Interactive Video Training System *mt*
IVU In-View Update (SW Function) *fu*
IVV Independent Validation and Verification [also IV&V] *hc*
IW Interceptor Warning *hc*
IWA Individual Work Authorization *pbl, ah, jn, fu, 8, 17, sb, H, pl, ie5.11.12, lt, jdr*
IWARS Installation Worldwide Ammunitions Reporting System [USA] *mt*
IWC Intermediate Work Center *fu*
IWCS Integrated Weapon Control System *wp*
IWCS Integrated Wideband Communications System *wp, hc*
IWCS Interceptor Weapon Control System *hc*
IWD Integrated Weapon Display *fu*
IWD Intermediate Water Depth *hc*
IWDS Imager Wideband Data Software *oe*
IWDS Improved Weapon Delivery System *hc*
IWDS Integrated Warning and Display System *mt*
IWG Group / Intersystem Working Group {?} *mt*
IWG IEMATS Working *mt*
IWG Intercept Word Generator *fu*
IWG Interface Working Group *mt, hc*
IWG Interprogram Working Group *tb*
IWG Investigator Working Group *hy*
IWGDMGC Interagency Working GRoup on Data Management for Global Change *hy*
IWGN Input Waveguide Network *jdr*
IWIS Interceptor Weapons Instructor School *wp*
IWM Incremental Warehouse Methodology *em*
IWM Integrated Woz Machine *em, ct*
IWP In service Work Plan *mt*
IWPTB Integrated Weather Product Test Bed *dm*
IWR Interceptor Warning Receiver *hc*
IWS Integrated Weapon System *mt, wp*
IWS Integrated Wiring System *fu*
IWS Intelligent Work Station *mt, wp*
IWSM Integrated Weapon Support Management *wp, kf*
IWST Integrated Weapon System Training *wp*
IWTS I/E Warhead Test Set Program *fu*
IX (IX-64 Wolverine) (IX-81 Sable) [training carrier] *mt*
IX Ion Exchanger *wp*
IX [unclassified auxiliary] [US Navy] *mt*
IXC Inter Exchange Carrier [also IEC] *mt, ct, em, kg, hi*
IXC Interexchange Channel *do93*
IXD Index Duplicate *hi*
IXSS [unclassified auxiliary submarine] [US Navy] *mt*
IXTR (IXTR versus Kp and Delay Slope) *smiv*
IXYS {a San Jose silicon electronics company} *is26.1.55*

J

J Joint [used to refer to various directorates, J-3 for Operations, J-2 for Intelligence, J-4 for Logistics, J-6 for Communications, etc.] [USA DOD] *mt*
J-BADGE Japan Base Air Defense Ground Environment *fu*
J-FACT Joint Flight Acceptance Test *hc*
J-SAK Joint Second echelon Attack *mt*
J-SEAD Joint Suppression Of Enemy Air Defenses [USA DOD] *mt*
J-STARS Joint Surveillance Target Attack Radar System *mt*
J-T Joule-Thompson [cryostat] *17, sb*
J-TENS Joint-Service Tactical Exploitation of National Systems *kf*
J/FET Junction/Field-Effect Transistor *fu*
J/N Jammer-to-Noise (Ratio) *fu*
J/S Jammer (Jamming) to Signal [power ratio] *mt, fu, uh, wp, sd, hc, jdr, kf*
JA/ATT Joint Airborne / Air Transportability Training *mt*
JAA Joint Airworthiness Authorities *mt*
JAAD Justification, Approval, and Acquisition Documentation *mt*
JAAF Joint Army-Air Force *hc*
JAAG Judge Advocate General Legal Service Office *dt*
JAAP Joint Airborne Advance Party *mt*
JAASPO Joint Army-Air Force Special Project Office *hc*
JAAT Joint Air Attack Team *mt*
JAC Joint Analysis Center *mt*
JACC Joint Airborne Command Center *mt*
JACC Joint Airborne Communications Center *mt*
JACC/CP Joint Airborne Command Center / Command Post [JACKPOT] *mt*
JACG Joint Aeronautical Commanders Group *wp*
JACUDI Japan Computer Usage Development Institute *hi*
JACWA Joint Allied Command Western Approaches *fu*
JAD Joint Analyses Directorate [OJCS] *tb, mt*
JAD Joint Application Design *pk, kg*
JAD Joint Application Development *kf*
JADE Japanese Air Defense Environment *wp*
JADE Joint Allied Defense Experiment *st*
JADE [a non-acoustic antisubmarine warfare program, TRW] *kb*
JADMAG Joint Aeronautics Depot Maintenance Action Group *dt*
JADO Joint Air Defense Operations *mt*
JADREP Joint Resource Assessment Damage Report *mt*
JADS Joint-Service ASLBM Defense Study *fu*
JAE Jump if Above or Equal *pk, kg*
JAEW Japan Airborne Early Warning [system] *fu, hc*
JAFE Joint Advanced Fighter Engine *wp*
JAG Joint Analyses Group [TACS / TADS] *mt*
JAI Joint Administrative Instruction *mt*
JAIEG Joint Atomic Information Exchange Group *dt, wp*
JAIS Japan Air Intelligence System [PACAF] *mt*
JAISPAC Joint Area Information System Pacific *mt*
JAL Japan Air Lines *aw130.13.15, hc, wp*
JAM Jammer Analysis Model *fu*
JAMES Joint Automated Message Editing Software *mt*
JAMPAC Jamming Package *wp*
JAMPS JINTACCS Automated Message Preparation (Processing) System *fu, mt*
JAMS Japanese Management Society *wp*
JAN Jet Aircraft Noise *wp*
JAN Joint Army-Navy [USA] *mt, pl, wp, hc, nt, fu, oe*
JANAF Joint Army, Navy, Air Force *wp, fu*
JANAIR Joint Army Navy Aircraft Instrumentation Research *wp, hc*
JANAP Joint Army Navy Air Force Procedure *wp*
JANAP Joint Army, Navy, Air Force Publication *mt, sd, hc, kf, fu*
JANE Joint Air Force Navy Experiment *wp*
JANTX (minimum PMP quality levels *Semiconductors - JANTX >*
JANUS [force model, Livermore Labs] *mt*
JANUS [series of RV deployment experiments] *17, sb*
JAO Joint Area of Operations *mt*
JAPC Joint Air Photo Center *wp*
JAPO Joint Area Petroleum Officer *mt*
JAPSS Joint Automated Planning Support System *mt*
JAR Java Archive File [format] *pk, kg*
JAR Joint Airworthiness Requirements [European] *mt, aw118.14.31*
JARM Jammer, Artillery, Radar Missile *wp*
JARPA Jam-Resistant Phased Array *hc*
JARS Jamming Aircraft and Radar Simulation [model] *wp*
JASAR Jittered And Swept Active Radar *hc*
JASC Joint Actions Steering Committee *mt*
JASIN Joint Air Sea Interaction [Experiment] *ac2.7-13*
JASORS Joint Advanced Special Operations Radio System *mt*
JASS Jumps Automated Supplemental System [USA] *mt*
JAST Joint Advanced Strike Technology [USA] *mt*
JAT Jam Angle Tracking *hc*
JATCC Joint Air Traffic Control Center *fu*
JATF Joint Amphibious Task Force *mt*
JATO Jet Assisted Take Off. [more correct Rocket Assisted Take Off, or RATO] *mt, hc, 0, wp, 13, af72.5.142*

JAVA Jammer Azimuth Versus Amplitude *hc*
JAWF Jet Augmented Wing Flap *wp*
JAWOP Joint Automated Weather Observation Program *ag*
JAWS Jamming And Warning System *wp*
JAWS Joint Air Force Systems Command War Game System *mt*
JAWS Joint Airport Weather Studies *fu, ag*
JAWS Joint All Weather Seeker *hc*
JAWS Joint Automated Wiring System *fu*
JB Jet powered Bomb [USA] *mt*
JB Junction Box *0*
JBE Jump if Below or Equal *pk, kg*
JBG Jagd Bomber Geschwader [fighter bomber wing; as in JBG 31 Boelcke, Germany] *mt*
JBIS Journal of the British Interplanetary Society *sf31.1.10*
JBS Japanese Broadcast Satellite *sm11-5, hc*
JBUSDC Joint Brazil - U. S. Defense Commission *dt*
JC Jump if Carry set *pk, kg*
JCA Jamming Control Authority *mt*
JCAAD Joint Counter Air / Air Defense *mt*
JCAP Joint Conventional Ammunition Panel Or Program *wp*
JCC Joint Coordination Center *mt*
JCCA Joint Conex Control Agency *dt*
JCCB Joint Configuration Control Board *mt, fu*
JCCC Joint Communications Control Center *mt*
JCCCCM Joint Command, Control and Communications Countermeasures *mt*
JCCSOC Joint CCC Staff and Operations Course [AFSC*] *mt*
JCCEP Joint Crisis Communications Exercise Program *mt*
JCCG Joint / Combined Coordinating Group *mt*
JCCRS Joint Contingency Construction Requirements System *mt*
JCCSA Joint Communication Contingency Station Assets [US] *mt*
JCC[E] Joint Coordination Center [Europe] *mt*
JCDSI Joint Continental Defense Systems Integration *hc*
JCE Jet Control Electronics *gg80, hc*
JCEC Joint Chiefs Electronic Committee *hc*
JCEOI Joint Communications - Electronics Operations Instructions *mt*
JCEWS Joint Command, Control, and Electronic Warfare School *mt*
JCEWS Joint Commanders Electronic Warfare Staff *mt*
JCF Joint Communications Facility *mt*
JCG Joint Commanders Group *wp*
JCG Joint Coordinating Group *wp*
JCIOC Joint Counterintelligence Operations Center *mt*
JCIP JTIDS Class I Project *fu*
JCIS Joint Command Information Systems [Korea] *mt*
JCL Joint Logistics Commanders *mt*
JCLIC Joint Center for Low Intensity Conflict *mt*
JCLOT Joint Closed Loop Operational Test *0*
JCMC Joint Crisis Management Capability *mt, hc, wp*
JCMEC Joint Captured Materiel Exploitation Center [USA DOD] *mt*
JCMP Joint Cruise Missile Project *hc, dt, fu*
JCMPO Joint Cruise Missile Project Office *mt, dt*
JCMSPO Joint Cruise Missile System Program Office *hc*
JCN Job Control Number *mt*
JCOC Joint Command Operations Center *mt, fu*
JCOP JCS Concept and Objectives Paper *mt*
JCPES Joint Contingency Planning and Execution Support *mt*
JCRLCMP Joint Computer Resource Life Cycle Management Program *mt*
JCS Japanese Communications Satellite *sm11-5*
JCS Joint Chiefs of Staff [USA DOD] *mt, fu, 5, 12, 13, ph, sb, w9, af72.5.170, jb, nt, dt, wp, tb, as, jdr, kf*
JCS-ACA Joint Chiefs of Staff (JCS) Automatic Conference Arranger *mt*
JCSAN Joint Chiefs of Staff Alert(ing) Network [voice] *mt, fu, hc*
JCSAT Japan Communications Satellite [Company] *hn49.9.1, aw130.11.29*
JCSE Joint Communications Support Element [MacDill AFB] *wp, mt*
JCSEA Joint Chiefs of Staff (JCS) Emergency Actions *mt*
JCSM Joint Chiefs of Staff (JCS) Memorandum *jb, kf, mt*
JCSMC Joint Chiefs of Staff (JCS) Message Center *mt*
JCSP Joint Chiefs of Staff Publication *mt*
JCSS Joint Communications Support Squadron *mt*
JCS_NICA Joint Chiefs of Staff NICA Support [HQ USAF] *mt*
JDA Japan Defense Agency [also known as NDA] *aw130.13.22, hc*
JDA Joint Deployment Agency *mt, wp*
JDA Joint Duty Assignment [USA DOD] *mt*
JDAL Joint Duty Assignment List [USA DOD] *mt*
JDC Jet Deflection Control *wp*
JDC Joint Deployment Community *mt*
JDCRB Joint Document Change Review Board *fu*
JDEC Joint Documents Exploitation Center *mt*
JDES JTIDS Development and Evaluation Site *hc, fu*
JDHQ-SV-W Joint Defense Headquarters Services Washington *dt*
JDI Java Driver Interface *em*
JDISS Joint Deployable Intelligence Support System [GCCS] [USA DOD] *mt*
JDL Joint Directors Of Laboratories *wp*

JDMAG Joint Depot Maintenance Analysis Group *fu, wp*
JDMC JOPES Development Management Center *mt*
JDP JTIDS Discrete Panel *fu*
JDRP JTIDS Data Reduction Program *fu*
JDS Joint Deployment System *mt, wp*
JDSIP JDS Interface Processor [JDA] *mt*
JDSIR JDS Incident Reporting [REDCOM/JDA] *mt*
JDSQP JDS Query Processor [JDA] *mt*
JDSSC Joint Data Systems Support Center [DCA] [formerly CCTC] *dt, mt*
JDSUP JDS Update Processor [JDA] *mt*
JDT Joint Development Team *mt, wp*
JDW Jane's Defence Weekly *wp*
JE Jump if Equal *pk, kg*
JEA Joint Endeavor Agreement [NASA] *wl, mt*
JEC Joint Economic Committee *wp*
JECG Joint Exercise Control Group *mt*
JED Joint Exercise Division [J-3] *mt*
JEDEC Joint Electron Devices Engineering Council *ct, fu, em, kg, is26.1.76c, hi*
JEEP Joint Emergency Evacuation Program (Plan) *mt, 12*
JEEP Joint Environmental Evaluation Program *fu*
JEFF/TDMA Judiciously Efficient Fixed Frame TDMA *fu*
JEIDA Japan(ese) Electronics Industry Development Association *kg, hi*
JEIM Jet Engine Intermediate Maintenance *mt*
JEM Japanese Experiment Module [for Space Station Freedom] *mi, aa27.4.b48, wl, mb*
JEM Jet Engine Modulation *hc, ful, pm*
JEM Joint Exercise Manual [JCS] *mt*
JEOF Joint Exercise Observation File [USREDCOM] *mt*
JEP Joint Experiments Program *ac3.2-20*
JEPDS Jet Exhaust Powered Decontamination System [chemical warfare] *wp*
JERS Japan Environmental Research Satellite *aa26.10.10*
JERS Japanese (Japan's) Earth Resources Satellite *wp, hy, hg, ge*
JES Job Entry (Sub) System *fu, ct, kg, em*
JESAP Jet Engine Smoke Abatement Program *wp*
JESS Joint Exercise Support System *mt*
JET Jam Exceeds Threshold *hc*
JET JDS Evaluation Team *mt*
JET-X Joint European Telescope for X-rays on SXG *cu*
JETDS Joint Electronics Type Data System *fu, hc*
JETDS Joint Electronics Type Designation System *mt*
JETGLD (GTS Goldstone (JETGLD)) *0*
JETS Jammer Technique Simulation *hc*
JETS Joint Enroute and Terminal System [Canada] *hc, fu*
JEVA Jammer Evaluation Versus Amplitude *hc*
JEWC Joint Electronic Warfare Center *mt, sr9-82, hs, jb, tb*
JEWOC Joint Electronic Warfare Orientation Course *mt*
JEWSOC Joint Electronic Warfare Staff Officer Course [AFSC*] *mt*
JEZ Joint Engagement Zone [USA DOD] *mt*
JFAAD Joint Forward Area Air Defense *hc, fu*
JFAADS Joint Forward Area Air Defense System *hc*
JFACC Joint Force Air Component Commander [USA DOD] *mt*
JFACC Joint Forces Air Command/Control *kf*
JFACT Joint Flight Acceptance Composite Test *0*
JFAST Joint Flow and Analysis System for Transportation [GCCS] [USA] *mt*
JFC Joint Force Commander [USA DOD] *mt*
JFCC Joint Force Fires Coordinator *mt*
JFCC Joint Frequency Coordination Committee *mt*
JFDP Joint Force Development Process *mt, wp*
JFET Junction Field Effect Transistor *hc, kg, fu, wp*
JFFSC Joint Force Fire Support Coordinator *mt*
JFIDS J-5 Force Structure Information Display System *mt*
JFK John Fitzgerald Kennedy International Airport, New York *aw130.11.6, mj41.5.48*
JFLCC Joint Force Land Component Commander [USA DOD] *mt*
JFM Joint Force Memorandum *mt, wp, hc*
JFMCC Joint Force Maritime Component Commander [USA DOD] *mt*
JFMIP Joint Financial Management Improvement Project *wp*
JFMSES Joint Frequency Management and Spectrum Engineering System [PACOM] *mt*
JFP Joint Frequency Panel [USMCEP] *mt*
JFS Jet Fuel Starter *wp*
JFS Journaled File System [IBM] *kg, hi*
JFSL JFS Log *hi*
JFSOCC Joint Force Special Operations Component Commander [USA DOD] *mt*
JFV Jobs For Veterans [National Committee] *dt*
JG Jagd Geschwader [fighter wing, as in JG 71 Richthofen, JG 74 Molders] [Germany] *mt*
JG Jump if Greater *pk, kg*
JGD JOPES Global Dictionary *mt*
JGE Jump if Greater or Equal *pk, kg*
JGOFS Joint Global Ocean Flux Studies *hg, ge*
JGR Journal of Geophysical Research *mi*

JGSDF Japanese Ground Self-Defense Forces *hc, fu*
JHMET Joint Healthcare Management Engineering Team *af72.5.109*
JHN Japan Heli-Network *aw130.6.56*
JHSU Joint Helicopter Support Unit *mt*
JHU [Hughes Aircraft Company customer for ALBIS] *hc*
JIAFS Joint Institute for Advancement of Flight Sciences *aa26.11.55*
JIAWG Joint Integrated Avionics Working Group *rm8.2.5*
JIC Joint Industrial Counsel *fu*
JIC Joint Industry Conference *fu*
JIC Joint Information Center *mt*
JIC Joint Intelligence Center [USA DOD] *mt*
JIC Joint Intelligence Committee *wp, kf*
JIC Joint Intelligence Curriculum *mt*
JICST Japan Information Center for Science and Technology *mt*
JIDL JFAAD Instrumentation Data Link *fu*
JIEO Joint Interoperability [and] Engineering Organization *mt*
JIEP Joint Intelligence Estimate For Planning *mt, wp*
JIES Joint Interoperability Evaluation System [JTCCCA] *fu, hc, mt*
JIF Joint Interrogation Facility *mt*
JIF Joint Interrogation Vehicle [USA DOD] *mt*
JIFDATS Joint Services In-Flight Data Transmission System *hc*
JILA Joint Institute for Laboratory Astrophysics *mi*
JILE Joint Intelligence Liaison Element *mt*
JILO Joint Information Liaison Office *wp*
JILSMT Joint Integrated Logistics Support Management Team *nt*
JILSP Joint Integrated Logistic Support Plan *nt*
JIMA JDS Interface Method of Access [JDA] *mt*
JIMPACS Joint Improved Multimission Payload Aerial Surveillance Combat Survivable [system] *aw130.22.55*
JINTACCS Joint Integrated Tactical Command and Control System *fu, wp*
JINTACCS Joint Interoperability of Tactical Command and Control Systems *mt*
JIOP Joint Interface Operating Procedure *fu*
JIP Job Improvement Plan *wp*
JIP Joint Identifier Program [OJCS] *mt*
JIP Joint Interface Plan *hc*
JIPC Joint Imagery Production Complex *mt*
JIPD JINTACCS Interoperability Planning Document *mt*
JIPG Joint Interoperability Planners Group *mt*
JIR JOPES Information Requirements *mt*
JIS Jamming Influence Sector *fu*
JIS Japan Industrial Standards *hi*
JIS Joint Interoperability System *mt*
JIS JOPS Interim System *mt*
JISR Joint Information Search and Retrieval [OJCS] *mt*
JISS Japan Intelligence Support System *mt*
JITC Joint Interoperability Test Center *kf, mt*
JITC Joint Interoperability Test Command *kf*
JITF Joint Interface Task Force *tb, mt*
JITS JINTACCS Interoperability Test System *fu*
JITS Joint Interface Test System *mt, fu*
JIU JTIDS Interface Unit *fu*
JJ Josephson Junction *st*
JJD Josephson Junction Device *fu*
JKMS Joint Key Management System *mt*
JLC Joint Logistic Commander *mt, fu, wp*
JLREID Joint Long Range Estimation Intelligence Document *mt*
JLRPG Joint Long Range Proving Ground [later CCAFS] *ci.85, ai2*
JLRSA Joint Long Range Strategic Assessment *mt*
JLRSS Joint Long Range Strategic Study [USA] *fu, wp, mt*
JLSP Joint Logistics Support Plan *mt*
JLWG Joint Lethality Working Group *st*
JM Joint Mission *mt*
JMA Japan Meteorological Agency *hc, bl3-1.155*
JMA Joint Mission Application [WIS] *mt*
JMA Joint Mobilization Augmentation *mt*
JMAC Joint Munitions Allocation Committee *wp*
JMAS Joint Mission Application Software *mt*
JMC Joint Maritime Course *mt*
JMC Joint Message Center *wp*
JMC Joint Movement Center *mt*
JMCC Joint Movement Control Center *mt*
JMCIS Joint Maritime Command Information System [GCCS] [USA] *mt*
JMD JOPES Management Division [OJCS/J-3] *mt*
JMDS Joint Manpower Data System [OJCS] *mt*
JMEM-SO Joint Munitions Effectiveness Manual-Special Operations [USA DOD] *mt*
JMENS Joint Mission Element Need Statement *mt*
JMFU Joint Force Meteorological And Oceanographic Forecast Unit [USA DOD] *mt*
JMH Joint Mission Hardware [WIS] *mt*
JMHC Joint Mission Hardware Contractor [WIS] *mt*
JMMC Joint Military Medical Command *mt, af72.5.71*
JMO Joint Force Meteorological And Oceanographic Officer [USA DOD] *mt*
JMO Joint Maritime Operations *mt*

JMOA Joint Memorandum Of Agreement *mt, nt*
JMP Joint Manpower Program [USA DOD] *wp, mt*
JMP Joint Mission Processor [WIS] *mt*
JMPAB Joint Materiel Priorities [and] Allocation Board [JCS] *wp, mt*
JMPE Joint Mission Processing Environment [WIS] *mt*
JMPE Joint Mission Processing Equipment [WIS] *mt*
JMPO Joint MILSTAR Program Office *mt*
JMPP Joint Munitions Production Panel *wp*
JMRO Joint Medical Regulating Office *mt*
JMS Joint Mission Software [WIS] *mt*
JMSDF Japan(ese) Maritime Self Defense Force *mt, fu, hc*
JMSNS Justification of Major System New Start *mt, hc, wp, st*
JMSPO Joint Meteorological System Project Office *hc*
JMSW Joint Mission Software Subsystem [WIS] *mt*
JMSWG Joint Message Standard Working Group [JTIDS] *mt*
JMT Joint Management Team *fu, wp*
JMTB Joint Military Transportation Board *fu*
JMTG Joint Military Task Group *wp*
JMTSS Joint Multiple Channel Trunking and Switching System *mt*
JMUSDC Joint Mexican - US Defense Commission *dt*
JMVA Joint Meritorious Unit Award *mt*
JN Job Number *kf*
JNA Jump if Not Above *pk, kg*
JNACC Joint Nuclear Accident Coordinating Center *mt*
JNAE Jump if Not Above or Equal *pk, kg*
JNAT Jacchia-Nicolet Atmospheric (model) *sa12205.2-14*
JNB Jump if Not Below *pk, kg*
JNBE Jump if Not Below or Equal *pk, kg*
JNCPO National Climate Program Office *ag*
JNDI Java Naming Directory Interface *ct, em*
JNET Japanese Network *ct*
JNG Jump if Not Greater *pk, kg*
JNGE Jump if Not Greater or Equal *pk, kg*
JNIDS Joint National Intelligence Distribution Staff *fu*
JNLE Jump if Not Less or Equal *pk, kg*
JNO Jump if No Overflow *pk, kg*
JNP Jump if No Parity *pk, kg*
JNR Jammer to Noise Ratio *hc*
JNS Jump if No Sign *pk, kg*
JNWEB Joint New Weapons and Equipment Board *ci.36*
JOA Joint Operating Agency *mt*
JOA Joint Operations Area *mt*
JOC Joint Operational Community *mt*
JOC Joint Operations Center *mt, fu, hc*
JOCAR Joint Communications Allocation Requirement *mt*
JOCAS Job Order Cost Accounting System *wp*
JOCF Joint Operations Capability File *mt*
JOCS Job Order Control System *fu*
JOCS Joint Operational Climatological Support *mt*
JOD Joint Operations Division *mt*
JOD Journal of Development *mt*
JOG Joint Operations Graphics *hc, mt*
JOG Joint Operations Group *dt*
JOI Joint Oceanographic Institutions, [Inc.] *hy*
JOIN Joint Optical Information Network *hc*
JOISTS Joint Operational Interface Simulation Training System *mt*
JOOD Junior Officer Of the Deck [US Navy] [USA] *mt*
JOOW Junior Officer Of the Watch [US Navy] [USA] *mt*
JOP Job Opportunities Program *sc3.4.8*
JOP Joint Operating Procedure *mt, fu, wp*
JOP Jupiter Orbiter Probe [later known as Galileo] *hc, ci.806*
JOPES Joint Operation Planning and Execution System [USA DOD] *mt*
JOPES Joint Operational Planning and Execution System *fu, wp*
JOPES_ROLSTK JOPES Rolling Stock Summary Report *mt*
JOPREP Joint Operational Reporting System [JCS] *mt*
JOPS Joint Operations Planning System *mt*
JOPSREP JOPS Reporting System [JRS] *mt*
JOPSRSS JOPS Rolling Stock Sum System [USAF] *mt*
JOPS_III Joint Operations Planning Software Support System *mt*
JOPS_ROLSTK JOPS Rolling Stock Summary Report [MAC] *mt*
JOR Joint Operational Requirement *mt, fu*
JOR Joint Operations Room *fu*
JORAC Joint Operations Radar Airspace Control [Muenster, GE] *mt*
JOSAF Joint Operations Support Activity Frankfurt *-10*
JOSDEPS Joint Strategic Defense Concept Plan *mt*
JOSE Joint Optics Structures Experiment *hc*
JOSS Johnniac Open Shop System [Rand Corp.] *pk, kg*
JOSS Joint Overseas Switchboard Switch *mt*
JOT&E Joint Operational Test and Evaluation *mt, hc*
JOTR Joint Operational Technical Review *wp*
JOTS Joint Operational Tactical System *mt, fu, kf*
JOTS Joint Operational Telephone System *mt*
JOVIAL Jule's Own Version of the International Algorithmic Language [computer programming language] *ct, ro, fu, wp*
JOVIAL Jules Own Version of the International Algebraic Language [a computer language] *mt*

JP Jet Petroleum *wp*
JP Jet Propulsion [may be used to designate different grades of jet fuel] *w9, mt*
JP Joint Publication [USA DOD] *mt*
JPA Job Performance Aid *fu*
JPAM Joint Program Assessment Memorandum *mt, wp*
JPAO Joint Public Affairs Office *mt*
JPAPS Job Performance Aids Production System *mt*
JPATS Joint Primary Aircraft Trainer System [a basic trainer to replace the T-37 and T-34C, USA] *wp, mt*
JPB Joint Planning Board *wp*
JPCD Joint Policy Coordinating Group {?} *fu*
JPCG Joint Policy Coordinated Group *mt*
JPDS Joint Petroleum [POL] Data System [JCS] *mt*
JPE Jump if Parity Even *pk, kg*
JPEC Joint Planning and Execution Course *mt*
JPEC JOPES Planning and Execution Community *mt*
JPEDB Joint Planning and Execution Data Base *mt*
JPEG Joint Photographers (Photographic) Experts Group [still image file on net] *jr, ct, kg, hi, du*
JPEG Joint Planning and Execution Graphics *mt*
JPESC Joint Planning and Execution Steering Committee *mt*
JPF Joint Frequency Panel *mt*
JPG Joint Photographic [experts] Group [also JPEG] *hi*
JPG JOPES Project Group *mt*
JPGM Joint Planning Graphics Module *mt*
JPL Jet Propulsion Laboratory [operated by CIT, Pasadena, California] *sc4.11.2, 4, 13, 15, 16, pl, aa26.12.b3, sf31.1.1, es55.89, aw130.11.49, hc, wp, so, ns, mb, no, ep, hy, re, tb, wl, jdr, hg, kf, mb, ge, kg, mi, fu, oe, mt*
JPM Joint Program Manager *mt, nt*
JPMO Joint Program Management Office *mt, dt*
JPMR Joint Program Management Review *wp*
JPO Joint Program (Project) Office(r) *mt, fu, wp, aa26.11.10, af72.5.27, ts4-6, nt, tb, fu*
JPO Jump if Parity Odd *pk, kg*
JPOC Joint Planning Orientation Course *mt*
JPOP Japanese Polar Orbiting Platform *hy, hg, ge*
JPP Joint Planning Process *mt*
JPPSOWA Joint Personal Property Shipping Office *dt*
JPR Job Performance Requirement *kf*
JPT Jetpipe Temperature *mt*
JPT Joint Participation Team *fu*
JPTDS Junior Participating Tactical Data System *fu, hc*
JPTM Joint Procedures Training Manual *mt*
JPU JTIDS Participant Unit *fu*
JPWG Joint Projects Working Group *mt*
JP_COMNET Joint Pacific Command Teletype Network *mt##*
JP_VOICE Joint Pacific Command Control Voice Network *mt##*
JQS Job Qualification Standard *mt*
JQT Job Qualification Training *kf*
JR Jam Resistance *hc*
JRA Joint Rear Area *mt*
JRAC Joint Rear Area Coordinator [USA DOD] *mt*
JRADS Joint Resource Assessment Data Base System [JCS] *mt*
JRC Joint Reconnaissance Center *mt, 10, wp, dt*
JRC Joint Reporting Center *mt*
JRCC Joint Rescue Coordination Center *mt, fu*
JRCI Jamming Radar Coverage Indicator *fu*
JRD Japan Reconfiguration and Digitization *mt*
JRDB Joint Research and Development Board *ci.36*
JRDL Jam Resistant Data Link *hc, fu*
JRDOD Joint Research and Development Objectives Document *mt, wp, hc, fu*
JRE Joint Readiness Exercise *mt*
JRIM Joint Requirements Integration Manager [OJCS] *mt*
JRMB Joint Requirements [and] Management Board [OSD] *wp, mt*
JRMET Joint Reliability and Maintainability Evaluation Team *mt, wp, hc, kf*
JROC Joint Requirements Oversight Council *mt, wp, kf*
JRS Job Routing and Standards *fu*
JRS Joint Reporting Structure [JCS Pub 6] *mt*
JRSC Jam Resistant Secure Communication(s) *mt, wp, kf*
JRTC Joint Readiness Training Center [TRADOC] *wp, mt*
JRTOC Joint Rear Tactical Operations Center [USA DOD] *mt*
JRU JTIDS Relay Unit *fu*
JRX Joint Readiness Exercise *mt*
JS Joint Services *wp*
JS Jump if Sign *pk, kg*
JSA Jammer System Analysis *wp*
JSAM Joint Security Assistance Memorandum *mt, wp*
JSAP Joint Security Assistance Planning *mt*
JSARC Joint Search and Rescue Center *hc, fu*
JSASS Japan Society for Aeronautical and Space Sciences *aa26.10.b4*
JSAT Joint System Acceptance Test *mt*
JSAVLA Joint Services Advanced Vertical Lift Aircraft *wp*
JSC Joint Spectral Center *jdr*

JSC Joint Steering Committee *mt*
JSCAMPS Joint Service Common Airframe Multipurpose System *aa26.11.50*
JSCAN Joint Chiefs of Staff Alerting Network *jb*
JSCCB Joint Services Configuration Control Board *wp*
JSCE Joint Services Communications Element *mt*
JSCP Joint Strategic Capabilities Plan [USA] *mt, wp, jb, fu*
JSCS Joint Strategic Connectivity Staff *mt*
JSD Jam Strobe Detector *fu*
JSDF Japanese Self Defense Forces *mt*
JSDOP Joint Strategic Defense Operations Plan *mt*
JSE JOPES Support Element *mt, fu*
JSEAD Joint Suppression of Enemy Air Defenses *mt*
JSIC Joint Space Intelligence Center *go1.2-1, jb, cp, jmj, as*
JSIDS Joint Service Intrusion Detection System *wp*
JSIIDS Joint Services Interior Intrusion Detection System *hc*
JSIL Joint-Stars System Integration Lab *fu*
JSIPS Joint Services Imagery Processing System *mt, aw126.20.125, pf.f88.26, wp*
JSIPS Joint Strategic Integration Planning Staff *mt*
JSIT JDS Information Trace *mt*
JSL Job Shop Labor *-8*
JSLC Jiuquan Satellite Launch Center [China] *ai2*
JSNPE Joint Strategic Nuclear Planning Element *mt*
JSO Joint Service Office *dt, hc*
JSO Joint Specialty Officer / joint specialist [USA DOD] *mt*
JSOA Joint Special Operations Agency [OJCS] *dt, mt*
JSOA Joint Special Operations Area [DOD USA] *mt*
JSOACC Joint Special Operations Air Component Commander [USA DOD] *mt*
JSOC Joint Special Operations Command (Commander) *mt*
JSOF Joint Special Operations Force *mt*
JSOP Joint Strategic Objectives Plan [replaced by JSPD] [USA] *wp, fu, mt*
JSOR Joint Services Operational Requirement *mt, hc, st, fu*
JSOR Joint Services Organizational Requirements *wp*
JSOR Joint Statement of Requirement *mt*
JSOR Joint Systems Operational Requirements [USA] *mt*
JSORD Joint Systems Operational Requirements Document [requirements published by the USAF and USN for the JPATS program] [USA] *mt*
JSOSE Joint Special Operations Support Element [USREDCOM] *mt*
JSOTF Joint Special Operations Task Force [USA DOD] *mt*
JSP Jam Strobe Processor *fu*
JSP Joint Support Plan *mt*
JSPD Joint Strategic Planning Document *mt, wp, tb*
JSPDSA Joint Strategic Planning Document [and] Supporting Analyses *wp, mt*
JSPO Joint Systems Program Office [NEXRAD] *ag, hc, fu*
JSPOC Joint ASAT Planning and Operations Center *go3.3*
JSPS Joint Strategic Planning System [USA DOD] *wp, mt*
JSR Joint Status Review *wp*
JSRAAM Joint Short Range Air to Air Missile *hc*
JSRC Joint Services Review Committee *hc*
JSRR Jamming to Signal Ratio Required *mt*
JSRT Joint Short Range Technology *wp*
JSS Joint Services [Imaging] Seeker *hc*
JSS Joint Surveillance Strategy *jdr, kf*
JSS Joint Surveillance System *mt, hc, hn41.18.1, 12, go1.4.1, wp, tb, fu*
JSSC Joint Space Surveillance Center *go1.2-1*
JSSEE Joint Service Software-Engineering Environment *mt*
JSSG Jamming Signal Source Generator *hc*
JSSIS Joint Staff Support Information System [OJCS] *mt*
JSSPO Joint Surveillance System Project Office *fu*
JST Japanese Standard Time *pk, kg*
JSTARS Joint Surveillance and Target Acquisition Radar (Reconnaissance) System *wp, uc*
JSTARS Joint Surveillance Tracking (and Target) Attack Radar System [USA] *mt, fu, kf, hc, tb*
JSTC Joint Services Test Command *hc*
JSTP Joint System Test Program *mt, wp*
JSTPS Joint Strategic Target Planning Staff *mt, wp, tb*
JSTRC Joint Services Telecommunications Requirements Contract *mt*
JSWG Joint (Navy/Contractor) Safety Working Group *fu*
JSWG Joint Science Working Group *so*
JT JTIDS Terminal *fu*
JT&E Joint Test and Evaluation *mt, ph*
JTA Joint Tactical Architecture *kf*
JTAC Joint Technical Advisory Committee *wp*
JTACMS Joint Tactical Cruise Missile System [USAF/USA] *wp, hc, mt*
JTAG Joint Test Action Group *is26.1.51*
JTAGS Joint Tactical Ground Station *kf*
JTAMCS Joint Tactical Cruise Missile System {cf. JTACMS} *ph*
JTAMCS Joint Tactical Missile System {cf. JTACMS} *dh*
JTAO Joint Tactical Air Officer *fu*
JTAO Joint Tactical Air Operations *mt*
JTAW Joint Tactical Autonomous Weapons *wp*

JTB Joint Targeting Board *mt*
JTB Joint Transportation Board [JCS] *wp, mt*
JTC Joint Technical Committee *ch*
JTC Joint Technology Center *wp*
JTCB Joint Targeting Coordination Board [USA DOD] *mt*
JTCC Joint Tactical Command and Control *mt*
JTCC Joint Test Coordinating Committee *wp*
JTCCC-CDBS Joint Tactical Command, Control, and Communications - Central Data Base System *mt*
JTCCCA Joint Tactical Command, Control, and Communications Agency [later JIEO] *tb, mt*
JTCG Joint Technical Coordinating Group *mt, wp*
JTCG-ME Joint Technical Coordinating Group for Munitions Effectiveness [USA DOD] *mt*
JTCG/AS Joint Technical Coordinating Group for Aircraft Survivability *mt*
JTCG/AS Joint Technical Coordination Group on Aircraft Survivability *wp*
JTCG/EW Joint Technical Coordinating Group For Electronic Warfare *mt*
JTCG/ME Joint Technical Coordination Group for Munitions Effectiveness [USA DOD] *mt*
JTCO Joint Tactical Communications Office *mt*
JTD Joint Tables of Distribution [Manning Document] *mt*
JTD Joint Test Director *mt*
JTDE Joint Technology Demonstrator Engine [USA] *wp, mt*
JTDP Joint Technical Development Plan *wp*
JTE Joint Technical Evaluation *wp*
JTE Joint Test Element *mt*
JTECH Japanese Technology Evaluation Database *mt*
JTENS Joint Tactical Exploitation of National Systems *mt*
JTF Joint Tactical Fusion *fu*
JTF Joint Task Force [as in JTF-7 Automation Support] [USREDCOM] *mt, fu, hc, wp, kf, 12*
JTF Joint Test Force *ts17-11*
JTF-AK Joint Task Force - Alaska *mt*
JTF-AL Joint Task Force - Aleutians *mt*
JTF/LOCE Joint Tactical Division / Limited Operational Capability Europe *mt*
JTFAK Joint Task Force Alaska *-12*
JTFP Joint Tactical Fusion Program *mt, ph, wp*
JTFPMO Joint Tactical Fusion Program Management Office *dt*
JTFS Joint Tactical Fusion System *tb*
JTFTB-A Joint Tactical Fusion Test Bed-A *fu*
JTG Joint Test Group *fu*
JTIDS Joint Tactical Information Distribution System *fu, hn41.18.1, ph, hq5.2.5, hc, hc, af72.5.139, ts17-3, aw130.22.77, wp, pm, tb, kf*
JTIDS Joint Tactical Information Distribution System [US] *wp*
JTIDS Joint Tactical Integrated Display System [a tactical display that combines information from the aircraft's own equipment with information from an AWACS, ground stations or ships] [UK/USA] *mt*
JTIDS Joint Tactical Interoperable Data System *pf.f88.10*
JTIP Joint Technology Insertion Program *wp*
JTIRS Japanese Technical Information Research Service *mt*
JTL Joint Target List *mt*
JTLS Joint Theater Level Simulation (Simulator) *mt, wp*
JTMD Joint Theater Missile Defense *mt, wp*
JTMLS Joint Tactical Microwave Landing System *mt*
JTMPO Joint Tactical Missile Project Office *st*
JTN Joint Targeting Network *mt*
JTP Joint Theatre Plan *fu*
JTP Jolt Transaction Protocol *em*
JTR Joint Travel Regulation *mt*
JTRB Joint Telecommunications Resource Board *mt*
JTS Java Transaction Services *ct, em*
JTS Joint Training Squadron [USA] *mt*
JTSA Joint Technical Support Activity *dt*
JTSC Joint Technical Steering Committee *hc*
JTSG Joint Targeting Steering Group *mt*
JTSH Joint Threat Simulator Handbook *mt*
JTTP Joint Tactics, Techniques and Procedures [USA] *wp, mt*
JTWC Joint Typhoon Warning Center *pl*
JU Joint Use *wp*
JU JTIDS Unit *fu*
JUA Joint Usage Agreement *mt*
JUD Jam Until Destroyed *mt*
JUF Joint Users File *mt*
JUG JOPES Users Group *mt*
JUG JOVIAL-Ada User's Group *fu*
JUGHEAD Jonzy's Universal Gopher Hierarchy Excavation And Display [Internet] *kg, hi*
JUH-MTF Joint User Handbook for Message Text Formats *mt*
JULLS Joint Universal Lessons Learned System *mt*
JUMPS Joint Uniform Military Pay System [DOD] *wp, mt*
JUNET Japan Unix Network *hi*
JUSMAAG Joint U. S. Military Assistance Advisory Group *hc*
JUSMAG Joint U. S. Military Advisory Group *wp*

JUSMG Joint U. S. Military Group *hc*
JUSMMAT Joint U. S. Military Mission Aid and Training *hc*
JUSSC Joint United States Strategic Committee *wp*
JUTCPS Joint Uniform Telephone Communications Precedence System *mt*
JUWTF Joint Unconventional Warfare Task Force *mt*
JV Journal Voucher [cost transfer] *fu, ah, 8, hc, jdr, pbl*
JVIDS Joint Visually Integrated Display System *mt*
JVX Joint Vertical Lift Airlift [CV-22] *mt*
JVX Joint [Services Advanced] Vertical [Takeoff and Landing {aircraft}], Experimental *ph, hc, af72.5.145, wp, uc*
JWC Joint Warfare Center [USREDCOM] *mt*
JWDSC JOPES / WIS Data Standardization Committee *mt*
JWG Joint Working Group *mt, wp, aa26.12.34, ci.860, so, st*
JWICS Joint Worldwide Intelligence Communications System [USA DOD] *mt*
JWIDS Joint Worldwide Interoperability Demonstration System *mt*
JWIS Joint WWMCCS Information System *mt*
JWR Joint War Room *mt, wp*
JX Jammer {transmitter} *wp*
JZ Jump if Zero *pk, kg*

K

K [in operations date codes, K-day is the basic date for the introduction of a convoy system on any particular convoy lane. Compare D-day; M-day] [USA DOD] *mt*
K/J Keyboard/Joystick *fu*
K/P Keyboard/Printer *fu*
KA Kill Assessment (Analysis) *tb, st, kf, fu*
KAAO Korean Air Area of Operations *mt*
KABLE Kennedy Space Center Atmospheric Boundary Layer Experiment *wp*
KAC Key Access Code *mt*
KADIZ Korean Air Defense Identification Zone *fu*
KADS Korean Air Defense Sector *mt*
KADS Korean Air Defense System *mt*
KAF Kuwait Air Force *fu*
KAI Kurzweil Applied Intelligence *hi*
KAIS Korean Air Intelligence System *mt*
KALCC Korean Airlift Control Center *mt*
KAMO Korean Airlift Management Office *mt*
KAO Kuiper Airborne Observatory *st, mb, mi*
KAPSE Kernel ADA Program(ming) Support Environment *fu, mt*
KARIC Korean All-Resource Intelligence Center *fu*
KaSA Ka-band Single Access *vv*
KaSAR Ka-band Single Access Return *vv*
KASC Knowledge Availability Systems Center *ci.1005*
KASMS Korean Air Support Management System *mt*
KASS Kuwait Automated Support System *fu*
KATC Korean Air Traffic Control *fu*
KATE Knowledge Based Automatic Test Equipment *tb*
KATO KAW ASAT Targeting Optimizer *tb*
KB Konstruktorskoye Byuro [Design Bureau] [USSR] *mt, ai2*
KB/S Kilobits per Second *mt, tt*
KB/SEC KiloBit(s) per Second *vv, kf*
KBA Killed By Air *wp*
KBE Knowledge Based Engineering *pbl, jdr, ah, jn*
KBI Kill Before Intercept *hc*
KBIT Knowledge Based Intelligent Tracking *wp*
KBL Kill Before Launch *hc*
KBOR Kills Before Ordnance Release *fu*
KBPS Kilobits Per Second *bo, 4, so, mb, vv, kf, kg, ct, uh, hi, oe*
KBPS Kilobytes Per Second *wp, wl, em, kg, hi*
KBS Knowledge Based System *wp, fu*
KBSA Knowledge Based Software Assistant *hc, wp*
KBTS Ku Band Transmitter System *hc*
KBU Keyboard Unit *fu*
KC [designator, KC-10 is Extender, KC-135 is DOD Stratotanker] [USA] *wp, mt*
KCC KREMS Control Center *-17*
KCI Keyword Code Index *fu*
KCMX Keyset Central Multiplexer *fu*
KCOIC Korean Combat Operations Intelligence Center *hc, mt*
KCP Keyer Control Panel *fu*
KCP KG Control Panel *fu*
KCRCC Korean Combined Rescue Coordination Center *mt*
KCSS Korean Combat Support System *mt*
KD Kilo Digit *mt*
KDC Key Distribution Center *mt*
KDD K-band DG-1 Demodulator *-4*
KDF Kongelige Danske Flygvabnet ["Royal Danish Air Force"] [Denmark] *mt*
KDI KDI Score [supplier of AMRAAM thermal batteries] *-17*
KDM Key Decision Memorandum *tb*
KDPP Keyboard/Display/Printer/Punch [Hughes] *hc*

KDT Key Definition Table *pk, kg*
KDT Keyboard Display Terminal *mt*
KDU Keyboard Display Unit *mt, fu*
KE Kinetic Energy *mt, jb, wp, nt, tb, st, as*
KEAS Knots Estimated Air Speed *wp*
KED Kill Enhancement Device *nt, jmj, tb*
KED Knife Edge Diffraction *fu*
KEE Knowledge Engineering Environment *fu*
KEFIR Key Findings Reporter [GTE] *pk, kg*
KELV KKV ESM Launch Vehicle *5, 17, sb*
KEM Kinetic Energy Missile *hc, wp*
KEMA [Netherlands national utility company, Arnhem] *is26.1.58*
KERMA Kinetic Energy Released in Materials *wp*
KERMIT [Mainframe/Micro, Micro/Micro Protocol, named after a frog character in a television series] *mt*
KES Kwajalein [Missile Range] Engagement Simulation *st*
KESFF Kinetic Energy Self-Forging Fragments *st*
KESLOC Kilo [thousands of] Equivalent Source Lines Of Code *kf*
KEV Kilo Electron Volts *wp, mb, eo, oe*
KEV Kinetic Energy Vehicle *tb, st*
KEW Kinetic Energy Weapon [later SPI or SPICV] *mt, 1. 5, wp, 17, sb, hc, ts4-8, jb, st, fu*
KEWB Kinetic Experimental Water Boiler *hc*
KEWC Kinetic Energy Weapon, Chemically propelled *tb*
KEWE Kinetic Energy Weapon, Electromagnetically propelled *tb*
KEWG Kinetic Energy Weapon, Ground *tb*
KEWO Kinetic Energy Weapon, Orbital [later SBI] *tb*
KEYBBE Foreign language KEYBoard program - Belgium *pk, kg*
KEYBBR Foreign language KEYBoard program - Brazil *pk, kg*
KEYBCF Foreign language KEYBoard program - Canadian-French *pk, kg*
KEYBCZ Foreign language KEYBoard program - Czechoslovakia[Czech] *pk, kg*
KEYBDK Foreign language KEYBoard program - Denmark *pk, kg*
KEYBFR Foreign language KEYBoard program - France *pk, kg*
KEYBGR Foreign language KEYBoard program - Germany *pk, kg*
KEYBHU Foreign language KEYBoard program - Hungary *pk, kg*
KEYBIT Foreign language KEYBoard program - Italy *pk, kg*
KEYBLA Foreign language KEYBoard program - Latin America *pk, kg*
KEYBNL Foreign language KEYBoard program - Netherlands *pk, kg*
KEYBNO Foreign language KEYBoard program - Norway *pk, kg*
KEYBPL Foreign language KEYBoard program - Poland *pk, kg*
KEYBPO Foreign language KEYBoard program - Portugal *pk, kg*
KEYBSF Foreign language KEYBoard program - Swiss-French *pk, kg*
KEYBSG Foreign language KEYBoard program - Swiss-German *pk, kg*
KEYBSL Foreign language KEYBoard program - Czechoslovakia [Slovak] *pk, kg*
KEYBSP Foreign language KEYBoard program - Spain *pk, kg*
KEYBSU Foreign language KEYBoard program - Finland *pk, kg*
KEYBSV Foreign language KEYBoard program - Sweden *pk, kg*
KEYBUK Foreign language KEYBoard program - United Kingdom *pk, kg*
KEYBUS Foreign language KEYBoard program - United States *pk, kg*
KEYBYU Keyboard program [foreign language] for Yugoslavia *pk, kg*
KEYSTREAM Output of Crypto Device *mt*
KFL Key Facilities List *mt*
KFLOPS Thousand Floating Point Operations Per Second *fu*
KFOS Korean Fragmentary Order System *mt*
KFP Korean Fighter Program [South Korea] *mt*
KG Key Generator [Crypto] *fu, mt*
KG Kilo Gauss *wp*
KGB Komitet Gosudarstvennoy Bezopastny [Committee for State Security] [secret police of the former Soviet Union] *mt, su, fu, ph, 10, hc, w9, wp, tb*
KGPS Kilograms Per Second *w9*
KGSG KELV Ground Support [working] Group *-17*
KHA Killed by Hostile Action *wp*
KHB Kennedy Handbook *mi*
KHI Kawasaki Heavy Industries *hc*
KHILS Kinetic Kill Vehicle Hardware In the Loop Simulation *5, 17, sb, hc, cp*
KHIT Kinetic Kill Vehicle Hover Integrated Test [Station] *-17*
KIA Killed In Action [USA DOD] *mt, w9, wp, af72.5.13*
KIADS Korean Integrated Air Defense System *hc*
KIAS Knots Indicated Air Speed *mt, aw129.10.38*
KIF Knowledge Interchange Format *kg, ct*
KIM Kinetic Impact Munition *wp*
KIP Korean Industrial Participation *hc*
KIPS Kilo (thousand) Instructions Per Second *wp, fu*
KIPS Organic-Rankine Kilowatt Isotope Power System *ci.898*
KIS Knowbot Information Service [Internet] *pk, kg*
KISS Korean Intelligence Support System *mt, hc*
KIT KAPSE Interface Team *mt, fu*
KITE Kuiper Infrared Technology Experiment *hc*
KITIA KAPSE Interface Team for Industry and Academia *fu, mt*
KITSIM Kit Simulation *fu*
KIWI (series of AEC reactors) {1959} *ci.896, hc*

KKV Kinetic Kill Vehicle *5, wp, 17, sb, hc, tb, st, fu*
KKV-RET Retrodirective Antenna Emulation *-17*
KKVWS Kinetic Kill Vehicle Weapon System *17, sb, tb*
KLA Klystron Amplifier *fu, wp*
KLC Kodiak Launch Complex *ai2*
KLIZ Korean Limited Identification Zone *fu*
KLM Koninklijke Luchtvaart Maatschappij [(Netherlands) Royal Air Lines] *aa26.10.49, aw129.14.111, wp*
KLO Klystron Oscillator *fu*
KLu Koninklijke Luchtmacht ["Royal Air Force"] [Netherlands] *mt*
KMIP Key Management Interface Processor *tb*
KMPS Kilometers Per Second *w9*
KMR Kwajalein Missile Range [later called USAKA] *5, 17, hc, sb, wp, tb, st, fu*
KMU [guidance unit for guided bombs, USA] *mt*
KNL Kongelige Norske Luftforsvaret ["Royal Norwegian Air Force"] *mt*
KNOBS Knowledge Based System [also KBS] [RADC] *mt, wp, tb*
KNP Korean National Police *mt*
KNU Knowledge Utility *ct, em*
KOA Korean Operation Area *mt*
KOC Key Operational Capability *mt*
KOH Potassium Hydroxide *wp, oe*
KOPS Kilo [thousands of] Operations Per Second *hc, sd, wp, fu, hi*
KORBX [an AT&T system for handling large logistics problems] *aw130.11.57*
KOZ Keep Out Zone *oe*
KP (SPICE design parameters . . . such as . . . KP) *ne*
KP/V Keypunch/Verifier *fu*
KPI Kernel Programming Interface *pk, kg*
KPL Key Personnel Locator *mt*
KPNO Kitt Peak National Observatory *no, mb, mi*
KPSM Klystron Power Supply Modulator *fu, 0*
KPT Key Parameter Threshold *kf*
KQML Knowledge Query and Manipulation Language *pk, kg*
KRCTR Kit Requisition Configuration Traceability Record *ep*
KREEP Potassium, Rare Earth Elements, and Phosphorus [-rich rocks] *eo*
KREMS Kiernan Reentry Measurement Site (System) [a complex of KMR radars] *17, st*
KRESS [see CRESS, e.g. Combat Readiness Evaluation Simulation System] *st*
KRF Krypton Fluoride *tb*
KRPL Krylataya Raketa Podvodnykh Lodok [SLCM] [USSR] *mt*
KRS Knowledge based Replanning System *mt*
KRS Knowledge Retrieval System *pk, kg*
KRTR Kit Requisition and Traceability Record *gp, pl, ep, jdr, fu*
KRU Korabl Upravleniye [command ship] [USSR] *mt*
KRVB Krylataya Raketa Vozdushnogo Bazirovaniya [ALCM] [USSR] *mt*
KRZB Krylataya Raketa Zemnogo Bazirovaniya [GLCM] [USSR] *mt*
KS Kit Short *fu*
KS [Directorate of Mission Support, AFDSDC] *mt*
KS-87 [reconnaissance camera, RF-4C] *wp*
KS/SEC Kilosymbol(s) per Second *vv*
KSA K-band Single Access *2, 4*
KSAF KSA Forward *-4*
KSANG Kansas Air National Guard *-12*
KSAR KSA Return *-4*
KSAS KSA Shuttle *-4*
KSC Kagoshima Space Center [Japan] *ai2*
KSC Kaman Sciences Corporation *st*
KSC Kennedy Space Center [NASA] *mt, ci.569, 2, 4, 9, sf31.1.9, es55.42, ws4.10, wp, ns, mb, re, tb, wl, jdr, mb, kf, vv, mi, oe, ai2*
KSH Key Strokes per Hour [also KSPH] *hi*
KSH Korn Shell [program] [Unix] *kg, hi*
KSLOC Kilo [thousands of] Source Lines Of Code *kf*
KSM Keyboard Switch Module *fu*
KSN Kit Serial Number *mt*
KSOS Kernalized Secure Operating System *mt, fu*
KSPH Key Strokes Per Hour [also KSH] *pk, kg*
KSPM Key Strokes Per Minute *hi*
KSPS Kilo Symbols Per Second *re*
KSR Keyboard Send/Receive *mt, kg, fu, hi*
KSRP Keyboard Send/Receive Printer *fu*
KSUB Submarine Broadcast *mt*
KT Key Telephone *fu*
KTA Keep Terminal Alive *uh*
KTA Korean Telecommunication Authority *mt*
KTACS Korean Tactical Air Control System *mt, hc, fu*
KTB [Cretaceous Tertiary Boundary (from German)] *mi*
KTE Key Telephone Equipment *fu*
KTO Kuwait Theater of Operations *mt*
KTR Kwajalein Test Range *kf*
KTS Key Telephone System *fu*
KTTC Keesler Technical Training Center *mt*
KU Koreblnyi Uchebno [shipboard instructional, as in Su-27KU] [USSR] *mt*

KuSA Ku-band Single Access *vv*
KuSAR Ku-band Single Access Return *vv*
KUTLAS KU-Band Tactical Lightweight Avionics System *hc*
KV Kill Vehicle *nt, sb, jmj, tb, st, fu*
KV Kongsburg Vaapenfabrikk [a Scandinavian weapons manufacturer] *fu*
KVA Kill Vehicle Assembly *-17*
KVA Kilo Volt Ampere *mt, aw118.18.70, kf*
KVAN Korean Voice Alerting Network *mt*
KVDT Keyboard Video Display Terminal *mt, fu*
KVDU Keyboard/Visual Display Unit *mt*
KVETS KKV Validation and Evaluation Test System [integrated] *17, sb*
KVI Key Validity Interval *fu*
KWCC Korean Weapons Control Center *fu*
KWCC KTACS Warning and Control Center *mt*
KWE Kingdom-Wide Exercise *fu*
KWIC Key Word In Context *mt, wp, hc, fu*
KWN Korean Wideband Network *mt*
KWNX Keyword Index Subsystem *mt*
KWOC Key Work Out of Context *mt, hc, fu*

L

L [chemical warfare codename for lewisite] *wp*
L [code for time in an operations plan] (L-hour) [USA DOD] *mt*
L&I Launch and Impact [radar missile report] *fu, mt*
L&MM Logistics and Materiel Management *mt*
L&OG Logistics and Operations Group [USA] *mt*
L&TH Lethality and Target Hardness *tb*
L-COM Logistics Composite Model [USAF] *mt*
L-SAT Large [multipurpose] Satellite [European] *sc3.3.4*
L/C Launch Complex *0, tb*
L/C Letter Contract *fu, hc*
L/C Load Crew *mt*
L/D Length to Diameter *fu*
L/D Lift / Drag [ratio] *fu*
L/F Leader/Follower *dh, fu*
L/G {Lieutenant General} *af72.5.31*
L/I Laser Imager *st*
L/I Letter of Intent *hc*
L/L Long Lead *gp*
L/O Layout *hc*
L/O Length to Outside [diameter ratio for missile] *st*
L/O Liftoff *oe*
L/O [Lift to Drag ratio for airplane] *st*
L/PS Laser/Peritelescope System *hc*
L/PSA Logic Power Supply Assembly *jdr*
L/R Liaison and Reporting *fu*
L/R Locus of Radius *fu*
L/V Launch Vehicle *jdr, oe*
L/W Length to Width [Ratio] *fu*
L2F Layer Two Forwarding [Cisco] *kg, em*
L2TP Layer Two Tunneling Protocol *kg, em*
LA Laser Altimeter *re*
LA Lead Angle *fu*
LA Line Adapter *fu, hi*
LA Line Array *fu*
LA Log Analyzer *mt*
LA Logistics Assistance *mt*
LA Look Angle *tb*
LAA Limited Access Authorization *jdr, su*
LAAAS Low Altitude Airfield Attack Systems *hc*
LAADS Low Altitude Air Defense System *hc, fu*
LAAFB Los Angeles Air Force Base *kf*
LAAFS Los Angeles Air Force Station *mt, hc, pl*
LAAM Light Anti Aircraft Missile *hc*
LAAMP Lode Advanced Active Mirror Program *hc*
LAAMS Land Armaments Movement Model *mt*
LAANG Louisiana Air National Guard *-12*
LAAS Low Altitude Airfield Attack System *mt*
LAAS Low Altitude Alerting System *mt, wp*
LAAT Laser Augmented Airborne TOW *hn41.20.4, hc*
LAAV Light Airborne ASW Vehicle *hc*
LAAW Lightweight Assault Antitank Weapon *mt*
LAB Land Air Battle *mt, wp*
LAB Low Altitude Bombing *wp*
LABC Local Automatic Brightness Control *hc*
LABIL Light Aircraft Binary Information Link *fu*
LABP Lethal Aid for Bomber Penetration *hc*
LABS Low Altitude Bombing System [tossing nuclear bombs in a half-loop, returning to low altitude on the opposite direction] [USA] *mt*
LAC Large Anechoic Chamber *wp*
LAC Large Area Counter [on Ginga] *cu*
LAC Lead Angle Compensator *hc*
LAC Leading Air Craftsman *mt*
LAC Local Area Controller *mt*

LAC Local Area Coverage [NOAA] *ge, pl, hg, oe*
LACE Landline Air defense Communications Encryption *fu*
LACE Laser (Airborne) Communications Experiment *hc, wp*
LACE Liquid Air Cycle Engine *hc, ci.255*
LACE Low powered Atmospheric Compensation Experiment *ro*
LACI Land Air Campaign Initiative [USAREUR] *mt*
LACIE Large Area Crop Inventory Experiment *ac1.2-6, tb*
LACP Large Area Coverage Processor *hc*
LAD Laser Acquisition and Direction *hc*
LAD Laser Acquisition Device *hc, wp*
LAD Latest Arrival Date [USA DOD] *mt*
LAD Light Area Defense *fu*
LAD Low Altitude Dispenser *mt*
LADAR Laser Detection And Ranging [Laser Radar] *hc, wp, tb, st, fu*
LADD Lens Antenna Deployment Demonstration *hc*
LADD Low Altitude Drogue Delivery *hc*
LADD Low Altitude Drogue Delivery [dropping nuclear bombs that are retarded by a parachute and have a time fuse] [USA] *mt*
LADDA Layered Device Driver Architecture [Microsoft] *kg, hi*
LADIR Low-Cost Arrays for Detection of Infrared *hc, wp*
LADMIS Low Altitude Air Defense Missile *hc*
LADR Low Altitude Detection Radar *fu*
LADS Laser Air Defense Study *hc*
LADS Light Area Defense System *fu, hc*
LADS Lightweight Air Defense System *hc, uc*
LADS Low Altitude Defense System *hc*
LADSR Low Altitude Defense System Radar *hc*
LADT Local Area Data Transport *hi*
LADTAC Light Area Defense Technical Assistance Control *hc*
LAE Left Arithmetic Element *fu*
LAE Liquid Apogee Engine *jn*
LAEU Launching Adapter Electronic Unit *hc*
LAF List of Affected Files *fu*
LAF Load Alleviation Function *aw126.20.44*
LAFACTS Laser And FLIR Augmented Cobra TOW Sight *hc*
LAFATCS Laser And FLIR Augmented TOW Cobra Subsystem *hc*
LAFS Lightweight Airborne FLIR System *hc*
LAFTS Laser And FLIR Test Sets *hc*
LAG Logical Application Group *mt*
LAGC Instantaneous Automatic Gain Control *fu*
LAGEOS Laser Geodynamics Satellite [measures tectonic plate shifts] *ci.133, hy*
LAH Large Area Hybrid *fu*
LAH Light Attack Helicopter *hc, mt*
LAHETS Laser Heterodyne Sensor [carbon dioxide] *hc*
LAHIVE Low Altitude High Velocity *hc*
LAID Latin American Industrial Development *hc*
LAINS Low Altitude Inertial Navigation System *hc*
LAIRS Laser Imaging and Ranging System *hc*
LAIS Logistics Attrition Information System [TAC] *mt*
LAK Look Alike disk *mt*
LAL Lower Address Line *fu*
LALL Longest Allowed Lobe Length *pk, kg*
LALOC Laser Locator *wp*
LAM Liquid Apogee Motor *js17.5, cl, jdr, kf, vv, he, uh, jn*
LAM Local Area Missile *hc*
LAM Logistics Attrition Model [SAC] *mt*
LAM Long Aerial Mine [an explosive charge, towed behind an aircraft on a 600m line, in hope that enemy aircraft would hit it] [UK] *mt*
LAM Low Altitude Missile *hc*
LAM Lunar Excursion Module [also LEM] *wp*
LAMA Local Automatic Message Accounting *fu*
LAMARS Large Amplitude Multimode Aerospace Research Simulator *hc, wp*
LAMB Low Altitude Multi Burst *st*
LAMBDA Large Aperture Marine Basic Data Array *fu*
LAMFD Linear Analog Matched Filter Device *fu*
LAMINAR Low Altitude, Mapping, Interception, Navigation *hc*
LAMM Land Armament Manpower and Material Data Base [NATO] *mt*
LAMMR Large Antenna Multichannel (Multifrequency) Microwave Radiometer *hc, ac2.7-21*
LAMODX (Look Angle Module (Software Program)) *jb*
LAMP Large [LODE] Advanced Mirror Program *17, wp, sb, ro, fu*
LAMP Laser Modulation Program *hc*
LAMP LIDS Automatic Modulation Program *hc*
LAMP Logistic Assessment Methodology Prototype *wp*
LAMP Low Altitude Manned Penetration [later AMPSS] *hc*
LAMPS Light Airborne Multiple Purpose System [helicopter] [LAMPS I was the SH-2D Seasprite, LAMPS II is the SH-60B Seahawk] [USA] *mt, ph, hc, af72.5.15, pf.f88.25, pf1.2, fu*
LAMPS Logistics Assessment of Modifications Program *mt*
LAMRL Logistic Area Material Readiness List *wp*
LAMS Local Area Missile System *mt*
LAMS Local Asset Management System *hc*
LAN Launch After NUDET
LAN Local Area Network *mt, aw130.13.61, hc, wp, es55.51, ts4-39, tb, pbl, tm, kf, kg, nu, fu, ct, em, ah, hi, oe, tt, y2k, ai1*

LANA 50th Anniversary of Naval Aviation [L means 50 in Roman numerals] [USA] *mt*
LANA Low Altitude Night Attack system [podded system carrier by the A-7, USA] *mt, af72.5.136, wp*
LANAC Lawyers' Alliance for Nuclear Arms Control *-11*
LANACS Local Area Network Asynchronous Connection Server *kg, hi*
LANCC LAN Communications Controller *mt*
LANCC Local Area Network Control Center *mt*
LANCC/SM LAN Control Center and Security Monitor *mt*
LANCE (the 110 km range LANCE missile) *nd74.445.39*
LANDP LAN Distributed Platform *pk, kg*
LANDSAT Land Satellite [NASA's remote sensing, survey or resources] *ci.1, 13, hc, hy, hg, ge, fu*
LANDSOUTH Allied Land Forces Southern Europe [NATO] *mt*
LANDSOUTHEAST Allied Land Forces Southeastern Europe [NATO] *mt*
LANDSS Lightweight Advanced Night / Day Surveillance System [a surveillance RPV program] [USA] *mt*
LANE Local Area Network Emulation [provide interoperability of LAN and ATM] *kg, nu, ct, em*
LANFOX Local Area Network Fiber Optic Transceiver *hc*
LANL Los Alamos National Laboratory *pl, wp, ts4-14, tb, st, mb, cu, mi*
LANM Local Area Network Management *hi*
LANS Local Area Network Server *hi*
LANT Atlantic area [Command] *mt, kf, fu*
LANTCOM Atlantic Command [US Unified Command] *mt, 12, jb, kf*
LANTDAC Atlantic Command Defense Analysis Center *mt*
LANTFLT Atlantic Fleet *mt*
LANTIRN Low Altitude Navigation and Targeting Infra Red [System] for Night [equipment carried by aircraft configured for night attacks] [USA] *mt, 17, sb, hc, hn44, aw129.10.13, af72.5.11, wp, fu*
LAO Logistics Assistance Office *fu*
LAOD Los Angeles Ordnance District [US Army] *hc*
LAP Large Scale Advanced Propeller *aa26.10.17*
LAP Launch Assist Platform [K-1 first stage] *ai2*
LAP Launcher Adaptable Platform *hc*
LAP Launcher Avionics Package *mt, wp*
LAP Line (Link) Access Protocol *jdr, ric, hi*
LAP Link Access Procedure *fu*
LAP Load, Assemble, and Pack *fu*
LAP Low Altitude Performance *wp*
LAP Low Altitude Program *hc*
LAP-B Link Access Procedure (Protocol) - Balanced *mt, ct, oe*
LAPB Link Access Procedure Balanced [protocol] *kg, em, fu, hi*
LAPD Link Access Procedure (Protocol) Direct [protocol] *mt, kg, hi*
LAPE Low Altitude Parachute Extraction [system] *mt*
LAPES Low Altitude Parachute Extraction System [USA] *hc, mt*
LAPG Lightweight Antenna Positioner (Positioning) Group *ric, jdr*
LAPM Link Access Procedure for Modems *ct, em, kg, hi*
LAPS Low Altitude Proximity Sensor *wp*
LAPSS Large Area Pulsed Solar Simulator [Spectrosun] *hc*
LAPT Local Apparent Time *fu*
LAPTAD Laser Application for Precision Tracking/Advanced Discrimination *hc*
LAPTS Lens Array Polarization Transmitter System *fu*
LAR Launch Acceptability (Acceptable) Region *fu, hc, kf*
LAR Limited Access Required *hc, kf*
LAR Load Access Rights *pk, kg*
LAR Local Acquisition Radar *fu, hc, wp, kf*
LAR Logistics Assessment Review *mt, wp, kf*
LARA Light Attack Reconnaissance Aircraft *hc*
LARA Local Acquisition Radar *wp*
LARA Low Altitude Radar Altimeter *wp*
LARAM Line-Addressable RAM *fu*
LARC Langley Research Center [NASA, Hampton, VA] *ac1.2-6, tb, wl, mi*
LARC/TPI (the new polymer is similar to . . . LARC/TPI) *aa26.12.76*
LARF Low Altitude Radar Fuzing *fu*
LARS Laminar Rate Sensor *hc*
LARS Laser Aided Rocket System *hc*
LARS Laser Angular Rate Sensor *hc, wp*
LARS Low Altitude Radar System *wp*
LARTS LOGAIR Real Time Terminal System [AFLC] *wp, mt*
LARVA Low Altitude Research Vehicular Advancements *hc*
LAS Land Analysis Software *ns*
LAS Land Analysis System *ns*
LAS Landsat Assessment System *bl3-3.8*
LAS Launch Auxiliary System *wp*
LAS Light Activated Switch *fu*
LAS Loral AeroSys *oe*
LAS Low Altitude Satellite *hc*
LASA LIDAR Atmospheric Sounder and Altimeter *hy*
LASAM Laser [Anti Tank] Semi Active Missile *hc, wp*
LASAR Logic Automated Stimulus and Response *fu, hc*
LASAR LORAN Aided Search and Rescue *hc*
LASAT Laser Antisatellite Satellite *sd*
LASCA Large Area Solar Cell Array *hc*

LASCR Light Activated Semiconductor (Silicon) Controlled Rectifier *wp, fu*
LASE Laser Atmospheric Sensing Experiment *hy*
LASE Logistic Asset Support Estimate *wp*
LASER [a device using] Light Amplification by Stimulated Emission of Radiation [for its operation] "laser" *mt, hc, kg, wp, tb, hi*
LASERCOM LASER Communication *tb*
LASGAM Laser Semi Active Guided Anti Tank Missile *hc*
LASH Laser Anti Tank Semi-Active Homing [Missile] *hc, wp*
LASHAR Laser System Hardware *hc*
LASHE Altitude Simultaneous HAWK Engagement *fu*
LASINT Laser Intelligence [USA DOD] *mt*
LASL Los Alamos Scientific Laboratory [USA] *hc, mt*
LASO Low Altitude Search Option *fu*
LASORT Laser Airborne Simulated Optical Range Tester *hc*
LASP Low Altitude [Space] Surveillance Platform *10, hc, wp*
LASR Low Altitude Surveillance Radar *hn44.11.1, hc, fu*
LASRM Low Altitude Short Range Missile *hc*
LASRM Low Altitude Supersonic Ramjet Missile *hc*
LASS Land Applications Satellite System *ac3.7-21*
LASS Large Area (Amplitude) Space Simulator *aa26.11.26, wl*
LASS Laser Scanning System *hc*
LASS Launch Area Support Ship [US Navy] *mt, aw130.13.19*
LASS Logistics Activities Status System [JCS] *mt*
LASSM Line Amplifier and Super Sync Mixer *fu*
LASSO (the LASSO experiment for international clock synchronisation) *es55.45*
LAST Laser Aided Search and Track *hc*
LAST Laser SAW Trimming *fu*
LAST Low Altitude Supersonic Target *wp*
LASTport Local Area Storage Transport [protocol] [DEC] *pk, kg*
LASV Low Altitude Supersonic Vehicle [later SLAM] *hc*
LAT Laser Acquisition and Tracking *wp*
LAT Local Access Terminal *kg, hi*
LAT Local Area Transport [DEC] *kg, ct, em*
LATA Local Access and Transport Area *mt, ct, kg, em, hi*
LATAR Laser Augmented Target Acquisition and Recognition [System] *hc, wp*
LATF Low Altitude Tactical Formation *wp*
LATH Long-wavelength Above The Horizon *tb*
LATIS Lightweight Airborne Thermal Imaging System *wp*
LATR Low Altitude Threat Radar *wp*
LATRS Logistic Air Terminal Reporting System *wp*
LATS Lightweight Antenna Terminal Seeker *hc*
LATS Logic Automatic Test Set *fu*
LATS LWIR Advanced Seeker Technology *hc*
LAU Launcher Unit *mt*
LAV Light(weight) Armored Vehicle *mt, uc, hc, wp*
LAV Load Average *ct, em*
LAVC Local Area VAX Cluster *kg, hi*
LAW Light Antitank [rocket] Weapon *aw118.18.69, hc, wp*
LAW Light Armor Weapon *fu*
LAW Light Assault Weapon *wp*
LAWDS LORAN Assisted Weapon Delivery System *hc*
LAWN Local Area Wireless Network *hi*
LAWS Laser Atmospheric Wind Sounder [EOS] *ge, kg, wp, aw130.11.50, hy, hg*
LAWS Laser Weapon Simulator *hc*
LAWS Low Altitude Warning System *mt, wp*
LAX [Los Angeles International Airport] *sc2.7.2, kf, ai2*
LAZ Launchpoint Azimuth *kf*
LB Labor Burden *-8*
LB Line Block *mt*
LB Line Buffer *fu*
LB Local Battery *mt*
LB Low Band *fu*
LBA Linear Boom Actuator *jp*
LBA Logical Block Address (Addressing) *kg, em, ct*
LBA Luftfahrt Bundes Amt ["Federal Air Travel Office"] [Germany] *mt*
LBC Launcher Battery Charger *hc*
LBCT Launch Base Cargo Transporter *oe*
LBD Laser Beam Directors [RADC] *mt*
LBD Lifting Body Development *hc*
LBDS Laser Beam Diagnostic Scanner *hc*
LBEF Land-Based Evaluation Facility *fu*
LBF Linear Beamformer *fu*
LBF Pounds Force *oe*
LBG Local Battle Group *tb*
LBH Light Battlefield Helicopter *mt*
LBI Land Based Interceptor *wp*
LBI Long Baseline Interferometer *wp*
LBIR Laser Beam Image Recorder *ac3.3-21*
LBITS Land-Based Integration and Test Site *fu*
LBITTS Land-Based Integration Test/ Trainer System *fu*
LBL Low Brightness Laser *hc*
LBM Load Buffer Memory *fu*

LBM Local Battle Manager *tb*
LBM Pounds Mass *vv*
LBNP Lower Body Negative Pressure *ci.664*
LBO Launch to Burnout *kf*
LBO Line Build Out *kf*
LBPMR L-Band Pushbroom Microwave Radiometer *hy*
LBPO Lifting Body Program Office *wp*
LBR Laser Beam Recorder *ggIII.1.25*
LBR Laser Beam Rider *hc*
LBR Librarian *pk, kg*
LBRG Laser Beam Riding Guidance *wp*
LBRM Laser Beam Rider Missile *hc*
LBS Laser Bombing System *hc*
LBS Line Block Simulator *mt*
LBS Liquid Bipropellant (Sub)System *3, lt, jdr*
LBSR Lightweight Battlefield Surveillance Radar *hc*
LBT Launch Base Test *ggIII.1.31*
LBT Logical Bus Technology *hi*
LBT Low Band Pass Transformer *hc*
LBTP Launch Base Test Plan *-17*
LBTS Land-Based Test Site (Station) *fu*
LBX Local Bus Accelerator *pk, kg*
LC Inductance Capacitance *fu*
LC Landing Craft *w9*
LC Launch Complex *9, oe, ai2*
LC Launch Contractor *oe*
LC Letter Contract *fu, wp*
LC Limited Capability *jb*
LC Line of Contact *wp*
LC Liquid Chromatography *wp*
LC Liquid Crystal *wp*
LC Load Compensating *fu*
LC Logistic Command *wp*
LC Logistics Center *fu*
LC Low Component *kf*
LC20DL Low Cost 20-MHz Data Link *fu*
LCA Landing Craft, Assault *mt*
LCA Level Compensation Attenuator *uh*
LCA Light Combat Aircraft [indigenous fighter aircraft, under development in India] *wp, mt*
LCA Logic Cell Array *fu*
LCA Logistics Control Activity [USA] *mt*
LCA Lotus Communications Architecture [Lotus] *kg, hi*
LCA Low Cost Aircraft *wp*
LCAC Landing Craft, Air Cushion(ed) [US Navy] *wp, mt, hc, aw118.18.37, aa27.4.b12, ct*
LCAJDL Low Cost Anti Jam Data Link *fu*
LCAMIMS Logistics Capability Assessment Models Information Management System [HQ USAF] *mt*
LCAO Linear Combination of Atomic Orbitals *wp*
LCAP Linear Control [Systems] Analysis Program *id4142-10/831*
LCAP Low Cost Accurate Programmer *hc*
LCAR Low Cost Attack Radar *hc*
LCAT Laser Communications Airborne Testbed *hc, wp*
LCATS Large Capacity Automated Telecommunications System [USAF] *mt*
LCATS Laser Communications Airborne Test Set *hc*
LCB Least Common Bit *wp*
LCB Logic Control Block *wp*
LCB Longitudinal (Position of) Center of Buoyancy *fu*
LCC Amphibious Command Ship [USN ship classification] *fu*
LCC Land Component Commander [USA] *mt*
LCC Launch Command and Control *hc*
LCC Launch Control Center *mt, fu, hc, cp, tb, kf, ai2*
LCC Leadless Chip Carrier *kg, fu, ct, em, hn44.7.6, md 2.1.3*
LCC Life Cycle Cost(ing) *mt, fu, ah, 5, wp, 16, 17, ic2-5, sb, hc, go3.5.2, nt, cp, dm, tb, st, jdr, kf*
LCC Limited Capability Configuration *wp*
LCC Local Console Controller *-4*
LCC Logistics Coordination / Control Center *mt*
LCC {Landing Craft, Command} [amphibious command ship, a naval ship from which a commander exercises control in amphibious operations, USA DOD] *mt*
LCCA Life Cycle Cost Analysis *wp, fu*
LCCC Logistics Command, Control, and Communications *mt*
LCCE Life Cycle Cost Estimate *mt, wp*
LCCM Life Cycle Cost Model *hc, fu*
LCCP Launcher Captain's Control Panel *hc*
LCCP Life Cycle Cost Plan *jdr*
LCCS Life Cycle Contractor Support *mt*
LCD Life Cycle Documentation *fu*
LCD Liquid Crystal Display [hardware] *mt, fu, kg, wp, aw130.13.79, 6, hc, is26.1.61, mj31.5.30, jdr, pbl, tm, ct, em, ah, jn, hi, ai1*
LCD Load Classification Group *mt*
LCD Load Control Device *fu*
LCD Logistics Communication Division *mt*
LCD Lowest Common Denominator *kg, cr, w9*

LCDC Liquid Crystal Display Console *hi*
LCDDS Leased Circuit Digital Data Service *hi*
LCDTL Load Compensated Diode Transistor Logic *fu*
LCE Launch Complex Equipment *fu, 0*
LCE Launch Control Equipment *hc*
LCE Load Circuit Efficiency *fu*
LCE Logistics Capability Estimator *mt*
LCEHV Low Cost Expendable Harassment Vehicle [RPV development program, USA] *mt*
LCES Lightweight Communications Equipment Subsystem *hc*
LCF Last Control Frame *fu*
LCF Launch Control Facility *fu, hc*
LCF Low Cost Fiber *kg, hi*
LCFDU Laser Countermeasure Frequency Double Unit *hc*
LCG Lasar Carrier Grading *fu*
LCG Longitudinal [position of] Center of Gravity *fu*
LCH Light Combat Helicopter *mt*
LCI Landing Craft, Infantry [US Navy] *mt*
LCI Low Cost Inertial *hc*
LCIGS Low Cost Inertial Guidance System *wp*
LCK Library Construction Kit [Microsoft FoxPro] *pk, kg*
LCL Lower Confidence Limit *wp*
LCL Lower Control Limit *fu*
LCLA Liquid Crystal Light Amplifier *fu*
LCLM Low Cost Lightweight Missile [possible Sidewinder replacement, USA] *mt*
LCLV Liquid Crystal Light Valve *hc, 17, fu*
LCM Laboratory Contract Manager *mt*
LCM Land Combat Missile *hc*
LCM Landing Craft, Mechanized [units] [US Navy] *fu, mt*
LCM Laser Countermeasure *wp*
LCM Least (Lowest) Common Multiple *cr, w9, wp*
LCM Life Cycle Management *mt, wp*
LCM Line Control Module *mt*
LCM Logistics Capability Model [USAFE] *mt*
LCM Low Component Module *kf*
LCM Low Cruise Missile *fu*
LCMCC Loosely Coupled Multiprocessing Computer Complex *fu*
LCMD Laser Countermeasures Materials Development *st*
LCMM Life Cycle Management Model *mt*
LCMMD Laser Countermeasures Material Development *hc*
LCMS Laser Cavity Mode Spacing *hc*
LCMS Logistics Capability Measurement System [USAF] *mt*
LCMU Load Current Monitoring Unit *-8*
LCMWG Life Cycle Management Working Group *mt*
LCN Load Classification Number *mt*
LCN Local Communication Network *fu*
LCN Local Computer Network *mt*
LCN Logical Channel Number *fu*
LCN Logistic Control Number *wp*
LCOD Last Cutoff Date *hc*
LCOM Logistics Composite Model [a computer simulation] *af72.5.109*
LCOS Launch Checkout Stations *0*
LCOS Lead Computing Optical Sight *hc*
LCOSS Lead Computing Optical Sight System *wp*
LCP Landing Craft, Personnel [US Navy] *mt*
LCP Left [hand] Circular Polarization *wp, 2, 4, fu, oe*
LCP Line Control Protocol *jdr, ric, kg*
LCP Link Control Procedure *fu*
LCP Liquid Crystal Polymer *aa26.11.42*
LCP Logistic Capability Plan *wp*
LCP Low Cost Processor *hc*
LCPVC Life Cycle Present Value Costs *mt*
LCR Laboratoire Central de Recherche [(French) "Central Research Laboratory"] *wp*
LCR Launch Control Rack *hc*
LCR Launch Control Room *oe*
LCR Line Control Register *pk, kg*
LCR Low Pass Coaxial Relay *hc*
LCRU Lunar Communication Relay Unit *fu*
LCS Labor Collection System *fu*
LCS Laser Communication System *wp*
LCS Laser Cross Section *wp*
LCS Launch Control Shelter *0*
LCS Launch Control System *mt*
LCS Life Cycle Support *mt*
LCS Life Cycle Survivability *fu*
LCS Line of sight Control Subsystem *kf*
LCS Logistics Control System *fu*
LCS Lower Cost Seeker *af72.5.147*
LCSA Life Cycle Support Activity *fu*
LCSE Low Component Systems Engineering *kf*
LCSMM Life Cycle System Management Model *st, fu*
LCSS Land Combat Support System *hc*
LCSS Life Cycle Support Software *fu*
LCT Landing Craft, Tank [US Navy] *mt*

LCT Local Civil Time *w9, fu*
LCT Low Cost Terminal *mt*
LCT50 [vapor inhalation dosage lethal at 0.5 probability] *wp*
LCTAR Le Centre Thompson d' Applications Radars *wp*
LCU Landing Craft, Utility [US Navy] *wp, mt*
LCU Laser Coding Unit *hc*
LCU Last Cluster Used *pk, kg*
LCU Launch Control Unit *jmj*
LCU Line Converter Unit *hc*
LCU Line of sight Control Unit *kf*
LCU Local Control Unit *fu*
LCV Labor Cost Variance *hc*
LCV Low Cost Visual [simulation system] *aw129.21.161*
LCVP Landing Craft, Vehicles and Personnel [US Navy] *mt*
LCW Local Communications Network *fu*
LD Laser Designator *17, hc*
LD Laser Diode *ful. wp*
LD Lethal Dose *w9, wp*
LD Loaded Deployability posture *mt*
LD Logic Driver *fu*
LD Logistics Demonstration *fu*
LD-CELP Low-Delay Code Excited Linear Prediction *ct*
LD/SD Look Down/Shoot Down *wp*
LDA Landing Distance Available *mt*
LDA Laser Doppler Anemometer *wp*
LDA Line Driving Amplifier *fu*
LDA Local Distribution Accesses *hc*
LDA Locate Drum Address *fu*
LDA Logical Device Address *kg, hi*
LDA Logistics Design Appraisal *fu*
LDAP Lightweight Directory Access Protocol [IBM] *ct, kg, em*
LDAS Lightning Data Acquisition System *oe*
LDAU Laboratory Data Adapter Unit *hc*
LDB Limited Data Block *fu*
LDB Local Database *fu*
LDC Land Defense of CONUS *mt*
LDC Laydown Code Development *st*
LDC Less (Lower) Developed Country *hc, w9*
LDC Line Drop Compensator *fu*
LDC Local Distribution Center Telephone *mt*
LDC Lotus Development Corporation *kg, hi*
LDCELP Low Delay Code Exciting Linear Prediction *em*
LDCON Load Device Connected *fu*
LDEF Long Duration Exposure Facility [RADC] *mt, hc, wp, ci.600, aw129.1.49, mb, mi*
LDF Lesson Development Folder *fu*
LDG Logistics Data Gateway *mt*
LDGP Low Drag, General Purpose [bomb] *mt*
LDHILS Laser Designator Hardware In the Loop Simulation *-17*
LDI Long Dwell Imager *pbl*
LDIN Lean In Lighting System *fu*
LDIN Load Device In *fu*
LDIP Logistic Data Improvement Program *fu*
LDL Labor Distribution Log *fu*
LDLAN Long Distance Local Area Network *hi*
LDM Limited Distance Modem *hi*
LDM Linear Drive Mechanism, [a.k.a. nuller gimbal mechanism] *jdr*
LDM Logical Database Model *fu*
LDM Long Distance Modem *kg, hi*
LDME Laser Distance Measuring Equipment *hc*
LDMS Laser Distance Measuring Subsystems *hc*
LDMX Local Digital Message Exchange *mt*
LDNS Lightweight Doppler Navigation System *fu*
LDO Limited Duty Officer [USA, US Navy] *mt*
LDOE Longitude Drift/Orbit Eccentricity *-3*
LDP Language Data Processing *fu*
LDP Logistics Data Package *hc*
LDP Logistics Data Processing *hc*
LDR Large Deployable Reflector *hc, tb, wl*
LDR Laser Designator Receiver *hc*
LDR Low Data Rate *wl, jdr, kf, fu*
LDRA Low Data Rate Auxiliary *hc*
LDRI Low Data Rate Input *fu*
LDRM Laser Designator Rangefinder Module *hc*
LDS Laser Detection System *17, sb*
LDS Laser Device Subsystem *hc*
LDS Launch Detection Satellite *sd*
LDS Launch Display System *oe*
LDS Layered Defense System *hc, tb, st*
LDS Lethal Defense [Sub] System *wp, 17, sb, hc*
LDS Lightweight Decontamination System [chemical warfare] *wp*
LDS Local Distribution System *mt*
LDS Logic Development System *fu*
LDS Logistics Data System *mt*
LDSC Layered Defense Systems Countermeasures *hc*
LDSD Look Down Shoot Down [Radar] *wp, hc*

LDSRA Logistic Doctrine, Systems and Readiness Agency *wp*
LDSS Laser Designator Seeker System *hc*
LDT Linear Differential Transformer *fu*
LDT Local Descriptor Table *pk, kg*
LDTS Laser Designator Targeting System *hc*
LDTS Laser Designator Tracker System *hc*
LDV Laser Doppler Velocimeter [also LVD] *wp, aa26.11.b30, ai1*
LDVS Logistics Data Validation System [JCS] *mt*
LDXL Large Diameter eXtended Length *ai2*
LE Lafayette Escadrille *af72.5.58*
LE LAN (Local Area Network) Emulation *hi*
LE Lead (Leading) Edge *hc, w9, lt, oe*
LE Less or Equal *pk, kg*
LE Low Energy *cu*
LE Lupus Erythematosus *wp*
LE-ARP LAN Emulation - Address Resolution Protocol *hi*
LEA Launch Enable Alarm *wp*
LEA Launcher Electronics Assembly *hc*
LEA Law Enforcement Agency *mt*
LEA Limited Exclusion Area *mt*
LEA Line Equalizing Amplifier *fu*
LEA Load Effective Address *pk, kg*
LEA Logistics Evaluation Agency [USA] *st, mt*
LEA Long Endurance Aircraft *wp*
LEAD Low Cost Encryption Authentication Devices *mt*
LEAF Law Enforcement Access Field *pk, kg*
LEAMS Law Enforcement Automated Management Subsystem *mt*
LEAP Lightweight Endo {?} Atmospheric Projectile *fu*
LEAP Lightweight Exo-Atmospheric Projectile *hc, 5, sb, aa26.10.36, rm8.2.5, st, wp, tb*
LEAPS Laser Engineering and Application Prototype Systems *wp*
LEAR Logistic Evaluation And Review *wp*
LEASAT Leased Satellite [for US Navy communications] *mt, hc, 16, ph, wp, tb, uh*
LEASATCOM Leased Satellite (LEASAT) Communications *mt*
LEC LAN [Local Area Network] Emulation Client *kg, ct, em*
LEC LANTCOM Electronic Intelligence [ELINT] Center Local Exchange *mt*
LEC Local Exchange Carrier (Company) *kg, ct, hi*
LECC Layered ECC *hi*
LECS LAN Emulation Configuration Server *ct, em, hi*
LED Lethality Enhancing Device *sb, wp*
LED Light Emitting Diode [hardware] *mt, gp, kg, wp, 6, hc, af72.5.171, aw129.21.16, is26.1.60, tb, jdr, pbl, tm, ct, em, fu, hi, ai1*
LED Logical Error Detection *wp*
LED Low Endo Defense *st*
LEDC Logistics Executive Development Course *mt*
LEDD Light Emitting Diode Display *hi*
LEDI Low Endo-Atmospheric Defense Interceptor (Interception) *hc, fu, 5, sb, wp, tb*
LEDS Low Endo-atmospheric Defense System *tb*
LEED Low Energy Electron Diffraction *hc, wp*
LEF Laser Excited Fluorescence *wp*
LEI Laboratory Engineering Instruction *ep*
LEI Laser Enhanced Ionization *wp*
LEI Link Eleven Improvements *mt*
LEID Low Energy Ion Detector *fu*
LEL Link, Embed and Launch to edit [Lotus] *pk, kg*
LEL Low Energy Laser *hc*
LEL Lower Explosive Limit *oe*
LELU Launch Enable Logic Unit *fu*
LeLv Lentolaivue [Finland] *mt*
LEM Language Extension Module *pk, kg*
LEM Launch Escape Monitor *wp*
LEM Logical End of Media *sp5*
LEM Logistics Element Manager *mt, fu*
LEM Lunar Excursion Module [a.k.a. LM, in US Apollo spacecraft] "lem" *mt, mi, hc, wp, hn44.12.1, 0, ep, mb*
LEMARS Linear Elastic Matrix Analysis Routine *hc*
LEMAS Large Element Multiple Aperture System *hc*
LEMDE [an analog to digital conversion program] *hc*
LEMS Low Electromagnetic Signature Targets *wp*
LEN Line Equipment Number *ct, em*
LEN Low Entry Networking *pk, kg*
LENNK Low Endoatmospheric Non Nuclear Kill *st*
LENSCE Limited Enemy Situation Correlation Element [USAF] *mt*
LEO Launch and Early Orbit *jdr*
LEO Low Earth Orbit *ci.826, 14, 16, 17, hc, aa26.11.33, es55.58, nt, ep, ce, hy, re, tb, st, wl, as, jdr, mb, kf, vv, mi, oe, ai2*
LEOA Launch, Early Orbit Anomaly *kf*
LEOMA Laser Electro Optic Measurement Alignment *wp*
LEOP Launch and Early Operations (Orbit) Phase *hc, sf31.1.20, es55.23*
LEOS Large Erectable Optical System *hc*
LEOS Left Edge Of Scan *oe*
LEOT Launch and Early Orbit Testing *kf*
LEP Laser Eye Protection *hc, wp*
LEP Logistics Excellence Program *mt*

LEP Lowest Effective Power *fu*
LEP Lowest Evaluated Price *mt, wp*
LEPC Low Energy Proportional Counter [on SXG] *cu*
LEPT Long Endurance Patrolling Torpedo *hc*
LER Leading Edge Radius *fu*
LERC Lewis Research Center [Ohio, NASA] *wp, ci.93, 4, tb, wl, mi, oe*
LERTCON Alert Condition system [JCS] *jb, mt*
LERTS Laboratoire d'Etudes et de Recherches en TÈlÈdetection Spatiale [(French) Laboratory for Studies and Research in Remote Sensing from Space] *ge, hg*
LERX Leading Edge Root eXtension *mt*
LES LAN Emulation Server *ct*
LES Laser Engagement System *hc*
LES Launch Escape System *wp*
LES Leading Edge Simulation [TAC] *mt*
LES Leave and Earnings Statement *mt*
LES Lincoln [Laboratory] Experimental Satellite [often followed by sequence number, as in (LES 8/9)] *mt, sd, hc, ro*
LES Link Evaluation Simulation *fu*
LES Local Area Network Emulation Server *kg, em*
LES Logistics Element Specialist *fu*
LES Loop Error Signal *fu*
LES LOX [liquid oxygen] Expert System *ws4.10*
LES/BUS LAN Emulation Server / Broadcast and Unknown Server *em*
LESA Lunar Exploration System for Apollo *hc*
LESAR Linear Elastic Structural Analysis Routine *hc*
LESEM Low Energy Scanning Electron Microscope *hc*
LESO Lifting Engineering Stop Order [a document] *ep, jdr, tm*
LESS Leading Edge Structural Subsystem *ci.971*
LEST Large Earth based Solar Telescope *mi*
LET Large Earth Terminal *fu*
LET Leading Edge Track (Tracker) (Tracking) *fu, hc, wp*
LET Linear Energy Transfer *fu, hc*
LET Low Energy Telescope *es55.98*
LET [Czechoslovakian aircraft manufacturer] *aw130.6.32*
LETR Life Evaluation Test Report *jdr*
LETS Leading Edge Tracker Seeker *hc*
LETS Linear Energy Transfer System *hc*
LETS Lunar Experiments Telemetry System *hc*
LEU Launcher Electronics Unit *hc*
LEV Lunar Excursion Vehicle *hc*
LEWDD Light Weight Early Warning Detection Device *wp*
LEWWG Land Electronic Warfare Working Group *mt*
LEX Leading Edge eXtension *mt*
LEXSWG Lunar Exploration Science Working Group *eo, re*
LEZ [Laser Rangefinder for German Leopard Tank] *hc*
LF Landing Force *mt*
LF Late Finish *jdr*
LF Launch Facility *mt, hc, tb*
LF Learning Factor *fu*
LF Line Feed *fu, kg, ct, em, wp, do242, hi*
LF Line Flow *fu*
LF Low Fidelity *tb*
LF Low Frequency Synthesizer *mt*
LF Low Frequency [30 - 300 kHz] *fu, aw118.14.65, 12, w9, wp, af72.5.139, tb, kf, nu, hi, mt, ai1*
LFA Low Frequency Active *mt*
LFA Low Frequency Array *fu, kb*
LFAT Live Fire Acceptance Test *fu*
LFATDS Lightweight Field Artillery Tactical Data System *hc*
LFC Laminar Flow Control *wp*
LFC Large Format Camera *hy*
LFC Low Frequency Correction *fu*
LFCM Low Flyer Cruise Missile *fu*
LFF Logistics Factors File [WWMCCS] *mt*
LFI Last File Indicator *pk, kg*
LFICS Landing Force Integrated Communications System *mt, fu*
LFL Lower Frequency Limit *hc*
LFLID Long Focal Length Imaging Demonstration *hc*
LFM Large Function Modules *hc*
LFM Linear Frequency Modulation *hc, st, fu*
LFMOP Linear Frequency Modulation On Pulse *wp*
LFMR Low Frequency Microwave Radiometer *hc, 16*
LFN Long File Name *kg, em*
LFO Low Frequency Oscillator *fu, hi*
LFOC Landing Force Operations Center [USMC] *mt*
LFRD Lot Fraction Reliability Definition *fu*
LFRED Liquid Fuel Ramjet Engine Demonstrator [study] *hc*
LFS Loop Feedback Signal *fu*
LFS Low Frequency Synchronization *hc*
LFSA List of Frequently Seen Acronyms *mi*
LFSMS Logistics Force Structure Management Support {?} *hc*
LFSMS Logistics Force Structure Management System *hc*
LFSP Landing Force Support Party [USMC] *mt*
LFSPT Long Form System Performance Test *pl, hc*
LFT Limited Functional Test *jdr*

LFT Live Fire Test *fu*
LFT Low Function Terminal [IBM] *pk, kg*
LFTS Low Frequency Test Station *hc*
LFU Laser Firing Unit *hc*
LFU Least Frequently Used *pk, kg*
LG Landing Gear *wp*
LG Light Gun *fu*
LG Local Ground *tb*
LG Loop Gain *fu*
LGA La Guardia Airport *mj31.5.48*
LGA Light Gun Amplifier *fu*
LGA Low Gain Antenna *jp, so, re, mb, mi*
LGAF Light Ground Attack Fighter *hc*
LGB Laser Guided Bomb [USA] *wp, mt*
LGBM Local Ground Based Battle Manager *tb*
LGC Longitude Grid Control *re*
LGD Large Group Display *mt, fu*
LGDT Load Global Descriptor Table *pk, kg*
LGE Laser Guidance Element *hc*
LGM Launched from silo, Guided, ground-to-ground Missile [USA] (LGM-118A Peacekeeper ICBM, LGM-25 Titan II, LGM-30 Minuteman) *mt*
LGM Little Green Men *mi*
LGMS Laser Ground Mapping System *hc*
LGP Laser Guided Projectile *mt, wp*
LGP Low G Projectile *st*
LGWD Large Group Wall Display *mt*
LH Labor Hour *hc*
LH Lewisite mustard mix [chemical warfare] *wp*
LH Light Helicopter *mt*
LH Load High [command] *hi*
LH Long Haul *hc*
LHA Amphibious Assault Ship [General Purpose USN ship classification] *fu*
LHA Landing Helicopter Assault *hc*
LHA Lower Hour Angle *0*
LHA {Landing, Helicopter, Assault} [general purpose amphibious assault ship, carrier for helicopters and VTOL aircraft, combined with loading dock for amphibious operations, USA] *mt*
LHB Linear History Buffer *jdr*
LHC Left Hand Circular [polarization] *gg24, 4, jdr, vv, oe*
LHCN Long Haul Communication Network *mt*
LHCP Left Hand Circularly Polarized (Polarization) *gp, 16, uh, fu*
LHCRC Lap Held Computer Requirements Contract *mt*
LHCS Long-Haul Communications System *fu*
LHD Amphibious Assault Ship *wp*
LHD [amphibious assault ship, an improved LHA, circa 1990] [US Navy] *mt*
LHDLC Logical High Level Data Lead Control *mt*
LHE Liquid Helium *wp*
LHEA Laboratory for High Energy Astrophysics (GSFC, Code 660) *cu*
LHIA Logistics Horizontal Integration Analysis *mt*
LHITA Long Haul Information Transfer Architecture *mt*
LHM Laser Hardened Materials *wp*
LHMEL Laser Hardened Material Evaluation Laboratory *wp*
LHR Low Hop Rate *fu, jdr, uh*
LHS Laser Heterodyne Sounder [Shuttle] *hg, ge*
LHS Left Hand Side *ai1*
LHS Lightweight Hydraulic System *hc*
LHTEC (the LHTEC ... engine ... was designed to power the LHX) *aw130.11.6*
LHTSCT Light, Helicopter-Transportable Satellite Communications Terminal *fu*
LHX Light Helicopter eXperimental [the RAH-66 Comanche, USA] *mt*
LHX Lightweight Helicopter, Experimental *fu, aa26.12.1, wp, ph, hc, aw130.11.6, hn49.3.1*
LI Length Indicator *fu*
LI List Item *ct*
LI Local Intraconnect *mt*
LIB Light Infantry Brigade *mt*
LIBSM Low Inertia Beam Steering Minor *hc*
LIC Laser Induced Chemistry *wp*
LIC Line Interface Coupler [IBM] *pk, kg*
LIC Linear Integrated Circuit *fu, ai1*
LIC Low Intensity Conflict [USA DOD] *mt, wp, nd74.445.16*
LICALM LORAN Inertial Command Air Launched Missile *hc*
LICK Lightweight Communications Kit *mt*
LICS Lotus International Character Set [LDC] *kg, hi*
LICT Laser Imaging Component Technology *hc*
LID Laser Illuminator Detector *wp*
LID Laser Image Display *wp*
LID Leadless Inverted Device *fu*
LID Lift Improvement Device [on the AV-8 Harrier, two strikes on the underside of the fuselage and a retractable ventral dam, improving flight for vertical take off, USA] *mt*
LID Light Infantry Division [USA] *mt*
LID Locked In Device *fu*
LIDAR Laser Induced Differential Absorption Radar *hc, wp*

LIDAR Laser Light Detection And Ranging "lidar" *mt*
LIDAR Light Detection And Ranging [light radar] *fu, ge, hc, pl, 16, aw129.14.0, wp, hy, eo, hg*
LIDS Laser Illumination Detection System *wp*
LIDS Laser Infrared Demonstration System *wp*
LIDS Laser Instrumentation Detection System *hc*
LIDS Laser IRCM Demonstrator System *hc*
LIDT Load Interrupt Descriptor Table *pk, kg*
LIEP Large Internet Exchange Packet [Novell] *kg, hi*
LIF Laser Induced Fluorescence *wp*
LIF Logistics Intelligence File [DARCOM] *fu, mt*
LIF Low Insertion Force *ct, em, kg*
LIFE Laser Induced Fluorescence Emission *wp*
LIFE Likelihood Function Estimation *wp*
LIFMOP Linearly Frequency Modulated Pulse *fu*
LIFO Last In, First Out *fu, w9, wp, ct, kg, em, hi*
LIFRAM Liquid Fueled Ramjet [Engine] *hc, wp*
LIFSUM Airlift Summary *mt*
LIFT Lead In Fighter Training *af72.5.144*
LIFT Logistics Improvement Facility Technology *mt*
LIIB Lithium Ion Battery *hi*
LIL Lunar International Laboratory *wp*
LILO Last In, Last Out *pk, kg*
LILO Linux Loader *ct, em*
LIM Link Interface Module *fu*
LIM Liquid Injection Molding *hc, wp*
LIM [silo-launched SAM; as in LIM-49 Nike Zeus / Spartan, USA] *mt*
LIM/EMS Lotus Intel Microsoft / Expanded Memory Specification *hi*
LIMA Lotus Intel Microsoft AST *kg, hi*
LIMDIS Limited Distribution *wp*
LIMDO Light Intensity Modulation Direct Overwrite *pk, kg*
LIMFAC Limiting Factor *mt*
LIMM Light Intensity Modulation Method *pk, kg*
LIMPES Logistics Interactive Mobilization / Planning and Execution System [USAF] *mt*
LIMS Laser Ionization Mass Spectrometer *wp*
LIMS Library Information Management System *pk, kg*
LIMS Limb Infrared Monitoring (Monitor) of the Stratosphere [NIM-BUS] *ac2.8-4, hc, hy, hg, ge*
LIMSS Logistics Information Management Support System [USAF] *wp, mt*
LIN Line Item Number *mt, st*
LIN Liquid Nitrogen *wp*
LINAC Linear [Induction] Accelerator *fu, 17, wp, sb, hc, tb, st*
LINAS Laser Integrated Navigation/Attack System *wp*
LINCOLN Laser Input Consolidation [program] *hc*
LINK Data Link *mt*
LINK [cost planning and budget calendarization program] *hc, gb, 8*
LINS Laser Inertial Navigation System *wp*
LINUX [operating system named for Linus Torvalds] *pk, kg*
LIP LAN Interface Processing *fu*
LIP Large Internet Packet *kg, hi*
LIP Laser Induced Plasma *wp*
LIP Lunar Impact Probe *hc*
LIPAS Laser Induced Photoacoustic Spectroscopy *wp*
LIPPS [Naval Research Lab Spacecraft] *sc4.3.6*
LIPS Laser Image Processing Scanner *wp, ro*
LIPS Laser Intensity Profile System *hc*
LIPS Logical Inferences Per Second *kg, wp*
LIPS Logical Instructions Per Second *hi*
LIR Logic Information Release *fu*
LIRD Launch Interface Requirements Document *vv*
LIRIS Loral Infrared and Imaging Systems *oe*
LIRL Low Intensity Runway Lights *fu*
LIRLY Load Indicating Relay *fu*
LIS Laboratory Integrated System *hc*
LIS Land Information System *wp*
LIS Laser Isotope Separation *hc, wp*
LIS Lightning Imaging Sensor [EOS] *hy*
LIS Lithium Ion Storage *ct, em*
LIS Logistics Information System *mt*
LIS Loop Input Signal *fu*
LIS/CIV Logistics Information System / Critical Item Visibility system [USMC] *mt*
LISA Linear System Analysis *fu*
LISA Linked Index Sequential Access *hi*
LISA Logistics Information Systems Analysis [USAF] *mt*
LISD Logistics Information Systems Division [AFLC] *mt*
LISE Laser Integrated Space Experiment *wp*
LISE Low Intensity Special Event *kf*
LISI Industrial Security, International [Office of] *hc*
LISN Line Impedance Stabilization Network *jdr*
LISP List Processing (Processor) [a computer programming language] *mt, fu, kg, ct, em, wp, go1.5.3, tb, hi*
LISS Laser Illuminator Subsystem *hc*
LISS Linear Imaging Scanning System *ac3.2-20*

LIST Low Intensity Static Target Report *kf*
LIT Light Intra theater Transport *hc*
LIT Local Information Transfer *mt*
LITA Local Information Transfer Architecture *mt*
LITE Laser Illuminator Targeting Equipment *hc*
LITE Laser Infrared Tracking Experiment *st*
LITE Lead Integration and Test Engineer *jn*
LITESAT {Light Weight Satellite} *rg*
LITHCOA Lithium Corporation Of America *pl*
LITS Local Information Transfer System *mt*
LITTS Land-based Integration Test/ Trainer System *fu*
LITVC Liquid Injection Thrust Vector Control *ai2*
LIU LAN Interface Unit *fu*
LJJ Long Josephson Junction *st*
LKA [attack cargo ship] [USA DOD] *mt*
LKdo Luftflotten Kommando [Germany] *mt*
LKE Lockheed Khrunichev Energia *ai2*
LKG Loop Key Generator *mt*
LL Land Line [US] *tb, mt*
LL Lincoln Laboratories *oe*
LL Lincoln Libraries *tb*
LL Little Link *mt*
LL Local Line *hi*
LL Long Lead *fu, tb, kf, jn*
LL Long Line [NATO] *mt*
LL Low Level *mt, fu, bl2-7a.23, 0*
LL(N)L Lawrence Livermore National Laboratory *17, sb*
LLA Launch Load Accelerometer *jn*
LLA Low Level Attack *fu*
LLAD Low Level Air Defense *mt, hc, wp*
LLC Limited Life Components *tb*
LLC Line Land Control *mt*
LLC Local Link Control *ct, em*
LLC Logical Link Control *mt, kg, fu, ct, em, hi*
LLCCA Low Life Cycle Cost Avionics *hc*
LLCSC Logistics Life Cycle Support Cost *fu*
LLCSC Lower Level Computer Software Component *mt, fu, ep*
LLD Laser Locator Designator *hc*
LLD Long Line Driver *fu*
LLD Low Level Detector *fu*
LLD Lower Level Discriminator *pl*
LLD/R Laser Locator Designator/Rangefinder *hc*
LLDT Load Local Descriptor Table *pk, kg*
LLF Low Level Formatting (Format) *ct, em, kg, hi*
LLG LAN-LAN Gateway *fu*
LLGB Launch and Leave Guided Bomb *mt*
LLGB Low Level [Laser] Guided Bomb [also LLLGB] *mt*
LLH Light Lift Helicopter *wp*
LLH Linked-List Histogrammers *17, sb*
LLI Long Lead time Item *fu, hc, tb*
LLIL Long Lead Items List *fu, hc*
LLL Lawrence Livermore Laboratory *mt*
LLL Low Level Language *wp*
LLL Low Level Logic *fu*
LLLGB Low Level Laser Guided Bomb *mt, dh*
LLLTV Low Light Level Television [camera] *mt, hc, ac3.2-23, af72.5.137, wp, tb*
LLM Long Lead Material *fu*
LLNL Lawrence Livermore National Laboratory *mt, mi, wp, ts4-20, hc, tb, st, mb, pbl*
LLP Long Line Program *mt*
LLPI Link Level Protocol Interpreter *mt*
LLR Lunar Laser Ranging *ns*
LLRES Load Limiting Resistor *fu*
LLRP Low Level Reference Point *fu*
LLRV Lunar Landing Research Vehicle [USA] *hc, 0, mt*
LLS Laser Line Scanner *hc*
LLSI Low Level Serial Interface *fu*
LLSV Lunar Logistics Support/Service Vehicle *hc*
LLT Long Lead Time *fu, st*
LLTI Long Lead Time Items *hc*
LLTIL Long Lead Time Items List *hc, fu*
LLTR Low Level Transit Route [USA] *mt*
LLTV Low Light Television *mt, hc*
LLTV Lunar Landing Training Vehicle [USA] *mt, hc, ci.938*
LLV Lockheed Launch Vehicle [precursor to LMLV and Athena] *ai2*
LLWAS Low Level Wind Shear Alert System *fu*
LM Large Momentum *sd*
LM Late Midcourse [phase] *tb*
LM Lightning Mapper *oe*
LM List of Materials *hc, fu, wp*
LM Load Master *mt*
LM Load Module *hi*
LM Logistics Module *wl*
LM Lunar Module [a.k.a. LEM] [the actual moon landing craft in US Apollo spacecraft] *mt, mi, hc, ci.172, w9, mb*

LM/X LAN Manager / UNIX *hi*
LMAE Lunar Module Ascent Engine *ci.172*
LMAV Laser Maverick [program] *hc*
LMB Lower Memory Block *em*
LMBCS Lotus Multi Byte Character Set [Lotus] *pk, kg*
LMC Large Magellanic Cloud *mi*
LMC Late Mid Course *tb*
LMC Least Material Condition *fu*
LMC Lockheed Martin Corporation *jdr*
LMCA Laboratory Material Control Activity *wp*
LMCA Logistic Material Control Activity *wp*
LMCM Land Mine Countermeasures *hc*
LMCS Local Monitor and Control Subsystem *fu*
LMD Lead Military Department *mt*
LMD Logistics Maintenance Data *hc*
LMDC Leadership and Management Development Center [AU] *mt*
LMDC Leadership Management Development Center *wp*
LMDE Lunar Module Descent Engine *ci.172*
LMDS Lightweight Modular (Module) Display System *fu, hc*
LMDS Local Multipoint Distribution Service (System) *ric, kg, nu, ct, em, jdr*
LME Liquid Metal Embrittlement *wp*
LMF Language Media Format *mt*
LMG Light Machine Gun *w9*
LMG Liquid Methane Gas *wp*
LMI Local Management Interface *pk, kg*
LMI Logistics Management Institute *mt*
LMIF LOGMOD / MANPER Interface *mt*
LMIS Logistics Management Information System [USMC] *tb, mt*
LMLV Lockheed Martin Launch Vehicle [precursor to Athena] *ai2*
LMMF Local Maintenance and Management Facilities *wp*
LMMS Logistics Maintenance Management System *mt*
LMO Lens Modulated Oscillator *fu*
LMOF Local Media Output Format *mt*
LMP LANTCOM Modernization Program *mt*
LMR Land Mobile Radio *mt*
LMR Long Message Recovery *mt*
LMR Lowest Maximum Range *fu*
LMRA Land Mobile Radio Architecture *mt*
LMS Laser Mass Spectrometry *wp*
LMS Least Mean Square *fu*
LMS Lightning Mapper Sensor *hy*
LMS Local Maintenance (Monitoring) Station *fu*
LMS Logistics Management System *mt, wp*
LMS Lotus Messaging Switch [Lotus] *pk, kg*
LMSC Lockheed Missiles and Space Company *kf, hc, sc3.3.1, 0, tb, st, jdr, kf, oe*
LMSC Logistics Management Systems Center *mt*
LMSD [Hughes Aircraft Company customer for independent landing monitor radar system] *hc*
LMSS Land Mobile Satellite System *sc2.8.5*
LMSW Load Machine Status Word *pk, kg*
LMT Local Mean Time *fu, w9*
LMT Logic Module Tester *hc*
LMTR Laser Marker and Target Ranger *mt*
LMU LAN Management Utilities [IBM] *pk, kg*
LMU Line Modem Unit *mt*
LMU Line Monitoring Unit *fu*
LMU Local Monitoring Unit *fu*
LN (Call us for A.N.D.s, LN and NSA) *aw129.21.4*
LN Load Number *pk, kg*
LN Logistics Needs *mt*
LN Lotus Notes *pk, kg*
LNA Low Noise Amplifier *mt, hc, csl.7, 0, 4, 16, 17, sb, pl, jdr, kf, fu, ct, uh, tt*
LNA/DC Low Noise Amplifier/Downconverter *hc*
LNB Low Noise Block Deconverter *ct*
LNC Low Noise Converter *dn, kf*
LNG Liquified Natural Gas *wp, aa26.8.8, w9*
LNKPSC Link Performance Assessment Program *mt*
LNM LCN Network Management *fu*
LNNI LAN [Emulation] Network-to-Network Interface *ct, em*
LNO Liaison Officer [USA] *fu, mt*
LNO Limited Nuclear Option *mt*
LNP (LNP Engineering Plastics) *aa26.11.42*
LNP Local Number Portability *y2k*
LNR Line Number *mt*
LNR Low Noise Receiver *gs1-5.76, hc, kf*
LO Launch Officer *uh*
LO Launch Operations *jdr*
LO Liaison Officer [USAF] *hc, mt*
LO Lift Off [launch] *jn*
LO Liquid Oxygen *wp*
LO Local Oscillator *fu, uh, wp, gg158, 4, pl, mj31.5.26, lt, jdr, oe*
LO Lock On *fu*
LO Low Observable [technology] *mt*

LO Lubrication Order *fu*
LO Lunar Observer *eo, re*
LOA Launch On Assessment *mt, tb*
LOA Letter Of Agreement *mt, wp, tb, fu, st*
LOA Letter Of Authorization *st*
LOA Letter of Offer and Acceptance *mt, wp*
LOA Local Ocean Area *mt*
LOA Local Oscillator Amplifier *jdr*
LOAA Local Oscillator Amplifier Assembly *jdr*
LOAC Low Accuracy *fu*
LOAD Laser Optoacoustic Detection *wp*
LOAD Low Altitude Defense [program] *st, mt*
LOADS Low Altitude Defense System *hc, tb*
LOAL Lock On After Launch *uc, hc*
LOAP Lightweight Directory Access Protocol *pk, kg*
LOB Launch Operations Building *0*
LOB Limited Operating Base *fu, wp*
LOB Line Of Balance *mt, hc*
LOB Line Of Bearing *mt, fu*
LOBAR Long Baseline Radar *fu, ro*
LOBAT Long Base Line Correlation And Tracking *wp*
LOBL Lock On Before Launch *hc*
LOBO Lobe On Receive Only [radar] *wp*
LOC Launch Operations Center *0*
LOC Letter Of Consent *tb, jdr, su*
LOC Limitation Of Cost *hc*
LOC Limited Operational Capability *mt, fu, hc, jb*
LOC Line Of Code [software] *mt, kg, 17, sb, sr9-82, tb, jdr, kf*
LOC Line(s) Of Communication [USA] *mt, wp, pbl*
LOC Logistics Operations Center [AFLC] *wp, mt*
LOC Loop On line Control *kg, hi*
LOC Lunar Observer Camera *re*
LOCAP Low Combat Air Patrol [USA] *mt*
LOCAT Low Altitude Clear Air Turbulence *wp*
LOCC Launch Operations Control Center *hc, tb*
LOCCAP Transportation Plans LOCs Capacities and Capabilities [PACOM] *mt*
LOCE Limited Operational Capability for Europe *mt, ph*
LOCI (Constant Intermodulation LOCI Measure) *mj31.5.52*
LOCIST Low Cost Improved Sensors Technology *hc*
LOCMEM Local Memory *fu*
LOCO Low Cost Optint [Receiver] *hc*
LOCP Local Operator's Control Panel *hc*
LOCPOD Low Cost Powered Dispenser *mt*
LOCS Librascope Operations Control System *hc*
LOCS Local Optical Clean up System *hc*
LOCS Local Optical Correction System *hc, 17*
LOD Launch Operations Directorate *0*
LOD Line Of Departure *fu*
LODE Large Optics Demonstration Experiment *mt, ro, fu, hc, wp, 14, 17, sb, tb, st*
LODSB Load String Byte *pk, kg*
LODSC Logistics Operations Decision Support Center *mt*
LODTM Large Optics Diamond Turning Machine *tb*
LOE Level Of Effort *mt, 17, fu, wp, hc, st, jdr, kf*
LOEP List Of Effective Pages *fu*
LOF Line Of Flight *wp*
LOF Local Oscillator Frequency *fu*
LOFAAD Lower Altitude Forward Area Air Defense *hc*
LOFAADS Low Level Forward Area Air Defense System *mt*
LOFAR Low Frequency Acquisition and Ranging *fu*
LOFFIRS Low Cost Fire/Forget Infrared Seeker *hc*
LOFP Launch Operations and Flight Plan *oe*
LOFT Level Of Fault Tolerance *fu*
LOFTI Low Frequency Trans Ionosphere {an early satellite} *le6*
LOG-CCC Logistics Command, Control, and Communications [EDS] *mt*
LOG-CCCI Logistics Command, Control, Communications, and Intelligence [AFLC] *mt*
LOGAI Logical Addressing and Interoperability *mt*
LOGAIR Logistic Airlift *wp*
LOGAIRNET Logistic Air Network *wp*
LOGCAP Logistics Civil Augmentation Program *mt, wp*
LOGCEN Logistics Center [USA] *mt*
LOGCOMNET [AFLC teletypewriter communications system] *mt*
LOGCON Medical Logistics and Contingency Planning system [US Navy] *mt*
LOGDESMAP Logistics Data Element Dictionary [DOD] *mt*
LOGDET Logistics Detachment *mt*
LOGDET Logistics Detail *mt*
LOGEN Aircraft Load Generator [MAC*] *mt*
LOGEX Logistical Exercise *wp*
LOGFAC Logistics Feasibility Analysis Capability [COMPES] [USAF] *mt*
LOGFACREP Logistics Factors Report [JRS] *mt*
LOGFOR Logistics Force Packaging System [COMPES [USAF] *mt*

LOGFTC Logarithmic Fast Time Constant *fu*
LOGICA Logistics Operations Graphical Integration Capability Assessment System *mt*
LOGICON Logicon, Inc. *tb*
LOGMAP Logistic Master Plan [NATO-US] *mt*
LOGMARS Logistics Applications of Automated Marking and Reading Symbols *mt, aw129.21.84, wp*
LOGMARS Logistics Marking and Reading Symbology *mt*
LOGMDS Logistics Management Data System [COMPES] [USAF] *mt*
LOGMIS Logistics Management Information System *mt*
LOGMOD Logistics Module *mt*
LOGMOD-B Logistics Module - Base Level [COMPES] [USAF] *mt*
LOGMOD-M Logistics Module - Major Command Level [COMPES] [USAF] *mt*
LOGNET Logistic Network *wp*
LOGNET Logistics Data Network [USA] *mt*
LOGNEW Logistics Network *mt*
LOGO Limit(ation) Of Government Obligation *fu, hc, mt*
LOGO [a programming language] *kg, hi*
LOGPLAN Logistics Planning Module [COMPRES] [USAF] *mt*
LOGPLAN Logistics Plans Generation Subsystem *mt*
LOGRAM Logistics Readiness Assessment Model *mt*
LOGREADI Logistics Production / Maintenance/Readiness *mt*
LOGREDI Logistics Readiness [HQ USAF] *mt*
LOGREP Logistics Report [COMUSK] *mt*
LOGREQ Logistics Requirement (Report) *fu*
LOGSACS Logistics Structure and Composition System [USA] *mt*
LOGSAFE Logistics Sustainment Analysis and Feasibility Estimator [GCCS] [USA] *mt*
LOGSARC Logistics System Acquisition Review Council *wp*
LOGSIM Logic Simulator [Program] *hc*
LOGSIM Logistic Simulation [model] *wp*
LOGSTAR Logistics Status Report *mt*
LOGSTAT Logistics Status [monitoring subsystem] *mt*
LOGSUM Logistics Summary Data [USAF] *mt*
LOGTIES Logistic Technology Initiatives For Existing Systems *wp*
LOGWARS Logistic Wartime Automated Readiness System *mt*
LOH Light(weight) Observation Helicopter [USA] *fu, hc, mt*
LOI Lunar Orbit Insertion *re*
LOL Loss Of Learning *mt*
LOL Loss Of Lock *oe*
LOL Lost On Line *fu*
LOL Lower (Low) Order Language *fu, wp*
LOLA Lunar Orbit Landing Approach *hc*
LOLADS Low Altitude Laser Air Defense System *hc*
LOLOS Logon/Logoff Service *fu*
LOM List Of Material *mt*
LOM Living Operating Module *hc*
LOM Low Order Memory *fu*
LOMA Large Order Matrix Arithmetic *hc*
LOMAD Low to Medium Air Defense [Missile] *uc*
LOMEZ Low Altitude Missile Engagement Zone [USA DOD] *mt*
LOMMCA Logistics Operations Manpower and Materiel Cost Analysis *mt*
LOMUX Low Frequency Multiplexing *mt*
LONEX Laboratory Office Network Experiment *mt*
LONGIN Logistics Installation File *mt*
LONS Laboratory Office Network System [AFSC*] *hc, mt*
LONS Local On Line Network System [AFSC*] *wp, mt*
LONS Logistic Office Network System *wp*
LOOKING_GLASS [USAF EC-135C command post (SAC)] *wp*
LOOMS Laser Optical O&M Support *hc*
LOOPE Loop while Equal *pk, kg*
LOOPNE Loop while Not Equal *pk, kg*
LOOPNZ Loop while Not Zero *pk, kg*
LOOPZ Loop while Zero *pk, kg*
LOP Letter Of Proposal *wp*
LOP Line Of Position *fu, hc*
LOP Local Operating Procedure *fu*
LOPAIR Long Path InfraRed [alarm system] *hc, fu*
LOPAR Low Power Acquisition Radar *fu, wp*
LOQG Locked Oscillator Quadrature Grid *fu*
LOR Launch and Orbit Raising *oe*
LOR Level Of Repair *mt, fu, wp*
LOR Low frequency Omnidirectional Range *wp*
LOR Lunar Orbit Rendezvous *ci.386, 13*
LORA Level Of Repair Analysis *mt, hc, wp, fu*
LORA-HOJ Long Range Home On Jam *hc*
LORAD Long Range Active Detection [USA] *fu, mt*
LORAH Long Range Area Homing [Interceptor Study] *hc*
LORAL (LORAL Elec.) [Hughes Aircraft Company customer] *hc*
LORAN Long Range [Aids to] Navigation [system] [a long-range radio navigation position fixing system using the time difference of reception of pulse type transmissions from two or more fixed stations] *mt, fu, wp, ro, ci.97, 12, hc, af72.5.144*
LORAR Level Of Repair Analysis Report *fu*
LOREC Long Range Earth Current Communications *fu*

LOREOR Long Range Electro Optical Reconnaissance *wp*
LOREORS Long Range Electro Optical Reconnaissance System *hc, wp*
LORI Logistics Operation Readiness Inspection *mt*
LORO Lobe On Receive Only *mt, fu, wp*
LOROP Long Range Oblique Photography *mt, hc, wp*
LORS LEM (Lunar) Optical Rendezvous System *hc, hn44.12.2*
LORS Lunar Orbiting Reconnaissance System *hc*
LORTID Long Range Target Identification *hc*
LORU Launch Order Responded Unit *fu*
LORV Low Observable Reentry Vehicle *hc, wp*
LOS Land Observation Satellite *un*
LOS Launch and Orbital Support *hc*
LOS Line Of Sight *mt, fu, bl1-5.7, wp, 1, 5, 8, 16, 17, sb, hc, w9, jb, jmj, lt, bf, tb, st, kf, oe, ai1*
LOS Loss Of Signal *fu, 4, tb, wl, uh, oe*
LOSACA Liaison Officer to Supreme Allied Commander Atlantic [U. S.] *dt*
LOSE Line Of Sight Equipment *hc*
LOSE [a communications jamming system] *hc*
LOSM Lunar Orbital Survey Missions *hc*
LOSMAD Line Of Sight Missile for Air Defense *hc*
LOSR Line Of Sight Range *tb*
LOSREP Loss Report [Aircraft/Aircrew] *mt*
LOSS Launch Operations Support Services *-16*
LOSS Lunar Orbit Space Station *hc*
LOT (Poland's LOT national airline) *aa26.12.1*
LOT Lock On and Tracking *hc*
LOT Low Observables Technology *mt, wp*
LOT Lumisis Operations Team *re*
LOTAS Large Optical Tracker-Aerospace *hc*
LOTAWS Laser Obstacle Terrain Avoidance Warning System *hc*
LOTEX Life Of Type Extension *mt*
LOTRANS [a laser program] *hc*
LOTS LEM Optical Tracker System *hc*
LOTS Logistics Over The Shore *mt, wp*
LOUIS Logical On Line User Inquiry System [USAF] *mt*
LOVA Low Vulnerability Ammunition *wp*
LOW Launch On Warning *11, tb*
LOW Link Order Wire *hi*
LOWG Launch Operations Working Group *-17*
LOWTRAN Low Resolution Transmission [model] *wp*
LOX Liquid OXygen *mt, ac2.5-2, mi, wp, 13, tb, mb, ai1, ai2*
LP Launch Platform *ai2*
LP Launch Point *tb, kf*
LP Launch Procedure *jdr*
LP Learning Prototype *mt*
LP Limited Production *mt*
LP Line Printer *fu, oe*
LP Linear Polarization *gp, 16*
LP Linear Programming *fu, hc*
LP Liquid Propane *wp*
LP Liquid Propellant *wp, ai2*
LP Low Pass *fu*
LP Low Point *fu*
LP Low Pressure *w9, mt*
LPA Link Pack Area *fu*
LPA Low Power Amplifier *fu, hc*
LPA [amphibious transport, US Navy] *mt*
LPAPS Low Power Amplifier Power Supply *fu*
LPAR Large Phased Array Radar *mt, ps204*
LPAR Logic Programming and Automated Reasoning *pk, kg*
LPB Launch Pad Building *0*
LPC Line Printer Controller *fu*
LPC Linear Predictive Coding *mt, fu*
LPC Local Procedure Call *kg, ct, em*
LPC Logical Processing Capabilities *-4*
LPCW Long Pulse Continuous Wave *fu*
LPD Landing Platform Dock [amphibious transport dock, a ship designed to transport and land troops, equipment, and supplies by means of embarked landing craft, amphibious vehicles, and helicopters] [USA DOD] *mt*
LPD Landing Ship Dock *fu*
LPD Launch Point Determination *hc, kf*
LPD Line Printer Daemon [protocol] [Berkley] *pk, kg*
LPD Low Probability of Detection *uh, mt, fu*
LPE Launch Point Estimate *kf*
LPE Leading Payload Engineer *jn*
LPE Liquid Phase Epitaxy *fu*
LPE Low Probability of Exploitation *fu*
LPF Large Payload Fairing *kf, oe, ai2*
LPF Laser Plotting Facility *fu*
LPF Low Pass Filter *fu, csIII.10, lt, jdr, uh*
LPFL Low Pass Filter *fu*
LPG Language Processor Group *fu*
LPG Laser Pattern Generator *fu*
LPG Liquified Petroleum Gas *wp, aa26.8.8, w9, ai1*

LPH Amphibious Helicopter Assault Ship (USN ship classification) *fu*
LPH Landing Platform Helicopter ship [amphibious assault ship, includes a number of former CVs of the Essex class and CVEs of the Anzio and Commencement Bay class] [USA] *mt*
LPI Lines Per Inch *kg, wp, hi*
LPI Low Probability of Intercept *mt, hc, sa, wp, jdr, fu, uh, ric*
LPIR Low Probability of Intercept Radar *fu, hc*
LPL Lotus Programming Language [Lotus 1-2-3] [LDC] *pk, kg*
LPL Low Power Laser *hc*
LPLV Large Payload Life Vehicle *tb*
LPM Laser Pointing Mirror *hc*
LPM Lines Per Minute *mt, fu, kg, muc, hi, oe*
LPM Linked Page Memory *fu*
LPM Liquid Propulsion Module *jrm*
LPM Low Power Model *fu*
LPMEDLEY [CIA cryptonym for an operation in support of the NSA] *-10*
LPMES Logistics Performance Measurement and Evaluation System *mt*
LPMO LOAD Project Manager's Office *st*
LPMS Logistic Program Management System *wp*
LPN Long Part Number *fu*
LPN Lumped Parameter Network *hc*
LPO La Palma Observatory *mi*
LPO Leading Petty Officer [US Navy] [USA] *fu, mt*
LPO Low Power Output *fu*
LPO Lunar Polar Orbiter *hc*
LPP Launch Point Prediction *mt*
LPP Learning Prototype Phase *mt*
LPPS Logistics Plan for Pre-Operational Support *fu*
LPR Line Printer Remote *pk, kg*
LPR Long Playing Rocket [meaning *satellite*]
LPR Low cost Packet Radio *fu*
LPRE Launch Prediction *jb*
LPRF Low Power Radio Frequency *wp*
LPRF Low Pulse Repetition Frequency [mode] *hc, fu*
LPS Large Processing Station *mt, kf*
LPS Launch Processing System [Shuttle] *mt, ci.943, tb*
LPS Launch Protection System *tb*
LPS Lines Per Second *hi*
LPS Liquid Propulsion [Sub] System *gp, 16, jn*
LPS Logistics Planning Support *mt*
LPS Low Power Schottky *pk, kg*
LPSD Logically Passive Self-Dual *fu*
LPSO Laboratory Procurement Support (Supply) Office [Army] *fu, hc*
LPSS Amphibious Transport Submarine [US Navy] *mt*
LPST Latest Possible Start Time *mt*
LPT Laboratory Point and Tracking *hc*
LPT Legal Path Table *fu*
LPT Limited Performance Test *oe*
LPT Line Printer [usually parallel port, identified by a number] *kg, hi*
LPT Liquid Pressure Transducer *pbl*
LPT Local Printer Terminal *ct, em*
LPTD Launch Point and Time Determination *kf*
LPTV Large Payload Test Vehicle *hc*
LPTV Low Power Television *dn*
LPU Line Printer Unit *mt*
LPV Launch Preparation Van *ai2*
LPV Long Period Variable *mi*
LPWG Logistics Planning Working Group *mt*
LQ Letter Quality *kg, hi*
LQ Link Quality *fu*
LQA Link Quality Array *fu*
LQA Living Quarters Allowance *wp*
LQG Linear Quadratic Gaussian *ai1*
LQG [a label on a flowchart] *bl1-6.29*
LQLSD Liquid Crystal Large Screen Display *hc*
LQM Link Quality Monitoring [protocol] *pk, kg*
LQR Linear Quadratic Regulator *ai1*
LQU Link Quality Analysis *mt*
LR Laser Rangefinder *tb*
LR Lethal Radius *wp*
LR Letter of Requirements *mt*
LR Letter Requirement *st, tb*
LR Linear Regression *wp*
LR Link Register *pk, kg*
LR Load Ratio *fu*
LR Load Resistor *fu*
LR Long Range *fu*
LR Longley-Rice [radio propagation loss prediction model] *fu*
LR Low Rate *bl1-5.87*
LR Lunar Rover *re*
LRA Laser Retroreflector Array *ac2.7.21*
LRA Light Replaceable Assemblies *hc*
LRA Line Receiving Amplifier *fu*
LRA Long Range Aircraft *mt*
LRA Long Range Aviation [Soviet] *fu, mt, tb*

LRA Lowest Replaceable Assembly *fu*
LRAACA Long Range [Air] Antisubmarine-warfare Capable Aircraft *aa26.12.1, aw129.21.148, wp*
LRAAM Long Range Air to Air Missile [Study] *hc, wp*
LRAAS Long Range Airborne Air to Surface [Weapon System] *hc*
LRAF Long Range Aviation Forces *mt*
LRAM Local Random Access Memory *fu*
LRAM Lr {Long Range}Attack Missile *wp*
LRAP Long Range Acoustic Propagation *fu*
LRAPIS Long Range Advanced Piloted Interceptor System *hc*
LRARFS Long Range Airlift Requirements Forecast System *mt*
LRAT Large Radar Array Technology [RADC] *tb, mt*
LRAT Long Range Anti Tank *hc*
LRB Liquid Rocket Booster *hc, mb, mi, ai2*
LRB Lang Runlength Beacon *fu*
LRBA Laboratoirre de Recherches Balistiques et Aerodynamiques *crl*
LRBM Long Range Ballistic Missile *mt*
LRC Langly Research Center [NASA] *hc, sm18-57*
LRC Learning Resource Center *mt*
LRC Lesser Regional Conflict *mt, kf*
LRC Line Replaceable Components *hc*
LRC Load Ratio Control *fu*
LRC Local Register Cache *kg, hi*
LRC Logistics Readiness Center *wp, mt*
LRC Long Range Cargo *mt*
LRC Longitudinal Redundancy Check *fu, do263, jdr, kg*
LRC-TCC Logistics Readiness Center-Transportation Control Center [USAFE] *mt*
LRCA Long Range Combat Aircraft [the B-1B, USA] *wp, mt*
LRCA-STS Long Range Combat Aircraft System Test Station *hc*
LRCC Longitudinal Redundancy Check Character *hi*
LRCCM Long Range Conventional Cruise Missile *wp*
LRCD Lunar Rock Coring Device *hc*
LRCM Long Range Cruise Missile *tb*
LRCSOW Long Range Conventional Stand Off Weapon [also LRCSW] *wp*
LRCSS Logistics Resource Control and Support System [MAC] *mt*
LRCSW Long Range Conventional Standoff Weapon [also LRCSOW] *wp*
LRD Lightning and Radio [emission] Detector *es55.28*
LRD Long Range Data *fu*
LRD Long Range Designator *hc*
LRD Low Rate Demodulator *-4*
LRD/T Long Range Designator/Transceiver *hc*
LRDMM Long Range Dual Mission Missile [a ship-launched missile program, intended against both ships and aircraft, USA] *mt*
LRDS Laser Ranging and Designation System *wp*
LRE Laser Ranger Experiment [USA] *mt*
LRE Latest Revised Estimate [contractor estimate] *mt, jdr, kf, fu*
LRE Low Rate Encoding *jgo*
LRECL Logical Record Length *mt*
LREP Light Replica Decay *tb*
LRES Long Range Earth Sensor *0*
LRF Laser Range Finder *mt, hc, wp, pm, fu*
LRF Laser Resonance Fluorescence *wp*
LRF Lesson Reference File *hc*
LRFAX Low Resolution Facsimile [for GMS] *pl*
LRFD Low Resolution Fault Dictionary [program] *hc*
LRFOP Limited-Rate Follow-On Production *fu*
LRFOP Low Rate Follow On Production *fu*
LRGB Long Range Glide Bomb *mt*
LRHIT Long Range Homing Interceptor *fu*
LRI Left/Right Indicator *fu*
LRI Limited Range Intercept *mt, fu*
LRI Long Range Interceptor *hc, wp*
LRI Long Range International [CRAF] *mt*
LRI Lowest Replaceable Item *fu*
LRIC Long Run Incremental Costing *do35*
LRINF Long Range Intermediate Nuclear Forces [Soviet] *mt*
LRIP Local and Remote Interface Processing *fu*
LRIP Long Range Initial Production *hc*
LRIP Low Rate Industrial Production *wp*
LRIP Low Rate Initial Production *mt, fu, hc, hq5.3.1, nt, jmj, st, jdr, kf*
LRIR Limb Radiance Inversion Radiometer *ac2.8-3*
LRIRR Low Resolution Infrared Radiometer *fu*
LRIX Long Range Interceptor Experimental [a Mach 3+ fighter, F-108 Rapier project, canceled, USA] *mt*
LRL Laser Relay Link *hc*
LRL Lawrence Radiation Laboratory *pl*
LRL Least Recently Loaded *pk, kg*
LRL Lunar Receiving Laboratory [at Johnson Space Center] *ci.83*
LRM Labor Reimbursable Material *hc*
LRM Language Reference Manual *pk, kg*
LRM Layer Reference Model *hi*
LRM Least Recently used Master *em, kg*
LRM Line Replaceable Module *wp, tb*
LRM Linear Radiating Module *hc*

LRM Low Rate Multiplexer *uh*
LRM Lunar Reconnaissance Mission *hc*
LRMP Long Range Maritime Patrol [USA] *fu, mt*
LRMPA Long Range Maritime Patrol Aircraft *hc*
LRMS Low Resolution Mass Spectrometry *wp*
LRMTS Laser Range Meter and Target Seeker [targeting equipment fitted to the Jaguar and Harrier] [UK] *wp*
LRMTS Laser Rangefinder and Marked Target Seeker *wp*
LRNC List of Specifications {?} *fu*
LRNC Long Reference Number Code *fu*
LRNTF Long Range Intermediate Nuclear Forces {?} *tb*
LRO Long Range Objective *mt, fu*
LROS Long Range Optical System [space based active HEL surveillance system] *wp*
LRP Lang Recovery Package *hc*
LRP Logistics Recovery Program *mt*
LRP Long Range Interceptor *wp*
LRP Low Rate Production *hc, nt, fu*
LRPA Laser Radar Power Amplifier *hc*
LRPA Long Range Patrol Aircraft *mt, hc*
LRPE Long Range Procurement Estimate *hc*
LRPG Long Range Proving Ground *hc*
LRPPD Long Range Planning Purpose Document *hc*
LRPS Laser Reliability Prediction Study *hc*
LRPV Light [weight] RPV *wp*
LRR Laser Radiation Receiver *hc*
LRR Launch Readiness Review *17, sb, re, jdr, oe*
LRR Long Range Radar *mt, hc, wp, fu*
LRR Loop Regenerative Repeater *do181*
LRRDA Long Range Research, Development, and Acquisition *st*
LRRDAP Long Range Research, Development, and Acquisition Plan *mt*
LRS Laser Radiometer System *hc*
LRS Laser Raman Scattering *wp*
LRS Laser Range [Sub] System *hc*
LRS Laser Rangefinder Set *hc*
LRS Laser Raster Scanner *hc*
LRS Long Ranging Subsystem *hc*
LRSA Long Range Strategic Aircraft *tb*
LRSC Lowest Replaceable Software Component *fu*
LRSCA Large Retractable Solar Cell Array *hc*
LRSI Low temperature Reusable Surface Insulation *ci.xxii*
LRSLBM Long Range Sea Launched Ballistic Missile *tb*
LRSOM Long Range Stand Off Missile *hc, mt*
LRSR Long Range Storage Requirements [PACOM] *mt*
LRSS Logistics Readiness Simulation System [TAC] *mt*
LRSS Long Range Survey System *hc*
LRT Launch Readiness Test *gp*
LRT Link Readiness Test *-4*
LRTF LANTIRN [eye-level] Remote Test Facility *hc*
LRTNF Long Range Theater Nuclear Forces *mt*
LRU Laser Rangefinder Unit *hc*
LRU Least (Lowest) Replaceable Unit *fu*
LRU Least Recently Used *pk, kg*
LRU Line Replaceable (Replacement) Unit *mt, fu, hc, wp, bo, aw129.14.99, nt, tb, jdr, kf*
LRV Last Received Value *-3*
LRV Last Recorded (Reported) Value *jdr, kf, uh, jn*
LRV Lunar Roving Vehicle *hc*
LRW (LRW Associates . . . have worked diligently) *mj31.5.40*
LRWR Long-Range Warning Receiver *fu*
LS Landing Ship [US Navy] *mt*
LS Landing System *wp*
LS Late Start *jdr*
LS Launch Site *jdr*
LS Lead Service *wp*
LS Least Significant *hi*
LS Logistics Summary [USAREUR] *mt*
LS Low Speed *hi*
LSA Labor Surplus Area *kf*
LSA LAN and SCSI Adapter [IBM] *kg, hi*
LSA Large Search Area *fu*
LSA Large Space Antenna *wl*
LSA Laser Seeker Assembly *hc*
LSA Launch Services Agreement *oe*
LSA Launcher Structure Assembly *hc*
LSA Light Source Assembly *hc*
LSA Line Sharing Adapter *pk, kg*
LSA Local Security Authority [Microsoft] *pk, kg*
LSA Logistics Support Analysis *fu, mt, hc, wp, 17, sb, go3.6, nt, dm, tb, st, jdr, kf*
LSA Logistics Systems Applications *mt*
LSA Low Sidelobe Antenna *hc*
LSAA Low Sidelobe Antenna Array *hc*
LSABA Low Sidelobe Agile Beam Antenna *fu*
LSACN Logistic Support Analysis Control Numbers *fu*
LSAM Logistics Support Analysis Modeling *fu*

LSAO Logistics Systems Analysis Office *dt*
LSAP Logistics Support Analysis Plan *mt, wp, fu, 17, sb, kf*
LSAPI License Services Application Program Interface *kg, hi*
LSAR Logistics Support Analysis Record (Requirements) *mt, hc, fu, hn43.9.3, wp, st, kf*
LSAT Ladar Satellite *tb*
LSAT Large Scale Applications Test *ac1.2-21*
LSAT Large Staring Array Technology *kf*
LSAT Logistics Shelter, Air Transportable *hc*
LSB Launch and Service Building *ai2*
LSB Launch Support Building *oe*
LSB Least Significant Bit *gg64, he, uh, kg, wp, 4, 9, hc, st, jdr, kf, fu, hi, oe, ai1*
LSB Least Significant Byte *hi*
LSB List of Successful Bidders *wp*
LSB Low Speed Buffer *fu*
LSB Lower Sideband *mt, fu*
LSBGA Last Sortie Before Ground Alert *mt*
LSC Large Sized Combatant *hc*
LSC Least Significant Character *kg, hi*
LSC Littleton System Center *tb*
LSC Load Standardization Crew *mt*
LSC Logistics Support Center *hc, kf*
LSC Logistics Support Cost *wp, fu*
LSC Low-Speed Concentrator *fu*
LSCB Launching Section Control Box *fu*
LSCD Large Screen Color Display *mt, hc*
LSCD Launch System Control Document *oe*
LSCM Low Slow Cruise Missile *fu*
LSCPS Logistics Support Concept of Pre-Operational Support *hc*
LSCRC Lapheld Small Computer Requirements Contract *mt*
LSCSIM Logistics Support Cost Simulation Model *fu*
LSCSS Limited Scale Command and Control System *hc*
LSCU Lightning System Control Unit *oe*
LSD Landing Ship, Dock [USA DOD] *wp, mt*
LSD Large Scale Display *hc*
LSD Large Screen Display *mt, fu, hc, tb*
LSD Least Significant Digit *fu, kg, 4, hi*
LSD Low Speed Data *kf*
LSD Lowest Significant Dose *wp*
LSDC Large Screen Display Controller *fu*
LSDS Large Screen Display System *mt*
LSDT Local Sidereal Time *fu*
LSE Language Sensitive Editor *oe*
LSE Launch System Engineering *jdr*
LSE Launcher Servo Electronics *hc*
LSE Lead Software Engineer *kf*
LSE Lead System Engineer *fu*
LSE Least Squares Estimation *fu*
LSE Life Support Equipment *wp*
LSE Logistics System Engineering *fu*
LSE Loop Stress Error *jdr*
LSEAD Lethal Sead *wp*
LSEDIT Language-Sensitive Editor *fu*
LSEP Limited System Evaluation Program *hc*
LSEP Lunar Surface Experiment Package *hc*
LSF Line Spread Function *wp*
LSFT Launch Site Functional Test *jn*
LSG Linear Sequence Generator *fu*
LSGE Large Screen Graphic Equipment *fu*
LSH Least Significant Half *jdr, uh*
LSHI Large-Scale Hybrid Integration *fu*
LSI Landing Ship, Infantry [US Navy] *mt*
LSI Large Scale Integrated [circuit] *sb, uh, ah*
LSI Large Scale Integration [of circuit elements] *kg, aa26.12.17, fu, wp, gg161, 6, 17, sb, pl, hc, rm8.1.1, ep, tb, jdr, pbl, tm, hi, oe, mt*
LSI Launch Systems Integration *pbl, ah, jn*
LSI Lead System Integrator *jdr*
LSI-M-CAPS Large Scale Integrated - Mobile CAPS *mt*
LSIC Large Scale Integrated Circuit *hc, ep, jdr, fu*
LSIC Lockheed System Integration Contractor *st*
LSIR Limb Scanning Infrared Radiometer *hc*
LSIR Low Ship Impact Ranging *fu*
LSL Link Support Layer *pk, kg*
LSL Load Segment Limit *pk, kg*
LSL Local Sequence List *fu*
LSLC Loop Side Lobe Canceler *hc*
LSM Landing Ship, Medium [rockets] [US Navy] *mt*
LSM Local Summary Messages *kf*
LSM Logistics Support Manager (Management) *wp, mt*
LSMI Logistics Support Management Information *mt, wp*
LSMIS Logistics Support Management Information System *mt*
LSMP Logistics Support and Mobilization Plan *mt*
LSMP Logistics Systems Modernization Program [DLA] *mt*
LSMS Logistics Support Management System *fu*
LSO Landing Safety Officer [US Navy] *mt*

LSOP Launch Site Operations Plan *oe*
LSP Large Screen Projector *fu*
LSP Launch Services Provider *jdr*
LSP Logistics Support Plan *mt, fu, wp, tb*
LSP Logistics Support Priorities *mt*
LSP Low Speed Printer *mt, fu*
LSP Lunar Survey Probe *hc*
LSPCC Logistics Support Planning and Control Center *tb*
LSPES Logistics System Planning and Execution System *mt*
LSPI Laboratory Service Process Instructors *hc*
LSPPS Logistics Support Plan for Pre-operational Support *fu, hc, tb*
LSPR Logistics Support Property Representative *hc*
LSPT Launch Site Performance Test *9, uh*
LSPT Logistics Support Planning Team *hc*
LSPTP Low Speed Paper Tape Punch *mt*
LSQF Least Squares Fit *hc*
LSR Line Stock Requisition *fu*
LSR Load Shifting Resistor *fu*
LSR Local Senior Representative *fu*
LSR Local Standard of Rest *mi*
LSR Logistic Support Requirements *wp*
LSR Logistics Service Representatives *hc*
LSRC Logistics System Review Committee *mt*
LSRD Launch Site Requirements Document *gp*
LSRD Logistic Support Readiness Date *wp*
LSRF Logistics Support Resource Funds *mt*
LSS Large Space System *hc*
LSS Laser Sizing Study *tb*
LSS Launch Status Summarizer *fu*
LSS Launch System Integrator *jdr, ric*
LSS Lightning Sensor System *aw129.1.26*
LSS Local Subscriber Switch *mt*
LSS Logistics Support Station *mt*
LSS Lunar Survey System *hc*
LSSA Logistics System Support Activity [USA] *mt*
LSSC Launch Support Services Contracts *mt*
LSSD Level Sensitive Scan Design *fu*
LSSO Launch Site Safety Office *oe*
LSSO Launch Site System Operations *gp*
LSSP Launch Site Safety Plan *oe*
LSSP Launch Site Support Plan *gp, 4, oe*
LSSS Laboratory Support Signal Switching System *fu*
LSSS Lightweight SHF SATCOM System *mt*
LSST Launch Site Support Team *oe*
LSST List of Specifications and Standards *fu*
LST Landing Ship, Tank [USA DOD] *w9, mt*
LST Large Scale Telescope *hc*
LST Laser Spot Tracker *hc*
LST Least Squares Tracking *hc*
LST Lightweight Satellite Terminal *mt*
LST Local Satellite Time *uh*
LST Local Sidereal Time *w9*
LST Local Solar Time *oe*
LST Local Spacecraft Time *jdr*
LST Local Standard Time *fu*
LST Loop Stress Transient *jdr*
LST Lunar Surface Transponder *hc*
LSTDM Low Speed Time Division Multiplex(er) *hc, mt*
LSTP Local Solar Time at Perogee *oe*
LSTP Logistic Systems Training Program *wp*
LSTP Low Speed Tape Processor *gp, 16, oe*
LSTR Launch System Test Rack *gp*
LSTS Low Speed Tape System *oe*
LSTTL Low Power Schottky Transistor-Transistor Logic *fu*
LSTTY Low Speed Teletypewriter *mt*
LSU Landing Ship, Utility [US Navy] *mt*
LSU Large Shelterized Unit *fu*
LSU Line Select Unit *fu*
LSV Large Scale Vehicle *pf.f88.17*
LSV Logistic Support Vessel *wp*
LSVD Large Screen Visual Display *mt*
LSW Large Scale Walleye *hc*
LSWD Large Screen Wall Display *mt*
LT Kueytebabt {?} *mt*
LT Landing Time *mt*
LT Lead Time *fu*
LT Learning Terminal *fu*
LT Light Terminal [Satellite Ground] *mt*
LT Line Termination *ct, em*
LT Line Traffic Coordinator *mt*
LT Link Terminal *mt*
LT-SURV Lethality-Survivability Simulator *-17*
LTA LAM Thrust Accelerometer *he*
LTA Lighter Than Air *mt, wp, aa26.12.28*
LTA Long Term Agreement *kf*
LTA Long Time Averaging *fu*

LTAA License and Technical Assistance Agreement *fu, hc*
LTAC Linear Thruster Attitude Control *jdr, uh*
LTB Limited Track Block *fu*
LTB Line Termination Buffer *mt*
LTBO Linear Time Base Oscillator *fu*
LTC (a catering company in D,sseldorf, operated by LTC) *aw130.13.93*
LTC Large Temperature Chamber *fu*
LTC Level and Timing Controller *hc*
LTC Lieutenant Colonel *mt, fr, w9, ts7-1*
LTC Load Tap Changing *fu*
LTC Longitudinal Time Constant *fu*
LTCC Low Temperature Cofired Ceramic *md2.1.2*
LTD Large Text Display *fu*
LTD Laser Target Designator *mt, fu, hc, wp*
LTD Live Test Demonstration *mt*
LTD Local Data Terminal {?} *-4*
LTD/R Laser Target Designator/Rangefinder (Ranger) *hc, hn49.8.1*
LTDP Long Term Defense Program [NATO] *ph, mt*
LTDR Laser Target Designator Rangefinder [Subsystem] *hc*
LTDR Laser Target Designator Receiver *hc*
LTDS Laser Target Designation Set *fu, hn41.25.1*
LTDS Laser Target Designator System *hc*
LTDS Laser Tracking and Discrimination Study *fu, hc*
LTE Launch To Eject *hc*
LTE Local Thermodynamic Equilibrium *pl*
LTE [a subdivision of LTU operating in Spain] *aw130.13.93*
LTER Long Term Ecological Research [Site] *hy*
LTF Lead The Force [program] *hc*
LTF Lightning Training Flight [UK] *mt*
LTFCS Laser Tank Fire Control System *hc*
LTFT Long Term Full-time Training *wp*
LTG Lieutenant General *w9*
LTG Luft Transport Geschwader [air transport wing] [Germany] *mt*
LTH Lethality [and] Target Hardening *5, 17, sb, wp, tb, st*
LTI (LTU . . . established an investment company called LTI) *aw130.13.92*
LTI Logistic Technical Information *mt*
LTI Long Time Integration *fu*
LTJ Look-Through Jamming *fu*
LTJG Lieutenant, Junior Grade *w9*
LTL Listing Time Limit *fu*
LTM Laser Target Marker *hc*
LTM Line Type Modulation *fu*
LTM Local Transverse Mercator *fu*
LTMS Lightning Transient Measurement System *oe*
LTO Low Temperature Oxide *fu*
LTOC Lowest Total Overall Cost *mt*
LTOD Linear Time Of Day *fu*
LTOE Living TOE *fu*
LTP Life Test Procedure *hc, jdr*
LTP Lunar Transient Phenomenon *mi*
LTPD Lot Tolerance Percent Defective *fu, jdr*
LTPP Long Term Planning Project *wp*
LTQ Low Torque *fu*
LTR Laser Target Rangefinder *hc*
LTR Left To Right *pk, kg*
LTR Load Task Register *pk, kg*
LTR Low Thermal Regime *st*
LTRS Laser Target Recognition System *hc*
LTS Laser Target Simulator *hc*
LTS Laser Tracking System *hc*
LTS Launch Telemetry Station *w9*
LTS Launch Tracking System *w9*
LTS Lethality Test System *tb, st*
LTS Levothyroxine Sodium *wp*
LTS Light Traffic Station *hc*
LTS Logical Time Slot *fu*
LTS Low Temperature Superconductor *wp*
LTT Laser Thermal Tagger *tb*
LTT Low Thrust Thruster *jdr, jn*
LTTAID Long Tank Thrust Augmented Improved Thor *ci.xvii*
LTTAT Long Tank Thrust Augmented Thor *ci.xvii*
LTTC Lowry Technical Training Center *hc*
LTTE Link Terminal Terminating Equipment *mt*
LTU (the German charter airline LTU) *aw130.13.93*
LTU Laser Tracking Unit *hc*
LTU Line Terminating Unit *mt, fu*
LTU Logic Test Unit *fu*
LTUM Line Terminating Unit Module *mt*
LTV Laser Doppler Velocimetry *te23, 11, 22*
LTV Launch Test Vehicle *tb*
LTV Linear Velocity Transducer *st*
LTV Ling Temco Vought, [LTV Aerospace and Defense Company, Dallas, Texas] *hc, jb, tb, st, ci.195, aw130.11.11, aa26.11.16, af72.5.28, hc, jmj*

LU LGBM Utility *tb*
LU Link Unavailable *mt*
LU Logical Unit *ct, em, kg, fu, hi*
LUA Launch Under Attack *mt, hc, tb*
LUA Logical Unit Application [interface] *pk, kg*
LUF Lowest Usable Frequency *fu*
LUHF Lowest Usable High Frequency *fu*
LUI Library User Interface *fu*
LUI Local User Input *pk, kg*
LUIS Library User Information Service *pk, kg*
LUMS Land Use Management System [USMC] *mt*
LUN Logical Unit Number *ct, em, kg*
LUNI LAN User-to-Network Interface *ct*
LUPI Laser Unequal Path Interferometer *hc*
LUSAM Lumped Shell Analysis Methods *hc*
LUT Local User Terminal *aw129.14.45, oe*
LUT Look Up Table *fu, kg*
LV Latch Valve *gg37, jn, oe*
LV Launch Vehicle *ci.193, 16, 17, jdr, kf, vv, he, uh, jn*
LV Logical Volume [IBM] *pk, kg*
LV Low Voltage *fu*
LVA Large Vertical Aperture *fu*
LVA Launch Vehicle Adapter *kf, jn*
LVA Log Video Amplifier *fu*
LVA Lucht Vaart Afdeling ["Air Corps", name of home based Air Force before and during WWII] [Netherlands] *mt*
LVC Launch Vehicle Contingency *oe*
LVC Launch Vehicle Contractor *kf, he, jn*
LVC Low Voltage Controller *he*
LVC Low Voltage Converter *pbl*
LVCD Least Voltage Coincidence Detector *fu*
LVCLS Launch Vehicle Component {Level} Simulation *hc*
LVCU Latch Valve Control Unit *oe*
LVD Laser Doppler Velocimeter [also LDV] *aa26.11.55*
LVDE Large Volume Data Exchange *mt*
LVDT Linear Variable Differential Transformer *fu, hc*
LVDT Linear Variable Displacement [Differential] Transducer *17, sb*
LVE Lead Verification Engineer *jn*
LVFEL Low Voltage Free Electron Laser *wp*
LVI Launch Vehicle Integration *jdr, kf*
LVI Launch Vehicle Interface *jn*
LVIWG Launch Vehicle Integration Working Group *hc*
LVLH Local Vertical, Local Horizontal *wl*
LVM Layered Virtual Machine *fu*
LVM Logical Volume Management [IBM] *pk, kg*
LVMP Launch Vehicle Mission Particular [hardware] *oe*
LVP Low Voltage Protection *fu*
LVPC low voltage Power Controller *he*
LVPO Launch Vehicle Project Office *oe*
LVPS Low Voltage Power Supply *ac10.4-5, hc, fu*
LVRE Low Voltage Release Effect *fu*
LVRJ Low Volume Ram Jet *hc*
LVRLD Load Variable Request to Load Device *fu*
LVS Launch Vehicle System *dm, oe*
LVS Logistics Vehicle System *mt*
LVS Low Velocity Scanning *fu*
LVSS Laser Vector Scoring System *hc*
LVT Landing Vehicle, Tracked [US Navy] *uc, w9, mt*
LVTE Landing Vehicle, Tracked, Engineer [as in LVTE-l for model 1] [USA DOD] *mt*
LVTR Low VHF Transmitter Receiver *fu*
LVTX Landing Vehicle, Tracked, Experimental *hc, uc*
LW Logistics Wing *fu*
LW Long Wavelength (Wave) *uc*
LW Lucas Weinschel [Incorporated] *hq5.2.7*
LW-3D Light Weight Three Dimensional Radar *hc*
LW3-D Lightweight Three-Dimensional [Radar] [also LW-3D] *fu*
LWB Logistics Workbench Project *fu*
LWDB Land Warfare Data Base *wp*
LWDM Light Weight Dogfight Missile *hc*
LWDS Local Weather Dissemination Systems *mt*
LWELJ Long Wavelength Expendable Laser Jammer *hc*
LWESS Lightweight Weapons Engagement Scoring System *hc*
LWF Light Weight Fighter [program that created the F-16 and F-17, USA] *mt*
LWHSS Lightweight Honeycomb Sandwich Structure *hc*
LWIR Long Wavelength Infra Red [probe] *mt, fu, 16, 17, sb, hc, jb, nt, wp, tb, st, pbl, kf*
LWIRJ Long Wavelength Infrared Jammer *hc*
LWL Land Warfare Laboratories [US Army] *hc*
LWL Liquid Water, Land *ac7.5-12*
LWL Long Wavelength [Laser] *uc, bf*
LWLD Lightweight Laser Designator *hc*
LWLOJS Long Wavelength Optical Jamming Simulator *hc*
LWO Limited Warfare Office (Officer) *hc*
LWO Liquid Water, Ocean *ac7.5-12*

LWOP Lease With Option to Purchase *mt*
LWOP Leave Without Pay *wp*
LWP Leave Without Pay *pbl*
LWR Laser Warning Receiver *mt*
LWRL Light Weight Rail Launcher *hc*
LWRM Light Weight Radar Missile *hc*
LWS Laser Warning System *mt*
LWS Laser Weapon System *fu*
LWSD Large Wall Screen Display *mt*
LWSF Light Weight Strike Fighter *hc*
LWTP Light Weight Tracking Pedestal *hc*
LX Low power Duplex *hi*
LX/LSD Liquid-Crystal Large-Screen Display *fu*
LYRIC Language for Your Remote Instruction by Computer *wp*
LZ Landing Zone *mt, w9, wp, fu*
LZA Local Zenith Angle *kf*
LZC Landing Zone Construction [dropping blast bombs to clear an area] [USA] *mt*
LZCO Landing Zone Control Officer *mt*
LZP Landing Zone Preparation [suppressing enemy action to make landing of aircraft or helicopters possible, USA] *mt*
LZT Local Zone Time *fu*
LZW Lempel, Ziv, Walsh (Welch) [a compression algorithm] *ct, kg, em, him, mt*

M

M [code applied to designations for armament systems and equipment, such as tanks, used by the US Army] [USA] *mt*
M&A Maintenance and Administration *fu*
M&C Monitor and Control *oe*
M&DOD Mission and Data Operations Directorate *2, 4, oe*
M&DOR Mission and Data Operations Review *oe*
M&E Machinery and Equipment *fu, hc*
M&IO Movements and Identification Officer *fu*
M&LC Mission and Launch Control *st*
M&O Maintenance and Operations *hc, 4*
M&P Materials and Processes *aw129.21.153, fu, ah, ep, jdr, pbl, tm, jn*
M&PE Material and Process Engineering *fu*
M&PL Materials and Processes List *gp*
M&R Maintenance and Repair *fu*
M&RA Manpower and Reserve Affairs *dt*
M&S Maintainability and Supportability *rm2.2.4*
M&S Maintenance and Supply *fu*
M&T Models and Tools *fu*
M&T Monitor and Test *gp, oe*
M&TC Mission and Traffic Control *hc*
M&TE Measuring and Test Equipment *jdr*
M-3D Mobile 3-Dimension *fu*
M-DAY Mobilization Day *wp*
M-DEMO Maintenance Demonstration *kf*
M-FSK {(?)..Frequency Shift Keying} [a waveform] *sd*
M-SPEC NAVSEA Modular Specification System *fu*
M-X Multiple-Warhead Experimental [missile] *uc*
M/A Microanalytical *ne*
M/A_COM (Federal Express has awarded M/A COM DCC the . . . contract) [Burlington, Massachusetts] *sc4.4.4, aw130.22.77*
M/C Mid Course *0, kf*
M/G {Major General} *af72.5.28*
M/P Manpower *hc*
M/S Mail Server *kf*
M/S Mail Station *fu, sc2.7.4, oe*
M/S Main/Standby *fu*
M/S Margin of Safety *jn*
M/S Meters per Second *kf*
M/SEC Meter(s) per Second *vv*
MA Maintenance Action *wp*
MA Maintenance Agency *mt*
MA Maintenance Aid *fu*
MA Manufacturing Agreement *fu*
MA Manufacturing Authorization *hc, 8*
MA Memory Address *fu*
MA Mercury Arc *fu*
MA Message Assembler *fu*
MA Missile Armed *hc*
MA Mission Abort *mt*
MA Mission Accomplished *wp*
MA Mission Assignment *wl*
MA Mission Assurance *oe*
MA Mobilization AugmentÈe *mt*
MA Multiple Access *2, 4, wl, vv*
MA/PS {Mathematics and Physical Science} *aa27.4.60*
MAA Military Advisory Assistance *hc*
MAA Mission Area Analysis *mt, hc, fu, wp, st, kf*
MAAC Minimum Aft Axial Clearance *sa12474.32*

MAAG Military Assistance Advisory Group [USA DOD] *mt, fu, su, hc, wp, tb, jdr*
MAAMA [Hughes Aircraft Company customer for airborne intercept missile] *hc*
MAAR Maintenance Action Arrival Rate *hc*
MAAS Machine Assisted Assembly System *hc*
MAAS Manpower Allocations and Accounting System *mt*
MAAST Multiple Application Addressable Secure Television *dn*
MAATE Multiple Application Automatic Test Equipment *hc*
MAB Marine Amphibious Brigade *12, fu, wp, ph, hc, mt*
MAB Missile Assembly Building *17, ai2*
MABS Marine Air Base Squadron [USMC] *mt*
MAC Machine Aided Cognition *0*
MAC Madrid Automated Center *hc*
MAC Maintenance Action Code *fu*
MAC Maintenance Allocation Chart *fu*
MAC Mandatory Access Control *tb*
MAC Maneuver Area Command [USA] *mt*
MAC Manufacturing Artwork Check *fu*
MAC Manufacturing Assembly Code *fu*
MAC Marine Artillery Consultant *hc*
MAC Master Aperture Card [microfilm copy] *ah, jn, ep, jdr, pbl, tm*
MAC Maximum Allowable Concentration *wp*
MAC Mean Aerodynamic Chord *mt*
MAC Media (Medium) Access Control *mt, kg, nu, fu, ct, em, hi*
MAC Memory Access Controller *fu*
MAC Microfilm Aperture Card *fu*
MAC Microprogrammed Ada Compiler *fu*
MAC Military Airlift Command [formerly MATS] [USA] * *mt, fu, 12, 17, hc, w9, af72.5.31, dt, wp, tb*
MAC Mobile Automated Correlator *hc*
MAC Months (Month) After Contract [award] *fu, wp, tb*
MAC Multiple Access Computer *0*
MAC Multiple Access Controller *mt*
MAC Multiple Array Correlation *wp*
MAC Multiplexed Analog Component *dn*
MAC-TCU Maintenance Alignment Cart-TCU *un*
MAC/RAN Measurement Analysis Corporation/Random Data *hc*
MACA Military Air Clearance Authority *mt*
MACA Missile Assembly and Checkout Area *fu*
MACA Modular Automatic Conferencing Arranger *mt*
MACA Months After Contract Award *fu*
MACAF MAC* numbered Air Force [21st, 22nd, and 23rd] *mt*
MACALT MAC* Alternate Headquarters *mt*
MACARMS MAC* Aircrew Resources Management System [MAC*] *mt*
MACARNET MAC* Airlift Recovery Network *mt*
MACB Missile Assembly and Checkout Building *fu*
MACCONNET MAC* Contingency Network *mt*
MACCS Marine Air Command and Control System *mt*
MACCS Marine Aviation (Air) Command and Control System [formerly MTDS] *fu, hc*
MACCSU MAC* Accelerated Command Center System Upgrade [MAC*] *mt*
MACE Maneuvering Attack Concept Evaluation *hc*
MACE Mechanical Antenna / Array Control Electronics *hc*
MACE Military Airlift Capability Estimator *mt*
MACE Military Airlift Center Europe [obsolete] *mt*
MACE Mission Adaptive Combat Ensemble *wp*
MACEW Military Airlift Command Electronic Warfare *hc*
MACG Military Air Control Group *-12*
MACH Multilayer Actuator Head [Epson] *pk, kg*
MACI Military Adaptation of Commercial Instruments (Items) *ph, fu, st*
MACIMS Military Airlift Command Integrated Management System *mt*
MACIPS MAC Information Processing System *mt*
MACIPS Military Air Command Information Processing System *wp*
MACISIN MAC Information Systems Internetting *mt*
MACL Memory Access Control Logic *em*
MACLO Military Airlift Command Liaison Officer *mt*
MACLOGFOR MAC* Logistics Force Packing System [MAC*] *mt*
MACM Multiple Architecture Cost Model [TRW] *-17*
MACMA MAC* Minicomputer Acquisition *mt*
MACMIP MAC* Mainframe Internetting Project *mt*
MACOM Major Command [US Army] *fu, mt, st*
MACOPS MAC* Operational Phone Systems *mt*
MACP Multiple Access Command and Pilot *-4*
MACR MAC* Regulation *mt*
MACR Materiel Acquisition Control Record *mt, wp*
MACRO MAC* Resource Optimization *mt*
MACRO Monopoles, Astrophysics, and Cosmic Ray Observatory *mi*
MACS Manned Air Combat Simulator *wp*
MACS Marine Air Control Squadron *mt*
MACS Medium Altitude Communication Satellite *hc*
MACS Modular Attitude Control [Sub] System *sd, hc*
MACS Multiple Access Communications System *mt, fu*
MACSAM MAC* System Architecture Modernization *mt*

MACSO Military Airlift Command Support Office *mt*
MACU Material Cost per Unit *wp*
MACV Military Assistance Command in Vietnam [USA] *hc, mt*
MACWACC MAC* Washington Area Computer Center *mt*
MAD Magnetic Airborne Detector [detects submarines by measuring the changes in the magnetic field of the earth] *mt*
MAD Magnetic Anomaly Detector (Detection) [detects submarines by measuring the changes in the magnetic field of the earth] *fu, hc, mt*
MAD Maintenance and Diagnostics *fu*
MAD Mean Absolute Deviation *hc*
MAD Message Address Directory *mt*
MAD Mission Area Deficiency *st*
MAD Multifunction Antenna Development *hc*
MAD Multiple Aperture Device *fu, hc*
MAD Mutual(ly) Assured Destruction *fu, mt, 11, wp, 13, 14, w9, tb*
MADA Multiple Access Demand Assignment *mt*
MADA Multiple Access Discrete Address *hc*
MADAC Madrid Air Defense Automated Center *hc, fu*
MADAN Multimission Attitude Determination/Autonomous Navigation *hc*
MADAR Malfunction Analysis Detection And Recording (Reporting) *mt, af72.5.140, wp, fu*
MADARS Malfunction Detection, Analysis and Recording System *wp*
MADCAP Mission Assessment and Design Capability [Program] [model] *hc, wp*
MADCAP Mosaic Array Data Compression and Processing *kf*
MADD Missile Aerosurface Development Device *hc*
MADDIDA Magnetic Drum Digital Differential Analyzer *fu*
MADE Manufacturing and Automated Design Engineering *pk, kg*
MADGE Malaysian Air Defense Ground Environment *fu*
MADGE Microwave Aircraft Digital Guidance Equipment. A landing-aid *mt*
MADI Master Data [submittal] Index *hc*
MADIZ Military Air Defense Identification Zone *wp*
MADP Materiel Acquisition Decision Process *mt, st*
MADRE Magnetic Drum Receiving Equipment *fu*
MADRE [high power, long range, over the horizon radar] *hc*
MADREC Malfunction Detection and Recording *wp*
MADS Maintenance And Diagnostic Subsystem *wp*
MADS Manned Airborne Defense Station *hc*
MADS Mars Atmosphere Density Sensor *hc*
MADS Militarized Advanced Disk System *fu*
MADS Mission Area Deficiency Statement *st*
MADS Mobile Air Defense System *hc*
MADS Mobile Airborne Data System *hc*
MADT Malaysian Air Defense Trainer *fu*
MADT Maximum Allowable Down Time *fu*
MADT Microalloy Diffused Transistor *fu*
MADYMO Mathematical Dynamic Modeling [TNO] *pk, kg*
MAE Mean Area of Effectiveness *hc*
MAE Mechanical and Aerospace Engineering [department at IIT] *aa26.11.52*
MAEC Manufacturing Analysis of Engineering Change *fu, hc*
MAF MA Forward *-4*
MAF Marine Amphibious Force [USMC] *mt, 12, wp, fu*
MAF Multiple Access Forward *vv*
MAFD Minimum Acquisition Flux Density *fu*
MAFEE Mission and Flight Essential Equipment *fu*
MAFF Ministry of Agriculture, Forest, and Fisheries [Japan] *ac3.3-23*
MAFFS Modular Airborne (Aerial) Fire Fighting System [installed in ANG C-130s to assist in firefighting] [USA] *mt, af72.5.35*
MAFIA Missile Auxiliaries Firing Interlock Assembly *hc*
MAFIS Mobile Area Field Instrumentation System [USA] *mt*
MAFLIR Modified Advanced Forward Looking Infrared *hc, wp*
MAFR Major Frame *ggIII.1.54*
MAFS Manned Aerospace Flight Simulator *ci.971*
MAG MA Antenna Group *vv*
MAG Marine Air Group [USA] *mt, 12, hc*
MAG Maritime Air Group [Canada] *mt*
MAG Marketing Action Group *st*
MAG Military Airlift Group [USA] *mt*
MAG/ER Magnetometer/Electron Reflectometer *re*
MAGE Mechanical Aerospace Ground Equipment *gp, jdr, pbl, tm, kf, ah, jn, oe*
MAGE Mechanical Assembly Ground Equipment *-4*
MAGE Multiple Access Ground Equipment *-4*
MAGGE Medium Altitude Gravity Gradient Experiment *hc*
MAGIC Maritime Air Ground Intelligence Center [PACFLT] *mt*
MAGIC Microcomputer Application of Graphics and Interactive Communication *wp*
MAGIC Missile Agile de Combat [(French) Agile Combat Missile] *wp*
MAGIIC Mobile Army Ground Imagery Interpretation Center [USA] *fu, mt*
MAGIS Marine Air Ground Intelligence System [USMC] *mt, hc, wp, fu*
MAGIS Megawatt Air to Ground Illuminating System *wp*
MAGISIAC Marine Air Ground Intelligence System Intelligence Analysis Center *mt*
MAGLAD Marksmanship And Gunnery Laser Device *hc*

MAGNA (MAGNA_8 fourth generation software) *mt*
MAGSAT Magnetic Field Satellite *hy*
MAGTF Marine Air Ground Task Force [USA DOD] *fu, mt*
MAH Milliampere-Hour *pk, kg*
MAI Management Action Indicator *mt*
MAI Manufacturing Assembly Instructions *fu*
MAI Military Assistance Institute *dt*
MAI Multiple Applications Interface *pk, kg*
MAID Magnetic Anti Intrusion Detector *mt*
MAID Manual Intervention and Display *hc*
MAID Modular AUTODIN Interface Device *mt*
MAIDA Multiple Attribute Identification Analysis *hc*
MAIDS Multiple purpose Automatic Inspection Diagnostic System *hc*
MAIL Multiple Aperture Interlinked *hc*
MAILS Multiple Antenna Instrument Landing System *wp*
MAIR Molecular Airborne Intercept Radar *hc*
MAIRS Medium Altitude Infrared System *ro*
MAIRS Military Airlift Integrated Reporting System [MAC*] *mt*
MAIRU Mobile Aircraft Instrument Repair Unit *wp*
MAIS MAC Air Intelligence System *hc*
MAIS Maintenance Information System *wp*
MAIS Major Automated Information Systems *mt*
MAIS Military Airlift Intelligence System [MAC*] *hc, mt*
MAISITE Modular Automatic Integrated System/Interoperability Test and Evaluation *hc*
MAISRC Major Automated Information System Acquisition Review Council *wp*
MAISRC Major Automated Information Systems Review Council *mt*
MAIT Missile Airframe Integration Technology *hc*
MAITS MILSTAR Automatic Intermediate-level Test Station *fu*
MAJB Multi-channel Anti-Jam Broadcast *jdr, uh*
MAJCOM Major Command [USAF] *mt, wp, af72.5.103, dm, st*
MAL Macro Assembly Language *hi*
MAL Materiel Allowance List *wp*
MAL Memory Access List *wp*
MALE Multiaperture Logic Element *fu*
MALLAR Manned Lunar Landing And Return *ci.971*
MALOR Mortar and Artillery Locating Radar [Army] *fu*
MALOS Maintenance And Logistics Space program *hc*
MALS Medium Intensity Approach Lights *fu*
MALS Miniature Air Launcher (Launched) System *mt, hc, sr9-82*
MALSR MALS with Runway Alignment Indicator Lights *fu*
MAM Maintain Airspace Model *fu*
MAM Maintenance Assistance Model *fu*
MAM Management Analysis Model *mt*
MAM Medium Altitude Missile *hc*
MAM Munitions Assessment Model [HQ USAF] *mt*
MAMA Maintenance And Malfunction Analysis *hc*
MAMB Mission Avionics Multiplex Bus *wp*
MAMOS Marine Automatic Meteorological Observing *hc*
MAMP Mission Area Materiel Plan *mt*
MAMP Mission Assurance Management Plan *oe*
MAMS Military Aircraft Marshalling System *hc*
MAMS Multispectral Atmospheric Mapping Sensor *hy*
MAMS-II Maintenance Activity Management System II [USA] *mt*
MAN Material Alert Notice *fu*
MAN Metropolitan Area Network *ct, em, kg, nu, hi*
MANDC (Manual Differential Correction (Software Program)) *jb*
MANFOR Manpower Force Packaging System [USAF] *mt*
MANIAC Mathematical Analyzer Numerical Integrator and Computer *pk, kg*
MANIET Manifest Update *mt*
MANOVA Multivariate Analysis Of the Variance *wp*
MANPAD Man Portable Air Defense *hc, fu*
MANPADS Man Portable Air Defense System [became FIM-92 Stinger] [USA] *hc, mt*
MANPER Manpower and Personnel Module *mt*
MANPER-B Manpower and Personnel Module - Base Level [COMPES] [USAF] *mt*
MANPER-M Manpower and Personnel Module - Major Command Level [COMPES] [USAF] *mt*
MANPLAWS Manportable Assault Laser Weapons System *hc*
MANPRINT Manpower and Personnel Integration *mt, fu, nt*
MANREQ Manpower Requirements System [USAF] *mt*
MANS Mission Analysis for Missile and NUDET Surveillance *mt*
MANTECH Manufacturing Technology [Army Program] *uc, 17, sb, mj31.5.20, wp, kf, fu*
MANTRAC Manual Angle Tracking Capability *fu*
MAN_TECH Manufacturing Technology [also MANTECH] *mj31.5.20*
MAO Major Attack Option *mt*
MAO Mono Amine Oxidase *w9, wp*
MAOS Minimum Aircraft Operating Strips *mt*
MAOT Maximum Allowable Operating Time *fu*
MAOTS Missile Auxiliary Output Testers *hc*
MAP Macro Assembly Program *wp*
MAP Main memory, Arithmetic unit, and Post-processor *hc*
MAP Maintenance Analysis Procedures *pk, kg*

MAP Maintenance and Administration Position *mt*
MAP Major Activities and Products *fu*
MAP Management Action Program [Hughes Aircraft Company] *pl*
MAP Manual Array Processing *fu*
MAP Manufacturing Automation Processor *em, ct*
MAP Manufacturing Automation Protocol *mt, fu, hc, kg, is26.1.46*
MAP Master Automation Plan *mt, tb*
MAP Measurement and Processing *fu*
MAP Memory Allocation Map *kg, hi*
MAP Message Acceptance Pulse *fu*
MAP Microcomputer Array Processor *fu*
MAP Middle Atmosphere Program *hg, ge*
MAP Midyear Appraisal and Planning *pbl*
MAP Military Aid Program *mt*
MAP Military Assistance Program [USA] *fu, mt*
MAP Milstar Advanced Processor *jdr*
MAP Milstar Alternate Processor [aka, resource controller] *jdr*
MAP Mission Activity Plan *wl*
MAP Mission Area Plans *wp*
MAP Mission Automation Plan *mt*
MAP Modern Aids to Planning *mt*
MAP Modular Arithmetic Processor *em*
MAP Modular Array Processor *fu*
MAP Multiple Aim Point *mt, tb*
MAP Municipal Airport *-12*
MAP Mutual Aid Program *tb*
MAP Mutual Assistance Program *wp, hc*
MAP Mutual Assistance Program [USA] *mt*
MAPAC Military Assistance Program Address Code *wp*
MAPAD Military Assistance Program Address Directory *mt, wp*
MAPAG Multiple Association Policy Advisory Group *ph*
MAPAR Materials And Processes Acceptance Requirement *hc*
MAPE Multimode (Display) Ancillary Programming Equipment *fu*
MAPES Management of Personnel Records [USMC] *mt*
MAPG Maximum Available Power Gain *fu*
MAPG Monograph Advisory Planning Group *hc*
MAPI Mail/Messaging Applications Programming Interface [Microsoft] *kg, hi*
MAPI Messaging Applications Programming Interface *ct, em*
MAPICS Manufacturing, Accounting and Production
MAPL Manufacturing Assembly Parts List *8, hc, fu*
MAPLE Multipurpose Long Endurance Aircraft [USAF] *hc*
MAPMIS Manpower and Personnel Management Information System [US Navy] *mt*
MAPP Modern Aids to Planning Program [OJCS] *mt, aw130.6.2*
MAPPER Maintaining, Preparing and Processing Executive Reports *mt*
MAPR Miniaturization and Automation of Personnel Record [USMC] *mt*
MAPS MAC* Automated Planning System [MAC*] *mt*
MAPS Managerial Administrative Problem Solving *hn43.18.2, 8*
MAPS Manpower and Personnel System *mt*
MAPS Measurement of Air Pollution from Satellite [Shuttle] *ge, hc, hy, hg*
MAPS Measurement of Atmospheric Pollutants from Space *ac1.7-26*
MAPS Missile Application Propulsion Study *hc*
MAPS Mobile Aerial Port Squadron *mt*
MAPS Mobility Analysis and Planning System [MTMC] *mt*
MAPS Modular Azimuth Position System *aw129.21.c2*
MAPS Multiple Application Pro-fan Studies *hc*
MAPS Multiservice Access Platforms *ct, em*
MAPS Multispectral Active Passive Scanner *hc*
MAPSE Minimal Ada Programming Support Environment *fu*
MAPSE Minimal APSE *mt*
MAPU Memory Allocation and Protection Unit *fu*
MAQ Maximum Authorized Quantity *wp*
MAR Management Assessment Report *wp*
MAR Manual Analysis Receiver *fu*
MAR Manufacturing Assessment Review *fu*
MAR Memory Address Register *3, 17, sb, fu*
MAR Minimally Attended Radar [SEEK_IGLOO] *fu, mt*
MAR Multi-function Array Radar [Nike X] [also MFAR] *hc, wp*
MAR Multiple Access Receiver *4, 17, sb*
MAR Multiple Access Return *vv*
MARAD Maritime Administration *mt*
MARAIRMED Maritime Air Force Mediterranean [USMC] *mt*
MARC MAC ALCE Reactions Communications *mt*
MARC Machine Readable Cataloging *fu*
MARC Material Accountability Recoverability Code *wp*
MARC Military Airlift Command/Airlift Control Element/Reaction Communication *fu*
MARC Mobile Area Repair Calibration *hc*
MARCCO Man's Role in Command and Control *mt, hc*
MARCEP Maintainabilty And Reliability Cost Effectiveness Program [model] *wp*
MARCOR Marine Corps Radar *hc*
MARCORPS Marine Corps *mt*
MARDAC Manpower Research and Data Analysis Center *mt*

MARDIS Modernized Army Research and Development Information System *mt, st*
MARDIV Marine Division *mt, fu*
MARECS Maritime European Communications Satellite *sc3.3.4, sf31.1.23*
MAREMIC Maintenance, Repair and Minor Construction *wp*
MAREMICS Maintenance, Repair and Minor Construction *mt*
MAREP Maritime Reporting System *mt*
MARFE Multiple Access Radio Frequency Equipment *-4*
MARFOR Marine Forces *mt*
MARISAT Marine Satellite System *mt*
MARISAT Maritime [telecommunications] Satellite [system] *fu, sm18-24, lt*
MARK 1B Military Affiliated Radio System [Designated AN/MSC-46 by Army SATCOM] *fu*
MARK 1B Transportable Satellite Communications Link Terminal *fu*
MARK IV Light Helicopter Transportable Satellite Communications Terminal *fu*
MARMIC Command Level Maintenance Repair and Minor Construction Program reporting system *mt*
MARP Manpower Requirements Plan *mt*
MARP Mating and Ranging Program [SAC] *mt*
MARR Manpower Authorization Requirement Review *mt*
MARRES Manual Radar Reconnaissance Exploitation Systems [ESD] *hc, wp, ro, mt*
MARS (modal characteristics of substructures ... computing) *gg20*
MARS Maintenance And Repair System *wp*
MARS Manned Astronomical Research Station *wp*
MARS Marine Corps Ammunition Reporting System [USMC] *mt*
MARS Material Analysis Review System *fu*
MARS Matrix Analysis of Redundant/Routine Structure *hc*
MARS Microprogrammable Accelerator for Rapid Simulations *is26.1.35*
MARS Mid Air Recovery / Retrieval System *mt*
MARS Military Affiliate Radio System [USA DOD] *hc, wp, mt*
MARS Military Affiliated Radio System *fu*
MARS Mission Maintenance And Reliability Simulation *wp*
MARS Mobile Air Defense Radar System *hc*
MARS Monitoring And Reporting System *fu*
MARS Multicast Address Resolution Server *ct, em*
MARS Multiple Artillery Rocket System *hc*
MARSAM Multiple Airborne Reconnaissance Sensor Assessment Model *wp*
MARSS Meteorological And Range Safety Support *wp*
MART A coded mask X-ray telescope on SXG *cu*
MART Mobile Automatic Radiating Tester *hc*
MART Modular Automated Remote Terminal *mt*
MARTS Master Radar Tracking System *hc*
MARV Maneuvering (Maneuverable) Reentry Vehicle [also MRV] *mt, uc, wp, hc, aw129.10.13, tb, st*
MARV Maneuvering Anti Radiation Vehicle *0*
MARV Mobile Armored Reconnaissance Vehicle *wp*
MARVEL Machine Assisted Realization of the Virtual Electronic Library [Library of Congress] *pk, kg*
MARVS Material Acquisition Requirements Validation System *mt*
MAS MA Subsystem *vv*
MAS Military Agency for Standardization [NATO] *wp, mt*
MAS Military Airlift Squadron [MAC*] *mt, af72.5.121*
MAS Mission Area Summary *hc*
MAS Mission Auxiliary Subsystem *hc*
MAS Mobile Access Structure *ai2*
MAS Multifunction Acquisition System *fu*
MAS Multiplex Applications Software *mt*
MAS Mutually Assured Survival *tb*
MAS/S Mechanical Arm Sub-System *hc*
MASA Modular Avionics System Architecture *wp*
MASAR Multimode Airborne Solid State Array Radar *wp*
MASC MAGTF Automated Services Center [USMC] *mt*
MASC Milstar Auxiliary (Ancillary) Support Center *jdr*
MASCDC Military Aircraft Storage and Disposition Center [at Davis-Monthan AFB, Arizona] [USA] *mt*
MASCOT Military Air Transportable Satellite Communications Terminal *wp*
MASDC Military Aircraft Storage and Disposition Center [AFLC] *mt*
MASDCMIS MASDC Management Information System *mt*
MASE Mission And Systems Engineering *re*
MASER MAterial Science Experimental Rocket *crl*
MASER Microwave Amplification by Stimulated Emission of Radiation [e.g, device that uses this for its operating principle] "maser" *fu, 0, wp, hc*
MASF Military Assistance Service Fund *wp*
MASF Mobile Aeromedical Staging Facilities *mt*
MASFC MAGTF All Source Fusion Center *mt*
MASH Mobile Army Surgical Hospital *mt, w9*
MASIIS Maintenance Analysis Structural Integrity Information System *mt*
MASINT Measurement And Signature (Signal) Intelligence [USA DOD] *kf, mt*
MASIS Management And Scientific Information System *wp*

MASKAP (we use MASKAP to first check design rule integrity) *ne*
MASNET Management Support Network [USA] *mt*
MASP Multiple Access Signal Processor *-4*
MASR Microwave Atmospheric Sounding Radiometer *hc*
MASR Miniature Airborne/Spaceborne Reconnaissance [Data Link] *wp*
MASR Multiple Antenna Surveillance Radar *wp*
MASS Manned Aircraft Surface to Surface *hc*
MASS Manned Aircraft versus Surface to Surface [missile] *hc*
MASS Maximum Availability and Support Subsystem [Parallan] *pk, kg*
MASS Medium Altitude Sensor System *fu*
MASS MICAP Asset Sourcing System *mt*
MASS Military Airlift Support Squadron [MAC*] *mt*
MASS Military Airlift Survivability Study *wp*
MASS Mobilization Automated Support System [FORSCOM] *mt*
MASS Multiple Array Serial Scan *hc*
MASSTER Mobile Army Sensor System Test, Evaluation, and Review *hc*
MAST Maintenance and Status Terminal *fu*
MAST Model Assembly Sterling Center *hc*
MASTACS Maneuvering Augmentation Simulation of Tactical Aerial Combat System *fu*
MASTER Mission Area Simulation To Evaluate Requirements [model] *wp*
MASTIFF [Israeli Tactical RECCE/Surveillance RPV Carried Radar] *wp*
MASW (Cmdr., 1605th MASW, 21st AF,) *af72.5.31*
MASW Mission Application Software [WIS] *mt*
MAT Maintenance Access Terminal *pk, kg*
MAT Microalloy Transistor *fu*
MAT Missile Adapter Testers *hc*
MAT Mission Allowable Temperature *oe*
MAT Mobilization Assistance Team *mt*
MAT Module Adaptive Threshold *kf*
MAT Monopulse Angle Tracker *hc*
MATB Missile Auxiliary Test Benches *hc*
MATC Missile Auxiliary Test Console *hc*
MATCALS Marine Air Traffic Control and Landing System [USMC] *mt, fu, hc, pf.f88.9*
MATCO Military Air Traffic Coordinating Office *mt, wp*
MATCON Microwave Aerospace Terminal Control *fu*
MATCU Military Air Traffic Coordinating Unit *mt*
MATDEV Materiel Developer *st*
MATE Manual Adaptive TMA Evaluation *fu*
MATE Modular AUTODIN Terminal Equipment *mt*
MATE Modular Automatic (Automated) Test Equipment *mt, fu, wp, hc, af72.5.77, is26.1.51, tb, kf*
MATE Modular Avionics Test Equipment *wp*
MATEAM Manufacturing Process Applications Team *hc*
MATER Magnetic Tape Event Recorder *hc*
MATES Mobilization and Training Equipment Site *mt*
MATH Modular Air Transportable Hospital *wp*
MATLID Multiple Frequency Atmospheric LIDAR *ge, hg*
MATMS Marine Aviation Training Management System [USMC] *mt*
MATOC Mobility Air Terminal Operations Center *mt*
MATP Missile Auxiliary Test Position *hc*
MATRA Mechanique Aviation TRAction *wp, crl*
MATRACS Military Air Traffic Control System *hc*
MATS Midcourse Airborne Target Signatures *hc*
MATS Military Air Transport Service (System) [USA] *hc, wp*
MATS Miniature Addressable Transceiver *hc*
MATS Missile Adapter Test Set *hc*
MATS Missile Auxiliary Test Set *hc*
MATS Monitor and Test System *fu*
MATS Multiparticipant Airbattle Training System *wp*
MATV Master Antenna Television *ct, dn*
MATV Multiple Axis Thrust Vectoring [an F-16 with a vectoring nozzle] [USA] *mt*
MATZ Military Air Traffic Zone *mt*
MAU Marine Amphibious Unit [USMC] *fu, 12, mt*
MAU Media Access Unit *kg, hi*
MAU Medium Attachment Unit *ct, em*
MAU Million Accounting Units *es55.11, ai2*
MAU Miscellaneous Armament Unit *wp*
MAU Missile Auxiliary Unit *hc*
MAU Multi Station Access Unit *ct*
MAU Multiple Aircraft Universal *hc*
MAU Multistation Access Unit *pk, kg*
MAV Mars Ascent Vehicle *re*
MAVAR Modulating Amplifier using Variable Reactance *fu*
MAVCS Multi Axis Vibration Control System *tb*
MAVDM Multiple Application VDM *kg, hi*
MAVERICK [TV guided air-to-surface missile] *wp*
MAVMS MACOM Automated Vehicle Management System [USA] *mt*
MAVS Manned Aerial (Airborne) Vehicle for Surveillance *wp, hc*
MAW Management Action Workshop *pbl, hn44.20.4, ah, jn*
MAW Marine Air(craft) Wing [USA] *fu, 12, mt*
MAW Mechanically Armed Warhead *st*

MAW Medium Assault Weapon [anti-tank] *hc*
MAW Military Air (Airlift) Wing [USA] *mt, wp, af72.5.120*
MAW Minor (Manufacturing) Assist Work [authorization or document] *hc, 8, pl, ep, jdr, pbl, tm, fu, su, ah, jn*
MAW Mission Adaptive Wing *mt*
MAW Multipurpose Assault Weapon *wp*
MAWG Mission Analysis Working Group *tb*
MAWS Missile Approach Warning System [UK] *mt*
MAWS Missile Attack Warning System *wp*
MAWTS Marine Aviation Weapons and Tactics Squadron [USMC] *mt*
MAX Message Automatic Exchange *mt*
MAXCMT Maximum Correction Maintenance Time *fu*
MAXI Modular Architecture for the Exchange of Intelligence [DODIIS] *mt*
MAXIE Magnetospheric Atmospheric X-ray Imaging Experiment [USA] *hg, ge*
MaxTTRF Maximum Time To Restore Function *kf*
MAYSAT [navigational satellite] *sd*
MB Main Base *mt*
MB Maintenance Bulletin *fu*
MB Maneuver Box *fu*
MB Manned Base *mi*
MB Mediciniae Baccaloreatus *wp*
MB Memory Buffer *fu*
MB Mid Byte *-3*
MB Milli Bar(s) *pl, oe*
MB Momentum Bias *sd*
MBA Main Battle Area *fu, mt*
MBA Master of Business Administration *wp, hn43.22.3, w9*
MBA Motor Bearing Assembly *bl1-3.6*
MBA Multiple Beam Antenna *mt*
MBA Multiple Beam Array [Antenna] *kja, tb, jdr*
MBAM Main Beam Avoidance Maneuver *hc*
MBASIC Microsoft BASIC [Microsoft] *kg, hi*
MBB Make Before Break *fu*
MBB Make-or-Buy Board *fu*
MBB Messerschmitt-Boelkow-Blohm [Germany] *fu, aa26.12.20, aw130.13.5, hc, sc2.7.3*
MBB Modular Building Block *mt, wp*
MBBG Main Bang Blanking Generator *hc*
MBBLS Thousands of Barrels *mt*
MBC Meteor Burst Communications *mt, ro, fu*
MBC Military Budget Committee *mt, hc*
MBC Missile Boresight Correlator *hc*
MBCP Mauler Battery Command Post *hc*
MBCP Missile Base Communications Processor *mt*
MBCS Meteor Burst Communication System *hc*
MBCS Multi-Byte Character Set [IBM] *pk, kg*
MBDC Modified Battery Discharge Controller *jdr*
MBDC Modular Battery Discharge Controller *uh*
MBDL Missile Battery Data Link [USA] *mt, fu, go3.1-1*
MBE Molecular Beam Epitaxy *fu, wp, is26.1.89*
MBFR Mutual and Balanced Force Reduction *mt, wp, 14*
MBFT Multi Band Frequency Translator *hc*
MBI Major Budget Issues *wp*
MBI Multi Beam Imaging [Radar] *so*
MBIC Memory Bus Interface Chip *oe*
MBIU Maintenance Bus Interface Unit *wp*
MBK Multiple Beam Klystron *fu*
MBL Medium Brightness Laser *hc*
MBM (MBM Technology, Ltd, Brighton, UK) *aw129.10.53*
MBM Magnetic Bubble Memory *fu*
MBM Manufacturing Bill of Material *fu*
MBN Material Breakdown Number *fu*
MBO Management by Objectives [a DOD program] *mt, fu, wp*
MBP Management By Planning [Policy] *ad*
MBPS Mega Bits Per Second *mt, ggIII.1.41, ic2-66, aa26.11.44, af72.5.171, pbl, kf, kg, hi, oe*
MBPS Mega Bytes Per Second *tb, wl, kg, hi*
MBR Marker Beacon Receiver *fu*
MBR Mars Balloon Relay *un*
MBR Master Boot Record *pk, kg*
MBR Mechanical Buffer Register *fu*
MBR Mezhkontenentalnaya Ballisticheskaya Raketa [ICBM] [USSR] *mt*
MBRRGChIN MBRs Razdelyayuchimi Golovnyym Chastyami Indiviualnogo Navedeniya [MIRV] [USSR] *mt*
MBRV Maneuverable Ballistic Re-Entry Vehicle *hc, 0*
MBS Magnetron Beam Switching *fu*
MBS Main "Bang" Suppressor *fu*
MBS Maintenance Bench Set *fu*
MBS Master Boot Sector *pk, kg*
MBS Maximum Burst Size *em, ct*
MBS Mission Bit Stream *mt*
MBS Modular Bremsstrahlung Source [a nuclear hardness test device] *-17*
MBSGM Multi Base Sortie Generation Model [HQ USAF] *mt*
MBSS Multi Band Staring Sensor *hc*

MBSTA Mobilization Station *mt*
MBT Main Battle Tank *mt, dh, hc*
MBT Maximum Burst Tolerance *ct, em*
MBU Materiel Business Unit [organization] *pbl, tm, ah, jn*
MC Magnetic Core *fu*
MC Main Coolant *fu*
MC Maintenance Console *fu*
MC Major Component *fu*
MC Major Critical [subcontract determination] *-17*
MC Manufacturing Convenience *fu*
MC Marine Corps *mt, dh, 12, dt*
MC Mean Chord *ai1*
MC Medical Corps [USA] *mt*
MC Message Center *mt*
MC Mid Course [phase] *fu, 17, sb, tb*
MC Military Characteristics *fu*
MC Military Committee *mt*
MC Military Computer *wp*
MC Mini Cartridge *ct, em*
MC Mission Capable [USA DOD] *mt*
MC Mission Commanders *fu*
MC Mission Complete *wp*
MC Mission Computer *fu*
MC Mission Control *fu, tb*
MC Mobility Control Center *mt*
MC Mode Change *fu*
MC Mode Control *fu*
MC Momentary Contact *fu*
MC Monitoring Center [DDN] *mt*
MC Multiple Contact *fu*
MC [designator, MC-130 is Combat Talon] *mt*
MC&G Mapping, Charting, and Geodesy [USA DOD] *mt*
MC&R Message Composition and Response *fu*
MC-PGA Metallized Ceramic - Pin Grid Array *pk, kg*
MC-QFP Metallized Ceramic - Quad Flat Pack *pk, kg*
MCA Manufacturing Change Analysis *hc, wp*
MCA Manufacturing Change Authorization *fu*
MCA Manufacturing Chemists Association *wp*
MCA Micro Channel (Adapter) Architecture [IBM] *kg, ct, em, hi*
MCA Military Construction, Army *mt*
MCA Missile Command Amplifier *hc*
MCA Monoclonal Antibodies *wp*
MCA Movement Control Agency *mt*
MCA Multiprocessor Communication Adapter *fu*
MCAD Mechanical Computer Aided Design *pk, kg*
MCAIR McDonnell [Douglas] *hc*
MCAM Manufacturing Cost Analysis Model *fu*
MCAP Major Command ADP Plan [obsolete] *mt*
MCAP Maximum Calling Area Procedure *mt*
MCAP Mobile Consolidated Aerial Port Subsystem *mt*
MCAR Military Construction Army Reserve *mt*
MCAR Multiple Channel Acoustic Relay [US Navy] *mt*
MCAS Marine Corps Air Station *mt, hc, 12, aw130.22.42*
MCAS Modular Component Assembly System *aw129.14.91*
MCATF Mechanized Combined Arms Task Force *mt*
MCATS Medium Capacity Automated Telecommunications System [USAF] *mt*
MCAWN Microcomputer Connectivity to AWN *mt*
MCB Materiel Contract Brief *fu*
MCB Memory Control Block *mt, kg*
MCB Mobile Construction Battalion [USA] *mt*
MCBD Multi Purpose Chemical Biological Decontaminant [chemical warfare] *wp*
MCBF Mean Cycles Between Failures *fu*
MCBIT Memory Card Burn In Tester *fu*
MCC Magnetometer Calibration Coil *jp*
MCC Maintenance Control Center *mt, fu, tb*
MCC Manual Control Console *fu*
MCC Manufacturing Change Control *fu*
MCC Master Control Center *lt*
MCC Master Control Console *fu*
MCC Message Class Code *-4*
MCC Microelectronics and Computer [Technology] Corporation *aa26.12.15, md2.1.1*
MCC Military Cooperation Community *dt*
MCC Mini Chip Carrier *fu*
MCC Mission Control Center *mt, hc, bl3-1.164, 4, 5, 15, 16, sb, pf1.22, jb, nt, cp, wp, tb, wl, jdr, kf, mb, mi, ah, uh, jn, pbl, oe*
MCC Mission Control Complex *hc, kf, ai2*
MCC Mission Critical Circuit *fu*
MCC Mission Critical Computer *jdr*
MCC Mobile Command Center [formerly GMCP] *ph, mt*
MCC Mobile Communications Center *mt*
MCC Mobile Control Center *fu*
MCC Mobility Control Center *mt*
MCC Monitor and Control Console *fu*

MCC Movement Control Center *mt*
MCCB MFR Change Control Board *fu*
MCCB MILSTAR Change Control Board *fu*
MCCB Multinational Configuration Control Board *wp*
MCCC Malaysian Command, Control and Communication *fu*
MCCC Marine Corps Command Center *mt*
MCCC Mission Control and Computing Center *gp, 16*
MCCC Mobile Consolidated Control Center *kf*
MCCD Mission Control Console Display *wp*
MCCD Multispectral Close Combat Decoys *hc*
MCCIMS Military Command and Control Information Management System *fu*
MCCISWG Military Command, Control Information Systems Working Group [NATO] *mt*
MCCM Multichannel Communications Module *fu*
MCCOR Motion Compensation Coherent On Receive *hc*
MCCP Military Consolidated Command Post *mt*
MCCP Multichip Cluster Package *fu*
MCCR Master Change Compliance Record *hc*
MCCR Mission Critical Computer Resources *mt, fu, hc, tb, jdr, kf*
MCCRES Marine Corps Combat Readiness Evaluation System [USMC] *mt*
MCCS Manpack Configuration Cable Set *fu*
MCCS Manufacturing Configuration Control System *fu*
MCCS Master Control Communication Station *hc*
MCCS Mission Critical Computer System *mt*
MCCS Mobile Command and Control Squadron *kf*
MCCS Mobile Command and Control System *hc*
MCCS Monitor and Control Console Suite *fu*
MCCSS Mobile Command Center Strategic System [OJCS] [formerly NEACP] *mt*
MCCW Monitor and Control Console Workstation *fu*
MCD Manipulative Communications Deception *mt*
MCD Manufacturing Control Document *oe*
MCD Master Control Drawing *fu*
MCD Medical Crew Director *mt*
MCD Microelectronics Circuits Division [of Hughes IEG] *hq5.3.5, hn48.13.8, md2.1.1*
MCD Minimum Cost Design *hc, wp*
MCD-ROM Multimedia CD-ROM *hi*
MCDAS Multiple Channel Data Acquisition System *hc*
MCDB Manufacturing Card Database *fu*
MCDC Mobilization Concepts Development Center *tb*
MCDD Multichannel Demux/Demod *rg*
MCDEC Marine Corps Development and Education Command *fu*
MCDN Marine Corps Data Network *mt*
MCDS Management Communications and Data System *wl*
MCDS Mission Critical Defense Systems *wp*
MCDS Multifunction CRT Display System *9, oe*
MCDT Maximum Corrective Down Time *fu*
MCDT Mean Corrective Down Time *hc*
MCE Matrix Computational Element *fu*
MCE Mid Course Early *tb*
MCE Missile Command Electronics *hc, nt*
MCE Mission Control Element *fu, mt, nt, as, jdr*
MCE Modular Control Element *fu, wp, pm*
MCE Modular Control Equipment [TAC] *kf, mt*
MCE Monitoring and Control Equipment *fu*
MCEAMS Marine Corps Expeditionary Aircraft Maintenance Shelter [USMC] *mt*
MCEB Military Comm Electronics Board *dt*
MCEB Military Communications Electronics Board [JCS] *mt*
MCES Major City Earth Stations *-9*
MCES Modular Command and Control Evaluation Structure *mt*
MCF Maintenance Control Facility *mt*
MCF Master Control Facility *ac3.2-21*
MCF Message Control Field *fu*
MCF Military Computer Family *hc, mt, fu*
MCF Mission Control Forecast *mt*
MCF Multilink Control Field *fu*
MCG Material Conservation Group *fu*
MCG Material Control Group *fu*
MCG Mid Course Guidance *0, tb, st*
MCG Millimeter wave Contrast Guidance *wp*
MCG Mission Control Group *fu*
MCG Movement Control Group *mt*
MCG Multifunction Cogenerator *hn50.8.4*
MCG Multiple Co-Generator *fu*
MCGA Monochrome and Color Graphics Adapter *wp*
MCGA Multi Color Graphics Array (Adapter) *kg, ct, em, hi*
MCGD Millimeter-wave Contrast Guidance Demonstration [Boeing] *hc, wp*
MCGS Microwave Command Guidance System [for drones] *fu, wp*
MCGW Millimeter-wave Contrast Guidance Weapon *wp*
MCHQ Marine Corps Headquarters *mt*
MCHS Microclimate Cooling / Heating System *hc*
MCI Mandatory Customer Inspection *fu, gp*

MCI Media Control Interface *ct, em, kg*
MCI Microwave Communications, Inc.[later renamed MCI Communications Corp.] *ct, kg, hn44.17.3*
MCI Mission Capability Inspection *mt*
MCIC Micro-Computer Information Center *-15*
MCIDAS Man-Computer Interactive Data Access (Analysis, Acquisition) System *kf, hn44.7.8, hy, oe*
MCIS Maintenance Control Information System *wp*
MCIS Management Control and Information System *fu*
MCL Management Control Limit *fu*
MCL Manufacturing Cost Level *fu, hu*
MCL Manufacturing Cost Line *jdr*
MCL Master Configuration List *hc*
MCL Maximum Contaminant Level *wp*
MCL Microsoft Compatibility Labs [Microsoft] *pk, kg*
MCL Mid Course Late *tb*
MCL Minimum Clear Length *wp*
MCL Mobilization Cross Leveling [USA] *mt*
MCLBA Marine Corps Logistics Base Albany, Georgia *mt*
MCLO Marine Corps Liaison Office *dt*
MCM Magnetic Core Memory *fu, wp*
MCM Micro(wave) Circuit Module *fu, hc*
MCM Mid Course Maneuver *0*
MCM Militarized Computer Module *fu*
MCM Military Committee Memorandum *mt*
MCM Mine Counter Measures Ship [US Navy] *mt, fu, wp, ph, hc*
MCM Multiple Chip Module *fu, kf, kg*
MCMCC Marine Corps Manpower Control Center *mt*
MCMM Manufacturing Control Material Management *fu*
MCMT Maximum Correction (Mean Corrective) Maintenance Time *fu*
MCMW Multichannel Microwave *fu*
MCN MACIMS Communication Network [MAC] *mt*
MCN Management Control Number *mt*
MCO Missile Control Officer *hc, wp*
MCO Mission Control Office *ac3.3-21*
MCOC Main Combined Operations Center *fu*
MCOG Maneuver Commands Group *0*
MCOT Missile Control Officers' Trainer *hc, hn44.17.12*
MCP Main Control Panel *hc*
MCP Maintenance Computer Program *fu*
MCP Maintenance Control Position *fu*
MCP Management and Control of Provisioning *mt*
MCP Marine Corps Capabilities Plan *mt*
MCP Master COMSEC Plan *mt*
MCP Master Control Program *mt, do422*
MCP Materiel Command Procedures *hc*
MCP Measure and Control Panel *gp*
MCP Meteorological Communications Package *hg, ge*
MCP Microchannel Plate *17, es55.4*
MCP Microwave Communication Products *fu*
MCP Military Construction Program (Plan) (Project) *mt, wp, hc*
MCP Minicommand Post *fu*
MCP Mission Control Processor *-17*
MCP Mobile Launch Plaform *-9*
MCP Modular Communications Programs *kf*
MCP Monitor and Control Position *fu*
MCPC Multiple Channel Per Carrier *ce*
MCPE Modular Collective Protection Equipment [chemical warfare, Army] *fu, wp*
MCPL Mandatory Components Parts List *hc*
MCPO Master Chief Petty Officer *w9*
MCPU Maintenance Central Processing Unit *fu*
MCPU MILSTAR Communications Processor Unit *fu*
MCR Manpower Change Request *mt*
MCR Master Change Record *fu*
MCR Microwave Cloud Radiometer *hy*
MCR Military Compact Reactor *fu*
MCR Minimum Cell Rate *ct*
MCR Minor Clarification Request *kf*
MCR Mission Control Room *oe*
MCR Modem Control Register *pk, kg*
MCRC Master Control and Reporting Center *mt, fu, wp*
MCRL Master Cross Reference Library *mt*
MCRL Master Cross Reference List *fu*
MCRSCMS Marine Corps Reserve Support Center Management System *mt*
MCS Maintenance Control System [USA] *mt*
MCS Maintenance Cost System [USAF] *mt*
MCS Management Control System *mt, fu*
MCS Maneuver Control System *fu, mt, hc, aw130.11.57*
MCS Maritime Communications System *sc3.1.3*
MCS Master Control Station *fu, hc, ab.13*
MCS Material Control System *fu*
MCS Micro Computer System; *sb*
MCS Microprocessor Control System *17, sb*
MCS Microwave Carrier Supply *fu*

MCS Millimeter Wave Contrast Seekers *wp*
MCS Mine Countermeasures Support *hc*
MCS Mission Control Segment *jdr*
MCS Mission Control Station *kf*
MCS Mobile Calibration Station *N75-27202-485*
MCS Module Control Station *es55.71*
MCS Monitor and Control Subsystem *fu*
MCS Monthly Cost Summary *-4*
MCS Motion Compensation System *hc*
MCS Multiple Channel System *dn*
MCS/SPALS Mode and Carrier Selection / Shipment Planning and Address Labeling System *wp*
MCSB Mission Control Station Backup *kf*
MCSF Marine Corps Security Force *mt*
MCSF Mobile Cryptologic Support Facility [NSA] *mt*
MCSMAW Marine Corps Shoulder launched Multiple purpose Assault Weapon *hc*
MCSP Mission Completion Success Probability *wp*
MCSPS Monitor and Control Subsystem Processor Suite *fu*
MCSR Materiel Condition Status Reporting system [USA] *mt*
MCSS Magnetron Compensator Signal Simulator *hc*
MCSS MATE Control and Support Software *fu*
MCSS Mid Course Surveillance System *tb*
MCSS Military Communications Satellite Station *hc, wp*
MCSS Military Communications Satellite System *fu*
MCSS Missile Checkout System Selector *fu*
MCSS Missile Control Sub System *hc*
MCSSC Multi-Color Spin-Scan Camera *hc*
MCSST Multichannel Sea Surface Temperature *ge, hg*
MCT Marine Corps Terminal *fu*
MCT Maximum Corrective Time *mt*
MCT Mean Corrective Time *mt, fu*
MCT Medium Chain Triglycerides *wp*
MCT Mercury Cadmium Telluride *wp, tb, kf*
MCT Mission Control Team *uh, jdr*
MCT Mobile Communications Terminal *mt*
MCT Module and Cell Tester *hc*
MCT Movement Control Team *mt*
MCT Multichannel Transponder *fu*
MCTI Motion Compensated Target Identification *hc*
MCTL Militarily Critical Technologies List [DOD] *fu, wp, mt*
MCTR Message Center [also MSGCEN] *dt*
MCTS Manufacturing Card Test System *fu*
MCTSSA Maritime (Marine) Corps Tactical Systems Support Activity *fu, mt*
MCTV Maintenance Console Television *fu*
MCU Manual Control Unit *hc*
MCU Master Control Unit *fu*
MCU Memory Controller Unit *mt*
MCU MicroController Unit *pk, kg*
MCU Microprogrammed Control Unit *mt*
MCU Module Control Unit *fu*
MCU Multiple Chip Unit [DEC] *pk, kg*
MCU Multipoint Control Unit *pk, kg*
MCU-2P [Improved chemical warfare mask] *wp*
MCVDU Microprocessor-Controlled Video Display Unit *fu*
MCVS Multi-Crew Visual System *hc*
MCVS Multichannel Video Switcher *fu*
MCW Minimum Clear Width *wp*
MCW Modulated Continuous Wave *mt, hc, fu*
MCWG Milstar Communications Working Group *jdr*
MCXO Microcomputer Compensated Crystal Oscillator *hc*
MD Magnetic Disk *hi*
MD Main Drum *fu*
MD Maintainability Demonstration *hc*
MD Management Document *fu*
MD Manual Data *fu*
MD Manufacturing Days *fu*
MD Manufacturing Directive *fu*
MD Manufacturing Division *fu, hn42.3.3*
MD Maritime Defense *mt*
MD Marshaled Deployability posture *mt*
MD Material Developer *mt*
MD Material Distribution *fu*
MD Mean Deviation *fu*
MD Micrometeoroid Detector *re*
MD Military District [Soviet] *-12*
MD Miss Distance *tb*
MD Missile Defense *mt, kf*
MD Modification Drawing *fu*
MD Monochromatic Display *fu*
MD Monochrome Display *pk, kg*
MD Motor Driver *jdr*
MD [chemical warfare codename for methyldichloroarsine] *wp*
MD-NA Missile Defense-North America *kf*
MD-NMD Missile Defense-National Missile Defense *kf*

MD-T Missile Defense-Theater *kf*
MD/XMP Message Distribution / Xerox Message Processor *mt*
MDA Manufacturing Defects Analyzer *fu*
MDA McDonald, Dettwiley and Associates *fu, ac3.5-4, es55.44*
MDA McDonnell Douglas Aerospace *oe*
MDA Mechanically Despun Antenna *hc*
MDA Methylenedioxyamphetamine *wp*
MDA Military Damage Assessment *mt*
MDA Minimum Descent Altitude *aw129.10.73*
MDA Mission Data Assurance *jb*
MDA Monochrome Display Adapter *ct, em, wp, kg, hi*
MDA Multidimensional Analysis *pk, kg*
MDA Multiple Docking Adapter *ac2.3-3*
MDAA Mutual Defense Assistance Act *wp*
MDAC McDonnell-Douglas Astronautics Corporation *hc, gg31, 4, 9, 17, aa26.10.55, lt, tb, st*
MDAC Multiplying Digital to Analog Converter *hc*
MDAC-HB McDonnell-Douglas Astronautics Corporation, Huntington Beach *-17*
MDACA Medical Defense Against Chemical Agents [chemical warfare] *wp*
MDAO Mutual Defense Assistance Office *hc*
MDAP Major Defense Acquisition Program *kf*
MDAP Military Department Aid Program *hc*
MDAP Mutual Defense Air Program *hc, tb*
MDAP Mutual Defense Assistance Program *mt, fu*
MDAR Malfunction Detection Analysis and Recording *wp*
MDAS Missile Defense Architecture Simulation *kf*
MDAS Multispectral Data Analysis System *ac3.2-23*
MDAY Manufacturing Day *fu*
MDB Manufacturing Database *fu*
MDB Manufacturing Division Bulletin *fu*
MDB Master Data Base *wl*
MDB Microelectronic Data Bank *hc*
MDB Multiplex Data Bus *wp*
MDB Mutual Defense Board *mt*
MDBL Maintainability Design Baseline *aa26.12.16, hc*
MDC Maintenance Data Collection *mt, fu, wp*
MDC Maintenance Dependency Chart *hc, fu*
MDC Manual Designation Console *fu*
MDC Master Direction Center *mt*
MDC Master Document Control *se1.8.2*
MDC McAfee Development Center *ct*
MDC Message Display Console *mt*
MDC Message Distribution Center *mt*
MDC Miniature Detonating Cord [used to fragment canopies before ejection] *mt*
MDC Missile Display Conference *mt*
MDC Mission Director's Center *uh, jn, oe*
MDC Mission Duty Cycle *wp*
MDC Mobile Data Center *wp*
MDC Movement Designator Code *mt*
MDC Multifunction Display Console *fu*
MDCA Manufacturing Design Change Analysis *hc*
MDCC Make Department/Cost Center *fu*
MDCP Manufacturing Data Control Point *fu*
MDCS Maintenance Data Collection System *hc*
MDCS Management Data Communications System *fu*
MDCSR Mission Data Collection Summary Report *kf*
MDDBMS Multidimensional Data Base Management System *pk, kg*
MDDR Maintenance Design Data Report *hc*
MDE Message Distribution Element *mt*
MDE Mission Development Element *jdr*
MDE Motor Drive Electronics *gp, 8*
MDEC Marine Corps Development and Education Command *mt*
MDET Militarized Digital Element Tester *hc*
MDF Main Distributing (Distribution) Frame *fu, 0, hc, hi*
MDF Main Distribution Frame *mt*
MDF Maintenance Data Form *hc*
MDF Manpower Data File *mt*
MDF Manual Data Input *fu*
MDF Manual Direction Finder *fu*
MDF Master Data File *hc, wp*
MDF Master Document File *oe*
MDF Metric Data Facility *hc*
MDFR Master Data File Record *hc*
MDG Mission Data Generation (Generator) *mt, jb*
MDG Multiple Diffraction Gratings *hc*
MDG Multipurpose Display Group *hc*
MDGWS Modular Digital Guided Weapon System *hc*
MDI Magnetic Direction Indicator *fu*
MDI Manufacturing Division Instructions *-8*
MDI Mechanical Design Integration *jdr, kf*
MDI Memory Display Interface *pk, kg*
MDI Miss Distance Indicator *17, wp, hc, tb, st*
MDI Multiple Document Interface *kg, ct, em, hi*

MDI Multipurpose Display Indicator *hc*
MDIC Manchester Decoder and Interface Chip [AT&T] *pk, kg*
MDIDP Manufacturing Division Industrial Defense Plan *fu*
MDIF Manual Data Input Function *fu*
MDIS Maintenance and Diagnostic Information System *wp*
MDIU Manual Data Input Unit *fu*
MDK Multimedia Developers Kit [Microsoft] *kg, hi*
MDL Management Data List *fu*
MDL Maximum Data Load *fu*
MDL Microcomputer Development *fu*
MDL Microwave Development Laboratories [Natick, MD] *hq5.3.5*
MDL Mission Data Load *bo, jb, ai2*
MDL Multiuse Data Link *oe*
MDLM Mobile Depot Level Maintenance *kf*
MDLP Mobile Data Link Protocol *kg, hi*
MDM Magnetic Drum Memory *hc*
MDM Medium Depth Mine *uc*
MDM Metal-Dielectric-Metal *fu*
MDM Mission Data Message *mt, kf*
MDM Mission Data Module *hc*
MDM Mobile Depot Maintenance *fu, wp*
MDM Modulator or Demodulator *-2*
MDM Momentum Dump Mode *jdr, uh*
MDM Monomethylol Dimethyl *wp*
MDM Multiplexer / (or) Demultiplexer *gp, 2, 4, 9, 16, oe*
MDMS Maintenance Data Management System [USA] *mt*
MDOF Multiple Degree Of Freedom *17, sb, st*
MDP Maintenance Data Processing *hc*
MDP Mandatory Diversion Point *fu*
MDP Manufacturing Division Procedure *fu*
MDP Master Data Processing *hc*
MDP Mission Data Processing *kf*
MDP Multiplex / Demultiplex Processor *fu*
MDPA Master Data Processing Authorization *hf*
MDPS Main Data Processing System *fu*
MDPS Metric Data Processing System *hc*
MDPS Mission Data Preparations Systems *mt*
MDQO Manufacturing Division Quality Office *fu*
MDQS Management Data Query System *mt*
MDR Maintenance Data Recorder *hc*
MDR Master Data Record *fu*
MDR Material Deficiency Report *mt, wp, fu*
MDR Maximum Detection Range *fu*
MDR Medium Data Rate *mt, ah, pbl, jdr, kf, jn*
MDR Message Directed Relay *fu*
MDR Microwave Digital Radio *fu*
MDR Minimum Design Requirement *pk, kg*
MDR Mission Data Reduction *fu*
MDR Mission Dress Rehearsal *jdr*
MDRA Material Deficiency Report Analysis *mt*
MDRAM Multibank Dynamic Random Access Memory *ct, em*
MDRT Manufacturing Defect Review Team *fu*
MDS Maintenance Data System *fu*
MDS Manpower Data System [HQ USAF] *mt*
MDS Manufacturing Document Services *fu*
MDS Master Dimension Specification *fu*
MDS Meteoroid Detection Satellite *hc*
MDS Minimum Detectable (Discernible) Signal *fu*
MDS Minimum Discernible Signal *mt, wp*
MDS Miss Distance Sensor *hc*
MDS Missile Detection System *wp*
MDS Mission Design Series *wp*
MDS Mission, Design, and Series *mt*
MDS Mobilization, Deployment, and Sustainment *mt*
MDS Model Design Series *mt, fu*
MDS Modem Digital Switch *fu*
MDS Modular Display System [SPACECOM] *mt, sr9-82*
MDS Multipoint (Multichannel) Distribution Service *jr, hc, do*
MDSC Microprocessor Design Support Center *fu*
MDSC Modular Digital Scan Converter *hc, fu*
MDSI Modular Display Systems, Incorporated *aw130.13.79*
MDSS Maritime Pre Positioned Ships (MPS) Decision Support System [USMC] *mt*
MDSS Multiple Distribution Switching Subsystem *mt*
MDSSC McDonnell Douglas Space Systems Company *oe*
MDSV Manned Deep Space Vehicle *hc*
MDT Maintenance Down Time *wp, kf*
MDT Master Data Channel *fu*
MDT Master Data Tape *hc*
MDT Mean Down Time *mt, tb, fu*
MDT Mechanical Design Team *jr*
MDT Minimum Detectable Target *kf*
MDT Mission Design Team *-17*
MDT Monochromatic Display Terminal *fu*
MDT Mountain Daylight Time *kg, w9, wp, kf*
MDTA Manpower Development and Training Act *hc*

MDTD Minimum Detectable Temperature Difference *wp*
MDTP Main Display Touch Panel *fu*
MDTR Maximum Data Transfer Rate *hi*
MDTS MD-918/GRC Digital Troposcatter Subsystem *fu*
MDTS Megabit Digital Troposcatter Subsystem *hc*
MDTS Mission Data Transfer System *mt*
MDTT Mission Data Transfer Terminal *fu*
MDU Master Data Unit *oe, ai2*
MDU Memory Drum Unit *hc*
MDU Message Decoder Unit *mt, fu*
MDU Mission Data Update *fu*
MDU Multipurpose Display Unit *fu*
MDUC Meteorological Data Utilization Center [New Delhi, India] *ac3.2-19*
MDUS Medium Data Utilization Station *bl3-1.59, pbl*
MDW Military District of Washington [USA] *dt, mt*
MDWS Missile Detection and Warning System *mt*
MDX {McDonnell Douglas Experimental} [helicopter] *aw130.6.54*
MDY Month Day Year *kg, hi*
MDZ Maritime Defense Zone *mt*
MDZL Maritime Defense Zone Atlantic *mt*
MDZP Maritime Defense Zone Pacific *mt*
ME Maintenance Engineering / Engineer *hc*
ME Manufacturing Engineering *fu, 8, aa26.11.55*
ME Mechanical Engineer(ing) *ma16, w9*
ME Medium Energy *cu*
ME Mission Equipment *kf*
ME Mobility Equipment *kf, fu*
ME Month End *fu*
ME&F Manufacturing Engineering and Facilities (Dept.) *fu*
ME/DC Memory Expansion/Disk Control *fu*
MEA Maintenance Engineering Analysis *fu, wp*
MEA Major Emergency Actions *mt*
MEA Materials Experiment Assembly *ci.821*
MEA Mono Ethanol Amine *wp*
MEAC Management Estimated Actual Completion *fu*
MEAD Maintenance Engineering Analysis Data *hc*
MEADS Maintenance Engineering Analysis Data System *mt*
MEADS Maintenance Engineering Analysis Documentation Summary *fu*
MEADS Medium Extended Air Defense System [Corps SAM] *kf*
MEAI Modified Experiment ARPAT Interceptor *hc*
MEANG Maine Air National Guard *-12*
MEAP Maintenance Engineering Analysis Process *fu*
MEAR Maintenance Engineering Analysis Record *hc, fu*
MEAST Multi-Echelon Automatic Shop Tester *hc*
MEASURE Metrology Automated System for Universal Recall/Reporting *hc*
MEB Marine Expeditionary Brigade [USA] *fu, mt*
MEB Memory Expansion Board *pk, kg*
MEBS Management Evaluation Guides {?} *mt*
MEBU Minimum Essential Backup *mt*
MEBU Mission Essential Back Up *mt, sr9-82*
MEC Main Evaluation Center *fu*
MEC Master Events Controller *wp*
MEC Master Executive Control *hc*
MEC Message Evaluation Center *fu*
MEC Microsystems Engineering Corporation [Hoffman Estates, Illinois] *aw130.22.65*
MEC Military Essentiality Code *fu*
MEC Missile Event Conference *mt*
MEC Multiple Element Correlation *hc*
MECB Misso Espacial Completa Brasileira [Complete Space Mission, Brazil] *ai2*
MECCA Missile Environmental Computer Control Analysis *hc*
MECH Mechanized *mt*
MECL Minimum Essential Circuit List *mt*
MECL Motorola Emitter Coupled Logic *hc*
MECO Main Engine Cut Off (MECO2 is 2nd stage LV Main Engine Cut Off) *kf, uh, mi, gg17, 4, 17, sb, hc, pf1.25, lt, tb, mb, oe, ai2*
MECS Modular Embedded Computer Software *wp*
MECSIP Mechanical Subsystems and Equipment Integrity Program *wp*
MED Maintenance Engineering Data *fu*
MED Manipulative Electronic Deception *mt*
MED Manual Entry Device *fu*
MED Manufacturing Engineering Directive *oe*
MED Message Element Dictionary [JINTACCS] *mt*
MED Microelectronic Device *fu*
MED Microelectronics Division [Hughes] *kf*
MED Molecular Electronic Device *wp*
MED-WRAP Medical War Reserve Automated Process [USA] *mt*
MEDA Military Emergency Diversion Airfield *mt*
MEDA Multiplex Electronic Doppler Analyzer *hc*
MEDALS Military Engineering Data Asset Locator *mt*
MEDAT Multi-Element Discrete Angle Tracker *hc*
MEDCAT Medium Altitude Critical Atmospheric Turbulence *wp*
MEDCOM Mediterranean Communication System *mt*

MEDD Manual Entry Data Device *fu*
MEDEA Multiple-Discipline, Engineering, Design, Evaluation and Analysis *wp*
MEDFAD Medical Field Assistance Branch *mt*
MEDFLI Miniaturized ESM Direction Finding /Locating Intercept *hc*
MEDLARS Medical Literature Analysis And Retrieval System *wp*
MEDP Mission Essential Data Processing *mt*
MEDS Maintenance Engineering Data System *mt*
MEDS Management Engineering Data System [as in MEDS 70 for 70's] *mt*
MEDS Meteorological Data Systems *mt*
MEDS Multifunction Electronic Display Subsystem *ai2*
MEDUSA Concept To Deploy Missiles From Submarines and C-5a Transports *wp*
MEE Maintenance Engineering Evaluation *fu*
MEE Mission-Essential Equipment *fu*
MEECN Minimum Essential Emergency Communications Network *mt, ro, 12, wp, hc, go3.1.1, tb*
MEES Missile End Game Evaluation System *wp*
MEETS Minimum Engine Tracking System *mt*
MEF Marine Expeditionary Force [USA] *mt*
MEF Minimum Essential Facilities *mt*
MEF Minimum Essential Functions *mt*
MEFPAK Manpower and Equipment Force Packaging System [USAF] *mt*
MEFT Minimum Essential Functional Task *mt*
MEG MA Electronics Group *vv*
MEG Message Entry Generator *fu*
MEG Message Expediting Group *mt*
MEI Maine Electronics, Incorporated *hq5.3.5*
MEI Maintenance Engineering Inspection *hc*
MEI Major End Item *mt*
MEI Management Effectiveness Inspection *mt*
MEI Management Efficiency Inspection *wp*
MEI Manufacturing Engineering Instructions *gp*
MEI Method Engineering Instruction *fu*
MEIDS Military (Miniaturized) Electronic Information Delivery System [USA] *mt*
MEIR Minimum Essential Information Requirement *mt*
MEIS [airborne multispectral push-broom scanner] *ac3.5-7*
MEISER Minimum Essential Improvement in System Reliability *hc*
MEISR Minimum Essential Improvement in System Reliability [a 1967 upgrade of the F-106] [USA] *mt*
MEISTER Multimode Electronically Scanned Interdiction *hc*
MEITS Mission Essential / Effective Information Transmission System *mt*
MEIWIC Mapping, Earth Imagery, Weather Imagery Component *kf*
MEK Methyl Ethyl Ketone *wp*
MEL Manufacturing Engineering Laboratory [Hughes] *gp, 8, 17*
MEL Master Equipment List *fu, mt*
MEL Materials Evaluation Laboratory *fu*
MEL Medium Energy Laser *mt*
MEL Microelectronics Laboratory *fu*
MEL Minimum Equipment List *mt, aw129.21.120*
MELIOS Mini Eyesafe Laser Infrared Observation Set *hc, hn43.23.1*
MELT Microelectronics Technology *fu*
MELVA Military Electronic Light Valve *fu*
MEM Mars Excursion Module *hc*
MEM Maximum Entropy Method *fu*
MEM MicroElectricalMechanical *rih*
MEM Mission Effectiveness Model *ts4-17, cp, tb*
MEMDT Message Execution Matrix Display Task *mt*
MEMO Mission Essential Maintenance Only *hc*
MENA Mission Element Need Analysis *wp*
MENS Mission Element Need Statement *mt, fu, hc, wp, sd, kf*
MEO Manual Entry Operator *fu*
MEO Medium (Intermediate) Earth Orbit *hs, tb, jdr, ai2*
MEO Message Exchange Occurrence *mt*
MEOC Medium Earth Orbit, Circular *sd*
MEOM Medium Earth Orbit, Molniya *sd*
MEOP Maximum Expected Operating (Operation) Pressure *hc, gd2-32, jdr, jn, oe*
MEOSS {a German mapping satellite} *aa26.10.10*
MEOWS Multi-Mission Electro-Optical Weapon System *hc*
MEP Main Enable Plug *gp, oe*
MEP Maintenance Engineering Program *mt*
MEP Management Engineering Plan *mt*
MEP Management Engineering Program *af72.5.108*
MEP Manual Entry Panel *fu*
MEP Manufacturing Extension Program *y2k*
MEP Mean Effective Pressure *wp*
MEP Microcircuit Emulation Program *mt*
MEP Mission Equipment Package [USA] *mt, aw129.10.57, hn49.3.1*
MEPA Meteorological and Environmental Protection Agency *fu*
MEPCOM Military Entrance Processing Command [USA] *mt*
MEPED Medium Energy Proton and Electron Detector [GOES, TIROS] *no*

MEPIS Management Engineering Program Information System *mt*
MEPS Military Entrance Processing Stations *mt*
MEQPT Major Equipment ID Code *mt*
MER Merchant Ship *mt*
MER Message Error Rate *fu*
MER Multiple Ejector (Ejection) Rack *mt, wp*
MERA Microelectronic Radar *hc*
MERA Molecular Electronics for Radar Applications *hc*
MERADCOM Mobility Equipment Research and Development Command [US Army] *T.C, hc, dt, wp*
MERCAST Merchant Shop Broadcast System *mt*
MERCI Multimedia European Research Conferencing Integration *pk, kg*
MEREP Merchant Ship Report *mt*
MERI Moderate Resolution Imaging Spectrometer [ESA] *hy*
MERINT Merchant (ship) Intelligence *fu*
MERIS Medium Resolution Imaging Spectrometer [ESA] *hg, ge*
MERIT Military Exploitation of Reconnaissance and Intelligence Technology *mt*
MERS Mobile Emergency Response System [FEMA] *mt*
MERS Multiple Element Radiometer System *hc*
MERSHIP Merchant Shipping system [COMUSK] *mt*
MERT Mean Emergency Resupply Time *fu*
MERTER Multiple Ejector Rack/Triple Ejector Rack *hc*
MES Main Engine Start *uh, oe*
MES Major Equipment Supplier *hc*
MES Management Engineering Squadron [AFCC*] *mt*
MES Manufacturing Execution System *pk, kg*
MES Message Entry Subsystem *mt*
MES Methylsalicylate [chemical warfare Simulant] [also MS] *wp*
MES Minor Earth Stations *-9*
MES M^ssbauer Emission Spectroscopy *wp*
MESA Maximum Entropy Spectral Analysis *fu*
MESA Module [tool set] Evaluation Strategic Architectures *tb*
MESA Multiplexed Electronic Synthetic Array *hc*
MESAMAC Engine Status Accounting System [MAC] *mt*
MESAWS Marine Expeditionary Surface to Air Weapons System *hc*
MESDIS Message Distribution System [PACOM] *mt*
MESFET Metal [Epitaxial] Semiconductor Field Effect Transistor *fu, wp, 17, mj31.5.52*
MESH (one of the leaders of the MESH consortium, Matra is) *sc2.8.4*
MESI Modified, Exclusive, Shared and Invalid *em, kg*
MESL Minimum Essential Subsystem List *mt*
MESP Management Engineering Scheduling Program [USAFE] *mt*
MESS Military Equipment Support Subsystem [USA] *mt*
MESS Monitoring Equipment Shipboard Simulator *fu*
MESSR Multispectral Electronic Sel(f)-Scanning Radiometer *ge, hg, ac3.3-13*
MESYLATE Methansulfonate *wp*
MET Management Engineering Team *mt, wp*
MET Memory Enhancement Technology [Hewlett-Packard] *kg, hi*
MET Missile Early Warning System *fu*
MET Mission Elapsed Time (Magellan / IUS deploy during Orbit 5, MET 6 *18 >*
MET Mission Equipment Team *aw129.10.57, hn49.3.4*
MET Multiple Element Tracker *hc*
MET Multiple Emitter Transistor *fu*
MET Multiple Environment(al) Trainer *fu, hc*
META [two groups arranged so that there is another group between them] *wp*
METAL [German-English machine translator] *ro*
METEOR Meteorological Event Oriented Reporting System [USAFE] *mt*
METEOSAT Meteorological [European Space Agency] Satellite *ge, bl3-1.12, kf, oe*
METG Middle East Task Group *hc*
METGAP Mission Event Timeline Graphical Application *uh*
METL Mission Essential Task List *mt*
METOC Meteorological and Oceanographic [USA] *mt*
METRA Metal Radar *hc*
METRIC Multiple Echelon Technique Recoverable Item Control [RAND] *mt*
METS Mechanized Export Traffic System [MTMC] *mt*
METS Mobile Electronic Test Set *mt, wp*
METS Multiple Engined Training Squadron [UK] *mt*
METSAT Meteorological Satellite *gp, wp, 16, hy, oe*
METSATS Multi-Echelon Tester Standard Analog Test Set *hc*
METSII/ACI Military Export Traffic System II - Enhanced / Automated Carrier Interface [MTMC] *mt*
METT Mission, Enemy, Terrain, and Troops *mt*
METTT Mission, Enemy, Terrain, Troops and Time *mt*
METU Mobile Electronics Technical Unit *wp*
MEU Marine Expeditionary Unit [USA] *mt*
MEU Mission Essential Unit [US Navy] *mt*
MEU Multiple Event Upset (?) *jdr*
MEU[SOC] Marine Expeditionary Unit [Special Operations Capable] [USA DOD] *mt*
MEVA Million Electron Volt Accelerator [Hughes subsidiary] *fu*

MEVDP Modified Elemental Volume Dose Program *jdr*
MEW Marine Early Warning *hc*
MEW Microwave Early Warning *fu*
MEWA Missile Electronics Warfare Area *hc*
MEWES Mobile Electronic Warfare Environment Simulator *hc*
MEWETS Multiple Electronic Warfare Emitter Target System *wp*
MEWS Microwave Electronic Warfare System *wp*
MEWS Missile Early Warning Station *mt, wp, hc*
MEWS Mobile Electronic Warfare Simulator *hc*
MEWS Modular Electronic Warfare System *hc, fu*
MEWS Multisignal ECM System *hc*
MEWSG Maritime Electronic Warfare Support Group *mt*
MEWSG Multiple Service Electronic Warfare Support Group [NATO] *wp, mt*
MEWSS Mobile Electronic Warfare Support System [USMC] *mt*
MEXP Monitor Exercise Processor *fu*
MEZ Maritime Exclusion Zone *mt*
MEZ Missile Engagement Zone *mt, fu*
MF Main Frame *bo, oe*
MF Maintenance Facility *tb*
MF Mass Fraction *tb*
MF Matched Filter *fu*
MF Medium Frequency [0.03 to 3 MHz] *mt, nu, fu, 0, wp, w9, go1.4.1, tb*
MF Mega FLOPS *tb*
MF Methyl Formamide *wp*
MF Micro Film (Fiche) *fu, w9*
MF Middle Frequency *kf*
MF Modulated Frequency *ct*
MF Multi Frequency *fu*
MF Multiple Frequency Signaling *mt*
MF Multiple Function *mt*
MF Munitions Facility *tb*
MFA Materiel Fielding Agreement *mt, fu*
MFA Methyl Fluoro Acetate [chemical warfare] *wp*
MFAA Massive Fire Power Aircraft *wp*
MFAR Multi-Function Array Radar [also MAR] *hc*
MFASMR Multifrequency Aperture Synthesis Microwave Radiometer *sa830182P.24*
MFBARS Multifunction-Multiband Airborne Radio System [modular] *fu, hc, wp*
MFC Manual Frequency Control *fu*
MFC Message Formatting Committee *mt*
MFC Meteorological and oceanographic Forecast Center [USA DOD] *mt*
MFC Micro Functional Circuit *fu*
MFC Microsoft Foundation Class *em, kg*
MFC Missile Fire Control *mt*
MFC Mortar Fire Control *hc*
MFCC Missile Fire Control Computer *wp*
MFCL Microsoft Foundation Class Library *hi*
MFCRS Multifunction Flight Control Reference System *wp*
MFCS Microprocessor Flight Control System *af72.5.150*
MFCS Missile Fire Control System *fu*
MFD Magnetic Frequency Detector *fu*
MFD MagnetoFluiDynamics *ai1*
MFD Master File Directory *mt*
MFD Multiple Function Display *fu, hc, wp, mt*
MFDU Multifunction Display Unit *wp*
MFE Magnetic Field Explorer *hy*
MFE Manpower Force Element *mt*
MFECC Magnetic Fusion Energy Computing Center *tb*
MFEL Manpower Force Element Listing *mt*
MFFS Microsoft Flash File System [Microsoft] *kg, hi*
MFG Marine Flieger Geschwader (MFG 3 Graf Zeppelin) [Germany] *mt*
MFHBF Mean Flight Hours Between Failures *mt, wp, rm3.1.2*
MFHC Missile Flight Hazard Corridor *fu*
MFI Manufacturing Finance Instructions *fu*
MFID Main Frame Identification *oe*
MFIT Manual Fault Isolation Testing *hc*
MFL Maintenance Fault List *wp*
MFL Master Force List *mt*
MFLAD Multiflexible Analysis Display *fu*
MFLOPS Mega (Million) Floating Point Operations Per Second *mt, kg, fu, em, aw130.13.61, aa26.11.44, ie26.1.8d, tb, kf, hi*
MFM Modified Frequency Modulation *ct, kg, em, hi*
MFMBARS Multifunction Multiband Airborne Radio System *wp*
MFME MFM Encoding *hi*
MFMS Multi-CPU File Management System *fu*
MFO Multiple Facility Organization *su*
MFOX Multipurpose Fiber Optic Transceiver *hc*
MFP Major Force Program *mt, wp*
MFP Master Personnel File [USAF] *mt*
MFP Materiel Fielding Plan *mt, st, fu*
MFP Mean Free Path *wp, ai1*
MFP Minimum Facility Plan *hc*
MFP Multi Frequency Pulsing *fu*

MFP Multi Function Peripheral *em, kg*
MFP Multi Function Product *kg, hi*
MFP Multiple Purpose Facility *mt*
MFP [an input to a timing and control box] *ac7.2-15*
MFPA Monolithic Focal Plane Array *fu, hc*
MFPA Mosaic Focal Plane Array *hc*
MFPG Mechanical Failures Prevention Group *rf84.3.26*
MFPI Multifunction Peripheral Interface *pk, kg*
MFPS Materiel Fielding Plans *hc*
MFQF Multi File Query Filter *hi*
MFR Memorandum for the Record *wp, mt*
MFR Multi Function Radar *fu, wp, hn43.9.1, af72.5.86*
MFR Multi Function Receiver *hc, gd2-40, 4*
MFR Mutual Force Reduction *-14*
MFS Magnetic tape Field Search *pk, kg*
MFS Major Frame Synchronization *gp, 16*
MFS Memory File System *pk, kg*
MFS Modified Filing System [Revelation Technologies] *pk, kg*
MFS Monitored Frequency Synthesizer *fu*
MFS Multi Frequency Signature *hc, fu*
MFSA Missile Flight Safety Approval *hc*
MFSK Multi(ple) Frequency Shift Keying *kf, hc, fu, ric, jdr*
MFSK Multilevel Frequency Shift Keying *mt*
MFSO Missile Flight Safety Officer *hc, wp*
MFT Master File Table *pk, kg*
MFT Materiel Fielding Team *fu*
MFT Meter Fix Time *fu*
MFT Minimum Flexible Targeting *mt*
MFT Minuteman Flexible Targeting *wp*
MFT Multiprogramming with a Fixed number of Tasks *pk, kg*
MFTL Modular Field Test Laser *hc*
MFTP Multi-Cast File Transfer Protocol *ct, em*
MG Machine Gun *w9*
MG Major General *w9, tb*
MG Map Generation *fu*
MG Maschinen Gewehr ["machine gun"] [Germany] *mt*
MG Military Government *w9*
MG Missile Guidance *hc, tb*
MG Motor Generator *tb*
MG/R Message Generator / Recorder *mt*
MGA Medium Gain Antenna *re*
MGA Monochrome Graphics Adapter *kg, ct, em, hi*
MGAB Maintenance Ground Abort *wp*
MGABR Maintenance Ground Abort Rate *wp*
MGB Ministervo Gosudarstvennoi Bezopasnosti [(Russian) Ministry of State Security] *w9*
MGC Manual Gain Control *0, wp, fu, jn*
MGC Missile Guidance Computer *wp*
MGC Missile/Gun Coordinator *fu*
MGCO Mars Geoscience Climatology Orbiter *hc, sc3.9.3, 8*
MGE Maintenance Ground Equipment *hc, fu, 0*
MGE Missile Guidance Equipment/Element *hc*
MGE Modular GIS Environment *pk, kg*
MGES Maintenance Ground Equipment Section *hc*
MGG Memory Gate Generator *fu*
MGGB Modular Guided Glide Bomb [GBU-15[V] HOBOS, USA] *fu, wp, hc, mt*
MGI Mandatory Government Inspection *jdr*
MGM Master Group Multiplexer *mt*
MGM Mobile, Guided ground to ground Missile [USA] *mt*
MGP Meteorology/Geophysics Package *re*
MGP Multiplane Graphics Processor *fu*
MGR Military Grid Reference *fu*
MGR Mobile, Ground targeted Rocket [MGR-1 is Honest John, MGR-3 is Little John] [USA] *mt*
MGRS Military Grid Reference System *tb, fu*
MGS Message Generation System [as in MGS2] *mt*
MGS Missile Guidance Set *hc, tb*
MGS Mission Ground Station [DSPO] *pbl, mt*
MGS Mobile Ground Station *kf*
MGS Mobile Ground System [DSP] *mt*
MGS Modified Gram-Schmidt *fu*
MGS Moveable Ground Station *sc4.1.center, lt*
MGSE Mechanical Ground Support Equipment *gp, 4, jdr, kf, jn, oe*
MGSU Mobile Ground System Unit *kf*
MGT Mobile Ground Terminal *mt, wp, hc, sd, kf*
MGU Mid course Guidance Unit [for missiles] *mt*
MH Material Handling *fu*
MH/MA Manhours/Maintenance Action *wp*
MH/P&TE Material Handling/Packaging and Tool Engineering *fu*
MH/PE Material Handling/Packaging Engineering *fu*
MH/PP Material Handling/Parts Protection *fu*
MH/TS Manhours/Troubleshhoting Action *wp*
MHA Maintenance Hazards Analysis *fu*
MHC Minehunter, Coastal [US Navy] *mt*
MHCT Microsoft Hardware Compatibility Testing [Microsoft] *hi*

MHD Magneto Hydro Dynamics *fu, ai1*
MHD Material Handling Drawing *fu, aa26.12.25, ci.236, 0, w9, is26.1.76kk, wp, tp*
MHD Moving Head Disk *mt, fu*
MHDEMP Magneto Hydro Dynamic EMP *mt*
MHE Materiel Handling Equipment *mt, fu, wp*
MHE Message Handling Element *mt*
MHEOWS Multicolored High Energy Outgoing Wavefront Sampling *hc*
MHF Medium High Frequency *mt, fu*
MHF Mission History File *fu*
MHF Mixed Hydrazine Fuel *wp*
MHFARS Minuteman High Frequency Antenna Replacement System *mt*
MHFEP Multiple Host Front End Processor *mt*
MHI Material Handling Instruction *fu*
MHI Mitsubishi Heavy Industries [Japan] *hc*
MHPE Material Handling and Packaging Engineering *fu*
MHQ Maritime Self-Defense Force Headquarters *fu*
MHR Man Hours *wp*
MHR Missile Hazard Report *fu*
MHRS Magnetic Heading Reference System *wp*
MHS Mechanical Hardware System *fu*
MHS Message Handling Service (System) *mt, kg, ct, em, hi*
MHSV Multipurpose High Speed Vehicle *hc*
MHT Maximum Holding Time *csII.87*
MHV Miniature Homing Vehicle *mt, wp, sr9-82, 5, 11, 14, sb, nt, tb, st*
MHW Multi-Hundred Watt *ci.231*
MI Management Indicator *mt*
MI Management Interface *pk, kg*
MI Manual Input *fu*
MI Master Index *fu, hc, ep, jdr*
MI Master Item *fu*
MI Material Index *fu*
MI Microwave Imager *-16*
MI Military Intelligence *mt, fu, w9*
MI Minority Institution *nt*
MI Mode Indicate *ct, em*
MI Modification Instruction [Marine Corps] *hc, fu*
MI Modulation Index *oe*
MI Mutual Inductance *fu*
MI {Manpower and Installations} *dt*
MI [BN][CO] Military Intelligence [Battalion] [Company] *mt*
MI/MIC Mode Indicate/Mode Indicate Common *pk, kg*
MIA Management Information Analysis *mt*
MIA Missile Intelligence Agency *st*
MIA Missing In Action [USA DOD] *mt, w9, wp, af72.5.13*
MIAG Management Information Analysis Group *fu*
MIAMI Microwave Ice Accretion Measurement Instrument *hc*
MIANG Michigan Air National Guard *-12*
MIAP Military Incentive Analysis Program *wp*
MIARS Maintenance Information Automated Retrieval System [Navy] *hc, wp, fu*
MIAS Mobile Intelligence Analysis Center *wp*
MIB Management Information Base *kg, ct, em, hi*
MIB Military Intelligence Board [USA DOD] *mt*
MIB Minimum Impulse Bit *oe*
MIB Missile Interceptor Base *mt*
MIBARS Mechanized Infantry Battalion Air Reconnaissance Support *wp*
MIBK Methyl Iso Butyl Ketone *wp*
MIBL Masked Ion Beam Lithography *fu*
MIC Management Information Center *fu*
MIC Matched Indicator Code *hc*
MIC Media Interface Connector *fu*
MIC Memory In Cassette *em*
MIC Memory Interface Connection *hc*
MIC Microcomputer Information Center *sc4.9.3*
MIC Microelectronic (Monolithic) Integrated Circuit *fu*
MIC Microwave Integrated Circuit *mt, fu, pl, wp, gg171, hc, mj31.5.15, lt, ep, jn*
MIC Mobile Intelligence Center *mt*
MIC Mode Indicate - Common *ct*
MIC Monitoring, Identification and Correlation *fu*
MIC Multifield Identification Chip *-17*
MICA Modem ISDN Channel Aggregation *ct, em*
MICAD Multipurpose Integrated Chemical Agent Detector [chemical warfare] *wp*
MICAF Measuring Improved Capabilities of Army Forces *mt*
MICAM Micro Connection and Assembly Method [Hughes program] *fu, gg57, hc, lt*
MICAM Module Integrated Connection and Assembly Method *gp*
MICAP Mission Capability *mt, wp*
MICAP Mission Capability Analysis System [USAF] *mt*
MICAP Mission Impaired Capability Awaiting Parts *mt*
MICBM Mobile Inter Continental Ballistic Missile *hc*
MICD Mechanical Interface Control Drawing *pl*
MICE Micro-In-Circuit Emulator *fu*

MICE Missile Intercept Computer Evaluation [model] *wp*
MICE Morphological Image Complexity Evaluation *hc*
MICL Mid Infrared Chemical Laser *tb*
MICNS Modular Integrated Communications and Navigation System *hc, mt*
MICo Military Intelligence Company [USA] *mt*
MICOM Missile Command [Huntsville, AL, US Army] *mt, fu, hn41.18.4, hc, bf, st, kf*
MICOS Multi Functional Infrared Coherent Optical Sensor *wp*
MICR Magnetic Ink Character Recognition *mt, w9*
MICRA Microelectronic Integrated Circuit Replaceable *hc*
MICRAD Microwave Radiometric Reconnaissance System *wp*
MICRAD Microwave Radiometry *hc*
MICROFIX [near term intelligence/electronic warfare microprocessor program] *wp*
MICROFX [intelligence data processing set] [FORSCOM] *mt*
MICROMUSE [A COTS software company] *kf*
MICRONAV [Hughes Aircraft Company customer for microwave landing system] *hc*
MICS Macro Interpretive Commands *pk, kg*
MICS Management Information and Control System *hc*
MICS Materiel Intransit Control System [MTMC] *mt*
MICSLP Maintenance Inventory Control Stock Levels Projection *mt*
MICV Mechanized Infantry Combat Vehicle *mt, hc, wp*
MID Military Intelligence Detachment *mt*
MID Military Intelligence Division *-10*
MID Multiple Individually Designated *hc*
MIDAR Microwave Detection and Ranging *fu*
MIDAS Detect Missile Launching Satellite *0*
MIDAS Maintenance Integrated Data Access System *mt*
MIDAS Manufacturing Information Distribution and Acquisition System *fu, vc25.un.un*
MIDAS Mine Detection and Avoidance Sonar *hc*
MIDAS Missile Defense Alarm System *wp*
MIDAS Model for Intertheater Deployment and Scheduling [USAF] *mt*
MIDAS Modified Integration Digital Analog Simulator *hc*
MIDAS Munich Image Data Analysis System *es55.51, ns*
MIDATS Modular Intermediate Depot Automatic Test System *af72.5.78*
MIDB Major Item Data Base *mt*
MIDDS Meteorological Interactive Data Display System *hy*
MIDET Military Intelligence Detachment [Pentagon counterintelligence force] *dt*
MIDGET Modular Integrated Digital Equipment Tester *wp*
MIDI Master Index of Drawing Information *fu*
MIDL Miniature Interoperable Data Link *wp*
MIDL Modular Interoperable Data Link *pf.f88.26*
MIDP Major Item Distribution Plan *mt*
MIDR Mosaicked Image Data Record *ct*
MIDS Management Information Data System *fu*
MIDS Management Information Display System [USA] *mt*
MIDS MPAC Information Data System *mt*
MIDS Multifunction(al) Information Distribution System [NATO] *mt, fu, wp, aw130.22.61*
MIDSIM Midcourse Simulation *tb*
MIE Monitor Inertial Electronics *hc*
MIEC Meteorological Information Extraction Center [Darmstadt] *pl*
MIEP Modifications/ Improvements Evolutionary Program *fu*
MIES Multi-Image (Imaging) Exploitation System *hc, wp*
MIF Management Information Format *kg, hi*
MIF Mobile Instrumentation Facility *wp*
MIF Modernization and Implementation Facility [Gunter AFS] *mt*
MIFASS Maintenance Integration Fire And Support System *hc*
MIFASS Marine Integrated Fire and Air Support System [USA] *hc, fu, mt*
MIFR Minor Frame *bl*
MIG Metal In Gap *ct, pk, kg*
MIG Metal Inert Gas (Welding) *wp*
MIG Mikoyan and Gurevich *wp*
MIG Mikoyan Gurevich [designator for aircraft designed by or named after] "mig" [USSR] *mt*
MIG Military Interface Group *fu*
MIG Miniature Integrating Gyro *hc*
MIGCAP MIG Combat Air Patrol *mt*
MIGS Miniature Infrared Guidance Sensor *hc, wp*
MIHT Moscow Institute of Heat Technology *ai2*
MII Master Item Identification *fu*
MII Modular Image Interpretation *wp*
MIICS Master Item Identification Control System *wp*
MIIDS Military Intelligence Integrated Data System *mt*
MIJI Meaconing, Interference, Jamming and Intrusion *tb, mt*
MIJO Missile Joint Optimization [Study] *hc*
MIK Methyl Isobutyl Ketone *wp*
MIK Missile Installation Kit *hc*
MIK Montazhno Ispytatelnyj Korpus [assembly test complex] *ai2*
MIL Machine Interface Layer [Go Corporation] *pk, kg*
MIL Man In the Loop *tb, st*
MIL Micro Instruction Level *fu*

MIL Military In the Loop *tb*
MIL-HDBK Military Handbook [a document] *mt, tb, fu*
MIL-I Military Instruction *nt*
MIL-SPEC Military Specification [also MILSPEC] [a document] *fu, hc, ep, hi, ai1*
MIL-STD Military Standard [a document] *fu, 16, hc, aa26.12.65, es55.6, aw130.11.56, nt, mb, tb, kf, vv, oe, ai1*
MILA Merritt Island Launch Annex (Area) [Florida STDN Tracking Station] *gp, 0, 4, oe*
MILAP Maintenance Information Logically Analyzed and Presented *mt*
MILCAP Military Standard Contract Administration Procedures [DOD] *mt*
MILCOM Military Communication *fu*
MILCOMSAT Military Communications Satellite *mt, hc, wp*
MILCON Military Construction program [US Navy] *mt, fu, uc, wp, hc, dm, st, kf*
MILD Minipro Interactive Load Debug *fu*
MILDEP Military Department *mt*
MILES Magnetic Intrusion Line Sensor *mt*
MILES Merisel's Information and Logistical Efficiency System *ct*
MILES Multiple Integrated Laser Engagement System *mt, hn44.20.1, hc*
MILESTONE [stage in program development, ~_0 program initiation/mission need decision, ~1 concept demonstration/validation decision, ~2 full scale development decision, ~3 full rate production decision, ~4 operational readiness and support review, ~5 major upgrade/ *wp*
MILFAC Military Facilities File *mt*
MILID Military Identification *fu*
MILIRAD Millimeter Wave Radar Fuze *wp*
MILIROS Mini-Laser Infrared Observation Set *hc*
MILL_RACE [high-explosive test to simulate airblast effects for 1 kt nuclear blast] *wp*
MILNET Military Network [DDN] [an operational packet switchin network] *mt, ct, 1, tb, hi*
MILO Microphone Locator *hc*
MILOGS Marine Integrated Logistic System *mt, fu*
MILPAY Military Pay *mt*
MILPER Military Personnel [also MILPERS] *st*
MILPERCEN Military Personnel Center [US Army] *mt*
MILPERS Military Personnel [also MILPER] *wp*
MILREP Military Representative *mt*
MILS Military Standard Logistic System [DOD] *wp, mt*
MILS Missile Impact Location System *hc*
MILSATCOM Military Satellite Communications *mt, cp, wp, kf*
MILSCAP Military Standard Contract Administration Procedures [DOD] *wp, fu, mt, hc*
MILSIMDS Military Standard Item Management Data Systems *wp*
MILSO Military Standards Logistics Systems Office *dt*
MILSPEC Military Specification [also MIL-SPEC] "milspec" *wp, mt*
MILSPETS Military Standard Petroleum System [DOD] *mt*
MILSTAAD Military Standard Activity Address Directory *hc*
MILSTAMP Military Standard Movement Procedures *mt*
MILSTAMP Military Standard Transportation And Movement Procedures [DOD] *mt, fu, hc, wp*
MILSTAR Military Satellite Tracking And Reconnaissance *tb*
MILSTAR Military Strategic and Tactical Relay Satellite [EHF satellite program] "milstar" *mt, kf, hc, ph, wp, tb, jdr, oe*
MILSTART (such CCCI programs as SDI, GWEN, WIS, MILSTART) *is26.1.76cc0*
MILSTD Military Standard [also MS] *mt, 16, wp, bo*
MILSTEP Military Supply and Transportation Evaluation Procedures [DOD] *mt*
MILSTRAP Military Standard Transaction Reporting and Accounting Procedures [DOD] *mt, wp*
MILSTRIP Military Standard Requisition(ing) and Issuing Procedures [USA] *mt, fu, wp, tb*
MILSVC Military Services *mt*
MILTRACS Military Air Traffic Control System *mt*
MILVAN [military owned demountable container, conforming to US and international standards, operated in a centrally controlled fleet for movement of military cargo] [USA DOD] *mt*
MIL_BASOPS Military Base Operations *fu*
MIL_STD Military Standard [also MIL-STD] *wp, ep*
MIL_TRACS Military Air Traffic Control System *fu*
MIL_VIG Military Vigilance [NATO] *mt*
MIM Metal Insulator Metal [screen] *pk, kg*
MIM Mobile SAM [USA] {?} *mt*
MIM Multiplex Interface Module *hc*
MIMD Multiple Instruction Multiple Data stream [processor] *kg, wp, is26.1.30, tb, hi*
MIME Multipurpose Internet Mail Extension [encoding and decoding protocol for attached, non-ASCII files] *kg, ct, em, nu, hi, du*
MIMI Magnetospheric Imaging Instrument *es55.26*
MIMI Missile Intensive Maintenance Item *mt*
MIMIC (2 rigid body MIMIC simulation) *id4142-10/831*
MIMIC Computer Language For Simulation *wp*
MIMIC Microwave and Millimeter-wave Monolithic Integrated Circuit Program *hc, aa26.12.72, hq6.2.7, aw130.22.77*

MIMIC [computer program - circuit design] *hc*
MIMIS Major Item Management Information System [USA] *mt*
MIMIS Municipal Improvement Management Information System *hc*
MIMMS Marine [Corps] Integrated Maintenance Management System *fu, mt*
MIMO Multi Input / Multi Output *ai1*
MIMOSA Mission Modes and Systems Analysis *wp, hc*
MIMR Multifrequency Imaging Microwave Radiometer *ge, hg*
MIMS Mobile Information System *mt*
MIMS Multiple Independent Maneuvering Submissile (Submunitions) *hc, wp*
MIN Mobile Identification Number *ct, jdr, kg, em*
MINDM Miniature Dyna Metric Model *mt*
MINET Movement Information Network [USEUCOM] *mt*
MINGEL Martin Integrated Neutral Graphics Engineering Language *fu*
MINI-MEG Mini Message Entry Generator *hc*
MINI-STE Mini-Special Test Equipment *hc*
MINI-TAC Mini Terminal Access Controller *mt*
MINIAPS Miniature-Accessory Power Supply *fu*
MINIBAC Miniature Battlefield Computer *fu*
MINIPRO Hughes Programmable Signal Miniprocessor (HSP-1032) *fu*
MINIS Mini Intraport Network Information System *mt*
MINOD Miniaturized Infrared Night Observation Device *hc*
MINSET Minimum Set *fu*
MINT Management Information Network for Training [ATC] *mt*
MINTSS MAC* Intelligence Support System *mt*
MINX Multimedia Information Network Exchange *hi*
MIO Management Information Office *wp*
MIO Management Integration Office *mt*
MIO Manual Inputs Operator *fu*
MIO Military Industrial Organization *hc*
MIO Movement Identification Officer *fu*
MIOD Message Input/Output Device *hc, fu*
MIOP Message Input/Output Procedure *fu*
MIOT Manual Input/Output Terminal *fu*
MIP Magnetic Index Pulse *gg61*
MIP Mainframe Internetting Project *mt*
MIP Maintenance Index Page *fu*
MIP Mandatory Inspection Point *17, wp, sb, jn, oe*
MIP Manual Input Processing *fu*
MIP Master Index Pulse *gp, 3, 9, pl, lt*
MIP Material Improvement Project *hc, wp*
MIP Mechanical Integrated Product *fu*
MIP Message Interface Processor *mt, kf*
MIP Military Improvement Program *mt*
MIP Minimum Impulse Pulse *fu*
MIP Missile Impact Prediction (Predictor) *fu, mt*
MIP Mission Integration Panel *wl*
MIP Model Installation Program *mt*
MIPA Monoisopropanolamine *wp*
MIPAS Michelson Interferometer for Passive Atmospheric Soundings [FRG] *ge, hg*
MIPB Material Improvement Project Board *mt*
MIPC Master Index Pulse Corrected *pl*
MIPDS Minuteman Instrumented Payload Delivery System *hc*
MIPE Modular Information Processing Equipment *fu*
MIPG Master Index Pulse Generator *bl2-6.20, 9, lt*
MIPIR Multi Mission Imagery Photo Interpretation Report *wp*
MIPL Manufacturing Indentured Parts List *fu, gp*
MIPL Master Indentured Parts List *hc*
MIPL Multimission Image Processing Laboratory (Library) *ct, em*
MIPR Master Index Pulse Reference *3, 8*
MIPR Master Inspection Program and Record *fu*
MIPR Military Interagency Procurement Requisition *mt*
MIPR Military Interdepartmental Purchase Request [DD 448] *mt, hc, ro, wp*
MIPRG Master Index Pulse Reference Generator *3, 8*
MIPS Maintenance Index Pages *hc*
MIPS Marine Integrated Personnel system [USMC] *fu, mt*
MIPS Master Installment Purchase System *wp*
MIPS Master Integrated Program Schedule *kf*
MIPS Mechanical Impulse Pyroshock Simulator *jdr*
MIPS Mechanical Integrated Products System *fu*
MIPS Mechanically Induced Pyrotechnic Simulator *jdr*
MIPS Millions of Instructions Per Second *mt, fu, em, ct, aw130.13.61, 4, 17, sb, pl, aa26.12.65, w9, pf1.10, pf.f88.27, wp, re, tb, kf, hi*
MIPS Mission and Information Planning System *ns*
MIPS Multimission Image Processing *re*
MIPS Multimission Image Processing Subsystem *em, ct*
MIPSL Manufacturing Indentured Parts Summary List *gp, sc9.6.4, jdr*
MIR Material Inspection and Receiving report [also MIRR] *hc*
MIR Material Inspection Report *fu*
MIR Mechanical Information Release *fu*
MIR Memory Input Register *fu*
MIR Mid Infrared *wp*
MIRA Miniature Infrared Alarm *hc*

MIRACL Mid (Medium) Infrared Advanced Chemical Laser *ro, fu, hc, aa27.4.1, af72.5.18, hn49.11.1, tb, wp*
MIRACLE Mid Infrared Advanced Chemical Laser, Experimental *hn44.14.1*
MIRADCOM Missile Research and Development Command [Army] *fu*
MIRAGE Microelectronic Indicator for Radar Ground Equipment *ro*
MIRAGE [Upa-56 Ground Radar Display System] *wp*
MIRC Missile In Range Computer *hc*
MIRCLE Mid-Infrared Chemical Laser *sb*
MIRCOM Missile Materiel Readiness Command [Army] *fu*
MIRG Management Information and Report Generator *fu*
MIRID Mobile Infrared Inspection and Diagnostic [System] *hc*
MIRL Medium-Intensity Runway Lights *fu*
MIRLTD Mid-Infrared Laser Target Designator *hc*
MIROS Modulation Inducing Retrodirective Optical *hc*
MIRR Material Inspection and Receiving Report [also MIR] *fu, hc, wp*
MIRS Management Information and Retrieval System *wp*
MIRS Mid-Infrared Source *hc*
MIRS Millimeter Wave Instrumentation Radar Station *hc*
MIRS Modular Integrated Radar System *uc, hc*
MIRS Multipurpose Infrared Sensor *hc*
MIRSE Multipurpose Imaging Radiometer Spectrometer *hc*
MirSIP Mirage Safety Improvement Programme [Belgium] *mt*
MIRU Missile Inertial Reference Unit *hc*
MIRV Multiple Independently Targeted Reentry Vehicle [a payload for a ballistic missile consisting of multiple, independently guided warheads or decoys] *mt, fu, ca.8, 13, 14, hc, af72.5.145, wp, tb, st, kf*
MIS Maintenance Information System *mt*
MIS Management Information Services [system] [an organization] *hc, ms3.1.2, 15, sc8.5.3, ep*
MIS Management Information System *mt, fu, kg, em, wp, tb, st, wl, oe*
MIS Manual Input Station *fu*
MIS Material Inspection Service *wp*
MIS Metal Insulated Semiconductor (Silicon) *fu, hc*
MIS Meteorological Impact Statement *fu*
MIS Microwave Imager and Sounder *16, pl*
MIS Missile Interim Specification *fu, hc*
MIS Missile Squadron [USA] *mt*
MIS Motor Inert Storage *ai2*
MIS Multimedia Information Sources [Internet] *pk, kg*
MIS Multiplexer Interface Subunit *fu*
MIS/INAS MIS for Industrial Naval Air Stations [US Navy] *mt*
MISCAP Mission Capability Statement *mt*
MISD Multiple Instruction Single Data *fu*
MISER Microwave Space Electronics Relay *hc*
MISER Minicomputer for Scientific and Engineering Requirements *fu*
MISICC Management Information System Input to Command and Control [USA] *mt*
MISM MAJCOM Information System Manager *mt*
MISMO Maintenance Interservice Support Management Office *fu, wp*
MISMO MAJCOM Information Systems Management Office [ASPO] *mt*
MISO Maintenance Interservice [Support] Officer *wp, fu*
MISP MAJCOM Information Systems Plan [USAF] *mt*
MISPE Monopulse Information Signal Processing Element *hc*
MISR Major Item Status Report *wp*
MISR Mobile Intercept-Resistant Radio *fu*
MISR Multi-angle Imaging Spectro-Radiometer [EOS] *hy*
MISREP Mission Report [Joint Tactical Air Reconnaissance / Surveillance] [USA DOD] *fu, mt*
MISS Man In Space Simulation *hc*
MISS Man In Space Soonest *hc, ci.1062*
MISS Manpower Information Systems Support [USMC] *mt*
MISS Microwave Imager Sensor Study *hc*
MISS Mobile Instrumentation Support System *jmj*
MISS Modular Infrared Sensor System *hc*
MISS-IDA Meteorology Interactive Software System for Image Data Analysis *pl*
MIST Modularized Interoperable Surface Terminal *fu*
MIST Mosaic Infrared Sensor Technology *hc, wp, tb*
MISTC Multiple Information Set Tracking Correlator System [SDI] *mt*
MISTEC Microprocessor Support, Training, Evaluation, and Critique *fu*
MISTEC Mosaic Infrared Seeker T{est} and E{valuation} Chamber *st*
MISTER Mobile Integrated System Trainer Evaluation *hc*
MISTER Mobile Integrated System Trainer, Evaluator, and Recorder *fu*
MISTI Miniature Sensor Technology Integration *kf*
MISTIC Missile System Target Illuminated Control *hc*
MISTIC Multiple Information Set Tracking Correlation *hc*
MISTIR Multifunction Imaging Search/Track Infrared *hc*
MISTR Management of Items Subject To Repair *mt, wp, hc, wp*
MISTRAM Missile Trajectory Measurement System *wp*
MISTY Missile System for Tactical Telephony *hc*
MISTY_JADE [DNA project to investigate nuclear explosion effects] *wp*
MISTY_RAIN [DNA project to assess radiation impact on satellites] *wp*
MISVAL Missile Evaluation *hc*
MIT Manufacturing, Inspection and Test [flow chart] *oe*

MIT Massachusetts Institute of Technology *aa26.12.25, hc, hn44.19.2, wp, ns, tb*
MIT Materiel Improvement Team *wp*
MIT {Missile Interface Test} [see ALARM] *hc*
MIT/LL Massachusetts Institute of Technology Lincoln Laboratories *uh, 17, st*
MIT/MAT Missile Interface Test/Missile Auxiliary Test *hc*
MITAS Missile Threat Analysis Simulator *ts4-17, tb*
MITB Missile Interface Test Bench *hc*
MITE Master Instrumentation Timing Equipment *fu*
MITE Missile Integration Terminal Equipment *hc, fu*
MITE Mission Training and Evaluation *hc*
MITE Multiple Input Terminal Equipment *fu*
MITEC Machine Intelligent Technical Controller *mt*
MITI Ministry of International Trade and Industry [Japan] *ge, ac3.3-23, aw129.14.38, hg*
MITP Master Integration and Test Plan *fu*
MITRE Massachusetts Institute of Technology Research and Engineering [a corporation] *fu, mt, wp, tb*
MITS MAC* Intra Theater Transmission System *mt*
MITS Management Information Telecommunication System *fu*
MITS Military Air Lift Command Imagery Transmission System *wp*
MITS MILSTAR Intermediate Test Station *fu*
MITS Missile Interface Test Set *hc, jdr*
MITS Mission Integrated Transparency System *wp*
MITSY Memory Interface Test System *hc*
MIU MAGE Interface Unit *-4*
MIU Memory (Message) Interface Unit *fu*
MIU Message Injection Unit *fu*
MIU Midas Interface Unit *fu*
MIU Multifunction Interface Unit *fu*
MIU Multiplexing Interface Unit *mt*
MIVP Mission Interface Verification Plan *oe*
MIW Mine Warfare *mt*
MIWG MAC Interoperability Working Group *mt*
MIX Member Information Exchange *pk, kg*
MJCOM Major Command [USA] *mt*
MJCS Joint Chiefs of Staff Memorandum *mt*
MJCS Memorandum of the Joint Chiefs of Staff *dm*
MJEMS Modular Jet Engine Management Simulator *mt*
MJF Major Frame [structure] *uh*
MJNPE Mobile Joint Nuclear Planning Element *mt*
MJPO Milstar Joint Program Office *jdr*
MJT Maintenance Job Tracking system *wp, mt*
MJU Mariner/Jupiter/Uranus *hc*
MK Maschinen Kanone [Germany] *mt*
MK Modification Kit *fu*
MKP Modem Kernel Processor *fu*
MKR Modification Kit Requirement *fu*
MKS Meter, Kilogram, Second *wp, ai1*
MKV Miniature Kill Vehicle *wp, tb, st, jmj*
MKV Multiple Kill Vehicle *wp, tb*
ML Machine Language *pk, kg*
ML Major Lobe *fu*
ML Master Library *fu*
ML Materials Laboratory [of AFWAL] *tc*
ML Materiel List *wp*
ML Maximum Likelihood *fu*
ML Meta Language *pk, kg*
ML Minelayer [ship] [US Navy] *mt*
ML Mobile Launcher *ai2*
ML/PWB Multilayer Printed Wiring Board *fu*
MLA Microwave Link Analyzer *-4*
MLA Multispectral Linear Array *hc, ac2.2-23*
MLA/SWIR Multispectral Linear Array Short Wave Infrared *wp*
MLAN Midas Local Area Network *fu*
MLAPI Multilingual Application Programming Interface *kg, hi*
MLB Message Link Buffer *mt*
MLB Multilayer Board *fu*
MLBS Multi Layer Boards *hc*
MLC Main Lobe Cancellation *wp*
MLC Main Lobe Clutter *hc*
MLC Management Level Chart *wp*
MLC Management Level Code *wp*
MLC Maneuver Load Control *aa26.8.22*
MLC Military Liaison Committee *dt*
MLC Milstar Leadership Council *jdr*
MLC Multilayer Ceramic *pk, kg*
MLC Multilevel Cell [program] [Internet] *pk, kg*
MLC Multiline Communications Controller [Honeywell] *mt*
MLCAD Maintenance and Logistics Factors in Computer Aided Design *mt*
MLCB Multilayer Circuit Board *fu*
MLCC Modular Life Cycle Cost *wp*
MLD MAJCOM Level Data *mt*
MLD Marine Luchtvaart Dienst [(Netherlands) Naval Air Service] *mt*

MLD Master Layout Duplicate *fu*
MLD Maximum Likelihood Detector *fu*
MLD Mean Level Detector *hc, fu*
MLD Median Lethal Dose *w9*
MLD Minimum Lethal Dose *w9, wp*
MLD Minimum Line of Detection *fu*
MLD Missile Launch Detector (Detection) *hc, wp*
MLDT Mean Logistics Down Time *fu*
MLE Maximum Likelihood Estimate *wp*
MLE Mesoscale Lightning Experiment *ws4.10*
MLE Missile Launch Envelope *hc, wp*
MLEP Multi-purpose Long Endurance Plane *hc*
MLF Maintenance Level Function *hc*
MLF Medium Low Frequency *fu*
MLF Multi Lateral Force [NATO] *hc, w9*
MLF Multilateral Force *fu*
MLG Main Landing Gear *wp*
MLGW Maximum Landing Gross Weight *wp*
MLI Machine Language Instruction *tb*
MLI Minimum Line of Interception *fu*
MLI Multi(ple) Layer Insulation *uh, gd2-12, pl, lt, wl, jdr, kf, jn, oe*
MLID Multiple Link Interface Driver *kg, hi*
MLIM Microminiature Low Level Multicoders
MLLDI Multi-Level Laser Designator Illuminator *hc*
MLLP Mainstream Lunar Landing Program *hc*
MLMS Multipurpose Lightweight Missile System [an air launched development of Stinger, intended as a self-defense weapon for helicopters and small aircraft] [USA] *mt*
MLN Message Ledger Number *mt*
MLO Master Layout Original *fu*
MLO Military Liaison Officer *mt*
MLOC Million Lines Of Code *tb*
MLP Mobile Launch Platform *oe, ai2*
MLP Multi Level Precedence *mt*
MLP MultiLink Procedure *fu*
MLPI Maximum Likelihood Parameter Identification *wp*
MLPP Multi Level Precedence (Priority) and Preemption *fu, mt*
MLR Main Line of Resistance *hc*
MLR Modular Laser Rangefinder *hc*
MLR Monodisperse Latex Reactor *sc4.6.6*
MLR Multichannel Linear Recording *em*
MLR Multiple Linear Regression *wp*
MLRF Mini Laser Range Finder *hc*
MLRS Multiple Launch Rocket System *mt, fu, hc, wp, ph, af72.5.28, nd74.445.30, pm, kf*
MLS Master Log Sheet *fu*
MLS Microtel Learning Services *fu*
MLS Microwave Landing System *fu, mt, hc, wp, aw129.14.115*
MLS Microwave Limb Sounder [UARS, JPL; EOS] *ge, hg, hy, pbl*
MLS Missile Launch System *tb*
MLS Month Long (Length) Select *uh, jdr*
MLS Multiple Level Security (Secure) *mt, fu, kf, wp, tb*
MLS SEED Month Length Select *ric*
MLSA Modified Launch Services Agreement *wl*
MLSF Mobile Logistics Support Force [US Navy] *wp, mt*
MLSI Multi-Level Large Scale Integrated/Integration *hc*
MLS_DDN Multi Level Secure Defense Data Network *mt*
MLT Mean Level Tracker *hc*
MLTAA Manufacturing License and Technical Assistance Agreement *fu*
MLU Memory Load(ing) Unit *fu, bo*
MLU Mid Life Upgrade *af72.5.136, mt*
MLU Modular Logic Unit *fu*
MLU [Japanese magnetic levitation train] *is26.1.63*
MLV Medium Launch Vehicle *16, ic2-63, dm, kf, mt*
MLV Medium Lift Vehicle *tb*
MLV Memory Loader / Verifier *mt*
MLVT Missile Launch Vehicle Transporter *tb*
MLW Maximum Landing Weight *mt*
MLWIR Medium Long Wave Infrared *kf*
MM Maintenance Manual *fu, mt*
MM Man-Month *fu, hc, cca, kf*
MM Manufacturing Methods [and Technology] *hc*
MM Mariner Mars *hc*
MM Mass Memory *hc, tb, jdr*
MM Materiel Management *mt*
MM Micro Mole *wp*
MM Microwave Module *fu*
MM Middle Marker *fu*
MM Minute-Man [missile] *sd, tb, st, kf*
MM Mission Management *fu*
MM Multi Mission [Surveillance System] *tb*
MM&T Manufacturing Methods and Technology *st, fu*
MM/PCR Manufacturing Management Production Capability Review *fu*
MMA Martin Marietta Aerospace [Company] *-4*
MMA Microcomputer Managers Association *pk, kg*
MMA Momentum Management Assembly *hc*

MMA Monmethyl Aniline *wp*
MMAC Material Management Aggregation Code *hc, wp*
MMAD Millimeter Wave Active Decoy *hc*
MMAIDS Martin Marietta Aerospace Integrated Data Systems *fu*
MMAX Maximum Repair Time *fu*
MMAXCT Maximum Corrective Time *mt*
MMB Modified Monobit *fu*
MMB Multi Mission Bus *sc3.10.1, 17, kf*
MMBF Mean Miles Between Failure *wp*
MMBR Multipurpose Supersonic Beamrider *hc*
MMC Martin Marietta Corporation *hc, sa12474.ep*
MMC Massachusetts Microelectronics Center *is26.1.86*
MMC Matched Memory Cycle *pk, kg*
MMC Material Management Center [USA] *fu, mt*
MMC Material Management Code *fu, hc, wp*
MMC Maximum Material Condition *fu*
MMC Metal Matrix Composite *wp, aa26.8.40, hc, tb, st*
MMC Microcomputer Marketing Council *pk, kg*
MMC Mirror Motion Compensation *oe*
MMC Movements Monitoring Center *mt*
MMC MultiMode Coupler *jdr*
MMCD Multi Media Compact Disk *kg, hi*
MMCIAC Metal Matrix Composite Information Analysis Center *hc*
MMCS Missile and Munitions Center and School [US Army] *hn44.11.8*
MMCSRS MACOM Materiel Condition Status Reporting System [USA] *mt*
MMCX Multimedia Communication Exchange [Lucent Technologies] *pk, kg*
MMD Manufacturing Management Directive *fu*
MMD Master Modification Drawing *fu*
MMD Master Monitor Display [system study] *hc*
MMD Materiel Management Directive *hc*
MMD Mean Mission Duration *hc, jdr, kf, pbl*
MMD Moving Map Display *hc*
MMD Multimode Display *fu*
MMDC Master Message Display Console *mt*
MMDMTI Multi Mode Display Moving Target Indicator *hc*
MMDS Martin Marietta Data Systems *tb*
MMDS Militarized Magnetic Drug {Drum} System *hc*
MMDS Multichannel Memorandum Distribution Facility *mt*
MMDS Multipoint Multichannel Distribution Service *ct, ric, jdr, em*
MME Master Mode Entry *mt*
MME Military Message Experiment *mt*
MMEC Materiel Maintenance Engineering Center *wp*
MMES Maritime Mobile Earth Station *hc*
MMES Martin Marietta Energy Systems *mt*
MMF Machine Master File *hc*
MMF Magneto Motive Force [mmf also used] *hc, w9, fu, ai1*
MMF Man-Machine Model *fu*
MMF Mass Memory Module *fu*
MMF Master Message File *mt*
MMF Missile Maintenance Facility *tb*
MMF Mobile Maintenance Facility *hc*
MMF Multi Message Fusion *kf*
MMF Multi Mode Fiber *fu*
MMFCS Micro Modulator Fire Control System *hc*
MMFL Mobile Maintenance Facility Liaison *hc*
MMFP Modular Maverick FLIR Pod *hc*
MMG Multi Mode Guidance *hc, fu*
MMGT Microwave Magneto-Transconductance *mj31.5.52*
MMH Maintenance Man-Hours *fu, hc, wp*
MMH Mono Methyl Hydrazine [CH3N2H3] *mi, uh, mb, kf, jn, ac3.2-17, 16, 17, lt, wp, oe, ai2*
MMH/FH Maintenance Man Hours per Flight Hour *mt, fu, hc, wp*
MMH/MA Maintenance Man-Hours per Maintenance Action *fu*
MMH/OH Maintenance Man Hours per Operating Hour *mt, fu, hc*
MMH/UH Maintenance Man-Hours per Utilization Hour *hc*
MMHS Mechanized Material Handling System *wp*
MMI Man Machine Interface *mt, fu, kg, sr9-82, 1, 16, wp, pf1.11, go1.1, tb, oe*
MMI Minuteman I *17, sb*
MMI&CS Martin Marietta Information and Communication Systems *tb*
MMIC Microwave Monolithic Integrated Circuits *mt, fu, ro, wp, ep, jdr, kf, jn, aw130.13.8, hc, 17, hq5.3.5, af72.5.95, sc9.6.3, 19*
MMIC Millimeter / Microwave Integrated Circuit *mt*
MMICS Maintenance Management Information and Control System [USAF] *wp, tb, mt*
MMIIS Materiel Management Integrated Information System *wp*
MMIR Multispectral Microwave Imaging Radiometer *wp*
MMIS Materials Manager Information System *pk, kg*
MMITI Martin Marietta Information Technology Institute *tb*
MML Mass Memory Loader *fu*
MMLAN Multi Media Local Area Network *hc*
MMLRF Micro Modular Laser Range Finder *hc*
MMLS Mobile Microwave Landing System *mt*
MMM Maintenance Material Management [also called 3-M system] *wp, fu*

MMM Mariner Mission to Mars *hc*
MMM Mars Mission Module *hc*
MMM Meaningful Measures of Merit *mt*
MMM Minnesota Mining and Manufacturing [Company] *-19*
MMM Mode Management Module *mt*
MMM Multi Mission Missile *hc, wp*
MMM/UH Maintenance Man-Minutes per Utilization Hour *hc*
MMM65 Mast Mounted M65 *hc*
MMMC MAJCOM Manpower Management Computer [also M3C] [USAF] *mt*
MMMDCS Multinational Maintenance Management Data Collection System *mt*
MMMIC Monolithic Microwave and Millimeter Wave Integrated Circuit [also M3IC] *mt*
MMMR Medical Materiel Mission Reserve *wp*
MMMS MAJCOM Manpower Management System [also M3S] [USAF] *mt*
MMMS Medical Materiel Management System [USAF] *mt*
MMMS-OL Medical Materiel Management System On Line *mt*
MMN Maintenance Module Node *wp*
MMO MIPR Management Office *wp*
MMO Mission Management Office *ac3.7-33*
Mmo Maximum operating Mach number *mt*
MMOC Multimissions Operations Center *hy*
MMP Maintenance and Modernization Program *wp*
MMP Maintenance Management Program *mt*
MMP Master Mobilization Plan *mt, wp*
MMP Medical Mission Planner *mt*
MMP MEECN Master Plan *mt*
MMP MEECN Message Processing Mode *mt*
MMP Military Microprocessor *hc*
MMPC Mobilization Materiel Procurement Capability *wp*
MMPF Microgravity and Materials Processing Facility *wl*
MMPM Marine Management Programming Model [USMC] *mt*
MMPM Multi Media Presentation Manager *pk, kg*
MMPS Manpower Mobilization Planning System [USMC] *mt*
MMPS Manual Message Preparation System [Consolidated] *hc*
MMPS MEECN Message Processing System *mt*
MMR Maintenance Management Report *mt*
MMR Maintenance Management Review *wp*
MMR Manufacturing Method Report *fu*
MMR Mobilization Material Requirement *wp*
MMR Modified Modified Read *hi*
MMR Monthly Management Review *17, wp*
MMR Multi Mission Radar [later FARMAR] *hc*
MMR Multi Mode Radar *hc, wp*
MMRB Materiel Management Review Board *wp*
MMRBM Mobile Medium Range Ballistic Missile *fu, hc, 0*
MMRC Materials and Mechanics Research Center [US Army] *tc*
MMRD Multi Mode Radar Display *hc*
MMRDDP Multi Mode Radar Digital Doppler Processor *hc*
MMRFS Multi Mode Radar Feasibility Study *hc*
MMRP Missile Master Replacement Program *hc*
MMRPV Multi Mission Remotely Piloted Vehicle *wp*
MMS Main Motor Start *fu*
MMS Maintenance Management System *fu*
MMS Mass Memory System *hc*
MMS Mast Mounted Sight [for helicopters] *hc, mt*
MMS Material Management System *fu*
MMS Meteorological Measuring System *ag*
MMS Minimum Manned Satellite *ci.1062*
MMS Modular Multispectral Scanner *ci.940*
MMS Movement Monitoring System [EUCOM] *mt*
MMS Multimission Modular Spacecraft [NASA] *hc, mt, wp, sc2.7.2, ep, hy, ai1*
MMS Multiple Microprocessor System *hc*
MMS Munitions Maintenance Squadron *mt*
MMS [a time bias (FE)] *sb*
MMSA Multiple Mission Surveillance Aircraft [USA] *mt*
MMSC Missile Monitor Signal Conditioner *fu*
MMSC Multi Mode Sonar Console *hc*
MMSC Multimode Sonar Console *fu*
MMSCV Manned Military System Capability Vehicle *hc*
MMSE Minimum Mean Square Error *fu*
MMSE Molecular Monitor Shuttle Experiment *hc*
MMSE Multi Mission Support Equipment *bo*
MMSR Master Materiel Support Record *hc, wp*
MMSS Maritime Mobile Satellite Service *wp*
MMSS Multi Mission System Study *hc*
MMSS Multimode Scanning System *fu*
MMSSF Man Machine System Simulation Facility *hc*
MMST Multi Mode Storage Tube *fu, hc*
MMT Manufacturing Methods and Technology *wp, fu*
MMT Master Message Terminal *mt*
MMT Mean Maintenance Time *kf*
MMT Missile Maintenance Trainer *hc*

MMT Mission Maintenance Technician *wp*
MMT Mobile Maintenance Team *fu*
MMT Multi Mode Tonotron *hc*
MMT Multi Mode Transponder *2, 4*
MMT Multimode Terminal *fu*
MMT Multiple Mirror Telescope *mi*
MMTE Manufacturing Methods Technology Engineering *hc*
MMTPS Maintenance Materiel Transaction Processing System [TAC] *mt*
MMTS Multimode Transmission System *fu*
MMTTRF Mean Maintenance Time To Restore Function *kf*
MMU Main Memory Unit *mt*
MMU Manned Maneuvering Unit *mi, aa26.12.26, ci.611, tb, wl, jdr, mb*
MMU Mass Memory Unit *9, jdr, fu, pbl*
MMU Memory Management Unit *mt, kg, em, fu, hi*
MMV Mobile Maintenance Van *fu*
MMW Millimeter Wave *mt, fu, uc, wp, hc, aa26.11.4, af72.5.8, aw130.22.14, tb, st, jdr*
MMW Multi Mega Watt *tb*
MMWA/ACN Millimeter Wave Armor/Air Covert Net *fu*
MMWIFE Millimeter Wave Integrated Front End *hc*
MMWR Millimeter Wave Radar *mt*
MMX Matrix Manipulation Extensions [Intel] *pk, kg*
MMX Multimedia Extension *ct, em, kg*
MN Magnetic North *w9*
MN Material Need *fu*
MN Merchant Navy *wp*
MN Message Number *0*
MNA Mission Need Analysis *kf*
MNANG Minnesota Air National Guard *-12*
MNC Major NATO Command(er) *fu, mt*
MNC Multinet Conflict *fu*
MNC Multinet Control *fu*
MNCS Master Net Control Station *mt*
MND Ministry of National Defense *hc*
MNDO Modified Neglect of Differential Overlap *wp*
Mne Mach number Never to be Exceeded *mt*
MNF Minor Frame *jdr, uh*
MNFP Multi Nation Fighter Program *wp, af72.5.136*
MNFR Malfunction Nonconformance Failure Report *oe*
MNFS Minor Frame Synchronization *bl1-5.134*
MNG Multi-Net Gateway *mt*
MNM Multinet model *fu*
MNOS Metal Nitride Oxide [on] Silicon [semiconductor] *kg, fu, hc, wp*
MNP Microcom Networking Protocol *wp, kg, ct, hi*
MNPO Multi National Program Office *st*
MNPWR Manpower Standard Application Program *mt*
MNRAS Monthly Notices of the Royal Astronomical Society *mi*
MNS Mission Need Statement *mt, wp, kf*
MNS Multi Net Simulation *fu*
MNSL Maintenance Non Significant Lists *hc*
MNTVC Movable Nozzle Thrust Vector Control *ai2*
MNV Mine Neutralization Vehicle *uc*
MO Magneto Optical [disk drive] *ct, em, kg, hi*
MO Mail Order *w9, wp*
MO Maintenance Officer *wp*
MO Manual Output *fu*
MO Manufacturing Option *fu*
MO Mars Observer [mission or spacecraft] *eo, re*
MO Master Oscillator *mt, fu, uh, jdr*
MO Modification Order *wp*
MO Molecular Orbital *wp*
MO Money Order *w9, wp*
MO&DA Mission Operations and Data Analysis *re*
MO&DS Mission Operations and Data System [NASA] *mt*
MO&DSD Mission Operations and Data Systems Directorate *oe*
MOA Make On Assembly *fu*
MOA Matrix Output Amplifier *fu*
MOA Memorandum Of Agreement *mt, fu, uh, 4, wp, hc, af72.5.13, tb, jdr, kf, oe*
MOA Method Of Accomplishment *wp*
MOA Military Operations Area *fu*
MOA Ministry Of Aviation [Germany or UK] *hc*
MOAMA Mobile Air Materiel Area *wp*
MOANG Missouri Air National Guard *-12*
MOAT Missile Off Aircraft Test *hc*
MOAT Mission Opportunities For Airship Technology *wp*
MOATRS MAJCOM On Line Aerospace Vehicle Training Report System [USAF] *mt*
MOATS Message Originating and Terminating Station *hc, fu*
MOB Main Operating Base [CMAH + MGS] *fu, 12, wp, hc, tb, kf*
MOB Main Operations Base [USA DOD] *mt*
MOB Memory Order Buffer *kg, hi*
MOB Military Order of Battle *mt*
MOB Missile Order of Battle *mt*
MOB Mobile Operating Base *tb*
MOBCC Mobilization Command and Control [USA] *mt*

MOBCON Mobilization Construction Plan *mt*
MOBDET Mobility Deployment Planning System [PACAF] *mt*
MOBERS Mobilization Equipment Redistribution System [USA] *mt*
MOBIDIC Mobile Digital Computer *wp*
MOBILAB Mobile Rapid Photo Processing Laboratory *wp*
MOBLAS Mobile Laser *ci.133*
MOBPERS Mobilization Personnel Processing System [USA] *mt*
MOBPLANS Mobility Plans *mt*
MOBREM Mobilization Base Requirements Model [USA] *mt*
MOBS Multiple Orbit Bombardment Systems *tb*
MOBS Multiple Orbital Bombardment System *mt*
MOBSCOPE Mobilization Shipments Configured for Operations, Planning and Execution *mt*
MOBSS Mobile Support Squadron *mt*
MOBTDA Mobilization Table of Distribution and Allowances *mt*
MOBTRAN Mobilization and Deployment Transportability [MTMC] *mt*
MOC Magnetic Optic Converter *fu*
MOC Maintenance Operations Center *mt*
MOC Mars Observer Camera [on Mars Observer] *eo, re, mb, mi*
MOC Master Operations Central *hc*
MOC Materiel Obligation Code *wp*
MOC Measures Of Correlation *fu*
MOC Milstar Operations Center *jdr*
MOC Minimum Operating Capability *mt*
MOC Missile Operations Center *go1.2-1, jb*
MOC Mission Operations Computer *pf1.24*
MOC Mobile Operations Center *fu*
MOCA Merisel Open Computing Alliance *ct*
MOCAS Mechanization Of Contract Administration Sources *mt*
MOCAT Measurement and Observation of Clear Air Turbulence *wp*
MOCC Meteosat Operations Control Centre *es55.45*
MOCC Mission Operations Control Center *pl*
MOCOCO Movement Coordinating Committee *mt*
MOCP Mission Operational Computer Program *mt*
MOCR Mission Operations Control Room *oe*
MOCS Missile Operational Communications Systems *wp*
MOCS Mission Operations and Control Subsystem [at GSFC] *ac2.7-11*
MOCS Multichannel Ocean Color Scanner *hy*
MOCTR Multichannel Optical Communications and Tracking Receiver *ro*
MOCVD Metallo Organic Chemical Vapor Deposition *wp*
MOD MAC Operational Directive *mt*
MOD Magneto Optical Drive *hi*
MOD Maintenance Of Deception *st*
MOD Maintenance Of Defense *st*
MOD Masters Of Deception *ct*
MOD Military Orbital Development *hc*
MOD Ministry Of Defense [Germany, United Kingdom, etc.] *mt, fu, hc, kf*
MOD Miscellaneous Obligation Document *wp*
MOD Mission Operating Directive *mt*
MOD-PE Ministry Of Defence - Procurement Executive (U.K.) *fu*
MODA Ministry Of Defense and Aviation *fu*
MODAPS Modal Data Acquisition and Processing System *hc*
MODAS Maintenance and Operational Data Access (Sub)System *mt, wp*
MODAS Model for Intertheater Deployment by Air and Sea [OSD] *mt*
MODB Mission Operations Data Base *oe*
MODBSC Modular Boresight Computer *hc*
MODCAP Material, ODC {Other Direct Costs}, Analysis and Printout *hc*
MODELS Modernization Of the Defense Logistics Standard System [DLA] *mt*
MODEM Modulation-Demodulation *wp*
MODEM MOdulator / DEModulator *mt, ct, fu, kg, em, csIII.30, 6, hc, tb, hi, tt*
MODES Mode Optimization and Delivery Estimation System [JDS] *mt*
MODES Mode Select Beacon System *fu*
Mode_A Key Auto-Key Encryption and Decryption *fu*
Mode_B Cipher Text Auto-Key Encryption and Decryption *fu*
Mode_C Altitude Reporting Mode of Secondary Radar *fu*
Mode_C Long-Cycle Operation *fu*
Mode_S Discrete Interrogation Beacon and Data Link *fu*
Mode_S Mode Select Beacon System *fu*
MODFET Modulating Field Effect Transistor *fu*
MODILS Modular Microwave Instrument Landing System *wp*
MODIS Moderate Resolution Imaging Spectrometer (Spectroradiometer) [EOS] *kf, aw130.11.49, hy*
MODIS-N Moderate Resolution Imaging Spectrometer-Nadir (GSFC) *ge, hg*
MODIS-T Moderate Resolution Imaging Spectrometer-Tilt (GSFC) *hg, ge*
MODMATS Modified Missile Auxiliary Test Set *hc*
MODS Military Orbital Development System *hc*
MODS Mobilization Planning Data System [MTMC] *mt*
MODU Master Oscillator Distribution Unit *jdr*
MODWORS Modification Work Order/Status System *mt*
MOE Measure Of Effectiveness *mt, fu, sd, wp, tb, st, jdr, kf*

MOE Measure Of Efficiency *tb*
MOE Motor Overload Electronics *oe*
MOF Ministry Of Finance, [Japan] *bl3-1.155*
MOFFS Multi-Megabit Operation Flexible Frame Synchronizer *hc*
MOFS Maintenance Of Flying Skills *wp*
MOG Master Oscillator Group *uh*
MOG Material Ordering Guide *wp*
MOG Maximum [Aircraft] On the Ground *mt*
MOG Micro Optic Gyro *17, sb*
MOG-R Master Oscillator Group-Rubidium *jdr*
MOG-X Master Oscillator Group-Crystal *jdr*
MOHATS Mobile Overland Hauling And Transport System *wp*
MOHLL Machine Oriented High Level Language *pk, kg*
MOI Maintenance Operating Instruction *fu*
MOI Mars Orbit Insertion *ma20, mb, re*
MOI Master Oscillator Input *jdr*
MOI Mission Oriented Items *mt*
MOI Moment Of Inertia *gd2-30, pl, jdr, kf, jn, oe*
MOIA Mission Oriented Item Activity *mt*
MOIC Missile Ordnance Inhibit Circuit *hc*
MOIS Missile Operations Intercommunications System *0*
MOL Machine Oriented Language *fu*
MOL Manned Orbiting Laboratory *fu, mi, hc, ci.209, 13, aw, w9, mb, ai2*
MOL Maximum Order Limitation *mt*
MOL Microsoft Open License *ct*
MOL Middle Of Life *3, jn*
MOL Motor Over Load [sounder/imager scanner motor] *oe*
MOLA Mars Observer Laser Altimeter [on Mars Observer] *eo, re, mb, mi*
MOLAB Mobile Lunar Laboratory *hc*
MOLAP Multidimensional On-Line Analytical Processing *pk, kg*
MOLECAP Molecular Contamination Analysis Program *kf*
MOLINK Moscow Link [Moscow-Washington Direct Communications Link, a.k.a Hotline] *mt, wp*
MOLS Machine Oriented Language Lines *mt*
MOLS Multiple Object Location System *hc*
MOLT Manually Operated Lift Trailer *wp*
MOM Many On Many *wp*
MOM Measure of Merit *fu, wp, tb, st*
MOM Message Oriented Middleware *em*
MOM Military Official Mail *wp*
MOM MIS Modification Module *mt*
MOM Mission Operations Manager *gp, oe*
MOM Modified Operational Missile *tb*
MOMAR Modern Mobile Army *hc*
MOMTD Metal On Metal Tunnel Diode *wp*
MOMV Manned Orbital Maneuvering Vehicle *mi*
MON Memorandum Of Negotiation *jdr*
MON Mixed Oxides of Nitrogen *oe*
MON Motor Octane Number *wp*
MON-3 Mixed Oxides of Nitrogen [nitrogen tetroxide with 3% nitric oxide] *jdr, oe, ai2*
MON-3 Nitrogen Tetroxide [oxidizer] *uh, kf*
MONEX Monsoon Experiment *pl*
MOO MUD (Multi User Dimensions) Object Oriented [Internet] *kg, hi*
MOOG (MOOG, Inc. . . . East Aurora, NY 14052) *jmj*
MOOSE Milstar On Orbit Support Equipment *jr*
MOOSSE Manned Orbital Oceanographic Survey System *hc*
MOP (the testing of the three MOP spacecraft) [METEOSAT] *es55.45*
MOP Machine Operating Processing *tb*
MOP Maintenance Operations Protocol *pk, kg*
MOP Measure(ment) Of Performance *fu, kf, tb*
MOP Memorandum Of Policy [JCS] *tb, mt*
MOP Message Operating Processing *tb*
MOP Message Output Processor *mt*
MOP METEOSAT Operational Program [EUMETSAT] *hg, ge*
MOP Mission Operations Plan(ning) *0, oe*
MOP Mobility Operating Procedure *mt*
MOP Modulation on Pulse *fu*
MOP Monthly Obligation Plan *mt*
MOPA Master Oscillator Power Amplifier *mt, fu, hc, wp, tb*
MOPAR Management Of Post Attack Resources *wp*
MOPAR Master Oscillator Power Amplifier Radar *fu*
MOPE Multiple Object Parameter Estimation *hc*
MOPITT Measurements Of Pollution In The Troposphere *hy*
MOPLN Mission Orbit Planning *tb*
MOPM Office Of the Product Manager *dt*
MOPMS Modular Pack Mine System *dh, fu, hc*
MOPP Material Operations and Parts Procurement *ep*
MOPP Mission Objective Protective Posture *mt*
MOPP Mission Ordered Protective Posture *fu*
MOPP Mission Oriented Protective Posture [chemical warfare] *wp*
MOPR Mission Operations Planning Review *gp, oe*
MOPS Message Output Processing System *mt*
MOPS Million Operations Per Second *fr, 17, tb, hi*
MOPS Missile Operations Paging System *0*

MOPS Mission Operations *jdr*
MOPT Mission Operations Planning Team *oe*
MOPTAR Multiple Object Phase Tracking and Ranging *fu*
MOR Mars Orbital Rendezvous *hc*
MOR Memory Output Register *fu*
MOR Military Operational Requirement *1, tb*
MOR Mission Operations Review *gp, 17, sb, oe*
MORD Mission Operations Requirement Document *jdr*
MORL Manned Orbital Research Laboratory *wp, hc, ci.1065, 0*
MORR Mission Operations Readiness Review *oe*
MORS Military Operations Research Society *mt, wp, tb*
MORSA Movement Requirements for Staff Planning and Special Studies Applications [JCS] *mt*
MORT Master Operational Recording Tape *fu*
MORT MOSART Related Technology *hc*
MOS Magneto Optic Storage *pk, kg*
MOS Manned Orbital Station *mt, wp*
MOS Marine Observation Satellite [Japan] *ge, ac3.3-1, hy, hg*
MOS Mass On Line Storage *mt*
MOS Memorandum Of Support *mt*
MOS Metal Oxide Semiconductor (Silicon) *mt, fu, ct, em, kg, wp, hc, rm1.2.1, 6, w9, is26.1.55, tb, hi*
MOS Military Occupational Skill [i.e. the task one is trained for] *mt*
MOS Military Occupational Specialty [USA/USMC] *fu, hc, st, mt*
MOS Minimum Operating Strip *mt, wp*
MOS Mission Operations Software *bo*
MOS Mission Operations Support *oe*
MOS Mission Operations System *so, mb, re*
MOS Mode Of Shipment *wp*
MOS Model Output Statistics *ag*
MOS Monolithic Oxide Silicon *mt*
MOS Multi Orbit Satellite *kf*
MOSAIC Multiple user, On line System for Automated Information Communication [a non-proprietary software tool using hypertext links to navigate and retrieve data from the Internet] *cu, mt*
MOSAIDS MOS Automated Integral Design System *fu*
MOSAR Modulation Scan Array Radar *wp*
MOSART Monolithic Signal Processor and Detector Array *hc*
MOSC Mission Operations System Center *re*
MOSC Modified Operations Summary Console *fu*
MOSFET Metal Oxide Semiconductor Field Effect Transistor *fu, kg, em, he, uh, hc, hm, 6, 17, sb, th4, w9, wp, jn, hi*
MOSIS Metal Oxide Semiconductor Implementation System *em*
MOSIS Monolithic Silicon Detector *hc*
MOSIS MOS Implementation Service [DARPA] *1, tb*
MOSK Multimode Operations Support Network *ic1-4, rg*
MOSP Mission Operations Support Plan *oe*
MOSPF Multicast Open Shortest Path First *em*
MOSS Maintenance Operating (Operations) Support Set *mt, wp*
MOSS Manned Orbital Space Station *0, hc*
MOSS MIME Object Security Services *em*
MOSS Mission Operation Support Services *hc*
MOSS Mission Operations Support Satellites *lka*
MOSS Mobility Support Sets *wp*
MOSS [NATO code name for Soviet TU-26 aircraft] *wp*
MOST Metal Oxide Semiconductor Transistor *fu, em*
MOST Missile On Shipboard Test *hc*
MOST Mission Operation Support Team *oe*
MOST Mobile Optical Surveillance Tracker *hc*
MOST Mobile Sonar Technology *hc*
MOST Mosaic Optical Sensor Technology *wp*
MOSTT Mosaic Optical Sensor Technology Testbed *tb*
MOT Maximum Operating Time *fu, wp*
MOT Ministry Of Telecommunications *hc*
MOT Ministry Of Transport, [Japan] *bl3-1.155*
MOT Ministry Of Transportation (Canada) *fu*
MOT&E Multinational Operational Test and Evaluation [NATO] *mt*
MOTAM Moving Target Attack Missile *wp*
MOTARDES Moving Target Detection System *wp*
MOTD Maintenance and Operating Technical Data *hc*
MOTD Maintenance and Operation Technical Data *fu*
MOTD Message Of The Day *pk, kg*
MOTE [NATO code name for Soviet MBR-2 reconnaissance aircraft] *wp*
MOTIF Maui Optical Tracking and Identification Facility *hc, ps204, jb, tb*
MOTP Minuteman Operational Targeting Program *hc*
MOTS Military Off The Shelf *tb*
MOTS Module Test Sets *hc*
MOTS Multiple OPLAN Tasking Summary [AFCC*] *mt*
MOTU Mobile Operation Training Unit *hc*
MOTU Mobile Technical Units *fu*
MOTV Manned Orbital Transfer Vehicle *mi*
MOTWU Man On The Way Up *wp*
MOU Memorandum Of Understanding *mt, ge, fu, wp, ci.570, hc, es55.92, af72.5.25, 19, hy, tb, wl, hg, kf, oe, y2k*
MOU Mobile Observer Unit *fu*
MOUSE Minimum Orbital Unmanned Satellite of the Earth *ci.40*

MOUT Military Operations on Urban Terrain *mt*
MOV Metal Oxide Varistor *fu, kg*
MOVLAS Manually Operated Visual Landing Aid System *wp*
MOVPE Metal Oxide Vapor Phase Epoxy *jn*
MOVREP Movement Reporting [US Navy] *mt*
MOVS Move String *pk, kg*
MOVTAS Modified Visual Target Acquisition System *wp*
MOWG Mission Operations Working Group *oe*
MOXE [All sky monitor on SXG] *cu*
MP Main Processor *fu*
MP Maintenance Plan *fu*
MP Maintenance Planning *mt*
MP Maintenance Processor *fu*
MP Man Pack *fu*
MP Manpower and Personnel *mt*
MP Manpower Plan *mt*
MP Manual Proportional [attitude control mode] *ci.349*
MP Manufacturing Plan *fu*
MP Mass Properties *jdr, jn*
MP Massively Parallel [processing] *kg, hi*
MP Master Pattern *fu, oe*
MP Materiel Procedure *fu*
MP Melting Point *fu, wp, w9, ai1*
MP Message Processing (Processor) *fu, kf, mt*
MP Micro Processor *kf, fu*
MP Military Police (Policeman) *mt, 12, w9*
MP Minuteman Platform *hc*
MP Mobilization Plan *mt*
MP Modulation Parameter *fu*
MP Multi Pole *fu*
MP Multi Purpose *oe*
MP Multiple Paths *fu*
MP Multiple Processor *mt, kg*
MP&T Manpower, Personnel and Training *fu*
MPA Maintenance Planning Analysis *fu*
MPA Man Powered Aircraft *mt*
MPA Manufacturing Product Assurance *fu*
MPA Maritime Patrol Aircraft *mt, ph, wp*
MPA Material Procurement Administrator *fu*
MPA Materiel Parts Administrator *fu*
MPA Materiel Project Administrator *hc*
MPA Medium Performance Amplifier *hc*
MPA Medium Power Amplifier *fu*
MPA Microwave Power Amplifier *uh*
MPA Mission Peculiar Adapter *oe*
MPA Modification Proposal Analysis *wp*
MPA Multimode Processing Array *rg*
MPA Multiple Payload Adapter *ai2*
MPAA Material Program Administration Analyst *fu*
MPAC Materials and Parts Availability Control *mt*
MPAD Manpower Personnel Assignment Document *mt*
MPAR Maintainability Program Analysis Report *hc*
MPASS Mobile Processing And Support System *wp*
MPB Manufacturing Project Bulletin *fu*
MPBI Multiple Post Boost Intercept [Study] *hc*
MPBRS Maintenance Production/Backlog Reporting System [USA] *mt*
MPC Magnetic Program Card *fu*
MPC Main Processor-Correlator *fu*
MPC Mandatory Product Control *fu, hc*
MPC Manufacturing Planning and Control *sc4.2.4*
MPC MAPL Planning Change *fu*
MPC Materiel Program Code *8, fu, wp, hc*
MPC Materiel Program Costs *hc*
MPC Message Processing Center [USAF] *hc, fu, wp, mt*
MPC Micro Programmable Computer *fu*
MPC Military Personnel Center *mt*
MPC Military Property Custodian *wp*
MPC Milstar Payload Computer *jdr*
MPC Minor Planets Circular *mi*
MPC Missile Practice Camp *mt*
MPC Mobile Processing Center *fu*
MPC Modification Planning Center *fu*
MPC Monitoring Proportional Counter [Einstein Observatory (HEAO-2)] *cu*
MPC Most Probable Cost *kf, ah, pbl*
MPC Multimedia Personal Computer *ct, kg, em*
MPC Multipath Channel *pk, kg*
MPC Multiple Processor Computer *hc*
MPC Multiple Purpose Communications *fu*
MPC Multipurpose Console (Command) *fu*
MPCAG Military Parts Control Advisory Group *mt, fu, hc*
MPCASS Modernization Parts Control Automated Support System *mt*
MPCCC Multimedia PC Compatibility Specifications [also MPC3] *hi*
MPCI Multi-Purpose Custom Interface *oe*
MPCL Multi-Purpose Chemical Laser *wp*
MPCMP Mass Properties Control and Management Plan *hc*

MPCRB Materials and Processes Change Review Board *fu*
MPCS Mission Planning and Control Station [Software] *pk, kg*
MPCS Multimedia PC System *hi*
MPCTNS Manportable Common Thermal Night Sight *hc*
MPD Magneto Plasma Dynamics *wp, hc, js, ai1*
MPD Microwave Power Devices (Company) *fu*
MPD Microwave Products Division [of IEG at Hughes Aircraft Company] *fu, hn44.9.10, pl*
MPD Mini Port Driver *ct, em*
MPD Missile Purchase Description *hc*
MPD Modification Program Directive *wp, fu*
MPD Multiple Purpose Display *fu, hc*
MPDD Meteoroid Penetration Detector Development *hc*
MPDG Multi-Purpose Display Generator *hc*
MPDR Mobile Pulse Doppler Radar *fu*
MPDS Message Processing and Distribution System *mt*
MPDT Mean Preventive Down Time *hc*
MPDT Message Processing Data Terminal *mt*
MPDV Modulation Parameter Data Vectors *fu*
MPE Maintainability Project Engineer *fu*
MPE Manual Plot Extractor *fu*
MPE Manufacturing Producibility Engineering *fu*
MPE Mass Properties Engineer *jdr*
MPE Max Planck Institute [Federal Republic of Germany] *ns*
MPE Maximum Permissible Exposure [to radiation] *hc*
MPE Mission Peculiar Equipment *wl*
MPE Mission Planning Element *jdr*
MPE Modern Programming Environment *hc*
MPE Multiple Programming Executive [HP] *pk, kg*
MPE Multiple Proximate Events *kf*
MPEG Motion (Moving) Picture Experts Group [type of net video file] *ct, jdr, kg, hi, du*
MPER Material in Process Engineering Request *hc*
MPES Maximum Performance Ejection Seat *hc*
MPF Main Processor Facility *fu*
MPF Medium Payload Fairing *ai2*
MPF Message Processing Function *fu*
MPF Mission Planning Facility *mt*
MPF Mission Planning Forecast *mt*
MPF Multipurpose Facility *kf*
MPFCS Multi Purpose Fire Control System *hc*
MPG Max Planck Gesellschaft *crl*
MPG Message Processing Group *fu*
MPG Microwave Pulse Generator *fu*
MPG Miles Per Gallon *wp*
MPG Mission Planners Guide *kf*
MPG Mobile Protected Gun *mt, fu*
MPG Multi Channel Pulse Generator *hc*
MPG-NT Mobile Protected Gun, Near Term *uc*
MPGP Multiphase Graphics Processor *fu*
MPGS Mobile Protected Gun System *hc, uc*
MPH Miles Per Hour *wp*
MPI Machine Performance Integration *mt*
MPI Magnetic Particle Inspection *fu*
MPI Mean Point of Impact *fu, wp*
MPI Message Passing Interface *em, kg*
MPI Multi Photon Ionization *wp*
MPI Multiprecision Integer *pk, kg*
MPIC Message Processing Interrupt Count *fu*
MPIF Microprocessor Interface *fu*
MPIK Manpack Installation Kit *fu*
MPIP Maintenance Posture Improve Program *mt*
MPIR Missile Precision Instrumentation Radar *fu*
MPK Maintenance Parts Kit *fu*
MPK Maliy Protivolodochniy Korabl [small anti submarine ship] [USSR] *mt*
MPL MADS Program Language *wp*
MPL Maintenance Parts List *hc, fu, wp*
MPL Mandatory Parts List *mt*
MPL Master Parts List *hc*
MPL Materials and Processes Laboratory *hd1.89.y26, ep*
MPL Maximum Procurement Level *wp*
MPL Mixer/Preamplifier/Local [oscillator] *pl*
MPL Multiple Pulsed Laser *hc*
MPL Multischedule Private Line *do38*
MPLCC Most Probable Life Cycle Cost *wp*
MPLM Machine Prepared List of Materials *hc, fu*
MPLM Mini Pressurized Logistics Module *ls4, 5, 38*
MPLO Mill Postal Liaison Office *dt*
MPLS Multi-Protocol Label Switching *rih*
MPM Manufacturing Program Manger *fu*
MPM Matched Port Module *jdr*
MPM Medical Planning Module [JOPS] *mt*
MPM Message Passing Library [IBM] *pk, kg*
MPM Message Processing Module *hc, fu*
MPM Meters Per Minute *w9*

MPM Microframe Program Manager *jdr, kf*
MPM Movement Planning Module *mt*
MPM Multi Purpose Missile *hc*
MPM Multiport Memory *fu*
MPMD Multiple Processor/Multiple Data *pk, kg*
MPMIS Military Personnel Management Information System *mt*
MPMIS Military Police Management Information System [USA] *mt*
MPMP Master Program Management Plan *mt*
MPN Manufacturer's Part Number *fu*
MPN Military Personnel, Navy *fu*
MPN Mobility Position Number *mt*
MPN Most Probable Number *fu*
MPN-XX Mobile Precision Number-XX [next generation mobile ground radar] *mt*
MPNB Master Project Notebook *fu*
MPO Master Purchase Order *fu*
MPO Material Project Office *hc*
MPO Maximum Power Output *fu*
MPO Military Post Office *wp*
MPO Mission Payload Operator *fu*
MPO Mission Planning Officer *fu*
MPO Model Purchase Order *hc*
MPOA Multi-Protocol over Asynchronous-Transfer-Mode *ct, em, kg*
MPP Maintainability Program Plan *hc, fu*
MPP Massively Parallel Processing (Processor) *em, kg, ns, pbl, hi*
MPP Matrix Plotting Program *fu*
MPP Maximum Performance Penalty *hc*
MPP Medical Planning Program [JOPS] *mt*
MPP Message Posting Protocol *pk, kg*
MPP Message Processing Program *pk, kg*
MPP Mission Payload Package *wp*
MPP Modern Programming Practices *hd1.89.y30*
MPP Most Probable Position *fu*
MPPCS Manpower and Personnel Contingency Support System *mt*
MPPL Mobilization Procurement Planning List *mt*
MPPT Message Preparation Processing Task *mt*
MPQ Minimum Procurement Quantity *fu*
MPR Manufacturing Producibility Review *fu*
MPR Master Personnel Records [USAF] *wp, mt*
MPR Material Purchase Requisition *hc*
MPR Material Purge Report *fu*
MPR Maximum Performance Reward *hc*
MPR Medium Power RADAR *fu, mt*
MPR Microwave Precipitation Sounder *hy*
MPR Milepost Review *nt*
MPR Monopulse Radar *fu*
MPR Monthly Progress Review *tb*
MPR Multi Protocol Router [Novell] *pk, kg*
MPR Multipart Repeater *pk, kg*
MPR Multiport Register *fu*
MPRF Medium PRF [mode] *hc*
MPRL Master Parts Reference List *wp*
MPRO Machine Processing [section of NSA's production organization] -*10*
MPRR Mobility Personnel Resource Roster *mt*
MPS Main Power Source *fu*
MPS Maintenance Processing Subsystem *fu*
MPS Management of Parts Shortages *mt*
MPS Manpower Planning System *mt*
MPS Manufacturing Production Schedule *oe*
MPS Maritime Prepositioning Ships [USA DOD] *mt*
MPS Master Phasing Schedule -*17*
MPS Master Production Schedule *fu*
MPS Master Program Schedule *fu, hc*
MPS Materials Processing in Space *ci.627, wl*
MPS Materiel Planning Study *wp*
MPS Memory Processor Switch *fu*
MPS Message Processing Subsystem *mt*
MPS Meters Per Second *w9, so*
MPS Mission Payload Subsystem *fu*
MPS Mission Planning System [TAF] *mt*
MPS Mission Processing Subsystem *hn49.3.4*
MPS Mobile Pointing Station *0*
MPS Mobile Power Stand *gp, jn*
MPS Moteur Propergol Solide [Ariane 5 solid motor] *ai2*
MPS Multi Processor Specification *ct, em*
MPS Multiple Protection Structures *st*
MPS Multiple Protective Shelter *st*
MPS Multiple Protective Structure *mt*
MPS Multiprocessor Specification *pk, kg*
MPS Multipurpose Sonar *fu*
MPS Muzzle Position Sensor *hc*
MPSA Military Postal Service Agency *dt*
MPSG Maintenance Philosophy Steering Group *fu*
MPSK Multilevel Phase Shift Keying *mt*
MPSK Multiple Phase Shift Keying *spix*

MPSOC Multi Purpose Satellite Operations Center [Offutt AFB, NE] -*16*
MPSP Micro Programmable Scene Processor *hc*
MPSP Modular Programmable Signal Processor *hc*
MPSR Manufacturing Program Status Report *fu*
MPSR Military Petroleum Stocks Report *mt*
MPSS Master Production Scheduling System *fu*
MPSS Message Processing Sub System *jb*
MPT Manpower, Personnel and Training *mt, wp*
MPT Mean Preventive maintenance Time *fu*
MPT Message Processing Time *mt*
MPT Mission Planning Terminal *2, 4*
MPTDS Multi-Purpose Tactical Display Set *hc*
MPTE Manual Peculiar Test Equipment *hc*
MPTIS Manpower, Personnel and Training Integration System *wp*
MPTN Multiple Protocol Transport (Transfer) Network *kg, hi*
MPTR Mobile Position Tracking Radar *fu*
MPTS Mobile Photographic Tracking Station *wp*
MPTS Multi-Purpose Test Set *hc*
MPTS Multiple Protocol Transport (Transfer) Services *kg, hi*
MPTT Management Practices Tools and Techniques *fu*
MPU Maintenance Panel Unit *fu*
MPU Maintenance Processor Unit *fu*
MPU Manpack Unit *fu*
MPU Message Processor Unit *mt, fu, jdr*
MPU Micro Processor Unit *fu, kg, hi*
MPU Microphone Preamplifier Unit *hc*
MPU Mobility Processing Unit *mt*
MPUIK Manpack Unit Installation Kit *fu*
MPWS Mobile Protected Weapons System *hc*
MQ Multiplier-Quotient *hc*
MQA Manufacturing Quality Assurance *fu*
MQE Manufacturing Quality Engineering *fu*
MQF Multiquery Facility *fu*
MQI Message Queue Interface *pk, kg*
MQS Maximum Query Sequence *uh*
MQT Mission Qualification Training *mt*
MR Magnetic Field Radiation *fu*
MR Magneto Resistive [thin film] *kg, ct, em, hi*
MR Maintenance Request *fu*
MR Malfunction Report *fu, oe*
MR Management Reserve *mt, fu, jdr*
MR Management Review *dm*
MR Maneuver Radius *st*
MR Manufacturing Request [document] *ah, fu, gp, 8, pl, jdr, pbl, tm, hc, jn*
MR Manufacturing Requisition *hc*
MR Manufacturing Responsibility *fu*
MR Map Reference *w9*
MR MAPL Report *fu*
MR Maritime Reconnaissance *mt*
MR Material Request *pl*
MR Material Requisition *fu*
MR Material Review *fu, pl, ep*
MR Medium Range *hc, fu, tb*
MR Message Re-registration *fu*
MR Message Repeat *fu*
MR Microminiature Relay *fu*
MR Microwave Radiometer *hy*
MR Mission Ready *mt*
MR Mission Reliability *kf*
MR Mission Requirements *0*
MR Mixture Ratio [oxidizer/fuel] *gp, oe*
MR Modem Ready *pk, kg*
MR Movement Region *mt*
MR-RPV Mid Range Remotely Piloted Vehicle *hc*
MRA Main Response Axis *fu*
MRA Material Review Action *fu*
MRA Material Review Action (Activity) *17, hc, ep, jdr, pbl, tm, jn*
MRA Maximum Response Axis *fu*
MRA Mission Readiness Availability *fu*
MRA Mission Ready and Available *mt, wp*
MRA Module Rack Assembly *hc*
MRAAM Medium Range Air to Air Missile *mt, wp*
MRAD Microwave Radiometer *re*
MRAPCON Mobile Radar Approach Control *fu*
MRAR Mishap Risk Assessment Report *kf*
MRASM Medium Range Air to Surface Missile [tactical, non-nuclear development of the AGM-109 Tomahawk cruise missile, BGM-109, USA] *hc, ph, fu, mt*
MRB MADS Resource Board *wp*
MRB Materials Review Board [organization] *ah, fu, gp, 0, 8, 17, sb, pl, hc, nt, lt, wp, ep, jdr, pbl, tm, kf, jn, oe*
MRB Modification Review Board *hc*
MRB Multiple Role Bomber [USA] *mt*
MRBF Main Rounds Before Failure [number indicating the reliability of a gun] *mt*

MRBF Mean Ranging Between Failure *hc*
MRBM Medium Range Ballistic Missile *mt, wp, ac3.4-3, 13, 14, tb, kf*
MRBS Medium Range Bit Synchronizer *-4*
MRC Maintenance Requirement Card *mt, hc, fu*
MRC Major Regional Conflict *kf*
MRC Manufacturing Record Change *fu*
MRC Manufacturing Record Code *fu*
MRC Manufacturing Reference Code *fu*
MRC Materiel Release Confirmation *wp*
MRC Media Resource Controller *hc*
MRC Media Resource Controller (USAF Program) *fu*
MRC Mission Research Corporation *kf*
MRC Monthly Recurring Charge *mt*
MRC Motorized Rifle Company *wp*
MRC Movement Report Center [USA] *mt*
MRCA Multiple Role Combat Aircraft [the Panavia Tornado] *fu, wp, hc, mt*
MRCC Movement Reports Control Center [US Navy] *mt*
MRCF Memory Read Correction Flip-Flop *fu*
MRCF Microsoft Realtime Compression Format *kg, hi*
MRCI Microsoft Realtime Compression Interface *pk, kg*
MRD Material Requirement Document *mt*
MRD Materials Required Data *mt*
MRD Medium Rate Demodulator *-4*
MRD Message Route Determination *fu*
MRD Minimum Reactive Display *fu*
MRD Mission Requirements Document *wl*
MRD Modification Record Drawing *fu*
MRD Motorized Rifle Division *-12*
MRDA Mission Requirements Definition and Analysis [SDI Study] *mt, go4.2, tb*
MRDB Materiel Returns Data Base [USA] *mt*
MRDB Mission Requirements Data Base *wl*
MRDFS Manpack Radio Direction Finding System [USA/USMC] *mt*
MRDS Maintenance Requirements Data Systems *mt*
MRE Mean Radical Error *hc*
MRES Material Requirements Estimation System *wp*
MRF Maintenance Replacement Factor *wp*
MRF Meteorological Radar Facility *hc*
MRF Meteorological Research Flight [UK] *mt*
MRF Module Release Form *gp*
MRF Multiple Role Fighter *mt*
MRG Management Requirement General *hc*
MRG Material Review Group *wp*
MRG Movement Requirements Generator [JOPS] *mt*
MRG [Electro-] Magnetic Rail Gun *tb*
MRHT Magneto Resistive Head Technology *hi*
MRI Magnetic Resonance Image (Imaging) *wp, is26.1.68, kg*
MRI Midwest Research Institute *ci.978*
MRI MILSTRIP Routing Identifier *wp*
MRI Miscellaneous Radar Input *fu*
MRI Monopulse Resolution Improvement *hc*
MRIL Master Repairable Items List *fu*
MRIR Medium Resolution Infrared Radiometer *hc, ac2.8-2*
MRIR Modification and Retrofit Installation Report *fu*
MRIT Mobile Remote Intelligence Terminal *mt, wp*
MRJE Multileaving Remote Job Entry *fu*
MRK Maliy Raketniy Korabl [small rocket ship] [USSR] *mt*
MRL Maintenance Requirement List *mt*
MRL Malibu Research Laboratory [Hughes] *ge, hg, hc*
MRL Master Requirements List *mt*
MRL Material Requirements List *hc, fu*
MRL MCO Rapid Lock-on *hc*
MRL Minimum Risk Level [USA DOD] *mt*
MRL Multiple Rocket Launcher *mt, wp*
MRLC Maintenance Repair Level Code *hc*
MRM Maintenance Reporting and Management *mt*
MRM Materiel Readiness and Modernization *mt*
MRM Most Recently Used Master *kg, em*
MRMS Mobile Remote Manipulator System *wl*
MRMU Mobile Remote Manipulating Unit *hc*
MRND Maintenance Required Not Developed *fu*
MRO Maintenance Repair and Operating [supplies] *hc, fu*
MRO Materiel Release Order *fu, hc, wp*
MRO Memory Readout *jdr, uh*
MRO Movement Report Office [USA] *mt*
MRO Multiple Region Operation *pk, kg*
MROC Multiple Command Required Operational Capability *mt, lka, go3.5.1, nt, jmj, as*
MRP Maintenance Repair Part *hc*
MRP Manpower Requirements and Personnel *mt*
MRP Manual Reporting Post *hc, fu*
MRP Manufacturing Requirements (Resources) Planning *fu*
MRP Material Requirements Planning *kg, wp, fu*
MRP Mid Range Plan *mt*
MRP Militarized Reconfigurable Processor (US Army) *fu*

MRP Mission Requirements Package *wp*
MRP Mobile Repair Party *mt*
MRP Modification Refurbishment Phase *fu*
MRP Multimode Radar Processor *hc*
MRP Multiple Rate Processor *mt*
MRPD Medium Repetition-rate Pulsed Doppler *wp*
MRPL Main Ring Path Length *pk, kg*
MRR Manual Rating Request *fu*
MRR Manufacturing Readiness Review *pl, ep, jdr, jn, oe*
MRR Material Receiving Report *fu*
MRR Material Release Point *mt*
MRR Material Requisition and Receiver *fu*
MRR Material Review Report *oe*
MRR Medium Range Radar *fu*
MRR Minimum Risk Route [USA DOD] *mt*
MRR Mission Readiness Review *17, fu, oe*
MRR Mission Requirements Review *re*
MRR Multi Role Radar *hc, wp*
MRRC Mechanical Reliability Research Center *hc*
MRRC Military Requirements Review Committee *fu*
MRRL Materiel Repair Requirements List *wp*
MRRM Medical Resupply Requirements Module [USAF] *mt*
MRRR Mobility Requirements / Resource Roster *mt*
MRS Maintenance Repair and Service *wp*
MRS Maintenance Requirement Substantiated *fu*
MRS Management Reporting System *hc*
MRS Master Radar Site *hc*
MRS Master Ranging Station *pl, pbl*
MRS Media Recognition System *pk, kg*
MRS Medical Readiness System *mt*
MRS Medical Resupply Set Model [USA] *mt*
MRS Medium Range SAR *hc*
MRS Military Radar Station *fu, hc*
MRS Minimum Raid Size *fu*
MRS Minimum Residual Shutdown *uh, oe, ai2*
MRS Mission Radio System [USSOUTHCOM] *mt*
MRS Multiple Representation System *tb*
MRS Multispectral Resource Scanner *ac5-i*
MRSA Materiel Readiness Support Activity [USA] *fu, mt*
MRSE Microwave Remote Sensing Experiment *ac3.7-17*
MRSL Marconi Radar Systems, Ltd. *fu*
MRSP Multimode Radar Signal Processor *hc*
MRSR Mars Rover Sample Return [mission] *mi, eo, re, mb*
MRSRM Mars Rover and Sample Return Mission *mi*
MRSSS Multilateration Radar Surveillance and Strike System *fu*
MRSV Maneuvering Recoverable Space Vehicle *hc*
MRT Maintenance Recovery Team *mt*
MRT Material Requirements Transfer *fu*
MRT Mean Repair Time *kg, kf*
MRT Mean Resolvable Temperature *wp*
MRT Miniature Receiver Terminal *mt, wp*
MRT Minimum Resolvable Temperature *hc, bf*
MRT Multiple Radar Tracking *fu*
MRTB Metallic Return Transfer Breaker *hc*
MRTB Missile Research Test Building *9, oe*
MRTCS Minaturized Real Time Computational System *wp*
MRTD Minimum Resolvable Temperature Difference *hc*
MRTF/A Manual Radar Terrain Following/Avoidance *hc*
MRTFB Major Range and Test Facility Base *wp, mt*
MRTS Master Radar Tracking Station *hc*
MRTT Mission Readiness Test Team *oe*
MRTT Modular Record Traffic Terminal *mt*
MRU Maximum Receive Unit *pk, kg*
MRU Message Retransmission Unit *fu*
MRU Microwave Relay Unit *fu*
MRU Military Radar Unit *mt*
MRU Minimum Replacement Unit *wp*
MRU Mobile Radio Unit *fu, wp*
MRUASTAS Medium Range Unmanned Aerial Surveillance and Target Acquisition System *fu*
MRV Maneuvering Reentry Vehicle *hc, wp*
MRV Message Receipt and Validation *fu*
MRV Multiple Reentry Vehicle *mt, fu, 13, tb, st, kf*
MRVLP Maneuvering Reentry Vehicle for Low Level Penetration *hc, wp*
MRW Maximum Ramp Weight *mt*
MRWG Mission Requirements Working Group *wl*
MS Magnetic Storage *fu*
MS Mail Station [also M/S] *fu, aa26.11.51*
MS Management Science *hi*
MS Management Systems *ms3.1.1*
MS Margin of Safety *gp, fu, 17, sb*
MS Mass Spectrometry *wp*
MS Mass Storage *wp*
MS Master Station *fu*
MS Materiel Supplement *fu*
MS Memory System *pk, kg*

MS Message Store *pk, kg*
MS Methyl Salicylate [chemical warfare Simulant] *wp*
MS Microprogram Store *fu*
MS Microsoft System(s) *ct*
MS Military Standard *hc*
MS Mine Sweeper [ship] [US Navy] *mt*
MS Mission Simulator *wp*
MS Mission Support *mt*
MS Mobilization Station *mt*
MS Modem Sharing *do64*
MS Monitor(ing) Station *sd, jdr*
MS Monitoring Subsystem *fu*
MS Motor Ship *w9*
MS Multiple Satellite *jdr*
MS {Mission Specialist} *ws4.10*
MS-DOS Microsoft Disk Operating System *fu, ct, kg, hi*
MS/EMI Mission Sequence and Electromagnetic Interference *hc*
MS/NAS Mil Spec / National Aircraft Standard *oe*
MSA Maintenance Support Activities *mt*
MSA Major Subcontract Administrator *fu*
MSA Major System Acquisition *mt, wp*
MSA Management System [USA] *mt*
MSA Medium Service Area *jdr*
MSA Microgravity Science and Applications *wl*
MSA Microwave System Analyzer *-4*
MSA Minimum Safe Altitude *mt, fu*
MSA Mission Support Analyst *-2*
MSA Mission Support Area *so, mb, re*
MSA Mobile Subscriber Access *mt*
MSA Mutual Security Administration *hc*
MSACM Microsoft Audio Compression Manager *kg, hi*
MSALT Mean Sea Level Altitude *wp*
MSAM Mobile Surface to Air Missile *uc*
MSAMS Medium-Range Surface-to-Air Missile System *fu*
MSAMS Mobile (Medium altitude) Surface to Air Missile System *hc, hn43.9.6, pm*
MSAN Microwave Steerable Null Antenna Array *hc*
MSAP Military Security Assistance Program *mt*
MSAPPS Microsoft Applications *hi*
MSAR Manpower Standards Application Routine *mt*
MSAR Microwave Spectrometer And Radiometer *es55.26*
MSARC Marine Systems Acquisition Review Council [USMC] *mt*
MSAS Manpower Standards Application System [AFMEA] *mt*
MSAT Maritime Satellite *jdr*
MSAT Mobile Satellite *kf, hi*
MSATC Mobile Satellite Communicators *hi*
MSATNET Mobile Satellite Network *hi*
MSAU Multi-Station Access Unit *pk, kg*
MSAV Microsoft Anti Virus *pk, kg*
MSAW {Minimum Safe Altitude Warning *mt, fu*
MSB Main Support Base *wp*
MSB Memory Storage Buffer *wp*
MSB Mine Sweeping Boat *mt*
MSB Mobile Support Base *fu*
MSB Most Significant Bit *he, fu, uh, kg, gp, wp, hc, gg64, 4, 9, jdr, kf, hi, oe*
MSB Most Significant Byte *hi*
MSBF Mean Swaps Between Failures *pk, kg*
MSBS Mer-Sol Balistique Strategique [French SLBM] *12, crl*
MSC Major Subordinate Command [NATO] *fu, mt*
MSC Make Span Code *fu*
MSC Manned Spacecraft *hc*
MSC Manned Spacecraft Center [NASA] *fu, hc, ci.103*
MSC Manning Schedule Changes *uh*
MSC Manual Spacecraft Center *0*
MSC Material Support Concept *mt*
MSC Message Switching Center *mt*
MSC Message Switching Concentration *do306*
MSC Microwave Carrier Supply *fu*
MSC Microwave Semiconductors Corporation *gg142, fu*
MSC Milestone Schedule Chart *wp*
MSC Military Sealift Command [USA DOD] *fu, dt, mt*
MSC Mobile Satellite Communication *mt*
MSC Modified Stirling Cycle *hc*
MSC Modular Spacecraft Computer *aa26.12.51, hc*
MSC-IMIS Military Sealift Command Integrated Management Information System [MSC] *mt*
MSCD Military Support to Civil Defense *mt*
MSCIT Medium Speed Communications Interface Terminal *mt*
MSCLNO Major Subordinate Command Liaison Officer *fu*
MSCN Manned Satellite Communication Network *0*
MSCP Mean Spherical Candlepower *fu*
MSCPAC Military Sealift Command, Pacific *fu*
MSCPLANT Military Sealift Command, Atlantic *fu*
MSCR Multipoint Strip Chart Recorder *hc*
MSCS Manned Spaceflight Control Squadron [AF Space Command] *mt*
MSCS Manual SHORAD Control System *fu*

MSCS Mobile Service Communications Satellite *hc*
MSCS Multi Service Communications System *mt*
MSCS Multiple Satellite Communications System *ic2-26*
MSD Management Systems Designers *mt*
MSD Mass Storage Device *mt, kg, wp, fu*
MSD Material Support Date *hc*
MSD Method of Steepest Descent *hc*
MSD Microelectronics Systems Division *fu, hn43.21.6*
MSD Microsoft Diagnostics *ct, hi*
MSD Microsoft System Diagnostics [Microsoft] *kg, hi*
MSD Minimum Safe Distance *wp*
MSD Mission Sensor Data *pl*
MSD Most Significant Digit *kg, fu*
MSD Multi Sensor Display *hc*
MSD Multiple Sensor Discrimination *tb*
MSDB Mission Schedule Data Base *mt*
MSDE Maintainability Simulation and Demonstration Equipment *fu*
MSDF Multi-Sensor Data Fusion *tb*
MSDG Multiple Sensor Display Group *hc*
MSDP Maintenance Systems Development Program *mt*
MSDR Multi Spectral Scanner Data Redundancy *hc*
MSDR Multiplexed Streaming Data Request *pk, kg*
MSDS Manpower Standards Development System [AMEA] *mt*
MSDS Material Safety Data Sheet *kf*
MSDS Microsoft Developer Support [Microsoft] *pk, kg*
MSDS Multispectral Scanner and Data System *wp*
MSE (MSE, which operates CDIF under contract) *aa26.12.25*
MSE Maintenance Support Equipment *fu*
MSE Manned Spaceflight Engineer *sf31.1.26*
MSE Manufacturing Support Equipment *wp*
MSE Materials Science and Engineering *wp*
MSE Mean Square Error *sp39, fu, wp*
MSE Mechanical Systems Engineer *jdr*
MSE Military Specification Exception *fu*
MSE Mission Support Element *jdr*
MSE Mobile Subscriber Equipment [TRI-TAC] *fu, hc, pm, kf, mt*
MSE Modernization Support Environment *mt*
MSEC Message Security *fu, wp*
MSEL Master Scenario Events List *mt*
MSEO Mission Systems Engineering Organization *jn*
MSER Multiple Stores Ejector Rack *hc*
MSF Mission Support Facility *mt*
MSF Multiple Sensor Fusion *tb*
MSF Munitions Storage Facility *mt*
MSFC Marshall Space Flight Center [Huntsville, AL; NASA] *fu, mi, hc, ci.98, 0, 4, wp, ns, no, tb, wl, mb, oe, ai2*
MSFN Manned Space Flight Network [stations] *hc, ci.ix*
MSFR Minimum Security Function Requirements [IBM] *pk, kg*
MSFS Maintenance Shop Floor System [USA] *mt*
MSFTP {Minitrack System Flight Tracking..(?)..} *ac2.9-10*
MSG Mapper Sweep Generator *fu*
MSG Master Sergeant *w9*
MSG METEOSAT Second Generation [ESA] *hg, ge*
MSG Microwave Signal Generator *wp*
MSG Miscellaneous Simulation Generation *fu*
MSG-ROUTE Message Routing Emulation *-17*
MSGCEN Message Center *mt, wp*
MSGID Message Identifier Set *mt*
MSGT Master Sergeant *mt, wp*
MSG_STDS Message Standards *fu*
MSH MAC Standard Header *mt*
MSH Metastable Helium *mt*
MSH Mine Sweeper and Hunter [Ship] *uc*
MSH Mine Sweeper, coastal [US Navy] *mt*
MSH Most Significant Half *jdr, uh*
MSI Maintenance Significant Item *fu*
MSI Management System Indicator *mt, fu*
MSI Manufacturing Standing Instructions *pl*
MSI Master Scheduled Item *fu*
MSI Medium Scale Integration *mt, fu, hc, wp, 6, tb, kg, hi, ai1*
MSI Message Start Interval *fu*
MSI Milestone I *nt*
MSI Multi Sensor Integration *fu, wp, tb*
MSI Multi Spectral Imaging (Imager) *gp, 16*
MSI Multiple Source Integration *mt*
MSI Multispectral Imagery *mt*
MSI Multispectral Scan Imaging *gd2-8*
MSI Multispectral Spectral Imaging *oe*
MSIA Multispectral Image Analyzer *wp*
MSIC Medium-Scale Integrated Circuit *fu*
MSIC Missile and Space Intelligence Center [USA] *jb, tb, kf, mt*
MSID Master Station Identification *fu*
MSID Multisource Identification *fu*
MSIP Multinational Staged Improvement Program [NATO] *mt*
MSIP Multiple Stage Improvement Program [an upgrade program for the F-15, USA] *mt, fu, hc, hn43.6.8, hq5.3.5, af72.5.135, wp*

MSIS Marine Safety Information System *hc*
MSIS Mass Spectrometer Incoherent Scatter [atmosphere model] *ns*
MSIT Missile Subsystem Integration Technology *hc*
MSI_TTL Medium-Scale Integration Transistor-Transistor Logic *fu*
MSK Minimal (Minimum) Shift Keying *mt, fu, sp661, ce*
MSK Minimum Spares Kit *mt, wp*
MSK Mission Support Kit *mt*
MSL Maintenance Supply Liaison *mt*
MSL Map Specification Library *pk, kg*
MSL Master Support List *fu*
MSL Mean Sea Level *mt, fu, wp, ac10.1-1, hc, w9, aw129.10.38, st, ai1*
MSL Microwave Landing System {?} *mt*
MSL Microwave Systems Laboratory *-19*
MSL Mirrored Server Link *pk, kg*
MSLP Mean Sea Level Pressure *fu*
MSLP Medium Speed Line Printer *mt*
MSLS Modular Site Location System *hc*
MSLS Multi Spectral Line Scan [Sensor] *hc*
MSM Medical Sourcing Model [DLA] *mt*
MSM MEECN System Model *mt*
MSM Microwave Switch Matrix *ep*
MSM Mission Support Manager *2, 4*
MSM Mission Support Model *mt*
MSM Mission Systems Manager *oe*
MSMLC Medium Speed Multiline Controller *mt*
MSMP Multispectral Measurement Program *-17*
MSMS Microsoft System Management Server *hi*
MSMS Microwave Self Mixing Sensor *hc*
MSN Manufacturing Sequence Number *fu*
MSN Message Switching Network *fu*
MSNET Microsoft Network *hi*
MSO Management Systems Operations *-8*
MSO Mandatory Scramble Order *mt*
MSO Maritime Staff Office *hc*
MSO Message Start Opportunity *fu*
MSO Military Satellite Communications Systems Office [DCA] *mt*
MSO Mine Sweeper, Ocean [US Navy] *hc, mt*
MSO Multiple Systems Operators *ct, em, kg*
MSOCC Multi Satellite Operations Control Center [GSFC] *gp, 2, 4, 16*
MSOCC Multiple Satellite Operations Control Center [NASA] *mt*
MSOGS Molecular Sieve Oxygen Generation System *mt, wp*
MSOP Mission Support Operations Plan *kf*
MSOS Mass Storage Operating System *fu, wp*
MSOW Modular Stand Off Weapon [NATO] [USA] *mt, wp, hc, aw129.21.12*
MSP Mach Sweep Programmer [automatic wing sweep control system of the F-14 Tomcat] [USA] *mt*
MSP Maintenance Support Plan *mt, wp*
MSP Manpower Support Plan *mt*
MSP Manufacturing Support Program *fu*
MSP Manufacturing Systems Products [an IBM operation, Boca Raton, FL] *is26.1.54*
MSP Mass Storage Processor *mt*
MSP Medium-Speed Printer *fu*
MSP Milstar Standard Processor *jdr*
MSP Missile Status Panel *hc*
MSP Mission Support Plan *mt, fu, wp*
MSP Mission Support Processing *kf*
MSP Modular Switching Peripheral *is26.1.76g*
MSP Movement Support Plans *17, sb*
MSP Multi-Sensor Processing *hc*
MSP Multicamera Synthesizing Projector *ac2.8-2*
MSPB Merit System Protection Board *wp*
MSPF Maritime Special Purpose Force [USA DOD] *mt*
MSPL Military Space Parts List *jr*
MSPS Mega Samples Per Second *kf*
MSPS Mobilization Stationing and Planning System *mt*
MSPS Multi Spectral Point Scanner *hc*
MSPSP Missile Systems Prelaunch Safety Package *jdr, kf, vv, oe*
MSPU Manned Simulated Participating Unit *fu*
MSPU Master Signal Processor Unit *fu*
MSR Material Status Record *fu*
MSR Mechanically Steered Radar *ps204*
MSR Microwave Scanning Radiometer *ac3.3-13*
MSR Minimum Sustaining Rate *fu*
MSR Missile Simulation Round *hc*
MSR Missile Site Radar *fu, hc, tb*
MSR Mission Success Review *jdr*
MSR Monthly Status Review *gp, 16, oe*
MSRA Module Screening and Repair Activity *fu*
MSRDPS Missile Site Radar Data Processing System *mt*
MSRL Mobile Secondary Reference Laboratory *hc*
MSRP Multi Stage Retrofit Program *wp*
MSRS Military System Requirement Specification *st*
MSRT Mobile Subscriber Radio Telephone Terminal *mt*
MSS Maintenance Support Schedule *mt*

MSS Management Support Services *hc*
MSS Management Support System *mt*
MSS Mass Storage System *fu, kg, hi*
MSS MEECN System Simulator *mt*
MSS Message Source Synchronization *fu*
MSS Message Support Subsystem *mt*
MSS Message Switching (Switched) System *fu, 2, oe*
MSS Midcourse Surveillance System *hc*
MSS Military Supply Standard *fu*
MSS Missile Sensor Study *hc*
MSS Mission Support System *mt, uh, wp, jdr*
MSS Mobile Satellite Service (Services) *wp, tt*
MSS Mobile Satellite System *-19*
MSS Mobile Subscriber Subsystem *hc*
MSS Mobility Support Squadron *mt*
MSS Mobilization Stations *mt*
MSS Modularized Simulation System *hc*
MSS Moored Sonobuoy Program *hc*
MSS Moored Surveillance System *fu*
MSS Multiple Satellite System [DARPA] *1, tb, mt*
MSS Multiprotocol Switched Services [IBM] *pk, kg*
MSS Multispectral Scanner System *hc, hn43.21.4, 16, ep, hy, hg, ge*
MSS&P Maintenance Support Software and Procedures *fu*
MSSBR Multi purpose Supersonic Beamrider [Missile] *hc*
MSSC MAJCOM Spare Support List *mt*
MSSC Multi-Spectral Scanning Camera *hc*
MSSCC Multicolor Spin Scan Cloud Camera *gp*
MSSG MAU Service Support Group [USMC] *mt*
MSSM Mars Spinning Support Module *hc*
MSSP Maintenance Standards, Schedule and Procedures (Manual) *fu*
MSSS Manned Space Station Simulator *fu*
MSSS Marine Corps Standard Supply System [also M3S] [USMC] *mt*
MSSS Midcourse Surveillance Satellite Study *hc*
MSSTM Military Space Systems Technology Model *mt*
MSSW Mission Support Software *mt*
MST Main Satellite Thruster *oe*
MST Maintenance Support Team *fu*
MST Management Survey Team *wp*
MST Master Station Trainer *fu*
MST Mean Solar Time *fu*
MST Mini Star Tracker *kf*
MST Missile Simulation Test *hc, jdr*
MST Mission Simulation Test *sb*
MST Mission Support Team *mt*
MST Mixed Simulation Test (Testing) *gp, lt, ep, kf, jn*
MST Mobile Service Tower *bo, hc, oe, ai2*
MST Monolithic Systems Technology *wp*
MST Mountain Standard Time *kg. wp, aw130.13.25, w9, kf*
MST Multi Sensor Track *tb*
MST&E Multiservice Test and Evaluation *mt*
MSTART Missile System To Attack Relocatable Targets *wp*
MSTC Multi-Spectral Target Cueing *hc*
MSTC Multisensor Tracking and Correlation *fu*
MSTC Multisource Track Control *fu*
MSTI Miniature Space Technology Investigation *kf*
MSTM Military Standard Transportation and Movement *mt*
MSTM Multisource Track Management *fu*
MSTP Multimission Software Transmission Project *ct, em*
MSTRS Miniaturized Satellite Threat Reporting System *kf*
MSTS Military Sea Transportation Service *hc, 10, w9*
MSTS Minuteman System Test Station *hc*
MSTS Missile Subsystem Test Set *hc*
MSTS Munitions/Submunitions Tracking System *hc*
MSTT Multi-Spectral Tracking Telescope *hc*
MSTV Multi-Spectral Television *hc*
MSU Magnetic Storage Unit *fu*
MSU Maintenance Station Unit *hc*
MSU Mass Storage Unit *mt, wp*
MSU Message Switching Unit *mt, fu*
MSU Microwave Sounding Unit [NOAA] *pl, aw129.14.43, hy, hg, ge*
MSU Modular Switching Unit *fu*
MSUS Monitoring Subsystem User's Simulator *fu*
MSV Magnetically Supported Vehicle *wp*
MSVD Message Status Visual Display *mt*
MSVDP Message Status Visual Display Panel *mt*
MSW Machine Status Word *pk, kg*
MSW Magnetostatic Wave *fu*
MSW Mission Software *fu*
MSWDS Multi-Sensor Weapon Delivery System *hc*
MSWG Message Standards Working Group [NATO] *mt*
MSX Midcourse Space Experiment *kf*
MT Magnetic Tape *fu*
MT Magnetic Torquer [Torque] *jn*
MT Maintenance Trailer *hc*
MT Management Target *fu*
MT Management Threshold *mt*

MT Maneuverable Target *fu*
MT Manufacturing Technology *fu, mt*
MT Manufacturing Type *fu*
MT Mean Time *fu*
MT Measurement Ton [40 cubic feet] *mt*
MT Medium [Satellite Ground] Terminal *mt*
MT Mission Trainer *wp*
MT Morskoy Tralshchik [seagoing minesweeper] [USSR] *mt*
MTA (MTA, Inc. . . . Huntsville, AL 35805) *jmj*
MTA MAC* Transportation Authorization *mt*
MTA Mail Transfer Agent *ct*
MTA Maintenance Task Analysis *fu*
MTA Major Trading Area *ct*
MTA Mass Transit Agencies *y2k*
MTA Measurement Tolerance Allowed *-4*
MTA Message Transfer Agent *ct, kg*
MTA Military Training Area *mt*
MTA Multiple Terminal Access *pk, kg*
MTAC Main Target Allocation Center *fu*
MTACC Marine Tactical Command and Control System [USA] *mt*
MTACCS Marine Corps Tactical CC System *mt*
MTACCS Marine Tactical Command and Control System *fu, hc*
MTAD Manufacturing Training and Development System *fu*
MTADS Man Transportable Air Defense System *hc*
MTAE Multiple Time Around Echoes *hc*
MTAE Multiple Time Around Elimination *hc*
MTAP Multifunction Target Acquisition Processor *hc*
MTAR Mission/Task Analysis Report *kf*
MTAS Microwave Transistor Amplifiers *-4*
MTAS Millimetric Target Acquisition System *mt*
MTAS Modular Target Acquisition System *mt, wp*
MTAS Multimission Target Acquisition System *fu*
MTAS Multiple mission TAS Radar System *hc*
MTAS Multiple Sensor Target Acquisition System [USA] *wp, hc, mt*
MTASS Machine Transferable AN 10 YK 20 Support Software *mt*
MTASS Machine-Transportable AN/UYK Support Software *fu*
MTB Maintenance Test Bench *fu*
MTBA Master Test Bed Aircraft *hc*
MTBB Mean Time Between Breakdowns *pk, kg*
MTBCF Mean Time Between Critical Failures *mt, fu, hc, 17, tb, as, kf*
MTBCMA Mean Time Between Corrective Maintenance Action *mt, as*
MTBD Mean Time Between Demand *kf*
MTBDE Mean Time Between {Downing} Events *fu, kf*
MTBEF Mean Time Between Equipment Failures *hc*
MTBF Mean Time Between (Before) Failures *wp, hc, rm3.1.1, 0, 4, 16, 17, aw129.21.120, sb, af72.5.92, mj31.5.2, tb, st, kf, kg, fu, ct, em, hi, oe*
MTBF Mean Time Between Failures *mt*
MTBFI Mean Time Between Failure, Instantaneous *fu*
MTBFMI Mean Time Between False Maintenance Indicators *tb*
MTBI Mean Time Between Incidents *fu*
MTBJ Mean Time Between Jams *pk, kg*
MTBM Mean Time Between Maintenance *mt, fu, wp, kf*
MTBMA Mean Time Between Maintenance Actions *fu, wp, hc*
MTBMR Mean Time Between Maintenance Requests *fu*
MTBO Mean Time Between Outages *mt*
MTBOMF Mean Time Between Operational Mission Failure *fu*
MTBPE Mean Time Between Program Errors *fu*
MTBPS Mean Time Between Program Stops *fu*
MTBR Mean Time Between Removals *mt, hc, fu*
MTBR Mean Time Between Repairs *fu*
MTBSF Mean Time Between System Failures *hc*
MTBSP Mobilization Troop Basic Stationing Plan *mt*
MTBUM Mean Time Between Unscheduled Maintenance *wp*
MTBUMA Mean Time Between Unscheduled Maintenance Actions *hc*
MTBUR Mean Time Between Unscheduled Removal *wp*
MTC Magnetic Tape Controller *mt, fu*
MTC Maintenance Training Center *fu*
MTC Man Tended Capability *mi*
MTC Maneuver Training Command *mt*
MTC Marine Traffic Control *fu*
MTC Master Tape Control *fu*
MTC Memory Test Computer *fu*
MTC Message Traffic Control *fu*
MTC Missile Threat Conference *mt*
MTC Mission and Traffic Control *fu*
MTC Monitor Test Console *mt*
MTCC Marine Traffic Control Center *fu*
MTCC Modular Tactical Communications Center [TRI-TAC] *hc, mt*
MTCC Modular Tactical Control Center *mt*
MTCF Mean Time to Catastrophic Failure *fu, wp*
MTCIP Manual Technical Control Improvement Program *mt*
MTCN Management Task Control Number *mt*
MTCR Missile Technology Control Regime *ai2*
MTCS Marine Traffic Control System *fu*
MTCS Mean Time to Cycle Slip *-4*

MTCU Mobile Temperature Conditioning Units *hc*
MTD Magnetic Tape Drive *hi*
MTD Maintenance Task Distribution *fu*
MTD Maneuvering Technology Demonstrator [program] *aa26.12.b22, aw130.22.44, wp*
MTD Manufacturing and Test Documentation *gp*
MTD Manufacturing Technology Division *wp*
MTD Maximum Tolerated Dose *wp*
MTD Microelectronics Technology Department *fu*
MTD Minimal Toxic Dose *wp*
MTD Moscow Time Daylight *pk, kg*
MTD Moving Target Detection (Detector) *hc, mt*
MTD Multiple Target Deception *fu*
MTD Multiple Target Discrimination *wp*
MTDD Master Time Division Demultiplexer *mt*
MTDL Maintenance Task Description List *hc*
MTDM Master Time Division Multiplexer *mt*
MTDS Marine Tactical Data System [USMC] *hc, mt*
MTDT Mean Time to Detect Fault *fu*
MTE Maintenance Test Equipment *fu*
MTE Manual Test Equipment *dm*
MTE Maximum Tracking Error *fu*
MTE Measuring and Test Equipment *fu*
MTE Mesosphere-Thermosphere Explorer *hy*
MTE Multisystem Test Equipment *fu, hc*
MTEPP Master Test and Evaluation Program Plan *fu*
MTF Mean Time to Failure *hc*
MTF Medical Treatment Facility *mt*
MTF Message Test Format(ting) [JCS Publication 25] *fu, mt*
MTF Microsoft Tape Format *kg, hi*
MTF Missile Track File *tb*
MTF Mississippi Test Facility *ci.100*
MTF Modulation (Modulated) Transfer Function *kg, hc, bl3-2.23, wp, dm, st, pbl, hi, oe, fu, ai1*
MTFA Modulation Transfer Function Area *wp*
MTFF Man Tended Free Flier *aw126.20.66, es55.46*
MTFS Medical Treatment Facility System *mt*
MTFT Mean Time of Functional Test *fu, hc*
MTG Multiple Trigger Generator *fu*
MTH Magnetic Tape Handler *fu*
MTH Magnetic Tape Holder *mt*
MTI Manufacturing Technology Initiative *aw126.20.22*
MTI Moving Target Indication (Indicator) *mt, ro, fu, hc, aw118.14.5, ic2-42, 17, sb, wp, pbl*
MTI Multiple Terminal Interface *fu*
MTI Multispectral Thermal Imaging [System] *kf*
MTIP Moving Target Indication Processor *hc*
MTIS Maintenance Task Information System *wp*
MTIT Mean Time of Fault Isolation Test *fu*
MTL Manufacturing and Technology Laboratory *wl*
MTL Master Tape Loading *fu*
MTL Merged Transistor Logic *fu*
MTL Minimum Triggering Level *fu*
MTL Mobile Transport Launcher *fu*
MTLCCS Mediterranean Tactical Logistic Command and Control System [US Navy] *mt*
MTLSTE MILSTAR Terminal-Level System Test Equipment *fu*
MTM Methods Time Measurement *mt*
MTM Military Training Mission *hc*
MTM Million Ton Miles *wp*
MTM Mission Table Management *mt*
MTM Multiple Terminal Manager *mt*
MTM-UAS Motion Time Measurement System-Universal Analyzing System *fu*
MTM/D Million Ton Miles Per Day [MAC*] *wp, mt*
MTMC Military Traffic Management Command [formerly MTMTS] [USA DOD] *mt, fu, hc, dt, wp, tb, st, jdr, su*
MTMC Multithreshold Multicriteria *fu*
MTMC-TTCE MTMC Transportation Terminal Command Europe [USA] *mt*
MTMCEA Military Traffic Management Command, Eastern Area *mt*
MTMCTEA Military Traffic Management Command Transportation Engineering Agency *mt*
MTMCWA Military Traffic Management Command, Western Area *mt*
MTMG Microcomputer Test Message Generator *fu*
MTMR Military Traffic Management Regulation *mt*
MTMT Master Target Magnetic Tape *kf*
MTMTS (MTMC
MTMTS Military Traffic Management and Terminal Service (currently MTMC) *su*
MTNS Manportable Thermal Night Sight *hc*
MTO Mediterranean Theater of Operations *w9*
MTO Message To Observer *fu*
MTO Mid Term Objective *wp*
MTOE Modified Table of Organization and Equipment [USA] *mt*
MTOGW Maximum Take Off Gross Weight *mt, wp*
MTON Measurement Ton [40 cu ft] *mt*

MTOW Maximum Take Off Weight *wp, mt*
MTP Magnetic Tape Processors *mt*
MTP Maintenance Test Package *wp*
MTP Manufacturing Technology Program *hc*
MTP Master Test Plan *fu, wp*
MTP Materiel Transfer Plan *mt, st*
MTP Message Translator Processor *fu*
MTP Mission Test Plan *wp*
MTP Multi Megawatt Terrestrial Power Plant *mt*
MTP Multiline Telephone Panel *fu*
MTPE Mission To Planet Earth *hy*
MTPS Maintenance Test Packages *hc*
MTPSI Master Test Program Set Index *hc*
MTPT Mean Time of Performance Test *fu*
MTR Magnetic Tape Recorder *fu*
MTR Maintenance Trouble Report *fu*
MTR Major Threat Region *kf*
MTR Mandatory Turn-in Repairable *fu*
MTR Materials Technology Report *oe*
MTR Maximum Report Time *mt*
MTR Mean Time to Restore *fu*
MTR Missile Tracking Radar *fu, mt, hc, wp*
MTRAT Maverick Target Recognition and Acquisition Trainer *hc*
MTRE Missile Test Readiness Equipment *hc*
MTRS Multimode Tactical Radar Simulator *hc*
MTS (leader in dynamic testing, MTS . . . Minneapolis, Minnesota) *aa26.11.10*
MTS Magnetic Tape Synchronizer *fu*
MTS Maintenance Training Set *hc*
MTS Marine Tactical System *fu*
MTS Masked Target Sensor *pm*
MTS Master Timing Source *fu*
MTS Material Test Specification *fu*
MTS Member of the Technical Staff *fu, ma16, 8, 15, sb, hc, sc9.6.2*
MTS Message Transfer Service/System *pk, kg*
MTS Microsoft Transaction Server *em*
MTS Microwave Temperature Sounder *hy*
MTS Military Training Standard *wp*
MTS Mobile Test Stand *-9*
MTS Mobile Tracking Station *0, wp*
MTS Mobile Training Set *jdr, mt*
MTS Module Test Set *fu*
MTS Module Test Station *hc*
MTS Moscow Time Standard *pk, kg*
MTS Multichannel Television Sound *pk, kg*
MTS Multiline Telephone Set *fu*
MTSAT Multifunction Transport Satellite *pbl*
MTSO Mobile Telephone Switching Office *hi*
MTSP Microelectronic Technology Support Program *hn50.9.1*
MTSS Military Test Space Station *hc*
MTST Magnetic Tape Selectric Typewriter [IBM] *hc, fu*
MTST Maneuvering Target Statistical Tracker *fu*
MTSU Multiple Transport Synchronization Unit *fu*
MTT Magnetic Tape Transport *fu*
MTT Medium Thrust Thruster *jn*
MTT Mobile Training Team [USA DOD] *wp, mt*
MTT Moving Target Tracking (Tracker) *hc, wp*
MTT Multi Transaction Timer *pk, kg*
MTT Multiple Target Tracking *mt*
MTT-S Microwave Theory and Techniques Society *mj31.5.45*
MTTD Mean Time To Detect *wp*
MTTD Mean Time To Diagnose *pk, kg*
MTTE Magnetic Tape Terminal Equipment *fu*
MTTF Mean Time To Failure *kg, fu, hi*
MTTFF Mean Time to First Failure *fu*
MTTI Magnetic Tape Transport Interface *fu*
MTTM Mean Time to Maintain *fu*
MTTR Mean Time To Repair (Restore) (Recover) *mt, ct, em, kg, fu, wp, tb, as, rm1.2.3, 4, 17, sb, hc, oe*
MTTRCF Mean Time to Restore Critical Function *kf*
MTTRF Mean Time To Restore Function *kf*
MTTRS Mean Time To Restore System *tb, as, kf*
MTTS Magnetic Tape Terminal Station *mt*
MTTS Mean Time to Trouble Shoot *hc*
MTTS Moving Target Tracking Study *hc*
MTTT Mean Time to Test *fu*
MTTT Mobile Threat Tracking Testbed *ro*
MTTY Maintenance Module Teletype *mt*
MTU Magnetic Tape Unit *mt, fu*
MTU Magnetometer Test Unit *gp, 16*
MTU Maintenance Training Unit *hc*
MTU Master Timing Unit *fu, jdr*
MTU Maximum Transmission Unit *pk, kg*
MTU Microwave Transmission Unit *ac10.4-4*
MTU Missile Training Unit *hc*
MTU Mobile Test Unit *hc, st*

MTU Mobile Training Unit *fu*
MTU Motoren und Turbinen Union [Germany] *aa26.12.59*
MTU Multiplex Terminal Unit *fu, hc*
MTU Multiplexed Telemetry Unit *jn*
MTV Miniature Test Vehicle *st*
MTVAL Master Tape Validation *fu*
MTW Mobile Training Wing *wp*
MTWR Micro-Thrust Water Rocket *hc*
MTX Motor Transport *mt*
MU Maintenance Unit [UK] *mt*
MU Master Unit *fu*
MU Master Unit [PLRS] *mt*
MU Memory Unit *fu, hc, hi*
MU Midcourse Update *tb*
MU Mobile Unit *fu*
MUA Mail User Agent *ct, em, kg*
MUA Materials Usage Agreement *oe*
MUCH-FET Multi-Channel Field Effect Transistor *hc*
MUCO Materiel Utilization Control Office *wp*
MUD Multi User Dialogue [Internet] *hi, kg*
MUD Multi-User Dimension/Domain [Internet] *kg, hi*
MUDR Multidetail Drawing *fu*
MUDS Man-hour Use Data System *mt*
MUDS Multiple Usage Data Sheet *hc, fu*
MUF Maximum Usable Frequency *mt, fu*
MUF Modeling Uncertainty Factor *jdr, ric*
MUFFLIR Multiple Function Forward Looking Infrared *hc*
MUFFS Multiple Frequency Firing System *hc*
MUG Microcomputer Users' Group *wp*
MUID Master Unit Identifier *fu*
MUL Master Urgency List *mt*
MUL Material Utilization List *jdr*
MULAH Multilayer Heating Model *-17*
MULE Modular Universal Laser Equipment *mt, fu, hc, hn41.20.6, hq5.3.5, wp*
MULE Multiple Use Laser Equipment *hc*
MULTACK Multiple Target Attack *wp*
MULTACK Multiple Target Attack System Integration and Simulation *hc*
MULTACKS Multi Attack System *hc*
MULTAM Multiple Target Assessment Model [AFSC*] *wp, mt*
MULTEWS Multiple Target Electronic Warfare System *hc, fu*
MULTI-SDF Multiple Signal Direction Finder *hc*
MULTICS Multiplexed Information and Computing Service [USAF] *ct, em, ro, mt*
MULTIPLEX [combine multiple circuits into one data stream] *mt*
MULTOTS Multiple Unit Link Test and Operational Training System *mt*
MUM Maximum Useful Magnification *wp*
MUMMS Marine Corps Unified Material Management System [USMC] *hc, mt*
MUNEX Munitions Allocation and Expenditure Tracking System *mt, wp*
MUNIREP Air Munitions Report *mt*
MUNSCS 7361 MUNSS Small Computer System *mt*
MUNSS Munition Storage Site *mt*
MUNSS Munition Support Squadron *12, mt*
MUNSTOR Munitions Storage Facilities [USAFE] *mt*
MUOS Mobile User Objective System *rih*
MUP Multi-User/Multi-Project *mt*
MUPO Maximum Undistorted Power Output *wp*
MUR Manpower Utilization Report *wp*
MURD Multi-Use Ranger/Designator *hc*
MUS Major User Stations *hc*
MUS Manned Underwater Station *fu*
MUS Mission Unique Software *jdr, uh*
MUS Multiuser System *fu*
MUSARC Major US Army Reserve Command *mt*
MUSB Mobile Unit Support Base *fu*
MUSE Monitor of Ultraviolet Solar Energy *hc, ac2.8-2*
MUSE Multi User Shared Environment *kg, hi*
MUSE Multiple Sub-Nyquist Encoding *is26.1.59*
MUSIC Multi-Spectral Infrared Camera [a DARPA program] *kb*
MUSIC Multi-System Integrated Control [a study into thrust-vectoring] [USA] *mt*
MUST MAC* UHF Satellite Terminal *mt*
MUST Mobile Undersea Systems Test [Laboratory] *aw129.14.0*
MUST [air transportable field hospital project] *hc*
MUSTRAC Multiple Target Steerable Telemetry Tracker *hc*
MUT Mean Up Time *fu*
MUT Monitor Under Test *pk, kg*
MUTE Multiplied Unit for Transmission Elimination [US Navy] *mt*
MUTES Multiple Threat Emitter System *mt*
MUTS Manual Unit Test Set *hc*
MUTT Multiuse Tactical Terminal *fu*
MUTT/JEFF Multi-Use Tactical Terminal/Judiciously Efficient Fixed Frame *hc*
MUX Multiplexer (Multiplexing) *hc, wp, sp661, 8, tb, jdr, kf, ct, em, uh, fu, kg, hi, oe*

MUX/DEMUX Multiplexer/Demultiplexer *mt*
MV Merchant Vessel [US Navy] *mt*
MV Miniature Vehicle *gr, wp, jb, tb*
MV Multi Vibrator *hc*
MV Muzzle Velocity *fu*
MV {Data General's MV model 10000 computer, which runs SEL-DADS} *pr*
MVA Manufactured Valued Asses {?} *tb*
MVA Minimum Vectoring Altitude *fu*
MVAT Multiple Vandal Assembly Terminal *hc*
MVB Microsoft Visual Basic *hi*
MVB Multimedia Viewer Book *pk, kg*
MVC Manual Volume Control *fu*
MVC Multimedia Viewer Compiler *pk, kg*
MVCP Management Visibility and Control of Provisioning *mt*
MVD Ministersvo Vnutrennikh Del [(Russian) Ministry of Internal Affairs] *w9*
MVDM Multiple Virtual DOS Machines *kg, hi*
MVE Muzzle Velocity Error *fu*
MVGA Monochrome Video Graphics Array *pk, kg*
MVIP Multi-Vendor Integration Protocol *pk, kg*
MVL Metal Vapor Laser *hc*
MVM Multivehicle Module *fu*
MVO Military Vehicles Operation [a unit of General Motors] *aw130.11.75*
MVP Magnetic Vector Potential *fu*
MVP Modular Voice Processor *ct, em*
MVP Multimedia Video Processor *pk, kg*
MVR Mission Voice Radio *fu*
MVS Magnetic Voltage Stabilizer *fu*
MVS Military Vehicle Systems *aw118.18.4*
MVS Miniature Vehicle Sensor *fu, hc, 17*
MVS Modularized Vehicle Simulation *tb*
MVS Multiple Virtual Storage *fu, hd, wp, is26.1.89, ct, em, kg, hi*
MVS Multiple Virtual Storage Operating System [IBM] *mt*
MVS Multiple Virtual System *mt*
MVS/ESA Multiple Virtual Storage / Enterprise Systems Architecture *ct, em*
MVS/SP Multiple Virtual Storage / System Product *ct, em*
MVS/TSO Multiple Virtual Storage / Time Sharing Option *ct, em*
MVS/XA Multiple Virtual Storage / Extended Architecture [IBM] *ct, em, fu, mt*
MVT Multiprogramming with a Variable number of Tasks *pk, kg*
MVTOD Message Validation and Time Of Day *fu*
MVTS Mating Verification Test Set *fu*
MW Medium Wave *fu*
MW Mehrzweck Waffe [(German) "multipurpose weapon"] *mt*
MW Micro Wave *mt, jgo, mj31.5.22*
MW Mid Wave *oe*
MW Missile Warning *mt, kf*
MW Missile Wing [USA] *mt*
MW Momentum Wheel *oe*
MW-NA Missile Warning - North America *kf*
MW-ROW Missile Warning - Rest Of World *kf*
MW-T Missile Warning-Theater *kf*
MWA Manufacturing Work Authorization *fu*
MWA Millimeter Wave Assembly *jdr*
MWA Minimum Warning Attack *mt*
MWA Momentum Wheel Assembly *pl, cl, jdr, kf, uh, jn, oe*
MWAP Minimum Warning Attack Plan *mt*
MWARS MACOM Worldwide Ammunition Reporting System [USA] *mt*
MWBP Missile Warning Bypass *mt*
MWC Missile Warning Center [SPACECMD] *mt, wp, sr9-82, hs, jb, jmj, tb, kf*
MWC Multiwarfare Control *fu*
MWCCG Multiwarfare Coordination Control Group *fu*
MWCE Millimeter Wave Communications Experiment *hc*
MWCF Memory Write Correction Flip-Flop *fu*
MWCS Mobile Weapons Control System *hc*
MWD Magnetostatic Wave Device *wp*
MWDS Missile Warning and Display System [NORAD] *mt*
MWEM Major Work Element Manager *gp*
MWG MANPRINT Working Group *fu*
MWI Message Waiting Indicator *ct, em*
MWIR Medium (Middle) Wavelength Infrared *hc, 39, 17, sb, tb, kf*
MWL Milliwatt Logic *fu*
MWN Message Waiting Notification *ct, em*
MWO Missile Warning Office *mt*
MWO Modification Work Order *fu, hc*
MWP Momentum Wheel Platform *cl, jdr, he, jn*
MWPA Momentum Wheel Platform Assembly *jn*
MWPC Multi Wire Proportional Chamber *hc*
MWR Mean Width Ratio *fu*
MWRMQAS Medical War Reserve Materiel Quality Assurance Subsystem *mt*
MWS Manned Weapons Station *hc*
MWS Missile Warning Squadron *mt, 12*

MWS Missile Warning System *wp*
MWS Multiple Weapons Simulator *fu*
MWSC Missile Warning Systems Conference *mt*
MWSG Marine Wing Support Group [USMC] *mt*
MWSOC Missile Warning and Space Operations Center *mt*
MWSS Marine Wing Support Squadron [USMC] *mt*
MWSS Millimeter Wave Signal Source *hc*
MWT Millimeter Wave Technology *wp*
MWT Missile Warning-Theater *kf*
MWTGM Millimeter Wave Terminal Guided Missile *wp*
MWTS Millimeter Wave Transmitting Subsystem *hc*
MWTTY Missile Warning Teletypewriter *mt*
MWU Millimeter Wave Unit *jdr*
MWV Maximum Working Voltage *fu*
MWWP Mine Warfare Working Party *mt*
MWWU Marine Wing Weapons Unit *-12*
MX Mail Exchanger [Internet] *kg, hi*
MX Multiplexer [also MUX] *hc*
MX Mutually Exclusive *dn*
MX {Multiple-warhead} Experimental [US advanced intercontinental ballistic missile, later dubbed "Peacekeeper"] *mt, ro, aw130.13.18, hc, 11, 14, ph, aa26.12.72, st, wp*
MXP MDR Crosslink Processor *jdr*
MXPO MX Project Office *tb, st*
MXS Microsoft Exchange Server *kg, hi*
MXSP Maintenance Support Plan *mt*
MY Multi Year [a type of sea ice] *ac*
MYE Man Year Equivalent *wp*
MYP Multiple Year Procurement *mt, hc, wp, uc, st, fu*
MYR Mean Year *wp*
MYSTIC_LINK [secure, jam resistant digital information distribution system] *wp*
MYSTIC_STAR [Presidential Support Air Ground Communications System] *mt*
MZ Mashinostroitelny Zavod [machine building plant] *ai2*
MZE Multifunctional Zone Evaluation *hm, hn49.11.12, sc9.6.4, jdr*
MZFW Maximum Zero Fuel Weight *mt*
MZR Multiple Zone Recording *kg, hi*

N

N [code for date in operations plan, as in N-day] [USA DOD] *mt*
N-AMPS Narrow band - Advanced Mobile Phone Service [Motorola] *hi*
N-HADR Norwegian-Hughes Air Defense Radar Program *fu*
N-ISDN Narrowband ISDN *pk, kg*
N-ISDN National ISDN *pk, kg*
N-LOS Non-Line Of Sight *aw130.13.67*
N-SITE Near-term System Integration Test and Evaluation *st*
N/A Next Assembly *fu*
N/B Noise Power/Unit Bandwidth *0*
N/C No Change *fu*
N/C Not Critical *tb*
N/C Numerically Controlled *fu, hd*
N/I Non Interlaced *kg, hi*
N/N Noise to Noise *mt*
NA National Archives *wp*
NA Next Assembly *fu*
NA No Action *0*
NA North America *w9, kf*
NA Not Applicable *w9, wp, tb, kf, vv, he, fu*
NA Not Available *fu, ci.217, 0, 2, w9*
NA Nuclear Attack *kf*
NA Numerical Aperture *wp, fu*
NAA National Aeronautical Administration *0*
NAA Neutron Activation Analysis *wp*
NAA North American Aviation [company, later merged to North American Rockwell] *ci.410*
NAA North Atlantic Assembly [NATO] *mt*
NAADM North American Defense Modernization [NORAD] *mt*
NAAFI Navy, Army and Air Force Institutes [USA] *fu, mt*
NAAG NATO Army Armaments Group *wp*
NAAR Night Air to Air Refueling *mt*
NAAS Naval Auxiliary Air Station [USA] *mt*
NAAS NORAD Attack Alert System *mt*
NAAW NATO Anti Air Warfare *hc*
NAAWS NATO Anti Air Warfare System *wp*
NABS NATO Air Base SATCOM [ESD] *hc, mt*
NAC NASA Advisory Council (Committee) *ci.645, hy, wl*
NAC National Agency Check *mt, hc, wp, tb, su*
NAC Naval Air (Avionics) Center *tc, wp*
NAC Naval Air Command [UK] *mt*
NAC Network Access Component/Centers *mt*
NAC Network Adapter Card *pk, kg*
NAC NMCS Automated Control *mt*
NAC North Atlantic Council [NATO] *mt*

NAC Number of Azimuth Cells *fu*
NAC/DPC North Atlantic Council / Defense Planning Committee [NATO] *mt*
NACA National Advisory Committee on Aeronautics [later became NASA] *mt, mi, ci.xiii, 13, mb, aa27.4.59, aw130.13.36*
NACC National Air Communications Council [NATO] *mt*
NACD National Association of Computer Dealers *pk, kg*
NACE National Association of Corrosion Engineers *aw130.13.104*
NACE NMCS Automated Control Executive *mt*
NACES Navy Aircrew Common Ejection Seat [USA] *mt*
NACF Naval Air Combat Fighter [the F-18 program, USA] *mt*
NACISA NATO Communications and Information Systems Agency [NATO] *fu, wp, mt*
NACISC NATO Communications and Information Systems Committee *mt*
NACISO NATO Communications and Information Systems Organization *mt*
NACK Negative Acknowledgement [character] *csIII.21, tb, ct, jdr*
NACL Nippon Aviotronics Co., Ltd. *fu*
NACM Network Authentication Control Module *mt*
NACO National Association of Countries *y2k*
NACOA National Advisory Committee on Oceans and Atmosphere *sc3.6.4*
NACOM National Alert Communications *mt*
NACORE International {?} Association of Corporate Real Estate Executives *hn4*
NACP NORAD/CONAD Airborne Command Post *mt*
NACS National Advisory Committee on Semiconductors *pk, kg*
NACS Network Assisted Coordinated Science *ns*
NACSEM National Communications Security/Emanations *fu*
NACSI National COMSEC Instruction *mt, nd74.445.11*
NACSIM National Communications Security Information Memorandum *kf, fu, tb*
NACSM National Communications Security Information Memoranda *mt*
NAD Naval Air Department [UK] *mt*
NAD Nicotinamide Adenine Dinucleotide *wp*
NADACS National Air Defense and Airspace Control System *aw130.6.34*
NADB Network Accessible Data Base *mt*
NADC NATO Air Defense Committee *mt*
NADC Naval Air Defense Center *kf*
NADC Naval Air Development Center [US Navy] *hc, fu, tc, mt*
NADEC Navy Decision Center *mt*
NADGE NATO Air Defense Ground Environment [NATO] *mt, hc, fu, wp, hn41.18.4, 12, tb*
NADGECO NADGE Company [an international consortium] *fu*
NADGEMO NADGE Management Office *fu*
NADH North American Digital Hierarchy *hi*
NADIN National Airspace Data Interchange Network *mt, fu*
NADIS NATO Defense Information System *mt, wp*
NADO Navy Accounts Disbursing Office *dt*
NADP Navy Advanced Development Plans *hc*
NADP Nicotinamide Adenine Dinucleotide Phosphate *wp*
NADS NATO Air Defense System *mt*
NADS North Atlantic Defense System *mt, wp*
NADSS Navy Area Defense Supplement to Safeguard *fu*
NAE National Academy of Engineering *ci.873, aa26.12.b4*
NAE Not Above or Equal *pk, kg*
NAEC Naval Air Engineering Center *mt, fu*
NAECON National Aerospace and Electronics Conference *aw118.14.4*
NAEDS Nonaqueous Equipment Decontamination System [chemical warfare] *wp*
NAEGIS NADGE Airborne Early Warning Ground Environment Integrated Segment *fu*
NAEGIS NATO Air Environment Ground Information System *hc, go3.1.3*
NAEGIS NATO Airborne Early warning and Ground control Integration Segment *mt*
NAEW NATO Airborne Early Warning [NATO] *fu, mt*
NAEWF NATO Airborne Early Warning Force [NATO] *mt*
NAF Naval Air Facility [USA] *mt*
NAF Naval Avionics Facility *fu*
NAF Non Appropriated Funds *wp, mt*
NAF Numbered Air Force [USAF] *mt*
NAFAG NATO Air Force Armaments Group *mt*
NAFAX National Facsimile *mt*
NAFB Norton Air Force Base *-17*
NAFC Navy Accounting and Financing Center *dt*
NAFCR NAF Computer Replacement [SAC] *mt*
NAFEC National Aviation Facilities Experimental Center [Pomona, NJ, USA] *wp, mt*
NAFI [Hughes Aircraft Company customer for deep submergence rescue vehicle] *hc*
NAFISS Non Appropriated Funds Information Standard System [USA] *mt*
NAFLIR Navigation Forward Looking Infrared *hc*
NAFMIS Non Appropriated Funds Management Information System [USAF] *wp, mt*

NAFTA North American Free Trade Agreement *hi*
NAFTOC NORAD Automated Forward Tell Output to Canada *mt*
NAG Naval (Navy) Astronautics Group *mt, jb, no*
NAG Nulling Antenna Group *jdr*
NAG Numerical Algorithm Group *tb*
NAI Next Assembly Indicator *fu*
NAI No Airborne Intercept *wp*
NAI Nonautomatic Initiation *fu*
NAIAD Nerve Agent Immobilized Enzyme Alarm and Detector [chemical warfare] *wp*
NAIC National Air Intelligence Center *kf*
NAIC Nuclear Accident / Incident Control *mt*
NAICO Nuclear Accident / Incident Control Officer *mt*
NAIF Navigation Ancillary Information Facility *re*
NAILS National Airspace Integrated Logistics Support *fu*
NAIPS National Aeronautical Information Processing System *fu*
NAIPS NMCS Automated Input Processing System *mt*
NAIR Naval Air Systems Command Headquarters *dt*
NAL National Aerospace Laboratory [Japan] *aw129.14.38*
NALC Naval Aviation Logistics Command *fu*
NALCOMIS Naval Aviation Logistics Command Management Information System *mt*
NALCOP NORAD Alternate Command Post *mt*
NALDA Naval Aviation Logistic Data Analysis System *mt*
NALE Naval, Amphibious Liaison Element *mt*
NALLA National Long Lines Agency *mt*
NALLADS Norway (Norwegian) Army Low Level Air Defense System *hc, fu*
NALTOACS Naval Laboratories Technical Office Automation and Communications System *mt*
NAM Naval Armaments Model [DIA] *mt*
NAM Net Assignment Map *fu*
NAM Number Assignment Module *kg, jdr*
NAMC [regional aircraft manufacturer] *aw129.21.97*
NAMFI NATO Missile Firing Installation [a firing ground for missiles at Akrotiri, Crete] *mt*
NAMLS Naval Automated Medical Logistics System *mt*
NAMMA NATO MRCA development and procurement Management Agency *mt*
NAMMOS Navy Manpower Mobilization System *mt*
NAMP Naval Aviation Maintenance Procedures *fu*
NAMPS Narrow band Analog Mobile Phone Service [Motorola] *pk, kg*
NAMS NATO AWACS Maintenance System *mt*
NAMSA NATO Maintenance and Supply Agency *mt, hn43.22.12, fu, wp*
NAMSO Navy Maintenance Support Office *mt*
NAMT Naval Air Maintenance Trainer *hc*
NANS Night Attack Navigation System *hc*
NAOC National Airborne Operations Center [USAF E-4B aircraft, 747, formerly known as NEACP] *mt*
NAOMI NADIN/AUTODIN Mode-1 Interface *mt*
NAOMIS Navy Materiel Transportation Office Operations and Management Information System *mt*
NAP Naval Air Plan *mt*
NAP Network Access Point *ct, em, kg, hi*
NAP Non Nuclear Armament Plan *wp*
NAP Not Authorized POMCUS *mt*
NAP Nuclear Auxiliary Power *wp*
NAP Nulling Algorithm Processor *jdr*
NAP Nulling Antenna Processor *jdr*
NAP(E) Nulling Antenna Processor (Emulator) *jdr*
NAPA National Academy of Public Administration *ci.xxvii*
NAPC Naval Air Project Coordinator *dt*
NAPC Naval Air Propulsion Test Center *tc*
NAPLDS North American Presentation Level Autocol Syntax *mt*
NAPLPS North American Presentation Level Protocol Syntax [graphics] *kg, fu, hi*
NAPMA NATO AEW C&C Program Management Agency *fu*
NAPMA NATO AWACS Program Management Agency *mt*
NAPO NATO AWACS Project Office *fu*
NAPPS Navy Automated Publishing and Printing System *mt*
NAPR NATO Armaments Planning Review *mt*
NAPRS National Airspace Performance Reporting System *fu*
NAPS Navy Automated Publications System *wp*
NAPS Night Aerial Photographic System *wp*
NAR National Airspace Review *fu*
NAR Negotiated As Required *fu*
NAR Non-Advocacy Review *hy*
NARA Navy Appellate Review Activity *dt*
NARBS Night/Day Angle Rate Bombing System *hc*
NARCOG Narcotics Coordination Group *-10*
NARDAC Naval (Navy) Regional Data Automation Center *dt, mt*
NARF (NATS
NARF Naval Air Reserve Force [USA] *mt*
NARF Naval Air Rework Facility *mt*
Narkomaviaprom [State Commissariat for the Aviation Industry, USSR] *mt*
NARM Navy Resources Model *mt*

NARS Non-nuclear kill Avionics Requirements Study *-17*
NAS NADGE Analysis System *fu*
NAS National Academy of Sciences *mt, fu, wp, ci.555, 0, 13, w9, tb*
NAS National Aerospace Standard *fu, oe*
NAS National Aircraft Standard *fu*
NAS National Airspace (Aerospace) System (Spec) *aa26.12.35, ag, hc, is26.1.62, wp, tb, jdr, y2k*
NAS National Airspace System [FAA] *fu, mt*
NAS Naval Air Squadron [USA] *mt*
NAS Naval Air Station *mt, fu, 12, w9, af72.5.28, ai2*
NAS Naval Air Transport Service *mt*
NAS Naval Audit Service *dt*
NAS Network Access Switch *mt*
NAS Network Application Support [DEC] *pk, kg*
NAS NORAD Alert System *mt*
NAS North Arabian Sea *mt*
NAS Numerical Aerodynamic Simulation [NASA] *mt*
NASA National Aeronautics and Space Administration (Agency) [organization, USA] "nasa" *hc, 0, 2, 4, 8, 9, 12, 13, 14, 17, sb, aw130.13.25, aa26.12.3, w9, wp, so, ns, mb, no, ce, hy, re, tb, st, wl, pbl, hg, mb, kf, vv, ah, su, ge, kg, mi, fu, tm, oe, mt, y2k, crl, ai2*
NASAMS Norwegian Advanced Surface to Air Missile System *hc, hq6.2.7, hn50.8.1*
NASAP Navy Alcohol Safety Action Program *dt*
NASARR North American Search And Range Radar [radar of the F-104] [USA] *mt*
NASC National Aeronautics and Space Council *ci.867, 13*
NASC Naval Air Systems Command [US Navy] *tc, mt*
NASCOM NASA Communications [systems] [center, network of lines linking STDN sites] *mt, hc, 0, 2, 4, mb, hy, ac2.7-11, oe*
NASCOM Naval Air Station Communication *mt*
NASCOP NASCOM Operating Procedures *oe*
NASD NMCS Applications Software Description *mt*
NASDA National Space Development Agency [Japan] *hn44.17.1, aa27.4.49, hq6.2.1, ep, ce, hy, wl, hg, mb, fu, ge, mi, ai2*
NASEE NAVAIR Software Engineering Environment *aw130.11.57*
NASI NetWare Asynchronous Services Interface [Novell] *pk, kg*
NASIRC NASA Automated Systems Internet Response Capability *pk, kg*
NASKER NASA Ames Kernel [benchmark] *pk, kg*
NASM National Air and Space Museum *mi*
NASMO NATO Starfighter Management Office [NATO] *mt*
NASNET National Airspace System Network *fu*
NASP National Aero Space Plane (also known as the X-30) [USA] *mt, mi, aw130.13.45, aa26.12.13, af72.5.27, wp, mb, ai1*
NASP National Airspace System Plan (Program) [FAA] *mt, aa26.12.35, tb, fu*
NASP NATO Atomic Supply Point *mt*
NASRS Non Available Status Reporting System [USA] *mt*
NASTRAN NASA Structural Analysis [a computer program] *fu, aa26.12.51, ci.1018, 16, 17*
NASWS Naval Anti Submarine Warfare Systems *tc*
NAT Not Air Transportable *mt, wp*
NATC Naval Air Test Center [at Patuxent, Maryland, USA] *tc, hc, fu, mt*
NATC Naval Air Training Command [USA] *mt*
NATCA National Air Traffic Controllers' Association *aw130.13.30, y2k*
NATCOM National Communication Center *fu*
NATF Naval Air Test Facility *tc*
NATF Navy Advanced Tactical Fighter [USA] *mt, aa26.8.43*
NATFAX National Weather Facsimile Network *mt*
NATIC Naval Air Technical Training Center *wp*
NATO North Atlantic Treaty Organization [created in 1949] "nato" *mt, aw130.13.10, hc, ac7.1-39, 11, 12, 13, 14, w9, dt, wp, pm, tb, st, fu, su, hi, ai2*
NATOPS Naval Air Training and Operating Procedures Standardization *mt*
NATS NARF Automatic Test Station *hc*
NATS National Air Traffic Control Simulator *fu*
NATS NMCS AUTODIN Terminal Subsystem *mt*
NATS North Atlantic Track System *mt*
NATSF Naval Air Technical Services Facility *fu*
NAU Naval Administrative Unit *dt*
NAU Network Addressable Unit *ct, em, kg*
NAU Northern Arizona University *is26.1.87*
NAUHF Northern Area Ultra High Frequency radio system [GREEN_PINE] *mt*
NAV/MOL Naval Manned Orbital Laboratory *hc*
NAVADS Navy Automated Transportation Documentation System *mt*
NAVAGLOBE Long Distance Radio Navigation System *ro*
NAVAID Navigational Aid *pl, wp, fu*
NAVAIDS Navigational Aids *mt*
NAVAIR Naval Air [Systems Command] *mt, hc, tc, aw130.11.57, wp, fu*
NAVAIR-SYSCOM Naval Air Systems Command *kf*
NAVAIRDEVCEN Naval Air Development Center *fu*
NAVAIRMIS Naval Aviation LOGCEN MIS [Industrial NAIR Station Workload Control] *mt*
NAVALT Naval Alteration [USA] *mt*
NAVARHO Navigation and Radio Homing [a long distance, ground-based navigation system] *fu, ro*

NAVASTROGRU Navy Astronautics Group *jb*
NAVC Naval Audio Visual Center *dt*
NAVCAMS Naval Communications Area Master Station *mt, fu, 12, uh*
NAVCAS Navigation and Collision Avoidance System *fu*
NAVCO-SSACT Naval Command System Support Activity *mt*
NAVCOMM Naval Communications Command *fu*
NAVCOMMSTA Naval Communications Station *mt*
NAVCOMPARS Naval Communications Processing and Routing System *mt*
NAVCOMSTA Naval Communications Station *mt, fu*
NAVCOMTELSTA Naval Computer and Telecommunications Station *mt*
NAVCON Navigation/Control System *fu*
NAVDAC Naval Data Automation Command *mt*
NAVDF Naval Direction Finding Network *mt*
NAVELEX Naval Electronic Systems Command [US Navy] *mt, fu, hc, sc8.3.4, tc*
NAVEONIC Nautical Auto Vessel Environmental Ocean Naval Information Communications System *hc*
NAVEUR Naval Forces, Europe *mt*
NAVEX Navigation Exercise *mt*
NAVFAC Naval Facilities Engineering Command *fu*
NAVFACENGCOM Naval Facilities Engineering Command *fu*
NAVFLIR Navigational Forward Looking Infrared *hc*
NAVILCO Navy International Logistics Control Office *-18*
NAVINTCOM Naval Intelligence Command *mt*
NAVLIS Navy Logistics Information System *fu*
NAVMACS Navy Modular Automated Communications System *mt*
NAVMACS {equipment on board destroyers} *pf1.4*
NAVMAT Navy Material Command [later ONAS] *mt, rm1.1.2, aw130.11.27, nt, fu*
NAVMECDET Navy Manpower Engineering Center Det. *dt*
NAVMEDCOM Naval Medical Command *mt, dt*
NAVMIC Naval Military Intelligence Center *mt*
NAVMOD Navy Model *mt*
NAVOBSY Naval Observatory *dt*
NAVOC Navy Operations Center *fu*
NAVOCEANO Naval Oceanographic Office [also NOC] *ag, dm*
NAVOPTINCEN Naval Operations Intelligence Center *jb*
NAVORDSTA Naval Ordnance Station *fu*
NAVPERS Naval Personnel Command *fu*
NAVPRO Naval Plant Representative's Office *fu*
NAVRADSTA Naval Radio Station *mt*
NAVSAT Navigation Satellite *hc, mj31.5.9, fu*
NAVSAT Navy Satellite System [USA] *mt*
NAVSEA Naval Sea Systems Command [replaced NAVSHIPS and NAVORD] *mt, fu, hc, tc*
NAVSEC Naval Ship Engineering Center *fu*
NAVSECGRU Navy Security Group *mt*
NAVSIM Navigation Simulation *fu*
NAVSOC Naval Satellite Operations Center *jdr, uh*
NAVSOC Naval Special Warfare Special Operations Component [USA DOD] *mt*
NAVSOF Naval Special Warfare Forces [USA DOD] *mt*
NAVSOUTH Naval Forces, Southern Region *mt*
NAVSPACECOM Naval Space Command *sc4.1.5, 16, go2.4, jb, kf, uh*
NAVSPASUR Naval Space Surveillance system [also NSSS] *mt, hc, 12, ps204, go2.3.2.1, jb, kf*
NAVSPECWAR Navy Special Warfare *wp*
NAVSPOC Naval Space Operations Center *jb, cp*
NAVSTAR Navigation [Satellite] System using Timing And Ranging [USAF/USN global navigation system] [later GPS] *mt, 12, 14, ci.571, wp, tb, fu*
NAVSUP Naval Supply Systems Command *mt, fu*
NAVTACSTANS {Naval Tactical Data Systems Standards} *mt*
NAVTAG Naval Tactical Analysis Game *fu*
NAVTAS Navigation and Target Acquisition System *hc*
NAVTELCOM Naval Telecommunications Command *mt, jb*
NAVWASS Navigation and Weapon Aiming Sub System *mt*
NAVWESA Naval Weapons Engineering Support Activity *fu*
NAVWESTOCEANCEN Naval Western Oceanography Center [Pearl Harbor, Hawaii] *ag*
NAVWPNCEN Naval Weapons Center *mt*
NAWAS National Warning System [FEMA] *ag, mt*
NAWC Naval Air Warfare Center [USA] *mt, heh2, 21, 1*
NAWDEX National Weather Data Exchange [system of the OWDC] *ac1.5-14*
NAWG NTB Architecture Working Group *tb*
NAWP National Aviation Weather Processing *fu*
NAWS Negotiate About Window Size *pk, kg*
NAWS NORAD Alert (Attack) Warning System *fu, wp*
NB Narrow Band *mt, fu, wp, kf, hi*
NB Narrow Beam [satellite] *mt, jdr, ggIII.2.378*
NBAA National Business Aircraft Association [USA] *mt, aw129.10.82*
NBAE New Business Acquisition Expenditures *tb*
NBB Number of Bytes of Binary *ct, em*
NBC Narrow Body CRAF *mt*

NBC Navy Branch Clinic *dt*
NBC Nuclear, Biological, and Chemical [weapons] [as in the phrase NBC defense] [USA] *mt, fu, ph, hc, af72.5.43, wp, pm*
NBC/MMT Nuclear Biological Contamination/Manufacturing Methods Technology *hc*
NBCWRS Nuclear, Biological, Chemical Warfare Reporting System *mt*
NBCWRS Nuclear/Biological/Chemical Warning and Reporting System *mt*
NBDFX Narrowband Dicke-Fix *fu*
NBE Not Below or Equal *pk, kg*
NBEUI Network BIOS Extended User Interface *em*
NBFM Narrow Band Frequency Modulation *wp, jdr*
NBG Naval Beach Group *mt*
NBHF Narrowband High Frequency Communications System *hc*
NBI Niels Bohr Institute [Copenhagen, Denmark] *sm*
NBI Non Battle Injury *mt*
NBM Narrowband Modem *fu*
NBMB National Bus Military Bureau *dt*
NBMR [NBMR-3 is the requirement formulated in 1960-1961 for a supersonic VTOL strike fighter] [NATO] *mt*
NBP No Baseband Processing *ce*
NBS Narrow Band Service *kf*
NBS National Bureau of Standards [later NIST] *mt, wp, ci.879, 2, 4, hc, w9, mj31.5.47, no, tb, mb, fu, mi, kg, oe*
NBS Navy Broadcasting Service *dt*
NBS Numeric Backspace *pk, kg*
NBSCU Narrowband Signal Conversion Units *fu*
NBST Narrow Band Secure Terminal *mt, wp*
NBST Narrow Band Subscriber Terminal *mt*
NBSV Narrow Band Secure Voice *fu, mt*
NBSVS Narrow Band Secure Voice System *wp*
NBTU Narrow Band Trunking Unit *mt*
NC Narrowbeam Coverage *mt*
NC Native Course *fu*
NC NATO Confidential *hc*
NC Navigation Controller *fu*
NC Network Computer *ct, kg, em*
NC Network Control *mt*
NC Network Controller *fu*
NC No Carry (Not Carried) *kg, fu*
NC No Change *fu*
NC Noise Correlation *fu*
NC Normally Closed *fu, gp, wp, hi, oe*
NC Numerical Control *mt, fu, kg, 6*
NC/STRC North Carolina Science and Technology Research Center *ci.1005*
NCA National Command Authority [USA DOD] *mt, fu, sr9-82, 5, 12, sb, go1.2-1, jb, nt, cp, wp, tb, as, kf*
NCA Network Communications Adapter *kg, hi*
NCA Network Computing Architecture *ct, em*
NCA Non Contractual Authorization *fu, 8, hc*
NCA Northern Communications Area *mt*
NCAA NATO Civil Air Augmentation *mt*
NCAA Non Nuclear Consumables Annual Analysis (system) *mt, wp*
NCAC National COMSEC Assessment Center *tb*
NCAD Northrop Computer Aided Design *ep, jdr*
NCAP National COMPUSEC Assessment Program *tb*
NCAPS Naval Control and Protection of Shipping [LANTFLT] *mt*
NCAR National Center for Atmospheric Research *wp, ac1.7-17, no, hy, tb*
NCARAI Navy Center for Applied Research in Artificial Intelligence [NRL] *mt*
NCB Net Control Block *fu*
NCB Nuclear Contingency Branch *mt*
NCBC National Codification Bureau Code *fu*
NCC NASA Communications Control(ler) *wl, hi*
NCC National Climate Center *ci.882*
NCC National Command Center *fu, 17, sb*
NCC National Coordinating Center [NCS/DCA] *mt*
NCC Netscape Communications Corporation *hi*
NCC Network Computer Complex *fu*
NCC Network Control Center [EOS] *mt, ct, em, kg, 2, 4, hy, fu, oe*
NCC Non Cancellable Commitment *wp*
NCC NORAD Command Center *mt, kf*
NCC NORAD Control Center *mt*
NCC NTB Control Center *tb*
NCCDPC NATO Command, Control and Information System and ADP Committee *mt*
NCCDS NCC Data System *-2*
NCCE Novell Collaborative Computing Environment *hi*
NCCF Network Communications Control Facility *mt*
NCCIS NATO Command, Control and Information System *mt*
NCCMC Navy Civilian Career Management Center *dt*
NCCOSC Naval Command, Control, and Ocean Surveillance Center *uh*
NCCR NCC Representative *-2*
NCCS Navy Command and Control System *mt, tb, fu*
NCCS Network Communications and Control Stations *fu, jdr*

NCCS NORAD Command and Control System *mt*
NCCS Nuclear Command and Control System *mt*
NCD Net Control Device *mt*
NCD Network Computing Device *pk, kg*
NCD No Change Drawing *fu*
NCD Numerical Concept Data *fu*
NCDC National Climatic Data Center *ag*
NCDF Nested-Cup Dipole Feed *fu*
NCDRS NMCS Crisis Data Retrieval System *mt*
NCDS NASA Climate Data System [formerly PCDS] *ns*
NCE Network Communications Engineer *2, 4*
NCE Network Control Element *mt*
NCE New, Changed, Extended *jdr*
NCE Nomadic Computing Environment *hi*
NCEB NATO Communications Electronics Board *mt*
NCEL Naval Civil Engineering Laboratory [Port Hueneme, CA] *tc, mt*
NCEN Network Compatibility Engineer *-2*
NCEPA National Civil Emergency Planning Authorities *mt*
NCEUR National Security Agency Europe *mt*
NCF Netware Command File *ct, em*
NCF Network Control Facility *mt*
NCF Noise Combining Factor *jdr*
NCF NSSDC Computer Facility *ns*
NCF Nuclear Capable Forces *mt*
NCFPO Northern California Field Procurement Office *hq5.3.5*
NCFSK Non Coherent Frequency Shift Keying *mt*
NCGA National Computer Graphics Association *kg, fu*
NCI Net Control Initialization *fu*
NCI New Concepts Initiative *wp*
NCIC National Cartographic Information Center *fu*
NCIC National Crime Information Center (Computer) [FBI] *kg, ct, mt*
NCID Non Cooperative Identification *wp*
NCIS NATO Common Interface (Interoperability) Standards *go3.1-2, fu*
NCISC Naval Counter Intelligence Support Center *dt*
NCM Network Configuration Management *mt*
NCM Node Controller Module *ct*
NCM Non Corrosive Metal *fu*
NCMA National Contract Management Association *hn41.17.1*
NCMB NATO Configuration Management Board *mt*
NCMC National Critical Materials Council *mt*
NCMC NORAD Cheyenne Mountain Complex (Center) *mt, fu, wp, sr9-82, 12, tb, kf*
NCMO NAVAIDS Communications Management Office *mt*
NCMOS N-channel Complementary Metal Oxide Semiconductor *em*
NCMP NATO Configuration Management Plan *mt*
NCMP Navy Capabilities and Mobilization Plan *mt*
NCMR Non Conforming Material Report [document] *ah, 8, 17, fu, sb, ep, jdr, pbl, tm, jn*
NCMS National Classification Management Society *st*
NCMT Numerical Control for Machine Tools *pk, kg*
NCO Net Control Operator *mt*
NCO Non Commissioned Officer *mt, fu, wp, af72.5.72*
NCO Numerically Controlled Oscillator *fu, jdr*
NCOA Noncommissioned Officer's Association *fu*
NCOC NORAD Combat Operations Center *mt, wp*
NCOIC Non Commissioned Officer In Charge *mt, kf*
NCOS Network Computer Operating System *pk, kg*
NCOSE National Council on Systems Engineering *pbl, jn*
NCP National Climate Program *ag*
NCP National Command Post *tb, kf*
NCP Navy Capabilities Plan *mt*
NCP Network (Netware) Core Protocol *em, ct, kg*
NCP Network Computer Processor *fu*
NCP Network Control Processor *kg, fu*
NCP Network Control Program [IBM] *mt, em, ct, kg*
NCP Network Control Protocol *kg, csII.80, tb, hi*
NCP Noise Cover Pulse *fu*
NCP NORAD Command Post *mt, sd, wp, jb, kf*
NCP Not Copy Protected *pk, kg*
NCP Nutation Control Processor *-8*
NCPB Navy Council of Personnel Boards *dt*
NCPC Naval Civilian Personnel Command *dt*
NCPS Netware Cross Platform Services *ct, em*
NCPS Nuclear Contingency Planning System [part of GWS] *mt*
NCR National Capital Region *mt, dt*
NCR No Calibration Required *fu*
NCR No Carbon Required *wp*
NCR Non Conformance Report *tb, oe*
NCR Notice of Change Revision *fu*
NCRC Navigational Aids [NAVAIDS] Communications Readiness Center [USAFE] *mt*
NCS National Communications System [USA DOD] *mt, aa26.12.54, dt, y2k*
NCS National Cryptographic School *-10*
NCS Naval Communications Station *mt*
NCS Net Control Station [USA DOD] *pm, mt*

NCS Network Computer System *ns*
NCS Network Control Segment *hc*
NCS Network Control Station *fu*
NCS Network Control Subsystem *mt*
NCS NORAD Computer Segment *jb*
NCS NORAD Computer System *mt, sr9-82, jb*
NCS Nutation Control System *ai2*
NCS-R NCS Replacement [CCPDS, CCPDS-R] *mt*
NCS-R NORAD Computer System Replacement *dh*
NCSA National Center for Supercomputing Applications *ct, em, kg, hi*
NCSA National Computer Security Association *ct*
NCSC National Communications Security Committee *mt, tb*
NCSC National Computer Security Center *mt, wp, tb, fu, kg, ct*
NCSC National Computer Security Council *mt*
NCSI Network Communications Services Interface [Network Products Corp.] *kg, hi*
NCSIU Net Control Station Interface Unit *fu*
NCSL Naval Coastal Systems Laboratory *tc*
NCSORG Navy Control of Shipping Organization *mt*
NCSR National Committee on Space Research [India] *ai2*
NCSS National Computer Security System *tb*
NCSS Naval Control Shipping System [LANTFLT] *mt*
NCSSA Naval Command Systems Support Activity *dt*
NCSSC Naval Command Systems Support Center *dt, fu*
NCT NATO Comparative Test *wp*
NCT Network Control Terminal [JRSC] *mt*
NCTAF Nuclear Communications Task Force *mt*
NCTAMS Naval Computer and Telecommunication Area Master Station *mt, uh*
NCTC Naval Computer and Telecommunications Command *mt*
NCTE Network Channel Terminating Equipment *mt*
NCTI/R Non Cooperative Target Identification / Recognition *mt*
NCTID Non Cooperative Target ID *fu*
NCTR Non Cooperative Target Recognition *fu, hc, wp, uc, ic2-13, ts11-17, pm*
NCU Naval Communications Unit [Washington] *dt*
NCU Net Control Unit *fu*
NCU Nozzle Control Unit *-17*
NCUID Net Control Unit Identifier *fu*
NCURSRV Naval Courier Service *dt*
NCW Naval Coastal Warfare [USA DOD] *mt*
NCW Normal Control Word *fu*
NCWC Naval Coastal Warfare Commander [USA DOD] *mt*
ND Naval Dispensary *dt*
ND Network[s] Director[ate] *2, 4*
ND Nondestructive Count *0*
ND Normal Deployability Posture *mt*
ND Norsk Data *fu*
ND Null Depth *fu*
NDA National Defense Area [USA DOD] *mt*
NDA Non Disclosure Agreement *mt*
NDAA Non Developmental Airlift Aircraft [an off the shelf transport as a cheaper alternative for the C-17] [USA] *mt*
NDAC Naval Data Automation Command *dt*
NDAC Nuclear Defense Affairs Committee *mt*
NDADS NSSDC Data Archive and Distribution System *cu*
NDANG North Dakota Air National Guard *-12*
NDB Non Directional Beacon *fu, kg, mt*
NDB Nuclear Depth Bomb *fu*
NDBDM Navy Department Board of Decorations and Medals *dt*
NDBO NOAA Data Buoy Office *ag*
NDC National Distribution Circuit *ag*
NDC NATO Defense College *mt, wp*
NDC Newport Design Center [Hughes Aircraft Company] *hn44.18.1*
NDC Node Dispatch Controller *fu*
NDC Nuclear Detector Circuit *mt*
NDC Number of Doppler Cells *fu*
NDCC National Defense Command Center Canada *mt*
NDCS Network Data Control System *mt*
NDD Network Delivery Device *mt*
NDDK Network Device Development Kit [Microsoft] *kg, hi*
NDE Non Destructive Evaluation *hc, wp, aa26.8.37, tb, jdr, fu*
NDEA National Defense Education Act *-13*
NDEF Not to be Defined *pk, kg*
NDEL Non Deliverable *kf*
NDEO Naval Disability Evaluation Office *dt*
NDER National Defense Executive Reserve *mt*
NDEV Non Developmental *kf*
NDEW Neutral Particle Directed Energy Weapon *-17*
NDEW Nuclear Directed Energy Weapon *mt, hc, wp, tb, st*
NDEW-G Nuclear Directed Energy Weapon - Ground *tb*
NDF Neutral Density Filter *bl3-2.12*
NDHC National Defense Headquarters, Canada *mt*
NDI Non Destructive Inspection *fu, hc, aw130.11.53c, tb*
NDI Non Development(al) Item [an off the shelf product] *mt, fu, hc, wp, kf, 18, pf.f88.9, tb, st*

NDIAG Norton Diagnostics *ct*
NDIC National Defense Intelligence Center [Canada] *mt*
NDIS Network (Device) Driver Interface Specifications *kg, ct, em, hi*
NDL Network Data Language *mt*
NDL Network Database Language *mt*
NDL Network Description Language *fu*
NDMP Network Data Management Protocol *ct, em*
NDN National Diffusion Network *fu*
NDN Nodal Data Network *mt*
NDN Non Delivery Notice *pk, kg*
NDOC National Defense Operations Center [Ottawa] *mt, kf, wp, jb*
NDP National Disclosure Policy *mt*
NDP Neutral Density Profile *pbl*
NDP Numeric Data Processor *kg, fu*
NDPC National Disclosure Policy Committee *dt*
NDPF NASA Data Processing Facility [Greenbelt, Maryland] *ac*
NDR Network Data Reduction *fu*
NDR Network Data Representation *pk, kg*
NDRE Norwegian Defense Research Establishment [the FFI] [Norway] *fu, mt, crl*
NDRF National Defense Reserve Fleet [USA DOD] *ph, wp, mt*
NDRG NATO Defense Research Group *mt*
NDRO Non Destructive Readout [aerospace memory] *kg, fu, hc*
NDRT Naval Design Review Team *fu*
NDS Navigation Development Satellite *N75-27202-465*
NDS NetWare Directory Service [Novell] *pk, kg*
NDS Network Development System *fu*
NDS Network Directory Services *em*
NDS Non Developmental Software *jdr*
NDS NPIC Data System [DIA] *mt*
NDS NUDET (Nuclear Detonation) Detection System *mt, 5, ph, sb, cp, wp, jdr, kf*
NDSC Network for the Detection of Stratospheric Change *hy*
NDSN NASA Deep Space Network *jdr*
NDT Network Design Tool *do452*
NDT Non Destructive Test(ing) *mt, hc, wp, tb, st, jdr, fu*
NDTC Non Destructive Test Laboratory {?} *-9*
NDTI Non Destructive Testing and Inspection *wp, kf*
NDTM National Defense Transportation Model *mt*
NDTP Network Development Test Plan *-4*
NDU National Defense University *mt, dt*
NDU Navigational Display Unit *fu*
NDV NASP Derived Vehicle *mb, wp, mi*
NDVI Normalized Difference Vegetation Index *hg, ge*
NDW Naval District Washington *dt*
NDW Normal Data Word *fu*
NE Noise Equivalent *ac1.2-13*
NE Nuclear Energy *mt, wp*
NEA Nitrogen Enriched Air *wp*
NEA North East Asia *wp*
NEACP National Emergency Airborne Command Post. [USAF E-4B aircraft, 747, later known as NAOC] *mt, fu, 12, wp, af72.5.139, jb, dt, tb*
NEACP_TTY NEACP Teletypewriter Network *mt*
NEARNET New England Academic and Research Network *ct*
NEAT NCR Electronic Autocoding Technique *wp*
NEAT New Equipment Advisory Team *hc*
NEATS North East Air Traffic System *fu*
NEAWS NORAD/ADCOM Emergency Airborne Warning System *mt*
NEB Noise Equivalent Bandwith *wp*
NEBS New Equipment Building System *fu*
NEC National Electrical Code *fu, wp*
NEC Navy Enlisted Classification *fu*
NEC Nippon Electric Company (Corporation) *fu, kg, sc2.7.3, pl, aw129.21.13, is26.1.33, pbl*
NEC Noise Equivalent Carriers *kf*
NEC Northern European Command [NATO] *fu, mt*
NECC Net Entry Control Channel *fu*
NECCCIS Northern European CCC Information System *tb*
NECOS Network Coordination Station *mt*
NECPA National Emergency Command Post Afloat *wp*
NECS National Electrical Code Standards *fu*
NECS Navy Embedded Computer System *hc*
NECT Net Entry Control Terminal *fu*
NEC_CCIS Northern European Command CCIS *fu*
NED NASA Extragalactic Database [NASA] *kg, hi*
NED Network Engineering Division *-4*
NED Northrop Electronics Division *-17*
NED Nuclear Event Detected *fu*
NEDBAG NATO Emitter Database Advisory Group *mt*
NEDBAS NATO Emitter Database *mt*
NEDN Naval (Navy) Environmental Data Network *mt, ag*
NEDR Noise Equivalent Delta Radiance *oe*
NEDRES National Environmental Data Referral Service *ag*
NEDS Naval Environmental Data System *mt*
NEDS Naval Environmental Display Station *ag*
NEDSA NAVSEA Engineering Drawing Support Activity *mt*

NEDT Noise Equivalent Differential (Delta) Temperature *pl, bf, pbl, oe*
NEE North Enter Earth [Pulse] *-9*
NEEC Net Entry Entering Channel *fu*
NEEO Naval Electronics Engineering Office *fu*
NEET Nonlinear Estimation for Exoatmospheric Trajectories *hc*
NEF NASA Extended FORTRAN *mt*
NEFD Noise Equivalent Flux Density *tb, fu, kf*
NEFMA NATO EFA/Eurofighter Management Agency [NATO] *mt*
NEFMA NATO European Fighter Management Agency *hc*
NEFPO New England Field Procurement Office *hq5.3.5*
NEHRP National Earthquake Hazards Reduction Program [FEMA] *mt*
NEI Noise Equivalent Irradiance *bf, kf*
NEI Non Explosive Initiator *-4*
NEI Nuclear Energy Institute *y2k*
NEL Naval Electronics Laboratory *fu, hc*
NELC Naval Electronics Laboratory Center *fu, hc*
NEMA National Electrical Manufacturers Association *fu, wp*
NEMA National Emergency Management Agency *lat*
NEMP Nuclear Electro Magnetic Pulse *hc, wp, fu*
NEMS National Emergency Management System [FEMA] *wp, mt*
NEMS Navigation and Environmental Monitoring System *hy*
NEMS Nimbus E Microwave Spectrometer *ac2.8-3*
NEN {Noise Equivalent Radiance} *pl, oe*
NEO Non Combatant Evacuation Operation *mt*
NEOCS Navy Enlisted Occupational Classification System *fu*
NEODF Naval Explosive Ordnance Disposal Facility *tc*
NEOS National Earth Orientation Service *hy*
NEP Network Entry Point *kg, hi*
NEP Noise Equivalent Power *pl, wp, kf, fu*
NEPA National Environmental Policy Act *nt, kf, oe*
NEPA National Environmental Protection Act *mt*
NEPD Noise Equivalent Power Density *fu*
NEPP Nuclear Effects Post Processor *st*
NEPRF Navy Environmental Prediction Research Facility [Monterey, California] *pl*
NEQ Noise Equivalent Flux *kf*
NER Net Entry Request *fu*
NER Noise Equivalent Radiance *oe*
NERAC New England Research Applications Center *ci.1005*
NERC National Environmental Research Centre [United Kingdom] *hy*
NERC North American Electric Reliability Council *y2k*
NERD Neuro Evolutionary Rostral Developer *wp*
NEREM (Granger's paper, NEREM Record, vol. 13 (1971)) *do41*
NEREP Nuclear Execution and Reporting *mt*
NERF Navy Emitter Reference File *mt*
NERV Nuclear Emulsion Recovery Program *0*
NERVA Nuclear Engine for Rocket Vehicle Application *ci.226, mb, mi*
NES National Education Supercomputer *pk, kg*
NES Net Entry Signals *fu*
NES Net Entry Solicitation *fu*
NES Network Encryption System *kf*
NES Noise Equivalent Signal *wp*
NESAC Naval Engineering Support and Analysis Center *fu*
NESC National Electrical Safety Code *fu*
NESC National {Environmental} Satellite Center [ESSA] [later NESS] *pl, hg, ge*
NESC Naval Electronic Systems Command *tc, tb*
NESCO Network and Schedule Committee *fu*
NESDIS National Environmental Satellite Data and Information Service [NOAA] *16, gp, aw129.10.76, jb, no, pr, hy, hg, ge, oe, ls4, 5, 16*
NESEA Naval Electronics System Engineering Activity *fu*
NESEC Naval Electronic Systems Engineering Center *mt, fu*
NESO NAVAIR Engineering Support Office *mt*
NESP Navy EHF SATCOM (Satellite) Program *fu, mt*
NESS National Environmental Satellite Service *gd1-4, 16, hc, no*
NESS Navy EHF SATCOM System *mt*
NESS NMCS Environmental Support System [OJCS] *mt*
NESSEC Naval Electronics Systems Security Engineering Center *dt*
NESSIE [a thermal imaging system] *hc*
NESSUS Nonlinear Evaluation of Stochastic Structures Under Stress *aa26.11.38*
NEST Novell Embedded Systems Technology *kg, hi*
NEST Nuclear Effects Support Team *wp*
NEST Nuclear Energy Search Team *mt*
NESTOR Neutron Source Thermal Reactor *wp*
NESTOR [older family of tactical secure voice equipment, e.g., KY28] *mt*
NET New Equipment Training *hc, fu*
NET NIMBUS Experiment Team *ac2.8-12*
NET Noise Equivalent Target *fu, kf*
NETADM Network Administrator *hi*
NetBEUI NetBIOS Extended User Interface [IBM] *kg, ct, hi*
NetBIOS Network Basic Input/Output System [IBM] *kg, ct, hi*
NETCAP Network Capability *mt*
NETCOS NASA-ESRO Traffic Control Satellite *hc*
NETD Noise Equivalent Temperature Difference *pl, wp*

NETMON Network Monitor *pk, kg*
NETS National Emergency Telecommunications Service *mt*
NETS Nationwide Emergency Telecommunications System [NCS] *mt*
NETS New Threats Simulator *hc, wp*
NETSIM Network Simulation Language *fu*
NETT New Equipment Training Team *fu*
NET[P] New Equipment Training [Plan] *mt*
NEVADA Net Energy Verification And Determination Analyzer *un*
NEW Net Explosive Weight *mt, wp*
NEWAC NATO Electronic Warfare Advisory Committee *mt*
NEWMAG NATO Electronic Warfare Management Assessment Group *mt*
NEWRADS Nuclear Explosion Warning and Radiological Data System [USA] *wp, mt*
NEWS Naval Electronic Warfare Simulator *fu*
NEWS NetWare Early Warning System [Frye Computer] *pk, kg*
NEWS Network Extensible Window System *pk, kg*
NEWS Novell Electronic Webcasting Service *ct*
NEX National Exchange {Incorporated} *sc4.2.1*
NEX Navy Exchange [USA] *mt*
NEXRAD Next [Generation Weather] Radar *mt, ph, wp, fu*
NEXT Near End Crosstalk *pk, kg*
NF Near Field *fu*
NF No Failure *fu*
NF Noise Factor *fu*
NF Noise Figure *fu, 4, pl, lt, jdr*
NF Noise Frequency *fu*
NFAC National Foreign Assessment Center [DCI] *mt*
NFAC Naval Facilities Engineering Command [Headquarters] *dt*
NFADB Naval Facilities Automated Data Base *mt*
NFARS NORAD Forward Automated Reporting System *mt*
NFC Node Finish Controller *fu*
NFCC Navy Finance Center, Cleveland *dt*
NFCS Nuclear Force Communications Satellite *mt*
NFD No Foreign Dissemination *wp*
NFDC National Flight Data Center *fu*
NFE Network Front End *mt, hc, tb*
NFGF Naval Flying Grading Flight [UK] *mt*
NFH NATO Frigate Helicopter [NATO] *mt*
NFIB National Foreign Intelligence Board *mt*
NFIC National Foreign Intelligence Community *mt*
NFID Next Free ID *fu*
NFIP National Foreign Intelligence Program *mt, uc, wp*
NFL New Foreign Launch *jb, nt, cp*
NFM North Finding Module *hc*
NFMC Not Fully Mission Capable *mt, wp*
NFMCS Not Fully Mission Capable Supply *mt*
NFMRAD Null Filter Mobile Radar *hc*
NFO Naval Flight Officer [USA] *mt*
NFOIO Navy Fleet Operational Intelligence Office *dt*
NFOV Narrow Field Of View *wp, kf*
NFP Navy Flag Plot [Washington, DC] *mt*
NFPA National Fire Protection Association *nt*
NFPPE Nuclear Force Policy, Planning and Execution *mt*
NFR Needline Failure Report *fu*
NFS NASA FAR Supplement *16, oe*
NFS Navy Facilities System *mt*
NFS Network File System *kg, tb, ct, em, cu, hi*
NFSG National Field Support Group *fu*
NFSO Navy Fuel Supply Office *dt*
NFST National Fleet Surgical Team [USA] *mt*
NFSW Nonfunctional Software *fu*
NFT Network File Transfer *hi*
NFT Norsk Forsvarsteknologi A/S [a Norwegian company] *hn50.8.1*
NFTI Naval Firefighters Thermal Imager *hc*
NFWS Naval Fighter Weapons School [Top Gun] *mt*
NFZ No Fly Zone *mt*
NFZ Nuclear Free Zone *wp*
NG National Guard *mt, w9, wp, af72.5.151*
NGB National Guard Bureau [USA] *dt, mt*
NGC National Gas Association *y2k*
NGC New General Catalog *mi*
NGCADMM Next Generation Computer Aided Design *mt*
NGCC National Guard Computer Center *dt*
NGDA Nordihydroguaiaretic Acid *wp*
NGDC National Geophysical Data Center *no, pr*
NGE Noise Generation Equipment *-4*
NGE Not Greater or Equal *pk, kg*
NGF Naval Gun Fire *fu*
NGFS Naval Gun Fire Support [USA DOD] *mt*
NGMH New Generation Military Hospital *mt*
NGNF Next Generation Navy Fighter *hc*
NGO Non Governmental Organization *pbl, ai2*
NGP NaviGation Processor *ai2*
NGP Next Generation Processor *hn43.7.5*
NGPC Node Group Primitive Constant *fu*

NGR National Guard Regulation *mt*
NGS NSSDC Graphics System *ns*
NGSA National Gas Supply Association *y2k*
NGSDC National Geophysical and Solar Terrestrial Data Center [Boulder, CO] *ci.882, no*
NGSO Non Geostationary Orbit *jr*
NGSP National Geodetic Satellite Program *ac2.4-1*
NGST Next Generation Space Telescope *te23, 11, 22*
NGT NASA Ground Terminal [colocated with TDRSS/WSGT] *-2*
NGT Next Generation Trainer [a replacement for the T-37, USA] *ph, wp, hc, mt*
NGTE National Gas Turbine Establishment [established 1946 at Farnborough, UK] *mt*
NH&S Nuclear Hardening and Survivability *mt*
NHA Next Higher Assembly *fu*
NHANG New Hampshire Air National Guard *-12*
NHB NASA Handbook *gp, 16, mb, vv, fu, oe*
NHC National Hurricane Center [Miami, Florida] *pl, wp, kg*
NHC Naval Historical Center *dt*
NHCP National HUMINT Collection Plan *wp*
NHELTR National High Energy Laser Test Range *ag*
NHEP Nuclear Hardness Evaluation Procedures Program *wp*
NHF Naval Historical Foundation *dt*
NHF Nodding Height Finder *fu*
NHOS Nuclear Hardened Optical Sensor *st*
NHPP Non-Homogeneous Poisson Process *fu*
NHR National Handwriting Recognition *pk, kg*
NHRP Next Hop Resolution Protocol *em, ct*
NHRP Non Hierarchical Routing Protocol *em, ct*
NHS New Hampshire Station *hc*
NHZL (signal paths to transmitter input are NHZL-CLK) *bo*
NI Natural Intelligence *mt*
NI Noise Index *fu*
NIA Next Instruction Address *pk, kg*
NIAC National Information and Analysis Center *wp*
NIAC Naval Intelligence Automation Center *mt*
NIAG NATO Industrial Advisory Group *wp*
NIB Noise Investigation Bureau [studied German radar emissions] [Allied] *mt*
NIB Non Interference Basis *fu*
NIC National Intelligence Community *tb*
NIC Naval Intelligence Command *mt, fu, dt*
NIC Nearly Instantaneous Companding *hc, ce*
NIC Negative Impedance Converter *fu*
NIC Network Information Center [DDN] *mt, em, ct, ns, kg, hi, du*
NIC Network Interface Card *kg, em, ct, hi*
NIC Newly Industrializing Country *-14*
NIC North Integration Cell *jdr*
NIC Not in Contract *fu, tb*
NIC Numerically Intensive Computing *pk, kg*
NICB National Industrial Conference Board *fu*
NICIM Negative Ion Chemical Ionization Mass Spectroscopy *wp*
NICKA [codeword, nickname and exercise term system] *mt*
NICMIS Navy Integrated Command / Management Information System *mt*
NICMOS Near Infrared Camera / Multi Object Spectrometer [HST upgrade] *mi*
NICN Navy Item Control Number *fu*
NICOLAS Network Information Center On Line Aid System [NASA] *kg, hi*
NICP National Inventory Control Point *mt*
NICP Network Interface Computer Program *fu*
NICRAD Naval-Industry Cooperative Research and Development *fu*
NICRAD Navy/Industry Cooperative Agreement Document *hc*
NICS NATO Integrated Communications System *mt, fu, wp*
NICSMA NATO Integrated Communications System Management Agency *mt*
NICSOI NICS Operating Instruction *mt*
NICSS Nellis Integrated Communications Switching System *hc*
NICS_COA NICS Control Operating Authority *mt*
NID Network Interface Device *mt*
NID New Interactive Display [NEC] *pk, kg*
NID Next ID *pk, kg*
NID Non Interactive Display *mt*
NIDC Net Interface Digital Computer *fu*
NIDMS NMCS Information for Decision Makers System [OJCS] *mt*
NIDP Network Interface Data Processor *fu*
NIDS National Intelligence Display System *wp*
NIDS NMCS Information [and] Display System [OJCS] *mt*
NIE National Institute of Education *ci.901*
NIE National Intelligence Estimate *mt, wp, af72.5.107*
NIF Navy Industrial Fund *mt*
NIGHT_OWL [program to develop infrared seeker for air-to-surface missiles] *wp*
NIGHT_WATCH [LLTV Mod On EC-135 for night surveillance] *wp*
NIH National Institutes of Health *wp, ci.900, w9*
NIH Nickel Hydrogen (Hydroxide) {?} *kf, jn, hi*

NII National Information Infrastructure *ct, em, kg*
NII Nauchno Issledovatelsky Institut [Scientific Research Institute] *ai2*
NIIN National Item Identification Number *fu*
NIIRS National Imagery Interpretation Rating Scale *wp*
NII_V-VS Nauchno Issledovatelsky Institut V-VS [Scientific Research Institute of the V-VS] [USSR] *mt*
NIL Noninspectable Items List *fu*
NIL Nuclear Induced Lightning *wp*
NIM Network Installation Management [IBM] *pk, kg*
NIM NORAD Intelligence Memorandum *wp*
NIMH Nickel Metal Hydride (Hydroxide) [batteries] *ct, em, kg, hi*
NIMMS NAVAIR Industrial Material Management System [Navy] *mt*
NIMPHE Nuclear Isotope Monopropellant Hydrazine Engine *0*
NIMROD [U. K. maritime reconnaissance aircraft] *wp*
NIMS NASA Interface Monitoring System *2, 4*
NIMS Near Infrared Mapping Spectrometer [on Galileo] *jp, mb, ct, mi*
NIMS Noiseless Integral Magnetic Scanners *hc*
NIMSC Nonconsumable Item Materiel Support Code *wp*
NIMSR Nonconsumable Item Materiel Support Request *wp*
NIO Native Input/Output *pk, kg*
NIOSH National Institute for Occupational Safety and Health *wp*
NIP NATO Interoperability Plan *mt, wp*
NIP Network Interface Processor *fu*
NIPD NATO Interoperability Planning Document *mt*
NIPIR Nuclear Immediate Photo Interpretation Report *mt*
NIPS Naval Intelligence Processing System *mt, fu*
NIPS Network I/Os Per Second *kg, hi*
NIPS NMCS Information Processing System [OJCS] *wp, mt*
NIPSSA Naval Intelligence Processing Systems Support Activity *dt*
NIPWG NTG Integrated Planning Working Group *tb*
NIR Near InfraRed *mb, wp, mi*
NIR Network Information Retrieval *pk, kg*
NIRFOOPS Nonscanning InfraRed Focal plane Options Study *hc*
NIRS Near Infrared Spectrometer *es55.26*
NIS NATO Identification System *mt, ph, wp, aw130.22.61*
NIS NATO Information System *fu*
NIS NATO Infrastructure System *mt*
NIS Naval Investigative Service *mt, hc, aw129.1.17, tb, su*
NIS Network Information System [DCSO] *ct, em, kg, hi, mt*
NIS NMCS Information System [OJCS] *mt*
NISC National Intelligence Space Center *jb*
NISC Naval Intelligence Support Center *mt, dt, st, kf, fu*
NISDN Narrowband ISDN [see ISDN] *nu, hi*
NISE Naval In Service Engineering *ai2*
NISHQ Naval Investigation Service Headquarters *dt*
NISIE NATO Integrated System for Information Exchange *mt*
NISO National Individual Standing Offer *hc*
NISO National Information Standards Organization *pk, kg*
NISO Naval Investigative Service Office *dt*
NISP Networked Information Services Project *pk, kg*
NISPOM National Industrial Security Program Operating Manual *kf*
NIST National Institute of Standards and Technology [formerly NBS] *mt, wp, aw132.10.6, no, mb, mi, kg, em, oe, y2k*
NIST National Intelligence Support Team [USA DOD] *mt*
NIT Network Integrity Test *mt*
NITADS NATO Interoperability of Tactical Automated Data Systems *mt*
NITC National Information Technology Center *pk, kg*
NITE-OP Night Imaging Through Electro Optics *mt*
NITF National Imagery Transmission Format *mt*
NITO/W National Intelligence Tasking Office, Warning and Crisis Management *dt*
NITS National Institute of Standards and Technology [also NIST] *wp*
NIU NASCOM Interface Unit *-4*
NIU Navigation Interface Unit *fu*
NIU Navy Satellite Control Station Interface Unit *uh*
NIU Network Interface Unit *mt, fu, kg*
NIU NSCS Interface Unit *jdr*
NIVR Netherlands Agency for Aerospace Programmes *hg, ge*
NJANG New Jersey Air National Guard *-12*
NJE Network Job Entry [protocol] [IBM] *fu, kg*
NJP Network Job Processing *fu*
NK Nuclear Kill *st*
NKDS Navy Key Distribution System *hc*
NKSRS Non-Kernel Security Related Software *tb*
NKVD Narodnyi Komissariat Vnutrennikh Del [(Russian) People's Commisariat of Internal Affairs] *w9*
NLCC NORTHAG Logistics Command Center [NATO] *mt*
NLDM Network Logical Data Manager *mt*
NLDP National Launch Development Program *mi*
NLE North Leave Earth [pulse] *-9*
NLE Not Less or Equal *pk, kg*
NLG Nose Landing Gear *wp*
NLI Natural Language Interface *mt*
NLL Night Low Level *mt*
NLM NetWare Loadable Module *kg, ct, em, hi*
NLO Nonlinear Optics *wp*

NLOS Non Line Of Sight *fu, mt*
NLP Natural Language Processing *ct, em, kg*
NLP Non Linear Programming *wp*
NLQP Natural Language Query Processor *ns*
NLR Noise Load Radio *fu*
NLRP NMCS Long Range Plan *mt*
NLS Network Logic Schedule *fu*
NLS oNLine Service (System) *em, ct*
NLSFUNC National Language Support Function *pk, kg*
NLSO Naval Legal Service Office *dt*
NLUF National Laser Users Facility *wp*
NLUT Normalization LookUp Table *oe*
NLV National Language Version [IBM] *pk, kg*
NM Nautical Mile *mt, fu, sd, wp, hc, w9, tb, wl, oe*
NM Net(work) Management *mt, fu, hi*
NM Newton Meter *jdr, uh, oe*
NM North Mark *fu*
NMA Naval Mutual Aid Association *dt*
NMA Next Manufacturing Assembly *fu*
NMADT Net Management Algorithm Development Tool *fu*
NMB (Ramtron licensed its technology to NMB Semiconductor Co., Tokyo) *is26.1.48*
NMC National Meteorological (Meteorology) Center *mt, ci.881, hy, oe*
NMC Naval Materiel Command *fu, tc, dt*
NMC Naval Medical Command *mt*
NMC Naval Missile Center *fu*
NMC Network Monitoring Center *mt*
NMC Not Mission Capable *mt, wp*
NMCC National Military Command Center [JCS] *mt, fu, wp, sr9-82, jb, cp, dt, tb, kf*
NMCC Network Management and Control Center *fu*
NMCC/MC NMCC / Message Center *mt*
NMCCS Navy Manpower Configuration Control System *mt*
NMCEC Navy Marine Corps Exhibit Center *dt*
NMCM Not Mission Capable, Maintenance [USA DOD] *mt*
NMCNCR Naval Medical Command National Capital Region *dt*
NMCS National Military Command *fu*
NMCS National Military Command Segment (System) [JCS, USA DOD] *mt, wp, hs, jb*
NMCS Network Monitoring and Control System *fu*
NMCS Not Mission Capable, Supply [USA DOD] *mt*
NMCSA Naval Material Command Support Activity *dt*
NMCSSC National Military Command System Support Center *mt, dt*
NMD NASA Master Directory *hy*
NMD National Missile Defense *kf*
NMD/TMD National Missile Defense/Theater Missile Defense *kf*
NMDAS Navy Manpower Data Accounting System *mt*
NMDF NATO Main Distribution Frame *mt*
NMDN Nonconformance Material Disposition Notice *fu*
NMDS Network and Mission Operations Support [NASA] {?} *mt*
NMDSC Naval Medical Data Services Center *dt*
NME Navy Materiel Establishment *mt*
NMEC National Meteorological and Environmental Center *fu*
NMEP NMCS Management Engineering Plan *mt*
NMET Naval Mobile Environmental Team [USA DOD] *mt*
NMFS National Marine Fisheries Service *ci.882*
NMI NASA Management Instruction *ci.116, 2, oe*
NMI Nautical Mile *mt, tb, kf, vv, uh*
NMI Non Maskable Interrupt *ct, kg, fu, em, hi*
NMIBT New Material Introductory Briefing Team *fu*
NMIC National Military Intelligence Center [JCS] *kf, mt*
NMIC Navy Management Information System *mt*
NMIC-SS National Military Intelligence Center Support System *mt*
NMIMT New Mexico Institute of Mining and Technology *st*
NMIST National Military Intelligence Support Team *mt*
NMITT New Material Introductory Training Team *fu*
NMJIC National Military Joint Intelligence Center *mt*
NMM NetWare Management Map [NetWare] *kg, hi*
NMMIS Naval Materiel Management Information System *mt*
NMMPS National Military Message Processor System *mt*
NMMW Near Millimeter Wave *fu, hc*
NMO Net Management Operator *fu*
NMOS Negative channel Metal Oxide Semiconductor *kg, em, fu*
NMP National Maintenance Point *mt, fu*
NMP Net Management Processor *fu*
NMP Network Management Protocol [AT&T] *kg, hi*
NMP NMCS Master Plan *mt*
NMPC Naval Military Personnel Command [USA] *dt, mt*
NMPS Naval Military Pay System *dt*
NMPS Normal Mode Phase Sequencer *jdr, uh*
NMR National Military Representative [to SHAPE] *mt*
NMR New Mobile Radar *fu*
NMR New Mobile RAPCON *hc*
NMR No Maintenance Required *fu*
NMR Nonconforming Material Report *pl*
NMRA Not Mission Ready and Available *mt*

NMRI Naval Medical Research Institute *dt*
NMRO Normal Memory Readout *jdr, uh*
NMRS Navy Manpower Requirement System *mt*
NMS Network Management System [Novell] *kg, hi*
NMS Newton Meter Second *oe*
NMS Noise Monitoring System *sn2.10, ma40*
NMSS NASCOM Manual Scheduling System *-2*
NMSST Navy Manpower Shore Survey Team *dt*
NMT Nordic Mobile Telephone *hi*
NMU {North Michigan University} *aw129.10.76*
NMWG Network Management Working Group *fu*
NMX Not Multiplexed *gs1-5.93*
NN Network Node *pk, kg*
NN Non Nuclear *wp*
NNA Neutral Non-Allied *wp*
NNAG NATO Naval Armaments Group *mt*
NNBIS National Narcotics Border Interdiction System *wp*
NNC Not Normally Calibrated *oe*
NNCS NICS Network Control System *mt*
NNDC National Naval Dental Center *dt*
NNEMP Non Nuclear Electromagnetic Pulse *fu*
NNI Network Node Interface *em, ct*
NNI Network to Network Interface *em, ct, kg*
NNK Non Nuclear Kill *hc, wp, tb, st*
NNOR Non Nuclear Ordnance Requirements Model [US Navy] *mt*
NNPRB NOAA-NASA Program Review Board *oe*
NNS Near Net Shape *17, sb*
NNSC NSF Network Service Center *hi*
NNTP Network News Transfer Protocol [Internet] *kg, ct, em, hi*
NNTRP National Nuclear Test Readiness Program *wp*
NNWDB National Nuclear Weapons Data Base *mt*
NNWSI Nevada Nuclear Waste Storage Investigation *st*
NO Nitric Oxide *oe*
NO Normally Open *gp, wp, fu, hi, oe*
NOA New Obligational (Obligation) Authority *wp, oe*
NOA [scientific test aerodrome, replaced by the NII_V-VS in 1926] [USSR] *mt*
NOAA National Oceanic (Oceanographic) and Atmospheric Administration [formerly ESSA] [USA] *aa26.12.26, sc3.1.4, 16, aw129.10.76, pl, hc, wp, ns no pr ep hy wl jdr, hg, mb, ge, mi, fu, oe, mt, ai2*
NOAD Navy Officers, Accounts Office, [U. S]. *dt*
NOAH Norwegian Adapted HAWK *pm, fu*
NOAO National Optical Astronomy Observatories *mi*
NOB Naval Order of Battle *mt*
NOB Nuclear Order of Battle *wp*
NOBS Naval Operations Bases System *mt*
NOC Naval Oceanographic Office [U. S.] *dt*
NOC Naval Operations Center [MSN] *fu, mt*
NOC Network Operations Center [WIN] *mt, ct, kg, em, hi*
NOCC Network Operations Control Center *ac3.2-19, 4, re, oe*
NOCGI Night Only Computer Generated Image *wp*
NOCMIS NOC Management Information System [WIN] *mt*
NOD Network Operations Division *-4*
NOD Numerical Optical Disk *hi*
NOD-LR Night Observation Device-Long Range [also NODLR] *hc*
NODC National Oceanographic (Oceanography) Data Center [NOAA] *ag, hy, hg, ge*
NODDS NOAA Oceanographic Data Distribution System *mt*
NODECA Norwegian Defense Communications Agency *fu*
NODIS NSSDC Online Data and Information Services *ns*
NODLR Night Observation Device Long Range [also NOD-LR] *wp*
NODOZ Nuclear Offense/Defense Operational Zone *st*
NODS NASA Ocean Data System [JPL] *hg, hy, ge*
NOE Nap(e) Of the Earth *hc, wp, mt*
NOESS National Operational Environmental Satellite System *oe*
NOFORN Not (No) Foreign Nationals [a release restriction] *mt, wp, tb, pbl, kf, su*
NOG NSA Pacific Operations Group *-10*
NOGS Night Observation Gunship [System] *mt, hc*
NOI Not Otherwise Identified *fu*
NOIC Naval Oceanic Intelligence Center *kf*
NOIC Naval Operational Information Center [US Navy] *mt*
NOISE National Organization to Insure a Sound-controlled Environment *aw130.6.60*
NOISH National Institute of Occupational Safety and Health *fu*
NOIWAD National Operations and Intelligence Watch Officer's Advisory *mt*
NOIWON National Operations Intelligence Watch Officers Network [OJCS] *mt*
NOL Naval Ordnance Laboratory [USA] *fu, hc, wp, mt*
NOLDS Naval Ordnance Laboratory Data Service *dt*
NOLS Nuclear Ordnance Logistics Systems *mt*
NOMAC Noise Modulation and Correlation *mt*
NOMAD Navy Oceanographic and Meteorological Automatic Device [USA] *mt*
NOMIS Naval Ordnance Management Information System *mt*

NOMS Nuclear Operations Monitoring Systems *mt*
NOMSS National Operational Meteorological Satellite System *ci.880, oe*
NOMSS Naval Oceanographic and Meteorological Support System *mt, ag*
NONA [Hughes Aircraft Company customer for NESS] *hc*
NOP Naval Operations [Office of the Chief of Staff] *dt*
NOP No Operation *pk, kg*
NOP Nuclear Operating Plan [NATO] *wp, mt*
NOPAC Network OPAC *pk, kg*
NOPF Naval Oceanographic Processing Facility *fu*
NOPLAN No Operational Plan published [JOPS] *mt*
NOPS Nimbus Observation Processing System *ac2.8-12*
NOR Network Operations Representative *-2*
NOR Not Operationally Ready *fu*
NOR Notice Of Revision *mt, fu, tb, jdr*
NORAD North American Air (Aerospace) Defense [Command] *sc3.1.5, 5, 12, 14, sb, w9, aa27.4.12, af72.5.67, ps201, jb, wp, ns, tb, jdr, mb, kf, fu, mi, ct, hi, oe*
NORAD North Atlantic Aerospace Defense Command [created in 1957, USA, Canada] "Norad" *mt*
NORAD_OPSTAR North American Air Defense Command Operational Status Reporting System *ro*
NORCCIS Norwegian Command, Control, and Information System *fu, hc*
NORF National Offense Reserve Fleet *ph*
NORM Not Operational Ready, Maintenance *fu, mt*
NORMEDS Northern Meteorological Data System *no*
NORPAX North Pacific Experiment *pl*
NORS Not Operational Ready, Supply *fu, mt*
NORTHAG Northern Army Group [Central Europe, NATO] *fu, mt*
NOR_ADMIN NORAD Administrative Telephone Network *mt*
NOR_ALERT NORAD Voice Alert Network *mt*
NOR_COMNET NORAD Teletypewriter Network *mt*
NOR_TAC NORAD HQ ADC Tactical Telephone Network *mt*
NOR_WARN NORAD Warning Network *mt*
NOS National Ocean Survey [USA] *mt, ci.882*
NOS Network Operating System *ct, kg, em, is26.1.89, tb, hi*
NOS Night Observation Surveillance [aircraft] *mt, wp*
NOS Nominal Operations Sequence [test] *oe*
NOS Not Otherwise Specified *fu, w9*
NOS NPS Operating System *fu*
NOSC NATO Operations Support Cell / Center *mt*
NOSC Naval Ocean Systems Center [formerly Naval Electronics Laboratory] *mt, fu, 12, md2.1.1, hc, no, tb*
NOSIC Naval (Navy) Ocean Surveillance Information Center *mt, dt*
NOSP Network Operations Support Plan *2, 4, oe*
NOSS NAS Operational Support System *fu*
NOSS National Oceanic (Oceanographic) Satellite System *hc, ac1.6-23, 14, hy*
NOSS National Orbital Space Station *0*
NOSS Network Operations Support Specialist *-4*
NOTAM Notice To Airmen [Military] [USA DOD] *mt, fu, ag, af72.5.169, aw130.11.72*
NOTAMS Notice to Airmen and Sailors *mt, pf1.27*
NOTAR No Tail Rotor [helicopter without tail rotor, in which contra-torque is achieved by the air flow through a slit in the tail boom] *mt*
NOTCC Network Operational / Technical Coordination Center [DCA] *mt*
NOTEMPS Nontemporary Storage of household goods System *mt*
NOTS Naval Ordnance Test Station [China Lake, California] *ci.183, fu*
NOTS NMCS Operations Team Support Subsystem *mt*
NOV Night Only Visual *wp*
NOVRAM Nonvolatile Random Access Memory *fu*
NOX Oxides of Nitrogen *wp*
NP New [work item] Proposal *tt*
NP North Panel *oe*
NP Not Planned *jdr*
NP Not Provided *wp*
NPA Network Printer Alliance *pk, kg*
NPA Numbering Plan Area (North American Numbering Plan) *ct, hi*
NPA [an airline, subsidiary or West Air, Pasco, Washington] *aw129.21.31*
NPB Neutral (Neutron) Particle Beam *mt, dh, wp, 14, 17, hc, tb, st, tb*
NPC NASA Procurement Circular *fu*
NPC National Programming Center (for Germany) *fu*
NPC NATO Programming Center [Belgium] *fu, hc*
NPC Naval Photographic Center *fu*
NPC Network Parameter Control *ct, em*
NPD Navy Procurement Directive *fu*
NPDNET Nordic Public Data Network *hi*
NPDS National Military Command System Processing and Display System *mt*
NPDS NMCC Processing and Display System [JCS] *mt*
NPDU Nested Protocol Data Unit *hi*
NPE Naval Preliminary Examination (Evaluation) [testing of equipment by the US Navy before procurement] *hc, mt*
NPE Nuclear Planning and Execution *mt*
NPES Nuclear Planning and Execution System *mt, wp*

NPETOFF Naval Petroleum Office *dt*
NPF NAVSTAR Processing Facility *ai2*
NPFC Naval Publications and Forms Center *fu*
NPG Needline Participation Group *fu*
NPG Net Participation Group *fu*
NPG Non unit Personnel Generator *mt*
NPG Nuclear Planning Group [NATO] *mt, nd74.445.39*
NPGA National Propane Gas Association *y2k*
NPGS Naval Postgraduate School *mt, fu*
NPIAS National Plan of Integrated Airport Systems *fu*
NPIC National Photographic Interpretation Center [USA] *mt, 10, wp, jb, tb*
NPIN Negative-Positive-Intrinsic-Negative *fu*
NPIP Navigation Port Interface Processing *fu*
NPK Nitrogen, Phosphorous, Potassium *wp*
NPL National Physics Laboratory *fu*
NPL Non Procedural Language *pk, kg*
NPLG Navy Program Language Group *dt*
NPLO NATO Production Logistics Organization *hc, tb, su*
NPM Network Performance Monitor *mt*
NPM Next Page Memory *fu*
NPMA NATO Project Management Agency *mt*
NPN N-type P-type N-type (Negative-Positive-Negative) [detail of design of transistor] *0, mj31.5.60, fu, wp*
NPN Non Protein Nitrogen *wp, w9*
NPN Notes Public Network *ct*
NPO Nauchno Proizvodstvennoe Ob'edinenie [Scientific Production Association] *ai2*
NPO Negative, Positive, Zero *fu*
NPO Nuclear Propulsion Office *ci.108, hc*
NPO/LA [Hughes Aircraft Company customer for project GENIE] *hc*
NPOC National Points Of Contact *ac3.7-12*
NPOC Naval Polar Oceanography Center *mt*
NPOESS National Polar-orbiting Operational Environmental Satellite System *jdr, pbl, ls4, 5, 16*
NPOLA Naval Purchasing Office, Los Angeles *fu*
NPOP NASA Polar Orbiting Platform *hy*
NPPS Navy Publication and Printing Service *dt, mt*
NPPSA Naval Personnel Program Support Activity *dt*
NPPSO Navy Publications and Printing Services Office [Washington] *dt, fu*
NPR Network Process Engineering *ct*
NPR New Purchase Requisition *fu*
NPR Noise to Power Ratio *fu, 7, ce, pbl, oe*
NPRA National Petroleum and Refiners Association *y2k*
NPRDC Naval (Naval) Personnel Research and Development Center *mt, tc, fu, dt*
NPRIS Nuclear Planning and Resource Information System *mt*
NPRM Notice of Proposed Rule Making *aw129.21.76*
NPRO Naval Representative Offices *dt*
NPS Naval Postgraduate School [Monterey, California] *mt, kb, fu*
NPS Navy Primary Standards *fu*
NPS Night Pilotage System *hc*
NPS Novell Productivity Specialist *pk, kg*
NPS NTDS Peripheral Simulator *fu*
NPSS NASA Packet Switched System *ns, hy*
NPT Navy Pointer and Tracker *hc, 17, fu*
NPT Non Proliferation Treaty *mt*
NPTN National Public Telecomputing Network *pk, kg*
NPU Natural Processing Unit *pk, kg*
NPV Name Plate Value *fu*
NPX Numeric Processor Extension *pk, kg*
NQAA National Quality Assurance Agency *fu*
NQAR National Quality Assurance Representative *fu*
NQCC Network Quality Control Center *mt*
NQL Network Query Language *mt*
NQL NORD Query Language *fu*
NQR Nuclear Quadrupole Resonance *wp*
NQS Network Queuing System [Cray] *pk, kg*
NR NATO Restricted *hc*
NR Negative Resistance *fu*
NR Noise Ratio *fu*
NR Non Reactive *fu*
NR Non Recurring [costs] *sd, jn*
NR Non-conformance Report *pl*
NR Not Required *sd*
NRA NASA Research Announcement *19, hy, cu*
NRA Network Reconfiguration Algorithm *fu*
NRaD Naval Command and Control and Ocean Surveillance Center, Research {?} *uh*
NRALCC Northern Region Airlift Control Center *mt*
NRALD Northern Region Airlift Division *mt*
NRAO National Radio Astronomy Observatory *mb, ns, mi*
NRC National Regulatory Commission *y2k*
NRC National Research Council [Canada] *mt, kg, wp, ci.555, w9, hy, tb*
NRC Navy Recruiting Command *dt*
NRC Nichols Research Corporation *tb, st*

NRC Non unit Related Cargo *mt*
NRC Nonrecurring Recoupment Charge *wp*
NRC NORAD Readiness Center *mt, sr9-82*
NRC Nuclear Regulatory Commission *mt, su, hc, wp, pl, w9, st*
NRC Number of Range Cells *fu*
NRCC Naval Regional Contracting Center *mt, dt, hc*
NRCC NORAD Region Control Center *mt*
NRCO (NRCO LB) [Hughes Aircraft Company customer] *hc*
NRCP Non Reciprocal Circular Polarizer *fu*
NRCS National Radio Communications System *fu*
NRCS Normalized Radar Cross Section *ac9.3-1*
NRD Noise Recognition Differential *fu*
NRDC Navy Relief Society, DC Auxiliary *dt*
NRDC {Nuclear Rocket Development Center} *ci.109*
NRE Non Recurring Engineering *mt, pbl, jdr, kf, ah, fu, jn*
NRE Non Recurring Expense *-18*
NREB Navy Reserve Evaluation *dt*
NREC Navy Recruiting Exhibit Center *dt*
NRECA National Rural Electric Cooperative Association *y2k*
NREN National Research and Education Network *pk, kg*
NRF Naval Reserve Force *mt, 12*
NRF Non Relevant Failure *fu*
NRFC Navy Regional Finance Center *dt*
NRFI Not Ready For Issue *fu*
NRI Net Radio Interface *mt*
NRIC Network Reliability and Interoperability Council *y2k*
NRIFSD Non-Recoverable In-Flight Shut-Down *wp*
NRKM Non Radar Keyboard Multiplexer *fu*
NRL Naval Research Laboratory [Washington, D. C.] *mt, fu, uh, ci.6, 12, 13, 17, sb, mj31.5.28, dt, wp, no, tb, st, kf, oe*
NRL NUWEP Reconnaissance List *mt*
NRL/USRD Naval Research Laboratory/Under-water Sound Reference Division *fu*
NRLA Network Repair Level Analysis *fu, hc*
NRM Network Resource Manager *fu*
NRMC Naval Regional Medical Clinic *dt*
NRN Novell Remote Network *ct*
NRO National Reconnaissance Office [USA DOD] *10, mb, mi, mt, ai2*
NROFF New Runoff [a document processing program for UNIX systems] *fu*
NROSS Naval (Navy) Remote Ocean Sensing System [USA] *mt, hc, sc3.6.4, af72.5.149, tb, hg, ge*
NROTC Naval Reserve Officer Training Corps [USA] *mt*
NRP Non unit Related Personnel *mt*
NRP North Reference Pulse *fu*
NRPO Naval Regional Procurement Office *fu, hc*
NRRC Naval Reserve Readiness Command *dt*
NRRF Naval Radio Receiving Facility *mt*
NRS Naval Radio Station *mt*
NRS Naval Relief Society *dt*
NRS Navy Reporting Structure *mt*
NRSA National Remote Sensing Agency [India] *ac3.2-1*
NRT Near Real Time *mt, sd, wp, pbl*
NRT Network Readiness Test *-4*
NRT Non Real Time *0, fu, ct, em*
NRTC National Rural Telecommunications Cooperative *rih*
NRTC Naval Reserve Training Center *dt*
NRTC Northrop Research and Technology Center *17, sb*
NRTEM Near Real Time Exploitation Module *wp*
NRTF Navy Radio Transmitting Facility *hc*
NRTI Naval Revolutionary Technology Initiative *hc*
NRTM Near Real Time Module *wp*
NRTS Near Real Time Reconnaissance *wp*
NRTS Non Removable Temperature Sensors *he, jn*
NRTS Not Repairable (at) This Station *mt, wp*
NRU Noise Reduction Unit *fu*
NRV Nuclear Reentry Vehicle *kf*
NRX NERVA Reactor-Experimental *ci.227*
NRZ Non Return to Zero *mt, gg52, 0, 2, 4, lt, vv, fu, kg, oe*
NRZ-L Non Return to Zero - Level *he, 4, jdr, kf, vv, oe*
NRZ-L Non Return to Zero - Logic *uh*
NRZ-M Non Return to Zero - Mark *sm3-12, 4, vv*
NRZ-S Non Return to Zero - Space *2, 4, vv, he, oe*
NRZC Non Return to Zero Change *fu*
NRZI Non Return to Zero Inverted *pk, kg*
NRZM Non Return To Zero-Mark *fu*
NS National Semiconductor [as in NS16550 UART] *pk, kg*
NS NATO Secret *hc*
NS Network Supervisor *pk, kg*
NS Neutron Spectrometer *re*
NS Neutron Star *mi*
NS Non Selectively *fu*
NS Non Stop *pk, kg*
NS Nuclear Ship *w9*
NS Nutation Synchronous (mode) *ggIII.2.371*
NS&MP Navy Support and Mobilization Plan *mt*

NSA National Security Agency [USA] *mt, fu, mi, ct, su, ro, uh, hc, wp, 12, 16, 17, ph, w9, jb, tb, jdr, mb, kf, kg, jn, oe*
NSA New Shipborne (Shipboard) Aircraft *hc, pf1.9*
NSA/CSS National Security Agency / Central Security Service *mt*
NSAC National Space Activities Council *ac3.3-3*
NSAP Network Services Access Point *kg, nu, fu, em*
NSAPAC National Security Agency, Pacific *-10*
NSAPI Netscape Server Application Programming Interface *em, kg*
NSARC Navy System Acquisition Review Council *mt, fu, hc*
NSASAB National Security Agency Scientific Advisory Board *-10*
NSB Narrow Spot Beam (antenna) *jdr*
NSC National Security Council [USA DOD] *mt, fu, wp, ci.1117, 13, w9, af72.5.31, hd, dt jdr*
NSC NATO Supply Center *mt*
NSC Naval (Navy) Supply Center *fu, mt*
NSC Network Support Center [DODIIS] *mt*
NSC Network Support Committee *-2*
NSC Norwegian Space Center *crl*
NSCAT NASA Scatterometer [NASA, NROSS] *ge, hg, hc, hy*
NSCAT Navy Scatterometer *hc*
NSCC N-Site Control Center *st*
NSCCA Nuclear Safety Cross Check Analysis *mt*
NSCI NASA System Control Interface *-4*
NSCI NASCOM System Control Interface *-2*
NSCID National Security Council Intelligence Directive *-10*
NSCM NATO Supply Codes for Manufacturers *fu*
NSCN Navy Satellite Control Network *uh*
NSCS Navy Satellite Control Station *jdr, uh*
NSCSES NAVSEA Combat System Engineering Station *fu*
NSD National Security Division *wp*
NSD Navy Supply Depot *fu*
NSD NMCS Support Division *mt*
NSDA Non Self Deployment Aircraft *mt*
NSDC National Space Development Center [Japan] *ac3.3-3*
NSDD National Security Decision Directive *hc, wp, ph, tb*
NSDD National Security Defense Directive *mt*
NSDF Nondedicated Software Development Facility *fu*
NSDM National Security Decision Memorandum *mt*
NSDS Navy Source Data System *mt*
NSDSA Naval Sea Data Support Activity *fu*
NSDSSO NASA Science Data Systems Standards Office *ns*
NSE Not Separately Estimated *kf*
NSEA Naval Sea Systems Command Headquarters *dt*
NSEC Naval Ship Engineering Center *tc*
NSEC Nonspecific Emitter Correlation *fu*
NSEN Network Simulations Engineer *-2*
NSEN NMD System Engineering Notebook *kf*
NSEP National Security and Emergency Preparedness *wp, mt*
NSERC Natural Sciences and Engineering Research Council [Canada] *aa26.12.95, kg, hi*
NSES National Security Electronic Surveillance *-10*
NSF National Science Foundation *mt, fu, kg, mi, su, wp, ac1.7-14, 13, hc, w9, ns, no, hy, tb, mb, hi, y2k*
NSF National Support Facility *hc*
NSFNET National Science Foundation Network *kg, ct, hi*
NSFO Navy Standard Fuel Oil [USA] *mt*
NSFS Naval Surface Fire Support [USA DOD] *mt*
NSG Naval Security Group [USA] *10, mt*
NSGA Naval Security Group Activity *mt*
NSI NASA Science Internet *kg, hy, hi*
NSI NASA Standard Initiator *gp, lt, he, uh, jn, oe*
NSI National Space Institute *ci.1246*
NSI Not Seriously Injured [USA DOD] *mt*
NSIA National Security Industrial Association *hn17.2, nd74.445.27, fu*
NSIC Nuclear Safety Information Center [NRC] *mt*
NSIDC National Snow and Ice Data Center [University of Colorado, USA] *ge, hg*
NSIWG NATO SATCOM Interoperability Working Group *mt*
NSL Network Software Laboratory *mt*
NSLOOKUP Name Server Lookup [Unix] *pk, kg*
NSM Narrow Band Sensor Monitor *wp*
NSM Netscape Server Manager [Netscape] *pk, kg*
NSM Network Support Manager *2, 4, oe*
NSM Network Systems Management *ct, em*
NSM NMCS Simulation Model *mt*
NSMTFS NATO Standard Message Text Formatting System *mt*
NSN NASA Science Network *ns, hy*
NSN National Stock Number *mt, wp, tb, fu, oe*
NSN {a type of night light marking} *af72.5.43*
NSNF Non Strategic Nuclear Forces *mt, ph, wp*
NSO National Solar Observatory (Observatories) *no, pr, mb, mi*
NSO Navy Staff Offices *dt*
NSO Non SIOP Options *wp*
NSO Nuclear Safety Office Non-SIOP Option *mt*
NSOC National SIGINT Operations Center *fu*
NSOC National Signals Intelligence Operations Center *mt*

NSOF Navy Status Of Forces File *mt*
NSOR NATO Statement of Operational Requirement *mt*
NSOW Naval Statement Of Work *dm*
NSP NASA Support Plan *2, 4, oe*
NSP National Service Provider *ct, em*
NSP Native Signal Processing [Intel] *kg, hi*
NSP Navy Special Projects *dt*
NSP Network Service Provider *pk, kg*
NSP Network Services Protocol *do426*
NSP Nonstandard Part *fu*
NSP Not Separately Priced *fu, wp, tb*
NSPAR Non Standard Part Approval Request *gp, 15, ie5.11.8, jdr, fu, oe*
NSPS Nonsynchronous Pulse Suppression *fu*
NSPSC Naval Supply Systems Command *tc*
NSPT NASA / SOCC Processed Telemetry *oe*
NSR Needline Service Request *fu*
NSRB National Security Resources Board *dt*
NSRDC Naval Ship Research and Development Center [David W. Taylor] [USA] *mt, tc, hc, dt, fu*
NSRL National SIGINT Requirements List *mt, wp*
NSRL National Strategic Reconnaissance List *mt*
NSRR NATO Southern Region Radar *fu*
NSS National Strategic Stockpile *mt*
NSS National Supply System *wp*
NSS Navy Secondary Standards *fu*
NSS Network System Simulator *tb*
NSS Non Self Sustaining *mt*
NSS Non Subscriber Site *go3.1-1*
NSSC NASA Standard Spacecraft Computer *hc, kg*
NSSC Naval Sea Systems Command *tc*
NSSC NORAD Space Surveillance Center *mt, wp*
NSSD National Security Study Directive *tb*
NSSDC National Space Science Data Center [at Goddard SFC, NASA] *cu, ge, mi, pl, ns, no, hy, tb, wl, mb, hg*
NSSF NAS Simulation Support Facility *fu*
NSSFC National Severe Storms Forecast Center [Kansas City, MO] *pl, oe*
NSSK North South Station Keeping *jdr, oe, ls4, 5, 28*
NSSL National Severe Storms Laboratory *ag*
NSSMS NATO Sea Sparrow Missile System [USA] *fu, mt*
NSSPO Navy Strategic Systems Projects Office *tc*
NSSPS New Space Signals Processing Stations *hc*
NSSS Naval Space Surveillance System [also NAVSPASUR] *0*
NST (Near-earth NST) *gg164*
NST [Nuclear and Peace Talks] *wp*
NSTAC National Security Telecommunications Advisory Committee *tb, mt, y2k*
NSTC National Science and Technology Council *pk, kg*
NSTDB National Strategic Target Data Base *mt*
NSTI NASA Simulation Traffic Interface *-4*
NSTIS Navy Standard Technical Information Systems *mt*
NSTL National Software Testing Laboratories, Inc. [US] *kg, hi*
NSTL National Space Technology Laboratories [NASA] *mt, ci.70, hy, wl*
NSTL National Strategic Targets List *mt*
NSTP National Security Telecommunication Policy *mt*
NSTS National Space Transportation System [the Shuttle] *16, tb, wl*
NSU Network Service / Interface Unit *mt*
NSUP Naval Supply [Systems Command Headquarters] *dt*
NSVN NATO Secure Voice Network [NICS] *wp, mt*
NSVS NATO Secure Voice System *mt*
NSW Naval Special Warfare [USA DOD] *mt*
NSWC Naval Surface Warfare (Weapons) Center [USA] *mt, fu, 12, tc, sd, hc, jb, dt, tb, st*
NSWCDD Naval Surface Warfare Center Dahlgren Division [US Navy] *mt*
NSWG Naval Special Warfare Group [USA DOD] *mt*
NSWP Non Soviet Warsaw Pact *mt, wp*
NSWSES Naval Ship Weapon Systems Engineering Station *fu*
NSWTE Naval Special Warfare Task Element [USA DOD] *mt*
NSWTG Naval Special Warfare Task Group [USA DOD] *mt*
NSWTG/TU Naval Special Warfare Task Group / Unit [USA DOD] *mt*
NSWTU Naval Special Warfare Task Unit [USA DOD] *mt*
NSWU Naval Special Warfare Unit [USA DOD] *mt*
NT Network Termination (Terminator) *nu, kg, hi*
NT Noise Temperature *pl*
NT Normal Threat *mt*
NTA National Telecommunications Agency *hi*
NTA Nitrilo Triacetic Acid *wp*
NTAP Navy Occupational Task Analysis Program *fu*
NTAS New Technology Advanced Server [Microsoft] *kg, hi*
NTAS NORAD Tactical AUTOVON (Automatic Voice Network) System *jb, mt*
NTAS Norwegian Tracking Adjunct System *hc*
NTB National Test Bed [SDI] [ESD] *mt, 1, 5, 16, wp, 17, sb, hc, ts.5, tb, st*
NTB Network Test Bed *mt*
NTBIC National Test Bed Integration Contract *tb*

NTBJPO National Test Bed Joint Program Office *tb, kf*
NTBSIM National Test Bed Simulation *tb*
NTC National Territorial Command *mt*
NTC National Training Center *wp*
NTC Naval Telecommunications Command *mt*
NTC Negative Temperature Coefficient *fu*
NTC Network Terminal Concentrator *mt*
NTCC Naval Telecommunications Center *mt*
NTCF NATO Technical Control Facility *mt*
NTCOC Naval Telecommunications Command Operations Center *mt, sc4.1.center*
NTCS Naval Telecommunications and Computer System *mt*
NTCSA Navy Tactical Command System Afloat *mt*
NTD Naval Tactical Data *fu*
NTDA Navy Tactical Doctrine Production Activity *dt*
NTDB Naval Tactical Data System *tb*
NTDC Naval Training Device Center *fu*
NTDN Navy Terminal Data Node *uh*
NTDO Naval Technical Data Office [USA] *mt*
NTDS Naval Tactical Data System [a data processing system fitted to the E-2C Hawkeye, USA] *fu, mt*
NTDS Naval Tactical Display System *hc, pf1.4*
NTDT Non-Time-Dependent Target *st*
NTE Net Time Entry *fu*
NTE Not To Exceed *mt, fu, nt, wp, tb, jdr, kf, oe*
NTEC Naval Training Equipment Center *tc, hc*
NTEP NTDS Executive Program *fu*
NTF National Test Facility [ESD] *mt, wp, tb, jdr, kf, ts4-26, st*
NTF Needham Test Facility *fu*
NTF New Tactical Fighter *hc*
NTF No Trouble Found *kg, hi*
NTF/SPC National Test Facility/Special Program Center *kf*
NTFMRS Navy Total Force Manpower Requirements System *mt*
NTFS New Technology File System *kg, ct, em, hi*
NTI Near Term Improvements *mt*
NTI Near Term Initiative *wp*
NTI New Technology, Incorporated *st*
NTIA National Telecommunications and Information Administration *ci.879, 2, ce, jdr, fu, kg*
NTIA National Telecommunications Information Agency [DOC] *mt*
NTIAC Nondestructive Testing Information Analysis Center *hc*
NTIPP Navy Technical Information Presentation Program *fu*
NTIPS Navy Technical Information Presentation System *mt, fu, hc*
NTIPS Navy Technical Information Processing System *hc*
NTIS National Technical Information Service *kg, fu, wp, ci.895, rg, tb*
NTISA Naval (Navy) Tactical Interoperability Support Activity *fu, mt*
NTISSC National Telecommunications and Information Systems Security Committee *tb, mt*
NTISSI National Telecommunications and Information System Security Information *kf*
NTLM NT LAN Manager *em*
NTM National Technical Means [formerly National Assets] [of surveillance] *14, 17, wp, tb*
NTM Network Terminal Manager [Honeywell] *mt*
NTM Network Test Manager *-2*
NTM Network Transaction Manager *fu*
NTMP NMCS Technical Management Plan *mt*
NTO Nitrogen Tetroxide [oxidizer, N2O4] *uh, 16, oe*
NTP Navy Telecommunications Publications *mt*
NTP Near Term Penetrator [USA] *mt*
NTP Network Terminal Protocol *hi*
NTP Network Time Protocol *ct, em*
NTP Normal Temperature and Pressure *w9, wp*
NTPD Normal Temperature and Pressure Dry *wp*
NTPF Near Term Prepositioning Force *mt*
NTPI Nuclear Weapons Test Proficiency Inspection *fu*
NTPS Naval Test Pilots School [Patuxent River NAWC] [USA] *mt*
NTPS Near Term Prepositioned Ships *mt*
NTR Network Time Reference *fu*
NTR Nine Thousand Remote *mt*
NTR Non Tactical Radio *mt*
NTR Nuclear Thermal Rocket (Rocketry) *mi*
NTRAS NT Remote Access Services [Microsoft] *pk, kg*
NTRID Netted Telephone Radio Interface *mt*
NTRS Navy Tactical Reconnaissance System *hc*
NTS National Technical Services *fu*
NTS National Technical Systems [Fullerton, California] *aw130.22.87*
NTS National Testing Services *fu*
NTS National Traffic System *mt*
NTS Naval Telecommunications System *mt*
NTS Navigation Technology Satellite *tk*
NTS Net Time Station *fu*
NTS Nevada Test Site *tb*
NTS Not To Scale *fu, wp*
NTSA Naval Tactical Support Activity *fu*
NTSA NetWare Telephony Services Architecture [Novell] *kg, hi*

NTSAC Net Time Station Authentication Code *fu*
NTSB National Transportation Safety Board *aw126.20.31, aa27.4.8, fu*
NTSC Navy Training Systems Center *mt*
NTSMP Naval Telecommunications System Master Plan *mt*
NTSS National Time Sensitive System *fu*
NTT New Technology Telescope *mi*
NTT Nippon Telephone and Telegraph [Company] *sc4.10.3, is26.1.43*
NTT Numbered Test Trunk *ct*
NTTF Network Test and Training Facility [GSFC] *ci.940, 2, 4*
NTTPC Nippon Telephone and Telegraph Public Corporation *fu*
NTTR Non Tactical Telecommunications Requirements *mt*
NTTS NATO Terrestrial Transmission System *mt*
NTU National Technological University *aw130.6.7*
NTU Naval Toxicology Unit *dt*
NTU Network Terminating Unit *hi*
NTU New Threat Upgrade *fu, wp*
NTV Next Traffic Variable *fu*
NTWR New Threat Warning Receiver *uc*
NTWS New Threat Warning System *mt, wp, hc*
NTWS Non Track While Scan *fu*
NU Network Utility *mt*
NUC Navy Unit Commendation [USA] *mt*
NUC Non Uniformity Compensation *sb*
NUC Non Uniformity Correction [circuit] *sb*
NUCAP Nuclear Capabilities [data base] *mt, wp*
NUCEWA Nuclear Weapons Availability *mt*
NUCINT Nuclear Intelligence [USA DOD] *wp, mt*
NUCMATTS Nuclear Materials Transportation Tracking System [DOE] *mt*
NUCSTAT Nuclear Operational Status Report *mt*
NUCWA Nuclear Weapons Accounting [GCCS] [USA] *mt*
NUCWAL Nuclear Weapons Allocation Logistics *mt*
NUCWEP Nuclear Weapons *mt*
NUDAC Nuclear Data Center *wp*
NUDAP Nuclear Detonation Data Point *mt*
NUDET Nuclear Detonation [report] *mt, fu*
NUDET Nuclear [Detonation] Detection *hc, wp, tb, kf*
NUDETS Nuclear Detonation detection and reporting System [USA DOD] *mt*
NUDIS NUDET and Damage Information Summary *mt*
NUI Network User Identification *em, ct, kg*
NUI Network User Interface *em, kg*
NUI Notebook User Interface [Go Corporation] *pk, kg*
NUICCS NORAD/USSPACECOM Integrated Command and Control System *kf*
NUMA Non Uniform Memory Access *ct, em, kg*
NUOWG NTB User Operations Working Group *tb*
NURADS Netted Universal Radar System *hc*
NURBS Non Uniform Rational B-Spline *hi*
NURC Non Unit Related Cargo *mt*
NUREP Nuclear Reporting *mt*
NURP Non Unit Related Personnel *mt*
NUS No Upper Stage *ai2*
NUSC Naval Underwater Systems Center *12, tc, hc, ie26.1.8b, fu*
NUSI NASCOM User Simulation Interface [WSGT to NGT] *-2*
NUSSE Nonuniform Simple Surface Evaporation [chemical warfare model] *wp*
NUT Network Utility File Transfer *mt*
NUTEX Nuclear Tactical Exercise *wp*
NUTI NASA User Traffic Interface *-4*
NUTI NASCOM User Traffic Interface (WSGT to NGT) *-2*
NUWAX Nuclear Weapon Accident [Exercise] *mt, ph*
NUWC Naval Undersea Warfare Center *fu*
NUWEAMP Nuclear Weapons Employment and Acquisition Master Plan *sd*
NUWEP Nuclear Weapons Employment Procedures / Policy *mt*
NUWES Naval Undersea Warfare Engineering Station *fu*
NUWES Naval Underwater Weapons Evaluation Station *-12*
NV No Overflow *pk, kg*
NV Non Volatile [memory] *fu*
NV-EOL Night Vision - Electro Optics Laboratory *hc*
NVA North Vietnamese Army *mt, 10*
NVDET Network Virtual Data Entry Terminal *mt*
NVEOL Night Vision and Electro Optics Laboratory [of ERADCOM] *tc*
NVG Night Vision Goggles *mt, wp, aw130.13.23, hc, fu*
NVI Near Vertical Incident *hc*
NVIS Near Vertical Incidence Skywave *fu*
NVM Non Volatile Memory *pk, kg*
NVOD Near-Video On Demand *pk, kg*
NVP Network Voice Protocol [research-oriented] *csII.60*
NVP Nominal Velocity of Propagation *pk, kg*
NVPS Night Vision Pilotage System [USA] *mt, aw129.10.58, hn49.3.4*
NVR No Voltage Release *fu*
NVR Non Volatile Residue *tr, 4, he, oe*
NVRAM Non Volatile Random Access Memory *ct, em, kg, fu, hi*
NVS Night Vision Sight *hc*

NVS Night Vision System *mt, wp*
NVT Network Virtual Terminal *mt, kg*
NVT Novell Virtual Terminal [Novell] *kg, hi*
NW Nuclear Warfare *wp*
NWA {North West Airlines} *aw130.13.11*
NWC National War College [NDU] *dt, wp, mt*
NWC Naval War College *mt*
NWC Naval Weapons Center [USA] *12, wp, SCG830182P.15, hc fu, mt*
NWCG National Wire Gage *fu*
NWD Navigation Weapon Delivery *wp*
NWDS Network Wide Directory System *hi*
NWE Nuclear Weapons Effect *mt, wp, tb, st*
NWEP Nuclear Weapons Effect Panel *wp*
NWES Naval Weapons Engineering Support Activity *dt*
NWFZ Nuclear Weapon Free Zone *-14*
NWG Network Working Group *ct, em*
NWGS Naval Warfare Gaming System *fu*
NWIS Navy WWMCCS Information System *mt*
NWL Naval Weapons Laboratory *17, wp, tb*
NWMP Nuclear Weapons Master Plan *mt*
NWNCS Netway Net Control Station *fu*
NWO Negation Weapons Officer *jb*
NWOO NATO Wartime Oil Organization [NATO] *mt*
NWP Naval (Navy) Warfare Publication *mt, fu*
NWP Numerical Weather Prediction *hy, oe*
NWRP Nuclear Weapons Release Procedures *mt*
NWS National Weather Service *mt, wp, ci.880, 16, pr, fu, oe*
NWS Naval Weapons Station *12, fu*
NWS NetWare Web Server [Netware] *pk, kg*
NWS North Warning System *mt, hc, wp*
NWS Nuclear Weapon Storage *mt, wp*
NWSB Nuclear Warfare Status Branch *mt*
NWSC National Weather Satellite Center *ci.880*
NWSC National Weather Service Center *wp*
NWSC Naval Weapons Support Center [Crane, Indiana] *-12*
NWSC Naval Weather Service Command *wp*
NWSCS North Warning Satellite Communication System *fu*
NWSO Naval Weapons Services Office *dt*
NWSS Navy WWMCCS Software Standardization *mt*
NWTDB Naval Warfare Tactical Data Base *mt*
NWTP NMCS WWMCCS Transition Plan *mt*
NWWCSS Navy Worldwide Command Support System *fu*
NX Non-Expendable *wp*
NXT NASA X-band Transponder *mi*
NYANG New York Air National Guard *-12*
NYNEX New York - New England Exchange *ct*
NYSE New York Stock Exchange *aw130.13.11, rm1.2.1, w9*
NYSERNET New York State Education Research Network *ct*
NYU New York University *is26.2.67*
NZ Neutral Zone *mt*
NZAMP New Zealand Airways Modernisation Project *fu*
NZPO New Zealand Project Office *fu*
NZRAF Royal New Zealand Air Force *aw130.6.47*

O

O&A Orbit and Attitude *gp, 16, oe*
O&C Operation and Checkout [a facility at ELS] *bo, fu, oe*
O&C Operations and Control *fu*
O&I Organizational and Intermediate *wp*
O&M Operation(s) and Maintenance *mt, fu, hc, wp, gp, 16, tb, st, jdr, kf, oe*
O&M Outline and Mounting [drawing] *fu, 8, ep, jdr, jn*
O&MA Operation and Maintenance, Army *st*
O&MS Operations and Maintenance Supervisor *fu*
O&O Operational and Organizational *nt, st, fu*
O&S Operations and Support *hc, wp, 16, 17, sb, ts18-18, st, kf, fu*
O&S Operations and Sustainment *kf*
O&SHA Operating and Support Hazard Analysis *fu*
O&ST Order and Ship Time *wp*
O&T Operations and Test *kf*
O&TE Operational Test and Evaluation *kf*
O-CLS Organizational Contractor Logistics Support *kf*
O/A Operations / Administration *mt*
O/A Orbit/Attitude *pl*
O/D Operational Difficulty *0*
O/F Oxidizer to Fuel ratio *ai2*
O/H On Hand *mt*
O/H Over Head *jdr*
O/I/D Organizational/Intermediate/Depot *kf*
O/L Operational Location *fu*
O/M Operations and Maintenance [also O&M] *mt*
O/O Owner/Operator *jb*
O/P Output *jdr*
O/P Output Power *bl1-5.9*

O/S Operations and Support *mt*
O/V Over Voltage *jn*
OA Obligational Authority *wp*
OA Offensive Avionics *fu*
OA Office Automation *mt, fu*
OA Operational Area *mt*
OA Operational Assessment *jdr, kf*
OA Operational Availability *kf*
OA Operations Analysis *mt*
OA Optical Adjunct *tb, st*
OA Optical Augmentation *hc*
OA Orbit Analyst *uh, jdr*
OA Output Acknowledge *oe*
OA&M Operations Administration and Maintainance *ct*
OAATM Office of the Assistant for Automation *dt*
OAB One-to-All Broadcast *pk, kg*
OAB Outer Air Battle *dh*
OABM Outer Air Battle Missile *hc, dh*
OABWS Outer Air Battle Weapon System *fu*
OAC Observation, Analysis and Control *fu*
OAC Operator Access Console *mt*
OAD Open Architecture Driver [Bernoulli] *pk, kg*
OAD Operational Availability Date *wp*
OAD Orbit and Attitude Determination *oe*
OADR Originating Agency's Determination Required *mt, wp, jdr, kf, fu, su*
OADSS Office Assignment Decision Support System [USMC] *mt*
OAFB Offutt Air Force Base *jdr*
OAFDS Orbiter Aft Flight Deck Simulator *jdr*
OAG Official Airline Guide *pk, kg*
OAG Online Air Guide *pk, kg*
OAG Open Applications Group *ct, em*
OAI Open Applications Interface *pk, kg*
OAISN OJCS Automated Information System Network *mt*
OALC Ogden Air Logistics Center [AFLC] *wp, mt*
OAM Operations, administration, and maintenance *nu*
OAM Operator Assistance Menu *mt*
OAM Orbit Adjust Module *ai2*
OAMP Optical Airborne (Aircraft) Measurement Program *hc, fr, tb, st*
OAMS Optical Angular Motion Sensor *wp*
OAMS Orbital Attitude and Maneuvering System *ci.370*
OAO Orbital Analysis Officer *uh*
OAO Orbiting Astronomical Observatory *mi, hc, gs1-1.17, 0, es55.12, mb*
OAP Offset Aim(ing) Point *hc, mt*
OAP Operator Assistance Position *fu*
OAP Optical Adjunct Program *st*
OAP Organizational Assessment Package *wp*
OAR Off Axis Rejection *kf*
OAR Office of Aerospace Research *wp*
OARS On-Focal-Plane Array Signal Processing *hc*
OARS Order Analysis Reporting System *fu*
OART Office of Advanced Research and Technology [NASA] *ci.643, 0*
OAS Offensive Air Support *mt, fu, ph, wp*
OAS Offensive Avionics System [B-52s] *mt, wp, af72.5.134*
OAS One to All Scatter *pk, kg*
OAS Orbiter Aerofight Simulator *hc*
OAS Orbiter Avionics Simulator *jdr*
OAS Organization of American States *fu, w9, wp*
OAS Other ADGE Site *go3.1-1*
OASA Office of the Assistant Secretary of the Army *dt*
OASC Office Automation Support Center *mt*
OASD Office of the Assistant Secretary of Defense [OSD] *wp, tb, mt*
OASD(C) Office of the Assistant Secretary of Defense (Comptroller) *hc*
OASD(PA) Office of The Assistant Secretary of Defense (Public Affairs) *su, hc, tb*
OASD/ISP Office of the Assistant Secretary of Defense/International Security Policy *tb*
OASD[PA] Office of the Assistant Secretary of Defense [Public Affairs] *mt*
OASF Office Automation System Facility [Honeywell] *mt*
OASIS Ocean All Source Information System *wp*
OASIS Operational Analysis Strategic Interaction Simulator [ICBMs] *mt*
OASIS Operational Application of Special Intelligence Systems [US Navy] *wp, fu, hc, mt*
OASIS Operations Analysis Strategic Interactions Simulation [model] *wp*
OASIS Optimized Air to Surface Infrared Seeker *wp*
OASMA Offensive Air Support Mission Analysis *wp*
OASN Office of the Assistant Secretary, Navy *dt*
OAST Office of Aeronautics and Space Technology [NASA] *sc4.6.6, aa27.4.b45, aw129.21.158, wl, wp*
OASYS Obstacle Avoidance System *hc*
OAT Outside Air Temperature *mt*
OATS Operational Assessment Technical Support *fu*
OATS Optical Augmentation Target Screen *hc*
OATS Orbit and Attitude Tracking Subsystem *gp, 16, oe*
OAU Organization of African Unity *w9, wp*

OAWC Overseas Air Weapons Control *mt*
OAY Outstanding Airmen of the Year *mt*
OB Operating Budget *wp*
OB Operational Base [USAF] *tb*
OB Order of Battle (OB1KB is Order of Battle Version 1 Knowledge Based) *mt, fu, wp, tb*
OB Output Buffer *fu*
OB Outside Board [structure] *jn*
OBA Operating Budget Authority *mt, wp*
OBA Optical Barrel Assembly *gp, 16, lt*
OBA Oxygen Breathing Apparatus *mt, wp*
OBAD Operating Budget Authority Document *wp*
OBC On Board Computer *4, wp, 16, oe, ai2*
OBC Optical Bar Camera *wp*
OBD Off Board Drone *fu*
OBD Omni Bearing Distance *fu*
OBD On Board Diagnostics *pk, kg*
OBDB Order of Battle Data Base *fu*
OBDH On Board Data Handling *es55.60*
OBE Overcome By Events *mt, jdr*
OBE Overtaken By Events *wp*
OBEWS On Board Electronic Warfare Simulator *wp*
OBEX Object Exchange [Borland] *kg, hi*
OBI Omnibearing Indicator *fu*
OBIFCO On Board In Flight Checkout *hc*
OBIGGS On Board Inert Gas Generating System [C-17] *mt*
OBIGS On Board Inert Gas Generating System *wp*
OBIO Order of Battle Inputs Operator *fu*
OBL Operating Budget Ledger *wp*
OBOE Offensive Built Operational Environment *st*
OBOGS On Board Oxygen Generating System [USA] *mt, wp, aa26.8.25*
OBOS On Line Back Order System *fu*
OBP On board Processor *he*
OBRC Operating Budget Review Committee *wp*
OBRP On Board Repair Parts *fu*
OBS Omnibearing Selector *fu*
OBS Onboard Shuttle *tb*
OBS Optical Bypass Switch *fu*
OBS Organization(al) Breakdown Structure *mt, jdr*
OBS OSIS Baseline (Sub)System [US Navy] *tb, mt*
OBS Output Batch Scaling *fu*
OBSS Off Board Sensor Systems *wp*
OBSS Operations Briefing Support System *hc*
OBT Officers Basic Training *wp*
OBT On-Board Trainer *fu*
OBU Ocean Information System Baseline Upgrade [US Navy] *mt*
OBU OSIS Baseline Upgrade [US Navy] *mt*
OC Officer Candidate *w9*
OC Officer Commanding *mt*
OC Operand Cache *fu*
OC Operating Characteristics *fu*
OC Operational Capability *mt*
OC Operational Concept *mt*
OC Operational Condition *mt*
OC Operations Center (Central) *mt, fu, kf, tb*
OC Operations Communications *mt*
OC Operator Code *fu*
OC Optical Carrier *kg, nu*
OC Output Controller *-2*
OC-ALC Oklahoma City Air Logistics Center *mt*
OC-N Office of the Comptroller, Navy *dt*
OCA Obstacle Clearance Altitude *mt*
OCA Oceanic Control Area *mt*
OCA Offensive Counter Air [USA] *fu, wp, mt*
OCA Operational Control Authority *mt, wp*
OCA Orbital Carrier Aircraft *ai2*
OCA Original Classification Authority *mt*
OCAC Operations Control and Analysis Center *mt, wp*
OCAL On Line Cryptanalytic Aid Language *wp*
OCALA Oklahoma City Logistic Center *wp*
OCALC Oklahoma City Air Logistic Center *fu*
OCAS On-Line Cryptanaytic Aid System *wp*
OCB Operations Coordinating Board *-13*
OCBB Operating Cost Board Budget *wp*
OCC Office of the Comptroller of the Currency *y2k*
OCC Operations (Operational) Control Center *mt, hn44.14.1, 15, lt, ep*
OCC Operations Command Center *hc*
OCC Operator Circuit Control *fu*
OCC Operator Control Console *fu, jn*
OCCM Optical Counter Counter-Measure *hc*
OCD Office of Civil Defense *fu*
OCD Operational Concept Document *mt, fu, uh, wp*
OCD Ordnance Classification of Defects *fu*
OCDM Office of Civil and Defense Mobilization *w9*
OCE Office, Chief of Engineers *mt*
OCE Open Collaboration (Collaborative) Environment *ct, em, kg, hi*

OCE Orbital Computations Engineer *-2*
OCEANAV Oceanographer of the Navy *dt*
OCF Objects Components Framework [Borland] *kg, hi*
OCG Operations Control Group *fu*
OCH Obstacle Clearance Height *mt*
OCI Ocean Color Imager [NASA] *hg, hy, ge*
OCI Organizational Conflict of Interest *st*
OCIS Organized Crime Information Systems *ct*
OCJCS Office of the Chairman of the Joint Chiefs of Staff *dt*
OCL Open Class Library [IBM] *hi*
OCL Operation (Operator) Control Language *pk, kg*
OCLI Optical Coating Laboratories, Incorporated *17, kf*
OCM Ocean Color Monitor [ESA] *hg, ac3.7-23, ge*
OCM On Condition Maintenance *wp*
OCM Optical Countermeasure *hc, wp*
OCMM-N Office of Civilian Manpower Management - Navy *dt*
OCN Operations Control Number *wp*
OCNI Optimal Communications, Navigation and Identification *wp*
OCO Operational Capabilities Objective *wp*
OCONUS Outside CONUS (Contiguous United States) [USA] *kf, mt*
OCP Ocean Surveillance Product {?} *mt*
OCP Office of Commercial Programs [NASA] *wl*
OCP Operational Capability Plan *mt*
OCP Operational Computer (Control) Program *fu, hc*
OCP Out of Commission Parts *wp*
OCP Overcharge Protection *gp*
OCP Overcurrent Protection *jdr*
OCPCSB Operational Computer Program Configuration Sub Board *mt*
OCPO Office of Civilian Personnel Operations *mt*
OCPO Office of Computer Processing Operations *wp*
OCR Office of Corollary Responsibility *mt, wp*
OCR Operational Change Report *mt*
OCR Operational Concept Review *17, sb*
OCR Optical Character Reader (Recognition) *mt, fu, w9, kg, wp, ct, em, hi*
OCR Overcurrent Relay *fu*
OCRD Office, Chief of Research and Development [Army] *fu*
OCRE Optical Character Recognition Equipment *mt*
OCRS Operational Change Reporting System [DNA] *mt*
OCS Ocean Color Scanner *hy*
OCS Office of Communication Systems *wp*
OCS Officer Candidate School *mt, w9*
OCS On Card Sequencer *pk, kg*
OCS Operations Central System *fu*
OCS Operator Control Segment *jdr*
OCS Optical Character Scanner *wp*
OCS Optical Contrast Seeker *hc*
OCS Optical Control System *wp*
OCSA Office, Chief of Staff, Army *mt*
OCSM Operations Central System Modernization *fu*
OCSOT Overall Combat System Operability Test *fu*
OCST Office of Commercial Space Transportation *mi*
OCTC Operators Console Transfer Channel *mt*
OCTS Ocean Color and Temperature Scanner [Japan] *hg, ge*
OCTS Ocean Color and Temperature Sensor *hy*
OCU Office Channel Unit *do186*
OCU Operational Capability Update [USA] *mt*
OCU Operational Capability Upgrade *af72.5.135*
OCU Operational Conversion Unit [UK] *mt*
OCU Operator Console Unit *fu*
OCVCXO Oven Controlled Voltage Controlled Crystal Oscillator *jdr*
OCWG Operations Concept Working Group *kf*
OCZ Operational Control Zone *wl*
OD Officer of the Day *w9*
OD Operational (Operations) Directive *0, 17, tb, kf*
OD Operations Director *fu*
OD Orbit Determination *sd, tb*
OD Ordnance Data *fu*
OD Outside Diameter (Dimension) *w9, oe, ai1*
OD Over Dose *wp*
ODA Office Document Architecture *kg, hi*
ODA Omni Deployment Actuator *cl, jdr, kf, uh, he, jn*
ODA Other Design Activity *fu*
ODA/ODIF Office Document Architecture / Office Document Interchange Format *mt*
ODAC Other Design Activity Control *fu*
ODAL Open Discrepancy Action Log *jdr*
ODALS Omnidirectional Airport Lighting Support *fu*
ODAPI Open Database Application Program(ming) Interface [Borland] *kg, hi*
ODAPS Oceanic Display And Planning System *fu*
ODAPS OGE Data Acquisition and Patching Systems *oe*
ODAS Oceanographic Data Acquisition System *hy*
ODB Operational Data Base *wp, oe*
ODBC Object oriented Data Base Connectivity *pk, kg*
ODBC Open Data Base Connectivity [Microsoft] *kg, em, hi*

ODBC Open DataBase Compliant *em, ct*
ODBMS Object-oriented Database Management System [also OODBMS] *kg, hi*
ODC Office of Defense Cooperation *hc, wp, su*
ODC Open Database Connectivity *hi*
ODC Operator Display Console *fu*
ODC Other Direct Charges *fu, 16, ep, jdr, ah, pbl, tm*
ODC Other Direct Costs *mo, 8, hc, 15, kf, fu, jn*
ODC Ozone Depleting Chemical (Chemicals) *tm, jdr*
ODCARP Operational Data Collection, Analysis and Reporting Program *hc*
ODCM Office of the Director of Civilian Marksmanship *dt*
ODCR Operations Deputies Conference Room [NMCC] *mt*
ODCSLOG Offices Deputy Chief of Staff for Logistics *mt*
ODCSOPS Office, Deputy Chief of Staff for Operations and Plans *mt*
ODCSPER Office, Deputy Chief of Staff for Personnel *mt*
ODD Offboard Deception Device *mt*
ODD On Demand Dialing *hi*
ODD Optical Downconverter Demultiplexer *hc*
ODDD Operational Data Dictionary / Directory *mt*
ODDL Onboard Digital Data Load *bo*
ODDR&E [see ODDRE] *un*
ODDRE Office of the Director of Defense Research and Engineering *aw130.13.104, ci.160*
ODE Optical Discrimination Evaluation *st*
ODE Ordinary Differential Equation *ai1*
ODEMS Operational Design and Management System *fu*
ODF Orbit Determination Facility *-4*
ODI Open Data link Interface [Novell] *ct, em, kg, hi*
ODI Open Device Interconnect [NetWare] *kg, hi*
ODID Operator Display Interaction Device *fu*
ODISS Optical Digital Image Storage System *wp*
ODIT Optical Dynamic Interaction Test *oe*
ODJS Office of the Director of the Joint Staff *dt*
ODL Open Discrepancy List *lt*
ODL Operational Design Life *jn*
ODL ORACLE Data Load *mt*
ODM Object Data Manager [IBM] *pk, kg*
ODM Office of Defense Mobilization *-13*
ODM Office of the Division Manager *fu*
ODM Operational Data Message *-4*
ODM Operational Data Model *fu*
ODM Operational Development Model *hc*
ODM Output Data Monitor *fu*
ODMA Open Document Management API *kg, hi*
ODN Operations Data Network *oe*
ODN Out Dial Notification *ct, em*
ODN Own Doppler Nullifier *fu*
ODP Object Dependent Processor *kf*
ODP Open Distributed Processing *kg, hi*
ODP Orbit Determination Program *0*
ODP Output Data Processor *fu*
ODR Officer of Defense Resources *mt*
ODR Omni Direction Range *fu*
ODR Operational Design Review *fu*
ODR Output Data Request *oe*
ODRC Orbiter Data Reduction Complex *pf1.23*
ODRN Orbiting Data Relay Network *hc*
ODS Ocean Data System [NASA] *hg, ge*
ODS Open Data Services *pk, kg*
ODS Open Data Services [Microsoft] *hi*
ODS Ozone Depleting Substance *kf*
ODSI Open Directory Service Interface *ct, kg, em*
ODTC Office of Defense Trade Controls *rih*
ODTP Operational Demonstration and Test Program *fu*
ODUS Ozone Dynamics Ultraviolet Spectrometer [EOS] *pbl*
ODUSN Office of the Deputy Under Secretary of the Navy *dt*
ODW Output Data Word *fu*
ODWG Operational Design Working Group *fu*
ODZ Outer Defense Zone *mt*
OE Office of Education *ci.901*
OEA Oxygen Enriched Air *wp*
OEC Other Early Capability *ts4-41, tb*
OECD Organization for Economic Cooperation and Development *14, w9, y2k*
OECS Organization of Eastern Caribbean States *mt*
OED Operational Effectiveness Demonstration *mt, fu*
OED Operational Evaluation Demonstration *ph, wp*
OEF Operator Evaluation Function *fu*
OEG Operations Evaluation Group [Navy] *fu*
OEIC Opto Electronic Integrated Circuits *hi*
OEM Original Equipment Manufacturer *mt, fu, kg, ct, em, hc, wp, pl, aw130.11.75, is26.1.c2, kf, hi*
OEP Office of Emergency Preparedness *mt*
OEP Operand Execution Pipeline *pk, kg*
OEP&GR Office of Employment Policy and Grievance Review *dt*

OER Officer Effectiveness Report *mt*
OES Office Evaluation System *mt*
OESS Organizational Effectiveness Survey System [USA] *mt*
OET Office of Emergency Transportation *mt*
OEU Operational Evaluation Unit [UK] *mt*
OEX Orbiter Experiments *wp*
OF Operating Frequency *fu*
OF Operational Functionality *mt*
OF Over Frequency *fu*
OF Overflow Flag *pk, kg*
OF Oxygen Free *wp*
OFA Office of Flight Assurance *oe*
OFA Optics/FPA/ASP (sensor model in payload testbed) *kf*
OFB Output Feedback [mode] *pk, kg*
OFCATS Optical Fiber Cable Assembly Automatic Test System *hc*
OFCO Office of the Federal Coordinating Officer *mt*
OFD Operations Flow Diagram *fu*
OFDM Orthogonal Frequency Division Multiplexing *ct, em*
OFET Teledetection Education Office, [France] *ac3.6-7*
OFFET Offline Equipment Test *fu*
OFFS Optical Fiber Field Sensor *wp*
OFG Optical Frequency Generator *fu*
OFI Opportunity For Improvement *wp*
OFM Ordnance Field Manual *wp*
OFMT Output Format for Numbers *pk, kg*
OFP Operational Flight Program *fu, wp*
OFP Oscilloscope Face Plane *fu*
OFPP Office of Federal Procurement Policy *fu, hc, wp, nd744.445.57, jmj*
OFR Overfrequency Relay *fu*
OFS Object File System [Microsoft] *kg, hi*
OFS Operational Fixed Service *dn*
OFS Operational Flight Software *bo*
OFS Optical Fiber Sensor *wp*
OFS Orbiter Functional Simulator *jdr*
OFS Output Field Separator *pk, kg*
OFT Operational Flight Trainer *mt, hc, wp*
OFT Orbital Flight Test *ac1.4-21, 2, 4, lt, wl*
OG Officer of the Guard *w9*
OG Operations Group [USA] *mt*
OG Order Generator *fu*
OGA Option Generation Aid *mt*
OGA Other Government Activity *fu*
OGC Office General Counsel *dt*
OGC Other Government Costs *16, kf*
OGC-N Office of the General Counsel Navy *dt*
OGE On Gimbal Electronics *17, sb*
OGE Operations (Operational) Ground Equipment *fu, gp, 0, af72.5.145, oe*
OGE Out of Ground Effect *mt*
OGIP Office of Guest Investigator Programs [GSFC, Code 668] *cu*
OGL Open GL *hi*
OGMA Officinas Gerais de Material Aeronautica ["General Offices of Aeronautical Materiel", Portugal] *mt*
OGO Orbiting Geophysical Observatory *ac2.4-1, 0*
OGPU [Soviet secret police, also NKVD and KGB] *-13*
OGR Officer Grade Requirements *wp*
OGS Overseas Ground Station *mt, wp, kf*
OH Operating Hours *wp, jj*
OH Over Head *oe*
OHA Operating (Operational, Operations) Hazard(s) Analysis *fu, kf, oe*
OHA Outside Helix Angle *fu*
OHC [Observatoire de Haute Provence] *hy*
OHF O'Neill Hull Form *aa27.4.b14*
OHFR Objective High Frequency Radio *fu, hc*
OHM Output Hybrid Matrix *jr*
OHQ Originating Headquarters *mt*
OHS Open Heterogeneous System *fu*
OHSPC Oil and Hazardous Substance Pollution Contingency *mt*
OHT Over the Horizon Targeting *fu*
OI Office of Information *mt*
OI Operating Instructions *mt, wp, ep, fu*
OI Operational Instrumentation *2, 4*
OI Operations Intelligence *mt*
OI Operator Interface *fu*
OI-N Office of Information - Navy *dt*
OIA Operations Intelligence Automation [PACAF] *mt*
OIC Officer In Charge *mt, wp*
OIC Orbit Injection Correction *jdr*
OICA OPLAN Implementation Capabilities Report *mt*
OICC Operational Intelligence Crisis Center *mt*
OICCD Oblique Imaging Charged Couple Device *wp*
OICR Operation, Implementation, Capabilities Report [AFCC*] *mt*
OICS Operational Information Collection System *fu*
OIDL Object Interface Definition Language *pk, kg*
OIDS Operational Information Display System *fu*

OIFC Oil-Insulated Fan-Cooled *fu*
OIG Office of the Inspector General [also DAIG] *dt, y2k*
OIG Operations Interface Group *mt*
OIIBS Operator Interface Information Base Subsystem *fu*
OIM Organizational Intermediate Maintenance *mt, wp*
OIM {Orbit Insertion Maneuver} *ma20*
OIMS [All Union Society for the Study of Interplanetary Communications] *-13*
OIP OGE Input Simulator *oe*
OIP Operational Incentive Plan [TRW] *kf*
OIP Optical Improvement Program *hc*
OIP Output Intercept Point *jdr*
OIPD Operations Interface Procedure Document *-4*
OIR Operational and Information Requirements *mt*
OIRM Office of Information Resources Management *mt, wp*
OIS Office Information System [JCS] *kg, fu, mt*
OIS Operation Instruction Sheet *fu*
OIS Operational (Operation) Instruction Sheet [a document] *ah, gp, pl, lt, ep, jdr, pbl, tm, jn*
OIS Orbit Injection [Sub] System *16, pl, kf*
OIS Orbit Insertion System *kf*
OIS/IGP Office Information System Intelligent Gateway Processor *mt*
OISE Office Information System Equipment *mt*
OISI Office of Industrial Security, International *su, tb*
OISS Operational Intelligence Support System *mt, wp*
OIST Office of Innovative Science and Technology *tb*
OIT Office Software Development and Information Technology [GSA] *mt*
OIT Open, Inspect and Test *fu*
OIT Operational Instruction Title *pl*
OITC Officer In Tactical Command *mt*
OIU Operator Interface Unit *fu*
OIUC Optical Infrared Ultraviolet Communications *wp*
OIUS Optical Infrared Ultraviolet Surveillance *wp*
OIWG Operations Intelligence Working Group *mt*
OIWG Operations Interface Working Group *-4*
OJCS Office (Organization) of the Joint Chiefs of Staff *mt, fu, ph, af72.5.31, wp*
OJCS Office of Joint Computer Support *fu*
OJP Outer Join Processor *fu*
OJT On the Job Training *mt, wp, sc4.9.5, hc, w9, kf, fu, oe*
OKB Osoboye Konstruktorskoye Buro [Special Design Bureau] *ai2*
OKB [Design Bureau, USSR] *mt*
OKEAN [Oceanographic satellite series, USSR] *ge, hg*
OL On Line *wp*
OL Open Loop *fu*
OL Operating Location *mt, 12*
OL Operations Leader *pbl, jdr, ah, jn*
OL Ordered List *ct*
OLA (specialized agencies are OLA at the Pentagon, which) *af72.5.109*
OLA On Line AUTODIN [US Navy] *mt*
OLA-N Office of Legislative Liaison Affairs - Navy *dt*
OLAP On Line Analytical Processing *kg, ct, em, hi*
OLB Outer Lead Bonded *fu*
OLC Operating Location Clerk *mt*
OLC Ordnance Load Code *fu*
OLCMS On Line Cargo Movement System [HQ USAF] *mt*
OLD Off-Line Diagnostics *fu*
OLDS On Line Display (Data) System *wp, no*
OLD_DIPPER [Photographic Target Selection Program] *wp*
OLD_GRIND [Photography of Strategic Air Command Bases] *wp*
OLD_TOWN [Photographic Exploitation and Processing Program] *wp*
OLE Object Linking and Embedding [Microsoft] *ct, em, kg, hi*
OLE Optical Logic Etalon *aa26.10.30*
OLEASS Organic Long Endurance Airborne Area Surveillance System *hc*
OLF Orbital Launch Facilities *wp*
OLGA On Line Guitar Archive *pk, kg*
OLI Optical Line Interface [AT&T] *pk, kg*
OLIPS Open Literature Information Processing System *wp*
OLLS Operational Logistics Support Summary [US Navy] *mt*
OLMC Output Logic Macrocell *pk, kg*
OLOGS Open Loop Oxygen Generating System *wp*
OLPARS On Line Pattern Analysis and Recognition System *wp*
OLPM On-Line Performance Monitor(ing) *fu*
OLR Off Line Recovery *mt*
OLR Office Loop Regenerator *do181*
OLRP On Line Report Processor *mt*
OLS Operational Line scan System [primary sensor in DMSP] *16, ac7.5-4, pl, wp*
OLS Optical Linescan Subsystem *-16*
OLSCA Orientation Linkage for a Solar Cell Array *hc*
OLSP Off Line Simulation Program *fu*
OLSP On Line Service Provider *pk, kg*
OLT Off-Line Test *fu*
OLT On-Line Test *fu*

OLTP On Line Transaction Processor (Processing) *mt*
OLTP On Line Transaction Protocol *kg, em, hi*
OLVIMS On Line Vehicle Interactive Management System *mt*
OM Object Manager *pk, kg*
OM Office Manager *hi*
OM Operations Manual [for a computer] *mt*
OM Operations Module *mt, fu*
OM Operator's Manual *fu*
OM Optical Microscope *re*
OM Option Module *fu*
OM Organo Metallic *wp*
OM Outer Marker *fu*
OM&E Operations, Maintenance And Engineering *kf*
OMA Object Management Architecture *kg, em*
OMA Operation and Maintenance, Army *mt*
OMA Operations Management Application *aa26.11.21*
OMA Organizational Maintenance Activity *wp*
OMAC Old Man's Aircraft Company *aa26.10.24, aw130.11.25*
OMAG Osobaya Morskaya Aviatsionna Grupa [Independent Naval Aviation Group] [USSR] *mt*
OMAR Objectives and Media Analysis Report *kf*
OMAR Ozone Monitoring And Research *wp*
OMAT Optical Measurement Control System *wp*
OMB Office of Management and Budget *mt, fu, mi, wp, ci.141, hc, w9, aw129.10.4, hy, tb, mb, oe, y2k, ai2*
OMC Occupational Measurement Center *mt*
OMC Office of Munitions Control [hardware export approval by the U.S. State Department] *fu, bl3-1.142*
OMCFP Optimized MAC* Computer Flight Plan *mt*
OMCS Operations and Maintenance Communications System *tb*
OME Open Messaging Environment [protocol] *kg, hi*
OMEI Other Major Equipment Items *mt*
OMEW Office of Missile Electronic Warfare [USA] *wp, mt*
OMF Object Module Format [Microsoft] *kg, hi*
OMF Open Media Framework *pk, kg*
OMF Open Message Format *kg, hi*
OMF Organization Master File [USA] *mt*
OMFCU Outboard Message Format Conventional Unit [NAVELEX] *mt*
OMG Object Management Group *em, kg, hi*
OMG Operational Maneuver Group *mt, wp*
OMI Office and Maintenance Instruction *oe*
OMI Office of Management Information [Navy] *dt, wp*
OMI Open Messaging Interface [Lotus] *kg, hi*
OMI Operator Message Input *mt*
OMIS Operational Multi-Spectral Imager Suite *dm*
OMMEN On-Board Multi-Function Maintenance Evaluation Network *wp*
OMNI Mobilization and Deployment Data Base [FORSCOM] *mt*
OMNI Omnidirectional Antenna *uh, vv*
OMNICS Office of the Management National Communications *dt*
OMNS Open Network Management System *pk, kg*
OMO Operator Message Output *mt*
OMP Organometallic Polymers *wp*
OMP Output Message Processing *fu*
OMPA Octa Methyl Pyrophosphor Amide *wp*
OMR Operation and Maintenance Report *fu*
OMR Operation Maintenance Request *fu*
OMR Optical Mark Reader *mt*
OMR Optical Mark Recognition *kg, ct, em*
OMR Output Message Report *mt*
OMRS Output Mechanical Redundancy Switching *jdr*
OMS Object space Mirror Sensor *kf*
OMS On board Maintenance System *aw126.20.36*
OMS Optimum Mode Selection *fu*
OMS Orbit Mode Software *-16*
OMS Orbital Maneuvering [Sub] System *mi, ci.xxii, 9, aw129.14.18, ws4.10, gp, 16, dm, wl, mb, ai2*
OMS Orbiter Mechanical Simulator *jdr*
OMS Ordnance Management System [US Navy] *mt*
OMS Organizational Maintenance Squadron (Shop) *wp, fu*
OMS Output Multiplex Synchronizer *fu*
OMSF Office of Manned Space Flight [NASA] *ci.643*
OMSRADS Optimum (Optimal) Mix of Short Range Air Defense Systems *hc, wp*
OMT Object Modeling Technique *hi*
OMT Ortho Mode Transition *uh*
OMT Orthogonal Mode Transducer *csl.7, 4, pl, jdr*
OMT Other Military Target *mt, tb, st*
OMTF Optical Modulation Transfer Function *gp*
OMTPE Office of Mission To Planet Earth *oe*
OMTS Organizational Maintenance Test Set *wp*
OMTU Organizational Maintenance Trainer Unit *hc*
OMU Orbital Maneuvering Unit *wl*
OMUX Output Multiplexer *jdr, uh, oe*
OMV Orbital Maneuvering Vehicle *wp, aa26.12.26, sc4.3.3, 16, 17, sb, hc, dm, tb, wl*
ONA Open Network Architecture *mt*

ONAS Office of Naval Acquisition Support *mt, dt*
ONC Open Network Computing [Sun] *kg, hi*
ONCON Operational Control *fu*
ONDE Office of Naval Disability Evaluation *dt*
ONDS Open Network Distribution Services [IBM] *pk, kg*
ONE Open Network Environment *ct, em, kg*
ONE/R One [time and revisions as] Required *tb, kf*
ONERA Office National d'Etudes et de Recherches Aerospatiales *aa26.11.b29, crl*
ONI Office of Naval Intelligence *-10*
ONLEXCRT Online Exercise Critique system [JCS] *mt*
ONM Office of Naval Material *fu*
ONMS Open Network Management System *ct, em, hi*
ONNI Office of Naval Narcotics Intelligence *-10*
ONO Office of Naval Operations *mt*
ONPG Operations Nuclear Planning Group *mt*
ONR Office of Naval Research [USA] *mt, fu, tc, wp, 1, aa26.11.50, hc, dt, hy, tb, st*
ONSCOST [a logistics computer program] *hc*
ONT Office of Naval Technology *mt, tc*
ONU Optical Networking (Network) Unit *ct, kg, em*
OO Object Orientation (Oriented) *em, kf*
OO-ALC Ogden Air Logistics Center *mt*
OOA Office of Ocean Affairs *dt*
OOAD Object Oriented Analysis and Design *hi*
OOALC [Hughes Aircraft Company customer for automated technical order system] *hc*
OOAMA [Hughes Aircraft Company customer for EW simulator modification] *hc*
OOB Order Of Battle *fu, kf*
OOB Out Of Band *jdr*
OOC Office Of the Chairman *jdr*
OOD Object Oriented Design *mt, fu, hi*
OOD Officer Of the Day *wp*
OOD Officer Of the Deck [US Navy] [USA] *mt*
OODB Object Oriented Data Base *kg, hi*
OODBMS Object Oriented Data Base Management System [also ODBMS] *hi*
OODEPS Owners, Officers, Directors, Partners, Regents, Trustees, or Executive Personnel *su*
OODMS Object Oriented Database Management System *pk, kg*
OOFS Object Oriented File System *hi*
OOG Object Oriented Graphics *hi*
OOH Orbital Operations Handbook *jdr, kf, uh*
OOK On Off Keying *mt, ce*
OOL Object Oriented Language *pk, kg*
OOL Out Of Limits *oe*
OOMS On Orbit Maintenance/Servicing *tb*
OOOS Object Oriented Operating System *kg, hi*
OOP Object Oriented Programming *em, wp*
OOP Office Of the President *ah, pbl, jdr, jn*
OOPL Object Oriented Programming Language *kg, hi*
OOS Object Oriented System *kg, hi*
OOS Off line Operating Simulator *pk, kg*
OOS On Orbit Support *jdr*
OOS Orbit to Orbit unmanned System *ci.263*
OOS Out Of Scope *jn*
OOS Out Of Specification *jdr*
OOT Object Oriented Technologies *jdr, kg*
OOT On Orbit Test *jdr, uh*
OOTF On Orbit Test Facility *jdr*
OOUI Object Oriented User Interface *kg, hi*
OOW Officer Of the Watch [US Navy] [USA] *mt*
OP Observation Post *fu, w9*
OP Operational Performance *mt*
OP Operational Profile *fu*
OP Operational Program *fu*
OP Operator Code *fu*
OP Ordnance Publication *fu*
OP Output Processor *mt*
OPA Office of Program Appraisal - Navy *dt*
OPA Optoelectronic Pulse Amplifier *fu*
OPA Other Procurement Army *mt*
OPAC Online Public Access Catalog [Internet] *kg, hi*
OPADEC Optical Particle Decoy *hc, wp*
OPAQUE Optical Atmospheric Quantities in Europe [model] *wp*
OPAT OGE Preliminary Acceptance Test *oe*
OPBS Orbiter Payload Bay Simulator *jdr*
OPC Optical Phase Conjunction *tb*
OPC Organic (Optical) Photoconductor *pk, kg*
OPC Organic Photoconducting Cartridge *ct, em*
OPCC Offutt Processing and Correlation Center *mt, wp*
OPCODE Operational Code *pk, kg*
OPCOM Operational Command *mt, go3.2, jb*
OPCOM Operations Communications *fu*
OPCOM Optical Communications *-19*

OPCON Operational Control [USA DOD] *wp, as, kf, mt*
OPCONCTR Operations Control Center *mt*
OPD Optical Path Difference *wp, st*
OPD Optical Proximity Detector *wp*
OPDAR Optical Detection and Ranging *fu*
OPDEC Operational Deception [US Navy] *fu, wp, mt*
OPDEP Operations Deputy *mt*
OPDOC Operational Documentation [USA DOD] *mt*
OPDS Officer Professional Development Seminar *mt*
OPDS Offshore POL Discharge System *wp*
OPEC Organization of Petroleum Exporting Countries *14, wp, w9*
OPED Operator Position Entry Device *fu*
OPEI Office of Public Education and Information [JPL] *0*
OPEN Open Protocol Enhanced Network *mt*
OPEN Origin of Plasma in the Earth's Neighborhood *hc*
OPEVAL Operational Evaluation *mt, fu, hc*
OPF Official Personnel Folder *wp*
OPF Orbiter Processing Facility *bo, mi, 9, jdr, mb, ai2*
OPFAC Operational Facility *fu*
OPFCO Operational Functional Checkout *fu*
OPFOR Opposing Force *wp*
OPG Operations Planners Group [JCS] *mt*
OPGEN Operation Plan Generation [PACFLT] *mt*
OPGEN Operational General [Message Format] *fu*
OPGEN Operations General Message *mt*
OPI Office of Primary Interest *fu*
OPI Open Prepress Interface *pk, kg*
OPI Operator Interaction *fu*
OPI Orbiter Payload Interface *-4*
OPIC Overseas Processing and Interpretation Center *mt*
OPINDOC Operational Indoctrination *hc*
OPINE Operation In a Nuclear Environment *st*
OPINTEL Operations Intelligence *fu*
OPIT Interministerial Teledetection Pilot Operation [France] *ac3.6-7*
OPL Organizer Programming Language *mt*
OPLAN Operation(al) Plan [USA DOD] *kf, oe, mt*
OPLANS Operations Plans *fu, tb*
OPLAN_REV Operation Plan Review [JCS] *mt*
OPLE Omega Position Locator [and Location] Equipment *hc, gp, 16*
OPM Office of Personnel Management [OSD] *mt, wp, tb, y2k*
OPM Office of Program Maintenance *fu*
OPM Operating Procedures Manual *fu*
OPM Operator Performance Metrics *kf*
OPM Orbital Parameters Message *uh*
OPMET Operational Meteorological *fu*
OPMS Officer Personnel Management System *mt, wp*
OPNAV Office of the Chief of Naval Operations [Navy Staff] *fu, mt*
OPNAVINST Operational Navy Instruction *mt*
OPNL_RPTS Operational Reports *mt*
OPORD Operation Order [USA DOD] *mt*
OPORD Operational Orders [executable part of OPLAN] *kf*
OPORDS Operations Orders *fu*
OPP Oriented Poly Propane *wp*
OPPLAN Operation Plan [USA] *mt*
OPR Office(r) of Primary (Prime) Responsibility *mt, fu, sb, wp, hc, tb, kf*
OPR Optical Page Reader *mt*
OPR Optical Pattern Recognition *wp*
OPR Orbiter Payload Recorder *-2*
OPRC Offsite Production Review Committee *fu*
OPREP Operational Event / Incident {Report} *mt*
OPREP Operational Report *mt*
OPREPS Operational Reporting System *mt*
OPRES Operations Research *wp*
OPROM Optical Programmable ROM *hi*
OPS On Line Provisioning System *fu*
OPSA Output Power Steering Assembly *fu*
OPSAN Operations Analysis System [PACAF] *mt*
OPSCAP Operational Status Capability *kf*
OPSCOM Operational Secure Communications *mt*
OPSCOMM Operational Communications *mt*
OPSDEPS Operations Deputies *mt*
OPSEC Operations Security [USA DOD] *mt, fu, su, 16, wp, 17, af72.5.75, nt, tb, st, jdr, kf*
OPSEC_M&A Operations Security Management and Analysis section [USA] *mt*
OPSG Operation Planning Steering Group *mt*
OPSIM Officer Planning and Simulation Model *wp*
OPSIM Operation Simulation *fu*
OPSIT Operational Situation *fu*
OPSMOD Operational Validation *mt*
OPSMOD Operations Planning Module [COMPES] [USAF] *mt*
OPSNOTE Operations Note *mt*
OPSORD Operations Order *mt*
OPSS On Pad Spacecraft Simulator *oe*
OPSTAT Operations Status Report [NATO] *mt*
OPSU Oxidizer and Pressurant Servicing Unit *oe*

OPSU Oxidizer Propulsion Servicing Unit *oe*
OPSUM Operational Summary *mt*
OPSYS Operating System *tb*
OPT Open Protocol Technology *kg, ct, em*
OPT Operational Parameter Translation *fu*
OPT Optimized Production Technology *fu*
OPTAC Optical Target Acquisition and Cueing *wp*
OPTADS Operations Tactical Data System [USA] *wp, mt*
OPTAR Optical Automatic Ranging *fu*
OPTEVFOR Operational Technical Evaluation Force *fu*
OPTEVFOR Operational Test and Evaluation Force [USA] *mt*
OPTINT Optical Intelligence *fu, wp*
OPTIR Optical Infrared *wp*
OPTRAJ Optimal Trajectory *fu*
OPTRAK Optical Tracking and Ranging Kit *wp*
OPUS Octal Program Updating System *pk, kg*
OPUS Officer Planning Utilization System [USMC] *mt*
OPWG Output Waveguide *jdr*
OQPSK Offset QPSK *oe*
OR Operational Readiness *mt, wp*
OR Operational Recording *fu*
OR Operational Requirement [UK] *fu, mt*
OR Operationally Ready *mt, fu*
OR Operations Requirement [a document] *0, a2, 4, 17*
OR Operations Research *mt, hc, w9, st*
OR Output Register *fu*
OR Outside Radius *fu*
OR/SA Operations Research/Systems Analysis *st*
ORA Operations Research Analyst *wp*
ORA Overtime Request/Authorization *fu*
ORACLE Optimized (Optimum) Reliability and Component Life Estimator *ro, wp*
ORACLE Ordnance Rapid Area Clearance System *wp*
ORACLE Oversight of Resources And Capability for Logistics Effectiveness Relational Data *mt*
ORADS Optical Ranging and Detection System *hc, wp*
ORATMS Off Route Antitank Mine System *fu*
ORB Object Request Broker *kg, em*
ORB Operational Review Board *fu*
ORB Orbit generation/magnetic calculation [program] *17, sb*
ORBA Object Request Broker Architecture *hi*
ORBATTOA Order of Battle - Transfer Of Authority *mt*
ORBITSIM Orbit Simulation [program] *-19*
ORBX [system of three privately owned communications satellites] *hn51.16.1*
ORCA Operations Requirements Continuity Assessment *mt*
ORCHIDEE Observatoire Radar Coherent HeliportÈ d'Investigation Des Elements Ennemis [a pulse-doppler radar carried on a helicopter and used to track moving ground targets] [France] *wp*
ORCON Originator Control [of information release] *mt*
ORCS On line Remote Compile System *mt*
ORD Operational Readiness Demonstration (Date) *jr, fu*
ORD Operations (Operational) Requirements Document *17, kf*
ORD Orbital Requirements Document *uh, 4*
ORDALT Ordnance Alteration *fu*
ORDBMS Object-Relational Data Base Management System *em*
ORDCIT Ordnance [Department of Army] and California Institute of Technology *ci.167, 0*
ORDET Orbit Determination [Group] *0*
ORDIR Omnirange Digital Radar *fu*
ORDLIS Ordnance Logistics Information System *fu*
ORDVAC Ordnance Variable Automatic Computer *pk, kg*
ORE Operational (Operations) Readiness Exercise *mt, wp*
ORE Operational Readiness Evaluation [USA] *mt*
ORESTES [intrabattle group tactical HF circuit] *mt*
OREX Orbital ReEntry vehicle *ai2*
ORF Operational Readiness Fleet *mt*
ORF Operational Readiness Float *fu*
ORFEUS Orbiting and Retrievable Far and Extreme Ultraviolet Spectrometer *mi*
ORI Online Retrieval Interface *pk, kg*
ORI Operational Readiness Inspection [USA] *mt, fu, wp, af72.5.107, tb*
ORICS Optical Ranging, Identification [friend or foe] and Communications System *wp, hc*
ORIT Operational Readiness Inspection Test *wp*
ORJETS On Line Remote Job Entry Terminal System *mt*
ORL Operations Recording List *fu*
ORLA Optimum Repair Level Analysis *mt, wp, tb, fu*
ORLRAPS Online Remedial Action Progress System [JCS] *mt*
ORMAS Operational Resource Management Assessment Plan *mt*
ORMS Operating Resource Management System *ct, em*
ORNL Oak Ridge National Laboratory *mt, tb*
ORO Operations Research Office *fu*
ORP Operational and Recording Program *fu*
ORP Operational Readiness Platform [RAF, UK] *mt*
ORP Orbital Radiation Program *17, sb*
ORR Operational Radar Replacement *mt*

ORR Operations Readiness Review -2
ORR Outside Receiving Reports *fu*
ORS Offensive Radar System *hc, wp*
ORS Offset Radiation Source *jdr, uh, jn*
ORS Operational Reports Section *mt*
ORS Operational Research Section [Allied] *mt*
ORS Optical Rendezvous System *hc*
ORS Output Record Separator *pk, kg*
ORSA Operations Research and System Analysis *wp*
ORSA Operations Research Systems Analysis *mt*
ORSL Optional Recording Specification List *fu*
ORT Operational Readiness Test *fu, wp*
ORT Operational Readiness Training *mt*
ORT Operational Review Team *fu*
ORT Origin Routing Table *fu*
ORT Overland Radar Technology (system) *fu, wp*
ORT-L Operations Readiness Test-Launch *re*
ORTS Operational Readiness Test System *wp*
ORU Orbital Replacement Unit *aa26.11.25, tb, wl*
ORUS Orbital Replacement Units *mt*
ORV Operational Range Vessels *0*
OS Ocean Surveillance *ic2-6*
OS Operating System (OS/2 is Operating System/2 [IBM]) *mt, pk, kg, fu, kg, ct, em, wp, tb, hi*
OS Operations Specialist *fu*
OS Ordered Statistics *fu*
OS Ordinary Seaman *w9*
OS Organizational Support *mt*
OS Output Slice *jdr*
OS Over Seas *mt*
OS Own Ship *fu*
OS&M Operational Support and Maintenance *mt*
OS/CMP Operational Support / Configuration Management Plan *mt*
OS/E Operating System / Environment *kg, hi*
OSA Office of the Secretary of the Army *mt, dt*
OSA Open System (Scripting) Architecture *kg, hi*
OSA Operational Support Aircraft *wp*
OSA Operational Support Airlift / Aircraft [USA] *mt*
OSA Optical Society of America *fu*
OSA Output Switch Assembly *jdr*
OSAF Office of the Secretary of the Air Force *mt*
OSAM Overflow Sequential Access Method *hi*
OSAP ORACLE Stress Analysis Program *hc*
OSAP Otdelnii Smeshannii Aviatsonii Polk [Independent Mixed Aviation Regiment] [USSR] *mt*
OSAR Optical Storage And Retrieval *wp*
OSASN Office of the Special Assistant to the Secretary of the Navy *dt*
OSAT Operational System Acceptance Test *fu*
OSATRMS Operational Resource Tracking and Management System *mt*
OSC On Scene Commander *mt*
OSC Operational System Control *mt*
OSC Operations Summary Console *fu*
OSC Operations Support Center *mt, wp*
OSC Optical Signatures Code *st*
OSC Orbital Sciences (Systems) Corporation *mi, sc3.1.4, aw129.21.13, mb, ai2*
OSC Organizational Supply Code *wp*
OSCAR Operations Support Center (OSC) Automated Reports system [USAFE] *mt*
OSCAR Orbiting Satellite Carrying Amateur Radio *mt, mi*
OSCD Operations Support Computing Division -2
OSCF Operations Support Computing Facility *2, 4*
OSCFAR Ordered Statistic Constant False Alarm Rate *fu*
OSCP On Site Computer Programmer *mt*
OSD Office of the Secretary of Defense *mt, fu, su, ca.4, 13, hc, af72.5.31, nd74.445.46, ts.6, dt, wp, tb, st, jdr, kf*
OSD Operational Sequence Diagram *fu*
OSD Operational Support Directive *wp*
OSD Operational System Demonstration *fu*
OSD Operational Systems Development *wp*
OSD Over Seas Duty *wp*
OSDIT Office of Software Development and Information Technology [GSA] *mt*
OSDMP O,S-Diethylmethylphosphonate [chemical warfare simulant] *wp*
OSDP On board Signal and Data Processor *kf*
OSDS Own-Ship Data Set *fu*
OSE Operational Support Equipment *fu, 0, wp*
OSF Office of Space Flight [NASA] *2, 4, wl, oe*
OSF Open Software Foundation *ct, em, kg, is26.1.38, hi*
OSF Open System Foundation *kf*
OSF Operating System Function *fu*
OSF Operational Support Facility [WIS] *jdr, mt*
OSF Ordnance Storage Facility *wp*
OSFD Operational Sequence Flow Diagram *fu*
OSG Operations Summary Group (was called NTDS III at Hughes) *fu*
OSGP Operations Support Group Prototype [PACFLT] *mt*

OSHA Occupational Safety and Health Administration *mt, fu, wp, hn48.13.8, 14, 16, ph, aw129.14.27, w9, nd74.445.4, tb, jdr, kf*
OSHA Operational Safety and Health Act *tb*
OSHF Operational Summaries and Histories Function *fu*
OSI Office of Scientific Information *wp*
OSI Office of Special Investigations [USAF] [U. S. Dept. of Justice] *mt, su, hc, aa26.11.6, aw130.11.24, wp, tb*
OSI Office of Strategic Information *wp*
OSI Open Standards Interconnection *em, ct*
OSI Open Systems Interconnection [defines 7 layers of communications *physical, data link, network, transport, session, presentation, application]*
OSI/RM Open System Interconnection / Reference Model *mt*
OSIA On Site Inspection Agency *hn48.13.1, aw129.10.17*
OSID Operational System Interface Document -4
OSIFA Operational Sequence and Information Flow Analysis *mt*
OSINET Open System Interconnection Network *mt*
OSINT Open Source Intelligence [USA DOD] *mt*
OSIP Operational Satellite Improvement Program *hy, oe*
OSIRM Open Systems Interconnection Reference Model *nu, fu*
OSIS Ocean Surveillance Information System [US Navy] *mt, fu, wp, aw118.18.37, pf1.4, tb*
OSIS Operating Systems Installation Support *mt*
OSL Orbiting Space Laboratory *wp*
OSM OS Specific Module *em*
OSM Ownship Surveillance Management *fu*
OSMR Operational Software Maintenance Report *oe*
OSMV One-Shot Multivibrator *fu*
OSN Office of the Secretary of the Navy *dt*
OSO Offensive System Operator *aw129.10.72*
OSO Office of Satellite Operations *oe*
OSO Office of Space Operations *re*
OSO Orbiting Solar Observatory *cu, hn43.21.4, 16, 17, hc, ep*
OSP Objective and Strategies Plan *mt*
OSP Off line Support *fu*
OSP Office of Scientific Personnel *wp*
OSP On board Signal Processor *kf*
OSP On Screen Programming *pk, kg*
OSP On Site Programmer *mt*
OSP Optical Storage Processor *pk, kg*
OSP Orbital Suborbital Program *ai2*
OSP Orbital Support Plan *uh*
OSP Orbiting Standards Platform *hc*
OSPE Organizational Spare Parts and Equipment *wp*
OSPF Open Shortest Path First *ct, em, kg*
OSPO On-Site Program Office *fu*
OSQL Object Structured Query Language *kg, hi*
OSR OEM System Release *ct, em, kg*
OSR Office of Scientific Research [USAF] *fu, wp*
OSR On-Site Representative *fu*
OSR Operational Scanning Recognition *fu*
OSR Operational Support Requirement *fu, mt, wp*
OSR Optical Surface (Solar) Reflector *jdr, oe*
OSR Output Shift Register *fu*
OSR {a receiver on INSAT-1} *ac3.2-16*
OSRD Office for Scientific Research and Development *ci.36, 13*
OSRI Originating Station Routing Indicator *mt*
OSRPC Optical Solar Reflector Protective Covers *oe*
OSS Office of Secret Services [USA] *mt*
OSS Office of Space Sciences [NASA] *ci.643, wp*
OSS Office of Space Station [NASA] *wl*
OSS Office of Space Systems *wp*
OSS Office of Strategic Services *10, w9*
OSS Operating System Set *fu*
OSS Operation Support Services *kf*
OSS Operational (Operations) Support System *fu, mt*
OSS Operational Storage Site *mt, wp*
OSS Operational Support Squadron [USA] *mt*
OSS Optical Surveillance System *mt*
OSS Orbital Stabilization System *hc*
OSS Order Status System *fu*
OSSA Ocean Surveillance Aircraft *wp*
OSSA Office of Space Science and Applications [NASA headquarters] *hc, ci.643, hy, re, wl, mb, mi, oe*
OSSE Oriented Scintillation Spectrometer Experiment [on CGRO] *mi*
OSSN Originating Station Serial Number *mt*
OSSV Ognichenie i Sokrashchenie Stratigicheskykh Vooruzhenii [START] [USSR] *mt*
OST Office of Science and Technology *ci.869*
OSTA Office of Space and Terrestrial Applications [NASA] *sc4.6.6, hg, ge*
OSTDS Office of Space Tracking and Data Systems [NASA] *ci.xli, 2, 4, wl*
OSTF Off Site Test Facility [SPACECMD] *mt*
OSTF Operational System Test Facility *0*
OSTF Optical Sensor Test Facilities *st*
OSTG Ordered Statistics Threshold Generator *fu*

OSTP Office of Science and Technology Policy *mt, ci.869, wp, fu*
OSU Oxidizer Service Unit *oe*
OSUS Ocean Surveillance [US Navy] *mt*
OSV Ogranichenie Stratigicheskykh Vooruzhenii [SALT] [USSR] *mt*
OSV On Site Verification *fu*
OSV Orbital Servicing Vehicle [Japan] *wl*
OT Object Technology *pk, kg*
OT Offensive Threat *5, sb*
OT On Trajectory *tb*
OT Open Transport *ct, em*
OT Operational Test(ing) [and evaluation] *mt, fu, ph, hc, tb, st, kf*
OT Operations Team *mt*
OT Operator Technician *fu*
OT Optical Tracker *wp*
OT&A Operational Test and Acceptance *mt*
OT&E Operational Test and Evaluation *mt, fu, aw126.20.93, af72.5.54, jmj, tb, st, kf*
OT/DT Operational and Development Testing *fu*
OTA Operation Triggered Architecture *pk, kg*
OTA Operational Test Organization *kf*
OTA Operator Task Analysis *fu*
OTA Optical Telescope Assembly [on HST] *mi*
OTA Over Target Authorization (budget term) *fu*
OTAN {Organization du TraitÈ de l'Atlantique Nord} [(French) NATO] *mt*
OTAR Over The Air Rekey *mt, jdr, uh, fu*
OTAS Observers Target Acquisition System *wp*
OTC Officer in Tactical Command *mt, fu, tb*
OTC Operational Test Center *wp*
OTC Operational Training Capability *mt*
OTC Overseas Telecommunications Commission *mt, wp*
OTC [Officer Training School] *wp*
OTCCC Open Type Control Circuit Contact *fu*
OTCIXS Office(r) in Tactical Command Information Exchange (Sub)System *fu, mt*
OTD Optical Technology Development *st*
OTDA Office of Tracking and Data Acquisition *2, 4*
OTDB Operations Tasking Data Base [PACAF] *mt*
OTDR Optical Time Domain Reflectometer (Reflectometry) *fu, kg, hc, wp*
OTE Operational Test and Evaluation [also OT&E] *wp*
OTEA Operational Test and Evaluation Agency [US Army] *wp, st, mt*
OTEC Ocean Thermal Energy Conversion *hc, wp*
OTEC Operational Test and Evaluation Center *mt*
OTEF Operational Test and Evaluation Facility *hc*
OTEP (OTEP - a Vision for the Future) *aa26.11.b48*
OTEP Operational Test and Evaluation Plan *wp*
OTES Operational Test and Evaluation Squadron [AFCC*] *mt*
OTF Open Token Foundation *pk, kg*
OTF Operational Test Flight *tb*
OTF Optical Transfer Function *wp*
OTG OPTEVFOR Tactics Guide [US Navy] *mt*
OTH Over The Horizon *mt, pbl, ro, fu, 12, 24, bm132, wp, hc, is26.1.66, pf.f88.16, kf*
OTH-B Over The Horizon Backscatter [also OTHB] *mt, ro, fu, 12, 16, aw, ph, af72.5.150, tb, kf*
OTH-B_WCRS Over The Horizon Backscatter West Coast Radar System *mt*
OTH-R Over The Horizon Radar [also OTHR] *ph*
OTH-T Over The Horizon Targeting *mt, uc, pf1.4, fu*
OTHB Over The Horizon Backscatter *hc, wp, mb, mi*
OTHR Over The Horizon Radar *wp, fu, aw130.6.34, tb*
OTI Ocean Technology, Incorporated *fu*
OTIIA Operational Test IIA *hc*
OTIP Offense Technology Interaction Program *tb*
OTM Object Transaction Monitor *em*
OTM Operational Test Models *mt*
OTM Orbit Trim Maneuver *mi*
OTP Office of Telecommunications Policy *fu*
OTP One-Time Programmable *pk, kg*
OTP Operating Temperature Range *wp*
OTP Operations (Operational) Test Plan *0, wp*
OTP Orbital Test Plans and Procedures *uh*
OTP Orbital Test Plans/Procedures *jdr*
OTP Ozone Trend Panel [NASA] *ge, hg*
OTPP Operational Transfer Point Plan *mt*
OTPS Operation(al) Test Program Sets [F/A-18] *mt, hc*
OTR Operational Trouble Report *fu*
OTR Optical Tracking *wp*
OTRA Oversea Theater Requisitioning Authority *wp*
OTRAG Orbital Transport und Raketen AktienGesellschaft *sc3.12.4, 14, crl*
OTRS Operational Testing Readiness Statement *st*
OTS Office of Thrift Supervision *y2k*
OTS Officer's Training School [ATC] *w9, mt*
OTS Operational Test Site *mt*
OTS Optimal Thruster Section *jdr, uh, jn*

OTS Orbital Test Satellite *sc2.8.3, hc, sf31.1.16*
OTS Orbital Tracking System *oe*
OTS Ordnance Test Set *oe*
OTSG Operational TADIL Standards Group *fu*
OTSR Optics Technology for Standoff Reconnaissance *fu, hc*
OTSR Optimum Track Ship Routing *hy*
OTSS Operational Telecommunications Switching System *mt*
OTU Operational Training Unit *wp*
OTU Oxidizer Transfer Unit *oe*
OTV Operational Television *oe*
OTV Orbital Transfer Vehicle *mi, wp, sc4.1.4, 16, 17, aa26.11.24, ce, dm, tb, wl, mb*
OTWG Operational Test Working Group *mt*
OU Organizational Unit *ct, em*
OUADP Operational Utility of ADP *mt*
OUE Operational Utility Evaluation *mt*
OUIC Operational Unit Identification Code *mt*
OURS Open Users Recommended Solutions *kg, ct, em*
OUSA Office of the Under Secretary of the Army *dt*
OUSD Office of the Under Secretary of Defense *wp*
OUSDR&E Office of the Under Secretary of Defense for Research and Engineering *dt*
OUSDRE Office of the Under Secretary of Defense for Research and Engineering *12, hc, st*
OUSN Office of the Under Secretary of the Navy *dt*
OUT Operational Utilization Test *wp*
OUTS Output String *pk, kg*
OV Open Ventilated *fu*
OV Operational Verification *mt*
OV Orbiting (Orbital) Vehicle *pl, wp, fu, mi, ai2*
OV Orientation Visit *mt*
OV [designator, OV-10 for Bronco] [USA DOD] *mt*
OV [prefix for small observation aircraft] *wp*
OVAL Object based Virtual Application Language *pk, kg*
OVE On Vehicle Equipment *fu*
OVE Otdelnii Vertoletnii Eskadrilya ["independent helicopter squadron"] [USSR] *mt*
OVERVIEW [POM Spares Forecasting System] *mt*
OVMS Open VMS *hi*
OVO Orbiting Volcanological Observatory *hy*
OVP On-orbit Verification Review *oe*
OVP Otdelnii Vertoletnii Polk ["independent helicopter regiment"] [USSR] *mt*
OVP Over Voltage Protection *fu, jdr*
OVT Operational Validation Test *mt*
OVTR Operational Video Tape Recorder *wp*
OW Oil immersed Water cooled *wp*
OW Order Wire *mt, fu, kf*
OWDC Office of Water Data Coordination [of the USGS] *ac1.5-14*
OWDM Optical Wavelength Division Multiplexer *fu*
OWE Operating Weight Empty *mt*
OWF Optimum Working Frequency *fu*
OWG Operation Working Group *oe*
OWGN Output Wave Guide Network *jdr*
OWL Object Windows Library [Borland] *hi*
OWL One Watt Linear *hc, hq6.2.7*
OWL Open Windows Library [Borland] *hi*
OWL/D Optical Warning and Location Designator *hc*
OWM Office of Weights and Measures *wp*
OWM Order Wire Modem *fu*
OWO On Work Order *wp*
OWRM Office of Weather Research and Modification *un*
OWRM Other War Reserve Material *mt*
OWRMR Other War Reserve Materiel Requirements *mt*
OWS Operator Work Station *mt*
OWS Orbital Workshop *ac2.3-3, hc*
OWS Overload Warning System *wp*
OXO Orbiting X-Ray Observatory *wp*
OXRO Orbiting X-Ray Observatory *wp*
OXS Objective X-ray Crystal Spectrometer *cu*
OXS Oxygen Sensor *wp*
OYSTER Optical Yardsticks Toward Error Reduction [a Hughes system] *17, kf*

P

P {Production} [code for date, P-day is the date at which the rate of production of an item for military consumption equals the rate at which the item is required by the Armed Forces [USA] [DOD] *mt*
P&A Parameter and Assembly *fu*
P&A Program and Acquisition *st*
P&A Programming and Analysis *mt*
P&ES-W Personnel and Employment Service - Washington *dt*
P&F Program and Financial Plan *mt*
P&IC Production and Inventory Control *fu*
P&L Power and Lighting *fu*

P&MD Parts and Materials Document *oe*
P&O Positioning and Orientation *bl2-9.15*
P&P Planning and Programming *mt, af72.5.31*
P&P Procurement and Production *mt*
P&R Planning and Resources *mt*
P&R Plans and Requirements *mt*
P&RF Personnel and Reserve Force *dt*
P&S Personnel and Security [REDCOM/JDA] *mt*
P&T Pointing and Tracking *st*
P&T Posts and Telegraph [Department, Ministry of Communications, India] *ac3.2-15*
P&TF Prototype and Test Facility *mt*
P&W {Pratt and Whitney} *af72.5.82*
P&Y Pitch and Yaw [steering law coefficients] *oe*
P(R) Packet Receive Sequence Number *fu*
P(S) Packet Send Sequence Number *fu*
P-3 Orion [USA DOD] *mt*
P-DAT Portable Digital Access Transceiver *hc*
P-DIL Prereleased Design Integration Laboratory *ep*
P-MAIL Paper Mail *pk, kg*
P-P Peak to Peak *gp, 16, fu, he*
P-S Parallel-to-Serial *fu*
P-SRAM Pseudo-Static Random Access Memory *kg, hi*
P-V Peak-To-Valley *fu*
P/A Polar To Analog *fu*
P/A Product Assurance *fu*
P/A Programmer/ Analyst *fu*
P/D (P/D amplifier and sync. demod) *ac7.2-15*
P/DU Processing or Distribution Unit *gp, 16*
P/N Part Number *mt, 8, fu, oe*
P/N Pseudo Noise *0*
P/O Purchase Order *fu*
P/P Parts Protection *fu*
P/R Primary/Redundant *jdr*
P/S Phase Shift *ac7.2-15*
P/S Power Supply *jdr*
P/SR&A Propulsion/Structures Research and Analysis *st*
P/S_CP Platoon/Section Command Post *fu*
P/T Part Time *pk, kg*
P/TR Program/Technical Review *fu*
PA Penetration Aid *kf*
PA Performance Adjustment *oe*
PA Performance Analysis [group] *0*
PA Performance Appraisal *ah, pbl*
PA Permanently Assigned [Channel] *dmi*
PA Power Amplifier *gg149, 2, 4, w9, tb, wl, fu, oe*
PA Privacy Act *mt, wp*
PA Probability of Arrival *mt*
PA Problem Analysis *hi*
PA Problem Area *0*
PA Process Anomaly *jdr, pbl, tm, jn*
PA Process Average *fu*
PA Procurement Authorization *mt, tb*
PA Procuring Authority *fu*
PA Product Assurance [an organization] *mt, sc4.7.center, 16, 17, sb, pl, nt, ep, st, jdr, pbl, tm, fu, ah, jn, oe*
PA Program Access *fu*
PA Program Analysis [and evaluation] *hc*
PA Program Authorization *mt, wp*
PA Prologue Auditor *hd1.89.y30*
PA Provisional Acceptance *mt*
PA Public Affairs *mt, dt*
PA Pulse Amplifier *fu*
PA Pulse Amplitude *fu*
PA Purchase Agreement *fu, jdr*
PA&E Program Analysis and Evaluation *mt, tb*
PA/E&A Product Assurance, Engineering and Audit *fu*
PA/EDS Product Assurance, Engineering Data Support *fu*
PA/MD Product Assurance, Manufacturing Division *fu*
PA/PO Product Assurance Programs Office *fu*
PA/POE Product Assurance Programs Office Engineers *fu*
PA/RA Product Assurance, Reliability Assurance *fu*
PA/SQS Product Assurance Software Quality Support *fu*
PAA Passive Acoustic Analysis *hc*
PAA Payload Attach Assembly *ai2*
PAA Primary Aircraft Assigned *wp*
PAA Primary Aircraft Authorization *wp*
PAA Primary Authorized Aircraft *wp*
PAANG Pennsylvania Air National Guard *-12*
PAAS Passive Active All Weather System *wp*
PAAS Passive Active Attack System *hc*
PAATS Precision Approach Area Tracking System *wp*
PAAW Preliminary Announcement of Anticipated Waiver *oe*
PAAWS Precision Advanced All Weather Strike *hc*
PABA Para Amino Benzoic Acid *wp*
PABA Post Air Battle Analysis *fu*

PABX Private Automatic Branch Exchange [telephone] *mt, do 331, fu, kg, ct, hi*
PAC (PAC air takes over this route from . . . Eastern Express) *aw130.13.15*
PAC Pacific Air Command *wp*
PAC Pacific Area *mt*
PAC Parachute and Cable [rocket] [type L was a 2 inch rocket launched to 1000ft, dispensing a 600ft cable with a parachute at either end and an explosive charge at one. Types J, K had other altitudes and cable lengths] [UK] *mt*
PAC Personal Authenticator Card [DD Form 1833] *mt*
PAC Personnel Attendance Concept *fu*
PAC Phase/Amplitude Control *jdr*
PAC Planning Advisory Committee *mt*
PAC Post Award Conference *tb*
PAC Primary Address Code *fu*
PACAF Pacific Air Forces [USAF] *mt, fu, hn44.9.10, wp, af72.5.35, hc*
PACAIDS PACOM Crisis Action Information Distribution System *mt*
PACAMS Pacific Aircrew Management system *mt*
PACBAR Pacific Barrier [space surveillance radar system] *mt, ps204, wp*
PACC Pacific Airlift Control Center *mt, wp*
PACCA Airfield Capabilities Application [PACOM] *mt*
PACCAT Pacific Area Command and Control AUTODIN Terminal *mt*
PACCOMMAREA Pacific Communications Area [AFCC*] *mt*
PACCS Post Attack Command and Control System [USAF] [later known as Strategic Command and Control System, SCACS] [USA] *mt, wp, tb*
PACDIGS Pacific Digital Graphics System *mt*
PACE Pacific Airlift Center *mt*
PACE Performance Analysis and Continuous Evaluation *fu*
PACE Performance and Cost Evaluation *fu*
PACE Perigee Augmentation Control Electronics *gp, 16*
PACE Priority Access Control Enabled [3Com] *pk, kg*
PACE Processing and Classification of Enlistees [ATC] *mt*
PACE Program Acquisition Cost Estimate *mt*
PACE Program for Airport Efficiency *aw129.21.98*
PACE Programmable Aerospace Checkout Equipment *tb*
PACE Programmable Aerospace Control Equipment *oe*
PACER Program Assisted Console Evaluation and Review *mt, ro*
PACERS Peacetime Analysis of the Combat Effectiveness of Repairable Spares [TAC*] *mt*
PACER_ACE [support of prototype aircraft] *wp*
PACER_ACQUIRE [AFLC Management System Acquisition] [AFCL] *mt*
PACER_ANGLER [Kc-135a conversion program] *wp*
PACER_ASH [resupply destroyed assets] *wp*
PACER_BASS [movement of non-DOD cargo] *wp*
PACER_BELL [an excess inventory reduction program] *wp*
PACER_BOX [battle damaged or crashed aircraft requirement program] *wp*
PACER_BRIGHT [F-4 assets identification program] *wp*
PACER_CHECK [KC-135 supply support program] *wp*
PACER_CLAM [F-15 age management program] *wp*
PACER_CRAFT [special aircraft for Iran mission] *wp*
PACER_DAMAGE [C-141B recovery] *wp*
PACER_DART [F106 conversion to F-4d] *wp*
PACER_GOOSE [sealift resupply of Greenland bases] *wp*
PACER_HILL [F-4c support program] *wp*
PACER_LINK [Ec-135 modified with EMP hardened systems] *wp*
PACER_PATTERN [F-15 avionics support] *wp*
PACER_PUP [C-5 support and depot maintenance program] *wp*
PACER_ROCK [F-4c support] *wp*
PACER_SACK [simulator for air to air combat] *wp*
PACER_STABILIZER [assets move to support grounded F-15-S] *wp*
PACER_WAX [AWACS DDT&E support] *wp*
PACER_WING [C-5 wing mod] *wp*
PACFLT Pacific Fleet [US Navy] *12, mt*
PACHEM Point Area Chemical Effects Model [chemical warfare] *wp*
PACIR Propulsion, Aerodynamic, Controls Integrated Research [USA] *mt*
PACM Passive Access Control Module *mt*
PACM Pulse Amplitude Code Modulation *fu*
PACMEDS Pacific Meteorological Distribution Systems *mt*
PACMS Pacific Crisis Management System *mt*
PACOM Pacific Command *mt, 12, wp, jb, pm, kf*
PACOR Passive Correlation and Ranging *fu*
PACOSS Passive and Active Control Of Space Structure *hc*
PACR Perimeter Acquisition Radar *fu*
PACR Planned Activity Change Request *oe*
PACS Passenger Automated Check In System [MAC*] *mt*
PACS Pentagon Automated Communications System [USA] *mt*
PACS Personal Access Communications System *jr, heh2.21.4*
PACS Picture Archiving and Communication System *kg, cm16.8.9*
PACS Plotter and Combat System *fu*
PACS Plotting and Computing System *fu*
PACS Pointing and Control System *wl*
PACS Programmable Armament Control Set (System) *hc, wp*
PACS [horizontal plotter combat summary display] *hc*

PACS-L Public Access Computer Systems List [Internet] *kg, hi*
PACSBB Phased Array Concept Study and Brass Board *hc*
PACT Passive Active Correlation Techniques *hc*
PACT Portable Automatic Calibration Tracker *hc*
PACTAIS Pacific Theater Air Intelligence System [PACAF] *mt*
PACTIDS Pacific TAC Intelligence Data System [PACOM] *mt*
PACVI Product Assurance Configuration Verification Index *fu*
PACWRAC PACOM WWMCCS Regional ADP Center *mt*
PAC_II CCSC Project Management System *mt*
PAD Packet Assemble(r) / Disassemble(r) *tb, fu, ct, kg, em, hi*
PAD Packet Assembly / Disassembly *mt*
PAD Port Air Defense *fu*
PAD Power Amplifier Driver *fu*
PAD Preferential Adaptive Defense *tb*
PAD Product Assembly Department *fu*
PAD Product Assurance Directive *gp*
PAD Program Action Directive *mt, wp*
PAD Program and Acquisition Division *st*
PAD Project Approval (Authorization) Document *pl, re, oe*
PAD Propellant Actuated Device *wp*
PAD Property Accountability Document *fu, sc8.5.4, 15, jdr*
PAD [fixed attenuator] *bl2-9.15*
PADA Physical Disability Agency, [US Army] *dt*
PADAR Passive [Airborne] Detection And Ranging *fu, wp*
PADDS Procurement Automated Data and Document System [USA] *mt*
PADEX Passive Dumping on a Beam Expandable [Telescope] *hc*
PADGE Philippine Air Defense Ground Environment *fu*
PADLOC Passive Detection and Location Of Countermeasures *fu*
PADM Product Assurance Directives Manual *gp*
PADMS Producibility based Automated Design and Manufacturing System *fu*
PADRE Portable Automatic Data Recording Equipment *fu*
PADS Passive Active Data Simulation *fu*
PADS Passive Advanced Sonobuoy *hc*
PADS Pen Application Development System [Slate Corporation] *pk, kg*
PADS Penetration Aids Deployment System *wp*
PADS Position and Azimuth Determining System *fu*
PADT Preliminary Aircraft Design Technology *wp*
PAE Phase Angle Error *fu*
PAE Polyarylene Ether *aa26.8.40*
PAE Precision Attack Enhancement *hc*
PAE Program Analysis and Evaluation *mt*
PAE&A Product Assurance Engineering and Audit *fu*
PAEDS Product Assurance Engineering Data Support *fu*
PAETS Product Assurance Estimating Techniques System *gb*
PAF Payload Attach Fitting *bl1-2.170, 9, lt, oe, ai2*
PAF Precision Processing and Archiving Facility [ESA] *ge, hg*
PAF Preprocessing and Archiving Facility *es55.44*
PAFAM Performance And Failure Assessment Monitor *wp*
PAFB Patrick Air Force Base [Cocoa Beach, Florida] *0, wp*
PAFB Peterson Air Force Base *kf*
PAFCCN PACAF Command and Control Network *mt*
PAFCONET PACAF Teletypewriter Network *mt*
PAFDEF PACAF Defense Network *mt*
PAFSC Primary Air Force Specialty Code *mt*
PAFVONET PACAF Command and Control Voice Network *mt*
PAG Parts Advisory Group *fu*
PAG Platform Applications Group *fu*
PAG Prototype Antenna Group *fu*
PAGEL Priced Aerospace Ground Equipment List *mt, fu*
PAGEOS Passive Geodetic Orbiting Satellite *ci.950*
PAGES Print And Graphics Express Station *tb*
PAGR Projected Annual Growth Rate *hi*
PAH Panzer Abwehr Hubschrauber [(German) anti tank helicopter] *mt*
PAHOC [German] Patriot / Hawk Operations Center [later SAMOC] *hc*
PAI (The editorial office (SSD/PAI)) {an organizational division of SSD} *an31.1.2*
PAI Primary Aircraft Inventory *mt, wp*
PAI Product Assurance Instruction *gp*
PAI Provisional Acceptance Inspection *mt*
PAIH Public Access Internet Host [Internet] *kg, hi*
PAILS Projectile Airburst and Impact Locating System *hc, wp*
PAINT Post Attack Intelligence *hc*
PAIP Product Assurance Implementation Plan *gp, oe*
PAIR Phased Array Instrumentation Radar *wp*
PAIRS Program For the Analysis of Infrared Spectra *wp*
PAIRTECH Passive Autonomous Infrared [Sensor] Technology *hc*
PAIS Program Analysis and Information Support *st*
PAIS Prototype Advanced Indications System *wp*
PAIS Public Access Internet Site [Internet] *pk, kg*
PAIT PLRS/ADDS Improved Terminal *fu*
PAL Hughes Process Access Library *kf*
PAL Paradox Applications Language [Borland] *kg, hi*
PAL Penetration Aid Launcher *wp*
PAL Permissible (Permissive) Action Link *fu, 14, wp*
PAL Permissible Access Links *tb*

PAL Permission Active Link [device to trigger nuclear bombs from the cockpit] [USA] *mt*
PAL Phase Alteration Standard *ct*
PAL Phase Alternation Line [a type of videocassette] *fu, dn, cs1.20, aw129.10.60, ct, kg*
PAL Phased Array Lens *un*
PAL Prescribed Action Link *fu*
PAL Program Assembler Language *mt*
PAL Programmable (Programmed) Array Logic *fu, kg, ct, em, hi, oe*
PAL Programming Assembly Language *kg, hi*
PAL Propulseur d' Appoint Liquid [Ariane 4 liquid rocket booster] *ai2*
PALACE [nickname for HQ USAF programs connected with personnel] *wp*
PALACE_DOG [TDY contingency program] *wp*
PALACE_SENTINEL [intelligence officer career management] *wp*
PALADIN [US Army's anti air weapon system] *hc*
PALC Plasma Addressed Liquid Crystal [display] *pk, kg*
PALCC Pacific Airlift Control Center [PACAF] *mt*
PALCD Plasma Addressed Liquid Crystal Display *hi*
PALDEM PAL Demonstration *tb*
PALM Perigee or Apogee Liquid Motor *gp, 16*
PALS Photo Area and Location System *wp*
PALS Portable Airfield Light Set *wp*
PALS Portable Airfield Lighting System *mt*
PALS Principles of the Alphabet Literacy System *pk, kg*
PALT Procurement Acquisition Lead Times *mt*
PALT Procurement Administrative Lead Time *wp*
PALTT Preflight Actual Launch Time Trajectory *oe*
PAM Payload Assist Module *mi, hn44.4.2, 17, wp, aw, af72.5.149, lt, ep, ce, wl, jdr, mb*
PAM Perigee Assist Motor (Module) *gp, 9, 16*
PAM Peripheral Adapter Module *fu*
PAM Power Amplifier Module *fu*
PAM Pre Amplifier Module *fu*
PAM Procurement and Acquisition Management *mt*
PAM Product Assurance Manager *17, sb*
PAM Program Analysis Memorandum *mt*
PAM Pulse Amplitude Modulated (Modulation) *fu, mt, kg, bl1-5.122, 0, wp, 4, aa26.11.44, ce, jdr, hi, ai1*
PAM Pyridine Aldoxime Methiodide [chemical warfare nerve agent antidote] *wp*
PAM-D Payload Assist Module, Delta-class *mi, oe, ai2*
PAMCS Panel on Air Space Management and Control System [ACCS] *mt*
PAMD Product Assurance Manufacturing Division *fu*
PAMELA Process Abstraction Method for Embedded Large Applications *mt*
PAMIS PSYOP Automated Manpower Information System *mt*
PAML Program Authorized Materials List *oe*
PAMO Product Assurance Management Office *fu*
PAMOATE [4,4'-methylenbis93-hydrox2-2-naphtoate] *wp*
PAMPL Program Approved Materials and Processes List *jdr*
PAMRI Peripheral Adapter Module Replacement Item *fu*
PAMS Pacific AUTODIN Multiplex System *mt*
PAMS Patent Application Management System *wp*
PAMS PMEL Automated Management System *mt*
PAMS Position/Altitude Measuring System *hc*
PAN Peroxy Acyl Nitrate *ac1.7-13, aa26.12.76*
PAN Poly Acrylo Nitrile *kf*
PAN Primary Alerting Network *mt*
PANEX Pantheistic Executive Program *fu*
PAO Primary Action Office *wp*
PAO Program Action Officer *mt*
PAO Property Accountability Order *jdr*
PAO Property Accounting Office *mt*
PAO Property Action Order *sc8.5.4*
PAO Public Affairs Office(r) *mt*
PAOC Pacific Air Operations Center [PACAF] *mt*
PAOC Post Award Orientation Conference *dm*
PAOTS Power And Ordnance Test Set *-4*
PAP Packet level Procedure *kg, hi*
PAP Password Authentication Protocol *ct, kg, em, hi*
PAP Property Administration Procedure *fu*
PAP Propulseur d' Appoint Poudre [Ariane 4 solid rocket booster] *ai2*
PAP [solid propellant strap on booster for Ariane] *es55.83*
PAPA [nickname for Alaskan Air Command programs] *wp*
PAPE Product Assurance Project Engineer *fu*
PAPI Precision Approach Path Indicator *mt, aw118.18.89, fu*
PAPL Preliminary Advance Parts List *fu*
PAPL Program Approved Parts List *vv*
PAPL Program Authorized Parts List *oe*
PAPL/S Program Approved Parts List and Specifications *jdr*
PAPM Product Assurance Program Manager *oe*
PAPM Pulse Amplitude and Phase Modulation *hc*
PAPO Product Assurance Program Office *fu*
PAPOE Product Assurance Programs Office Engineer *fu*
PAPP Product Assurance Program Plan *fu, nt*

PAPRL Program Approved Parts Requirement List *oe*
PAPS Periodic Armaments Planning System [NATO] *hc, mt*
PAPS Portable ADA Programming System *mt*
PAQ Pending Action Queue *mt*
PAR Page Address Register *-3*
PAR Panzer Abwehr Regiment ["anti tank regiment"] [Germany] *mt*
PAR Peak to Average Ratio *mt*
PAR Performance Analysis and Review *wp*
PAR Perimeter Acquisition Radar *mt, fu, wp, tb, st*
PAR Personal Animation Recorder *pk, kg*
PAR Phased Array Radar *mt, 12, wp, sd, tb*
PAR Photosynthetically Active Radiation *ge, hg*
PAR Planning Activity Report *mt*
PAR Population / Personnel At Risk *mt*
PAR Positive Acknowledgement with Retransmission *mt, fu*
PAR Precision Approach Radar *mt, fu, hc, wp, tb*
PAR Preferential Arrival Route *fu*
PAR Preliminary Analysis Review *mt*
PAR Product Assurance Requirements *fu, pl, jdr*
PATHS Program Acquisition Request *mt*
PAR Program Acquisition Review *hc*
PAR Program ADP Requirement *mt*
PAR Program Analysis and Review *fu*
PAR Program Assessment Review *fu, hc, wp*
PAR Projected Automation Requirement *mt*
PARA [two groups arranged in reflexion (180 deg opposite)] *wp*
PARABEN Parahydroxybenzoate *wp*
PARAMP Parametric Amplifier *wp*
PARC Palo Alto Research Center [XEROX] *kg, ct, sm, hi*
PARC Progressive Aircraft Recoditioning Cycle *wp*
PARCS Perimeter Acquisition Radar [attack] Characterization System *mt, hc, 12, sd, jb, wp, tb*
PARD Phased Array Radar Detection / Track *mt*
PARD Pilot Airborne Recovery Device *wp*
PARDOP Passive Ranging Doppler *fu*
PARDPS PAR Data Processing System *mt*
PARDS Phased Array Radar Detection System *wp*
PARIS Passive/Active Radar Identification System *hc*
PARL Program Applicable Requirements List *fu*
PARM Post Attack Resource Management *wp*
PARM Primary Alignment Reference Mirror *gp, 16*
PARMIS Planning and Resource Management Information System *mt*
PARP Preengineered AUTOVON Restoral Plan Pacific *mt*
PARPRO Peacetime Aerial Reconnaissance Program *mt*
PARR Program Analysis and Resource Review *mt*
PARROT Position Adjustable Radar Range and Orientation Transponder *hc*
PARS Parachute Altitude Recognition System *wp*
PARS Performance Analysis Reports System *mt*
PARS Procurement Action Reporting System *wp*
PARS Property Accountability Record System *wp*
PARTS Parts Analysis and Review Technique for Spares *mt*
PARTS Parts Automated Repairable Tracking System [USAF] *mt*
PAS Payload Assist Stage *-17*
PAS Penetration Aids System *tb*
PAS Performance Analysis System *jdr*
PAS Performance Assessment System *hc*
PAS Personnel Accounting Symbol *mt*
PAS Personnel Accounting System *mt*
PAS Planning and Strategy *fu*
PAS Pneumatic Actuation System *rm8.1.3*
PAS Primary Alert System *tb*
PAS Primary Alerting System *mt*
PAS Procurement Action System *wp*
PAS Propellant Acquisition System *ai2*
PAS Publicly Available Submitter *ct, em*
PASA Pay Administration Standards and Appeals *dt*
PASARS Podded Advanced Synthetic Aperture Radar System *hc*
PASC Pacific Area Standards Conference *fu*
PASCALS Passive Automatic Carrier Landing System *hc*
PASCOM Passive Communicator *hc*
PASLA Programmable, Asynchronous, Single-Line Adapter *fu*
PASS Parked Aircraft Sentry System *wp*
PASS Personal Access Satellite System *-19*
PASS Photo Interpretation Analyst Support System *wp*
PASS POCC Applications Software Support [Space Telescope] *hc*
PASS Pointing And Stabilization Subsystem *sd*
PASS Procurement Automated Source System *wp, st*
PASTRAM Passenger Traffic Management System [MTMC] *mt*
PASU Preliminary Approval for Service Use *fu*
PAT Passive Angle Track(ing) *fu, mt*
PAT Performance Analysis and Test *kf*
PAT Perigee Augmentation Timer *uh, gp, jn*
PAT Peripheral Allocation Table *hi*
PAT Portable Analog Tester *fu*
PAT Preliminary Acceptance Trial (Test) [Navy] *gp, fu, oe*

PAT Product Assurance Test *fu*
PAT Production Acceptance Test *fu, mt*
PAT Program Acceptance Test *fu*
PAT Pulse Ablative Thruster *js12.643*
PAT&E Production Acceptance Test and Evaluation *fu, st*
PAT/ARM Passive Angle Tracking / Anti Radiation Missile system [fitted to some A-6B Intruders for operations against SAM* sites in Vietnam] [USA] *mt*
PATAT Passive Acquisition/Tracking of Airborne Targets *hc*
PATCH [nickname for HQ USAF programs connected with supply] *wp*
PATCH_BUGGIE [program for medical supply for chemical warfare] *wp*
PATCO Professional Air Traffic Controllers' Organization *aw126.20.32, aa26.8.6*
PATE Production Acceptance Test and Evaluation *wp*
PATE Programmable ATEC Terminal Element *mt*
PATEC Portable Automatic Test Equipment Calibrator *fu*
PATHFINDER Passive Thermal Forward-Looking Infrared for Navigation, Detection and Enhanced Resolution *aw87sep21.101*
PATHS Precursor Above The Horizon Sensor *hn49.11.3, ie5.11.8*
PATN Promotional Port Access Telephone Number *pk, kg*
PATR Portable Acoustic Tracking Range *hc*
PATRAN (PATRAN serves as an interactive graphics postprocessor) *hd1.89.y17*
PATRIC Pattern Recognition Interpretation and Correlation *fu, wp*
PATRIOT [an anti tactical ballistic missile system of the US Army] *mt, nd74.445.18*
PATS Parts Approval Tracking System *fu*
PATS Precision Acquisition Tracking System *hc*
PATS Precision Angular Tracking System *ro*
PATS Primary Aircraft Training System *wp*
PATS Prototype Automated Telecommunications System *mt*
PATWAS Pilots' Automatic Telephone Weather Answering Service *fu, ag*
PATWING Patrol Wing *-12*
PAU Power Amplifier Unit *fu*
PAV Position And Velocity *fu*
PAVE Position And Velocity Extraction *fu*
PAVE Precision Acquisition of Vehicle Entry *hc, 12, sd, jb*
PAVE Principles and Applications of Value Engineering *mt*
PAVE [USAF programs connected to (night) avionics, target designators, vision] *wp*
PAVE_Aegis (the Lockheed AC-130 PAVE_Aegis carries a 105mm cannon) [USA] *mt*
PAVE_Arm [program to create missiles that homed on the emissions of aircraft] [USA] *mt*
PAVE_Arrow [two F-4Ds were modified to locate heat sources on the ground, using modified AIM-9 seekers] [USA] *mt*
PAVE_BOW [night avionics for O-2 aircraft] *wp*
PAVE_BRAZO [Eglin AFB program] *hc*
PAVE_CAP [a Wright Patterson AFB palletized night attack system program] *hc*
PAVE_CLAW [program to destroy armor and fixed targets] *wp*
PAVE_Claw [the GPU-5/A 30mm gun pod, the gun is the GAU-13/A, a four-barreled version of the GAU-8 Avenger of the A-10, USA] *mt*
PAVE_Coin [development program for a low cost, high performance close support aircraft, undertaken during the Vietnam war, USA] *mt*
PAVE_CRICKET [mini drone jammer] *wp*
PAVE_CUPS [computer updating of SAC C2 systems] *wp*
PAVE_DELTA [air to air missile warhead lethality program] *wp*
PAVE_Deuce [conversion of F-102 for use as manned or unmanned target, USA] *mt*
PAVE_Fire [pod with a low light level television camera and a laser ranger, to be used in night attacks, unsuccessful] [USA] *mt*
PAVE_GAMMA [aircraft engine radar signal feturn recognition (F-15 APG-63 radar)] *wp*
PAVE_Gat [Martin B-57G night attack aircraft equipped with a triple-barrel 20mm gun] [USA] *mt*
PAVE_HAWK [night vision on drones program] *wp*
PAVE_Hawk [MH-60G rescue helicopter, USA] *mt*
PAVE_Knife Guidance pod for laser-guided bombs. AN/AVQ-10 [USA] *mt*
PAVE_Light AN/AVQ-9 laser designator [USA] *mt*
PAVE_Low [MH-53J helicopter] [USA] *mt*
PAVE_MACE [program to develop mobile ground target destruction] *wp*
PAVE_MINT [EW countermeasures for the F-15 and B-52g] *wp*
PAVE_MOVER [moving target indicator radar] *wp*
PAVE_Mover [side looking radar for targeting missiles to enemy forces] [USA] *mt*
PAVE_Nail [modification of the OV-10 for night precision attacks] [USA] *mt*
PAVE_Nickel [radar reconnaissance flights of the RB-57F at the border of the Warsaw pact territory] [USA] *mt*
PAVE_Onyx [tactical electronic reconnaissance sensor, Litton AN/ALQ-125 Terec] [USA] *mt*
PAVE_PACE [21st century avionics design and demonstration program] *wp*
PAVE_PAT [fuel-air explosives connected program] *wp*
PAVE_Pat II [BLU-76 fuel-air explosive submunition bomb] [USA] *mt*

PAVE_PAWS Precision (Perimeter) Acquisition of Vehicle Entry [and] Phased Array Warning System [an early warning system for submarine launched missiles, USA] *mt, wp, tb, st*

PAVE_PENNY [Wright Patterson AFB advanced target identification laser program to track ground designated targets] *wp, hc*

PAVE_Penny [laser designation system for missiles, also known as TISL, USA] *mt*

PAVE_Phantom [modification of the F-4D Phantom II with AN/ARN-92 LORAN navigation equipment, USA] *mt*

PAVE_PILAR [advanced tactical fighter integration] *wp*

PAVE_PILLAR [a second generation avionics standardization R&D initiative] *dh, hc*

PAVE_PINE [over the horizon radar propagation program] *wp*

PAVE_POINTER [automatic target surveying system using laser designator] *wp*

PAVE_PRISM [evaluation of active radar guidance for the Aim-9l (Sidewinder)] *wp*

PAVE_Pronto [AC-130A gunship, USA] *mt*

PAVE_ROCK [warhead for 2.75" rocket to penetrate aircraft shelters] *wp*

PAVE_SCOPE [program to integrate target acquisition aids to fighter aircraft] *wp*

PAVE_SHIELD [infrared suppression program] *wp*

PAVE_Spectre [AC-130 gunship, USA] *mt*

PAVE_SPIKE [laser pod for the F-4 to acquire and designate targets] *wp*

PAVE_Spike [AN/AVQ-23 or AN/ASQ-152[V] laser designation system, USA] *mt*

PAVE_SPLINTER [part of PAVE_PILLAR] *wp*

PAVE_Spot [equipment for the coordination of air strikes, installed in USA] *mt*

PAVE_Storm [KMU-421/B guidance unit, a member of the PAVE_WAY family, fitted to cluster bombs, USA] *mt*

PAVE_Strike [defense suppression program, USA] *mt*

PAVE_Sword [AN/AVQ-11 precision attack sensor. A modified AIM-9 seeker head, used to track the targets designated by the AN/AVQ-12 Pave Spot laser designator, USA] *mt*

PAVE_TACK [IR night fision system (FLIR) and laser target designator pod] *mt, wp*

PAVE_TIGER [communications jamming mini-RPV] [mini-drone against C3 terminals] *dh, wp*

PAVE_WAY [a Wright Patterson AFB laser guided bomb program] *mt, hc, wp*

PAVS Passive Aided Visual Sensor *hc*

PAVT Position And Velocity Tracking *fu, wp*

PAW Peachtree Accounting for Windows *ct*

PAWLS Passive Artillery Weapons Locating System *hc*

PAWOS Portable Automatic Weather Observing Station *hc*

PAWS Personal Atlas Workstation *fu*

PAWS Phased Array Warning System *12, hc, sd, jb, wp*

PAWS Portable ASAS/ENSCE Work Station *mt*

PAWS Programmable ASAS/ENSCE Workstation *fu*

PAWS Prototype Analyst Work Station *mt*

PAX Portable Archive Exchange [Unix] *kg, hi*

PAX Private Automatic Exchange *ct, em, hi*

PAX_DOC Passenger Documentation *mt*

PA_IPT Performance Assurance IPT *kf*

PB Parallel Binary *-4*

PB Partial Band *jdr*

PB Particle Beam *hc, wp*

PB Patrol Boat (USN ship classification) *fu*

PB Pikuriyushi Bombardirovshchik [dive bomber] [USSR] *mt*

PB Post Boost *tb*

PB [chemical warfare codename for phenyldichloroarsine] *wp*

PB/BP Passive Beamformer/Broad-band Processor *fu*

PB/L Product Baseline *fu*

PBA Provincetown Boston Airline *aw129.21.15*

PBAN {Poly Butadiene Acryl Nitrite} (propellant will be 0.12 PBAN) *ci.223*

PBB Passive Broadband *fu*

PBB Programmable Building Block *fu*

PBBP Passive Beamformer Broadband Processor *hc*

PBBS Public Bulletin Board System [SESC] *pr*

PBC Pipelined Burst Cache *hi*

PBC Program Breakdown Code *tb*

PBD Production Buy Decision *mt*

PBD Program Budget Decision *mt*

PBD Program Budget Directive (Decision) *wp, tb*

PBDC Post Boost Detection System [also PBDS] *tb*

PBDS Post Boost Detection System [also PBDC] *tb*

PBG Program and Budget Guidance *mt*

PBGC Pension Benefit Guarantee Corporation *sc8.9.3, pbl, ah*

PBI Post Boost Intercept *17, sb, st*

PBI Push Button Indicator *mt*

PBIT [In Flight BIT] *rm2.3.4*

PBM Patrol Boat Multi-mission *hc*

PBMR [PLDS data set] *hg, ge*

PBP Post Boost Phase *17, sb*

PBR Patrol Boat, River [USN ship classification] *mt, fu*

PBR Post Boost Regime *st*

PBR Program Budgetary Review *mt*

PBR Project Business Review *fu*

PBRC Program Budget Review Committee *mt*

PBS Portable Base Station *ct*

PBS Production Base Support *wp*

PBS Public Broadcasting Service *dn, wp, w9*

PBS/IS Public Buildings Service / Information System *mt*

PBT Polybenzothiazole *wp*

PBU Passive Beamforming Unit [built by Hughes for AN/BQQ-5] *fu*

PBU Product Business Unit [an organization] *ah, pbl. jdr, tm, jn*

PBV Post Boost Vehicle *mt, fr, wp, 5, 17, sb, tb, st, kf*

PBVCO Post Boost Vehicle Cut-Off *tb*

PBVI Post Boost Vehicle Ignition *tb*

PBW Particle Beam Weapon *mt, wp, tb, st*

PBW Parts By Weight *wp*

PBX Plastic Bonded Explosive *wp*

PBX Private Branch Exchange *mt, kg, nu, fu, ct, em, csIII.58, wp, w9, hi, y2k*

PBX Private Business Exchange *jdr*

PC Parts Catalog *fu*

PC Patrol Corvette / Patrol Craft [US Navy] *mt*

PC Payload Control *kf*

PC Personal Computer [IBM compatible] *fu, ct, em, kg, ah, aa26.12.19, sv1.9.2, 8, w9, aw130.11.76, hn49.11.8, wp, pr, tb, wl, jdr, pbl, tm, kf, hi, oe, mt*

PC Photo Conductor (Conductive) *fu, gp, 17, sb, kf*

PC Planning Controller *fu*

PC Power Conversion (Converter) *fu, jdr*

PC Prime Contractor *fu, st*

PC Printed Circuit *fu, kg, wp, tb, hi*

PC Probability of Collection *kf*

PC Procurement Code *fu*

PC Procurement Cost *fu*

PC Production Control *mt, fu*

PC Production Coordination *fu*

PC Professional Computer *mt*

PC Program Control *wp, jdr*

PC Program Counter *fu, kg*

PC Proposed Change *fu*

PC Pulsating Current *fu*

PC Pulse Compression *fu, wp*

PC&E Program Control and Evaluation *fu*

PC-DOS Personal Computer - Disk Operating System [IBM] *kg, hi*

PC-I/O Program Controlled I/O *pk, kg*

PC-III Personnel Concept III *mt*

PC/G Parity Checker - Generator *fu*

PC/LT Procurement Code/Lead Time *fu*

PC/T Portable Loader/Tester *fu*

PCA 2-Pyrrolidone-5-Carboxylic Acid *wp*

PCA Pacific Communications Area [AFCC*] *mt*

PCA Percent Coverage Area *jdr*

PCA Performance and Coverage Analyzer *ct, em*

PCA Permanent Change of Assignment *mt, wp*

PCA Physical Configuration Audit *mt, fu, uh, pl, wp, hc, tb. jdr, kf, oe*

PCA Polar Cap Absorption *mt, 16, pr*

PCA Positive Control Airspace *fu*

PCA Pre Contractual Agreement *jdr*

PCA Pre Contractual Authorization *fu, 8, 15, hc, jdr, pbl*

PCA Product Configuration Audit *fu, jdr*

PCA Program Controls Analyst *jdr*

PCA Project (Program) Controls Administrator *fu, jdr*

PCA Project Control Analyst *-8*

PCA Propellant Control Assembly *jp*

PCAC Primary Control and Analysis Center *mt*

PCACIAS Personal Computer Automated Calibration Interval Analysis System *pk, kg*

PCAD (the following design capture systems *PCAD, >*

PCAM Punch Card Accounting Machine *fu*

PCAM Punched Card Accounting Machine *mt*

PCAS Personnel Cost Accounting System *mt*

PCAS Production Cost Analysis Summary *fu*

PCASO Pilot Classification And Screening Operation *wp*

PCASS Parts Control Automated Support System *mt*

PCB Parts Control Board *fu*

PCB Plenum Chamber Burning [the burning of fuel in the forward nozzles of the Rolls Royce Pegasus engine] [UK] *mt*

PCB Polychlorinated Biphenyl *wp*

PCB Power Circuit Breaker *fu*

PCB Pre Commit Boost *st*

PCB Printed Circuit Board *mt, fu, gp, kg, ah, wp, hd, 17, hc, ep, tb, jdr, pbl, tm, jn, hi, oe*

PCB Program Control Block *pk, kg*

PCBC Plain Cipher Block Chaining *pk, kg*

PCC PC Cluster Controller *mt*

PCC Primary Command Center *mt*

PCC Production Compression Capability *wp*
PCC Production Control Coordination *fu*
PCC Program Configuration Control *oe*
PCC Program Controlled Clock *fu*
PCC Program Coordinating Committee *ci.899*
PCC Programmable Communications Controller *fu*
PCC Provisioning Control Code *fu*
PCCADS Panoramic Cockpit Controls and Displays Systems [Program] *hc, wp*
PCCB Program Change Control Board *fu*
PCCIE Power Conditioning and Continuation Interfacing Equipment *mt*
PCCIP President's Commission on Critical Infrastructure Protection *y2k*
PCCN Provisioning Contract Control Number *fu, wp*
PCCS PACOM Command and Control System *mt*
PCCS Personnel Command and Control System *mt*
PCCS PLRS Communications Control Software *fu*
PCCS Portable Computers and Communications Services *hi*
PCCU Power Conditioning and Cover Control Unit *sd*
PCD Pacific Communications Division *mt*
PCD Parts Control Drawing *fu*
PCD Performance Commence(ment) Date *jk, jn*
PCD Photo Compact Disk *kg, hi*
PCD Power Control and Distribution *fu*
PCD Precision Course Direction *jj*
PCD Product Change Directory *jn*
PCD Product Configuration Documentation *jdr*
PCD Program Change Decision *mt, wp, tb*
PCD Program Coding Description *fu*
PCDB Project Comparative Database *kf*
PCDMO Program Configuration and Data Management Office *jdr*
PCDOS Personal Computer Disk Operating System *ct, em*
PCDP Pilot Control and Display Panel *fu*
PCDS Pilot Climate Data System [NASA] *hg, ge*
PCDU Power Control and Distribution Unit *jdr*
PCE PLRS Communications Enhancement *fu*
PCE Positional Control Equipment *fu*
PCE Positioner Control Electronics *jdr*
PCE Preliminary Closure Estimate *mt*
PCE Professional Continuing Education *mt, af72.5.72*
PCE Program Cost Estimate *wp*
PCESP Program Communication Electronic Support Program [AFCC*] *mt*
PCF Parallel Computing Forum *is26.1.38*
PCF Patrol Craft, Fast [USN ship classification] *fu*
PCF Personal Computing Facility *fu*
PCF Pounds per Cubic Foot *oe*
PCF Pulse to Cycle Fraction *fu*
PCFIS PACAF Combat Fuels Information System [PACAF] *mt*
PCG Power Connector Group *fu*
PCH Parts Cost History *fu*
PCH Patrol Craft, Hydrofoil (USN ship classification) *fu*
PCH Production Control Handbook *fu*
PCI Periodic Convolutional Interleaving *2, 4*
PCI Peripheral Command Indicator *fu*
PCI Peripheral Component Interface (Interconnect) *kg, hi*
PCI Personal Computer Interconnect *ct*
PCI Personal Computer Interface *em*
PCI Physical (Product) Configuration Identification *fu*
PCI Platform/Control Interface *fu*
PCI Process Control Instruction *fu*
PCI Procurement Control Identifier *fu*
PCI Product Configuration Identification *mt, jdr*
PCI Program Controlled Interruption *em*
PCI Project Code Identifier *fu*
PCI Project Control Identifier *fu*
PCIC PC-Card Interrupt Controller *pk, kg*
PCIT Production Control Inspection Test *fu*
PCL Parts Complement List *pl*
PCL Passive Coherent Location *hc*
PCL Performance (Process) Capability Limit *fu*
PCL Personnel [security] Clearance *tb, su*
PCL Positive Control Launch *mt, fu*
PCL Post Conference List *fu*
PCL Printer Command (Control) Language [HP] *kg, hi*
PCL Process Control Language *pk, kg*
PCL Process Control Levels *fu*
PCL Product Cleanliness level *he*
PCL Pulse Compression Line *fu*
PCLS Passive Coherent Location System *hc*
PCM Parametric Cost Model *16, kf*
PCM Passive Countermeasure *wp*
PCM Photometric Calibration Mirror *jp*
PCM Plane Change Maneuver *sa122052.1.1.2*
PCM Program Configuration Manager *oe*
PCM Pulse Code Modulation *mt, sp662, fu, kg, ct, he, uh, hc, gp, 4, 9, 17, aa26.11.44, af72.5.171, lt, wp, ep, ce, tb, wl, pbl, jdr, kf, hi, oe, ai1, ai2*

PCMAS Portable Computer Based Maintenance System *mt*
PCMC PCI, Cache, Memory Controller [Intel] *pk, kg*
PCMCIA Personal Computer Memory Card International Association *kg, ct, em, hi*
PCMD Pulse Code Modulation, Digital *fu*
PCMD Pulse Command *jdr, uh*
PCMF Parts Common Master File *fu*
PCMI President's Council on Management Improvement *wp*
PCMIM Personal Computer Media Interface Module *pk, kg*
PCMMU Pulse Code Modulation Master Unit *-4*
PCMO Program Configuration (Change) Management Office [an organization] *ep, pbl, tm, fu, ah, jn*
PCMS Program Configuration Management System *fu*
PCN Personal Computer Network *pk, kg*
PCNFS Personal Computer Network File System *kg, hi*
PCO Point of Control and Observation *pk, kg*
PCO Principal Contracting Officer *mt, tb, fu*
PCO Procurement Change Order *wp*
PCO Procuring (Procurement) Contracting Officer *mt, su, 17, wp, 18, sb, hc, nt, tb, kf*
PCO Program Contracting Officer *mt*
PCO Publications Control Officer *fu*
PCOL Procuring Contracting Officer Letter *wp*
PCOM Probability of Communication *fu*
PCP Parts Control Program *fu*
PCP Payload Control Processor *he*
PCP Peak Cell Rate *ct*
PCP PhenCyclohexylPiperidine (Phenylcyclidine) *wp*
PCP Pilot's Control Panel *hc, wp*
PCP Platoon Command Post *fu*
PCP Potential Contractor Program *fu*
PCP Power Control Panel *fu, gp, 9*
PCP Program Change Proposal *mt, fu, wp*
PCP Program Control Plan *mt*
PCP Programmable Data Processors *mt*
PCP Pulse Comparator *fu*
PCP Purge Control Panel *oe*
PCPDS Personnel Data Support Center [US Army] *dt*
PCPP Parts Control Program Plan *mt*
PCPS Portable Collective Protection Shelter [chemical warfare] *wp*
PCR Payload Certification Review *wl*
PCR Payload Changeout Room *bo, 9, jdr, oe*
PCR Payload Chargeout {?} Room *gp*
PCR Planning Change Request *fu*
PCR Post Change Request *fu*
PCR Problem Change Request *kf*
PCR Problem Correction Report *fu*
PCR Program Change Report *fu*
PCR Program Change Request *mt, fu, wp, tb*
PCR Program Control Room *0*
PCR Projected Communications Requirement *mt*
PCR Pulse Compression Radar *wp*
PCR Pulse Compression Ratio *fu, sd*
PCRB Program (Project) Change Review Board *fu*
PCS Patchable Control Store *pk, kg*
PCS Payload Cable Set *kf*
PCS Permanent Change [of] Station *mt, wp, tb, fu*
PCS Personal Communication System (Services) *em, ct, kg, hi*
PCS Personal Conferencing Specification *pk, kg*
PCS Physical Control Space *mt*
PCS Planning Control Sheet *pk, kg*
PCS Plastic Clad Silica [cable] *jr, jdr*
PCS Power Conditioning System *fu*
PCS Power Conversion System *fu*
PCS Primary Coolant System *fu*
PCS Print Contrast Signal *pk, kg*
PCS Printed Circuit Substrate *fu*
PCS Probability of Command Shutdown *ai2*
PCS Process (Project) Control System *pk, kg*
PCS Production Control Specialist *fu*
PCS Production Control System *fu*
PCS Program Counter Store *pk, kg*
PCS Propulsion Control System *wp*
PCS Protected Cable System *mt*
PCS Proxy Cache Server *em, ct*
PCSMS PACAF Combat Supplies Management Subsystem [PACAF] *mt*
PCSO Presidential Communications Support Office *mt*
PCSP Program Communications Electronic Support Program [USAF/PACAF] *mt*
PCT Photometric Calibration Target *jp*
PCT Photon Induced Charge Transfer *wp*
PCT Private Communications Technology *pk, kg*
PCT Propulsion Component Technology *wp*
PCTA Plastic Cased Telescoped Ammunition *mt*
PCTC Pentagon Consolidated Telecommunications Center *mt*
PCTE Portable Common Tool Environment *mt*

PCTM Pulse Count Modulation *fu*
PCU Payload Control Unit *kf*
PCU Portable Computer Unit *fu*
PCU Power Conditioner Unit *oe*
PCU Power Control Unit *mt, fu, oe*
PCU Power Converter (Conversion) Unit *fu, wl*
PCU Printer Control Unit *fu*
PCU Process Control Unit *oe*
PCU Program Control Unit *fu, 4, 17, sb, bf*
PCU Pyro Control Unit *jdr*
PCV Pollution Control Valve *wp*
PCVI Product Configuration Verification Index *fu*
PCWBS Preliminary Contract Work Breakdown Structure *kf, fu*
PCWO Printed Circuit Work Order *fu*
PCX PC paintbrush X *hi*
PCZ Physical Control Zone *jdr*
PD Passive Detection *fu*
PD Patent Disclosure *pbl, ah, jn*
PD Perceptive Diagnostics *jdr*
PD Phase change Dual *kg, ct*
PD Planning Directive *mt*
PD Point Defense *fu*
PD Position Description *wp, oe*
PD Power Distribution *fu*
PD Power Divider *fu*
PD Preliminary Design *fu, tb*
PD Preliminary Draft *tb*
PD Presidential Directive *mt, ci.871*
PD Primary Demonstration *fu*
PD Prime Driver *fu*
PD Priority Directive *mt*
PD Probability of Damage *mt, wp*
PD Probability of Destruction *mt*
PD Probability of Detection *mt, fu, kf, wp, tb, jdr*
PD Procedures Description *mt*
PD Procurement Directive *fu*
PD Procurement Drawing *tb*
PD Program (Project) Director *mt, wp, tb*
PD Program Description *fu*
PD Program Document *mt*
PD Project (Program) Directive *wp, tb*
PD Prototype Demonstration *tb*
PD Pulse Doppler *mt, fu, hc*
PD Pulse Duration *fu*
PD Purchase Description *fu*
PD [chemical warfare codename for] Phenyl Dichloroarsine *wp*
PD&C Plan, Direct, and Control [also PDC] *fu*
PD-V Project Definition - Validation *jdr*
PD/U Processing Distribution Unit *oe*
PDA (PDA Engineering, Costa Mesa, California) *aa26.11.52*
PDA Percent Defective Allowable *bl2-9.48, lt*
PDA Personal Digital Assistant *ct, kg, em, hi*
PDA PIN diode Attenuator *4, uh*
PDA Post Delivery Availability *fu*
PDA Power Drive Amplifier *jdr*
PDA Predicted Drift Angle *w9*
PDA Preliminary Design Audit *-4*
PDA Principal Development Activity [US Navy] *mt, fu*
PDA Principal Development Agency *wp*
PDA Procurement Defense Agency *mt, wp*
PDA Program Departure Authorization *fu*
PDA Proprietary Disclosure Agreement *jdr*
PDAF Playback Data Acquisition Facility *ac3.7-33*
PDAP Programmable Digital Auto Pilot *hc*
PDAR Preferential Departure and Arrival Route *fu*
PDB Partial Data Block *fu*
PDB Picture Data Base *fu*
PDB Project Development Brochure *tb*
PDC Payload Design Center *pbl*
PDC Peripheral Device Controller *fu*
PDC Platform Despin Controller *3, 8*
PDC Primary Development Contract *mt*
PDC Primary Domain Controller *ct, kg, em*
PDC Probability of Data Collection *kf*
PDC Product Data Center *pl*
PDC Professional Development Center [AFCEA] *mt*
PDC Program Coding Description *fu*
PDC Program Design Change *oe*
PDC Program Development Center *fu*
PDC Programmable Display Controller *fu*
PDC Programming, Design and Construction System *mt*
PDC Proteus Digital Channel *fu*
PDCALC Computer Code To Determine Probability of Damage *wp*
PDCU Power Distribution and Control Unit *fu*
PDD Physical Device Driver *pk, kg*
PDD Product Definition Data *fu*

PDD Product Delivery Device *mt*
PDD Program Description Document *fu, tb*
PDD Proposal Due Date *mt*
PDDI Product Definition Data Interface *mt*
PDDS Program Data Distribution System *fu*
PDE Partial Differential Equation *ai1*
PDE Printer Description Extension [a file name extension] *pk, kg*
PDE Product Design Engineering *sd*
PDE Propulsion Driver Electronics *jp*
PDEC Program Development Edit Center *fu*
PDES Product Data Exchange Standard *mt*
PDES Product Definition Exchange Specification *is26.1.37*
PDES Product Description Exchange Standard *mt*
PDF Package Definition File *pk, kg*
PDF Patient Data Form *mt*
PDF Portable Data Format [Adobe] *oe*
PDF Portable Document File *em, kg*
PDF Portable Document Format *jr, kg, pbl, hi*
PDF Precision Direction Finding (Finder) *mt, fu*
PDF Probability Density Function *fu, csIII.9, wp, aa26.11.37, kf, ai1*
PDF Processor Defined Function *pk, kg*
PDF Program Data Form *st*
PDF Program Development Facility *pk, kg*
PDF Project Data Format *2, 4*
PDG Program Decision Group *wp*
PDG Prototype Development Group *fu*
PDI Payload Data Interleaver *2, 4, 9, lt, oe*
PDI Process Definition Interface *fu*
PDI Pyronetics Devices, Incorporated *hq6.2.2*
PDIAL Public Dialup Internet Access List [Internet] *kg, hi*
PDIP Program Development Increment Package *mt*
PDIR Peripheral Data Set Information Record *fu*
PDIS Product Definition Information System *fu*
PDIT Product Development Improvement Team *sn2.8*
PDK Plasma Display and Keyboard *fu*
PDKU Plasma Display and Keyboard Unit *fu*
PDL Page Description Language *pk, kg*
PDL Parameter Data Load *gp*
PDL Parts Deletion List *fu*
PDL Permanent Duty Location *mt, wp*
PDL Physical Data Language *fu*
PDL Position Description Language *fu*
PDL Process Definition List *fu*
PDL Processor Data Load *gd2-41, oe*
PDL Program Description Language *pk, kg*
PDL Program Design Language [ADA] *mt, fu, kg, 17, wp, sb, hc, tb, hi, oe*
PDLM Periodic Depot Level Maintenance *wp*
PDM Planned Depot Maintenance *mt*
PDM Precedence Diagramming Method *jdr*
PDM Procedure Diagramming Method *fu*
PDM Product Data Management *ah, kg, pbl, jdr, kf, jn, tm*
PDM Program Decision Memorandum *mt, hc, dh, wp, tb, st, kf*
PDM Program Development Manual *wp*
PDM Programmed Depot Maintenance *mt, fu*
PDM Project Data Manual *wp*
PDM Propellant Dispersion Munition *wp*
PDM Propellant Distribution Module *-4*
PDM Pulse Duration Modulation *mt, fu, 0, 4, oe*
PDM Pursuit Deterrent Munition *dh*
PDMFM Pulse Duration Modulation/Frequency Modulation *fu*
PDMO Program Data Management Office *jn, pbl, tm*
PDMP Project Data Management Plan *hy, cu*
PDMS Point Defense Missile System [SAM* system for ships] [USA] *wp, mt*
PDMS Product Data Management System *hc, fu, cd*
PDMT Provisioning Data on Magnetic Tape *fu*
PDN Public Data Network *mt, ct, kg, em, wp, hi*
PDO Portable Distributed Object *kg, em*
PDO Procedure Description Overview *mt*
PDO Product Data Operations *jdr, pbl, ah, jn*
PDO Project Data Operations *jn*
PDO Publications Distribution Office *mt*
PDOS Publishing Distribution Office System *mt*
PDP Plasma Display Panel *pk, kg*
PDP Power Distribution Panel *fu*
PDP Program Decision Package *wp, mt*
PDP Program Definition Phase *mt, fu*
PDP Program Development Plan *fu, 0*
PDP Programmed (Programmable) Data Processor [DEC] *0, kg, ct, em, hi*
PDPA Project Data Processing Authorization *hf*
PDPIF Post Delivery Performance Incentive Fee *jdr*
PDPS Project Data Processing Subsystem *ac2.7-11*
PDR Parts Data Record *fu*
PDR Parts Discrepancy Record *fu*

PDR Periscope Detection Radar *mt*
PDR Phase Derived Range *fu*
PDR Preferential Departure Route *fu*
PDR Preliminary Design Review [a meeting] *mt, ah, fu, hc, wp, gp, 4, 16, 17, sb, pl, nt, lt, ep, dm, re, tb, st, wl, as, jdr, pbl, tm, kf, vv, jn, oe*
PDR Priority Data Reduction *fu*
PDR Processed Data Records *ac*
PDR Processed Data Relay *oe*
PDR Production Design Review *fu*
PDR Program Drum Recording *fu*
PDRC Program Development Review Committee *mt*
PDRD Program Definition and Requirements Document *wl*
PDRL Preliminary (Proposal) Data Requirements List *fu*
PDRR Program Definition Risk Reduction *ls4, 5, 16*
PDRS Payload Deployment and Retrieval System *ci.628*
PDRS Program Document Reporting System *fu*
PDS Pacific Distribution System [PACAF] *mt*
PDS Packet Data Switch *fu*
PDS Packet Driver Specification *kg, hi*
PDS Partitioned (Partition) DataSet *ct, em, kg*
PDS Passenger Documentation System *mt*
PDS Passive Detection System *wp*
PDS Patient Distribution System [USAF] *mt*
PDS Payload Data Subsystem *mb, re*
PDS Personnel Data System [USAF] *mt*
PDS Planetary Data System *ns, re, mb, mi, kg, em, ct*
PDS Portable Document Software *pk, kg*
PDS Positional Data System [US Navy] *mt*
PDS Premise Distribution System *ct, em*
PDS Processing and Display System *mt, kf*
PDS Processor Direct Slot [Macintosh] *kg, hi*
PDS Product Data Services *ah, pbl, tm, jn*
PDS Product Definition System *fu*
PDS Product Design Standard [a document] *ep, jdr, fu*
PDS Product Development System *ah, pbl, tm, jn*
PDS Product Distribution System *fu*
PDS Program Design Specification *fu*
PDS Program Distribution System [USAF] *mt*
PDS Proprietary Data Summary *fu*
PDS Protected Distribution System *mt, kf*
PDS Provisioning Data System *fu*
PDS Purchasing Department Specification *fu*
PDSC PACOM Data Systems Center *mt*
PDSMS Point Defense Surface Missile Systems *hc*
PDSP PACOM Data System Program *mt*
PDSS Post Deployment Software Support *mt, fu, st, kf*
PDSS Post Development and Software Support *pk, kg*
PDSS Procurement Decision Support System *ep*
PDSSC Post Deployment Software Support Center *st*
PDST Payload Design Steering Team *jdr*
PDST Personnel Data System for Training [USAF] *mt*
PDT Pacific Daylight Time *kg, sc7.8.1, 0, w9, kf*
PDT Performance Diagnostic Tool [IBM] *kg, hi*
PDT Processor Diagnostic Test *hc*
PDT Product Development Team *jdr*
PDT Programmable Drive Table *pk, kg*
PDT Provisioning Technical Documentation *fu*
PDT&TAC Per Diem Travel and Transportation Allowance Committee *dt*
PDU Philippines Digital Upgrade *mt*
PDU Pilot Display Unit *wp*
PDU Plasma Display Unit *fu*
PDU Plug Distribution Unit *pk, kg*
PDU Power Distribution Unit *fu, jn, bo, jdr, kf, he, ah, uh, jn*
PDU Protocol Data Unit *mt, ct, fu, nu, em*
PDVF Payload Design Verification Facility *hc, 4*
PDW Personal Defense Weapon *wp*
PDW Preliminary Design Walkthrough *kf*
PDW Pulse Description Word *fu*
PDWE Pulse Detonation Wave Engine [uses explosions of fuel rather than continuous burning of fuel] *mt*
PE Parity Even *pk, kg*
PE Passive Element *fu*
PE Percent Bit Error Probability *mt*
PE Percent Error *dm*
PE Peripheral Equipment *fu*
PE Perkin Elmer *17, sb*
PE Permanent Echo *fu*
PE Potential Energy *fu, ai1*
PE Primary Echo *fu*
PE Probable Error *fu, w9, wp*
PE Processing Element *fu, kg*
PE Procurement Executive [UK] *mt*
PE Product Effectiveness *jdr, fu*
PE Product Enhancement *jdr*
PE Program Element *mt, fu, wp, tb, st*

PE Project Engineer *fu, wp, tb, jdr*
PE Propellant Excess [sum of FPR, LVC, PA and PM] *oe*
PE Protect Enable *pk, kg*
PE(M) Program Element (Monitor) *hc*
PE/CP Process Engineering Control Point *fu*
PE/L Probability of Early/Late *jdr*
PE/L Probe Early/Late *jdr*
PEA Parity Error Alarm *fu*
PEA Patterson Experimental Array *fu*
PEA Pitched Earth Acquisition *oe*
PEA Pocket Ethernet Adapter *kg, hi*
PEA Production Engineering Analysis *fu*
PEACE [nickname for HQ USAF programs connected with military aid or sale] *wp*
PEACE_ALPS [F-5 FMS for Switzerland] *wp*
PEACE_DIAMOND [F-4e FMS for Turkey] *wp*
PEACE_DRUM [F-5e FMS for Kenya] *wp*
PEACE_EAGLE [F-15 sale for Japan] *wp*
PEACE_FOX [F-15 FMS for Israel] *wp*
PEACE_HAWK [F-5 FMS for Saudi Arabia] *wp*
PEACE_JEWEL [T-38 FMS to Turkey] *wp*
PEACE_LADY [An/Trn-26 FMS for Taiwan] *wp*
PEACE_MARBLE [potential F-16 FMS to Israel] *wp*
PEACE_ORO [A/T-37 FMS for Colombia] *wp*
PEACE_PEARL [FMS fire control system for the Chinese F-811] *wp*
PEACE_PULSE [radar FMS to Saudi Arabia] *wp*
PEACE_RAMA [F-5e FMS for Thailand] *wp*
PEACE_SHIELD Royal Saudi Air Force C3 System *fu*
PEACE_SUN [F-15 simulator support] *wp*
PEACE_VOICE [MRC-108 radio FMS to Korea] *wp*
PEACE_WINGS [program to modernize Canadian interceptor fleet] *wp*
PEAM Personal Electronic Aid for Maintenance *mt*
PEAM Portable Electronic Aid to Maintenance *wp*
PEAS/SOAS PSYOP Effects and Analysis/SOAS *mt*
PEATS Programmable Environmental and Automatic Test Systems *hc*
PEB Performance Evaluation Board *oe*
PEB Power Electronics Box [ref. SXI] *oe*
PEC Passive Equipment Cabinet *fu*
PEC Performance Evaluation Categories *jdr*
PEC Photoelectric Cell *fu, kf*
PEC Preliminary Engineering Change *oe*
PEC Program Element Code *mt, wp*
PECC Civilian Personnel Center Army *dt*
PECI Productivity Enhanced Capital Investment *mt*
PECM Passive ECM *mt, wp*
PECOS Personnel Contamination Sensor [chemical warfare] *wp*
PECP Preliminary Engineering Change Proposal *mt, fu*
PECP Process Engineering Control Point *fu*
PECS Pennsylvania Educational Communications System *hn43.15.8*
PECS Portable Environmental Control System *wp*
PECVD Plasma Enhanced Chemical Vapor Deposition *wp*
PED Probable Error Deflection *fu*
PED Processing, Exploitation and Dissemination *wp*
PED Program Element Description *wp*
PED Program Entry Device *fu*
PED Purchase Early Development *mt*
PEDA Power Entry Distribution Assembly *fu*
PEDRS Process Evaluation and Defect Reporting System *pl*
PEDS Program Element Descriptive Summary *mt, wp*
PEE Proof and Experimental Establishment [at Foulness, UK] *mt*
PEEC Personnel Emergency Estimator Capability *mt*
PEEK Poly Ether Ether Ketone *wp, aa26.10.42, aw129.21.13*
PEEL Programmable Electrically Erasable Logic *hi*
PEFF Portable Executable File Format *hi*
PEG Polyethylene Glycol *wp*
PEG Power Electronics Group *vv*
PEG Program Evaluation Group *fu, mt, wp*
PEG Programmable Event Generators *fu*
PEG Pulse Enable Gate *fu*
PEGO Product Effectiveness Group Office *fu*
PEI Process Editor Interface *fu*
PEI Process Engineering Instruction *fu*
PEJ Premolded Expansion Joint *fu*
PEL Picture Element [pixel] *fu, kg, hi*
PEL Pulse Expansion Line *fu*
PELISS Precision Emitter Location Strike System *hc*
PELJ Pyrotechnical Expendable Laser Jammer *hc*
PELS Parity Error Logic Strobe *fu*
PELSS Precision Emitter Locator Strike System *mt*
PEM Pacific Exploratory Mission *hy*
PEM Particle Environment Monitor [UARS] *ge, hg*
PEM Payload Ejection Mechanism *wp*
PEM Performance Evaluation Missile *aw130.13.19*
PEM Privacy Enhanced Mail [Internet] *kg, hi*
PEM Product Engineering Measures *fu*
PEM Product Error Message *ct, em*

PEM Program Element Monitor [also PE(M)] *mt, hc, wp, tb*
PEM Project Engineering Manager *fu*
PEMARS Procurement of Equipment, Missiles Army Management And Reporting System *wp*
PEMF Pulsating Electromagnetic Field *wp*
PEMRC Program Executive Management Review Committee *mt*
PEMS Performance Evaluation and Measurement System *fu, sc8.5.3, 8, 15, jdr*
PEN Penetrator Rod *re*
PENAID Penetration Aid *mt, wp, tb*
PENAIDS Penetration Aids *uc, st*
PENSAR Penetrating Synthetic Aperture Radar *wp*
PEO Polar Earth Orbit *kf*
PEO Program Executive Officer *wp, kf*
PEO-SCS Program Executive Officer for Space, Communications and Sensors *uh*
PEOC Presidential Emergency Operations Room *mt*
PEP Packet Exchange Protocol *pk, kg*
PEP Packetized Ensemble Protocol [Telebit] *kg, hi*
PEP Peak Envelope Power *jdr, mt, fu*
PEP Performance Evaluation and Prediction *wp*
PEP Performance Evaluation Program *mt*
PEP Planar Epitaxial Passivated *fu*
PEP Polynomial Error Protection *oe*
PEP Preliminary Evaluation Plan *mt*
PEP Process Enhancement Program *wp*
PEP Procurement Evaluation Panel *wp*
PEP Producibility (Production) Engineering and Planning *ph, st, nt, wp*
PEP Producibility Engineering and Planning *fu*
PEP Product Engineering and Production *fu*
PEP Production Engineering Program *fu*
PEP Productivity Enhancement Program *mt*
PEP Project Element Plan *mt*
PEP Pyrotechnic Enable Plug *gp, oe*
PEP [Producibility and Production Engineering] *nt*
PEPCI Providing Emergency Presidential Communications Interface *mt*
PEPE Parallel Element Processing Ensemble *fu, hc*
PEPE Perkin Elmer Performance Evaluation *mt*
PEPP Producibility Engineering and Production Planning *tb*
PEPS Photovoltaic Electric Power Systems *hc, wp*
PEPSI Parameter Extraction Program for Silicon on Insulator *hc*
PEPSOC Portable ESA Package for Synchronous Orbit Control *es55.86*
PEPSY Precision Earth Pointing System *hc, wp*
PEQ Position Extrapolation Quality *fu*
PEQUA Production Equipment Agency *wp*
PER Personnel Resource [Status] *kf*
PER Pert Event Report *wp*
PER Pre Environmental Review *gp, oe*
PER Process Engineering Request *fu*
PER Productivity Evaluation Review *fu*
PER Program Element Review *wp*
PERA Planning and Engineering for Repairs and Alterations *hc, wp*
PERCOM Personnel Command *dt*
PERG Production Equipment Redistribution Group *dt*
PERL Practical Extraction and Report(ing) Language [Unix] *kg, hi*
PERL Prepositioned Equipment Requirements List *mt*
PERMARS Personnel Management Reports System [MACOM, Military, USA] *mt*
PERMIS Preference Management Information System *mt*
PERS ACD Personnel subsystem [PACAF] *mt*
PERS Product Effectiveness Requirements Summary *fu*
PERSAS Personnel Accounting System [USAFE] *wp, mt*
PERSCO Personnel Support For Contingency Operations [USAF] *mt*
PERSCOM Personnel Command *mt*
PERSDEP Personnel Deployment Report *mt*
PERSEC Personnel Security *kf*
PERSID Personnel Seismic Intruder Detector *wp*
PERSINSCOM Personnel Information System Command [USA] *mt*
PERSITREP Personnel Situation Report [USA] *mt*
PERSNET Personnel Communications Network *mt*
PERSTAT Personnel Status Report *mt*
PERT Performance (Program) Evaluation and Review Technique *fu, kg*
PERT Performance Evaluation and Review (Reporting) Technique *8, ep, tb*
PERT Program Evaluation and Review Team *st, jdr*
PERT Program Evaluation and Review Technique (Technology) *mt, hc, 0, 13, dt, wp, hi*
PERT Project Evaluation Review Technique [schedule network] *-4*
PERUMTEL [procuring agency for Indonesian national communications satellite] *sc4.7.1*
PES PACOM Exercise Schedule *mt*
PES Peripheral Equipment Simulator *fu*
PES Personal Earth Stations *sc8.5.1*
PES Photoelectric Scanner *fu*
PES Positioning Error Signal *pk, kg*
PES Post Ejection Sequencer *gp, jdr*
PES Power Electronics Subsystem *kf*

PES Processor Enhancement Socket *pk, kg*
PES Program Element Summary *wp*
PESA Provisioning Engineering Support Agency *fu*
PESA Pyroelectric Earth Sensor Assembly *kf*
PESAM Penetration Survivability Assessment Model *wp*
PESL Plessey Electronics Systems, Ltd. *fu*
PESO Product Engineering Services Office [DOD] *dt*
PESQ Product Effectiveness Software Quality *fu*
PEST Pesticide Evaluation Summary Tabulation System [USAF] *mt*
PET Partitioner for Exhaustive Testing *fu*
PET Performance Evaluation Test *-4*
PET Phase Elapsed Time *-4*
PET Polyethylene Terephtalate *wp*
PET Positron Emission Tomography *wp, is26.1.68*
PET Print Enhancement Technology [Compaq] *kg, hi*
PET Process Evaluation Tool *jdr, kf*
PET Project Engineering Team *fu*
PET Proposal Evaluation Team *tb*
PET Prototype Evaluation Testing *fu*
PETA Parabolic Expandable Truss Antenna *sd*
PETE Portable Environmental Test Equipment *fu*
PETN Pentaerythritol Tetranitrate *wp*
PETNAIS Pacific Essential Telecommunications Network and Information System [PACAF] *mt*
PETR Post Environmental Test Review *gp, oe*
PETROL_RAM Petroleum Resource Automated Management *mt*
PETS Photographic Equipment Test System *wp*
PEWS Portable EDS Workstation *mt*
PEWSX PEWS Q-D Subsystem *mt*
PF Patrol Frigate [US Navy] *mt*
PF Potential for Failure *fu*
PF Power Factor *mt, fu, w9*
PF Probability of Failure *wp*
PF Probability of Fool *tb*
PF Probability of Fratricide *tb*
PF Program Function *ct, em, fu*
PF Programmed Function [key on a computer keyboard] *-8*
PF Pulse Frequency *fu*
PF2 Project Forecast II [USAF HQ AFSC R&D look into the future] *wp*
PFA Participating Field Activity *fu*
PFA Peachtree First Accounting *ct*
PFA Pixel Flow Architecture *hi*
PFA Popular Flying Association *mt*
PFA Probability of False Alarm *mt, fu, kf, wp, jdr*
PFB Preformed Beam *fu*
PFC Private First Class *w9*
PFC Production Flow Council *jn*
PFCS Primary Flight Control System *mt, wp*
PFD Particle Flux Detector *0*
PFD Phase Frequency Detector *fu*
PFD Phase Front Distortion *fu*
PFD Power Fault Detector *fu*
PFD Power Flux Density *gp, 2, oe*
PFD Primary Flight Display *fu*
PFD Probability of False Detection *kf*
PFE Programmer's File Editor *pk, kg*
PFE Purchaser Furnished Equipment *fu*
PFF Planning Factors File [JOPS] *fu*
PFIAB President's Foreign Intelligence Advisory Board *-10*
PFIS MAJCOM Personnel Feedback Information System [MAC] *mt*
PFK Perimeter Function Key *fu*
PFK Program Function Key *mt*
PFL Planned Fire List *fu*
PFM Multimission Platform [for the SPOT program] *ac3.7-27*
PFM Patient Flow Model [USA] *mt*
PFM Power Factor Meter *fu*
PFM Prefaulted Module *fu*
PFM Proto Flight Model *bl3-1.2, es55.43, oe*
PFM Pulse Frequency Modulation *mt, fu, 0, wp*
PFMDS PACAF Force Management Display System *mt*
PFMF Printer Font Metric File *hi*
PFN Pulse Forming Network *fu, hc*
PFP Partnership For Peace [NATO] *mt*
PFP Portable Firing Panel *fu*
PFP Purchaser Furnished Property *fu*
PFPD Process Flight Plan Data *fu*
PFR Part Failure Rate *fu*
PFR Path Failure Report *fu*
PFR Power Fail / Restart *kg, fu*
PFR Problem or Failure Report *gp*
PFR Program Financial Review *mt, wp*
PFRB Project Failure Review Board *fu*
PFRT Preliminary Flight Rating Test *fu, wp*
PFS Payload Feasibility Study *re*
PFSDR Parts Failure Service Difficulties Report *wp*
PFTA Payload Flight Test Article *aw118.18.18*

PG Participation Group *fu*
PG Patrol Gunboard [US Navy ship classification] *mt, fu*
PG Planning Guidance *mt*
PG Prediction Gate *fu*
PG Pressure Gage *fu*
PG Project Group *mt*
PG Proving Ground *wp*
PG Pulse Generator *fu*
PGA Pin Grid Array *em, ct, fu, kg, hi*
PGA Professional Graphics Adapter *em, ct, kg, hi*
PGA Programmable Gate Array *hi, ai1*
PGAPL Preliminary Group Assembly Parts List *fu*
PGB Patrol Gun Boat *hc*
PGBC Pension Benefit Guaranty Corporation [also PBGC] *sc8.9.3*
PGC Policy Guidance Council *wp*
PGC Program Generation Center *fu, jb*
PGC Program Group Control [Microsoft] *pk, kg*
PGDF Product Generation and Distribution Facility *ac2.2-16*
PGH Patrol Gunboard, Hydrofoil [US Navy ship classification] *mt, fu*
PGHM Payload Ground Handling Mechanism (Mechanical) *gp, 9, jdr, oe*
PGI Parameter Group Identifier *fu*
PGIP Predicted Ground Impact Point *kf*
PGL Professional Graphics Language *hi*
PGM Planning Guidance Memorandum [SECAF/CSAF] *mt*
PGM Precision Guided Missile (PGM-11 is Redstone) *mt, wp*
PGM Precision Guided Munition *mt, fu, hc, dh, 14*
PGM Processing Graph Methodology *fu*
PGNS Primary Guidance and Navigation Section *ci.392*
PGO Periodic Grating Oscillator *fu*
PGOC Payload Ground Operations Control [NASA] *mt*
PGOWG Payload Ground Operations Working Group *jdr*
PGP Pretty Good Privacy [an encryption program] *kg, ct, em, hi, du*
PGPSIU PLRS-GPS Interface Unit *fu*
PGRF Pulse Group Repetition Frequency *fu*
PGRI Pulse Group Repetition Interval *fu*
PGRV Precision Guided Reentry Vehicle *hc, wp*
PGS Papergram System *fu*
PGS PLSS Ground Station *wp*
PGS Power Generation Subsystem *wl*
PGS Program Generation System *fu*
PGS Propellant Gauging System *jdr, jn*
PGSC Payload and General Support Computer *ws4.10*
PGSE Peculiar Ground Support Equipment *wp*
PGSE Pre released GSE *jdr*
PGSE Processing Graph Support Environment *fu*
PGU Power Generator Unit *kf*
PH Probability of Hit *wp*
PHA Preliminary Hazard Analysis *17, tb, fu, jn*
PHA Pulse Height Analyzer *pl*
PHAROS Phased Array Radar for Overland Surveillance (System) *hc, wp*
PHAROS Planning Handling And Radar Operating System *hc*
PHASE_IV [base level data automation program] [USAF] *mt*
PHC Payload Heater Control *jn*
PHD Philosophiae Doctor [(Latin) doctor of philosphy] *wp, ai1*
PHELM Pulsed High Energy Laser countermeasures *wp*
PHF Payload Handling Fixture *jdr*
PHH (PHH Corp.) [an airlines services company] *aw130.2.49*
PHI Position Homing Indicator *fu*
PHIGS Programmers' Hierarchical Interactive Graphics System (Standards) *mt, fu, kg, hi*
PHJ PLRS/JTIDS Hybrid [program] *hn44.7.4*
PHL Preliminary Hazard List *fu, tb*
PHM Patrol Craft, Hydrofoil, Missile armed [USN ship classification] *mt, fu*
PHO Production Hold Order *fu*
PHOTINT Photographic Intelligence [USA] [DOD] *mt*
PHOTOINT Photographic Intelligence *wp*
PHOTONICS Photo Electronics *wp*
PHOTOX Photochemical Vapor Deposition Process *hc*
PHP Planetary Horizontal Platform *0*
PHQ Peacetime Headquarters [NATO] *mt*
PHQPP Preliminary Hardware Quality Program Plan *fu*
PHR Payload Hazard Report *gp, oe*
PHS Personal Handyphone System *pk, kg*
PHS&T Packaging, Handling, Storage, and Transportation *fu, kf*
PHSL Program Hardware/Software List *jdr*
PHT Packaging, Handling and Transportability (Transportation) *mt, hc, wp*
PHT Port Hold Time *mt*
PHTS Packaging, Handling, Transportation, Storage [ILS] *mt*
PHY Physical Layer Protocol *hi*
PHYSEC Physical Security *kf*
PHYSNET Physics Network [also known as HEPNET] *ns*
PI Parameter Identifier *fu*

PI Payload Interrogator [Orbiter] *-4*
PI Performance (Product) Improvement *fu*
PI Performance Improvement *hn44.23.9, 8, 15, sc9.6.1, jdr*
PI Performance Index *fu*
PI Periodic Inspection *mt*
PI Photo Interpreter *mt*
PI Photographic Intelligence *mt*
PI Plan Information *mt*
PI Point of Impact *wp, mt, fu*
PI Point of Intersection *fu*
PI Power Incentive [a fee] *jdr*
PI Practice Interception *mt*
PI Predicted Impact *mt*
PI Prime (Primary) Item *fu, tb*
PI Principal Investigator *pl, wp, so, ns, re, wl, jdr, mb, cu, mi*
PI Process Instructions *fu*
PI Program (Programmed) Instruction [a directive] *fu, hc, gp, 15, sb, w9, ep, jdr, pbl, tm*
PI Program Instruction [directive] *ah, jn*
PI Program Interruptor (Interruption) *fu, kg*
PI Program Introduction [document] *2, 17, st*
PI Proportional Integral [Controller, AOCS] *oe*
PI Protocol Interpreter *mt*
PIA Peripheral Interface Adapter *pk, kg*
PIA Pointed Instrument Assembly *pl*
PIA Propellant Isolation Assembly *jp*
PIAP Process Improvement Action Plan *kf*
PIAS PLSS Intelligence Augmentation System *mt*
PIAT Projector, Infantry, Anti Tank. [WWII anti tank weapon, a shoulder-held spigot mortar] [UK] *mt*
PIB Polar Ionosphere Beacon *fu*
PIB Pulse Interference Blanker *fu*
PIC Pacific Imagery Processing and Interpretation Center *mt*
PIC Parent Indicator Code *fu*
PIC Payload Integration Contract *tb*
PIC Personal Intelligent Communicator *pk, kg*
PIC Personnel Investigations Center *su, tb*
PIC Photo Interpretation Console *mt*
PIC Physical Interface Card *fu*
PIC Pilot In Command *mt*
PIC Polyethylene Insulated Conductor *fu*
PIC Preferred Inter exchange Carrier *em, ct*
PIC Primary Inter exchange Carrier *em, ct*
PIC Priority Interrupt Controller *pk, kg*
PIC Process Incentive Contracting *fu*
PIC Processor Interface Control *wp*
PIC Programmable (Program) Interrupt Controller *em, ct, kg*
PIC Programmed Instruction Configuration *fu*
PIC Pyrotechnic Initiator Controller *hc*
PIC [occurs in Hughes Aircraft Company customer list] *sa12474.ep*
PICA Primary Inventory Control Activity *wp*
PICA Process Improvement Cycle Analysis *fu*
PICE Programmable Integrated Control Equipment *fu*
PICES Programmed Interactive Cost Estimating System [model] *nt*
PICF Principal Investigator Computing Facilities *hy*
PICIM Positive Ion Chemical Ionization Mass [Spectroscopy] *wp*
PICN Preliminary Interface Change Notice *fu*
PICO Preinstallation, Installation, and Checkout *hc*
PICON Process Intelligent Controller *wp*
PICP Program Support Inventory Control Point *fu*
PICS Perform Intertask Communications Services *fu*
PICS Platform for Internet Content Selection *pk, kg, du*
PICS Pulsed Image Converter System *wp*
PID Parameter Identification *oe*
PID Plan Identification Number [USA DOD] *mt*
PID Position Identifier *mt*
PID Priority Identification *oe*
PID Process Identification Number [also PIN] *kg, hi*
PID Processed Instrument Data *oe*
PID Program Identifier *mt, wp*
PID Program Introduction Document [also PI] *wp, kf*
PID Proportional, Integral, Derivative (Differential) [gain] *kg, aa26.11.b9, wl*
PIDA Power Interface/Distribution Assembly *fu*
PIDA Proprietary Information Disclosure Agreement *kf*
PIDAS Portable Instant Display and Analysis Spectrometer *hy*
PIDDP Planetary Instrument Definition and Development Program *so*
PIDP Programmable Indicator Data Processor *mt, fu, hc*
PIDP_II Second Generation PIDP Hardware *fu*
PIDS Primary (Prime) Item Development Specification *hc, fu, wp*
PIE Performance Improvement Effectiveness *fu*
PIE Photo Image Enhancement *hc*
PIE Priority Interrupt Expander *fu*
PIE Pulse Interference Elimination *fu, wp*
PIECES (PIECES (software program)) *jb*
PIER Procedures for Internet Enterprise Renumbering *pk, kg*

PIES Post Installation Engineering Support [GIMTACS] *oe*
PIF Photo Interpretation Facility [USA] *mt*
PIF Picture Interchange Format (File) *kg, hi*
PIF PolyIsocyanurate Foam *ai2*
PIF Productivity Investment Fund *mt, wp*
PIF Program Information File *ct, kg, hi*
PIF Provisions of Industrial Facilities *st*
PIFRS Prototype Increase Frequency Reporting System [OJCS] *mt*
PIG Pendulous Integrating Gyro *hc*
PIG Product Information Guide *ct, em*
PIGA Pendulous Integrating Gyro Accelerometer *hc*
PIGS Passive Infrared Guidance System *wp*
PIH Parallel Interrupt Handler *fu*
PII Pershing II [a missile] *jmj*
PII Pre Installation Investigation *fu*
PII Procurement Instrument Identification *fu*
PII Program Integrated Information *pk, kg*
PIIN Procurement Instruction ID Number *fu*
PIL Parts Identification List *gp*
PIL Preferred Item List *mt*
PILOT Phased Integrated Laser Optics Technology *wp*
PILOT Piloted Lowspeed Tests [Martin-Marietta X-24, low speed in comparison with the X-23 PRIME vehicle] [USA] *mt*
PILOT Programmed Inquiry Learning Or Teaching *pk, kg*
PilotACE Pilot Automatic Computing Engine *pk, kg*
PIM Passive Intermodulation *ep, pbl, jdr, kf, jn, uh, he, ep, pbl, jdr, kf*
PIM Personal Information Manager *em, ct, hi, pk, kg*
PIM Planning Interchange Meeting *tb*
PIM Plug In Module *wp*
PIM Position and Intended Movement *fu*
PIM Position of Intended Movement *mt*
PIM Pretrained Individual Manpower *wp*
PIM Primary Interface Module *pk, kg*
PIM Process Interface Module *fu*
PIM Program Instruction Manual *fu*
PIM Programmatic Interchange Meetings *mt*
PIM Protocol Independent Multicast *em*
PIM Pulse Interval Modulation *fu, wp*
PIMO Presentation of Information for Maintenance and Operation *fu*
PIMS Phase Interference Modulation System *wp*
PIMS Procurement Information Management Services [DLA] *mt*
PIMS Program Interface for Multiplex System *hc*
PIN Particle Inclusion Noise (diodes) *gg148, mj31.5.32*
PIN Particle Induced Noise *mt*
PIN Personnel Increment Number [USA DOD] *mt*
PIN Plan Identification Number *mt*
PIN Position Indicator *fu*
PIN Positive, Intrinsic, Negative [device] *fu*
PIN Process Identification Number [Unix] *pk, kg*
PIN Project ID Number *fu*
PIND Particle Impact Noise Detection *fu, jdr, oe*
PIND Particle Inclusion Noise Detection *gp*
PINE Passive Infrared Night Equipment *hc*
PINE Program for Internet News and Email *pk, kg*
PINES PACAF Interim Nation Exploitation Segment *hc*
PING Packet Internet Gopher *ct, em, kg, hi*
PING_PONG [short range, front line reconnaissance] *wp*
PINPOINT [advanced version of the An/Apr-25 Radar homing and warning system] *wp*
PINS Point In Space *jb*
PINS Precise Integration Navigation System *hc*
PINS Product Information System *fu, sc4.2.4, 15, ep, jdr*
PINT (sample incoming inspection and test including PINT) *bl2-9.45*
PIO Parallel Input/Output *pk, kg*
PIO Peripheral Input / Output *fu, hi*
PIO Pilot Induced Oscillations *wp*
PIO Processor Input/Output *pk, kg*
PIO Programmable Input/Output *fu*
PIO Programmed Input / Output *ct, em, kg*
PIO Provisioned Item Order *fu*
PIO Public Information Officer *mt, wp, af72.5.14, oe*
PIOA Process Input/Output Activity *fu*
PIOC Programmable Input / Output Controller *fu*
PIOCS Physical Input / Output Control System *hi*
PIOM Programmable Input/Output Module *fu*
PIP Parallel Image Processing *wp*
PIP Parts Improvement Program *bl2-9.52*
PIP Payload Integration Plan *gp, 4, wl, jdr, oe*
PIP Performance Improvement Program *sa816391B, ig21.12.17*
PIP Performance Improvement Proposal [a document] *ep, fu*
PIP Picture In Picture *is26.1.60, kg, hi*
PIP Planning Integration Process *mt*
PIP Predicted Impact Point *17, sb, tb*
PIP Predicted Intercept Point *fu*
PIP Preliminary Installation Plan *fu*
PIP Primary Injection Point *rih*

PIP Problem Isolation Procedure *pk, kg*
PIP Product Improvement Program *hc, st*
PIP Product Integration Program *mt*
PIP Program Implementation Plan *mt, wp*
PIP Program Improvement Plan *wp*
PIP Program Integration Plan *gp*
PIP Programmable Interconnect Point *pk, kg*
PIP Programmable Interface Processor *tb*
PIPA Pulsed Integrating Pendulous Accelerometer *hc*
PIPD Planning Input for Program Development *mt, wp*
PIPEX Public Internet Protocol Exchange *hi*
PIPF Previously Identified Pattern Failures *fu*
PIPO Parallel In, Parallel Out *kg, hi*
PIPOR Program for International Polar Oceans Research *hg, ge*
PIPPS Publication Information Processing and Printing System *mt*
PIPS Passive Infrared Personnel Sensor *hc, hm*
PIPS PCAC Information Processing System [DIA/NSA] *mt*
PIPS Plans Integration Partition System [JCS] *mt*
PIPS Portable Interactive Planning system [SAC] *mt*
PIR Payload Integration Review *bo*
PIR Photo Interpretation Report *wp*
PIR Primary Intelligence Requirements *mt*
PIR Procurement Information Reporting *wp*
PIR Procurement Initiation Request *fu*
PIR Protein Identification Resource *wp*
PIR Pulse Interference Rejection *fu*
PIRAD Passive Infrared Detector *wp*
PIRD Program Instrumentation Requirements Document *0*
PIRLS Probe Infrared Laser Spectrometer *es55.28*
PIRN Preliminary Interface Review (Revision) Notice *4, fu*
PIRS Passive InfraRed Sensor *wp*
PIRS Pulsed InfraRed System *wp*
PIRSA Passive InfraRed Situation Awareness *wp*
PIRT Precision Infrared Tracking *fu*
PIRT Production Inspection Reliability Test *fu*
PIS Platform Integration Simulator *fu*
PISAB Pulse Interference Separation and Blanking *fu*
PISD Pacific Information Systems Division [AFCC*] *mt*
PISO Parallel-In, Serial-Out *fu, ma72*
PIT Payload Integration and Test *kf*
PIT Peripheral Input Tape *fu*
PIT Preliminary Integration Test *fu*
PIT Process Implementation Team *pbl, ah, jn*
PIT Programmable Interval Timer *fu, kg*
PITMOS Post Mode Optimizing System *mt*
PITR Platform Interface Testing Remote Chassis *fu*
PIU Peripheral Interface Unit *fu*
PIU PLRS Interface Unit *fu*
PIU Plug In Unit *fu*
PIU Power Interface Unit *ma83, 9, lt, jdr*
PIU Programmable Interface Unit *fu*
PIV Particle Image Velocimetry *te23, 11, 22*
PIV Post Indicator Valve *fu*
PIWG Permanent Interoperability Working Group *mt*
PJBD Permanent Joint Board of Defense [Canada - U. S.] *dt*
PJH PLRS / JTIDS Hybrid system *mt*
PJH PLRS/JTIDS Hybrid *fu*
PJHI PJH interface *fu*
PK Peace Keeper [a missile] *kf*
PK Probability of Kill *mt, fu, wp, tb, st*
PKC Public Key Cryptography *em*
PKG-POL Packaged POL *mt*
PKM Perigee Kick Maneuver *sa122052.1.1.2*
PKM Perigee Kick Motor *sc2.12.2, 9, lt, jdr, he, uh, jn, oe*
PKO Peace Keeping Operations *wp*
PKR Protivolodochniy Kreyser [anti-submarine cruiser] [USSR] *mt*
PKS Perigee Kick Stage *bl1-2.162, lt*
PKSS Probability of Kill Single Shot [also SSPK] *wp, st*
PL Parting Line *fu*
PL Parts List *fu, jdr, oe*
PL Phase Line *fu*
PL Physical Layer *hi*
PL Podvodnaya Lodka [patrol submarine] [USSR] *mt*
PL Position Location *fu*
PL Probability of Leakage *tb*
PL Product Line *fu*
PL Programming Language [as in PL1] *wp, hi, mt*
PL Proportional Limit *fu*
PL Provisioning List *fu*
PL-DRO Phase Locked Dielectric Resonator Oscillator *jdr*
PL/M Programming Language for Micros *kg, hi*
PLA Atomnaya Podvodnaya Lodka [nuclear attack submarine] [USSR] *mt*
PLA Payload Adapter *oe*
PLA People's Liberation Army [Chinese] *-12*
PLA Plain Language Addressee *mt*

PLA Port Link Assignment *fu*
PLA Programmable Logic Array *mt, 6, kg, fu, hi, ai1*
PLAAR Packaged Liquid Air Augmented Rocket *hc*
PLAC Preliminary Load Analysis Cycle *kf, jn*
PLACE Position Locating and Checking of Entries *fu*
PLACE Position Location and Aircraft Communication Equipment *hc*
PLACE Post LANDSAT-D Advanced Concept Evaluation *hc*
PLAD Plain Language Address Directory *mt*
PLADS Plain Language Address System *mt*
PLAN Parts Logistic Analysis Network *wp, jdr*
PLAP Atomnaya Protivolodochnaya Podlodka [nuclear hunter-killer submarine] [USSR] *mt*
PLARB Atomnaya Podlodka Raketnaya Ballisticheskaya [nuclear ballistic missile submarine] [USSR] *mt*
PLARK Atomnaya Podlodka Raketnaya Krilataya [nuclear cruise missile submarine] [USSR] *mt*
PLASI Pulse Light Approach Slope Indicator *fu*
PLAT Pilot Landing Aid Television *hc*
PLATO Program (Programmed) Logic for Automated Teaching Operations *hc, wp, kg*
PLB Payload Bay *9, oe*
PLC Power Line Carrier *fu*
PLC Preliminary Loads Cycle *kf*
PLC Product Line Code *fu*
PLC Product Line Controller *fu*
PLC Programmable Logic Controller *pk, kg*
PLC Propellant Loading Cart *oe*
PLC Public Limited Company *ai1*
PLCC Plastic Leaded Chip Carrier *em, em*
PLCC Plastic Leadless Chip Carrier *pk, kg*
PLCP Payload Leak Contingency Plan *oe*
PLCP Physical Layer Convergence Protocol *ct, em*
PLCR Post Launch Checkout Report *gp*
PLCRB Product Line Change Review Board *fu*
PLD Personnel Lowering Device *wp*
PLD Product Line Document *pbl, jdr, tm, ah, jn*
PLD Program Load Disc *mt*
PLD Programmable Logic Device *kg, wp*
PLDS Pilot Land Data System [NASA] *kg, ns, hy, hg, ge*
PLE Public Local Exchange *ct, em*
PLET Project Labor Estimation Tool *kf*
PLF Payload Fairing *jdr, he, ai2*
PLF Payload Launch Fairing *jn*
PLF Proof Load Factor *jdr*
PLFA Primary Level Field Activity *mt*
PLFPF Payload Fairing Processing Facility *ai2*
PLG Parts List Generator *fu*
PLI Phase Lock Injection *jdr*
PLI Private Line Interface *mt*
PLIM Propellant Lining Installation Metal *hn44.17.3*
PLIR Position Location Information Report *fu*
PLISN Provisioning List Item Sequence Number *fu*
PLK Podlodka Raketnaya Krilataya [cruise missile submarine] [USSR] *mt*
PLL Phase Lock (Locked) Loop *kg, fu, hi, gp, ct, 0, 2, 4, 7, 16, mj31.5.23, lt, ep, jdr, oe*
PLL Postscript Level Language *hi*
PLL Prescribed Load List *mt, fu*
PLLATS Portable Lightweight LAFS Azimuth Test Set *hc*
PLM Post Launch Maneuver *fu*
PLN Program Line Number *mt*
PLO Palestine Liberation Organization *10, w9, wp, nd74.445.17*
PLO Phase Locked Oscillator *mj31.5.75, fu, hi*
PLOO Pacific Launch Operations Office *ci.92*
PLP Packet Level Procedure *ct, em*
PLP Procedural Language Processor *mt*
PLP Product Line Processor *fu*
PLP Propagation Loss Processing *fu*
PLPS Presentation Level Protocol Standards *hi*
PLRACTA Position, Location, Reporting and Control of Tactical Aircraft *hc, wp, fu*
PLRS Position Location (Locating and) Reporting System [USMC] *mt, wp, hn42.4.8, hq5.2.5, aa26.10.13, hc, tb, fu*
PLS Payload Specification *kf*
PLS Personal Locator System *wp*
PLS Plasma Spectrometer *es55.26*
PLS Plasma Subsystem *jp*
PLS Pre Launch Survivability *mt, wp*
PLS Preoperational Logistics Support *fu*
PLS Primary Link Station *pk, kg*
PLS Product Line Simulator *fu*
PLS Product Line Specialist *fu*
PLSC PACAF Logistics Support Center *mt*
PLSF Post Loop Spur Filter (PLL) *jdr*
PLSO Phonetic Letter Spell Out *fu*
PLSO Phonetic Letters Spelled Out *mt*
PLSS Portable Life Support System *w9, mb, mi*

PLSS Precision (Position) Location and Strike System [USAF] *mt, fu, ph, hc, nd74.445.30, wp*
PLT Parts List Transmittal *fu*
PLT Pilot *mt, ws4.10*
PLT Post Launch Testing (Test) *oe*
PLT Procurement Lead Time *fu*
PLT Production Lead Time *fu*
PLTT Post Launch Test Team *oe*
PLTT Propulsion Leak Test Tooling *oe*
PLU Position Location Uncertainty *st*
PLU Preservation of Location Uncertainty *hc*
PLU Program Load Unit *fu*
PLUC (PLUC control panel . . . of CM, Apollo Program) *ci.388*
PLUS Precision LIDAR Utility Study *hc*
PLUTO Pipe Line Under The Ocean *wp*
PLV Personnel Launch Vehicle *ci.1056*
PLV Production Level Video *pk, kg*
PLVP Project Line Video Processor *fu*
PLWS Pulsed Laser Weapon System *wp*
PM Patrol Missile Boat [US Navy] *mt*
PM Performance Measurement *wp, jdr*
PM Performance Monitoring *fu*
PM Perigee Motor *oe*
PM Periodic Maintenance *fu*
PM Phase Modulation *mt, fu, bl1-5.38, wp, 0, 2, 4, 16, mj31.5.30, jdr, vv, oe, ai1*
PM Plovuchaya Masterskaya [floating workshop] [USSR] *mt*
PM Point Mugu [site] *fu*
PM Presentation Manager [IBM] *kg, hi*
PM Pressurized Module *mi*
PM Preventive Maintenance *mt, fu, 2, 4, tb, kg, hi*
PM Procedure Memory *mt*
PM Process Manager *fu, kg*
PM Producibility Manual *ep*
PM Product Management *mt*
PM Product Manager *st*
PM Product Monitor *oe*
PM Program Manager (Management) *mt, fu, kf, wp, 17, nt, jmj, tb, jdr, hi*
PM Program Memorandum *mt*
PM Program Memory *fu*
PM Project Manager *mt, gp, oe*
PM Project Monitor *oe*
PM Propulsion Module *gr, oe*
PM Pulse Modulation *fu, lt*
PM {Photo Multiplier} *es55.6*
PM&P Parts, Materials and Processes *jdr, kf, oe*
PM-TRADE Project Manager, Training Devices [Office of] *tc*
PM/FL Performance Monitoring/Fault Location *fu*
PM/ST Performance Monitoring/Self-Test *fu*
PMA Polymethylacrylate *wp*
PMA Preliminary Mission Analysis *oe*
PMA Principal Maintenance Activity *fu*
PMA Property Movement Authorization *-4*
PMAA Petroleum Marketers Association of America *y2k*
PMAB Program Manager's Advisory Board *fu*
PMAC Preliminary Maintenance Allocation Chart *fu*
PMAD Portable Maintenance Aid Device [USA] *mt*
PMAD Power Management And Distribution *wl*
PMAG Program Management Advisory Group *mt*
PMAG Program Management Assistance Group *wp*
PMALS Prototype Miniature Air Launch (Launched) System *uc, hc*
PMAPS Predirected Munitions Automated Planning System [USAFE] *wp, mt*
PMAST Process Management, Assessment and Standard Tools [Team] *kf, jdr*
PMB Performance Measurement Baseline *17, sb, jdr, kf, oe*
PMBX Private Manual Branch Exchange *hi*
PMC Partial Mission Capability *wp*
PMC Partial Mission Capable [USA DOD] *mt*
PMC Permanently Manned Capability *mi*
PMC Port Management Center [MAC] *mt*
PMC Precision Measurement Work Cell *hc*
PMC Procurement Method Code *fu*
PMC Program Management Charter *mt*
PMC Program Management Control *wp*
PMC Program Management Council *heh2.21.7*
PMC Program Manager Charter *mt*
PMC Project Management Chart *fu*
PMC Publications Media Center [an organization] *ah, jdr, pbl, tm, jn*
PMCC Port Management Control Center *mt*
PMCM Partial Mission Capable, Maintenance [USA DOD] *mt*
PMCS Partial Mission Capable, Supply [USA DOD] *mt*
PMCS Preventive Maintenance Checks and Services *fu*
PMC_[B,M,S] Partially Mission Capable [Both, Maintenance, Supply] *mt*
PMD Packet Mode Data *pk, kg*

PMD Program Management Directive (Direction) *mt, hc, fu, wp, tb, kf*
PMD Program Management Document *mt*
PMD Program Management Documentation *st*
PMD Propellant Management Device *sd, jdr, kf, vv, he, uh, jn, oe*
PMDC Program Management Development Course *heh2.21.7*
PMDP Project Manager Development Program *mt*
PME Payload Mission Engineering *jr*
PME Precision Measurement (Measuring) Equipment *mt, fu, wp, hc*
PME Prime Mission Equipment *mt, hc, wp, dm, tb, fu*
PME Professional Military Education *mt, af72.5.72*
PMEL Precision Measurement (Measuring) Equipment Laboratory *mt, fu, hc, wp, tb, kf*
PMF Perigee Maneuver Firing *oe*
PMF Perigee Motor Firing *uh, oe*
PMF Principal Management Facility *su, hc, jdr*
PMF Program Management Facility *tb*
PMG Prediction Marker Generator *fu*
PMG Program Maintenance Group *mt*
PMGL Precision Measuring Equipment Lab *tb*
PMGM Program Manager Guidance Memorandum *mt*
PMH Petroleum Movement History [MSC] *mt*
PMH Prime Mission Hardware *hc*
PMHL Payload Module Hardline *uh*
PMI Permanent Manufacturing Information *fu*
PMI Preventive Maintenance Inspection *mt, wp*
PMI Preventive Maintenance Instructions *fu, hc*
PMI Program Management Instruction *fu*
PMI Purchased Material Inspection *oe*
PMIM Production Management Information Module *fu*
PMIR Pressure Modulated Infrared Radiometer *re, mi*
PMIRR Pressure Modulated InfraRed Radiometer [on Mars Observer] [also PMIR] *mi*
PMIS Personnel Management Information System [USAF] *wp, mt*
PMIS Production Management Information System [DMMIS] [AFLC] *mt*
PMIS Publications (Production) Management Information System *fu*
PMISP Personnel Management Information System Plan [USAF] *mt*
PML Processor Maintenance Logic *fu*
PML Program (Project) Master Library *fu*
PMLCD Passive Matrix LCD *hi*
PMM Polymethylmethacrylate *wp*
PMM Pulse Mode Multiplex *fu*
PMMA Poly Methyl Methacrylate *fu, wp*
PMMC Permanent Magnet Movable Coil *fu*
PMMR Passive Multichannel Microwave Radiometer *ac7.4-1*
PMMU Paged Memory Management Unit *kg, hi*
PMO Program Management Office *mt, 16, wp, tb, kf, fu, oe*
PMOC Prototype Mission Operations Center [SPACECMD] *mt, sr9-82, hs, go3.2, jb, tb*
PMOS Positive channel Metal Oxide Semiconductor *fu, 6, em, kg, hi*
PMP Parallel Microprogrammed Processor *hc*
PMP Parts, Materials and Processes [hardware] *ah, he, gp, 5, 16, 17, sb, pl, nt, ep, tb, pbl, tm, kf, jn*
PMP Performance Monitoring Program *mt*
PMP Portable Maintenance Panel *fu*
PMP Precision Mounting Platform *-16*
PMP Prime Mission Product *mt*
PMP Program Management Directive *mt*
PMP Program Management Plan *mt*
PMP Program Master Plan *wp*
PMP Project (Program) Management Plan *fu, 0, wp, tb*
PMP Property Management Procedure *fu*
PMPCB Parts, Materials and Processes Control Board *gp, 17, sb, pl, lt, ep, jdr, tm, kf, jn*
PMPCC Parts, Materials or Process Change Coordination *fu*
PMPSL Parts, Materials and Processes Selection List *17, sb*
PMPSR Payload Multi Purpose Support Room *oe*
PMR Pacific Missile Range *fu, ci.91, 0, sb*
PMR Performance Measurement Report *-4*
PMR Preliminary Material Review *oe*
PMR Pressure Modulated (Modulator) Radiometer *ac2.8-3, pl*
PMR Primary Mission Readiness *mt*
PMR Private Mobile Radio *hi*
PMR Problem Management Report [IBM] *pk, kg*
PMR Program Management Responsibilities *mt*
PMR Program Management Review *mt, fu, wp, tb, kf*
PMR Program Manager's Representative *fu*
PMR Program Modification Request *mt*
PMR Proton Magnetic Resonance *wp*
PMR Provisioning Master Record [USA] *fu, mt*
PMRB Preliminary Material Review Board *fu*
PMRF Pacific Missile Range Facility *12, fu*
PMRM Periodic Maintenance Requirements Manual *fu*
PMRR Prepositioned Materiel Receipt Record *wp*
PMRT Program Management Responsibility Transfer *mt, fu, wp*
PMRTD Program / Project Management Responsibility Date *mt*

PMRTP Program Management Responsibility Transfer Plan *mt*
PMS PACOM Munitions Storage *mt*
PMS Payload Management Subsystem *jdr*
PMS Pedestal Mounted Stinger *pm*
PMS Performance Measurement System *fu, gp, 4, 16, 17, ep, kf, oe*
PMS Performance Monitoring System *mt, wp*
PMS Pipeline Management System [USAF] *mt*
PMS Planned (Preventive) Maintenance System *fu*
PMS Planned Maintenance Support *mt*
PMS Policy Management System *pk, kg*
PMS Production (Product) Management System *fu, bl3-3.8*
PMS Production Management Support *mt*
PMS Project (Program) Master Schedule *fu, jdr*
PMS Propulsion Module Subsystem *so*
PMSR Performance Measurement Status Report *oe*
PMSS Program Manager's (Management) Support System *mt, wp*
PMSV Pilot to Metro Services *mt*
PMT Phase Modulated Transmitter *-4*
PMT Photo Multiplier Tube *fu, gp, 16, 17, mb, mi, oe*
PMT Product Management Team *jn*
PMT Product Maturity Test *fu*
PMT Program Management Tool *fu*
PMT Program Master Tape *fu*
PMTC Pacific Missile Test Center *fu, 12, tc, hc, tb*
PMTC PLRS Message Traffic Control *fu*
PMTC Point Mugu Test Center [missile test center of the US Navy] *mt*
PMTP Program Management Transfer Point *mt*
PMTPD Program Management Transfer Point Date *mt*
PMTPM Program Management Transfer Point Milestone *mt*
PMTPP Program Management Transfer Point Plan *mt*
PMU Portable Memory Unit *fu*
PMU Precision Mockup *-17*
PMW 146 UHF F/O Program Office *uh*
PMWC Preventive Maintenance Workcard *fu*
PMX Private Manual Exchange *hi*
PN Part Number [a designator] *bo, ah, wp, ep, jdr, pbl, tm, jn, hi*
PN Positive-Negative *fu*
PN Primary Node *mt*
PN Processing Node *pk, kg*
PN Pseudo(random) Noise *mt, fu, spxi, 2, 4, 17, vv, oe*
PN Public Network *y2k*
PN/FH (spread-spectrum PN/FH technique) *sd*
PNA Programmable Network Access *pk, kg*
PNAF Primary Nuclear Airlift Force *-12*
PNB Passive Narrow Band *fu*
PNB Producibility Notebook *fu*
PNB Programmer's Notebook *fu*
PNC Preferred Noise Criteria *fu*
PNCP PLRS Net Control Processor *fu*
PNCTR Passive Non Cooperative Target Recognition *hc, fu*
PNDI Proven Non-Developmental Items *kf*
PNETS Public Non Exclusive Telecom Services *hi*
PNFE Prototype Network Front End *mt*
PNG Portable Network Graphics *pk, kg*
PNG Pseudo Noise Generator *fu*
PNI Packet Network Intercommunications *hi*
PNIP Positive-Negative-Intrinsic-Positive *fu*
PNL Pacific Northwest Laboratory *tb*
PNM Price Negotiation Memorandum *wp*
PNMP PLRS Net Management Processor *fu*
PNMP Preliminary NMCS Master Plan *mt*
PNNI Private Network to Network Interface *ct, em*
PNOHS Precision Null Operating Horizon Sensor *jdr*
PNP Positive-Negative-Positive [transistor] *fu, wp, bl2-9.46, mj31.5.60*
PNP Preliminary Network Plan *-2*
PNPN Postive-Negative-Positive-Negative *fu*
PNS Parabolized Navier-Stokes *aa26.12.b8*
PNS Part Number Sequence *fu*
PNS Primary North-South Stationkeeping *-3*
PNSS Pseudonoise Spread Spectrum *mt*
PNVS Pilot Night Vision Sensor *hc, wp*
PNW Protracted Nuclear Warfighting *uc*
PO Parity Odd *pk, kg*
PO Petty Officer *w9*
PO Power Oscillator *fu*
PO Production Order *fu*
PO Program (Project) Office *mt, fu, tb, st*
PO Program Office(r) *mt*
PO Proizvodstvennoe Ob'edinenie [Production Association] *ai2*
PO Project Organization *fu*
PO Purchase Order [a document] *bl2-9.50, ah, fu, wp, 8, 16, 17, w9, ep, tb, jdr, pbl, tm, jn*
PO Purchasing Office *mt*
POA Power Open Association *ct, em*
POA&MS Plan Of Action and Milestone Schedule *fu*
POBAL Powered Balloon *wp*

POC Parts Ordering Control *fu*
POC Point Of Contact *fu, mt, jmj, wp, tb, st, jdr, pbl, kf, kg, ah, uh, jn*
POC Privately Owned Conveyance *wp*
POC Process Owner Councils *jdr*
POC Productional Operational Capability *0*
POC Proof-Of-Concept *fu, sb, 19, jdr*
POC-ET Proof Of Concept Experimental Testbed *mt*
POC/POT Proof Of Concept / Proof Of Technology *-5*
POCA Point Of Closest Approach *st*
POCC Payload Operations Control (Coordination) Center *ci.940, 9, es55.71, tb, wl, jdr, oe*
POCC Project Operations Control Center *ci.141, 2, 4, mb*
POCL Paid Out Cable Length *fu*
POD Port Of Debarkation [USA DOD] *mt*
POD Power On Diagnostics *fu*
POD Preliminary Orbit Determination *ai2*
POD Private Orientation Device *hc*
POD Probability Of Detention *mt*
POD Products Operations Division [Hughes Aircraft Company] *hn43.22.8*
POD Proposed Operational Dates *mt*
PODA [an approach to] Pulse transmission On Demand Assignment *csII.82*
PODS Pilot Ocean Data System [JPL] *hg, ge*
POE Part Of the Equipment *fu*
POE Port Of Embarkation [USA DOD] *17, w9, mt*
POE Port Of Entry *w9, wp*
POE Power Open Environment *pk, kg*
POE Program Office Estimate *kf*
POEM European Polar Platform *hg, ge*
POEM Photonic Opto Electronic Material *is26.1.76u*
POEM Polar Orbiting Earth (observation) Mission *ls4, 5, 44*
POEMS Polar Orbit Earth Observation System [ESA] *hg, ge*
POEMS Positron Electron Magnet Spectrometer [EOS] *hy*
POES Polar-orbiting Operational Environmental Satellite [NOAA] *hg, ge, aw129.10.76, hy, ls4, 5, 16*
POET (see POPO) *tb*
POET Primer Oscillator Expendable Transponder *mt*
POETS Product Operations Estimating Techniques System *gb, 8*
POEWGP WIN Operational Experiments Working Group *mt*
POFA Performance and Operability Functional Analysis *fu*
POFA Programmed Operational and Functional Appraisal [program] *wp*
POGO (critical ascent conditions are the MECO/POGO and liftoff) *gg17*
POGO Polar Orbiting Geophysical Observatory *hc, le28*
POH Performance Overhead *jdr, pbl*
POI Period Of Interest *mt, kf*
POI Probability Of Intercept *wp*
POI Program Of Instruction *mt, fu*
POIS Performance-Oriented Instructional System *fu*
POL Petroleum Oil and Lubricants *fu, hc, wp, pm, tb, st*
POL Petroleum, Oils, and Lubricants [broad term that includes all petroleum products used by U. S. Armed Forces] *mt*
POL Problem Oriented Language *pk, kg*
POL Procedure Oriented Language *fu*
POL/MIL Political/Military *mt*
POLAMCO [a Polish manufacturing firm] *sc3.2.6*
POLAR (A POLAR code simulation of SPEAR I shows) *aa26.12.24*
POLCAP Petroleum Capability *mt*
POLDER Polarization and Directionality of the Earth's Reflectances [CNES] *ge, hg*
POLEX Polar Experiment *pl*
POLREP PACOM Petroleum Report [PACOM] *mt*
POLT Program Optimizing Low-thrust Trajectory *hc*
POM (a POM/AM/AM telecommand system) *ac3.2-6*
POM Preparation for Overseas Movement *mt*
POM Program Objectives Memorandum *mt, ph, wp, hc, nt, tb, st*
POM Project Operations and Milestones *mt*
POM Proof Of Manufacturing *rf84.3.23, fu*
POMC Period Of Minimum Change *mt*
POMCUS Prepositioned Overseas Materiel Configured to Unit Sets [USA] *mt*
POMCUS Prepositioning Of Materiel Configured to Unit Sets *wp, pm, aj89.9.40*
POMINS Portable Mine Neutralization System *mt, hc*
POMO Production Oriented Maintenance Organization *mt, fu*
POMP (Sorts and Routes Obs (Software Program)) *jb*
POMSS Prepositioned Operational Material Storage Site *mt*
POMWG Program Objective Memorandum Working Group *mt*
PON Program Office Need *jn*
PON Project Originator Number *mt*
POOD Provisioning Order Obligation Document *fu*
POP Period Of Performance *mt, tb*
POP Perpendicular to Orbit Plane *wl*
POP Point Of Presence *mt, em, ct, kg*
POP Post Office Protocol *em, kg, ct, hi, du*
POP Preliminary Operating Plan *fu*
POP Product Operations Procedures *fu*

POP Program Operating Plan *tb, wl, oe*
POPF Pop Flags *pk, kg*
POPIM Product Operations Planning Instruction Manual [a document] *ep*
POPO Phase One Program Office *tb*
POPR Power On Preset Reset *jr*
POPS Principles of Operation Simulator *fu*
POPS Pyrotechnic Optical Plume Simulators *wp*
POPSAT Precise Orbit Positioning Spacecraft *sc3.4.6*
POR Power On Request *gp*
POR Power On Reset *he, gp, kg, uh*
POR Purchase Order Request *wp*
PORL Playback Operational Recording List *fu*
PORT Photo-Optical Recorder Tracker *fu*
PORT Programmable Organization for Reciprocal Transactions *fu*
PORTER Performance Oriented Tracking of Equipment Repair *wp*
PORTS Portable Imagery Telecommunications System *wp*
PORTS Ports Characteristics File [WWMCCS] *mt*
PORTS Preliminary Operations Requirements Test System *tb*
POS Peacetime Operating Stocks *mt, wp*
POS Portable Oxygen Subsytem *wp*
POS Positional Accuracy *fu*
POS Pre Operational Support *kf*
POS Primary Operating Stocks [USA DOD] *mt*
POS Program Operational Summary *fu*
POS Programmable Object Select *pk, kg*
POS Programmable Option Select *ct, em*
POS Proof Of Station *fu*
POSC Primary Operations System Control *mt*
POSE Portable Operating System Environment *mt*
POSE Portable Operating System Extension *hi*
POSF Ports Of Support File *mt*
POSIT Position Report *mt*
POSIX Portable Operating System Interface for [Unix] [Computer Environments] *mt, kg, kf, aw130.13.63, hi*
POSNAV Position Navigational Assistance *fu*
POSS Positioner Subsystem *jdr*
POSS Preliminary Operational Safety Study *fu*
POSS Prototype Optical Surveillance Subsystem *mt*
POSS Prototype Optical Surveillance System *wp*
POSS Purchase Order Status System *fu*
POST Passive Optical Seeker Technique *wp*
POST Point Of Sale Terminal *hi*
POST Point Of Sale Transaction *hc*
POST Portable Optical Sensor Tester *hc, st*
POST Power On Self Test *ct, kg, em, hi*
POST Proposed Optical Stinger [a FIM-92 version with a new seeker, USA] *mt*
POST Prototype Ocean Surveillance Terminal [US Navy] *mt*
POSTNET Postal Numeric Encoding Technique *pk, kg*
POT Preproduction Operational Test *mt*
POTS Purchase Of Telephones and Systems *mt*
POTT Patriot Operator Tactics Trainer *aw130.13.87*
POTV Personnel Orbital Transfer Vehicle *ci.1056*
POV Peak Operating Voltage *fu*
POW Prisoner Of War *mt, wp, ie5.11.3, w9*
POWER Performance Optimization With Enhanced RISC *kg, hi*
POWERPC Performance Optimization With Enhanced RISC-Performance Computing *kg, hi*
POWG Payload Operations Working Group *oe*
PP Panel Point *fu*
PP Parallel Port *pk, kg*
PP Parcel Post *wp*
PP Parts Protection *fu*
PP Peak to Peak *-2*
PP Penetration Point *fu*
PP Physics Package [RMO] *jdr*
PP Planning Package *fu, kf*
PP Point to Point *mt, fu, hi*
PP Precision Platform *-16*
PP Program Package *fu*
PP Project Plan *oe*
PP&C Procurement Planning and Control *fu*
PP&P Preservation, Packaging and Packing *fu*
PPA PACAF Product Performance Agreement Center [ASD] *mt*
PPA Pastanovchik-Pamech Aktivnii [active jammer, as in Mi-18PPA Hip-J] [USSR] *mt*
PPA Performance Planning and Appraisal *ah, pbl, jn*
PPA Phenyl Propanol Amine *wp*
PPA Platform Pointing Angle *-3*
PPA Program, Project, and Activity *mt*
PPA Pulse-by-Pulse Antenna *fu*
PPAM Program Product Assurance Manager *17, sb*
PPAP Procurement Product Assurance Plan *fu*
PPAR Program Page Address Register *-3*
PPB Parts Per Billion *fu*

PPB Product Producibility Board *fu*
PPB Provisioning Parts Breakdown *fu*
PPBES Planning, Programming, Budgeting, and Execution (Executing) System *fu, nt*
PPBG Preliminary Program and Budget Guidance *mt*
PPBS Planning, Programming and Budgeting System *mt, wp, fu, 1, 13, ph, hc, tb, st, kf*
PPC Personnel Planning Conference *fu*
PPC Personnel Program Coordinator *fu*
PPC Petroleum Planning Committee *mt*
PPC Phased Provisioning Code *fu*
PPC Polar Path Compass *wp*
PPC Precipitating Particle Counter *-16*
PPC Precision Plotting Center *fu*
PPC Production Planning and Control *mt*
PPC Programmable Peripheral Controller *fu*
PPC Pulse to Pulse Correlation *wp*
PPCA Preliminary Physical Configuration Audit *mt*
PPCAS Production Program Cost Analysis Summary *fu*
PPD Penetration Protection Device *fu*
PPD Policy and Procedures Directive *fu*
PPD Pre-Prototype Demonstration *st*
PPD Program Package Document *fu*
PPDB Point Positioning Data Base *mt*
PPDF Portable Postscript Document Format *hi*
PPDL Point to Point Data Link *hi*
PPDU Payload Power Distribution Unit *he, kf, jn*
PPDU Presentation Protocol Data Unit *fu*
PPDU Propulsion Power Distribution Unit *cl, uh, jn*
PPE PAVE_PAWS East *mt*
PPE Pre Production Evaluation *mt*
PPE Premodulation Processing Equipment *fu*
PPE Problem Program Evaluation *fu*
PPF Pattern Propagation Factor *fu*
PPF Payload Processing Facility *uh, 9, jdr, oe, ai2*
PPF Polar Platform *ls4, 5, 44*
PPF Precision Plotting Facility *fu*
PPFS Product Performance Feedback System *mt*
PPG Programmable Pulse Generator *ai2*
PPG Propylene Glycol *wp*
PPG Pulses Per Group *mt*
PPGA Plastic Pin Grid Array *pk, kg*
PPGM Planning, Programming, and Guidance Memorandum *mt*
PPH Pulses Per Hour *fu*
PPI (PPI gauge) [international loading gauge] [USA DOD] *mt*
PPI Pictorial Position Indicator *fu*
PPI Plan Position Indicator *mt, fu, ai1*
PPI Planned Position Indicator *hn42.1.6, hc*
PPI Preplanned Product Improvements *mt, tb*
PPI Programmable Peripheral Interface *fu, ct, em*
PPI Pulses Per Inch *hi*
PPIF Photographic Processing and Interpretation Facility *mt, wp*
PPIO Priced Provisioned Item Order *fu*
PPIP Program Protection Implementation Plan *kf*
PPIRG PAVE_PAWS Interface Requirements Group *mt*
PPIU Probe Power Interface Unit *jp*
PPK Planned Probability of Kill *fu*
PPL Preferred Parts List *fu, gp, oe*
PPL Preferred Products List *wp*
PPL Preliminary Parts List *fu*
PPL Private Pilots License *mt*
PPL Provisioning Parts List *fu, hc*
PPLAN Programming Plan *mt*
PPLI Precise Participant Location and Identification *fu*
PPLI Precise Position Location and Identification *fu*
PPLI Provisioning Parts List Index *fu*
PPLIT Precise Participant Location Identification Tracking *fu*
PPLL Programmable Phase Lock Loop *jdr*
PPM Pages Per Minute *em, kg, hi*
PPM Parts Per Million *fu, gp, wp, 16, 17, st, jdr, hg, ge, oe*
PPM Periodic Permanent Magnet *fu*
PPM Pilot Performance Measurement *wp*
PPM Planned Preventive Maintenance *wp*
PPM Platform Pointing Mode *-3*
PPM Positions and Proper Motions [catalog] *mi*
PPM Preliminary Producibility Model *fu*
PPM Presidential Review Memorandum *mt*
PPM Principal Period of Maintenance *mt*
PPM Program Plans Manager *fu*
PPM Pulse Position Modulation *mt, fu, 0, wp, tb, jdr*
PPM Pulses per Minute *fu, wp*
PPMMS Polaris / Poseidon Material Management System [US Navy] *mt*
PPMP Preliminary Program Management Plan *fu*
PPMS Personal Property Movement and Storage System [USAF] *mt*
PPMS Program Planning Management System *-17*
PPN Peer to Peer Networking *hi*

PPN Project-Programmer Number *ct, em*
PPO Plant Protection Officer *su*
PPO Preferred Provider Organization *sc8.9.3*
PPO Principal Period of Operation *wp*
PPO Program Planning Objectives *wp*
PPO Purchaser Project Office *fu*
PPP Peak Pulse Power *fu*
PPP Personnel Performance Profile *fu*
PPP Plans, Progress, and Problems *tb*
PPP Point to Point Protocol [Internet protocol for dial up connection] *nu, kg, ct, em, ric, jdr, hi, du*
PPP Policy, Procedures, and Practice *tb*
PPP Pre Planned Product [Improvement] *-5*
PPP Preposition Procurement Package *mt*
PPP Production Process Procedure *fu*
PPP Program Protection Plan *kf*
PPP Programmable Power Processor *hc*
PPPI Planned Product Performance Improvement *jhm*
PPPI Potential Preplanned Product Improvement *st*
PPPI Pre Planned Product Improvement [also P^3I] *mt, jdr, kf, fu, 17, ph, sb, hc, wp, st*
PPPIP Pre Planned Product Improvement Program [also P3IP] *mt*
PPPO Personal Property Processing Office *mt*
PPPPI Pre Planned Product and Process Improvement *kf*
PPPSL Proposed Program Parts Selection List *fu*
PPQ Polyphenylquinoxaline *aa26.8.40*
PPR Photopolarimeter Radiometer *hc, jp*
PPR Pre Procurement Request *jdr*
PPR Pre Publication Review *mt*
PPR Preliminary Purchase Request *gp*
PPR Prior Permission Required *mt*
PPR Program Planning Review *wp*
PPR Program Progress Review *fu*
PPRNS Pulse Phased Radio Navigation System [Soviet] *-12*
PPRR Program Pre Release Reviews *mt*
PPRS Promotions and Placements Referral System *wp*
PPS Packets Per Second *kg, hi*
PPS Parallel Processing System *wp*
PPS Parts Population Summary *fu*
PPS Ping Pong Switch *jdr*
PPS Power Personal Systems [computing] *pk, kg*
PPS Precise Positioning Service [GPS] *jdr*
PPS Preproduction Sensor *fu*
PPS Primary Power Supply *gp*
PPS Program Performance Schedule *fu*
PPS Program Performance Specification *fu*
PPS Proposal Pricing System *mt*
PPS Provisioning Performance Schedule *fu*
PPS Pulses Per Second *mt, fu, ct, wp, tb, hi, oe*
PPS [Data Processing Station] *wp*
PPSL Program Parts Selection List *mt, fu, hc, 17, sb, tb, jdr*
PPSN Public Packet Switched Network *hi*
PPSO Personal Property Shipping Office *mt*
PPSR Parts Protection Stock Request *fu*
PPSU Propellant and Pressurant Servicing Unit *oe*
PPT Packed Pixel Technology *hi*
PPT Parts Per Trillion *wp, kf*
PPT Ping Pong Tray *jdr*
PPT Port Processing Time *mt*
PPT Product Positioning Time *fu*
PPT Pulsed Plasma Thrusters *js12.643*
PPTP Point to Point Tunneling Protocol *kg, em*
PPTS Propellant Pressurant Test Set *oe*
PPU Post Processor Unit *fu*
PPU Power Processing (Processor) Unit *hc, jn*
PPU Projection Plotting Unit *fu*
PPU Propulsion Power Unit *jn*
PPW PAVE_PAWS West *mt*
PPWSM Post Production Weapon System Management *mt*
PQ Planetary Quarantine *ma10*
PQ Preventive Quality *fu*
PQ Proto Qualification *jdr*
PQA Preliminary Quantitative Analysis *st*
PQA Product Quality Assurance *tb*
PQAE Program Quality Assurance Engineer *gp, lt*
PQAM Purchaser Quality Assurance Manager *17, sb*
PQE Project Quality Engineer(ing) *fu, oe*
PQFP Plastic Quad Flat Pack (Package) *kg, ct, em*
PQI Process Quality Indicator *fu, ig21.12.18*
PQO Project Quality Office *fu*
PQPP Project Quality Program Planning *fu*
PQR Prequalification Review *ep*
PQR Program Quality Requirement(s) [a document] *fu, pl, gp, 17, sb, ep, jdr, pbl, tm, jn*
PQS Personnel Qualification Standards *fu*
PQT Performance Qualification Testing *fu*

PQT Preliminary (Production) Qualification Test *mt, fu, hc, wp, tb, kf, 10*
PQT Prototype Qualification Test *fu, bf, jdr*
PQT-C Pre Qualification Test Contractor *fu*
PQV Preliminary Qualification Verification *fu*
PR Packaging Review *oe*
PR Packet Radio *fu*
PR Passive Ranging *wp*
PR Pattern Recognition *wp*
PR Photo Reconnaissance *mt*
PR Preliminary Review *mt, fu, wp*
PR Pressure Ratio *fu*
PR Problem Report *oe*
PR Procurement Request *jdr*
PR Procurement Requirement *fu*
PR Program Requirements *0*
PR Programs and Requirements [Deputy Chief of Staff for] *mt*
PR Programs and Resources [directorate of] *mt*
PR Pseudo Random *fu*
PR Public Relations *mt, wp, sc3.6.1, w9*
PR Pulse Rate *mt, fu*
PR Purchase Request (Requisition) [a document] *mt, ah, fu, gp, 8, wp, ep, tb, jdr, pbl, tm, jn*
PR&R Project Review and Reporting *-4*
PRA Parts Review Activity *fu*
PRA Performance Risk Assessment *kf*
PRA Photon Research Associates *17, sb, kf*
PRA Planetary Radio Astronomy *ns*
PRACSA Public Remote Access Computer Standards Association *pk, kg*
PRADS Product Reliability Assurance Data System *fu*
PRAG Performance Risk Assessment Group *kf*
PRAM Parallel Random Access Memory *em, kg*
PRAM Parameter Random Access Memory *pk, kg*
PRAM Productivity, Reliability, Availability, and Maintainability *mt, hc, wp*
PRAM Programmable Random Access Memory *fu, em, hi*
PRAM Programmed Symbols *em, ct*
PRAMS Passenger Reservation and Manifesting System [MAC*] *mt*
PRANG Puerto Rico Air National Guard *-12*
PRARE Precision Range And Rate Experiment - [Extended Version, ERS] *hg, ge*
PRAT Product Reliability Acceptance Test *fu*
PRAW [Naval Personnel Research Activity] *dt*
PRB Performance Review Board *wp*
PRB Program Review Board [DCA] *tb, mt*
PRC Passenger Reservation Center *mt*
PRC People's Republic of China [e.g. communist governed mainland China] *mt, fu, wp, ci.xxix, aw124.21.21, w9, hn50.8.3, kf*
PRC Personnel Readiness Center *mt*
PRC Physical Review Council, [Navy] *dt*
PRC Planning Research Corporation [of Kentron] *aw130.13.79, 17*
PRC Point of Reverse Curve *fu*
PRC Policy Review Committee [NSC] *mt*
PRC Presidential Review Committee *ci.1117*
PRC Procurement Review Committee *fu*
PRC Program Review Committee *mt*
PRC Program Review Council *wp*
PRC Programmable Radar Controller *fu*
PRC Project Review Committee *fu*
PRC/OS Programmable Radar Controller/Operator Station *fu*
PRCB Program Requirements Change Board *wl*
PRCP Programmed Communication Electronic Support Program *mt*
PRD (PRD Electronics Division, Harris Corporation) *aw118.18.6*
PRD Product Requirements Document *mt*
PRD Program Requirements Document *uh, wp, 0, 2, 4, 5, 17, sb, tb, st, jdr, kf, oe*
PRDA Program Research and Development Announcement *mt, fu, wp, tb, hc, ic1-1, cp*
PRDAVX Predictive Avoidance [software program] *jb*
PRDL Personnel Research and Development Lab *dt*
PRDR Preproduction Reliability Design Review *rm1.2.3*
PRE Performance Requirements Envelope *wl*
PRE Program Reliability Engineer *17, sb*
PRE Project Reliability Engineer *fu*
Pre-EMD Pre-Engineering and Manufacturing Development *kf*
PRECAT Precision Card and Assembly Tester *fu*
PREDICT Corporate Reliability Prediction Program *fu*
PRELOGE Preliminary Logistics Evolution *mt*
PRELORTR Precision Long Range Tracking Radar *fu*
PREMPT Program for Electromagnetic Pulse Testing *mt*
PREMSS Photo Reconnaissance and Exploitation Management Support System *wp*
PREP Power PC Reference Platform [IBM/Motorola] *kg, hi*
PREP Pre Edit Processor [WWMCCS] *mt*
PREP/RAMP Pre Edit Processor / Report And Message Processor [WWMCCS] *mt*
PREPAS Precise Personnel Assignment System [USMC] *mt*
PREPO Pre Positioning *wp*

PREPROD Pre Production *wp*
PRES Program Reporting and Evaluation System *wp*
PRES Promotion Recommendation Forms *wp*
PRESSURS Pre-Strike Surveillance and Reconnaissance System *-16*
PRF Personnel Resources File *mt*
PRF Pulse Recurrence Frequency *mt*
PRF Pulse Repetition Frequency *mt, fu, kg, wp, ac2.7-5, hc, tb, st*
PRG Problem Review Group *kf*
PRG Program Review Group *mt, wp*
PRHA Parts Radiation Hardness Assurance *jdr*
PRI Primary Rate Interface *ct, em, kg, hi*
PRI Pulse (Ping) Repetition Interval *wp, fu*
PRI Pulse Rate Indicator *fu*
PRI Pulse Recurrence Interval *mt*
PRICE Program Review of Information for Costing and Evaluation *wp, go3.5.1*
PRICE Programmed Review of Information for Costing and Estimation *mt*
PRICE-S [a software costing model] [RCA] *mt*
PRIDE Performance Readiness Initiatives Decision Evaluation System [AFLC] *mt*
PRIDE Production Responsibility in Design Engineering *fu*
PRIDE Program To Revitalize Industrial Defense Efficiency *wp*
PRIME Precision Recovery Including Maneuvering Entry [Martin X-23 lifting-body entry vehicle, USA] *mt*
PRIMES Preflight Integration of Munitions and Electronic Systems *wp*
PRIMIR Product Improvement Management Information Report *mt, st*
PRIMOS Prime Operating System *ct, em*
PRIMS Provisioning Information Management System *fu*
PRIO Sequencing Program *0*
PRIP PSP Radar Improvement Program *hc*
PRIRODA [METEOR series satellite] *ge*
PRIS Personnel Resource Impact System *mt*
PRISM Photo Refractive Information Storage Material *pk, kg*
PRISM Prioritized Requirements, Impacts and Schedule Milestones *mt, wp*
PRISS Post Deployment Real Item Interactive Simulator/Driver System *fu*
PRL Playback Recording List *fu*
PRL Program Requirements List *jdr*
PRM Policy and Requirements Manual *kf*
PRM Presidential Review Memorandum *mt, ci.871*
PRM Programmers Reference Manual *fu*
PRM Project Review Meeting *fu*
PRM Pulse Rate Modulation *fu*
PRM Pulse Repetition Per Minute *mt*
PRMC Pacific Regional Monitoring Center [DDN] *mt*
PRMD Private Management Domain [X.400] *kg, hi*
PRMIS Printing Resources Management Information System *mt*
PRML Partial Response Maximum Likelihood *ct, kg, em, hi*
PRN Process Re-engineering Network *jdr*
PRN Program Revision Number *mt*
PRN Pseudo Random Noise *mt, fu, uh, ab.13, 4, jdr, kf*
PRN Pseudo Random Test {?} *wp*
PRN Pulse Ranging Navigation *fu*
PRNET Packet Radio Network *mt, fu*
PRO Projection Readout *fu*
PRO Protivo Raketnaya Oborona [ABM] [USSR] *mt*
PRO Public Records Office [UK] *mt*
PROBE Program Optimization and Budget Evaluation *wp*
PROBE [a discriminating sensor flown out with or ahead of weapons] [later GSTS] [Pop-up IR Sensor] *17, st*
PROCAS Process Oriented Contract Administrative Services *kf*
PROCURE Command Procurement System [USAF] *mt*
PRODAT [mobile land vehicle communications program] *sf31.1.23, es55.101*
PRODSEC Product Security *dm*
PROFIS Professional Officer Filler Information System *mt*
PROFIT Planner's Remote Offline Form Image Tape *mt*
PROFIT Propulsion / Flight Control Integration Technology [USA] *mt, wp*
PROFS Professional Office System [IBM Electronic Unit Product] *fu, mt, ct, kg, tb, em*
PROFS Prototype Regional Observing and Forecasting Service *es55.51*
PROLOG Programming In Logic [an artificial intelligence language] *fu, wp, kg, hi*
PROM Programmable Read Only Memory [ADP] *fu, ah, ct, em, he, uh, kg, gg63, 6, 8, 16, pl, hc, lt, wp, ep, tb, jdr, pbl, tm, kf, vv, hi, oe, mt, ai1*
PROMACX Production Management and Control System *mt*
PROMIS Polar Regions and Outer Magnetosphere International Study *ns, hy*
PROMIS Procurement Management Information System [USAF] *mt*
PROMMIS Programmed Restructuring of Maintenance Management Information System *mt*
PROMPT Proposal Management Pricing Tool *jdr*
PROMT Programmable Miniature Message Terminal *hc*
PROPIN Proprietary Information *mt, wp*

PROS ROSAT (and X-ray) analysis system developed within IRAF *cu*
PROVORG Providing Organization *mt*
PROWLER Programmable Robot With Logical Enemy Response *wp*
PRP (HP Graphics Language or Auto CAD PRP format) *is26.1.74*
PRP Personnel Reliability Program *mt, hc, tb, su*
PRP Pipelined Resampling Processor *wp*
PRP Pulse Repetition Period *fu, wp*
PRPL Priced Repair Parts List *fu*
PRR Parts Replacement Requisition *fu*
PRR Prelaunch Readiness Review *gp, oe*
PRR Preliminary Requirements Review *mt, re, tb, wl*
PRR Production Readiness Review *mt, fu, hc, wp, tb, jdr*
PRR Program Readiness Review *fu*
PRR Pulse Repetition Rate *fu, wp, hi*
PRRS PACAF Readiness Reporting System [PACAF] *mt*
PRS Performance Reporting System *fu*
PRS Personnel Readiness System *mt*
PRS Portable Radar Simulator *wp*
PRS Production Reporting System *mt*
PRS Project Reporting System *fu*
PRS Provisioning Requirements Summary *fu*
PRSS Polystatic Radar Satellite System *ic2-25*
PRST Probability Ratio Sequential Test *fu*
PRT Personnel Rapid Transit *fu*
PRT Phase Reversal Transducer *fu*
PRT Physical Readiness Training *wp*
PRT Platinum Resistance Temperature *oe*
PRT Production Reliability Test *fu*
PRT Project Readiness Test *-4*
PRT Pulse Recurrence Time *mt*
PRT Pulse Repetition Time *fu, jdr*
PRT Pulsed Radar Transmitter *wp*
PRTSC Print Screen *pk, kg*
PRU Personnel Readiness Unit *mt*
PRU Photographic Reconnaissance Unit [UK] *wp, mt*
PRU Primary Replaceable Unit *fu*
PRV Peak Reverse (inverse) Voltage *fu*
PRWS Plasma and Radio Wave Spectrometer *es55.26*
PS Packaging Specification *fu*
PS Packet Switch(ing) *hi*
PS Participating Scientist *so*
PS Passazhirskii i Syaz [passenger and liaison / communications, as in Mi-8PS] [USSR] *mt*
PS Payload Specialist *sf31.1.29*
PS Payload Station *jdr*
PS Personal System [a line of IBM products] *ct, em, hi*
PS Physical Sequence (Sequential) *em, ct*
PS Pilot Simulator *fu*
PS Point of Switch *fu*
PS Potentiometer Synchro *fu*
PS Power Supply *fu, pl, w9, jdr, hi*
PS Primary Sensor *fu*
PS Probability of Survival *mt*
PS Process Specification *fu*
PS Procurement Specification *fu*
PS Product Specification *fu*
PS Program Sequencer *fu*
PS Program Specification *mt, fu*
PS Program Support *kf*
PS Programmed Symbols *mt*
PS Proportional Spacing *kg, hi*
PS Propulsion Subsystem *vv*
PS Pulse Sensor *fu*
PS [chemical warfare codename for trichloronitromethane (an irritant)] *wp*
PS&C Program Systems and Controls *fu*
PS&L Power Switching and Logic *0*
PS/2 Programming System 2 [IBM] *pk, kg*
PS/PC Physical Security/Pilferage Code *fu*
PSA Pacific Southwest Airlines *hn43.23.2, aw129.21.92*
PSA Passive Situation Awareness *hc*
PSA Perimeter Switch Assembly *fu*
PSA Pitch Sun Acquisition *jdr, uh*
PSA Post-Shakedown Availability *fu*
PSA Preliminary (Provisional) Site Acceptance *fu*
PSA Preliminary Satellite Assembly *jdr*
PSA Problem Statement Analyzer (Analysis) *mt, fu, tb*
PSA Project Support Agreement *mt*
PSA Provisional Site Acceptance *mt*
PSAA Pacific Special Activities Area *mt*
PSAA Payload Support Access Assembly *jdr*
PSAC President's Science Advisory Committee *ci.xxi, 13*
PSACEI Propulsion Subsystem for Attitude Control and Engine Ignition *ai2*
PSAM Probabilistic Structural Analysis Methodology *aa26.11.36*
PSAP Public Safety Answering Points *y2k*

PSAT Pluribus Satellite [interface message processor] *csII.96*
PSAT Predicted Site Acquisition Table *-2*
PSB Program Status Book *mt*
PSC Personal Super Computer *pk, kg*
PSC Platform Support Center [EOS] *hy*
PSC Polar Stratospheric Cloud *hy*
PSC Principal Subordinate Command [NATO] *fu, mt*
PSC Print Server Command *pk, kg*
PSC Product Service Center *pk, kg*
PSC Programming Support Center *fu*
PSC Protoflight Spacecraft Cryocooler *kf*
PSCDR Preliminary Software Critical Design Review *fu*
PSCF [Industrial] Personnel Security Clearance Files *hc, tb*
PSCN Program Support Communications Network *mt, ns, hy*
PSCN Proposed Specification Change Notice *fu*
PSCN/R Production Schedule Completion Notice/Report *fu*
PSCRC Portable Small Computer Requirements Contract *mt*
PSCU Power Supply Control Unit *fu*
PSD Performance and Security Document *mt*
PSD Phase Sensitive Demodulator *fu*
PSD Port Sharing Device *do68*
PSD Power Spectral Density *fu, wp, gg19, 16, gp, he, oe*
PSD Power Structure Density {probably "Spectral" ?} *st*
PSD Power Supply Data *fu*
PSD Prepare Simulation Data *jb*
PSD Program System Description *mt*
PSD Pulse Shape Discriminator *fu, wp*
PSDE [a telecommunications program of ESA] *es55.43*
PSDL Process Data Sensitivity Label *tb*
PSDN Packet Switched Data Network *ct, em, fu, kg, hi*
PSDS Packet Switched Data Service *kg, hi*
PSE Packet Switching Exchange *hi*
PSE Payload Service Equipment *bo*
PSE Peculiar Support Equipment *mt, 16, bo, wp, dm, tb, kf, fu*
PSE Physical Security Equipment *st*
PSE Power Switching Electronics *-4*
PSE Program Support Environment *fu*
PSE Program System Engineering *jdr, kf*
PSERVER Print Server [NetWare] *kg, hi*
PSF Permanent Swap File *kg, hi*
PSF Point Spread Function *pbl, mb, wp, mi*
PSF Pounds per Square Foot *jdr*
PSF Programming Support Facility *fu*
PSFK Permanent Software Function Key *fu*
PSFL Propulsion System Firing Laboratory *jn*
PSG Platoon Sergeant *w9*
PSG Power Sources Group *vv*
PSG Program Steering Group *oe*
PSG Project Science Group *so*
PSGP Power Supply and Grid Pulser *fu*
PSI (PSI and PSI-2 sun pulses) *gg67*
PSI Per Square Inch *tb*
PSI Pounds per Square Inch [a unit of pressure] *mt, 13, w9, ws4.10, nt, wp, tb, jdr, kf, fu, uh, hi, oe*
PSI Pressure System Incorporated [a company] *jn*
PSIA Pounds per Square Inch Absolute *fu, vv, uh, oe*
PSIC Planning and Systems Integration Center [DCA] *mt*
PSID PostScript Image Data *pk, kg*
PSIG Pounds [force] per Square Inch Gauge *fu, jdr, uh, oe*
PSIL Preferred Speech Interference Level *fu*
PSIU Programmable Signal Interface Unit *fu*
PSK Phase Shift Key(ing) (Keyed) *mt, gp, 0, 2, 4, 16, af72.5.171, lt, ep, ce, jdr, kf, vv, uh, fu, hi, oe*
PSKR Pogranichniy Storozhevoy Korabl [border patrol ship] [USSR] *mt*
PSL Power Systems Laboratories [at Hughes] *ep*
PSL Primary Standards Laboratory *fu, jdr*
PSL Problem Statement Language *mt, fu, wp, tb*
PSL Processor Stack Local *oe*
PSL Program Supply Logic unit *mt*
PSL Program Support Letter *mt*
PSL Program Support Library *mt, fu, tb*
PSL/PSA Problem Statement Language / Problem Statement Analysis (Analyzer) *fu, mt*
PSLU Power Supply Logic Unit *mt*
PSLV Polar Satellite Launch Vehicle [India] *ac3.2-3*
PSM Personnel Systems Manager *mt*
PSM Power Supply Module *oe*
PSM Printing Systems Manager *pk, kg*
PSM Project Support Manager *oe*
PSM Purchaser Supplied Materiel *fu*
PSMTA Power Supply Modulator Test Adapter *hc*
PSN Packet Switching (Switched) Network *mt, fu, ns, kg, hi*
PSN Packet Switching Node *mt*
PSN Pseudorandom Noise *jdr*
PSN Public Switch(ed) Network *mt, fu, y2k*
PSNAP PLRS Steerable Null Antenna Processor *fu, hc*

PSNG Programmable Signal and Noise Generator *fu*
PSNR Peak Signal-to-Noise Ratio *fu*
PSO Pilot Simulator Operator *fu*
PSO Port Services Office *fu*
PSO Program Security Office *pbl*
PSO Program Support Operations *kf*
PSOC Preliminary System Operational Concept *mt, hc*
PSOS Probably Secure Operating System *hc*
PSP Packet Switching Processor *hi*
PSP Payload Signal Processor [Orbiter] *-4*
PSP Payload Support Platform *kf*
PSP Personal Software Products [group] [IBM] *pk, kg*
PSP Phased Support Plan *mt, hc*
PSP Pierced Steel Planking [surface for improvised runways] [USA] *mt*
PSP Pilot Simulator Position *fu*
PSP Priced Spare Parts *wp*
PSP Primary Supply Point *mt*
PSP Priority Strike Plan [NATO] *mt*
PSP Program (Project) Support Plan *0, 2, 4, 17, oe*
PSP Program Segment Prefix *pk, kg*
PSP Program Support Plans *mt*
PSP Programmable Signal Processor *fu, hc, hn42.4.3, wp*
PSPA Plant Site Property Administrator *17, sb*
PSPC [ROSAT] Position Sensitive Proportional Counter *cu, ns*
PSPDN Packet Switched Public Data Network *kg, nu, hi*
PSPL Priced Spare Parts List *fu*
PSPP Proposed System Package Plan *fu*
PSQ Personnel Security Questionnaire *hc, tb, jdr, su, fu*
PSR Performance Summary Report *wp*
PSR Personal Service Radio *mt*
PSR Poly Static Radar *ic2-17*
PSR Post Strike Reconnaissance *mt*
PSR Pre Shipment Review *bl3-1.142, gp, re, jn, oe*
PSR Primary Surveillance Radar *fu*
PSR Program Status Report (Review) *mt, 16, 17, sb, dm, st, jdr, fu, pbl*
PSRAM Pseudo Static Random Access Memory *em*
PSRB Pre Ship Review Board *kf*
PSRR Power Supply Rejection Ratio *fu*
PSRS Program Status Reporting System *fu*
PSRT Passive Satellite Research Terminal *fu, hc*
PSRX Position Situation Report [software program] *jb*
PSS Packet Switching Services *hi*
PSS Payload Support Structure *kf*
PSS Personnel Security System *44.8.8*
PSS Personnel Support Services *fu*
PSS Portable Simulation System *4, oe*
PSS Portable Space Simulator *oe*
PSS Precision Sun Sensor *gp, 16*
PSS Premature Separation System *-17*
PSS Probe Support Subsystem *so*
PSS Product Structure Subsystem *fu*
PSS Program Status Summary *fu*
PSS Project Summary Sheet *mt*
PSS Propulsion Support System *wp*
PSS Protective Security Service *fu, hc, su*
PSSA Pyroelectric Solid State Array *hc*
PSSC Preliminary System Security Concept *kf*
PSSC Public Service Satellite Consortium *-19*
PSSG Protocol Standards Steering Group *mt*
PSSK Probability of Single Shot Kill *mt, fu*
PSSR Program Status Summary Report *fu*
PST Pacific Standard Time *sc2.11.1, 0, w9, wp, kf, kg*
PST Pad Service Tower *ai2*
PST Pair Selected Ternary *fu, sp662*
PST Pre Shipment Test *oe*
PST Preliminary System Test *fu*
PSTE Personnel Subsystem Test and Evaluation *fu*
PSTE Production Special Test Equipment *jdr*
PSTN Public Switched Telephone Network *mt, kg, nu, fu, ct, tb, hi, tt*
PSTP Protocol Standards Technical Panel *mt*
PSTS Portable Subsystem Test Set *fu*
PSU Packet Switching Unit *hi*
PSU Power Supply Unit *mt, em, kg*
PSU Pyrotechnic Switching Unit *jp, lt*
PSV Preliminary Site Visit *mt*
PSVP Pilot Secure Voice Project [NATO] *mt*
PSVT Pilot Secure Voice Terminal [NATO] *mt*
PSW Payload System Weight *oe*
PSW Program Status Word *fu, kg, oe*
PSW Program Support Workstation *tb*
PSWBS Project Summary Work Breakdown Structure *fu, nt*
PSWC Payload Systems Weight Capability *ai2*
PSYOP Psychological Operations [military] *fu, wp, mt*
PSYWAR Psychological Warfare [USA DOD] *mt*
PT Page Table *pk, kg*
PT Personnel and Training *mt*

PT Plain Text *mt, fu*
PT PLRS Terminal *fu*
PT Point of Tangency *fu*
PT Pointer / Tracker *fu*
PT Position Tracking *fu*
PT Potential Transformer *fu*
PT Precision Tracker *fu*
PT Pressure Transducer *jn*
PT Product (Part) Type *fu*
PT Proficiency Training *kf*
PT Project Technology *wp*
PT Pulse Time *fu*
PT Pulse Transformer *fu*
PT Test Plan *mt*
PT Torpedo Boat [US Navy] *mt*
PTA Point of Total Assumption [a budget term] *fu, jdr*
PTA Programmed Time of Arrival *wp*
PTA Propfan Test Assessment *aa26.10.17, aw130.6.31*
PTA Proposed Technical Approach *mt*
PTAB (July 26 *Board/PTAB Annual Program Review* >
PTAR Personnel Task Analysis Report *fu*
PTB Payload Testbed *kf*
PTBT Partial Test Ban Treaty *-14*
PTC Peace Through Confrontation *mt*
PTC Pentagon Telecommunications Center [USA] *mt*
PTC Phase Transfer Catalysis *wp*
PTC Positive Target Control *fu*
PTC Positive Temperature Coefficient *fu*
PTC Primary Technical Control *mt*
PTC Programming Training Center *fu*
PTC Prototype Console *fu*
PTCR Pad Terminal Connection Room *-9*
PTCS Pentagon Telecommunication Center System [USA] *mt*
PTD Parallel Transfer Disk (Drive) *fu, kg*
PTD Post-Tuning Drift *fu*
PTD Power Transient Detector *fu*
PTD Program Task Description *fu*
PTD Provisioning Technical Documentation *mt, fu, hc*
PTDP Preliminary Technical Development Plan *fu*
PTF Fast Patrol Craft [USN ship classification] *fu*
PTF Patch and Test Facility *mt*
PTF Patch and Test Frame *fu*
PTF Payload Test Facility, Vandenberg AFB, California *-16*
PTF Phase Transfer Function *wp*
PTF Problem Temporary/Trouble Fix *pk, kg*
PTF Programmable Transversal Filter *hc*
PTF Prototype and Test Facility *oe*
PTFE Poly Tetra Fluoro Ethylene *he, wp*
PTG Torpedo Boat, Guided missile armament [US Navy] *mt*
PTH Plated-Through Hole *fu*
PTH Power Transient Hybrid *fu*
PTI Personnel Transaction Identifier *mt*
PTI Product Test Interface *fu*
PTI Production Test Instructions *fu*
PTI Programmed Test Identifier *fu*
PTL Primary Target Line *fu, pm*
PTM Patch, Test, and Monitor *fu*
PTM Performance Test Model *wp*
PTM Phase Time Modulation *hi*
PTM Pilot Training Missile *mt, hc*
PTM Proof Test Model *bl1-9.9, 0*
PTM Propellant-pressurant Tank Module *-4*
PTM Pulse Time Modulation *fu*
PTM Torpedo Boat, Missile-armed [US Navy] *mt*
PTML PNPN Transistor Magnetic Logic *fu*
PTMOP Preliminary Technical Manual Organization Plan *fu*
PTMUX Pulse-Time Multiplex *fu*
PTN Public Telephone Network *mt*
PTO Patent and Trademark Office [DOC] *wp, mt*
PTO Power Takeoff *fu*
PTO Preliminary Technical Order *fu, hc, tb*
PTO Primary Test Organization *wp*
PTO Public Telephone Operator *hi*
PTOC Platoon Tactical Operations Center *pm*
PTOC Position, Time and Organization Channels *fu*
PTOL Peace Time Operating Level *wp*
PTOP Point To Point *fu*
PTP Patch and Test Panel *mt*
PTP Peer To Peer *em*
PTP Point To Point *mt, uh, jdr*
PTP Probability To Penetrate *mt*
PTP Programmable Telemetry Processor [programmable data formatter] *oe*
PTP Protoqualification Test Procedure *jdr*
PTPB Partial Time/Partial Band Jamming *fu*
PTR Paper Tape Reader *fu*

PTR Position Track Radar *fu*
PTR Precision Tracking Radar *fu*
PTR Program Trouble Report *mt, fu, kf*
PTR Pulse Transmission Rate *fu*
PTRD Payload Test Requirements Document *jn*
PTRRS Pipeline Time Recording and Reporting Systems *mt*
PTS PLRS Test Set *fu*
PTS Pneumatic Test Set *oe*
PTS Portable Test Set (also PTU, Portable Test Unit) *fu*
PTS Pressure and Temperature Sensor *fu*
PTS Probe Track System *fu*
PTS Program Test System *fu*
PTSA (attending UCLA on a PTSA scholarship) *hn43.22.3*
PTSF Payload Trunnion Support Fitting *jdr*
PTSP Prototype Console Demonstration Support *fu*
PTSR Preliminary Technical Survey Report *mt*
PTT Parallel Tasking Technology *hi*
PTT Part Task Trainer [USA] *hc, mt*
PTT Platform Transmitter Terminal *-16*
PTT Postal, Telephone, and Telegraph [National Administrations] *mt, fu, csII.66, es55.67, ce*
PTTC Petroleum Technology Transfer Council *y2k*
PTU PLRS Test Unit *fu*
PTU Portable Test Unit *fu, oe*
PTU Power Transfer Unit *bo*
PTV Propulsion Technology Validation *hc*
PTV Propulsion Test Vehicle *fu, st*
PTZ (PTZ camera) [a component of an OMV] *aa26.11.28*
PTZ Predesignated Target Zones *fu*
PU Participating Unit [TACS/TADS] *fu, mt*
PU Peripheral Unit *fu*
PU Physical Unit *ct, em*
PU Plutonium *wp*
PU Power Unit *fu*
PU Processor Unit *fu*
PU Production Unit *fu*
PUAR Pulse Acquisition Radar *fu*
PUBDIS Publishing Distribution System *mt*
PUC Presidential Unit Citation [USA] *mt*
PUC Public Utilities Commission *mt, wp, y2k*
PUD Physical Unit Directory *mt*
PUF Planning Units File *mt*
PUFFS Passive Underwater Fire Control Feasibility Study *fu*
PUJ Pop-Up Jammer *fu*
PUJT Programmable Unijunction Transistor *fu*
PULSAR Pulsating Star *wp*
PUMA PACAF Unit Munitions Allocation System [PACAF] *mt*
PUN Physical Unit Number *ct, kg, em*
PUP PARC Universal Packet [protocol] *kg, ct, hi*
PUP Performance Upgrade (Update) Program *af72.5.134, wp*
PUP Pop Up Point *wp*
PUP Power Upgrade Program *wp*
PUP Product Update Program *mt*
PURE POMCUS Unit Residual Equipment [USA] *mt*
PUS Processor Upgrade Socket *pk, kg*
PUSHA Push All Registers *pk, kg*
PUSHF Push Flags *pk, kg*
PUT Programmable Unijunction Transistor *wp*
PV Pioneer Venus [a space mission] *sc3.12.1, ep*
PV Plan View *jdr*
PV Present Value *jdr, kf*
PV Privileged Information *mt*
PV Pulemot Vozdushny [machine gun, air, PV-1 7.62mm 780rds/m] [USSR] *mt*
PVA Perigee Velocity Augmentation *jn, oe*
PVAC Position Velocity, Altitude Correction *fu*
PVAL Poly Vinyl Alcohol *wp*
PVAS Primary Voice Alerting System [NORAD] *mt*
PVB Performance Verification Board *oe*
PVC Permanent Virtual Circuit *em, ct, fu, kg, hi*
PVC Permanent Virtual Connection *em, ct*
PVC Poly Vinyl Chloride *w9, wp, kg*
PVD Plan-View Display *fu*
PVDF Polyvinylidene Fluoride *wp*
PVI Pilot Vehicle Interface *wp*
PVM Parallel Virtual Machine *kg, em, hi*
PVM Pass-through Virtual Machine [protocol] [IBM] *pk, kg*
PVO Pioneer Venus Orbiter *mi*
PVOR Precision Very High Frequency Omnidirectional Range *wp, fu*
PVOS Protivo Vozdushnoy Oborony Strany [air defense of the homeland] [USSR] *mt*
PVP Parallel Vector Processing *pk, kg*
PVP Performance verification Plan *he*
PVP Poly Vinyl Pyrrolidone *wp*
PVR Precision Voltage Reference *fu*
PVR Product Verification Review *mt, wp*

PVS Parallel Visualization Server *pk, kg*
PVSD Programmable Variable Storage Device *fu*
PVT Performance Verification Test *fu*
PVT Pressure, Volume, Temperature *w9, wp, fu*
PVU Performance Verification Unit *fu*
PVU Production Verification Unit *hc*
PVWA Planned Value of Work Accomplished *fu*
PW Passing Window *fu*
PW Pilot Wire *fu*
PW Prisoner of War *w9*
PW Probability of Warning *kf*
PW Prototype Workstation *fu*
PW Pulse Width *mt, fu, wp*
PWA (Calgary based PWA Corp. owns Canadian Airlines International) *aw130.13.90*
PWA Printed Wiring Assembly *mt, rm7.1.3, hc, fu*
PWA Project Work Authorization *fu, 4*
PWASS Precision Wide Area Surveillance System *wp*
PWB Printed Wire (Wiring) Board *mt, ah, uh, kg, gp, 8, hc, ie5.11.6, lt, wp, ep, jdr, pbl, tm, kf, jn, oe*
PWB Printed Wiring *fu*
PWB Procedure Work Block *oe*
PWBA {Printed Wiring Board Assembly} *hd1.89.y27*
PWBD Printed Wiring Board Documentation *fu*
PWBS Program (Project) Work Breakdown Structure *mt, kf, fu*
PWC Public Work Center *mt*
PWC Pulse Width Coded *fu*
PWC/MIS Public Work Centers / MIS [US Navy] *mt*
PWD Print Working Directory [Unix] *kg, hi*
PWD Pulse Width Discriminator *fu*
PWDS Protected Wire Distribution System *mt*
PWDT Processor Watchdog Timer *jdr, uh*
PWDTS Peacekeeper Warranty Data Tracking System *mt*
PWE Pulse Width Encoder *fu*
PWF Personnel Working File *mt*
PWGS Pacific Weather Graphics System *mt*
PWHQ Peace War Headquarters [Europe] *kf*
PWHQ Primary War Headquarters *mt*
PWI Process Work Instruction *fu*
PWI Proximity Warning Indicator *fu*
PWIC Prisoner of War Information Center *mt*
PWIN Prototype WWMCCS Intercomputer Network *mt*
PWIS Prisoners of War Information System *mt*
PWM Pulse Width Modulated (Modulator) (Modulation) *fu, csIII.7, 0, 17, pl, jdr, hi, oe*
PWO Property Work Order *jdr*
PWP Program Work Plan *fu*
PWPF Pulse Width Pulse Frequency *oe*
PWR Passive Warning Receiver [UK] *mt*
PWRMR Prepositioned War Reserve Materiel Requirement *mt*
PWRMS Prepositioned War Reserve Materiel Stocks *wp, mt*
PWRR Prepositioned War Reserve Requirement *wp*
PWRS Prepositioned War Reserve Stock *mt, wp*
PWS Peer Web Services *ct, em*
PWS Plasma Wave Subsystem *jp*
PWS Program Word Structure *fu*
PWS Programmable Work Station *mt*
PWS Proximity Warning System *wp*
PWS Public Water System *y2k*
PWSCS Programmable Workstation Communication Services [IBM] *pk, kg*
PWW Planar Wing Weapon [a canceled alternative for CWW, USA] *mt, wp*
PX Post Exchange *mt, w9*
PY Prior Year *mt*
PZT Lead Zirconium Titanate *is26.1.48*
PZT Piezoelectric Transducer *te23, 11, 40*
PZT Piezoelectric Zirconate Titanate *fu*

Q

Q Quadrature [channel] [phase] *fu, vv*
Q Quadrillion BTU [also known as quads] *wp*
Q Qualification *jdr*
Q [chemical warfare codename for 1,2-di(chloroethylthio)ethane (sesquimustard)] *wp*
Q&A Query and Answer *fu*
Q&A Question and Answer *pbl, sc2.8.6, fu, ah, hi*
Q&E Querying and Editing *hi*
Q&RA Quality and Reliability Assurance *oe*
Q/R Query/Response *mt*
QA Quality Assurance [an organization] *mt, fu, ah, kg, gp, 8, 16, 17, sb, hq5.3.5, pl, hc, nt, wp, ep, tb, st, jdr, pbl, tm, kf, jn, hi, oe*
QA Quick Acting *fu*
QA/RSR Quality Status/Reliability Summary Report *fu*

QACP Quality Assurance Control Point *fu*
QAD Quality Assurance Directive *fu*
QADI Quality Assurance Departmental Instructions *oe*
QADR Quality Assurance Deficiency Report *fu*
QADS Quality Assurance Data Support *fu*
QAE Quality Assurance Engineer *fu*
QAE Quality Assurance Evaluator *mt, wp*
QAEST Quality Assurance Estimating [program] *gb*
QAGO Quality Assurance Group Office *fu*
QAI Quality Assurance Inspection *fu*
QAI Quality Assurance Institute *mt*
QAM Quadrature Amplitude Modulation *jdr, ct, fu, kg*
QAM Quality Assurance Manager *mt*
QAO Quality Assurance Office [Naval Weapons] *dt, wp*
QAP Quality Assurance Plan *mt, hc, wp, tb, jn*
QAP Quality Assurance Procedure [a directive] *pbl, jdr, tm, ah, fu*
QAP Quality Assurance Program *mt*
QAP Quality Assurance Provision *tb*
QAPO Quality Assurance Project Office *jdr*
QAPP Quality Assurance Program Plan *fu, jdr, oe*
QAQMS Quality Assurance Quality Method Sheet *fu*
QAR Quality Analysis Request *fu*
QAR Quality Assurance and Reliability *pl*
QAR Quality Assurance Report *wp*
QAR Quality Assurance Representative *mt, fu*
QAR Quality Assurance Requirements *pl*
QAR Quality Audit Report *fu*
QARC Quality Assurance Representative's Correspondence *fu*
QART Quality Assurance Routing Tag *fu*
QASIMRR Quality Assurance Source Inspection Material Report *oe*
QASK Quadrature Amplitude Shift Keying *spx*
QASP Quality Assurance Support Plan *mt*
QASR Quality Assurance Supplier Report (Record) *8, 17, sb, jdr*
QASR Quality Assurance Supply Record *gp*
QAT Quality and Testing *pl*
QAT Quick roll estimate and Attitude Trim *-3*
QATIP Quality Assurance Test and Inspection Procedure *fu*
QATS Quality Assurance Test System *mt*
QAVC Quiet Automatic Volume Control *fu*
QB Quick Break *fu*
QBE Query By Example *ct, em, kg, wp, hi*
QBF Query By Form *kg, hi*
QBIC Query By Image Content [IBM] *pk, kg*
QBL Quasi Band Limited *fu*
QBSV Quiescent Beam Steering Vector *jdr*
QC Quality Circle *fu, sc4.7.7, 8, ig21.12.17*
QC Quality Control *mt, kg, fu, gp, 8, 16, 17, sb, hq5.3.5, hc, w9, wp, tb, st, jdr, oe*
QCAL Qualified Contractor Access List *mt*
QCATS Quality Corrective Action Tracking System *fu*
QCHR Quality Control History Record (Report) *fu, gp, 17, sb, pl, lt, ep, jdr, kf*
QCI Quality Conformance Inspection *fu, oe*
QCI Quality Control Inspection *jdr*
QCI/T Quality Conformance Inspection and Test *fu*
QCIC Quality Control/Assurance and Inspection *hc*
QCIF Quarter Common Interchange Format *hi*
QCL Quality Characteristics List *fu*
QCM Quality Control Monitor *oe*
QCM Quartz Crystal Microbalance *gp, oe*
QCOI Quality Control Operating Instructions *pl*
QCR Quality Control Report *mt*
QCRI Qualification Cross-Reference Index *fu*
QCRR Quality Control Requirement Report *0*
QCS Quality Control Screening *fu*
QCS Quality Control System *fu*
QCS Quality Cost Summary *fu*
QCS Query Control Station [Army] *fu, hc*
QCSP Quality Control System Plan *fu*
QD Quadrature Detector *fu*
QD Qwerty/Dvorak *pk, kg*
QDC Quiet Day Curve *pr*
QDM Quantitative Debris Monitoring *aw129.21.92*
QDN Query Direct Number *ct, em*
QDOS Quick and Dirty Operating System *pk, kg*
QDR Qualification Design Review *fu*
QDR Quality Deficiency Report *fu, wp, mt*
QDR Quality Discrepancy Record *oe*
QDRI Qualitative Development Requirements Information *fu*
QDRS Quality Data Reporting System *fu*
QDSB Quadrature Double[d] Sideband Modulation *2, 4*
QE Quadrant Elevation *0, fu*
QE Quality Elevation *fu*
QE Quality Engineer(ing) *wp, fu, oe*
QEC Quick Engine Change *wp*
QED Quality Engineering Directive *oe*

QED Quod Erat Demonstrandum [(Latin) which was to be demonstrated] *w9, wp, ai1*
QEMM Quarterdeck Expanded Memory Manager [Quarterdeck Corp.] *ct, kg, hi*
QEPL Quality Engineering Planning List *fu*
QETW Quality Engineering Test Worksheet *fu*
QF Query Filter *hi*
QF Quick Firing *w9, fu*
QF Quick Fix *fu*
QFA Quick File Access *pk, kg*
QFC Quality Flash Copy *fu*
QFC Quaternary Function Channel *fu*
QFD Quality Function Deployment [a process] *fu, ah, jn, rm7.1.1, hn49.12.8, ad, ig21.12.17, jdr, pbl, tm, kf*
QFE [atmospheric pressure at aerodrome elevation] *mt*
QFI Qualified Flying Instructor [RAF, UK] *mt*
QFI Quality Form Instructions *fu*
QFM Quantized Frequency Modulation *mt, fu*
QFP Quad Flat Pack *kg, hi*
QGV Quantized Gate Video *fu*
QHR Quality History Record *fu*
QI Quality Indicator *fu*
QI Quality Instruction *oe*
QIC Quality Information using Cycle time [Hewlett-Packard] *pk, kg*
QIC Quarter Inch Cartridge *ct, em, kg*
QIC Quarter Inch Committee *ct*
QICS Quality Inspection Comment Sheet *fu*
QICSC Quarter Inch Cartridge Standards Committee *hi*
QICW Quarter Inch Cartridge Wide *hi*
QID Queue ID *fu*
QIFM Quadrature Intermediate Frequency Mixer (Modulator) *fu*
QIO Queue Input / Output *ct, em*
QIP Quality Improvement Program *nd74.445.26*
QIP Quality Indicator Program *kf*
QIS Quality Information System *fu, hc*
QL Quick Look *mt, fu, 0*
QLCI Quick Look Capabilities Index *mt*
QLD Quick Look Diagnostic *jdr*
QLIT Quick Look Intermediate Tape *fu*
QLP Query Language Processor *mt*
QM Quadrature Modulation *fu*
QM Qualification Model *oe*
QM Quality Manual *oe*
QM Quartermaster *mt, w9*
QMC Quality Management Committee *hn49.12.8*
QMC Quartermaster Corps *w9*
QMCS Quality Monitoring and Control System *wp*
QMDO Qualified Military Development Objective *hc*
QMDR Quality Assurance Deficiency Report *fu*
QME Quadrature Modulator Exciter *kf*
QMG Quartermaster General *w9*
QML Qualified Manufacturers List *is26.1.70*
QMQB Quick Make, Quick Break *fu*
QMR Qualitative Materiel Requirement *fu*
QMR Quarterly Management Report *ah, pbl, jn*
QMR Quarterly Management Review *jdr*
QMRD Quad Medium Rate Demodulator *jdr*
QMS Quality Management System *fu*
QMS Quality Method Sheet *fu*
QNE [altimeter reading at airfield when subscale is set to 1013.2 millibars] *mt*
QNH [altimeter subscale setting to destination airfield altitude] *mt*
QNM Queuing Network Model *mt*
QO Quality Operations (QO-Mfg Quality Operations - Manufacturing) *fu*
QOD Quick Opening Device *fu*
QOQI Quality Operations Quality Instructions *fu*
QOR Qualitative Operational Requirement *mt, fu*
QOS Quality Of Service [ATM performance parameters, delays, throughput, etc] *nu, kg, fu, ct, rih*
QOSC Quality Of Service Class [ATM Forum definition of QOS parameters] *nu*
QOT&E Qualification, Operational Test and Evaluation *mt*
QP Quality Practice *fu*
QPG Quantum Phase Gate *pk, kg*
QPI Quality Performance Indicator *fu*
QPI Quality Program Instruction *oe*
QPL Qualified Parts List *jdr*
QPL Qualified Products List *fu, nt*
QPL Quality Parts List *5, sb, hc, mj31.5.6*
QPM Quality Practice Manual *ep*
QPM Quantitative Process Management *kf*
QPM Quantized Phase Modulation *fu*
QPMIS Quantitative Process Management Information System *kf*
QPP Quality Planning Profile *oe*
QPP Quality Program Plan *fu, hc*
QPP Quantized Pulse Position *fu*

QPPM Quantized Pulse-Position Modulation *fu*
QPR Qualitative Performance Requirements *mt*
QPR Quarterly Progress Report *mt*
QPR Quarterly Progress Review *mt, fu*
QPSK Quadrature (Quadrative {?}) Phase Shift Keying (Keyed) *mt, kf, vv, fu, ct, kg, hi, oe, tt*
QPSK Quadriphase (Quaternary Phase) Shift Key[ing] *sp662, 2, 4, 19, ce*
QPT Quality and Production Tool *hc*
QQPRI Qualitative and Quantitative Personnel Requirements Information *mt, st, fu*
QR Quality Record *oe*
QR Quality Representative *fu*
QR Quick Response *fu*
QR Quiet Radar *fu, hc*
QRA Quality Reliability Assurance *wp*
QRA Quick Reaction Alert *-12*
QRA Quick Reaction Alert [NATO] *mt*
QRC Quick Reaction Capability *mt, fu, wp, st*
QRCG Quasi-Random Code Generator *fu*
QRC_CAC Quick Reaction Center Combat Analysis Center [USA] *mt*
QRF Quick Reaction File *mt*
QRI Qualitative Requirements Information *fu, hc*
QRL Quick Reference List *mt*
QRO Quality and Reliability Officer *fu*
QRO Quick Reaction Order *hc*
QRP Quick Reaction Packages *mt*
QRR Qualitative Research Requirement [for nuclear weapons effect] *st*
QRS Quality Requirements Summary *fu*
QS Quarter Set *fu*
QSAM Queued Sequential Access Method *fu*
QSD Quality Systems Division *fu*
QSM Queue Statue Memory *fu*
QSO Quasi Stellar Object *mb, wp, mi*
QSPOD Quality Systems and Product Operations Division *fu*
QSR Quarterly Summary Report *0*
QSR Query and Reporting Process [Honeywell] *mt*
QSR Quick Strike Reconnaissance *mt, wp*
QSRA Quiet Short-Haul Research Aircraft *aw118.18.15*
QSTOL Quiet Short Take-Off and Landing *wp*
QSW Quality/Software *pl*
QT Qualification Test *tb, oe*
QT Quart *wp*
QT Quick Turn *mt*
QT Quiet Torpedo *fu*
QT&E Qualification Test and Evaluation *mt*
QTA Qualification Test Article *fu*
QTA Quick Turn Area *mt*
QTAM Queued Terminal Access Method *hi*
QTC Quicktime Conferencing [Macintosh] *pk, kg*
QTM Quadratic Texture Map *hi*
QTP Qualification Test Procedure *jdr*
QTP Qualification Training Package *mt*
QTR Qualification Temperature Range *oe*
QTR Quarter [calendar] *-3*
QTRLY Quarterly *tb*
QTW Quick-Time for Windows *ct*
QTY Quantity *wp, jdr*
QUASAR Quasi Stellar [star-resembling object] *wp*
QUASER Quantum Amplification by Stimulated Emission of Radiation *fu, wp*
QUEL Query Language *hd1.89.y18*
QUEST Quality and Efficiency in Simulation Techniques *ne*
QUEST Quantitative Understanding of Explosive Stimulus Transfer *fu*
QUEST Query, Update, Enter, Search, and Time-share *fu*
QUI Quito [Ecuador, STDN site] *ac, 4*
QUICC Quad Integrated Communications Controller [Motorola] *hi*
QUICK Quick Reacting General War Gaming System *mt*
QUICK_LOOK [border surveillance radar on the Ov-1c] *wp*
QUINCE [prototype Chinese-to-English machine translator] *ro*
QUP Quantity Per Unit Package *fu*
QV Quo Vide [which see] *wp*
QVI Quality Verification Inspection *mt*
QVT Quality Verification Testing *fu*
QWERTY [designation for the standard typewriter keyboard] *fu, pf.f88.11*
QWEST Quantum Well Envelope State Transition *aa26.10.30*
QXI Queue Executive Interface *hi*

R

R&A Reliability and Availability *mt*
R&A Review and Analysis *mt*
R&D Research and Development *aa26.12.20, aw130.13.56, hc, pl, ac1.2-21, 0, 12, 13, 14, 15, w9, es55.2, af72.5.35, dt, tb, st, wl, hg, kg, gem ful, ro, hi, oe, mt*

R&E Research and Engineering [office of] *fu, 10*
R&E Research and Experimentation *fu*
R&M Reliability and Maintainability [also R/M] *mt, hc, rm2.3.3, 17, sb, aa26.12.66, nd74.445.26, kf*
R&O Repair and Overhaul *fu*
R&QA Reliability and Quality Assurance [also RQA] *gp*
R&R Remove and Replace *fu*
R&RR Range and Range Rate [also RRR] *ac3.3-17, 2, 4*
R&SU Repair and Salvage Unit [sometimes mobile technical units that followed WWII RAF units] [UK] *mt*
R&T Research and Technology *ci.111, tb, fu*
R(accel) Range Acceleration *vv*
R(vel) Range Velocity *vv*
R-C Resistance-Capacitance [also RC] *0*
R-DIL Released Design Integration Laboratory *ep*
R-HOUR Release Hour *mt*
R-IL Repeat-Incorrect Length *fu*
R-NETS Requirement Networks *fu*
R-O Read Only *pk, kg*
R-PCM Remote Pulse Code Modulators *ai2*
R-PE Repeat-Parity Error *fu*
R-sw R-switch *uh*
R-TDPC Real Time Data Processing Center *fu*
R/A {Radar} Altimeter *0*
R/ASR Revisions As Required *fu, kf*
R/B Red / Black *mt*
R/E Receiver/Exciter *fu*
R/I Receiving Inspection *fu*
R/J Rotary Joint *bl1-5.99*
R/M Reliability and Maintainability *rm2.1.2, tb, fu*
R/M/A Reliability / Maintainability / Availability *fu*
R/M/S Reliability / Maintainability / Safety *fu*
R/M/T Reliability / Maintainability / Testability *fu, mt*
R/O Receive Only *mt, fu*
R/O Roll Over *gp, kf*
R/PA Receive / Processor Assembly *sd*
R/R Readout and Relay *fu*
R/R Recorder / Reproduced *fu*
R/R Remove and Replace *fu*
R/R/R Reassignment / Retargeting / Reconstitution *mt*
R/RE Recording / Reproducing Equipment *fu*
R/RR Rework / Retest Request *fu*
R/S Receive / Store *fu*
R/T Radio Telephone *fu*
R/T Real Time *tb*
R/T Receive(r) / Transmit(ter) *mt, fu, kg, hi*
R/W Re Work *jdr*
R/W Read / Write *fu, jdr, kg, hi*
R/Y Roll / Yaw *oe*
R/Y [message traffic type designation, R is Genser, Y is DSSCS] *mt*
RA Radar Airport *fu*
RA Radar Altimeter *fu, ac3.7-23, wp, hg, ge*
RA Random Access *ac3.2-19*
RA Range Acceleration *vv*
RA Read Amplifier *fu*
RA Regular Army *w9*
RA Release Authorization *fu*
RA Reliability Analysis *wp*
RA Requesting Agency *oe*
RA Requirements Analysis *mt, kf, fu*
RA Responsible Activity *fu*
RA Return Authorization *pk, kg*
RA Right Ascension *0, w9*
RA/FA Report of Accounting and Finance *mt*
RAA Regional Airline Association *aw129.21.29*
RAA Resource Apportionment Aid *mt*
RAA Responsibility, Authority, and Accountability *kf*
RAAA Reliability Allocation, Assessment, and Analysis *jdr*
RAAF Royal Australian Air Force *mt, w9, aw130.6.4, wp*
RAAFAS Report of Accounting and Finance Activities [USAFE] *mt*
RAAFB Royal Australian Air Force Base *aw130.6.41*
RAAM Remote Artillery-Delivered Antitank Mine *fu*
RAAMS Remote Anti-Armor Mine System *fu*
RAAN Right Ascension [of the] Ascending Node *4, jdr, vv, uh*
RAAP Rapid Application of Air Power [USAF] *ro, mt*
RAAS Robotic Antenna Assembly System *hc*
RAAWS Radar Altimeter and Altitude Warning System Radar [RAdio Detection and Ranging, detection, location and tracking of aircraft by reflected radio waves] *mt*
RAB RAID Advisory Board *hi*
RAB Readiness Assessment Board [USAF] *mt*
RAC Radar Area Correlation *mt*
RAC Radar Azimuth Converter *fu*
RAC Radiometric Area Correlation *hc*
RAC Rapid Action Changes *fu*
RAC Recombinant DNA Advisory Committee *wp*

RAC Reflective Array Compressor *fu*
RAC Reliability Analysis Center *wp*
RAC/SMC Radar Azimuth Converter/Ship's Motion Converter *fu*
RACC Repairable Asset Control Center *mt*
RACE Research in Advanced Communications for Europe *mt*
RACES Radio Amateur Civil Emergency Services *mt*
RACF Resource Access Control Facility *mt, fu, wp, kg*
RACPAS Radar Coverage and Penetration Analysis System *hc, fu*
RACS Rocket Attitude Control System *hc*
RAD Radar Augmentation System *hc*
RAD Radio Science [an instrument] *es55.26*
RAD Rapid Application Development *mt, ct, em, kg, hi*
RAD Recruiting Aids Division *dt*
RAD Remote Antenna Driver *ct*
RAD Required Availability Data *mt*
RAD Required Availability Date *wp*
RAD Requirements Action Document [USAF] *mt*
RAD Requirements Allocation Document *kf, jdr*
RAD Roentgen Absorbed Dose [unit of ionizing radiation] *un, jdr, uh*
RAD TP Radiation Test Point *jdr*
RAD/RASP Remote Antenna Driver / Remote Antenna Signal Processor *ct, em*
RADA Random Access Discrete Address *fu*
RADA Receive Antenna Deployment Actuator *cl, jdr, uh, he*
RADA Reflector Antenna Deployment Actuator *jn*
RADAC Radar Area Correlator [guidance system] *-11*
RADAG Radar Area Correlation Guidance *wp*
RADAL Radio Detection and Location *fu*
RADALON [weather resistant paint] *ro*
RADAM (project RADAM made a radar imaging survey of Brazil) *ac1.4-12*
RADAM Radar Detection of Agitated Metal *hc*
RADAR Radio Detection and Ranging [a radio detection device that provides information on range, azimuth and / or elevation of objects] "radar" *mt, hg, ge, fu, ct, wp, eo*
RADAR Rights, Availabilities, Distribution Analysis and Reporting [Disney] *pk, kg*
RADAREXREP Radar Exploitation Report [a formatted statement of the results of a tactical radar imagery reconnaissance mission, including the interpretation of the sensor imagery] [USA] *mt*
RADARSAT Radar Satellite [Canadian mapping satellite] *aa26.10.10, hy, ge*
RADAS Random Access Discrete Address System *fu*
RADAT Radar Data Transmission *fu*
RADAY Radio Day *mt*
RADB Routing Arbiter Data Base *pk, kg*
RADBN Radio Battalion [USMC] *mt*
RADC Rome Air Defense Center *fu*
RADC Rome Air Development Center [Rome, New York] [USAF] *mt, ro, aw130.13.51, hc, rm1.1.1, 1, 5, 12, 17, ic2-63, sb, wp, tb, kf*
RADCAL Radar Calibration *mt*
RADCM Radar Counter Measure [also RCM] *wp*
RADC_East RADC's Directorates at Hanscom AFB, Massachusetts *ro*
RADE Research And Development Division [within NSA] *-10*
RADEC Radiation Detection *kf*
RADEC Radiation Detection Capability *mt*
RADEM Random Access Delta Modulation *fu*
RADES Radar Evaluation Squadron [84th RADES—ACC, based at Hill AFB, Utah, USA] *mt*
RADES-MR Radar Evaluation Squadron Minicomputer Replacement *mt*
RADEX Radar Data Extractor *fu, wp*
RADF Radio Failure *fu*
RADFAC Radiation Interchange Factor *hc*
RADHAZ Radiation Hazard *fu, wp*
RADIAC Radiation Detection, Indication, And Computation *wp, fu*
RADIAC Radioactivity Detection, Indication, And Computation *mt*
RADID Radar Information Display *hc*
RADIL Region Airborne Digital Information Link *hc*
RADIL ROCC / AWACS Digital Information Link *mt*
RADINT Radar Intelligence [USA DOD] *tb, 10, mt*
RADINT Radiation Intelligence *wp, fu*
RADIST Radar Distance Indicator *fu*
RADIT Radar Derived Image Transmitter *hc*
RADLO Radar Liaison Officer *mt*
RADM Rear Admiral *w9*
RADMON Radiological Monitoring *fu*
RADOME Radar Dome *wp*
RADS Radar Alphanumeric Display System *fu*
RADS Radar Squadron *-12*
RADS Rapid Area Distribution Support *wp*
RADSCAT Radiometer/Scatterometer *ac7.3-1*
RADSL Rate Adaptive Digital Subscriber Line [variation of ADSL] *nu, ct, em, kg*
RADVS Radar Altimeter and Doppler Velocity Sensor *sm18-3, 0, wp, fu*
RADX Requirements Analysis and Data Extraction *fu*
RAE (the RAE table of earth satellites, 1957 - 1986) *sf31.1.24*
RAE Radio Assignment Equipment *fu*

RAE Radio Astronomy Explorer *ci.738, hc*
RAE Ranger Azimuth Elevation *0*
RAE Royal Aircraft Establishment [later Royal Aerospace Establishment, UK] *mt, crl*
RAES Royal Aeronautical Society [UK] *mt*
RAF Royal Air Force [United Kingdom] *mt, fu, 12, w9, aa27.4.59, af72.5.35, wp*
RAF-T Reconnaissance Attack Fighter Trainer *mt*
RAFAN Royal Air Force Airmove Network [UK] *mt*
RAFAS Report of Accounting and Finance Activities System *mt*
RAFAX Radar Facsimile Transmission *fu*
RAFG RAF Germany [UK] *mt*
RAFKADPS RAF Kemble ADP Support *mt*
RAFS Rubidium Atomic Frequency Standard *jdr*
RAFTEC Real Time Aircraft Flight Test Evaluation Center *aw130.6.41*
RAFTS Reconnaissance, Attack, Fighter Training System [USAF] *wp, mt*
RAFUS Radio Frequency Usage Reporting System *mt*
RAFVR RAF Volunteer Reserve [created in 1936, UK] *mt*
RAG Replenishment Air Group [USA] *mt*
RAG Requirements Assessment Group [USAF] *wp, mt*
RAG Row Address Generator *pk, kg*
RAHAM Radiation Hazard Meter *wp*
RAI Random Access and Inquiry *fu*
RAIAM Random Access Indestructive Advanced Memory *fu*
RAID Ram Air Inflatable Decelerator *wp*
RAID Recallable Airborne Infrared Display *hc*
RAID Redundant Array of Inexpensive Drives (Disks) *ct, kg, em, hi*
RAID Remote Access Interactive Debugger *fu*
RAIDS Radar Airborne Intrusion Detection System *wp*
RAIDS Real time AUTODIN Interface Distribution System *mt*
RAIL Runway Alignment Indicator Lights *fu, wp*
RAIS Redundant Arrays of Inexpensive Systems *pk, kg*
RAKR Atomnaya Raketniy Kreyser [nuclear rocket cruiser] [USSR] *mt*
RALACS Radar Low-Altitude Control System *fu*
RALS Remote Augmentation Lift System *mt*
RALU Register [and] Arithmetic Logic Unit *fu, kg*
RAM (overview of RAM accelerator technology) *aa26.12.b25*
RAM Radar Absorbing (Absorption, Absorptive, Absorbent) Material *mt, hc, wp, tb, jdr, fu, ai1*
RAM Radiation Attenuation Measurement *fu*
RAM Raid Assessment Mode *hc*
RAM Ramenskoye [aircraft first identified at Ramenskoye airfield, USSR, were temporarily labeled RAM-x by NATO, were _x_ is a sequence letter. RAM-K is Su-27, RAM-L is MIG-29]
RAM Random Access Memory "ram" *mt, kg, em, ct, he, su, uh, ah, hc, gg161, 6, 8, 15, 16, pl, w9, lt, wp, ep, tb, st, jdr, pbl, tm, kf, fu, hi, oe, ai1*
RAM Readiness Assessment Module [WSMIS] [AFLC] *mt*
RAM Redeye Air Launched Missile *hc*
RAM Relative Absorption Mass *ai1*
RAM Reliability, Availabilit,y and Maintainability *mt, fu, rm3.1.2, nt, wp, jdr*
RAM Remote Access Modem *em, ct*
RAM Responsibility Assignment Matrix *fu*
RAM Role Accountability Matrix *jdr*
RAM Rolling Airframe Missile *fu, hc, uc, ci.179*
RAM {a boom requiring pointing, part of Mars orbiter} *ma40*
RAM/ACE Radar Aided Mission/Aircrew Capability Exploration *hc*
RAM/LOG Reliability/Availability/Maintainability and Logistics *hc*
RAMA Reliability, Availability, and Maintainability Analysis *fu*
RAMAC RAID Architecture with Multilevel Adaptive Cache *em*
RAMAN Raman Cell (gas) [improve beam quality] {?} *tb*
RAMAS Reliability, Availability, Maintainability Analysis System [AFLC] *mt*
RAMBO Rapid Analytical Model of BMD Optimization *tb*
RAMCAD Reliability And Maintainability in Computer Aided Design [USA] *mt, aa26.12.66, wp*
RAMD Reliability, Availability, Maintainability, Durability *wp*
RAMDAC Random Access Memory Digital to Analog Converter *em, kg, hi*
RAMER Reliability / Availability / Maintainability Event Report *fu*
RAMIS Rapid Access Management Information System *wp*
RAMIS Reliability and Maintainability Information System *mt*
RAMIT Rate Aided Manually Initiated Tracking *fu*
RAMM Recording Ammeter *fu*
RAMMS Radiating Millimeter Source *fu, hc*
RAMOS Remote Automated Meteorological Observing Station *pl*
RAMP Radar Modernization Project *fu*
RAMP Radiation Airborne Measurement Program *wp*
RAMP Rapid Acquisition of Manufactured Parts *mt*
RAMP Reliability And Maintainability Program *wp*
RAMP Remote Access Maintenance Protocol [Internet] *kg, hi*
RAMP Report and Message Processor *mt*
RAMP Risk Assessment Management Program *tb*
RAMPART Radar Advanced Measurement Program For Analysis of Reentry Techniques *wp*

RAMPS Recipe and Menu Pricing System [USAF] *mt*
RAMPS Resource Allocation and Multiproject Scheduling *wp*
RAMS Racal Avionics Management Systems *aw118.18.78*
RAMS Range Measurement System [TFWC] *mt*
RAMS Recruiting Automated Management System [USAF] *mt*
RAMS Reliability And Maintainability Symposium *rm7.1.6*
RAMS Reproduction Administrative Management System *mt*
RAMS Reprographics Automated Management System [USAF] *mt*
RAMSTAT Recovery Airfield Monitoring and Status *wp*
RAMTIP Reliability And Maintainability Insertion Program *wp*
RAN Regional Area Network *fu*
RAN Reporting Account Number *wp*
RAN Request for Authority to Negotiate *wp*
RAN Royal Australian Navy *hc*
RANCID Real and Not Corrected Input Data *fu*
RAND Request Activity and Name Directory *ns*
RAND Research And Development [a corporation in California] *ci.37, 13*
RANDAT Random Data *fu*
RANGEX Ranging Exercises *fu*
RANN Research Applied to National Needs *hc, wp*
RAOC Rear Area Operations Center *mt*
RAOC Regional Air Operation Center *mt*
RAP (High energy laser windows (RAP)) *hn50.9.5*
RAP Rapid Application Prototyping *pk, kg*
RAP Readiness Action Proposal *wp*
RAP Recognized Air Picture *fu, pm*
RAP Reliability Assurance Program *pl*
RAP Remedial Action Project *mt*
RAP Remote Advisory Panel [Hahn AFB] *mt*
RAP Risk Abatement Plan *fu*
RAP Risk Aversion Plan *jdr*
RAP Rocket Assisted Projectile *hc*
RAPCOE Random Access Programming and Checkout Equipment *fu*
RAPCON Radar Approach Control *mt, fu, wp*
RAPCONS Radar Approach Controls *wp*
RAPID Radar and Power Improvement Program *hc*
RAPID Random Access Provision for Inbound Demands *fu*
RAPID Rapid Access Programmed (Program) Information Display *fu, hc*
RAPID Real Time Acquisition and Processing of In-Flight Data *wp*
RAPID Retrieval through Automated Publication and Information Digest *wp*
RAPIDE Relocatable Army Processors for Intelligence Data, Europe [USAREUR] *mt*
RAPIDS Random Access Personnel Information System *wp*
RAPIDS Real time Automated Personnel Identification System *mt*
RAPIDSIM Rapid Intertheatre Deployment Simulator [JCS] *mt*
RAPIER Rapid Emergency Reconstitution *mt, jb*
RAPIER Rapid Emergency Relocation *mt, fu*
RAPLOC Rapid Acoustic Passive Localizer *hc*
RAPPI Random Access Plan Position Indicator *fu, mt*
RAPPORT_III ECM System *mt*
RAPS Radar Absorbing Primary Structure *wp*
RAPS Radar Proficiency Simulator *mt*
RAPSE Report on Active and Planned Spacecraft and Experiments *ns*
RAR Radio Acoustic Ranging *fu*
RAR Reliability Assurance Requirements *pl*
RAR Revise As Required *wp*
RARC Regional Administrative Radio Conference *sc3.5.7, ce*
RARP Reverse Address Resolution Protocol *ct, kg, em*
RAS Radar Absorbing Structures *hc*
RAS Radar Altimeter Science *es55.28*
RAS Radar Augmentation System *wp*
RAS Radar Automatic System *fu*
RAS Random Access Storage *pk, kg*
RAS Reader Admission System [British Library] *pk, kg*
RAS Rectified Air Speed *mt*
RAS Reliability, Availability and Serviceability *kg, hi*
RAS Remote Access Server (Service) *ct, kg, em, hi*
RAS Remote Active Spectrometer [Dial] *hc, wp*
RAS Requirements Allocation Sheet *mt, fu, tb*
RAS Row Address Select *pk, kg*
RAS Royal Astronomical Society *mi*
RAS/STADES Record Association System / Standard Data Elements System [WWMCCS] *mt*
RASAPI Remote Access Service Application Programming Interface [Microsoft] *kg, hi*
RASC Regional AUTODIN Switching Center *mt*
RASC Royal Astronomical Society of Canada *mi*
RASCAL Random Access Secure Communications Anti Jam Link *wp*
RASER Radio [Frequency] Amplification By Simulated Emission of Radiation *fu*
RASGA Radar Antenna-System Group Alternate (Patriot) *fu*
RASP Receive Active Signal Processor *fu*
RASP Recognized Air and Sea Picture *fu*
RASP Remote Antenna Signal Processor *ct, em*

RASS Random Access Storage Set *fu*
RASSR Reliable Advanced Solid State Phased Array Radar *wp*
RASTA Radiant Special Test Apparatus *wp*
RASTAC Random Access Storage and Control *fu*
RASV Reusable Aerodynamic Space Vehicle *wp*
RAT Ram Air Turbine *mt, wp*
RAT Reliability and Availability Test *oe*
RAT Reliability Assurance Test *wp*
RAT Remote Access Terminal *fu*
RAT Resource Allocation Tool *fu*
RAT Resource Availability Table *fu*
RAT RISC Architecture Technology *hi*
RATAN Radar and Television Aid to Navigation *mt, fu*
RATCC Radar Approach Control [Navy equivalent to TRACON] *fu*
RATCF_DAIR Radar Air Traffic Control Facility Direct Altitude and Identity Readout *mt*
RATD Report Aboard Trained Date *fu*
RATEL Radio Telephone *wp*
RATFOR Rational FORTRAN *hi*
RATO Rocket Assisted Take Off *mt, hc, fu, wp*
RATOG Rocket Assisted Take Off Gear *wp*
RATS Radar Acquisition and Tracking System *hc, wp*
RATS Radar Altimeter Target Simulator *wp*
RATS Rapid Area Transportation Support *wp*
RATS Remote Alarm Transmission System *hc*
RATS Retrieval And Transmission System *mt*
RATS Roll Axis Tracking System *wp*
RATSCAT Radar Target Scatter [upgrade] *hc, wp*
RATSCAT Radar Targeting and Scattering *st*
RATSCAT [radar reflectivity measurement range] *ro*
RATT Radio Teletype(writer) *fu, mt*
RATTLER [shoulder-fired antitank weapon (DRAGON follow-on)] *uc*
RAU Radio Access Unit *mt*
RAU Registered Arithmetic Unit *fu*
RAuxAF Royal Auxiliary Air Force [UK] *mt*
RAV Restricted Availability *fu*
RAV Robotic Air Vehicle *wp*
RAVE Radar Acquisition Visual (Tracking) Equipment *fu*
RAVE Rendering Acceleration Virtual Engine *pk, kg*
RAVE Research Aircraft Visual Environment *hc*
RAVEN Ranging And Velocity Navigation *wp, fu*
RAVEN [nickname for EF111 aircraft, USAF] *mt*
RAVES Rapid Aerospace Vehicle Evaluation System *wp*
RAWIN Radar Wind sounding [USA] *mt*
RAWINSONDE Radar Wind Sounding and Radiosonde [USA] *mt*
RAWS Radar Altitude Warning System *mt*
RAWS Radar Attack and Warning System *mt*
RAWS Remote Area Weather Station *fu, hc*
RAZEL Range Azimuth and Elevation *fu*
RAZON Range and Azimuth Only *fu*
RB Radar Battalion *fu*
RB Reconnaissance Bomber *wp*
RB Report Back *fu*
RB Revision Block *fu*
RB/ER Reduced Blast, Enhanced Radiation *ph*
RBB Residential Broadband *ct*
RBBS Remote Bulletin Board System *kg, hi*
RBC Relocatable Binary Code *mt*
RBC Route Broadcast Channel *fu*
RBCC Radar Battalion Control Center *fu*
RBCS Remote Bar Code System *pk, kg*
RBD Reliability Block Diagram *fu*
RBDE Radar Bright Display Equipment *fu*
RBECS Revised Battlefield Electronic CEOI System *mt*
RBF Remote Batch Facility [Honeywell] *mt*
RBFIAS Rule Based Force Integration Analysis System *fu*
RBGM Real Beam Ground Map *hc*
RBM Red Ballistic Missile *tb*
RBM Regional Battle Manager *tb*
RBN Radio Beacon *fu*
RBNS Research Board for National Security *-13*
RBO Rubidium Oscillator *jdr*
RBOC Rapid Blooming Offboard Chaff *fu*
RBP Reactive Bed Plasma *wp*
RBP Relative Breakpoint Parameter *fu*
RBP Report Back Processor *jdr*
RBRP (contact . . . RADC/RBRP, Griffiss AFB, N.Y.) *is26.1.72*
RBS Radar Bomb Scoring *mt, 12*
RBS Radar Bomb Site *mt*
RBS Radar Bombing Score *fu*
RBS Remote Batch Station *mt*
RBSN Regional Basic Synoptic Networks *pl*
RBT Remote Batch Terminal *mt*
RBV Return Beacon Vidicon [LANDSAT] *hg, ge, ac2.2-5*
RC Radar Controller *fu*
RC Radiant Cooler *oe*

RC Radio Control *mt*
RC Range Correction *fu*
RC Rate of Change *fu*
RC Reaction Control *9, vv*
RC Receive Component *-4*
RC Reconnaissance Capability *fu*
RC Record Change [for Correction] *fu*
RC Reduced Capability *fu*
RC Regulatory Commission *fu*
RC Relationship Code *fu*
RC Remote Control *fu*
RC Reserve Component *mt*
RC Reset Channel *fu*
RC Resistance - Capacitance *fu, wp, ai1*
RC Resistor - Capacitor (filter) *sp662, 0, w9*
RC Resource Center [SPACECMD] *mt*
RC Resource Controller *jdr*
RC Response Cell *mt*
RC Responsibility Center *wp*
RC/CC Responsibility Center/Cost Center *wp*
RCA Radio Corporation of America *mt, aa26.12.28, sc2.7.3, aw, hc, jb, tb, ro*
RCA Regional [battle manager] Control Algorithm *tb*
RCA Request for Corrective Action *fu*
RCA Riot Control Agent [chemical warfare] *mt, wp*
RCA Royal Canadian Air Force *wp*
RCADS Radar Control and Display Subsystem *fu*
RCAF Royal Canadian Air Force *w9, wp*
RCAG Remote Communications Air/Ground Facility *fu*
RCAG Remote-Controlled Air/Ground (Radio) *fu*
RCAM Refueling Capability Assessment Model [HQ USAF] *mt*
RCAM Remotely Controlled Attack Missile *hc*
RCAS REDCOM Command and Control System [US] *mt*
RCAS Reserve Component Automated Systems [USA] [formerly CAMIS] *mt*
RCA_PRICE Cost estimating system developed by RCA *fu*
RCB Remotely Controlled Boat *wp*
RCB Requirements Change Board *wl*
RCC Radar Control Computer *fu*
RCC Radio Communications Coordinator *mt*
RCC Range Control Complex *hc*
RCC Redundant Code Correction *fu*
RCC Regional Computer Center *mt*
RCC Regional Control Center *fu, 12*
RCC Reinforced Carbon - Carbon *ci.xxii*
RCC Remote Command and Control *wp*
RCC Remote Communications Center *mt*
RCC Remote Computer Center *fu*
RCC Remote Control Center *mt*
RCC Rescue Coordination Center [USA] *fu, mt*
RCC RPV Control Center *fu*
RCCC Regional Communications Control Center *fu*
RCCP Rough Cut Capacity Planning *jdr*
RCD Research and Acquisition Communications Division *mt*
RCD Reverse Current Device *fu*
RCDS Radar Control and Display Subsystem *fu*
RCE Remote Control Equipment *fu*
RCF Range Control Facility *hc*
RCF Recovery Control Facility *fu*
RCF Regional Control Facility *st*
RCF Remote Call Forwarding *ct, hi*
RCF Remote Communications Facility *fu*
RCF Remote Concentration Facility [Honeywell] *mt*
RCG Reference Concept Group *wl*
RCH Route Characteristics File *mt*
RCI Radar Coverage Indicator *fu*
RCI Reliability Critical Item *fu, nt*
RCI Resonant Cavity Interferometer *wp*
RCI Rodent Cage Interface [for SLS mission] *mi*
RCIRS Recoverable Consumption Item Requirements System [AFLC] *mt*
RCIU Radio Control Interface Unit *fu*
RCL Rotate Carry Left *pk, kg*
RCLL Runway Center Line Lights *fu*
RCM Radar Counter Measures *mt, fu, 0, wp*
RCM Radio Counter Measures *mt*
RCM Reliability Centered Maintenance *mt, fu, wp, st*
RCM Requirements Correlation Matrix *wp, kf*
RCME Rockwell - Collins Middle East *fu*
RCMS Remote Control and Monitoring System *fu*
RCMS Runway Configuration Management System *fu*
RCN Royal Canadian Navy *w9*
RCO Range Control Officer *mt*
RCO Record Communications Operator *mt*
RCO Regional Coordination Office *fu*
RCO Remote Communications Outlet *fu*

RCO Remote Control Oscillator *fu*
RCOC Region Communications Operations Centers [DCA] *mt*
RCP Radar Control Panel *fu*
RCP Radar/ Communications Processor *fu*
RCP Register Clock Pulse *bl2-7b.42*
RCP Remote Control Panel *pk, kg*
RCP Remote Copy [Internet] *kg, hi*
RCP Request Change Proposal *mt*
RCP Restore Cursor Position *pk, kg*
RCP Right [hand] Circular Polarization *0, 2, 4, wp*
RCPAC Reserve Component Personnel and Administration Center [USA] *mt*
RCR Record Correction Request *fu*
RCR Rotate Carry Right *pk, kg*
RCR Runway Condition Rating (Reading) *mt*
RCRI Requirement Cross-Reference Index *fu*
RCS Radar Cross Section *mt, fu, hc, aw129.21.12, jb, kb, wp, tb, st*
RCS Reaction Control (Sub)System *gs2-2.48, 4, 9, 17, sb, aa26.11.26, es55.60, nt, lt, ce, wl, jdr, mb, vv, he, mi, ah, uh, jn, oe, ai2*
RCS Rearward Communications System *mt*
RCS Records Communications Switching System *pk, kg*
RCS Regional Control Station *mt*
RCS Remote Channel Selector *fu*
RCS Remote Control System *fu*
RCS Reports Control System *mt*
RCS Revision Control System [Unix] *kg, fu*
RCS Roll Control System *ai2*
RCT Radar Control Terminal *fu*
RCT Radiometric Calibration Target *jp*
RCT Repair Cycle Time *mt, fu, wp*
RCTEC Reduced Cost Turbine Engine Concept *wp*
RCTL Resistor, Capacitor Transistor Logic *wp, fu*
RCU Radar Command Unit *0*
RCU Remote Command Unit *kgl*
RCU Remote Control Unit *fu, ai2*
RCU Reserve Component Unit *mt*
RCU RMDU Control Unit *-4*
RCV Reaction Control Valves [puffer jets using compressed air, used to control VTOL aircraft in the hover mode]
RCV Remotely Controlled Vehicle *fu, N75-27202-483*
RCW Request Control Word *fu*
RCZ Rear Combat Zone *mt*
RD Radar Data *fu*
RD Radiation Detection *fu*
RD Raketnay Dvigatel [rocket motor] *ai2*
RD Random Drift *fu*
RD Receive Data [also RXD] *fu, 4, kg, hi*
RD Register Drive *fu*
RD Remove Directory *pk, kg*
RD Requirements Document [for data] *mt*
RD Restart Definition *fu*
RD Restricted Data *mt, sy9, hc, tb, jdr, kf, su*
RD Root Diameter *fu*
RD Rough Draft *fu*
RD&A Research, Development, and Acquisition *ph, hc*
RD&D Research, Development and Dissemination *fu*
RD&E Research, Development, and Engineering *mt*
RD&T Research, Development, and Testing *12, ac1.2-21*
RD/GT Reliability Development and Growth Test *fu*
RD/PAT Reliability Demonstration/Pro-duction Acceptance Tests *fu*
RD/W Request for Deviation/Waiver *fu*
RDA Radio Direct Access *fu*
RDA Recommended Daily (Dietary) Allowance *w9, wp*
RDA Reduced Diameter Array *fu*
RDA Reflective Dot Array *fu*
RDA Reflector Deployment Actuator *cl, jn*
RDA Reliability Design Analysis *wp*
RDA Remote Database Access *pk, kg*
RDA Requirements Definition and Analysis *fu*
RDA Research and Development Associates *tb*
RDA Research and Development, Army *fu*
RDA Research, Development, and Acquisition *mt, uc, dt, st*
RDAC (ergonomic study initiated at RSG's RDAC laboratory) *ne*
RDACS Remote Data Acquisition System *ci.1027*
RDAF Real Time Data Acquisition Facility *ac3.7-33*
RDAF Royal Danish Air Force *fu*
RDAPS Real Time Data Acquisition and Process System *hc*
RDAS Radar Data Acquisition System *fu*
RDAT Research and Development Acceptance Test *fu*
RDATA Raw Imaging and Sounding Data *gp*
RDB Radar Data Block *fu*
RDB Receive Data Buffer *fu, kg*
RDB Relational Data Base *mt, kg, hi*
RDB Requirements Data Bank *wp, st*
RDBMS Relational Data Base Management System *fu, em, tb, kg, hi*
RDBW Request Data Block Word *fu*

RDC Radar Data Converter *fu*
RDC Radar Display Console *fu*
RDC Radar Distribution Center *fu*
RDC Reference Designation Code *fu*
RDC Regional Distribution Center *fu*
RDCC Research and Development Computer Complex *fu*
RDCE Reply Detector and Code Extractor *fu*
RDCF Relay Drone Control Facility *fu*
RDCLK Received Timing Clock *pk, kg*
RDCU Recording Data Control Unit *kf*
RDD Radar Data Display *fu*
RDD Radar Data Distribution *fu*
RDD Remote Data Device *fu*
RDD Remote Display Device *fu*
RDD Required Delivery Date [USA DOD] *fu, wp, mt*
RDD Requirements Driven Development *jdr*
RDD Return Data Delay *-4*
RDDC Radar Data Distribution Controller *fu*
RDDDD Relational Design Data Dictionary and Directory *fu*
RDDM Return Data Delay Measurement *-4*
RDDS Radar Data Distribution Switchboard *fu*
RDDT&E Research, Development, Design, Test, and Evaluation *ci.1055*
RDE Radar Data Extractor *fu*
RDEX Readiness Exercise *mt*
RDF Radar Direction Finder (Finding) *fu, w9, ct*
RDF Radial Distribution Function *wp*
RDF Radio Direction Finding (Finder) [acronym for early versions of RADAR, UK] *mt*
RDF Rapid Deployment Force *mt, fu, ph, hc*
RDF Research and Development Facility *fu*
RDFCS Reconfigurable Digital Flight Control System *fu*
RDG Radar Display Group *fu*
RDG Rate Delay Generator *jdr*
RDG Resolver Differential Generator *fu*
RDGT Reliability Development and Growth Test *fu, hc*
RDI Reconnaissance, Detection and Identification *hc, wp*
RDJTF Rapid Deployment Joint Task Force [later U.S. CENTCOM] *dt, wp, ro, mt*
RDL Requirements Definition Language *fu*
RDLRP Research and Development Long-Range Plan *fu*
RDM Radar Data Monitor *fu*
RDM Radiometer Deployment Mechanism *pl*
RDM Reflective Dual Mode *lt*
RDM Reliability, Dependability, and Maintainability *kf*
RDM Remotely Delivered Mine *fu*
RDM Rendezvous and Docking Module *re*
RDM Report Display Manager *ct, em*
RDM Requirements Definition Methodology *fu*
RDM Requirements Development Manager *jdr*
RDM/BURP Radar Data Monitoring/Backup Radar Processing *fu*
RDMC Radar Data Monitoring Console *fu*
RDMF Rapid Deployable Medical Facility *mt*
RDMS Rapid Deployable Mobile System [ESC*] *mt*
RDMSS Rapidly Deployable Mobile SIGINT System *mt, wp*
RDMU Radar-Drift Measuring Unit *fu*
RDN Receiving Discrepancy Notice *fu*
RDN Relative Distinguished Name *ct, em*
RDN Reliability Discrepancy Notice *oe*
RDO Read Only *hi*
RDO Reconnaissance Duty Officer *mt*
RDO Redistribution Order *mt, wp*
RDO Remote Data Object *pk, kg*
RDO Research and Development Objectives *wp*
RDOC Reference Designation Overflow Code *fu*
RDP Radar Data Processing (Processor) *fu, hc, wp*
RDP Rapid Deployment Planning *mt*
RDP Relay Driver Pulse *uh*
RDP Reliable Datagram Protocol *ct, em, kg*
RDP Remote Data Processing *fu*
RDPC Regional Data Processing Center *mt*
RDPI Remote Data Processing Installation [later RIPC] *mt*
RDPM Reflector Deployment Positioning Mechanism *cl, jdr, ric*
RDPS Radar Data Processing System *fu*
RDQ Radar Data Quality *fu*
RDR Remote Data Readout *fu*
RDR Report Distribution Request *-8*
RDR Revision Difference Reconciliation *oe*
RDRAM Rambus Dynamic Random Access Memory *kg, hi*
RDRM Reliability Data Request Memorandum *fu*
RDRS Radar Data Reduction System *fu*
RDS Radar Defense System *fu*
RDS Relational Database Systems *fu*
RDSR Receiver Data Service Request *kg, hi*
RDSS Radio Determination Satellite Service *wp*
RDSS Rapidly Deployable Surveillance System *wp*
RDT Radar Display Terminal *fu*

RDT Receiving Day Tag *fu*
RDT Reliability Demonstration Test *wp, fu*
RDT Remote Data Terminal *fu*
RDT Remote Data Transmitter *wp, jdr, ric*
RDT&E Research, Development, Test(ing), and Evaluation *mt, 12, aw130.13.104, ph, hc, ci.604, wp, dm, tb, st, jdr, kf, fu*
RDTE Research, Development, Test, and Evaluation [also RDT&E] *nt, wp*
RDTM Rated Distribution Training Management *mt*
RDTO Receive Data Transfer Offset [IBM] *pk, kg*
RDU Radar Display Unit *fu*
RDU RADEC Data Unit *mt*
RDU Radiation Detector (Detection) Unit *cl, jdr, uh, kf*
RDU Remote Data Unit *ai2*
RDU Remote Decoder Unit *sc3.6.2, jdr*
RDVS Remote Document Viewing System *fu, hc*
RDW Request Data Word *fu*
RDW Request for Deviation or Waiver [a document] *17, pl, ep, jdr, tm, he, oe*
RDW Request for Deviation/Waiv-er *fu*
RDX Directorate of Program Installation *mt*
RDX Research Depart. Explosive [cyclonite/cyclotrimethylen-etrinitramine] *wp*
RDYLD Ready to Load Data *mt*
RE Radiated Emission *fu, oe*
RE Radiative Equilibrium *pl*
RE Receiver-Exciter *fu*
RE Recurring Cost *kf*
RE Recurring Expense *-18*
RE Reliability Engineer *fu*
RE Reporting Element *fu*
RE Reporting Subsystem [WIS] *mt*
RE Resident Engineer *fu*
RE Responsible Engineer [an individual] *fu, ep, jdr, pbl, tm, jn*
RE [Hughes Aircraft Company customer for multiplex interface module] *hc*
REA Request for Equitable Adjustment *mt, oe*
REA Responsible Engineering Activity *gs2-2.48, 15, sb, lt, ep, bf, jdr, pbl, tm, ah, he, jn*
REA Responsible Engineering Authority *fu, ah, he, hc, hn43.22.4, 8, 17, pl, jdr, pbl, tm, jn*
REA Rocket Engine Assembly *hc*
REA Roll Earth Acquisition *oe*
REACT Radar Electro-optical Area Correlator Tracker *wp*
REACT Resource Allocation and Control Technique *wp*
REACTS Radiation Emergency Assistance Center Training Site *mt*
REACTS Readiness Exercise and Crisis Transportation System *mt*
READ Radar Echo Augmentation Device *fu, wp*
READ Readability Assessment Device *fu*
READGE Republic of Ecuador ADGE *fu*
REALP Real Property Inventory [USAF] *mt*
REAMS Resources Evaluation And Management System *wp*
REBE Recovery Beacon Evaluation *fu*
REBM Radio Electronic Battle Management *mt*
REC Radar Elevation Converter *fu*
REC Radio Electronic Combat [translation of the Russian terminology for ECM] *fu, mt*
REC Reconstitutable Enduring Communication *mt*
REC Regional Evaluation Center *fu*
REC Repeat Error Counter *fu*
REC Request for Engineering Change *fu, wp*
REC Research and Engineering Center *fu*
RECA Residual Capability Assessment *mt*
RECAS Residual Capability Assessment System [part of GWS] *mt*
RECC Regional Emergency Communications Coordinator [GSA] *mt*
RECCCI Reconstitutable Enduring Communications Command, Control, and Intelligence *mt*
RECCEXREP Reconnaissance Exploitation Report [USA DOD] *mt*
RECON Reconnaissance Information System [WWMCCS] *mt*
RECON Reconnaissance [System] *wp, mt*
RECREP Reconnaissance Report *mt*
RECS Radar Embedded Computer System *wp*
RECS Radio Electronic Combat Support [FSU] [NATO] *mt*
RECS Rear Echelon HF COMINT Systems *mt*
RECS Reconfigurable EC System *wp*
RECS Reseller Electronic Communication System *ct*
RECSHIP Receiving Ship [USA] *mt*
RECSTA Receiving Station [USA] *mt*
RED Remote Entry Device *fu*
RED Risk Estimate Distance [safe distance to friendly forces to use weapons] *mt*
REDAP Reentrant Data Processing *fu*
REDARC Reentry Data Acquisition Range Contractor *-17*
REDCAP Real time Electromagnetic Digitally Controlled Analyzer and Processor *wp*
REDCOM Readiness Command [part of US Unified Command] *mt, 12, jb*

REDCON Readiness Condition [USA] *mt*
REDHAT Radar Exploration Development for High-value target Acquisition/Track. *hc*
REDOX Reduction Monitor *st*
REDOX Reduction/Oxidation *wp, ai1*
REDSS RTADS Electronic Delivery and Status System *fu*
RED_FLAG [training program at Nellis AFB] *wp*
RED_HORSE Rapid Engineer Deployable, Heavy Operational Repair Squadron, Engineer [USAF] [USA] *mt*
RED_MILL [forward scatter and surveillance system for the over-the-horizon radar] *wp*
RED_SPIKE [an infrared spectral band where aircraft engine emissions peak] *wp*
REEDM Rocket Exhaust Effluent Dispersion Model *wp*
REES Radar Electromagnetic Environment Simulator *wp*
REFILES Reference Files [JCS Standard] [JOPS] *mt*
REFTRA Refresher Training *fu*
REGAL Rigid Epoxy Glass Acrylic Laminate *pk, kg*
REGENCY_NET [code name for USEUCOM HF/SSB communications system] *mt*
REI Request Engineering Information *fu*
REI Research Equipment, Inc. *tb*
REIL Runway End Identifier Lights *fu*
REIS Reconstitutable and Enduring Intelligence System *mt*
REJ Reject *jdr*
RELACS Radar Emission Location Attack Control System *hc*
RELNAV Relative Navigation *fu*
RELOC Relocation File *mt*
RELROK Releasable to Republic Of Korea *mt*
RELSAT Reliability for Satellite *hc*
RELSE Relative State Estimator *-3*
RELSECT Relative Sector *pk, kg*
REL_COPAN USAF HQ ADVANCE Network *mt*
REM Radiation Effects Monitor *es55.99, bo*
REM Rat Enclosure Module [for SLS mission] *mi*
REM Release Escape Mechanism *wp*
REM Reliability, Environmental, and Maintainability *fu*
REM Ring Error Monitor *pk, kg*
REM Roentgen Equivalent, Mammal [quantity of ionizing radiation which produces a physiological effect equivalent to that of 1 roentgen of X-rays if absorbed by man or mammals] *mt*
REMAB Remote Marshaling Base [PACOM] *mt*
REMAD Remote Magnetic Anomaly Detection *fu*
REMAR Reentry Effectiveness Measurements Array Radar *fu*
REMARC Resource Management Accounting Control System *mt*
REMBASS Remotely Monitored Battlefield Sensor System [USA] *hc, fu, mt*
REMC Resin-Encapsulated Mica Capacitor *fu*
REMIDS Remote Minefield Detection System *hc*
REMIDS Remote Minefield Identification and Display System *fu, hc*
REMIS Reliability and Maintainability Information System *mt, hc, wp*
REMOB Remote Observation *ct, em*
REMP Research, Engineering, Mathematics, and Physics [division in NSA] *-10*
REMS Registered Equipment Management System *wp*
REMUS [German computer controlled standardized measurement test system for weapon system electronics] *wp*
RENO Radiant Energy Network Option *gp*
RENS Reconnaissance, Electronic Warfare and Nala [intelligence] System *wp*
REO Readiness and Evaluation Office *fu*
REO Regenerated Electrical Output *fu*
REO Responsible Engineering Organization *fu*
REOS Right End Of Scan *oe*
REP Range Error Probable *wp*
REP Range-Error Probability *fu*
REP Rendezvous Evaluation Pod *ci.372*
REP Repeat Reports *mt*
REP Request for Programming *0*
REP Reserve Enlistment Program [USA] *mt*
REP Resource Estimating Procedure *mt*
REP Roentgen Equivalent Physical *fu, wp*
REPCAT Report Corrective Action Taken *mt*
REPE Repeat while Equal *pk, kg*
REPGEN Report Generator [FLOGEN] [IMAPS] *mt*
REPNE Repeat while Not Equal *pk, kg*
REPNZ Repeat while Not Zero *pk, kg*
REPOL Reporting Emergency Petroleum, Oils, and Lubricants *mt*
REPPAC Repetitively Pulsed Plasma Accelerator *fu*
REPRO Regional Programs Office *fu*
REPSHIP Report of Shipment *-18*
REPSHIPS Report of Shipments *wp*
REPZ Repeat while Zero *pk, kg*
REQ/ACK Request/Acknowledge *fu*
REQCAP Requirement Capability [USA] *mt*
REQID Requirement Identification (Identifier) *jr, jdr*
REQMT Requirement *wp*

RER Residual Error Rate *do1*
RERL Residual Equivalent Return Loss *fu*
RES Radar Environment Simulator *fu, hc*
RES Reading Ease Score *fu*
RES Remote Execution Service *pk, kg*
RESC Reconstitutable Enduring Satellite Communications System *mt*
RESCAP Rescue Combat Air Patrol [USA] *mt*
RESCU Resource Cost and Utilization System [U.S. Marine Corps] *mt*
RESEP Re Entry Systems Environmental Protection *wp*
RESER Re Entry Systems Evaluation Radar *hc, wp, fu*
RESG Research Engineering Standing Group *dt*
RESINS Redundant Strapdown Inertial Navigation System *ai2*
RESR Rotating Electronically Scanned Radar *fu*
RESS Radar Environmental Simulator System *fu*
REST Real Time Embedded Systems Testbed *wp*
RESTA Reconnaissance, Surveillance, Target Acquisition [also RSTA] *wp*
RESTORE Rapid Evaluation System to Repair Equipment *fu*
RESURS [Earth resources satellite series, USSR] *hg, ge*
RET Radar Equipment Trailer *fu*
RET Resolution Enhancement Technology [HP] *kg, hi*
RETA Rigid Explosive Transfer Assembly *oe*
RETAWS Real Time All Weather Surveillance Sensor *hc*
RETMA (is available in a RETMA enclosure) *hd1.89.y13*
RETS Remote Target System *hc*
REVEL Reverberation Elimination *fu*
REVIC Revised Intermediate COCOMO *kf*
REVOCON Remote Volume Control *fu*
REVS Requirement Engineering and Validation System *fu*
REWSON Reconnaissance, Electronic Warfare, Special Operations and Naval [Intelligence Processing System] *wp, fu*
REX Readiness Exercise *mt*
REX Receiver Exciter *fu*
REX Relocatable Executable *pk, kg*
REXEC Remote Executable (Execution) *ct, kg, em*
REXX Restructured Extended Executor [language] [IBM] *kg, ct, em, hi*
RF Radar Flight *hc*
RF Radio Frequency *hc, sp662, 0, 4, 5, 7, 9, 17, w9, sb, pl, aa26.11.11, es55.22, af72.5.44, md2.1.3, wp, pm, ep, ce, dm, eo, re, tb, st, wl, jdr, mb, kf, vv, kg, mi, he, uh, ct, fu, hi, oe, mt, ai1, ai2*
RF Range Finder *fu*
RF Rapid Finder *fu*
RF Rapid Fire *tb*
RF Reconnaissance Fighter *wp*
RF Relevant Failure *fu*
RF Response Force *tb*
RF Risk Factor *fu*
RF [designation, RF-4 is Phantom II] [USA DOD] *mt*
RFA Radio Frequency Amplifier *fu*
RFA Radio Frequency Assembly *fu*
RFA Radio Frequency Authorization *0*
RFA Request for Action *fu, wp, tb*
RFA Request For Addition [a form] *he, pl*
RFA Request For Authorization *jdr*
RFB (RFB research and development in WIG vehicles) *aa27.4.b13*
RFC Radio Frequency Choke *fu*
RFC Radio Frequency Combiner *fu*
RFC Radio Frequency Correlator *jdr*
RFC Reply File Conversion *fu*
RFC Request For Comment [Internet] *kg, ct, hi, du*
RFC River Forecast Center *un*
RFC Royal Flying Corps [WWI air corps of the British Army, UK] *mt, af72.5.58*
RFCA Radiator Flow Control Assembly *ci.970*
RFCS RF Control System *fu*
RFD Request For Deviation *fu, jdr*
RFD Request For Discussion [Internet] *kg, hi*
RFDB RED FLAG Data Base [TFWC] *mt*
RFEG Radio Frequency Environmental Generator *hc*
RFF Remote File Facility [Honeywell] *mt*
RFF Request For Funds *mt*
RFG Receive Format Generator *fu*
RFG Reference Frequency Generator *jdr*
RFGG Reference Frequency Generation Group *jdr*
RFI Radio Frequency Interference *ggIII.1.67, 0, 2, 4, 7, hc, aa26.12.76, mj31.5.25, jb, tb, wp, wl, jdr, mb, ct, mi, kg, fu, he, hi, oe, mt, tt*
RFI Ready For Issue *fu, wp*
RFI Representative of a Foreign Interest *tb, jdr, su*
RFI Requests For Information *fu, ct, lka, cb5.15.89, wp, mt*
RFI/CA Request For Investigation or Corrective Action *gp, fu*
RFIA Radio Frequency Interface Assembly *fu*
RFIC Radio Frequency Interface Chip *wp*
RFID Request For Implementation Date *wp*
RFIS Radar Warning Receiver/Fire Control Interface Software *hc, wp*
RFIU RF Interface Unit *fu*
RFM Radio Frequency Measurement *fu*

RFM Radio Frequency Module *fu*
RFM Reactive Factor Meter *fu*
RFMU Radio Frequency Memory Unit *fu*
RFN Request For Nomenclature *fu*
RFNA Red Fuming Nitric Acid [nitric acid with a stabilizer added, an easily storable oxidizer used in rocket engines] *mb, ci.168, fu, mi, mt*
RFNM Request For Next Message *do404, ct*
RFNSI Request for New Stores Item *fu*
RFO Radio Frequency Oscillator *fu*
RFO Rework Fabrication Order *fu*
RFOG Resonance (Resonant) Fiber Optic Gyro *17, sb, st*
RFP Request For Proposal [a document] *ah, kg, ge, fu, su, sc4.11.4, 0, 8, 15, 16, 17, sb, pl, hc, af72.5.144, aw130.11.77, wp, ep, re, tb, st, wl, jdr, pbl, hg, tm, kf, jn, oe, mt*
RFP RF Processor *fu*
RFPA Request For Plan of Action *fu*
RFPD Request For Proposal Data *fu*
RFPP Repetitive Flight Plans Program *fu*
RFPT Roland Field Proficiency Trainer *hc*
RFPU Radio Frequency Processor (Processing) Unit *4, jdr*
RFQ Radio Frequency Quadrupoles *tb*
RFQ Request For [price] Quotation (Quote) [a document] *ah, su, 8, 17, sb, hc, wp, ep, tb, st, wl, jdr, pbl, tm, kf, fu, kg, jn, mt*
RFR Radio Frequency Radiation *mt*
RFS Radio Frequency Subsystem *so, jdr*
RFS Remote File Service *em, ct*
RFS Remote File Sharing *pk, kg*
RFS Remote File System *em, ct, kg*
RFS Request For Service *mt*
RFS Rubidium Frequency Standard *jdr*
RFS/ECM Radio Frequency Surveillance/Electronics Countermeasures *wp RFS*
RFSP Replacement Flight Strip Printers *fu*
RFSS Radio Frequency Subsystem *jdr*
RFSS Radio Frequency Surveillance System *wp*
RFSS Red Flag Scheduling System *ro*
RFT (Rich Text Format) {?} *pk, kg*
RFT Request For Tender [a document] *pbl, tm, fu, jn*
RFT Revisable Form Text *pk, kg*
RFTP Radio Frequency Test Position *fu*
RFTP Request For Technical Proposal *mt*
RFTT RF Test Target *fu*
RFU Radio Frequency Unit *fu, jdr, uh*
RFU Ready For Use *mt*
RFU Reserved for Future Use *pk, kg*
RFW Radio Frequency Weapons *mt, wp*
RFW Request for Waiver *fu, jdr*
RFX Reconnaissance Fighter, Experimental [tactical experimental reconnaissance aircraft] *uc, wp*
RF_PITS Radio Frequency Portable Input Terminal System *wp*
RG Reception Good *fu*
RG Reconnaissance Group [USA] *mt*
RG Red, Green *fu*
RG Reference Guide *hi*
RG Replay Group *fu*
RG Reset Gate *fu*
RGA Residual Gas Analysis *jdr*
RGA Residual Gas Analyzer *oe*
RGB Red Green Blue [color model] *em, ct, wp, fu, kg, hi*
RGC Receiver Gain Control *fu*
RGD Range Gate Deception *wp*
RGF Remote Ground Facility *uh*
RGG Rotating Gravity Gradiometer *hc*
RGM [ship launched, Guided ground-to-ground Missile] [USA] *mt*
RGON Remote Geophysical Observing Network *no*
RGP Rate Gyro Package *wp*
RGPD Range-Gated Pulse Doppler *fu*
RGPI Range-Gate Pull-In *fu*
RGPO Range-Gate Pull-Off *fu*
RGS Rate Gyro System *wp*
RGS Readiness Groups [USA] *mt*
RGS Receiving Ground Station *hy*
RGS Relay Ground Station *kf*
RGS Remote Ground Station *uh, jdr*
RGS Report Generation Subsystem *fu*
RGS-A Relay Ground Station-Atlantic *kf*
RGS-A/EGS Relay Ground Station-Atlantic/European Ground Station *kf*
RGS-P Relay Ground Station-Pacific *kf*
RGS-P/SRP Relay Ground Station-Pacific/Survivable Relay Processor *kf*
RGSE Released Ground Support Equipment *jdr*
RGT Radioisotope Thermoelectric Generator *wp*
RGT Reliability Growth Test(ing) *fu*
RGT Remote Ground Terminal *fu*
RGU Radar Graphics Unit *hc*
RGU Rate Gyro Unit *ai2*
RGWS Radar Guidance Weapon System *wp, hc*

RH Range-Height *fu*
RH Relative Humidity *fu, 0, wp, jdr, oe*
RH Right Hand *fu, w9, aw130.22.30*
RH Rockwell Hardness *fu*
RH [USA DOD designator, RH-53 is Sea Stallion] *mt*
RH-CMOS Radiation Hardened - Complimentary Metal Oxide Semiconductor *kf*
RHA Rolled, Hardened Armor *wp*
RHAW Radar Homing And Warning system [used for strikes at enemy radar sites, AN/APR-25] [USA] *fu, wp, mt*
RHAWS Radar Homing And Warning System *wp*
RHB Radar Homing Bomb *wp*
RHC Regiment d'HélicoptÈres de Combat ["combat helicopter regiment", France] *mt*
RHC Right Handed Circular [polarization] *gd2-21, 2, 4, vv*
RHCP Right Hand Circular Polarization *pbl, kf, jdr, uh, fu, oe*
RHCP Right Handed (Hand) Circularly Polarized (Polarization) *ggIII.1.13, 16, jdr*
RHD Removable Hard Disk *mt*
RHI Radar Height Indicator *mt*
RHI Range Height Indicator *fu*
RHI Risk Hazard Index *fu*
RHIMS Red Horse Information Management System [USAF] *mt*
RHM Radiation Hardened Memory (RAM) *uh, jdr, kf*
RHOGI Radar Homing Guidance Investigation *hc*
RHS Right Hand Side *ai1*
RHWR Radar Homing and Warning Receiver *mt*
RHWS Radar Homing and Warning System *mt*
RI International Relative [sunspot number] *pr*
RI Radar Input *fu*
RI Radio Interference *hi*
RI Range Instrumentation *fu*
RI Receiving Inspection *fu*
RI Redundant Item *fu*
RI Referential Integrity *pk, kg*
RI Reflective Insulation *fu*
RI Replaceable Item *fu*
RI Require Identification *fu*
RI Responsible Individual *fu*
RI Ring Indicate *pk, kg*
RI Routing Indicator *mt, fu*
RI [Hughes Aircraft Company customer for airborne target augmentor] *hc*
RI&T Receiving Inspection and Test (also RI/T) *fu*
RI/TAL Receiving Inspection/Test Activity Log *fu*
RIA Radio Immuno Assay *w9, wp*
RIA Range Insensitive Axis *tb*
RIA RDPS Interface Application *fu*
RIA [Hughes Aircraft Company customer for advanced forward area defense system] *hc*
RIACS Research Institute for Advanced Computer Science *mi*
RIAS Research Institute for Advanced Studies *wp*
RIATAS Requirements Identification And Technology Assessment Summary *wp*
RIB Radar Input Buffer *fu*
RIB Recoverable Item Breakdown *mt, fu*
RIBS Readiness In Base Service *mt, af72.5.121*
RIC Radar Integration Control *fu*
RIC Radar Intercept Control *wp*
RIC Reconnaissance Intelligence Centre [UK] *mt*
RIC Requirement Identity Code *mt*
RIC Resource Identification Code *mt*
RIC Routing Indicator Code *mt*
RICMO Radar Input and Countermeasures Officer *fu*
RICO Radar Inputs Control Officer *fu*
RICS Range Instrumentation for Control and Scoring *hc*
RID Radar Input Drum *fu*
RID Raw Instrument Data *oe*
RID Reset Inhibit Drum *fu*
RID Retrofit Installation Drawing *fu*
RID Review Item Disposition *mt*
RIDB Readiness Integrated Data Base [USA] *mt*
RIDE Reading Impact Difficulty Estimate *fu*
RIDS Readiness Improvement Data System [WWMCCS] *mt*
RIE Radar Interface Equipment *fu*
RIF Reduction In Force *mt, ci.71, wp*
RIFCA Redundant Inertial Flight Control Assembly *ai2*
RIFF Raster Image File Format *hi*
RIFF Resource Interchange File Format [Microsoft] *kg, hi*
RIFI Radio Interference Field Intensity *fu*
RIFT Reactor in Flight Test *ci.229*
RIG Radar Image Generator *wp*
RIG Radar Inputs Group *fu*
RIG Raster Image Generator *fu*
RII Request for Intelligence Information *mt*
RIIP Roll Isolated Inertial Platform *hc*
RIIXS Remote Interrogation Information Exchange Subsystem *mt*

RIL Radio Interference Level *fu*
RIL Repairable Items List *fu*
RIM Radar Input Mapper *fu*
RIM Radiant Intensity Measurement *fu*
RIM Relational Information Management *fu*
RIM Remote Installation and Maintenance [Microsoft] *pk, kg*
RIM Rotating Interface Mechanism *gp*
RIM Runway Incursion Management *aw130.22.28*
RIM [ship launched SAM, USA] *mt*
RIMCS Reparable Item Movement Control Center *wp*
RIME RelayNet International Message Exchange *kg, hi*
RIMP Remote Input Message Processor *mt*
RIMS Records Information Management System [USAF] *mt*
RIMS Replacement Inertial Measurement System *aw129.1.57*
RIMS Rescue Information Management System [MAC*] *mt*
RIMS Responsive Interdict Missile Systems *sb*
RIMS Rim Inertial Measuring System *hc*
RIMS Roll Stabilized Inertial Measurement System *-17*
RIMSTOP Retail Inventory Management and Stockage Policy *mt*
RIMU Redundant Inertial Measurement Unit *pl, ai2*
RIN Relative Intensity Noise *fu*
RIN Remote Intelligent Network *mt*
RIN Routing Instruction Note *wp*
RING_TAIL [fragmentation bomblet] *wp*
RINT Radiation Intelligence [also RADINT] [unintentional] [USA DOD] *wp, mt*
RIO Radar Intercept Officer [the second crewmember in US Navy two-seat fighters] *mt, hn48.13.5, hc*
RIOD Remote Input/Output Device *fu*
RIOT Real Time Input/Output Transducer *fu*
RIP Radar Improvement Program *hc*
RIP Radar Interface Processor *fu*
RIP Rapid Installation Procedure *fu*
RIP Raster Image Processor *kg, hi*
RIP Received Isotropic Power *pbl, jdr, kf, ah*
RIP Reconnaissance Information Point *wp*
RIP Remote Imaging Protocol *kg, hi*
RIP Report In Point *fu*
RIP Report on Individual Personnel *mt*
RIP Routing Information Protocol [Novell] *ct, em, kg, hi*
RIP/M Radar Improvement Plan - Mediterranean [also called SRAS] *fu*
RIPC Remote Information Processing Center *mt*
RIPEM Riordan's Internet Privacy Enhanced Mail *pk, kg*
RIPL Reconnaissance and Interdiction Planning Line *mt*
RIPS Range [Control Center] Integration and Processor System *hc*
RIPS Raster Image Processing System *pk, kg*
RIPS Reconnaissance Information System [JCS] *mt*
RIQAP Reduced Inspection Quality Assurance Program *fu*
RIS Radiometric Imaging System *hc*
RIS Range Information System [TFWC] *mt*
RIS Resonance Ionization Spectroscopy *wp*
RIS Retroreflector in Space [Japan, Environmental Agency] *ge, hg*
RIS [ELS facility] *bo*
RISA Radioimmunosorbent Assay *wp*
RISC Reduced Instruction Set Code *em*
RISC Reduced Instruction Set Computer *mt*
RISC Reduced Instruction Set Computer (Computing) *em, ct, aa26.12.65, aw129.10.53, is26.1.28, pf.f88.27, wp, tb, kf, fu, kg, hi*
RISD Research and acquisition Information Systems Division [AFSC*] *mt*
RISE Readiness Improvement through Systems Engineering [ESD] *wp, mt*
RISF Resupply and Replacement Intermediate Storage File *mt*
RISO Range Instrumentation Systems Office *wp*
RISOP Red Integrated Strategic Offensive Plan *mt*
RISOP Russian Integrated Strategic Operating Plan *sd, tb, kf*
RISTA Reconnaissance, Intelligence, Surveillance, and Target Acquisition *uc*
RIT Raw Input Thread *pk, kg*
RIT Readiness Initiative Team *mt*
RIT Receiving Inspection and Test *bl2-9.50, ep, jdr, tm*
RIT Red Interface Terminal *mt*
RIT Roland Institutional Trainer *fu*
RITES Radiometric Imaging Thermal Energy Sensor *hc*
RIU Radar Interface Unit *fu, wp*
RIU Remote Intercom Unit *fu*
RIU Remote Interface Unit *mt, bl3-2.27, jdr*
RIV Radio Influence Voltage *fu*
RIVER_WAGON [C-130 airborne warning capability program] *wp*
RIVET [nickname for HQ USAF programs usually for special aircraft missions] *wp*
RIVET_AMBER [Classified project, previously called Project 863] *fu*
RIVET_CARD [special C-135b SAC aircraft] *wp*
RIVET_CLAMP [special C-130E aircraft] *wp*
RIVET_DANDY [special C-135 SAC aircraft] *wp*
RIVET_DOCTOR [special mission C-130e aircraft for TAC] *wp*
RIVET_EAGLE [special mod Rf-4c for Pac] *wp*

RIVET_FLARE [C-97 training aircraft for Pac] *wp*
RIVET_GUN [to repair and return battle damaged aircraft to active inventory] *wp*
RIVET_JOINT [ELINT collector on-board Rc-135] *wp*
RIVET_REDSKIN [improved Slar for Rf-4c] *wp*
RIW Reliability Improvement Warranty *fu, wp*
RIXT Remote Information Exchange Terminal *mt, hc*
RJ Rivet Joint [as in RJ_III] *mt*
RJAF Royal Jordanian Air Force *fu*
RJE Remote Job Entry *mt, ct, em, fu, kg, hc, hd1.89.y17*
RJET Remote Job Entry Terminal *mt*
RJO/USAF [Hughes Aircraft Company customer for LRTF] *hc*
RJP Remote Job Processing *fu*
RKA Raketniy Kater [rocket cutter] [USSR] *mt*
RKA Rossiiskoye Kosmicheskoye Agentstvo [Russian Space Agency] *ai2*
RKM Radar Keyboard Multiplexer *fu*
RKO {a US communications corporation} *sc3.6.4*
RKR Raketniy Kreyser [rocket cruiser] [USSR] *mt*
RKS Remote Key Set *fu*
RKVAM Recording Kilovolt-Ammeter *fu*
RL Precipitation {Rain ?}over Land *ac7.5-12*
RL Receive Location *mt*
RL Resistance Inductance *fu*
RL Rocket Launcher *mt*
RLA Radar Line Adapter *fu*
RLA Repair Level Analysis *mt, fu, hc, wp, kf*
RLADD Radar Low Angle Drogue Delivery *fu*
RLAF Royal Laotian Air Force *mt*
RLAT Radiation Lot Acceptance Test(ing) *jdr, he, kf*
RLB Relocatable Binary Code *mt*
RLC Random Logic Controller *fu*
RLC Resistance-Inductance-Capacitance *fu*
RLCC Remote Launch Control Center *ai2*
RLCE Ringdown Line Compatibility Equipment *fu*
RLCM Remote Line Control Module *mt*
RLD (the Civil Aviation Department of the Netherlands (RLD)) *aw130.11.64*
RLD Ready to Load Date [USA DOD] *mt*
RLD Received Line Detect *pk, kg*
RLE Run Length Encoding (Encoded) *ct, kg, em, hi*
RLG Ring Laser Gyro *17, sb, hc, af72.5.81, aw130.11.c1, wp*
RLI Right-Left Indicator *fu*
RLL Run Length Limited *ct, em, kg, hi*
RLM Reichs Luftfahrt Ministerium [imperial air travel ministry] [Germany] *mt*
RLN Remote LAN Node [DCA] *kg, hi*
RLOGIN Remote Login *ct, kg*
RLP Remote Line Printer *mt*
RLPROP Real Property Inventory Accounting System [USAF] *mt*
RLRC Regional Logistic Readiness Center *mt*
RLS Radar Line of Sight *fu*
RLS Radio Lokatisonnaya Stantsiya [RADAR] [USSR] *mt*
RLSD Received Line Signal Detected *pk, kg*
RLSI Ridiculously Large Scale Integration *kg, hi*
RLSTK Rolling Stock Summary System *mt*
RLSV Reentry Launch System Vehicle *-17*
RLT Regimental Landing Team [USMC] *mt*
RLT Relocatable Target *wp*
RLTS Return to Launch Site *ws4.10, pf1.25*
RLU Remote Load Unit *fu*
RLV Reusable Launch Vehicle *ai2*
RM Deputy Commander for Resources *mt*
RM Rate Multiplier *-4*
RM Raw Material *fu*
RM Redundancy Manager *jdr*
RM Reeling Machine *fu*
RM Reference Manual *fu*
RM Reference Model *fu*
RM Reject Message *mt*
RM Relationship Manager *pbl, ah, jn*
RM Reliability Manager *fu*
RM Remedial Maintenance *fu, tb*
RM Remote Monitoring *fu*
RM Reset Mode *pk, kg*
RM Resource Maintenance *mt*
RM Resource Management (Manager) *mt, ct, em, af72.5.100, tb, jdr*
RM Risk Manager *fu, jdr*
RM Rocket Motor *jdr*
RM&D Reliability, Maintainability, and Dependability *kf*
RM&QA Redundancy Management and Quality Assurance *rm1.1.3, nd74.445.36*
RM&S Reliability, Maintainability, and Supportability *mt*
RM/EOS Range Marks/End Of Sweep *fu*
RMA Reliability, Maintainability, and Availability [also R/M/A] *fu, tb, kf, mt*

RMA Responsible Manufacturing Activity *jdr, pbl, tm, ah, jn, fu*
RMA Responsible Manufacturing Authority *hc, sc9.6.4, pl, gp, 8, lt4, ep, jdr*
RMA Return Material Authorization *pk, kg*
RMA Return Merchandise Authorization *ct, hi*
RMA Return to Manufacturer Authorization *pk, kg*
RMA Rosin Mildly Activated *fu*
RMAF Royal Malaysian Air Force *fu, hc*
RMAPS Radar Material Acquisition and Processing System *hc*
RMAS Reliability, Maintainability, Availability, and Safety *fu*
RMB Requirements Management Board *mt*
RMB Risk Management Board *fu, kf*
RMC Regional Meteorological Center *pl*
RMC Remote Multiplexer Connector *fu*
RMC Resource Management Center *mt*
RMC Returned to Military Control [USA DOD] *mt*
RMCS Reeling Machine Control System *fu*
RMCS Regional Master Control Station *fu*
RMCS Remote Monitoring and Control Station *fu*
RMCU Reeling Machine Control Unit *fu*
RMDC Radar Monitor Display Console *fu*
RMDIR Remove Directory *pk, kg*
RMDU Remote Multiplexer/Demultiplexer Unit *-4*
RME Relay Mirror Experiment *aw129.1.3, ro*
RMED Remote Manual Entry Device *fu*
RMF Resource Management Facility *tb*
RMF Resource Measurement Facility *mt*
RMG Resource Management Group *mt, af72.5.127*
RMGF Radar Mapper Gap Filler *fu*
RMI Radio Magnetic Indicator *fu*
RMI Remote Memory Interface *oe*
RMI Remote Method Invocation *em, kg*
RMI/HSI Radio Magnetic Indicator/Horizontal Situation Indicator *hc*
RMIS Reprographics Management Information System *mt*
RMIS Resource Management Information System *mt*
RMISTB Resource Management Information System Test Bed [MAC*] *mt*
RMIU Reeling Machine Interface Unit *fu*
RML Radar Microwave Link *fu, wp*
RML Remote Microwave Link *mt*
RMLR Radar Mapper, Long Range *fu*
RMLR Radar Microwave Link Repeater *wp*
RMM Radar Map Matching *fu*
RMM Read-Mostly Memory *fu*
RMM Remote Maintenance Monitoring *fu*
RMMC Remote Maintenance Monitoring and Control *fu*
RMMO Risk Measurement and Management Organization *fu*
RMMP Reliability and Maintainability Management Plan *mt, wp*
RMMP Risk Measurement and Management Plan *fu*
RMMS Remote Maintenance Monitoring (Sub)System *fu, hc*
RMMT Rail Movements Management Team [USA] *mt*
RMN Return Material Negotiation *fu*
RMNET Reuter's Money Network *hi*
RMO Requirements Management Office *mt*
RMO Resource Management Office (mt, dt)
RMO Rubidium Master Oscillator *jdr*
RMON Remote Monitor (Monitoring) *kg, hi*
RMOS Refractory Metal Oxide Semiconductor *fu*
RMP Rate Measuring Package *ac2.8-2*
RMP Remote Maintenance Processor [IBM] *pk, kg*
RMP Requirements Management Plan *mt*
RMP Reserve Mobility Plan [HQ USAF] *mt*
RMP Resource Management Plan *mt*
RMP Risk Management Plan *kf*
RMPE Reliability / Maintainability Project Engineer *fu*
RMR Reliability, Maintainability, and Recovery *mt*
RMR Request for Material Review *oe*
RMS Radar Manual System *fu*
RMS Radiation and Meteoroid Satellite *ci.969*
RMS Reconnaissance Management System *hc*
RMS Record Management Service *fu, ct, kg*
RMS Recording and Monitoring System *fu*
RMS Redundancy Management Subsystem *jdr*
RMS Reliability, Maintainability, and Safety *fu*
RMS Reliability, Maintainability, Supportability *wp*
RMS Remote Maintenance Subsystem *fu*
RMS Remote Manipulator System *mi, ci.467, aa26.11.20, ws4.10, wl, jdr, mb*
RMS Remote Monitoring Subsystem *fu*
RMS Resource Management System *mt*
RMS Resources Management Study *dt*
RMS Root Mean Square *mt, fu, kg, he, sp664, 2, 5, 17, sb, es55.6, wp, so, st, kf, vv, oe, ai1*
RMSC Remote Monitoring Subsystem Concentrator *fu*
RMSDS Reserve Merchant Ship Defense System *hc*
RMSE Root Mean Square Error *dm*

RMSM Root Mean Square Misconvergence *hi*
RMSP Root Mean Square Pincushioning *hi*
RMT Radio Monitor Technician *fu*
RMTC Regional Military Training Center *wp*
RMTP Reliable Multicast Tranport Protocol *em*
RMU Radar Monitoring Unit *hc*
RMW Read Modify Write *kg, hi*
RMWG Risk Management Working Group *kf*
RN Radio Navigation *fu*
RN Read News [Internet] *kg, hi*
RN Reference Noise *fu*
RN Regency Net *mt*
RN Relay Node *mt*
RN Revision Notice [a document] *ah, fu, ep, jj, jdr, pbl, tm, jn, oe*
RN Royal Navy [United Kingdom] *12, w9, af72.5.58*
RNAF Royal Norwegian Air Force *mt*
RNAS Royal Naval Air Service [UK] *mt*
RNAS Royal Navy Air Station [United Kingdom] *mt, 12*
RNAV Area Navigation *fu*
RNAV [navigation system that makes use of airway beacons] *mt*
RNCC Reference Number Category Code *fu*
RNDES Range NTDS Display Emulator Subsystem *fu*
RNG Random Number Generator *pk, kg*
RNG Range *mt*
RNG Ranging *bl3-1.167*
RNGC Revised New General Catalog *mi*
RNGM Radar Navigation Ground Map *hc*
RNHF Royal Navy Historical Flight *mt*
RNII [Jet Scientific Research Institute] *-13*
RNIU Reverse NASCOM Interface Unit *-4*
RNO Regional Nuclear Options *mt*
RNOAF Royal Norwegian Air Force *fu*
RNP Regional Network Provider *kg, hi*
RNP Remote Network Printer *mt*
RNP Remote Network Processor *mt*
RNR Receive Not Ready *do378, fu*
RNS Radar Netting System *fu, hc*
RNS Residue Number System *fu*
RNS Reusable Nuclear Shuttle *ci.972*
RNSC Reference Number Status Code *fu*
RNSP Residual Nuclear Strike Plan *mt*
RNTDS Restructured Naval Tactical Data System *fu*
RNVC Reference Number Variation Code *fu*
RNX Restricted Numeric Exchange *hi*
RNZAF Royal New Zealand Air Force [also NZRAF] *mt, w9*
RO Precipitation {Rain ?} over Ocean *ac7.5-12*
RO Radio Operator *fu*
RO Receive Only *fu, wp, ct, em*
RO Relay Optics *oe*
RO Remote Operations *fu*
RO Research Objective *mt, wp*
RO Reverse Osmosis *wp*
RO/RO Roll On / Roll Off [ship] *ph, mt*
ROA Restricted Operation Area *fu*
ROA Return On Assets *aw130.22.66*
ROADRUNNER [a MICOM laser system] *hc*
ROAMA Rome Air Material Area *ro*
ROB Radar Order of Battle *fu, wp*
ROB Re Order Buffer [Intel] *kg, hi*
ROB Red Order of Battle *tb*
ROBAT Robotic Obstacle Breaching Assault Tank [mine clearing] *hc, dh*
ROBCO Readiness Objective Code [UNITREP] *mt*
ROBO Remote Office/Branch Office *nu*
ROBY Results, Output, Benefits, or Yield *mt*
ROC Rate Of Climb *mt*
ROC Receiver Operating Characteristics *fu*
ROC Receiver Operational Capability *fu*
ROC Reconnaissance Operations Center *mt*
ROC Regional Operating Center [NATO] *mt, fu*
ROC Republic Of China *hc*
ROC Required Operational Capability *mt, fu, sd, hc, wp, tb, st*
ROC Reserve Officer Candidate [USA] *mt*
ROCC Region(al) Operations Control Center [NORAD] *fu, mt, 12, hc, hn41.18.1, af72.5.74, hs, wp, tb*
ROCLANT Regional Operations Center Atlantic [Norfolk] *mt*
ROCN Republic Of China Navy *fu*
ROCOZ Rocket Ozone Correlation Program *hy*
ROCS Range Only Correlation System *wp*
ROCS Range Operations Control System *hc*
ROD Report Of Discrepancy *wp, fu*
ROD Required Operational Date *mt, fu*
ROE Rules Of Engagement [USA DOD] *fu, tb, mt*
ROF Rate Of Fire *tb*
ROF Royal Ordnance Factory [United Kingdom] *-12*
ROF Runway Operating Facility *fu*
ROFL Rolling On Floor Laughing *du*

ROFT Rapid Optical Fabrication Techniques *st*
ROH Regular Overhaul *mt, fu, hc*
ROH Restricted Overhaul *mt*
ROHAS Robotized Wire-Harness Assembly System *fu*
ROI Return On Investment *mt, dh, w9, jdr, fu, ls4, 5, 27*
ROIC Readout Integrated Circuit *kf*
ROID Report Of Item Discrepancy *fu*
ROJ Range On Jamming *fu*
ROK Republic Of Korea *mt, ph, 10, 12, wp, fu*
ROKA Republic Of Korea Army *mt, fu*
ROKAF Republic Of Korea Air Force [South Korea] *fu, mt*
ROKMC Republic Of Korea Marine Corps *mt*
ROKUS Republic Of Korea / United States *mt*
ROL Rotate Left *pk, kg*
ROLAND [a MICOM European missile] *hc*
ROLAP Relational OnLine Analytical Processing *em, kg*
ROM Read Only Memory *gg161, 6, 8, 16, pl, hc, w9, nd74.445.5, lt, wp, ep, tb, st, jdr, pbl, tm, kf, kg, ct, em, su, uh, ah, fu, hi, oe, ai1*
ROM Read Only Memory "rom" *mt*
ROM Rough Order of Magnitude *hn43.23.8, 8, hc, go3.5.1, wp, tb, st, jdr, pbl, tm, kf, fu, ah, jn, oe*
ROMAC Range Operation Monitoring and Control (System) *fu*
ROMAD Radio Operator, Maintenance and Driver *mt*
ROMANS Range Only Multiple Aircraft Navigation System *ro*
ROMAN_CANDLE [jelled fuel cartridge jettisoned for infrared decoy] *wp*
ROME Remotely Operated Mobile Excavator *wp*
ROMS Remote Ocean Surface Measuring System *un*
RON Remain Over Night *mt, wp*
RON Research Octane Number *wp*
RONA Return on Net Assets [financial] *ah, pbl, jdr, tm, jn*
RONT Radar On Target [also ROT] *wp*
ROOM Real Time Object-Oriented Modeling *pk, kg*
ROOT Relaxation Oscillator, Optically Tuned *fu*
ROP Raster Operation *pk, kg*
ROP Receive Only Printer *mt, fu*
ROP Recommended Operating Procedure *uh, jdr, jn*
ROP Reorder Point *wp, fu*
ROP Report Out Point *fu*
ROP RISC Operation *kg, hi*
ROP Run Of Paper *w9*
ROPACI Rocket Propulsion Analysis Capability Implementation *hc*
ROPES Route and Penetration Evaluation *mt*
ROPP Receive Only Page Printer *fu*
ROPS Readiness, Operations and Planning Systems Division [J-3] *mt*
ROR Range Only Radar *mt, fu, wp*
ROR Rate Of Return *wp*
ROR Repair of Repairables *fu*
ROR ROSAT Observation Request *ns*
ROR Rotate Right *pk, kg*
RORO Read Only Read Only *hi*
RORO Roll On, Roll Off *mt*
RORSAT Radar Ocean Reconnaissance Satellite [Soviet] *mt, fu, 14, sd, aj89.9.40, jmj, wp, tb*
RORW Read Only Read and Write *hi*
ROS Read Only Storage *kg, hi*
ROS Real time Operating System *fu*
ROS Reduced Operating Status *wp*
ROS Reduced Operational Status [USA DOD] *mt*
ROS Resident Operating System *hi*
ROS [USA STDN ground station in] Rosman, [North Carolina] *ac2.7-10, 4*
ROSAT ROentgen SATellite:(mi, cu)
ROSE Remote Optical Sensing of the Environment *wp*
ROSS RAND Object [Oriented] Simulation System *tb, mt*
ROT Radar On Target *fu, wp*
ROT Range Only Track *fu*
ROT Running Object Table *pk, kg*
ROTAB Rotating Table *msm*
ROTC Reserve Officers' Training Corps [USA] *mt, hn43.5.3, w9, af72.5.14, wp*
ROTH Relocatable Over the Horizon [radar] *dh*
ROTHR Relocatable Over The Horizon Radar [US Navy] *mt*
ROTI Recording Optical Tracking Instrument *fu*
ROTP Receive-Only Tape Perforator *fu*
ROTS Range On Target Signal *fu*
ROTV Reusable Orbital Transfer Vehicle *wp*
ROUS Rodents Of Unusual Size [mythical {?}] *mi*
ROV (an optimized ROV intervention system) *aa27.4.b11*
ROVER Remotely Operated Vehicle for Emplacement and Reconnaissance *wp*
ROVER Rocket Vehicle Research *ci.896*
ROVN Received on Voucher Number *fu*
ROWPS Reverse Osmosis Water Purification System *wp*
ROWPU Reverse Osmosis Water Purification Unit *wp*
ROWS Radar Ocean Wave Spectrometer *hy*
ROZ Restricted Operating Zone *mt*

RP Radar Post *fu*
RP Radar Processor *fu*
RP Receiver Processor *fu*
RP Recommended Procedure *nt*
RP Recurrence Period *fu*
RP Reference Publication *wl*
RP Reference Pulse *fu*
RP Relative Performance *mt*
RP Relay Platform *-17*
RP Reliability Plan *-17*
RP Rendezvous Point *fu*
RP Reporting Post *mt, fu*
RP Response Planning *mt*
RP Restoration Priority *mt*
RP Restricted Project *fu*
RP Rocket Projectile [UK] *mt*
RP Rocket Propellant *wp, wl, ai2*
RP Rocket Propelled *mt*
RP Route Package [USA] *mt*
RP&E Receive, Process and Exploit *wp*
RP-1 Rocket Propellant [high-grade kerosene] *ai2*
RPA Radar Performance Analyzer *fu*
RPA Replacement Part Analysis *fu*
RPA Route Planning Aid *fu*
RPA {instrument that is part of Mars orbiter} *ma40*
RPAM Recruiting Policy Analysis Model *mt*
RPBM Real Property Building Manager *wp*
RPC Remote Position Control *fu*
RPC Remote Power Control *fu*
RPC Remote Procedure Call *ct, em, kg, hi*
RPC Repair Parts Cost *wp*
RPC Reparable Processing Center *mt*
RPC/XDR Remote Procedure Call/External Data Reference *tb*
RPCA Remotely Programmable Conference Arranger *mt*
RPD Radar Planning Device *fu*
RPD Renewal Parts Data *fu*
RPD Revolutions Per Day *id4142-10/831*
RPDM Reflector Positioning Deployment Mechanism *jn*
RPDS Radar Data Processing System *fu*
RPE Registered Professional Engineer *wp*
RPE Reliability Project Engineer *fu*
RPE Rocket Propulsion Establishment *crl*
RPED Receive, Process, Exploit, and Dissemination *hc*
RPF Radiometer Performance Factor *fu*
RPF Real Property Facilities *fu*
RPF Remote Processor Function *mt*
RPF Report Format *mt*
RPF Report Program Facility *mt*
RPFO Resupply Planning Factors Office *mt*
RPG Radar Processor Group *fu*
RPG Rapid Isotope Power Generator *wp*
RPG Receiver Processing Group *mt*
RPG Report Program Generator [a programming language] *kg, fu, wp*
RPG Research Planning Group *wp*
RPG Research Planning Guide *mt*
RPG Role Playing Game *ct*
RPG Rounds Per Gun *mt*
RPI Radar Precipitation Integrator *fu*
RPI Rated Position Identifier *fu*
RPI Rensselaer Polytechnic Institute *ag, is26.1.87*
RPI Representative Present Implementation *fu*
RPI Restoration Priority Indication *mt*
RPIE Real Property Installed Equipment *mt, tb, fu*
RPIE Real Property Inventory Account System *mt*
RPIFS Real Property Industrial Fund System [USAF] *mt*
RPJ Random Pulse Jamming *wp*
RPL Remote Program Load *em*
RPL Repair Parts List *wp*
RPL Requested Privilege Level *pk, kg*
RPL Resident Programming Language *pk, kg*
RPL Rocket Propulsion Laboratory [USAF] *17, sb, tb*
RPL RPV-IT Program Library *fu*
RPM Radar Performance Monitor *fu*
RPM Reflector Positioner (Positioning) Mechanism *he, cl, jn*
RPM Reliability Program Management *fu*
RPM Remote Performance Monitoring *fu*
RPM Revenue Passenger Mile *aw129.21.95*
RPM Revolutions Per Minute *fu, gp, 16, w9, af72.5.174, sn2.4, rm8.1.4, wp, tb, jdr, vv, jn, gp, 16, w9, af72.5.174, sn2.4, rm8.1.4, wp, tb, jdr, vv, jn, hi, oe, mt*
RPMA/MIS Real Property Maintenance Activities/MIS [US Navy] *mt*
RPMB Remotely Piloted Mini Blimp [a small airship for surveillance purposes] [USA] *mt*
RPMC Remote Performance Monitoring Control *fu*
RPN Radar Post North (NATO) *fu*
RPN Real Page Number *pk, kg*

RPN Reverse Polish Notation *kg, hi*
RPO Resident Project Office *fu*
RPO Responsible Property Officer *mt*
RPP Red Patch Panel *mt*
RPPP Repair Parts Program Plan *fu*
RPPROM Reprogrammable Programmable Read Only Memory *kg, hi*
RPQ Request for Price Quotation *fu, kg*
RPR Reduced Packet Radio *fu*
RPRA Remotely Piloted Research Aircraft [teleguided aircraft to test the oblique wing concept] [USA] *mt*
RPRINTER Remote Printer [NetWare] *pk, kg*
RPRV Remotely Piloted Research Vehicle *mt, wp*
RPS Rapid Prototyping System *fu*
RPS Redundant Power Supply *gp*
RPS Remote Processing Station *mt*
RPS Revolutions Per Second *fu, w9, wp, hi*
RPSL Repair Parts Stockage List *fu*
RPSP Real Time Programmable Signal Processor *fu*
RPSTL Repair Parts and Special Tools List *fu*
RPT Receive Plain Text *fu*
RPT Resident Provisioning Team *fu*
RPTF Red Patch and Test Facility *mt*
RPTOR Reporting Organization *mt*
RPU Remotely Piloted Vehicle *oe*
RPV Remotely Piloted Vehicle *mt, fu, ci.74, 14, aw129.1.2, hc, aa26.11.9, af72.5.142, wp, tb, st*
RPV-IT Remotely Piloted Vehicle - Institutional Trainer *fu*
RQ Ready Queue *fu*
RQ Relational Query *mt, hi*
RQ Repeat Request *do256*
RQA Reliability and Quality Assurance [also R&QA] *gp, 16*
RQBE Relational Query By Example [Fox Pro] *kg, hi*
RQC Radar Quality Control *fu*
RQE Reply Queue Element *hi*
RQF Rescue Flight [USA] *mt*
RQG Rescue Group [USA] *mt*
RQS Rescue Squadron [USA] *mt*
RQT Reliability Qualification Test(ing) *fu, hc, wp*
RQUERY RISOP Query *mt*
RQW Rescue Wing [USA] *mt*
RR Radar Regiment *fu*
RR Rapid Rectilinear *fu*
RR Rapid Response *fu*
RR Readiness Reports *mt*
RR Readiness Review *pl*
RR Real Reality *kg, hi*
RR Receive Ready *fu, do378*
RR Receiving Report *ep, fu*
RR Recoilless Rifle *wp*
RR Recurrence Rate *fu*
RR Relief Radii *fu*
RR Remove and Replace *wp*
RR Report(ing) Responsibility *fu, pm*
RR Requirements Review *mt*
RR Respiration Rate *wp*
RR Revision Record *fu*
RR Rocket Research *id4133.10/92*
RRA Regular Reports of Audit *mt*
RRC Radio Relay Control *fu*
RRC Rapid Reaction Corps [NATO] *mt*
RRC Relative Rate Controller *-8*
RRC Remote Rekey Controller *fu*
RRD Radar Resolved Doppler *fu*
RRDC Risk Reduction Data Collection [a SBKEWS phase I task] *-17*
RRDCCC Rapid Reaction Deployable Command and Control Center [also R2DCCC] *mt*
RRE Remote Rekey Equipment *fu*
RRF Rapid Reaction Force [NATO] *mt*
RRF Ready Reserve Fleet *mt*
RRF Ready Reserve Force [USA DOD] *ph, wp, mt*
RRF Receive Reject Filter *jn*
RRF Regional Relay / Radio Facility *mt*
RRF Regional Repair Facility *fu*
RRG Requirements Review Group [USAF] *kf, mt*
RRG Research Review Group *wp*
RRH Relay Radio Hardware *jp*
RRH Remote Radar Head *fu*
RRI Range Rate Indicator *fu*
RRIP Required Receiver Input Power *jdr*
RRIS Remote Radar Integration Station *fu, wp*
RRITA Rapid Response Intra Theater Airlifter [USA] *mt*
RRM Rapid Relational Modeling *tb*
RRM Red Resource Monitoring *mt*
RRM Relative Rate Mode *-3*
RRO Random Roundoff *fu*
RRO Remote Readout *fu*

RRP Reconnaissance Reference Point *mt*
RRP Remote Radar Post *fu*
RRP Remote Reporting Post *fu*
RRP RSAF Readiness Plan *fu*
RRPL Recommended Repair Parts List *tb*
RRPP Rapid Retargeting and Pointing Platform *tb*
RRR Exceedingly Rare *wp*
RRR Range and Range Rate [also R&RR] *wp*
RRR Range Readiness Review *jn*
RRR Rapid Runway Repair *mt, af72.5.103, wp*
RRR Reduced Residual Radiation *wp*
RRR Remove, Repair and Replace *mt, wp*
RRR Resistor Reactor Rectifier *wp*
RRR Retrieval, Repair, and Refurbishment *-16*
RRR Rework/Retest Request *fu*
RRR Risk Reduction Report *fu*
RRRV Rate of Rise of Recovery Voltage *fu*
RRS Radio Relay Site *mt*
RRS Readiness Reporting System [FEMA] *mt*
RRS Ready Reserve Status *mt*
RRS Remote Radiation Sensor (Signal) *jdr, ric*
RRS Resupply Requirement System [HQ USAF] *mt*
RRSB Record/Reproduce Switchboard *fu*
RRT Radar Readiness Technologies *hc*
RRT Relative Rate Toggle *gp, 8a*
RRU Record/Reproduce Unit *fu*
RRV Robotic Research Vehicle *hc*
RRW Relative Risk Weighting *kf*
RRWDS Radar Remote Weather Display System *fu*
RS Radiated Susceptibility *fu*
RS Range Scale *fu*
RS Range Selector *fu*
RS Range Surveillance *fu*
RS Range System *oe*
RS Range, Slant *tb*
RS Ranger Subsystem *hc*
RS Receive/Store *fu*
RS Recommended Standard *wp, kg*
RS Reconfiguration Subsystem *fu*
RS Reconnaissance and Strike *wp*
RS Reconnaissance Squadron [USA] *mt*
RS Record Separator *fu, do242, kg*
RS Red Switch *mt*
RS Reed-Solomon *fu*
RS Remoting Subsystem *mt*
RS Remove Strip *fu*
RS Request to Send *kg, hi*
RS Research Summary *0*
RS [an infrared] Reconnaissance System *wp*
RS&I Ready Stock and Issue *fu*
RSA Radar Signature Analysis *wp*
RSA Rate Sensor Assembly *-8*
RSA Responsible Specification Activity *fu*
RSA Rivest, Shamir and Adleman [encryption] *ct, em, hi*
RSA [included in the RSA spacecraft to STS Interface Control Doc] *sa122474.5*
RSAC RAND Strategy Assessment Center *mt*
RSAC Recreational Software Advisory Council *pk, kg*
RSACG Royal Saudi Arabian Coast Guard *fu*
RSADS RSA *Data Security*
RSAF Royal Saudi Air Force [Saudi Arabia] *fu, kf, mt*
RSAFF Royal Saudi Arabian Frontier Force *fu*
RSAM Relative Sequential Access Method *hi*
RSAS RAND Strategic Assessment System *mt*
RSC Radar Signal Converter *fu*
RSC Radio (Radar) Set Control *fu*
RSC Runway Surface Condition *mt*
RSCAAL Remote Sensing Chemical Agent Alarm [chemical warfare] *wp*
RSCP Radar Signal and Control Processor *fu*
RSCS Rate Stabilization Control System *ci.349*
RSCS Remote Spooling Communications System *pk, kg*
RSD Route Server Daemon *pk, kg*
RSDM Radar System Design Methodology *fu*
RSDT Remote Station Data Terminals *wp*
RSDW Request Single Data Word *fu*
RSE Radar Signal, ECCM *hc*
RSE Raman Scattering Experiment *hc*
RSE Rotor State Estimator *-3*
RSER Reentry System Evaluation Radar *fu*
RSEXEC Resource Sharing Executive *ct, em*
RSF Refresh Scrub Flip-Flop *fu*
RSFP Rivet Switch Follow on Program *mt*
RSFSR Russian Soviet Federated Socialist Republic *w9*
RSG Radar Systems Group [Hughes Aircraft Company] *fu, hc, hq5.3.5, sc2.7.1, 8, 15, 16, sb, ep*
RSG Recording Specification Generator *fu*

RSGF [a truck-pulled item of Pershing-II ground support equipment] *jmj*
RSGMD Radar Systems Group Manufacturing Division *hn44.18.5, 8*
RSGMM Reduced Sensor Geometric Math Model *pl*
RSH Remote SHell *em, ct, kg*
RSH Restricted SHell *em, ct*
RSHCLK Receive Shaped Clock (from the KG Synchronizer) *fu*
RSI Radar Sensitivity Improvement *wp*
RSI Radiation Status Indicator *fu*
RSI Rainbow Satellite, Incorporated *sc3.5.7*
RSI Rationalization, Standardization, and Interoperability [NATO] *mt, hc, wp, st*
RSI Ready Stock and Issue *fu*
RSI Reflected Signal Indicator *fu*
RSI Reusable Surface Insulation *hc*
RSIC Radiation Shielding Information Database [DOE] *mt*
RSIC Redstone Scientific Information Center *st*
RSIL Radar Subsystem Integration Lab *fu*
RSIP Radar System Improvement Program *af72.5.139*
RSIP Reusable Software Implementation Program *fu*
RSL Received Signal Level *fu*
RSL Request-and-Status Link *pk, kg*
RSL Requirements Specifications Language *tb*
RSL Requirements Statement Language *fu*
RSLC Red Signal Line Conducted *fu*
RSLF Royal Saudi Land Forces *fu*
RSLI Reply Sidelobe Inhibit *fu*
RSLP Reentry System Launch Program *17, sb*
RSLP Rocket Systems Launch Program *ai2*
RSLS Receiving Sidelobe Suppression *fu*
RSLV Reentry Systems Launch Vehicle *st*
RSM Real Storage Management *hi*
RSM Removable Storage Media *fu*
RSM Rubidium [master oscillator] Substitution Module *jdr, ric*
RSM {an organization within Hughes Aircraft Company} *hn50.8.5*
RSMA Regional Satellite Monitoring Agency *-14*
RSMT Resident Subcontract Management Team *fu*
RSN Radio SuperNova *mi*
RSN Real Soon Now [vaporware] *mi, du*
RSNF Royal Saudi Naval Forces *fu*
RSNS Registration Services Network Solutions *hi*
RSO Reconnaissance System Officer *wp*
RSO Reconnaissance System Operator *mt*
RSO Resident Space Object *5, 17, sb, jb, cp, kf*
RSO Runway Supervisory Officer *mt, aw130.11.71*
RSOF Required Special Optional Features *mt*
RSP Radar Signal Processor *fu*
RSP Rapid Solidification Processing *hc*
RSP Red Switch Program *mt*
RSP Render Safe Procedure *mt, fu*
RSP Required Space Character *pk, kg*
RSPL Recommended Spare Parts List *mt, fu*
RSPP Remote Sensing Preparatory Program *ac3.7-35*
RSPS Requantization and Sensor Performance Summing *fu*
RSPX Remote Sequenced Packet Exchange *pk, kg*
RSR Radar Service Request *fu*
RSR Rain Scatterometer and Radiometer *hg, ge*
RSR Resource Status Review *oe*
RSRC Resource System Review Committee *mt*
RSRE Royal Signals and Radar Establishment [UK] *mt*
RSRM Redesigned Solid Rocket Motor *ai2*
RSRP Radar for Southern Regions and Portugal *hc, go2.3.2.2, fu*
RSS Radar Scenario Simulator *fu*
RSS Radio Science Subsystem [a candidate facility instrument] *so*
RSS Radio Set Simulator *fu*
RSS Ready Supply Store *fu*
RSS Real Time Support System *fu*
RSS Relaxed Static Stability *mt*
RSS Reliability, Supportability, and Survivability *mt*
RSS Remote Slave Station *fu*
RSS Remote Subsystem *fu*
RSS Root Sum of Squares (Square Sum) *fu, jn, he, uh, kf, vv, wp, 16, sb, 17, gp, st, jdr, oe, ai1*
RSS Rosette Scan Seeker *wp*
RSS Rotating Service Structure *gg190, 4, 9, lt, jdr, oe, ai2*
RSS Routing and Switching System *hi*
RSS Runway Supervisory Sets *mt*
RSSC Regional System Support Center *fu*
RSSC Routine Support Schedule Changes *-4*
RSSI Receive Signal Strength Indicator *jdr*
RSSPS Remote Sensor System for Physical Security *hc*
RSSU Radar Signal Simulation Unit *fu*
RSSU Reentry System Substitute Unit *fu*
RST Radar Start *fu*
RST Remote Switching Terminal *mt*
RSTA Reconnaissance, Surveillance, and Target Acquisition [also RESTA] *mt, ph, nd74.445.29, wp*

RSTAA Reconnaissance, Surveillance, and Target Acquisition Aircraft *wp*
RSTMM Reduced Sensor Thermal Math Model *pl*
RSTN Radio Solar Telescopic (Telescope) Network *no, mt*
RSTN Regional Seismic Test Network *-12*
RSTS Resource Sharing and Time Sharing [Digital] *kg, hi*
RSTS/E Resource Sharing Time Sharing System / Extended *mt*
RSTU Radar Self-Test Unit *fu*
RSU Remote Station Unit *fu*
RSU Remote Switching Units *mt*
RSU Runway Supervisory Unit *mt*
RSVP Resource Reservation Protocol *ct, em, kg*
RSX Real time resource Sharing Executive *kg, hi*
RT Radar Tracking *fu*
RT Radar Trigger *fu*
RT Radio / Telephone *fu, w9*
RT Range Tracking *fu*
RT Real Time *fu, ct, em, kg, ac3.7-33, 0, 9, hi*
RT Receive Timing *-4*
RT Receiver/ Transmitter *fu*
RT Relocatable Target [also RLT] *mt, wp*
RT Relocatable Terminal *kf*
RT Remote Terminal *fu, he, wp*
RT Repeat Time *tb*
RT Report Test *mt*
RT Report Type *fu*
RT Reydovoy Tralshchik ["roadstead minesweeper"] [USSR] *mt*
RT RISC Technology *kg, hi*
RT Room Temperature *w9, jdr*
RT Run Time *pk, kg*
RTA Real Time Attitude *oe*
RTA Records Transfer Authorization *id3/18/83*
RTA Remote Technical Advisor *jr*
RTA Request for Technical Action *fu*
RTA Residual Threat Assessment *mt*
RTA Responsible Test Agency *wp*
RTA Responsive Threat Assessment *hc*
RTAC Real Time Adaptive Control *mt, fu*
RTACS Real Time Adaptive Central System *mt, hc*
RTADS Real Time Attitude Display System *oe*
RTADS Royal Thai Air Defense System *fu, hc*
RTAF Royal Thai Air Force *fu*
RTAM Remote Telecommunications Access Method *hi*
RTAM Remote Terminal Access Method *kg, hi*
RTAPS Real Time Analysis Processing System *hc*
RTAS Real Time Advisory System *mt*
RTASC Real Time Air/Ground Advisory Control *hc*
RTASS Remote Tactical Airborne SIGINT System [ESC*] *mt*
RTASTK (routine situation report (software program)) *jb*
RTB Return To Base [USA DOD] *fu, wp, mt*
RTBS Radar Test Bench Set *fu*
RTBTL Radar Tracking and Beacon Tracking Level *fu*
RTC Radiometric Test Chamber *kf*
RTC Real Time Channel *mt*
RTC Real Time Clock *fu, em, uh, ct, kg, wp, hi*
RTC Real Time Control [command] *fu, kf*
RTC Remote Terminal Concentrator *mt*
RTCC Real Time Computer Complex *ci.954*
RTCC Real Time Mission Control *17, sb*
RTCG Real Time Control Group *fu*
RTCI Real Time Clock Interrupt *uh, jn*
RTCMM Requirement Tracking Configuration Management Matrix *wp*
RTCON Real Time Connect *mt*
RTCP Real Time Transport Control Protocol *ct, em*
RTCU Real Time Control Unit *fu*
RTCU Remote Telemetry and Command Unit *jr, he, kf*
RTCU Remote Terminal Control Unit *fu*
RTCU/SCP Remote Telemetry and Command Unit/Spacecraft Processor *he*
RTD Radioisotope Thermionic Diode *0*
RTD Range Time Decoder *fu*
RTD Rapid Transit District [Southern California] *sc2.8.4*
RTD Research and Technique Development *hc*
RTD Resistance Temperature Detector *fu*
RTD Returns To Duty *mt*
RTDA Records Transfer and Disposal Authorization *id3/18/83*
RTDD Real Time Data Distribution *fu*
RTDD Remote Timing and Data Distribution *fu*
RTDDC Real Time Digital Data Correction *fu*
RTDL Read Tegas Data Language *fu*
RTDM Real Time Data Migration *pk, kg*
RTDS Real Time Display System *hc, wp*
RTDU Real Time Data Unit *gp*
RTE Real Time Environment *fu*
RTE Real Time Executive *fu, tb*
RTE Remote Terminal Emulator *mt, wp*

RTE Remote Terminal Equipment *wp*
RTE Return-To-Enable *fu*
RTE Run Time Environment *mt*
RTE Run-to-Enable *fu*
RTECS Registry of Toxic Effects of Chemical Substances *wp*
RTEL Reverse Telnet [Internet] *kg, hi*
RTEM Radar Tracking Error Measurement *fu, wp*
RTF Radar Training Facility *fu*
RTF Radio Telephone *mt*
RTF Readiness Training Facility *fu*
RTF Reconnaissance Technical Flight [USAF] *mt*
RTF Remote Task Force *mt*
RTF Remote Tracking Facility *kf*
RTF Rich Text Format [Microsoft] *ct, kg, hi*
RTG Radioactive Thermal *jr*
RTG Radioisotope Thermoelectric (Thermal) Generator [also RGT] *jr, mi, wp, ci.897, so, ce, re, jdr, mb*
RTG Reconnaissance Technical Group *mt*
RTG Remote Tracking Ground-station *kf*
RTG [ELS storage building] *bo*
RTGP Real Time GOES Processor *oe*
RTH Regional Telecommunication Hub *pl*
RTI Real Time Interface *he, hi*
RTI Real Time Interrupt *uh, gp, jdr*
RTI Relational Technology, Incorporated *hd1.89.y18*
RTI Remote Terminal Interface *aw130.22.115*
RTI Routing Indicator *mt*
RTIC Radar Target Imaging and Classification *hc*
RTIF RAMP Test and Integration Facility *mt*
RTIP Radar Target Identification Point *fu*
RTIP Real Time Interactive Processor *mt*
RTL Radar Threshold Limit *fu*
RTL Register Transfer Language *pk, kg*
RTL Register Transfer Level *fu, kg*
RTL Research and Technology Laboratories [of AVSCOM] *tc*
RTL Resistor Transfer Logic *mt*
RTL Resistor Transistor Logic *kg, fu, 6, is26.1.76v*
RTL Right To Left *pk, kg*
RTL Run Time Library *ct, em, kg, hi*
RTLS Return To Launch Site [Shuttle abort plan] *mi, ci.607, 9, mb*
RTM Radar Target Material *fu*
RTM Rapid Tuning Magnetron *fu*
RTM Real Time Monitor *fu, wp, hi*
RTM Receiver / Transmitter Modulator *fu*
RTM Requirements Traceability Matrix *mt, hc, fu*
RTM Response Time Measurement *hc*
RTM Response Time Monitor *pk, kg*
RTM Running Time Meter *fu*
RTM Running Time Monitor *oe*
RTM Runtime Manager [Borland] *pk, kg*
RTMP Remote Terminal with Multi-Protocol *aw130.22.115*
RTMP Routing Table Maintenance Protocol *em, kg*
RTO Responsible Test Organization *mt, wp*
RTOC Rear Area Tactical Operations Center *mt*
RTOD Random Time Of Day *fu*
RTOK Retest OK *wp*
RTOL Reduced Takeoff and Landing *wp*
RTOM Real Time Option Module *fu*
RTOS Real Time Operating System [MAC] *mt, fu, kg, em, jdr, hi*
RTOT Range Track On Target *fu*
RTOV Real Time Operation Validation *fu*
RTP Range Time Processor *jn*
RTP Rapid Transport Protocol *pk, kg*
RTP Real Time Processing *hc*
RTP Real Time Protocol *pk, kg*
RTP Real Time Transport Protocol *ct, em*
RTP Request for Technical Proposal *fu*
RTPF Real Time Processing Facility *ac3.7-33*
RTPJH Real Time PJH *fu*
RTPLRS Real Time PLRS *fu*
RTPS Remote Tape Processing System *mt*
RTQC Real Time Quality Control *fu*
RTQM Requirements Traceability Qualification Matrix *tb*
RTR Real Time Raster *wp*
RTR Remote Terminal with internal RAM *aw130.22.115*
RTR Remote Transmitter/ Receiver *fu*
RTR Requirement Testability Report *fu*
RTR Resident Technical Representative *fu*
RTR Rotating Threat Recognizer *fu*
RTS Radar Terrain Sensor *hc*
RTS Radar Test System *hn43.17.8, hc*
RTS Radar Tracking Station *fu*
RTS Radar Tracking System *fu*
RTS Range Tracking Station *5, sb*
RTS Ready To Send *mt, fu*
RTS Real Time System *hi*

RTS Reconnaissance Technical Squadron *mt*
RTS Remote Takeover System *pk, kg*
RTS Remote Terminal for Stores *aw130.22.115*
RTS Remote Terminal System *mt*
RTS Remote Tracking Station [of AFSCF] *ric, hc, 4, 16, jb, dm, kf, jdr, uh*
RTS Remote Transmission Station *go3.3*
RTS Request to Send *fu, do202, kf, ct, em*
RTS Return to Supplier *fu*
RTS-II Radar Tracking Station Type II *fu*
RTS/CTS Request To Send / Clear To Send *hi*
RTSP Real Time Signal Processor *fu*
RTSP Real Time Streaming Protocol *ct, em, kg*
RTSS Red Telephone Switching System *aw129.14.103*
RTSS Request To Send {?} [usually RTS] *pk, kg*
RTSS Reserve Training Support System [US Navy] *mt*
RTSS Run Time Support System *fu*
RTSSF Real Time Simulation Support Facility *hc*
RTT Radiation Tracking Transducer *fu*
RTT Real Time Telemetry *gp*
RTT Remote Transmit Unit *fu*
RTT Required Track Time *fu*
RTT Round Trip Timing *fu*
RTTI RTT Interrogate *fu*
RTTI Run Time Type Information *pk, kg*
RTTM Round Trip Timing Message *fu*
RTTR Round Trip Timing Reply *fu*
RTTY Radio Teletype(writer) [communications] *mt, fu, ct, kg, hi*
RTU Real Time UNIX *is26.1.c2*
RTU Receive/ Transmit Unit *fu*
RTU Remote Telemetry Unit *sd, jdr*
RTU Remote Terminal Units *4, y2k*
RTU Replacement Training Unit *mt, af72.5.123*
RTU Reserve Training Unit *wp*
RTU Retransmission Unit *mt*
RTV (there were also insulating materials called RTV) (the shelf RTV weighs 0.77 lb) *ci.413, gg32*
RTV Real Time Television *fu*
RTV Real Time Verification *hc*
RTV Real Time Video *kg, hi*
RTV Return to Vendor *fu, lt, wp, ep, jdr*
RTV Room Temperature Vulcanized (Vulcanizer, Vulcanizing, Vulcanization) *gp, lt, wp, jdr, jn*
RTVM Requirements Traceability Verification Matrix *tb*
RTVS Real Time Velocimeter System *hc*
RTWS Raw Tape Write Submodule *fu*
RTXT Rich Text *hi*
RU Range User *fu*
RU Reconnaissance Utility [as in RU-21 for small, propeller aircraft] *wp*
RU Relay Unit *fu*
RU Remote Unit *gp*
RU Reporting Unit [TACS / TADS] *fu, mt*
RU Reproducing Unit *fu*
RU Request Unit *ct, em*
RU Response Unit *ct*
RUI Replot User Interface *fu*
RUIC Reporting Unit Identification Code *mt*
RUM Resource and Unit Monitoring *mt*
RUMM Radar Upgrade Modernization Modification *hc*
RUMS Remote Update Message Service *fu*
RUP Remote User's Package [JDS] *mt*
RUPS Recorder Utility Processor System *oe*
RUR Reference Update Review *wl*
RUR Resource Utilization Report [DSDO]. *mt*
RUR [ship-launched, anti submarine rocket] [the torpedo is dropped by the missile at the end of flight] (RUR-5 ASROC) [USA] *mt*
RURPOP Rural Population file *mt*
RUS Rural Utilities Service *y2k*
RUS [air defense radar set; RUS-1 and RUS-2 were put in service in May 1941, USSR] *mt*
RUX Remote Unit Crosslinks *jdr*
RV Range Velocity *vv*
RV Rated Voltage *fu*
RV Ratio to Velocity *tb*
RV Reentry Vehicle *mt, fu, ca.7, 5, 11, 14, 17, sb, hc, wp, tb, st, kf*
RV Routine Verification *-4*
RV&H Radiation Vulnerability and Hardening [committee] *rm6.4.3*
RVA Reactive Voltampere meter *fu*
RVA Receiver Vulnerability Analysis *fu*
RVA Relative Virtual Address *pk, kg*
RVAO Reentry Vehicle Associated Object *17, tb, st*
RVARM Recording Varimeter *fu*
RVC Radar Video Converter *fu*
RVCF [AF] Remote Vehicle Checkout Facility *4, hc*
RVD Remote Virtual Disk *ct, em*
RVDS Radar Video Distribution System *fu*

RVE Routine Verification Equipment *-4*
RVI Reverse Interrupt *pk, kg*
RVM Reactive Voltmeter *fu*
RVM Requirements Verification Matrix *kf*
RVN Regional Voice Network [USSOUTHCOM] *mt*
RVP Radar Video Processor *fu, hc*
RVR Runway Visual Range *mt, fu*
RVS Routine Verification SHO *-4*
RVSN Rakentye Voiska Strategityesko Naznatseniya [strategic missile force] [USSR] *mt*
RVSP Radar VHSIC Signal Processor *fu*
RVV Runway Visibility Value *fu*
RW Radioactive Waste *mt*
RW Radiological Warfare *w9*
RW Reaction Wheel *kf, oe*
RW Reconnaissance Wing [USA] *mt*
RW Rivet Workforce *mt*
RW Rotary Wing *fu*
RWA Reaction Wheel Assembly *4, jdr, he, kf, oe*
RWC Read, Write, Computer *fu*
RWC Regional Warning Center *no*
RWCEP Real World Crisis Evaluation Plan *mt*
RWH Radar Warning and Homing *wp*
RWI Radio Wire Integration *fu, mt*
RWM Read / Write Memory *kg, fu, hi*
RWM Real World Model *fu*
RWO Ramstein Warning Office *mt*
RWO Reproduction Work Order *fu*
RWP Real Time Weather Processor *fu*
RWP Recurring Work Program *mt*
RWR Radar Warning Receiver *mt, fu, hc, mj31.5.18, ae151.52.460, wp*
RWR Rear Warning Radar *mt*
RWRO Read and Write Read Only *hi*
RWRS Radar Warning Receiver System *hc*
RWRW Read and Write Read and Write *hi*
RWRW Rescue and Weather Reconnaissance Wing *mt*
RWS Range While Search *hc, wp*
RWS Remote Workstation *fu, bl1-8.23*
RWV Read, Write, Verify *0*
RX Resolver Transmitter *fu*
RX-TX {Receive Switch-Transmit Switch} *gg147*
RXD Receive Data *kg, hi*
RXKGD Receive KG Variable Data *fu*
RZ Recovery Zone [USA DOD] *mt*
RZ Return to Zero *fu, gd2-26, 4, 9, jdr*

S

S&A Safe and Arm *ggIII.1.31, 9, lt*
S&A Safing and Arming *fu*
S&BMS-W Space and Building Management Service - Washington *dt*
S&CG Space and Communications Group [of Hughes Aircraft Company] [also SCG] *fu, hc, 8*
S&E Science and Engineering *fu*
S&EG Space and Electronics Group [of TRW] *kf*
S&G Sargent and Greenleaf, [Incorporated] *se24.09.84.1*
S&L (Office of the Assistant Secretary of the Navy (S&L)) *nd74.445.36*
S&M (purge system (S&M)) *bo*
S&M Scheduling and Movement [GCCS] [USA] *mt*
S&P Standard and Poor's Corporation *aw130.11.9*
S&PO Subcontract and Production Operations *fu*
S&R Search and Replace *hi*
S&R Search and Rescue [System] [also SAR] *ge, hy, hg*
S&SS Surveillance and Sensor Systems *fu*
S&T Science and Technology *mt, ph, hc*
S&T Simulation and Training *fu*
S&TE Support and Test Equipment *fu, kf*
S&TG Space and Technology Group [of TRW] *sc2.8.4*
S&TI Scientific and Technical Intelligence *hc*
S&TS Systems and Technology Symposium [DARPA] *kb*
S&W Surveillance and Warning *mt, dh*
S&WC Surveillance and Warning Center *mt*
S-CDMA Synchronous Code-Division Multiple Access *pk, kg*
S-HDSL Single-Line - High-bit-rate Digital Subscriber Line *ct*
S-HTTP Secure Hypertext Transfer Protocol *kg, hi*
S-MIME Secure MIME *pk, kg*
S-VHS Super Video Home System *kg, hi*
S-W Shick Wolverton *fu*
S/A Safe and Arm *jdr, oe*
S/A Sanders Associates *fu*
S/A Services and Agencies *mt*
S/A Solar Array *jdr*
S/B Spot Beam *jdr*
S/C+N Signal-to-Clutter + Noise [ratio] *fu*
S/CR Signal to Clutter Ratio *st*

S/DB Synchronizer / Data Buffer *gd3-2, oe*
S/EWCC Signals Intelligence / Electronic Warfare Coordination Center [USMC] *mt*
S/F Software / Firmware *fu*
S/F Store and Forward *mt*
S/G Space / Ground *jdr*
S/H Sample and Hold *fu, kg, gg161, hi*
S/I Signal-to-Interference Ratio *fu*
S/I Speaker / Interrupt *mt*
S/J Signal to Jamming (Jammer) ratio *fu, wp, mt*
S/L Sea Level *mt*
S/L Shelf Life *fu*
S/L Submittal Letter *jdr*
S/M Strength Member *fu*
S/MP Signal / Message Processor *fu*
S/MTD STOL/Maneuvering Technology Demonstrator *af72.5.135, aw130.22.44*
S/N Serial Number [also SN] *fu, 8, tb, jdr*
S/N Signal to Noise [ratio] [also SNR] *mt, kg, fu, ac3.3-14, 0, tb, pbl, jdr, hi, oe*
S/N Summary Number *-8*
S/P Series / Parallel *fu*
S/P Sound Powered *fu*
S/PP Subsystem Project Plans *mt*
S/R Send / Receive *fu*
S/R Set / Reset *fu*
S/R Shift Register *jdr*
S/R Sub Reflector *jdr*
S/RD Secret / Restricted Data *mt*
S/S Ship to Shore *mt*
S/S Shop Stock *fu*
S/S Space Segment *-4*
S/S Sub System *tb*
S/SDB Small / Small Disadvantaged Business *tb*
S/SEE System / Software Engineering Environment *kf*
S/STE Special / System Test Equipment *oe*
S/T Search / Track *fu*
S/T Short Ton [2000 lbs] *mt*
S/TK Sectors per Track *kg, hi*
S/U Set / Use *fu*
S/V Space Vehicle *17, sb, jdr*
S/V Survivability / Vulnerability *fu, nt, tb, jdr*
S/VP System Integration and Verification Plan *kf*
S/W Software *mt, fu, kg, 4, 17, bo, sb, pl, jb, wp, tb, kf, hi*
S/WI&T Software Integration and Test *fu*
S/WIL Software-In-the-Loop *tb*
SA Seaman Apprentice *w9*
SA Security Architecture *mt*
SA Security Assistance *mt*
SA Select(ive) Availability *jdr, kg*
SA Senior ADLO *fu*
SA Sensor Amplifier *fu*
SA Separation Assurance *fu*
SA Service Availability *mt*
SA Significant Accomplishment *kf*
SA Single Access *2, 4, vv*
SA Site Activation *kf*
SA Situation Assessment *mt, tb*
SA Situation Awareness *mt*
SA Solar Array *jn, oe*
SA Space Available *mt*
SA Special Access *hc*
SA Spectrum Analyzer *fu*
SA Standalone *fu*
SA Storage Analyzer *mt*
SA Strike Assessment *mt*
SA Structural Assembly *kf*
SA Sun Acquisition *oe*
SA Supplemental Agreement *mt, fu*
SA Support Agency *oe*
SA Surface Attack *mt*
SA Surface to Air [missile] *fu, mt*
SA Swept Audio *fu*
SA System Administrator *fu*
SA System Architecture *jdr*
SA Systems Analysis *wp, tb, st*
SA [chemical warfare codename for arsine (arsenic hydride)] *wp*
SA-OCRE Stand Alone Optical Character Recognition Equipment *mt*
SA/BM System Analysis / Battle Management *st*
SA/TT Solar Array Trim Tab *oe*
SAA Science And Applications *wl*
SAA Service Aspects and Applications *em, ct*
SAA Single Access Antenna *vv*
SAA Single Agency Action *mt*
SAA South Atlantic Anomaly *mi, pl, mb, kf*
SAA Systems Application Architecture *em, ct, kg, hi*

SAAA Security Assistance Automation Army *mt*
SAAA Simulated Air to Air Attack [REDCOM / JDA] *mt*
SAAC Science and Applications Advisory Committee *wl*
SAAC Security Assistance Accounting Center *wp*
SAAC Simulator for Air to Air Combat *hc*
SAACONS Standard Army Automated Contracting System *mt*
SAADTA SA-ALC Damage Tolerance Analysis [AFLC] *mt*
SAAF South African Air Force [South Africa]
SAAFR Standard use Army Aircraft Flight Route [USA DOD] *mt*
SAAG Science and Applications Advocacy Group *wl*
SAAHS Stability Augmentation and Altitude Hold System *mt*
SAALC San Antonio Air Logistics Center *mt, hc*
SAALC San Antonio Air Logistics Command *hc*
SAAM (SAAM cargo specialist) *af72.5.27*
SAAM Special Assignment Airlift Mission [MAC*] *wp, mt*
SAAM Surface to Air Anti Missile *mt*
SAAMA [Hughes Aircraft Company customer for IMAT] *hc*
SAAMII Solar Array Assembly Machine II *gp*
SAAR Search And Acquisition Radar *fu*
SAAS Semiconductor Array Acquisition System *hc*
SAAS Standard Army Ammunition System *mt*
SAASS Semi Automated Acquisition Subsystem *fu*
SAAW Spin Axis Attitude and Wobble *uh*
SAAX Science and Applications Missions *wl*
SAB Satellite Assembly Building *bo, 9, jdr*
SAB Science Advisory Board *wp*
SAB Shuttle Assembly Building *ai2*
SAB Space Applications Board [of National Academy of Engineering] *ac1.1-1*
SAB Support Activities Building *-10*
SABA Small Agile Battlefield Aircraft *mt, wp*
SABCA [Hughes Aircraft Company customer for an electro-optical fire control system] *hc*
SABE Shipboard Armament Boresight Equipment *hn42.1.3, hc*
SABENA SocietÈ Anonyme Belge d'Exploitation de la Navigation
SABER (a new circuit /system simulation tool called SABER) *aa26.12.66*
SABER Simplified Acquisition of Base Engineering Requirements *af72.5.104*
SABER [nickname for HQ USAF programs usually to evaluate new systems] *wp*
SABER_EVOLVE [evaluation of FYDP SAC potential] *wp*
SABER_FIRE [overall TAC evaluation] *wp*
SABER_GRAND [trade-off in tactical warfare] *wp*
SABER_HEAT [B-1 infrared study] *wp*
SABER_INDIA [reconnaissance capabilities evaluation] *wp*
SABER_JET [air to air evaluation] *wp*
SABER_LAUNCH [pre-launch survivability of AWACS evaluation] *wp*
SABER_LIFT [TAC airlift capability evaluation] *wp*
SABER_SILO [ICBM survival evaluation] *wp*
SABER_VALUE [cost effectiveness methodology for force analysis] *wp*
SABIR [a type of kill vehicle] *jmj*
SABLE Structure And Behavior Linking Environment *fu*
SABM Set Asynchronous Balanced Mode *fu*
SABM {a technology directorate within USASDC} *jmj*
SABMIS Sea-Based Antiballistic Missile Intercept System *hc, fu*
SABO Sense Amplifier Blocking Oscillator *fu*
SABOC SAM Battalion Operations Center *fu*
SABOC SAM Brigade Operations Center *fu*
SABRE Secured Airborne Radar Equipment *wp*
SABRE Support Airlift By Remote Entry *mt*
SABS Shared Aperture Boresight Sensor *hc*
SAC Senate Appropriations Committee *mt, hc, wp*
SAC Shared Aperture Component *hc*
SAC Signal Analysis Center *fu*
SAC Single Access Compartment *-4*
SAC Single Attachment Concentrator *pk, kg*
SAC Slanted Array Compressor *fu*
SAC Sonar Azimuth Converter *fu*
SAC Space Applications Center [Ahmedabad, India] *ac3.2-2*
SAC Special Access Clearance *mt*
SAC Special Area Code *hi*
SAC Spin Axis Controller *-3*
SAC Strategic Air Command [later renamed US Strategic Command, USSTRATCOM] [USA] * *mt, ro, hc, ci.570, 11, 12, 13, 14, w9, af72.5.47, jb, wp, tb, st, jdr, kf*
SAC Structural Assembly Compartment *kf*
SAC Sun Angle Counter *pl*
SAC Switching Amplifier Compensation *0*
SAC/LGL [SAC / LGL contingency planning and management system] [SAC*] *mt*
SACAPS SAC Adaptive Planning System *wp*
SACARMS SAC* Aircrew Resource Management System [SAC] *mt*
SACAW Strike Avionics Concepts for Advanced Weapons *hc*
SACC San Antonio Contracting Center *mt*
SACC Science and Applications Computer Center *ac2.8-12*
SACC Semiautomated Area Control Center *fu*

SACC Shore ASW Command Centers *mt*
SACC Supporting Arms Control Center *fu*
SACC Supporting Arms Coordination Center [USMC] *mt*
SACCA Strategic Air Command Communications Area *mt*
SACCOM Strategic Air Command Communications *wp*
SACCOP SAC* Continuity of Operations Plan *mt*
SACCS SAC Airborne Command and Control System *-12*
SACCS_DTS SAC* Automated Command and Control System Data Transmission Subsystem [formerly SACDIN] *mt*
SACDEF Strategic Avionics Crewstation Design Evaluation Facility *hc*
SACDIN Strategic Air Command Digital Information Network [formerly SATIN] *mt, 12, hc, wp, tb*
SACE Semi Automatic Checkout Equipment *fu*
SACE Shore Based Acceptance Checkout Equipment *fu*
SACEUR Supreme Allied Commander, Europe [NATO Command] *mt, fu, 12, wp*
SACINTNET Strategic Air Command Intelligence Network *mt, pf.f88.27*
SACLANT Supreme Allied Commander, Atlantic [NATO Command] *mt, fu, 12, dt, wp*
SACLOS Semi Active Command Line Of Sight guidance *mt*
SACLOS Semi Automatic Command to Line Of Sight *mt*
SACM SAC* Manual *mt*
SACMPC Systems Acquisition Career Management Program for Civilians *wp*
SACNET Secure Automatic Communications Network *mt*
SACOM Surveillance for Aircraft and Cruise Missile [RADC] *mt*
SACOP SAC* Continuity of Operations Plan *mt*
SACOPS Strategic Air Command Operations Planning [computer] System *hc, wp*
SACOS Strategic Air Combat Operations Staff *mt*
SACP SAC* Pamphlet *mt*
SACP Standalone Control Panel *fu*
SACPAS SAC* Primary Alerting System *mt*
SACR SAC* Regulation *mt*
SACRED Simplification And Cost Reduction Program *hc*
SACRIN Strategic Air Command Readiness Information Network *mt*
SACS Structuring And Composition System [USA] *mt*
SACS Synchronous Altitude Communications Satellite *hc*
SACSA Special Assistant for Counter-Insurgency and Special Activities *mt*
SACSO SAC* Security Office *mt*
SACT Standalone Console Trainer *fu*
SACTELNET SAC* Telephone Network *mt*
SACTTYNET SAC* Teletypewriter Network *mt*
SACU Synchronizer/Acquisition/Calibrate Unit *ac10.4-4*
SACWARNS Strategic {Air Command} Warning System *mt*
SAC_ADA SAC* Airborne Data Automation *mt*
SAC_ADVON SAC* Advanced Echelon *mt*
SAC_APS SAC* Adaptive Planning System *mt*
SAC_ECMP SAC* Engine Condition Monitoring Program [SAC] *mt*
SAD Satellite Attitude Dynamics *sa12474.13*
SAD Special Activities Division *mt*
SAD Standalone Diagnostics *fu*
SAD State of the Art Document *fu*
SAD Status Advisory Display *hc*
SAD Supplementary Alteration Drawing *fu*
SAD System Allocation Document *fu*
SADA Semi-Automatic Defense Area [Spain] [Sistema Semiautomatico de Defensa Aerea] *fu, hc*
SADA Solar Array Drive Assembly *4, oe*
SADANG Acoustics Detection and Analysis New Generation *fu*
SADARM Search (Sense) (Seek) and Destroy Armor [fire and forget] Munition [USA] *fu, mt, hc, uc, nd74.445.14, wp*
SADAS Semi Automatic Data Acquisition System *fu*
SADC Sector Air Defense Commander *fu*
SADE Sensor Algorithm Development Environment *wp*
SADE Silicide Array Development and Evaluation *hc*
SADE Solar Array Drive Electronics *jdr, oe*
SADEC Spin Axis Declination *uh*
SADF System Architecture Development Facility *tb*
SADIE Scanning Analog-to-Digital Input Equipment *fu*
SADIFS SAC* Automated Deployable Intelligence Fusion System *mt*
SADL Synchronous Data Link Control [Racal-Vadic] *pk, kg*
SADM System Acquisition Decision Memorandum *st*
SADMT SDI Data Flow Moduling {Modeling?} Technique *tb*
SADOC Standby Air Defense Operations Center *fu*
SADR Satellite Anomaly Detection and Resolution *jdr*
SADR Secure Acoustic Data Relay *fu*
SADRAD Synthetic Aperture Dual Frequency Radar *wp*
SADRAM Seek And Destroy Radar Assisted Mission *wp*
SADS Semi-Automatic Air Defense System *fu*
SADS Simulated Air Defense System [Model III] *hc*
SADS Special Area Defense Study (Lockheed) *fu*
SADS Submarine Active Detection Sensor *hc*
SADS Submarine Active Detection Sonar *fu*
SADSC San Antonio Data Services Center [later 2nd ISG] *mt*
SADT Structured Analysis and Design Technique *mt, fu*

SAE Shaft Angle Encoder *pl*
SAE Society of Automotive Engineers *fu, aw126.20.7, hc, aa26.12.77, hn49.9.3, wp*
SAEC Shaft Angle Encoder Corrected *pl*
SAEC Support Analysis of Engineering Change *fu*
SAEF Spacecraft Assembly and Encapsulation Facility *pl, sf31.1.9, aw130.22.40, jdr, ai2*
SAEL (Grumman's SAEL aeroservoelastic analysis code) *aa26.11.6*
SAEOC Second Alternate Emergency Operations Center [GSA] *mt*
SAES Scanning Auger Electron Spectroscopy *wp*
SAES System Analysis and Evaluation Study (Boeing) *fu*
SAEWS Ship's Advanced Electronic Warfare System *fu*
SAF Safe Arming and Fuze *-17*
SAF Secretary of the Air Force [also SAFOS] *mt, wp, kf*
SAF Sequence Active Flag *fu*
SAF Soviet Air Force *fu*
SAF Spacecraft Assembly Facility *gp, 0, jdr*
SAF Spanish Air Force *fu*
SAF Spec Airfield File [REDCOM/JDA] *mt*
SAF Special Access Facility *tb*
SAF Special Applications File *mt*
SAF Strategic Army Force *mt*
SAF-TE SCSI Accessed Fault Tolerance Enclosures *hi*
SAF/AL Secretary AF / Acquisition and Logistics *mt*
SAF/AQ Secretary of the Air Force, Acquisition *kf*
SAF/FM Secretary AF / Financial Management *mt*
SAF/FMD Deputy Assistant Secretary AF [Information Systems Management] *mt*
SAF/LLP Secretary of the Air Force, Legislative Liaison Programs Division *kf*
SAFC Sector Air Force Commander *fu*
SAFE San Andreas Fault Experiment *ac2.6-9*
SAFE Secure Analyst File Environment *mt*
SAFE Security And Freedom through Encryption *ct, em*
SAFE Selected Area For Evasion [USA] *mt*
SAFE Self Adjusting Focal Plane *hc*
SAFE Support to Analysts File Environment [DIA] *mt*
SAFE Supportability Analysis Forecasting and Evaluation *mt*
SAFE System Analyst File Environment *mt*
SAFE [nickname for HQ USAF programs usually connected with security] *wp*
SAFE SIDE [intrusion detection equipment] *ro*
SAFEGUARD [Army antiballistic missile program] *-17*
SAFENET Survivable Adaptable Fault tolerant Embedded local area Network *pf1.10*
SAFENET Survivable Adaptable Fiber Optic Embedded Network *mt*
SAFF Safe-Arm-Fuse and Fire *st*
SAFIRE Spectroscopy in the Atmosphere using Far Infrared Emission *hy*
SAFM Satellite Anticipated Failure Modes *jdr*
SAFOC Semi-Automated Flight Operations Center *hc, fu*
SAFOC DOE SAFOC Design Of Experiments *fu*
SAFOS Secretary of the Air Force [also SAF] *dt*
SAFS Simulated Aircraft Flight Station *fu*
SAFSEA Safeguard System Evaluation Agency *fu, hc*
SAFSS [a commercial satellite] *sd*
SAFT Subsequent Article Factory Test *fu*
SAG Science Analysis Group *0*
SAG Single Access Antenna Group *vv*
SAG Sub Activity Group *mt*
SAG SUBACS Advisory Group *fu*
SAG Surface Action Group *mt*
SAG Systems Analysis Group *mt*
SAGA Solar Array Gain Augmentation [for HST] *mi*
SAGA Studies, Analysis, and Gaming Agency [OJCS] *dt, mt*
SAGE Semi Automatic Ground Environment [USA] *mt, ro, fu, hc, hn41.18.1, 12*
SAGE Simulation for Air and Ground Engagements [model] *wp*
SAGE Stratospheric Aerosol and Gas Experiment [NASA, AEM-II] *ge, ac6-i, hy, hg*
SAGEM [aerospace company, France] *es55.44*
SAGITTAR_EXP [Kirtland AFB program] *hc*
SAH Semi Active Homing *wp, tb, kf*
SAHRS Standard Attitude and Heading Reference System *mt, aw130.11.c2, wp*
SAI Sanders Associates Inc. *fu*
SAI Science Applications, Incorporated [Hughes Aircraft Company customer for JFAADS] *fu, ci.824, tb, hc*
SAI Supplier Assembly Instruction *fu*
SAI Swales and Associates Inc. *oe*
SAIC Science Applications International Corporation [Washington, D. C.] *mt, cp, jmj, re, st*
SAIC Shanghai Aviation Industrial Corporation *aw126.20.31*
SAID Solar Array Isolation Diode (assembly) *jdr*
SAIDS Space Analysis Intervention Display System *hc*
SAIE Special Acceptance and Inspection Equipment *fu, hc*
SAIF Spec Automated Installation File [REDCOM / JDA] *mt*
SAIF Standard Avionics Integrated Fuze *wp*

SAIK Saudi Arabian Intra Kingdom *fu*
SAIL Shuttle Avionics Integration Laboratory *pf1.23, wl*
SAILS Simplified Aircraft Instrument Landing System *wp*
SAILS Standard Army Intermediate Level Supply system *fu, mt*
SAIMS Selected Acquisition Information and Management System *fu*
SAIMS Supersonic Airborne Infrared Measurement System *wp*
SAINT Satellite Inspection Technique *-14*
SAINT Satellite Interceptor [a USA DOD system designed to demonstrate the feasibility of intercepting, inspecting, and reporting on the characteristics of satellites in orbit] *mt, 0, 11, sd*
SAINT Symbolic Automatic Integrator *kg, hi*
SAINT Systems Analysis Integrated Networks of Tasks *wp*
SAIP Spares Acquisition Integrated with Production *fu, hc, wp*
SAIRS Standard Advanced Infrared System *wp*
SAIRS Synthesized Advanced Infrared Sensor *hc*
SAIS Science and Applications Information System *hy*
SAIS Standard Airborne Instrumentation System *hc*
SAIS Standard Automated Information Systems *mt*
SAISS Subcommittee on Automated Information System Security *tb*
SAJB Single Channel Anti Jam Broadcast *jdr, uh*
SAL Shift Arithmetic Left *pk, kg*
SAL Shipboard Allowance List *fu*
SAL Strategic ALCM Launcher [the Rockwell B-1] [USA] *wp, mt*
SAL Strategic Arms Limitation *mt, wp, tb*
SAL System Activity Log *jdr*
SAL Systems Analysis Laboratories [Hughes Aircraft Company] *sa12474.1*
SAL Systems Applications Laboratory *oe*
SAL-GP Semi-Active Laser-Guided Projectile *uc*
SALCC Satellite Airlift Control Center *mt*
SALDS Stabilized Airborne Laser Designator System *hc*
SALF Short Approach Lighting System w/Sequenced Flashing Lights *fu*
SALH Semi Active Laser Homing [USA] *mt*
SALID Solar and Lunar Illumination Data [PACAF] *mt*
SALIS Security Assistance Logistics Information System [ILC] *mt*
SALM Strategic Air Launched Missile *wp*
SALS Short-Approach Lighting System *fu*
SALT (used the SALT . . . program to simulate an entire graphics system) *hn50.8.4*
SALT Simulation Analysis for Logic and Timing *fu*
SALT Strategic Arms Limitations Treaty (Talks) "salt" *fu, mt, ci.159, 11, 13, 14, 17, w9, wp, tb*
SALTY [nickname for Military Airlift Command programs] *wp*
SALTY_CHASE [a USAF chemical warfare exercise at Hahn AFB, 1987] *wp*
SALTY_DEMO [USAFE chemical warfare exercise] *wp*
SALTY_NATION [large USAFE chemical warfare exercise] *wp*
SAM Scanning Auger Microscopy *wp*
SAM School of Aerospace Medicine *wp*
SAM Security Accounts Manager [Microsoft] *pk, kg*
SAM Self Assembling Material *wp*
SAM Sequential Access Method *kg, wp, hi*
SAM Serial Access Memory *pk, kg*
SAM Single Application Mode [Microsoft] *pk, kg*
SAM Solar Array Model *jn*
SAM Space Available Mail *mt*
SAM Special Air Mission [USA] *mt, af72.5.114*
SAM Spotbeam Antenna Mechanism *jn*
SAM Standard Avionics Module (Hughes) *fu*
SAM Station Acquisition Maneuver *oe*
SAM Stationing Analysis Model [USA] *mt*
SAM Steerable Antenna Mechanism *jn*
SAM Stratospheric Aerosol Monitor (Measurement) [NIMBUS-7] *ge, ac2.8-5, hy, hg*
SAM Surface to Air Missile * *mt, fu, 0, 12, 17, ph, sb, hc, w9, kb, wp, tb, st, kf*
SAM Sustainability Assessment Module [WSMIS] [AFLC] *mt*
SAM Synchronous Amplitude Modulation *fu*
SAM-D Surface to Air Missile Development *hc*
SAM-D Surface-to-Air Missile - Developmental (formerly AADS-70s) *fu*
SAM-X Next generation SAM *fu*
SAM/ADA Surface-to-Air Missile/Air Defense Artillery *fu*
SAMAS Security Assistance Manpower Accounting System *wp*
SAME Society of American Military Engineers *wp*
SAMF Special Air Mission Fleet [C-140] *wp*
SAMFARS Stand Alone Automated Message Formatting *mt*
SAMH Standard Automated Message Handling *mt*
SAMHS Standard Automated Message Handling System *mt*
SAMI System Acquisition Management Inspection *mt, af72.5.107, wp*
SAMICS Systems Application of Millimeter Wave Contrast Seeker *wp*
SAMIR Satellite Microwave Radiometer *ac3.2-9*
SAMIS Security Assistance Management Information System [AFLC] *mt*
SAMIU SAM Interface Unit *fu*
SAMM Semi Autonomous Mission Module *hc*
SAMMS Standard Automated Material Management System [DLA] *mt*
SAMO Secretary of the Army Management Office *dt*

SAMOB Surface to Air Missile Order of Battle *mt*
SAMOC Surface to Air Operations Center [formerly PAHOC] *hc, go3.1-2*
SAMOS Photographic Satellite *0*
SAMP Single Acquisition Management Plan *kf*
SAMP Sum Amplifier *fu*
SAMPE Society of Aerospace Material and Process Engineers *0, aw130.22.136*
SAMPEX Solar Anomalous and Magnetospheric Particle EXplorer *mi*
SAMRAAM Surface Advanced Medium Range Air to Air Missile *hc*
SAMRAM Surface to Air Medium Range [Active] Missile *fu, hc*
SAMREP SAM Activity Mission Report *fu*
SAMS Standard Army Maintenance System *mt, fu*
SAMS Standard Automated Maintenance System [SAC] *mt*
SAMS Stratospheric and Mesospheric Sounder [NIMBUS-7] *ge, ac2.8-5, hy, hg*
SAMS Surface to Air Missile System *hc*
SAMSIM SAM* Simulation *mt*
SAMSO Space And Missile Systems Organization [US Air Force, later Space Division] *mt, fu, ro, hc, ac7.5-1, 4, 12, 17, wp, tb, oe*
SAMSON Active Augmentation System Program *hc*
SAMSON Special Avionics Mission Strap-on Now [avionics pod] *mt*
SAMSS Space Assembly Maintenance and Servicing Study [AFSC*] [TRW] *mt*
SAMT Simulated Aircraft Maintenance Trainer *wp*
SAMTEC Space And Missile Test Center [USAF] *mt, ci.569, hc, wp, tb*
SAMTO Space And Missile Test Organization [AFSC*] *5, sb, tb, mt*
SAMWAR Surface to Air Missile Warning Radar *hc*
SAM_BN_CDR SAM Battalion Commander *fu*
SAN Satellite Access Node *ric, jdr*
SAN System Advisory Notices *mt*
SAN System Area Network *em, kg*
SANATOR [Norwegian designed steam chemical warfare decontamination system, M-17] *wp*
SANDAC [Sandia airborne computer guidance system] *sb*
SANE Standard Apple Numeric Environment *kg, hi*
SANG Saudi Arabian National Guard *fu*
SANIRS [a second generation focal plane array system] *17, sb*
SAO SAM Assignment Officer *fu*
SAO Security Assistance Office *mt*
SAO Selected Attack Option *mt*
SAO Smithsonian Astrophysical Observatory *mi, ac2.6-5, ns, mb*
SAO Special Activities Office *mt*
SAO System Acquisition Officer *wp*
SAOC Singapore Air Operations Center *fu*
SAOC Submarine Air Optical Communications *hc*
SAOCC Sector Air Operations Coordination Centers *mt*
SAOD Shared Aperture Optical Device *hc*
SAOEU Strike / Attack Operational Evaluation Unit [UK] *mt*
SAOM Solar Array Orientation Mechanism [also SOAM] *hc*
SAOT SAM Assignment Officer Technician *fu*
SAP Satellite Autonomy Program *hc*
SAP Second Audio Program *pk, kg*
SAP Security Assistance Program *mt, wp*
SAP Semi Armor Piercing *mt, fu*
SAP Service Access Point [DEC] *kg, nu*
SAP Service Access Protocol *mt*
SAP Service Advertising Protocol *ct, kg, em, hi*
SAP Site Activation Plan *kf*
SAP Special Access Program [organization] [USA DOD] *mt, ah, su, jn, pbl, jdr, tm, kf*
SAP Special Application Program *fu*
SAP Structural Analysis Program *hi*
SAP Symbolic Assembly Program *pk, kg*
SAPAS Semiautomated Population Analysis System *hc*
SAPBB Systolic Array Processor Brassboard *fu*
SAPE Survivable Adaptive Planning Experiment *mt, ro*
SAPI Service Access Protocol Interpreter *mt*
SAPI Speech Application Program Interface [Microsoft] *pk, kg*
SAPM Software Associate Project Manager *fu*
SAPO Subordinate Area Petroleum Officer *mt*
SAPPHIRE Synthetic Aperture Precision Processor High Reliability *wp*
SAR Safety Action Request *fu*
SAR Safety Assessment Report *fu*
SAR Search And Rescue [System] [also S&R] [USA DOD] *mt, fu, ge, mi, gp, hg, mb, oe*
SAR Segmentation And Reassembly *ct, kg, em*
SAR Selected Acquisition Report *mt, fu, hc, wp, tb, st*
SAR Selected Acquisition Review *mt*
SAR Semi Active Radar *wp, mt*
SAR Sensor Algorithm Research *hc*
SAR Shift Arithmetic Right *pk, kg*
SAR Sonar Analog Receiver *fu*
SAR Special Access Required *mt, fu, sy9, wp, kf*
SAR Specially Authorized Representative *-18*
SAR Storage Address Register *fu*
SAR Subsequent Application Review *mt, fu, tb, kf*

SAR Successive Approximation Register *pk, kg*
SAR Synthetic Airborne Radar *-14*
SAR Synthetic Aperture Radar *mt, fu, hc, kg, ge, mi, hn44.2.1, af72.5.139, wp, so, ep, hy, tb, mb, hg*
SAR Synthetic Array Radar *mt*
SAR System Acceptance Review *jn*
SAR System Acquisition Report *mt*
SAR System Analysis Recorder *fu*
SARA Satellite pour Astronomie Radio Amateur *mi*
SARA Spin Axis Right Ascension *uh*
SARA Station Aerostransportable Reconnaissance Aerienne *wp*
SARA System Architect's Apprentice *fu*
SARA Systems Accounting and Resource Analysis [software] *wp*
SARAC Steerable Array for Radar And Communications *ro, wp*
SARAH Semi Automatic Range, Azimuth, Height (system) *fu*
SARAH Standard Remote to AUTODIN Host *mt*
SARAH System Analysis and Resource Accounting for Honeywell [also SARA-H] (SARAH-Lite is a miniaturized administrative version of SARAH) * *mt*
SARB Spacecraft Anomaly Review Board *oe*
SARBE Search And Rescue Beacon Equipment [UK] *mt*
SARC Symantec Antivirus Research Center *ct, em*
SARC System Acquisition Review Council [also SARCS] *fu, wp, tb*
SARCALM Synthetic Array Radar Command Air Launched Missile *hc*
SARCALM Synthetic Array Radar Data *fu*
SARCS System Acquisition Review Council [also SARC] *ph*
SARDAB Search and Aerospace Rescue Data Base [MAC*] *mt*
SARDO Search And Rescue Deputy Office *mt*
SAREX Search and Rescue Exercise *mi*
SAREX Shuttle Amateur Radio Experiment *mi*
SARF Semi-Automatic Reconstruction Facility *fu*
SARG Synthetic Aperture Radar Guidance *wp*
SARGOS {space based data acquisition system, France} *ac3.6-1*
SARGUN Synthetic Array Pointing for a Gun *hc*
SARH Semi Active Radar Homing [a missile guidance system that steers on the reflection of the radar beam transmitted by the aircraft] *mt*
SARIN [isopropyl methyl phosphonofluoridate, a chemical warfare nerve agent] *wp*
SARIPS Search and Rescue Information Processing System *mt*
SARM SAR to Seeker *hc*
SARM Search And Rescue Memory *oe*
SARM Secondary Alignment Reference Mirror *gp*
SARM Set Asynchronous Response Mode *fu*
SAROP Synthetic Aperture Radar Operator Performance *hc*
SARP Ship Alteration and Repair Package [US Navy] *mt*
SARP Standards And Recommended Practices *aw129.21.137*
SARPMA San Antonio Real Property Maintenance Agency *mt, wp*
SARPS Synthetic Aperture Radar Processing System *wp*
SARS System Architecture Requirements Specification *mt*
SARSAT Search And Rescue Satellite Aided Tracking [system] *mt, sc3.3.3, 11, 19, tb, oe*
SARSS Standard Army Retail Supply System *mt*
SARTF Search and Rescue Task Force *mt*
SAS Safety Analysis Summary *tb*
SAS Sales Accounting System *pk, kg*
SAS Scan Actuator Subassembly *jp*
SAS Scandinavian Airlines System *aw129.14.110, wp*
SAS Sealed Authentication System *mt*
SAS Segment Arrival Storage *ai2*
SAS Semi-Automation System *fu*
SAS Sensor Assembly Simulator *oe*
SAS SGL/SA Antenna Subsystem *vv*
SAS Shoot, Assess, Shoot *tb*
SAS Single Attached Station *pk, kg*
SAS Single Audio System *fu, hc, kg*
SAS Small Astronomy Satellite *ci.735*
SAS Solar Array Simulator *tb*
SAS Solar Array System *hc, ep*
SAS Space Activity Suit *mi*
SAS Space Adaptation Syndrome *mi*
SAS Space Asset Support *dm*
SAS Sparse Array System *ic2-19*
SAS Speaker Authentication System *hc*
SAS Special Air Service [UK] *mt*
SAS Special Ammunition Site / Storage *mt*
SAS Stability Augmentation System [equipment installed in the B-52 to increase controllability in turbulence] [USA] *wp, mt*
SAS Standalone Simulator *fu*
SAS Statistical Analysis System [OJCS] *kg, 8, mt*
SAS Studies and Analysis Staff *dt*
SAS Sun Angle Sensor *oe*
SAS Supply Analysis System [USAFE] *mt*
SAS Survival Avionics System *hc, aw118.14.65, wp*
SAS System Activity Schedule *kf*
SAS System Architecture Study *tb*
SASC Semi Automatic Switch Center *mt*
SASC Senate Armed Services Committee *mt, hc, fu, af72.5.18, wp*

SASCS SHF Antenna Sidelobe Canceller System *hc*
SASE Special ARPA Space Equipment *hc*
SASF Spacing and Separation Feasibility *fu*
SASF System Architecture Study Facility *tb*
SASI Ships and Air Systems Integration Project Office *hc*
SASI Shugart Associates System Interface [SCSI originated from SASI] *ct, em, kg*
SASLS/P Synthetic Aperture Side Look Sonar/Processor *hc*
SASM Single Axis Stepper Motor [mechanism] *ric, kf, jdr*
SASM Special Assistant for Strategic Mobility *mt*
SASO Saudi Arabian Standard Organization *mt*
SASP SJCS Administrative Support Programs [OJCS] *mt*
SASR Spare Available Space Required System *mt*
SASS SAC Aircraft Security System *wp*
SASS Schedule Assessment Simulation System *tb*
SASS Seasat A [radar] Scatterometer System *ac2.7-1*
SASS Semi Automatic Switchboard (Switching) System *mt, wp*
SASS Shared Aperture Surveillance System *hc*
SASS SIOP Additive Spare Support *mt*
SASS Small Aerostat Surveillance System *hc*
SASS Space Assets Support System *tb*
SASS Spectrally Agile Sensor System *wp*
SASS Strategic Airborne Surveillance System *aw126.20.126*
SASS Suspended Array Surveillance System *fu*
SASS [SEASAT] *ge*
SASSETS Space and Strategic Engineering Estimating Technique System *gb*
SASSI Synthetic Aperture Seeker System Investigation *hc*
SAST Senior Air Surveillance Technician *fu*
SAST Shanghai Academy of Spaceflight Technology *ai2*
SASWC SACLANT Anti Submarine Warfare Center [NATO] *mt*
SASWC Ship Antisubmarine Warfare Coordinator *fu*
SAT SAM Assignment Technician *fu*
SAT Search, Acquisition, and Track *fu*
SAT Security Alert Team *mt, tb*
SAT Select At Test *oe*
SAT Service Acceptance Test *-4*
SAT Shared Aperture Technology *-17*
SAT Site Acceptance Test *jdr, jn, fu*
SAT Situational Awareness Terminal *mt*
SAT Software Analysis Team *wp*
SAT Sonar Analog Transmitter *fu*
SAT Split Array Tracker *fu*
SAT Subroutine Address Table *fu*
SAT Surveillance Acquisition and Tracking *tb, st*
SAT Synthetic Aperture Telescope *mi*
SAT System Acceptance Test *fu*
SAT System Analysis Test *fu*
SAT System Approach to Training *fu*
SAT Systems Acceptance Test *mt*
SAT IV Standard AUTODIN Terminal IV System *mt*
SATA Societè Antillaise de Transports AÈriens *aw118.18.23*
SATAF Site Activation Task Force *mt, fu, af72.5.153, wp, kf*
SATAN Security Administrator's Tool for Analyzing Networks *kg, ct, em, hi*
SATAR Satellite Aerospace Research *0*
SATB Standard Avionics Test Bed *wp*
SATC Simulation And Test Capability *st*
SATCOL [Satellite for Colombia] *sc2.7.3*
SATCOM Satellite Communications [system] [1974] *mt, fu, id4142-10/831, ph, af72.5.142, aw129.21.162, mj31.5.43, wp, tb, st, kf, oe*
SATCOMA Satellite Communications Agency [USA] *fu, mt*
SATCOMFAC Satellite Communications Facility *fu*
SATE Special Automatic Test Equipment *hc*
SATEC Semi Automatic Technical Center *hc*
SATEX Satellite Signal External Analysis *hc*
SATFA Security Assistance Training Field Activity *hc*
SATGEN Semi-Automatic Test Generation *fu*
SATGEN Sequential Array Test Generation *hc*
SATIN SAC* Automated Total Information Network [see SACCS DTS] *mt*
SATIN Survivability Augmentation for Transport aIrcraft Now! [an ECM system for transport aircraft] ["SATAN" would have been logical, but has objectionable connotations] [USA] *mt*
SATIRE Semi Automatic Technical Information Retrieval *0*
SATKA Surveillance, Acquisition, Tracking, and Kill Assessment *mt, fu, ro, fr, 1, 5, sb, ts4-8, nt, cp, wp, tb, st*
SATLID Single Frequency Atmospheric LIDAR *hg, ge*
SATMOS Service d'Archivage et a Traitement Meteorologique des Observations Satellitares [France] *ge, hg*
SATNAV Satellite Navigation *mt, fu*
SATNET Satellite Network *ct, em*
SATO Scheduled Airlines Ticket (Traffic) Office *mt, af72.5.119, wp*
SATP Service Acceptance Test Plan *-4*
SATPLT Satellite Plot [software program] *jb*
SATR Search And Tracking Radar *hc*
SATRAN Satellite Reconnaissance Advance Notice *mt, hs, jb, wp*

SATRAN Satellite Transmission *mt*
SATS Scout Launched ATS *ci.969*
SATS Short Airfield for Tactical Support *mt*
SATS Southwest Asia Telecommunications System *mt*
SATS Standardized Automatic Test System *hn43.22.12*
SATS System Acceptance Test Station *bf*
SATS System Architecture and Trade Study *st*
SATSIM [DSP satellite command verification simulation] *kf*
SATSS Seventh Army Tactical Automatic Switching System *hc*
SATTDE Solar Array Trim Tab Drive Electronics *oe*
SATTIRE Study of Advanced Technology Targets for Intelligence Research *hc*
SATVUL Satellite Vulnerability *fu*
SAU Search and Attack Unit *mt*
SAU Signal Analyzer Unit [part of GOES EPS] *gp, oe*
SAU Strap Around Unit *pl*
SAU Surface Attack Unit *mt*
SAV Service Appointed Validator *mt*
SAVA Standard Army Vectronics Architecture *hc*
SAVAR Standard Army Validation and Reconciliation System *mt*
SAVDM Single Application Virtual DOS Machine *kg, hi*
SAVE Situation Analysis and Vulnerability Estimate *mt, wp*
SAVI Space Active Vibration Isolation *hc*
SAW Satellite Attack Warning *sd, tb*
SAW Selectively Aimable Warhead *st*
SAW Spin Axis Attitude and Wobble control *uh*
SAW Standoff Attack Weapon *hc, dh, wp*
SAW Surface Acoustic Wave [detection system] *gg139, hc, hd1.89.y20, rg, mj31.5.15, jdr, oe, ai1*
SAW Surface Acoustic Waveguide *wp*
SAW Surveillance Acoustic Wave (device) *fu*
SAW/V Surveillance and Warning / Verification *mt*
SAWACS Soviet Union Airborne Warning And Control System *hc*
SAWAF SAW Adaptive Filter *fu*
SAWB Systems Analysis Workbench *fu*
SAWC Surveillance and Weapons Control *fu*
SAWDL SAW Delay Line *fu*
SAWE Simulation of Area Weapons Effects *hc*
SAWG Satellite Anomaly Working Group *kf*
SAWG SED Architecture Working Group *tb*
SAWPS Stand Alone Ware Plan System [SAC] *mt*
SAWS Satellite Attack Warning System *jb*
SAWS Silent Air Warning System *hc*
SAWS Silent Attack Warning System *mt, hc, wp*
SAWS Spaceborne Advanced Warning System *wp*
SAWS Submarine Acoustic Warfare System *fu, pf.f88.17*
SAWS Surveillance and Warning System *wp, fu*
SAWVS Satellite Attack Warning and Verification System *mt*
SB Serial Binary *fu*
SB Solid Booster *tb*
SB Sound Blaster [Creative Labs] *pk, kg*
SB Sound Board *pk, kg*
SB Space Based *wp, st*
SB Spot Beam [antenna] *uh, jdr*
SB Standby Base *mt*
SB Supply Bulletin *fu*
SB Synchronization Bit *fu*
SB System Base [as in SB3 for System Base 3] *mt*
SB [SBIR System IPT] *kf*
SBA S-Band Antenna *oe*
SBA Scene Balance Algorithms [Kodak] *pk, kg*
SBA Spin Bearing Assembly *jp*
SBA Spot Beam Antenna *uh*
SBA Standard Beam Approach *fu*
SBAC Society of British Aerospace Companies *aw124.21*
SBAM Single Base Aircraft Maintenance Model *mt*
SBAMS Space Based Anti-Missile Satellite *tb*
SBAP System Block Analysis Program *fu*
SBASS Space Based Atmospheric Surveillance System *mt, wp*
SBAW Shallow Bulk Acoustic Wave *fu*
SBB Subtract with Borrow *pk, kg*
SBB System Building Block *fu*
SBC Single Board Computer *fu, kg*
SBC Space Based Communication *mt*
SBCA Satellite Business and Communications Association *rih*
SBCS Single Byte Character Set *pk, kg*
SBCS Survivable BRAAT Communications System *af72.5.19*
SBD Smart Battery Data *hi*
SBDL Submarine B Data Link *fu*
SBEO Space Based Electro Optical [sensor] *ett*
SBES S-Band Excitor Systems *hc*
SBEV Space Based Interceptor [instead of KEW] {?} *tb*
SBEWS Space Based Early Warning System *kf*
SBF Short Backfire *uh*
SBF Simulation Build Function *fu*
SBF Stream Builder and Formatter *fu*

SBF Survivable Backup Facility *kf*
SBHP Single Bore Heat Pipe *jn*
SBHRG Space Based Hypervelocity Rail Gun *wp, tb*
SBI Serial Bus Interface *fu*
SBI Short Baseline Interferometer *wp*
SBI Sound Blaster Instrument.SC *Script [file name extension]*
SBI Space Based Interceptor *mt, aw129.14.25, cp, wp, tb*
SBI Special Background Investigation *mt, 10, wp, tb, jdr*
SBI Synchronous Backplane Interconnect *fu*
SBICV Space Based Interceptor Carrier Vehicle *tb*
SBIR Small Business Innovation Research [DOD] *mt, nd74.445.6, wp, tb, ai2*
SBIR Space Based Infrared *jdr*
SBIRS Space Based Infrared System *kf*
SBIRSSim SBIRS Simulation *kf*
SBKEW Space Based Kinetic Energy Weapon *hc, tb*
SBKEWS Space Based Kinetic Energy Weapon System *5, 17, sb*
SBKKV Space Based Kinetic Kill Vehicle [SDI] *mt, 17, sb, wp, st*
SBKV Space Based Kill Vehicle *tb*
SBL Space Based Laser *mt, fu, fr, 5, 17, sb, hc, wp, tb, st*
SBLADAR Space Based LADAR *st*
SBLC Standard Base Level Computer *mt*
SBLC/S Standard Base Level Computer / Systems *mt*
SBM Satellite Broadcast Manager *jdr*
SBM Space Battle Manager *tb*
SBM Strategic Ballistic Missile *17, sb*
SBM System Battle Manager *fu*
SBMC System Battle Management Controller *hc*
SBN Structure-Borne Noise *fu*
SBNPB Space Based Neutral Particle Beam *hc, wp, tb*
SBO Subcarrier Oscillator *jdr*
SBORAT Shipboard Operational Readiness Assistance Training *fu*
SBP Sedimentary Basins Project *ns*
SBP Sonobuoy Processing *fu*
SBPB Space Based Particle Beam *17, sb, wp, tb, st*
SBPMCCC Space Battle Program Management Command, Control, and Communications *tb*
SBR Signal to Background Ratio *ai1*
SBR Space Based Radar [RADC] *mt, fu, 5, ic2-5, sb, hc, rg, cp, wp, tb, st*
SBR Special Boat Squadron [USA DOD] *mt*
SBR Stacked Beam Radar *fu*
SBR/IR Space Based Radar/Infrared *dh*
SBRC Santa Barbara Research Center [division of Hughes Aircraft Company] *ge, fu, hc, pl, hq5.3.5, sc3.1.6, 17, sb, hn, th4, ep, jdr, hg, kf*
SBRCEA {Santa Barbara Research Center Employees' Association} *hn49.12.4*
SBRCP [Hughes Aircraft Company customer for portable digital access transceiver P-DAT] *hc*
SBRS Sonic Boom Reporting System *mt*
SBS Satellite Business Systems [commercial COMSAT] *mt, fu, sc2.7.2, 17, hc, jb, lt, ce*
SBS Semiconductor Bilateral Switch *fu*
SBS Senior Battle Staff *mt*
SBS Silicon Bilateral Switch *wp*
SBS Space Based Surveillance *fu*
SBS Stimulated Brillouin Scattering *wp*
SBS Submarine Broadcast System *mt*
SBS Support Battle Staff *mt*
SBS Surface Barrier Transistor *fu*
SBSA Small Business Set Aside *mt*
SBSS Space Based Surveillance System *sc3.4.7, hc, tb, st*
SBSS Standard Base Supply System [USAF] *mt, hc, wp, kf*
SBT Shore Based Trainer *fu*
SBT Solar Beam Test *oe*
SBT/LBTF Share-Based Trainer/Land-Based Test Facility *fu*
SBTS Sistema Brasileiro de TelecomunicaÁoes por Satelite [Brazilian System for Telecommunications by Satellite] *sc2.8.1*
SBTWR Space Based Threat Warning Receiver *hc*
SBU Service Business Unit *jdr*
SBU Special Boat Unit [USA DOD] *mt*
SBUV Solar Backscatter Ultraviolet [spectrometer, radiometer, NOAA, NIMBUS-7] *ge, hc, ac1.6-13, aa26.10.33, aw129.14.43, hy, hg*
SBWAS Space Based Wide Area Surveillance *mt, kf*
SBX S Band Transponder *fu*
SBX Slow Burning eXplosive [bomb using dust explosions that are more effective against buildings than conventional explosions] [USA] *mt*
SC Directorate of Systems Control or Systems Code *mt*
SC Saturable Core *fu*
SC Sector Controller *fu*
SC Segment Control [computer] *-4*
SC Semi Conductor *st*
SC Serial Command *-3*
SC Shoe Cove [Canada, station] *ac3.5-5*
SC Signal Conditioner *fu*
SC Simulation Center *st*
SC Single Contact *fu*
SC Solar Cell *fu*

SC Sonar Controller *fu*
SC Source Control *tb*
SC Spacecraft [also S/C] *bo, 4, jdr, kf, oe*
SC State and Country Code *mt*
SC Status Code *fu*
SC Steering Committee *fu*
SC Stored Command *0*
SC Sub Committee *tt*
SC Subcutaneous *wp*
SC Surveillance and Control *fu*
SC Surveillance Channel *fu*
SC Switching Center *mt*
SC System Clock *fu*
SC System Code *fu*
SC System Concept *5, 17, sb*
SC System Contractor *st*
SC System Controller *mt, fu*
SC&A Security Certification and Accreditation [also SC and A] *mt*
SC&CU Signal Conditioning and Control Unit *mt*
SC&D Stock Control and Distribution system [AFLC] *wp, mt*
SC/CU Signal Conditioning and Control Unit *hc*
SC/DDS Sensor Control/Data Display System *hc*
SCA Satellite Control Authority *uh, kf*
SCA SCG CITE Authorization [a document] *pl, ep*
SCA Scope Change Agreement *jn*
SCA Scope Change Authorization *jdr*
SCA Self Controlled Assembly *-9*
SCA Sensor Chip Assembly *kf*
SCA Sequence Control Assembly *lt*
SCA Service Cryptologic Agency *mt, 10*
SCA Shuttle Carrier Aircraft *mi, ci.461, mb*
SCA Sneak Circuit Analysis *fu*
SCA Software Configuration Audit *fu*
SCA Source Code Analyzer *fu*
SCA Spanish Civil Aviation *fu*
SCA Specification Compliance Analysis *fu*
SCA Stabilization Control Amplifier *hc*
SCA Sub Contract Administrator *17, sb, fu*
SCA Switching Center AUTOVON *mt*
SCA Systme de Controle d'Attitude [attitude control system] *ai2*
SCAD Subsonic Cruise Armed Decoy [the first designation of the AGM-86 program] *mt, fu, hc, wp*
SCADA Supervisory Control And Data Acquisition *kg, fu, hi, y2k*
SCADAR Scatter Detection and Ranging *fu*
SCADC Standard Central Air Data Computer *wp*
SCADS Scanning Celestial Altitude Determination System *hc*
SCADS Supplemental Corridor Air Defense System *wp*
SCALD Structured Computer Aided Logic Design *fu*
SCALO Scanning Local Oscillator *fu*
SCALP Scenario Determined, Computer Assisted Logistics [USA] *mt*
SCAM SCSI Configured (Configuration) Automatically *kg, hi*
SCAM Subsonic Cruise Attack Missile *hc*
SCAM Support Capability Assessment Model *mt*
SCAMA Signaling, Conferencing, and Monitoring Arrangement *0*
SCAMA Switching, Conferencing, and Monitoring Arrangement *oe*
SCAMP Single Channel Anti jam Man Portable *kf*
SCAMP Standard Configuration and Modification Program *mt*
SCAMPI [network of leased wideband T1 carriers] *mt*
SCAMS Scanning Microwave Spectrometer *ac2.8-3*
SCAMS Security Combat Automated Management Subsystem *mt*
SCAN Selective Current Aerospace Notices *hn43.22.6*
SCAN Shipboard Communications Area Network [US Navy] *mt, af72.5.44*
SCAN Silent Communications Alarm Network *ci.1024*
SCAN Switched Circuit Automatic Network *mt*
SCANA Self Contained Adverse [weather] Night Attack *hc, wp*
SCANFAR Scanning Fixed Array Radar (designation for Hughes AN/SPS-32, 33 radar-computer system) *fu*
SCANG South Carolina Air National Guard *-12*
SCANS Status Configuration and Alarm Notification System *mt, hc*
SCANSCAT [advanced scatterometer for studies in meteorology and oceanography, EOS] *hy*
SCANTS Ship Control and Navigation Training System *hc*
SCAP Silent Compact Auxiliary Power *fu*
SCAP Sneak Circuit Analysis Plan *jdr*
SCAPE Self Contained Atmospheric Protection (Protective) Ensemble *9, jdr, oe*
SCAPE System Capability And Performance Evaluation *fu*
SCAQMD South Coast Air Quality Management District *hn48.20.2*
SCAR (Observations by Nimbus, GOES . . . SCAR, SLAR . . .) *aw130.11.51*
SCAR Satellite Communications Applications Research *-19*
SCAR Strike Control And Reconnaissance *wp*
SCAR Supplier Corrective Action Request [a document] *fu, gp, 17, sb, ep, jdr, tm, kf*
SCARAB [Reconnaissance RPV} *wp*
SCARF Sidelooking Coherent All Range Focuser *fu*

SCARS Status, Control, Alerting, and Reporting System [SCARS II is a SCARS upgrade] [NATO] *fu, ro, mt*
SCAS Scan String *pk, kg*
SCAS Stability and Control Augmentation System *mt, aw118.18.73*
SCAT Scatterometer [SEASAT] [nadir looking radar wind field scatterometer] *ac1.6-13, so, ge, hg*
SCAT School and College Ability Test *w9, wp*
SCAT Scout ATtack [a version of LHX] [USA] *mt*
SCAT Scout [and] Attack [helicopter, US Army] *dh, fu*
SCAT Secure CATCOMS *mt*
SCAT Sequential Component Automatic Testing *fu*
SCAT Single Contractor Aviation Training [program] *aw130.13.5*
SCAT Spacecraft Assemblies Transfer *ai2*
SCAT Supersonic Commercial Air Transport *w9, wp*
SCAT Surface Controlled Avalanche Transistor *fu*
SCAT System Control of AUTODIN Traffic *mt*
SCATANA Security Control of Air Traffic and Air Navigation(al) Aids *fu, mt*
SCATHA Spacecraft Charging At High Altitudes *hc, ci.894*
SCATR Synchronous Communication And Tracking Relay System *hc*
SCATS Self Contained Automatic Test System *hc*
SCATT [Wind Scatterometer, ESA] *hy*
SCAWG Systems Control Architecture Working Group *mt*
SCB SPO Control Board [WIS] *mt*
SCB Subsystem Control Block [IBM] *pk, kg*
SCC SAM Control Center *fu*
SCC Satellite Control Center *ac3.2-20, kf, jn*
SCC Sector Command Center *fu*
SCC Secure Communications Controller *mt*
SCC Sensor Correlator Center *fu*
SCC Sequence Control Code *-3*
SCC Serial Communications Controller *kg, fu, hi*
SCC Serial Controller Chip *pk, kg*
SCC SIE Control Center *st*
SCC Simulation and Computing Center *fu*
SCC Single Conductor Cable *fu*
SCC Space Command and Control *cp*
SCC Space Computation Center *rva, hs*
SCC Spacecraft (Satellite) Control Center *4, jdr*
SCC SPADOC Computation Center *mt, sd, jb*
SCC Special Computation Center *mt*
SCC Special Coordination Committee [NSC] *mt*
SCC Specialized Common Carriers *do80, hi*
SCC Specialized Communication Carriers *mt*
SCC Standing Consultative Committee (Commission) [set up by SALT I] *11, tb*
SCC Stress Corrosion Cracking *jn*
SCC Sub Control Center *fu*
SCC Switching Control Center *ct, em*
SCC Synchronous Channel Check [IBM] *kg, hi*
SCC System Control Center *mt*
SCC System Control Console (Center) *fu*
SCC/SOC Sector Command Center/Sector Operation Center *fu*
SCCB Software Change Control Board *kf*
SCCB Software Configuration Control Board *mt, tb, fu, oe*
SCCB Sub-Contractor Configuration Control Board [organization] *tm*
SCCC Surveillance, Command, Control, and Communications *mt*
SCCC System Control and Computing Component *-4*
SCCC System Control Console Controller *fu*
SCCCC Systems for Command, Control, Communications and Computers *mt*
SCCCCTAP System for Command, Control, Communications, and Computer Technology Application Program [SSC] [also SC4TAP] *mt*
SCCCS Ship Command Control Communication System *hc*
SCCE Satellite Configuration Control Element *mt*
SCCISA Standard Command and Control Information Systems Architecture *mt*
SCCM Self Checking Computer Module *wp*
SCCMT Software Change and Configuration Management Tool *hc*
SCCN Secure Command and Control Net [USCG] *mt*
SCCO Special Circuit Control Officer *mt*
SCCS SAC* Command Conference System *mt*
SCCS Singapore Command and Control System *fu*
SCCS Software Change Control System *kf*
SCCS Software Configuration Control System *fu*
SCCS Source Code Control System *fu, kg, hd, tb, kf*
SCCS Standard Communications Computer Summary *mt*
SCCS Standard Cubic Centimeter per Second *oe*
SCCS Switching Control Center System *ct, em*
SCCSB Software Computer Configuration Control Sub Board *mt*
SCCT System Control Console Terminal *fu*
SCD S-Band Cassegrain Diplexer *0*
SCD Silicon Capacitor Diodes *mt*
SCD Source Control Document *jn*
SCD Source Control Drawing *fu, 17, jdr, pbl, tm, jn, oe*
SCD Specification Control Drawing *fu, oe, pl*
SCD Standard Color Display *pk, kg*

SCD Stock Control and Distribution *wp*
SCD Strategic Communications Division *jb, mt*
SCD System Clarification Document *kf*
SCD System Concept Diagram *fu*
SCDC Sampled Channel Doppler Corrector *-4*
SCDC Strategic Concepts Development Center *tb*
SCDIN Strategic Air Command Digital Information Network [also SACDIN] *wp*
SCDL Surveillance and Control Data Link *fu*
SCDPD Scheduled Power Down *fu*
SCDR Software Critical Design Review *fu*
SCDR Subcontract Cost Data Report *jdr*
SCDR System-level Critical Design Review *kf*
SCDRL Subcontract[or] Data Requirements List *fu, ah, gp, 17, ep, pbl, tm*
SCE S Class [equipment] Equivalent *-17*
SCE Satellite Control Equipment *jn*
SCE Saturated Calomel Electrode *wp*
SCE Service Creation Environment *ct, em*
SCE Signal Conditioning Equipment *fu*
SCE Software Capability Evaluation *kf*
SCE Spacecraft Command Encoder *-4*
SCE System Command Element *fu*
SCE System Cost Estimate *mt*
SCEC Small Computer Engineering Center [USA] *mt*
SCENARIO Exercise Scenario [Schedule] System [JCS] *mt*
SCEO Strategic Connectivity Engineering Office *mt*
SCEPC Senior Civil Emergency Planning Committee *mt*
SCEPS Stored Chemical Energy Propulsion System *hc, fu*
SCEPTRE System for Circuit Evaluation of Transient Radiation Effects *fu*
SCER SC Engineering Request [a document] *ep*
SCERT Systems and Computer Evaluation and Review Technique *mt, wp*
SCF Satellite Checkout Facility *0*
SCF Satellite Control Facility [Sunnyvale, CA] *mt, ah, ci.892, 4, 5, sb, hc, dm, tb, jn*
SCF Signal Combining Factor *jdr*
SCF Simulation Control Function *fu*
SCF Specification Change Form *kf*
SCF Standard Cubic Feet *oe*
SCF Switched Capacitor Filter *fu*
SCF System Control Facility [Honeywell] *pk, kg, mt*
SCFEL Standard Cryptologic Facility Equipment List *mt*
SCFH Standard Cubic Feet per Hour *oe*
SCFIS Satellite Control Facility Interface System *hc*
SCFM Standard Cubic Feet per Minute *oe*
SCFS Satellite Control Facility Support *jdr*
SCG Security Classification Guide *mt, wp, tb, kf*
SCG Sensor Converter Group *fu*
SCG Space [and] Communications Group [Hughes Aircraft Company] [also S&CG] *fu, ge, hq5.3.5, hn41.16.2, 15, 17, sb, pl, ep, hg*
SCG Stored Command Generator *fu*
SCG Structured Chart Graphics *fu*
SCG/DQM Structure Chart Graphics/Design Quality Measure (System) *fu*
SCGMG Single Gimbal Control Moment Gyro *sd*
SCGMS Space and Communications Group Material Specification [a document] *ep, tm*
SCGPS Space and Communications Group Process Specification [a document] *ep, tm*
SCGS Space and Communications Group Standard [a document] *tm*
SCGWI Space and Communications Group Work Instruction [a document] *ep, tm*
SCHS School Squadron *mt*
SCI SC Instruction [a Hughes directive] *tm*
SCI Science Citation Index *wp, fu*
SCI Science Institute [Space Telescope] *no*
SCI Scientific Calculations Incorporated *fu*
SCI Secure Compartmental Information *se24.0, tb*
SCI Sensitive Compartmented Information [USA DOD] *mt, su, wp, kf*
SCI Sequence Complete Interrupt *fu*
SCI Special Compartmented Intelligence *mt*
SCI Strategic Computing Initiative *hc, tb*
SCI System Configuration Index *fu*
SCI Systems Control Incorporated *mt*
SCI [an avionics/military electronics company] (biotelemetry techniques by SCI Systems, Inc., Huntsville, Alabama) *ci.1013, aw130.22.49, jmj*
SCIAC SIGINT Correlation and Analysis Capability *wp*
SCIAMACHY Scanning Imaging Absorption Spectrometer For Atmospheric Cartography [ESA] *ge*
SCID Small Column Insulated Delay *lt*
SCID Spacecraft Identifier *oe*
SCIF Sensitive Compartmented Information Facility *mt, tb, jdr, kf*
SCIF Signal to Clutter Improvement Factor *fu*
SCIF System Certification and Integration Facility *fu*
SCIM Supersonic Cruise Intercept Missile *wp*

SCIMITAR System for Countering Interdiction Missiles and Target Acquisition Radars *mt*
SCINET Sensitive Compartmented Information Network *mt, wp*
SCIP Site Configuration and Installation Plan [NICS] *mt*
SCIP Systems Control Improvement Program *mt*
SCIPMIS Standard Civilian Personnel Management Information System [USA] *mt*
SCIRS Sensor Contributor Integrated Requirements System *hc*
SCIS Survivable Communications Integration System [NORAD] *mt, af72.5.44, wp, kf*
SCIT System Concept and Integrated Technology *mt, 17, sb*
SCITNET SOUTHCOM Intel Teletype Network *mt*
SCITS Single Channel Injection Transponder System *mt*
SCIU Satellite Command Interface Unit *jn*
SCL Separately Controlled Link *mt*
SCL Standard Conventional Load *mt*
SCLM Software Configuration and Library Manager (Management) *ct, kg, em*
SCLSC Ship Configuration and Logistics Support Control *mt*
SCM S-Band Cassegrain Monopulse [cone assembly] *hc*
SCM Samarium Cobalt Magnet *wp*
SCM Satellite Communications Module *mt*
SCM Saturated Clutter Map *fu*
SCM Scientific Computing Manual *fu*
SCM SDI Control Module *ts7-24*
SCM Software Configuration Management *fu, em, kg, jmj, wp, kf, oe*
SCM Space Cargo Modified [two C-5As, modified to carry Space Shuttle payloads] [USA] *mt*
SCM Spectrum Control Module *fu*
SCM Spectrum Control Monitor *fu*
SCM Station Class Mark *pk, kg*
SCM Strategic Cruise Missile *wp*
SCM Surface Contamination Monitor [chemical warfare] *wp*
SCM Synchronous Communication Multiplexer *gp, oe*
SCM System Configuration Manager *fu*
SCM System Control and Monitor *ag*
SCM [S-Band monopulse feedhorn and bridge system] *0*
SCMD Serial Command *jdr, uh*
SCML Software Configuration Management Library *kf*
SCMP Software Configuration Management Plan *mt, fu, wp, kf*
SCMP System Console Monitor Printer *mt*
SCMR Surface Composition Mapping Radiometer *ac2.8-3*
SCMS Source Code Management System *fu*
SCMT Secure Clear Mode Timer *vv*
SCMU Small Community Master Unit *fu*
SCN Satellite Control Network [SPACEMD] *mt*
SCN Search Control Number *wp*
SCN Sensor Control Network *wp*
SCN Shipbuilding and Conversion Navy [USA] *mt*
SCN Site Communications Network *jn*
SCN Software Change Notice *fu*
SCN Specification Change Notice [a document] *mt, fu, ah, he, pl, hc, wp, ep, tb, jdr, pbl, tm, jn, oe*
SCNA Self Contained Night Attack *hc*
SCNS Self Contained Navigation System [NASA] *fu, af72.5.142*
SCO Santa Cruz Operation [software company] *ct, em, kg*
SCO State Coordinating Officer *mt*
SCO Subcarrier Oscillator *bl2-7b.5, 0, 4, fu, oe*
SCO System Control Officer *mt*
SCOAS Small Computer / Office Automation Services *mt*
SCOASB Small Computer / Office Automation Service Branch *mt*
SCOASO Small Computer / Office Automation Service Organization *mt*
SCOC Satellite Control and Operation Center *-9*
SCOC Sector Command Operations Center *fu*
SCOC System Control Operations Center [SPACECMD] *mt*
SCOCE Special Committee On Compromising Emanations *mt*
SCOD Sub Contract and Other Direct costs *mt*
SCODA Scan Coherent Doppler Attachment *fu*
SCODP Scan Conversion Object Description Page *hi*
SCOE Software Center Of Excellence *tb*
SCOMP Secure Communications Processor [Honeywell] *tb, mt*
SCOOP Study Confirmation Of Optical Phenomenology *hc*
SCOP System Concept Option Plan *mt*
SCOPA Survivable Concentrating Photovoltaic Array *aa26.12.47, hc*
SCOPE Schedule, Cost, Performance *wp*
SCOPE Simple Check Out Oriented Program Language *wp*
SCOPE Simple Communications Programming Environment [Hayes] *kg, hi*
SCOPE [nickname for HQ USAF programs usually connected with communications] *wp*
SCOPE_DIAL [base telecommunications modernization] [USAF] *mt*
SCOPE_SIGNAL [USAF program for world wide HF ground toair broadcast] *mt*
SCOR Scientific Committee on Ocean (Oceanic) Research *hg, pl, ge*
SCORE Signal Corps Experimental [US satellite, 1958] *gs2-1.2*
SCORE Spacecraft Operational Readiness Exhibit System *mt*
SCOSTEP Scientific Committee on Solar-Terrestrial Physics *ge, hg*

SCOT Satellite Communication Terminal *wp*
SCOT Single Channel Objective Terminal *wp, tb*
SCOT Special Command, Control and Communication [CCC] Operations and Testing [RADC] *mt*
SCOTS Shuttle Compatible Orbital Transfer Subsystem *sc4.7.3*
SCOTT Single Channel Objective Tactical Terminal *mt, hc, fu*
SCOUT Surface Controlled Oxide Unipolar Transistor *wp, fu*
SCP Satellite Control Processor *fu*
SCP Secure Conferencing Program *mt*
SCP Secure Conferencing Project *mt, wp*
SCP Service Control Point *ct, em*
SCP Signal Control Processor *hc*
SCP Simulation Computer Program *fu*
SCP Software Change Proposal *fu*
SCP Source Control Program *wp*
SCP Space and Communications Practice [Hughes directive] *tm*
SCP Spacecraft Control Processor *he, uh, jdr, cl, kf, vv, jn*
SCP Spherical Candlepower *fu*
SCP Standard Change Package *fu*
SCP Stored Command Processor *gg74, 8*
SCP Strategic Computing Program *mt*
SCP Subnet Communications Processor *mt*
SCP Subsystem Control Port *pk, kg*
SCP Support Center Pacific *mt*
SCP Support Computer Program *fu*
SCP Surveillance/ Communication Processor *fu*
SCP Surveyor Command Program *0*
SCP Survivable Command Post *mt*
SCP System Change Package *mt*
SCP System Concept Paper *mt, hc, wp, tb, st*
SCP System Control Program *pk, kg*
SCP [Computer Security Program Office] *mt*
SCPC Single Channel Per Carrier *mt, fu, ct, csIII.2, 7, lt, ce*
SCPC Single Channel Per Connection *em*
SCPE Simplified Collective Protection Equipment [chemical warfare] *wp*
SCPI Session Control Programmable Interface *mt*
SCPLRS Small Community PLRS *fu*
SCPO Senior Chief Petty Officer *w9*
SCPPL System Configuration Provisioning Parts List *fu*
SCPS Survivable Collective Protection Shelters [chemical warfare] *wp*
SCPS Survivable Collective Protective System *af72.5.99*
SCPS-M Survivable Collective Protection System - Medical [AFLC] *mt*
SCR Schedule Change Request *jdr*
SCR Science Confirmation Review *re*
SCR Selective Chopper Radiometer *ac2.8-3*
SCR Semiconductor (Silicon) Controlled Rectifier *fu, kg, 0, 6, wp, hi*
SCR Sidelobe Clutter [power] Ratio *ic2-50*
SCR Software Change Request *kf, jn, oe*
SCR Software Cost Reduction *fu*
SCR Special Contract Requirement *kf*
SCR Specification Clarification Request *fu*
SCR Specification Compliance Review *fu*
SCR Stolen Computer Registry *hi*
SCR Strip Chart Recorder [display] *uh, gp, oe*
SCR Surface Contouring Radar *hy*
SCR Sustained Cell Rate *ct, em*
SCR System (Software) Change Request *fu*
SCR System Circuit Register *mt*
SCR System Concept Review *gp, fu, oe*
SCRA Single Channel Radio Access *mt*
SCRAM Spares Control, Release And Monitoring *wp*
SCRAM Supersonic Combustion Ramjet *wp*
SCRAMJET Supersonic Combustion Ramjet *wp, tb*
SCRAT Safety Criteria Retrieval and Tabulation *fu*
SCRB Software Change Review Board *fu, hd1.89.y30*
SCRB Software Configuration Review Board *wp*
SCRB SPO Control Review Board *mt*
SCRC Small Computer Requirements Contract *mt*
SCRG Stable Coordinate Reference Guidance *hc*
SCRS Scalable Cluster of RISC Systems *pk, kg*
SCRS Sorption Compressor Refrigerator System *hc*
SCS Safety and Crashworthiness System [General Motors] *hc*
SCS Satellite Control Satellite *hc, lt*
SCS Satellite Control Site *sc4.1.center*
SCS Satellite Control Station *jdr*
SCS Sea Control Ship *fu*
SCS Security Control Station *mt*
SCS Self-Contained System *fu*
SCS Semiconductor Controlled Switch *fu*
SCS Sensors and Communications Systems *ah*
SCS Ship Control Station *fu*
SCS Simulation Control Subsystem *fu*
SCS Single-Channel Complex *fu*
SCS Small Computer System *fu*
SCS Society for Computer Simulation *aa26.8.b34*
SCS Spacecraft Simulator [DSP] *kf, oe*

SCS Stabilization and Control [Sub] System *ci.390, fu*
SCS Standard Communications Computer System *mt*
SCS Standardized Citation Score *fu*
SCS Submarine Combat System *fu*
SCS Support Center Switch *fu*
SCS System Control Set *fu*
SCSA Signal Computing System Architecture [Dialogic] *kg, hi*
SCSB Small Computer Support Bulletin *mt*
SCSC Schedule Control System Criteria *-8*
SCSC South Carolina State College *hn49.14.3*
SCSC Strategic Conventional Standoff Capability *dh, wp*
SCSD Systems Control and Support Directorate *mt*
SCSE Spacecraft Simulation Equipment *kf*
SCSI SCSI Parallel Interface *pk, kg*
SCSI Small Computer Standard Interface *em*
SCSI Small Computer System(s) Interface *em, ct, wp, fu, kg, is26.1.50, hi*
SCSMO Systems Control Software Management Office *mt*
SCSRSS Standard Command Supply Review System - SAILS [USA] *mt*
SCSS Ship's Combat System Simulation *fu*
SCSS Space Station Communication System Simulation *hc*
SCSW Single Conductor Shielded Wire *jdr*
SCSW Systems Communications Software *mt*
SCT Satellite Communications Terminal *mt*
SCT Schmidt-Cassegrain Telescope *mi*
SCT Secretarìa de Communicaciones y Transportes [Secretariat for Communications and Transports] [Mexico] *sc2.11.1*
SCT Sector Control Technician *fu*
SCT Shuang Cheng Tzu [satellite launch site, China] *ac3.4-4*
SCT Single Channel Transponder *mt, hc, wp, fu*
SCT Software Certification Test *fu*
SCT Surface Charge Transistor *fu*
SCT System Compatibility Test *mt*
SCT System Confidence Test *fu*
SCT System Control Team *tb*
SCTC Small Computer Technical Centers *mt*
SCTC Software Development and Training Center *fu*
SCTC Spacecraft Thermal Cycle (Cycling) *uh, jdr*
SCTC System Console Transfer Channel *mt*
SCTIS SCT [EAM] Injection Subsystem *hc*
SCTIS Single Channel Transponder Injector System *mt*
SCTR Signal Corps Technical Requirement *fu*
SCTV Spacecraft Thermal Vacuum [a test] *uh, 9, lt, ep, jdr, kf, jn*
SCTV Spun Component of Spacecraft Thermal Vacuum *-8*
SCU Scan Converter Unit *fu*
SCU Scanner Control Unit *fu*
SCU Sensor Control Unit *wp*
SCU Service Control Unit *-4*
SCU Signal Conditioner (Conditioning) Unit *fu, bo, 9, aw130.11.71, lt, oe*
SCU Slew Control Unit *fu*
SCU STAJ Control Unit *fu*
SCU Standalone Computer Unit *fu*
SCU System Control Unit *mt, fu*
SCUBA Self Contained Underwater Breathing Apparatus *0, wp, pbl*
SCUD Subsonic Cruise Unarmed Decoy *wp*
SCUD [Soviet army level mid range battlefield missile] *wp, 12*
SCV Subclutter Visibility *fu*
SCVM Shuttle Command and Voice Multiplexer *wp*
SCVR Super Crystal Video Receiver *fu*
SCWBS Subcontractor Contract Work Breakdown Structure *fu*
SCWG Satellite Communications Working Group [NATO] *mt*
SCWG Security Certification Working Group *mt*
SD Sample Data *fu*
SD Scan Duration *mt*
SD Scoping Document *fu*
SD Secondary Demonstration *fu*
SD Send Data *fu, kg*
SD Senior Director *fu*
SD Sensor Data *oe*
SD Serial Decimal *-4*
SD Situation Display *fu*
SD Smart Dispenser *ai2*
SD Software Development directorate *mt*
SD Space Division [of U. S. AFSC] *hc, sc2.8.2, 17, sb, pl, wp, tb, jdr*
SD Specification Document *fu*
SD Spectral Distribution *fu*
SD Standard Deviation *fu, ac2.4-5, w9, wp, tb*
SD Strategic Defense *tb*
SD Subcontract Directive *tb*
SD Super Density *pk, kg*
SD Supplier Directly [list] *sb*
SD Supplier Directory *fu*
SD Sweep Driver *fu*
SD System Data *fu*
SD System Description *mt*

SD&IC System Design and Integration Contract (a Hughes contract with NACISA) *fu*
SD-ROM Super Density ROM *pk, kg*
SD/SDM System Design/Static and Dynamic Modeling *fu*
SDA Ship Design Agent *fu*
SDA Software Disk Array *pk, kg*
SDA Source Data Analysis *fu*
SDA Source Data Automation *kg, fu*
SDA Strategic Defense Architecture *tb*
SDA Subcontract Data Administrator *fu*
SDA System Display Architecture [Digital] *kg, hi*
SDA SystÈme de Detection AeroportÉe [airborne detection system in the Boeing E-3F Sentry aircraft] [France] *mt*
SDACC Source Data Automation for Command and Control *hc*
SDAE Source Data Automation Equipment *hc*
SDAF Special Defense Acquisition Fund *fu*
SDAIP System Description, Analysis and Implementation Plan *oe*
SDANG South Dakota Air National Guard *-12*
SDAPS Space Defense and Asset Protection Study *hs*
SDAR System Design Analysis Report *jb*
SDAS System Design and Analysis Support *mt*
SDAT Spacecraft Data Analysis Team *0*
SDB Serial Data Bus *fu*
SDB Small Disadvantaged Business *hq6.2.1, kf*
SDB Source Data Base *oe*
SDB Standby Dispersal Base *mt*
SDB Survivability Data Base *st*
SDB Symbolic Debugger [Unix] *kg, hi*
SDB Synchronous Data Buffer *pl*
SDBC Serial Data Bus Controller *fu*
SDC Sector Duty Center *fu*
SDC Secure Data Communications *fu*
SDC Signal Data Converter *fu, hn49.3.1*
SDC Silicon Detector Company *oe*
SDC Situation (Status) Display Console *fu*
SDC Situation Display Converter *fu*
SDC Slave Display Console *mt*
SDC Software Development Center [USA] *fu, mt*
SDC Source Data Collection *fu*
SDC Space Data Corporation *17, sb*
SDC Strategic Defense Command [US Army] *mt, ts4-14, wp, tb, st*
SDC Strategic Direction Center [SACLANT] *mt*
SDC Structural Data Collector *aw130.11.71*
SDC Structural Design Criteria *jdr*
SDC Systems Development Corporation *mt, fu*
SDCC Strategic Defense Command Center *5, 17, sb, tb*
SDCCC Standard Desktop Computer Companion Contact *mt*
SDCD Super Density Compact Disk *hi*
SDCE Signal Data Conversion Equipment *fu*
SDCE Software Development Capability Evaluation *kf*
SDCOM Strategic Defense Command [also SDC] *tb*
SDCR System Design Change Request *tb*
SDCS Situation Display Console System *wp*
SDCS Survivable DSCS Control Segment *mt*
SDCT System Design Certification Test *fu*
SDD Software Description Database *hi*
SDD Software Description Database [Internet] *pk, kg*
SDD Software Design Document *fu, pl, wp, jdr, oe*
SDD System Definition Document *tb*
SDD System Description Document *kf, fu, jn*
SDD System Design and Development *fu*
SDD System Design Document *fu*
SDD Systems Development Documentation *-15*
SDDBMS Secure Distributed Database Management System *mt*
SDDD Software Detail(ed) Design Document *mt, hc, tb, ah, pbl, jdr, jn, fu*
SDDE Secure Digital Data Equipment *fu*
SDDL Software Design Documentation Language *fu*
SDDM Secretary of Defense Decision Memorandum *mt, hc, st*
SDDM Standard Design Drawing Manual *fu*
SDDN Secure Data Defense Network *mt*
SDDP Shipboard Data Display Processor *fu*
SDDR Subcontractor Detail Design Review *fu*
SDDS Sensor Data Distribution System *fu*
SDDS Software Design and Develop-ment Segment *fu*
SDDU Simplex Data Distribution Unit *fu*
SDE SADA Drive Electronics *-4*
SDE Software Design Engineer *fu*
SDE Software Development Environment *sd, tb, kf*
SDE Standard Data Element *mt*
SDE Standard Data Entry *mt*
SDEM Source Data Entry Module *hc*
SDEX Standard Executive *fu*
SDF Satellite Database Facility *jr*
SDF Software Development Facility *mt, jr*
SDF Software Development File *mt, 17, tb, kf, fu*

SDF Software Development Folder *jr, jdr*
SDF Space Delimited File *pk, kg*
SDF Space Delimited Format *kg, hi*
SDF Standard Distance File *mt*
SDF Supplementary Data File *so*
SDF System Development Facility *mt, wp, fu*
SDF System Development File *fu*
SDFOV Simultaneous Dual Field Of View *hc*
SDFP Sensor Dependent Front-End Processing *kf*
SDG Safe Data Group *fu*
SDG Simulated Data Generator *fu*
SDG Situation Display Generator *fu*
SDG Standard Design Guide *fu*
SDGN System Design Guidance Notebook *fu*
SDGP Simulated Data Generator Program *fu*
SDH Synchronous Digital Hierarchy [signal interface for high speed digital fiber transmission] *nu, ct, em, kg, hi*
SDHS Satellite Data Handling System [USAF] *mt, hc, pl, pl*
SDHS-COMNET SDHS Communications Network *mt*
SDI Selective Dissemination of Information *pk, kg*
SDI Sensor Data Interface *oe*
SDI Serial Data In(terface) *fu, hi*
SDI Ships Drawing Index *fu*
SDI Single Document Interface *kg, hi*
SDI Software Development Interface [Mosaic] *kg, hi*
SDI Space Defense Initiative *ro*
SDI Specially Designated Interceptor *fu*
SDI State [unemployment] Disability Insurance *hn25.11.2*
SDI Strategic Defense Initiative [an anti ICBM defense project, USA] *hc, sc4.9.3, 1, 5, 11, 14, 17, ic2-19, sb, th4, aw129.14.25, aa26.12.32, af72.5.150, wp, tb, st, mb, kf, mi, fu, mt*
SDI Supplier (Subcontractor) Developed Item *fu*
SDI/SAS Strategic Defense Initiative System Architecture Study *sb*
SDIA Special Design Inspection Aids *fu*
SDIAC Strategic Defense Initiative Advisory Council *wp, tb*
SDIAS Strategic Defense Initiative Architecture Study *17, sb*
SDII Strategic Defense Initiative Institute *wp*
SDIN Special Defense Intelligence Notice *mt*
SDINET Strategic Defense Initiative Network *tb*
SDIO Strategic Defense Initiative Organization (Office) [later BMDO] *mt, kf, mi, aw130.13.104, hc, 1, 11, 17, sb, aa26.10.36, nt, dt, wp, tb, st, mb*
SDIP Software Development Integrity Program *wp*
SDIP Strategic Defense Initiative Program *tb*
SDIS Satellite Data Insertion Device *hc*
SDIS Surveillance and Display Interface System *mt*
SDISM Strategic Defense Initiative Simulation *tb*
SDIT Software Design and Integration Test *fu*
SDK Software Developer's (Development) Kit [Microsoft] *em, kg, hi*
SDK Sredniy Desantniy Korabl [medium landing ship] [USSR] *mt*
SDL Sensor Down Link *oe*
SDL Serial Data Link *fu*
SDL Shielded Data Link *ct, em*
SDL Software Design Library *fu*
SDL Software Development Laboratory *fu*
SDL Software Development Library *mt, kf*
SDL Space Disturbance Laboratory *gp*
SDL Structured Design Language *fu*
SDL System Design Language *fu, hi*
SDLC Synchronous Data Link Control [protocol] *ct, em, fu, kg, csIII.48, wp, tb, hi*
SDLM Standard Depot Level Maintenance [US Navy] *hc, aw126.20.100*
SDLS Satellite Data Link Standard *uh, kf*
SDLS Space Data Link Standard *tb*
SDLS Standard Data Link System *-17*
SDM Secure Digital Modem *mt*
SDM Semantic Data Model *fu*
SDM Site Defense of Minuteman [an ABM research program, USA] *mt*
SDM System Design Memorandum *fu*
SDM System Design Model *fu*
SDM System Development Methodology *fu*
SDM System Development Multitasking *pk, kg*
SDME Software [System] Development and Maintenance Environment Software *mt*
SDMMC Software Development, Modification and Maintenance Contract *mt*
SDMP System Development Master Plan *fu*
SDMS SCSI Device Management System [NCR] *kg, hi*
SDMS Shipboard Data Multiplex System *fu*
SDMS Spatial Data Management System [US Navy] *mt*
SDMT Silicone Deformable Mirror Technology *fu*
SDMX Space Division Matrix *mt*
SDN Secret Document Number *sy14, hc, fu, su, uh*
SDN Software Defined Network [AT&T] *mt, kg, hi*
SDN System Development Notification *mt*
SDNS Secure Data Network Service *fu, kg*
SDO Screening and Debuffing Optimization *fu*

SDO Senior Duty Officer *mt*
SDP Serial Data Processor *fu*
SDP Shipboard Data Processor *fu*
SDP Signal Data Processor *kf*
SDP Silent Discharge Plasma *wp*
SDP Software Development Plan *mt, fu, pl, hc, wp, jdr, kf*
SDP Solar Drum Positioner *ma78, lt, jdr*
SDP Status Display Panel *fu*
SDP Storage and Distribution Point *wp*
SDP Strategic Defense Program *fu*
SDP Student Device Processor *fu*
SDP Synchronous Distributed Processor *fu*
SDP System Data Processor *fu*
SDP System Decision Package [Paper] *mt*
SDP System Design Process *jn, pbl, ah*
SDP System Development Plan *mt*
SDPE Special Design Protective Equipment *fu*
SDPF Sensor Data Processing Facility *-4*
SDPF Situation Display Processing Function *fu*
SDPF Spacelab Data Processing Facility *-2*
SDPL Sensor Data Processing Laboratory *mt, kf*
SDPP Software Development Project Plan *fu*
SDPS SAR Data Processing Subsystem *ac2.7-11*
SDPSK Symmetric Differential Phase Shift Key *jdr*
SDPVGS Scanning Digitizer and Prospective View Generating System *hc*
SDR Sensor Data Record *ge, ac7.5-9, pl, dm, hg, pbl*
SDR Site Defense Radar *tb*
SDR Software Discrepancy Report *oe*
SDR Solid Ducted Rocket *hc*
SDR Sonar Data Recorder *fu*
SDR Sponsor-Designated Representative *fu*
SDR Streaming Data Request *pk, kg*
SDR Surface Duct Receiver *fu*
SDR System (Subsystem) Design Review [Board] *mt, wp, 4, pl, hc, gp, sb, ts4-13, nt, ep, dm, tb, st, wl, jdr, kf*
SDR System Design Report *fu*
SDR System Development Requirement [USA] *mt*
SDRA Software Development Resources Architecture *mt*
SDRAM Synchronous Dynamic Random Access Memory *kg, em, hi*
SDRC Structural Dynamics Research Company *jn*
SDRL Subcontract(or) Data Requirements List *fu, kf, jdr, jn, oe*
SDRL Subcontractor Deliverable Requirements List *ep*
SDRM Spares Detailed Requirements Maintenance *fu*
SDRS {Software Documentation Requirements Specification} *hd1.89.y15*
SDS Satellite Data System *mt, fu, sc2.8.2, 14, 17, sb, lt, tb, kf*
SDS Satellite Defense System *fu, 17, sb, hc, tb*
SDS SIGINT Dissemination System *fu*
SDS Software Development System (Support) *fu*
SDS Spacecraft Design Specification *0*
SDS Standard Depot System [USA] *mt*
SDS Strategic Defense System *mt, tb*
SDS Supplier Data Sheet *fu*
SDS Surface Distribution System [USA] *mt*
SDS Surveillance Direction System *fu*
SDS SYSOPS Distribution System *pk, kg*
SDS System Demonstration Scenario *-4*
SDSC San Diego Supercomputer Center *ct, em*
SDSD Satellite Data Services Division [of NOAA] *ns*
SDSF Software Development Support Facility *fu*
SDSF System (Spool) Display and Search Facility *ct, em*
SDSL Single-Line Digital Subscriber Line *em, kg*
SDSP Software Development Support Plan *mt*
SDSP Space Defense System Program *mt, jb*
SDSS Shuttle Data Select Switch *-2*
SDSS Software Development Support System *hc*
SDSS [one of two major SDI architecture evaluation models] *17, sb*
SDT Satellite Data Transmission *hc*
SDT Scientific Data Translator *0*
SDT Semiconductor Division Technology Center *fu*
SDT Senior Director Technician *fu*
SDT Simulated Data Tape *fu*
SDT Soldier Data Tag [USA] *mt*
SDT Standalone Display Unit *fu*
SDT Status Determination Test *fu*
SDT System Design Tool *hc*
SDT System Diagnostic Test *fu*
SDTC Software (System) Development and Training Center *fu*
SDTF Software Development and Test Facility *kf*
SDTL System Development and Test Laboratory *hc*
SDTN Synchronous Digital Transmission Network *mt*
SDTP Signal Detection and Time of Arrival Processor *fu*
SDU Secure Data Unit *mt, hc, fu*
SDU Service Data Unit *nu*
SDU Signal Distribution Unit *4, mj31.5.30*

SDU Simulator Demonstrator Unit *hc*
SDU Slave Display Unit *mt*
SDU Spun Driver Unit *gp*
SDU Squib Driver Unit *he, uh, cl, jdr, kf, jn*
SDUC SDU Controller *fu*
SDUC Secondary Data Utilization Center *ac3.2-19*
SDUS Small Data Utilization Station *bl3-1.58, pbl*
SDUSN SDU Serial Number *fu*
SDV Shuttle Derived Vehicle *tb*
SDV Slowed Down Video *fu*
SDV Switched Digital Video [AT&T] *kg, hi*
SDVD System Design Verification Demonstration *fu*
SDVS Software Design and Verification System *bo*
SDVS State Delta Verification System *tb*
SDVT System Design Verification Test *fu*
SDW Single Data Word *fu*
SDWA Single Data Word Available *fu*
SDX Satellite Data Exchange *hi*
SDX Storage Data Acceleration *em, kg*
SDZ Space Defense Zone *sd*
SD_STB Streaming Data Strobe [IBM] *hi, kg*
SE Secondary Emitter *fu*
SE Shielding Effectiveness *fu*
SE Simulation Equipment *-4*
SE Single Event Transient *he*
SE Software Engineering *mt, fu, hi*
SE Source Engineering *jdr*
SE Spacecraft Element *jdr*
SE Special Equipment *fu*
SE Specialty Engineering *fu*
SE Support Equipment *mt, fu, 17, sb, hc, wp, tb, jdr, kf*
SE Switched Ethernet *hi*
SE System Exercises *fu*
SE Systems Effectiveness *pl, fu*
SE Systems Engineer(ing) *fu, 17, sb, pl, nt, jdr, kf, oe*
SE [High Component Systems Engineering (IPT Identifier)] *kf*
SE&I System Engineering And Integration *fu, wl*
SE&T System Engineering and Test *fu*
SE/TA Systems Engineering/Technical Analysis *-17*
SEA (served with distinction in SEA in 1972-73) *af72.5.136*
SEA Science and Engineering Associates *fu*
SEA Senior Enlisted Advisor *mt*
SEA Software Engineering Activity *fu*
SEA South East Asia *wp, ro*
SEA Standard Extended Attribute [OS/2] *kg, hi*
SEA Sun Elevation Angle *uh*
SEA Systems Engineering Analysis *fu*
SEA Systems Evaluation Area *fu*
SEA [satellite for Earth observation, India] [also SEO] *ac3-i*
SEAADSA Sea Automated Data Systems Activity [US Navy] *mt*
SEABEE Sea Barge System *mt*
SEAC Standards Eastern Automatic Calculator [name of first computer to use transistors, built by National Bureau of Standards] [Also see SWAC] *pk, kg*
SEACOP Strategic Sealift Contingency Planning System [MCS] *mt*
SEACP System Exercise and Computer Program *fu*
SEAD Suppression of Enemy Air Defenses [USA DOD] *wp, mt*
SEADCUGNAV SEA Data Communications User Group *mt*
SEADS Shuttle Reentry Air Data System *ci.971*
SEAFARER Surface ELF Antenna for Addressing Remotely EMP Receivers *mt*
SEAL Screening External Access Link [Digital-DEC] *pk, kg*
SEAL Sea, Air, Land [USA DOD] *wp, mt*
SEAL Secure Electronic Authorization Laboratory *em, ct*
SEAL Segmentation and Reassembly Layer [protocol] *pk, kg*
SEAL Severe Environmental Air Launch *hc*
SEAL Simple and Efficient Adaption Layer *em, ct*
SEAM Sidewinder Expanded Acquisition Mode [an upgrade of the guidance systems for the AIM-9] [USA] *hc, mt*
SEAN Strapdown ESG Aerospace Navigation *hc*
SEAO Systems Engineering, Analysis and Operations *pbl, jn, ah*
SEAORS Southwest Asia Operational Requirements *hc*
SEAP Specialized Emphasis Area Plan *wp*
SEAR System Engineering Analysis Report *wp*
SEAS Submarine Expert Analysis System *pf1.10*
SEAS Systems Engineering Analysis Support *hc*
SEAS Systems Engineering Analysis System *tb*
SEAS Systems, Engineering, and Analysis Support *mt*
SEASAT Sea Satellite [for marine and weather observations] [NASA] *fu, ge, ac2.3-10, ns, hy, hg*
SEASTAG SEATO Standardization Agreement *mt*
SEASTRAT Sealift Strategic Contingency Planning System [MSC] *mt*
SEAT Site Equipment Acceptance Testing *fu*
SEATECS Software Engineering Automation for Tactical Embedded Computer *fu*
SEATO Southeast Asia Treaty Organization [now defunct] *fu, su, hc, w9, wp, tb*

SEAWIFS [NASA] *ge*
SEA_LITE [control and pointer/tracker system for high energy laser beam] *hn4, 17, hc*
SEA_PHOENIX [basic point defense ship missile system] *hc*
SEA_SPARROW [basic point defense surface missile system] *hc*
SEB Source Evaluation Board *mt, re, tb, wl, jdr*
SEB/SEGR Single Event Burnout Gate Rupture *he*
SEBM Scanning Electron Beam Microscope *fu*
SEBS Submarine Emergency Buoyancy System [USA] *mt*
SEC Satellite Education Center *jgo*
SEC Security Clearance *mt*
SEC Sensor and Equipment Controller *st*
SEC Single Edge Contact *em*
SEC Single Engine Centaur *ai2*
SEC Single Error Correction *kg, sp65, hi*
SEC Space Environment Center [Boulder, Colorado] *oe*
SEC Space Event Conference *mt*
SEC Spacecraft Equipment Converter *-4*
SEC Standard Evaluation Circuit *is26.1.70*
SECAM Sequentiel Couleur Avec (‡) MÈmoire [(French) Sequential Color With Memory] *ct, em, dn, kg, hi*
SECAN Communications Security and Evaluation Agency *mt*
SECAS Ship Equipment Configuration Accounting System [US Navy] *hc, fu, mt*
SECC Survivable Enduring Command Center *mt*
SECDEF Secretary of Defense *mt, fu, hc, wp, tb, st, jdr*
SECGRU Security Group *mt*
SECNAV Secretary of the Navy [USA] *mt*
SECO Second Engine Cut Off [of a 3 stage launch vehicle] *jn, oe*
SECO Sequential Coding *fu*
SECO Sustainer Engine Cut Off [Atlas] *0, uh, fu, oe*
SECOMP Secure Enroute Communications Package *mt*
SECOR Sequential Collation Of Range *fu, le37*
SECORD Secure Cord Switch Board *mt*
SECP Systems Engineering Common Process *kf*
SECPLN Security Plan *mt*
SECSTATE Secretary of State *mt*
SED Sensor Evolution(ary) Development *mt, hc, kf*
SED Simulative Electronic Deception *mt*
SED Software Engineering Division *fu, hd1.89.y30*
SED System Environment Document *fu*
SED [Electronics firm, Saskatoon, Saskatchewan, Canada] *hn41.16.2*
SEDD Systems Engineering and Development Division *fu*
SEDE Systems Engineering Design Environment *fu*
SEDI Systems Effectiveness Department Instruction *fu*
SEDIS Site Exercise Data Injection System *mt*
SEDO Spacecraft Engineering Documentation Organization *jn*
SEDP System Engineering Development Proposal *wp*
SEDS Students for the Exploration and Development of Space *mi*
SEDS System Effectiveness Data System *hc, kf*
SEDSCAF Standard ELINT Data System Codes And Format *fu, wp*
SEE Secondary Electron Emission *wp*
SEE Single Event Effects [survivability] *he, jn*
SEE Software Engineering Environment *mt, fu, hc, hd1.89.y29, kf, fu*
SEE South Enter Earth Pulse *-9*
SEED Self Electro-Optical Effect Device *aa26.10.29*
SEED Supply of Essential Engineering Data *fu*
SEEDS Spacetrack Experiment Evaluation Design Study *hc*
SEEK [nicknames for USAF projects]
SEEK_BAT [EGLIN AFB program] *hc*
SEEK_BUS [AF communications system merged into JTIDS] *hc*
SEEK_BUS [high data rate, anti jam communications system] *mt*
SEEK_BUS [Hughes project for a digital communications system - TDMA for AWACS] *fu*
SEEK_FLEX [post 75 auto-TACC design study, ESD] *hc, fu*
SEEK_FROST [distant early warning system, ESD] *hc*
SEEK_IGLOO [Alaskan air defense and surveillance radar system] *fu, hc, 12*
SEEK_IGLOO [the FPS-39 radar system, a ground-based radar now replacing older systems in the DEW line] [USA] *mt*
SEEK_RAM [WPAFB countermeasures program] *hc*
SEEK_SCORE [ESD program] *hc*
SEEK_SENTRY [ESD advanced radar program] *hc*
SEEK_SPINNER [ground launched, low speed emitter-attack weapon system] *wp*
SEEK_STORM [ESD hurricane finding radar program] *hc*
SEEK_SWITCH [national Iranian communication system] *hc*
SEEK_TALK [jam resistant USAF communication system] *wp*
SEEK_TALK [RADC program] *hc*
SEEK_TELL [EGLIN AFB program] *hc*
SEELEY_TED [NSA program] *hc*
SEEM Secondary Electron Emission Mass Spectroscopy *wp*
SEEM Support Electronic Equipment Module *fu*
SEER Sensor Experimental Evaluation and Review *hc, 17*
SEER Software Evaluation and Estimation of Resources *fu, hd1.89.y30*
SEER System Evaluation and Estimation of Resources *kf*
SEESET System End-to-End Simulator / Emulator Testbed *17, sb*

SEF Satellite Evaluation Facilities *mt*
SEF Software Engineering Facility *fu*
SEG SGL Electronics Group *vv*
SEG Simulation Engineer(ing) Group *ts4-17, tb*
SEGR Single Event Gate rupture *he*
SEH Structured Exception Handling *pk, kg*
SEH Systems Engineering Handbook *fu*
SEI Second Engine Ignition *tb*
SEI Shippable End Item *fu*
SEI Software Engineering Institute *mt, jdr, kf, fu*
SEI Space Exploration Initiative *mi*
SEI Special Emphasis Item *af72.5.104*
SEI Special Experience Identifier *mt*
SEI Specific Emitter Identification *hc*
SEI Systems Engineering and Integration *fu, wp, tb*
SEID System Electrical Interface Definition *jdr*
SEIMSS System Engineering Integration Management Support Services *oe*
SEIP Standard Engineering Installation Package *mt*
SEIP Systems Engineering Implementation Plan *wp*
SEIP Systems Engineering Internship Program *msm*
SEIS Senior Executive Information System *mt*
SEIT Systems Engineering, Integration, and Test *fu, tb, kf*
SEIT_IPT Systems Engineering, Integration and Test IPT *kf*
SEL Single Event Latch-up *he, jdr*
SEL Space Environment Laboratory [Boulder, Colorado] *gp, no, pr, oe*
SEL Support Equipment List *fu*
SEL System Engineering Laboratories [division of Gould] *-17*
SEL System Engineering Language *fu*
SELA System Error Log Analyzer *mt*
SELCAL Selected Calling *fu*
SELCAL Selective Call *mt*
SELCH Selector Channel *fu*
SELDADS Space Environmental Laboratory Data Acquisition and Display System *no, pr*
SELETS System Engineering Laboratory Estimating Technique System *gb*
SELF Submarine Extremely Low Frequency [cfr. ELF] *mt*
SELOR Shipboard Emitter Location Group *mt*
SELREL Selective Release *mt*
SELRES Selective Reserves *mt*
SELRIP Selective Release Improvement Program *mt*
SELS Severe {?} Local Storms [forecasting] *oe*
SELSCAN Automatic Frequency Scanning / Channel Selecting *mt*
SELSIS Space Environment Laboratory Solar Imaging System *no, pr*
SELSYN Self Synchronous *fu*
SELVAX {the SPAN address of the SEL VAX computer} *pr*
SELVS Small Expendable Launch Vehicle Services *ai2*
SEM Satellite Ejection Mechanism *-8*
SEM Scanning Electron Microscope *fu, kg, bl2-3.32, hc, hd1.89.y19, wp, jdr, oe*
SEM Significant Events Message [REDCOM/JDA] *mt*
SEM Software Engineering Manager *fu*
SEM Software Estimation Model *kf*
SEM Space Environment Monitor [on GOES, TIROS] *ge, sc4.7.center, no, hy. jdr, hg, oe*
SEM Spiral Electron Multiplier *un*
SEM SQL Enterprise Manager *hi*
SEM Standard Electronic Module *fu, kg, hc, aw129.10.57, wp*
SEM Stray Energy Monitor *lt*
SEM System Engineering Management *fu*
SEM System Engineering Methodology *fu*
SEM System Error Message *ct, em*
SEM-E Standard Electronics Module - Extended *kf*
SEMA Special Electronic Mission Aircraft *mt, hc, uc*
SEMA-X Special Electronic Mission Aircraft {Experimental}[a twin turbofan VSTOL aircraft designed by Grumman] [USA] *mt*
SEMANOL Semantics Oriented Language *ro*
SEMAT Site Equipment Modification Acceptance Test *fu*
SEMATECH (E. D. Fett, SEMATECH, Austin, Texas) *kb*
SEMIS Solar Energy Monitor In Space *wp*
SEMMS Solar Electric Manned Mars Spacecraft *hc*
SEMP Standard Electronic Module Program *fu*
SEMP Standard Embarkation Management System [USMC] *mt*
SEMP System Engineering Management Plan *mt, ah, fu*
SEMP Systems Engineering Master Plan *pl, hc, nt, wp, dm, tb, st, jdr, kf*
SEMS Stray Energy Monitor System *hc*
SEMS Systems Engineering and Management Support *mt*
SEMS Systems Engineering Master Schedule *kf, jdr*
SEN Software Engineering Notebook *kf, fu*
SEN Space Engagement Node *jmj, as*
SEN System Evaluation Network *mt*
SENIOR [nickname for HQ USAF programs usually connected with logistic support] *wp*
SENIOR_CITIZEN [a classified reconnaissance program] *wp*
SENIOR_DAGGER [reconnaissance test and evaluation program connected with SR-71 (??)] *wp*

SENIOR_RUBY [ELINT collector Onboard U-2r] *wp*
SENIOR_SPEAR [COMINT on U-2r] *wp*
SENIOR_STRETCH [COMINT on U-2r] *wp*
SENIOR_YEAR [black program in intelligence and communications] *wp*
SENIT [French version of NTDS] *fu*
SENRAD Sensible Radar *hc*
SENSO (The SENSO shouts . . .) {an officer in charge of sensors} *pf1.3*
SENTAR Senior Engineering Technology Assessment Review [AFSC*] *wp, mt*
SENTINEL [nickname for HQ USAF programs usually connected with intelligence] *wp*
SENTINEL [small, unmanned Canadian reconnaissance helicopter drone] *wp*
SENTINEL_KEEP [a photographic processing program] *wp*
SENTINEL_LOCK [deployable geodetic photographic data base] *wp*
SENTINEL_MARK [a photographic processing program] *wp*
SENTINEL_PAY [program to update photo equipment (??)] *wp*
SENTINEL_SCORE [program to utilize and manage HUMINT personnel] *wp*
SENTINEL_SIGNAL [intelligence training program] *wp*
SENTRY (create data usable on the SENTRY) *ne*
SEO Satellite for Earth Observation [also SEA] *ac3.2-2*
SEO System Engineering Organization *fu*
SEO System Evaluation and Operations *kf*
SEO Systems Engineering Operations *jdr*
SEON Solar Electro Optical Newtwork [for meteorology observations, USAF] *mt, wp, no*
SEP Security Engineering Program *fu*
SEP Selective Employment Plan *mt*
SEP Service Entry Panel *fu*
SEP Societe Europeenne de Propulsion [European Propulsion Society] *ac3.7-4, crl*
SEP Software Enhancement Proposal *fu*
SEP Specific Excess Power *mt*
SEP Spherical Error Probable *mt, fu, tb, kf*
SEP Support Equipment Plan *hc, fu*
SEP System Engineering Plan *mt*
SEP System Engineering Process *fu*
SEPA System Evaluation and Performance Analysis [Honeywell] *mt*
SEPG Software Engineering Process Group *kf*
SEPN Software Engineering Procedures Notebook *fu*
SEPP Secure Encryption Payment Protocol *pk, kg*
SEPP Software Engineering Practices and Procedures *kf, fu*
SEPR Senior Enlisted Performance Report *mt*
SEPR Societe d'Etude de la Propulsion par Reaction *crl*
SEPS Solar Electric Propulsion Stage [System] *hc, ci.237*
SEPS System Engineering Productivity System *fu*
SEPW Strategic Earth Penetrating Weapon *wp*
SEP_BDE Separate Brigade *mt*
SER Signal Environment Range *fu*
SER Special Event Report *kf*
SER Symbol Error Rate *fu, 2, 4*
SERA Software Engineering Review and Audits *mt*
SERA Systems Engineering Requirements and Analysis *st*
SERC Science and Engineering Research Council *ns*
SERC Software Engineering Research Center *mt*
SERD Support Equipment Recommendation Data *mt, fu, hc, wp, tb*
SERD Support Equipment Recommendation Document *kf*
SERD Support Equipment Requirements Data *mt*
SERDS Support Equipment Requirements Data Sheet *kf*
SEREB SocietÈ pour l'Etude et la Realisation d'Engins Balistiques *ac3.6-3, crl*
SERF [solar simulator and vacuum chamber] *hc*
SERGE Space-Earth Resources Geophysical Experiment *hc*
SERI [Hughes Aircraft Company customer] *hc*
SERIES Satellite Emission Range Infrared Earth Surveying *tb*
SERL System Engineering Reference Library *mt*
SERMIS Support Equipment Rework MIS [US Navy] *mt*
SEROCC Southeast Region Operations Center *mt*
SERR Software Engineering Requirements Review *fu*
SERS Special Event Reporting System *kf*
SERS Standardized Electronic Repair Station *hc*
SERS Surface Enhanced Raman Spectroscopy *wp*
SERT Ship Electronic Readiness Team *mt*
SERT Space Electric Rocket Test *0*
SERVFOR Service Force [USA] *mt*
SES Section Experimental et de Servitude [France] *mt*
SES Senior Executive Service [government service] *wp, mt*
SES SocietÈ EuropÈenne des Satellites *wp*
SES Solid Earth Science *hy*
SES Standards Engineers Society *0*
SES Surface Effect Ship *mt, fu, hc, aa27.4.b10*
SES System Engineering Specification *mt, wp*
SES System Exercise Set *fu*
SESA Society for Experimental Stress Analysis *aw118.14.3*

SESAC System Effectiveness Simulator and Availability Computer *hc*
SESC Space Environment Services Center [Boulder, Colorado] *ci.882, no, oe*
SESD Space and Electronics Systems Division [later: FSEC] *ci.945*
SESI System Engineering and System Integration *mt*
SESL Space Environment Simulation Laboratory *wp*
SESP Satellite Engineering Support Project *oe*
SESS Space Environmental Sensor Suite *dm*
SEST Swedish ESO Submillimeter Telescope *mi*
SESTS Security Evaluation Space Transportation System *mt*
SET Secure Electronic Transactions *kg, ct, em*
SET Situational Emergency Training *wp*
SET Software Engineering Technology *pk, kg*
SET Solar Energy Thermionic [Conversion] *0*
SET System Effectiveness Test *fu*
SET System Embedded Training *fu*
SETA Scientific and Engineering Technical Assistance *hc, ts4-6, cp, wp*
SETA Systems Engineering Technical Advisor *tb*
SETA Systems Engineering [and] Technical Assistance *mt, kf, hc, wp, tb, st, ai2*
SETAC Systems Engineering [and] Technical Assistance Contract(or) *fu, hc, tb, st*
SETAD Secure Transmission of Acoustic Data *fu*
SETADS [a NAVAIR anti submarine warfare program] *hc*
SETAF Southern European Task Force [USA] *mt*
SETAG Software Engineering Technology Advancement Group *mt*
SETEXT Structure Enhanced Text [Internet] *kg, hi*
SETI Search for Extra Terrestrial Intelligence *fu, mi, hc, sf31.1.15, wp, mb*
SETS System Engineering and Technical Support *wp*
SETS System Environment and Threat Simulation *st*
SETS System Environmental Test Station *bf*
SETT Static Environment Test Tool *fu*
SEU Servo Electronics Unit *sd*
SEU Single Event Upset *es55.99, gp, pl, hc, he, uh, jdr, kf, vv, oe*
SEU Smallest Executable Unit *pk, kg*
SEU Stabilization Electronics Unit *hc*
SEU System Electronic Unit *hc*
SEV Servo Electronics Unit *kf*
SEV Stockpile Emergency Verification *mt*
SEVAC Secure Voice Access Console *mt*
SEVEN [nickname for HQ USAF programs usually connected with surveys] *wp*
SEVOC Secure Voice *mt*
SEVOCOM Secure Voice Communication [also SVC] *mt, wp*
SEW Shipboard Electronic Warfare *hc*
SEW Silicon Epitaxial Wafer *wp*
SEW Space and Electronic Warfare *mt*
SEW Special Effects Weapon *hc*
SEW Surface Electromagnetic Wave *wp*
SEWB System Engineering Workbench *fu*
SEWC Space and Electronic Warfare Commander *mt*
SEWCC Signals Intelligence Electronic Warfare Coordination Center [USMC] *mt*
SEWS Satellite Early Warning System *mt*
SEWS Satellite [Strategic] Early Warning System *17, sd, cp, wp, tb*
SEWS Scientific Engineering Work Station [AFSC*] *mt*
SEWS Shipborne Electronic Warfare Suites *af72.5.22*
SEWT Simulator for Electronic Warfare Training *wp*
SF Safety Factor *fu*
SF Sign Flag *pk, kg*
SF Single Frequency *mt, fu*
SF Space Flight *fu*
SF Special Forces [USA DOD] *wp, mt*
SF Square Foot *mt*
SF Standard Form *mt, tb*
SF Static Facility *fu*
SF Storage Facility *oe*
SF Sub Frame *oe*
SF Support Facility *mt*
SFA Sales Force Automation *ct, em*
SFA Security Facility Analysis *tb*
SFA Security Failure Analysis *fu*
SFA Silver Flag Alpha [TAC] *mt*
SFA Single Failure Analysis *fu*
SFA Spaceport Florida Authority *ai2*
SFA Squib Firing Assembly *0*
SFAD System Field Acceptance Demonstration *fu*
SFAF Standard Frequency Action Format *kf*
SFC Shop Floor Control *jdr*
SFC Space Forecast Center [Falcon AFB, Colorado] [MAC*] *mt, no*
SFC Spacecraft Final Checks *jn*
SFC Specific Fuel Consumption *mt, aa26.12.59, wp*
SFC Supercritical Fluid Chromatography *wp*
SFCF Standby Functional Check Flight Crew *mt*
SFCP Shore Fire Control Party [US Navy] *mt*

SFCP Strategy Force and Capabilities Plan [USAF] *mt*
SFCR Super Fire Control Radar *fu*
SFCS Survivable Flight Control System *wp*
SFD Saturated (Saturation) Flux Density *kf, jdr, uh*
SFDG Source File Dependency Graph *fu*
SFDR Subcontractor Functional Design Review *fu*
SFE Saudi Furnished Equipment *fu*
SFEL Standard Facility Equipment Lists *mt*
SFENA [Aerospatiale subsidiary, France] *aa27.4.14, aw129.21.62*
SFEO-RN Short Form EO-RN [a document] *ep, tm*
SFF Self Forging Fragments *st*
SFG Special Forces Group [USA DOD] *mt*
SFH Slow Frequency Hop *mt*
SFI Search For Information *kf*
SFI Space and Communications Group Instruction [a directive] {Short Form Instruction} *ep*
SFJ Stand Forward Jamming *wp*
SFK Soft Function Key *fu*
SFK Subfunction Key *fu*
SFL Sequence Flashing Lights *fu*
SFL Spacecraft Functional Test *kf*
SFL Support Facility Laboratory *fu*
SFM SDI (System) Force Manager *tb*
SFM Subcontractor Furnished Material *jdr*
SFMSF Software/Firmware Maintenance Support Facility *fu*
SFN Surveillance File Number *fu*
SFO Space Flight Operations *0*
SFOB Special Forces Operations (Operating) Base [USA DOD] *wp, mt*
SFOC Space Flight Operations Complex (Center) *0, so, re*
SFOD Space Flight Operations Director *0*
SFOF Space Flight Operations Facility *fu, ci.117, 0, wp, re*
SFOM Space Flight Operations Memorandum *0*
SFOP Space Flight Operations Plan *0*
SFOS Space Flight Operations System *0*
SFP Single Function Processor *mt*
SFP Straight Fixed Price (contract) *fu*
SFPA Secretary of the Army, Public Affairs Field Agencies *dt*
SFPPL Short-Form Provisioning Parts *fu*
SFQL Structured Full-text Query Language *kg, hi*
SFR System Functional Review *jdr, kf*
SFRA System Failure and Recovery Analysis *tb*
SFRD Secret, Formerly Restricted Data *mt*
SFRD System Functional Requirements Description [JRSC] *mt*
SFRP SAC Inflight Refueling Program [Non-AF] *mt*
SFRS Survivability Functional Requirements Specification *tb*
SFS Shoot, Fail, Shoot *tb*
SFS System File Server *pk, kg*
SFSAS Standard Fuel Savings Advisory System *wp*
SFSCN Short Form SCN [a document] *ep, tm*
SFSMA Strategic Future Systems Mission Analysis *wp*
SFSS Satellite Field Service Station *gd1-4, oe*
SFT Simplified Formation Trainer *wp*
SFT System Fault Tolerance *ct, kg, em, hi*
SFT System Functional Test *9, uh*
SFTA Software Fault Tree Analysis *fu*
SFTB (field support of the FSTB for the LHX multi-sensor fusion) *ie5.11.7*
SFTOA Systems and Feasibility Tradeoff Analyses *fu*
SFUL Summary Fault Unit *mt*
SFW Sensor Fused Weapon *mt, dh, wp*
SFWRR Stock Fund Ware Reserve Requirements [USAF] *mt*
SFX Sound Effect[s] *pk, kg*
SG Salary Grade *mc*
SG Sawtooth Generator *fu*
SG Secretary General [CNES] *ac3.6-3*
SG Sensor Group *vv*
SG Signal Generator *fu*
SG Sortie Generation *wp, fu*
SG Space to Ground *kf*
SG Sub Group *mt*
SG Sun Gate *0*
SG Surveillance Gateway *fu*
SGA Sensor Gimbal Assembly *sd, kf*
SGARS Second Generation Alerting and Reporting System *hc*
SGAS Solar Geophysical Activity Summary *no*
SGDB Satellite Global Data Base *pl*
SGDT Store Global Descriptor Table *pk, kg*
SGEL Satellite Ground Equipment Laboratory *sc4.9.5*
SGEMP Source Generated Electromagnetic Pulse *tb*
SGEMP Spacecraft Generated Electromagnetic Pulse *uh*
SGEMP System (Self) Generated Electromagnetic Pulse *mt, he, hc, 5, 11, 17, sb, wp, st, jdr, kf*
SGEN Signal Generator *pk, kg*
SGEN System Generator *pk, kg*
SGF Screen Generation Facility *mt*
SGG Supercooled Gravity Gradiometer *hy*

SGGE Supercooled Gravity Gradiometer Experiment *hy*
SGGM Supercooled Gravity Gradiometer Mission *hy*
SGI Silicon Graphics, Inc. *ct, em, kf, pbl, kg*
SGL Space Ground Link *4, vv*
SGLS Space Ground Link [Sub]System [AFSCF] *uh, bo, 4, tb, jdr, kf*
SGM Shaded Graphics Modeling *pk, kg*
SGM Sortie Generation Model [also SORGEN] *wp*
SGMI [US Army Medical Intelligence and Information Agency] *dt*
SGML Standard Generalized Markup Language *mt, fu, kf, kg, hi*
SGMTI/L Slow Ground Moving Target Indicator/Locator [radar] *wp*
SGO Support Government Operations *pbl*
SGP Structural Ground Point *oe*
SGR Set Graphics Rendition *pk, kg*
SGRAM Synchronous Graphics Random Access Memory *kg, em, hi*
SGRM SBIRS Ground Resource Manager *kf*
SGS Sensor Ground Station *mt*
SGS Simulation Generation Set *fu*
SGS Single Gridded Surface *jn*
SGS Software Generation System *fu*
SGS Spacecraft Gravity System *re*
SGT Satellite Ground Terminal *mt*
SGTS Second Generation Tactical Sensor *kf*
SH Southern Hemisphere *oe*
SH Spectral Hamming *fu*
SH [designator, SH-2F is Seasprite] [US Navy] *mt*
SHA Secure Hash Algorithm [NSA] *em, kg*
SHA Sidereal Hour Angle *0*
SHA System Hazard Analysis *fu*
SHAEF Supreme Headquarters Allied Expeditionary Forces [Allied] *mt*
SHAFTS Semi Hardened Facilities Tracking System [USAFE] *mt*
SHAG Simple (Simplified) High Accuracy Guidance *hc, wp*
SHALE Stand-Off High Altitude Long Endurance [aircraft] *wp*
SHAPE Spacecraft Health and Performance Evaluation *sc4.1.center*
SHAPE Supreme (Strategic) Headquarters, Allied Powers in Europe [NATO] *mt, fu, 12, hc, jb, dt, wp, st*
SHAPM Ship Acquisition Program Manager *fu*
SHAR Shell Archive *pk, kg*
SHAR Sriharikota Range [India] *ac3.2-3*
SHAREM SHIP ASW Readiness Effectiveness Measuring *fu*
SHARP Standard Hardware Acquisition and Reliability Program *mt*
SHARP Super High Accuracy / Resolution Processing [radar] *ro*
SHARP Surface Hugging Aid for Route Planning *mt*
SHC Special Handling Code *fu*
SHC Sun Hold Calibration *jdr, uh*
SHC Supreme High Command [Soviet] *mt*
SHCPO Secretariat Headquarters Civilian Personnel Office [Navy] *dt*
SHD Satellite Handling Dolly *oe*
SHD Special Handling Designator *mt*
SHDD Scenario Host Demonstration Device *fu*
SHDL SALT Hardware Description Language *fu*
SHDSL Single-Line - High-bit-rate Digital Subscriber Line *em*
SHE Support and Handling Equipment *kf*
SHEA Safety, Health, and Environmental Affairs *fu, hn48.13.8, hc, jdr, kf*
SHED Segmented Hypergraphic Editor *pk, kg*
SHEL Space based HEL *sd*
SHELF Super Hard Extremely Low Frequency *mt*
SHERL Scalable High Energy Raman Laser *hc*
SHF Super High Frequency [3 to 30 GHz] *mt, fo, uh, nu, fu, hc, sc3.1.5, 17, 0, w9, rg, wp, tb, jdr, pbl, tm, kf, hi*
SHFG Super High Frequency Group *jdr*
SHFS Supportable Hybrid Fighter Structures *wp*
SHFTG SHF Transmit Group *jdr*
SHG Segmented Hypergraphics *pk, kg*
SHIELD Sylvania High Intelligence Electronic Defense *wp*
SHILLELAGH [Redstone arsenal surface to surface missile system] *hc*
SHINPADS Ships Integrated Navigation Processing And Display System *fu*
SHIN_PADS Shipboard Integrated Processing And Display System *hc*
SHIP [Atlantic missile range tracking ship] *0*
SHIPALT Ship Alteration [USA] *mt*
SHIPALTS Ship Alterations *fu*
SHIPCAD/CAM Ship Computer Aided Design / Computer Aided Manufacturing *mt*
SHIRAN S-Band High-Accuracy Ranging and Navigation *fu*
ShKAS Shpitalny-Komaritsky Aviatsionny Skorostrelny [rapid firing aircraft gun designed by B. G. Shpitalny and I. A. Komaritsky. 7.62mm, 1800rpm, 10kg] [USSR] *mt*
SHL Shift Logical Left *pk, kg*
SHL Short Haul Link *fu*
SHM Safe Hold Mode *oe*
SHM Simple Harmonic Motion *ai1*
SHME Safe Hold Mode Electronics *oe*
SHMU Stabilized Head Mirror Unit *hc*
SHNC Sun Hold No Calibration *jdr, uh*
SHO Scheduling Operations [Message] [NCC to WSGT] *-2*
SHO [Service] Schedule Order *-4*

SHOC SHAPE Operations Center [NATO] *fu, mt*
SHOC Software/ Hardware Operations Control *fu*
SHOL Simulation Higher Order Language *wp*
SHOMADS Short to Medium Range Air Defense System *fu*
SHOPAIR Short Path Infrared *wp*
SHORAD Short Range Air Defense *mt, ph, hc, wp, pm*
SHORAD Short Range Radar *mt*
SHORADEZ Short Range Air Defense Engagement Zone [USA DOD] *mt*
SHORADS Short Range Air Defense System *hc, fu*
SHORAN Short Range Navigation *mt, hc, wp, fu*
SHOROC Shore Required Operational Capability *mt*
SHORT_ORD SAC Short Order Network *mt*
SHP Standard Hardware Program [module] *fu, hc*
SHP Sub Host Processor *mt*
SHP/AMD Standard Hardware Program/Assembly Module Drawing *fu*
SHPE Society of Hispanic Professional Engineers *hn48.20.3*
SHR Shift Logical Right *pk, kg*
SHRAM Sheet Random Access Memory *hi*
SHREWD Satellite Handbook for Remote Environment and Weather Determination *ac2.v*
SHRIKE Anti Radiation Missile [AGM-45] *mt*
SHRILS Short Pulse High Repetition Intensified Laser System *hc*
SHS System Hardware Support *fu*
SHSS System Hardware Support Subsegment *fu*
SHTC Short Time Constant *fu*
SHUP Silo Hardness Upgrade Program *wp*
ShVAK Shpitalny Vladimirov Aviatsionnaya Krupno Kalibernaya ["large caliber aircraft gun designed by B.G. Shpitalny and S. V. Vladimirov" 20mm, 800m/sec, 800rpm, 42kg] [USSR] *mt*
SI Information Systems *mt*
SI Selective Identification *mt*
SI Semi-Insulating *fu*
SI Sensitive Intelligence *mt*
SI Session Identifier *fu*
SI Shift In *fu, do242, kg*
SI Shipping Instruction *wp*
SI Source Index *ct, em, kg, wp*
SI Special Inspection *fu*
SI Special Intelligence *mt, wp, fu*
SI Specific Impulse [for rockets, thrust per unit of fuel per second.] *mt*
SI Supplementary Instructions *fu*
SI System Information *pk, kg*
SI System Integration *fu*
SI SystÉme International [(French) International System of units] *fu, sm1-1, w9, wp, wl, ai1*
SI Systems Integrator *pbl, hi*
SI&T Software Integration and Test *tb*
SI/SAO Special Intelligence / Special Actions Office *mt*
SI/SO Serial In / Serial Out *pk, kg*
SI/SO Shift In / Shift Out *pk, kg*
SIA Securities Industry Association *y2k*
SIA Sensor Interface Adapter *kf*
SIA Special Interface Adapter *mt*
SIA Standard Instrument Approach *fu*
SIA Standard Interface Adapter *hi*
SIA Surveillance Interface Adapter *fu*
SIAC [Securities Industries Communications Agency] *jgo*
SIADS Sensor Integration and Display Sharing *fu*
SIAM Self Initiated Anti-aircraft Missile *hc*
SIAM Signal Information and Monitoring *fu*
SIAM Society for Industrial and Applied Mathematics *wp*
SIAP Strategic Intelligence Architecture Program *mt*
SIAP Surveillance Integration Automation Project *wp*
SIAS Special Intelligence Analysis Study *hc*
SIAS Submarine Integrated Antenna System *fu*
SIAWS SAC Intelligence Advanced Workstation Software *mt*
SIB Ship Information Booklet *mt*
SIB Simulator Interface Box *fu*
SIB System Integration Board *wl*
SIBAS_DBMS SIBAS Database Management System [Norsk Data's CODASYL DBMS] *fu*
SIC Service Identification Code *-4*
SIC SHORADS Information Center *fu*
SIC Simulator Information Center *tb*
SIC Small Intelligence Computer *mt*
SIC South Integration Cell *mt*
SIC Standard Industrial Classification *mt, fu, tb, st, kf*
SIC Standard Industrial Code *jdr, kg*
SIC Standard Inlet Condition *oe*
SIC Subject Indicator Code *mt, fu*
SIC Synchronizer and Interface Control *fu*
SIC System Integration Contractor *fu, wp*
SICA Signal Intelligence Control and Analysis *hc*
SICAM Sortie Integration Capability Assessment Model [HQ USAF] *mt*
SICB SPO Interface Control Board *mt*

SICBM Small Intercontinental Ballistic Missile *mt, aw126.20.47, af72.5.18, wp, tb*
SICBM Strategic ICBM *fu*
SICC SATKA Integrated [Experiment] Control Center *tb*
SICC Service Item Control Center *wp*
SICC Standard Industrial Classification Code *hi*
SICCIM System Input from Command and Control to Information Management [USA] *mt*
SICDH Scientific Instruments Control and Data Handling *hc*
SICEA [air integrated airspace control system, Argentina] *hc*
SICM Scenario Input Correlation and Merge (Program) *fu*
SICP Subscriber Interface Computer Program *fu*
SICPS Standardized Integrated Command Post System *fu*
SICR Selected Item Configuration Record *fu*
SICT SBIRS Integrated Cost Tool *kf*
SID Sealift Information Data Base [MSC] *mt*
SID Secondary Image Dissemination *mt*
SID Security IDentifier *em, kg*
SID Security Interface Device *fu*
SID Seismic Intrusion Detector *wp*
SID Selected Item Drawing *jdr*
SID Selective Imagery Dissemination System *mt*
SID Self Improved Diagnostic *hc*
SID Serial Input Data *pk, kg*
SID Situation Display *wp*
SID Sound Interface Device *wp*
SID Standard Information Display *mt, fu, hc, sc3.4.8*
SID Standard Instrument Departure *mt, fu*
SID Station Identification [AT&T] *kg, hi*
SID Subscriber Identification *fu*
SID Sudden Ionospheric Disturbance *0, mi, pr, mb*
SID Symbolic Interactive Debugger *pk, kg*
SID System Integration and Development *fu*
SIDA Single Integrated Data Base {?}:(mt)
SIDA System Integration Design Aid *fu*
SIDAC Single Integrated Damage Assessment Capability [WWMCCS] *mt*
SIDAC Supportability Investment Decision Analysis Center *wp*
SIDAR Satellite Image Data Acquisition and Reduction *pl*
SIDC Subscriber Interface Digital *fu*
SIDD System Interface Description Document *kf*
SIDE Sensor Integrated Discrimination Experiment *wp*
SIDEARM Sidewinder Anti Radiation Missile *uc, wp*
SIDESWIPE Simulation of Defensive Weapon Intercept Program *-17*
SIDF Standard Interchange Data Form *hi*
SIDF System Independent Data Format *pk, kg*
SIDH System Identification for Home Systems *pk, kg*
SIDP Subscriber Interface Data Processor *fu*
SIDPERS Standard Installation / Division Personnel System [USA] *hc, mt*
SIDR Software Implementation Design Review *fu*
SIDR System Integration Design Review *re*
SIDS Satellite Imagery Dissemination System [part of AWS] *mt, pl*
SIDS Sensor Integration and Display System *wp*
SIDS Stellar Inertial Doppler System *hc, wp*
SIDT Store Interrupt Descriptor Table *pk, kg*
SIDTC Single Integration Development Test Cycle *bf*
SIDU Space Identification Utility *wp*
SIE SATKA Integrated (Integration) Experiment [RADC] *mt, cp, tb, st*
SIE Special Inspection Equipment *fu*
SIETP Software Integration and Evaluation Test Plan *tb*
SIF Selective Identification Feature *mt, ro, wp, fu*
SIF Standard Input Format *ct, em*
SIF System Integration Facility *fu*
SIFP SIF Processor *fu*
SIFS Supply Interface System *mt*
SIFT Software Implemented Fault Tolerance *fu*
SIFT Stanford Information Filtering Tool *pk, kg*
SIG Special Interest Group *w9, wp, kg, ct, em, hi*
SIG Special Investigative Group *dt*
SIG Systems Integration Group, [TRW] *kf*
SIG-EVENTS Significant Events message system *mt*
SIGADA Special Interest Group Ada *fu*
SIGCAT Special Interest Group on CD-ROM Applications and Technology *kg, hi*
SIGCEN Signal Center [USA] *fu, mt*
SIGFET Semi Insulated Gate FET *mj31.5.72*
SIGINT Signal Intelligence [as a rule, SIGINT is RADINT plus COMINT] *mt, fu, ph, hc, aw129.21.48, wp, tb, st, kf*
SIGLEX Special Interest Group on Lexicography *-10*
SIGMA Force Level and Maneuver Control System [USA] *mt*
SIGMA [CORADCOM program] *hc*
SIGMALOG Simulation and Gaming Method for Analyses and Logistics [USA] *mt*
SIGMET Significant Meteorological Information *fu*
SIGMOD Special Interest Group on Management Of Data *fu*
SIGN System Interaction Guidance Notebook *fu*

SIGSEC Signal Security *mt, fu, wp, tb*
SIGTRAN Special Interest Group on Translation *-10*
SIGVOICE Special Interest Group on Voice *-10*
SIG_ADA Special Interest Group on ADA *mt*
SII Seriously Ill or Injured [USA DOD] *mt*
SII Special Interest Item *mt*
SII Statement of Intelligence Interest *mt*
SIIG Site Installation / Implementation Guide *mt*
SIIP Site Installation and Integration Plan *mt*
SIITP System Implementation, Integration, and Test Plan *-4*
SIJ Stand In Jammer *mt*
SIL Site d'Integration des Lanceurs *ai2*
SIL Speech Interference Level *0, fu*
SIL System Interface Logic *-8*
SIL Systems Implementation Laboratory *fu*
SIL Systems Integration Laboratory *bo, 4, mt*
SILA Self Locking Incremental Linear Actuator *uh*
SILENT_FOX [CECOM program, formely MEDFLI] *hc*
SILENT_PARTNER [WPAFB program] *hc*
SILF System In the Loop Facility [KEW] *tb*
SILO Signal Location *mt*
SILTS Shuttle Infrared Leeside Temperature Sensing [system] *wp*
SILVER_BALL [ECM project for quick reaction capability] *wp*
SIM Scanning Ion Microspcope *wp*
SIM Sensor Input Module *wp*
SIM Sensor Interface Module *fu*
SIM Specially Instrumented Missile *hc*
SIM Subscriber Identity Module *ct, em*
SIM/VER Simulation and Verification *-4*
SIMA Shore Intermediate Maintenance Activity *fu, hn43.17.5*
SIMAS Ship Intermediate Maintenance Activities *mt*
SIMBAC System for Integrated Nuclear Battle Analysis Calculus *tb*
SIMBAD Set of Identifications, Measurements, and Bibliography for Astronomical Data *es55.66, ns*
SIMCE Simulation, Communication Electronics *mt*
SIMCOM Simulated Compiler *fu*
SIMCOM Singer-Link's Simulation Complex *tb*
SIMD Single Instruction Multiple Data-stream [processor] *kg, fu, hi, ai1*
SIMD Single Instruction Multiple Date *tb*
SIMDB Systems Integration and Management Data Base *mt*
SIMDE Simplified Method of Damage Estimation *mt*
SIMFAX SIMulation FACility [a Hughes HIL emulator for missile flights] *hc, sb*
SIMGEN Simulation Generation *fu*
SIMICOR Simultaneous Multiple Image Correlation *fu*
SIMM Single In line Memory Module *ct, kg, em, hi*
SIMM Symbolic Integrated Maintenance Manual [Navy] *hc, fu*
SIMMS Shipboard Integrated Maintenance Management System [US Navy] *mt*
SIMNET Large Scale Interactive Simulator Networking *mt*
SIMNET Simulation Network [model] *wp*
SIMO Special Items Management Office *wp*
SIMOX Separation by Implantation of Oxygen *wp*
SIMP Sensor Independent Mission Processor *kf*
SIMP Specific Impulse *fu*
SIMPLE Script application language Implementation *hi*
SIMPLRS Simulated PLRS (computer program) *fu*
SIMR System Integration Management Review *wp*
SIMS Secondary Ion Mass Spectrometry (Spectroscopy) *wp, jdr*
SIMS Services Information Management System [USAF] *mt*
SIMS Shuttle Imaging Microwave System *hc*
SIMS Simulation SHO *-4*
SIMS Single Item, Multi Source *wp*
SIMS Support Information Management System *mt*
SIMS-X Selected Item Management System - Expanded [USA] *mt*
SIMTEL Simulation and Teleprocessing *pk, kg*
SIMTRAIN Training Device and Simulation Device *hc*
SIMULA Simulation [language] *pk, kg*
SIMWG Site Installation Management Working Group *mt*
SIN Spanish International Network *sc2.10.6*
SINAD (measurements from .01% to 100% THD or 0 to 80 dB SINAD at baseband frequencies) *mj31.5.19*
SINAD Signal plus Noise And Distortion *fu, jdr*
SINC Space[craft] Integration Contract[or] *bo, 4*
SINCGARS Single Channel Ground and Airborne (Aerial) Radio (Sub)System *mt, fu, hc, ph, wp, pm*
SINDRE (ADGE systems in the . . . northern region (SINDRE)) *go3.4.1*
SINEWS Ships Integrated Electronic Warfare System *fu*
SINGER/LINK Singer Company, Link [Flight Simulation Division] *tb*
SINR Signal-to-Interference-plus-Noise Ratio *fu*
SINS (ADGE systems in the . . . Southern Region (SINS)) *go3.4.1*
SINS Ship's (Shipboard) Inertial Navigation System *mt, ci.131, fu*
SINS Ships In *0*
SINS Southern Improved NADGE Sites *fu*
SINTRAN_OS SINTRAN [III/VS, Norsk Data's Multi-Programming, Virtual Memory, Real Time] Operating System *fu*

SIO Senior Intelligence Officer *jdr*
SIO Serial Input / Output [communications driver] *wp, fu, kg*
SIO Signal Intelligence Officer [USMC] *mt*
SIO Site Intelligence Officer *mt*
SIO System Integration Office *mt, wp*
SIOH Supervision, Inspection and Overhead *wp*
SIOP Secure Identification Operating Procedure *wp*
SIOP Single Integrated Operations (Operational) Plan [war plan in case of a major war] [USA] *mt, fu, uc, 12, 13, hc, af72.5.142, wp, tb, kf*
SIOP Strategic Integrated Operational Plan *hc*
SIOP Structural Improvement of Operational Aircraft *wp*
SIOP-ESI SIOP-Extremely Sensitive Information *mt*
SIOP-MATE SIOP-Matrix Automation Through the Emergency Message Automatic Transmission System *mt*
SIP Scientific Instruction Processor [Honeywell] *mt*
SIP Segment Interface Protocol *mt*
SIP Simple Internet Protocol *hi*
SIP Single In line Package *ct, em, kg, 6, fu, hi*
SIP Site Installation Plan *fu*
SIP Software Instrumentation Package *mt*
SIP Special Inspection Plans *fu*
SIP Standard Interface Panel *-8*
SIP Standardization Instructor Pilots *aw129.14.13*
SIP Study and Implementation Plan *hc*
SIP Surveillance Interface Processor *fu*
SIP System Improvement Plan *mt*
SIP System Integration Plan *wp*
SIP Systems Implementation Plan *mt*
SIP Systems Improvement Program *mt*
SIP Systems Information Package *mt*
SIPC Simply Interactive Personal Computer [Microsoft] *pk, kg*
SIPC Simulated Information Processing Center *mt*
SIPD (PMT's and SIPD's) *bl3-2.27*
SIPG Special Intercept Priorities Group *-10*
SIPO Serial-In, Parallel-Out [a circuit configuration] *fu, kg, ma72, hi*
SIPO System Integration Project Office[s] *mt*
SIPP Single In-line Pin Package *kg, ct, em, hi*
SIPP Sodium Iron Pyrophosphate *wp*
SIPREP Special Intelligence Report *mt*
SIPRI Stockholm International Peace Research Institute *14, wp*
SIPRNET Secret Internet Protocol Router Network *kf*
SIPS Supplier In Plant Store *fu*
SIPSS (he is representative to the SIPSS program) *hn43.9.3*
SIPSS Smithsonian Institute Proprietary Security System *hc*
SIR Scientific Information Retrieval *wp*
SIR Selected Inertial Reference *-3*
SIR Semiannual Inventory Report *mt*
SIR Serial Infra Red [Hewlett-Packard] *kg, hi*
SIR Serious Incident Report *mt*
SIR Shuttle Imaging Radar *mi, ac1.2-14, aa26.10.10, wp, re, mb, jdr*
SIR Side Looking Radar *fu*
SIR Signal to Interference Ratio *wp, fu*
SIR Spaceborne Imaging Radar *hy*
SIR Supplier Information Report *-4*
SIR Supplier Information Request *jdr*
SIR Systems Integrated Receiver [receiver equipment in the tail pod of the EA-6B Prowler] [USA] *mt*
SIR Systems Integration Review *tb, wl*
SIRAS System Interface Requirements Analysis System *fu*
SIRCS Shipboard Intermediate Range Combat System *fu, hc*
SIRD Support Instrumentation Requirements Document *2, oe*
SIRD System Instrumentation Requirements Document *-4*
SIRDS Single-Image Random Dot Stereogram *pk, kg*
SIRE Satellite Infrared Experiment *hc, wp*
SIRE Space Imagery Receive and Exploitation *kf*
SIRE Space Infrared Experiment *ci.572, 17, kf*
SIRE Space Infrared [Sensor] *sb*
SIREWS Shipboard Infrared Electronic Warfare System [Navy] *hc, fu*
SIRIO [Italian Experimental Communications Satellite Program] *gs2-10.62*
SIRIS Shipboard Infrared Imaging System/Surveillance *hc*
SIRIUS Spaceborne Intensifier Radiometer Imaging Ultraviolet Spectroscope *hc*
SIRP Space Infrared Sensor Program *tb*
SIRR System Integration Readiness Review *re*
SIRRM Standard Infrared Radiation Model *-17*
SIRS Satellite Infrared Sounders *gp*
SIRS Satellite Infrared Spectrometer *ac2.8-2, wp*
SIRS Single Integrated Reporting System [DOD] *mt*
SIRS Standoff Infrared Sensor [system] *hc*
SIRS Staring Infrared Sensor *hc*
SIRS Staring Infrared Study *hc*
SIRS Status Inventory Reporting System *mt*
SIRS Strapdown Inertial Reference System *fu*
SIRTF Space [formerly Shuttle] InfraRed Telescope Facility *mi, mb, hc*
SIRU Spherical Inertial Reference Unit *he*

SIRVES SIGINT Requirements Validation and Evaluation Subcommittee *mt*
SIS SAM Interface Set *fu*
SIS Satellite Interceptor System *mt*
SIS Satellite Interconnect System *jdr*
SIS Scanning Imaging Spectroradiometer (Spectrometer) *ac1.4-19, wp*
SIS Science Information System *re*
SIS Selective Inquiry System *mt*
SIS Sensor Image Simulation *hc*
SIS Short Interval Scheduling *pl*
SIS Signal Intelligence Service *-10*
SIS Signal Interface Specification *fu*
SIS Site Installation Specification *fu*
SIS Software Integration Software *fu*
SIS Spacecraft Interface Subassembly *oe*
SIS Special Information System *mt*
SIS Stall Inhibitor System *aw129.10.72*
SIS Standard Information System *mt*
SIS Standard Instruction Set *fu*
SIS Submarine, Fleet Ballistic Missile [SSBN] Indications and Warning Support *mt*
SISC Standard Information Systems Center [AFCC*] *mt*
SISCR SISC Regulation *mt*
SISD Single Instruction Single Data *tb, fu*
SISD Strategic Information Systems Division [SAC] *mt*
SISEX Shuttle Imaging Spectrometer Experiment *wp*
SISM Standard System Systems Manager *mt*
SISMS Standard Integrated Support Management System *mt, fu*
SISO Single Input / Single Output *ai1*
SISP Standard Information Systems Panel *mt*
SISP Support Information System Panel [HQ USAF] *mt*
SISP System Interface Support Plan *mt*
SISPRE Societa Italiana Sviluppo Propulsione a REazione *crl*
SISR Standard Information Systems Review *mt*
SISS Surrogate Indirect Sighting System *hc*
SIT Select In Test *jdr*
SIT Silicon Intensifier Target [Tubes] *ci.960*
SIT Single Target Track *fu*
SIT Software Integration [and] Test *gp, fu, oe*
SIT Spaceborne Infrared Tracker *wp*
SIT Special Information Tones *pk, kg*
SIT Special Interest Target [USA DOD] *mt*
SIT Structurally Integrated Thruster *js10.503*
SIT System Integration Test(ing) *mt, fu, uh, wp*
SIT/TE System Integration and Test/Test Equipment *kf*
SITA International Society for Aeronautical Telecommunications *csII.65*
SITAP Simulation for Transportation Analysis and Planning [JCS] *mt*
SITCARD Card to Tape Activity *mt*
SITCEN NATO Situation Center *mt*
SITE Search Information Tape Equipment *fu*
SITE System Installation, Test, and Evaluation [personnel] *fu*
SITE Systems Integration, Test and Evaluation *-19*
SITGEM Space Intercept Tracking and Guidance Evaluation Model *-17*
SITP Shanghai Institute of Technical Physics *ac*
SITRB System Integration Test Review Board *mt*
SITREP Situation Report *mt, fu, wp*
SITS Satellite Integrated Test System *kf*
SITS Secure Imagery Transmission System *ro, wp*
SITS Single Integrated Telecommunications System *mt*
SITS System Integration Test Site *fu*
SITT System Integration and Test Terminal *wp*
SITVC Secondary Injection Thrust Vector Control *ai2*
SIU Sensor Interface Unit *fu*
SIU Signal Interface Unit *bo, 4, 9, lt*
SIU Station Interface Unit *fu*
SIU System Interface Unit *pk, kg*
SIV Special Interest Vehicle [USA] *mt*
SIVP SBIRS Integration and Verification Plan *kf*
SIVVP Software Independent Verification and Validation Plan *mt*
SIW Strategic Intelligence Wing [SAC] *mt*
SIWG Simulation Interoperability Working Group *tb*
SIWG Site Implementation Working Group *fu*
SIWG System Interface Working Group *fu*
SIXA SXG solid state Si(Li) detectors *cu*
SIXDOF Six Degrees Of Freedom [a simulation model] *st*
SIXS Satellite Information Exchange System *fu*
SJ Single Junction *jn*
SJ Support Jamming *wp*
SJAR Simultaneous Jam And Receive *hc*
SJNR Signal-to-Jammer-Noise Ratio *fu*
SK Automated Telecommunications Systems Directorate *mt*
SK Station Keeping *id4142-10/831, oe*
SKAPE [simulation] [one of two major SDI architecture evaluation models] *sb, tb*
SKAT Smart Knowledge Acquisition Tool *mt*

SKB Spetsializirovannoye Konstruktorckoye Byuro [Specialized Design Bureau] *ai2*
SKE Station Keeping Equipment *mt*
SKEET [low cost submunition for tanks with infrared sensor] *wp*
SKG Schnell Kampf Geschwader ["fast fighter-bomber unit"] [Germany] *mt*
SKI System Key Issues *st*
SKICAT Sky Image Cataloging and Analysis Tool [NASA] *hi*
SKIP Simple Key Management for Internet Protocols *ct, em, kg*
SKIPPER [Laser guided bomb-powered] *dh*
SKIPS COMP Kernel Interface Package [Honeywell] *mt*
SKR Saturn Kilometric Radiation *so*
SKR Storozhevoy Korabl [patrol ship] [USSR] *mt*
SKSS Station Keeping Sun Sensor *oe*
SKT Skill Knowledge Test *mt, fu*
SKU Stock Keeping Unit *kg, hi*
SKV Single Kill Vehicle *st*
SKW (SKW Corporation . . . Arlington, VA) *jmj*
SKYHOOK [USAF balloon program to pick up people from the ground] *wp*
SKYNET [United Kingdom Communications Satellite] *mt*
SKY_LITE [Air Force Space Division program] *hc*
SL Shelf Life (Code) *fu*
SL Shoemaker-Levy [comet] *mi*
SL Side Lobe *wp*
SL Space Lab *ci.488, mb*
SL SpaceLab:(mi)
SL Standard Label *fu*
SL Stop Lift *fu*
SL Systems Leader *ah, pbl, jn*
SLA Software Library Administrator *mt*
SLAB Subsonic Low Altitude Bomber [a requirement for a long-range bomber, new in 1961] [USA] *mt*
SLAC Stanford Linear Accelerator Center *tb*
SLAC Support List Allowance Card *mt, fu*
SLAD System Loads Analysis and Definition Document *fu*
SLAD System Logic and Algorithm Development *hc*
SLAFDR System-Level Assisted Fault Detection and Recovery *fu*
SLAM Semi Active Laser Guided Aerial Munition *wp*
SLAM Simulation Language for Alternative Modeling *mt, hc, tb*
SLAM Stand off Land Attack Missile [a derivative of the Harpoon missile, USA] *mt, hc, hn49.14.6, wp*
SLAM Strategic Lift Analysis Model [US Navy] *mt*
SLAM Strategic Low Altitude Missile *wp*
SLAM Submarine Launched Air (Attack) Missile *fu, wp*
SLAM Supersonic Low Altitude Missile *mt, wp*
SLAM Symbolic Language Adapted for Microcomputers *wp*
SLAMMR Side Looking Airborne Modular Multimission Radar [fitted to a maritime surveillance version of the Boeing 737] *mt, wp*
SLAP Strategic Laser Aircraft Program *hc*
SLAPC Selected Linear Area Pattern Control *hc*
SLAPS System Level Air-to-Air Performance Study *hc*
SLAR Side Looking Airborne Radar *mt, hc, fu, mi, ac1.4-19, 14, hc, af72.5.138, aw130.11.51, wp, mb*
SLAR Side Looking Array Radar *mt*
SLASH Self Limiting Activated Solution of Hypoclorite [chemical warfare decontamination solution] *wp*
SLASH Synthetic [Aperture] Land And Sea Homing *wp*
SLAT Supersonic Low-Altitude Target *fu*
SLAT Surface Launched Aerial (Air) Targeted [missile] *uc, wp*
SLATE Small Lightweight Transmission Equipment *fu*
SLATE System Level Automation Tool for Engineering *ah, pbl*
SLATEC Sandia, Los Alamos, Technical Exchange Committee [Air Force Weapons Laboratory] *tb*
SLB Side Lobe Blanking *mt, fu, wp*
SLBD SEA_LITE Beam Director *17, hc, af72.5.31*
SLBM Sea (Surface) Launched Ballistic Missile *mt, hc, tb*
SLBM Submarine Launched Ballistic Missile *fu, hc, ca, 1, 11, 12, 13, 14, 17, sb, w9, wp, st, kf, ai2*
SLBMAN SLBM Alerting Network *mt*
SLC Side Lobe Canceling *mt*
SLC Side Lobe Cancellation (Canceller) *wp, fu*
SLC Source Lines of Code *fu*
SLC Space Launch Complex *mi, oe, ai2*
SLC Strategic Laser Communication *hc*
SLC Submarine Laser Communication *17, hc*
SLC SX Low-power Cache *hi*
SLC-SAT Strategic Laser Communications Satellite *dh*
SLCAIR Strategic Laser Communication Airborne *hc*
SLCM Sea (Submarine) (Ship) Launched Cruise Missile *mt, fu, hc, 12, ph, ts12-6, wp, tb*
SLCSAT Submarine Laser Communications Satellite *tb*
SLCSE Software Life Cycle Support Environment *mt, wp*
SLD Satellite Launch Dispenser *mt*
SLD System-Level Diagnostics *fu*
SLDCOM Satellite Launch Dispenser Communications *mt*
SLDF Sierra Leone Defense Force [Sierra Leone] *mt*

SLDP System Level Diagnostics Program *fu*
SLDPF Spacelab Data Processing Facility *-2*
SLDT Store Local Descriptor Table *pk, kg*
SLE South Leave Earth [pulse] *-9*
SLE Systemic Lupus Erythematosus *w9, wp*
SLED Second Level Discriminator *fu*
SLED Ship Launched Electronic Decoy *mt*
SLED Single Large Expensive Disk *kg, ct, em*
SLED Space Laser Experimental Definition *hc*
SLEDGE Simulating Large Explosive Detonable Gas Experiment *wp*
SLEP Service Life Extension Program (Plan) [modernization program for carriers] [USA] *mt, fu, hc, aw126.20.100, af72.5.143, wp*
SLF Super Low Frequency [0.300-3KHZ] *mt*
SLFC Survivable Low Frequency Communications *wp*
SLFCS Survivable (Survivability) Low Frequency Communications System *mt, fu, 12, wp, tb*
SLFMS SAC* Logistics Force Management System [SAC*] *mt*
SLHC Standards for Long Haul Communications *mt*
SLI Sidelobe Inhibition *fu*
SLIC System Level Integration Circuit *pk, kg*
SLID System Level Interface Definition *kf*
SLIM Scanned LWIR Intrinsic Module *hc*
SLIM Simplified Logistics and Improved Maintenance [an upgrade for the F-106, USA] *mt*
SLIM Software Life Cycle Management *mt*
SLIM System Level Interface Manual [trainers] *fu*
SLIN Subline Item Number *mt*
SLIP Serial Line Interface Protocol *hi*
SLIP Serial Link Internet Protocol [Internet protocol for dial up connection] *ct, em, nu, kg, hi, du*
SLIR Stock Level Item Record *wp*
SLKT Survivability, Lethality and Key (Key) Technology *wp, nt, jmj, tb, st*
SLl Structural Language 1 *fu*
SLM Software Line Manager *fu*
SLM Spatial Light Modulator *hc, kg*
SLMM Submarine Launched Mobile Mine [USA] *fu, hc, mt*
SLO Service Level Objective *mt*
SLO Squadron Liaison Officer *mt*
SLO Sweeping Local Oscillator *fu*
SLOC Sea Lanes Of Communications *mt*
SLOC Sea Lines Of Communications *mt, wp*
SLOC Source Lines Of Code *fu, ric, kf, jdr*
SLOC Standard Lines Of Code {?}:(jdr)
SLOCS (interface simulations required about 750K SLOCS) {probably SLOC was meant} *ts10-11*
SLOE Special List Of Equipment *mt, 0*
SLOS Synthetic Line Of Sight *17, sb*
SLOSH Sea, Lake and Overland Surge from Hurricane [program] *pk, kg*
SLOW_HOPPER [Fort Monmouth program] *hc*
SLP Seaward Launch Point [USA DOD] *mt*
SLPPL Ship-Level Provisioning Parts List *fu*
SLR Satellite Laser Ranging *ns, hy*
SLR Side Looking Radar *mt, fu, hc, ro, wp*
SLR Side Looking Reconnaissance Radar *mt*
SLR Single Lens Reflex *w9, nd74.445.5, wp*
SLR Status Level Reports *-4*
SLR Stock Level Requirement *wp*
SLR Supplier Lot Record *fu*
SLR System Level Requirements *fu*
SLS Sea Level Static *fu*
SLS Secondary Landing Site *-9*
SLS Service Level Status *-4*
SLS Shoot Look Shoot *tb, st, fu*
SLS Sidelobe Suppression *wp, fu*
SLS Sodium Lauryl Sulfate *wp*
SLS Space Lab Simulation *tb*
SLS Space(lab) Life Sciences *mi*
SLS System Level Specification *fu*
SLS/DF Sidelobe Suppression/Dicke-Fix *fu*
SLSI Super Large Scale Integration *kg, hi*
SLSS Shuttle Launch Support System *-2*
SLSS Systems Library Subscription Service [IBM] *pk, kg*
SLSTO Survivability, Lethatlity, and Support Technologies Office *tb*
SLT Senior Leadership Team *ah*
SLT Spacecraft Local Time *oe*
SLT System Laboratory Test *fu*
SLU Spoken Language Understanding *pk, kg*
SLUFAE Surface Launched Unit, Fuel Air Explosive *hc*
SLV Satellite Launching (Launch) Vehicle *pl, ce*
SLV Source Likelihood Vector *fu*
SLV Space Launch Vehicle *tb, ai2*
SLV Standard Launch Vehicle *ci.948*
SLV Standardized Launch Vehicle [Atlas] *ai2*
SLVD System Level Validation Demonstration *hc*
SLVSV Section de Liaison et de Vol Sans VisibilitÈ [liaison and IFR training section of an Air Force squadron] [France] *mt*

SLW Space based Laser Weapon *wp*
SLW Store Logical Word *fu*
SLWL Straight-Line Wavelength *fu*
SLWS Submarine-Launched Weapon System *fu*
SLWT Super Light Weight Tank *ai2*
SM Deputy Chief of Staff, Maintenance and Modification *mt*
SM Deputy Commander for Maintenance and Modification *mt*
SM Security Monitor *mt*
SM Seeker Module *17, sb*
SM Segment Manager *tb*
SM Sergeant Major *w9*
SM Service Module [for Apollo program] *fu, ci.172*
SM Set Mode *pk, kg*
SM Shadowmask *fu*
SM Shared Memory *pk, kg*
SM Shielded Miniature *jdr*
SM Signal Message *fu*
SM Situation Monitoring *mt*
SM Software Manager *fu*
SM Soil Moisture *ac7.5-12*
SM Staff Memorandum [JCS] *mt*
SM Staff Months *kf*
SM Standard Missile *ph, hc*
SM Station Master *w9*
SM Statistical Multiplexer *-2*
SM Statute Miles *mt*
SM Strategic Missile *fu*
SM Synchronous Modem *hi*
SM System Manager [AFLC] *wp, tb, mt*
SM System Monitor [JOPS] *mt*
SM [Ground to ground missile] [USA] *mt*
SM {Ship}, Minelayer [US Navy] *mt*
SM&R Source, Maintenance and Recoverability *fu*
SM-ALC Sacramento Air Logistics Center [AFLC] *mt*
SM/GPC System Management/General Purpose Computer *-4*
SM/SW Situation Monitoring/Strategic Warning *mt*
SMA Satellite Monitoring Agency *-14*
SMA Sergeant Major of the Army *w9*
SMA Specialized Mission Aircraft *wp*
SMA [a coaxial cable connector] *aw129.10.7, mj31.5.6*
SMAB Solid Motor Assembly Building *bo, jdr*
SMAC Supervision Master Configuration and Control Console *fu*
SMACT Space Material and Component Testing *hc*
SMADTTA SM-ALC Damage Tolerance Analysis *mt*
SMALC Sacramento Air Logistics Center *fu*
SMALC Sacramento Air Logistics Command *hc*
SMAP Software Management and Assurance Program *oe*
SMARF Solid Motor Assembly and Readiness Facility *ai2*
SMARP SIOP Mating and Ranging Program [SAC] *mt*
SMART Satellite Maintenance And Repair Techniques *wp*
SMART Satellite Manual and Reference for Telecommunications *smi*
SMART Selected Methods of Attacking the Right Target *fu*
SMART Self Monitoring Analysis and Reporting Technology *pk, kg*
SMART Simulation Model for Allocation of Resources for Training [ATC] *mt*
SMART Software Metrics Accumulation and Reporting Tool *kf*
SMART Space Maintenance and Repair Terminal *hc*
SMART Spares Management Analysis Review Technique *mt*
SMART Store Management Armament Release Tester *hc*
SMART System (Spares) Management, Analysis, Reporting and Tracking *fu*
SMART System Management and Allocation of Resources Technique *wp*
SMART System Memory Array Tester *fu*
SMART Systems Management Analysis, Research and Test *wp*
SMART-T Secure, Mobile, Antijam, Reliable, Tactical Terminal *jdr, kf*
SMARTD SMART Drive *hi*
SMARTS Simulation Monitoring Analysis Reduction and Test System *hc, tb*
SMARTS Survivable Multimission Air Reconnaissance and Targeting System *wp*
SMAS Standard Materiel Accounting System *mt*
SMASH Sensor, Multiple Armament System, Helicopter *hc*
SMAT System Memory Array Tester *fu*
SMAT System Monitor and Test *fu*
SMATH Satellite Materials [and target] Hardening [program] *5, sb, hc, wp*
SMATV Satellite Master Antenna TV [systems] *dn*
SMAW Shoulder launched Multipurpose Assault Weapon *mt, hc, wp*
SMAW-HE Shoulder Launched Multipurpose Assault Weapon High Explosive *hc*
SMB Server Message Block [protocol] *ct, em, kg*
SMB [a coaxial cable connector] *mj31.5.6*
SMBUS System Management Bus *hi*
SMC Air Force Space and Missile Center *uh*
SMC Ship's Motion Converter *fu*
SMC Small Magellanic Cloud *mi*

SMC Space and Missile Center *jdr, kf*
SMC Spacecraft Motion Compensation *oe*
SMC Standard Mean Chord *ai1*
SMC Structures, Mechanisms, and Control *pl*
SMC Supervisor / Maintenance Console *fu*
SMC Supervisory Monitor and Control *fu*
SMC Surplus Material Certification *fu*
SMC Suspended Material Control *fu, gp, 17, sb, ep, jdr*
SMC System Management and Control *fu*
SMC System Monitor and Control *fu*
SMC System Monitoring Center [DDN] *mt*
SMC Systems, Man and Cybernetics [an IEEE Society] *is26.1.70*
SMC [a coaxial cable connector] *mj31.5.6*
SMCA Single Manager for Conventional Ammunition *mt, wp*
SMCC Sun Microsystems Computer Company *ct, em*
SMCC Surveyor Mission Control Center *0*
SMCC System Monitor Coordination Center *mt*
SMCC Systems Monitoring and Coordination Center *ag*
SMCH Standard Mission Cargo Harness *oe*
SMCH Standard Mixed Cargo Harness *gp, 9*
SMCM Sea Mine Countermeasures *hc*
SMCP Satellite Maintenance Computer Program *fu*
SMCS Satellite Mission (Master) Control Subsystem *jdr*
SMCS Software Maintenance Control System *fu*
SMD Ship Manning Document *fu*
SMD Ship Manpower Document *mt*
SMD Specification Maintenance Document *fu*
SMD Step Motor Drive *jdr*
SMD Surface Mounted Device *kg, hc*
SMDC Shielded Mild Detonatic Cord *aw130.11.71*
SMDD Shipboard Maintenance Delivery Device *hc*
SMDD Storage Module Disk Drive *mt*
SMDPS Strategic Mission Data Preparation System [B-52, B-1] *mt*
SMDR Station Message Detail Recording *pk, kg*
SMDR Subscriber (Station) Message Detail Recording *mt*
SMDR System Manager Data Report *fu*
SMDS Switched Multi Media Data Service [B-ISDN up to 45 Mbps] *nu, ct, em, kg, hi*
SMDS Switched Multi Megabit Data Service *hi*
SME Safe Mode Electronics *jr, jdr*
SME Satellite Mission Engineering *jr*
SME Solar Mesosphere Explorer [USA] *ge, mi, ac1.7-26, hy, mb, hg*
SME Subject Matter Expert *fu, kf, hn43.23.6, tb*
SME Surface Movement Element *mt*
SMEADO Selected Major Exploratory Advanced Development Objectives *hc*
SMEK Summary Message Enable Keyboard *fu*
SMER San Marco Equatorial Range *crl*
SMES Strategic Missile Evaluation Squadron *mt*
SMES Superconducting Magnetic Energy Storage *wp*
SMET Simulated Mission Endurance Test *wp*
SMEX SMall EXplorers *mi*
SMF Simplified Message Format *fu*
SMF Single Mode Fiber *fu, kg, hi*
SMF Software Maintenance Facility *fu, mt*
SMF Space Manufacturing Facility *ci.1033*
SMF System Maintenance Facility *fu*
SMF System Manager Facility [Compaq] *pk, kg*
SMF Systems Management Facility *mt*
SMI Sample Matrix Inverse *fu*
SMI Shared Memory Interface *oe*
SMI Solder Mask Image *fu*
SMI Soldier Machine Interface *st*
SMI Standard Measuring Instrument *fu*
SMI System Management Interrupt *ct, em, kg*
SMIDD Shipboard Maintenance Information Delivery Device *fu*
SMIDS Standard Maintenance Information Display System *hc*
SMIF Standard Mechanical Interface *pk, kg*
SMILS Sonobuoy Missile Impact Location System *af72.5.140, wp*
SMIME Secure Multipurpose Internet Mail Extensions *em*
SMIP Signal Message Interface Processor *fu*
SMIRR Shuttle Multispectral Infrared Radiometer *wp*
SMIS Safeguard Management Information System *fu*
SMIS Security Management Information System [FEMA] *mt*
SMIS Ships Management Information System [US Navy] *mt*
SMIT System Management Interface Tool [IBM] *pk, kg*
SMK Software Migration Kit [Microsoft] *kg, hi*
SML Software Maintenance Library *fu*
SML Standard Material List *fu*
SML Standard Meta Language *pk, kg*
SML Support Material List *fu*
SML Symbolic Machine Language *wp, fu*
SMLR Single Most-Likely Replacement *fu*
SMLS Shelterized Microwave Landing System *mt*
SMLV Standard Memory Load Verifier *mt*
SMM Software Maintenance Manual *oe*

SMM Solar Maximum Mission [satellite] [NASA] *mt, ge, mi, ci.752, 4, hc, ns, no, hy, mb, hg*
SMM System Management Mode *ct, kg, em*
SMMC System Maintenance and Monitor Console *fu*
SMMD Simplified Maintenance Manual Design *fu*
SMMDS Switched Multi Megabit Data Service *hi*
SMMP System MANPRINT Management Plan *nt*
SMMR Scanning Multi channel [frequency, spectral] Microwave Radiometer [NIMBUS-7, SEASAT] *ge, hg, hc, ac7.1-13, hy*
SMMS Support Maintenance Management System [USA] *mt*
SMMS Support Ships Material Management System [US Navy] *mt*
SMMS System Maintenance and Monitoring Station *fu*
SMMW Sub Millimeter Wave *hc, wp*
SMO Senior Meteorological and Oceanographic Officer [USA DOD] *mt*
SMO Senior METOC Officer [specifically, the METOC officer on a Unified Command staff] [USA] *mt*
SMO Shelter Management Office [ESD] *mt*
SMO Spares Management Organization *fu*
SMO Support Military Operations *pbl*
SMO Support to Military Operations *mt*
SMO Surveillance Management Officer *fu*
SMO System Management Office *kf*
SMOBC Solder Mask Over Bare Copper *pk, kg*
SMOLAN Single Mode Fiber Optic Local Area Network *wp*
SMOOS Shipboard Meteorological and Oceanographic Observing System *hc*
SMOR Standard Method Of Repair *oe*
SMOTE Simulation Of Turbofan Engine *wp*
SMP Safe Mode Processor *jdr*
SMP SBSS Master Plan *mt*
SMP Service Message Blocks *em*
SMP Signal / Message Processor *fu*
SMP Simple Management Protocol *pk, kg*
SMP Small Mechanical Parts *mt*
SMP Software Management Plan *oe*
SMP Software Modernization Plan *mt*
SMP Space Mine Payload *hs*
SMP Sub Motor Pool *mt*
SMP Symbolic Manipulation Program *kg, sm*
SMP Symmetric (Symmetrical) Multi Processing *kg, em, ct*
SMP Symmetric Multi processor *pk, kg*
SMP System Management Plan *mt*
SMPC Shared Memory Parallel Computer *pk, kg*
SMPS Switch(ing) Mode Power Supply *fu, kg*
SMPS Symmetric Multi Processing System *hi*
SMR (SMR Technologies — formerly a BF Goodrich unit) *aw129.10.54*
SMR Selective Message Router *mt, fu*
SMR Signature Measurement Radar *hc, wp, st*
SMR Software Maintenance Responsibility *mt*
SMR Source, Maintenance, (Maintainability) and Recoverability [also SM&R] *mt, kf, wp, fu*
SMR Special Management Review *mt*
SMR Specialized Mobile Radio *hi*
SMRAM System Management Random Access Memory *kg, em, hi*
SMRC Solid Motor for Roll Control *ai2*
SMRC Source Maintenance Recoverability Code *mt*
SMRI Secure Message Routing Indicator *mt*
SMRS Subcontractor Management and Reporting System *fu*
SMRT Software Maintenance Responsibility Transfer *mt*
SMS SBIRS Mission Simulator *kf*
SMS Scientific Memory System *fu*
SMS Selective Mass Storage *wp*
SMS Senior Master Sergeant *mt*
SMS Service Management Systems *ct, em*
SMS Shuttle Mission Simulation *fu*
SMS Shuttle Mission Simulator *hc, pf1.23, tb*
SMS Small Messaging System *pk, kg*
SMS Small Meteorological Satellite *oe*
SMS Spares Management System *fu*
SMS Standard Modular System *fu*
SMS Storage Management Services [NetWare] *pk, kg*
SMS Storage Management [Sub] System *ct, kg, em, hi*
SMS Stores Management Set (System) *hc, wp*
SMS Surface Missile System *fu*
SMS SURTASS Mission Supervisor *fu*
SMS Synchronous Meteorological Satellite [NOAA] *ge, gd1-0, hc, hg, fu, oe*
SMS System Maintenance Monitoring Support *wp*
SMS Systems Management Server [Microsoft] *pk, kg, em, hi*
SMSC Standard Multi-user Small Computer *mt*
SMSCRC Standard Multi-user Small Computer Requirements Contract [Project 251] *mt*
SMSD Scheduled Maintenance Service Data *fu*
SMSF Software Maintenance Support Facility *fu*
SMSGT Senior Master Sergeant *mt, wp*
SMSJ Solid Motor Side Jet *ai2*
SMSS Sealift Management Support System [MSC] *mt*

SMSS Station Management Sub System *lt*
SMSS Status Monitoring Sub System *-9*
SMSS Survivable Missile Surveillance System *wp*
SMSW Store Machine Status Word *pk, kg*
SMT Special Maintenance Team *mt*
SMT Station Management [protocol] *pk, kg*
SMT Subcontract Management Team *-17*
SMT Surface Mount Technology *mt, fu, kg, hi*
SMTD STOL / Maneuvering Technology Demonstrator [an F-15 with canards and two-dimensional thrust-vectoring nozzles] [USA] *mt*
SMTG Structural / Mechanical Technical Group *oe*
SMTI Selective Moving Target Indicator *mt*
SMTP Simple Mail Transfer (Transport) Protocol [program] [e-mail over IP] *mt, tb, ct, em, nu, kg, hi, du*
SMTP Smart Maintenance Test Panel *fu*
SMTS Senior Member of the Technical Staff *sb, fu*
SMTS Simulated Maintenance Trainer Systems *fu*
SMTS Space and Missile Training Squadron [USA] *mt*
SMTS Space [and] Missile Tracking System *kf, jdr*
SMTS_FDV Space Missile Tracking System Flight Demonstration Vehicle *kf*
SMU Secure Mobile Unit *mt*
SMU Self Maneuvering Unit *ci.970*
SMU Special Mission Unit [USA DOD] *mt*
SMU Subsystem Management Unit *ai2*
SMU Switch Matrix Unit *fu*
SMU System Management Utility *pk, kg*
SMU {Southern Methodist University} *aw130.6.73*
SMUD Standoff Munitions Dispenser *mt*
SMUR Spectrum Management Update and Review System [PACAF] *mt*
SMUT Sensor Module Unit Test *oe*
SMW Strategic Missile Wing [SAC*, USA] *12, mt*
SN Self Noise *fu*
SN Sequence Number *mt*
SN Serial Number [also S/N] [a designator] *0, kg, ah, ep, pbl, tm, hi, oe*
SN Signal Node *hi*
SN Space Network *vv*
SN Statement of Need *st*
SN Stock Number *wp*
SN Summary Number *hc, jdr*
SN SuperNova [e.g., SN1987A] *mi*
SN2 Solid Nitrogen *oe*
SNA Soviet Naval Aviation *12, wp*
SNA Systems Network Architecture [IBM] *mt, do348, is26.1.43, wp, tb, ct, em, fu, kg, hi*
SNAC Summary Number and Calendarization [form] *fu, hc, lt*
SNADS Systems Network Architecture Distributed (Distribution) Services *ct, mt*
SNAF Soviet Naval Air Force *fu*
SNAFU Situation Normal, All Fouled Up *wp*
SNAME Society of Naval Architects and Marine Engineers *mt*
SNAP Secure Network Application Processor *tb*
SNAP Shipboard Nontactical ADP (Automated Data Processing) Program [USA] *mt, pf1.26*
SNAP Spacecraft Nuclear Auxiliary Power [Unit] *0*
SNAP Standard Network Access Protocol *hc, do425*
SNAP Steerable Null Antenna Processor *mt*
SNAP Stochastic Network Analysis Program *fu*
SNAP Strategic Nuclear Attack Planning [model] *wp*
SNAP Sub-Network Access Protocol *ct*
SNAP Systems for Nuclear Auxiliary Power *wp, tb, ci.229*
SNAPII Shipboard Nontactical Automated Data Processing II [USA] *mt*
SNAPS Strategic Nuclear Attack Planning System *mt*
SNAPS Switched Network Automatic Profile System *mt*
SNAPSHOT [Wright Patterson AFB program] *hc*
SNARS Switched Network Automatic Routing System *mt, hc*
SNB Supportability Notebook *fu*
SNC Site Nodal Controller *mt*
SNC Stock Number Code *wp*
SNC Summary Number Calendarization *-8*
SNCOA Senior Non Commissioned Officer Academy *af72.5.72*
SNCR Standard Notes Change Request *fu*
SNCS Satellite Network Control Station *mt*
SND Standard Naval Display *fu*
SNDV Strategic Nuclear Delivery Vehicles [USSR] *wp, mt*
SNECMA (SNECMA of France, . . . and Fiat . . . participate in the engine program) *aa26.12.59, af72.5.142*
SNEP Saudi Naval Expansion Program *hc, dt, fu*
SNEWS Secure News Server [Internet] *pk, kg*
SNF Short range Nuclear Forces *ph, nd74.445.39, wp*
SNF System Noise Factor *fu*
SNFE Standard Network Front End *mt*
SNI (the . . . system was developed by . . . SNI Aerospatiale, France) *hn44.23.4*
SNI Sequence Number Indicator *fu*
SNI Standard Network Interface *mt*
SNID Standard Network Interface Device *mt*

SNIPE SDI Network Interface Processing Element *hc, tb*
SNL Sandia National Laboratory *tb, st*
SNL Space Network List *jdr*
SNL Standard Nomenclature List *fu*
SNLE Sous-Marin NuclÈaire Lanceur d'Engins [(French) nuclear powered submarine, BM armed] *-12*
SNM System Notification Message *mt*
SNMC Software Network Management Council *kf*
SNMP Sensor Network Management Processor *fu, ct, kf, kg*
SNMP Simple Network Management Protocol *hi*
SNOBOL String Oriented Symbolic Language [a progamming language] *kg, hi*
SNOE Smart Noise Equipment *mt*
SNOI Signal Not Of Interest *fu*
SNOPPY_RECEIVER [Wright Patterson AFB program] *hc*
SNORT Supersonic Naval Ordnance Research Track *wp*
SNOWCAT Support of Nuclear Operations with Conventional Air Tactics *mt*
SNOWTIME [SAC/NORAD electronic warfare exercises] *wp*
SNP Serial Number/Password [Omen Technology] *pk, kg*
SNP Subnet Protocol *mt*
SNPO Space Nuclear Propulsion Office *ci.108*
SNR Short Notice Requirement *mt*
SNR Signal to Noise [power] Ratio [usually in dB] [also S/N] *uh, fu, mi, kg, gs1-5.A2, 0, 2, 7, wp, tb, st, jdr, mb, kf, hi, oe, mt, ai1*
SNR Super Nova Remnant *mi, es55.31, mb*
SNRM Set Normal Response Mode *fu*
SNS SBIRS Network Simulation *kf*
SNS Sensor Netting Station *fu*
SNS Site Navigation System *wp*
SNS Somatic Nervous System *wp*
SNSB Senior Navy Steering Board *mt*
SNSL Stock Number Sequence Listing *fu*
SNTK Special Need To Know *wp*
SNU Solar Neutrino Units *mi, mb, pbl*
SNUD Stock Number User Directory *mt, wp*
SNUMB System Number *mt*
SNWSC Space and Naval Warfare Systems Command *mt, tb*
SO Scheduling Operator *-2*
SO Security Office *dt*
SO Shift Out *fu, do242*
SO Shipping Order *wp*
SO Shop Order *fu*
SO Slow Operate *fu*
SO Spares Order *fu*
SO Special Operations [USA DOD] *mt*
SO Special Options *mt*
SO Spurious Outputs *-2*
SO Stock Order *fu*
SO Stop Order *fu*
SO Surveillance Officer *fu*
SO Synchronous Orbit *oe*
SO [system Of Systems Engineering (IPT identifier)] *kf*
SO-DIMM Small Outline DIMM *pk, kg*
SO-J Small Outline J-lead *ct, em*
SOA S-band Omni Antenna *oe*
SOA Separate Operating Agency *mt, wp*
SOA Special Operating Agency *mt, af72.5.5*
SOA Special Operational (Operations) Aircraft *hc, jmj, wp*
SOA Speed Of Advance *fu*
SOA Start Of Authority *pk, kg*
SOA State Of the Art *wp, st, fu, hi*
SOA Status Of Actions system [PACAF] *mt*
SOACS Sub-to-Air Laser Communications System *hc*
SOAL Scanning Optical Augmentation Locator *hc*
SOAM Solar Array Orientation Mechanism [also SAOM] *hc*
SOAP Spectroscopic Oil Analysis Program *mt, wp*
SOAP Systems Operations Analysis Program *0*
SOAR Simplified Off Axis Rejection *is26.1.76h*
SOAR Simulation Of Airlift Resources *wp*
SOAR Software for Optical Archival and Retrieval *ns*
SOAR Spacecraft Operation Anomaly Report *oe*
SOARDS Stand-Off Airborne Radar Demonstrator System *wp*
SOARS Satellite On Board Attack Reporting System *mt, hc, wp*
SOARS Shuttle Operation Automated Reporting System *wp*
SOAS Special Operations ADP System [OJCS] *mt*
SOAW Stand Off Attack Weapon *hc*
SOB Science Operations Board *0*
SOB Space Order of Battle *jb*
SOB Start Of Business
SOBS Spares Order Balance System *fu*
SOBT Submarine On Board Training [USA] *mt*
SOC Satellite Operations Center *mt, wp, pbl*
SOC Satellite Operations Complex *hc*
SOC Section (Sector) Operations Center *fu, 12*
SOC Section Operations Center *mt*

SOC Sector Operations Center *mt*
SOC Secure Operator Console *mt*
SOC Security Operations Center [US Navy] *mt*
SOC Service and Overhaul Change *fu*
SOC Simulation Operation Center *oe*
SOC Simulations Operations Center [GSFC] *2, 4*
SOC Space (Spacecraft) Operations Center *wp, jdr, kf*
SOC Special Operations Capable *mt*
SOC Special Operations Command [USA DOD] *mt*
SOC Squadron Operations Center *mt*
SOC State Of Charge *jdr, uh, vv, jnd*
SOC Statement Of Compatibility *mt*
SOC Statement of [Operational] Capability *sd, wp*
SOC Station Operation Conductor *0*
SOC Struck Off Charge [UK] *mt*
SOC SURTASS Operations Center *fu*
SOC System On a Chip *pk, kg*
SOC System Operational Concept *mt*
SOC System Operations Center *pbl*
SOC System Ownership Costing *wp*
SOC Systems Operational Concept *kf, wp*
SOCC Satellite Operations Control Center [NOAA] *gd3-0, oe*
SOCC Sector Operations Control Center [NORAD] *mt*
SOCC Submarine Operations Command Center [US Navy] *mt*
SOCC Subordinate Operations / Command Center *mt*
SOCCT Special Operations Combat Control Team [USA DOD] *mt*
SOCD Source Control Drawing *fu*
SOCEUR Special Operations Command Europe [NATO] *mt*
SOCKS Socket Secure [server] [Internet] *kg, hi*
SOCM Stand-Off Cluster Munition *hc*
SOCN Source Control Number *fu*
SOCOLS Source Control Library System *fu*
SOCOM Special Operations Command *mt, wp*
SOCR Stand alone Optical Character Recognition *mt*
SOCRATES Special Operations Command Research Analysis and Target (Threat) Evaluation System *mt, wp*
SOCS SAC Operational Communications System *mt*
SOCS SAC Operational Conference System *mt*
SOCS SPADCCS Owner/Operator Communications System *jb*
SOCSE Special Operations Communications Supply Element [USA] *mt*
SOD Secretary Of Defense *jdr*
SOD Standard Operational Design *fu*
SODA Sight Operational Data Acquisition *wp*
SODA Source Data Automation *mt*
SODAR Sound Detection and Ranging *fu*
SODART X-ray telescope on SXG *cu*
SODD Spacecraft Operations Database Document *uh, jdr*
SODO Senior Operations Duty Officer *mt*
SODSIM Strategic Offensive/Defensive Simulation *ts4-17*
SOE Sequence Of Events *jn, oe*
SOE Special Operations Executive [UK] *mt*
SOE Standard Operating Environment *pk, kg*
SOE Status Of Equipment *mt*
SOF Safety Of Flight *hc*
SOF Special Operations Forces [USA DOD] *hc, wp, mt*
SOF Status Of Forces *tb, mt*
SOF Strategic Offensive Force *mt, tb, fu, st*
SOF Supervisor Of Flying *mt, aw130.11.71*
SOF Sustainability Of Forces [JCS] *mt*
SOFA Status Of Forces Agreement *mt, tb*
SOFADS/SOAS Special Operating Foreign Analysis Data System/SOAS [JCS] *mt*
SOFAR Sound(ing), Fixing, And Ranging *fu, mt*
SOFAS Survivable Optical Forward Acquisition Sensor [system] *hc, uc*
SOFIA Stratospheric Observatory For Infrared Astronomy *mi*
SOFNET Solar Observing and Forecasting Network *wp*
SOFT Site Operational and Functional Test *fu*
SOFTCON Software Configuration System, WWMCCS Site [JDSSC] *mt*
SOFT_TALK [secure ground-air-ground digital voice communications] *wp*
SOG Satellite Operations Group *jb*
SOG Special Operations Group [USA] *mt, af72.5.55*
SOGS Science Operation Ground System *hc*
SOH Satellite Operations Handbook *oe*
SOH Suborbital Operations Handbook *-17*
SOHO Small Office / Home Office *nu, kg, ct, em, hi*
SOHO Solar Heliospheric Observatory [joint U.S./European] *mi, aw130.13.34, mb*
SOI (arrival at the moon's SOI occurs prior to apogee) *bm342*
SOI Saturn Orbit Insertion *so*
SOI Signal Of Interest *fu*
SOI Signal Operating Instruction [NATO] *mt*
SOI Silicon On Insulator *hc, wp, jdr, kf*
SOI Space Object Identification *mt, fu, ro, sd, hc, jb, cp, wp*
SOI Start Of Installation *mt*

SOI Stratigicheskaya Oboronitelnaya Initsiativa [SDI] [USSR] *mt*
SOI Synchronous Orbit Injection *-4*
SOIC Space Operational (Operations) Intelligence Center *mt, wp*
SOIC Submarine Officer's Indoctrination Course *fu*
SOID Ship's Ordnance Infrared Decoy *fu*
SOIF Strategic Operation and Integration Functions *lka*
SOIR Stand Off Infrared *hc*
SOJ Stand Off Jamming (Jammers) *mt, wp, fu*
SOJS Stand Off [anti-missile] Jammer System *uc*
SOL Sequence Order List *mt*
SOL Simulation Oriented Language *hc, kg, wp*
SOL Start Of Line *oe*
SOL System Operation Log *fu*
SOLAR Solar / Lunar Calculator [WWMCCS] *mt*
SOLARS SAC On Line Analysis Retrieval System [SAC] *mt*
SOLD Simulation Of Logic Display [Hughes] *fu*
SOLE Society Of Logistics Engineers *hc, aa26.8.b7*
SOLID Software Lifecycle Development *fu*
SOLID Spatial Offset Location Investigation and Development *hc*
SOLIS SIGINT On Line Information System [DIA] *mt*
SOLL Special Operations Low Level *mt, hc, af72.5.143*
SOLRAD Solar Radiation [a study program of NRL] *ci.735, pbl*
SOLSTICE Solar Stellar Irradiance (Inter) Comparison Experiment [UARS] *hy, hg, ge*
SOLV Solenoid Valve *fu*
SOM Spacecraft Operations Manual *pl, oe*
SOM Stand Off Missile *mt, wp, fu*
SOM Start Of Message *mt, sp5, 0, kg, fu*
SOM Support Operating Master *fu*
SOM System Object Model [IBM] *kg, hi*
SOM System Operator's Manual *fu*
SOM-H Start Of Message, High (Precedence) *fu*
SOM-L Start Of Message, Low (Precedence) *fu*
SOM/SOH Spacecraft Operations Manual / Handbook *gp*
SOMA Second Order Multiple Access *fu*
SOMAN [Pinacolyl Methyl Phosphonofluoridate, chemical warfare nerve agent, Gd] *wp*
SOMI Start Of Message In *mt*
SOMP Start Of Message Priority *fu*
SOMPF Special Operations Mission Planning Folder [USA DOD] *mt*
SOMS Synchronous Operational Meteorological Satellite *hc*
SOMSA Space Object Measurement and Signal Analysis [RADC] *mt*
SON Southern (South) Norway *fu*
SON State of Operational Need *hc*
SON Statement Of Need *mt, hc*
SON Statement of Operational Needs *mt, sd, wp, kf*
SONAC Sonar Nacelle *fu*
SONAR Sound Navigation And Ranging [sonic device used primarily for the detection and location of underwater objects] [USA DOD] "sonar" *fu, mt*
SONCM&D SONAR Countermeasures and Deception *fu*
SONCR SONAR Control Room *fu*
SONET Synchronous Optical Network [North American standard for high speed optical transmission] *nu, kg, ct, em*
SONIC SAC On Line Interactive Controller *mt*
SONMET Special Operations Naval Mobile Environment Team [USA DOD] *mt*
SOO Senior Operation Officer *mt*
SOO Statement Of Objectives *kf, pbl*
SOON Solar Observing Optical Network [USAF] *12, pl, no, pr*
SOP Senior Officer Present [USA] *mt*
SOP Ship Operational Program *fu*
SOP Special Observing Period *pl*
SOP Standing Operating Procedure *mt, ah, kg, w9, jdr, pbl, jn, fu, oe*
SOP Strategic Objectives Plan *fu*
SOP System Operation Plan *0*
SOPA Senior Officer Present Abroad *fu*
SOPA Senior Officer Present Afloat [USA] *mt*
SOPC Science Operations and Planning Computer *so*
SOPC Shuttle Operations and Planning Complex *mt, hc, tb*
SOPRANO [surface to target to missile path] *hc*
SOPS Satellite Operations Squadron *kf, uh*
SOPS Space Operations Squadron *uh*
SOQ System Optical Quality *hc*
SOQAS Statement Of Quality and Support *fu*
SOQS System Optical Quality Study *hc*
SOR Specific Operational Requirement *fu, wp*
SOR Specific Operational Requirement [SOR-49-2 was requirement for the RF-105 derivative of the Thunderchief] [USA] *mt*
SOR Squadron Operations Room *mt*
SOR Stand Off Range *st*
SOR Statement Of Requirement *mt, wp*
SOR Stimulus-Operation-Response *fu*
SOR Struck Off Records *mt*
SOR Surface Obscuration Ratio *he*
SOR/J Set-On Receiver/Jammer *hc*
SORA Scheduled Overtime Request Authorization *15, jdr*

SORD Submerged Ordnance Recovery Device [USA] *mt*
SORD System (Statement of) Operational Requirements Document *wp, as, kf*
SORD System and Operations Requirements Document *oe*
SORD System Operational Requirements Document *mt*
SORGEN Sortie Generation (Model) *wp*
SORO Scan On Receive Only *wp*
SORS SIGINT Overhead Reconnaissance Subcommittee *mt*
SORT Simulated Optical Range Target *hc*
SORT Simulated Optical Range Tester *wp*
SORT Standard Operator Response Time *fu*
SORT Structures for Orbiting Radio Telescope *hc*
SORTE Summary Of Radiation-Tolerant Electronics *fu*
SORTS Status Of Resources and Training System [JCS] [formerly UNI-TREP] *mt*
SOS Sealift Operating System [MSC] *mt*
SOS Silicon On Sapphire *hn43.16.3, sb, th4, hc, fu, wp, jdr, kf, kg, hi*
SOS Software Obsolescence Study *mt*
SOS Sophisticated Operating System *pk, kg*
SOS Source Of Supply *wp*
SOS Space Object Surveillance *kf*
SOS Space Operations Simulator *tb*
SOS Special Operations Squadron [USA] *mt, af72.5.121, wp*
SOS Speed Of Service *mt, fu*
SOS Squadron Officers School *mt, af72.5.55, wp*
SOS Stabilized Sight *mt*
SOS Standards and Open Systems *pk, kg*
SOS Submarine Operating System *fu*
SOS System Of Systems *kf*
SOS System Operation Specialist *fu*
SOS System Operational Specification *fu*
SOS Systems Operation Station *fu*
SOS [internationally recognized distress signal, in Morse code] *mt, hn25.11.5, wp*
SOSE System Of Systems Engineering *kf*
SOSG Senior Officer Steering Group [USAFE] *mt*
SOSI Space Object Surveillance and Identification *kf*
SOSI System Of Systems Integration *kf*
SOSIP Space Object Surveillance and Identification Program *wp*
SOSO Synchronous Orbiting Solar Observatory *hc*
SOSS Satellite Optical Surveillance Station *wp*
SOSS Soviet Ocean Surveillance System *mt, wp, fu*
SOSTEL Solid State Electronic Logic *fu, wp*
SOSTF SPADCO Off Site Test Facility *mt*
SOSUS Sound Surveillance Undersea (Underwater) System [US Navy] *fu, mt, 12, hc, pf1.4, kb, tb, st*
SOT Single Offset Tactic *fu*
SOT Solar Optical Telescope *hc*
SOT State Of Technology *kf*
SOTA SIGINT Operational Tasking Authority *mt*
SOTA State Of The Art [also SOA] *kg, tb, st, hi*
SOTACA State Of the Art Contingency Analysis *mt*
SOTAS Stand Off Target Acquisition System *hc, fu, mt*
SOTAS Standoff Target Acquisition and Surveillance *wp*
SOTAS State Of The Art Survey *fu*
SOTFA Special Operations Task Force Atlantic *mt*
SOTFE Support Operations Task Force Europe *mt*
SOTI System Operability Testing Inspection *fu*
SOTS Sonar Operator Training System *hc*
SOTWG SACDIN Offutt Transition Working Group *mt*
SOU Sensor Optical Unit *kf*
SOU Sounder Mode [of a radar] *so*
SOUP Spectrum Orbit Utilization Program *-19*
SOUTHAG Southern Army Group [NATO] *fu, mt*
SOUTHCOM Southern Command [US] *jb, mt*
SOVN Shipped On Voucher Number *fu*
SOVT System Operational Verification Test *fu*
SOW Scope Of Work *nt, st*
SOW Special Operations Wing [USA DOD] *wp, mt*
SOW Stand Off Weapon *mt, ph, wp*
SOW Statement Of Work [a document] *mt, fu, gp, 17, sb, pl, hc, lt, wp, ep, dm, tb, st, wl, as, jdr, pbl, tm, kf, jn, ah, oe*
SOWG Spacecraft Operations Working Group *oe*
SOWT Special Operations Weather Team *mt*
SOWT/TE Special Operations Weather Team / Tactical Element [USA DOD] *mt*
SOX Sound Exchange *pk, kg*
SP Sample Part *fu*
SP Scalable POWER/parallel {?} *hi*
SP Sector Processor *fu*
SP Security Police *mt, wp*
SP Security Police Squadron *dt*
SP Self Propelled *mt, w9, wp*
SP Sequence Phase *fu*
SP Service Pack [IBM] *pk, kg*
SP Shore Patrol (Patrolman) [USA] *mt, w9*

SP Shore Police *w9*
SP Short Persistence *fu*
SP Short Pulse *fu*
SP Signal Processing *kf*
SP Signal Processor *17, sb, hc, fu*
SP Signal Pulse *fu*
SP Single Pole *fu, w9*
SP Sound Powered *fu*
SP South Panel *oe*
SP Spacecraft Package *uh*
SP Spare Part *fu*
SP Special Performance *aw130.6.54*
SP Special Perturbations *jb*
SP Special Purpose *fu*
SP Spurious PM *-2*
SP Stack Pointer *ct, em, wp, kg*
SP Standard and Performance *mt*
SP Stellar Polarimeter *pl*
SP Structured Programming *fu*
SP Subcontracting Plan [for small businesses, small disadvantaged businesses, etc]. *kf*
SP Sun Presence *oe*
SP System Printer *fu*
SP System Product *pk, kg*
SP/MR Supplier's Problem/Material Report *fu*
SPA S-Band Power Amplifier *fu*
SPA Selectable Power Adapter *fu*
SPA Sensor Performance Analysis *wp*
SPA Skill Performance Aid *fu, hc*
SPA Socioeconomic Programs Administrator *hq5.2.2*
SPA Software Publishers Association *kg, hi*
SPA Southwest Procurement Agency (Army; formerly LAPD) *fu*
SPA Space Processing Applications *hc*
SPA Spaces Per Aircraft *wp*
SPA Special Power Adapter *fu*
SPA Strategy and Policy Assessment *mt*
SPA Sudden Phase Anomaly *fu*
SPA Survivable Penetration and Attack *hc*
SPA System Performance Assessment *mt*
SPAA Spacecraft Performance Analysis Area *0*
SPAAG Self Propelled Anti Aircraft Gun *mt*
SPABAT Space Battle [simulator program] *-17*
SPAC Space Program Advisory Council *ci.645*
SPAC Spacecraft Performance Analysis and Command *0*
SPACC Space Command Center [USSPACECOM] *mt*
SPACC Space Control Center [USSPACECOM Command Center] *jb, cp, as*
SPACC Space Crisis Center *kf*
SPACE Society for Private And Commercial Earth [stations] *dn*
SPACECMD Air Force Space Command *mt*
SPACECOM Space Command [US Air Force] *mt, 12, sd, wp, tb*
SPACETRACK [USAF portion of NORAD space tracking and detection] *mt, wp*
SPACEWARN [world warning agency for satellites] *ns*
SPAD Software Product Assurance Directorate *fu*
SPAD System Performance Assessment Document *jdr*
SPADATS Space Detection and Tracking System [NORAD] [also SPADOTS] *mt, sd, hs, wp, tb*
SPADCCS Space Defense Command and Control System [SPACECMD] *mt, sd, hc, hs, go1.2, jb, cp, tb*
SPADE Settable Pneumatic Altitude Detection *wp*
SPADE Single Channel Demand Assignment Technique *spix*
SPADE Spare Parts Analysis, Documentation and Evaluation *wp*
SPADEF Space Defense Simulation *st*
SPADOC Space and Air Defense Command *fu*
SPADOC Space Defense Operations Center [SPACECMD] *mt, sr9-82, 17, hc, jb, nt, wp, tb, as, jdr, kf*
SPADOTS Space Detection and Tracking System [also SPADATS] *hc*
SPADS Satellite Data Processing and Display System *pl*
SPADS Shuttle Problem Action Data System *fu*
SPADS Smart Parachute Airdrop System *hc*
SPADS Staff Planning and Decision Support System [USAREUR/DNA] *mt*
SPADS Stinger-Post Advanced Decoy System *hc*
SPAFB [Hughes Aircraft Company customer for AWARE] *hc*
SPAIS South Pacific Air Intelligence System [PACAF] *mt*
SPALS Shipment Planning and Address Labeling System *wp*
SPAM Sensor POST Analytic Model *st*
SPAM Space Point and Alignment Mirror *hc*
SPAM [Scanning Photoacoustic Spectroscopy] *wp*
SPAN Space Physics [and] Analysis Network [NASA] *ge, mi, kg, es55.50, ns, no, pr, hg, mb*
SPANS Spectral Analysis Processing System *ro*
SPANS Standardization Potential Across Navigation *hc*
SPAN_NIC SPAN Network Information Center *ns*
SPAP Secure Password Authentication Protocol *hi*
SPAR (SPAR Aerospace, another TRW subcontractor) *aa26.11.28*

SPAR Solid State Phased Array Radar *wp*
SPAR Space Available / Required system [MAC*] *mt*
SPAR Space Processing Application Rocket *ci.819, hc*
SPAR Space Radar *wp*
SPAR Special Progressive Aircraft Re-Work *wp*
SPAR Standard Payload Assurance Requirements [NASA] *oe*
SPAR Super Precision Approach Radar *fu*
SPAR Supplier Produce Assurance Requirements [Ford Aerospace] *oe*
SPAR Supplier Product Assurance Reporting *fu*
SPARC Scalable (Scalar) Processor Architecture [Sun] *ct, kg, jdr, em, hi*
SPARC Standards Planning and Requirements Committee *mt*
SPARCS SPARC Server *hi*
SPARK Solid Propellant, Advanced Ramjet, Kinetic energy [a small anti tank missile] [USA] *mt*
SPARS Spacecraft Precision and Attitude Reference System *hc, wp, fu*
SPARS Supplier Product Assurance Reporting System *fu*
SPARTA (SPARTA, Inc. . . . Huntsville, AL 35805) *jmj*
SPARTIS Space Related Tactical Intelligence Systems *hc*
SPARX Space Applications Research Experiment *sc4.1.4*
SPAS Security Police Automation System [USAF] *mt*
SPAS Shuttle Pallet Satellite *ci.631*
SPAS Skill Performance Aids Program *mt*
SPASM System Performance and Activity Software Monitor *wp*
SPASUR Space Surveillance [Network] [System] [also SSS] [US Navy] *bm132, mt*
SPAT Silicon Precision Alloy Transistor *fu*
SPAT Single Performance Assessment Test *kf*
SPATOPS Space Training and Operations Procedure Standard *hc*
SPAWAR Space And Naval Warfare Systems Command *mt, hc, uh, kb, dt*
SPAWAR Space Warfare, NAVELEX *fu*
SPAWARCOM Space and Naval Warfare Systems Command [US Navy] *mt*
SPB Supplier Packaging Bulletin *fu*
SPBS Standard Property Book Systems *mt*
SPC Satellite Planning Center *fu*
SPC Satellite Processing Center *mt*
SPC Science Programme Committee [of ESA] *sf31.1.9, es55.8, so*
SPC Senior Project Consultant *fu*
SPC Shuttle Processing Contractor *mt*
SPC Small Peripheral Controller *pk, kg*
SPC Software Productivity Center *mt*
SPC Specific Fuel Consumption *mt*
SPC Standard Parts Committee *fu*
SPC Statistical Process Control *fu, kg, ad, rm7.1.1, 8, nd74.445.26, is26.1.70, wp, ig21.12.17, hn50.8.4, jdr, kf, jn, oe*
SPC Stored Commands Controller *0*
SPC Stored Program Command *jdr*
SPC Supplier Packaging Code *fu*
SPC Switch and Panel Controller *fu*
SPCC Ship Parts Control Center [Mechanicsburg, PA] *fu, mt*
SPCC Southern Pacific Communications Corporation *sc2.11.3*
SPCCA Silver Plated Copper Clad Aluminum *jn*
SPCD Space Communications Division *mt*
SPCDS Small Permanent Communications Display System *wp*
SPCL Spectrum Cellular Corporation *pk, kg*
SPCR Software Problem/Change Report *fu*
SPCS Short Ported Coaxial System *fu*
SPCS Stored Program Controlled Switch *pk, kg*
SPCU Satellite Processing Center Upgrade *pl*
SPD Solar Panel Drive *-9*
SPD Summary Plan Description *pbl*
SPD System Planning Document *0*
SPD System Program Directive *mt*
SPD System Program Director *fu*
SPDIF Sony Phillips Digital Interface Format *hi*
SPDM Special Purpose Dextrous Manipulator:(mi)
SPDP Shelter Power Distribution Panel *fu*
SPDR Software Preliminary Design Review *fu*
SPDR System level Preliminary Design Review *kf*
SPDR System Program Director's Review *mt*
SPDT Single Pole Double Throw *mt, fu, gp, mj31.5.57, lt, hi, oe*
SPDT Supplementary Provisioning Technical Documentation *fu*
SPDU Session Protocol Data Unit *fu*
SPE Signal Processing Electronics *kf*
SPE Signal Processing Element *mt, fu*
SPE Space tracking and Pointing Experiment *mt*
SPE Special Purpose Equipment *hc*
SPE Static Phase Error *gg158, 0, oe*
SPE Stored Program Element *fu*
SPE System Project Engineer *fu*
SPE/AADC Signal Processing Element/All Applications Digital Computer *hc*
SPEAR Signal Processing Evaluation and Reporting *mt*
SPEAR Signal Processing, Evaluation, Alert and Report *fu, wp*
SPEAR Space Power Experiments Aboard Rockets *aa26.12.24*
SPEAR [Illuminator X-band ground target system] *hc*

SPEC Systems Performance Evaluation Cooperative *pk, kg*
SPECAT Special Category *mt*
SPECM Spectrally Pure Electromagnetic (Electronic) Countermeasures *fu, hc*
SPECREQ Special Request *fu*
SPECTRUM Simulation Package for Evaluation by Computer Techniques of Readiness, Utilization, and Maintenance *mt*
SPEE Signal Processing Evaluation Environment *tb*
SPEED Signal Processing in Evaluated Electronic Devices *fu*
SPEEDEX System wide Project for Electronic Equipment at Depots [USA] *mt*
SPEF Sensor Performance Evaluation Function *fu*
SPELDA Structure Porteuse Externe pour Lancement Doubles Ariane *es55.75, ai2*
SPELS Single Platform Emitter Location Study *hc*
SPELTA Structure Porteuse Externe pour Lancement Triple Adriane *ai2*
SPETE Special-Purpose Electronic Tools and Equipment *fu*
SPEW Small Platform Electronic Warfare *hc*
SPEW Special Purpose Electronic Warfare *fu*
SPEWX Special-Purpose Electronic Warfare Transmitters *fu*
SPF Shortest Path First *pk, kg*
SPF SIDPERS Personnel File [USA] *mt*
SPF Single Point Failure [a phenomenon] *ah, ep, jdr, pbl, tm, vv, jn*
SPF Software Production Facility *pf1.23, tb*
SPF Space Power Facility *ci.95*
SPF Space Processing Facility *ci.821*
SPF Standard Plume Flowfield [computer model] *-17*
SPF Strategic Projection Force *dh, ph*
SPF Structure Programming Facility *jdr*
SPF Sun Protection Factor *wp*
SPF Superplastic Forming Process *aw129.1.25*
SPF System Programming Facility *pk, kg*
SPFA Single Point Failure Analysis *ep*
SPFM Single Point Failure Mode *pl, jdr*
SPG Space Group [USA] *mt*
SPG System Protection Guide *kf*
SPGN Signal Processing Graph Notation *fu*
SPHINX (technique in computer voice recognition . . . called SPHINX) *aa26.12.16*
SPI Schedule Performance Index *mt, jdr, kf, fu*
SPI SCSI Parallel Interface *hi*
SPI Security Parameters Index *pk, kg*
SPI Service Provider Interface *pk, kg*
SPI Special Position Identification (Indicator) *fu*
SPI Supervised Parallel Input *-4*
SPIBS Satellite Positive Ion Beam System *hc*
SPICE Satellite Personal Information Communications Equipment *ne, th4, hc*
SPICE Simulation Program with Integrated Circuit Emphasis *fu, md2.1.1*
SPID Service Profile Identification *ct, em, kg*
SPID Service Provider Identifier (Identification) *kg, hi*
SPIDPO Shuttle Payload Integration Development Program Office *-4*
SPIE (SPIE Vol. 322. Picosecond Laser and Applications] *vc25.3.21*
SPIE International Society for Optical Engineering *hn43.22.3*
SPIF Satellite Processing and Integration Facility *kf*
SPIF Shuttle Payload Integration Facility *bo, jdr*
SPIF Shuttle Payload Interface Facility *-4*
SPIFIN SCI PACOM Intelligence File *mt*
SPIIN Supplemental Procurement Instrument Identification Number *fu*
SPIKE Science Planning Intelligent Knowledge-Based Environment [STScI] *pk, kg*
SPIKE [an AI scheduling tool used on HST, ASCA and XTE] *cu*
SPILS Spin Prevention and Incidence Limiting System *mt*
SPILS Stall Protection and Incidence Limiting System *mt*
SPIN Space Intercept *14, wp*
SPIN Special Instruction Referenced in the FRAG *mt*
SPINE Shared Project Information Network *kf*
SPINE Shipboard (Submarine) Passive Infrared Night Equipment *hc*
SPINS Special Instructions *mt*
SPINTCOM Special Intelligence Communications *-10*
SPINTCOM Special Intelligence Communications Handling System *mt*
SPINTCOMM Special Intelligence Communications Network *mt*
SPIRE Space Based Infrared Experiment *17, tb*
SPIREP Special Intelligence Report *mt*
SPIREP Spot Intelligence Report *mt, fu*
SPIRIT Support of IR Tech-Interferometer Telescope *hc*
SPISD Space Information Systems Division [SPACECOM] *mt*
SPIT Site Performance and Integration Test *mt*
SPITBOL Speedy Implementation of SNOBOL *hi*
SPL Security Problem Log *fu*
SPL Signal Processing Laboratory *fu*
SPL Software Program Listing *fu*
SPL Sonic (Sound) Pressure Level *fu, sb, 0, kf*
SPL Source Pressure Level *fu*
SPL Spare Parts List *fu, 15, 17, sb*
SPL Subsystem Parts List *jn*
SPL Summary Parts List *ms3.1.2, 8, pl, ep, jdr, fu*

SPL System Programming Language [HP] *pk, kg*
SPL/1 Signal Processing Language, No. 1 (SPL-1 also seen) *fu*
SPLAT System Pre-Launch Acceptance Test [ground] *oe*
SPLICE Stock Point Logistics Integrated Communications Environment [US Navy] *hc, mt*
SPLLATM Special Purpose Low Lethality Anti Terrorist Munition *wp*
SPLTS Special Private Line Telephone System *mt*
SPM Secure Process Manager *fu*
SPM Security Protection Module [Honeywell] *mt*
SPM Skill Projection Model [USAF] *mt*
SPM Software Project Manager *kf*
SPM Solar Proton Monitor *hg, ge*
SPM Solid Propellant Motor *jdr*
SPM Space Products Manufacturing *-8*
SPM Stochastic Planning Model [TRW] *17, sb*
SPM Subproject Manager *fu*
SPM Sun Pointing Mode *jdr, uh*
SPM Support Program Manager *8, fu*
SPM System Performance Measure *hc*
SPM System Performance Monitor [IBM] *pk, kg, fu*
SPM System Planning Manual *mt*
SPM System Program Manager *wp, mt*
SPMO Software Projects Management Office *fu*
SPMO SURTASS Program Management Office *fu*
SPMR Supplier Problem Material Report *fu*
SPMSS Standard Production Management Scheduling System *fu*
SPN Shipment Performance Notice *mt*
SPN SURTASS Part Number *fu*
SPN Switched Public Network *hi*
SPO Special Position Operator *mt*
SPO Special Projects Office [US Navy] *mt, fu, hc, tb*
SPO Subpurchase Order *fu*
SPO Supervised Parallel Output *-4*
SPO System Performance Officer *fu*
SPO System Program(s) Office *mt, 17, bo, sb, hc, wp, dm, tb, kf, oe*
SPO System Project (Program) Office *fu*
SPO System Project Office *hc*
SPO System Project Organization *mt*
SPOC Shuttle Payload Operations Contractor *jdr*
SPOC Single Point Of Contact *jdr*
SPOC Space Operations Center *as*
SPOD Sea Port Of Debarkation *mt*
SPOD Surface Port Of Debarkation *mt*
SPOE Sea Port Of Embarkation *mt*
SPOE Surface Port Of Embarkation *mt*
SPOER Special Operational ELINT Requirements *fu*
SPOILER Short Pulsed Optical Initiated Laser *hc*
SPOOL Simultaneous Peripheral Operations On line *fu, kg, hi*
SPOT Satellite Positioning and Tracking *w9*
SPOT Shared Product Object Tree [IBM] *pk, kg*
SPOT Small Portable Terminal *fu, hc*
SPOT Software Process Oversight Team *kf*
SPOT SystËme Probatoire d'Observation de la Terre (Satellite Pour l') [(French) a system for earth SPOT *System Performance and Operations Test*
SPP Security Policy and Procedures *mt*
SPP Security Practices and Procedures *fu*
SPP Sequenced Packet Protocol *kg, hi*
SPP Serial / Parallel Port *fu*
SPP Signal Processing Program *is26.1.76h*
SPP Solid Propellant Performance *st*
SPP Standard Parallel Port *ct, em*
SPP Standard Practice Procedures [a security manual] *su*
SPP Standard Practice [and] Procedure *hc, tb, jdr*
SPP Standard Practices and Procedures *mt, fu*
SPP Subproject Program Plan *mt*
SPP Sun Pointing Platform *oe*
SPP SWARM Post Processor *tb*
SPP System Package Program *mt, fu*
SPP System Project Plan *mt*
SPPD Signal Processor Packaging Design *hc*
SPPE Senior Publications Project Engineer *fu*
SPPOT Shipboard Propulsion Plant Operator Training *hc*
SPPS Scaleable Power Parallel System [IBM] *kg, hi*
SPPS Supplier Pack Packaging Specification *fu*
SPR Secretary of Air Force Program Review *mt*
SPR Silicon Power Rectifier *fu*
SPR Simplified Practice Recommendation *fu*
SPR Software (System) Problem Reports [AF Form 1775] *mt*
SPR Software Problem Report *gp, 4, jb, wp, oe*
SPR Space and Communications Group Practice [a directive] *wp*
SPR Spare Part Requisition *fu*
SPR Special Purpose Register *pk, kg*
SPR Standard Procurement Requirements *fu*
SPR Statistical Pattern Recognition *pk, kg*
SPR Strategic Petroleum Reserve *mt, wp*

SPR Supplier Problem Report *fu*
SPR System Performance Record *fu*
SPR System Problem Report *kf, fu*
SPRAA Strategic Plans and Resource Analysis Agency [OJCS] *wp, dt, mt*
SPRAM Special Purpose Recoverables Authorized to Maintenance *mt*
SPRD Science Policy Research Division *ci.xli*
SPREAD Systems Programming, Research, Engineering and Development [IBM] *pk, kg*
SPRF Sandia Pulsed Reactor Facility [a DNA test device] *-17*
SPROB Solid Propellant Space Booster Plant *ai2*
SPROM Switched PROM *hi*
SPRP Strategic Petroleum Reserve Program *wp*
SPRS Single Passenger Reservation System *mt*
SPRS Special Purpose Reconnaissance System *wp*
SPRT Sequential Probability Ratio Test *fu*
SPS (assigned to the 377th SPS during Tet in 1968) *af72.5.13*
SPS Samples Per Second *fu*
SPS Satellite Power Stations *ci.x*
SPS Satellite Power Systems *hc*
SPS Scalable POWER/parallel Server *hi*
SPS Security Policy Squadron *mt*
SPS Sense Point Scanner *mt*
SPS Sensor Processing System *oe*
SPS Series-Parallel-Series *fu*
SPS Service Propulsion System *ci.390*
SPS Simplified Processing Station *mt, hc, wp*
SPS Simulation Preparation System *fu*
SPS Software Product Specification *mt, fu, pl, ep*
SPS Solar Panel Simulator *jn*
SPS Solar Panel Substrate *gg10*
SPS Solar Power Satellite *mt, mi*
SPS Solid Propellant Subsystem *lt*
SPS Special Processing Slice *jdr*
SPS Standard Positioning Service *fu*
SPS Standby Power Supply *ct, em*
SPS Standby Power System *pk, kg*
SPS Statement of Prior Submission *fu*
SPS Supplier Packaging Specifications *fu*
SPS Survival Power Source *wp*
SPS Symbols Per Second *-4*
SPS System Performance Specialist *fu*
SPS System Performance Specification *kf*
SPSLV [variant of PSLV, more powerful] *ac3.2-3*
SPSP Site Preparation Support Plan *mt*
SPSR System Performance Status Report *fu*
SPSS Statistical Package for the Social Sciences *pk, kg*
SPSS Statistical Processing Social Sciences *tb*
SPST Single Pole Single Throw *fu, mj31.5.57*
SPSURV Space Surveillance SRBS *tb*
SPSW Support Software *mt*
SPT Sectors Per Track *kg, hi*
SPT Software Process Team [formerly SWG] *kf*
SPT Special Performance Test *es55.43*
SPT Stationary Plasma Thruster *ls4, 5, 28*
SPT System Performance Test(ing) [an event] *ah, gs2-13.14, 9, 8, ep, jdr, pbl, tm, jn, fu*
SPTD Supplementary Provisioning Technical Documentation *tb, fu*
SPTE Special Purpose Test Equipment *fu*
SPTS Special Purpose Test Sets *hc*
SPTT Single-Pole Triple-Throw *fu*
SPTW Support Wing [USA] *mt*
SPU Signal Processor (Processing) Unit *pl, fu*
SPU STAJ Processor Unit *fu*
SPUD Set-Up Program Utility Demonstration Program [a test environment for Set-up software] *fu*
SPUR System for Predetermining Unit Requirement [MTMC] *mt*
SPUR [an Air Force program for nuclear energy conversion] *ci.232*
SPW Self Protection Weapon *uc, hc, wp*
SPW Signal Processing Work [system] *oe*
SPW Space Wing [USA] *uh, mt*
SPX Sequenced Packet Exchange *ct, em, kg, hi*
SQ Software Quality *fu*
SQA Software Quality Assurance *mt, fu, 17, sb, nt, wp, tb, kf, oe*
SQADCS Software Quality Assurance Document Comment Sheets *fu*
SQAE Senior Quality Assurance Evaluator *mt*
SQALR Supplier Quality Acceptance Lot Record *fu*
SQAM Software Quality Assurance Method *fu*
SQAMS {Software ?}Quality Assurance Method Sheets *fu*
SQAP Software Quality Assurance Plan (Program) *mt, fu, wp, tb*
SQAP Supplementary Quality Assurance Provisions *fu*
SQAR Supplier Quality Assurance Report *oe*
SQC Shop Quantity Control *fu*
SQC Software Quality Control *fu*
SQC Statistical Quality Control *mt, fu, 8, ad, ig21.12.17, jdr, oe*
SQD Software Quality Director *fu*

SQDR Software Quality Deficiency Report *fu*
SQE Signal Quality Error [test] *pk, kg*
SQE Software Quality Engineer *kf, fu, oe*
SQE Software Quality Engineering *oe*
SQE Software Quality Evaluation *mt*
SQEP Software Quality Evaluation Plan *mt, fu, hc, jdr*
SQHR Source Quality History Record *pl*
SQL Sequel Query Language *fu*
SQL Software Quality Laboratory *fu*
SQL Software Quality Language *mt*
SQL Standard Query Language *mt, kf*
SQL Structured Query Language [the ANSI standard database language] *mt, hg, ge, fu, wp, ns, tb, nu, ct, em, cu, hi*
SQL/DS Structured Query Language / Data System [IBM] *mt, kg, hi*
SQLL SQL Link *hi*
SQLS SQL Server *hi*
SQMD Squadron Manpower Document *mt*
SQMS Software Quality Management System *oe*
SQN Sequence Number *fu*
SQP Software Quality Plan *kf*
SQPN Staggered Q channel to PN clock modulation *-2*
SQPN Staggered Quadriphase Pseudorandom Noise *4, vv*
SQPR Supplier Quality Performance Report *17, sb*
SQPSK Staggered Quadriphase Shift Key(ing) (Keyed) (modulation) *sp662, 2, 4, vv*
SQR Special Qualifications/Requirements *mt*
SQR Supplier Quality Rating *fu*
SQRT Square Root *pk, kg*
SQS/VIS Supplier Quality System/Vendor Information System *fu*
SQT Skill Qualification Test *fu, hc*
SQUEL Standard Query Language *hc*
SQUID Superconducting Quantum Interference Device *wp, st*
SQ_OC Squadron Operations Center *mt*
SR Sample Return *re*
SR Saturable Reactor *fu*
SR Scanning Radiometer *ge, ac1.6-13, hg*
SR Seaman Recruit *w9*
SR Search Radar *fu*
SR Selective Ringing *fu*
SR Selenium Rectifier *fu*
SR Service Report *hc*
SR Shift Register *fu, kg*
SR Shipping Request *fu*
SR Short Range *fu*
SR Slip Ring *fu*
SR Slow Release *fu*
SR Software Repository *fu*
SR Solid Rocket *ai1*
SR Sortie Rate *mt, wp*
SR Sound Ranging *fu*
SR Space Required *mt*
SR Special Reconnaissance [as in SR-71, the "Blackbird" airplane] [USAF] *mt*
SR Special Relativity *mi*
SR Speech Recognition *hi*
SR Steradian *wp, tb, oe*
SR Stores Requisition *fu*
SR Strategic Reconnaissance [USA] *wp, mt*
SR Study Requirement *fu*
SR Summary Requisition (Report) *fu*
SR Supply Requisition *oe*
SR Surveillance Radar *fu*
SR System Requirements *fu*
SR&QA Safety, Reliability, and Quality Assurance *wl*
SRA (former name of Ericson Radio Systems AB, Sweden) *vc25.3.18*
SRA Selected Repair Availability *hc*
SRA Selected Restricted Availability *mt, fu*
SRA Shop (Ship) Replaceable Assembly *fu*
SRA Slip Ring Assembly *9, lt*
SRA Software Requirements Analysis *kf, fu*
SRA Special(ized) Repair Activity *mt, wp, fu*
SRA Standard Repair Authorization *fu*
SRA Standard Requirements Code *mt*
SRA Supply Readiness Assessment [PACAF] *mt*
SRA Surveillance Radar Approach *mt*
SRA System Requirement Analysis *jdr*
SRA Systems Research and Applications Corporation *mt, fu*
SrA Senior Airman, E-4 [USAF] [USA] *mt*
SRAAM Short Range Air to Air Missile *mt, hc*
SRAC Sonar Radar Azimuth Converter *fu*
SRAD Systems Requirements Allocation Document *-4*
SRAG Semi-active Radar Anti-air Guidance *hc*
SRALCC Southern Region Airlift Control Center *mt*
SRALD Southern Region Airlift Division *mt*
SRAM Safety Reliability, Availability, and Maintainability *hc*
SRAM Shadow Random Access Memory *kg, hi*

SRAM Short Range Attack Missile [AGM-69, a nuclear-headed missile carried by the B-52 and FB-111] [USA] *mt, aw118.18.14, 12, 14, hc, af72.5.134, wp, tb*
SRAM Short-Range Attack Missile *fu*
SRAM Static Random Access Memory *mt, kf, fu, ct, em, kg, wp, hi*
SRAN Stock Record Account Number *mt, wp*
SRAP Sensor Receiver And Processor *fu*
SRAPI Speech Recognition API *pk, kg*
SRARM Short Range Anti Radiation Missile *mt, hc*
SRAS Southern Region Added Sites [NADGE, also called RIP/M] *hc, fu*
SRAT Short Range Advanced Technology *-17*
SRAT Short Range Applied Technology *wp*
SRATS Short Range Aircraft Tracking System *wp*
SRAW Short Range Anti-tank Weapon [also STRAW] *hc, wp*
SRB Short Runlength Beacon *wp*
SRB Solid Rocket Booster *mi, sc3.6.1, hc, ws4.10, af72.5.26, wp, jdr, mb, ai1, ai2*
SRB Source Route Bridge *pk, kg*
SRB Specification Review Board *oe*
SRB Surface Radiation Budget *hy*
SRBDM Short Range Bomber Defense Missile *hc*
SRBM Short Range Ballistic Missile *mt, 12, wp, tb, kf*
SRBOC Super Rapid Blooming Offboard Chaff [US Navy] *mt*
SRBS Skeletal Reference Baseline Simulation *tb*
SRBS Solid Rocket Boosters *-9*
SRC Sample Return Capsule *re*
SRC Science Research Council *crl*
SRC Scientific Requirements Specification *ge, hg*
SRC Shipment Release Code *wp*
SRC Solar Rejection Coating *kf*
SRC Source Code [Hughes Aircraft Company] *lc*
SRC Source Recoverability Code *fu*
SRC Spiral Redundancy Check *fu*
SRC Standard Requirement Code *mt*
SRC Stock Record Card *fu*
SRC Strategic Reconnaissance Center [SAC] *mt*
SRC Supercomputing Research Center *mt*
SRC Supply Readiness Center *mt*
SRC Survival and Recovery Cell *mt*
SRC Survival Recovery Center *mt*
SRC System Review Committee *mt*
SRCF Subregional Control Facility *mt*
SRCL Standard Requirements Check List *fu*
SRD Screen Reader System *pk, kg*
SRD Software Requirements Document *uh*
SRD Standard Reporting Designator *mt*
SRD Step Recovery Diode *fu, gg154, jdr*
SRD System Requirements Document *fu, mt, es55.72, wp, tb, kf*
SRDB System Requirements Data Base *fu*
SRDRAM Self-Refreshed DRAM *pk, kg*
SRE (Litton Industries provided SRE and SREC series connectors) *ci.954*
SRE Standards for Robot Exclusion *em*
SRE Surveillance Radar Element *fu*
SREC (Litton Industries provided SRE and SREC series connectors) *ci.954*
SREM Software Requirements Engineering Methodology *wp, fu*
SREMP Source Region Electromagnetic Pulse *mt, tb, st*
SREP Sales Representative *hi*
SRF Secure Reserve Force *mt, tb*
SRF Strategic Reserve Force *mt*
SRF Strategic Rocket Forces [Soviet] *fu, 12, tb*
SRF Summary Reference File [JOPS] *mt*
SRF System Relevant Failure *fu*
SRGB Sustained RGB [values] *pk, kg*
SRHIT Short Range Homing Intercept Technology [see FLAGE] *hc, st, tb*
SRHQ Sub Regional Headquarters [British Command Centers] *-12*
SRI Short Range International *mt*
SRI Southern Research Institute *wp*
SRI Standard Repair Instruction Information *fu*
SRI Standing Request for Information *fu*
SRI Stanford Research Institute [Menlo Park, California] *fu, hc, ie26.1.9, wp*
SRI Surveillance, Reconnaissance, and Intelligence *mt*
SRI Systems Research Institute [IBM Corporation] *do.v*
SRIMP Software Requirement Integrate Modeling Program *fu*
SRINF Shorter Range Intermediate Range Nuclear Forces *wp*
SRIS {Station for the Reception of Images from SPOT} *ac3.6-26*
SRL (SRL Defense Electronics Systems . . . {Incorporated}) *aa26.11.46, hc*
SRL Satellite Research Laboratory [NOAA] *hg, ge*
SRL Section Reacteur LÈger [light jet section] [France] *mt*
SRL Shuttle Radar Lab *hy*
SRL Single Rail Launcher *hc*
SRL Standard Recording List *fu*
SRLR Sample Return with Local Rover *re*

SRM Security Reference Monitor *pk, kg*
SRM Short Range [Nuclear] Missile *aw130.22.23*
SRM Solid Rocket Motor *mi, ci.963, 4, hc, jdr, mb, ai2*
SRMMOP Solid Rocket Motor Multiple Options Program *cp*
SRMON Super R Monitor (R Monitoring) *hi*
SRMU Solid Rocket Motor Upgrade *jdr, ai2*
SRM_Bde_OC Short Range Missile Brigade Operations Center *fu*
SRN Simulation (System) Reference Number *fu*
SRN Study Request Notice *fu*
SRNP Secure Remote Network Processor *mt*
SRO Sample Return Orbiter *re*
SRO Sharable and Read Only *pk, kg*
SRO Site Reconnaissance Orbiter *re*
SRO Superintendent of Range Operations *0*
SRO System Review Office *2, 4, oe*
SROSS Stretched Rohini-class Satellites [India} *ai2*
SRP Salt River Project [Arizona] *aw130.22.79*
SRP Sealift Readiness Program *mt*
SRP Sealift Reserve Program *mt*
SRP Search Radar Processor *fu*
SRP Seaward Recovery Point [USA DOD] *mt*
SRP Sensor Reporting Post *hc*
SRP Sensor Reporting Post [later called AN/TYC-6] *fu*
SRP SIOP Reconnaissance Plan *mt, wp*
SRP Solicitation Review Panel *mt*
SRP Sonobuoy Referenced Position *fu*
SRP Standard Repair Procedure *fu*
SRP Strategic Reconnaissance Plan *mt*
SRP Student Response Processor *fu*
SRP Suggested Retail Price *kg, hi*
SRP Surveillance Radar Processor *fu*
SRP Survivable Relay Processor *kf*
SRPI Server-Requester Programming Interface *pk, kg*
SRQ Service Request *kg, wp*
SRR Scrap, Rework, and Repair *nd74.445.26*
SRR Serially Reusable Resource *pk, kg*
SRR Short Range Radar *mt, fu*
SRR Software (Specification) Requirements Review *fu*
SRR Survivability, Recovery and Reconstitution *mt*
SRR System Readiness Review *jn*
SRR System Requirements Review *mt, hc, 17, sb, nt, wp, dm, tb, st, wl, jdr, kf, jn*
SRR [MX Extended Range Radar] *hc*
SRRE Satellite Range and Range Extraction *dm*
SRRH System Requirements Review HEO *kf*
SRRR Scrap, Rework, Repair and Retest *jdr*
SRRS Stimulated Rotational Raman Scattering *hc*
SRS (SRS . . . Huntsville, AL 35806) *jmj*
SRS Satellite Readout Station *mt, kf*
SRS Satellite Relay Station *kf*
SRS Segment Ready Storage *ai2*
SRS Shock Response Spectrum *he, jdr*
SRS Simulator Reconfiguration System *tb*
SRS Situation Reporting System *mt*
SRS Software Requirements Specification *mt, pl, hc, ep, jdr, kf, fu, uh, oe*
SRS Sound Retrieval System *kg, hn49.9.1, hd1.89.y27*
SRS Specification Requirement Sheet *fu*
SRS Speech Recognition System *hi*
SRS Stimulated Raman Scattering *wp*
SRS Strategic Reconnaissance Squadron [USAF] [USA] *mt*
SRS Supplier Rating System *fu*
SRS Surveillance Radar System *fu*
SRS Synchronous Relay Satellite *hc*
SRS System Requirement Specification (Studies) *fu*
SRS System Requirements Study [BMD] *tb, st*
SRS/IRS Software Requirement Specification / Interface Requirements *fu*
SRSCC Simulated Remote Station Control Center *mt*
SRSSS Satellite Readout Station Subsystem *kf*
SRSU Satellite Readout Station Upgrade *hc, kf*
SRT Safety Review Team *jdr*
SRT Scheduled Return Time *mt*
SRT Shipboard Repeatability Test *fu*
SRT Single Radar Tracking *fu*
SRT STAJ Receiver Transmitter *fu*
SRT Standard Remote Terminals *mt*
SRT Station Readiness Test *-4*
SRT Strategic Relocatable Target *mt, ic2-6, go3.1.1, wp, tb*
SRT System Readiness Test *fu*
SRT Systems Requirements Test *tb*
SRT&E Systems Requirements Test and Evaluation *tb*
SRT/AMME Standard Remote Terminal/Automated Multi-Media Message Exchange *mt*
SRTA Strategic Relocatable Target Attack *wp*
SRTC Search Radar Terrain Clearance *fu*
SRTC Solid Rockets Technical Committee *aa27.4.b51*

SRTP Simulation Raid Tape (generation) Program *fu*
SRTS Strategic Reconnaissance Training Squadron [USA] *mt*
SRU Secondary Reference Unit *fu*
SRU Secondary Replaceable Unit *fu*
SRU Ship Replaceable Unit *wp, fu*
SRU Shop Replaceable Unit *fu, hc, wp, mt*
SRU Spares Replacement Units *hc*
SRU Stratospheric Sounding Unit [also SSU] *aw129.14.43*
SRU System Resource Unit *nt*
SRV Short Range Viewer *hc*
SRV Shuttle Derived Vehicle *hc*
SRV Surveillance Restoral Vehicle *mt*
SRV Swirl Recovery Vanes *aa26.10.20*
SRW Strategic Reconnaissance Wing [USA] *12, mt*
SRWBR Short Range Wideband Radio *mt*
SRZ Special Rules Zone *mt*
SR[A] Staff Requirement [Air] [UK] *mt*
SS Attack Submarine *wp*
SS San Salvador [AMR 5] *0*
SS Satellite Switched *hi*
SS Sea State (usually with number, as SS0 or SS2) *fu*
SS Sector Suite *fu*
SS Self Sustaining *mt*
SS Session Service *fu*
SS Ship Shore *mt*
SS Signal Strength *fu*
SS Signaling System [a channel signaling interface] *ct, em, nu, hi*
SS Simulation Supervisor *fu*
SS Simulator System *mt*
SS Single Satellite *jdr*
SS Single Sided *kg, hi*
SS Slow Scan *oe*
SS Slow Swept *fu*
SS Solid State *ep, fu*
SS Space Segment *pbl, kf*
SS Space Station *wl*
SS Space Systems *dt*
SS Spin Synchronous *ggIII.2.371*
SS Split Stream *do64*
SS Spread Spectrum *jdr*
SS Stack Segment *fu, kg, wp*
SS Standard Sequential Delivery *mt*
SS Star Sequence *oe*
SS Steam Ship *w9, wp*
SS Structure Subsystem *jdr*
SS Submarine [USN ship classification] *mt, fu*
SS Subsystem *fu, kf, oe*
SS Summer Solstice *kf, oe*
SS Summing Selector *fu*
SS Sun Sensor *kf*
SS Supply Support *mt, fu*
SS Support Segment *hc*
SS Support Systems *fu, hc, hn44.18.4*
SS Surface to Surface *17, sb, tb*
SS Surveillance (System) Supervisor *fu*
SS System Specification *mt, fu*
SS System Support [IPT identifier] *kf*
SS Systems Support Deputy Chief of Staff *mt*
SS Systems Support Deputy Commander *mt*
SS [Soviet designator for various ICBMs, e.g. SS-18, carrying 10 RVs, or shorter range SS-19, SS-20, SS-24, SS-25, SS-13, etc.] *wp*
SS&D Synchronization Separator and Digitizer *fu*
SS&S Surveillance and Sensors Systems *fu*
SS/L Space Systems / Loral *jdr, oe*
SS/SSS System Subsystem Specification / Signal Simulator *mt*
SSA S-band Single Access *2, 4, vv*
SSA Sector Space Averaging *fu*
SSA Selective Service Administration *wp, mt*
SSA Sequential Shunt Assembly *oe*
SSA Serial Storage Architecture *ct, kg, em, hi*
SSA Ship System Architecture *hc*
SSA Signal Security Agency *-10*
SSA Small Search Area *fu*
SSA Social Security Administration *w9, wp*
SSA Software Safety Analysis *fu*
SSA Software Support Activity *fu*
SSA Software Support Agency *mt*
SSA Solar Scatter Angle *kf*
SSA Solid-State Amplifier *hc, fu*
SSA Source Selection Authority *mt, fu, wp, tb, kf*
SSA Space System Architecture *tb*
SSA Staff Support Agency [USA] *mt*
SSA Static Sensitive Assemblies *fu*
SSA Structured Systems Analysis *fu*
SSA Sun Sensor Assembly *3, jn, he, oe*
SSA System Security Administrator *fu*

SSA [Cargo Submarine] [US Navy] *mt*
SSA {Diesel Powered Auxiliary Submarine} *wp*
SSAA Short Span Attitude Adjustment *oe*
SSAA Space Science Analysis Area *0*
SSAC Source Selection Advisory Council (Committee) *wp, mt*
SSAC Space Science Advisory Committee [of ESA] *so, es55.9*
SSAC Space Science Analysis and Command *0*
SSACV (SSACV Icebreaking LNG/Oil Tanker) {? Air Cushion Vehicle} *aa27.4.b14*
SSAF S-band Single Access Forward *4, vv*
SSAG Source Selection Advisory Group *fu*
SSALF Simplified Short-Approach Light Facility *fu*
SSALR Simplified Short-Approach Lighting (System with) Runway [alignment] *fu*
SSALS Simplified Short-Approach Light System *fu*
SSAN {Submarine Ship, Auxiliary, Nuclear powered}:(wp)
SSAO Sr. SAM Assignment Officer *fu*
SSAP Source Service Access Point *pk, kg*
SSAR S-band Single Access Return *4, vv*
SSARC Service System Acquisition Review Council *hc*
SSAS Signal Security Assessment Study *hc*
SSAS SSA Shuttle *-4*
SSASC Space Science and Applications Steering Committee *so*
SSATS Surface Ship Advanced Tactical Sonar *hc*
SSB Ship, Submarine Ballistic [strategic non-nuclear powered] *12, ph*
SSB Single Side Band [amplitude modulation] *mt, fu, ct, em, bl1-5.44, 0, 2, pl, hc, wp, ce, jdr*
SSB Software Sub Board *fu*
SSB Solid Strap-on Boosters *ai2*
SSB Source Selection Board *wp, fu*
SSB Space Sciences Board [of U. S. Academy of Sciences] *ci.718*
SSB Sub System Bus *wp*
SSB Submarine, Ballistic Missile [US Navy] *mt*
SSB/KEW Small Space Booster/Kinetic Energy Weapon *hc*
SSBD Silicon Surface Barrier Detector *pl*
SSBD Solar Sail Boom Drive *oe*
SSBI Single Scope Background Investigation *jdr*
SSBN Submarine (Steam) Ship, Ballistic missile armed, Nuclear powered [USN ship classification] *mt, 11, 12, hc, dh, pf1.16, pf.f88.15, wp, tb, fu*
SSBN-X Strategic Submarine, Ballistic, Nuclear-Experimental *uc*
SSBO Single-Swing Blocking Oscillator *fu*
SSBS Sol-Sol Balistique Strategique [(French) land based strategic ballistic missile] *12, crl*
SSBUV Shuttle Solar Backscatter Ultraviolet Spectrometer *hy*
SSC Samara Space Center *ai2*
SSC Satellite Service Calculator [COMSAT] *csl.i*
SSC Satellite Shipping Container *oe*
SSC Satellite Situation Center *ns*
SSC Satellite Switching Center *hc*
SSC Senior Sector Controller *fu*
SSC Simulation Support Center *tb*
SSC Skill Speciality Code *fu*
SSC Software Support Center *fu*
SSC Solid State Circuit *fu*
SSC Soliders' Support Center *dt*
SSC Space Surveillance Center *jb, tb, as*
SSC Space Surveillance Center [SPACECOM] *jb, tb, as, mt*
SSC Standard Systems Center [at Gunter Annex-Maxwell AFB, Alabama, later known as the Standard Systems Group, SSG] *mt*
SSC Static Sensitive Component *fu*
SSC Subdecoder Switch Controller *jdr*
SSC Superconducting Super Collider *fu*
SSC Supply (System) Support Center *fu*
SSC Supply Systems Command [USA] *mt*
SSC SURTASS Support Center *fu*
SSC Swedish Space Corporation *crl*
SSC System Simulation Center *st*
SSC System Status and Control *fu*
SSC System Support Contractor *mt*
SSCB Space Station Control Board *wl*
SSCC Special Security Communications Center [US Navy] *mt*
SSCC System Support Computer Complex *fu*
SSCF Signal Strength Center Frequency *fu*
SSCL Superconducting Super Collider Laboratory *fu*
SSCM Spectral Sonar Countermeasures *hc*
SSCN Scientific Satellite Communication Network *0*
SSCOP Service Specified Convergence Protocol *ct, em*
SSCP Small Self Contained Payload *ci.601*
SSCP System Services Control Point *do400, kg*
SSCR Standard Systems Center Regulation *mt*
SSCS Shipboard Satellite Communications Set *fu*
SSCS Spread Spectrum Command System *fu*
SSCS Spread Spectrum Communications System *wp*
SSCS Standard Satellite Control System *kf*
SSCSAP Standard Small Computer Software Acquisition *mt*
SSCSC Support Service Computer Software Component *kf*

SSD Section de Soutien de Dugny [Dugny maintenance station] [France] *mt*
SSD Security Sentinel Device *fu*
SSD Solid State Detector *gp, fu, oe*
SSD Solid State Disk *ct, kg, em, fu*
SSD Space Systems Division [USAF] *fu, an31.7.9, hc, dm, jdr, oe*
SSD Static Sensitive Device *fu*
SSD Static Sensitive Device [hardware] *ah, ep, jdr, pbl, tm, jn*
SSD Strategic Systems Division *fu*
SSD Supplementary Supplier Data *fu*
SSD Support Systems Division *-8*
SSD System Support Division *wp*
SSD Systems Support Database *mt*
SSDB Spin System Distribution Box *9, lt*
SSDD System / Segment Design Document *pl, fu, dm, kf*
SSDF Satellite Software Development Facility *jr*
SSDMS Space Station Data Management System *wl*
SSDN Survivable Sensor Data Network *kf*
SSDR Sub System Design Review *st, fu*
SSDR Subsystem Development Requirement *fu*
SSDS Silo Self Defense System *wp*
SSDS Small (Surface) Ship Data System *fu*
SSDS Solid State Driver Stage *hc*
SSDSG Special State Defense Study Group *dt*
SSE Fly By Wire [Sweden] *mt*
SSE Satellite to Satellite Experiment *ac2.4-6*
SSE Senior Staff Engineer *fu*
SSE Software Support Environment *aa26.8.b25, wp*
SSE Software Support Equipment *fu*
SSE Solid State Electronics *fu*
SSE Source Surveillance Engineer *hq5.2.4*
SSE Spacecraft Systems Engineer *oe*
SSE Special Support Equipment *fu*
SSE Sum of Squares due to Error *wp*
SSE Sustaining System Engineering *fu*
SSE System Safety Engineering *fu*
SSE System Security Engineering *nt, kf*
SSE System Support Environment *fu*
SSE/EWE SIGINT Support Element/Electronic Warfare Element *fu*
SSEB Source Selection Evaluation Board *mt, fu, wp, st*
SSEC Selective Sequence Electronic Calculator [IBM] *pk, kg*
SSEC Solar System Exploration Committee *hc, so*
SSEC Source Selection Evaluation Committee *wp*
SSEC Space Science Engineering Center [University of Wisconsin] *pl*
SSED Software Systems Engineering Directorate *mt*
SSED Space and Strategic Systems Engineering Division *hn43.10.8*
SSEG Source Selection Evaluation Group *mt*
SSEG System Safety Engineering Group *fu*
SSEP Source Selection Evaluation Plan *mt*
SSEP System Safety Engineering Plan *fu, wp*
SSEP System Security Engineering Plan *kf*
SSES Ship Signal Exploitation Space *mt*
SSES Ship Signal Exploitation System *tb*
SSES Ship System Engineering Standard *fu, hc*
SSES Ship's Signal Exploitation Space *fu*
SSET Source Selection Evaluation Team *mt, kf*
SSEU Segment String Element Update *fu*
SSF Software Support Facility *wp, mt*
SSF Sonar Simulation Facility *fu*
SSF Space Station Freedom *re, mi*
SSF Switch Supervisory Function *mt*
SSF System Support Facility *mt, fu*
SSFD Solid State Floppy Disk *ct, em*
SSFL Steady State Fermi Level *fu*
SSFM Single Sideband Frequency Modulation *fu*
SSG Cruise Missile Attack Submarine *wp*
SSG Search Signal Generator *fu*
SSG Secure Sequence Generator *fu*
SSG Ship, Submarine, Guided missile armed [USN ship classification] *wp, fu*
SSG Special Study Group *st*
SSG Staff Sergeant *w9*
SSG Submarine, Guided missile [USA DOD] *mt*
SSG System Safety Group *mt, fu*
SSGA System Suppport Gate Array *pk, kg*
SSGB Subscale Blind Grating *hc*
SSGM Strategic Scene Generation Model *kf*
SSGN Ship, Submarine, Guided missile armed, Nuclear powered [USA DOD] *mt, ph, wp, fu*
SSGS Spacecraft Support Ground System *oe*
SSgt Staff Sergeant *mt*
SSHA System Safety Hazards Analysis *mt*
SSHAR System Safety Hazard Analysis Report *fu*
SSI Secondary Search Inhibit *fu*
SSI Server Side Includes *pk, kg*
SSI Similar Source Integration *fu*

SSI Single Source Integrator *fu*
SSI Single System Image *pk, kg*
SSI Small Scale Integration *mt, kg, fu, hi*
SSI Small Signal Integration *kg, jdr*
SSI Solid State Imaging (Imager) [subsystem] [on Galileo] *mi, jp, es55.26, mi*
SSI Space Segment Integrator *pl*
SSI Space Services, Incorporated *sc3.6.4*
SSI Space Studies Institute *mi*
SSI Space Systems, Incorporated *sc3.10.4*
SSI Spaceport Systems International *ai2*
SSI Specialty Skill Identifier *mt*
SSI Standard Subject Identification Code *fu*
SSI Supply System Item *mt*
SSIC Standard Subject Identification Code *mt*
SSID Supplemental Structural Inspection Document *aw129.21.75*
SSILS Solid State Instrument Landing System *mt*
SSIP Ship Support Improvement Program *hc*
SSIP Surveillance Sensor Interface Processing *fu*
SSIPS/LDS Ship Support Improvement Projects / Log Data System [US Navy] *mt*
SSIS Space Station Information System *19, hy, wl*
SSIU Subsystem Interface Unit *fu*
SSIXS Submarine Satellite Information Exchange Subsystem [US Navy] *mt*
SSJ Self Screen Jamming *fu*
SSJ Self Screening Jammer *wp, pm*
SSJ Single-Sideband Jammer *fu*
SSK Submarine Ship, Hunter Killer [US Navy] *mt*
SSKN Submarine Ship, Hunter-Killer, Nuclear powered [US Navy] *mt*
SSKP Single Shot Kill Probability *mt*
SSL Secondary Standards Laboratory *fu*
SSL Secure Socket Layer *em, kg, hi, du*
SSL Space Sciences Laboratory [at MSFC, Aerospace Corporation, etc.] *pl*
SSL Space Simulation Laboratories [facility, Hughes Aircraft Company] *kf, sc3.12.2, 8, 15, ep, jdr, pbl, jn, tm, ah*
SSL Space Space Link *vv*
SSL Suspended Strip Line *mj31.5.68*
SSL System Specification Language *fu*
SSL Systems Support Laboratory *fu*
SSLD Single Seat Laser Designator *hc*
SSLSM Single Service Logistics Support Manager [AFLC] *mt*
SSLV Standard Small Launch Vehicle *ai2*
SSM Second Surface Mirror *jdr*
SSM Sensor System Manager *tb*
SSM Ship System Manual *fu*
SSM Shop Standards Manual *fu*
SSM Single Sideband Modulation *fu*
SSM Software Segment Manager *fu*
SSM Software Subcontract Monitor *fu*
SSM Solid-State Materials *fu*
SSM Solid-State Modulators *fu*
SSM Special Support Manager *wp*
SSM Spread Spectrum Modulation *mt*
SSM Staff Sergeant Major *w9*
SSM Steady-State Model *fu*
SSM Stress Screening Model *fu*
SSM Super Spin Mode *-3*
SSM Surface to Surface Missile *mt, hc, 12, kf, fu*
SSM System Support Manager *fu*
SSM/I Special Sensor Microwave Imager *hc, sc3.11.2, pl, hy*
SSM/T Special Sensor Microwave Temperature [sounder] *pl*
SSMA Solid State Modulator Assembly *jn*
SSMA Spread Spectrum Multiple Access *mt*
SSMCC Space Shuttle Mission Control Center *2, 4*
SSMD Silicon Stud Mounted Diode *fu*
SSME Space Shuttle Main Engine *mi, ci.xxii, 9, ws4.10, tb, mb, te23, 11, 18, ai2*
SSME Spread Spectrum Modulation Equipment *mt*
SSMI Special Sensor Microwave / Imager [DMSP] *hg, ge*
SSMIA Senior Manager's Information Analysis *mt*
SSMIS Special Sensor Microwave Imager [and/or Sounder] [MIS + UAS] *pl, wp*
SSMP Software Subcontractor Monitoring Plan *fu*
SSMP Source Selection and Message Processing *fu*
SSMP Surface Sonar Modernization Program *fu*
SSMP System Security Management Plan *kf*
SSMS Spacecraft and Station Management Subsystem *sc3.2.2*
SSMS Spark Source Mass Spectroscopy *wp*
SSMS Standard Survivable Message Set *mt, kf*
SSMTC Sary Shagan Missile Test Center [USSR] *tb*
SSN Ship, Submersible, Nuclear [nuclear powered attack submarine] *uc, pf1.28, wp, st*
SSN Sortie Sequence Number *mt*
SSN Space Surveillance Network *mt, go2.1, lka, jb, nt, cp, as*
SSN Submarine Ship, Nuclear powered} [USN ship classification] *fu, mt*

SSN Sun Spot Number *mt*
SSN Supplier Serial Number *fu*
SSN Surface to Surface Naval *tb*
SSNX Submarine Ship, Nuclear powered, Experimental [new in 1985] *dh*
SSO Source Selection Office *mt*
SSO Space Shuttle Orbiter *-4*
SSO Space Systems Operation *oe*
SSO Spares Shipping Order *fu*
SSO Special Security Office(r) *kf, mt*
SSO Spurious Shut Off *jdr*
SSO Standard System Support Office [Gunter AFS] *mt*
SSO Steady State Oscillation *fu*
SSO Submarine Ship, Oiler [US Navy] *mt*
SSO Subsistence Support Officer *mt*
SSO Sun Synchronous Orbit *ai2*
SSO System Security Officer *su*
SSOEG Satellite Systems Operations Evaluation Group *mt*
SSOF Standby Supervisor Of Flying *mt*
SSOP Satellite Systems Operations Plan *wp*
SSOS Severe Storm Observation Spacecraft *hc*
SSP Security Support Plan *mt*
SSP Sensor Suite Positioner *cl, he, jn*
SSP Site Security Practices *kf*
SSP Software Support Package *fu*
SSP Solid State Power *pl*
SSP Sonar Signal Processor *fu*
SSP Source Selection Plan *mt*
SSP Space Station Program *wl*
SSP Special Sensor Processing *pl*
SSP Standard Switch Panel *gp, lt, oe*
SSP Strategic Strike Plan *mt*
SSP Sub Satellite Point *jdr*
SSP Supervisory Status Panel *fu*
SSP Surface Science Package *es55.28*
SSP Surveillance System Performance *kf*
SSP System Safety Plan *fu*
SSP System Security Plan *17, fu*
SSP System Status Panel *fu*
SSP [vehicle transport submarine, US Navy] *mt*
SSP&C Software Standards Procedures and Controls *oe*
SSPA Solid State Power Amplifier *ct, em, uh, sc2.7.2, 8, 17, sb, rg, ep, jdr, kf, vv, tt*
SSPC Solar Sail Protective Cover *oe*
SSPC Solid State Power Controller *ci.973, wp*
SSPD Solid State Products Division [Hughes Aircraft Company] *fu, hn43.9.2, hc*
SSPE Sonar Signal Processing Facility *fu*
SSPE Space Station Program Element *wl*
SSPF Space Station Processing Facility *mi*
SSPI Security Service Provider Interface *em, kg*
SSPIS SBIRS System Phenomenology Impact Study *kf*
SSPK Single Shot Probability of Kill [also PKSS] *tb*
SSPM Software Standards and Procedures Manual *mt, kf, fu*
SSPM Solid State Photo Multiplier *st*
SSPO Strategic Systems Program Office [US Navy] *mt*
SSPO Strategic Systems Project Office [US Navy] *hc, nt*
SSPP System Safety Program Plan *fu, 17, nt, dm, kf, oe*
SSPS Satellite Solar Power Station *ci.874*
SSPS Second Stage Propulsion System *ai2*
SSPS Sky Survey Prototype System *mi*
SSPT Surveillance System Performance Testbed *kf*
SSQ Auxiliary Submarine *wp*
SSQD Single Sided Quad Density *hi*
SSR SACEUR Strategic Reserve *mt*
SSR Schedule Status Report *8, wp*
SSR Secondary Surveillance Radar *mt, hc, wp, fu*
SSR Software Service Report *oe*
SSR Software Specification Review *mt, fu, kf, 17, pl, tb*
SSR Solid State Receiver *fu*
SSR Spread Spectrum Relay *fu*
SSR Submarine Ship, Radar picket [US Navy] *mt*
SSR Sum of Squares due to Regression *wp*
SSR Supplier Surveillance Record *fu*
SSR Supply Support Request *mt*
SSR Support System Requirement *mt*
SSR System Software Review *mt*
SSR System Specification Review *mt*
SSR System Status Review *fu*
SSR System Support Resources *sd, sb*
SSR Systems Service Request *fu*
SSRB Source Selection Review Board *kf*
SSRC Standard Software Requirements Contract *mt*
SSRM Solid Strap-on Rocket Motor *ai2*
SSRM Spin Stabilized Rocket Motor (Spinning Solid Rocket Motor) *9, lt*
SSRMS Space Station Remote Manipulator System *mi*
SSRN Submarine Ship, Radar Picket, Nuclear powered [US Navy] *mt*

SSRO Sector Scan Receive Only *wp*
SSRP Secondary Surveillance Radar Processor *fu*
SSRP Simple Server Redundancy Protocol [Cisco] *pk, kg*
SSRR Subsystem Requirements Review *fu*
SSRS Standard Service Request Specification *mt*
SSRT Single Stage Rocket Technology *mi*
SSS Einstein Observatory (HEAO-2) Solid State Spectrometer *cu*
SSS Sealift Support System [MSC] *mt*
SSS Selective Service System *mt, w9*
SSS Semiconductor Serial Storage *mt*
SSS Ship Maintenance Support System *mt*
SSS Ship Sensor Switchboard *fu*
SSS Silicon Symmetrical Switch *fu*
SSS Single Sensor Satellite *kf*
SSS Single Strip Scan *hi*
SSS Small Scientific Satellite *pl*
SSS Software Subsystem *jdr*
SSS Software Support Services *hd1.89.y30*
SSS Software Support Set *fu*
SSS Solid State Spectrometer *pl*
SSS Solid State Switching *fu*
SSS Sortie Support System *hc*
SSS Source Selection Sensitive *mt*
SSS Space Surveillance System [also SPASUR] *lka, tb*
SSS Spares Status System *fu*
SSS SRHIT System Simulation *st*
SSS Staff Summary Sheet *mt*
SSS Standoff Surgical Strike *hc*
SSS Station Satellite System *mt*
SSS Stationary Service Structure *bl1-2.258*
SSS Stellar Sensor Subsystem [MMC Technology Center] *lka*
SSS Storage Serviceability Standard *mt, fu*
SSS Strategic Satellite System *rva, mt*
SSS Sub System Specification *mt*
SSS Summer Solstice Simulator *oe*
SSS Surveillance and Sensor Systems *fu*
SSS Synergistic Strike System *wp*
SSS Synthesizer Select Switch *jdr*
SSS System / Segment Specification *mt*
SSS System Software Support *fu*
SSS System/ Segment Specification *fu, pl, ep*
SSSB System Source Selection Board *mt*
SSSC Single Sideband Suppressed Carrier *jdr*
SSSC Space Science Steering Committee *ci.719*
SSSC Space Station Support Center *wl*
SSSC Surface Sub-surface Surveillance Center *tb*
SSSD Secretariat Services Support Division [Navy] *dt*
SSSD Space and Strategic Systems Division [Hughes EDSG] *hq5.2.4*
SSSD Surface Ship Systems Division [Hughes GSG] *hq6.2.2*
SSSD Surveillance and Sensor Systems Division [Hughes EDSG] *fu, hq5.2.4*
SSSE SBKEW System Simulator Emulator *17, sb, 5*
SSSG Space System Support Group *kf*
SSSMP Surface Ship Sonar Modernization Program *fu, hc*
SSSN Satellite System Service Note *jn*
SSSO Standard Systems Support Office *mt*
SSSP Satellite Systems Survivability Program *wp*
SSSP Source Selection Support Plan *mt*
SSSP System Source Selection Procedure *fu*
SSST Single Subscriber Terminal [also SST] *mt*
SST Sea Surface Temperature *ge, hg, hy, pbl*
SST Single Subscriber Terminal *mt, fu, hc*
SST Site Survey Team *wp*
SST Source Selection Team *mt*
SST Spectroscopic Survey Telescope *mi*
SST Spread Spectrum Technology *pk, kg*
SST Sum of Squares Total *wp*
SST Super Serial Technology *hi*
SST Super Sonic Transport *mt, mi, fu, ci.1245, aa26.12.51, aw129.21.2, wp, mb, ai1*
SST Surveillance Supervisor Technician *fu*
SST System Supported Training *fu*
SST Systems Services and Technology *pk, kg*
SST Training Submarine *wp*
SSTD Supplementary Spares Technical Documentation *kf*
SSTDMA (mixed business communications in SSTDMA) {Shared Service Time Division Multiple Access} *sc3.3.3*
SSTDPS Specialized Short Term Development Planning Support *wp*
SSTDRSS Shared Service Tracking [and] Data Relay Satellite System *-4*
SSTEREC Supersonic Tactical Electronic Reconnaissance Program *wp*
SSTF Ship/Shore Trailer Facility *fu*
SSTI Small Spacecraft Technology Initiative *kf*
SSTIX Small Ship Teletypewriter Information Exchange system *mt*
SSTL Survey Satellite Technology, Ltd. *ls4, 5, 45*
SSTM Single Service Tanker Manager [SAC] *mt*
SSTM Single Service Training Manager [ATC] *mt*

SSTO Single Stage To Orbit *mi, wp, tb, mb*
SSTOL Super(sonic) Short Take Off and Landing *mt*
SSTS Space based Surveillance [and] Tracking [Satellite] System *mt, fu, tb, hc, 5, sb, aw129.14.25, cp, wp, tb, st, jdr, kf*
SSTSD Space and Secure Telecommunications Systems Division *ci.962*
SSTSS Strategic System Test Support Study *wp*
SSTV Slow Scan TV *mt*
SSTV Submarine Shock Test Vehicle *fu*
SSU SBC Support CSCI *fu*
SSU Sensor Simulation Unit *fu*
SSU Sequential Shunt Unit *oe*
SSU Stratospheric Sounding Unit [also SRU] *pl, hy*
SSU System Support Unit *mt*
SSULI Space Segment Ultraviolet Limb Imager *pbl*
SSUN Soviet Submarine of Unknown Function *wp*
SSUPS Solid State Uninterruptible Power Source [USAF] *mt*
SSUPS Station Static UPS *mt*
SSURADS Shipboard Surveillance Radar System *hc*
SSUS Spinning Solid Upper Stage *bl1-2.162, hc*
SSUSI Space Segment Ultraviolet Spectrographic Imager *pbl*
SSV Space Shuttle Vehicle *gp, 4, mb*
SSV Support System Verification *kf*
SSVR Subsystem Validation Review *wp*
SSW Support Software *fu*
SSW Surface Strike Warfare *fu*
SSW Synchronizing Switch *fu*
SSW System Software *mt*
SSWA Static Safe Work Area *fu*
SSWG System Safety Working Group *kf*
SSWG System Security Working Group *kf*
SSWP Space Station Work Package *wl*
SSWS Static Safe Work Station *fu*
SS_LORAN Skywave Synchronized Long-Range Navigation *fu*
ST Schmitt Trigger *fu*
ST Schuler Tulling *fu*
ST Self Test *wp, hi*
ST Service Test *mt, fu*
ST Ship Trial *fu*
ST Single Throw *fu, w9*
ST Small Terminal (Satellite) *fu*
ST Space Telescope *tb*
ST Space Tug *ci.262*
ST Special Tooling *fu, 18, hc, tb, jdr*
ST Specification Tree *fu*
ST Status Code *fu*
ST Surface Temperature *pl*
ST&D Strategic Trade and Disclosure [OASD(ISA)] *dt*
ST&E Security Test and Evaluation *mt, kf*
ST&E System Test and Evaluation *fu*
ST&ETR Security Test and Evaluation Test Report *mt*
ST&TE Special Tools and Test Equipment *fu*
ST-IPT System Test [and evaluation] IPT *kf*
ST/BOA System Timing, Buffer Output ASIC *jdr*
ST/ECF Space Telescope/European Coordinating Facility *ns*
ST/STE Special Tooling and Special Test Equipment *fu*
STA Science and Technology Agency [Japan] *bl3-1.154, hy*
STA Sensor Telescope Assembly *kf*
STA Service Terminal Arrangement *mt*
STA Shuttle Training Aircraft *hc*
STA Spanning Tree Algorithm *ct, em, kg*
STA Structural Test Article *ci.462*
STA System Test Analysis *fu*
STAAP SAR Targeting Accuracy Analysis Program *hc*
STAAS Surveillance and Target Acquisition Aircraft System *hc, wp*
STAC Satellite Tracking Angle Calculations [PACAF] *mt*
STAC Supersonic Transport Aircraft Committee. Created in 1956 [UK] *mt*
STACCS Standard Theater Army CC System *mt*
STACS Stand off Targeting and Classification System *hc*
STADAC Station Data Acquisition and Control System *hc*
STADAN Space Tracking and Data Acquisition Network *hc, ci.117*
STADD Ship Towed Acoustic Deception Device *mt*
STADS Swedish Transportable Air Defense Station *hc*
STAE Second Time Around Echo *fu*
STAF Systems Technical Applications Facility *mt*
STAFAST Standard Frequency and Standard Time *mt*
STAG Safeguard Threat Action Generation *fu*
STAG Strategy and Tactics Analysis Group [US Army] *hc, dt*
STAGEX (Sequential Processing (Software Program)) *jb*
STAGG Small Turbine Advanced Gas Generator *fu*
STAIRS Storage And Information Retrieval System *hi*
STAJ Short Term Anti Jam *hc, fu*
STALO Stabilized (Stable) Local Oscillator *mt, gp, fu*
STALOC Small Target and Low Optical Contrast *hc*
STALOC Station Logistics *hc*
STALOG System to Automate Logistics *mt*

STAMIS Standard Army Management Information System *fu, wp*
STAMMIS Standard Army Multiple Command Management Information System [USA] *st, mt*
STAMO Stabilized Master Oscillator *fu*
STAMP Standard Air Munitions Package *mt*
STAMP Strategic Tactical Attack Modeling Process *tb*
STANAG Standard NATO Agreement *mt*
STANAG Standardization Agreement [NATO] *mt, wp, fu*
STANAV-SFORLANT Standing Naval Force Atlantic *mt*
STANAVFORLANT Standing Naval Force, North Atlantic [NATO] *mt*
STANAVFORMED Standing Naval Force Mediterranean [NATO] *mt*
STANCIB State Army Navy Communications Intelligence Board *-10*
STANEVAL Standardization and Evaluation *mt*
STANFINS Standard Finance System [USA] *mt*
STANO Search Track And Night Operations *hc*
STANO Surveillance, Target Acquisition, and Night Observation *hc*
STANO Surveillance, Target Acquisition, and Night Operations *mt*
STAP Strategic Target Aim Point *mt*
STAPL Ship Tethered Aerial Platform [a small shipboard RPV of auto-giro design] [USA] *mt*
STAPRS Submarine Towed Array Passive Ranging System *hc*
STAR Satellite Telecommunications with Automatic Routing *hc*
STAR Satellite Telemetry Automatic Reduction *ci.123*
STAR Scientific and Technical Aerospace Reports [NASA] *hn43.22.6*
STAR Self Defining Text Archival *pk, kg*
STAR Self Test And Repair *wp*
STAR Ship-deployable Tactical Airborne Remotely [piloted vehicle] *hc*
STAR Significant Technical (Technology) Achievement[s] [in (and)] Research *5, 17, sb, tb, st*
STAR Simultaneous Transmitting and Receiving [radar] *fu, hc, wp*
STAR Stall Avoidance and Recovery *wp*
STAR Standard Terminal Arrival Route *fu*
STAR Staring Array *kf*
STAR Storage And Retrieval system (STAR 1100 tape library system) *wp, mt*
STAR Stress Testing to Achieve Reliability *fu*
STAR Support Tools Analysis Report *fu*
STAR System Testbed for Avionics Research *wp*
STAR System Threat Analysis Report *fu*
STAR System Threat Assessment Report *mt, hc, nt, jmj, wp, as, kf*
STARC State Area Command [USA] *mt*
STARC State Army Reserve Center *mt*
STARCAT Space Telescope Archive and Catalog *ns*
STARCIPS Standard Army Civilian Payroll System *mt*
STARFIARS Standard Army Financial Inventory and Reporting System *mt*
STARLETTE [satellite fitted with laser reflectors for earth studies] *ac3.6-4*
STARNET Standard Army Network *mt*
STARNET Sustaining Base Army Network *mt*
STARPAHC Space Technology Applied to Rural Papago Advanced Health Care *ci.95*
STARS Small Tethered Aerostat Relocatable System [USCG] *mt*
STARS Software Technology for Adaptable Reliable Systems [DOD] *mt, fu, hc, dh, aa26.12.15, pf1.17, wp, tb*
STARS SPACECOM Threat Assessment Review System *hc*
STARS Stellar Tracking Attitude Reference System *hc*
STARS Strategic Tactical And Relay Satellite *jdr*
STARS Submarine Towed Array Radar System *hc*
STARS Surveillance and Target (Tracking) Attack Radar System *mt, hc, hn43.7.5, aw129.1.3, af72.5.139*
STARS Surveillance, Target Acquisition and Reporting System *wp*
STARS Synchronized Time Automated Reporting System *wp*
START Strategic Arms Reduction Talks (Treaty) "start" *mt, fu, 11, 14, ph, aa26.12.73, aw130.6.23, tb, wp, kf, ai2*
STARTEX Start of Exercise *mt*
STARTLE Surveillance and Target Acquisition Radar Location and Engagement *hc*
STASS Submarine Towed Array Sonar System *fu*
STAT Simulation Test and Analysis Tool *tb*
STAT SWARM Post Processor Utility *tb*
STATDM Statistical Time Division Multiplexing *do305*
STATPRO Stationing Program, USAREUR *mt*
STATUS Staring Array Technology for Utilization in Space *hc*
STAVKA Supreme Command Staff [USSR] *mt*
STB Super Tropical Bleach [Ca(OCl)2 - chemical warfare decontaminant] *wp*
STBS Simulation Test Bed Subsystem *fu*
STC Satellite Television Corporation [part of COMSAT General] *hc, sc3.2.7, ce*
STC Satellite Test Center [of AFSCF] *mt, sd, 4, hc, wp, tb, jdr*
STC Satellite Test Conductor *jdr*
STC Scientific and Technical Computing *mt*
STC Security Transition Component *mt*
STC Sensitivity Time Control *mt, fu*
STC Set Carry Flag *pk, kg*
STC SHAPE Technical Center [NATO] *fu, mt*

STC Short Time Constant *mt*
STC Slow Time Constant *fu*
STC Software Technology Center *mt*
STC Space Technology Center [Kirkland AFB, Maryland] *mt, sc2.7.2, 5, sb, st*
STC Space Threat Conference *mt*
STC Space Tracking Center *mt*
STC Standard Telephone and Cables [Australian firm] *sc3.9.3*
STC Storage Transfer Controller *fu*
STC Stored Time Command *0*
STC Strike Command [UK] *mt*
STC Supplemental Type Certificate [USA] *mt*
STC System Test Complex *gp, oe*
STC System Test Console *fu*
STC System Test Equipment *ggiii*
STCC Space Transportation Control Center *dm*
STCC System Technical Coordination Center *fu*
STCS SOCC Telemetry and Command System *oe*
STCW System Time Code Word *fu*
STD (STD Research) [in a list of corporate donors] *aa26.11.b5*
STD Set Direction Flag *pk, kg*
STD Software Technical Director *fu*
STD Software Test Description *mt, fu, pl, hc, jdr*
STD Software Test Document *jn*
STD Statistical Target Detector *fu*
STD Survivability Technology Development *st*
STD System Technology Demonstrator *wp*
STDA StreetTalk Directory Assistance [Banyan] *kg, hi*
STDAUX Standard Auxiliary *pk, kg*
STDB Software / Firmware Trend Data Bank *fu*
STDB Strategic Target Data Base *mt*
STDB [database interface layer which insulates user software from vendor dependencies, developed at ST ScI.] *cu*
STDERR Standard Error *pk, kg*
STDIN Standard Input *pk, kg*
STDIO.H Standard Input / Output Header [C Programming Language] *pk, kg*
STDL Submarine Tactical Data Link *fu*
STDM Statistical Time Division Multiplexer [also STDMX] *pk, kg*
STDM Synchronous Time Division Multiplexing *do305*
STDMX Statistical Time Division Multiplexer *hi*
STDN Spacecraft (Spaceflight) Tracking and Data Network *vv, hc, sc4.1.2, 2, 4, oe*
STDOUT Standard Output *pk, kg*
STDPRN Standard Printer *pk, kg*
STDR System Test Discrepancy Report *gp, 17, sb*
STDS Space Tracking and Data Systems *-4*
STE Satellite Test Equipment *oe*
STE Satellite Tracking Element *mt*
STE Search Target Extractor *fu*
STE Simplified Test Equipment *hc*
STE Site Test Equipment *fu*
STE Special Test Equipment *mt, fu, ah, uh, 18, pl, hc, wp, ep, tb, jdr, pbl, tm, jn*
STE Subscriber Terminal Equipment *do181, hi*
STE System Test Equipment *ah, uh, gg33, 9, 17, sb, hc, lt, ep, jdr, pbl, tm, kf, jn*
STE Systems Technology Equipment *st*
STE/ICE Simplified Test Equipment/Internal Combustion Engines *hc*
STEA System Test Equipment Assembly [Surveyor] *0, hc*
STED Standard Technical Equipment Division [within NSA] *-10*
STEDMIS Ships Technical Data Management Information System *fu*
STEEL [DOD program] *hc*
STEER [RADC precision navigation system] *hc*
STEL Short Term Exposure Limit *oe*
STEM Simulation Training Evaluation Module *hc*
STEM Sonar Test Emulator *fu*
STEM System Trainer and Exercise Module [ESD] *hc, fu, mt*
STEM System Training Evaluation Module *fu*
STEM Systems Telecommunications Engineering Manager *mt*
STEP Safety Test Engineering Program *fu*
STEP Scientific and Technical Exploitation Program *wp*
STEP Service Test and Evaluation Program *fu*
STEP Software Technology Evaluation Panel *fu*
STEP Software Test and Evaluation Project *fu*
STEP Space Technology Experiments Platform *wl*
STEP Space Tracking Experiment Program *5, 17, sb*
STEP Standard for The Exchange of Product [model data] *kg, hi*
STEP Stripes for Exceptional Performers [USAF] *mt, af72.5.109*
STEP Surveillance and Tracking Experiment Program *mt, wp, tb*
STEP Symbolic Tape Edit Program *hc*
STEP System Test Equipment Program *fu*
STEPS Simulation and Training Equipment Planning Sources *hc*
STERF Special Test Equipment Repair Facility *hc*
STERT Special Test Equipment and Repair Techniques *fu*
STESA Static Earth Sensor Assembly *uh, jdr*

STESA Static Thermofile Earth Sensor Assembly *jn*
STEX Stimulation Test and Exercise *mt*
STF Satellite Tracking Facility *wp*
STF Software (System) Test Facility *fu*
STF Store and Forward *fu*
STG See To the Ground *kf*
STG Space Task Group *ci.xxi*
STG Synthetic Terrain Generator *wp*
STGP Simulation Tape Generation Program *fu*
STGS Second TDRSS Ground Station *tb*
STGT Second TDRSS Ground Terminal *aa26.10.19*
STI Scientific and Technical Information *fu*
STI Set Interrupt Flag *pk, kg*
STI Short Time Integration *fu*
STI Speech Transmission Index *wp*
STI Stanford Telecommunications, Inc. [Reston, VA 22090] *jmj, tb*
STIC Scientific and Technical Intelligence Committee *mt*
STICS Scalable Transportable Intelligence Communications System *mt*
STID Standard Identification Number *mt*
STIG Secondary Target Interface [working] Group *-17*
STIMS Shared Technical Information Management System *mt*
STINFO Scientific and Technical Information program *mt, fu, hc, wp, tb, jdr, kf*
STINGER [Eglin AFB low cost air to surface missile program] *hc*
STINGS Stellar Inertial Guidance System *hc*
STIP Scientific and Technical Information Program *st*
STIRS System Technical Information Request / Statement *oe*
STIS SORAK Tactical Information System *fu*
STIS Space Telescope Imaging Spectrometer [to replace FOC and GHRS] *mi*
STJ Slip and Temperature Jump [as in nozzle velocities] *js7.463*
STK Satellite Tool Kit [registered trademark] *te23, 11, 18*
STKFND Stock Fund and Equipment Requirements System [USAF] *mt*
STL Software Technology Laboratory *fu*
STL Source Tailoring Language *fu*
STL Space Technology Laboratory, [Incorporated] *ci.157, 0*
STL Standard Template Library *kg, hi*
STLA Simulated Thruster Load Assembly *oe*
STLC Satellite Transponder Leasing Corporation *hn48.20.1*
STLDD Software Top Level Design Document *mt, fu, hc, tb*
STM Service Test Model *fu*
STM Significant Technical Milestone *wp*
STM Structural Test Model *gp, ep, oe*
STM Synchronous Transfer Mode *ct, em, hi*
STMP Shuttle Target Measurement Program *hc*
STN SAC Telephone Network *mt*
STN Source Track Number *fu*
STN Super Twist Nematics [displays] *kg, hi*
STO Science and Technology Objectives *st*
STO Short Take Off *af72.5.145*
STO Subcontractor Task Order *mt*
STO Sun Transit Outage *jdr*
STO Supersynchronous Transfer Orbit *oe*
STO Systems Test Objectives *0*
STOF Standard Tactical Operations Facility [USAREUR] *mt*
STOG Science and Technology Objectives Guide *fu*
STOIC Stratospheric Ozone Intercomparison Campaign *hy*
STOL Short Take Off and Landing *mt, aw126.20.21, hc, aa26.12.2, w9, af72.5.135, wp, ai1*
STOL Systems Test and Operations Language *oe*
STON Short Ton *mt*
STONE Small Target Optimization Study *hc*
STOP SCOMP Trusted Operating Program [Honeywell] *mt*
STOP Sequential Thematic Organization of Publications (Proposals) *hc, fu*
STOP Sequential Topicizing Of Proposals *fu*
STOP Stabilized Optical Pointing *hc*
STOP Storyboarding Of Proposals *fu*
STOP Strategic Orbit Point *mt*
STORET [digital data storage system of the Environmental Protection Agency] *ac1.5-15*
STORM Sensor Tank Off Route Mine *hc*
STORM Storm Scale Operational and Research Meteorology *hy*
STOS Store String *pk, kg*
STOT Scheduled Time Over Target *fu*
STOVL Short Take Off and Vertical Landing *mt, aa26.12.34, aw130.22.30, wp, ai1*
STOVL Standard Take Off and Vertical Landing *ai1*
STP Science and Technology Program *un*
STP Scientifically Treated Petroleum *wp*
STP Search Target Processor *fu*
STP Secure Transfer Protocol *pk, kg*
STP Select Test Procedure *jdr*
STP Shielded Twisted Pair [cable] *em, ct, kg, nu, hi*
STP Signal Transfer Point *em, mt, pk, kg*
STP Signal Transfer Protocol *ct*

STP Software Test Plan *mt, fu, pl, hc, tb, jdr*
STP Software Test Procedure *fu*
STP Solar Terrestrial Physics *ns*
STP Space Test Program *ci.571, hc, dm, hy, mt*
STP Standard Temperature and Pressure *0, w9, wp, fu, ai1*
STP Stream Protocol *mt*
STP Synchronized Transaction Processing *pk, kg*
STP System Technology Program *hc, st, fu*
STP System Test Plan *mt, fu, 17, sb*
STP System Training Plan *kf*
STP System Transition Plan *mt*
STPI/LARC [inexpensive random] Thermoplastic Poly-Imide *aa26.12.76*
STPM Satellite Test Procedure Module *oe*
STPO System Technology Project Office *st*
STPR Software Test Procedure Report *fu*
STPR Software Test Procedures *mt, hc*
STR Software Tasks Requirements *-4*
STR Software Test Report *fu, mt, pl, hc, jdr*
STR Software Trouble Report *fu, kf*
STR Sonar Transmitter/ Receiver *fu*
STR Special Test Request *ep, uh, jdr*
STR Store Task Register *pk, kg*
STR Strategic Training Range [Strategic Air Command bombing, navigation training site] *-12*
STR SURTASS Ship Trainer *fu*
STR Synchronous Transmitter Receiver *pk, kg*
STR System (Senior) Technical Representative *fu*
STR System Test Report *fu*
STR System Trouble Report *fu*
STR Systems Technology Radar *hc, st*
STR/PTR System Trouble Report / Program Trouble Report *fu*
STRAC Strategic Army Corps *fu*
STRACS Small Transportable Communications Station *hc*
STRAD Strategic Air Division *oe*
STRADEM Strategic Defense Effectiveness Model *-17*
STRAF Strategic Army Forces *mt*
STRAG System Training and Readiness Assessment Group *fu*
STRAM Synchronous Transmit Receive Access Method *hi*
STRAP Straight-Through Repeater Antenna Performance *hc*
STRAPP Standard Tank, Rack, Adapter, and Pylon Package *mt*
STRATCOM Strategic Command [US] *kf, mt*
STRATMOB Strategic Mobility Support System [MSC] *mt*
STRATOPS Strategic Operations *mt*
STRATPRO Strategic Programs [OSD] *mt*
STRATUS Strategic Mobility Support System [MSC] *mt*
STRAT_FAX Strategic Facsimile Network [USAF] *mt*
STRAT_SIM Strategic Simulator *tb*
STRAW Short Range Anti Tank Weapon [also SRAW] *wp*
STRAW Simultaneous Tape Read and Write *fu*
STRB Specification and Test Review Board *oe*
STRC Strategic Target Range Complex [SAC*] *mt*
STRC Strategic Training Range Complex [Strategic Air Command] *-12*
STRD Systems Test Requirements Document *jn*
STREAMLINER [Special Intelligence Message Switching and Terminal System] *mt*
STREP System Technology Reentry Experiment Program *st*
STRES Simulator Training Requirements and Effectiveness Study [report] *hc, wp*
STRESS Structural Engineering System Solver [a programming language] *kg, hi*
STRICOM Strike Command [Army] *fu*
STRIKEFOR-SOUTH Striking and Support Forces, Southern Europe, Naval [NATO] *mt*
STRIM Societe Technique de Recherches Industrielles et Mecaniques *crl*
STRIPS S&T Reporting and Information Processing System *hc*
STRR System Test Readiness Review *re*
STRUDL Structural Design Language [programming language] *pk, kg*
STRV Space Technology Research Vehicle *kf*
STS Satellite Test Station *tb*
STS Satellite Tracking Station *oe*
STS Scheduled Truck Service [AFLC] *mt*
STS Shuttle Transport System *mi*
STS Signal Tracking Subsystem *fu*
STS Simulator Test Station *fu*
STS Space Transportation System [the NASA Space Shuttle] *mt, uh, mi, sc4.11.1, 2, 4, 5, 9, 13, 14, 17, sb, aa26.12.44, sf31.1.1, ws4.10, ep, dm, re, tb, wl, jdr, mb, mit, oe, ai1, ai2*
STS Special Test System *fu*
STS Specialty Training Standard *mt*
STS Stockpile to Target Sequence *mt*
STS Structure/Thermal Subsystem *vv*
STS Subcommittee on Telecommunications Security *tb*
STS Synchronous Time Sharing *fu*
STS Synchronous Transport Signal *hi*
STS System Tactical Support *hc*
STS System Technical Support *mt*

STS System Technology Support *hc*
STS System Test Set *wp*
STS System Test Support *fu*
STS System Timing Slice *jdr*
STSA Secure Telephone System Architecture [USAF] *mt*
STSC Software Technology Support Center *mt*
STSC [satellite control and test earth station, Fucino] *sf31.1.22*
STSCC Standing Transportation Support Coordinating Committee [NATO] *mt*
STScI Space Telescope Science Institute:(mi, kg)
STSDAS Space Telescope Science Data Analysis System *mi*
STSOC Space Transportation System Operations Contract *pf1.20*
STSP Secure Telephone Systems Program *mt*
STSP Solar Terrestrial Science Programme *es55.42*
STT Secure Transaction Technology [Microsoft] *kg, hi*
STT Shore Targeting Terminal *mt, fu*
STT Site Transition Team *mt*
STTAS Ships Tactical Threat Analysis System *hc*
STTC Sheppard Technical Training Center *mt*
STTC Space Tracking, Telemetry, and Command [simulator] *kf*
STTE Special (Specific purpose) Tools and Test Equipment *mt, fu*
STTF System Technology Test Facility *st*
STTL Schottky Transistor-Transistor Logic *fu*
STTM Stabilized Tracking Tripod Module *hc*
STTO Start Taxi Take Off *mt*
STU Secure(d) Telephone Unit *mt, kf, 17, ph, jb, tb*
STU Sensor Transmitter Unit *fu*
STU Special Test Unit *fu*
STU Star Tracker Unit *kf*
STU Submerged Trainable Unit *fu*
STU System Timing Unit *fu*
STUD Ship Towed Underwater Device (Detector) *hc, fu*
STUD-D Ship Towed Underwater Detector-Depressed *hc*
STUIII Secure Telephone Unit Third Generation *mt*
STV Solar Thermal Vacuum *gg35*
STV Standard Tactical Vehicle [USA] *mt*
STV Steerable Low-Light TV [system fitted in the turrets under the nose of the B-52] [USA] *mt*
STV Structural Test Vehicle *ric, 4, jdr*
STV Subscription Television *dn, w9*
STV Surveillance Television *wp*
STVP System Test and Verification Plan *pl, jdr*
STVS System Test and Verification Software *jdr, ric*
STVS System Test Verification Support *fu*
STVT Solar Thermal Vacuum Test *-9*
STVTS Submarine Tactical Visual Training System *hc*
STW (STW-2 Synchronous Telecommunications and Broadcasting Satellite, PRC) *mt*
STW Strike Warfare *fu*
STWA Short Trailing Wire Antenna *mt*
STWC Strike Warfare Commander (Coordinator) *fu*
STZ Surface Turbulence Zone *aa27.4.b11*
SU Sole User *fu*
SU Supporting Unit [TACS/TADS] *mt*
SU [Soviet designator , e.g. (SU-126 for Moss) (SU-15/21 for Flagon) (SU-24 for Fencer) (SU-25 for Frogfoot) (SU-27 for Flanker) (SU-7/17 for Fitter) *mt*
SU IPT Support IPT *kf*
SU/SO Start-Up/Start-Over *fu*
SUADPS-RT Shipboard Uniform ADP System Real Time [US Navy] *mt*
SUAEWICS Soviet Union Airborne Early Warning and Interceptor Control System *wp*
SUAV Study of Unmanned Air Vehicles *wp*
SUAWACS Soviet Union Airborne Warning And Control System *wp, mt*
SUBACS Submarine Advanced Combat System [US Navy] *mt, fu, uc, hc*
SUBAD Submarine Air Defense *fu*
SUBCAP Submarine Combat Air Patrol *mt*
SUBCOM Subordinate Command *mt*
SUBDEVRON Submarine Development Squadron *-12*
SUBGRU Submarine Group *-12*
SUBLANT Submarine Force Atlantic [US Navy] *mt*
SUBLIB Subroutine Library *ct, em*
SUBOPAUTH Submarine Operating Authority *mt*
SUBPAC Submarine Force, Pacific [US Navy] *mt*
SUBROC Submarine [launched] Rocket [USA DOD] *mt, aa26.12.39, 12, ph, wp*
SUBSIST Subsistence [USAREUR] *mt*
SUCAP Surface Combat Air Patrol [US Navy] *mt*
SUCCESS Synthesized UHF CC and Exploitation Subsystem *mt*
SUCOC Succession Of Command *mt*
SUCONV Off-Line Set-up Conversion Program *fu*
SUCP Software Utility Computer Program *fu*
SUDC Set-Up Data Conversion *fu*
SUDEX Supplier Document Exhibit *fu*
SUDS Set-Up Data Supplement *fu*
SUDT Silicon Unilateral Diffused Transistor *fu*
SUDU Spin Up Dispenser Unit *st*

SUE Second Units Equipped *fu*
SUE Stupid User Error *ct, em*
SUI SWARM User Interface *tb*
SUKLO Senior United Kingdom Liaison Officer *-10*
SUKS Set-Up Controls System; Set-up Conversion Software *fu*
SUM Surface to Underwater Missile *fu, wp*
SUM System Check and Utility Master *fu*
SUM System Users' Manual [ALERT] *kf, fu*
SUM-SPARE Spare Engine Report *mt*
SUMA Satellite Universal Modulation Adapter *mt*
SUMMA (SUMMA Technology, Inc. . . . Huntsville, AL 35805) *jmj*
SUMS Spectrum Utilization Management System *fu*
SUN (Sun Microsystems, Inc.) *pk, kg*
SUN Stanford University Networks *ct, em*
SUNDANCE [NOSC program] *hc*
SUNRAYCER [GM solar powered car] *hc*
SUNY State University of New York *aa26.11.b7*
SUP Standard Unit of Processing *mt*
SUP Support Services *kf*
SUPARCO Space Agency, Government of Pakistan *sc4.10.3*
SUPCOM Support Command *mt*
SUPER [survivable solar based power system] *hc*
SUPIDEN Support Identification [code] *-4*
SUPIR Supplementary Photo Interpretation Report *wp*
SUPPLOT Supplemental Plot *fu*
SUPS Static Uninterruptible Power Source *mt*
SUPT Specialized Undergraduate Pilot Training *wp*
SUR Small Unit Radio *mt*
SURE Sensor Update and Refurbishment Effort [for the RF-4B Phantom II, USA] *mt*
SURGE Supply User Reports Generator *mt*
SURIC Surface Ship Integrated Control *fu*
SUROM Start Up Read Only Memory *oe*
SURS Surface Export Cargo System [MTMC] *mt*
SURSAT Surveillance Satellite [Canada] *sc, wp*
SURSATCOM Survivable Satellite Communications *mt*
SURTAC Surveillance and Tactical (operation) *fu*
SURTAC Surveillance and Tactical Warning Network [NORAD] *mt*
SURTAC_TTY Surveillance and Tactical Action Teletype *mt*
SURTASS Surface Towed Array Surveillance System *mt*
SURTASS Surveillance Towed Array Sensor System *sc3.4.8, hc, kb*
SURTASS Surveillance Towed Array Sonar Segment *hc, fu*
SURVEYOR [lunar soft landing capsule] *hc*
SURVIAC Survivability and Vulnerability Information Analysis Center *hc, wp*
SURVSATCOM Survivable Satellite Communications *hc*
SUS Semiconductor Unilateral Switch *fu*
SUS Silicon Unilateral Switch *wp*
SUS Software Utility Services *fu*
SUS Speech Understanding System *wp*
SUS Support Set *fu*
SUSE System Utility and Survivability Evaluation *cp*
SUSIF Set-Up/Software Interface Form *fu*
SUSIM Solar Ultraviolet Spectral Irradiance Monitor [UARS] *ge, hg, hy*
SUSLO Senior United States Liaison Officer *-10*
SUT System Under Test *mt*
SUU Suspended Underwing Unit [a submunitions dispenser or gun pod, USA] *mt*
SUV Sole User Voice *fu*
SUW Surface to Under Water missile *mt*
SUW Surface Warfare *fu*
SV Schedule Variance *jdr, mt, fu*
SV Secure Voice *mt*
SV Space Vehicle *jdr, kf, uh*
SV Squib Valve *jdr, jn, uh*
SV Stretched Video *pl*
SV Surface Vehicle *fu*
SV [High Orbit Space Vehicle (IPT identifier)] *kf*
SV/GC Secure Voice and Graphics Conferencing *mt*
SVAR Space Vehicle Acquisition Requirements *jdr*
SVAS Stretched VISSR Atmospheric Sounder *gd2-6*
SVC Secure Voice Communication [also SEVOCOM] *mt, wp*
SVC Space Vector Corporation *-17*
SVC Supervisor Call *fu*
SVC Switched Virtual Circuit *ct, kg, em, hi*
SVCCN Secure Voice Command Control Network *mt*
SVCP Simulation Video Computer Program *fu*
SVCP Space Vehicle Control Processor *kf*
SVCS Software Version Control System *fu*
SVD Single Value Decomposition *ai1*
SVD Spacecraft Validation Document *jn*
SVDM Simultaneous Voice and Data Modem *hi*
SVE Spacecraft Vehicle Equipment *0*
SVED Space Vehicles Electronics Division [Hughes SCG] *hq5.2.4*
SVF Simple Vector Format *pk, kg*
SVFR Special Visual Flight Rules *fu*

SVGA Super Video Graphics Array (Adapter) *kg, ct, em, hi*
SVHS Satellite Vertical Handling Sling *oe*
SVI Spectral Vegetation Index *hy*
SVI&T Systems Verification, Integration and Test *jn*
SVID System V Interface Definition *hi*
SVID System V Interface Device *mt*
SVIK Space Vehicle Installation Kit *fu*
SVIMS Short VIMS *mt*
SVIP Secure Voice Improvement Program (Plan) [DOD] *wp, mt*
SVM State Variable Model *fu*
SVM System Validation Model *fu*
SVM System Virtual Machine [Microsoft] *kg, hi*
SVMST Space Vehicle Mixed Simulation Test *kf*
SVN Secure Voice Network *mt*
SVN Switched Virtual Network [IBM] *pk, kg*
SVP Secure Voice Program *mt*
SVP Software Validation Procedure *jdr*
SVP Sound Velocity Profile *fu*
SVPCS SBKEWS Validation Planning and Control System *sb*
SVPF Space Vehicle Processing Facility *ai2*
SVR System V Release [Number] [AT&T] *pk, kg*
SVR System Validation Review *wp, mt*
SVRAM Synchronous VRAM *hi*
SVS Section de Vol Sportif [France] *mt*
SVS Secure Voice Switch *rm1.2.1, fu*
SVS Secure Voice System *mt, fu, hc*
SVS Single Virtual Storage *hi*
SVS Space Vehicle Specification *kf*
SVS SSTS Validation Satellite *kf*
SVSI Secure Voice Systems Interface *mt*
SVT Sled Verification Test *st*
SVT System Verification Testing *fu*
SVTS Secure Video Teleconferencing System *mt*
SVTV Space Vehicle Thermal Vacuum Test *kf*
SVU Surface Vehicle Unit *fu*
SVUIK Surface Vehicle Unit Installation Kit *fu*
SVV System Verification and Validation *jdr*
SVWG Space Surveillance Working Group [NORAD] *mt*
SW Short Wavelength (Wave) *fu, uc, w9, wp, kf, oe*
SW Slow Walker *kf*
SW Space Wing *kf*
SW Spot Weld *fu*
SW Strategic Warning *mt*
SW Strategic Wing [USA, SAC*] *12, mt*
SW Surface Wind [speed] *ac7.5-11*
SW Switchband Wound *fu*
SWA Solar Wing Actuator *kf, uh, he, jn*
SWA Solar Wing Assembly *cl*
SWA South West Asia *mt, wp, kf*
SWAAT SDIO Working And Action Team *mt*
SWAC Standards Western Automatic Calculator [Mobile version of SEAC, built for deployment to White Sands] *pk, kg*
SWAD Southwestern Air Defense *fu*
SWAD Strategic Warning Assessment Development *hc*
SWADE Strategic Weapon Allocation and Damage Expectancy [model] *wp*
SWADE Surface Wave Dynamics Experiment *hy*
SWAIS Simple Wide Area Information Server [Internet] *kg, hi*
SWAMI Software Aided Multifont Input *hi*
SWAO Senior Weapons Assignment Officer *fu*
SWAOT Senior Weapons Assignment Officer Technician *fu*
SWAP Severe Weather Avoidance Plan *fu*
SWAP Standard WWMCCS Accounting Program [WWMCCS] *mt*
SWAR Software Acceptance Review *oe*
SWARM Space War Model *tb*
SWARM Space Weapon Assessment and Requirements Model *tb*
SWARM Strategic Warning and Response Model *tb, kf*
SWAS Submillimeter Wave Astronomy Satellite *mi*
SWAT Secure Wire Access Terminal *hc*
SWAT Software Acceptance Test *fu*
SWAT Software Analysis and Test [team] *mt*
SWAT Special Weapons and Tactics *w9*
SWATH (added resistance of SWATH models in uniform waves) *aa27.4.b12*
SWB Short Wheel Base *wp*
SWB Switch Board *mt, fu*
SWBS Subcontractor Work Breakdown Structure *fu*
SWC Scan With Compensation *fu*
SWC Ship Weapon Coordinator [US Navy] *mt*
SWC Special Weapons Center *wp*
SWCC SAC Warning Control Console *mt*
SWCDR Software Critical Design Review *oe*
SWCI Software Configuration Item *kf, fu*
SWCP SAC Wing Command Post *fu*
SWCR Software Concept Review *oe*
SWCS SAC Warning and Control System *mt*

SWD Solar Wing Drive(s) *he, uh, cl, jdr, kf, jn*
SWD Surface Wave Device *fu*
SWDC Sounder Wideband Data Collection *oe*
SWDE Software Development Environment *tb*
SWDS Sounder Wideband Diagnostic Software *oe*
SWECS Small Wind Energy Conversion System *hc*
SWEDAC (Swedish board for technical accreditation) *hi*
SWEE Software Engineering Environment *tb*
SWF Short Wave Fading *mi*
SWF Strategic Weapons Facilities [US Navy] *mt*
SWG Science Working Group *re*
SWG Security Working Group [NATO] *tb, mt*
SWG Software Working Group *kf*
SWG Space Wing *jb*
SWG Standard Wire Gage [British] *w9, fu*
SWG Stubs Wire Gage *fu*
SWH Significant Wave Height *ac10.1-6*
SWH Solar Wing Hinge *cl, jn*
SWHQ Static War Headquarters [NATO] *mt*
SWIL Software In the Loop *tb*
SWIM Super Woz Integrated Machine *ct, em*
SWIP Super Weight Improvement Program [USA] *mt*
SWIP Systems and Weapon Improvement Program [for the A-6 Intruder, USA] *mt*
SWIR Short Wave Infrared Radiometer [Japan] *ge, hg, fu*
SWIR Short Wave(length) Infrared *fu, ca.39, 17, sb, hc, wp, tb, st, pbl, kf*
SWIRLS Stratospheric Wind Infrared Limb Sounder [EOS] *hy*
SWISH Simple Web Indexing System for Humans *kg, hi*
SWL Short Wavelength *bf*
SWL Short Wavelength [high energy] Laser *dh, tb, st*
SWL Signals Warfare Laboratory [of ERADCOM, USA] *tc, mt*
SWL Strategic Weapons Launcher [USA] *hc, wp, mt*
SWN Switching Network *fu*
SWO Staff Watch Officer *fu*
SWO Staff Weather Officer *mt*
SWO Surface Warfare Officer [US Navy] *mt*
SWP Central Weather Processor *fu*
SWP Simple Web Printing *pk, kg*
SWP Solar Wing Positioner *he*
SWP Soviet Warsaw Pact *wp*
SWP Surface Warfare Plan *mt*
SWPA South West Pacific Area [Allied] *mt*
SWPA {instrument mounted on rotor of Mars orbiter} *ma40*
SWPDR Software Preliminary Design Review *oe*
SWPS Strategic War Planning System *mt, wp*
SWPT Software Engineering Process Team *kf*
SWR Sine Wave Response *wp*
SWR Standing Wave Ratio *fu, 0, mj31.5.54, jdr*
SWRQ Software Requirements Engineering *kf*
SWRR Software Requirements Review *oe*
SWRSC SAC WIS Requirements Steering Committee *mt*
SWS Sidewall Simulator *oe*
SWS Space-Based Warning System *kf*
SWS Swimmer Weapon System *hc*
SWSC Space Warning Systems Center *kf, as*
SWSI Single Wire Serial Interface [T&C digital] *he, jn*
SWSS Storm Warning Satellite System *hc*
SWT Science Working Team *es55.42*
SWT Software Tape *fu*
SWT Supersonic Wind Tunnel *0*
SWTL Surface Wave Transmission Line *fu*
SWTRR Software Test Readiness Review *oe*
SW_LOC Software Lines Of Code *kf*
SX Simplex *fu, hi*
SXG Spectrum X-Gamma mission *cu*
SXG-CF Spectrum X-Gamma Coordinating Facility *cu*
SXI Solar X-ray Imager *no, oe*
SXR Soft X-Ray *jdr*
SXRP Stellar X-Ray Polarimeter *cu*
SXTF Satellite X-ray Test Facility *hc*
SY System Release *mt*
SYC Single Year Contract *st*
SYDMIS Shipyard MIS [US Navy] *mt*
SYDP Six Year Defense Plan *mt*
SYERS Senior Year Electro-optical Relay System [Senior Year is the U-2R; SYERS is a CCD-camera] [USA] *mt*
SYLDA SystÉme de Lancement Double Ariane *sc3.7.3, es55.77, ai2*
SYLK Symbolic Link *kg, hi*
SYMAR System for Materiel Wartime Attrition and Replacement *wp*
SYN Synchronization Character *jdr*
SYN Synchronous Idle [a communications control character] *do364*
SYNC Synchronization Character *jdr*
SYNCOM [Geo-]Synchronous Communications [satellite] {1962} *fu, gs2-1.3, 0*
SYRUP System Resource Utilization Package *mt*
SYS (SYS . . . San Diego, California 92122) *jmj*

SYSADMIN System Administrator *pk, kg*
SYSCOM Synchronous Communications [satellite] *oe*
SYSCOM Systems Command [US Navy] *fu, mt*
SYSCON Systems Control [DCA] *fu, mt*
SYSCON [defense electronics contractor, Washington, D.C.] *aw129.10.2, jmj*
SYSDATA AFSC Data Tie Line Network *mt*
SYSGEN System Generation (Generator) *mt, kg, fu*
SYSLOG System Log *pk, kg*
SYSMOD System Modification *pk, kg*
SYSOP SYStem OPerator *em, wp, kg, ct, hi*
SYSOPS System Operations *pm*
SYSREQ System Request *kg, hi*
SYSTEM-80 Automatic Program Generator *mt*
SYSTO System Command Program Office *wp*
SYSTO System Officer *wp*
SYSTRAN System Analysis Translator *wp*
SYSVOICE AFSC Voice Tie Line Network *mt*

T

T [chemical warfare codename for 2,2-di(chloroethylthio)diethyleter] *wp*
T&C Telemetry and Command *he, uh, sc4.11.2, 4, 17, sb, lt, wl, jdr, kf, vv, oe*
T&C Terms and Conditions *jdr, fu*
T&C Timing and Control *fu*
T&D Test and Diagnostic *mt*
T&D Training and Development *mt*
T&D Transmission and Distribution *fu*
T&DA Tracking and Data Acquisition [also TDA] *fu, 2, 4*
T&DRE Tracking and Data Relay Experiment [also TDRE] *ac2.8-3*
T&E Test and Evaluation *mt, fu, 12, ph, hc, nt, cp, wp, tb, st, kf*
T&G Tracking and Guidance *fu*
T&H Transportation and Handling *tb*
T&M Technical and Management *fu*
T&M Time and Materials *mt, fu*
T&SA Task and Skill Analysis *fu*
T&SAR Task and Skill Analysis Report *fu*
T&TE Tools and Test Equipment *fu*
T&TEP Training and Training Equipment Plan *hc*
T&TW Test and Training Wing [USA] *mt*
T&V Test and Validation *tb*
T&V Thermal and Vibration *fu*
T&W Targeting and Weaponeering *mt*
T-AGOS Towing Auxiliary General Ocean Surveillance *fu*
T-CAN Transportation - Capability Assessment Net *mt*
T-CON Test Control area *fu*
T-HOST Transportable Host, WWMCCS [USCENTCOM] *mt*
T-MIL Transportation - Mini Information Library *mt*
T-MUX Terminal Multiplexer *mt*
T-R Transmit-Receive *0*
T-SCRC TEMPEST Small Computer Requirements Contract *mt*
T/A Traffic Analysis *-10*
T/B Thermal Balance *oe*
T/C Test Coupler [payload] *jn*
T/C Thermal Couple *oe*
T/C Thermal Cycle *jn*
T/CAS Threat Alert / Collision Avoidance System *aw129.14.15*
T/CCP Task / Contract Change Proposal *wp*
T/E Tables of Equipment *fu*
T/E Transport Erector *ai2*
T/F Time of Fail *fu*
T/FR Top of Frame *fu*
T/I TPFDD Interface *mt*
T/L Track and Listen *0*
T/O Tables of Organization *fu*
T/O&E Tables of Organization and Equipment (TOE also used) *fu*
T/R (antenna . . energy . . . is . . . directed through the T/R circulator) *ac9.3-3*
T/R Time of Rise *fu*
T/R Transformer / Rectifier *hc, rg*
T/R Transmit / (and) Receive *mt, kg, fu, hi*
T/R&G Transmit Receive and Guard *fu*
T/RIA Telemetry / Range Instrumented Aircraft [USA] *mt*
T/S Test Set *jdr, fu*
T/S Test Specification *fu*
T/V Thermal Vacuum *oe*
TA Table of Allowances *mt*
TA Target Acquisition *mt, wp*
TA Target Analysis *mt*
TA Task Assignment *st, fu*
TA Teaming Agreement *jdr*
TA Technical Analysis *tb*
TA Terminal Adapter *ct, em, hi*
TA Terminal Arrival *fu*

TA Terrain Avoidance *mt, hc, wp*
TA Test Anomaly *jdr, tm, uh*
TA Theater Army *mt*
TA Threat Assessment *tb*
TA Threat Avoidance *wp*
TA Time and Attendance *mt*
TA Towed Array *fu*
TA Track Association *fu*
TA Traffic Analysis *mt*
TA Transfer Aisle *jdr*
TA Transfer Authorization *fu*
TA Turnover Agreement *mt*
TA Type Approved *0*
TA&P Theater Assessments and Planning [OSD] *mt*
TA/CE Technical Analysis/Cost Estimate *mt*
TAA Tacan Analog Adapter *fu*
TAA Tactical Aerospace Assessment *wp*
TAA Tactical Air Army [FSU] *mt*
TAA Technical Advisor Analyst *kf*
TAA Technical Area Advisor *mt*
TAA Technical Assistance Agreement *hc, fu*
TAA Terminal Advanced Automation *fu*
TAA Total Active Aircraft [authorization] *mt*
TAA Total Army Analysis [USA] *mt*
TAAA Total Active Aircraft Authorization *mt*
TAAA Triaxial Accelerometer Assembly *jdr*
TAAATS The Advanced Australian Air Traffic System *hn51.16.1*
TAAC Theater Army Assistant Chief (TAAC of Staff) *mt*
TAACOM Theater Army Area Command *mt*
TAACS Tactical Air-to-Air Coupling System *hc*
TAACS Tactical Airlift Accomplishments and Commitments System [USAFE/MAC] *mt*
TAADS The Army Authorization Document(ation) System *mt, fu*
TAAED Tactical Airborne Advanced Expendable Decoy *wp*
TAAF Test, Analyze And Fix *mt, fu, wp*
TAAI Total Active Aircraft Inventory *mt*
TAALS Transportation Automated Address and Labeling System *wp*
TAAN Tactical Aircrew Alert(ing) Net(work) *wp, mt*
TAAO Time Available at Action Office *mt*
TAAP Towed Array Analysis Package *fu*
TAAR Target Area Analysis Radar *fu*
TAAS Terminal Advanced Automation System *fu*
TAAT Target Acquisition Aiming and Tracking System *hc*
TAB Tactical Air Base *mt, wp, fu*
TAB Tape Automated Bonding *fu, hc, md2.1.1, hd1.89.y4, hn49.12.4*
TAB Target Acquisition Battery *fu*
TAB Technical Abstract Bulletin [Department of Defense] *hn43.22.6*
TAB Terminally Aimed Bomb *hc*
TAB Time After Burst *5, sb*
TAB Total Allocated Budget *jdr*
TAB Toxogonin + Atropine + Benactizine [chemical warfare nerve agent antidote autoinjector] *wp*
TAB Training Aid Booklet *fu, mt*
TAB Transmit / Receive Assignment Buffer *fu*
TAB-VEE [hardened aircraft shelter half-round steel / concrete] *mt*
TABCASS Tactical Air Beacon Control and Surveillance Systems *hc*
TABES Technical And Business Exhibition and Symposium *aa27.4.44*
TABG Thresholded-Alpha-Beta-Gamma *fu*
TABM [auxiliary general missile ship] *ci.130*
TABP Towed Array Broadband Processor *fu*
TABR Targeted Area Background Report *mt*
TABSAC Targets And Backgrounds Signature Analysis Center *wp*
TABSTONE Target And Background Signal To Noise Emission *wp*
TABUN [N-Dimethylphosphoamido Cyanidate (chemical warfare nerve agent, also called Ga)] *wp*
TABWS Tactical Air Base Weather Station *fu, wp*
TABWS Tactical Air Base Weather System *mt*
TAC Tactical Air Command [USA] * *mt, fu, ro, aw130.13.71, 12, ph, hc, hn44.9.10, w9, af72.5.31, wp, tb, jdr, kf*
TAC Tactical Air Control *hc*
TAC Target Acquisition Console *fu*
TAC Target Allocation Center *fu*
TAC Target Assignment Command *tb*
TAC Technical Activities Committee *aa26.8.b4*
TAC Technical and Administrative Center *oe*
TAC Technology Applications Center *ci.1005*
TAC Telemetry Acquisition Console *oe*
TAC Terminal Access Controller *tb*
TAC Terminal Access Controller / Center *mt*
TAC Threat Adaptive Channel *fu*
TACOL Thruster Attitude Control *jdr, uh*
TAC Total Accumulated Cycle [tests] *aw130.11.32*
TAC Transistorized Automatic Control *fu*
TAC Translator Assembler Compiler *fu*
TAC Transportation Account Code *mt*
TAC ASA Tactical Support System for Army Security Agency *fu*

TAC-IROCS Tactical Intrusion Resistant Optical Fiber Communications System *hc*
TAC/TAD Tactical Air Control / Tactical Air Defense *fu*
TAC/TADS Tactical Air Control System / Tactical Air Defense System *wp, mt*
TAC/TASS Tactical Towed Array Sonar Program *hc*
TACA Tactical Airborne Controller Aircraft *wp*
TACACPS Theater Army Construction Automated Planning System *mt*
TACACS TAC* Access Control System [MILNET] *mt*
TACADS Tactical Automated Data Processing System *wp*
TACAIR Tactical Air *mt*
TACAIR Tactical Air Command *uc*
TACAIR Tactical Aircraft *fu*
TACAIRCG Tactical Air Control Group *mt*
TACALS TAC* Command and Control Alerting System *wp, mt*
TACAMO Airborne Strategic Communications System *ph*
TACAMO TAke Charge And Move Out [aircraft that relay communications with US Navy submarines] [USA] *fu, hc, wp, mt*
TACAN Tactical Air Navigation [also written tacan] *hc, wp, fu, ro*
TACAN Theoretic And Calculated Air Navigation *0, aw129.10.72, af72.5.92*
TACAP Tactical Air Command Aircraft Profiler [TAC*] *mt*
TACBEAMS TAC* Base Engineering Automated Management System *mt*
TACC Tactical Air Command Center *mt*
TACC Tactical Air Control Center *mt, wp, fu*
TACC Tactical And Control Center *pl, pbl*
TACC Taipei Area Control Center *fu*
TACC Taiwan Area Control Center *fu*
TACC Tanker Airlift Control Center, at Scott AFB [USA] *mt*
TACC Theater Army Command and Control [USA] *mt*
TACCAR Time Averaged Clutter Coherent Airborne Radar *fu, wp*
TACCIMS Theater Automated Command and Control Information Management System [Korea] *mt, hc, wp*
TACCIS Theater Army Command and Control Information System [USA] *mt*
TACCO Tactical Coordinator [the second crewmember in some US Navy aircraft] *pf1.9, mt*
TACCOMSIM Tactical Communications Simulator *hc*
TACCS Tactical Airborne Command Control and Surveillance *wp*
TACCS Tactical Army Combat Service Support Computer System *mt*
TACCS Tactical Command and Control System *fu*
TACCS-K Theater Automated Command and Control System Korea *fu, mt*
TACCSF Theater Air Command and Control Simulation Facility *ts17-1*
TACD Tactical Deception [USA] *mt*
TACDAR Tactical Dissemination and Reporting *kf*
TACDEN Tactical Data Entry *fu*
TACDEW Tactical Advanced Combat Direction and Electronic Warfare *wp*
TACDEW Tactical Director and Electronic Warfare *hc*
TACDEW Tactical Director of Electronic Warfare [Navy] *fu, hc*
TACE Technical Analysis and Cost Estimate *mt*
TACELINT Tactical Electronic Intelligence *wp, fu*
TACELINT Tactical ELINT *mt*
TACEVAL Tactical Evaluation [NATO] *fu, mt*
TACF Tactical Air Control Flight *mt*
TACFIRE Tactical Fire [Direction System] [Distribution System] [USA] *mt, hc, fu*
TACFITS Tactical Force Interaction Targeting and Strike (Study) *wp*
TACG (4450th TACG, Nellis AFB, Nevada) *af72.5.55*
TACG Tactical Air Command Guard *mt*
TACGRU Tactical Air Control Group [USA] *mt*
TACIDN Tactical Information Distribution Network *mt*
TACIES Tactical Image(ry) Exploitation System *mt, hc, wp*
TACIES Tactical Intelligence Information Exchange Subsystem [US Navy] *mt*
TACINTEL Tactical Intelligence [communication net] *fu*
TACIP The Army Command and Control Initiatives Program *fu*
TACIT_RAINBOW [autonomous loitering missile to search/attack radars] *wp*
TACJAM Tactical Jamming (Jammer) system [USA] *fu, mt*
TACJAM-A Tactical Jamming Army [mobile system for U.S.] *aw130.13.87*
TACM Tactical Air Command Manual *fu*
TACMAR Tactical Multifunction Array Radar *wp*
TACMET TAC* Forces Control Management System *mt*
TACMIP TACFIRE Modification Improvement Program *hc*
TACMIS Tactical Management Information System [USA] *mt*
TACMS Tactical Missile System [US Army] *mt, wp, nd74.445.30*
TACNET TAC* Force Control Management Network [TAC*] *mt*
TACOL Thinned Array Computed Lens *wp*
TACOM Tactical Communications *wp*
TACOM Tank and Automotive [Research and Development] Command [US Army] *mt, tc, hc*
TACOM Theater Army Command *mt*
TACOM_EWS Tactical Communications Electronic Warfare System *wp*
TACON Tactical Control [USA DOD] *mt*

TACOPS Tactical Air Combat Operations *mt*
TACOS Tactical Air Defense Computerized Operational Simulation [model] *wp*
TACOS Tactical Airborne Countermeasures Or Strike *wp*
TACOS Tropospheric Scatter Area Communication System *hc*
TACP Tactical Air Control Party [USA] *fu, mt*
TACP Tactical Air Control Post *wp*
TACP Theater Ammunition Control Point *mt*
TACPLAN Tactical Plan Generation *mt*
TACR TAC* Regulation *mt*
TACREP Tactical Report *mt, wp*
TACRON Tactical Air Control Squadron [USA] *mt*
TACS Tactical Air Command System *wp*
TACS Tactical Air Control System [USA] *hc, fu, wp, mt*
TACS Target Acquisition and Control System *mt*
TACS Telemetry And Command Station *lt*
TACS Theater Air Control System [USAF] *kf, mt*
TACS Thruster Attitude Control System *ci.432*
TACS Total Access Communication System *pk, kg*
TACS Tracking And Control Station *bl1-5.84, pbl*
TACS Treasury Automated Communications *hc*
TACS/TADS Tactical Air Control System/Tactical Air Defense System *fu*
TACSAT Tactical Satellite [Communications System] *mt, ro, fu, sc4.10.3, lt, wp*
TACSATCOM Tactical Satellite Communications *mt*
TACSI Tactical Air Control System Improvements *wp*
TACSIM Tactical Simulation *mt*
TACSIT Tactical Situation [of a US Navy battlegroup] *mt*
TACSM Tactical Air Combat Simulation Model *mt*
TACSO Tactical Air Command System Office *wp*
TACSS Tactical Army Combat Support System *fu*
TACT Tactical Air Control Team *fu*
TACT The Automated Controller Test *mt*
TACT Transonic Aircraft Technology [the modification of an F-111 to research supercritical wings, USA] *wp*
TACTAS Tactical Towed Array System (Sonar) [AN/SQR-18] *hc, fu*
TACTICS Tactical Information Control System *mt*
TACTRAGRU Tactical Training Group *fu*
TACTS Tactical Air(crew) Combat Training System *mt, hc, wp*
TACTS Telemetry Acquisition and Command Transmission System *oe*
TACW Tactical Air Control Wing *mt, af72.5.137*
TACWE Tactical Air Command Weather Element *wp*
TAC_CP Tactical Command Post *mt*
TAC_D&E Tactical Development and Evaluation *hc*
TAC_INTEL Tactical Intelligence *mt*
TAD Tactical Air Defense *tb*
TAD Target Acquisition Data *fu*
TAD Target Area Designation (Designator) *0, wp*
TAD Task Allocation Diagram *fu*
TAD Technical Area Director *0*
TAD Technology Area Description *wp*
TAD Technology Availability Date *wp*
TAD Temporary Duty [US Navy] *mt*
TAD Thrust Augmented Delta *sm18-9*
TAD Time Available for Delivery *fu*
TAD Trailing Arming Device *hc*
TAD Training Augmentation Device *kf*
TAD Transaction Application Domain *fu*
TADAS Tactical Air Defense Alerting System *wp*
TADB Terrain Analysis Data Base *mt*
TADC Tactical Air Direction Center *mt, fu*
TADCOM T.A.D. Communications Company [a joint Hughes-ITT venture] *fu, hn42.1.4*
TADDS Target Alert Data Display Set *fu*
TADGC Tactical Air Designation Grid [System] *wp*
TADIL Tactical Digital Information Link [USA DOD] *mt, fu, hc, af72.5.139, wp, kf*
TADIL-B TADIL type B [several other type letters are also in use] *fu*
TADIXS Tactical Digital (Data) Information Exchange (Sub)System [US Navy] *mt, kf, fu*
TADNET Test and Development Network [USA] *mt*
TADPOLE [system for engaging enemy targets with target acquisition and Laser design]. *wp*
TADS Tactical Advanced Digital Control System *wp*
TADS Tactical Air Defense System *fu, hc, wp*
TADS Tactical Automatic Digital Switches *hc*
TADS Target Acquisition and Detection System *mt*
TADS Target Acquisition [and] Designation System (Sight) [USA] *mt, hc, aw129.10.57*
TADS Target and Activity Display System *wp*
TADS Telemetry Analysis and Display System *fu*
TADS Tracking and Display System [also called AN/SYQ-8] *hc, fu*
TADSS Tactical Army Digital Switching System *hc*
TADSTAND Tactical Digital Standard *fu, hc*
TADWS Tactical Army Digital Wire Switch *hc*
TAE Transportable Applications Environment (Executive) *ct, ns, em*
TAEDP Total Army Equipment Distribution Plan [USA] *mt*

TAEG Training Analysis and Evaluation Group (USN) *fu*
TAER Time, Azimuth, Elevation, Range *wp*
TAERS Tactical Aircrew Eye/Respiratory Protection System [chemical warfare] *wp*
TAES Tactical Aeromedical Evacuation System *mt*
TAF Tactical Air Force *mt, pl, hc, wp*
TAF Time And Frequency *hi*
TAFB Travis Air Force Base [California] *wp*
TAFB Tyndall Air Force Base [Florida?] *wp*
TAFHUG Tactical Air Forces Tactical Host Upgrade *mt*
TAFIC Tactical Air Forces Intelligence Center *hc, wp*
TAFICS Turkish Armed Forces Integration Communication System *hc*
TAFIES Tactical Air Force Intelligence Exploitation System *wp*
TAFIG Tactical Air Forces Interoperability Group [TAF] *mt*
TAFIIS Tactical Air Forces Integrated Information System [TAF] *hc, wp, mt*
TAFIM Technical Architecture Framework for Information Management [GCCS] [USA] *kf, mt*
TAFMM Tactical Air Force Maintenance Management *wp*
TAFS Terminal Aerodrome Forecast *mt*
TAFSEG Tactical Air Force System Engineering Group *wp*
TAFSM Target Acquisition Fire Support Model *fu*
TAFSS Tailored Automatic Functional Support System *hc*
TAFTRAMS TAC* Automated Flying Training Management System [TAC*] *mt*
TAFTS Tailored Automatic Functional Test Stations *hc*
TAFWG Tactical Air Force Working Group *mt*
TAG T&C Antenna Group *vv*
TAG Tactical Airlift Group *mt, af72.5.121*
TAG Technical Advisory Group *wp, tb, st, ct, em*
TAG Technology Advocacy Group *wl*
TAG The Adjutant General *w9, st*
TAG Topic Area Guide *mt*
TAG Tracker Assignment Gates *fu*
TAGCEN The Adjutant General Center *dt*
TAGO The Adjutant General {Office} staff *mt*
TAGOS (TAGOS (SURTASS) "inherently stealthy". Small ships (low signature)) *kb*
TAGS Text And Graphics System *ws4.10*
TAH [hospital ship] [also AH] *wp*
TAHA Tapered Aperture Horn Antenna *hc*
TAHQ Theater Army Headquarters *mt*
TAI International Atomic Time [USA DOD] *mt*
TAI Tactical Air Interdiction *wp*
TAI Technology Applications, Inc. *fu*
TAI Total Aircraft Inventory *mt, wp*
TAID Thrust Augmented Improved Delta *sm18-9*
TAIES TDD Automated Information Exchange System *wp*
TAILRATS Tail Radar Threat Sensor *hc*
TAIRCG Tactical Air Control Group *mt*
TAIRCW (602d TAIRCW, Davis-Monthan AFB, Arizona) *af72.5.55*
TAIRCW Tactical Air Control Wing *mt*
TAIS Tactical Air Intelligence System *mt, wp*
TAKR Takticheskoye Avianosniy Kreyser [tactical aircraft-carrying cruiser] [USSR] *mt*
TAL Transatlantic (Transoceanic) Abort Landing [Shuttle abort plan] *mi, ws4.10*
TALAR Tactical Landing Approach Radar *mt*
TALAR Tactical Landing [Aid] Radar *wp, fu*
TALCE Transportable Airlift Control Element *mt*
TALCM Tactical Air Launched Cruise Missile *hc*
TALCM Tomahawk Air Launched Cruise Missile [the competition for ALCM, USA] *mt*
TALD Tactical Air Launched Decoy *wp*
TALD Tactical Airborne Laser Designator *mt*
TALI Tactical Analysis Logistics Information [TAF] [later LOGFAC] *mt*
TALI-WRM TALI-War Readiness Materials *mt*
TALO Tactical Airlift Liaison Officer *mt*
TALO Time After Lift Off *kf*
TALON Theater Application Launch On Notice *mt*
TALONS Tactical Avionics Low Level Navigation and Strike *wp*
TALONS Tactical Long Range Navigation System *wp*
TALON_GOLD [project to develop low energy laser for target acquisition and tracking] *wp*
TALS Transport Approach and Landing System *hc*
TALSO Theater Army Logistics System Office *mt*
TALT Tracking Altitude *fu*
TAM Tactical Air Meet *mt*
TAM Tactical Airlift Model *mt*
TAM Terminal Access Method *hi*
TAM Theater Airlift Manager *mt*
TAM Threat Assessment Model *17, sb*
TAM Tier Attrition Model *fu*
TAM Transceiver Antenna Module *fu*
TAMCA Theater Army Movement Control Agency [USA] *mt*
TAME Tactical Air to Air Mission (Missile) Evaluation Model [AFSC*] *wp, hc, mt*

TAME TEMPEST Advanced Multiplexer Equipment *mt*
TAMMC Theater Air Material Management Center *mt*
TAMMC Theater Army Materiel Management Center [USA] *mt*
TAMMIS Theater Army Medical Management Information System *mt*
TAMMS The Army Maintenance Management System *mt, fu*
TAMMS Transportation Movements Management System *mt*
TAMP Tactical Air Missile Program *mt*
TAMP Theater Army Maintenance Program *mt*
TAMPS Tactical Air(craft) Mission Planning System *fu, wp*
TAMS Technical Assistance and Management Services *wp*
TAMS Test And Monitoring System [US Navy] *wp, mt*
TAMS Theater Airlift Management System [MAC] *mt*
TAMU Texas Agricultural and Mechanical University *aa26.11.56z*
TAN Test Advisory Notice *mt*
TANC Thruster Active Nutation Control *uh, 3, jdr, kf, vv*
TANGO [command post for UNC/CFC/CGCC, Korea] *mt*
TANKBREAKER [MICOM missile system] *hc*
TANS Tactical Air(borne) Navigation System *mt, aw130.22.107*
TANSTAAFL There Ain't No Such Thing As A Free Lunch *hi*
TAO Tactical Action Officer [USA] *fu, mt*
TAO Tactical Air Operations *fu*
TAO Technical Analysis Office *fu*
TAO Telephony Application Object *hi*
TAOC Tactical Air(borne) Operations Center [USMC] *ph, wp, fu, mt*
TAOC Theater Air Operations Center *kf*
TAODB Tactical Air Operations Data Base *fu*
TAOM Tactical Air Operations Module *mt*
TAOR Tactical Area Of Responsibility [USA DOD] *fu, mt*
TAOS Target Area Optical System *hc*
TAOS Thrust Assisted Orbiter [Shuttle] System *ci.452, ai2*
TAP (a 32 TAP digitally controlled programmable transversal filter) *mj31.5.60*
TAP Tactical Armament Plan *hc*
TAP Tactical Assessment and Planning *fu*
TAP Target Analysis and Planning system *fu, mt*
TAP Target Approach Point *mt*
TAP Task Assignment Procedure *st*
TAP Technical Acceptance Plan *mt*
TAP Technical Assessment Program *fu*
TAP Technological Assistance Program *ct, em*
TAP Technology Area Plan *wp*
TAP Telelocator Alphanumeric Protocol *pk, kg*
TAP Thermal Analysis Program [Hughes Aircraft Company] *hc*
TAP Timing and Acquisition Processor *jdr*
TAP Towed Array Processor *fu*
TAP Tracking Analyzer (Accuracy) Program *fu*
TAP Transient Acoustic Processing *fu*
TAP Transition Analysis Program *fu*
TAPAC Tape Automatic Positioning and Control *fu*
TAPCIS The Access Program for the CompuServe Information Service *pk, kg*
TAPDB The Army Personnel Data Base [USA] *mt*
TAPE Tactical Air Power Evaluation *wp*
TAPE Technical Advisory Panel for Electronics *wp*
TAPER The Army Personnel roll up system *mt*
TAPES Tactical Automated Planning and Engineering System *mt*
TAPI Telephony Applications Program (Programming) Interface *kg, ct, em, hi*
TAPS Terrain Analysis Production System *mt, hc*
TAR Tactical Air Reconnaissance *mt, wp*
TAR Tactical Air Request *fu*
TAR Tape Archive [a UNIX command] *ct, em, hi*
TAR Technical Action Request *fu*
TAR Technical Activities Report *wp*
TAR Terrain Avoidance Radar *hc, wp, fu*
TAR Test Analysis Report *fu*
TAR Test Anomaly Report *jdr*
TAR Threat Assessment Report *mt, wp*
TAR Towed Array Receiver *fu*
TARADCOM [Hughes Aircraft Company customer for muzzle position sensor] *hc*
TARAN Tactical Attack Radar and Navigation *fu, hc*
TARCAP Target Combat Air Patrol [USA] *mt*
TARE Tactical Reconnaissance Evaluation *wp*
TARE Telegraph Automatic Relay Equipment [NATO] *mt*
TARES Tactical Reconnaissance Evaluation System [TAF] *mt*
TAREWS Tactical Air Reconnaissance and Electronic Warfare Support *wp*
TARF Target Report Form *fu*
TARG Telescoped Ammunition Revolver Gun *mt*
TARGET/DCP Theater Analysis and Replanning Graphical Execution Toolkit / Distributed Collaborative Planning [GCCS] [USA] *mt*
TARH Terminal Active Radar Guidance *mt*
TARMAC Tar covered Macadamized [broken stone base road or runway] *wp*
TARP Tactical Air Reconnaissance Pod *wp*
TARP Theater Army Rebuild Program *mt*

TARP Transmit Amplifier/Receive Preamplifier *fu*
TARPS Tactical Air(borne) Reconnaissance Pod System [USA] *mt, wp, hc, aw126.20.125, rih*
TARRS Tactical Air Reconnaissance Results Reporting System *mt*
TARS Tactical Aerial Reconnaissance and Surveillance needs *mt*
TARS Tactical Air Reconnaissance and Surveillance *mt, wp*
TARS Tactical Air Reconnaissance System *mt, hc*
TARS Terrain Analog Radar Simulator *fu*
TARS Tethered Aerostat Radar System *mt*
TARS Transportation Automated Routing System *mt*
TARS Turnaround Ranging Station *gd2-6, hc, pbl*
TARVAN Transportation Van *ai2*
TAS Tactical Air Support *wp*
TAS Tactical Airlift Squadron [USA] *mt, af72.5.121, wp*
TAS Target Acquisition System *mt, fu, hn43.12.1, hc, wp*
TAS Technical Area Specialist *mt*
TAS Tempelhof Automation System *fu*
TAS Terminal Access System *mt, tb*
TAS Terminate Active Sequence *fu*
TAS Test Access System *fu*
TAS Towed Array Sonar *hc, fu*
TAS Tracking Adjunct System *fu*
TAS Transatmospheric Aeronautical System *wp*
TAS True Air Speed *mt, fu, aw130.13.38*
TASA Task and Skill Analysis *fu*
TASA Terminal Area Support Aircraft *wp*
TASA Total Aircraft Status Accounting [System] *mt*
TASC Terminal Area Sequencing and Control *fu*
TASC The Analytic(al) Sciences Corporation [Huntsville, AL 35805] *mt, ts15-24, jmj, 19*
TASC Theater ADP Service Center *mt*
TASCAM Tactical Airlift System Comparative Analysis Model [MAC*] *mt*
TASCFORM Technique for Assessing Comparative Force Modernization *mt*
TASCS Tactical Air Support Control System *fu, hc, wp*
TASCS Training Analysis Support Computer System *mt*
TASE Tactical Air Support Element *mt, fu*
TASES Tactical Airborne Signal Exploitation Space (System) [US Navy] *mt*
TASG (110th TASG, Battle Creek ANGB, Michigan) *af72.5.55*
TASI Time Assignment Speech Interpolation *fu*
TASM Tactical Air to Surface Missile [UK] *kf, mt*
TASM Tactical Anti Ship Missile *dh*
TASM TLAM Anti Ship Missile *dh*
TASM Tomahawk Anti Ship(ping) Missile [USA] *ph, hc, mt*
TASM Turbo Assembler [Borland] *kg, hi*
TASMO Tactical Air Support of Maritime Operations *mt*
TASO Terminal Area Security Officers *mt*
TASO {an association defining SNR} *gs1-5.A4*
TASR Terminal Area Surveillance Radar *fu*
TASRAN Tactical Air Surveillance Radar Network *wp*
TASS Tactical Air Support Squadron [USA] *mt, wp, af72.5.55*
TASS Tactical Air to Surface System *wp, mt*
TASS Tactical Analysis Support System [TAC] *mt*
TASS Tactical Armed Surveillance System *fu, wp*
TASS Tactical Avionics System Simulator *wp*
TASS Technical Analytical Study Support *hc*
TASS Towed Array Sonar (Surveillance) System *fu*
TASS Traffic Analysis Systems Study *hc*
TASS Trojan Automatic Switching System *hc*
TASS [official Soviet news agency] *11, sf31.1.2*
TASSA Tactical Assessment of Survivable Stand Off Alternatives *wp*
TASSEL TASS Element *mt*
TASTG (549th TASTG, Patrick AFB, Florida) *af72.5.55*
TASTG Tactical Air Support Training Group *mt*
TASTS TOW Airborne System Test Set *hc*
TASVAL Tactical Aircraft Survivability Evaluation *wp*
TAT Tactical Airlift Technology *wp*
TAT Target Acquisition and Tracking *st, kf*
TAT Target Aircraft Transmitter *fu*
TAT Thrust Augmented Thor *ci.xvii*
TAT To Accompany Troops *mt*
TAT Touraine Air Transport [a French airline] *aw130.11.68*
TAT Trans Atlantic Telephone [cable] *-10*
TAT Transport AÈrien Transregional *aw129.10.13*
TAT Transportation and Ammunition Tracking *mt*
TATB Triaminotrinitrobenzene [explosive] *wp*
TATC Terminal Air Traffic Control *wp*
TATC Turkish Air Traffic Control *fu*
TATCE Terminal Air Traffic Control Element *fu*
TATCF Terminal Air Traffic Control Facilities *wp*
TATCS Turkish Air Traffic *fu*
TATEP Training and Training Equipment Plan *fu*
TATF Terminal Automation Test Facility *fu*
TATR Tactical Air Target Recommender *mt*

TATS Tactical Airlift Training Squadron *mt*
TATS Tactical Assessment Teams *mt*
TATS Target Acquisition and Tracking System *st*
TAU Thousand Astronomical Unit [mission] *mi*
TAU Transient Absorber Unit *jn*
TAUVEX [a UV images on SXG] *cu*
TAV Total Asset Visibility [GCCS] [USA] *mt*
TAV Trans Atmospheric Vehicle *mt, wp, hc, tb, jdr*
TAVC Test Area Validation Console *fu*
TAW Tactical Air(lift) Wing [USA] *mt, wp, af72.5.55*
TAWACS Tactical Airborne Warning And Control System *wp*
TAWAR Tactical All Weather Attack Requirement *wp*
TAWARS Tactical All Weather Attack Requirements Study *hc*
TAWC Tactical Air Warfare Center [Eglin AFB] *mt, af72.5.55, wp*
TAWCS Tactical Air Weapons Control System [called ADGE at Hughes since 1967] *hc, fu*
TAWDS Target Acquisition and Weapon Delivery System *fu, hc, wp*
TAWDS Terminal Aerial Weapon Delivery Simulation *wp*
TAWDS Transportable Automated Weather Distribution System *mt*
TAWMS Topographic All Weather Mapping System *wp*
TAWS Tactical All Weather Strike *hc*
TAXI Transparent Asynchronous Tranceiver / Receiver Interface *kg, hi*
TB Technical Bulletin *mt, nt, fu*
TB Technology Base *17, sb*
TB Terminal Board *fu*
TB Terminate Beacon *fu*
TB Test Bed *st*
TB Thermal Balance *jdr, kf*
TB Time of Bus impact *0*
TB Time-Bandwidth *fu*
TB Transec Blanking *jdr*
TBA Terminal Board Assembly *fu*
TBAP Tyazhyolo Bombardirovotschnii Aviatsonii Polk [heavy bomber regiment] [USSR] *mt*
TBC Time Based Corrector *jgo*
TBC Track Beam Control *fu*
TBC Trickle Battery Charge *gg102*
TBD Terrain Bounce Demonstration *hc*
TBE TB-16/BQ Equivalent [towed sonar array] *fu*
TBE Teledyne Brown Engineering *st*
TBE Towed Body Equivalent *hc*
TBEP Training Base Expansion Plan [TRADOC] *mt*
TBF Transmitter Beamformer *fu*
TBG (TBG Information Systems, Inc., Englewood, Colorado) *aw130.11.4*
TBGA Tape Ball Grid Array *pk, kg*
TBI Through Bulkhead Initiator *lt*
TBITS Treasury Board Information Technology Standard *hi*
TBL TEGAS Behavioral Language *fu*
TBL Turbulent Boundary Layer *fu*
TBLO Time Base Local Oscillator *jdr*
TBM Tactical Ballistic Missile *mt, fu, wp, tb, st, kf*
TBM Theater Ballistic Missile *kf*
TBM Theater Battle Management *mt*
TBM Training Baseline Materials *mt*
TBMW Tactical Ballistic Missile Warning *kf*
TBPS Terabits Per Second *ct*
TBPS TeraBytes Per Second *em*
TBR Test Burn-in requirement *he*
TBRRSS Tactical Battlefield Reconnaissance Relay/Surveillance System *hc*
TBS Talk Between Ships *w9*
TBU Tank Break Up *st*
TBU Tape Backup Unit *kg, hi*
TBX Terminal Box *fu*
TBX Towed Body (TB)-16 Extended *fu*
TBX [thin film display for towed array sonar] *hc*
TC Tape Core *fu*
TC Technical Circular *fu*
TC Technical Committee *aa26.8.b4, tt*
TC Technical Control *fu*
TC Temperature Cycling *jdr*
TC Test Center *fu*
TC Test Conductor *jdr*
TC Test Control *pk, kg*
TC Test Controller *uh*
TC Thermal Compensation *jdr*
TC Thermal Compressive *jdr*
TC Thermal Control *wl*
TC Thermal Cycle *he, kf*
TC Time Clock *hi*
TC Time of Capsule impact *0*
TC Training Center *kf*
TC Transfer Channel *mt*
TC Transit Corridor [NATO] *mt*
TC Transition Components [WIS] *mt*
TC Transmission Control *pk, kg*

TC Transmit Component *-4*
TC Transmit Controller *fu*
TC Transportation Coordinator *mt*
TC Tuning Curve *fu*
TC Type Classification *st*
TC {Trickle Charge} (Bus Current Slow TC) *ggIII.2.406*
TC&R Telemetry, Command and Ranging *jn, oe*
TC-ACCIS Transportation Coordinator's Automated Command and Control Information System [DCA] *mt*
TC/NC Tow Cable / Nose Cone *fu*
TCA Target Class Assessment *mt*
TCA Terminal Control Area *fu*
TCA Thrust Chamber Assembly *gg37*
TCA Traffic Control Area *wp*
TCAC Tactical Collection and Analysis Center *mt*
TCAC Tactical Command Analysis Center *fu*
TCACCIS Transportation Coordinator Automated Command and Control System *mt*
TCAC[D] Technical Control and Analysis Center [Division] [USA] *mt*
TCAE Technical Control and Analysis Element [USA] *mt*
TCAIMS Transportation Coordinator's Automated Information for Movements System *mt*
TCAM Tele Communications Access Method [IBM] *ct, em, do422*
TCAS Technical Control and Analysis Section [USA] *mt*
TCAS Threat Alert and Collision Avoidance System *fu*
TCAS Traffic Alert and Collision Avoidance System *mt, aw129.14.109, aa26.12.42, is26.1.62, wp*
TCATA TRADOC Combined Arms Test Activity [USA] *mt*
TCB Task Control Block *fu, uh, jdr*
TCB Tetrachlorobiphenyl *wp*
TCB Transmission Control Block *mt*
TCB Trusted Computer (Computing) Base *mt, fu, tb, kf*
TCBV Temperature Coefficient of Breakdown Voltage *fu*
TCC Tactical Command and Control study (TCC-21 for 21st Century version) [AFSC*] *mt*
TCC Tactical Control Console *fu*
TCC TADIL Communication Control *fu*
TCC Tape Chip Carrier *fu*
TCC Technical Cooperation Committee *ph*
TCC Telecommunication Center *mt*
TCC Telecommunications Control Center *mt*
TCC Terminal Communication Control *kf*
TCC Test Conductor Console *jdr*
TCC Timing Correction Command *fu*
TCC Traffic Control Center *mt*
TCC Transmission Control Codes *mt*
TCC Transportation Component Command [USA DOD] *mt*
TCC Transportation Control Code *wp*
TCC Transportation Coordination Center *mt*
TCC [Training MCC] *hc*
TCCC Tower Control Computer Complex *fu*
TCCCC Transportable Command and Control Computer Center [USA] *mt*
TCCD Tactical Command and Control Display *ts17-3*
TCCF Tactical Communications and Control Facility [TRI-TAC] *hc, fu, mt*
TCCP Technical Control Collection Processor *fu*
TCCP Telecommunications Command Control Projects *mt*
TCCS Transportation Command and Control System *mt*
TCC[A] Theater Communications Command [USA] *mt*
TCD Tactical Control Director *mt*
TCD Technical Compliance Designee *jdr*
TCD Terminal Countdown Demonstration *oe*
TCD Thermal Conductivity Detector *wp*
TCD Trident-Common Diagnostics *fu*
TCDD Tower Cab Digital Display *hc*
TCDIS Transportation Coordinator Deployment Information System [JDA] *mt*
TCE Terminal Control Element *mt*
TCE Thermal Coefficient of Expansion *fu, wp, md2.1.2*
TCEP Tris(Chloroethyl)Phosphite *wp*
TCF Tactical Combat Force *mt, wp*
TCF Technical Control Facility *mt, tb*
TCF Tower Control Facility *fu*
TCFC Transec Communications Front end Controller *mt*
TCG Tactical Control Group *mt*
TCG Target Constant Generation *mt*
TCG Teleport Communications Group *ct, em*
TCG Time Code Generator *mt*
TCG Time Constant Group *bl2-5.21*
TCG Track Control Group *mt*
TCGS Telemetry and Command Ground Station *jdr*
TCH Transfer in Channel *fu*
TCHTG Technical Training Group *mt*
TCI Target Class Identification *mt*
TCI Telemetry Control Interface *jn*
TCI Terminal Control Indicator *fu*

TCI Terrain Clearance Indicator *fu*
TCI Transpace Carriers, Incorporated [Delta launch services company] *sc4.3.5*
TCIO Task Control and Information Office *mt*
TCIP Technical Control Improvement Program *mt*
TCIP Technical Control System Improvement / Control Technician Improvement Plan *mt*
TCL TAE Command Language *ct, em*
TCL Thermal Control Louvers *jdr, oe*
TCL Tool Command Language *pk, kg*
TCL Transistor Coupled Logic *fu*
TCLBS Tropical Constant-Level Balloon System *pl*
TCM Tactical Cruise Missile *wp*
TCM TADIL Communications Management *fu*
TCM Technical Coordination Meeting *-4*
TCM Telemetry Code Modulation *fu*
TCM Terminal Control Mechanism *mt*
TCM Thermal Control Model *0*
TCM Time Compression Multiplexing *mt*
TCM Trajectory Correction Maneuver *ma22, mb, re*
TCM Transfer Control Module *fu*
TCM Trellis Coded Modulation *kg, ct, em, hi*
TCMD Transportation Control Movement Document *mt*
TCMMS Transport Canada Materiel Management System *fu*
TCMOD Trident Combat System Modification *fu*
TCMP Transportation Control and Movement Plan *mt*
TCMX Terminal Computer and Multiplexer *hi*
TCN Task Control Number *fu*
TCN Tele Communications Network
TCN Transportation Control Number *mt, wp*
TCNT Transportation Cooled Nose Tip *st*
TCO Tactical Combat Operation System [USMC] *mt*
TCO Tactical Combat Operations *fu*
TCO Tactical Control Officer *fu*
TCO Task Command Operation *fu*
TCO Technical Control Operator *mt*
TCO Telecommunications Certification Office *mt*
TCO Telecommunications Control Office *mt*
TCO Telephone Control Officer [USAF] *mt*
TCO Termination Contracting Office(r) *mt, wp, fu*
TCOC Technical Control Operator Cluster *fu*
TCOM [a division of Westinghouse] *aa26.12.28, aw124.21.13*
TCOSS Technical Control Operations Support System *mt*
TCOT Tactical Control Operations Teams *mt*
TCP Tactical Cryptologic (Cryptology) Program *wp, mt*
TCP Tape Carrier Packaging *em, ct*
TCP Task Change Proposal *tb, kf, mt, fu*
TCP Technical Change Proposal *wp*
TCP Technological Capabilities Panel *-13*
TCP Technology Control Plan *jdr*
TCP Technology Coordination Paper *wp*
TCP Telecommunications Protocol *mt*
TCP TEMPEST Control Program *17, sb*
TCP Test Control Processor *fu*
TCP Test Controller Program *fu*
TCP Thruster Control Processor *gp, 8*
TCP Time-limited Correlation Processing *wp*
TCP Top Command Pointer *fu*
TCP Traffic Control Post *wp*
TCP Training Certification Program *fu*
TCP Transmission Control Protocol [DOD Standard] *mt, cs, aw130.11.56, hn49.11.8, tb, jdr, pbl, em, ct, ah, fu, hi*
TCP Transport Control Protocol *mt, wp*
TCP Tricresyl Phosphate *wp*
TCP Trident Common Program *fu*
TCP Tube Control Panel *fu*
TCP/IP Transmission Control Protocol / Internet Protocol [DARPA] *mt, ct, em, kg, nu, is26.1.46, ns, kf, hi, oe, du*
TCPI To Complete Performance Index *fu*
TCPS Transportable Collective Protection Shelter [chemical warfare] *wp*
TCQAM Trellis Coded Quadrature Amplitude Modulation *ct, em*
TCR Task Change Request *kf*
TCR Technical Control Rebuild *mt*
TCR Temperature Coefficient of Resistance *wp, fu*
TCR Track Command Receiver *he*
TCR Traffic Control Radar *wp*
TCR Transfer Control Register *fu*
TCRP Tactical Command Readiness Program [USAREUR] *mt*
TCS Tactical Command System [US Navy] *mt*
TCS Tactical Computer System *mt, fu*
TCS Tactical Control Squadron *fu*
TCS Technical Control Station *fu*
TCS Telemetry and Command Simulator System [GIMTACS] *oe*
TCS Telemetry and Command Simulator [GOES simulation system] *oe*
TCS Telemetry and Command Subsystem [digital] *vv*

TCS Television Camera System [long distance TV cameras to aid in the visual identification of aircraft] [USA] *mt, rih*
TCS Television Control Sets *hc*
TCS Teradyne Connection Systems, [Incorporated] *hq5.3.5*
TCS Terminal Communication System *fu*
TCS Terminal Control System *fu*
TCS Theater Communications System *mt*
TCS Thermal Control System (Subsystem) *mi, lt, tb, wl, jdr, mb, kf, oe*
TCS Tower Communications System *fu*
TCS Troop Carrier Squadron [USA] *mt*
TCSE Thermal Control Surface Equipment *hc*
TCSEC Trusted Computer System Evaluation Criteria [DOD] *mt, fu, tb*
TCSP Technical Control Systems Program *mt*
TCSR Technical Counters and Response *tb*
TCSR Technical, Cost, and Schedule Realism *st*
TCSS Technical Control Software Support *mt*
TCSS Telecommunications Service System *mt*
TCSTS Tactical Communications Systems Technical Standards *mt*
TCT Tactical Computer Terminal *mt, fu*
TCT Test Center TCCC *fu*
TCT50 [threshold vapor inhalation dose with 50% probability] *wp*
TCTI Transport Canada Training Institute *fu*
TCTO Temporary Change Technical Order *tb*
TCTO Thermal Compliance Technical Order *hc*
TCTO Time Compliance Technical Order *mt, fu, wp, jdr*
TCTS Tactical Communications Systems Technical Standards *mt*
TCU TACAN Control Unit *fu*
TCU Tape Control Unit *fu*
TCU Target Collimator Unit *hc*
TCU Temperature Conditioning Unit *jdr*
TCU Terminal Control Unit *fu*
TCU Threshold Control Unit *fu*
TCU TOW Collimator Unit *hc*
TCU Transit Control Unit *mt*
TCU Transmitter (Transmission) Control Unit *mt, fu, hi*
TCU Transponder Control Unit *uh, jn*
TCU Transportable Computer Unit *fu*
TCU Transportation Control Unit *mt*
TCU Trunk Coupling Unit *fu*
TCU TWT Control Unit *fu*
TCUL Tape Changing Under Load *fu*
TCW Tactical Control Wing *mt, af72.5.74*
TCWG Tactical Communications Working Group *mt*
TCWG Tele-Communications Working Group *fu*
TCWS Tower Controller Workstation *fu*
TCXO Temperature Compensated Crystal Oscillator *fu, gg139, 2, 4, vv*
TCXO Temperature Controlled Crystal Oscillator *kf, oe*
TCZD Temperature-Compensated Zener Diode *fu*
TC_AIMS Transportation Coordinator's Automated Information for Movements System [NAVSUP] *mt*
TD Tank Destroyer *w9*
TD Tank Division [Soviet] *-12*
TD Tape Drive *hi*
TD Target Designation *hc*
TD Task Description *kf*
TD Technical Data *mt*
TD Technical Directive *nt, st*
TD Technical Director *fu*
TD Terminal Defense *st*
TD Terminal Departure *fu*
TD Test Data *jdr*
TD Test Director *fu, kf*
TD Threshold Detector *ggIII.1.68*
TD Total Dose *kf*
TD Training Deviceman *fu*
TD Transient Detect *fu*
TD Transmit Data *pk, kg*
TD Transmitter Distributor *mt, fu*
TD Tunnel Diode *wp*
TD/AE Target Detector/Azimuth Estimator *fu*
TD/MC Terminal Defense/Midcourse *fu*
TDA Table of Distribution and Authorization *mt*
TDA Tables of Distribution and Allowances *fu, mt*
TDA Tactical Decision Aid [AWS] *fu, mt*
TDA Tactical Display Area *fu*
TDA Technical Design Agency *fu*
TDA Technical Director Agent *fu*
TDA Telecommunications and Data Acquisition *re*
TDA Text Display Area *fu*
TDA Time Delay Amplifier *bl2-2.23*
TDA Time Delay Arrival *fu*
TDA Torque Drive Amplifier *kf*
TDA Tote Display Area *fu*
TDA Tracking and Data Acquisition [also T&DA] *wp*
TDA Transaction Data Area *fu*
TDA Tunnel Diode Amplifier *gs1-5.86, 4, wp, jdr, fu*

TDAC Technical Direction And Coordination *jdr*
TDAG Tracking Data Analysis Group *0*
TDAR Tactical Defense Alert Radar *wp*
TDAS TOW Data Acquisition System *hc*
TDAS Tracking and Data Acquisition Subsystem (Satellite) [NASA] *ac2.7-11, wl*
TDASS Tracking and Data Acquisition Satellite System *tb, wl*
TDB Technical Data Base *mt*
TDB Terminal Database *fu*
TDBOE Task Description/Basis Of Estimate *jdr*
TDC Tabular Display Console *fu*
TDC Tactical Data Center *kf*
TDC Target Designator Control *hc*
TDC Technical Development Center *mt*
TDC Test Data Center *17, fu*
TDC Test Director Console *fu*
TDC Time Delay Closing *fu*
TDC Time Dispersed Coding *fu*
TDC Time Distance Calculator [JCS] *mt*
TDC Torpedo Data Converter *fu*
TDC Track Data Coordinator *fu*
TDC TRIDENT Display Console *fu*
TDCA Tactical Deployment and Control Aircraft [USA] *mt*
TDCA Tactical Deployment Coordination Aircraft *mt*
TDCC Trident Digital Control Computer *hc*
TDCM Transistor Driver Core Memory *fu*
TDCP Transaction Data Communication Packet *fu*
TDCR Technical Data Contract Requirements *fu, wp*
TDCS Traffic Data Collection System *mt*
TDCSR Tactical Data Collection Summary Report *kf*
TDCT Technical Development Coordination Team *fu*
TDCTL Tunnel-Diode Charge Transformer Logic *fu*
TDD Target Detecting Device *hc, wp, fu*
TDD Target DGZ Designation *mt*
TDD Telecommunications Device for the Deaf *kg, hn43.15.4, wp*
TDD Test Definition Document *0*
TDD Testbed Description Document *fu*
TDD Top Down Design *fu*
TDD/SP Top Down Design and Structured Programming *mt*
TDDBMS Trusted Distributed Database Management System *fu*
TDDC Tactical Data Display Console *hc*
TDDL Time Division Data Line *fu*
TDDL Time Division Data Link *mt*
TDDS Tactical Data Display System *fu*
TDDS Tactical Data Dissemination System *kf*
TDDS Track Data Display System *hc*
TDDS TRAP Data Dissemination System *kf*
TDE Tactical Data Exploitation *kf*
TDECC Tactical Display and Engagement Control Console *fu*
TDEEM Theater Defense Effectiveness Evaluation Model *hs*
TDEP Tracking Data Editing Program *0*
TDF (business and television satellite, France) *sc3.2.7*
TDF Tactical Digital Facsimile *mt*
TDF Test Data File *kf*
TDF Test Development Folder *fu*
TDF Time Domain Filter *hc*
TDF Track Directional Filter *lt*
TDI Tactical Deception Indication *wp*
TDI Tank Miss Distance Indicator *hc*
TDI Target Data Inventory *mt, wp*
TDI Time Delay [and] Integration *gp, wp, tb, st, kf, oe*
TDI Top Down Implementation *fu*
TDI Topdown Design and Implementation *jdr*
TDI Transport Device Interface *pk, kg*
TDICL Technical Data Identification Check List *fu*
TDICMS Redesign of Technical Data / Configuration Management System *mt*
TDIMF Tactical Data Intercomputer Message Format *kf*
TDIN Time Dependent Interactive Network *fu*
TDIO Timing Data Input/Output *fu*
TDL Tapped Delay Line *fu*
TDL Technology Development Laboratory *wl*
TDL TEGAS Design Language *fu*
TDL Test Discrepancy Log *ep, jdr*
TDL Time Delay List *fu*
TDL Transducer Delay Line *fu*
TDL Transition Documentation Library *mt*
TDL Tunnel Diode Logic *fu*
TDLOA Training Device Letter Of Agreement *st*
TDLR Training Device Letter Requirement *st*
TDM Technical Document Management *pk, kg*
TDM Technology Development Manager *ah, pbl, jdr, jn*
TDM Technology Development Missions [also TDMX] *wl*
TDM Time Division Multiplexed *kf, uh, he, ct, em*
TDM Time Division Multiplexing (Multiplex) [also TDMX] *mt, kg, nu, fu, hc, bl1-5.80, 2, 4, 7, ce, jdr, vv*

TDMA Tape Direct Memory Access *hi*
TDMA Time Division Multiple(xing) Access *mt, fu, kg, ct, em, he, uh, hc, gs1-5.90, 17, mj31.5.33, 19, lt, wp, pm, ep, ce, st, jdr, hi, oe, tt*
TDMD Top Down Modular Design *fu*
TDMP Technology Development Master Plan *sb*
TDMP Technology Development Missions Polar [Canada] *wl*
TDMS Terminal Display Management System *pk, kg*
TDMS Time shared Data Management System *mt*
TDMX Technology Development Missions [also TDM] *wl*
TDMX Time Division Multiplexing [also TDM] *hi*
TDN Target Doppler Nullifier *fu*
TDO Tactical Deception Officer [TAF] *mt*
TDO Time Delay Opening *fu*
TDOA Time Difference Of Arrival *mt, fu*
TDOA Time Direction of Arrival *fu*
TDOP Total DCS Operations Plan [DCA] *mt*
TDP Tactical Data Processor *fu*
TDP Target Display Processor *fu*
TDP Technical Data Package *mt, wp, st, jdr, fu*
TDP Technical Development Plan *mt, fu*
TDP Technology Development Program *mt, wl*
TDP Telelocator Data Protocol *pk, kg*
TDP Threat Display and Projection *wp*
TDP Training Development Plan *mt*
TDPF Tabular Display Processing Function *fu*
TDR Technical Design Review *fu*
TDR Temperature Data Record *pbl*
TDR Terminal Defense Radar *tb*
TDR Test Discrepancy Report *4, oe*
TDR Time Dissemination Receiver *fu*
TDR Time Division Record *pl*
TDR Time Domain Reflectometer (Reflectometry) *fu, ct, em, kg*
TDR Tracking Data Relay *-4*
TDR Traffic Distribution Record *mt*
TDR Training Device Requirement *wp, st*
TDRE Tracking and Data Relay Experiment [also T&DRE] *2, 4*
TDRS Tactical Data Recording System *hc*
TDRS Telemetry Data Relay System *hc*
TDRS Tracking and Data Relay Satellite [NASA] *he, ge, mi, hn43.25.1, 2, aw129.14.18, is26.1.64, wp, no, ce, tb, wl, mb, hg, kf, vv, mt*
TDRSS Tracking and Data Relay Satellite System [NASA] *ge, mi, gg150, 2, 17, sb, af72.5.25, wp, ep, dm, hy, tb, wl, jdr, mb, hg, kf, vv, oe, mt, ai1, ai2*
TDRT Test Defect Reduction Team *fu*
TDS Tactical Data Station *wp*
TDS Tactical Data System [ESD] *mt*
TDS Tactical Display System *wp, tb, fu*
TDS Tape Data Selector *fu*
TDS Test Data Sheet *fu*
TDS Test Demonstration [Earth] Station *es55.43*
TDS Time Distribution Subsystem *fu*
TDS Track Data Simulator *fu*
TDS Transaction Driven System *mt*
TDS Transform Domain Signal Processor *fu*
TDS/CIC Tactical Data System/Combat Information Center *fu*
TDSC Tinker Data Services Center *mt*
TDSM Time Division Switching Matrix *mt*
TDSO Theater Data System Operators *kf*
TDSP Top Down Structured Programming *mt, fu*
TDSQ Training Development Squadron *mt*
TDSR Transmitter Data Service Request *pk, kg*
TDST Track Data Storage *fu*
TDT Target Designation Transmitter *fu*
TDT Time Dependent Target *st*
TDTC Test, Development, and Training Center *mt*
TDTL Tunnel-Diode Transistor Logic *fu*
TDtoDP Tablet Coordinates to Display Coordinates [converting] *pk, kg*
TDTS Technical Data and Training Systems *gp*
TDU Tabular Display Unit *fu*
TDU Target Detection Unit *fu*
TDU Threat Display Unit *mt*
TDV Tab Delimited Values *hi*
TDX Thermal Demand Transmitter *fu*
TDY Temporary Duty [USA] *mt, w9, af72.5.14, wp, tb, kf*
TDZ Touchdown Zone (Lights) *fu*
TE Tangent Elevation *fu*
TE Task Element *mt*
TE Technical Engineer *jdr*
TE Technical Evaluation *mt*
TE Terminal Electronics *fu*
TE Terminal Emulator *hi*
TE Terminal Equipment *mt, fu*
TE Test Engineer(ing) *fu*
TE Test Equipment *fu*
TE Text Editor *hi*
TE Thermal Element *fu*

TE Thermo Electric *fu, wp*
TE Trailing Edge *fu, oe*
TE Training Effectiveness *fu*
TE Training Equipment *fu*
TE Transient Event *kf*
TE Transverse Electric [mode] *mt, fu, sm10-2, mj31.5.60*
TE/2 Terminal Emulator/2 [Oberon] *kg, hi*
TE/R Trailing Edge Radius *fu*
TE/WA Threat Evaluation Weapon Assignment *fu*
TEA Tetra Ethyl Aluminum {?}:(wp)
TEA Tetra Ethyl Ammonium [ion] *wp*
TEA Torque Equilibrium Attitude *mi, mb, wl*
TEA Transverse Electric Discharge {?}:(sd)
TEA Transversely Excited Atmospheric [pressure laser] *wp, hy*
TEAC Tetra Ethyl Ammonium Chloride *wp*
TEAC {a manufacturer of video recorders} *aw129.10.46*
TEALE_AMBER [MICOM sensor program] *hc*
TEAL_AMBER [Malabar, Florida, surveillance system] *ps204*
TEAL_BLUE [DARPA program to compensate optical distortion due to air effects, at MOTIF] *ps204, wp*
TEAL_CAMEO [unmanned U-2 follow-on] *wp*
TEAL_DAWN [DARPA program to develop stealthy missile] *wp*
TEAL_JADE [USAF early warning study] *wp*
TEAL_ONYX [DARPA program for long range infrared technology] *wp*
TEAL_RUBY [program to develop a satellite to detect aircraft from space by infrared measurements, USA] *dh, hc, wp, mt*
TEAL_TORCH [DARPA program] *hc*
TEAM Technical-Engineer-Architect- Management (organization) *fu*
TEAM Terminal Expandable Added Memory *hi*
TEAMS Tactical EA6B Mission Planning System *mt*
TEAMS Telecommunications Equipment Accounting and Management System *cd*
TEAMS Test Evaluation And Monitoring System *hc*
TEAM_MATE [Comm Intercept System] [TRQ32V] [USA] *mt*
TEAM_PACK [Noncomm Intercept System] [MSQ103] [USA] *mt*
TEAM_SPIRIT [Exercise] [USA] *mt*
TEAS Threat Expert Analysis System *hc, wp*
TEASE Target Enter Angle Simulation Equipment *hc*
TEASE Tracking Errors and Simulation Evaluation *fu*
TEB Tactical Exploitation Battalion [USA] *mt*
TEB Thread Environment Block *pk, kg*
TEB Tone Encoded Burst *fu*
TEBS Target Echo Bistatic Subsystem *hc*
TEC Monitor Terminal Equipment Configuration *mt*
TEC Tactical ELINT Correlator *fu*
TEC Target Equivalent Carriers *kf*
TEC Technical Education Center (Corporate) *fu*
TEC Technical Extension Courses *hc*
TEC Temporary Engineering Change *fu*
TEC Test Equipment Center *jdr*
TEC Tokyo Electronics Corporation *pk, kg*
TEC Total [ionospheric] Electron Content *sd, pr, pbl, kf*
TEC Training Extension Course *hc, fu*
TECA Threat Evaluation Countermeasures Agent *wp*
TECDOC Technical Documentation [USA DOD] *mt*
TECEP Training Effectiveness and Cost Effectiveness Prediction *fu*
TECH Thermo-Electro Chemical *hc*
TECHBASE [6.1 + 6.2 projects and programs] *wp*
TECHEVAL Technical Evaluation *fu*
TECHMOD (Manufacturing) Technology Modernization *fu*
TECHNINT Technological Intelligence *wp*
TECH_MOD Technology Modernization *tb*
TECO Third Engine Cutoff *tb*
TECOM Test and Evaluation Command [US Army] *mt, fu, tc*
TECR Tactical Embedded Computer Resource *fu*
TECS Treasury Enforcement Communications System *mt*
TED Testability and Embedded Diagnostic *mt*
TED Threat Environment Description *kf, wp, tb*
TED Threshold Extension Demodulators *hc*
TED Threshold Extension Device *-7*
TED Total [particle] Energy Detector [on TIROS-N] *pl, noa*
TED Transfer(red) Electron Device *fu, hc, jdr*
TED Trunk Encryption Device *mt*
TEDA Triethylene Diamine *wp*
TEDREP Type Unit Equipment Detail Report [USA] *mt*
TEDS Tactical Electronic Decoy System (drone) *fu*
TEDS Tactical Expendable Drone System [USA] *wp, mt*
TEDSS Tactical Environmental Data Support System [TAC] *mt*
TEEAR Test Equipment Error Analysis Report *fu*
TEF Technical Evaluation Facility *ts14-1, tb*
TEF Transverse Electric Field *wp*
TEFF (spin to transverse inertia ratio, IZZ/TEFF) *gg13*
TEFSA Territorial Forces Effectiveness File *wp*
TEG Technical Evaluation Group *st*
TEG Test and Evaluation Group [USA] *mt*
TEGAS Test Generation and Simulation [System] *fu, at*

TEGATE (a design capture system . . . file of all circuits drawn up) *hm*
TEI Terminal Endpoint Identifier *pk, kg*
TEI Test Engineering Instruction *fu*
TEI Third Engine Ignition *tb*
TEI Trans Earth Injection *re*
TEKES [Finnish Technology Agency] *ge, hg*
TEL Tetra Ethyl Lead *wp*
TEL Training Equipment List *fu*
TEL Transporter Erector Launcher *mt, fu, 14, hc, wp, tb, st, kf*
TELAN USAF Intelligence and Communications Facility; Cryptology Studies *wp*
TELCOM Telecommunications Configuration System, WWMCCS Site *mt*
TELCOM Telemetry Communications *fu*
TELEDAC Telemetric Data Converter *fu*
TELEFAX Telecommunications Facsimile *mt*
TELEMAN Telephone Management System [USAF] *mt*
TELENET Telecommunications Network Protocol *mt*
TELENET Teletypewriter Network *mt*
TELER Telecommunication Requirements *st, mt*
TELERAN Television and Radar Navigation *fu*
TELESCOM Telemetry Surveillance Communications *fu*
TELESPO Computer Operator Console Training *fu*
TELEX Teletypewriter (Teleprinter) Exchange [automatic service] *mt, do80, wp, fu, hi*
TELINT Telemetry Intelligence [USA DOD] *10, mt*
TELMOD Telemetry Modeling *hc*
TELNET Telecommunications Network *mt, fu*
TELNET Terminal Emulation protocol Network *hi*
TELOP TELeconference OPerator *em, ct*
TELOPS Telemetry Online Processing System *ci.126, 4*
TELPAK Telecommunications Package *mt*
TELSTAR Trans-Atlantic [Television] Relay [AT&T satellite for] *gs2-1.3*
TEL[AR] Transporter Erector Launcher [And Radar] *mt*
TEM (photo of INTELSAT transmit TEM line) *bl2-3.47*
TEM Target Engagement Model *fu*
TEM Technical Exchange Meeting *jdr, kf*
TEM Teledyne Electro-Mechanisms, Incorporated *hq5.3.5*
TEM Terminal Electronics Module *fu*
TEM Test Equipment Manufacturing *fu*
TEM Transmission Electron Microscope (Microscopy) *jdr, fu*
TEM Transverse Electromagnetic [mode] *wp, fu*
TEML Turbo Editor Macro Language [Borland] *kg, hi*
TEMP Test [and] Evaluation Master Plan *mt, fu, ph, hc, jmj, wp, tb, st, kf*
TEMPEST Transient Electromagnetic {Pulse} Emanations Standard [a technique to ensure that electronics is free of signal radiation, or a program for control of compromising emanations] *fu, aw130.6.6, wp, kf, ct, em*
TEMPLAR Tactical Expert Mission Planner [model] [RADC] *mt, ro, hc, wp*
TEMS Technical Engineering and Management Support *mt*
TEMS Turbine Engine Monitoring System *mt*
TEN-ONE [classified air defense project] *fu*
TENCAP Tactical Exploitation of National [space programs'] Capabilities Program:(mt, fu, lka, tb, wp, kf)
TENIS Terminal Environment Network Information System *mt*
TENS Transcutaneous Electrical Nerve Stimulation *wp*
TEOA Test and Evaluation Objective Annex *mt*
TEOC Technical Objective Camera *wp*
TEOLPS Telemetry On Line Processing System *hc*
TEOT Target End Of Track *st*
TEOTI Target End Of Track Initiate *st*
TEP Tactical Earth Penetrator *wp*
TEP Tactical ELINT Processor *mt, wp*
TEP Technical Evaluation Program *mt*
TEP TEMPEST Endorsement Program *wp*
TEP Terminal Emulation Processor *mt, wp*
TEP Terrestrial Entry Point *kf*
TEP Tetraethyl Phosphate [chemical warfare simulant] *wp*
TEP Thermal Exchange Pump [Honeywell] *mt*
TEP Traffic Engineering Program *mt*
TEPI Training Equipment Planning Information *mt, fu*
TEPIAC Thermophysical/Electronic Properties Information Analysis Center *hc*
TEPP Tetraethyl Pyrophosphate *wp*
TER Terminal Engagement Radar *hc*
TER Thermal Eclipse Reading [Sony] *pk, kg*
TER Training Equipment Report *fu*
TER Triple Ejection Rack *mt*
TERA Terminal Effects Research and Analysis *st*
TERAG Transverse Electric Receive Antenna Group *tb*
TERB Time En Route Bulletin *wp*
TERCOM Terrain Contour Matching [navigation by comparison of the terrain with a stored map] *wp, hc, fu, mt*
TEREC Tactical Electronic Reconnaissance [USAF] *mt, fu, af72.5.138, wp*

TEREC_GP TEREC Ground Processor *mt*
TERF Terrain Following flight *mt*
TERLS Thumba Equatorial Rocket Launch Station [India] *ac3.2-1*
TERMPWR Terminator Power *kg, hi*
TERMS Terminal Management System [MTMC] *mt*
TERMS Test Equipment Reporting Management System *mt*
TERPE Tactical Electronic Reconnaissance Processing and Evaluation *fu, hc*
TERPES TAC* Electronic Reconnaissance Processing Evaluation System *wp, mt*
TERPROM Terrain Profile Matching *wp*
TERPS Terminal Instruments Procedures *mt, aw129.14.115*
TERRAIN Terrain Analysis File *mt*
TERS Tactical Electronic Reconnaissance System *wp*
TERS Tactical Event Reporting System *mt, kf*
TES Tactical Environment Simulator *mt*
TES Tactical Event System *kf*
TES Telephony Earth Station *jdr*
TES Test and Evaluation Squadron [USA] *mt*
TES Thermal Emission Spectrometer [on Mars Observer] *mi, re, mb*
TES Tropospheric Emission Spectrometer [EOS] *hy, pbl*
TESE Tactical Exercise Simulator and Evaluator (USMC) *fu*
TESLA Technical ELINT Support Location and Analysis *wp*
TESM Tactical Electronic Support Measures *wp*
TESS Tactical Electronic Surveillance System *lka*
TESS Tactical Environment(al) Support System *mt, fu, ag, hc*
TESS The Extended Simulation System [TAC*] *fu*
TEST Thesaurus of Engineering and Scientific Terms *fu*
TESWG TRAMCON Engineering Subworking Group *mt*
TET Technical Evaluation Test *mt*
TET Text Enhancement Technology *hi*
TET Trailing Edge Tracking *wp*
TET Turbine Entry Temperature *mt*
TETAP Test and Evaluation Tasking Plan *mt*
TEU Telemetry Encoder Unit *cl, ah, uh, jdr, jn*
TEU Tracker Electronics Unit *hc*
TEU Transducer Excitation Unit *hc*
TEV Tera Electron Volt *wp, eo*
TEW Tactical Electronic Warfare *wp*
TEWA Threat Evaluation and Weapon Assignment *pm, fu*
TEWES Tactical Electronic Warfare Environment Simulator [US Navy] *fu, mt*
TEWG Test and Evaluation Working Group *fu, wp*
TEWS Tactical Electronic Warfare System *mt, fu, hc, af72.5.135, ts17-3, wp*
TEX Text Executive Processor *mt*
TEXOUT TEGAS Extended Output Utility *at*
TEXSIM TEGAS Extended Simulator *at*
TEXTA Technical Extracts of Traffic *-10*
TEXUS Technologische Experimente Unter Schwerelosigkeit *crl*
TEZ Total Exclusion Zone *mt*
TF Task Force [USA] *12, w9, wp, mt*
TF TeleFax *tt*
TF Terrain Following *mt, hc*
TF Territorial Force *w9*
TF Thin Film *fu*
TF Total Float *jdr*
TF [fixed torques] *vv*
TFAT Torpedo Factory Acceptance Testing *fu*
TFC Tactical Fusion Center *mt*
TFCA Total Force Capability Assessment *mt*
TFCC Tactical Flag Command Center [US Navy] *fu, hc, tb, mt*
TFCC Tactical Fleet Control *mt*
TFCC/FDDS Tactical Flag Command Center / Flag Data Display System *mt*
TFCS Tank Fire Control System *hc*
TFCS Treasury Financial Communications System *mt*
TFD Tactical Fusion Division [USAF] *mt*
TFD Thin Film Deposition *wp*
TFDB TRW Fee Determination Board *jdr*
TFDD Text File Device Driver *pk, kg*
TFDP Tower Flight Data Processing *fu*
TFE Tetra Fluoro Ethylene *wp*
TFE Transportation Feasibility Estimator [JOPS] *mt*
TFE Turbo Fan Engine *aw130.13.27, af72.5.105*
TFEL Thin Film Electro Luminescent *kg, hc, hi*
TFF Tactical Fighter Force *hc*
TFF Transportable Field Facilities *fu*
TFFC {antisubmarine warfare equipment on board aircraft carrier} *pf1.4*
TFG Tactical Fighter Group [USAF] [USA] *mt, 12, af72.5.55, wp*
TFG Transmit Format Generator *fu*
TFG Transportation Facility Guide [MTMC] *mt*
TFH Thick Film Hybrid *fu*
TFIV Task Force IV *mt*
TFL Transponder Frequency Level *0*
TFMCC Traffic Flow Management Computer Center *fu*

TFMRDBS Total Force Manpower Requirements Data Base *mt*
TFP Time and Frequency Processing *mt*
TFPA Thermal Focal Plane Arrays *hc*
TFR Terrain Following Radar *mt, fu, hc, aw129.10.51, wp*
TFR Test Failure Report *oe*
TFR Training Facilities Report *fu*
TFR Trouble Failure Report *bf, oe*
TFR [Swedish air transport association] *aw130.22.77*
TFRD Test Facility Requirements Document *tb*
TFRG Time and Frequency Reference Group *17, jdr*
TFRS Time and Frequency Reference Subsystem *jdr*
TFS Tactical Fighter Squadron *af72.5.55, hn50.8.3*
TFS Tactical Fighter Squadron (136th TFS Rockys Raiders, 308th TFS Emerald Knights, 481th TFS Crusaders) [USA] *mt*
TFS Time Frequency Standard *fu*
TFS Top of Free Space *fu*
TFS Traffic Flow Security *mt*
TFS Trainer Flight Simulator *hc*
TFSUS Task Force on Scientific Used of the Space Station *wl*
TFT Thin Film Transistor [screens or displays] *fu, ct, em, kg, hi*
TFTAM Thin Film Transistor Active Matrix *hi*
TFTCD Thin Film Transistor Color Display *hi*
TFTD Thin Film Transistor Display *hi*
TFTG Tactical Fighter Training Group *mt*
TFTP Trivial File Transfer Protocol *ct, kg, em, hi*
TFTS Tactical Fighter Training Squadron [USA] *mt*
TFTS TOW Field Test Set *hc*
TFTU Time/Frequency Transfer Unit *fu*
TFTW Tactical Fighter Training Wing [USA] *12, mt*
TFU Tactical Forecast Unit *ag*
TFU Transmit Filter Unit *jdr*
TFW Tactical Fighter Wing (366th TFW 'Gunslingers' , 8th TFW Wolf Pack) [USA] *mt, af72.5.55, wp*
TFWC Tactical Fighter Weapons Center [NELLIS AFB] *mt, 12, hc, af72.5.94, wp*
TFX Tactical Fighter Experimental [became the F-111, USA] *mt, nd74.445.46*
TG Task Group *mt*
TG Terminal Guidance *fu*
TG Test Group [USA] *mt*
TG Torpedo Group *fu*
TG Tracking Group *fu*
TG Transaction Group *fu*
TG Transmission Group *mt*
TGA Telescope Gimbal Assembly *ie5.11.7*
TGA Thermogravimetric Analysis *wp*
TGAS Tacan Guidance Augmentation System *wp*
TGB Trunk Group Busy *fu*
TGC Transmit Gain Control *fu*
TGCR Tactical Generic Cable Replacement *mt*
TGD Target Generator Data *fu*
TGD Technical Guidance Directions *0*
TGD Thickened Gd [chemical warfare] *wp*
TGDS Tactical Graphics Display Study *fu*
TGDT Tactical Graphics Display Terminal *fu*
TGG Tactical Graphics Generator *fu*
TGG Third Generation Gyro *hc*
TGID Trunk Group Identification Number *ct, em*
TGIF Tactical Ground Intercept Facility *mt*
TGIF Transportable Ground Control Intercept Facility *wp*
TGIF Transportable Ground Intercept Facility *mt*
TGIF II Tactical Ground Intercept Facility [ESC*] *mt*
TGL Test Group Leader *fu*
TGLIP TOW Guidance Loop Improvement Program *bf*
TGM (equipment on the TGM Maverick missile) *af72.5.13*
TGM Trunk Group Multiplexer *mt*
TGP Trajectory Generation Program *fu*
TGPSG Tactical Global Positioning System Guidance *hc*
TGS Tactical Graphics System *fu*
TGS TDDS Steering Group *kf*
TGS Terminal Guidance Sensor *hc*
TGS Tri-Glycine Sulfate *ac2.8-7*
TGSE Telemetry Ground Support Equipment *fu*
TGSG Target [Launch Vehicle] Ground Support [working] Group *-17*
TGSM Terminally Guided Submunition *hc, wp, mt*
TGSS Tactical Ground Sensor System *hc*
TGV Train ‡ Grande Vitesse [(French) high speed train] *is26.1.63*
TGW Terminally Guided Weapon (Warhead) *mt, uc, hc, wp*
TH Track History *fu*
TH Transportation and Handling *mt*
THAAD Theater High Altitude Air (Area) Defense [USA] *kf, mt*
THADGE Thailand Air Defense Ground Equipment *hc*
THAIS Type Commanders, Headquarters Automated Information System [US Navy] *mt*
THAP Tactical High Altitude Penetrator [alleged Northrop technology demonstrator for the alleged stealth reconnaissance aircraft TR-3, 1981] [USA] *mt*

THC Target Homing Correlator *wp*
THC Tetrahydrocannabinol *wp*
THC Thermal Converter *fu*
THC Total HydroCarbon *wp, oe*
THD Thickened Sulfur Mustard [chemical warfare] *wp*
THD Total Harmonic Distortion *ct, em, fu, kg, mj31.5.19, hi*
THEP TOGA Heat Exchange Program *hy*
THIR Temperature Humidity Infrared Radiometer *hc, ac2.8-3*
THK Turk Hava Kuvvetleri [Turkish Air Force] *mt*
THM Tricode Hexaphase Modulation *rih*
THMT Tactical High Mobility Terminal *mt*
THOR Tandy High Performance Optical Recording *pk, kg*
THOR Tiered Hierarchy Overlayed Research *wp*
THOR Transistorized High Speed Operations Recorder *fu*
THOR [long-range ship to air missile] *uc*
THORNTON [NSA program] *hc*
THP Terminal to Host Protocol *mt*
THPC Tetrakis(Hydroxymethyl)Phosphonium Chloride *wp*
THR Transmit Holding Register *pk, kg*
THREATCON Threat Conditions, terrorist [existing in various cases denominated ALPHA, BRAVO, CHARLIE, DELTA] [USA DOD] *mt*
THS Terminal Homing Sonar *fu*
THS Test Host Substitute *mt*
THUMPER [Avco high energy Laser] *wp*
THURSIS The Human Roles in Space [NASA] *mt*
THV Threshold Limit Value *oe*
TI Tape Inverter *fu*
TI Target Identification *fu*
TI Technical Instruction *fu*
TI Technical Intelligence *kf*
TI Technical Interchange *mt, fu*
TI Technology Insertion *fu*
TI Technology Integration *tb*
TI Texas Instruments [a corporation] *kg, af72.5.86, aw130.11.27, hc, hi*
TI Thermal Imager *mt*
TI Time Index *fu*
TI Time to Intercept *tb*
TI Track Initiation *tb*
TI Track Initiator *fu*
TI Tracker-Imager *fu*
TI Transmission Identification *fu*
TI Trials Installation *mt*
TI/TD Technical Interchange/Technical Direction [meeting] *17, sb*
TIA Telecommunications Industry Association *kg, hi*
TIA The Internet Adapter *ct, em, kg*
TIA Trans Impedance Amplifier *pl*
TIA Type Inspection Authorization *aw118.18.23*
TIAC Technical Intelligence Analysis Center *kf*
TIALD Thermal Imaging Airborne Laser Designator *wp*
TIALD Thermal Imaging and Laser Designation [UK] *mt*
TIAP Theater Intelligence Architecture Program [DIA/AFIN] *wp, mt*
TIARA Tactical Intelligence And Related Activities [USA DOD] *uc, wp, kf, mt*
TIARRA Target Identification And Recognition Radar *wp*
TIAS Target Identification and Acquisition System [fitted to some A-6B Intruders, USA] *wp, mt*
TIAS Target Information and Acquisition System *hc*
TIB Target Intelligence Branch *mt*
TIB Technical Information Base *fu*
TIB Temporary Import Bond *fu*
TIBS Tactical Information Broadcast System *mt, kf*
TIC Tactical Information Coordinator *fu*
TIC Target Integration Center *hc*
TIC Target Intercept Computer *wp*
TIC Technical Information Center *tb*
TIC Technical Intelligence Center *kf*
TIC Technical Interface Committee *cu*
TIC Technical Interface Concept *mt, hc*
TIC Technical Internal Correspondence [a document] *ah, hf, ep, jdr, pbl, tm, jn*
TIC Telecommunications Information Center *mt*
TIC Television Input Converter *fu*
TIC Temperature Indicating Controller *fu*
TIC Theater Intelligence Center *mt*
TIC Token Ring Interface Coupler *ct, em*
TIC Transaction Identification Code *mt*
TIC Trial Intercept Calculation *fu*
TIC Troops In Contact [USA] *mt*
TICARRS Tactical Interim CMS and REMIS Reporting System *mt*
TICD Thermal Interface Control Drawing *pl*
TICM Thermal Imaging Common Module *wp*
TICOD Time-Compressed Display *fu*
TICODS Time-Compressed Display System [Navy] *fu*
TICP Tactical Information Correlation and Presentation *wp*
TICTAC Time Compression Tactical Communications *fu*
TID Tactical Information Display *hc*

TID Target Identification *pk, kg*
TID Target Identification Device *hc*
TID Technical Implementation Directive *mt, wp*
TID [tools for integrated diagnostics] *wp*
TIDAD Target Identification for Air Defense *hc*
TIDAR Time Delay Array Radar [techniques study] *hc*
TIDCS Transitional Integrated DCS *mt*
TIDL Test Instrumentation Data Link *fu*
TIDP Technical Interface Design Plan [TACS/TADS] *fu, mt*
TIDP Technical Interface Development Plan *fu*
TIDPTE Technical Interface Design Plan Test Edition *mt*
TIDS Tactical Information Distribution System [Army] *fu*
TIDS Terminal Information Display System *fu*
TIDY Teletype Integrated Display System *mt*
TIE Technical Integration and Evaluation *ci.942*
TIE Technology Integration (Integrated) Experiment *tb, st*
TIEDS Tactical Imagery Exploitation Demonstrator System *hc*
TIEN_WANG [a Taiwan program] *hc*
TIES Tactical Information Exchange System *hc, fu*
TIES Tactical Integrated Electronic System *hc*
TIES Time Independent Escape Sequence *pk, kg*
TIES Torpedo Instrumentation and Exercise System *fu*
TIES Transmission and Information Exchange System *wp*
TIF Tactical Interface Facility *mt*
TIF Technical Information File *wp*
TIF Telephone Interference Factor *fu*
TIF Test and Integration Facility *fu*
TIF Torpedo Integration Facility *fu*
TIF Trusted Interface Function *fu*
TIFF Tagged Image File Format *ct, em, jr, kg, hi*
TIFS Total In-Flight Simulator *hc*
TIG Tactical Intelligence Group *mt*
TIG Target Interface [working] Group *-17*
TIG Track Initiation and Grouping *mt*
TIG Transmission Identification Generator *mt*
TIG Tungsten Inert Gas *fu, gp, lt*
TIGER Thermal Infrared Ground Emission Radiometer [EOS] *hy*
TIGER Topologically Integrated Geographic Encoding and Referencing *pk, kg*
TIGHAR {?} International Group for Historic Aircraft Recovery *aa26.8.b5*
TII Technology Independent Interface *ct, em*
TIIAP Telecommunications and Information Infrastructure Assistance Program [NII] *pk, kg*
TIIC Tri-Service Industry Information Centers *fu*
TIICG Tactical Information Integration Control Group *fu*
TIIF Tactical Imagery Interpretation Facility *fu*
TIL Technology Insertion Laboratory *tb*
TIL Telecommunications Investigation Laboratory *0*
TIL Transportation Information Library *mt*
TILO Technical and Industrial Liaison Office *mt*
TILO Technical Industrial Liaison Office *hc, wp*
TILS TMDE Inventory List System *mt*
TIM TACTS Interface Module *hc*
TIM Target Identification Markers *mt*
TIM Theater Information Manager *jdr*
TIM Track Imitation *fu*
TIM Tracking Information Memorandum *0*
TIM Tracking Instruction Manual *0*
TIM Tracking Instrumentation Manual *0*
TIM Transmission Interface Module *fu*
TIMATION [Early name for US Navy concept for GPS] *N75-27202-465*
TIME Test and Integration Methodology Engineering *jdr*
TIMF TIBS Intercomputer Message Format *kf*
TIMI Technology Independent Machine Interface [IBM] *pk, kg*
TIMIT Time Interval *mt*
TIMP Test and Integration Master Plan *jdr, jn*
TIMS Tactical Information Management System *fu*
TIMS Technical Information Management System [USA] *mt*
TIMS Theater Information Management System *mt*
TIMS Thermal Image Management System *hn48.19.3, hd1.89.y5*
TIMS Thermal Infrared Multispectral Scanner *ma40, hy*
TIMS Thermal Ionization Mass Spectrometry *wp*
TIMS Trident Integrating Multiplexing Subsystem *hc*
TINA Telecommunication Information Networking Architecture *pk, kg*
TINS Thermal Imaging Navigation Set *mt*
TINS Thermal Imaging Navigation System *aa26.12.27, hc, hq6.2.7, wp*
TINTS Turret Integrated Night Thermal Sight *hc*
TIOP Tactical Interface Operational Procedures *mt*
TIOS Tactical Input / Output Simulator *fu*
TIP Target Impact Prediction *mt*
TIP Target Intelligence Package *wp*
TIP Technical Improvement Program *hc*
TIP Technical Investigation Program *hc*
TIP Technology Integration Program *wp*
TIP Technology Investment Plan *wp*

TIP Telecommunications Implementation Plan *mt*
TIP Terminal IMP *ct, em*
TIP Terminal Interface Processor [Telenet] *mt, fu, kg, do134*
TIP Theatre Injection Point *rih*
TIP Thermal Image Processor *hc*
TIP Tower Interface Processor *fu*
TIP Track Initiation and Prediction *fu*
TIP Tracking and Impact Prediction *mt, jb*
TIP Transit Improvement Program [Navy Satellite] *cca*
TIP Transition and Integration Plan [DSCS] *mt*
TIPA Triisopropanolamine *wp*
TIPI Tactical Image Processing Interpretation *mt*
TIPI Tactical Information Processing and Interpretation *mt, ro, fu, hc, wp*
TIPI-IIS Tactical Information Processing and Interpretation - Imagery Interpretation Segment *mt*
TIPIL Turn In Point Item List *mt*
TIPS Tactical Information Processing System *wp*
TIPS Technical Information Presentation Systems *hc*
TIPS Telemetry Integrated Processing System *hc*
TIPS Terminal Information Processing System *hc*
TIPS Terrain Intelligence Processing System *mt*
TIPS Thermal Image Periscope for Submarines *hc*
TIPS Transportation Information Processing System [MAC*] *mt*
TIR Tactical Intelligence, Reconnaissance *wp*
TIR Target Illuminating (Illumination) Radar *fu, mt*
TIR Terminal Imaging Radar *fu, 17, sb, hc, wp, tb, st*
TIR Test Incidence Report *oe*
TIR Thermal Imaging Radar *mt*
TIR Thermal Infrared *hy*
TIR Total Indicator Reading *fu*
TIR Total Internal Reflection *jhm*
TIR Tracking and Illumination Radar *mt*
TIR Transportation Item Reporters *mt*
TIRAS Task Identification and Resource Allocation System *mt*
TIRAS Technical Information Retrieval and Analysis System *wp*
TIRE Tank Infrared Elbow *hc*
TIREC Tactical Infrared Reconnaissance *hc*
TIROC Tactical Intrusion Resistant Optical Fiber Communications *hc*
TIROS Television Infrared Observation Satellites [NOAA] *mt*
TIROS Television [and] Infrared Observation Satellite [USA] *ge, mi, gd4-6, hc, wp, no, hy, mb, hg*
TIROS Television [and] Infrared Operational Satellite [USA] *oe*
TIRS Tactical Information Retrieval System *wp*
TIRS Target Infrared Simulator *hc*
TIRS Thermal Infrared Scanner *wp*
TIRSAG Tactical Intelligence, Reconnaissance and Surveillance Advisory Group *wp*
TIRSS Theater Intelligence, Reconnaissance and Surveillance Study *wp*
TIS TAC Information System *mt, wp*
TIS TACOPS Information System [TAC] *mt*
TIS Tactical Intelligence Squadron [USAF] *mt*
TIS Tactical Intelligence Support *mt*
TIS Target Information System *wp*
TIS Technical Information Support *tb*
TIS Technical Information System *fu*
TIS Technical Interface Specification *fu*
TIS Test Information Sheet *-17*
TIS Test Information Sheet (System) *fu*
TIS Thermal Imaging Sight *hc*
TIS Thermal Imaging [Sub] System *fu, 8, hc, wp*
TIS Tyazhely Istrebitel Soprovozdeniya [heavy escort fighter] [USSR] *mt*
TIS [International Atomic Time] *jdr, uh*
TISA TAC Information Systems Architecture *mt*
TISCA Technical Information Storage and Control Application *mt*
TISD Tactical Information Systems Division [TAC] *mt*
TISEO Target Identification System, Electro-Optical *hc, af72.5.134, wp*
TISL Target Indicator System Laser [the PAVE PENNY equipment, USA] *wp, mt*
TISS Tester [for] Intermediate Support System *wp*
TISS TEWS Intermediate Support System [AFSC*] *wp, mt*
TISS Transportation Information Subsystem *mt*
TISS Troop Issue Subsistence System *mt*
TISSS Tester Independent Support Software System *mt, ro, hc, wp*
TISURS Trainer Inventory Status Utilization Reporting System *mt*
TIT Theater Indoctrination Training [familiarization flights during the first combat operations] [USA] *mt*
TIT Thermal Imaging Technology *hc*
TIT Turbine Inlet Temperature *wp*
TITAN (TITAN Systems, Inc. . . . Huntsville, AL 35805) *jmj*
TITAS Thermal Imaging Tracking Aid System *wp*
TITE Technology and Innovations in Training and Education *nd74.445.59*
TITS Thermal Imaging Test Set *hc*
TIU Tape Identification Unit *fu*
TIU Target Insertion Unit *fu*
TIU TIBS Interface Unit *kf*

TIU Torpedo Interface Unit *fu*
TIU Trainer Interface Unit *fu*
TIU Transceiver Interface Unit *fu*
TIU Transec Interface Unit *jdr*
TIU Trusted Interface Unit *mt, fu*
TIWADS TIP Interface With Automated Data System *mt*
TIWG Test [and] Integration (Integrated) Working Group *fu, nt, st*
TJ Thermal Junction *oe*
TJC Trajectory Chart *0*
TJD Trajectory Diagram *0*
TJS Tactical Jamming System *wp*
TK Time of KELV [launch] *-17*
TK/TK Track to Track *kg, hi*
TKA Torpedniy Kater [torpedo cutter] [USSR] *mt*
TKB Tactical Keyboard *fu*
TKG Terminal Key Generators *mt*
TL Tetrahedron Lights *fu*
TL Threat Level *fu*
TL Transit Level [NATO] *mt*
TL Transmission Level *mt*
TLA Thinline Array *fu*
TLA Time Line Analysis *nt, jdr*
TLAHS Thinline Array Handling System *fu*
TLAM Tactical Land Attack Missile *dh*
TLAM TOMAHAWK Land Attack Missile [USA] *mt, hc, fu*
TLAM-C Tactical Land Attack Missile-Conventional *uc*
TLB Translation Look aside Buffer *kg, ct, em*
TLC Technology Leadership Council *pbl, jdr*
TLC Telecommunications Line Center *mt*
TLC Telecommunications Live Controller *mt*
TLC Thin Layer Chromatography *w9, wp*
TLCC Theater Logistics Coordination Center *mt*
TLCC Thinline Connectivity Capability *fu*
TLCD Tactical Low Cost Drone *wp*
TLCSC Top Level Computer Software Component *mt, fu, ep, tb, jdr*
TLD Technical Logistics Data *fu*
TLD Thermal Luminescence Detector *jdr*
TLD Top Level Design *fu*
TLD Top Level Domain *ct, em*
TLDD Top-Level Design Document *fu*
TLeLv Tiedustelu Lento Laivue [USSR] *mt*
TLF (housing . . . 182 transient (81 VOQ, 60 VAQ, 41 TLF) *af72.5.152*
TLGC Transfer Line Gas Chromatography *wp*
TLI Trans Lunar Insertion *-13*
TLI Transport Layer Interface *pk, kg*
TLIC Transaction Local Interface Controller *fu*
TLIP TOW Lethality Improvement Program *hc*
TLM Telemetry *uh, 4, 9, 17, bo, wl, jdr, kf, vv, jn, oe*
TLO Tool Loan Order *fu*
TLO Transfer to Low Orbit *re*
TLP Tactical Leadership Program. [NATO] *mt*
TLP Test Level Point *mt*
TLP Transmission Level Point *mt, 4, 7*
TLR Telecommunications Leasing Request *mt*
TLREOS Tactical Long Range Electro Optical System *hc*
TLRN Technical Logistics Reference Network *mt*
TLRS Transportable Laser Ranging System *ac2.6-7*
TLS Tactical Landing System *fu, wp*
TLS Telemetry Listing Submodule *fu*
TLS Terminal Loading Simulation *fu*
TLS Top Level Specification *tb*
TLSS Tactical Life Support System *wp*
TLSS Technical Library Services Section *fu*
TLT Terminate Loop Test *fu*
TLT Test Loop Translator *jn*
TLTA Thinline Towed Array *fu*
TLU Time of Last Update *0*
TLV Target Launch Vehicle *sb*
TLV Threshold Limit Valve *wp, fu, oe*
TM Tactical Memoranda *fu*
TM Tactical Missile *fu*
TM Task Memory *mt*
TM Tasking Memorandum *mt*
TM Technical Manual *mt, fu, ac9.2-10, hc, nt, tb, kf*
TM Technical Memo *oe*
TM Telemetry transmitter [also TXTR] *sb*
TM Test Modification *fu*
TM Test Module *oe*
TM Text Monitor *fu*
TM Theater Missile *kf*
TM Thematic Mapper [LANDSAT] *ge, hn43.21.4, ep, hy, hg*
TM Time and Material *jdr*
TM Tone Modulation *fu*
TM Track Monitor *fu*
TM Transaction Monitor *fu*
TM Transparent Message *jdr*

TM Transverse Magnetic [mode] *mt, fu, sm10-2, jdr*
TM [designator for missiles, TM-61 Matador, TM-76 Mace, USA] *mt*
TMA Target Motion Analysis *fu*
TMA Technical Manual Application *kf*
TMA Terminal Control Area *mt*
TMA Test Module Adapter *fu*
TMA Three Mirror Anastigmat *kf*
TMA Time and Motion Analysis *fu*
TMA Traffic Management Agency *mt*
TMAAF Tele-operated Mobile Anti-Armor Project *hc*
TMACS Training Management Control System [USA] *mt*
TMADS Technicians Maintenance Assessment and Diagnostic System *wp*
TMARP Tanker Mating And Ranging Program [SAC] *mt*
TMARS Technical Manual Acquisition Requirements System *fu*
TMAS Technical Meeting Advisory Service *nd74.445.8o*
TMAS Transaction Monitoring and Analysis System *mt*
TMB Technical Management Board *tt*
TMC Telecommunications Monitor Console *0*
TMC Traffic Management Coordinator *fu*
TMCA Transportation Management Control Agency *mt*
TMCC Traffic Management Computer Center *fu*
TMCCC Tactical Modular Command and Control Center *hc*
TMCF Transportable Maintenance Calibration Facility *hc*
TMCOMP Telemetry Computation *fu*
TMCR Technical Manual Contract Requirement *kf, fu*
TMCS Tactical Maintenance Control System *mt*
TMCS Telecommunications Management and Control Subsystems *mt*
TMCS Telemetry Controller System *hc*
TMCS Transition MCS *kf*
TMD Tactical Mission Data (USAF) *fu*
TMD Tactical Munition Dispenser *mt, hc, wp*
TMD Tagged Material Detector *wp*
TMD Theater Missile Defense *mt, st, kf, fu*
TMD Theoretical Maximum Density *fu*
TMDAPO {a project being managed within USASDC} *jmj*
TMDAS Theater Missile Defense Architecture Study *fu*
TMDE Test Measurement and Diagnostic Equipment *mt, fu, hc, dm, bf, st, kf*
TMDS Theater Missile Defense System *mt*
TMDS Threat Missile Detection System *hc*
TME Test and Measurement Equipment *kf*
TME Total Maintenance Events *wp*
TMEDA Tetramethylenediamine *wp*
TMEK Trifluoro-Methyl-Ethyl-Ketone *jdr*
TMF Time Management Facility [Honeywell] *mt*
TMF Trim Maneuver Firing *oe*
TMF Trim Motor Fire *oe*
TMG Test Message Generator *fu*
TMG Thermal Model Generator *oe*
TMGS Transportable Mobile Ground Station [USAF] *wp, mt*
TMI Test Method Instruction *fu*
TMI TRIMM Microwave Imager *pbl*
TMICS Transportation Management Information Control System *mt*
TMIS Technical Management Information System [NASA] *mt*
TMIS Technology Management Information System *st*
TMIS Training and Maintenance Information System *hc, fu*
TML Television Microwave Link *fu*
TML Teradyne Modeling Language *fu*
TML Total Mass Loss *he, oe*
TMLS Tactical Microwave Landing System *mt, wp*
TMM Thermal Mathematical (Math) Model *wl, jdr, jn, oe*
TMMA Tactical Mobility Mission Analysis *wp*
TMMH Total Maintenance Man Hours *wp*
TMMS TOW Mast Mounted Sight *hn44.22.1, hc*
TMN Time Management Networking *ct, em*
TMN True Mach Number *mt*
TMO Table Mountain Observatory *ns, no, hy*
TMO Tactical Materiel Operations [Hughes EDSG] *hq5.2.4*
TMO Theater Mission Operator *kf*
TMO Theater of Military Operations [Soviet] *mt*
TMO Track Management Officer *fu*
TMO Traffic Management Office *mt*
TMO Transportation Movement Office *mt*
TMO Transportation Officer [also TO, TRO] *wp*
TMOS Thermosetting *wp*
TMOS Transportation Management / Optempo System [US Navy] *mt*
TMP TAC* Mission Processor *mt*
TMP Tactical Mission Planning *fu*
TMP Technical Manual Plan *hc, fu*
TMP Telecommunications Management Plan *wp*
TMP Terminal Monitor Program *fu*
TMP Thermal Mounting Plate *fu*
TMP Traffic Management Processor *fu*
TMPC Theater Mission Planning Center *mt*
TMPS Tactical Mission Planning System *fu*

TMPS Theater Mission Planning System [SAC/JCS] *mt*
TMR Technology Management Review *wp*
TMR Thrust Measuring Rig [a crude VTOL craft, powered by two jet engines, whose efflux was directed downwards, UK] *mt*
TMR Triple Module (Modular) Redundancy *jdr, ric, hi*
TMR/TL Table of Manpower Requirements / Troops List system [USMC] *mt*
TMS Tactical Missile Squadron *mt*
TMS Technical Maintenance Service *fu*
TMS Telecommunications Monitor System *0*
TMS Telemetry Measuring [menuing of VAX software] System *oe*
TMS Test Measurement [Earth] Station *es55.43*
TMS Thematic Mapper Simulator *ge, hy, hg*
TMS Thermo Mechanical Shock *jdr*
TMS Traffic Management System *fu*
TMS Traffic Monitoring System *jn*
TMS Transaction Monitor System *fu*
TMSC Telemetry Subcommutator *-9*
TMT Training Management Team *fu*
TMT TRAMCON Master Terminal *mt*
TMTIS Technical Maintenance Terminal Information System *hc*
TMU Test Monitor Unit *oe*
TMU Thermal Mechanical Unit *pl*
TMU Timing Unit *jdr*
TMU Traffic Management Unit *fu*
TMVS Traffic Management Voice Switch *fu*
TMW Tactical Missile Warning *kf*
TMW Tactical Missile Wing *mt*
TMXO Tactical Miniature Crystal Oscillator *hc*
TN TDRSS Network *-2*
TN Terminal Node *hi*
TN Track Number *fu*
TN True North *w9*
TN Twisted Nematic *kg, hi*
TNA Thermal Neutron Analysis *wp*
TNAA Thermal Neutron Activation Analysis *wp*
TNANG Tennessee Air National Guard *-12*
TNAS TELCON Network Administration System *mt*
TNB Trusted Network Base *tb*
TNC Terminal Node Controller *kg, hi*
TNC Trinitrocellulose *wp*
TNC Trunk Node Control *mt*
TNC Trusted Network Control *fu*
TNCB Trusted Network Component Base *tb*
TNF Tactical Nuclear Base {?} *tb*
TNF Tactical Nuclear Forces *hc, ph, tb*
TNF Theater Nuclear Forces *mt*
TNFCCC Tactical Nuclear Force Command, Control and Communications *mt*
TNFIP Theater Nuclear Forces Improvement Plan *mt*
TNFSSS Theater Nuclear Forces Survivability, Security, and Safety *hc*
TNI Trusted Network Interpretation *fu*
TNIM Telecommunication, Navigation, and Information Management *-19*
TNL Target Nomination List *mt*
TNMCS Total Not Mission Capable - Supply *mt*
TNMCS Total Not Mission Capable for Supply *wp*
TNO The Netherlands Organization *pk, kg*
TNO Toegepast Natuurwetenschappelijk Onderzoek [Dutch Research Institute] *wp*
TNR Threshold to Noise Ratio *kf*
TNT (the Australian cargo company, TNT) *aw129.21.69*
TNT Tri Nitro Toluene [an explosive] (TNT equivalent is a measure of the energy released from the explosion of a given quantity of fissionable material, in terms of the amount of TNT which could release the same amount of energy when exploded) [USA] *mt, sm1-3, wp*
TNW Tactical Nuclear Warfare *hc, wp*
TNW Theater Nuclear War *mt*
TNW Theater Nuclear Weapons *mt*
TNWS Tactical Nuclear Weapons *mt*
TO Take Off *mt*
TO TDRSS Operations *-4*
TO Technical Officer *oe*
TO Technical Order *mt, hc, wp, tb, kf*
TO Time Zero *-4*
TO Transfer Orbit *oe*
TO Transportation Officer [also TMO, TRO] *mt, hc, tb*
TO Transverse Optical *wp*
TO&E Table of Organization and Equipment *mt*
TO&P Technical Objectives and Plan *mt*
TO&S Test Operations and Support *hn48.13.4*
TOA Table Of Allowances *mt*
TOA Task Order *mt*
TOA Task Ordering Agreement *hc*
TOA Time Of Arrival *mt, uh, gp, 8, wp, st, jdr, jn*
TOA Total Obligational Authority *mt, uc, wp, tb*
TOA Trade Off Analysis *mt, tb, st*

TOA Transfer Of Authority *mt*
TOA Transfer of Operational Authority *mt*
TOA Transportation Operating Agency *mt*
TOAME Three Object Angle Measuring Equipment *tb*
TOAP Time Of Arrival Processor *fu*
TOC Tactical Operations Center [USA] *mt, fu, wp, pm, kf*
TOC TAF OPS Center *mt*
TOC Technical Operations Center *jgo*
TOC Technical Order Compliance *mt, wp*
TOC Transfer Of Communication *fu*
TOC Transportable Operations Center *uh*
TOCA Tactical Operations Center Automated System [USA] *mt*
TOCC Tactical Operations Command Center *hs, go3.2*
TOCC TDRSS Operations Control Center *-4*
TOCP TCCC Operational Computer Program *fu*
TOCP Tri-Orthocresyl Phosphate *wp*
TOCSE Tactical Operations Center Support Element [USA] *mt*
TOD Technical Objective Directive *hc, wp*
TOD Technical Objective Documents *mt*
TOD Technical Order Data *kf*
TOD Theater Operations Director *kf*
TOD Trade Off Determination *mt, st*
TODA Take Off Distance Available *mt*
TODO Technical Order Distribution Office *wp*
TODR Take Off Distance Required *mt*
TODR Top Of Dynamic Range *uh*
TODS Tactical Optical Disk System *ro*
TODS Tactical Overlay Defense Simulation *st*
TODS Time Of Day Step *jdr*
TOE Table of Organization(al) and Equipment *mt, fu, wp*
TOE Tons of Oil Equivalent *-14*
TOES Telephone Order Entry System *kf*
TOF Time Of File *mt*
TOF Time Of Flight *mt, wp, tb, st, jdr*
TOFAC Technical Office Automation and Communication System [US Navy] *mt*
TOFMS Time Of Flight Mass Spectrometry *wp*
TOFU [a suite of tools developed at the OGIP to transform non-FITS information into FITS] *cu*
TOG Technical Operations Group [AFTAC] *-12*
TOGA Tropical Oceans Global Atmosphere [experiment] [program] *ge, hy, hg*
TOGI Troll-Oseberg Gas Injection *aa26.11.11, aw130.6.12, hn51.16.3*
TOGW Take Off Gross Weight *mt, wp*
TOI Track Of Interest [USA DOD] *mt*
TOJ Track On Jam *mt, wp*
TOL Tactical Operating Location *mt*
TOL Type Organization List *mt*
TOLCAT Takeoff and Landing Critical Athmospheric Turbulence *wp*
TOLD Take Off and Landing Data *mt*
TOLS TERMS-On Line System [MTMC] *mt*
TOLTS Total On Line Test System *mt*
TOM Target Object Map *kf*
TOM Technical Operations Manager *mt*
TOM Threat Object Map *17, fu*
TOM Tracking Operation Memorandum *0, sb*
TOM Transitioning Operations and Maintenance *hc*
TOMA Technical Order Management Agency *mt, kf*
TOMARR Total Manpower Requirements and Resources *mt*
TOMS Total Ozone Mapping Spectrometer [NIMBUS-7, NASA] *ge, mi, ac1.7-26, aa26.10.33, aw130.11.44, ns, hy, mb, hg*
TOO Target Of Opportunity *kf*
TOP Tactical Operations Planner *mt*
TOP Technical and Office Protocol *mt, kg, ct, em, is26.1.46*
TOP Technical Objectives and Plans *wp*
TOP Technical Operating Procedure *gp, 0*
TOP Time Sensitive Operation Planning Procedures *mt*
TOPCAP Total Objective Plan for Carrier Airmen Personnel *wp*
TOPEX Topographic Ocean Experiment [Satellite] [JPL] *sc3.10.2, ge, hc, hy, re, hg*
TOPINT Technical Operational Intelligence [USA DOD] *mt*
TOPREP Time-sensitive Operation Planning and Reporting *mt*
TOPS Technical Order Page Supplement(s) *tb*
TOPS Thermoelectric Outer Planet Spacecraft *hc*
TOPS Time Sharing Operating System *wp*
TOPS Toward Other Planetary Systems *mi*
TOPS Trade Observation Pairing System *hc*
TOPS TRADOC Obscuration Pairing System *hc*
TOPS Transistorized Operations Phone System *oe*
TOPS Transportation Operational Personal Property Standard System [DOD] *mt*
TOPSEC Top Secret *mt*
TOR Tactical Operation Room *mt*
TOR Tactical Operational Requirement *wp*
TOR Technical Operations (Operating) Report *17, hc, jdr, kf*
TOR Tentative Operational Requirement *mt*

TOR Terminal Outage Report *jdr*
TOR Terms Of Reference *mt, wp*
TOR Time Of Receipt *mt*
TORA Take Off Run Available *mt*
TORMS TAC Operations Resources Management System [TAC] *mt*
TORR Take Off Run Required *mt*
TOS Tactical Operation Simulation *st*
TOS Tactical Operations Squadron *mt*
TOS Tactical Operations System [USA] *hc, mt*
TOS Tape Operating System *hi*
TOS Technical Operations Squadron [AFTAC] *-12*
TOS Test of Operating Systems *jn*
TOS Thermal Oxidative Stability *kf*
TOS TIROS Operational Satellite *ge, ci.880, hg*
TOS TIROS Operational System *oe*
TOS Training Objective Statement *fu*
TOS Transfer Orbit Site {?} *jdr*
TOS Transfer Orbit Stage *sc4.4.4, re, jdr*
TOSCA TOTE System Computer Assisted *fu*
TOSH TOW Optical Sight Hardening *hc*
TOSI Technical On Site Inspection *hn48.13.1*
TOSLOG Tactical Operation Simulation [data recording] Logs *st*
TOSM Tenant Operations Systems Manager *mt*
TOSP Tailored Ocean Surveillance Product *mt*
TOSS Television Ordnance Scoring System *mt*
TOSS Transfer Orbit Sun Sensor *cl, jn*
TOSTE Transmitter Test Set *fu*
TOT Tactical Operations Target *ic2-6*
TOT Techniques Of Training *fu*
TOT Time Of Transmission *mt*
TOT Time On (Over) Target *mt, fu, w9, wp, st*
TOT Transfer Of Training *fu*
TOTE Textual Display (Text Or Table Entry) *fu*
TOTES Time Ordered Techniques Experimental System *hc*
TOTS TRW Orbital Test Station *kf*
TOT_BLOCK Time Over Target Block Time *mt*
TOU Transmission Operating Unit *mt*
TOV Tele Operated Vehicle *mt, wp*
TOVS TIROS Operational Vertical Sounder [NOAA] *ge, pl, aw129.14.43, hg*
TOW Take Off Weight *mt, aw129.10.39*
TOW Tube-launched, Optically-tracked, Wire-guided [an anti tank missile, probably the most numerous guided missile ever] [USA] *mt, fu, hc, hn41.16.8, 17, sb, wp*
TOWA Terrain and Obstacle Warning *fu*
TOWER [Fort Belvoir program] *hc*
TOWG Technical Order Working Group *mt*
TOW_COBRA Tube launched, Optically tracked, Wired guided COBRA *kf*
TP Tape Processor *oe*
TP Target Present *jdr*
TP Technical Proposal *gp*
TP Teleprocessing Monitor *mt*
TP Temporary Procedure *uh*
TP Terminal Processing *mt*
TP Test Parameter *kf*
TP Test Plan *mt, fu, wp*
TP Test Point *fu, jdr, he*
TP Test Port *oe*
TP Test Position *fu*
TP Test Procedure *fu, jdr, jn, oe*
TP Test Program *fu*
TP Theater Processor *kf*
TP Tie Plate *fu*
TP Time Pulse *fu*
TP Touch Panel *fu*
TP Tower Processor *fu*
TP Training Plan *fu*
TP Transaction Processing *em*
TP Transaction Processor [Honeywell] *mt*
TP Transition Plan *mt*
TP Translator Processor *fu*
TP Transport Protocol (TP4 Transport Protocol Class 4) *ct, em, mt*
TP Trial Period *kf*
TP True Position *fu*
TP True Profile *fu*
TP Twisted Pair *em, ct*
TP&TEP Training Program and Training Equipment Plan *fu*
TPA Tape Pulse Amplifier *fu*
TPA Target Planning Aide [RADC] *mt*
TPA Target Prioritization Aid *mt*
TPA Terminal Processor Assembly *mt*
TPA Third Party Application *ct, em*
TPA Tissue Plasminogen Activator *wp*
TPA Trajectory Prediction Accuracy *fu*
TPAM Technical Program Assessment Memorandum *wp*

TPAP Technical Performance Audit Plan *fu*
TPAP Transaction Processing Applications Program *mt*
TPAR Technology Producibility Assessment Review *kf*
TPAS Target Preparation and Analysis *fu*
TPAS Telemetry Pulse Analysis System *hc*
TPC Tactical Platform Coordination *fu*
TPC Technical Performance Criteria *wl*
TPC Technical Planning and Control *fu*
TPC Telecommunications Processing Console *jdr*
TPC Theater Processing Center *kf*
TPC Tower Position Console *fu*
TPC Transaction-Processing Performance Council *pk, kg*
TPD Technology Program Demonstration *tb*
TPD Terminal Protection Device *mt, jdr, he, fu*
TPD Test Procedure Document *fu*
TPD Test Profile Description *fu*
TPD Third Party Developer *ct, em*
TPD Threat Planning Document *mt, wp*
TPD Time Pulse Distributor *fu*
TPDA Test Point Data Analysis *fu*
TPDFS Teampack Direction Finding System [USMC] *mt*
TPDMP Training Program Development Master Plan *mt*
TPDU Transport Protocol Data Unit *fu*
TPE Tactical Parameter Estimator *kf*
TPE Technical Performance Evaluation *kf*
TPE Tracking and Pointing Experiment [SDI] *mt, 5, 17, sb, sf31.1.32, wp, tb*
TPE Transaction Processing Executive *mt*
TPEB Technical Proposal Evaluation Board *mt*
TPEW Technology Panel For Electronic Warfare *mt*
TPF Telemetry Processing Facility *2, 4*
TPF Terminal Presentation Facility *mt*
TPF Transactions Processing Facility [IBM] *pk, kg*
TPFDD Time Phased Force and Deployment Data [JOPES, GCCS] [USA DOD] *mt*
TPFDL Time-Phased Force and Deployment List [JOPES] *mt*
TPG Technology Planning Guide [ESD] *mt*
TPG TEREC Ground Processor *mt*
TPG Test Pattern Generator *gp, oe*
TPG Test Program Generation *fu*
TPG Timing Pulse Generator *fu*
TPI (the new polymer is similar to . . . LARC/TPI) *aa26.12.76*
TPI Tape Phase Inverter *fu*
TPI Target Position Indicator *fu*
TPI Technical Proposal Instructions *oe*
TPI Test Program Instructions *fu*
TPI Tracks Per Inch *ct, kg, em, hi*
TPI Training Planning Information *fu*
TPICK Telecommunication Plan for Improvement of Command in Korea [USA] *mt*
TPID Telecommunications Performance and Interface Document *-4*
TPIP THORNTON Product Improvement Program *fu*
TPIP Time Phased Implementation Plan *mt*
TPL Table Producing Language *pk, kg*
TPL Transaction Processing Language *pk, kg*
TPLOT Terra Plot [OJCS] *mt*
TPM Technical Parameter Monitor [a computer program] *-4*
TPM Technical Performance Measure(ment) *mt, fu, ah, hc, 17, sb, tb, pbl, jdr, kf, jn*
TPM Telemetry Processor Module *fu*
TPM Transactions Per Minute *pk, kg*
TPMCM Total Partial Mission Capable - Maintenance *mt*
TPMCS Total Partial Mission Capable - Supply *mt*
TPMR Technical Performance Measurement Report *fu*
TPMS Teleprocessing Management System *wp*
TPNET Teleprocessing Network *do452*
TPO Technical Planning Objective *wp*
TPO Technical Project Officer *mt, wp*
TPO Telecommunication Program Objective *mt*
TPO Test Plan Outline *mt*
TPO Transportation Packaging Order *tb*
TPO TRIMIS Program Office *mt, dt*
TPOP Telco Point Of Presence *mt*
TPORT Twisted pair Port Transceiver [AT&T] *kg, hi*
TPP Technical Performance Parameter *kf*
TPP Technical Program Plan *hc*
TPP Technology Program Plan *wp*
TPP Test Plan and Procedure *fu*
TPP Testability Program Plan *fu*
TPP Theater Planning Package *mt*
TPP Total Package Procurement *mt*
TPP Training Program Plan *fu*
TPP Trajectory Prediction Program *fu*
TPPC Total Package Procurement Concept *wp*
TPPC Total Production Program Cost *fu*
TPPD Twisted Pair - Physical-Media Dependent *ct, em*

TPPS Two-Pass Peak Shearing *fu*
TPPT Teflon Pulsed Plasma Thruster *ls4, 5, 28*
TPQ Threshold Planning Quantity *kf*
TPR Technical Progress Review *tb*
TPR Technical Proposal Requirements *fu*
TPR Test Problem Report *hc, tb, fu*
TPR Thermal Plastic Rubber *fu*
TPR Trained Personnel Requirements *mt*
TPR Transaction Processing Routine *mt*
TPR Traveling Purchase Requisition *fu*
TPS (a former military TPS instructor pilot) *aw130.6.69*
TPS Tactical Product Set *fu*
TPS Task Parameter Synthesizer *fu*
TPS Technical Performance Specifications *mt*
TPS Telemetry Processing System *0*
TPS Telenet Packet Switched (Switching) *cd, fu*
TPS Test Pilots School *mt*
TPS Test Program Set *mt, hc, wp, dm, fu*
TPS Thermal Protection System *hc, ci.17, aa26.12.b8, wp, mb, mi, ai2*
TPS Training Path System *fu*
TPS Transaction Processing System *pk, kg*
TPS Transactions Per Second *pk, kg*
TPS Transportation Protection Service *su, jdr*
TPSC Teleprocessing Services Center [Air Force] *dt*
TPSD Test Program Set Document *fu*
TPSF Training and Programming Support Facility *fu*
TPSN Troop Program Sequence Number [UNITREP] *mt*
TPSO Teleprocessing Services Center *mt*
TPSR Team Program Status Review *-4*
TPSSM Touch Pad Screen Selection Menu *fu*
TPT Training Planning Team *mt, kf*
TPT Tripicryltriazine *wp*
TPTA Transient Pressure Test Article *aw129.1.23*
TPTG Tuned-Plate Tuned-Grid *fu*
TPTRL Time Phased Transportation Requirements List [JOPS] *mt*
TPU Terminal Processor Unit *mt*
TPU Text Processing Unit *oe*
TPU Text Processing Utility *fu*
TPU Transceiver Processor Unit *fu*
TPUMF Total Package Unit Materiel Fielding [USA] *mt*
TPUS Troop Program Unit [USA] *mt*
TPWG Test Plan(ning) Working Group *mt, fu, hc, wp, tb, jdr*
TPX Military Beacon Decoder *fu*
TQ Total Quality *ig21.12.17*
TQ Track Quality *fu*
TQA Total Quality Assurance *fu*
TQCM Temperature-controlled Quartz Crystal Microbalance *oe*
TQFP Thin Quad Flat Pack *kg, hi*
TQM Total Quality Management *fu, kg, af72.5.68, pf1.14, nd74.445.24, nt, an31.13.4, wp, jdr, kf*
TQMS Total Quality Management System *fu*
TQS Total Quality System *fu, hn44.22.4*
TR Tactical Reconnaissance aircraft [TR-1 is a U-2 derivative] *mt*
TR Target Recognition *wp*
TR Technical Report *mt, fu, hn44.7.4, wp, tb, st*
TR Technical Review *17, sb, tb, kf*
TR Telephone Request *fu*
TR Terminal Ready *kg, hi*
TR Test Report *st, fu*
TR Theater Reserve *mt*
TR Threat Recognizer *fu*
TR Thrusted Replicas *tb*
TR Token Ring *hi*
TR Torque Ratio *oe*
TR Tracking Radar *mt, fu*
TR Transaction Register *mt*
TR Transformer Rectifier *fu*
TR Transmit / Receive *fu*
TR Transportation Request *mt*
TR Trouble Report *fu*
TR&C Telemetry, Ranging and Command *gg150*
TRA Technical Requirement Analysis *wp*
TRA Temporary Reserved Airspace *mt*
TRA Training Requirements Analysis *tb*
TRAC (... RSG employees who supported TRAC equipment during Desert Storm) *hn51.16.5*
TRAC Technical Report Awareness Circular *mt*
TRACAB Terminal Radar Approach Control in Tower Cab *fu*
TRACAL Traffic Approach Control And Landing *hc*
TRACALS Traffic Control and Landing System *mt*
TRACE Technical Report Analysis Condensation Evaluation *fu*
TRACE Technical Report Analysis, Condensation, Evaluation *hc*
TRACE Total Risk Assessing Cost Estimating (Estimate) *mt, fu, ph, tb, st*
TRACE Track Review and Correction Estimate *fu*
TRACE Tracking And Communications, Extraterrestrial *0*

TRACE Transport And Chemistry near the Equator [program] *hy*
TRACER Tropospheric Radiometer for Atmospheric Chemistry and Environmental Research [EOS] *hy*
TRACON Terminal Radar Approach Control *fu, aw129.21.143, pf.f88.26, wp*
TRACS Terminal Radar And Control System *hc*
TRACS Terminal Range Closure System *fu*
TRACS Transportation Requirements and Capabilities Simulation Model *mt*
TRAD Transportation Routing And Documentation *wp*
TRADE Training Services [USA] *mt*
TRADES Training Device System *hc*
TRADEX Target Resolution and Discrimination Experiment *bm139*
TRADIC Transistorized Airborne Digital Computer [Name of first computer to be entirely transistorized] *pk, kg*
TRADOC Training and Doctrine Command [US Army] *mt, fu, aw126.20.7, 12, ts17-4, hc, wp, tb, st*
TRAFDIST Traffic Distribution System *mt*
TRAFES [transportable remote sensing station] *es55.2*
TRAFFIC_JAM [COMJAM system TLQ17A, USA] *mt*
TRAILBLAZER [VHF communication ground direction finding system] *wp*
TRAILX (Ephemeris Generation (software program)) *jb*
TRAIL_BLAZER [comm intercept system TSQ114, USA] *mt*
TRAINERS [a subsystem of TISURS] *mt*
TRAIS Transportation Reporting and Inquiry System [MAC] *mt*
TRAJSIM Trajectory Simulation [Hughes space software] *-17*
TRAM Target Recognition and Attack Multisensor [a sensor package carried by the A-6 Intruder, USA] *mt, fu, hc, hn43.5.4, 8, hq5.3.5, wp*
TRAM Test Requirements Analysis Matrix *fu*
TRAM Test Requirements Analysis Model *hc*
TRAM Tracking Radar Automatic Monitoring *fu*
TRAMCOM Transmission Management and Control *mt*
TRAMCOM Transmission Monitoring and Control System *mt*
TRAMS Transportation (Traffic) Management System Component of DSATS *mt*
TRAMS Transportation And Movement Subsystem, UCCIS [USAREUR] *mt*
TRAN Transport Ship [US Navy] *mt*
TRANET Tracking Network [Navy Doppler] *ac2.6-7*
TRANSCOM Transport Command *fu*
TRANSCOM Transportation Command Management Information System [USAREUR] *mt*
TRANSCOM Transportation Command [USA] *mt*
TRANSCOM Transportation Communications [DOE] *mt*
TRANSEC Transmission Security *mt, fu, uh, go1.1, wp, tb, jdr, kf*
TRANSIT [navigation satellite system of the U.S. Navy] *ci.893, 12*
TRANSMOD Transportation Mobility Development *mt*
TRANSMOD Transportation Module [COMPES] *mt*
TRANSPAC French National Telecommunications System:(ge)
TRANSVANS Transportable Radar and Communications Simulator Vans [NATO] *mt*
TRAP Tactical Related Applications Program *mt, kf*
TRAP Tanks, Racks, Adapters and Pylons *mt, wp*
TRAP/MATSS Terminal Radiation Airborne Program/Midcourse A/B Target Sign St{atus} *hc*
TRAP/TRE TRAP and Tactical Receive Equipment *kf*
TRAPATT Trapped Plasma Avalanche Transit Time *hc, fu*
TRAR Training Requirements Analysis Report *kf*
TRASYS Thermal Radiation Analysis System *kf*
TRAT Torpedo Readiness Assistance Team *fu*
TRB Technical Review Board *oe*
TRB Test Review Board *mt, 4, kf*
TRB Training Recording and Playback *fu*
TRB Transportation Research Board *aw129.21.4*
TRC Tactical Reconnaissance Center *mt*
TRC Technical Repair Center *wp*
TRC Technology Repair Center *mt*
TRC TEREC Remote Terminal [also TRT] *mt, wp*
TRC Torrance Research Center *fu*
TRC Total Recurring Cost *fu*
TRC Transmission Release Code *mt*
TRCC Tracking Radar Central Control *mt*
TRCO Technical Representative of the Contracting Officer *mt*
TRCS Tactical Radio Communications System *hc*
TRCS Techniques for Radar Cross Section *wp*
TRD Technical Requirements Document *mt, 5, 17, sb, hc, nt, jmj, st, kf*
TRD Test Requirements Document *fu, wp, jdr*
TRDL Tactical Reconnaissance Data Link *hc, wp*
TRDTO Tracking Radar Data Takeoff *fu*
TRE Tactical Receive Equipment *mt, fu*
TRE Telecommunication Research Establishment [UK] *mt*
TRE Transient Radiation Effect *tb, fu*
TRE Transmission Resource *fu*
TREAD Troop Recognition And Detection *wp*
TREAT TRE Advanced Terminal *mt*
TREATS Trident Electronic Assembly Test Sets *hc*

TREC Target Recognition for Electronic Combat *mt*
TREC Tracking Radar Electronic Component *fu*
TREDS Tactical Reconnaissance Exploitation and Dissemination System [USAF] *wp, mt*
TREDS TR-1 Exploitation Development System [a data processing installation at Hahn AB for data acquired by the TR-1, USA] *mt*
TREE Transient Radiation Effects on Electronics *mt, fu, hc, 17, sb, tb, st, jdr, kf*
TREES Transient Radiation Effects to Electronic Systems *sd*
TREP Thrusted Replica [decoy] *tb, st*
TRESSCOM Technical Research Ship Special Communications *-10*
TRESTLE [EMP testing facility at Kirtland AFB] *wp*
TRETS Transportation, Energy, and Troop Support *mt*
TREWS Training Range Electronic Warfare Simulators *hc*
TREWS Trunk Radio Electronic Warfare Simulator *hc*
TRF Trident Refit Facility *fu*
TRF Tuned Radio Frequency *wp, fu*
TRG Tactical Reconnaissance Group *mt*
TRG Technical Review Group *mt, wp*
TRG Technology Research Group *kf*
TRI Tactical Reconnaissance Intelligence *fu*
TRI-TAC Tactical Reconnaissance Intelligence - Target Allocation Center *fu*
TRI-TAC Tri service Tactical communications system [e.g. joint Army, Navy, Air Force] *mt, hc, pf.f88.23*
TRIAD Tracking, Ranging, Interrogation, Analysis and Detection *fu*
TRIAD [strategic force of ICBMs + SLBMs + air breathing bombers] *wp*
TRIB Transfer Rate of Information Bits *do1*
TRIBLE Transaction Image Builder / Editor *mt*
TRIC Tracking Radar Input and Correlation *fu*
TRICCSMA Trident CCS Maintenance Activity *fu*
TRICOMS TRIAD Computer System *mt*
TRICON [special container for military transportation] *wp*
TRICS Transaction Identification Codes *mt*
TRID Track Identity *fu*
TRIDENT [Navy submarine-launched missile, successor to ULMS] *fu*
TRIES Tactical Radar Imagery Exploitation System *wp*
TRIGA Teaching, Research, Isotope General Atomic [reactor] *aw130.11.55*
TRIGAT Third Generation Anti Tank Missile *mt*
TRIGS TR-1 Ground Station *wp, mt*
TRILOG Tri-Service Logistics System *mt*
TRIM Time Related Instructional Management System [ATC] *mt*
TRIM Trails Roads Interdiction Mission *hc*
TRIM Trails, Roads, Interdiction Multisensor [USA] *mt*
TRIMIS Tri Service Medical Information System [DOD] *hc, mt*
TRIMS Transportation Integrated Management System *mt*
TRIS Transportation Reporting and Inquiry System [MAC*] *mt*
TRISP Time Resolved Infrared [absorption] Spectroscopy *wp*
TRITAC Tri-Service Tactical [communications program] *hc, ph*
TRL Target Recommendation List *mt*
TRL Transistor Resistor Logic *mt, fu, 0, mj31.5.47*
TRM Titan Radar Mapper *es55.26*
TRMM Tropical Rainfall Measurement (Measuring) Mission [Japan, USA] *ge, hc, hy, hg*
TRMS Telefunken Radar Mobile Search [antenna built by Hughes for AEG-Telefunken] *fu*
TRN Task Requirements Notice *tb*
TRN Terrain Reference Navigation *mt*
TRN Threaded Read News [Internet] *kg, hi*
TRN Token Ring Network *pk, kg*
TRN Track Reference Number *fu*
TRN Turn around actions *mt*
TRNA Token Ring Network Adapter *hi*
TRO Technical Reviewing Office *wp*
TRO Transportation Officer [also TO, TMO] *wp*
TROC Tactical Relay Operations Center *mt*
TRON The Real time Operating system Nucleus *pk, kg*
TRON Time Read Operations system Nucleus *hi*
TROO Transponder On/Off *fu*
TROOP Transportation Operational and Organizational Plan *mt*
TROPO Tropospheric Backscatter *tb*
TROPO Tropospheric Scatter [Radio] *mt*
TRORD Tracking and Orbit Determination *tb*
TROSCOM Troop Support Command [US Army] *mt, hc*
TRP Test Requirements Plan *fu*
TRP Track Return Processing *fu*
TRP Transec Processor *jdr*
TRPC Transaction Remote Procedure Call *ct, em*
TRR Tactical Range Recorder *fu*
TRR Target Ranging Radar *fu*
TRR Technical Readiness Review *mt, wp*
TRR Test and Reconfigure Redundancy *fu*
TRR Test Readiness Review *mt, fu, pl, es55.43, nt, st, jdr, kf, jn*
TRR Test Requirements Review *fu*
TRR Transition Readiness Review *jdr*
TRRB Test Readiness Review Board *mt*

TRRMS Training Requirements and Resource Management System [USMC] *mt*
TRRP Transportable Radar Post *fu*
TRRR Trilateration Range and Range Rate *hc, ac2.8*
TRS T&C RF Subsystem *vv*
TRS Tactical Reconnaissance Squadron [USA] *mt, wp*
TRS Tactical Reconnaissance System [TR-1] *wp, mt*
TRS Teleoperator Retrieval System *ci.xix*
TRS Terminal Reservation System *mt*
TRS Test Requirements Specification *fu*
TRS Tetrahedral Research Satellite *le38*
TRS Thermal Reference Source *kf*
TRS Token Ring Server *hi*
TRS Torpedo Room Simulator *fu*
TRS TOW Roof System *hc*
TRS Transportable Relay Station *wp*
TRSA Terminal Radar Service Area *mt*
TRSA Timesharing Retrieval System *mt*
TRSIURS Trainer Status, Inventory and Utilization Reporting System [USAF] *mt*
TRSM Token Ring Switching Module *hi*
TRSR Taxi and Runway Surveillance Radar *fu*
TRT TEREC Remote Terminal [also TRC] *mt, wp*
TRT Token Ring Technology *hi*
TRU Tester Replaceable Unit *fu*
TRU Transportable Radio Unit *fu*
TRUE Trident Refit Unique Equipment *fu*
TRUMP Technical Review and Update of Manuals and Publications [Navy] *fu*
TRV Tactical Robotics Vehicle *hc*
TRV Tower Restoral Vehicle *mt*
TRVM Test Requirement Verification Matrix *-4*
TRW Tactical Reconnaissance Wing [USA] *mt, wp, af72.5.55*
TRW Thompson-Ramo-Woolridge [an aerospace corporation] *fu, aw130.13.87, sc4.11.2, 4, 14, 17, sb, aa26.12.51, af72.5.19, hc, wp, tb, oe*
TR_GPT Target Recognizer Growth Provision Tester *hc*
TS Target Strength *fu*
TS Technical Study (Survey) *fu*
TS Telegraph System *fu*
TS Temperature Sensitive *fu*
TS Terminal Surveillance *tb*
TS Test Set *jdr*
TS Test Specification *fu, jn*
TS Test Squadron [USA] *mt*
TS Thermal Shock *jdr*
TS Thermoemission Spectrometer *hc*
TS Time Skewing *fu*
TS Time Standard *fu*
TS Time Switch *fu*
TS Top Secret *mt, kg, sy9, hc, tb, jdr*
TS Total Slack *jn*
TS Training Simulator *hc*
TS Type of Signaling *mt*
TS Type Specification *fu*
TS&A Transportation Shipping and Authorization *fu*
TS/ETI Temperature Sensing / Elapsed Time Indicator *fu*
TS/RD Top Secret / Restricted Data *mt*
TS/SI Top Secret / Sensitive Information *pk, kg*
TSA Tactical Situation Analysis *kf*
TSA Target Service Agent *pk, kg*
TSA Technical Support Alliance *pk, kg*
TSA Theater Situation Analyst *kf*
TSA Theater Storage Area *mt*
TSA Thomson Sintra ASM *fu*
TSA Time Series Analysis *wp*
TSA Time Slot Assignment *fu*
TSA Transportation Standardization Agency *dt*
TSA Troops Support Agency [USA] *mt*
TSAD Time Synchronization And Dissemination *fu*
TSAE Troop Support Agency Europe [USA] *mt*
TSAF Typical System Acquisition Flow *fu*
TsAGI Tsentralny Aerogidrodinamichesky Institut [Central Aero and Hydrodynamic Institute] [USSR] *mt*
TsAGI [Central Aero-Hydrodynamics Institute, USSR] *-13*
TSAPI Telephony Server (Services) Application Program Interface [AT&T, Novell] *kg, em, ct, hi*
TSAR Task and Skills Analysis Report *kf*
TSAR Theater Simulation of Airbase Resources [a model] [HQ USAF] *mt, wp*
TSAR Transmission Security Analysis Report *wp*
TSARINA TSAR Inputs using ADA [inputs to TSAR, model] *wp*
TSARS Tactical Situation Assessment and Response Strategy *hc*
TSB Termination Status Block *pk, kg*
TSB Twin Side Band *mt*
TSBAR Training System Basis Analysis Report *kf*
TsBIRP [Central Bureau for the Study of Rocket Problems, USSR] *-13*

TSC (DOT/TSC) [Hughes Aircraft Company customer for airport surface detection equipment] *hc*
TSC Tactical Support Center *mt, fu*
TSC Tape Station Conversion *mt*
TSC Target Signature Classification *tb*
TSC Task Set Chain *fu*
TSC Technical Supervisor's Console *fu*
TSC Technical System Center *fu*
TSC Telecommunications and Space Company [of Hughes] *jdr, tm*
TSC Thruster Spin (Stabilization) Control *jn*
TSC Transportation System Center *fu*
TSC Trident Support Complex *fu*
TSCA Test System Condition Assessment *fu*
TSCC TAAS Support Computer Complex *fu*
TSCIXS Tactical Support Center Information Exchange System *mt*
TSCLT Transportable Satellite Communications Link Terminal *fu*
TSCM Technical Surveillance Countermeasures *mt*
TSCO Target Selection Confusion of Operator [US Navy] *mt*
TSCO Top Secret Control Officer *mt, kf, su*
TSCT Tactical Support Center *mt*
TSCT Tandem Switch Center *mt*
TSCT Technical Support Contractor / Center *mt*
TSCT Transportable Satellite Communication Terminal *hc*
TSCT Transportation Systems Center [DOT] *mt*
TSCT [AF Small Computer] *mt*
TSD Tactical Situation Display [USA] *mt*
TSD Tactical Surveillance Demonstration *kf*
TSD Team Start Date / Test Start Date *mt*
TSD Technical Support Data *fu*
TSD Technology Support [Services] Division [of Hughes] *fu, gp, 8, ie5.11.8, ep, jdr, tm*
TSD Total Systems Design *hc*
TSD/MMS Tactical Situation Display/Moving Map System *hc*
TSDM Tracking Service Data Message *-4*
TSDU Target System Data Update *fu*
TSE Tactical Signals Exploitation *fu*
TSE Tactical Support Element *mt, fu*
TSE Tactical Support Equipment *wp*
TSE Test Support Equipment *4, 17, sb*
TSE The Semware Editor *pk, kg*
TSEA Training Subsystem Effectiveness Analysis *hc*
TSEC Telecommunications Security *mt*
TSEC Telecommunications Security [US] *mt, hc, fu*
TSEF Technical Simulation and Evaluation Facility *fu, hc*
TSEM Thermal [analysis of] Standard Electronic Modules *fu*
TSEM TOW System Evaluation Missile *hc*
TSEO Tactical Signals Exploitation Operator *fu*
TSEP Tactical Signals Exploitation Program *fu*
TSES Tactical Signals Exploitation System *fu*
TSFC Thrust Specific Fuel Consumption *mt*
TSG Time Signal Generator *fu*
TSG TMDE Support Group *mt*
TSGCEE Tri-Service Group on Communications and Electronic Equipment *mt*
TSGMS Test Set, Guided Missile System *hc*
TSGO Technical Services Group Office *fu*
TSGP Threat Scenario Generation Program *fu*
TSgt Technical Sergeant *mt*
TSI (TSI, Inc., St. Paul, Minnesota) *is26.1.18*
TSI Technical Studies Institute *fu*
TSI Terminal equipment System Initialization *mt*
TSI Time Slot Interchange *fu, mt*
TSI Time-Slot Index *fu*
TSI Time-Slot Interrupt *fu*
TsIAM Tsentralny Institut Aviatsionnove Motorostroeniya [Central Aeroengine Institute] [USSR] *mt*
TSIC Tri-Service Information Center *uc*
TSIM Trajectory Simulation *kf*
TSIMS Telemetry Simulation Submodule *fu*
TSIR Total System Integration Requirement *hc*
TSIR Total System Integration Responsibility *fu*
TSIS Total System Integration Simulator *hc*
TSISC Tactical System Interoperability and Support Center *hc*
TsKB Tsentralnoye Konstruktorskoye Byoro [Central Design Bureau] [USSR] *mt*
TSL Technical Support Laboratory *fu*
TSLC Taiyuan Satellite Launch Center *ai2*
TSM Technical Service Manager *fu*
TSM Technical System Manual *fu*
TSM Technical System Model *fu*
TSM Transportation Simulation Model *mt*
TSMA Theater of Strategic Military Action [Soviet] *mt*
TSMD Time Stress Measurement Device *wp*
TSMP Technical Supervisory Maintenance Position *fu*
TSMP Threat Simulator Master Plan *mt*
TSN Time-slot Set Number *fu*

TSNM Tactical Surveillance Net Management *fu*
TSO (many items are FAA certified and TSO'd.) *aw129.21.125*
TSO Technical Service Order *wp*
TSO Technical Standing Order *mt*
TSO Technical Support Order *mt*
TSO Telecommunications Service Order *mt*
TSO Test Site Officer *fu*
TSO Time Sharing Option [of a computer operating system] *mt, ct, em, kg, fu, hn44.13.8, 8, hd1.89.y14, wp, jdr*
TSO Time Since Overhaul *wp*
TSO Time Slot Over *fu*
TSO/E Time Sharing Option / Extension *ct, em, mt*
TSOR Tentative Specific Operational Requirement *mt, fu*
TSOS Time Sharing Operating System *hi*
TSOT Trident Sonar Operator Trainer *fu*
TSP Tactical Strike Program [NATO] *mt*
TSP Technical Support Package *ci.1010, te23, 11, 22*
TSP Technical Support Plan *mt*
TSP Technical Support Program *hc*
TSP Telecommunications Services Priority [DCA] *mt*
TSP Terminal Support Processor *mt*
TSP Time Space Processing *jb*
TSP Timely Spares Provision(ing) *mt, fu*
TSP TRAM Stabilizing Platform *hc*
TSP Transmission Simulation Program *fu*
TSP Turret Stabilized Platform *hn43.7.12*
TSP Twisted Shielded Pair [wire] *jdr, jn*
TSPEC Test Specification *fu*
TSPI Time and Space Positioning Information *af72.5.4*
TSPL Torpedo Summary Problem List *fu*
TSPR Total System Performance Responsibility *wp, jdr, kf*
TSPR Total System Provider Responsibility *jdr*
TSPS Traffic Service Position Station *fu*
TSPS Traffic Service Position System *ct, em*
TSR Tactical Search Range *fu*
TSR Technical Support Request *mt*
TSR Technical Support Requirement *mt, wp*
TSR Telecommunications Service Requests *mt*
TSRA Training System Requirements Analysis *kf*
TSRC TRW Senior Review Committee *jdr*
TSRL Turbine Systems Research Laboratory *fu*
TSRP Technical Support Real Property *fu*
TSRS Tracking and Data Relay Satellite *tb*
TSRT Transportable Standard Remote Terminal [USA] *mt*
TSS Tactical Shelter System program [later redesignated Transportable Shelter System to emphasize its role as a non-tactical system, USAF] *mt*
TSS Tactical Strike System *wp*
TSS Tactical Surveillance Sonobuoy [US Navy] *mt*
TSS Targeted Search System *mi*
TSS Task State Segment *pk, kg*
TSS Technical Support Services *kf*
TSS Telecommunications and Space Sector [of Hughes] *jdr*
TSS Telecommunications Security System [USA] *mt*
TSS Test Support System *-4*
TSS Tethered Satellite System *hy, wl, mb, mi*
TSS Theater Surveillance System *kf*
TSS Time Sharing System *mt, pk, kg*
TSS Time Slot Set *fu*
TSS Traffic Service Section *mt*
TSS TRIDENT Support System [US Navy] *mt*
TSSA TDRS [H,I,J] Spacecraft Subsystem Allocation [specification] *vv*
TSSAM TRANSEC Spreading and Associated Memories *fu*
TSSCR Total Software System Configuration Report *fu*
TSSE TOW Subsystem Support Equipment *hc*
TSSF Terminal System Support Facility *fu*
TSSG Training and Support Systems Group [of Hughes Aircraft Company] *hn49.8.3, hq6.2.7, rm8.2.6*
TsSKB Tsentralnoe Spetsializirovannoye KB [Central Specialized Design Bureau] *ai2*
TSSR Tropo Satellite Support Radio *mt*
TSSS Time Sensitive Support System *mt*
TSSS Top Secret System {System} [GCCS] [USA] *mt*
TSSW Technical Document Support Software *mt*
TSSW Technical Support Software *mt*
TST Tactical Surveillance Technology *wp*
TST Technical Support Team *mt, wp*
TSTI Typical Solar Terrestrial Instrumentation *hc*
TSTN Triple Super Twist(ed) Nematics [displays] *kg, hi*
TSTO Two Stage To Orbit [also 2STO] *mi*
TSTS Thermal Site Test Sets *hc*
TSTS Thermal System Test Sets *hc*
TSTS TOW Shop Test Set *hc*
TSTV Total System Tactical Validation *fu*
TSU Telescopic Sight Unit *hc*
TSURV Terminal Surveillance [SRBS] *tb*

TSVMS Tactical Secure Voice Management System *mt*
TSVP Tactical Secure Voice Program *mt*
TSWC Time Standard Work Center Code *fu*
TSWG Test Scenario Working Group *-4*
TSWG Topography Science Working Group *hy*
TT Tanker Transport *af72.5.71*
TT Technology Transition *wp*
TT Telemetry Trailer *0*
TT Teletype(writer) [usually TTY] *w9, fu*
TT Terminal Timing *-4*
TT Test Thoroughness *fu*
TT Timing and Telemetry *fu*
TT Total Time *fu*
TT Tracking Technician *fu*
TT Transition Terminal *fu*
TT Triple Thermoplastic *fu*
TT Turn Time *mt*
TT [total available static torque] *vv*
TT&C Telemetry Transmission and Control *dm*
TT&C Tracking, Telemetry and Command [Control] *uh, fu, ggix, 4, 17, sb, aa26.12.91, es55.7, jb, lt, ep, ce, tb, wl, jdr, kf, oe, tt*
TTA Technology Trends Assessment *tt*
TTA Test Target Array *fu*
TTA Training Task Analysis *fu*
TTA Transport Triggered Architecture *pk, kg*
TTA Treasury Telecommunications Architecture *mt*
TTA Trim Tab Assembly *oe*
TTAC Tracking, Telemetry And Command [subsystem] *jdr*
TTAE Total Time, Airframe and Engine *mt*
TTAI Tactical Target Aircraft Identification *fu*
TTAP Training Technology Applications Program [ATC] *hc, mt*
TTASP Target Tracking And State Prediction *st*
TTB Tank Test Bed *hc*
TTB Tanker Transport Bomber *mt*
TTBO Track To Burn Out *kf*
TTBO-FA Track To Burnout Focused Area *kf*
TTBO-MRC Track To Burnout Major Regional Conflict *kf*
TTBTS Tanker Transport Bomber Training System *wp*
TTC Tape to Card *fu*
TTC Technical Training Center [USA] *mt*
TTC Telecommunications Techniques Corporation [Germantown, Maryland] *is26.1.76gg*
TTC Telemetry, Timing, and Control *kf*
TTC Telemetry, Tracking and Command *ai2*
TTC Telemetry, Tracking and Control *ai2*
TTC Tracking, Telemetry and Control *wp*
TTC&M Tracking, Telemetry, Command and Monitoring *sc3.2.1, 9*
TTCCC Theater / Tactical Command, Control, and Communications [also T2CCC] *mt*
TTCE Transportation Terminal Center, Europe *mt*
TTCN Tree and Tabular Combined Notation *pk, kg*
TTCS Target Tracking and Control Systems *hc*
TTCS Transportation Transaction Communications System *wp*
TTE Technical Test Equipment *fu*
TTE Technical Training Engineers *fu*
TTE Thompson Tubes Electroniques *jn*
TTE Transportable Test Equipment *hc*
TTEP Training and Training Equipment Plan *fu, kf*
TTF Tagged Text File *hi*
TTF Tanker Task Force *mt*
TTF Threat Training Facility *mt*
TTF Trainer Test Facility *fu*
TTF Training Test and Ferry *mt*
TTF Trident Training Facility *fu*
TTG Tactical Training Group *mt*
TTG Test Target *fu*
TTG Test Target Generator *hc*
TTG Test Traffic Generator(s) [simulator] *2, 4*
TTGE Test Traffic Generation Equipment *-4*
TTGP Test-Tape Generation Program *fu*
TTI Tactical Target Identification *hc*
TTI Time to Intercept *fu*
TTL Transistor to Transistor Logic *mt, fu, kg, ct, em, uh, hc, gg76, 6, th4, jdr, oe*
TTM Thermal Test Model *0*
TTM Two Tone Modulation *fu*
TTMF Touchtone Multifrequency *fu*
TTNR Test Tone to Noise Ratio *gs1-5.A7*
TTNS TOW Thermal Night Sight *hc*
TTO Transmitter Turnoff *fu*
TTP Tactics, Techniques, Procedures *kf*
TTP Thermal Transfer Printing *pk, kg*
TTP Trim Tab Panel *oe*
TTP Trim Tab Positioner *oe*
TTPM Trim Tab Positioning Mechanism *oe*
TTPRR Trainer Test Procedures and Results Report *fu*

TTR Target Tracking Radar *mt, fu, hc, wp, kf*
TTR Teletype Translator *fu*
TTRS Taiwan Tracking Radar System *fu*
TTS Tank Thermal Sight *hc*
TTS Teletype Switching Subsystem *0*
TTS Test Timing Sequencer *fu*
TTS Test Tracking System *fu*
TTS Text To Speech *pk, kg*
TTS Thule Tracking Station *hc*
TTS Time Temperature Recorder *fu*
TTS Transaction Tracking System *kg, ct, em*
TTS Transportable Tracking Station *8, lt*
TTSD Test and Training Systems Division *hc*
TTSG Technical TADIL Standard Group *fu*
TTT Travan Tape Technology *hi*
TTT Trunk to Trunk Transfer *ct, em*
TTTE Tornado Trinational Training Establishment [set up at RAF Cottesmore, UK, in 1981] [NATO] *mt*
TTTS Tanker Transport Training System *af72.5.34, wp*
TTU Terminal Tester Unit *fu*
TTV (taken from the TTV schedule of August 7) *gs2-13.37*
TTV Tow Test Vehicle *fu*
TTW Tactical Training Wing [USA] *mt, 12, af72.5.55, wp*
TTW Technical Training Wing [USA] *mt*
TTWG Turnover and Transfer Working Group *mt*
TTY Teletype(writer) [also TT] *mt, ct, em, fu, kg, bl1-5.59, 2, 4, 0, w9, pf.f88.10, jb, kf, hi*
TTYNET Teletype Network *mt*
TU Tape Unit *hi*
TU Timing Unit *jdr*
TU Transmission Unit *w9, fu*
TU Tupolev [Soviet aircraft designer, e.g. TU-16 or Badger, TU-22 or Blinder, TU-26 or Backfire, TU-28/128 or Fiddler, TU-95/142 or Bear] *mt*
TUAF Turkish Air Force *mt*
TUAV Tactical Unmanned Air Vehicle *fu*
TUBA TCP and UDP with Big Addresses *hi*
TUCHA Type Unit Characteristics Data / File [JOPS] *mt*
TUCOWS The Ultimate Collection Of Winsock Software *pk, kg*
TUCP Tactical Unit Command Post *fu*
TUDE Teletype User Data Entry *mt, jb, kf*
TUDET Type Unit [Equipment] Detail *mt*
TUDM Tentara Udara Diraya Malaysia [Royal Malaysian Air Force] *mt*
TUE/LUE Trainer Unique Equipment/Lab Unique Equipment *fu*
TUF Terminal User Function *mt*
TUFI Tactical Unit Financial Interface [FORSCOM] *mt*
TUFMIS Tactical Unit Financial Management Information System [USA] *mt*
TUG Tactical UTC Generator [PACAF] *mt*
TUG Technical Users Group *mt*
TUI Tactical User oriented Intelligence system *mt*
TUI Text Based User Interface [WordPerfect] *pk, kg*
TUM Towed Unmanned [submersible] *hc*
TUMS Table Update and Management System [Stanford University] *pk, kg*
TUNA_SID [NOSC program] *hc*
TUO Technology Utilization Office *ci.1009*
TUOC Tactical Unit Operations Center *mt, fu*
TUS Transmission Unattended Site *mt*
TUSA Third US Army *mt*
TUSC Technology Use Studies Center *ci.1005*
TUSLOG The United States Logistic Group [Turkey] *mt*
TUSLOG Turkish-U.S. Logistics Group *-12*
TUT Telemetry Users Table *oe*
TUT Terminal Under Test *fu*
TUT Transistor Under Test *fu*
TUT Tube Under Test *fu*
TV Thermal Vacuum *he, bl2-3.80, jdr, kf, jn*
TV Thrust Vector (Vectoring) *bo, wp*
TV Thruster Valve *jn*
TV Transport Vehicle *fu*
TV [worst-case variable torques] *vv*
TV&V Technical Verification and Validation *fu*
TV/SC Television Scan Converter *fu*
TVA Target Value Analysis *mt*
TVA Thrust Vector Actuator *st*
TVA/ADF Target Value Analysis and Allocation and Distribution of Fire *mt*
TVAC Television Area Correlation *wp*
TVAS Target Velocity Acquisition System *st*
TVAT Thermal Vacuum Acceptance Test *rih*
TVC Threat Vector Control *-17*
TVC Thrust Vector Control [missiles] *hc, bo, jmj, dm, st, ai2*
TVC Transverse Curvature *js7.463*
TVC Trust Vector control *mt*
TVCF Test Vehicle Checkout Facility *oe*
TVCF Transportable Vehicle Checkout Facility *kf*

TVD Teatr Voennykh Deistvii [Theater of Operations] [USSR] *mt*
TVD Theater of Military Operations [Soviet] *mt, 12*
TVD Total Variation Diminishing *ai1*
TVDP Terminal Vector Display Unit *fu*
TVE Technical (Technology) Validation Experiment *wp, tb, st*
TVF Table of contents Verbosely from File [UNIX] *pk, kg*
TVFS Toronto Virtual File System [IBM] *pk, kg*
TVG Time-Varied Gain *fu*
TVI Television Interference *fu, kg, hi*
TVM Tachometer Voltmeter *fu*
TVM Tactical Video Mapping *mt*
TVM Test Verification Matrix *mt*
TVM TOW Visual Module *hc, ie5.11.6, bf*
TVM Track Via Missile *mt, wp*
TVM Transistor Voltmeter *fu*
TVOL Television On-Line *pk, kg*
TVOR Terminal VHF Omnirange *fu*
TVPC TOW Vehicle Power Condition *hc*
TVRI [Indonesian National Television Network] *hn44.3.5*
TVRO Television Receive Only *ct, em, hn41.22.1, mj31.5.66, ce, tt*
TVS Technology Validation Strategy *tb*
TVS Thermal Video System *hc, sn2.4, rm8.1.2*
TVTS TOW Verification Test Set *hc*
TW Tactical Warning *mt, wp, tb*
TW Test Wing [USA] *mt*
TW Threat Warning *fu*
TW Time Word *fu*
TW Training Wing [USA] *mt*
TW Travelling Wave *wp, fu*
TW {Tactical Wing} *af72.5.140*
TW/AA Tactical Warning and Attack Assessment [also TWAA] *mt, fu, jb, wp, tb, st, kf*
TW/AA Threat Warning and Attack Assessment *mt*
TW/AC Tactical Warning /Attack Characterization *mt*
TW/DECM Tactical Warning/ Deceptive Electronic Countermeasures *fu*
TW/SD Tactical Warning and Space Defense *mt*
TWA Tail Wire Antenna *wp*
TWA Time Weighted Average *hc, oe*
TWA Trailing Wire Antenna *mt*
TWA Trans World Airlines [Incorporated] *aw129.10.67, wp*
TWA Traveling Wave Amplifier *wp, fu*
TWAA Tactical Warning and Attack Assessment [also TW/AA] *wp*
TWAC Tactical Weather Analysis Center *mt, wp*
TWAES Tactical Warfare Analysis Evaluation System (USMC) *fu*
TWAIN Technology Without Any Interesting Name [connection between application and scanner software] *ct, em, kg*
TWAP Target Weapon Association Program *mt*
TWAT Traveling Wave Amplifier Tube *mt*
TWCS TOMAHAWK Weapon Control System *fu*
TWCS Tomahawk Weapons Control System *mt*
TWEB Transcribed Weather Broadcast *fu*
TWERLE Tropical Wind Energy Reference Level *ac2.8-3*
TWF Trident Weapons Facility *fu*
TWG Technical Working Group *mt, ph, oe*
TWG Test Working Group *mt, 4*
TWG Training Working Group *kf*
TWG Transfer Working Group *mt*
TWG Turnover Working Group *mt*
TWGSS Tank Weapon Gunnery Simulation System *hc*
TWI Threat Warning Indicator *mt*
TWIDS Threat Warning Information Display System *mt*
TWL Taxiway Lights *fu*
TWL Total Weight Loss *he*
TWM Traveling Wave Maser *fu, 0*
TWMBK Traveling Wave Multiple Beam Klystron *fu*
TWOS Time Warp Operating System *tb*
TWOS Tropical Wind Observing Ships *pl*
TWO_ATAF Second Allied Tactical Air Force *mt*
TWP Tactical Warfare Program *hc*
TWPS Traveling Wave Phase Shifter *hc*
TWR Tail Warning Radar *hc, wp*
TWR Threat Warning Receiver *mt, uc*
TWR ToWeR, aerodrome control *mt*
TWRAPS Traffic management Workload Reporting And Productivity System [USAF] *mt*
TWS Tactical Warning System *wp*
TWS Tactical Weapon Simulator *fu*
TWS Tactical Weather System *mt*
TWS Tactical Work Station *fu*
TWS Tail Warning System *wp*
TWS Thermal Weapon Sight *hc, aw129.10.12, kf*
TWS Threat Warning System *fu*
TWS Track While Scan [radar] *mt, fu, 0, hc, wp*
TWS-QR Track-While-Scan Quiet Radar *fu*
TWSB Twin Sideband *fu*
TWSC Time Standard Work Center *fu*

TWSR Track While Scan Radar *hc*
TWSRO Track While Scan On Receive Only *fu, wp*
TWT Traveling Wave Tube [wave generator used in Pulse-Doppler radars] *mt, fu, ro, he, uh, ah, hc, hn43.12.1, 0, 7, 9, 15, mj31.5.33, 19, lt, wp, ep, ce, hy, jdr*
TWTA Traveling Wave Tube Amplifier *mt, fu, ct, em, uh, sc2.7.2, 4, 7, 9, 15, 17, hc, lt, ep, ce, dm, jdr, kf*
TWTPS Traveling Wave Tube Power *fu*
TWTTY Tactical Warning Teletypewriter *mt*
TWU Tactical Weapons Unit [UK] *mt*
TWU Transport Workers Union *aw129.14.110*
TWX Joint Message Form *mt*
TWX Teletype Communications *su*
TWX Teletypewriter Exchange [Service] *fu, hc, do8, 0, w9, aw129.21.1, kg, hi*
TX Technology Exploitation *wp*
TX Transmit *mt, jdr, kf*
TX Transmitter [also XMTR] *fu, uh, gg129, 2, 4, wp*
TX/RX Transmitter/Receiver *fu*
TXA Terminal Exchange Area *hi*
TXA Track Crosstell Area *fu*
TXANG Texas Air National Guard *-12*
TXCOB Transmit Control Output, Block *fu*
TXCS Transmit, Control, Source *fu*
TXD Transmit Data *kg, hi*
TXEN Transmit Enable *fu*
TXKGD Transmit KG (Variable) Data *fu*
TXPNDR Transponder *uh, jdr*
TXT Text [format] *kg, hi*
TY Then Year *mt, kf*
TYCOA Type Configuration and Armament *fu*
TYMNET (the TYMNET access control system) *cd*
TYP Ten Year Plan *mt*
TYPEA Unit Type Code File System [WWMCCS] *mt*
TYPREP Type Unit Report *mt*
TYX (Va. based TYX Corp . . . specializes in productivity . . . software) *aw130.13.6*
TZ Time Zone [Unix] *pk, kg*
TZR Target Zero Range French-German Roland missile system *fu*

U

U Uluchshennyi [Improved] [USSR] *mt*
U Unity Matrix *0*
U Untersee [(German) "undersea", designator for submarine] [Germany] *mt*
U [designator for strategic reconnaissance aircraft, U-1 and U-2, USA] *mt*
U&S Unified and Specified *mt, jb*
U-GO [Japanese attack in Burma, against Imhpal and Kohima, cities in the border region of in India, 1944, that failed] *mt*
U-SCSI Ultra SCSI *hi*
U/A Using Assembly *fu*
U/C Up Converter *jdr*
U/I Unit of Issue *fu, tb*
U/I Urban / Industrial *tb, mt*
U/L Uplink *jdr*
U/M Unit of Measure *fu*
U/S Upper Stage *oe*
U/U Unencode / Undecode *hi*
U/UD U/U Decoding [UNIX] *hi*
U/UE U/U Encoding *hi*
U/V Under Voltage *jn*
UA Unnumbered Acnowledgement *fu*
UA User Agency [also known as " Contracting Officer"] *jdr, su*
UA User Agent *pk, kg*
UA User Area *pk, kg*
UA&A User Authentication and Authorization *mt*
UAA University of Alaska Anchorage *aw129.10.76*
UAAP USAREUR Automation Architecture Plan *mt*
UAC Unit Advisory Council *mt*
UAD Universal Attitude Determination *kf*
UADB Upper Air Data Base *pl*
UADPS-ICP Unified ADP System Inventory Control Points [US Navy] *mt*
UADPS-SP Unified ADP System Stock Points [US Navy] *mt*
UAE United Arab Emirates *w9, wp*
UAE Unrecoverable Application Error *ct, kg, em, hi*
UAF User Authorization Facility *mt*
UAH University of Alabama at Huntsville *st*
UAI Universal Application Interface *hi*
UAL Unit Authorization List *mt*
UAL United Air Lines *aw118.18.32, wp*
UAL Upper Address Lines *fu*
UAM Underwater to Air Missile *wp*
UAM User Authentication Method *pk, kg*

UAMHS User oriented Automated Message Handling System *mt*
UAMS USEUCOM Alert Management System *mt*
UAN User Access Network *fu*
UAP User Acceptance Program *mt*
UAQPSK Unbalanced Asynchronous Quadraphase Shift Keying *oe*
UAR United Arab Republic [Egypt] *10, w9, wp*
UARP Upper Atmosphere Research Program *hy*
UARRSI Universal Aerial Refueling Receptacle Slipway Installation *wp*
UARS Unmanned Air Reconnaissance System *wp*
UARS Upper Atmosphere Research Satellite [NASA] *mt, ge, mi, ac1.7-26, hc, ns, hy, mb, hg*
UART Universal Asynchronous Receive Transmit *mt*
UART Universal Asynchronous Receiver [/] Transmitter *jdr, ct, em, kg, hi*
UARTS Unmanned Air Reconnaissance and Targeting System *wp*
UARV Unmanned Air Reconnaissance Vehicles [USAF] *mt*
UAS Unified Antenna Structure *he, jn*
UAS University Air Squadron [UK] *mt*
UAS Upper Air Sounding [Sounder] *pl*
UAS User Attributes *fu*
UASC User Assignment Switch Control *-4*
UAV Unmanned Air (Aerial) Vehicle [USA DOD] *mt, fu, aa26.12.55, hc, 16, af72.5.35, aw130.6.43, kb, wp*
UAVM UAV Maneuver [USA] *mt*
UB Underwater Battery *fu*
UB Undistributed Budget *fu*
UB Usage Block *fu*
UBA Usable Band Width *oe*
UBD Utility Binary Dump *fu*
UBFC Underwater Battery Fire Control *fu*
UBG Underground Building *-10*
UBITRON Undulating Beam Interaction Electronic Tube *fu*
UBM Unpressurized Berthing Mechanism *mi*
UBR Unspecified Bit Rate *ct, em*
UBS Universal Breadboard Simulator [Harris] *-4*
UBV Ultraviolet Blue Visual *ai1*
UC Unit Cooler *fu*
UC Universal Console *fu*
UCC Uniform Code Council *wp*
UCC Uniform Commercial Code *jdr*
UCC Unit Control Center *mt*
UCCIS USAREUR Command and Control Information System [USA] *mt*
UCCS USEUCOM Command Center System *mt*
UCDC User Channel Doppler Corrector *-4*
UCE Utilization and Conversion of Energy *fu*
UCF Uniform Contract Format *mt*
UCFM Upconverter Frequency Multiplier *un*
UCG Underground Coal Gasification *wp*
UCI University of California at Irvine *fu*
UCI University of California, Irvine *hn41.19.5*
UCI Unsupervised Command Interface *-4*
UCI Utility Card Input *fu*
UCL Universal Communications Language *pk, kg*
UCL Upper Control Limit *fu*
UCLA University of California, [at] Los Angeles *fu, sc2.8.4, hn49.11.2, wp*
UCM Universal Cable Module *ct, em*
UCM User Command Message *fu*
UCM User Community Model *fu*
UCMJ Uniform Code of Military Justice [USA] *mt*
UCMS Unit Capability Measurement System [HQ USAF] *mt*
UCN Uniform Control Number *tb*
UCO Utility Compiler *fu*
UCP Unified Command Plan *mt, kf*
UCP Universal Communications Processor *fu*
UCP Utility Computer Program *fu*
UCR Unit Cost Report *mt*
UCR User Communicant Request *fu*
UCS Unicode Conversion Support *pk, kg*
UCS University Character Set *fu, kg*
UCS User Coordinate System *pk, kg*
UCSB University of California, Santa Barbara *id4091.2/1111*
UCSD Universal Communications Switching Device *fu*
UCSD University of California, San Diego *hn44.18.3*
UCT Uncorrelated Target *jb, cp, jmj*
UCT Universal Coordinated Time *pk, kg*
UD University of Dayton (Ohio) *wp*
UDB Undistributed Budget *fu, jdr*
UDB Unified Data Base [for acquisition logistics] *wp, mt*
UDC Universal Display Console *fu*
UDC Up-Down Converter *fu*
UDC User Defined Commands *pk, kg*
UDCT Unit Design Certification Test *fu*
UDDS User Data Distribution System *ac2.7-17*
UDE Universal Data Exchange *pk, kg*
UDE User Display Element *fu*

UDEC (UDEC Corporation, Waltham, Massachusetts) *ci.1023*
UDEC Universal Digital Electronic Computer [Burroughs] *pk, kg*
UDF Un Ducted Fan *mt, aw126.20.36, aa26.10.17*
UDF Unit Development Folder *4, fu, tb, oe*
UDF Universal Data Format *em*
UDF Universal Disk Format *em, kg, ct*
UDF User Defined Function *mt, pk, kg*
UDG User Defined Gateway *pk, kg*
UDI Unique Data Item (description) *fu*
UDI Unsupervised Data Interface *-4*
UDI Up-Down Indicator *fu*
UDI User Data Input *fu*
UDI/O User Data Input/Output *fu*
UDICON Universal Digital Communications Network *fu*
UDID Unique Data Item Description *wp*
UDITS Unified Diagnostic Test System *fu*
UDL Unclassified (Uniform) Data Link *fu*
UDL Unit Designation List *mt*
UDL Unit Detail Listing *fu*
UDM ULP Device Management *fu*
UDMH Unsymmetrical Di-Methyl Hydrazine [a propellant] *ci.169, es55.78, mb, mi, ai2*
UDO User Data Output *fu*
UDP Uplink/Downlink Processor *kf*
UDP User Datagram Protocol *mt, kg, ct, em, du*
UDR Urgent Data Request *fu*
UDS Universal Data Set *fu*
UDS Universal Documentation System *2, 4, oe*
UDS User Display Station *fu*
UDS User Display System *mt*
UDT Underwater Demolition Team *mt, hc*
UDT Uniform Data Transfer *pk, kg*
UDT Universal Data Transcriber *fu*
UDTI Universal Digital Transducer Indicator *fu*
UDU Unique Data Unit *fu*
UE Unit Equipage *mt*
UE Unit [of] Equipment *wp, fu*
UE User Equipment *mt*
UEFS User-Equipment Firmware Support *fu*
UEM User Ephemeris Message *uh, jdr*
UEP Undetected Error Probability *fu*
UEPR Unsatisfactory Equipment Performance Report *mt*
UER Unsatisfactory Equipment Report *fu*
UES UCCIS Engineer Subsystem [USA] *mt*
UF Ultrasonic Frequency *fu*
UF Unit of Fire *fu*
UFA Unnamed Fighter Aircraft *wp*
UFAS Unified File Access System *mt*
UFCS Underwater Fire Control System *fu*
UFD Unit Development Folder *17, sb*
UFD User File Directory *fu, jdr*
UFE Universal Front-End *fu*
UFF Universal File Format *jdr, ric*
UFI User Friendly Interface *mt*
UFIR Upper Flight Information Region *fu*
UFM USAFE Forms Mode [USAFE] *mt*
UFO UHF Follow On *mt, uh*
UFP Unit Focal Point *mt*
UFR Under Frequency Relay *fu*
UFS Unix File System *kg, hi*
UFT Unified File Transfer *mt*
UFV Unmanned Flight Vehicle *hc*
UGC Uppsala General Catalog *mi*
UGM Underwater launched, Guided ground-to-ground Missile [USA] *mt*
UGS Unattended Ground Sensor *wp*
UGS Uninterruptible Gas Supply *oe*
UGT Under Ground [nuclear] Test *mt, fu, st*
UH Unit Heater *fu*
UH Utility Helicopter *ph*
UH Utilization Hours *jj*
UH [helicopter designator, UH-1 Iroquois, UH-60 Black Hawk Helicopter, USA DOD] *mt*
UHB Ultra High Bypass [engine] *mt, aw126.20.34*
UHB User Home Base *es55.71*
UHF Ultra High Frequency [225 to 3000 Megahertz] *mt, ah, ro, uh, nu, mi, fu, hc, pl, sc3.5.5, 0, 12, 15, 17, aw129.14.109, w9, pf.f88.10, wp, ep, tb, jdr, pbl, tm, mb, kf, hi, oe, tt, ai1*
UHFDF UHF Direction Finding *fu*
UHFTT UHF Transition Terminal *fu*
UHF_F/O Ultra High Frequency Follow-On *uh, kf*
UHR Ultra High Resistance *fu*
UHR Ultra High Resolution *wp*
UHRBS Ultra High Rate Bit Synchronizer *-4*
UHRD Ultra High Rate Demodulator *-4*
UHRIR Ultra High Resolution Image Recorded *hc*
UHRR Ultra High Resolution Radar *hc*

UHSBL Ultra High Speed Bipolar Logic *hc*
UHU Unterstutzungs Hubschrauber [support helicopter, Germany] *mt*
UIC Unit Identification (Identity) Code *mt, fu, wp*
UIC Urban Industrial Complex *mt*
UIC User Identification Code *kg, ct, em*
UIC User Information Center *cd, hd1.89.y14*
UICC Universal Interface to Controller Computer *fu*
UICP Uniform Inventory Control Point *wp*
UID User Identifier *pk, kg*
UIMS User Interface Management System *pk, kg*
UIN [reverse NASCOM interface unit] *-4*
UIOD User I/0 Device *fu*
UIP Unmanned Instrument Platform *fu*
UIPT User Integrated Product Development Team *kf*
UIR User Interface Requirements *mt*
UIS User Interface Specification *mt*
UIT Ultraviolet Imaging Telescope [Astro package] *mi*
UIUC University of Illinois, Urbana-Champaign *aw130.11.75*
UJT Uni Junction Transistor *fu, 6, wp*
UK United Kingdom *mt, ge, hc, w9, es55.44, jb, wp, ns, tb, hg*
UKADGE United Kingdom Air Defense Ground Environment *mt, fu, sc4.9.6, 12, go3.4.1, tb*
UKADR United Kingdom Air Defense Region [UK] *mt*
UKADS United Kingdom Air Defense System *fu*
UKAIR United Kingdom Air Command and Control *mt*
UKAIR United Kingdom Air Forces [NATO] *mt*
UKIRT United Kingdom Infra Red Telescope *wp*
UKO Undesirable Known Object *fu*
UKSL UKADGE Systems Ltd. *fu*
UKST United Kingdom Schmidt Telescope *mi*
UKUSA United Kingdom-United States Agreement *-10*
UKV Udarnye Kosmicheskie Vooruzheniya [Space based offensive weapons] [USSR] *mt*
UKWMO United Kingdom Warning and Monitoring Organisation *-12*
UL UnderLay *st*
UL Underwriters' Laboratories *fu, hc, wp, kg, su, ct, em, hi*
UL Unordered List *ct, kg*
UL Up Load [stations] *kg, ab.13, jdr, hi*
UL Uplink *uh*
UL Usage List *fu*
ULA Uncommitted Logic Array *pk, kg*
ULA Unit Level Automation Program *mt*
ULA [STDN ground station, Fairbanks, Alaska] *ac2.7-10, 4*
ULA/ILA Unauthorized and Inadvertent Launch Analyses *fu*
ULAIDS Universal Locator Airborne Integrated Data System *hc*
ULANA Unified Local Area Network Architecture *mt, wp*
ULC Unit Level Code *mt*
ULC Unit Level Computer [USA] *mt*
ULCE Unified Life Cycle Engineering *wp*
ULCS Unit Level Circuit Switch [TRI-TAC] *fu, hc, mt*
ULD Upper Level Discriminator *pl*
ULDA Uniform Low Dispersion Archive *es55.42, ns*
ULE UnderLay Experiment *st*
ULF Ultra Low Frequency [below 30 Hertz] *0, fu, hc, aa26.11.b48, tb*
ULG Universal Logic Gate *fu*
ULI Up Link Interface *oe*
ULI Useful Longitude Increment *fu*
ULLS Unit Level Logistics System [USA] *mt*
ULM Ultra Light Motorized [aircraft] *mt*
ULMS Undersea (Underwater) Long- Range Missile System *fu*
ULMS Underwater Launched Missile System [later Trident] *hc*
ULMS Unit Level Message Switch *mt, hc*
ULN Unit Line Number [USA DOD] *mt*
ULN Universal Link Negotiation *pk, kg*
ULO Unmanned Launch Operations *ci.943*
ULP Upper Layer Protocol *mt, fu*
ULPC Up Link Power Control *jn*
ULR USAREUR Logistics Review *mt*
ULS Unit Level Switch *mt*
ULS Up Load Station *sd*
ULSA Ultra Low Sidelobe Antenna [USAF] *wp, mt*
ULSDS Ultra Large Screen Display System *mt*
ULSI Ultra Large Scale Integration *kg, fu, hi*
ULSP USAREUR Logistics Support Plan *mt*
ULT Underwater Location Transmitter *wp*
ULV Unmanned Launch Vehicle *wp*
UMA Unified Memory Architecture *pk, kg*
UMA Upper Memory Area *ct, em, hi*
UMB Upper Memory Block [LIM/AST] *ct, em, kg, hi*
UMC Unit Movement Coordinator [USA] *mt*
UMD Unit Manning Document *mt, kf*
UMD Unit Movement Data *mt*
UME Unformatted Message Element *hc, fu*
UMER Undetected Message Error Rate *fu*
UMLA Universal Multiple Line Adapter *mt*
UMMIPS Uniform Material Movement and Issue Priority System *mt*

UMMIS UCCIS Medical Management Information Subsystem [USAREUR] *mt*
UMN Unsatisfactory Material Notice *fu*
UMOP Unintentional Modulation On Pulse *fu*
UMPS Unit Mission Planning System [SAC] *mt*
UMR Unsatisfactory Material Report *fu, oe*
UMR Upper Maximum Range *fu*
UMR User Measurement Report *fu*
UMS Unified Message Switch *fu*
UMS Uniform Management System *fu*
UMS Utility Management System *wp*
UMSE Unit Maintenance Support Equipment *fu*
UMT Universal Mean Time *ai1*
UMT Universal Military Training *w9*
UMTS Universal Mobile Telecommunications System *rih*
UMV Unmanned Aerial Vehicle *fu*
UN United Nations *mt, ge, aa26.12.37, sc3.5.3, 13, w9, sf31.1.11, wp, hy, hg*
UNAAF Unified Action Armed Forces [USA DOD] *mt*
UNACE Universal Aircraft Com/Nav Evaluation *mt*
UNAS Universal Network Architecture Services *kf*
UNB Universal Navigation Beacon *fu*
UNC (UNC Incorporated) [aerospace, aviation, and defense company] *aw129.21.126*
UNC Unbalanced Normal Class *fu*
UNC Unencoded Netnews Collator [Unix] *pk, kg*
UNC Unified National Coarse *jdr*
UNC United Nations Command [Korea] *mt*
UNC Universal Naming Convention *ct, kg, em, hi*
UNCAIRCOMP United Nations Command Air Component [Korea] *mt*
UNCLE Unstable Resonator Conductively Cooled Laser Equipment *hc*
UNCLOS United Nations Convention on the Law Of the Sea *ci.1044, 14*
UNCOL Universal Computed Oriented Language *kg, hi*
UND Urgency of Need Designator *mt*
UNEP United Nations Environment (Environmental) Program *ge, pl, hg, hy*
UNESCO United Nations Educational Scientific Cultural Organization *ac3.1-42, w9, wp, hy*
UNF Unified National Fine *jdr*
UNFO Undergraduate Naval Flight Office *hc*
UNGA United Nations General Assembly *wp*
UNICOM Universal Integrated Communications [System] *kg, hi*
UNICORNS Uniformity Correction for Nonscanned System *hc*
UNICOS Universal Compiler FORTRAN compatible *pk, kg*
UNIDATA University Data Broadcast Project *hy*
UNIEF USEUCOM Nuclear Interface Element *mt*
UNIFET Unipolar Field-Effect Transistor *fu*
UNII Unlicensed National Information Infrastructure *pk, kg*
UNIMOP Unintentional Frequency Modulation Of (Transmitted) Pulse *fu*
UNISAT United Satellite, Limited [British communications satellite] *sc3.2.7*
UNITRACK Unit Tracking Function [NWSS] *mt*
UNITREP Unit Status and Identity Report [formerly FORSTAT] *mt*
UNITREPCSRC UNITREP System Chief of Staff's Readiness Chart [HQ USAF] *mt*
UNIVAC Universal Automatic Computer [Hughes Aircraft Company customer for TCDD Tower Cab Digital Display] *ct, kg, hc, em, hi*
UNIVAC University Automatic Computer *mt*
UNIX Uniplexed Information and Computer Systems *em*
UNMA Unified Network Management Architecture *kg, ct, em*
UNOLS Upgrade Nuclear Ordnance Logistics System *mt*
UNPROFOR United Nations Protection Force *mt*
UNPS Universal Power Supply *fu*
UNREP Underway Replenishment *mt, fu*
UNS Universal Navigation Corp *aw124.21.23*
UNSSD United Nations Special Session on Disarmament *-14*
UNT Undergraduate Navigation Training *mt*
UNT Unified Network Technology (USN program) *fu*
UOC Ultimate Operational Capability *0, fu*
UOES User Operational Evaluation System *kf*
UOL Underwater Object Locator *fu*
UOR Urgent Operational Requirements *mt*
UP Unrotated Projectile [a codename used for rockets, UK] *mt*
UP Unsolicited Proposal *mt*
UP Update Processor *mt*
UP Utility Program *fu*
UPAMS Unpriced Action Management System *mt*
UPC Unit Product Code *mt*
UPC Unit Production Cost *fu, wp*
UPC Universal Product Code *ct, kg, w9, wp, hi*
UPC UP Converter *jdr*
UPC Usage Parameter Control *ct, em*
UPC User Permit Code *mt*
UPD [designator for SLAR models] (UPD-4 is SLAR on Rf-4b; no Real Time, used by Marines) (UPD-6 is SLAR on Rf-4e; Real Time data link, used by Germany) (UPD-8 is SLAR on Rf-4c; long range, Real

Time data link) [USAF] (UPD-9 is similar to UPD-8 but no long *wp*
UPF User Productivity Facility *mt*
UPIR Uniform Photo Interpretation Report *fu*
UPL Universal Program Library *fu*
UPM User Profile Management [IBM] *pk, kg*
UPN Unique Project Number *oe*
UPN Unprocurable Part Notice *fu*
UPS Unattended Power Source *mt*
UPS Underwater Production System *hc*
UPS Uniform Procurement System *wp*
UPS Uniform Property System *cd, jdr*
UPS Uninterrupted Power System *sd*
UPS Uninterruptible Power Supply [System] *mt, tb, jdr, kf, fu, ct, em, kg, hi*
UPS Universal Polar Stereographic *fu*
UPS Upper Perigee Stage *hc*
UPS User Profile System *mt*
UPSTAGE Upper Stage Acceleration and Guidance Experiment *hc*
UPSTP Universal Power Supply Test Position *fu*
UPSTS Universal Power Supply Test Set *fu*
UPT Undergraduate Pilot Training [USA] *fu, wp, mt*
UPT Unit Performance Testing *jn*
UPT/UNT (career field . . . figures do not include UPT/UNT . . . students) *af72.5.46*
UQPSK Unbalanced QuadraPhase Shift Keying *2, 4, ric, jdr, oe*
UQT Unit Qualification Training *kf*
UR Universalnaya Raketa [universal rocket] *ai2*
URA Universities Research Association *fu*
URADS Unmanned Rendezvous and Docking Satellite *hc*
URC Uniform Resource Characteristics *pk, kg*
URC Uniform Resource Citation *pk, kg*
URD Unit Reference Designator [descriptor] *ep, jdr, pbl, tm, jn, ah*
URD User Requirements Document *mt*
URDB User Requirements Data Base [DDN] *mt*
URE User equivalent Range Error *jdr*
UREO UNIX RSCS Emulation Protocol *hi*
UREP Unix RSCS Emulation Protocol [protocol] *kg, hi*
URI Unfavorable Rotation Index *mt*
URI Uniform (Universal) Resource Identifier *pk, kg*
URI Uniform Resource Indicator *em*
URL Uniform (Universal) Resource Locator *kg, nu, jr, ct, em, hi, du*
URL User Reference Language *fu*
URN Uniform Resource Name / Number *kg, hi, du*
URN Unit Requirement Number *mt*
URO User Readout *fu*
UROS User Readout Simulator *hc*
URP Unit Record Processors *mt*
URPOP Urban Population File *mt*
URR Ultra Reliable Radar *mt, hc, wp*
URR User Rapid Report *fu*
URS User Requirement Specification *fu*
URSI International Union of Radio Science *mj31.5.15, ns*
URSIGRAM [message from URSI] *no*
URT Universal Radar Tracker *hc*
US Under Secretary *dt*
US Uniform System (Lens Marking) *fu*
US Unit Separator [information] *do242, fu, kg*
US United States *su, w9, sf31.1.2, tb, wl, kf, oe*
US Upper Stage *gs1-1.19*
USA United Space Alliance *ai2*
USA United States Army *mt, w9, wp, tb, fu, su*
USA United States of America *mt, ge, w9, aa26.11.b55, tb, hg*
USAAA US Army Audit Agency *mt*
USAAC United States Army Air Corps [USA] *mt*
USAADASCH United States Army Air Defense and Artillery School *st*
USAAF United States Army Air Force [USA] *mt*
USAAG US Army Artillery Group *mt*
USAAKA United States Army Kwajalein Atoll [formerly KMR] [also USAKA] *-17*
USAAS United States Army Air Service [created May, 1918, USA] *mt*
USACC United States Army Communications Command *fu, tc, st*
USACC US Army Communications Command *mt*
USACDC US Army Combat Development Command *mt*
USACE US Army Corps of Engineers *mt*
USACEAC United States Army Cost and Economic Analysis Center *nt*
USACEEIA-CONUS United States Army Communications Engineering Electronics Installation Agency - Continental United States *mt*
USACGSC US Army Command and General Staff College *mt*
USACSA US Army Communications System Agency *mt*
USACSG United States Army Command Support Group *mt*
USACSLA United States Army Communications Security Logistics Agency *mt*
USACSLA US Army Communications Security Logistics Activity *mt*
USACSSAA [Hughes Aircraft Company customer for Korean intelligence support system] *hc*
USACSSEC United States Army Computer Systems Support and Evaluation Command *mt*

USACTA United States Army Central TMDE Activity *st*
USADEA US Army Development and Employment Agency *fu*
USAEC [Hughes Aircraft Company customer for adaptive antenna control] *hc*
USAECOM US Army Electronics Command *fu*
USAEDS United States Atomic Energy Detection System *af72.5.128*
USAESEIA US Army Electronic Systems Engineering and Installation Activity *mt*
USAETL [Hughes Aircraft Company customer for SPARTIS space related tactical intelligence systems] {cfr. AETL} *hc*
USAEV United States Army Experimental Verification *tb*
USAF United States Air Force [created 1947, USA] *mt, fu, mi, su, aa26.12.9, aw130.13.5, sc2.7.1, 13, 16, pl, hc, w9, sf31.1.8, af72.5.5, nt, no, tb, st, wl, mb, kf, oe, ai2*
USAF/MP_EAB USAF Manpower Personnel Emergency Actions Book *mt*
USAFA United States Air Force Academy *mt, af72.5.9, wp*
USAFAD US Army Field Artillery Detachment *mt*
USAFADWC USAF Air Defense Weapons Center [TAC] *mt*
USAFA_MID US Air Force Academy Micros in the Dormitory *mt*
USAFE United States Air Forces in Europe *mt, fu, hn44.9.10, af72.5.47, wp*
USAFETAC USAF Environmental Technical Applications Center *ag*
USAFHRC United States Air Force Historical Research Center *mt, af72.5.47*
USAFLANT [Air Force Command, Langley AFB, Virginia] *af72.5.31*
USAFM Undetectable SAFM *jdr*
USAFOMC United States Air Force Occupational Measurement Center *af72.5.154*
USAFPP USAF Personnel Plan *mt*
USAFR United States Air Force Reserve *mt, ci.300*
USAFSAM USAF School of Aerospace Medicine *mt*
USAFSBSS USAF Standard Base Supply System *mt*
USAFSO United States Air Force Southern [Command] *wp*
USAFSS United States Air Force Security Service *mt, 10, hc, tb*
USAHEL United States Army Human Engineering Laboratory *st*
USAHSC United States Army Health Services Command *st*
USAICS US Army Intelligence Center and School *mt*
USAINTC United States Army Intelligence Command *hc*
USAISC United States Army Information Systems Command *mt, dt*
USAISEC US Army Information Systems Engineering Center [formerly USAISSSC] *mt*
USAISMA US Army Information Systems Management Activity *mt*
USAISSSC United States Army Information Systems Software Support Command *mt*
USAKA United States Army Kwajalein Atoll [formerly KMR] [also USAAKA] *sb*
USAM Unique Sequential Access Method *hi*
USAMMA United States Army Medical Materiel Agency *mt, st*
USAMMC US Army Maintenance Management Center *fu*
USAMRDC United States Army Medical Research and Development Command *st*
USAMSSA US Army Management Systems Support Agency *mt*
USAOTEA US Army Operational Test and Evaluation Agency *fu*
USAR US Army Reserve *mt*
USARADB US Army Air Defense Board *fu*
USARADCOM US Army Air Defense Command *fu*
USAREUR United States Army in Europe *mt, 12, fu*
USARI {US Army Research Institute} [Hughes Aircraft Company customer for computerized training system] *hc*
USARO United States Army Research Office *tc*
USARSPACE United States Army Space Command *jb*
USARSPACECOM {United States Army Space Command} [Kwajalein] *go2.4*
USARSPOC {United States Army Space Operations Center} *jmj*
USART Universal Synchronous - Asynchronous Receiver / Transmitter *fu, kg, hi*
USAS USA Standard *fu*
USASA United States Army Security Agency *hc, fu*
USASATCOMA US Army Satellite Communications Agency *mt*
USASCII United States of America Standard Code for Information Interchange *mt*
USASDC United States Army Strategic Defense Command *mt, 5, sb, th4, nt, jmj, tb, tb, st*
USASI United States American Standards Institute *fu*
USASTC US Army Signal Training Center *fu*
USASTRAT-COM United States Army Strategic Communications Command *hc*
USAT Unit Air Staff Tables *fu*
USAT User Satellite *vv*
USATD United States Army Training Devices *tc*
USATESAE United States Army Test Equipment Support Agency *mt*
USATSCH United States Army Transportation School *mt*
USA_SDC United States Army Strategic Defense Command *sb*
USA_SDG United States Army Space Defense Group *sb*
USB Unified S Band [transponder] *fu, ci.129, 2, 4*
USB Universal Serial Bus [Intel] *ct, em, kg*
USB Upper Side Band *mt, fu*

USB [on a map of NASA's Kennedy Space Center] *ci.85*
USBC Universal Serial Bus Connector *hi*
USBR United States Bureau of Reclamation *ac1.5-19*
USBRO United States Bases Requirements Overseas [WWMCCS] *mt*
USC Unit Source Listing [AFCC*] *mt*
USC United States Code *mt, su, ag, hc, af72.5.49, tb, wl*
USC University of Southern California *sc2.8.4*
USC User Spacecraft *-4*
USC/ISI University of Southern California/Information Sciences Institute *mt*
USCEEIA US Army Communications - Electronics Engineering Installation Agency *mt*
USCENTAF US Central Command Tactical Air Force *mt*
USCENTCOM United States Central Command *mt, ro*
USCG United States Coast Guard *mt, fu, su, ag, w9, hc, wp*
USCGNET United States Communications Grid Network *mt*
USCI United Satellite Communications, Incorporated *hn43.23.2*
USCIB United States Communications Intelligence Board *-10*
USCIB/IC USCIB Intelligence Committee *-10*
USCINCAFRED United States Commander In Chief, Air Force Readiness Command *mt*
USCINCARRED U.S. Commander in Chief, Army Forces Readiness Command *mt*
USCINCCENT United States Commander in Chief Central Command *mt, go3.1, wp*
USCINCEUR United States Commander In Chief, European Command *mt, wp*
USCINCLANT United States Commander In Chief, Atlantic Command *mt, go3.1, wp*
USCINCNORAD {United States CINC NORAD} *go3.1*
USCINCPAC United States Commander In Chief, Pacific Command *mt, go3.1, wp*
USCINCPACE [Commander In Chief, US Space Command] *wp*
USCINCRED United States Commander In Chief, Readiness Command *wp, mt*
USCINCSAC {United States CINC SAC} *go3.1*
USCINCSO United States Commander In Chief, Southern Command *mt*
USCINCSOCOM United States Commander In Chief, Southern Command *go3.1, wp*
USCINCSPACE United States Commander in Chief, Space [Command] *go2.3.2.1, jb, nt, jmj, as*
USCINCSPACE United States Space Command Commander in Chief *kf*
USCINCTRANSCOM {United States CINC Transportation Command} *go3.1*
USCM Unmanned Spacecraft Cost Model *kf*
USCS US Customs Service *mt*
USCS User Spacecraft Simulator *-4*
USCSB United States Communications Security Board *hc*
USD Under Secretary of Defense *hc, wp*
USDA Under Secretary of Defense for Acquisition *mt*
USDA United States Department of Agriculture *mt, ac1.2-6, w9, hy, y2k*
USDCFO US Defense Communications Field Office *mt*
USDEC Undesirable Signal Data Emanation *sd*
USDEL US Delegation *mt*
USDI (the USDI facility at Sioux Falls) *ac1.5-19*
USDP Under Secretary of Defense for Policy *mt, ph*
USDR&E Under Secretary of Defense for Research and Engineering [OSD] *mt*
USDRE Under Secretary of Defense for Research and Engineering *ph, tb*
USE UNIVAC Systems Exchange *mt*
USE/IT User System Evaluation and Integration Tool *fu*
USEC Underwater Systems Engineering Center *fu*
USEIT User System Evaluation and Integration Tool *mt*
USENET User Network [Internet] *ct, em, kg*
USERID User Identification *mt, pk, kg*
USES Universal Source Encoder for Space *kf*
USEUCOM United States European Command *wp, mt*
USFISC United States Foreign Intelligence Surveillance Court *-10*
USFJ United States Forces, Japan *mt*
USFK United States Forces, Korea *mt*
USFL United States Football League *hn43.11.6*
USFR USAF Reserve *wp*
USFS United States Forest Service *hy*
USFU Unglazed Structural Facing Units *fu*
USG United States Gage *fu*
USG United States Government *mt, fu, kf*
USGPO United States Government Printing Office *-12*
USGS United States Geological Survey *mt, fu, ac1.5-19, aw130.22.105, wp, ns, no, hy*
USI User System Interface *mt*
USIA United States Information Agency [later, International Communications Agency] *mt, su, 0, ci.913, 13, w9, wp, tb*
USIB United States Intelligence Board *hc*
USIC US International Carrier *mt*
USICA US International Communication Agency *mt*
USIDF United States Icelandic Defense Forces *mt*
USIS United States Identification System *hc*
USKAC United States Key Access Code *mt*

USL UNIX Systems Laboratories *ct, em, hi*
USLANTCOM United States Atlantic Command *mt, wp, af72.5.31*
USLO United States Liaison Officer *dt*
USM Underwater to Surface Missile *fu*
USM Unscheduled Maintenance *mt*
USM User Security Matrix *mt*
USM User Support Management *mt*
USMA {United States Marine Academy} (Cadet, USMA, ASAFA, or ROTC Contract/Scholarship) *af72.5.151*
USMC United States Marine Corps *mt, fu, su, ci.295, 16, hc, w9, af72.5.143, wp*
USMCC United States Mission Control Center *mt*
USMCEB United States Military Communications - Electronics Board *mt*
USML United States Microgravity Laboratory *aa26.12.46*
USML United States Munitions List *fu*
USMP United States Microgravity Payload *aa27.4.48, mb, mi*
USMR [Hughes Aircraft Company customer for TMCS telemetry controller system] *hc*
USMS United States Meteorological Service *wp*
USMTF United States Message Text Format(ting) *mt, kf*
USN United States Navy *mt, fu, su, ci.294, 16, pl, hc, w9, sf31.1.27, aa27.4.20, wp, tb, ai2*
USNATO US Mission to NATO *mt*
USNAVCENT US Naval Forces Central Command *mt*
USNAVEUR US Naval Forces Europe *mt*
USNI United States Naval Institute [USA] *mt*
USNMR US National Military Representative [SHAPE] *mt*
USNO United States Naval Observatory [Washington, D.C.] *-4*
USNR United States Navy Reserve [Ready] [USA] *mt*
USNS United States Naval Ship [DOD] *mt, aw130.13.19*
USNSTN United States Naval School, Transportation Management *mt*
USO Ultra Stable Oscillator *pl, re, vv*
USO United Services Organization *w9, an31.13.6, wp*
USOC Universal Service Ordering Code *pk, kg*
USOC User Support and Operations Centre *es55.71*
USP United States Pharmacopoeia *w9, wp*
USP Uplink Signal Processor *jdr, kf*
USPACOM United States Pacific Command *mt, wp*
USPS United States Postal Service *mt, w9, wp*
USQX [AN/USQ-104 MDI System] *-17*
USR U.S. Robotics [a corporation] *ct, kg, em, hi*
USREDCOM United States Readiness Command *mt, wp*
USRMCLO United States Representative NATO Military Committee Liaison Office *dt*
USRSDC United States ROSAT Science Data Center *ns*
USRT Universal Station Readiness Test *-4*
USRT Universal Synchronous Receiver / Transmitter *kg, hi*
USS United States Ship (Steamer) *mt, aw130.13.19, sm18-12, w9, af72.5.32, an31.13.8, wp*
USS United States Standards *fu*
USS User Service Support *-4*
USS User Source Synchronization *fu*
USS User Support Subsystem *mt*
USS [Deputy] Undersecretary [of the Air Force for] Space Systems *dt*
USSA User Supported Software Association [United Kingdom] *pk, kg*
USSB United States Satellite Broadcasting *sc3.11.5, hc*
USSC User Source Sync Channel *fu*
USSOCOM United States Special Operations Command *mt, kf*
USSOUTHCOM United States Southern Command *mt, wp*
USSP User Support Software Package *es55.42*
USSPACECOM United States Space Command *mt, af72.5.68, ts7-37, go1.2-1, wp, tb, as, kf*
USSPC United States Space Command *kf*
USSR Union of Soviet Socialist Republics *ge, ph, w9, sf31.1.1, es55.20, aw129.21.9, is26.1.70, wp, tb, st, hg*
USSS US SIGINT System *mt*
USSSI United States Satellite Services, Incorporated *sc3.2.3, 8*
USSTRATCOM United States Strategic Command *mt, kf*
UST UHF SATCOM Terminals [ESD] *mt*
UST Undergraduate Space Training *mt*
UST Underground Storage Tank *wp*
UST Undersea Technology *fu*
UST USAFE Sortie Tables [USAF] *mt*
USTF Uniformed Services Treatment Facilities *mt*
USTR United States Trade Representative *aw129.1.13*
USTRANSCOM United States Transportation Command *mt, ro, af72.5.140*
USTS Ultra High Frequency Satellite Terminal System [MAC] *hc, wp, mt*
USTS United States Telecommunications Security *tb*
USTTI United States Telecommunications Technology Institute *hn43.22.3*
USTV United Satellite Television *dn*
USUC US Unilateral Control *mt*
USUHS Uniformed Services University of the Health Sciences *dt*
USUTC United States Unified Transportation Command *mt*

USW Under Sea Warfare *fu, hc*
USWB United States Weather Bureau *mt*
USWest [telephone company that serves Colorado Springs area] *kf*
US_Roland [a short range, low altitude, all-weather, Army air defense artillery surface-to-air missile system which is based upon the FrancoGerman Roland III missile system] [USA DOD] *mt*
UT Umbilical Tower *oe, ai2*
UT Unit Test *fu*
UT Universal Time [a.k.a. GMT (Greenwich Mean Time) or Zulu Time] [see UTC] *mt, 0, w9, es55.97, wp, no, ce, mb, kf, mi, fu*
UT&E User Test and Evaluation *fu*
UTA (another Tracor/UTA Industries Super Guppy Transport) *aw129.21.13*
UTA University of Texas, Austin *aa26.11.53*
UTACCS USAREUR Tactical Command and Control System *mt*
UTAIN USAFE Tactical Air Intelligence Network *mt*
UTAIS USAFE Tactical Air Intelligence System *mt*
UTANG Utah Air National Guard *-12*
UTAPS Universal Tank-Automotive Prognostics System *hc*
UTC Unified Technology Center [of United Aircraft] *ci.219*
UTC Unified Transportation Command *mt, wp*
UTC Unit Time Coding *fu*
UTC Unit Type Code *mt*
UTC United Technologies Center *ci.968*
UTC United Technologies Corporation *aw124.21.20*
UTC Universal Time Code *mt, kf*
UTC Universal Time, Coordinated [a.k.a. Zulu] [subtly different from UT] *fu, mi, kg, uh, 2, 4, w9, wp, tb, jdr, oe, ai2*
UTC User Transmission Command *fu*
UTCE Uniform Task Cost Estimating *fu*
UTD Uniform Geometric Theory (of Refraction) *fu*
UTDC Unclassified Technical Document Centers *fu*
UTDF Universal Tracking Data Format *2, 4*
UTE Aircraft Utilization Rate *mt*
UTE Flying Hour Utilization *mt*
UTE Universal Terminal Emulator *kf*
UTE User Terminal Element *mt*
UTEELRAD Utilization of Enemy Electromagnetic Radiation *fu*
UTF Underground Test Facility *wp*
UTF Universal Tracking Format *-4*
UTFCS Universal Tank Fire Control System *hc*
UTH USAREUR Transportable Host *mt*
UTI Universal Text Interchange/Interface *pk, kg*
UTIC USAREUR Theater Intelligence Center *mt*
UTL (UTL Corporation) [an avionics/military electronics company] *aw130.22.49*
UTLFE User Traffic Link Forward Equipment *-4*
UTLRE User Traffic Link Return Equipment *-4*
UTM Universal Transverse Mercator (projection) *mt, fu, ac2.2-10, 0, pbl*
UTM Universal Turing Machine *sf31.1.15*
UTMC United Technologies Microelectronics Center *aw130.22.115*
UTP Unit Test Plan *fu*
UTP Unshielded Twisted Pair [cable] *ct, em, kg, nu, hi*
UTP USAREUR Theater Plan *mt*
UTR Unit Test Report *fu*
UTROF (in the UTROF experiment . . . photos . . . to assess . . . impact) *ac3.1-43*
UTS Universal Terminal System [usually followed by a number, as in UTS-20, etc.] *mt*
UTS Universal Test Station *fu*
UTS Universal Time Standards *fu*
UTTAS Utility Tactical Transport Aircraft System *fu*
UTTCo Utility Tactical Transport helicopter Company [USA] *mt*
UTTR Utah Test and Training Range *12, wp*
UTX Universal Transaction eXchange *em*
UUID Universal Unique Identifier *pk, kg*
UUM Underwater-to- Underwater Missile *fu*
UUM {Underwater to Underwater Missile} [submarine launched, anti submarine missile that is is launched underwater, breaks trough the surface, travels through air, and then drops its nuclear warhead into the water] [USA] *mt*
UUT Unit Under Test *pk, mt, fu, rm1.1.4, 17, hc, ie5.11.8, wp, jdr, oe*
UUV Unmanned Undersea Vehicle *mt, kb*
UV Ultraviolet [0.75 - 30 Phz] *mt, kg, ge, mi, fu, nu, ac2.2-17, 17, sb, aa26.12.45, w9, wp, eo, tb, st, mb, hg, kf, jn, hi, ai1*
UV Under Voltage *oe*
UV-VS Upravlenie Voenno-Vozdushnikh Sil [administration of the Air Force] [USSR] *mt*
UV/OL Undervoltage/Overload *ma80*
UVA University of Virginia *id4091.2/1111*
UVASER Ultraviolet Amplification by Stimulated Emission of Radiation *fu*
UVD Undervoltage Device *fu*
UVEPROM Ultraviolet Erasable PROM *fu*
UVHP (studies will include mounting the UVHP on spinning section) *ma9*
UVI Ultraviolet Imager *-16*
UVR Ultraviolet Radiometer *wp*

UVS Ultra Violet Sensor *-16*
UVS UltraViolet Spectrometer *mi, mb, jp*
UVSI UltraViolet Spectrometer Imager *es55.26*
UVSP Ultraviolet Spectrometer/Photometer *re*
UVUN Unique Variable Update Number *fu*
UW Unconventional Warfare [USA DOD] *wp, fu, mt*
UW University of Wisconsin *gd4-5*
UW-SCSI Ultra Wide SCSI *hi*
UWC Under Water Communications *mt*
UXART High Speed UART *hi*
UXB Unexploded Bomb *mt, w9, wp*
UXO Unexploded Ordnance [bomb, warhead, munitions, etc.] [USA] *wp, mt*

V

V [designator, as in V-22A Osprey] *mt*
V [stands for "squadron" in many US Navy acronyms, e.g. VA, VAQ, VBF, VC, VF, VFA, VFC, VMA, VMAQ, VMAT, VMB, VMCJ, VMF, VMFA, VMFAT, VMO, VQ, VS, VT, VXE] *mt*
V&H Vulnerability and Hardness (Hardening) *jdr, kf, st*
V&V Verification and Validation [also VV] *mt, em, hc, pl, tb, st, kf, kg, fu, oe*
V-EPUU VHSIC-Enhanced PLRS User Unit *fu*
V-MAN Vehicle Management *mt*
V-SMP VHSIC Signal Message Processor *fu*
V-VS Voenno-Vozdushniye Sily [Air force] [USSR] *mt*
V/HUD Vertical/Head Up Display *hc*
V/STOL Vertical / Short Take Off [and] Landing [aircraft] *mt, aa26.12.2, hc, w9, fu, ai1*
V/STOL Vertical / Standard Take Off [and] Landing [aircraft] *ai1*
VA Veterans' Administration *mt, hc, w9, hn49.9.4, af72.5.119, fu, y2k*
VA Voltage Variable Attenuator *fu*
VA Vulnerability Assessment *wp, tb*
VA {US Navy Squadron}, Attack *mt*
VAB Variable Action Button *fu*
VAB Vehicle Assembly Building [ELS facility, formerly Vertical Assembly Building] *bl1.2-238, jdr, mb, mi9, wl, ai2*
VAB Vertical Assembly Building *fu, 9, wl*
VAB Vilamor Air Base *fu*
VAC Value Added Carrier *dn*
VAC Variance At Completion *17, sb, jdr*
VAC Volts, Alternating Current *mt, hd, fu, ct, em, hi, ai2*
VACALES Variable Character Length Synchronous *fu*
VACE Versatile Automatic Control Equipment *hc*
VACI Video/Chroma and Analog Integrator *fu*
VACOSA Versatile Automatic Connector and Signal Assigner *fu*
VAD Value Added Dealer *em, kg, hi*
VAD Veterans Affairs Department *wp*
VAD Vulnerability Assessment Device *fu*
VAD&D Voluntary Accidental Death and Dismemberment [insurance] *hn49.8.2*
VADD Value Added Disk Driver *em, kg, hi*
VADER Vulnerability Assessment to Directed Energy Program *hc*
VADGE Venezuelan ADGE *fu*
VADM Vice Admiral *w9*
VADSL Very-High-Rate Asymmetric Digital Subscriber Line *em, kg*
VAFB Vandenberg Air Force Base [Lompoc, California] *hc, ci.568, 16, 17, tb, wl, mb, kf, mi, ai2*
VAG Vector Address Generator *fu*
VAID Variable Air Inlet Duct [USA] *mt*
VAL Vehicle Authorization List *wp*
VAL Voice Application Language *kg, em*
VAL {US Navy Squadron}, Attack, Light (VAL-4 'Black Ponies') [USA] *mt*
VALT VTOL Approach and Landing Technology *wp*
VALU Vector Arithmetic/Logic Unit *fu*
VAM Valve Actuation Model *jn*
VAM Virtual Access Method *kg, em, hi*
VAMOSC Visibility and Management of Operating and Support *mt*
VAMP VHSIC Avionics Modular Processor *wp*
VAN Value Added Network *mt, ct, em, kg*
VANCE Value-Added Networks for Computer Environments *fu*
VANGUARD [USAF long range planning program] *wp*
VAP Value Added Process *ct, em, kg*
VAP Vilivnoj Aviatsionnii Pribor ["dischargeable aviation system", a system of water tanks in the cargo hold of the Il-76 for firefighting tasks] [Russia] *mt*
VAPS Virtual Avionics Prototyping System *wp*
VAQ Visiting Airman Quarters *mt*
VAQ [US Navy tactical electronic warfare squadrons[(VAQ-13 'Scorpions', VAQ-34 'Flashbacks') *mt, af72.5.152*
VAR Variance Analysis Report *4, 17, jdr, fu*
VAR Visual Aural Range *w9*
VAR Volt Ampere Reactive *w9*
VARCS Vehicle Application and Replacement Categorical System *mt*

VARI Voice-channel Active Relay Indicator *fu*
VARISTOR Variable Resistor *wp, fu*
VARITRAN Variable Voltage Transformer *fu*
VARR Visual/Aural Radio Range *fu*
VAS Vehicle Allocation and Summary [MAC] *mt*
VAS Vertical Atmosphere Sounder [GOES] *hg, ge*
VAS VISSR Atmospheric Sounder *gd1-0, 16, hy, jdr, pbl, oe*
VASI Visual Approach Slope Indicator *mt, fu*
VAST Variable Array Storage Technology *em, kg*
VAST Versatile Avionics Shop Test *hc, fu*
VATE Versatile Automatic Test Equipment *hc, fu*
VATLS Visual Airborne Target Locator System *hc*
VATS Video Augmented Tracking System *hc*
VATTS Video Automatic Target Tracking System *hc*
VAW {US Navy Squadron}, Airborne early Warning (VAW-125 'Tigertails', VAW-127 'Seabats') *mt*
VAWX [follow-on carrier-based early warning aircraft] *dh*
VAX Virtual Address Extension [trade name for DEC minicomputers] *ct, em, aa26.12.65, aw130.11.61, ns, pr, tb, fu, oe*
VAX/VMS Virtual Address Extension / Virtual Memory System [DEC] *kg, em, hi*
VAXMATH Virtual Address Extension Library *tb*
VB Variable Block *kg, em*
VB Vertical Bomb [VB-1 and VB-2 were free-fall weapons with azimuth-only guidance, later VBs had complete guidance. VB-9 to VB-13 had annular wings. VB-13 was used on large scale in Korea] [USA] *mt*
VBE VESA BIOS Extension *em*
VBE [Voltage (beta) drop] *0*
VBE/AI VESA BIOS Extension / Audio Interface *em, kg, hi*
VBF {US Navy Squadron}, Bomber-Fighter *mt*
VBI Vertical Blanking Interface *ct, em*
VBITS Video Based Information Transmission System [FORSCOM] *mt*
VBNS Very High Speed Backbone Network Service [MCI, NSF] *kg, em, hi*
VBO Velocity at Burn-Out *tb*
VBR Variable Bit Rate *ct, em, kg, rih*
VBRUN Visual Basic Runtime *kg, hi*
VBX Visual Basic Extension *hi*
VBXS Visual Basic custom controls *hi*
VC Verify Code *fu*
VC Viet Cong *w9*
VC Voice Coil *fu*
VC Voltage Variable Capacitor *fu*
VC {US Navy Squadron}, Composite *mt*
VC {Velocity}, Cruising [a design parameter] *mt*
VCA Voltage Controlled Attenuator *jdr, uh*
VCASS Visually Coupled Airborne System Simulator *hc, wp*
VCB Vertical (Local of the) Center of Buoyancy *fu*
VCC Validation Count Code *fu*
VCC Variable Command Count *kf, jn*
VCC Variable Count Code *fu*
VCC Vehicle Command Count *jdr, uh, oe*
VCC Vehicle Command Counter *oe*
VCC Virtual Channel Connection *ct, em, kg, nu, hi*
VCD Valve Coil Driver *oe*
VCD Variable Center Distance *fu*
VCD Video Compact Disk *hi*
VCD Virtual Communications Driver *pk, kg, em, hi*
VCD Visual Class Designer *hi*
VCDA Valve Coil Driver Assembly *oe*
VCDE Valve Coil Driver Electronics *oe*
VCE Variable Cycle Engine *hc, wp*
VCE VSCS Console Equipment *fu*
VCG Verification Condition Generator *fu*
VCG Vertical (Location of the) Center of Gravity *fu*
VCHP Variable Conductance Heat Pipe *pl, jdr*
VCHS Visibly Clean Highly Sensitive *oe*
VCI Virtual Channel Identifier *em, ct*
VCI Virtual Circuit Identifier *kg, em*
VCID Virtual Channel Identification Code *oe*
VCJCS Vice Chairman, Joint Chiefs of Staff [OJCS] *mt*
VCL Vertical Center Line *fu*
VCL Visual Component Library *kg, em*
VCLK Variable Clock *fu*
VCLO Voltage Controlled Local Oscillator *hc*
VCLR Visual Control Logic Requirements *tb*
VCM Vibrating Coil Magnetometer *fu*
VCM Volatile Condensable Material *he, 4, oe*
VCN Voltage Conditioning Network *fu*
VCO Voice Communications Outlet *fu*
VCO Voltage Controlled Oscillator *fu, hc, sp663, 4, 7, 0, mj31.5.58, lt, wp, ep, hi, oe, te23, 11, 40*
VCOA Voltage Controlled Oscillator Amplifier *jdr*
VCOS Visual Caching Operating System [AT&T] *em, kg, hi*
VCOSS Vibration Control Of Space Structures *wp*
VCP Vector Coprocessor *wp*
VCP Voice Communications Panel *fu*

VCP Voluntary Cooperation Program *pl*
VCPH (Vapor Flow in the Condenser on VCPH) *aa27.4.b23*
VCPI Virtual Control Program Interface *kg, em, ct*
VCPS Velocity Control Propulsion Subsystem *hc*
VCPT VIsual Context Processor Technology *hi*
VCR Video Cassette Recorder *mt, sc2.10.3, aa26.11.39, wp, fu, kg, ct, em, hi*
VCRI Verification Cross Reference Index *tb, kf, fu, jn, he*
VCRM Verification Compliance Requirement Matrix *kf*
VCRM Verification Cross Reference Matrix *mt, jdr, oe*
VCS Validation Control System *17, fu*
VCS Voice Communications Switch (System) *fu*
VCSR Voltage Controlled Shift Register *fu*
VCT Voltage Controlled Transfer *fu*
VCXO Voltage Controlled Crystal Oscillator *hc, gg150, 2, 4, 0, jdr, fu*
VD Vertikalnye Dvigateli ['vertical engines', i.e. lift engines] [USSR] *mt*
VD Voltage Drop *fu*
VD {Velocity}, Dive, maximum *mt*
VDA Vacuum Deposited Aluminum *gp, lt, jdr, kf, jn*
VDA Vapor Deposited Aluminum *uh, oe*
VDA Video Distribution Amplifier *fu*
VDB Validation Data Base *-17*
VDB Video Distribution Buffer *fu*
VDC Video Data Console *gp*
VDC Video Display Controller *fu*
VDC Volts Direct Current *mt, ct, em, he, sc4.11.1, 17, kf, hi, oe, ai2*
VDCN Video Disk Controller *fu*
VDD Verified Design Document *fu*
VDD Version Description Document *mt, pl, hc, wp, tb, jdr, kf, fu, oe*
VDD Virtual Device Driver *kg, em*
VDDM Virtual Device Driver Manager *kg, em, hi*
VDDN Video Data Distribution Network *wp*
VDE Value Drive Electronics *-4*
VDE Video Display Editor *em, kg*
VDE Visual Development Environment *kg, em, hi*
VDEP Validation, Demonstration, and Evaluation Plan *fu*
VDET Voltage Detector *fu*
VDF VHF Direction Finding [station] *mt*
VDFG Variable Diode Function Generator *fu*
VDFM Voice Data Fax Modem *hi*
VDH Very Distant Host *fu*
VDHL Hardware Description Language *fu*
VDI Vertical Display Indicator *fu*
VDI Virtual Device Interface *fu, hi*
VDI Voice Data Integrator *wp*
VDIF Video Display Information File *hi*
VDISK Virtual Disk *kg, em, hi*
VDL Variable Data Label *fu*
VDL Vehicle Discrepancy Log (List) *gp, 8, ep, jdr*
VDL Video Data Link *wp*
VDM VAS Data Multiplexer *gg1*
VDM Virtual DOS Machine *em, kg, hi*
VDM VISR or DAS Multiplexer *gp*
VDM VISSR Digital Multiplexer *sc4.7.center, pbl*
VDM VSCS Display Module *fu*
VDMAD Virtual Direct Memory Access Device [Microsoft] *kg, em, hi*
VDP Vehicle Down for Parts *wp, mt*
VDP VHSIC Data Processor (Hughes') *fu*
VDPU Video Digital Processing Unit *fu*
VDR Variable Depression Receiver *fu*
VDS Variable Depth SONAR [US Navy] *fu, mt*
VDS Vehicular Data System *mt*
VDS Virtual DMA Services *pk, kg, em, hi*
VDS Voice Distribution Subsystem *fu*
VDSI Video Distribution Subsystem Interface [USSPACECOM] [USA] *mt*
VDSL Very high bit (data) rate Digital Subscriber Line *ct, kg, em*
VDT Variable Differential Transformer *17, sb*
VDT Video Data (Display) Terminal *fu*
VDT Video Dial Tone *em, ct*
VDT Video Display Terminal *w9, wp, em, pk, kg*
VDU Valve Driver Unit *uh, cl, jn*
VDU Versatile Display Unit *hc*
VDU Video Display Unit *fu, mt, kg, wp, em, ct*
VDU Video Distribution Unit *fu*
VDU Visual Display Unit *mt, em, kg, w9, ai1*
VDUC VAS Data Utilization Center *oe*
VDWT Variable Density Wind Tunnel *0*
VE Value Engineering *fu, hc, hn44.8.4, hq5.2.8, wp, ep, dm, tb, st, jdr, kf*
VE Vehicle Engineer *jdr*
VE Virtual Environment *is26.1.89, kf*
VE [o-ethyl S-2-diethylaminoethylethylphosphonothioate, chemical warfare nerve agent] *wp*
VEAS Vertical Expendable Array System *hc*
VEB Variable Elevation Beam *fu*
VEB Vehicle Equipment Bay *es55.77, ai2*

VECG Video Extractor [and] Control Group *hc, fu*
VECO Vernier Engine Cutoff *0, fu, oe*
VECP Value Engineering Change Proposal [a document] *mt, hc, hq5.2.8, nd74.445.27, aw130.6.12, rm8.1.1, wp, ep, st, jdr, fu*
VEDAR Visible Energy Detection and Ranging *fu*
VEEGA Venus-Earth-Earth Gravity Assist [Galileo flight path] *mi, aa27.4.b24*
VEGA Vettore Europeo di Generazione Avanzata *crl*
VEGA Video-7 Enhanced Graphics Adapter [Video 7, Inc.] *em, kg, hi*
VEGAS Verified Earth-shine Geometric Albedo and Solar *gp, 16*
VEI Value Engineered Indicator *fu*
VEI Value Engineering Incentive *fu, st*
VEID Vehicle Ephemeris Identifier *uh*
VELA [TRW nuclear burst detection satellite launched mid 1960's, as in VELA-5B, an all sky X-ray monitor] *17, cu*
VEMM Virtual Expanded Memory Manager *em, kg, hi*
VEMMI VErsatile MultiMedia Interface *kg, em*
VEN Variable Exhaust Nozzle *wp*
VEOS Versatile Electro Optical System *hc, hn49.3.8*
VEOS Virtual Environment Operating System *hi*
VEP Value Engineering Proposal *wp*
VEP Vehicle Evaluation Payload *ai2*
VEPR VE Program Requirement *fu*
VERDAN Versatile Differential Analyzer *fu*
VERDIN [VLF/LF Multiple Channel Broadcast System] [US Navy] *mt*
VERLORT Very Low Range Tracking *fu*
VERONICA Very Easy Rodent-Oriented Network-wide Index to Computerized Archives *ct, kg, em, hi*
VERR VERify Read access *em, kg*
VERTREP Vertical Replenishment [transport by helicopter of supplies to warships [USA] *mt*
VERW VERify Write access *em, kg*
VES Visitor Entry System *44.8.8*
VESA Video Electronics Standards Association *ct, kg, em, hi*
VETRONICS [ground vehicle electronics] *uc*
VEV Voice Excited Voicecoder *mt*
VF Variable Frequency *fu*
VF Vector Field *fu*
VF Verification Factor *fu*
VF Video Frequency *0, w9*
VF Virtual Floppy *em, kg, hi*
VF Voice Frequency *fu*
VF Voice Frequency [30-300 MHz] *w9, mt*
VF {US Navy Squadron}, Fighter *mt*
VFA Variable Gain Amplifier *fu*
VFA {US Navy Squadron}, Fighter Attack *mt*
VFAS Vertical Force Accounting System [USA] *mt*
VFAT Virtual File Allocation Table [Microsoft] *ct, kg, em, hi*
VFC Vector Function Chainer *em, ct*
VFC Version First Class [communications standard] *em, pk, kg*
VFC Video Feature Connector *em, ct*
VFC Voice Frequency Carrier *fu*
VFC {US Navy Squadron} Fighter Composite *mt*
VFCT Voice Frequency Carrier Telegraph *mt*
VFCTT Voice Frequency Carrier Teletype *fu*
VFD Vacuum Fluorescent Display *em, kg, hi*
VFDMIS Vertical Force Deployment Management Information System [USA] *mt*
VFDMIS Vertical Force Development Management and Information System *hc, dt*
VFDR Variable Flow Ducted Rocket *hc, wp*
VFDRE Variable Flow Ducted Rocket Engine *wp*
VFK Variable Function Key *mt, fu*
VFM Voice Fax Modem *hi*
VFMED Variable Format Message Entry Device *fu*
VFMP Vehicle Fleet Management Program [USA] *mt*
VFMX {CV} Fighter, Medium, {Experimental} [proposed new carrier based aircraft] *uc*
VFO Variable Frequency Oscillator *mt, 0, wp, fu*
VFP [US Navy photo reconnaissance squadron] *mt*
VFR Visual Flight Rules *mt, aw129.14.113, w9, af72.5.134, wp, fu*
VFS Variable Function Switch *fu*
VFS Verification of Flight Software *fu*
VFS Virtual File System *hi*
VFT Verification Flight Test *sc4.6.6*
VFT Voice Frequency Telegraph *fu*
VFTG Voice Frequency Telegraph Group *mt*
VFW Veterans of Foreign Wars *w9*
VFW Video For Windows *ct, em, kg, hi*
VFX [follow-on carrier-based light fighter/attack aircraft] *dh*
VG Variable Geometry *mt*
VGA Variable Gain Amplifier *fu*
VGA Video Graphics Adapter *em, hi*
VGA Video Graphics Array *em, ct, kg, is26.11.32, wp*
VGC Video Graphics Controller *em, kg, hi*
VGCA Voice Gate Circuit Adapter *hc*
VGK Supreme High Command [Soviet] *mt*

VGP Vehicle Ground Point *oe*
VGPI Visual Glide Path Indicator *fu*
VGPO Velocity Gate Pull-Off *fu*
VGS Velocity Gate Stealing *fu*
VGS Visual Glide Slope *fu*
VGSI Visual Glide Slope Indicator *fu*
VH Virtual Host *fu*
VH {Vertical-Horizontal} [polarizations] *un, fu*
VHA Variable Housing Allowance *an31.13.1*
VHDL VHSIC Hardware Description (Design) Language *em, jdr, hc, is26.1.37, aw130.22.136, kf, kg, fu, hi*
VHDR Very High Data Rate *-19*
VHDU Visual Handcopy Display Unit *mt*
VHF Very High Frequency [30 to 300 Megahertz] *mt, ro, nu, mi, fu, hc, bl3-1.24, 0, 4, 15, w9, pf.f88.10, aw130.22.107, lt, wp, ce, re, tb, jdr, mb, hi, oe, tt, ai1*
VHFDF Very High Frequency Direction *fu*
VHFS (CECOM VHFS) [Hughes Aircraft Company customer for AISAF] *hc*
VHOL Very High Order Language *wp*
VHPCM Very High Power Countermeasures *wp*
VHRR Very High Resolution Radar *wp*
VHRR Very High Resolution Radiometer *ac1.6-13, hc, hg, ge, oe*
VHS Very High Speed *kg, tb, em, fu*
VHS Video Home System *fu, kg, sc3.11.3, af72.5.13, hi*
VHS Virtual Host Storage *kg, em, hi*
VHSI Very High Scale Integration *hi*
VHSIC Very High Speed Integrated Circuit *hc, sc2.8.4, 1, 16, 1, sb, th4, aw130.13.8, aa26.12.65, af72.5.15, wp, hn50.8.1, ep, tb, wl, jdr, pbl, tm, kf, em, ro, ah, fu, kg, hi, oe, mt*
VHSR Verification Methods Summary Report *fu*
VI Video Integrator *fu*
VI Viscosity Index *w9, fu*
VI Visual Interactive [editor in UNIX] *em, kg*
VI Volume Indicator *fu, w9*
VI Vysotny Istrebitel [high altitude fighter] [USSR] *mt*
VIA VAX Information Architecture *fu*
VIA Virtual Interface Architecture *em*
VIABLE Vertical Installation Automated Baseline system [USA] *wp, mt*
VIB Vertical Integration Building [ELS facility] *bo, jdr, ai2*
VIC Vector In Command Model *mt*
VIC Vehicle Identification Code *-4*
VIC Visibility of Intransit Cargo [USA] *mt*
VIC VISSR Ingest Computer *gd3-10*
VIC [occurs on a map of NASA's KSC] *ci.85*
VICAR Video Image Communication And Retrieval *ct, ns, em*
VICC Variable-Increment Clutter Counter *fu*
VICC Visual Information Control Console *hi*
VICI Velocity Indicating Coherent Integrator *fu*
VID Virtual Image Display *hc*
VID Visual Identification *mt, wp, fu*
VIDAR Velocity Integration Detection and Ranging *fu*
VIDOC Visual Information Documentation [USA DOD] *mt*
VIDS Vehicle Integrated Defense System *wp*
VIDS VICAR Interactive Display Subsystem *ct*
VIDS Video Interactive Display Subsystem *em*
VIDS Virtual Image Display System *hc*
VIDS-DMS Vehicle Integrated Defense System - Data Management System *hc*
VIE VAS Interface Electronics *oe*
VIE Virtual Information Environment *kg, em*
VIF Virtual Interface *em, kg, hi*
VIF Virtual Interrupt Flag *kg, em*
VIFF Vectoring In Forward Flight [use of the vectoring jet nozzles of the Harrier to enhance maneuverability in combat] *mt*
VIGS Videodisc Gunnery Simulator *hc*
VILP Vector Impedance Locus Plotter *fu*
VIM Vendor Independent Mail *em, pk, kg*
VIM Vendor Independent Messaging *pk, kg*
VIM Vibration Isolation Module *fu, hc*
VIM Video Interface Module *em, pk, kg*
VIM Visual Instrument Mandatory *fu*
VIMC Vendor Independent Messaging Consortium *hi*
VIMIS Vertically Integrated Metal Insulator Semiconductor *wp*
VIMS Vehicle Integrated Management System [USAF] *mt*
VIMS Visible and Infrared Mapping Spectrometer *so, eo, re*
VINES Virtual Networking Software *ct, kg, em, hi*
VINES Virtual Networking System [Banyan] *hi*
VINSON Family of Tactical Secure Voice Equipment [E.G., KY58] *mt*
VIO Video Input/Output *kg, em*
VIO Virtual Input/Output *fu, kg, em*
VIP Variable Information Processing *kg, fu, em*
VIP VAS Image Processor *gp, 16, oe*
VIP Versatile Information Processor *hc*
VIP Video Image Processor *oe*
VIP Video Information Processor *mt*
VIP Video Information Provider *pk, kg, em*

VIP Virtual Interrupt Pending *em, kg*
VIP Visual Identification Point *fu*
VIP Visual Information Processor [Honeywell] *mt*
VIP VIsual Programmer (Programming) *kg, em*
VIP [a box in a diagram of Nimbus data flow] *ac2.8-13*
VIPER Variable Intensity Pulsed Effects Research *hc*
VIPER Verifiable Integrated Processor for Enhanced Reliability *em, kg, hi*
VIPS VAS Image Processing System *oe*
VIPS Video Image Processing System *hc*
VIRAM VATE Improved Reliability and Maintainability *hc*
VIRGS Visible InfraRed GOES System *oe*
VIRGS VISSR Image Registration and Gridding System *gd3-2*
VIRR Visible and Infrared Radiometer *ac2.7-1*
VIS Vehicular Intercommunications System *hc*
VIS Vendor Information System *kf, kg*
VIS Video Information System *em, kg*
VIS Visual Instruction Set [Sun] *pk, kg, em*
VIS Voice Information System *kg, em*
VISA VHSIC Imaging Sensor Algorithm *hc*
VISFOV Visibility/Field Of View *fu*
VISID Visual Identification *fu*
VISR Visible and Infrared Spinning Radiometer *gp, 16*
VISSD Vendor Information System Supplier Directory *fu*
VISSR Visible [and] Infrared Spin Scan Radiometer [GOES, GMS, Meteosat, Insat] *ge, hn43.21.4, hy, pbl, jdr, hg, oe*
VISTA Variable In-flight Simulator Test Aircraft *wp*
VISTA Variable stability In-flight Simulator Test Aircraft [a modified F-16] [USA] *mt*
VISTA Very Intelligent Search, Track and Acquisition *wp*
VISTA Very Intelligent Surveillance and Target Acquisition [System] *uc, hc*
VISTA [a program to simulate IC's in a computer] *hc*
VISTAS Very Intelligent Surveillance Target Acquisition System *hc*
VISTRAC Visual Target Acquisition [model] *wp*
VITA VME-bus International Trade Association *is26.1.71*
VITAL Vast Interface Test Application Language *fu*
VK {an organization within Hughes Aircraft Company} *hn50.8.5*
VKP Vozduzhnyi Komandnii Punkt [airborne command post, as in Mi-6VKP Hook-B] [USSR] *mt*
VKS [Russian Military Space Forces] *ai2*
VL Vertical Landing *mt*
VL Vertical Launch *dh*
VL Very Low *sd*
VL VESA Local *ct, em*
VL-BUS VESA Local-Bus [also VLB] *kg, hi*
VLA Very Large Array:(mi)
VLA Very Low Altitude *fu*
VLA Volume Licensing Agreement *ct, em*
VLAC Verification Load(s) Analysis Cycle *gp, 16, kf, jn*
VLAC Verify, Load, and Control *fu*
VLAC Very Large Aperture Camera *re*
VLAD Vertical Line Array Directional *uc*
VLADD Visual Low-Angle Drogue Delivery *fu*
VLAGES Very Low Altitude Gravity Extraction System [airdropping at low altitude and low speed, without parachutes] *mt*
VLAN Virtual Local Area Network *ct, kg, em*
VLASROC Vertical Launch Anti Submarine Rocket [US Navy] *mt*
VLB VESA Local Bus *ct, em*
VLBA Very Long Baseline [interferometry] Array *es55.20, mb, mi*
VLBC Visible Laser Beam Control *fu*
VLBI Very Long Baseline Interferometry *ac2.6-11, 12, es55.18, wp, ns, hy, tb, mb, mi, ai1*
VLBUS Vesa Local-Bus *em*
VLC Variable Load Control *fu*
VLC Verification Load Cycle *jdr, kf*
VLCR Variable Length Cavity Resonance *fu*
VLD Variable-Length Decoder *em, kg*
VLDB Very Large Data Base *hi*
VLDS Very Large Data Storage *fu*
VLE Vehicle Liaison Engineer *oe*
VLF Vehicle Launch Facility *0*
VLF Very Low Frequency [3-30 kHz] *mt, hc, sc3.5.3, 12, aw129.14.103, w9, aa26.11.b48, af72.5.139, wp, tb, mb, kf, nu, kg, mi, fu, ct, em, hi, oe, ai1*
VLFS Variable Low-Frequency Standard *fu*
VLIC Very Large Integrated Circuit *em*
VLIW Very Large (Long) Instruction Word *kg, em, hi*
VLLADWS Very Low Level Air Defense Weapon System *hc*
VLM Very Large Memory *hi*
VLM Virtual Loadable Module *ct, kg, em, hi*
VLMS Vertical Launch Missile System *fu*
VLO Very Low Observable *mt*
VLR Very Long Range *fu*
VLS Vandenberg Launch Site *-16*
VLS Vapor, Liquid, Solid *aa26.10.42*
VLS Veiculo LanÁador de Satelites [Vehicle for the Launching of Satellites (Brazilian)] *sc3.10.3, aw130.22.35*

VLS Vertical Launch System *cp, wp, tb, st, kf, fu*
VLSI Very Large Scale Integrated (Integration) *mt, hc, 1, 17, ph, sb, th4, aw129.1.61, rm8.2.1, wp, ep, re, tb, st, jdr, pbl, tm, kf, fu, kg, ah, ct, em, uh, hi, ai1*
VLSIC Very Large Scale Integrated Circuit *mt, ph, aa26.12.68, md2.1.1, wp*
VLSIPS Very Large Scale Immobilized Polymer Synthesis *em, kg, hi*
VLSISC Very Large Scale Integration Socketing Capability *mt*
VLT Variable List Table *em, kg*
VLT Very Large Telescope *mi*
VLVIP VHSIC LANTIRN Video Image Processor *hc*
VLW Very Long Wavelength *wp*
VLWIR Very Long Wave Infrared *kf*
VM Vector Magnitude *sd*
VM Velocity Modulation *fu*
VM Virtual Machine *mt, em, fu, ct, kg, hd, hi*
VM Virtual Memory *fu, em, ct, wp, tb, kg, hi*
VM [o-ethyl-s-2-diethylaminoethylmethylphophonothioate, a chemical warfare nerve agent] *wp*
VM/CMS Virtual Machine/Conversation Monitor System *hd, fu*
VM/SP Virtual Machine / System Product *mt*
VMA Virtual Memory Address *em, kg, hi*
VMA {US Navy Squadron}, Marine Corps, Attack *mt*
VMA-[AW] {US Navy Squadron}, Marine Corps, Attack, All Weather *mt*
VMAQ Marine Tactical Electronic Warfare Squadron *mt*
VMAS Versailles Project on Advanced Materials and Standards *tt*
VMAT {US Navy Squadron}, Marine Corps, Attack, Training *mt*
VMB Virtual Machine Boot *em, kg, hi*
VMB {US Navy Squadron}, Marine Corps, Bombardment *mt*
VMC Vehicle Mission Computer *ai2*
VMC VESA Media Channel *em, ct*
VMC Visual Meteorological Conditions [USA DOD] *mt, af72.5.145, aw129.21.135, fu*
VMCJ [USMC electronic warfare squadron, e.g, VMCJ-2 'Playboys'] [USA] *mt*
VMCP Virtual Machine Control Program *fu*
VMCS Vehicle Management Control System *wp*
VME Versa Module (Mode) Eurocard *em, kf, kg*
VME Versabus Motorola Eurocard *fu*
VME Virtual Memory Environment *kg, em, hi*
VMEbus Versabus Motorola Eurocard bus *fu, is26.1.50*
VMF Variable Message Format *fu*
VMF {US Navy Squadron}, Marine Corps, Fighter *mt*
VMFA {US Navy Squadron}, Marine Corps, Fighter Attack *mt*
VMFAT {US Navy Squadron}, Marine Corps, Fighter, Attack, Training *mt*
VMI Verified Milstar I *jdr*
VMI Virginia Military Institute *hn43.5.3*
VMIU Virtual Memory Interface Unit [Honeywell] *mt*
VMM Virtual Machine Manager *em, kg*
VMM Virtual Memory Manager *ct, em, kg, hi*
VMM VISSR Multiplexer-Modulator *bl3-1.26*
VMO Maximum permitted operating speed *mt*
VMO {US Navy Squadron}, Marine Corps, Observation [VMO-6 'Tomcats'] [USA] *mt*
VMOS Vertical Metal-Oxide Semiconductor *kg, em, fu*
VMOS Virtual Memory Operating System *fu, hi*
VMP Virtual Modem Protocol *kg, em*
VMRS Vessel Movement Reporting System *fu*
VMS Vehicle Management System *wp*
VMS Vertical Missile Simulator *fu*
VMS Vertical Motion Simulator *mi*
VMS Virtual Machine System *fu*
VMS Virtual Memory [operating] System *em, ct, hd1.89.y18, tb, kg, fu, em, ct, oe*
VMS Voice Message System *em, kg, hi*
VMSP Virtual Machine System Product *mt*
VMSR Verification Methods Summary Report *fu*
VMT Virtual Memory Technique *em, kg, hi*
VMX Voice Message Exchange [e.g. Voice Mail] *mt*
VMX Voice Message Network *hd*
VMX [follow-on carrier-based attack aircraft] *dh*
VN Van Nuys [Hughes site] *fu*
VN Virtual Network *em*
VNA Virtual Network Architecture *em, kg, hi*
VNAV Vertical Navigation *aw129.14.115*
VNE {Velocity} (speed) Never to Exceed [a structural limitation.] *mt*
VNIRP Visible and Near IR Polarimeter *hg, ge*
VNL Via Net Loss *ce*
VNTK Vulnerabilty Number for Thermonuclear Kill *wp*
VO Validation Office *mt*
VO Visual Object *hi*
VOAC Variable field of Off-Axis Coherent [system] *hc*
VOC Creative Voice [format] [Sound Blaster] *kg, hi*
VOC Voice Order Circuit *fu*
VOC Volatile Organic Compound [emission] *af72.5.180, hn50.8.i*
VOCODER Voice Coder *mt*

VOCODER Voice Operated Code *mt*
VOCOM Voice Communications *mt, fu*
VOD Vertical On Board Delivery *mt*
VOD Video On Demand *em, kg, hi*
VOD Virtual Optical Disk *ns*
VODAS Voice Operated Device Anti-Sing *fu*
VODAT Voice Operated Gain Adjusting Device *fu*
VOED Visual Observation Entry Device *fu*
VOG Vanguard Operations Group *ci.44*
VOGAD Voice-Operated Gain Adjusting Device *fu*
VOI Volume Of Interest *go3.1-1, fu*
VOIR Venus Orbiter Imaging Radar [superseded by VRM] *hn44.2.1, hc, mb, mi*
VOLAR Volunteer Army *w9*
VOLCANO [new land mine dispenser] *uc*
VOLID Volume Identification *fu*
VOLMET Routine broadcast of meteorological information *mt*
VOP Variation Of Price *fu*
VOP Visual Observation Post *fu*
VOPIO Visual Observation Post Input Operator *fu*
VOPO Visual Observation Post Operator *fu*
VOQ (housing . . . 182 transient (81 VOQ, 60 VAQ, 41 TLF)) *af72.5.152*
VOQ Visiting Officer Quarters *mt*
VOR Very high frequency Omnidirectional Radio range [an air navigational radio aid which uses phase comparison of a ground transmitted signal to determine bearing] [USA] *mt*
VOR Very high frequency, Omnidirectional (Radio) Range *hc, 0, w9, aw129.10.73, af72.5.142, wp*
VOR VHF Omnidirectional [Radio] Range *fu*
VOR Visions Of Reality *kg, em*
VOR Visual Observation Report *fu*
VOR Visual Operating Restrictions *mt*
VOR Voice Operated Relay *fu*
VORDME VHF Omnirange Distance Measuring Equipment *fu*
VORLOC Very High Frequency Omnirange Localizer *fu*
VORTAC VHF Omni Range and Tactical [Navigation] (Combined VOR and TACAN system colocated) *mt, wp, fu*
VOS Verbal Operating System *em, kg*
VOS Virtual Operating System *em, fu, hi*
VOS Voice Operating System *em, kg*
VOSIM Voice Simulation *hi*
VOT VHF Omnidirectional Range Test *fu*
VOX Voice Activated Transmitter *fu*
VOX Voice Extension *mt*
VOX Voice Operated Relay Circuit *mt*
VOX Voice Operator Transmit *fu*
VP Variable Pitch [propeller] *w9, mt*
VP Vendor Preference *fu*
VP Vertical Polarization *fu*
VP Vice President *hn25.11.2, 17, w9, aa26.22.b4, kf*
VP Video Processor *sb*
VP Virtual Path *em, ct, pk, kg*
VP Visotny Perekhvatchik [high altitude interceptor] [USSR] *mt*
VP [patrol squadron, a Navy unit flying maritime patrol aircraft] *12, pf1.7*
VP {US Navy Squadron}, Patrol [VP-17 'White Lightnings', VP-8 'Tigers'] *mt*
VPB Vertical Plotting Board *fu*
VPC Viewport Controller *fu*
VPC Virtual Path Connection *em, nu, ct*
VPCA Video Prelaunch Command Amplifier *fu*
VPCMS Vector Process Common Memory Software *oe*
VPD Variable Power Divider *bkr, jn*
VPD Vendor Promise Date *fu*
VPD Vital Product Data [IBM] *kg, em*
VPE Vapor Phase Epitaxial *fu*
VPE Video Port Extensions [Microsoft] *kg, em*
VPE Video Processing Equipment *fu*
VPE Virtual Programming Environment *em*
VPE Visual Programming Environment *pk, kg*
VPE VSCS Position Equipment *fu*
VPF Vehicle Processing Facility *ai2*
VPF Vertical Processing Facility *gp, 4, 9, lt, jdr, mb, mi, oe*
VPHD Vertical Payload Handling Device *jdr, oe*
VPHD Vertical Payload Handling Dolly *oe*
VPHS Vertical Payload Handling Service *-9*
VPI Vertical Payload Integration *pl*
VPI Virtual Path Identifier *em, ct*
VPIO Video Processor Input / Output *rm8.1.4*
VPIP Vinson/Parkhill Implementation Plan *mt*
VPISU (paper presented at the VPISU/AIAA symposium) *id4091.2/1111*
VPL Virtual Programming Language *em, kg*
VPM VHSIC Processing Module *wp*
VPM Vibration Per Minute *fu*
VPM Video Port Manager *em, kg*
VPN Vendor Part Number *fu*
VPN Virtual Page Number *em, kg*

VPN Virtual Private Network *kg, em, ct, tt*
VPOS Vector Processor Operating System *oe*
VPR Virtual PPI Reflectoscope *mt*
VPS Vanguard Planning Summary *mt*
VPS Variable Phase Shifter *bkr*
VPS Voice Processing System *mt, kg, em*
VPSC Vault, Process, Structure, Configuration *kg, em*
VPT Virtual Print(er) Technology [Dataproducts] *em, kg, hi*
VPT VLSIC Packaging Technology *md2.1.1*
VPU Vector Processing Unit *oe*
VPU Vibrator Power Unit *fu*
VPU Virtual Processing Unit *hi*
VPU Voltage Protection Unit *fu*
VPWPU SAFE Four Echelon Wartime Casualty System *mt*
VQ Shore Based Recon Aircraft [US Navy] *mt*
VQ Vector Quantification (em, ct)
VQ [US Navy electronic reconnaissance squadron] *mt*
VQA Vendor Quality Assurance *jdr*
VQU Video Quantizer Unit *fu*
VR Variability Reduction *hn49.12.8, ad*
VR Variable Voltage Rectifier *fu*
VR Virtual Reality *ct, em, kg, kf, hi*
VR Voltage Ratio *sm1-17*
VR Voltage Regulated (Regulator) *em, kg, fu, hi*
VR Volunteer Reserve *mt*
VRAD Verification Requirements Allocation Document *jdr*
VRAM Variable Random Access Memory *hi*
VRAM Video Random Access Memory *em, fu, kg, ct, hi*
VRB VHF Recovery Beacon *fu*
VRC Validation and Rule Checking *hi*
VRC Variability Reduction Committee *hn49.12.8*
VRC Vertical Redundancy Check *do262, jdr, fu, hi*
VRC Voice Response Controller *mt*
VRCI Variable Resistive Components Institute *fu*
VRD Vacuum Tube Relay Driver *fu*
VRDI Virtual Raster Display Interface *em, ct*
VRDK [air-turbocompressor reaction engine, also known as accelerator, USSR] *mt*
VRE Voltage Regulated Extended *em, kg*
VRF USN transport wing [USA] *mt*
VRF Visual Recording Facility *mt*
VRF Visual Reproduction Facility *ns*
VRFI Voice Reporting Fault Indicator *fu*
VRIC Vendor Request for Information or Change *jdr, fu*
VRL Vertical Reference Line *fu*
VRM Venus Radar Mapper [later called Magellan] *sc3.4.1, hc, ep, mb, fu, mi*
VRM Voltage Regulator Module *em, kg*
VRML Virtual Reality Modeling Language [originally named Virtual Reality Markup Language] *em, ct, kg, hi*
VROOMM Virtual Real Time Object Oriented Memory Manager [Borland] *em, kg, hi*
VRP Variability Reduction Program (Process) *rm7.1.1, aa26.12.66, nd74.445.26, ad, ig21.12.17*
VRPS Voltage Regulated Power Supply *fu*
VRS Voice Response System *ag, fu*
VRSA Voice Reporting Signal Assembly *fu*
VRSS Voice Relay Satellite System *hc*
VRT Voltage Reduction Technology *hi*
VRT Voltage Regulation Technology [Intel] *em, kg*
VRTX Virtual Real Time Executive *fu*
VRU Voice Response Unit *em, kg*
VRU Voltage Readout Unit *fu*
VS Validation Segment *hc*
VS Velocity Search *hc*
VS Video Services *sc2.10.3*
VS Video Switch *mt*
VS Virtual Storage *em, kg, hi*
VS Voltmeter Switch *fu*
VS [O-Ethyl-S-2-Diisopropylaminoethylethylphophonotioate, chemical warfare nerve agent] *wp*
VS {US Navy Squadron}, Submarine warfare (VS-21 'Fighting redtails', VS-30 'Diamondcutters') [USA] *mt*
VS/VD Virtual Source / Virtual Destination *ct*
VSA Validity and Statistical Analysis Subsystem *mt*
VSAM Virtual Sequential Access Method *mt*
VSAM Virtual Storage Access Method [IBM] *mt, wp, em, ct, fu, kg, hi*
VSAT Very Small Aperture Terminal *sc8.5.1, hn48.20.8, 19, kf, hj16.9.3, kg, em, ct, hi, tt*
VSB Vestigial Sideband *fu*
VSC Visual System Coordinator *fu*
VSC Voltage Saturated Capacitor *fu*
VSCA Vertical Scanning Colinear Array *gp*
VSCCS Voice Switch Communications Control System *fu*
VSCDP Visual System Component Development Program *hc*
VSCF Variable Speed Constant Frequency *aa26.10.43, fu*
VSCS Voice Switching and Control System *fu*

VSCU Vertical Sounding Control Unit *fu*
VSD Variable Slope Delta [modulation] *fu*
VSD Vertical Situation Display *hc*
VSDM Variable Slope Delta Modulation *sp663*
VSDN Very Survivable [sensor] Data Network *kf*
VSE Vehicle System Engineering *hc*
VSE Virtual Storage Extended *em, kg*
VSE Virtual System Extended *mt*
VSF Vertical Scanning Frequency *kg, em*
VSG Versatile Symbol Generator *hc*
VSG Visual Scene Generation *fu*
VSI Vertical Situation Indicator *wp*
VSI Vertical Speed Indicator *mt*
VSI Video Sweep Integrator *fu*
VSI Virtual Socket Interface *em, kg*
VSII Very Seriously Ill or Injured [USA DOD] *mt*
VSIO Virtual Serial Input Output *em, kg*
VSL Vehicle Locator System {?} *hc*
VSL Viscous Shock Layer *ai1*
VSLI Very Large Scale Integration *kg, em*
VSM Vestigial Sideband Modulation *fu*
VSM Virtual Shared Memory *kg, em*
VSM Virtual Storage Management *kg, em*
VSM Visual System Management (Manager) *kg, em, hi*
VSMF Visual Search Microfilm Facility *ep*
VSMF Visual Search Microfilm File *jdr*
VSMF Visual Standard Microfilm File *ep*
VSN Volume Serial Number *kg, em*
VSO Very Stable Oscillator *fu*
VSOS Virtual Storage Operating System *kg, em, hi*
VSP Virtual System Parameter *fu*
VSP Vendor Standard Parts *ric, jdr*
VSP VHSIC Signal Processor *fu*
VSP Video Select Panel *fu*
VSP Vision Service Plan *hm*
VSR Very Short Range *fu*
VSS Vector Scoring System *hc*
VSS Velocity Sensor Switch *fu*
VSS Video Signal Simulator *fu*
VSS Video Signal Switchboard *fu*
VSS Voice Switching System (Set) *fu*
VSS [a commercial satellite] *sd*
VSSC Vikram Sarabhai Space Center [Trivandrum, India] *ac3.2-2*
VST Vinson Subscriber Terminal *mt*
VSTAR Variable Search and Track Air (Array) [defense] Radar *hc, hn41.22.6, fu*
VSTAR-PT Variable Search and Track Array Radar - Precision Tracker *fu*
VSTOL Vertical or Short Take Off and Landing *mt, aa26.12.34, 12, hc, wp*
VSTT Variable Speed Training Target [competition for a target drone, won by the Beechcraft MQM-107, USA] *mt*
VSU Video Storage Unit *fu*
VSVD Virtual Source / Virtual Destination *em*
VSWR Voltage Standing Wave Ratio *mt, em, ct, gg148, 0, pl, mj31.5.3, lt, jdr, fu*
VSX [follow-on carrier-based ASW aircraft] *dh*
VSYNC Vertical Synchronization *kg, em, hi*
VT Variable Time [proximity] fuse *mt*
VT Video Tape *fu*
VT Video Terminal *hi*
VT Virtual Terminal *fu*
VT Virtual Tributary *ct, em*
VT Voice Technology *hi*
VT Voyenniy Transport [military transport ship] [USSR] *mt*
VT {US Navy Squadron}, Training *mt*
VTA Soviet Military Transport Aviation *mt*
VTA Vehicle Test Area *bo*
VTA Voenno-Transportnaya Aviatsiya [Military Transport Aviation] [USSR] *mt*
VTAADS Vertical The Army Authorization Document System *mt*
VTAM Virtual Telecommunications Access Method [IBM] *mt, kg, em, ct, do422, hi*
VTAM Virtual Terminal Access Method [IBM] *hi, mt*
VTANG Vermont Air National Guard *-12*
VTART Visual Target Angle and Range Trainer *hc*
VTAS Visual Target Acquisition System [a helmet sight, USA] *hc, mt*
VTC Video Tele-tape Conferencing *tb*
VTC Video Teleconference:(ric, kf, jdr)
VTC Video Teleconferencing Capability *hn50.8.4*
VTC Video Teleconferencing System [PACAF] *mt*
VTCA VHSIC Transmit Control Assembly *wp*
VTDRO Voltage Tuned Dielectric Resonator Oscillator *fu*
VTDU Video and Trigger Distribution Unit *fu*
VTHL Vertical Takeoff Horizontal Landing *mi*
VTIP Visual Target Identification Point *fu*
VTIR Visible and Thermal Infrared Radiometer *ac3.3-13*
VTKLPS Video Text, Keyboard and Light Pen Subsystem *fu*

VTL Variable Threshold Logic *fu*
VTL Verification Testing Laboratory *re*
VTM Vertical Test Methods *wp*
VTM Voltage Tunable Magnetron *fu*
VTMR Variance To Mean Ratio *wp*
VTNS Virtual Telecommunications Network Service *em, kg*
VTO (integral "spill proof" VTO eliminates . . . safety violations) *aw129.21.120*
VTO Voltage Tunable Oscillator *fu*
VTO {Vertical Take Off} *af72.5.145*
VTOL Vertical Take Off and Landing *mt, aa26.12.35, 14, aw130.13.101, wp, ai1*
VTP Vehicle Test Procedure *rih*
VTP Virtual Terminal Protocol *re*
VTPR Vertical Temperature Profile Radiometer *ag, hg, ge*
VTR Video Tape Recorder (Recording) *w9, 19, wp, fu, hi*
VTR Voice Tape Recorded *mt*
VTRAS Video Tape Recording Assemblies *hc*
VTS Vandenberg Tracking Station *hc, wp*
VTS Variable Tracking Strategy *mt*
VTS Vessel Traffic Services *fu*
VTS Video Transmission Services *tt*
VTSRS Voice Type Speech Recognition System *hi*
VTT Video Thermal Tracker *bf*
VTTG Variable Test Target Generator *fu*
VTU Variable Time Unit *fu*
VTVL Vertical Takeoff Vertical Landing *mi*
VTVM Vacuum Tube Volt Meter *0, fu*
VTX Videotax [system] *hd1.89.y30*
VTX [new Navy training aircraft system] *ph*
VTX-TS [trainer competition for the US Navy, won by the British Aerospace Hawk, known as T-45 in US Navy service] *mt*
VU Volume Unit *mt, w9*
VUE Visible / Ultraviolet Experiment *cp, kf*
VUI Video User Interface *em, kg*
VUP VAX Unit of Performance *em, kg, hi*
VUTS Verification Unit Test Set *fu*
VUV Vacuum Ultra Violet *tb*
VV Verification and Validation [also V&V] *wp*
VV Vocoded Voice *uh*
VV {Vertical-Vertical} [polarization of antenna beam] *ac3.7-24*
VV&A Validation, Verification, and Accreditation *kf*
VVI Vertical Velocity Indicator *aw129.10.73*
VVS-VMF Voenno-Vozdushniye Sily Voenno Morskovo Flota [Naval Air Force] *mt*
VVSRF [Russia] Voenno-Vozdushniye Sily Rossiskoi Federatsii [Air Forces of the Russian Federation] *mt*
VVT Voice View Technology *hi*
VWP Variable Width Pulse *fu*
VX [organophosphate chemical warfare nerve agent] *wp*
VX [US Navy test unit, VX-4 'Evaluators', VX-5 'Vampires'] *mt*
VXD Virtual Device Driver *ct*
VXE [US Navy development squadron, VXE-6 'Puckered Penguins'] [USA] *mt*
VXI (VME Extensions for Instrumentation) *is26.1.50*
VXO Variable Crystal Oscillator *fu*
VZPU Vozdushnyi Zapastnoi Punkt Upravlenya [Airborne Reserve Command Post. As in Mi-8VZPU 'Hip-D'] [USSR] *mt*

W

W [represents USCG ship in many acronyms, e.g. WSES, WAGB, WAGL, WHEC, WIX, WLB, WLM, WLR, WMEC, WPB, WSES] *met*
WA Western Australia *w9, fu*
WA Work Authorization [document] *tm*
WA Worst Accuracy *fu*
WAA Wartime Aircraft Activity *mt*
WAA Wide Aperture Array *fu*
WAACS Wartime Aircraft Activity Capability System [HQ USAF] *mt*
WAAF Womens Auxiliary Air Force [UK] *mt*
WAAM Wide Area Anti Armor (Tank) Munitions [a development program for Anti tank munitions, USA] *hc, mt*
WAAM WWMCCS Allocation and Assessment Model *mt*
WAANG Washington Air National Guard *-12*
WAARS Wartime Aircraft Activity Reporting System [USAF] *mt*
WAAS Wide Area Active Surveillance *hc*
WAAS Wide Area Augmentation System *kg, em, jdr*
WAATS Western ARNG Aviation Training Site [USA] *mt*
WAA_RET Wartime Aircraft Activity Reporting System Retrievals *mt*
WABNRES Worldwide Airborne Resources *mt*
WAC Weapon Aiming Computer *mt*
WAC Weapon Assignment Cycle *tb*
WAC Weeks After Contract [award] *fu*
WAC Wide Angle Camera *re*
WAC Without Attitude Control *ci.168*
WAC World Aeronautical Chart *mt*

WACC Washington Area Contracting Center [MAC*] *mt*
WACMS WWMCCS/WIS Automated Configuration Management System *mt*
WACO WWMCCS ADP Contracting Officer *mt*
WACOS WWMCCS ADP Continuity of Service *mt*
WAD Wash-Aqueduct Division [U. S. Army Engineers] *dt*
WAD Wireless Acoustic Device *hc*
WAD Work Authorization and Delegation *gp, 8, 15, H, pl, lt, jdr, pbl, tm, kf, fu, su, ah, jn*
WAD Work Authorization Document *17, sb, wp, oe*
WAD World Aviation Directory *aw129.10.83*
WAD WWMCCS ADP Directorate *mt*
WADC Wright Air Development Center [USA] *mt*
WADD Wright Air Development Division (USAF) *fu*
WADDS WWMCCS ADP Directory of Data Bases and Systems *mt*
WADE Wideband Archiving Data Equipment *oe*
WADS Wide Area Data Service *mt*
WAF Wiring Around Frame *fu*
WAF Women in the Air Force *ro*
WAFDS WWMCCS ADP Fiscal Data Systems *mt*
WAFTM Wide Area Far Term Munition *fu*
WAG Wild Assed Guess *pk*
WAGB [USCG Icebreaker, USA] *mt*
WAGE Weapons Aerospace Ground Equipment *mt*
WAGL [USCG Lighthouse Tender, USA] *mt*
WAIS Wide Area Information Search *em, ct*
WAIS Wide Area Information Server *kg, hi*
WAITS Wide Area Information Transfer Systems *em, kg, ct*
WALT Warning Assessment Logic Terminal *mt*
WAM Weapon Attack Manager *tb*
WAM Weapons Assessment Model *st*
WAM Weather Alert Message *fu*
WAM Wide Area Mine [US Army program] *fu*
WAM Workload Analysis Activity Report System {?} *mt*
WAMEX West African Monsoon Experiment *pl*
WAMOSCOPE Wave Modulated Oscilloscope *fu*
WAN Wide Area Network *mt, ns, tb, jdr, pbl, tm, kf, fu, nu, kg, ah, em, ct, hi, oe, tt, ai1*
WANC Wobble Action Nutation Control *jn*
WANL Westinghouse Astro Nuclear Laboratory *ci.972*
WANODI Wide Area Network ODI *hi*
WAO Weapons Assignment Officer *fu*
WAOSS Wide Angle Optoelectronic Stereo Scanner *em, ct*
WAPS Weighted Airman Promotion System [USAF] *mt, af72.5.109*
WARC World Administrative Radio Congress (Conference) *ci.885, is26.1.42, wp, fu*
WARCONFAC War Consumables Consumption Factor [USAF] *mt*
WARFTR (the WARFTR Institute) *aa26.8.b8*
WARM War(time) Reserve Mode [USA DOD] *wp, fu, mt*
WARMAPS Wartime Manpower Planning System [OSD] [USA DOD] *wp, mt*
WARMARS Wartime Manpower Requirements System [USAFE] *mt*
WARMG Water Resources Management Group *wp*
WARP Worldwide AUTODIN Restoral Plan *mt*
WARPAC Warsaw Pact *mt*
WARRS Wholesale And Retail Receiving/Shipping [system] *wp*
WARS Warfare Analysis and Research System *fu*
WARS Wartime Assessment and Requirements System [AFLC] *mt*
WARS Worldwide Ammunition Reporting System [USA] *mt*
WARSAMS WARSAW SAM Missile Simulation System *mt*
WARSL War Reserve Stockage List *mt*
WARSMART Simulation Model for Allocation of Resources for Training [ATC] *mt*
WAS War At Sea [US Navy] *mt*
WAS Warhead Add on Sensor *fu*
WAS Weapon Aiming System *mt*
WAS Weapons Alert System [NORAD] *mt*
WAS Web Application Server *em*
WAS Wide Angle Sensor *st*
WASER Wafer Scale Reconstruction *hc*
WASHFAX Washington Area High Speed Facsimile System *mt*
WASO WWMCCS ADP Security Officer *mt*
WASP War Air Service Program *mt*
WASP Weasel Attack Signal Processor *mt, wp*
WASP Weight Analysis Sounding Probe *ci.244*
WASP Wide Area Surveillance Program *fu*
WASP [a Hughes MSG air to surface missile] *-17*
WASPM Wide Area Side Penetrating Mine *dh*
WASPUM Wide Area [Anti-Armor] Mine *hc*
WASS Wide Angle Sun Sensor *jdr*
WASS Wide Angle Surveillance System *hc*
WASS Wide Area Surveillance System *mt, wp*
WASS WWMCCS ADP System Security *mt*
WASSM WWMCCS ADP System Security Manager *mt*
WASSO WWMCCS ADP System Security Officer *mt*
WAT Weapons Assignment Technician *fu*

WAT Weight, Altitude, Temperature [factors determining take off performance] *mt*
WAT Work Authorization Trail *fu*
WATASO WWMCCS ADP Terminal Area Security Officer *mt*
WATE Weapons Attack Evaluation [mode] *fu*
WATS Website Activity Tracking Statistics *em, ct*
WATS Wide Area Telecommunications Service *mt, sc4.3.6, w9, fu, kg, hi*
WATS Wide Area Telephone Service *em, ct*
WATS Wide Area Telephone System *mt*
WATSTORE [a digital data file of the OWDC] *ac1.5-15*
WAUXCP West Auxiliary Airborne Command Post [SAC] *mt*
WAV Windows Audio Video *hi*
WAWAS Washington Area Warning System *mt*
WAWS Washington Area Wideband System *mt*
WB Wide Band *mt, gp, 16, hc, jb, wp, kf, fu*
WB/WAAS Wartime Beddown / Wartime Aircraft Activity System [USAF] *mt*
WBC Wide Body Cargo *mt*
WBC Wide Body CRAF *mt*
WBCT Wideband Current Transformer *fu*
WBD Wide Band Data *4, kf, fu*
WBDFX Wideband Dicke-Fix *fu*
WBDI Wide Band Data Interface *-4*
WBDI Wide Band Data Interleaver *bo, 4*
WBDL Wideband Data Link *fu*
WBEM Web Based Enterprise Management *em, kg*
WBL Weighted Burg-Levinson *fu*
WBL Wide Band Limiting *wp*
WBLA Weighted Burg-Levinson Algorithm *fu*
WBM Wideband Modem *fu*
WBMS Work Breakdown Management Structure *mt*
WBN Weapon Battle Manager {?} *tb*
WBNL Wideband Noise Limiting *fu*
WBO Washington Branch Office *dt*
WBOMB Web BOMB *hi*
WBS Work Breakdown Structure *mt, hc, ms3.1.2, 4, 8, 15, 16, 17, nt, lt, wp, ep, dm, re, tb, st, wl, jdr, kf, fu, pbl, oe*
WBSC Wide Band Signal Channel radio *mt*
WBST Wideband Secure Terminal *mt*
WBST Wideband Subscriber Terminal *mt*
WBSV Wide Band Secure Voice *fu*
WBSVN Wideband Secure Voice Network *mt*
WBSVVS Wideband Secure Voice Via Satellite *mt*
WBT Wait Before Transmission *mt*
WBT Wide Band Tape *kf, oe*
WC WAD Coordinator *fu*
WC Weapon Carrier *fu*
WC Weapons Controller *mt, fu*
WC Weather Channel *fu*
WC {an organization within Hughes Aircraft Company} *hn50.8.5*
WCA Worst Case Analysis *mj31.5.24, ep, jn, oe*
WCAC WWMCCS Crisis Alerting Capability *mt*
WCAN WWMCCS Crisis Alerting Network *mt*
WCAS Workload Collection and Analysis System [USAF] *mt*
WCC Weapon Control Console *fu*
WCC Wideband Communications Capsule *mt*
WCCA Worst Case Circuit Analysis *he, jdr*
WCCC Warning, Command, Control and Communications *mt*
WCCM Wideband Command and Control Modem *hc, fu*
WCCS Washington Crisis Communications System *mt*
WCCS Wing Command and Control System [USAF] *mt*
WCDAS Wallops Command and Data Acquisition Station *oe*
WCDO War Consumables Distribution Objective *mt*
WCDOS War Consumable Distribution Objectives System *mt*
WCDR War Consumable Distribution Requirements *mt*
WCEMP Weapon Case EMP *jdr*
WCI White Consolidated Industries *aw118.18.39*
WCL Weapons Custody List *mt*
WCL Wireless Command Link *hc*
WCM Weapons Control Module *fu*
WCM Weapons Countermeasures and Mines [US Navy] *mt*
WCM Wired-Core Matrix *fu*
WCO Weapon Control Order *fu*
WCP Weapon Control Platform *5, sb, tb*
WCP Weather Communication Processor *fu*
WCP Wing Command Post *mt*
WCP Wireless Command Post *hc, fu*
WCP World Climate Program *pl, hg, ge*
WCP World Coordinate System *em*
WCRP World Climate Research Program *ag, hy, hg, ge*
WCRS War Consumable Reporting System [USAF/PACAF] *mt*
WCS Waste Control System *wl*
WCS Weapon Control System *mt, rm3.1.2, hc, wp*
WCS Wireless Communication Service(s) *em, ct*
WCS Workload Control system [TAC] *mt*
WCS Workstation Control and Synchronization *mt*

WCS World Coordinate System *pk, kg*
WCS Writable Control Store *fu*
WCSG WWMCCS Council Support Group *mt*
WCT Weapons Controller Technician *fu*
WD War Department *w9*
WD Weapons Director *mt, fu*
WD White Dwarf *mi*
WD Work Directives *fu*
WDA Wire Design Activity *fu*
WDAMS WIS Data Base Management System *mt*
WDBII World Data Base II *mt*
WDC Weapon Data Converter *fu*
WDC World Data Center *no, hg, ge*
WDC-A-R&S World Data Center A for Rockets and Satellites *ns*
WDCP WSPRS Data Collection Program *mt*
WDE Wheel Drive Electronics *jdr, uh, jn, oe*
WDL Weapons Data Link *hc, fu*
WDL Western Development Laboratories [was Ford Aerospace] *oe*
WDL Windows Driver Library [Microsoft] *kg, em, hi*
WDM Wave Division Multiplexing *em, ct*
WDM Wavelength Division Multiplexer (Multiplexing) *fu, kg, rih*
WDNS Weapons Delivery and Navigation System *mt, hc*
WDP Weapon Data Processor *fu*
WDR Waiver Deviation Request *oe*
WDR Wet Dry Rehearsal *oe*
WDRAM Windows Dynamic Random Access Memory *em, kg, hi*
WDS Weapon Direction System *fu*
WDS Wire Design System *fu*
WDSIL Data Automation Resource Management System {?}:(mt)
WDT Warning Display Terminal *mt*
WDT Watch Dog Timer *3, fu*
WDT Weapons Director Technician *fu*
WDU Wallops Distribution Unit *gp, oe*
WE (WE_177 is a family of nuclear weapons currently in use, UK) *mt*
WEA Weapons Expenditures Authorization *mt*
WEATHER Weather Satellite Tracking Parameter Generator *mt*
WEBNFS WEB Network File System *em*
WebNFS Web Network File System [Sun] *pk, kg*
WEC Westinghouse Electric Company [Corporation] *gp, pl, oe*
WECO Weather Current Operations *mt*
WECOM [Hughes Aircraft Company customer for GLAADS] *hc*
WED Western Europe Daylight *pk, kg*
WEFA (the WEFA Group, formed from the merger of Chase . . . and Wharton Econometrics) *is26.1.73*
WEFAX Weather Facsimile [GOES] *gd1-0, 16, hy, pbl, hg, ge, oe*
WEG Weapons Evaluation Group [USA] *mt*
WEGENER Working Group of European Geoscientists for the Establishment of Networks for Earthquake Research *hy*
WEIP Work Element Implementation Plan *gp*
WELL Whole Earth 'Lectronic Link [BBS] *kg, em, ct*
WELS Wideband Emitter Location System *fu*
WEM Work Element Manager *gp*
WEMA Wter Equipment Manufacturers Association *y2k*
WEO WWMCCS Evaluation Office *mt*
WEP WWMCCS Evaluation Program *mt*
WEPH Weapon Phenomenology *mt*
WES Warfare Environment Simulator *tb*
WES Warhead Electronics Subsystem *fu*
WES Warhead Exploder Section (Subsystem) *fu*
WES Waterways Experiment Station [US Army] *tc*
WES Weapon Entry Station *fu*
WES Western Europe Standard *pk, kg*
WES Wire Entry System *fu*
WES WWMCCS Entry System [FORSCOM] *mt*
WESCOM Weapons Effects on Satellite Communications *mt*
WESCON {proceedings of a communications conference, 1980} *sd*
WESRAC Western Research Applications Center *ci.1005*
WESS Weapon System Status *mt*
WESS Weapons Effect Signature Simulator *nd74.445.19*
WESS WWMCCS Environmental Support System [WWMCCS] *mt*
WESTAR {satellite serving Western Union} [first US domestic synchronous satellite] *mt, ma16, 4, lt*
WESTCOM Western Command [Pacific] [US Army] *mt*
WESTHEM Western Hemisphere *mt*
WESTLAND Western Atlantic *fu*
WESTLANT Western Atlantic Command [NATO] *-12*
WESTPAC Western Pacific *mt, kf*
WET Weapons Effectiveness Testing *hc, fu*
WET_EYE [500 lbs chemical warfare bomb] *wp*
WET_SNOW [SAC bombing method] *wp*
WEU Western European Union [a defense organization] *mt, 14, aa26.10.12, wp*
WEZ Weapon Engagement Zone [USA DOD] *mt*
WF Wave Form *tb*
WFA Wage Fixing Authority [DOD Technical Staff] *dt*
WFC Wallops Flight Center *ci.70*

WFC Wide Field Camera *ns*
WFC Wide Field Collimator *oe*
WFD Work Flow Diagram *jdr*
WFE Wave Front Error *tb*
WFF Wallops Flight Facility [NASA] *5, sb, hy, mb, mi, oe*
WFMU Weather and Fixed Map Unit *fu*
WFNA White Fuming Nitric Acid *ci.185*
WFOV Wide Field Of View *hc, wp, tb, st, kf*
WFPC Wide Field / Planetary Camera [on HST] *mi*
WFPCII [replacement for WFPC] *mi*
WFS Wideband Frequency Synthesizer *jdr*
WFSC Weather Facsimile Switching Center *ag*
WFT Warshot Fuel Tank *fu*
WFZ Weapons Free Zone *mt, fu*
WGHPF Waveguide Highpass Filter *fu*
WGM Weighted Guidelines Method *fu*
WGN White Gaussian Noise *fu*
WGS Weather Graphics System *mt*
WGS Work Group System *em, kg*
WGS World Geodetic System (WGS84 means 1984 datum of WGS) *kf, mt, sd, fu*
WGSC War Gaming and Simulation Center *tb*
WGSG Weapon System Vehicle Ground Support Working Group *-17*
WGT Waveguide Termination ?(2-8) *jdr*
WHA War Head:(wp)
WHA Weapon Hazard Analysis *fu*
WHAM West Hemisphere AUTODIN Management *mt*
WHC Workstation Host Connection *em, kg*
WHCA White House Communications Agency *mt, dt, fu*
WHDM Watthour Demand Meter *fu*
WHEC [USCG High Endurance Cutter, USA] *mt*
WHEP White House Emergency Plan *mt*
WHIP WSEM/Host Interface Protocol *fu*
WHM Watthour Meter *fu*
WHMO White House Military Office *mt*
WHNS Wartime Host Nation Support *mt, wp*
WHQ War Headquarters *fu*
WHT Walsh-Hadamard Transform *fu*
WI Wallops Island, [Virginia, CDAS site] *pl*
WI Weight Incentive [a fee] *jdr*
WIA Wounded In Action [USA DOD] *mt, w9, wp*
WIANG Wisconsin Air National Guard *-12*
WIAS Wholesale Inventory Audit System *wp*
WIC Warning Information Correlation *mt*
WICS Warehouse Inventory Control System *wp*
WICS Warning Information and Correlation System *mt*
WICS Worldwide Intelligence Communication System *mt*
WICU Weather Intercept Control Unit *mt*
WID Wire Identification *jdr*
WIDT Wideband Data Transmittal [system] *jdr*
WIF Weapons Integration Facility *fu*
WIFR WithIn Frame Registration *oe*
WIG (RFB research and development in WIG vehicles) *aa27.4.b13*
WIG Wing In Ground effect *mt*
WIGAS Wide Band Adaptive Ground Antenna *hc*
WIGE Wing In Ground Effect *mt*
WIJC Work Unit Code *fu*
WILD_WILD WEASEL [aircraft for lethal suppresion of enemy air defenses] *wp*
WIM Wideband Interconnection Module *mt*
WIMM Weapons Integrated Material Manager *mt*
WIMS Work Information Management System [USAF] *mt*
WIMS Worldwide Intra theater Mobility Study [OSD] *wp, mt*
WIMWSS Worldwide Integrated Management of Wholesale Subsistence Stock [DLA] *mt*
WIN Wireless Intelligent Network *jr*
WIN Worldwide Information Network *wp*
WIN WWMCCS Information Network *mt*
WIN WWMCCS Intercomputer Network *mt*
WINA WWMCCS Information Needs Analysis *mt*
WINCON Winter Convention [on military electronics] *hn43.5.3*
WINCS WWMCCS Intercomputer Network Communications Subsystem *mt*
WIND WWMCCS Intercomputer Network Director *mt*
WINDII Wind Imaging Interferometer [UARS, Canada, France] *hy, hg, ge*
WINDS Weather Information Network Display System *wp*
WINDSAT Wind Measurements Satellite *hg, ge*
WINFORUM Wireless Information Networks Forum *em, kg*
WINHEC WINdows Hardware Engineering Conference *em, ct*
WINS Windows Internet Name (Naming) Service [Microsoft] *kg, em, ct, hi*
WINSOCK Windows Open Systems Architecture [Microsoft] *pk, kg*
WINSOCK WINdows SOCKet *em, ct, hi*
WINSTATS WIN Statistics Systems [WWMCCS] *mt*
WINTEL Warning, Intelligence [material] {?} {WNINTEL ?} *wp*

WINTEL WINdows / InTEL *em, ct*
WINTEX/CIMEX Winter Exercise/Civilian - Military Exercise [NATO] *mt*
WINWORD Word For Windows [Microsoft] *pk, kg*
WIS Worldwide [Military Command and Control] Information System *uc, is26.1.76cc, tb, fu*
WIS WWMCCS Information System *mt*
WISDIM WWMCCS Dictionary for Information Management *mt*
WISE Weapon Improvement and Service Extension *hc*
WISE Weapon Integration System Engineering *fu*
WISE WordPerfect Information System Environment *em, kg*
WISNAS WIS Network Authentication Service *mt*
WISP Wartime Information Security Program *mt*
WISRA WIS Requirements Analysis *mt, fu*
WISSA Wholesale Interservice Supply Support [system] *wp*
WIS_CMS WIS Configuration Management System *mt*
WIS_JPM WIS Joint Program Manager *mt*
WIS_SPO WIS Systems Program Office [ESD] *mt*
WITS Washington Interagency Telecommunications System [GSA] *mt*
WITS WSGT Intelligent Terminal System [WWMCCS] *mt*
WIU Weapons Interface Unit *mt, fu*
WIX [USCG Training Cutter, USA] *mt*
WIYN Wisconsin / Indiana / Yale / NOAO [telescope] *mi*
WJ Watkins-Johnson [Company] *pl*
WL Water Line *w9, fu*
WL Wave Length *w9, wp, fu*
WL Weapons Laboratory *wp, fu*
WL Weiner-Levinson *fu*
WL White Light *fu*
WL Wire List *fu*
WLB [USCG Seagoing Buoy Tender, USA] *mt*
WLC Weapon Launch Console *fu*
WLC White Light Coronagraph *pl*
WLM [USCG Coastal Buoy Tender, USA] *mt*
WLP Wideband Link Processor *kf*
WLR Weapon Locating Radar *hc*
WLR [USCG River Buoy Tender, USA] *mt*
WLS Weapon Launch System *fu*
WLS Weighted Least Square *pbl*
WLT Weapons Load Training *mt*
WLTB Wire Line Termination Box *fu*
WLU Weapons Location Unit *fu*
WM Warfare Management *fu*
WMAPI Windows Messaging API *hi*
WMATC Washington Metropolitan Area Transit Command *dt*
WMC Workflow Management Coalition *em, kg*
WMC World Meteorological Center *pl*
WMD Weapon of Mass Destruction *kf*
WMEC [USCG Medium Endurance Cutter, USA] *mt*
WMO World Meteorological Organization *bl3-1.12, wp, no, hy, hg, fu, ge*
WMOP Weather Observation Master Plan *mt*
WMP War and Mobilization Plan [USAF] *mt*
WMP War Mobilization Plan *wp*
WMSC Weather Message Switching Center *mt, ag, fu*
WMSCR Weather Message Switch Center Replacement *fu*
WNFE WWMCCS Network Front End *mt*
WNIC Wide area Network Interface Co-processor *em, kg, hi*
WNIM Wide area Network Interface Module *em, kg*
WNINTEL Warning Notice Intelligence [sources and methods involved] *mt, tb, jdr, kf*
WNS Worldwide Navigation System *mt*
WO Warrant Officer *w9, fu*
WO Weapons Officer *fu*
WOA Wartime Operating Airfield *mt, wp*
WOC War Operations Center *wp*
WOC Wing Operations Center [USAF/NATO] *mt*
WOC Wing Operations Station *wp*
WOCE World Ocean Circulation Experiment *hy, hg, ge*
WOCWAR Wing Operation Center Wargame [model] *wp*
WOD Wind Over the Deck [of a carrier] *mt*
WOE Weapons Optical Effects [a code] *st*
WOG Weapon Order Generator *fu*
WOM Web Object Management *em*
WOM Write-Only Memory *fu*
WOMIS Wing Operations Management Information System [USAFE] *mt*
WOP Wing Operation Plan *mt*
WORAM Word Oriented Random Access Memory *fu*
WORCS Work Order Reporting and Communication System [USA] *mt*
WOS Weapon Order Subsystem *fu*
WOS Workstation Operating System *kg, em*
WOSA Windows Open Services (Systems) Architecture [Microsoft] *kg, em, hi*
WOSE Weather Officer in Space Experiment *sf31.1.30*
WOW Weight On Wheels *pl*

WOWS Wire Obstacle Warning System *hc*
WP Warsaw Pact *mt, fu, ph, hc, wp*
WP Weapon Platform *17, sb*
WP Weather Permitting *w9, wp*
WP White Phosphorus *mt, w9*
WP Word Processing (Processor) *mt, fu, w9, em, kg*
WP WordPerfect *kg, hi*
WP Work Package *mt, fu, hy, st, wl, jdr, kf*
WP Work Plan *mt*
WP Work Position *fu*
WP Working Paper *mt*
WP Write Protected *em, kg*
WPA Work Package Authorizations *fu*
WPAC West Pacific [Satellite] *mt*
WPAFB Wright Patterson Air Force Base [Dayton, Ohio] *hc, wp*
WPARR War Plans Additive Requirements Report *mt*
WPB Write Printer Binary *fu*
WPB [USCG Large Patrol Boat, USA] *mt*
WPC War Preparations Center *wp*
WPC Warrior Preparation Center [USAFE] *mt*
WPC Watts Per Candle *w9, fu*
WPC Word Processing Center *mt, hd*
WPD Work Package Directive *tb, st*
WPD Work Program Directive *5, sb*
WPD Write Printer Decimal *fu*
WPE Word Processing Equipment *mt*
WPF Warped Phase Front *fu*
WPHD Write Protect Hard Disk *kg, em, hi*
WPI Wholesale Price Index *wp*
WPIE Warning Plan Interface Equipment *mt*
WPIG Weapon system vehicle Payload Interface working Group *-17*
WPL (WPL, a leader in aerospace software engineering) *aw130.11.73*
WPL Wave Propagation Laboratory [NOAA] *ge, hg*
WPLID Wind Profile LIDAR *hg, ge*
WPM Words Per Minute *mt, kg, em, 0, w9, wp, oe*
WPMS Wideband Propagation Measurement System *hc*
WPOD Water Port Of Debarkation *wp*
WPOE Water Port Of Embarkation *wp*
WPOS Work Place Operating System *em, kg*
WPPS Work Package Planning Sheets *17, sb, fu*
WPREQ Request for Nuclear Reserve Weapons *mt*
WPRS War Powers Reporting System *mt*
WPS Word Processing System *mt, tb*
WPS Working Prototype System *mt*
WPS Workplace Shell [OS2] *kg, em, hi*
WPVM Windows Parallel Virtual Machine *kg, em*
WR Weapon Radius *wp*
WR Weather Reconnaissance *mt*
WR Western Range [formerly WTR] *mb, kf, mi*
WR-ALC Warner Robins Air Logistics Center [AFLC] *mt*
WRA Weapons Replaceable (Replacement) Assembly *mt, rm3.1.3, hc, fu*
WRAC Women's Royal Air Corps *w9*
WRADTA (upgrade of WR-ALC GO89 Damage Tolerance system) *mt*
WRAF Womens Royal Air Force [UK] *w9, mt*
WRALC Warner Robins Air Logistic Center [Georgia] *hc*
WRAM Windows Random Access Memory *em, kg, hi*
WRAMA Warner Robins Air Material Area [USA] *hc, mt*
WRAMC Walter Reed Army Medical Center *dt*
WRAMSOS WR-ALC Multiple System OFP Support System *mt*
WRAP War Reserve Automated Process *mt*
WRATE Warner Robins Automatic Test Equipment *wp*
WRATE Warning Receiver Automatic Test Equipment *fu*
WRB Web Request Broker *em*
WRBD Web Request Broker Dispatcher *em*
WRC Wartime Rate of Consumption *mt*
WRCS Weapon Release Computer Set *mt*
WRCS Weapon Release Computer System *aw118.18.99*
WRDC Wright Research and Development Center *wp, kb*
WRE Weapon Research Establishment *crl*
WRECS Weapon Radiation Effects on Communications Systems *mt*
WRF Woodbridge Research Facility *dt*
WRGS Western Relay Ground Station *kf*
WRM War Reserve Material [USAF] *mt, hc, wp*
WRMCMR WRM Control and Management Reporting [USAFE] *mt*
WRMLRSSL WRM Lists Requirements and Spares Supply List *mt*
WRMRS WRM Rating System *mt*
WRMS Web Reception Monitoring Service *hi*
WRN Weather Reference Number *fu*
WRNS Women's Royal Naval Service *w9*
WRO Western Regional Office *fu*
WROC War Room Operations Center *fu*
WRS War Reserve Stock *fu*
WRS Warning Radar System *hc*
WRS Weather Reconnaissance Squadron [USA] *mt*
WRS Wholesale Receiving System *wp*
WRSA War Reserve Stock for Allies *wp*

WRSK War Readiness Spares Kit [USAF] *mt, fu, wp*
WRSK War Reserve Support Kit *mt*
WRSK/BLSS WRSK / Base Level Self Sufficiency Spares *mt*
WRWC Wartime Requirements for War Consumables *mt*
WS Watch Supervisor *fu*
WS Weapon Specification *fu*
WS Weapon System *17, sb, jmj, fu*
WS Weasel Support *mt*
WS Weather Squadron [USA] *mt*
WS Winter Solstice *oe*
WS WordStar *pk, kg*
WS Work Station *ts4-40, kg, em, kf, oe*
WSA Weapon System Acquisition *wp*
WSA Weapons Storage Area *mt*
WSA Wide Service Area *jdr*
WSA WWMCCS System Architecture *mt*
WSA&E Warfare Systems Architecture and Engineering [US Navy] *fu, mt*
WSAB Weapon System Assessment Briefing *mt*
WSAFHWC Weather Support for AFHWC *mt*
WSAG Washington Special Action Group *mt*
WSAM Weapon System Assessment Model *mt*
WSAP Weapon System Acquisition Process *mt, wp, fu*
WSAPI Web Site Application Program Interface *em, kg*
WSAS WWMCCS Standard Application Software *mt*
WSAT Weapon System Alignment (Acceptance) Tests *fu*
WSB Weapons Services Branch *mt*
WSB Wide Spot Beam *jdr*
WSBCRCS Worldwide Stock Balance and Consumption Report Consolidation System *mt*
WSC War Support Center [Europe] *mt*
WSC Weapon Simulation Control *fu*
WSC Weapons System Controller *mt*
WSC White Sands Complex *vv*
WSC WIN Site Coordinator *mt*
WSC Wire Sounding Capability *hc*
WSCMO Weather Service Contract Meteorological Office *ag*
WSCS Weapon System Communications System *fu*
WSD Weapon System Development (Department) *fu*
WSDL Weapons System Data Link *mt*
WSE Weapon System Evaluator *hc*
WSE WWMCCS System Engineer [DCA] *mt*
WSEG Weapons Systems Evaluation Group *13, dt, fu*
WSEM Weapons Systems Evaluation Missile *hc*
WSEM Work Station Electronics Module *fu*
WSEO WWMCCS System Engineering Organization [DCA] *mt*
WSEP Weapon System Evaluation (Evaluator) Program *mt, hc, wp*
WSES [USCG Surface Effect Ship, USA] *mt*
WSESA Weapon System and Equipment Support Analysis *wp*
WSESRB Weapon System Explosive Safety Review Board *fu*
WSEU Work Station Electronics Unit *fu*
WSF Weapons Systems Fill *mt*
WSF Weather Support Force *mt*
WSF World Space Foundation [a Pasadena, California, corporation] *sc3.1.6*
WSFO Weather Service Forecast Office [U. S. National] *gd3-12*
WSGS White Sands Ground Station *-4*
WSGT White Sands Ground Terminal *2, 4, wl*
WSGT WWMCCS Standard Graphics Terminal [WWMCCS] *mt*
WSGTU White Sands Ground Terminal Upgrade *vv*
WSI Wafer Scale Integration *17, sb, wp*
WSIAO Weapon Support Improvement and Analysis Office [DOD] *dt*
WSIP Weapon Systems Improvement Program *mt*
WSL Wind Sock Lights *fu*
WSLO Weapon System Logistic Office *wp*
WSM Weapon Status Module *fu*
WSM Weapon System Model *mt*
WSMC Western Space and Missile Center [also known as WTR] [AFSC*] *mt, bl1-2.236, 12, jb, st, jdr, kf, oe*
WSMIS Weapon System Management Information System [AFLC] *wp, mt*
WSMO Weather Service Meteorological Office *ag*
WSMP Weapon System Master Plan *wp*
WSMR White Sands Missile Range [New Mexico, USA] *mt, hc, 0, 5, 12, 17, sb, nt, wp, st, mb, fu, mi*
WSN World Spaceflight News [ISSN 0737-8548] *ws4.10*
WSNGT White Sands NASA Ground Terminal *-4*
WSO (National) Weather Service Office *gd1-4*
WSO Warning Systems Operator [USSPACECOM / USSTRATCOM] *mt*
WSO Weapon System Operator *mt, wp*
WSO Weapon(s) Systems Officer *mt*
WSP Weapon Safety Program *fu*
WSP Weapons System Pouch *mt*
WSPAR Weapon System Program Assessment Review *mt*
WSPRS WIN SCF Performance Reporting System *mt*
WSPS Worldwide Standard Port System [MTMC] *mt*

WSR Weapon System Reliability *mt*
WSR Weapon System Review *fu*
WSR Weapons Status Report *mt*
WSS Warfare Support System *mt*
WSS Weapons System Security *mt*
WSS Wholesale Shipping System *wp*
WSS WWMCCS Standard System *mt*
WSSH White Sands Space Harbor *ws4.10*
WSSIB WWMCCS Standard System Information Base *mt*
WSSM Weapon System Support Model *mt*
WSSP Weapon System Support Program *wp*
WSSS Weapon Storage and Security System *mt*
WSSS WWMCCS Standard System Software *mt*
WST Weapon System Trainer *mt, wp*
WSTF White Sands Test Facility *ci.70, 4*
WSTS Weapons Systems Training System *hc*
WSU Washington State University *aa26.11.56*
WSU Weapon Simulation Unit *fu*
WSU Weather Support Unit *mt, sr9-82, hc, hs*
WSU Wright State University (Dayton, Ohio) *wp*
WSUMS WWDMS Scheduling and Management System *mt*
WSV Weapon System Vehicle *17, sb*
WSV Wideband Secure Voice *mt*
WSVN Worldwide Secure Voice Network *mt*
WT Wind Tunnel *0*
WTCA Water Terminal Clearance Authority *wp*
WTCV Weapons and Tracked Combat Vehicles [an Army program] *lka*
WTD Weapons Training Detachment *mt*
WTDOS Weapon Training Detachments Operating Spares *mt*
WTM WWMCCS Technical Manual *mt*
WTO Warsaw Treaty Organization *14, aw130.6.21*
WTO World Trade Organization *rih*
WTQD Work Transfer Quality Document *fu*
WTR Western Test Range *ci.1250, 4, 5, 12, 16, 17, sb, ps204, lt, wp, ep, dm, wl, jdr, mb, kf, mi, oe*
WTR Wideband Tape Recorder *oe*
WTR Work Transfer Request *fu*
WTS Washington Tactical Switch *mt*
WTT Weapons Tactics Trainer *hc*
WU Western Union *mt, sc3.6.3, 4*
WUC Work Unit Code *mt, wp*
WUCM Work Unit Code Manual *fu*
WUDO Western Union Defense Organization [precursor of NATO created in 1948 by the treaty of Brussels] *mt*
WUG WWMCCS (WIS) Users Group *mt*
WUIS Work Unit Information Summary *fu*
WUIS Work Unit Information System *mt*
WUPPE Wisconsin Ultraviolet PhotoPolarimter Experiment [Astro package] *mi*
WUSCI Western Union Space Communications, Incorporated *-4*
WV Wet Vegetation *ac7.1-37*
WV Working Voltage *fu*
WVD Weather Video Digitizer *fu*
WVM Wideband Video Modem *fu*
WVR Within Visual Range *hc*
WVS Women's Voluntary Services *w9*
WVT WES Validation Tester *fu*
WVU West Virginia University [Morgantown, WV] *aa27.4.62*
WW MCCS ADP Control Center *mt*
WW Weather Wing *mt*
WW Wild Weasel *mt, wp*
WW World War (WWII) *mt, w9, pf.f88.14, wp*
WWA Wild Weasel Augmentation *mt*
WWA World Warning Agency *no, pr*
WWABNCP World Wide Airborne Command Post *mt, dh, wp, fu*
WWABNRES WWMCCS Airborne Resources *wp*
WWAN Wireless Wide Area Network *ct, em*
WWANCP World Wide Airborne National Command Post *kf*
WWAS World Warning Agency for Satellites *ns*
WWB World Weather Building *ag*
WWC Wing Work Center *mt*
WWCC World-Wide Coordinate Conversion *fu*
WWDMS Worldwide Data Management System [WWMCCS] *mt*
WWDSA Worldwide Digital Systems Architecture *mt*
WWFO Wild Weasel Follow-On *wp*
WWI World War One [1914 to 1918] *mt, aa26.12.84, af72.5.11, aw129.21.70*
WWII World War Two [1939 to 1945] *mt, aa26.11.6, af72.5.11*
WWIMS Worldwide Indicator and Monitoring System *mt*
WWIMWSS Worldwide Management of Wholesale Subsistence Stock [DLA] *mt*
WWIS World Wide Information System [Internet] *kg, em, hi*
WWMCCS World Wide Military Command and Control System *mt, hc, 12, 17, ph, jb, dt, wp, kf, fu*
WWMCCSDPCE WWMCCS Data Processing Center, Europe *mt*
WWMCS World Wide Military Computer System *hc*

WWO Wing Weather Officer *mt*
WWOLS World Wide On Line System [DCA*] *mt*
WWP World Weather Program *ag*
WWP World Wide Procurement [system] *kf*
WWS WIS Workstation *mt*
WWSIM Welter Weight Simulator *kf*
WWSVA World Wide Secure Voice *mt, wp*
WWSVA Worldwide Secure Voice Architecture *mt*
WWSVCS Worldwide Secure Voice Conferencing System *mt*
WWTCIP Worldwide Technical Control Improvement Plan *mt*
WWV {call letters of the standard time and frequency radio station} *pr, no, wl*
WWW World Weather Watch *pl, oe*
WWW World Wide Web [a loose linkage of Internet sites which provide data and other services from around the world] [also W3] *cu, kg, nu, ct, em, hi, oe, du*
WWWA World Wide Web Applets [also W3A] *em, kg*
WWWC World Wide Web Consortium [also W3C] *em, kg, ct*
WWWW World Wide Web Worm *hi*

X

X Experimental *0*
X-Tell Cross Tell *fu*
X/L Crosslink *jdr, kf*
X2B heXadecimal to Binary *em, kg*
X2C heXadecimal to Character *em, kg*
X2D heXadecimal to Decimal *em, kg*
XA eXtended Architecture *em, kg, hi*
XA eXtended Attribute *em, kg*
XANADU [X-ray analysis system maintained at the HEASARC] *cu*
XAPIA X.400 Application Program Interface Association *em, kg, hi*
XDA XRS Drive Assembly *oe*
XDCR Transducer *uh, oe*
XDF Extended Density Format [IBM] *kg, em, hi*
XDIN Communications Subsystem of IMAPS *mt*
XDM {Experimental Demonstrator Model} *-17*
XDP Extended Data Processing *aa27.4.52*
XDR eXtended (External) Data Representation:(em, kg, ns)
XDS Exoatmospheric Defense System *st*
XED Transceiver / Encoder-Decoder *fu*
XF Crystal Filter *jdr*
XFCN eXternal Function *em, kg*
XFD Xenon Fill and Drain [valve] *jn*
XFER Transfer *jdr, vv*
XGA eXtended Graphics Adapter *em*
XGA Extended Graphics Array [IBM] *kg, ct, hi*
XGDS Exempt from the General Declassification Schedule *hc, fu*
XGRS X-ray and Gamma Ray Spectrometer *eo*
XI Exercise Indicator *fu*
XID eXchange IDentifier *em, kg*
XIE X-ray Imaging Equipment [EOS] *hy*
XILA Extended Incremental Linear Actuarot *uh*
XIMAGE [Image analysis program in XANADU] *cu*
XIOS eXtended Input / Output System *em, kg*
XIP Xenon Ion Propulsion *jdr*
XIPS Xenon Ion Propulsion System *ah, pbl, he, jn, ls4, 5, 28*
XLAC Crosslink Access Control [message] *jdr*
XLAT Translate *pk, kg*
XLAU Extended Limited AUTODIN Upgrade *mt*
XLNK Crosslink *jdr*
XLP Crosslink Processor *jdr*
XLS Crosslink Subsystem *jdr*
XLV Xenon Latch Valve *jn*
XM Experimental Model *0*
XMGM (XMGM-52) *mt*
XMIT Transmit *bl2-2.39, kg, oe*
XMIT [general purpose message transmission system, USAFE] *mt*
XML eXtensible Markup Language *em*
XMM Extended Memory Manager [LIM/AST] *kg, ct, em, hi*
XMM X-ray Multi Mirror [observatory] *es55.56, mb, mi*
XMO Crystal Master Oscillator *jdr*
XMS eXpanded Memory Service *em*
XMS Extended Memory Specifications [LIM/AST] *em, ct, hi, kg*
XMTL Transmittal *jdr*
XMTR Transmitter *gd2-21, 0, wp, jdr, uh, oe*
XN Execution Node *hi*
XO Executive Officer [USA] [slang] *mt*
XOFF Transmitter Off *pk, kg*
XON Transmitter On *pk, kg*
XON/XOFF Transmitter ON / OFF Telecommunications control protocol *hi*
XOR eXclusive Or [also EOR] *em, kg, hi*
XPC XIPs Power Controller *he*
XPD Exportable Protection Device *fu*
XPG X/Open Portability Guide *hi*

XPNDR Transponder *0, jdr, oe*
XPRM Xerox Print Resources Manager *kg, em*
XPS X-Ray Photoelectron Spectroscopy *wp*
XRD X-Ray Diffraction (Diffractometer) *fu, eo*
XRDA X-Ray Drive Assembly *oe*
XRDE X-Ray Drive Electronics *oe*
XREF Cross Reference [PEMS file] *mt, 8, wp*
XRF X-Ray Fluorescence [spectrometer] *eo*
XRI X-Ray Imager *gp, 16, oe*
XRL Extended Range Lance *hc*
XRL X-Ray Laser *hn48.20.1, wp, st*
XRONOS [Temporal analysis program in XANADU] *cu*
XRP X-Ray Positioner *oe*
XRP X-Ray Processor *oe*
XRPE X-Ray Processor Electronics *oe*
XRS X-Ray Sensor *gp, 16, oe*
XRT eXtensions for Real Time *em, kg*
XS Cross-Strap *gd2-21*
XSAR X-band Synthetic Aperture Radar *wp*
XSC Experimental System Contractor *st*
XSELECT [a High Level tool to manage the FTOOLS] *cu*
XSLC Xichang Satellite Launch Center *ai2*
XSMD Extended Storage Module Driver [interface] *kg, em, hi*
XSPEC [X-ray/gamma-ray spectral analysis package in XANADU] *cu*
XSS Crosslink Subsystem? *jdr*
XSSI eXtended Server Side Includes *em, kg*
XST Cross-Strapping *jdr*
XST Experimental Stealth Technology [a Lockheed technology demonstrator aircraft] [USA] *mt*
XT Extended Technology *hi*
XTASS Experimental Towed-Array Sonar System *fu*
XTCLK eXternal Transmit CLocK *em, kg*
XTE X-ray Timing Explorer *cu*
XTP Xenon [ion] Thruster Platform *jn*
XUTIL ISP Utility Program *mt*
XUV eXtreme UltraViolet *mi*
XY-R [Cartesian to polar coordinate conversion] *fu*

Y

Y Prototype [X Experimental]
Y/N YES or NO [request for decision] *pk, kg*
YAG Yttrium Aluminum Garnet [laser] *aa26.12.39, hn41.26.3, 17, aw129.21.25, wp, fu*
YAP Yield Analysis Pattern *hi*
YB Ytterbium *wp*
YDT Yukon Daylight Time *pk, kg*
YE Year End *fu*
YE Yield Enhancement *fu*
YEC (contact GPS program control (SAMSO/YEC)) *N75-27202-466*
YF Yei-ti Fa-dong-ji [Long March liquid engine] *ai2*
YIG Yttrium Iron Garnet *mj31.5.18, wp, fu*
YMS Young Micro Systems *ct, em*
YP Yield Point *fu*
YP [patrol craft, US Navy] *mt*
YPG Yuma Proving Ground [USA] *fu, mt*
YS Yield Strength *fu*
YSA Yaw Sun Acquisition *uh, jdr*
YSO Young Stellar Object *mi*
YSR [Seaplane Wrecking Derrick, US Navy] *mt*
YST Yukon Standard Time *pk, kg*
YTB [Large Harbor Tug, US Navy] *mt*
YTC Yield To Call *hi*
YTD Year To Date *jdr, fu, kg, em*

Z

Z-CAV Zoned Constant Angular Velocity *pk, kg*
ZAI Zero Administrative Initiative *ct, em*
ZAP Zone Acquisition Processor *hc*
ZAR Zeus Acquisition Radar [large pyramidal radar used with the Nike Zeus ABM missile] [USA] *mt*
ZBB Zero Base Budgeting *mt*
ZBR Zone Bit Recording [IBM] *kg, em*
ZCAV Zoned Constant Angular Velocity *em, hi*
ZCR Zero Crossing Rate *hi*
ZD Zenith Distance *w9*
ZD Zero Defects *fu*
ZD Ziff-Davis *ct, em*
ZDF Zero Doppler Filter *fu*
ZDL Zero Delay Lockout *em, kg*
ZDS Zenith Data Systems *em, kg*
ZEL Zero Launch system [a zero length launch system for F-100s that carried nuclear bombs] [USA] *mt*

ZF Zero Flag *ct, em*
ZFW Zero Fuel Weight *mt*
ZI Zone of Interior *mt, w9*
ZIF Zero Insertion Force [socket] *fu, ct, em, kg, hi*
ZIP Zig-zag In-line Package *ct, em, kg, hi*
ZIP Zodiacal Infrared Probe *hc*
ZIP Zone Improvement Plan [ZIPcode] *em, kg*
ZKA [communications signal to automatically seize the submarine broadcast network] *mt*
ZLG Zero-lock Laser Gyro *kf*
ZM Zagraditel' Minniy [mine layer] [USSR] *mt*
ZM Zero Momentum *sd*
ZM Zone Marker *mt*
ZO {Characteristic impedance} *jdr*
ZOE Zone Of Exclusion *2, 4*
ZORRO Zero Offset Rapid Reaction Ordnance *hc*
ZOT Zinc Ortho-Titanate *jdr*
ZPR Zero Power Resistance *hi*
ZPV Zoomed Port Video [Toshiba] *kg, em*
ZRE Zero Rate Error *hi*
ZSA Zero Set Amplifier *fu*
ZSL Zero Slot LAN *em, kg, hi*
ZST Zone Standard Time *0*
ZT Zone Time *fu*
ZTC Zero Temperature Coefficient *fu*
ZTO Zone Transportation Officer *mt*
ZWC Zero Word Count *hi*
ZXCFAR Zero Crossing Constant False Alarm Rate *fu*
ZYGO (testing (ZYGO interferometer, etc.)) {Zirconium Yttrium Garnet Oxide} *hd*

www.ingramcontent.com/pod-product-compliance
Lightning Source LLC
Chambersburg PA
CBHW08024030426
42334CB00023BA/2700